Smart Materials

Smart Materials

Edited by

Mel Schwartz

CRC Press
Taylor & Francis Group
Boca Raton London New York

CRC Press is an imprint of the
Taylor & Francis Group, an **informa** business

CRC Press
Taylor & Francis Group
6000 Broken Sound Parkway NW, Suite 300
Boca Raton, FL 33487-2742

10 9 8 7 6 5 4 3 2 1

International Standard Book Number-13: 978-1-4200-4372-3 (Hardcover)

Library of Congress Cataloging-in-Publication Data

Smart materials / [edited by] Mel Schwartz.
p. cm.
Includes bibliographical references and index.
ISBN 978-1-4200-4372-3 (alk. paper)
1. Smart materials. 2. Smart structures. I. Schwartz, Mel M. II. Title.

TA418.9.S62S48 2008
620.1'1--dc22 2008018721

Visit the Taylor & Francis Web site at
http://www.taylorandfrancis.com

and the CRC Press Web site at
http://www.crcpress.com

Dedication

To Carolyn and Anne-Marie who light up my life
every day of the year

Contents

Preface

Many "smart" materials were invented more than 30 years ago, but their development and improvement over the past three decades has led to new, more varied uses of these adaptable materials. Smart materials have properties that can be altered by temperature, moisture, electric or magnetic fields, pH, and stress. They can change shape and color, become stronger, or produce voltage as a result of external stimuli. Magnetostrictive materials—materials that expand when exposed to a magnetic field—were discovered in the 1800s, but have uses today as varied as automotive sensors for collision avoidance and vibration dampening for surgical tools.

Materials such as brass, nickel-titanium, and gold-cadmium are shape memory alloys (SMAs), which alter their shape in response to changes in temperature and then return to their original shape. In the past 30 years, SMAs have seen widespread use in applications such as miniature surgical tools that can twist and pull when a small amount of heat is applied, wires that expand and contract for use by dentists in straightening teeth, cell phone antennas that resist breaking, stents and bone plates that must fuse together or expand in order to stay in place, and flexible eyeglass frames.

"Smart materials," which find widespread applications today, should be recognized along with the familiar metals, plastics, ceramics, composites, powder metals, and specialty-type and multifunctional materials (SMAs, microelectromechanical systems [MEMS], functionally graded materials [FGMs], and nanomaterials). Three decades ago, it appeared as though smart materials would be the next step in engineering design.

By using smart materials instead of adding mass, engineers can endow structures with built-in responses to a myriad of contingencies. In their various forms, these materials can perform as actuators, which can adapt to their environments by changing characteristics such as shape and stiffness, or as sensors, which provide the actuators with information about structural and environmental changes.

All aircraft, whether military commercial, or privately owned, should be and are inspected and maintained on a regular basis. Intelligent research programs benefit this segment of the aerospace industry. Sensors introduced into current and future aircraft designs will help technicians and mechanics to provide a more sophisticated inspection technique for maintenance and repairs.

Smart materials are beginning to play an important role in civil engineering designs for dams, bridges, highways, and buildings. An example of a smart materials project sponsored by the U.S. Navy and Army Corps of Engineers is to remove corrosion that has damaged a Navy pier and install sheets of composite materials containing sensors. These sensors embedded throughout a concrete and composite structure can act like nerves, sensing when areas of the structure begin to degrade and alerting maintenance engineers to the need for repairs.

The automotive industry is also eager to incorporate intelligent materials technology. Currently, researchers are working on an industry-sponsored project to develop smart car seats that can identify primary occupants and adapt to their preferences for height, leg room, back support, and so forth.

More profound changes are looming in automotive design based on smart materials. For instance, the technology to enable cars to tell owners how much air pressure tires have, when oil changes are needed, and other maintenance information exists as of today. However, this technology is expensive, but developing solid state and smart materials technologies will bring costs down. There is a shift in the culture toward consumers being given more information and taking more responsibility for knowing when maintenance and repairs are needed. The future of stereophonic sound will be altered by another facet of smart materials research. The development of ultrahigh-fidelity stereo speakers using piezoelectric actuators, which expand and contract in thousandths of a second in response to applied voltage, is aimed at turning whole house walls or car interiors into speakers by imbedding them with the tiny actuators. Thus, 50 years from now, we would not need to install separate speakers in our homes and cars in an attempt to achieve maximum musical effects. Our cars and houses will offer built-in surround sound.

A new technology for implants that may improve construction or repair of bones in the face, skull, and jaw has been developed by researchers from the American Dental Association Foundation and the National Institute of Standards and Technology. The new technology provides a method for making scaffolds for bone tissue. The scaffold is seeded with the patient's own cells and is formed with a cement paste made of minerals also found in natural bone. The paste is mixed with beads of a natural polymer (made from seaweed) filled with bone cells. The paste is shaped or injected into a bone cavity and then allowed to harden with the encapsulated cells dispersed throughout the structure. The natural polymer beads gradually dissolve when exposed to the body's fluids, creating a scaffold that is filled by the now released bone cells.

The cement, a calcium phosphate material, is strengthened by adding chitosan, a biopolymer extracted from crustacean shells. The implant is further reinforced to about the same strength as spongy natural bone by covering it with several layers of a biodegradable fiber mesh already used in clinical practice.

Bone cells are very smart. They can tell the difference between materials that are bioactive compared to bioinert polymers. The material is so designed to be similar to mineral in bone so that cells readily attach to the scaffold.

In addition to creating pores in the hardened cement, the natural polymer beads protect the cells during the 30 min required for the cement to harden. Future experiments will develop methods for improving the material's mechanical properties by using smaller encapsulating beads that biodegrade at a predictable rate.

Other developments include a smart bandage, which has an in situ programmable medical device to treat wounds and burns, a smart pill for vitamin and nutraceuticals formulation that prevents cognitive decline in aging and amphiphilic polymeric materials where the polymers can be used in the development of physiologically stable, nonleaking, nonimmunogenic, safe, and efficacious targeted drug delivery systems.

Recent advances in nondestructive evaluation (NDE) sensor technologies, health monitoring, and life prediction models are revolutionizing component inspections and life management, and will significantly improve the reliability and airworthiness of aerospace systems. Several key advances in NDE, health, monitoring, and prognostics programs include emerging structural health monitoring technologies for aerospace applications, high-temperature health monitoring, advances in NDE technologies for measurement of subsurface residual stresses, computational methods and advanced NDE techniques, and materials damage assessment techniques and prognosis models for prediction of remaining useful life. The advancements described provide the technologies required for lowering maintenance costs, integrated damage prognosis and life prediction, enhancing reliability and safety, and improving the performance and operational efficiency of current and future aerospace systems.

Another area of focus is smart machines. For example, moving composite manufacturers out of hand layup and open molding into more environmentally acceptable closed-molding alternatives where a reliable source of preforms is a necessity for volume production in closed-mold processes, such as resin transfer molding (RTM), vacuum-assisted RTM (VARTM), and vacuum-assisted resin infusion molding (VARIM).

Large preforms are facilitated by the large-scale preformer (LSP). The LSP's computerized spray-up process requires no human contact from the time the tooling enters the preform manufacturing cell until it is ready to be demolded: spray-up, compression, cure of the binder, cooling, and demold operations are all accomplished robotically according to instructions preprogrammed into the system's software. LSP is a smart system that tells the user where every part is in the process at all times.

The LSP's inaugural application, and the largest preform produced thus far, was for fuel containment vessels, which were resin transfer molded. These vessels have been installed beneath gasoline pumps in gas stations to prevent contamination of soil if there is a leak in the pump-to-tank plumbing.

Composites are now used in monitoring systems where they are combined with other materials and sensors. An example is a composite "shape sensing mat" for use with a metallic riser system. The flexible mat, which incorporates fiber optics, wraps around a steel riser, and enables operators to monitor excessive bending and fatigue life during riser deployment on a dynamic positioned oil drilling ship moored in the Gulf of Mexico as well as smart downhole coiled tubing complete with power and data transmission capabilities for drilling or workover applications and oilfield pipelines.

Finally, machines can process simple commands, but they are not very good at figuring out complex orders or unstated common sense. Command a machine to "paint the computer case before you box it," or "provide power to the computer before you switch it on" and the machine may box the product before the case is dry, or plug and then unplug a computer before switching it on. The meaning of the word "before" is quite different in these two cases. Ontologists, who study and understand the thought process, hope to end the age of stupid machines. Ontologists, who have created some of the most advanced logic systems, plan to share their leading-edge concepts on such comprehensive ideas as time, space, and process. The promise to cooperate eventually could lead to software that will enable machines to interpret and act on commands with near human common sense.

Efforts to equip machines with artificial intelligence capacity have, up to now, been relatively rudimentary. Software programs might, for instance, give machines used to make furniture considerable "understanding" of terms and frames of reference used in the furniture business. But such collected knowledge is of limited use, and human operation is necessary at virtually every step in the manufacturing process. A machine that incorporates expanded frames of reference of such "higher ontologies" as space and cost might be able to make design and shipping decisions virtually on its own. The future is bright and with optimism will enable the leading ontologists throughout the world to continue this promising work.

The book contains many of the examples and of the aspects that I have discussed and the contributions and efforts of 60 experts in the various fields of smart materials and smart material systems. I hope the readers will appreciate the work of a multitude of scientists, educators, researchers, academia, and industry people who have made considerable innovative progress in bringing forth their endeavors.

Mel Schwartz

Editor

Mel Schwartz has degrees in metallurgy and engineering management and has studied law, metallurgical engineering, and education. His professional experience extends over 51 years. He has served as a metallurgist in the U.S. Bureau of Mines; as a metallurgist and producibility engineer in the U.S. Chemical Corps; as a technical manufacturing manager, chief R&D laboratory, research manufacturing engineering, and senior staff engineer in Martin-Marietta Corporation for 16 years; as a program director, manager and director of manufacturing for R&D, and chief metals researcher in Rohr Corp for 8 years; and as a staff engineer and specification specialist, chief metals and metal processes, and manager of manufacturing technology in Sikorsky Aircraft for 21 years. After retirement, Schwartz served as a consultant for many companies including Intel and Foster Wheeler, and is currently editor of *SAMPE Journal of Advanced Materials*.

Schwartz's professional awards and honors include Inventor Achievement Awards and Inventor of the Year at Martin-Marietta; C. Adams Award and Lecture and R.D. Thomas Memorial Award from AWS; first recipient of the G. Lubin Award and an elected fellow from *SAMPE*; an elected fellow and Engineer of the Year in CT from ASM; and the Jud Hall Award from SME.

Schwartz's other professional activities include his appointment to ASM technical committees (joining, composites and technical books, ceramics); manuscript board of review, *Journal of Metals Engineering* as peer reviewer; the Institute of Metals as well as *Welding Journal* as peer reviewer; U.S. leader of International Institute of Welding (IIW) Commission I (brazing and related processes) for 20 years and leader of IIW Commission IV (electron beam/laser and other specialized processes) for 18 years.

Schwartz owns five patents, the notable one being aluminum dip brazing paste commercially sold as Alumibraze. He has authored 16 books and over 100 technical papers and articles and is an internationally known lecturer in Europe, the Far East, and Canada. He has taught in U.S. Universities (San Diego State University, Yale University), ASM institutes, McGraw-Hill seminars, and in-house company courses.

Contributors

Ilkka Aaltio
Department of Materials Science and Engineering
Helsinki University of Technology
TKK, Finland

Yasuyuki Agari
Osaka Municipal Technical Research Institute
Osaka, Japan

Monica Barney
Nitinol Devices and Components
Freemont, California

Michelle Bartning
Cordis Advanced Medical Ventures
Freemont, California

John Bell
Queensland University of Technology
Brisbane, Queensland, Australia

Yves Bellouard
Mechanical Engineering Department
Eindhoven University of Technology
Eindhoven, The Netherlands

Davide Bernardini
Department of Structural and Geotechnical Engineering
University of Rome
Rome, Italy

Mahesh C. Bhardwaj
The Ultran Group
State College, Pennsylvania

Gopinath Bhimarasetti
Department of Chemical Engineering
University of Louisville
Louisville, Kentucky

John A. Brunk
National Nuclear Security Agency's Kansas City Plant
Kansas City, Missouri

J. David Carlson
Lord Corporation
Cary, North Carolina

Cai Chao
Institute of High Performance Computing
Singapore

Somali Chaterji
Weldon School of Biomedical Engineering
Purdue University
West Lafayette, Indiana

Seung-Bok Choi
Department of Mechanical Engineering
Inha University
Inchon, South Korea

Lu Chun
Institute of High Performance Computing
Singapore

Alison Roberts Cohan
Pacific Whale Foundation
Maui, Hawaii

Santoshrupa Dumpala
Department of Chemical Engineering
University of Louisville
Louisville, Kentucky

Hiroshi Eda
Intelligent Systems Engineering Department
Ibaraki University
Hitachi, Japan

Arthur J. Epstein
Department of Physics and Department of Chemistry
Ohio State University
Columbus, Ohio

Frank E. Filisko
The University of Michigan
Ann Arbor, Michigan

Yao Fu
Silverbrook Research Pty. Ltd.
St. Balmain, Sydney, Australia

Igor Yu Galaev
Lund University
Lund, Sweden

Muralidhar K. Ghantasala
Department of Mechanical and Aeronautical Engineering
Western Michigan University
Kalamazoo, Michigan

Suresh Gubbala
Department of Chemical Engineering
University of Louisville
Louisville, Kentucky

J.A. Güemes
Department of Aeronautics
Universidad Politécnica de Madrid
Madrid, Spain

Young-Min Han
Department of Mechanical Engineering
Inha University
Inchon, South Korea

Simo-Pekka Hannula
Department of Materials Science and Engineering
Helsinki University of Technology
TKK, Finland

Gabor Harsanyi
Department of Electronics Technology
Budapest University of Technology and Economics
Budapest, Hungary

Erol C. Harvey
Industrial Research Institute
Swinburne University of Technology
Hawthorne, Melbourne, Australia

James A. Harvey
Under the Bridge Consulting Inc.
Corvallis, Oregon

S. Hébert
Laboratoire Crismat
CNRS ENSICaen
Caen, France

Oleg Heczko
Department of Materials Science and Engineering
Helsinki University of Technology
TKK, Finland

Zheng Hui
Institute of High Performance Computing
Singapore

Jan Van Humbeeck
Catholic University of Leuven
Leuven, Belgium

Yukio Ito
The Pennsylvania State University
University Park, Pennsylvania

Andreas Janshoff
Johannes Gutenberg Universität
Mainz, Germany

Claire Jarry
Bio Syntech Canada, Inc.
Laval, Quebec, Canada

Martin Kendig
Teledyne Scientific Company
Thousand Oaks, California

A.L. Kholkin
Department of Ceramics and Glass Engineering and Center for Research in Ceramic and Composite Materials
University of Aveiro
Aveiro, Portugal

L.A. Kholkine
Department of Electrical and Computer Engineering
University of Porto
Porto, Portugal

Sung Won Kim
Department of Pharmaceutics
Purdue University
West Lafayette, Indiana

Patrick J. Kinlen
Crosslink
Fenton, Missouri

D.A. Kiselev
Department of Ceramics and Glass Engineering and Center for Research in Ceramic and Composite Materials.
University of Aveiro
Aveiro, Portugal

L.C. Klein
Rutgers University
Piscataway, New Jersey

Silpa Kona
Department of Electrical and Computer Engineering
University of Louisville
Louisville, Kentucky

Tatsuro Kosaka
Graduate School of Engineering
Osaka City University
Osaka, Japan

Joseph Kost
Department of Chemical Engineering
Ben-Gurion University
Beer Sheva, Israel

David Kranbuehl
Chemistry and Applied Science Departments
College of William and Mary
Williamsburg, Virginia

Il Keun Kwon
Department of Pharmaceutics
Purdue University
West Lafayette, Indiana

Manuel Laso
Laboratory of Non-Metallic Materials
Universidad Politécnica de Madrid
Madrid, Spain

Henry C.H. Li
School of Aerospace, Mechanical and Manufacturing Engineering
RMIT University
Fishermans Bend, Victoria, Australia

Jun Jie Li
School of Chemical Engineering and Technology
Tianjin University
Tianjin, China

Yi Li
School of Materials Science and Engineering
Georgia Institute of Technology
Atlanta, Georgia

Yuanchang Liang
Department of Mechanical Engineering
University of Washington
Seattle, Washington

A. Maignan
Laboratoire Crismat
CNRS ENSICaen
Caen, France

Dora Klara Makai
Department of Electronics Technology
Budapest University of Technology and Economics
Budapest, Hungary

Arumugam Manthiram
Materials Science and Engineering Program
University of Texas at Austin
Austin, Texas

B. de Marneffe
Active Structures Laboratory
Université Libre de Bruxelles
Brussels, Belgium

B. Mattiasson
Lund University
Lund, Sweden

Praveen Meduri
Department of Chemical Engineering
University of Louisville
Louisville, Kentucky

J.M. Menéndez
Composite Materials Technology Department
Airbus
Getafe (Madrid), Spain

Joel S. Miller
Department of Chemistry
University of Utah
Salt Lake City, Utah

Kyoung-sik Moon
School of Materials Science and Engineering
Georgia Institute of Technology
Atlanta, Georgia

Nezih Mrad
Department of National Defence
National Defence Headquarters
Ottowa, Ontario, Canada

K. Muthumani
Structural Dynamics Laboratory
Structural Engineering Research Centre, CSIR
Chennai, India

Hirotaka Ojima
Intelligent Systems Engineering Department
Ibaraki University
Hitachi, Japan

Christopher O. Oriakhi
Imaging and Printing Supplies
Hewlett-Packard Company
Corvallis, Orgeon

Kinam Park
Department of Biomedical Engineering
Purdue University
West Lafayette, Indiana

Thomas J. Pence
Department of Mechanical Engineering
Michigan State University
East Lansing, Michigan

André Preumont
Active Structures Laboratory
Université Libre de Bruxelles
Brussels, Belgium

S. Padma Priya
University of Mysore
Mandya, India

Jesse E. Purdy
Department of Psychology
Southwestern University
Georgetown, Texas

Richard E. Riman
Department of Materials Science and Engineering
Rutgers University
Piscataway, New Jersey

Bruno Rocha
Department of Mechanical Engineering
Instituto Superior Técnico
Lisbon, Portugal

A. Safari
Rutgers University
Piscataway, New Jersey

Johannes Schweiger
German Aerospace Society
Ban Heilbrunn, Germany

Matthew S. Shive
Bio Syntech Canada, Inc.
Laval, Quebec, Canada

Carlos Silva
Laboratory of Aeronautics
Portuguese Air Force Academy
Sintra, Portugal

Outi Söderberg
Department of Materials Science and Engineering
Helsinki University of Technology
TKK, Finland

David A. Spivak
Department of Chemistry
Louisiana State University
Baton Rouge, Louisiana

R. Sreekala
Structural Dynamics Laboratory
Structural Engineering Research Centre, CSIR
Chennai, India

Claudia Steinem
Georg-August Universität
Göttingen, Germany

Wojciech L. Suchanek
Sawyer Technical Materials, LLC
Eastlake, Ohio

Afzal Suleman
Department of Mechanical Engineering
University of Victoria
Victoria, British Columbia, Canada

Mahendra K. Sunkara
Department of Chemical Engineering
University of Louisville
Louisville, Kentucky

Minoru Taya
Department of Mechanical Engineering
University of Washington
Seattle, Washington

C.O. Too
Intelligent Polymer Research Institute
University of Wollongong
Wollongong, New South Wales, Australia

Kenji Uchino
Department of Electrical Engineering
The Pennsylvania State University
University Park, Pennsylvania

and

Micromechatronics Inc State College, Pennsylvania

Kumar Vedantham
Department of Pharmaceutics
Purdue University
West Lafayette, Indiana

A.G. Vedeshwar
Department of Physics and Astrophysics
University of Delhi
New Delhi, India

Aleksandra M. Vinogradov
Department of Mechanical and Industrial Engineering
Montana State University
Bozeman, Montana

Gordon G. Wallace
Intelligent Polymer Research Institute
University of Wollongong
Wollongong, New South Wales, Australia

Susan Williams
Hewlett Packard Company
Corvallis, Oregon

C.P. Wong
School of Materials Science and Engineering
Georgia Institute of Technology
Atlanta, Georgia

Tan Xiaoming
Institute of High Performance Computing
Singapore

Fang Lian Yao
School of Chemical Engineering and Technology
Tianjin University
Tianjin, China

Kang De Yao
Research Institute of Polymeric Materials
Tianjin University
Tianjin, China

Myung Jin Yim
School of Materials Science and Engineering
Georgia Institute of Technology
Atlanta, Georgia

Yu Ji Yin
Research Institute of Polymeric Materials
Tianjin University
Tianjin, China

Zhuqing Zhang
Hewlett-Packard Company
Corvallis, Oregon

1

Residual Stress in Thin Films

A.G. Vedeshwar
University of Delhi

1.1 Introduction

One of the most common and yet an important issue in thin films is the persistence of stress after the film growth, which is termed as the internal residual stress. Although the attempt to measure and understand the stress in films started as early as 1877 [1], the topic continues to be interesting and important even today with various innovative analyses and measurement techniques. A few early authoritative reviews [2,3] on this topic have been excellent sources of information, which has assisted in the further growth of knowledge. The residual stress depends on various factors like the method of growth, growth parameters, nature of substrates, and starting material and then after growth film processing, etc. Development of stress in films could be both advantageous and disadvantageous. A very well-known, advantageous effect is the formation of self-assembled quantum dots in heteroepitaxial growths caused by the development of stress due to the lattice mismatch between film and the substrate [4,5]. Thus, the quantum dot growth can be tailored or controlled in a desirable way by a proper choice of the extent of film–substrate lattice mismatch. This has already led to the successful development of various devices based on quantum dots. However, residual stress may be quite undesirable and disadvantageous in the fabrication of almost all other planar thin film electronic devices as it could lead to failure of the device, or else can modify the device performance undesirably, e.g., microelectronic or microelectromechanical systems (MEMS). Subsequently, the study of the residual stress effect and its elimination or minimization becomes necessary for such applications. Most of the properties of the materials are affected by the stress, either externally applied or the internal residual stress. Therefore, the study of the residual stress effect in thin films could be analogous to the externally applied pressure effect. This fact facilitates the possibility of studying the pressure-dependent physical properties without any actual externally applied high-pressure experiments. As the band structure depends crucially on the structure, the electronic properties of the semiconductors can be expected to exhibit interesting stress–strain dependence. The study of residual stress therefore seems to be of significant importance, although the fundamental mechanisms for its origin are far from being fully understood. Therefore, I will try to present the necessary and sufficient information on the topic in this limited review.

1.2 Models and Theoretical Background

Almost all films have stress and conceptually the total stress can be thought of having three major contributions, external, thermal and the intrinsic. It can be represented as [6]

$$\sigma = \sigma_{\text{external}} + \sigma_{\text{thermal}} + \sigma_{\text{intrinsic}} \tag{1.1}$$

as often distinguished in the literature. Intrinsic stresses are developed during the deposition process of the films. Thermal stress arises due to the difference between the thermal expansion coefficients of the substrate and the film material, and occurs especially during the cooling phase. External stresses are mainly considered as due to, for instance, oxidation and incorporation of impurities etc. Most of the formulas used in experimental determinations of film stress σ_f are the modifications of an equation first derived by Stoney in 1909 [7], which is given by

$$\sigma_{\text{f}} = \frac{E_{\text{S}} d_{\text{S}}^2}{6R(1-\nu_{\text{S}})d_{\text{f}}} \tag{1.2}$$

where

E_S and ν_S are Young's modulus and the Poisson ratio of the substrate, respectively
d_S and d_f are the thickness of substrate and film, respectively
R is the radius of curvature of the bending in the film–substrate system caused by the stress

The measurement of R will determine σ_f. In the literature, the $(1 - \nu_S)$ correction is frequently omitted. Most of the optical interferometric methods devised to measure R (or change in R) are based on the dependence of fringe-width on R.

The origin of intrinsic stress is not yet fully explained by any universal mechanism. However, Buckel [8] suggested and classified the processes leading to the generation of internal stress into the categories as: (1) differences in thermal expansion coefficient of film and substrate, (2) differences in the lattice spacing of single crystal substrates and the film during epitaxial growth, (3) variation of the interatomic spacing with crystal size, (4) adsorption or incorporation of atoms from residual gases or chemical reactions, (5) recrystallization processes, (6) microscopic voids and special arrangements of dislocation, and (7) phase transformations.

Large intrinsic tensile stresses observed in metal films has been explained by a model [9], which suggests that the annealing and shrinkage of the layer buried behind the advancing surface of the growing film leads to stress. Hoffman [10] tried to estimate the intrinsic residual stress using a model in which the isolated grains coalesce and form grain boundaries. Grains of radius R are assumed to be separated by a distance δ (constrained relaxation length), of the order of interatomic distances. This model uses a grain-boundary potential (similar to Moorse potential) having a minimum at equilibrium atomic separation. The tensile or compressive strain developed in the film is attributed to the situation of growth whether the interatomic distance is smaller or greater than the equilibrium distance, respectively. The intrinsic residual stress deduced by this model is given by

$$\sigma_{\text{intrinsic}} = \frac{E_f \delta P}{(1 - \nu_f) 2R} \tag{1.3}$$

where

P is the packing density of the film
E_f and ν_f are Young's modulus and Poisson ratio of the film, respectively

It should be noted that the term $E_f/(1 - \nu_f)$ replaces E_f because of the assumed biaxial nature of the stress. The parameter δ can be determined from the interaction potential between the concerned atoms, which is often not so easy. Recently, Nix and Clemens [11] attempted to improve this model of coalescence mechanism using a thermodynamical approach. They derived a relationship for the maximum value of the stress as

$$\sigma_{\text{intrinsic}}(\max) = \left[\frac{E_f (2\gamma_g - \gamma_{gb})}{(1 - \nu_f) R} \right]^{1/} \tag{1.4}$$

where γ_g and γ_{gb} are the surface tensions per unit area of the isolated grains and the grain boundary, respectively. Again, here too, the determination of the parameter γ_{gb} is quite difficult in this improved model. Another very recent report [12] attempted to explain the intrinsic residual stress in metal films using a model based on the size-dependent phase transformations of the nanograins via the size dependence of the melting temperature of nanoparticles assumed to be present at the early stage of film deposition. Similarly, a model for compressive stress generation in polycrystalline films during thin film growth is reported [13] in which the driving force is an increase in the surface chemical potential caused by the deposition of atoms from the vapor. The increase in surface chemical potential induces atoms to flow into the grain boundary and hence creates a compressive stress in the film. A number of other stress models [14–18] derived from these ideas have been invoked in the literature to explain both tensile and compressive stress in thin films, which will not be discussed further. The magnitude of intrinsic stresses in films may also be related to the microstructure of films, i.e., morphology, texture, grain size, etc.

Thermal effects can also contribute to film stress significantly. Films grown at elevated substrate temperatures and then cooled to ambient temperature will develop thermal stress. Similarly, the films either thermally cycled or cooled to cryogenic temperatures will also develop thermal stress. A biaxial stress can appear in films grown on substrates having different thermal expansion coefficients than the film, at a temperature T higher or lower than the substrate or deposition temperature T_S. The strain ε_T developed under such conditions is given by [3]

$$\varepsilon_T = (\alpha_f - \alpha_S) \Delta T \tag{1.5}$$

where $\Delta T = (T - T_S)$, α_s, and α_f are the thermal expansion coefficients of the substrate and film, respectively. Using Hooke's law, the relation between thermal stress and the elastic strain in the absence of any plastic deformation in the film–substrate structure during temperature change can be obtained as

$$\sigma_f(T) = \frac{E_f (\alpha_s - \alpha_f) \Delta T}{(1 - \nu_f)} \tag{1.6}$$

A third kind, termed as extrinsic stress, can be distinguished apart from thermal stress and intrinsic stress in the overall residual stresses in films. Various molecules can penetrate the open voids and pores present in a not so fully dense film and are adsorbed on pore walls. The interaction forces between adsorbed species, especially between polar species such as water molecules, can modify residual stresses. Hirsch [19] proposed a model to explain the origin of extrinsic stress based on the adsorption of polar species on pore walls.

1.3 Experimental Methods for the Measurement of Residual Stress

A wide variety of methods have been explored for measuring the stress in thin films [6,20]. They may broadly be categorized on the basis of the physical phenomena used as the technique such as diffraction (x-ray or electron diffraction), optical interferometry, electrical or electromechanical etc. However, a convenient and appropriate method may be chosen depending on the requirements like measurements in situ or ex situ, types of film–substrate material, level of accuracy etc. The interferometric methods are normally used to measure the extent of bending or deformation of a thin substrate (cantilever) from its equilibrium position caused by the film having stress deposited on it. The fringe width is proportional to the extent of bending or deformation and hence the stress can be determined. The nature of the biaxial stress such as balanced, unbalanced, or the different type of stress components, etc. can also be analyzed by the shape of the fringe pattern as has been reported for sputtered Mo films [21]. A tensile stress will bend the substrate so that the film surface is concave and the opposite (convex) happens with the compressive stress. A thin cantilever is most widely used as the substrate. A variety of interferometric techniques such as laser reflective interferometry [22–24], laser spot scanning interferometry [25], optical leverage with a laser beam [26,27], etc. have been developed. However, it should be noted that the stresses measured by cantilever techniques and x-ray diffraction (XRD) need not necessarily the same because XRD gives the strain in the crystal lattice of the grain (intragranular) while the cantilever method determines the stress due to grain boundaries between the grains (intergranular). There are a few other innovative modern techniques developed to meet specific requirements such as indentation [28,29] and laser spallation [30], strain gauges [31–33], Raman measurements [34–36], and so on.

X-ray or electron diffraction is one of the simplest yet most powerful methods for microstructural characterization. Therefore, more emphasis will be given to this technique here. The details can be found from the standard and widely referred books [37,38]. XRD data can also be used to determine either uniform or nonuniform residual stress in the films. The effect of strain, both uniform and nonuniform, on the direction of x-ray reflection causes the diffraction peak in a definite way. The uniform strain shifts the peak on either side of the unstrained 2θ to the lower side for tensile and to the higher side for compression. If a grain is given a uniform tensile strain at right angles to the reflecting planes (parallel to substrate plane), their spacing becomes larger than unstrained equilibrium spacing d^0 and the corresponding diffraction line shifts to lower angles but does not otherwise change. This line shift for the given (*hkl*) can be used to calculate the strain ε_z present on (*hkl*) along the *z*-axis in the film as

$$\varepsilon_z = \frac{\Delta d_{hkl}}{d^0_{hkl}} = \frac{d_{hkl} - d^0_{hkl}}{d^0_{hkl}} \tag{1.7}$$

The right hand side of Equation 1.7 will be positive for tensile stress and negative for a compressive stress. However, it should be noted that normally the d^0 value for the desired (*hkl*) is taken from powder data (ASTM or JCPDS) of the concerned material because for a good approximation, powder may be treated as stress free or having a negligible stress. Thus, strain for all (*hkl*) peaks of considerable intensity in the diffractogram can be determined and their average $\Delta d/d$ can be calculated for a randomly oriented polycrystalline film. The macrostress thus can be determined by multiplying the average strain $\Delta d/d$ by the elastic constant of the material. However, quite often, there will be a single most intense peak (*hkl*) in the diffractogram for (*hkl*)-oriented parallel to the substrate plane and the determination of strain is only for that (*hkl*). Similarly, using electron diffraction data, the *d*-spacings can be determined [39] and hence the strain. For a biaxial stress along the *x*- and *y*-axes ($\sigma_x \neq 0$, $\sigma_y \neq 0$), $\sigma_z = 0$ we have strain normal to the film surface given by [37,38]

$$\varepsilon_z = -\frac{\nu_f}{E_f}(\sigma_x + \sigma_y) \tag{1.8}$$

The above equation represents a contraction if σ_x and σ_y are tensile. Therefore, a positive ε_z determined by Equation 1.7 would imply the compressive biaxial stress. A component of biaxial stress in any desired direction on the *xy*-plane can be measured by tilting the sample by an angle ψ, the details of which may be found in Refs. [37,38]. This method is known as the $\sin^2\psi$ method in the modern literature. A nonuniform microstrain causes a broadening of the corresponding diffraction line. The relationship between the broadening produced and the nonuniformity of the strain can be obtained by differentiating Bragg's law. We therefore obtain

$$b = \Delta 2\theta = -2\frac{\Delta d}{d}\tan\theta \tag{1.9}$$

where *b* is the extra broadening, over and above the instrumental breadth of the line, due to a fractional variation in plane spacing, $\Delta d/d$. This equation allows the variation in strain, $\Delta d/d$, to be calculated from the observed broadening. The maximum strain thus found can be multiplied by the elastic modulus E_f to give the maximum stress present.

Although considerable progress has taken place both in theory and in experimental methods, the residual stress data in the literature show wide variations for various materials measured by different methods. Broad trends are visible in the published results and can be summarized as

1. Most of the metal films exhibit invariably tensile stress with a large magnitude (200–1100 MPa or 10^6 N/m^2) and less dependence on the nature of the substrate.
2. Refractory metals and metals having high melting points generally exhibit higher stresses (600–1200 MPa) than softer and lower melting point metals (20–100 MPa).
3. Normally, both compressive and tensile stresses are observed in nonmetallic films frequently with smaller magnitudes (7–700 MPa).

The crossover between the two types of stress is also observed. Data on some metals can be found in Refs. [3,9]. Similarly, early investigations on dielectric and optical coatings have been described in Refs. [3,40,41]. A recent report shows the relationship between the preferred orientations and stresses for some metals [42]. The stress behavior in a very important technological material, GaN, has been studied [43–45] and other nitrides also have been investigated [46–48]. These are only a few informative examples. However, there are a large numbers of reports on various materials, specific studies, experimental methods, analyses etc., which are beyond the scope and purpose of this chapter.

1.4 Residual Stress-Dependent Optical Properties of Some Layered Structured Semiconductors

It may be interesting to understand the origin and behavior of residual stress in films; however, it is more important to know how the residual stress affects the various properties of the film. In this section, some selected data will be presented for illustrating some of the points so far discussed. The layered structured semiconductors (like metal dihalides, chalcogenides, etc.) were chosen for this purpose because of their anisotropy along and across the layer. In a layered structure, the constituent atoms are bound by a strong covalent bonding within the *ab*-plane forming a sheet-like structure (layer) and these sheets are stacked along the third direction (*c*-axis) by a weak van der Waals bonding between these sheets to form a three-dimensional solid, thus leading to anisotropy along parallel and perpendicular to the *c*-axis. All the films presented in this section were grown by thermal evaporation using a molybdenum boat onto a glass substrate at room temperature (RT) or liquid nitrogen temperature (LNT) at a vacuum of about 10^{-6} Torr and their stoichiometry was well characterized and confirmed by XPS and EDAX. XRD in Bragg–Brentano focusing geometry was used for structural and residual stress analyses.

Apart from the inherent anisotropy, many metal iodides prefer certain (*hkl*)-oriented growth and a great affinity for crystallinity (even at LNT) in thin films. A typical case of lead iodide is illustrated in Figure 1.1 where the structural and strain analyses are summarized. Without any exception, PbI_2 films grow with (00*l*)-orientations (that is (00*l*) planes parallel or the *c*-axis perpendicular to the substrate plane) under any growth conditions including the low temperature. This is manifested by the appearance of only (00*l*) peaks in XRD as shown in Figure 1.1 even for ultrathin films. Films thinner than 20 nm were analyzed by transmission electron microscopy (TEM). Only representative micrographs are shown in the insets (a) and (b) for a 5 nm thick film. The *d*-spacings (with the Miller indices indicated on the

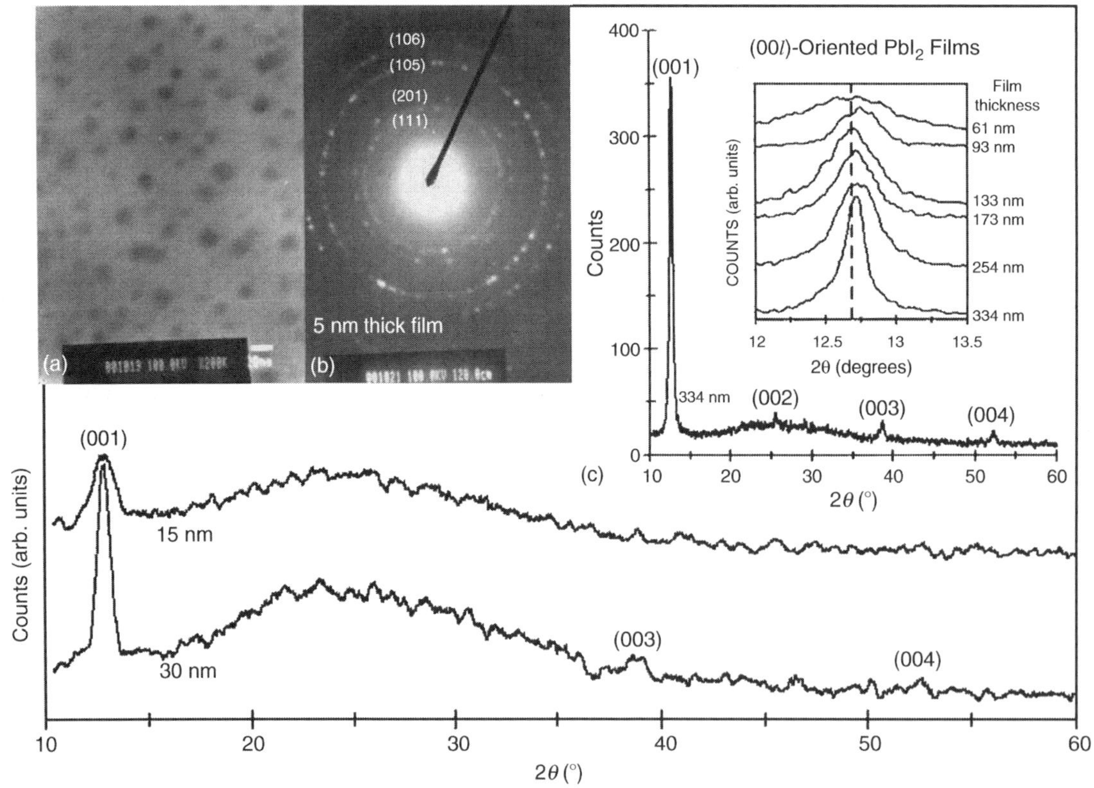

FIGURE 1.1 Typical XRD scan of (00*l*)-oriented ultrathin PbI_2 films for two film thicknesses as labeled in the figure. The inset (a) shows TEM for a 5 nm thick film and (b) is its electron diffraction. Inset (c) displays the XRD for very thick film and the inset within shows the shift of (001) peak from unstrained value (ASTM data indicated by the broken line) for various film thicknesses.

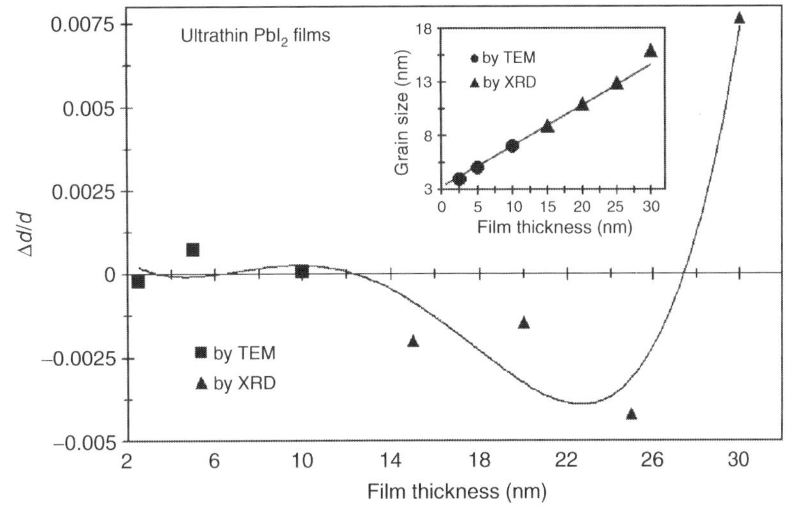

FIGURE 1.2 Development of strain ε_z in ultrathin PbI_2 films as a function of film thickness. Inset shows a linear grain size growth with film thickness. The method of strain and grain size determination is identified in the figure.

micrograph) calculated from the radius of the rings match quite well with ASTM card No. 07-0235 for 2H polytype of PbI_2. The unstrained value of d_{hkl} is taken from the above ASTM data for determining the strain. The (001) peak for thicker films are displayed in the inset within inset (c) depicting the strain, either tensile or compressive, in comparison with the ASTM value indicated by the broken vertical line. Strain is almost negligible in films thinner than 12 nm as depicted in Figure 1.2, which shows the development of strain with film thickness. The functional dependence of strain on grain size can be expected to be similar to that on film thickness because of the linear relationship between the two as shown in the inset. Qualitatively, this seems to be consistent with the coalescence model mentioned earlier as grains are well separated in ultrathin films (TEM in inset (a)) and start coalescing in thicker films, causing the biaxial stress responsible for the observed strain. However, quantitative estimation of the stress requires more information. Therefore, the development of biaxial stress can be expected, in general, for any film thicker than 10 nm. The important consequence and necessity of the strain analysis in ultrathin PbI_2 films is highlighted and demonstrated [49], especially while determining the quantum confinement contribution to the change in optical features. Results could lead to wrong conclusions if one ignores or overlooks the effect of strain on the optical properties. Results for thicker PbI_2 films are displayed in Figure 1.3 for various experimental parameters. The data shown are for films grown at RT and LNT. Interestingly, the behavior of strain is exactly opposite to each other in the two cases, most likely due to the different thermal stress contribution to the total stress. However, the variation of strain with film thickness, annealing temperature, and time as shown in Figure 1.3a, b, and c, respectively, may not be explained in a simple manner. Again, the qualitative explanations may have to be derived from the coalescence and grain boundary models with required modifications and inputs. For instance, the thickness dependence of strain requires the knowledge of grain size, grain density, and defect density, etc. for understanding the grain boundary structure. Switching from one type of biaxial stress to the other with film thickness could possibly be an explanation from their microstructure. That is, different σ_x and σ_y could lead to this kind of strain variation with film thickness. For example, strain is negligible for a 63 nm thick film, which indicates very small biaxial stress, possibly resulting from the opposite type of the two components. Similarly, it may be possible to account for the observed strain by the help of microstructural analyses. Annealing is normally carried out to facilitate the relaxation of the residual stress. However, it may not necessarily reduce the stress because the changing grain boundary, microstructure, defect density, etc. could increase the stress as well in some cases. Figure 1.3b displays the effect of annealing temperature on the strain. The wavy nature is difficult to explain just like the one with annealing duration as shown in Figure 1.3c. Nevertheless, data need to be analyzed more carefully using other possible macrostructural details obtained on the same sample for a better understanding of such behavior. The above discussion illustrates the typical intricacies inherent in the analyses of the residual stress origin. Such a detailed study is difficult to find in the literature.

The stress behavior differs for different materials. Another material, the zinc iodide has a different strain dependence on film thickness as shown in Figure 1.4. The strain in d_{104} is quite large compared to few other peaks of considerable intensity using XRD. ZnI_2 films are also not well oriented. Therefore, the average of strains of a few intense peaks is determined and plotted along with that of (104) in the figure. Although the comparative qualitative behavior of the two materials with film thickness is not very much different, their effect on optical properties is entirely different [50], which we will see little later. The increasing and saturating strain with the film thickness could simply be

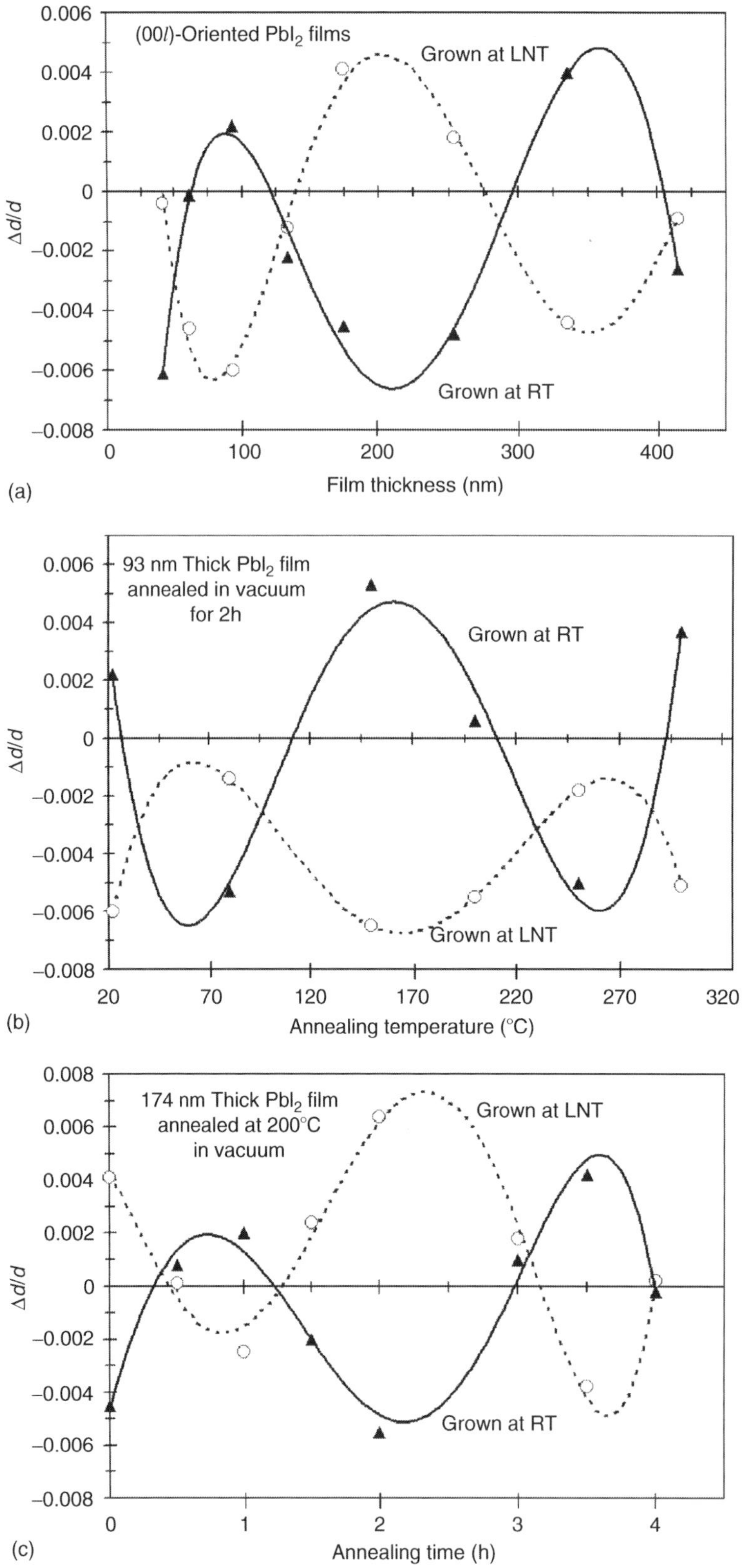

FIGURE 1.3 Residual strain in thicker PbI_2 films grown at RT and LNT as the function of (a) film thickness, (b) annealing temperature, and (c) annealing time.

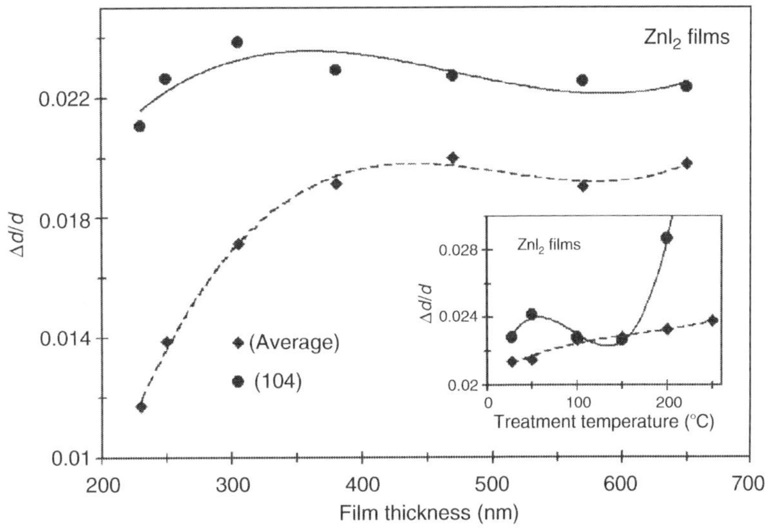

FIGURE 1.4 Average strain (average of few intense peaks of XRD) and strain in d_{104} as the function of thickness for ZnI_2 films. The inset shows the dependence of these quantities on annealing temperature (annealed in vacuum for 1 h).

understood to mimic the grain density behavior with film thickness [51,52]. The inset of Figure 1.4 shows the effect of vacuum annealing or heat treatment temperatures on the strain. Strain is found to increase with treatment in both the quantities, however, more dramatically for strain in (104). This increase could be due to the adsorption of residual gases present in the vacuum chamber maintained in the range 10^{-4}–10^{-5} Torr during annealing. The fluctuation in the vacuum level is due to degassing caused by the prolonged heating in the chamber. Another different type of strain dependence on film thickness is shown in Figure 1.5 for a (00*l*)-oriented CdI_2 film as well as (102) and (002)-oriented HgI_2 films. The typical thickness dependence of strain in CdI_2 is not surprising and can be explained by the increasing grain density with film thickness. Even the decreasing strain in CdI_2 films with annealing temperature as shown in the inset is very much like that one can ideally expect [53]. This opposite behavior compared to that of ZnI_2 in the inset of Figure 1.4 is mainly responsible for their basic microstructural difference, that is, CdI_2 films are completely oriented as opposed to randomly oriented ZnI_2 films. However, the increasing strain with substrate temperature as depicted in the inset is hard to explain by thermal stress alone because the substrate temperature

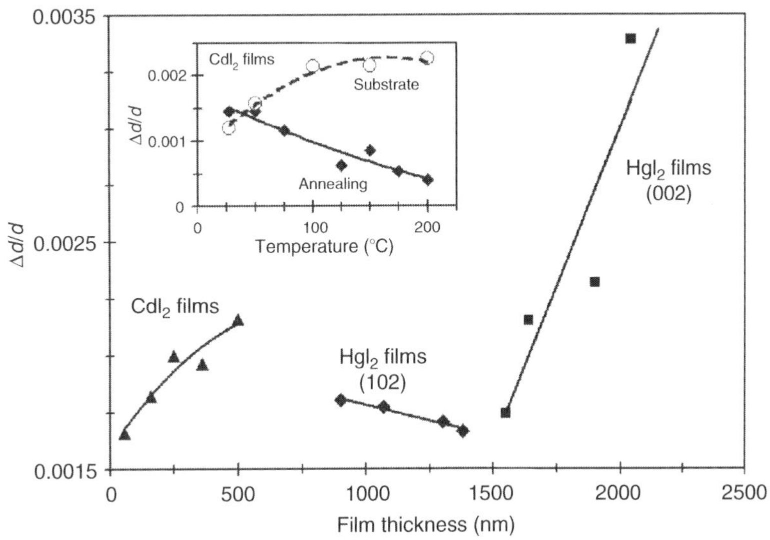

FIGURE 1.5 Film thickness dependence of strain for (00*l*)-oriented CdI_2 films and (102), (002)-oriented HgI_2 films. The inset shows the dependence of strain on annealing and substrate temperatures for (00*l*)-oriented CdI_2 films.

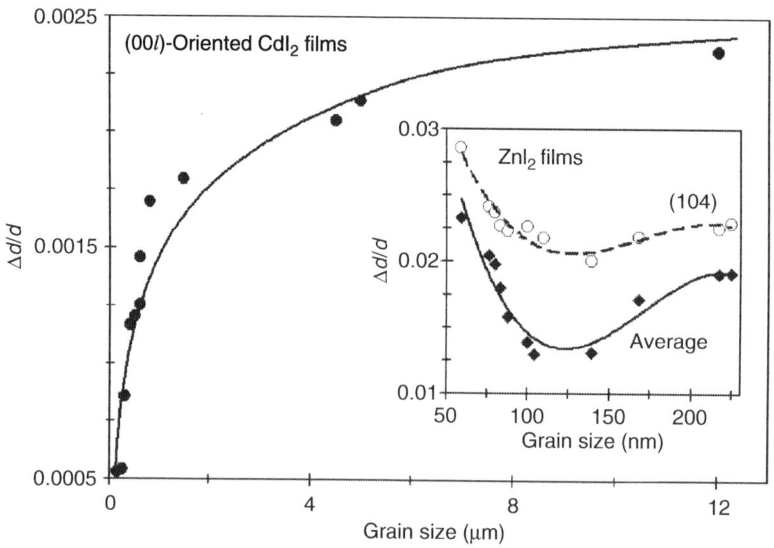

FIGURE 1.6 Dependence of strain on grain size for (00*l*)-oriented CdI_2 films. The inset shows the average strain and strain in d_{104} as a function of grain size for ZnI_2 films.

is known to affect the grain boundary structure as well. Another example in the figure, HgI_2, has a simple linear strain dependence on film thickness, although it differs substantially both in magnitude and slope for the two different growth orientations [54]. It will still be difficult to speculate or predict the origin and behavior of the stress with certainty in either completely or randomly oriented films of any material, especially, the dielectric and compound semiconductors etc. Nevertheless, the strain data will still be very useful in understanding its effect on the various physical properties of the material. The explicit dependence of strain on the grain size is illustrated for CdI_2 films in Figure 1.6 and for ZnI_2 films in the inset of the figure. The strain dependence of CdI_2 on the grain size seems to be qualitatively ideal and as expected according to the model of increasing grain density, especially in oriented films [53]. Normally, in such cases, the grain size varies linearly with film thickness and therefore a similar strain dependence on film thickness can be expected. Actually, this trend is indicated in the limited region of film thickness of CdI_2 in Figure 1.5. However, the grain size data shown in Figure 1.6 includes the results from film thickness, annealing, and substrate heating experiments. The more puzzling and strange result is the strain dependence of ZnI_2 shown in the inset where data for both average and the (104) plane are shown as discussed above. Any reasonable explanation of this trend is difficult. The unexpected behavior could well originate from the random orientation of grains and the minimum may be occurring for optimal close-packing grain size. It is only one more example illustrating the variation in diversified strain behavior. As we have seen so far from these limited examples, the origin and behavior of stress is fairly simple in some materials and are equally difficult and unpredictable for some other materials.

We will see how much essential and useful the stress analysis is in thin film study for understanding the affected physical properties. The above examples of strain for various materials will be extended to interpret the optical properties of a respective material. Analysis becomes fairly simple for oriented films like PbI_2. In *c*-axis oriented films, the strain along the *z*-axis resulting from the *xy*-plane biaxial stress can be treated as equivalent to the effect of uniaxial stress along the *z*-axis. This would certainly affect the band structure to a measurable extent in optical properties, at least in layer structured materials. Such an analysis is illustrated in Figure 1.7 for PbI_2. The various known excitonic peaks of PbI_2 observed in the optical absorption spectra are labeled in the inset of the figure for convenience. Both XRD and optical absorption measurements were done on the same sample for the reliability of the analyses. Excitonic peaks show a reasonably good and consistent linear dependence on strain, either positive or negative. Slightly different slopes of the peaks can be attributed to the extent of distortion of the atoms in the unit cell, which are responsible for the origin of the concerned peak. The further details of the analysis in view of the band structure of PbI_2 can be found in the recent report [55]. The important point here is the strain induced energy shift of the optical peaks and the magnitude of the shift compares with that due to quantum confinement. Therefore, the true picture of quantum confinement effect in PbI_2 is possible only if the strain contribution is separated [55]. Many such minute and important details could be obtained through a very careful stress analyses. Similar analyses for oriented CdI_2 and HgI_2 films are shown in Figure 1.8. Both these compounds did not show any excitonic peaks at RT and therefore, the optical band gap E_g was determined using standard method from their absorption spectra [53,54]. The strain dependence of E_g for CdI_2 is not linear and

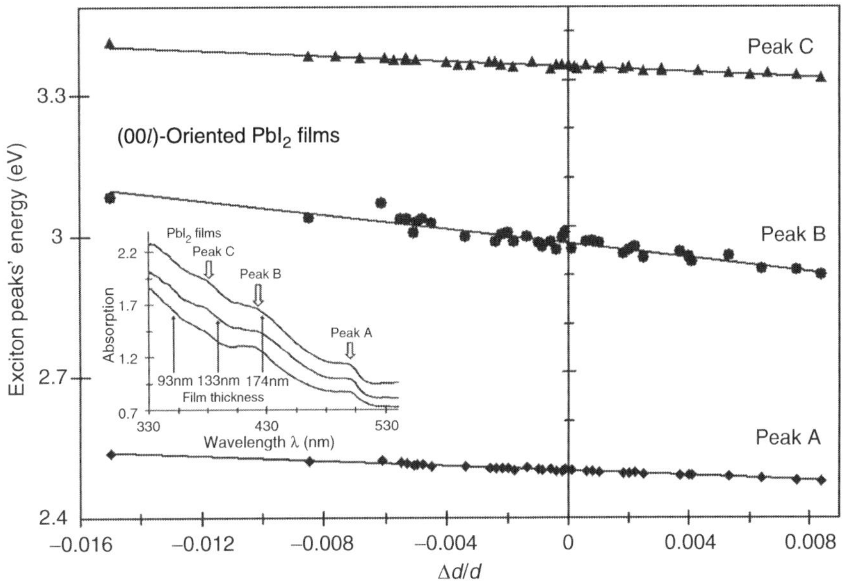

FIGURE 1.7 Strain dependence of three excitonic peaks of PbI_2 films. The three excitonic peaks are identified on the optical absorption spectra shown in the inset. The inset also depicts the shift in peaks for few film thicknesses due to strain.

unusual as can be seen from the figure. It suggests an important fact that there exists a threshold strain up to which E_g is slightly affected and decreases sharply beyond the threshold [56]. This threshold lies on the side of tensile strain and it increases with a small slope on the compressive side [57]. The compressive strain was produced in the sample by energetic ion irradiation [57]. The effect of strain on (102) and (002) oriented HgI_2 films is quite opposite both in the magnitude and the slope [54]. A few more examples of oriented growths are illustrated by XRD in Figure 1.9 for BaI_2 and SrI_2 films. As mentioned earlier, the presence of a very intense single peak indicates the (*hkl*)-orientation parallel to the substrate plane. Furthermore, such different orientations exhibit a different texture or morphology, which can be observed by scanning electron microscopy (SEM). Thus, both BaI_2 and SrI_2 films tend to grow in three different orientations under certain growth conditions as shown in Figure 1.9. For these two materials too, the strain dependence of E_g of respective orientation differs significantly from each other as shown in Figure 1.10. Here also, E_g was determined from their absorption spectra and no excitonic peak was visible at RT. These could be important results in understanding the band structure and probably can be verified by the density functional (DF) calculations. Finally, the

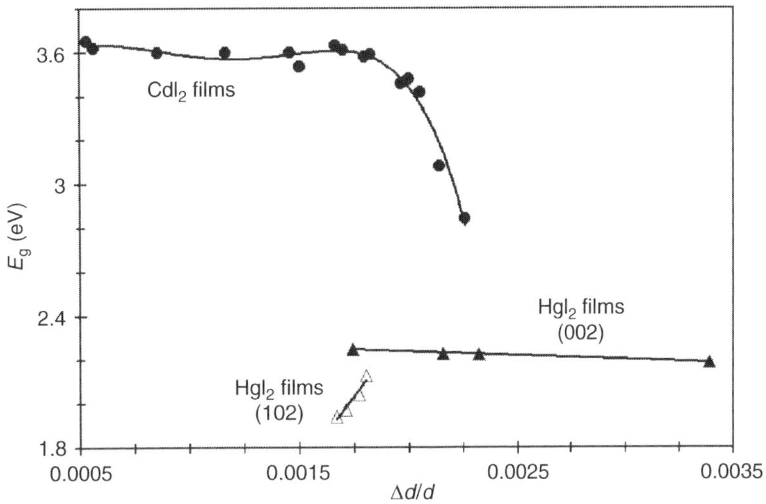

FIGURE 1.8 Strain-dependent optical band gap E_g for (00*l*)-oriented CdI_2 films and (102), (002)-oriented HgI_2 films.

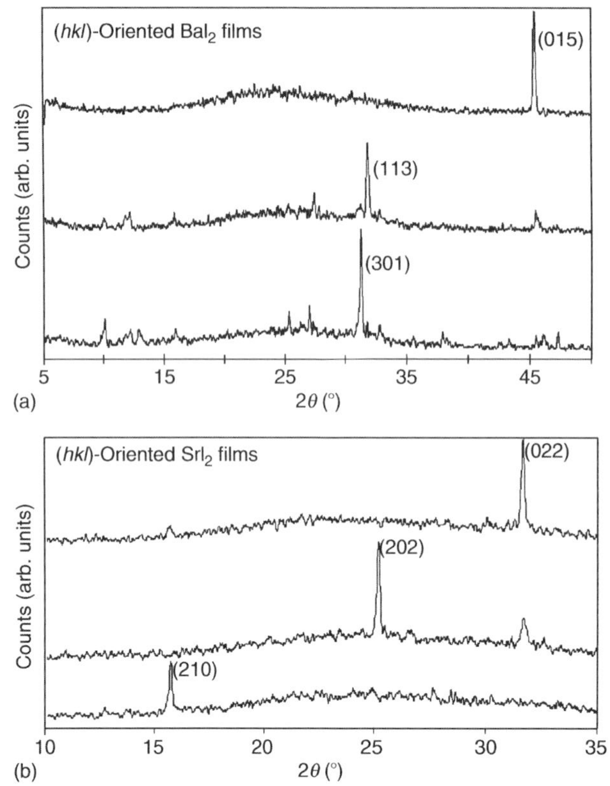

FIGURE 1.9 XRD scans depicting the various (*hkl*)-oriented film growth for (a) BaI_2 films and (b) SrI_2 films. The growth orientations are labeled on each scan.

example of randomly oriented polycrystalline ZnI_2 films is summarized in Figure 1.11. The absorption spectra of the various heat-treated samples are displayed in Figure 1.11a, which denotes various optical features. Peaks B and C are identified as excitonic peaks and the determination of E_g in such a case is described [50]. All these three quantities are to scale with the strain in d_{104} as shown in Figure 1.11b and do not show any systematic dependence on the average strain. In contrast, peak A does the opposite as shown in Figure 1.11c. This is the reason why the strain in (104) and the average strain are separately determined as shown

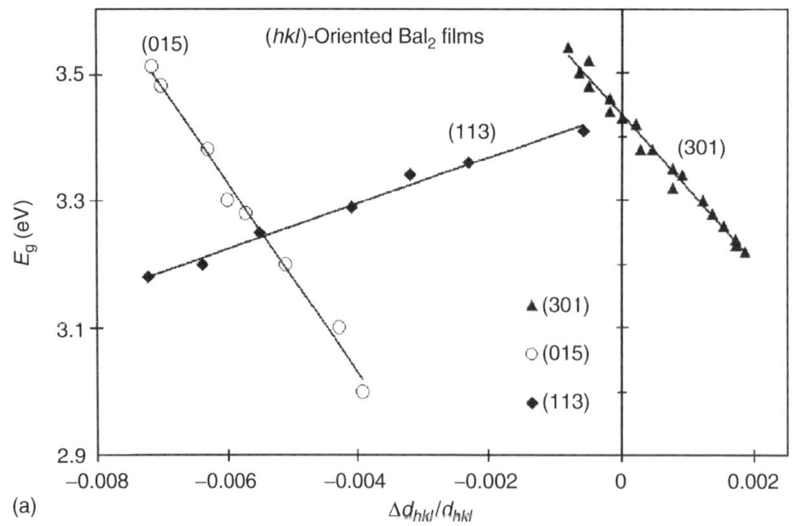

FIGURE 1.10 Strain dependence of optical band gap E_g for various (*hkl*)-orientations in (a) BaI_2 films

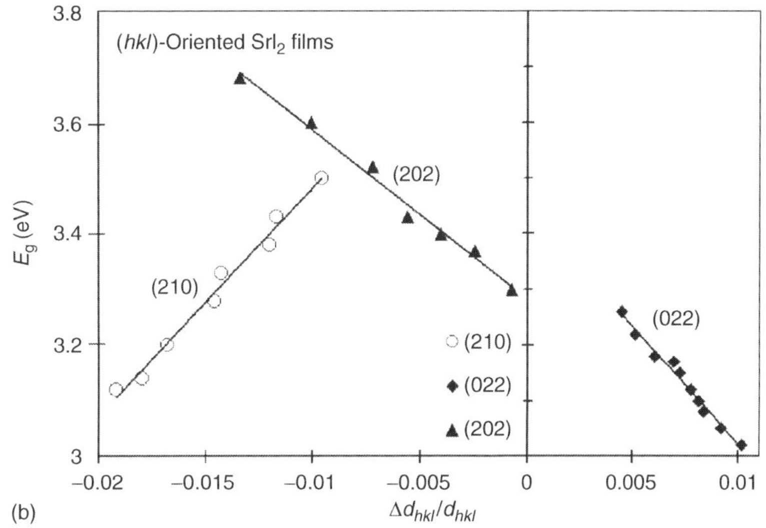

FIGURE 1.10 (continued) (b) SrI_2 films. The growth orientations are labeled in the figure.

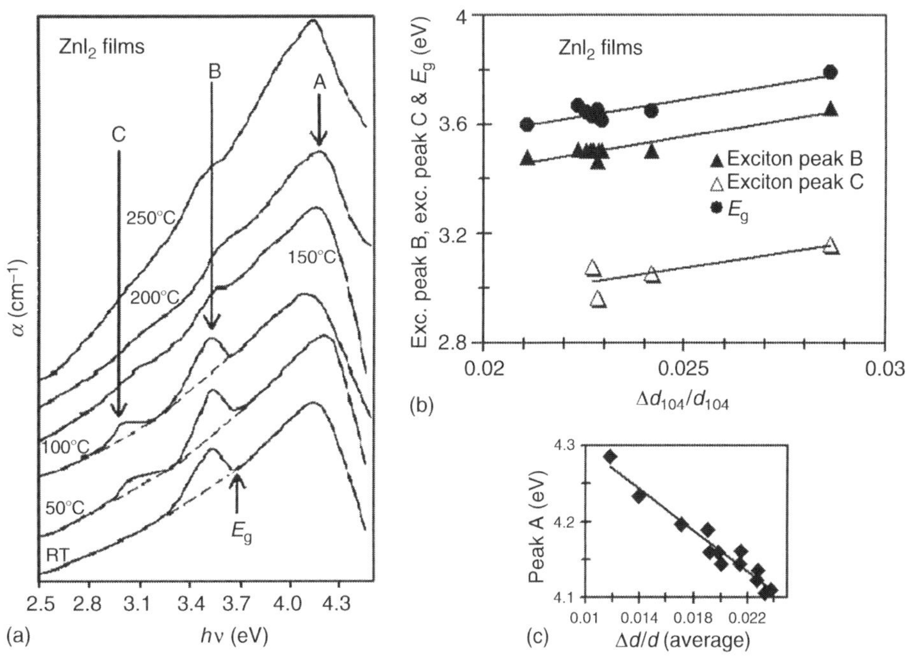

FIGURE 1.11 Summary of strain-dependent optical properties of ZnI_2 films: (a) Optical absorption spectra of 470 nm thick film for various annealing (in vacuum) temperatures as labeled on each spectrum. The various optical features like excitonic peaks B and C, peak A, and the band gap E_g are identified on the spectra, (b) dependence of peaks B, C, and E_g on the strain in d_{104}, and (c) the dependence of peak A on the average strain.

in Figure 1.4. Thus, the carefully carried out residual stress analysis along with other characterizations could help in better understanding the material.

These few diversified examples of optical properties analyzed in view of the stress demonstrate the importance of stress analyses in films. In fact, a number of other physical properties also depend on stress and their stress dependence can reveal many microscopic details. For instance, the Hall resistance of semiconductors, critical temperature of superconducting films, ferromagnetic resonance in ferromagnetic films, dielectric constant of dielectric films, and the likes can be expected to exhibit the strain dependence. Conversely, the shift in these relevant quantities may be used to estimate the order of magnitude of the stress present in the film provided the exact stress dependence of the concerned quantity is known. The linear stress dependence of an easily and conveniently measurable physical quantity of any material can qualify it as a smart material for strain gauge applications. The main reason for the wide data range of various physical quantities reported in the literature for the films of the same material could be due to ignoring the stress in the film. One such example is the optical band gap of a particular semiconductor, which mostly differs from one report to the other for thick films as much as in the range of 1 eV where stress is normally ignored. Without the knowledge of the influence of stress, it is difficult to determine the true intrinsic value of the band gap which is normally used as the reference for band structure calculations. Similar situations exist for a few of the other physical quantities as well as for which stress effect is quite significant.

1.5 Summary and Future Direction

It is a hard truth that any thin film is not completely free of residual stress. This fact makes us learn to live with stress by understanding its nature and behavior. Therefore, we practically we cannot afford to avoid or ignore the stress completely. An appropriate, affordable, and convenient method of stress analysis can be adopted out of several available choices from the literature. The information about the stress is already there in many of the routine structural characterization techniques, which have to be extracted by careful analyses as illustrated in this brief account. Indeed, a truly universal and quantitative theory for the internal residual stress in films is perhaps yet to be developed if there could be any. The entire data in the literature and their different possible explanations make it doubtful that such a unique theory valid for different film materials and different methods of deposition will ever emerge. The main stumbling blocks for such an attempt are some of the inherent problems and uncertainties of films and their growth processes. Film stoichiometry (or atomic composition), structural defects, grain-boundary structure, film–substrate interfacial properties etc. are some of the difficulties in describing them in terms of macroscopic elastic models. Nevertheless, the study of residual stress in thin films will continue to be important and necessary for exploiting them in various applications even if the origin cannot be traced convincingly. Sooner or later, the residual stress analyses in films will draw more attention and be consolidated in finding further importance and priority in materials research.

References

1. E. J. Mills, On electrostriction, *Proc. Roy. Soc. London* 26, 504–512, 1877.
2. R. W. Hoffman, The mechanical properties of thin films, in *The Use of Thin Films in Physical Investigations*, J. C. Anderson (ed.), Academic Press, New York, p. 261, 1966.
3. R. W. Hoffman, Mechanical properties of thin condensed films, in *Physics of Thin Films*, Vol. 3, G. Hass and R. E. Thun (eds.), Academic Press, New York, p. 211, 1966.
4. W. J. Parak, L. Manna, F. C. Simmel, D. Gerian, and P. Alivisatos, Quantum dots, in *Nanoparticles: From Theory to Application*, G. Schmid (ed.), Wiley-VCH Verlag, Weinheim, p. 27, 2004.
5. R. Notzel, Self-organized growth of quantum-dot structures, *Semicond. Sci. Technol.* 11, 1365–1379, 1996.
6. D. S. Campbell, Mechanical properties of thin films, in *Handbook of Thin Film Technology*, L. I. Maissel and R. Glang (eds.), McGraw-Hill, New York, 1970.
7. G. G. Stoney, The tension of metallic films deposited by electrolysis, *Proc. Roy. Soc. London* A82, 172–175, 1909.
8. W. Buckel, Internal stresses, *J. Vac. Sci. Tech.* 6, 606–609, 1969.
9. E. Klokholm and B. S. Berry, Intrinsic stress in evaporated metal films, *J. Electrochem. Soc.* 115, 823–826, 1968.
10. R. W. Hoffman, Stresses in thin films: The relevance of grain boundaries and impurities, *Thin Solid Films* 34, 185–190, 1976.
11. W. D. Nix and B. M. Clemens, Crystallite coalescence: A mechanism for intrinsic tensile stresses in thin films, *J. Mater. Res.* 14, 3467–3473, 1999.
12. G. Guisbiers, O. Van Overschelde, and M. Wautelet, Nanoparticulate origin of intrinsic residual stress in thin films, *Acta Mater.* 55, 3541–3546, 2007.
13. E. Chason, B. W. Sheldon, L. B. Freund, J. A. Floro, and S. J. Hearne, Origin of compressive residual stress in polycrystalline thin films, *Phys. Rev. Lett.* 88, 156103(1–4), 2002.
14. H. Windischmann, Intrinsic stress in sputter-deposited thin films, *Crit. Rev. Solid State Mater. Sci.* 17, 547–596, 1992.
15. Y. Pauleau, Generation and evolution of residual stresses in physical vapour-deposited thin films, *Vacuum* 61, 175–181, 2001.
16. C. A. Davis, A simple model for the formation of compressive stress in thin films by ion bombardment, *Thin Solid Films* 226, 30–34, 1993.
17. M. Itoh, M. Hori, and S. Nadahara, The origin of stress in sputter-deposited tungsten films for x-ray masks, *J. Vac. Sci. Technol.* B9, 149–153, 1991.
18. J. -D. Kamminga, T. H. de Keijser, R. Delhez, and E. J. Mittemeijer, A model for stress in thin layers induced by misfitting particles: An origin for growth stress, *Thin Solid Films* 317, 169–172, 1998.
19. E. H. Hirsch, Stress in porous thin films through adsorption of polar molecules, *J. Phys.* D13, 2081–2094, 1980.
20. D. A. Glecker and S. I. Shah, *Handbook of Thin Film Process Technology*, IOP, Bristol and Philadelphia, 1995.
21. R. E. Cuthrell, D. M. Mattox, C. R. Peeples, P. L. Dreike, and K. P. Lamppa, Residual stress anisotropy, stress control, and resistivity in post cathode magnetron sputter deposited molybdenum films, *J. Vac. Sci. Technol.* A6, 2914–2920, 1988.
22. C. H. Wu, W. H. Weber, T. J. Poter, and M. A. Tamor, Laser reflective interferometry for in situ monitoring of diamond film growth by chemical vapor deposition, *J. Appl. Phys.* 73, 2977–2982, 1993.
23. M. Sternheim, W. van Gelder, and A. W. Hartman, A laser interferometer system to monitor dry etching of patterned silicon, *J. Electrochem. Soc.* 130, 655–658, 1983.

24. J. Kempf, M. Nonnenmacher, and H. H. Wagner, Electron and ion beam induced heating effects in solids measured by laser interferometry, *Appl. Phys.* A56, 385–390, 1993.
25. J. Kempf, Optical in situ sputter rate measurements during ion sputtering, *Surf. Interf. Anal.* 4, 116–119, 1982.
26. G. J. Leusink, T. G. M. Oosterlaken, G. C. A. M. Janssen, and S. Radelaar, In situ sensitive measurement of stress in thin films, *Rev. Sci. Instrum.* 63, 3143–3146, 1992.
27. T. Aoki, Y. Nishikawa, and S. Kato, An improved optical lever technique for measuring film stress, *Jpn. J. Appl. Phys.* 28, 299–300, 1989.
28. S. Suresh and A. E. Giannakopoulos, A new method for estimating residual stresses by instrumented sharp indentation, *Acta Mater.* 46, 5755–5767, 1998.
29. M. Herrmann, N. Schwarzer, F. Richter, S. Frühauf, and S.E. Schulz, Determination of Young's modulus and yield stress of porous low-k materials by nanoindentation, *Surf. Coat. Technol.* 201, 4305–4310, 2006.
30. R. Ikeda, T. Uchiyama, H. Cho, T. Ogawa, and M. Takemoto, An advanced method for measuring the residual stress of deposited film utilizing laser spallation technique, *Sci. Technol. Adv. Mater.* 7, 90–96, 2006.
31. K. T. V. Grattan, B. T. Meggitt (eds.), *Optical Fiber Sensor Technology*, Chapman and Hall, London, 1995.
32. S. M. M. Quintero, W. G. Quirino, A. L. C. Triques, L. C. G. Valente, A. M. B. Braga, C. A. Achete, and M. Cremona, Thin film stress measurement by fiber optic strain gage, *Thin Solid Films* 494, 141–145, 2006.
33. A. Rizzo, L. Capodieci, D. Rizzo, and U. Galietti, Low cost technique for measuring in situ strain of nanostructures, *Mater. Sci. Eng.* C 25, 820–825, 2005.
34. J. K. Shin, C. S. Lee, K. R. Lee, and K. Y. Eun, Effect of residual stress on the Raman-spectrum analysis of tetrahedral amorphous carbon films, *Appl. Phys. Lett.* 78, 631–633, 2001.
35. A. C. Ferrari and J. Robertson, Interpretation of Raman spectra of disordered and amorphous carbon, *Phys. Rev.* B61, 14095–14107, 2000.
36. J. Zhu, J. Han, A. Liu, S. Meng, and C. Jiang, Mechanical properties and Raman characterization of amorphous diamond films as a function of film thickness, *Surf. Coat. Technol.* 201, 6667–6669, 2007.
37. B. D. Cullity, *Elements of X-ray Diffraction*, 2nd edn., Addison-Wesley, Reading, MA, 1978.
38. H. P. Klug and L. E. Alexander, *X-ray Diffraction Procedures for Polycrystalline and Amorphous Materials*, 2nd edn., John Wiley, New York, 1974.
39. D. Shindo and T. Oikawa, *Analytical Electron Microscopy for Materials Science*, Springer-Verlag, Tokyo, 2002.
40. A. E. Ennos, Stresses developed in optical film coatings, *Appl. Opt.* 5, 51–61, 1966.
41. H. K. Pulker, *Coatings on Glass*, Elsevier, Amsterdam, 1984.
42. J. Zhang, K. Xu, and V. Ji, Dependence of stresses on grain orientations in thin polycrystalline films on substrates: an explanation of the relationship between preferred orientations and stresses, *Appl. Surf. Sci.* 180, 1–5, 2001.
43. C. Kisielowski, J. Kruger, S. Ruvimov, T. Suski, J. W. Ager III, E. Jones, Z. Liliental-Weber, M. Rubin, E. R. Weber, M. D. Bremser, and R. F. Davis, Strain-related phenomena in GaN thin films, *Phys. Rev.* B54, 17745–17753, 1996.
44. R. F. Davis, S. Einfeldt, E. A. Preble, A. M. Roskowski, Z. J. Reitmeier, and P. Q. Miraglia, Gallium nitride and related materials: challenges in materials processing, *Acta Mater.* 51, 5961–5979, 2003.
45. E. Eiper, A. Hofmann, J. W. Gerlach, B. Rauschenbach, and J. Keckes, Anisotropic intrinsic and extrinsic stresses in epitaxial wurtzitic GaN thin film on γ-$LiAlO_2$ (1 0 0), *J. Cryst. Growth* 284, 561–566, 2005.
46. C. Sarioglu, The effect of anisotropy on residual stress values and modification of Serruys approach to residual stress calculations for coatings such as TiN, ZrN and HfN, *Surf. Coat. Technol.* 201, 707–717, 2006.
47. M. A. Moram, Z. H. Barber, C. J. Humphreys, T. B. Joyce, and P. R. Chalker, Young's modulus, Poisson's ratio, and residual stress and strain in (111)-oriented scandium nitride thin films on silicon, *J. Appl. Phys.* 100, 023514(1–6), 2006.
48. W. J. Zhang and S. Matsumoto, Investigations of crystallinity and residual stress of cubic boron nitride films by Raman spectroscopy, *Phys. Rev.* B63, 073201(1–4), 2001.
49. V. Gulia, A. G. Vedeshwar, and N. C. Mehra, Quantum dot-like behavior of ultrathin PbI_2 films, *Acta Mater.* 54, 3899–3905, 2006.
50. A. G. Vedeshwar and P. Tyagi, Excitonic absorption in ZnI_2 films, *J. Appl. Phys.* 100, 083522(1–6), 2006.
51. P. Tyagi and A. G. Vedeshwar, Optical properties of ZnI_2 films, *Phys. Rev.* B64, 245406(1–7), 2001.
52. P. Tyagi and A. G. Vedeshwar, Residual stress dependent optical properties of ZnI_2 films, *Phys. Stat. Sol.* (*a*) 191, 633–642, 2002.
53. P. Tyagi and A. G. Vedeshwar, Grain size dependent optical properties of CdI_2 films, *Eur. Phys. J.* AP19, 3–13, 2002.
54. P. Tyagi and A. G. Vedeshwar, Anisotropic optical band gap of (102) and (002) oriented films of red HgI_2, *Phys. Rev.* B63, 245315(1–6), 2001.
55. V. Gulia and A. G. Vedeshwar, Optical properties of PbI_2 films: Quantum confinement and residual stress effect, *Phys. Rev.* B75, 045409(1–6), 2007.
56. P. Tyagi and A. G. Vedeshwar, Effect of residual stress on the Optical properties of CdI_2 films, *Phys. Rev.* B66, 075422(1–8), 2002.
57. R. S. Rawat, P. Arun, A. G. Vedeshwar, P. Lee, and S. Lee, Effect of energetic ion irradiation on CdI_2 films, *J. Appl. Phys.* 95, 7725–7730, 2004.

2

Intelligent Synthesis of Smart Ceramic Materials

Wojciech L. Suchanek
Sawyer Technical Materials

Richard E. Riman
Rutgers University

2.1 Introduction

Ceramic materials are being used in nearly all advanced materials applications either as pure ceramics or as ceramic components of composites or devices with metallic or organic constituents. Growing demand from various industries to synthesize smart materials with more and more sophisticated features requires the use of smart starting materials, with very well-defined and controlled specific properties, such as size, shape (morphology), chemical composition and defect structure, surface functionalization, etc. This in turn requires use of advanced synthetic methods, which, in addition to producing superior product, should be also inexpensive and environmentally friendly. The low-temperature hydrothermal synthesis meets all of the above requirements; moreover, it is very well suited to produce both smart ceramic starting materials, such as powders or fibers, and already shaped products, such as bulk ceramic pieces, films or coatings and single crystals [1–6]. The number of scientific papers on hydrothermal synthesis of materials has been steadily increasing since 1989 and is currently on the level of hundreds per year [2,3]. The number of papers on hydrothermal synthesis of ceramic powders alone has nearly quadrupled between 2000 and 2004. Consequently, a large family of smart ceramics, primarily powders and coatings, has emerged that can be prepared under very mild hydrothermal conditions ($T < 200°C$, $P < 1.5$ MPa). This has generated lots of commercial interest in hydrothermal technology [1–6].

2.2 Hydrothermal Synthesis of Smart Ceramic Materials—An Overview

2.2.1 Process Definition

Hydrothermal synthesis is a process that utilizes single or heterogeneous phase reactions in aqueous media at elevated temperature ($T > 25°C$) and pressure ($P > 100$ kPa) to crystallize ceramic materials directly from solution [1,6]. Reactants used in hydrothermal synthesis are usually called precursors, which are administered in forms of solutions, gels, or suspensions. However, hydrothermal growth of single crystals requires in most cases use of solid nutrient, which recrystallizes during the growth process. Mineralizers are organic or inorganic additives that are used to control pH or enhance solubility. Other additives, also organic or inorganic, are used to serve other functions, such as controlling crystal morphology, chemical composition, particle dispersion, etc. Syntheses are usually conducted at autogeneous pressure, which corresponds to the saturated vapor pressure of the solution at the specified temperature and composition of the hydrothermal solution. Higher pressures up to over 500 MPa and temperatures over 1000°C [5] may be necessary to facilitate reactant dissolution and growth of certain types of ceramic materials, and are usually applied in single crystal growth. Nevertheless, mild conditions are preferred for commercial processes where temperatures are less than 350°C and pressures less than approximately 50 MPa. Intensive research has led to a better understanding of hydrothermal chemistry, which has significantly reduced the reaction time, temperature, and pressure for hydrothermal crystallization of ceramic materials, predominantly powders and coatings ($T < 200°C$, $P < 1.5\,MPa$) [2,4,5,7]. This breakthrough has made hydrothermal synthesis more economical since processes can be engineered using cost-effective and proven pressure reactor technology and methodologies already established by the chemical process industry.

2.2.2 Merits of Hydrothermal Synthesis of Ceramics

Hydrothermal synthesis offers many advantages over conventional and nonconventional ceramic synthetic methods. All forms of ceramics can be prepared with hydrothermal synthesis, such as powders, fibers, and single crystals, monolithic ceramic bodies, and coatings on metals, polymers, and ceramics (Figure 2.1). From the standpoint of ceramic powder production, there are far fewer time- and energy-consuming processing steps since high-temperature calcination, mixing, and milling steps are either not necessary or minimized. Moreover, the ability to precipitate already crystallized powders directly from solution regulates the rate and uniformity of nucleation, growth, and aging, which results in improved control of size and morphology of crystallites and significantly reduced aggregation levels, that is not possible with many other synthesis processes [8]. Figure 2.2 shows several examples of the varieties of morphologies and particle sizes possible with hydrothermal processing. The elimination or reduction of aggregates combined with narrow particle size distributions in the starting powders leads to optimized and

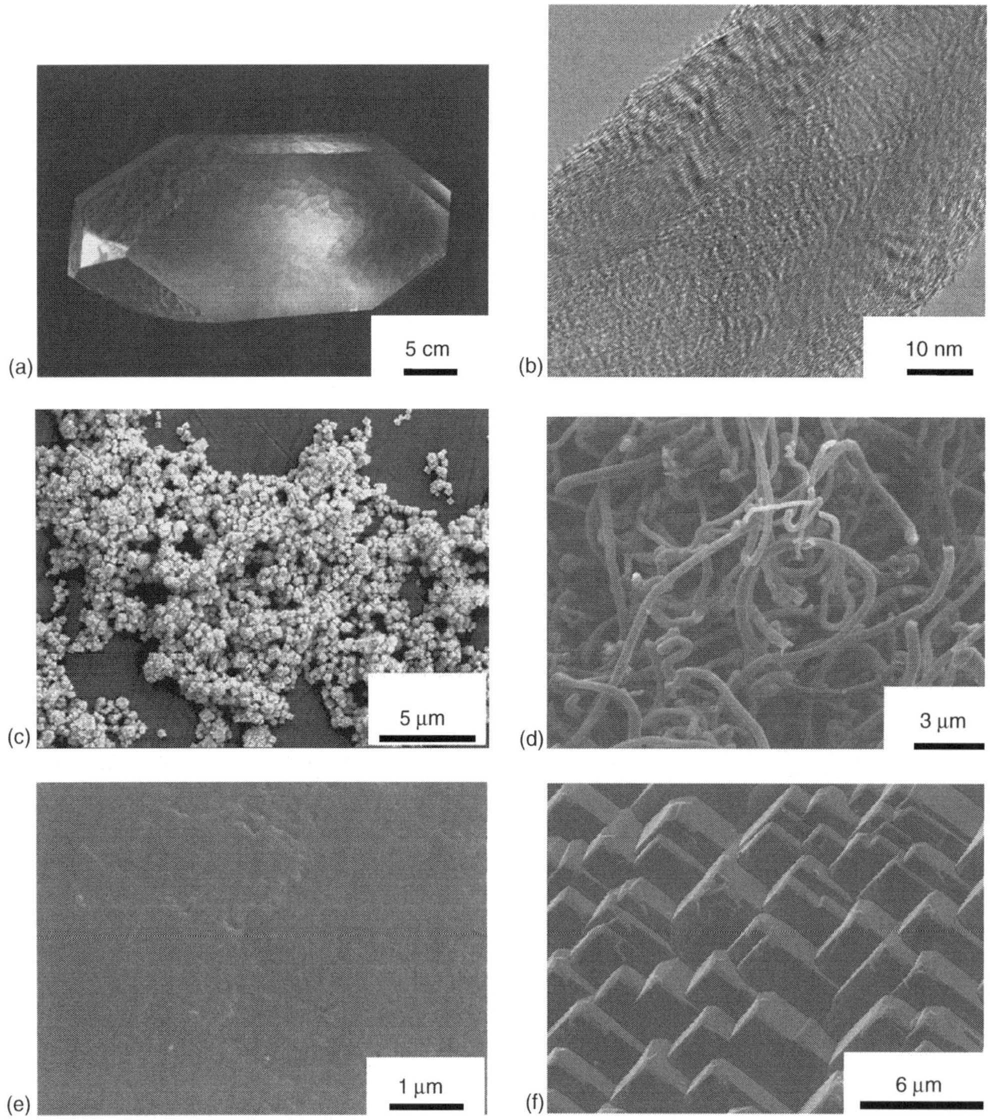

FIGURE 2.1 Examples of various forms of ceramic materials synthesized hydrothermally: (a) α-quartz single crystal, (b) carbon nanotube, (c) PZT powder, (d) carbon fibers, (e) epitaxial $KNbO_3$ film on $SrTiO_3$ wafer, and (f) epitaxial $KNbO_3$ film on $LiTaO_3$ wafer.

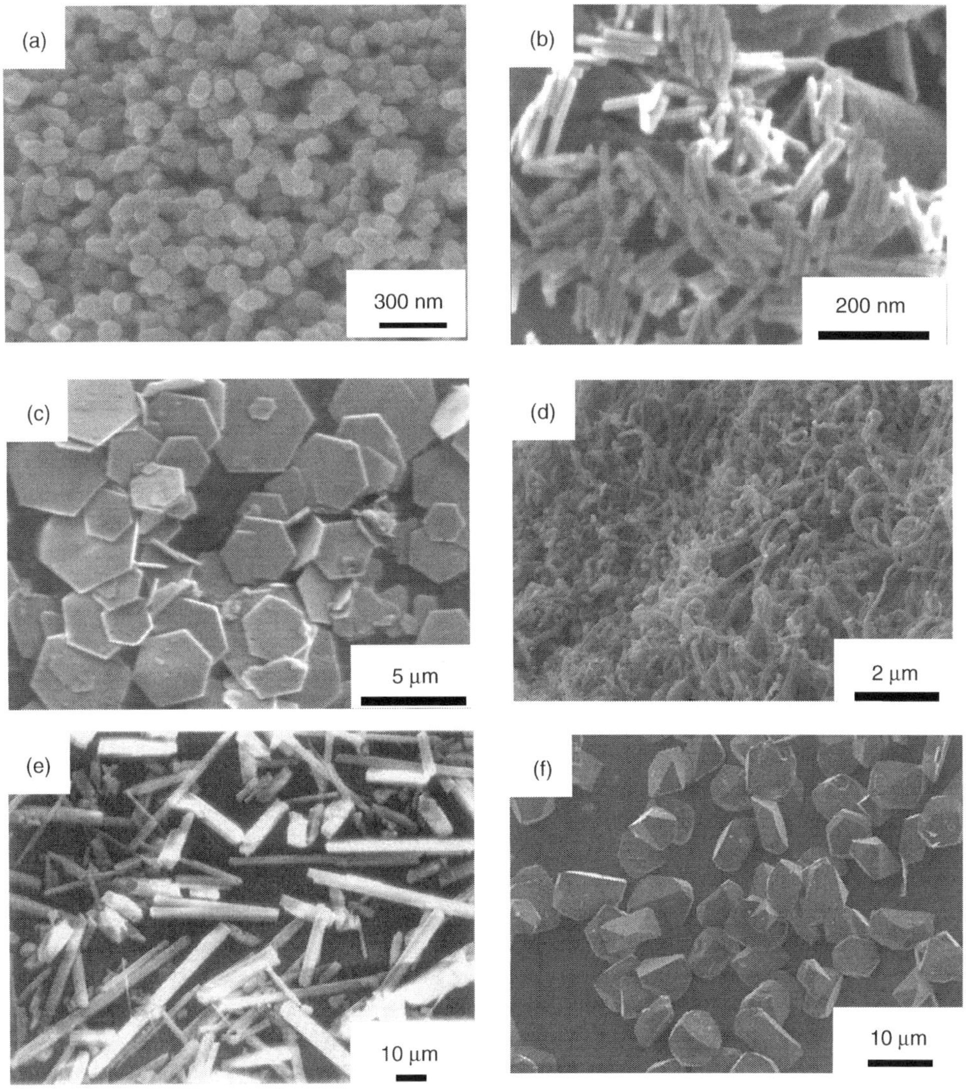

FIGURE 2.2 Examples of various sizes and morphologies of ceramic powders synthesized hydrothermally: (a) equiaxed nanosized ZnO, (b) nanosized hydroxyapatite needles, (c) $LiMnO_3$ platelets, (d) carbon nanotubes, 50–100 nm in diameter, several microns in length, (e) hydroxyapatite whiskers, and (f) equiaxed α-Al_2O_3.

reproducible properties of ceramics because of better microstructure control. From the standpoint of thin films (coatings), other methods such as physical vapor deposition, chemical vapor deposition, and sol–gel suffer from the disadvantage that they all require high-temperature processing to crystallize the ceramic phase. Thermally induced defects result, such as cracking, peeling, undesired reactions between the substrate and coating, and decomposition of the substrate material. In contrast, hydrothermal synthesis can be used to directly crystallize films on to substrate surfaces at low temperatures, thereby enabling new combinations of materials such as ceramic coatings on polymer substrates.

Hydrothermal processing can take place in a wide variety of combinations of aqueous and solvent mixture-based systems. Relative to solid-state processes, liquids accelerate diffusion, adsorption, reaction rate, and crystallization, especially under hydrothermal conditions [3,7]. However, unlike many advanced methods that can prepare a large variety of forms and chemical compounds, such as chemical vapor-based methods, the respective costs for instrumentation, energy, and precursors are far less for hydrothermal methods. Hydrothermal methods are more environmentally benign than many other synthesis methods, which can be attributed in part to energy-conserving low-processing temperatures, ability to recycle waste, and safe and convenient disposal of waste that cannot be recycled [3]. The low reaction temperatures also avoid other problems encountered with high-temperature processes, particularly during single crystal growth, for example, poor stoichiometry control due to volatilization of components or crystal cracking due to phase transformations taking place during cooling.

An important advantage of the hydrothermal synthesis is that the purity of hydrothermally synthesized materials significantly exceeds the purity of the starting materials. This is because hydrothermal crystallization is a purification process in itself, in which the growing crystals or crystallites reject impurities present in the growth environment. Materials synthesized under hydrothermal conditions often exhibit differences in point defects when compared to materials prepared by high-temperature synthesis methods. For instance, in barium titanate, hydroxyapatite, or α-quartz, water-related lattice defects are among the most common impurities and their concentration determines essential properties of these materials. The problem of water incorporation can be overcome by properly adjusting the synthesis conditions, use of water-blocking additives, or even nonaqueous solvents (solvothermal processing). Another important technological advantage of the hydrothermal technique is its capability for continuous materials production, which can be particularly useful in continuous fabrication of ceramic powders [9].

2.2.3 Chemical Compositions and Morphologies of Smart Ceramics

A great variety of ceramic materials have been synthesized by hydrothermal methods. Most common are oxide materials, both simple oxides, such as ZrO_2, TiO_2, SiO_2, ZnO, Fe_2O_3, Al_2O_3, CeO_2, SnO_2, Sb_2O_5, Co_3O_4, HfO_2, etc., and complex oxides, such as $BaTiO_3$, $SrTiO_3$, PZT, $PbTiO_3$, $KNbO_3$, $KTaO_3$, $LiNbO_3$, ferrites, apatites, tungstates, vanadates, molybdates, zeolites, etc., some of which are metastable compounds, which cannot be obtained using classical synthesis methods at high temperatures. Hydrothermal synthesis of a variety of oxide solid solutions and doped compositions is common. The hydrothermal technique is also well suited for nonoxides, such as pure elements (for example Si, Ge, Te, Ni, diamond, carbon nanotubes), selenides (CdSe, HgSe, $CoSe_2$, $NiSe_2$, $CsCuSe_4$), tellurides (CdTe, Bi_2Te_3, Cu_xTe_y, Ag_xTe_y), sulfides (CuS, ZnS, CdS, PbS, $PbSnS_3$), fluorides, nitrides (cubic BN, hexagonal BN), aresenides (InAs, GaAs), etc. [2,4,10–12].

Crystalline products with a specific chemical or phase composition can be usually synthesized hydrothermally in several different forms, such as single crystals, coatings, ceramic monoliths, or powders. Among them, the powders exhibit the largest variety of morphologies, such as equiaxed (for example cubes, spherical), elongated (fibers, whiskers, nanorods, nanotubes), platelets, nanoribbons, nanobelts, etc., with sizes ranging from a few nanometers to tens of microns (Figure 2.2). Core–shell particles and composite powders consisting of a mix of at least two different powders can be also prepared in one synthesis step. Some of the powders can even adopt nonequilibrium morphologies (Figure 2.4c and d).

2.2.4 Hydrothermal Hybrid Techniques

In order to additionally enhance the reaction kinetics or the ability to make new materials, a great amount of work has been done to hybridize the hydrothermal technique with microwaves (microwave–hydrothermal processing), electrochemistry (hydrothermal–electrochemical synthesis), ultrasound (hydrothermal–sonochemical synthesis), mechanochemistry (mechanochemical–hydrothermal synthesis), optical radiation (hydrothermal–photochemical synthesis), and hot-pressing (hydrothermal hot pressing), as reviewed elsewhere [1,3,5–7].

2.2.5 Industrial Production of Smart Ceramic Materials

Several hydrothermal technologies, primarily for the production of single crystals, such as α-quartz for frequency control and optical applications (Sawyer Technical Materials, Tokyo Denpa, NDK), ZnO for UV- and blue light-emitting devices (Tokyo Denpa), and $KTiOPO_4$ for nonlinear optical applications (Northrop Grumman-Synoptics), have already been developed that demonstrate the commercial potential of the hydrothermal method. The volume of the hydrothermal production of α-quartz single crystals is estimated at 3000 tons/year [2]. However, the largest potential growth area for commercialization is ceramic powder production. The widely used Bayer process uses hydrothermal methods to dissolve bauxite and subsequently precipitate aluminum hydroxide, which is later heat-treated at high temperature to crystallize as α-alumina. In 1989, the worldwide production rate was about 43 million tons/year. The production of perovskite-based dielectrics and zirconia-based structural ceramics is a promising growth area for hydrothermal methods [9]. Corporations such as Cabot Corporation, Sakai Chemical Company, Murata Industries, Ferro Corporation, Sawyer Technical Materials, and others have established commercial hydrothermal production processes for preparing ceramic powders.

2.3 Intelligent Control of Phase Assemblage

Thermodynamic modeling can be used to intelligently design a process to be thermodynamically favored using fundamental principles instead of the time-consuming Edisonian methods [1,13]. For a given hydrothermal synthesis of ceramic, the effects of precursor or additive concentrations, temperature, and pressure can be modeled to define the processing variable space over which the phases of interest are stable. The thermodynamic modeling can be accomplished using commercially available OLI computer software, which also contains a database of thermodynamic properties for many common systems [1,13].

2.3.1 Construction of a Thermodynamic Model

In order to construct a thermodynamic model, one has to know all possible species and independent chemical reactions occurring in the system being investigated. Even a simple hydrothermal system such as Ba–Ti–K–H_2O, which is used to synthesize $BaTiO_3$

ceramics, contains a significant number of species and independent chemical reactions. Determining the equilibrium concentrations for each of the species requires the use of automated computer-based solution algorithms. Stability diagrams concisely represent the thermodynamic state of multicomponent, multiphase aqueous systems in wide ranges of temperature and reagent (precursor) concentrations (Figure 2.3). Yield diagrams specify the synthesis conditions suitable for quantitative precipitation of the phase of interest (Figures 2.3b and 2.4a). Calculations of stability and yield diagrams are based on a thermodynamic

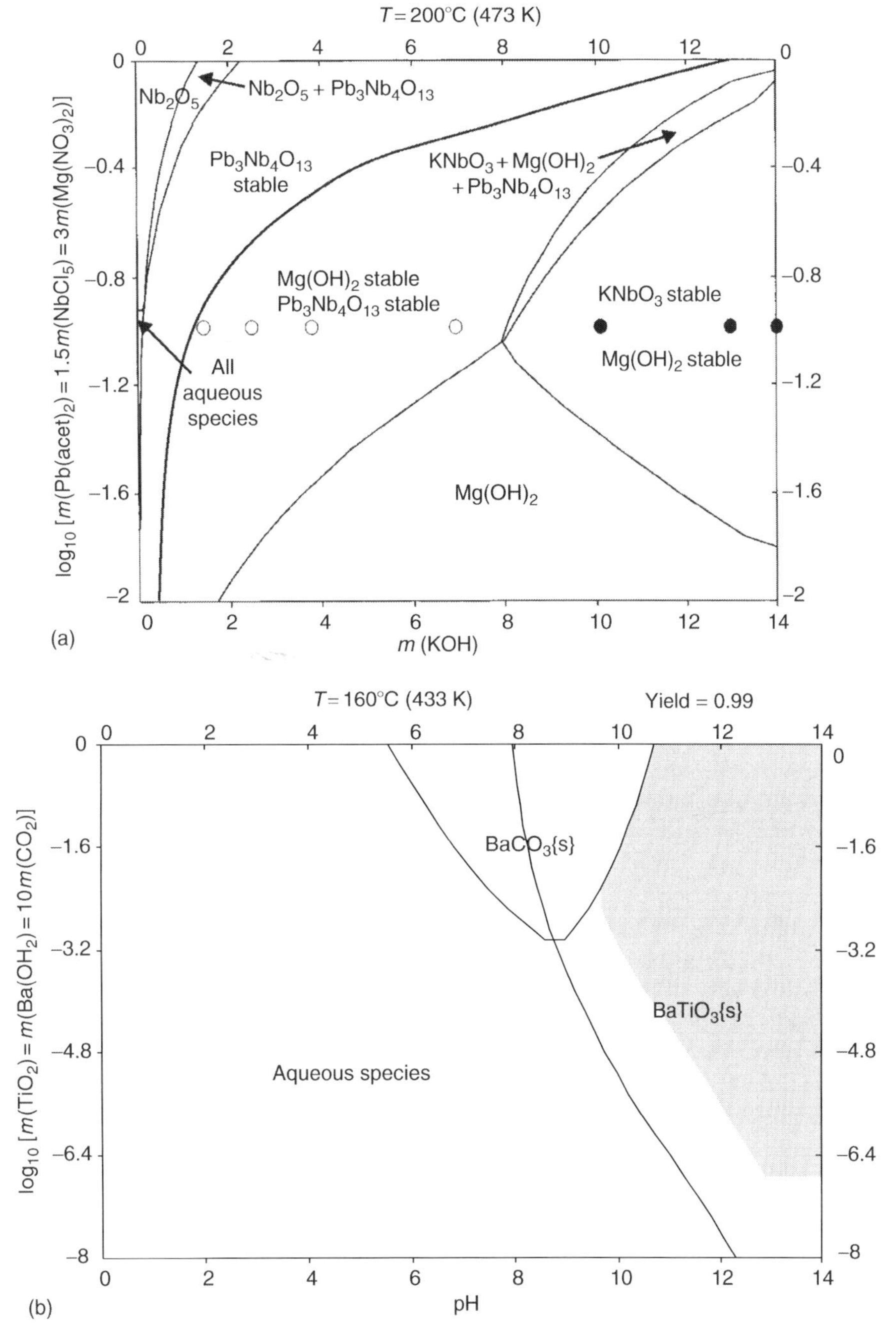

FIGURE 2.3 (a) Calculated stability diagram for the Pb–Mg–Nb–K–H_2O system at 200°C where input precursor concentration is plotted as a function of mineralizer (KOH) concentration. The symbols denote experimentally obtained phase assemblages corresponding to the reaction conditions specified by the equilibrium diagram: O $Mg(OH)_2$ + $Pb_3Nb_4O_{13}$, ● $KNbO_3$ + $Mg(OH)_2$. (b) Calculated stability/yield diagram for the Ba–Ti–CO_2–H_2O system at 160°C using $Ba(OH)_2$ and TiO_2 (rutile) when the amount of CO_2 in precursors is 0.1 times the amount of TiO_2 and the Ba/Ti ratio is equal to 1. The solid lines denote the incipient precipitation boundaries for $BaTiO_3$ and $BaCO_3$. The shaded area corresponds to the $BaTiO_3$ yield >99%.

FIGURE 2.4 Example of intelligent control of chemical composition, size, and morphology of PZT crystallites using thermodynamic and non-thermodynamic variables. (a) Calculated stability/yield diagram in Pb–Ti–Zr–K–H_2O system at 150°C showing stability field of the PZT phase with over 99% yield. (b) Control of crystallite size between 250 nm and 10 μm using simple agitation speed during hydrothermal synthesis and type of the precursor used. (c) and (d) Control of morphology during hydrothermal synthesis at 150°C using concentration of the TMAH mineralizer, which was (c) 0.5 m and (d) 1.0 m.

model that combines the Helgeson–Kirkham–Flowers–Tanger (HKFT) equation of state for standard-state properties of aqueous species with a nonideal solution model based on the activity coefficient expressions developed by Bromley and Pitzer, and modified by Zemaitis et al. For solid species, standard-state properties are used in conjunction with basic thermodynamic relationships. Fugacities of components in the gas phase are calculated from the Redlich–Kwong–Soave equation of state. A more detailed description of the thermodynamic model as well as citations that cover more of the fundamentals can be found elsewhere [1,13].

2.3.2 Methodology for Generating Stability and Yield Diagrams

First, the desired product and components of the hydrothermal system have to be defined. Thus, the identities of the precursors, mineralizers, and other additives needed for the synthesis of a required solid phase need to be specified. This information is used as input data, along with the range of reagent concentrations, temperature, and pressure specified by the user. Therefore, it is important that there is a data bank that is relevant for all the components in the system, which contains standard-state thermochemical properties and independent reaction sets for all species, HKFT equation of state parameters for aqueous species, and Redlich–Kwong–Soave equation of state parameters for gaseous species. The OLI software also stores binary parameters for ion–ion, ion–neutral, and neutral–neutral species interactions. It should be noted, however, that the standard-state properties and parameters are frequently obtained by regressing numerous kinds of thermodynamic data. These data include vapor pressures, osmotic coefficients, activity coefficients, enthalpies, and heat capacities of solutions, and solubilities and heat capacities of solids. When data are not available in the OLI data bank for the specified hydrothermal system, a private data bank must be constructed. The literature can be consulted for data, as well as several methods for estimating thermochemical data. The chemistry model generation step creates the species and reactions that are possible with the given components of the system. All possible combinations of ions, neutral complexes, and solids are considered in this step. The chemical speciation model is a set of equations, which contains chemical equilibrium equations, phase equilibrium equations, mass balance, and electroneutrality equations. Once the thermodynamic conditions are specified, the chemical speciation model is solved. Equilibrium concentrations of all species are calculated as a function of variables such as mineralizer concentration. This gives the equilibrium composition of a specific set of reaction conditions, but to understand the overall behavior of the system computations must be performed over a wide range of conditions. For this purpose, we specify the processing variables of interest. The OLI software offers a flexible choice of independent variables as x- and y-axes of stability and yield diagrams, which include precursor and mineralizer concentrations (Figures 2.3a and 2.4a), solution pH (Figure 2.3b) and temperature, in addition to the electrochemical potential, the latter being used for the simulation of corrosion. In the case of yield diagrams, the yield value (e.g., 99%, 99.9%, and 99.95%) of the desired material must be chosen. The upper temperature limit of the OLI software for creation of both stability and yield diagrams is around 300°C, which covers mild hydrothermal synthesis conditions for most ceramic materials.

2.3.3 Utilization of Thermodynamic Modeling

Calculated phase diagrams can perform many functions during the course of intelligent synthesis of ceramics. In addition to precisely defining the concentration–temperature–pressure processing variable space over which the phases of interest are stable, many different types of precursor systems and additives can be compared, and experiments can be designed to make materials that have never been previously prepared in hydrothermal solution [1,13]. Moreover, the thermodynamic modeling enables to compute chemical supersaturation in a given system. Thus, one can evaluate how the processing variables on the phase diagram influence supersaturation within a phase stability region. Once the supersaturation is known, by using conventional crystal growth models, one can identify when the process is dominated by nucleation and when by the growth rate, which can be utilized in morphology and size control, as demonstrated in the following section.

2.4 Intelligent Control of Crystal Size and Morphology

We will discuss this aspect of intelligent synthesis of ceramics using mostly powders and coatings as examples. These two forms of ceramics constitute the vast majority of ceramic materials synthesized under mild hydrothermal conditions and their size or morphology control is governed by principally the same rules. Control of single crystal growth has been reviewed elsewhere [2].

With the thermodynamic variables processing space well defined for the phase of interest, a range of conditions can be then explored to control reaction and crystallization kinetics for the purpose of developing a process suitable to produce the desired form of ceramics (e.g., particular size, morphology, aggregation level for powders, and microstructure, i.e., size and morphology of constituting crystallites for coatings). Thermodynamic processing variables, such as temperature, pH, concentrations of reactants, and additives, determine not only the processing space for a given material but also influence both reaction and crystallization kinetics. The phenomena that underlie the size and morphology or microstructure control using the thermodynamic variables are the overall nucleation and growth rates, which control crystal size and the competitive growth rates along principal crystallographic directions that control morphology. Size of the crystals can be thus controlled by varying temperature and concentration. Crystal morphology and size can be additionally affected by surfactants, which can adsorb on specific crystallographic faces and solvents, which adsorb similarly as well as regulate solubility. Unfortunately, changing the thermodynamic synthesis variables is constrained by the phase boundaries in a specific phase diagram. Therefore, it may or may not be possible to exploit all sizes, morphologies, and microstructures for a certain material by modifying only the thermodynamic variables. However, nonthermodynamic variables are also extremely important when operating in thermodynamically limited processing variable space. For instance, changing the stirring speed during synthesis of films or powder can change the crystallite size by orders of magnitude [14].

2.4.1 Thermodynamic Variables

The technique of using thermodynamic variables, such as temperature, concentrations of reactant, various additives or solvents, is the most widely used method to control size and morphology of crystals under hydrothermal conditions. It is because with increasing temperature and concentrations (i.e., supersaturation), both nucleation and growth rates significantly increase. It is widely believed that size of the crystals increases with increasing temperature or concentrations. However, this is true only when the growth rate dominates over the nucleation rate. If the nucleation rate dominates over the growth rate, the relationship will be opposite, i.e., size of the crystals will decrease with increasing temperature or concentrations, because there are so many particles formed rapidly and they have little time to grow. In extreme cases of very high nucleation rate, even amorphous materials can be synthesized. Nucleation- and growth-rate-controlled synthesis ranges differ for every material and have to be calculated or determined experimentally in each particular case.

Changing reactants concentration can be used not only to control size of the crystallites but also to control their shape. Fibers or cubes of lead zirconate titanate (PZT, chemical formula

$Pb(Zr_xTi_{1-x})O_3$, where $x = 0 - 1$) can form during the hydrothermal synthesis depending upon the concentration of tetramethylammonium hydroxide (TMAH) mineralizer (Figure 2.4c and d) [15].

The morphology of crystals can be furthermore controlled by the use of appropriate chemical additives or solvents present in the crystallization environment. This approach utilizes preferred adsorption of the additives or solvents on particular faces of the growing crystals, thus preventing growth in certain directions. Consequently, crystals of the same ceramic phase with various levels of aspect ratios, from platelets through whiskers, can be synthesized. A good example of using this approach is hydroxyapatite [16,17]. If chelating agents (lactic acid, EDTA, etc.) are used to prevent precipitation of amorphous precursor at ambient temperatures, precipitation at elevated temperatures usually yields larger crystals [16].

Use of templates, such as self-assembled organic molecules, emulsions, or nanotubes in order to direct the growth of the ceramic phase results in a variety of materials architectures including mesoporous materials, nanorods, and nanosized monodispersed powders [10]. These approaches, however, require removal of the template after the synthesis, which often involves high-temperature calcination.

Monodisperse ceramic particulate systems with minimized aggregation levels can also be synthesized from aqueous solutions without templates by precipitating first a small number of nuclei and subsequently reducing supersaturation level in order to prevent further nucleation and enable diffusion growth of the formed nuclei [18]. While the powder uniformity is excellent, it requires working with dilute reactant concentrations (e.g., $\approx$0.005 M) so that both supersaturation and nucleation are highly uniform and colloidal stability can be maintained [12].

2.4.2 Nonthermodynamic Variables

Nonthermodynamic synthesis variables, such as stirring rate and method of precursor preparation, can be essential in size and morphology control of crystals, while keeping the thermodynamic variables constant. Generally speaking, stirring during crystal growth leads to an increase in the probability of spontaneous nucleation, a decrease in supersaturation inhomogenities, and an increase of the growth rate. The final size of the crystals is then determined by a balance between the nucleation and growth rates. A good example of a very strong influence of nonthermodynamic variables on crystal size and morphology is PZT [14]. Concentration of the chemicals used in hydrothermal PZT synthesis was well within the calculated stability field of the PZT phase, which is shown in Figure 2.4a. Increasing the stirring speed alone from 200 to 1700 rpm has reduced the particle size of PZT crystals almost two orders of magnitude, from over 10 μm to a submicron range (Figure 2.4b). In this particular case, the nucleation rate seems to dominate over the growth rate, therefore the crystal size decreased with increasing stirring speed. At higher stirring speeds, the PZT particles exhibited better-defined cube habit and were more uniform in size, which is in line with reduction of supersaturation inhomogenities with increasing agitation rate. Use of different precursors impacted the crystal size as well (Figure 2.4b).

Stirring speed or velocity of the solution flow along the substrate can affect significantly the thickness, morphology, and chemical composition of films grown under hydrothermal conditions, as demonstrated using $BaTiO_3$, $SrTiO_3$, and PZT thin films [14]. Film thickness and morphology is also significantly impacted by the type and crystallographic orientation of the substrate used (Figure 2.1e and f).

2.5 Summary

Hydrothermal synthesis is environmentally friendly, energy-conserving low-temperature crystallization of all forms of smart materials directly from aqueous solutions. Hydrothermal crystallization affords excellent control of chemical composition (oxides, nonoxides, native elements, dopants, solid solutions), as well as shape, size, and morphology of ceramics. Understanding the physicochemical processes occurring in the aqueous solution is the key for intelligent engineering of hydrothermal synthesis of smart ceramics. Thermodynamic modeling is used to predict formation conditions for phases of interest. Subsequently, size, morphology, and agglomeration level of the synthesized ceramics can be controlled in wide ranges using both thermodynamic and nonthermodynamic (kinetic) variables. Such intelligent approach enables to achieve cost effective scale-up and commercial production of smart ceramics using the hydrothermal technology.

Acknowledgments

The authors wish to gratefully acknowledge the support of the Office of Naval Research, The National Institute of Health, OLI Systems, Inc., and Sawyer Technical Materials, LLC, for their generous support of the research cited in this manuscript.

References

1. Riman, R.E., Suchanek, W.L., and Lencka, M.M., Hydrothermal crystallization of ceramics, *Ann. Chim. Sci. Mat.*, 27, 15, 2002.
2. Byrappa, K. and Yoshimura, M., *Handbook of Hydrothermal Technology*, Noyes Publications/William Andrew Publishing LLC, Norwich, NY, 2001.
3. Yoshimura, M., Suchanek, W.L., and Byrappa, K., Soft solution processing: A strategy for one-step processing of advanced inorganic materials, *MRS Bull.*, 25, 17, 2000.
4. Sômiya, S., Ed., *Hydrothermal Reactions for Materials Science and Engineering. An Overview of Research in Japan*, Elsevier Science, London, 1989.
5. Roy, R. Accelerating the kinetics of low-temperature inorganic syntheses, *J. Solid State Chem.*, 111, 11, 1994.

6. Suchanek, W.L., Lencka, M.M., and Riman, R.E., Hydrothermal synthesis of ceramic materials, in *Aqueous Systems at Elevated Temperatures and Pressures: Physical Chemistry in Water, Steam, and Hydrothermal Solutions*, Palmer, D.A., Fernández-Prini, R., and Harvey, A.H., Eds., Elsevier, Amsterdam, 2004, Chapter 18.
7. Yoshimura, M. and Suchanek, W., In situ fabrication of morphology-controlled advanced ceramic materials by soft solution processing, *Solid State Ionics*, 98, 197, 1997.
8. Riman, R.E., in *High Performance Ceramics: Surface Chemistry in Processing Technology*, Pugh, R. and Bergström, L., Eds., Marcel-Dekker, New York, 1993, p. 29.
9. Dawson, W.J., Hydrothermal synthesis of advanced ceramic powders, *Ceram. Bull.*, 67, 1673, 1988.
10. Adair, J.H. and Suvaci, E., Morphological control of particles, *Curr. Opin. Colloid Interf. Sci.*, 5, 160, 2000.
11. Niesen, T.P. and DeGuire, M.R., Review: Deposition of ceramic thin films at low temperatures from aqueous solutions, *J. Electroceramics*, 6, 169, 2001.
12. Sugimoto, T., *Fine Particles, Synthesis, Characterization, and Mechanisms of Growth*, Marcel-Dekker, Inc., New York, 2000.
13. Lencka, M.M. and Riman, R.E., in *Encyclopedia of Smart Materials*, Vol. I, Schwartz, M., Ed., John Wiley, New York, 2002, pp. 568–580.
14. Suchanek, W.L. et al., Hydrothermal deposition of <001> oriented epitaxial $Pb(Zr,Ti)O_3$ films under varying hydrodynamic conditions, *Crystal Growth Design*, 5, 1715, 2005.
15. Cho, S.-B., Oledzka, M., and Riman, R.E., Hydrothermal synthesis of acicular lead zirconate titanate (PZT), *J. Crystal Growth*, 226, 313, 2001.
16. Suchanek, W. et al., Biocompatible whiskers with controlled morphology and stoichiometry, *J. Mater. Res.*, 10, 521, 1995.
17. Riman, R.E. et al., Solution synthesis of hydroxyapatite designer particulates, *Solid State Ionics*, 151, 393, 2002.
18. Matijevic, E., Monodispersed metal (hydrous) oxides—a fascinating field of colloid science, *Acc. Chem. Res.*, 14, 22, 1981.

3

Functionally Graded Polymer Blend

Yasuyuki Agari
Osaka Municipal Technical Research Institute

3.1 Introduction

Many reports have been published on functionally graded materials made of metals and ceramics [1]. These graded materials have characteristics of improved strength against thermal stress as well as excellent electromagnetic and optical properties. There have also been many reports on a functionally graded ceramic, which can be called a smart material. These ceramics showed a strong thermoelectric performance, which shifted with an increase in temperature. As a result, the thermoelectric performance was very high in a wide temperature gradient.

There have also been some reports on functionally graded polymeric materials [2–43]. These functionally graded polymeric materials can be classified into four types from currently used materials, as shown in Table 3.1. The graded structures may be classified into six types. However, reports on functionally graded polymer blends are somewhat limited [4–7,14–28], although there have been published reports on various types of blends. The functionally graded polymer blend has been considered to have a structure as shown in Figure 3.1. The blend has two different surfaces without an interface and can have both the advantages of a laminate and a homogenous blend.

We have devised a new method called the dissolution–diffusion method for preparing functionally graded polymer blends [4–7,24–28]. In this method, graded polymer blends are classified into two types. Graded polymer blends are usually prepared by three methods: surface inclination in the melt state [14,15], surface inclination in the solution method [16,17], and diffusion in the melt method [19–23]. The dissolution–diffusion method devised by us is only one of the methods that can be used in preparing both types of graded polymer blends. Our method has the following advantages in comparison with the other methods. The preparation time in our method is very short, and primarily is 100 times shorter than that of diffusion in the melt state method. The optimum condition can be easily determined because our method has many controllable factors. Furthermore, the chemical decomposition of molecules does not occur in the preparation by our method because the preparation is performed at a lower temperature. Therefore, we believe that our method is the most useful method.

In further work, we recently found a new preparation method (polymerization–diffusion method) by the polymerization of a monomer during diffusion using a macroazoinitiator [29,30].

TABLE 3.1 Various Types of Functionally Graded Polymeric Materials

Types of Used Materials	Structure	Preparation Method	Size of Dispersion Phase
Metal (or ceramic)/polymer	Composites	• Laminate method • Electric field method • Centrifugation method • Flame spraying method	Big
Polymer–polymer	Immiscible polymer blend	• Surface inclination in melt state method • Surface inclination in solution method • Dissolution–diffusion method	
	Miscible polymer blend	• Diffusion in melt method • Dissolution–diffusion method	Molecular order
Atom–atom (intramolecules)	Copolymer (random)	• Diffusion method of monomer in polymer gel	Atom order
	Copolymer (tapered)	• Living anion or radical polymerization method	
	Density of cross-linking	• Changing method of cross-linking concentration	
High-order structure (same polymer)		• Changing method of cross-linking temperature	Same atoms and molecules
	Crystal structure	• Injection mold method	

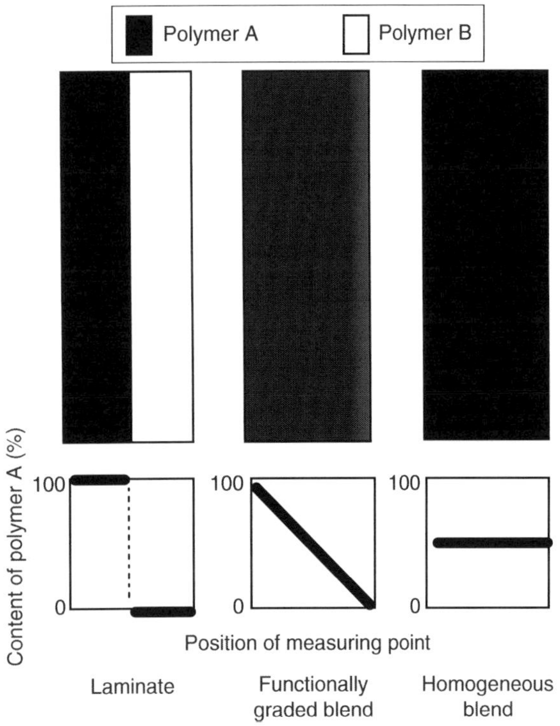

FIGURE 3.1 Schematic model of functionally graded blend.

It is useful for the preparation of an immiscible graded blend and can form a graded structure of similar molecular weight.

This chapter presents a detailed description of the preparation mechanism used in dissolution–diffusion and polymerization–diffusion methods to prepare functionally graded polymer blends. The chapter also explains how to determine the optimum condition for the several types of functionally graded polymer blends (polyvinyl chloride [PVC]/polymethacrylate, polymethyl methacrylate [PMMA], polyhexyl methacrylate [PHMA]) or polycaprolactone (PCL), polyethylene oxide (PEO)/poly(L-lactic acid) (PLLA), bisphenol A type polycarbonate (PC)/polystyrene (PS), PEO/polybutyl methacrylate (PBMA), and (PEO/$LiClO_4$)/PBMA used in characterizing graded structures for blends by measuring the Fourier transform infrared (FTIR) spectra, confocal Raman spectra, thermal behaviors around the glass transition temperature (T_g) by Differential Scanning Calorimetry (DSC) methods, by Scanning Electro-Microscopy–Energy Dispersive X-ray Spectrometry (SEM-EDX) observation, and by nuclear magnetic resonance (NMR), or Gel Permeated Chromatography (GPC) measurement. Furthermore, several types of functional properties are discussed especially for smart performance, which were caused by the graded structure. Finally, the prospects of functionally graded polymer blends and their applications are discussed.

3.2 Mechanism of the Preparation Method

3.2.1 Diffusion–Dissolution Method

The mechanism of formation of a graded structure by the diffusion–dissolution method is considered to be the following [27]. After a polymer B solution is poured onto a polymer A film in a glass petri dish, polymer A begins to dissolve and diffuse in the solution on the air side (Figure 3.2), but the diffusion is interrupted when all the solvent evaporates. Thus, a blend film is produced, which consists of a concentration of gradient of polymer A/polymer B in the thickness direction.

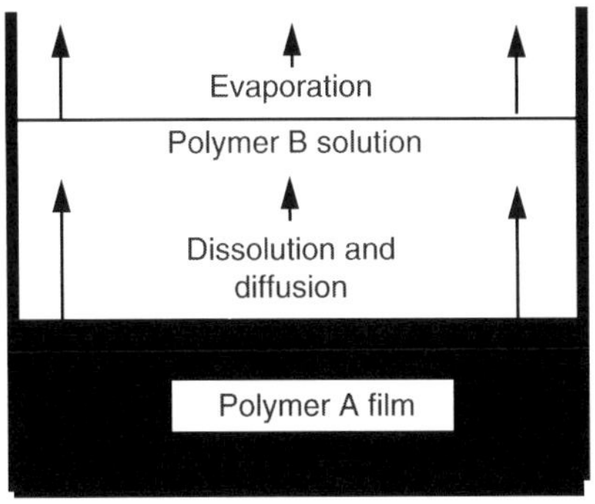

FIGURE 3.2 Schematic model of dissolution–diffusion method.

Following the steps of dissolution and diffusion of polymer A, the graded structures should be classified into three types (Figure 3.3).

First type: Polymer A begins to dissolve and then diffuses, but does not yet reach the air side surface of polymer B solution. The blend has three phases (polymer A, polymer B, and a thin graded structure).

Second type: Just when all of polymer A has dissolved, the diffusion frontier reaches up to the air side surface of polymer B solution. The blend has one graded phase from the surface to the other one, while the surfaces are composed of polymer A only or polymer B only.

Third type: After the dissolution and diffusion of polymer A reached up to the air side surface of polymer B solution, polymer A and polymer B molecules began to mix with each other and became miscibilized. The concentration gradient then began to disappear.

The formation of a concentration gradient should depend on (a) dissolution rate of polymer A in polymer B solution, (b) diffusion rate of polymer A in polymer B solution, and (c) interruption time for the diffusion due to the completion of solvent evaporation. Factors for controlling the above phenomena are (1) the type of solvent, (2) the casting temperature, (3) the molecular weight of polymer A, and (4) the amount of polymer B solution.

Until polymer A completely dissolves or reaches the surface of polymer B solution, i.e., in the formation of the first and second types of structure, the diffusion of polymer A in the polymer B solution is considered to obey Fick's second law (Equation 3.1) by the assumption of neglecting the evaporation of the solvent in polymer B solution during the diffusion:

$$\frac{\partial C_A}{\partial t} = D_{AB}\left(\frac{\partial^2 C_A}{\partial x^2}\right) \tag{3.1}$$

where
C_A is the concentration of polymer A
t is the passed time
x is the distance from the surface of polymer A sheet
D_{AB} is an apparent diffusion coefficient

The point where C_A approaches 1 and shifts to the petri glass side, thus carrying forward the dissolution of polymer A. Thus, by considering this effect and rearranging mathematically Equation 3.2 is obtained from Equation 3.1:

$$C_A = \text{erfc}\left(\frac{(x-b)}{2\sqrt{D_{AB}\,t}}\right)$$
$$\text{erfc}(x) = \left(\frac{2}{\sqrt{\pi}}\right)\int_x^{\infty} \exp\left(-\xi^2\right)d\xi \tag{3.2}$$

where b is the distance between the petri glass side surface and the other side of remainder of polymer A, which has not dissolved yet. Therefore, the gradient profile in the blend at t can be estimated by Equation 3.2.

Adaptability of Equation 3.2 to the experimental data was examined in the PVC/PMMA graded blend, which is given a detailed explanation in Section 3.3.1. The experimental data agreed approximately with the ones predicted by Equation 3.2, as shown in Figure 3.7. D_{AB} and b were obtained as 6.38 mm^2/s and 57 mm, respectively. The D_{AB} value was much larger than the value obtained by diffusion in melt state method, which implies that the dissolution–diffusion method is very useful.

Further, a thicker and a more excellent graded blend film was prepared by the multiple-step method, as illustrated in Figure 3.4. In this case, the graded blend was obtained by repeatedly preparing and changing the composition of the blend in the pouring solution.

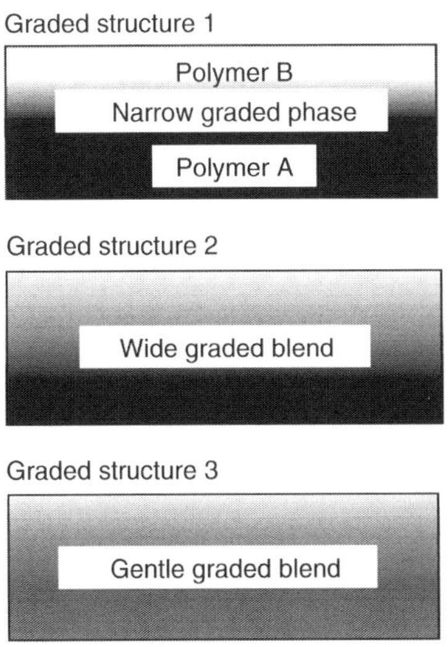

FIGURE 3.3 Schematic models of various types of graded structures.

3.2.2 Polymerization–Diffusion Method

The mechanism of formation of a graded structure by polymerization during the diffusion method is as follows [30]. After the polymer A film containing a macroazoinitiator, i.e., a radical type initiator, with oligomeric polymer A segment is cast on a filter, the

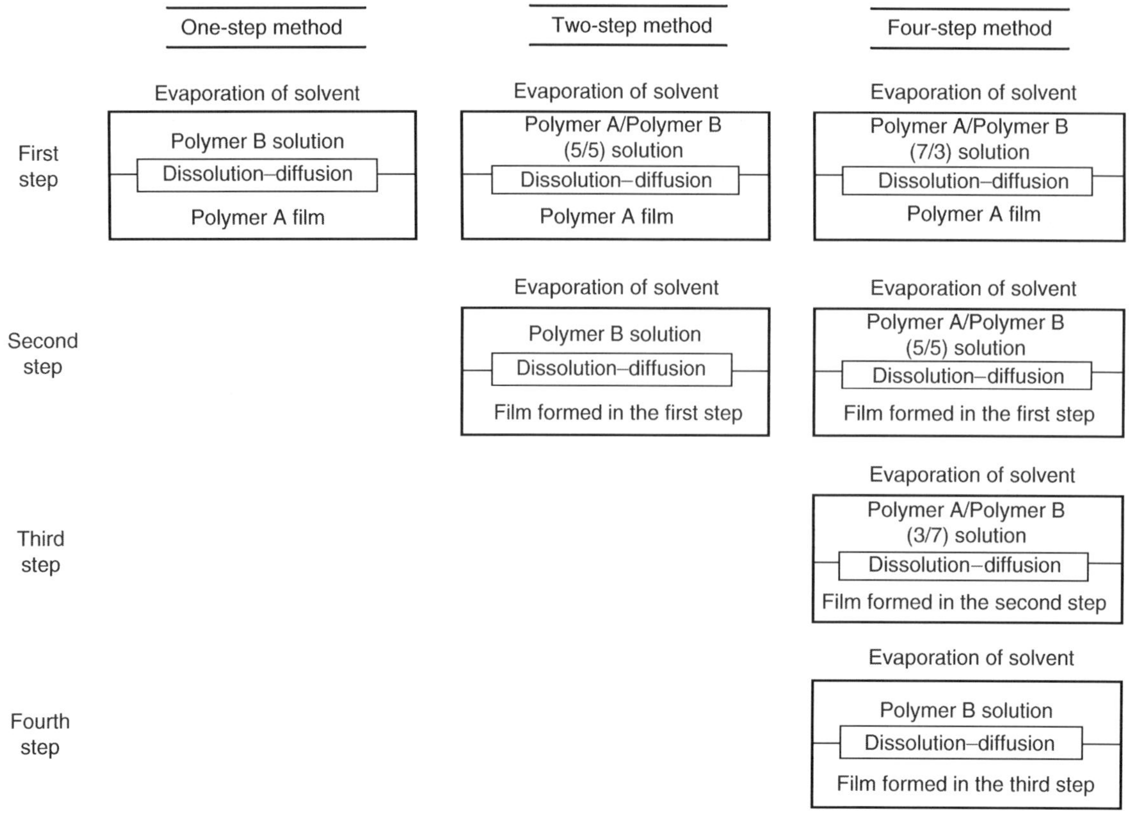

FIGURE 3.4 Schematic models of multiple steps method.

resulting laminate is put on a monomer B solution in a aluminum petri dish kept at a constant temperature (Figure 3.5). The monomer B begins to diffuse into polymer A film to the air side with polymerization. When the process is interrupted on the way by enough diffusion of the monomer B, a blend film is produced, which consists of a concentration gradient of polymer A/polymer B in the thickness direction.

The graded structures could be determined by the balance of polymerization and diffusion of monomer B. The temperature content of a macroazoinitiator, time, i.e., largely effected the balance. The temperature and the time enhanced both the polymerization and the diffusion, and the content of the macroazoinitiator enhanced only the polymerization. Ease of evaporation of *monomer B* enhanced only the diffusion.

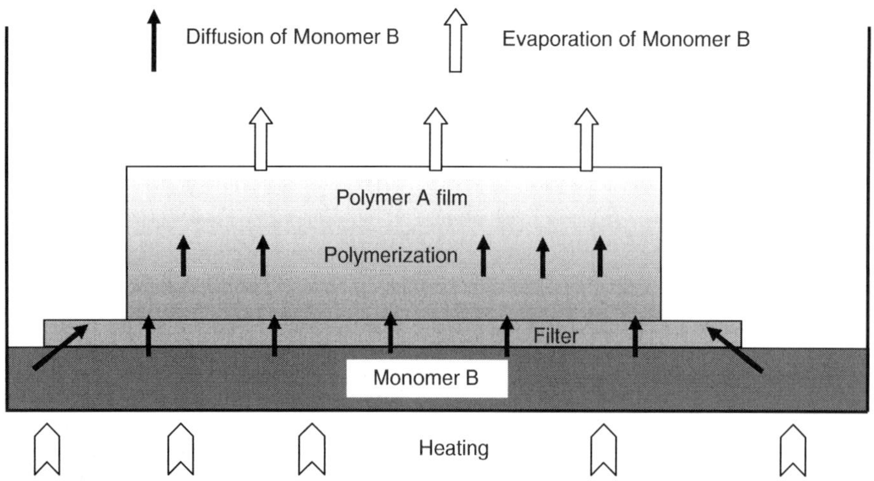

FIGURE 3.5 Schematic model of polymerization–diffusion method.

3.3 Preparation and Characterization of Several Types of Functionally Graded Polymer Blends

3.3.1 Amorphous Polymer/Amorphous Polymer Miscible Blend (Dissolution–Diffusion Method)

In the case of the PVC/PMMA system [6,7], samples were prepared by changing the four controllable conditions: (1) the type of solvent, (2) the casting temperature, (3) the molecular weight of the PVC, and (4) the amount of the PMMA solution. The graded structures of these samples were characterized by FTIR-ATR, Raman microscopic spectroscopy, and DSC methods. Here, FTIR-ATR was the spectroscopy, by which the IR spectra on the surface layer (about 1-10 μm of thickness) was measured. Figure 3.6 shows the graded structure of the samples in the direction of the thickness, measured by FTIR-ATR. In the similar blend with the graded structure 1, for a laminate, the PMMA content increased at 60% of the distance/thickness and was confirmed to have a thin graded layer (about 10%–20% of the distance/thickness). In the blend with the graded structure 3, the PMMA content was kept to about 50% in all distance ranges. However, in the blend with the graded structure 2, the PMMA content gradually increased in the range from 0% to 100% of the distance/thickness. Thus, it was found that this blend had an excellent wide concentration gradient. In this case, the PMMA content was estimated from the ratios of the absorption band intensities at 1728 cm^{-1} (the stretching of the carbonyl group in PMA) and 615 cm^{-1} (the stretching of C–Cl bond in the PVC). The change in the PMMA content in the thickness direction of the blended film was estimated by measuring the FTIR-ATR spectra on a sliced layer of the blended film.

The change in the PVC content of the graded blend was characterized by the confocal Raman spectroscopy method, similar to the FTIR-ATR method, as shown in Figure 3.7. The measurement of the Raman microscopic spectra was performed by measuring the Raman spectra at the focused point, which was shifted 10 by 10 mm, from one surface area to the other one. It could be confirmed that the blend had a comparatively thick layer of a graded structure phase. This method is considered to be significantly useful because an easy and detailed estimation can be made for a graded profile of a blend.

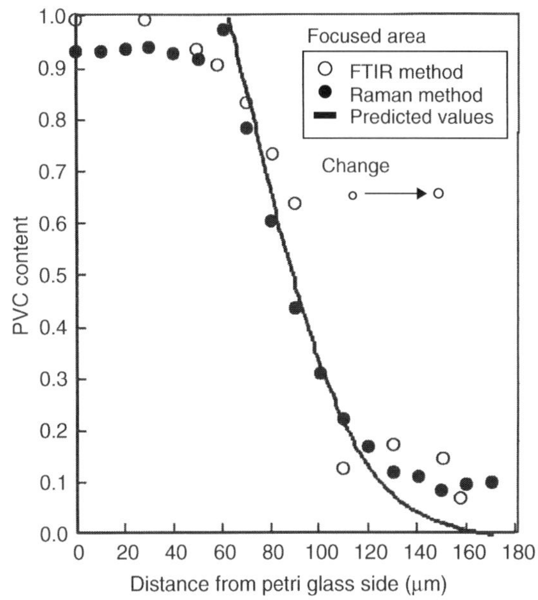

FIGURE 3.7 Change in the PVC content in the thickness direction of the PVC/PMMA blend.

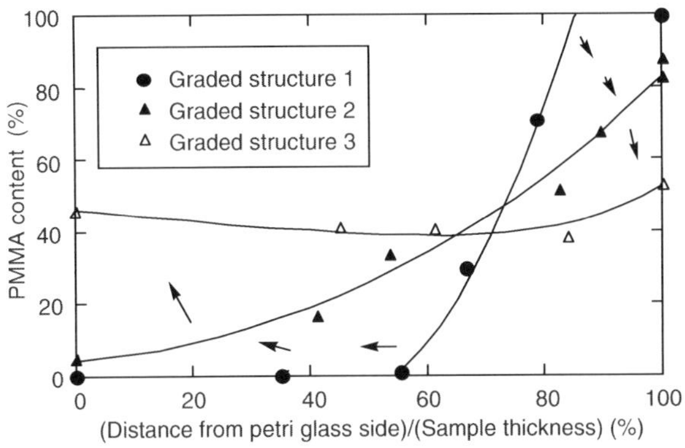

FIGURE 3.6 Change in PMMA content in the thickness direction of several types of the PVC/PMMA graded blends.

Furthermore, the graded structure was characterized by the DSC method. The DSC curve of the blend having a widely graded structure (graded structure 2), which shows a more gradual step around T_g than the others, is shown in Figure 3.8. Similarly, the structures of the samples, which were prepared in the several types of the conditions mentioned previously, were investigated and it was found that the typical optimum condition (molecular weight of PVC: $M_n = 35{,}600$, $M_w = 60{,}400$, type of solvent: tetrahydrofuran (THF)/toluene (5/1), volume of solvent: 0.23 ml/cm^2, temperature: 333 K) could be obtained.

In the case of PVC/PHMA system [7], the graded structure of the sample could not be estimated by the FTIR-ATR and the DSC methods because PHMA was very soft at room temperature. Thus, the graded structure was measured by the SEM-EDX method (Figure 3.9). The chlorine content in the sample increased gradually on the petri glass side and then it was confirmed to have a wide graded structure.

Finally, the structures of the samples, which were prepared in the several types of the previously mentioned conditions, were investigated. The typical optimum condition (molecular weight of PVC: $M_n = 35{,}600$, $M_w = 60{,}400$, type of solvent: MEK, volume of solvent: 0.37 ml/cm^2, temperature: 313 K) was obtained.

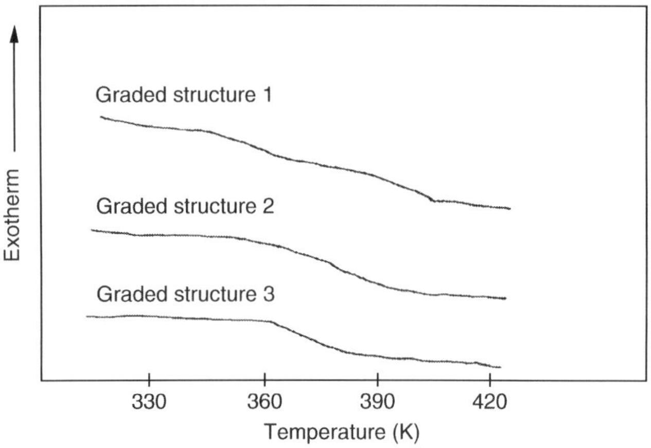

FIGURE 3.8 DSC curves of several types of PVC/PMMA blends.

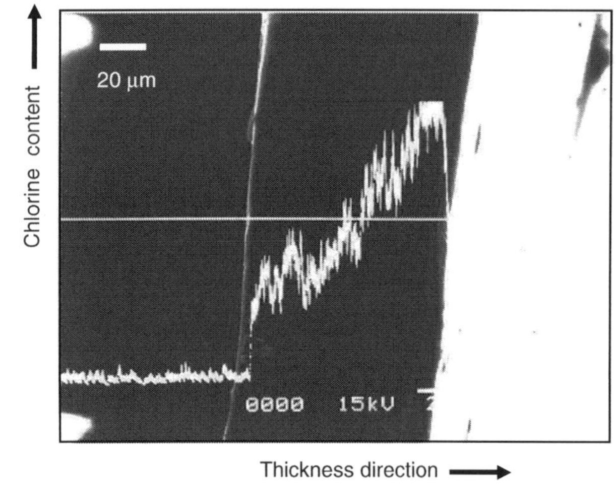

FIGURE 3.9 Chlorine content along the thickness of PVC/PHMA graded blend (×750).

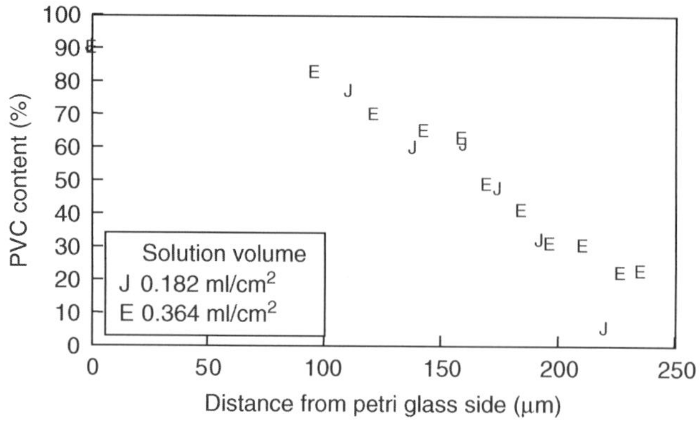

FIGURE 3.10 Graded structures of the PVC/PCL graded blends measured by the FTIR-ATR method.

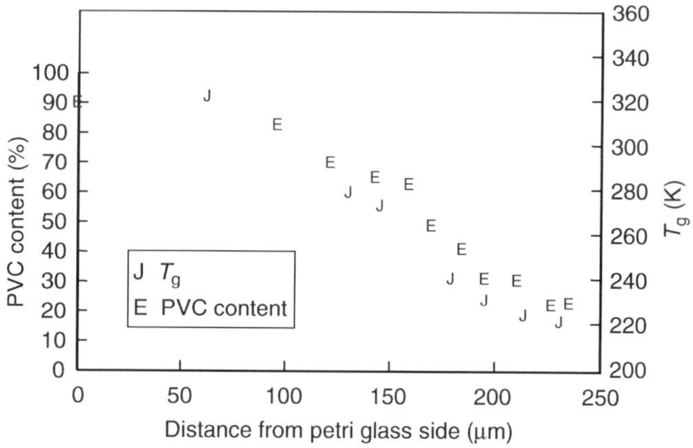

FIGURE 3.11 Graded structures of PVC/PCL graded blends characterized by the DSC method.

3.3.2 Amorphous Polymer/Crystalline Polymer Miscible Blend (Dissolution–Diffusion Method)

In the case of the PVC/PCL system [25], we obtained the optimum conditions when the graded polymer blend was prepared with a wider compositional gradient, similar to the PVC/PMMA system. Figure 3.10 shows the PVC content of the samples for the direction of thickness, measured by the FTIR-ATR method. PVC decreased at about 70 mm from the petri glass side and decreased gradually until the surface of the air side, that is, about 240 mm apart from the petri glass side in both cases of the solution volume.

Then, the change of T_g for the thickness direction of the blend film was characterized by the DSC method (Figure 3.11) in the case of 0.364 ml/cm^2 of solution volume. T_g decreased with the increase of the distance from the petri glass side, similar to the PVC content. Thus, the graded structure in the PVC content was confirmed by the graded profile in T_g.

Furthermore, the change of the PCL crystalline content was evaluated from that of the amount of the heat diffusion effect on the PCL crystalline, measured by the DSC method. The heat diffusion began to increase on the specimen after it was initially kept at 0 and then about 130 mm of the distance to the air side. Then, it increased immediately around about 180 mm. Thus, it was found that the graded structure in the PCL crystalline was formed in a distance of 130–240 mm. This means that the obtained graded PVC/PCL blend had both a gradient concentration of the PVC and a gradient content of the PCL crystalline, as shown in Figure 3.12. The PCL content was about 30% for a distance of 130 mm. This result corresponded to that of the PCL crystalline in a homogeneous PVC/PCL blend, which emerged in the case of the larger than 30% of the PCL [44]. It was further considered that the amorphous phase was made up of a miscible amorphous PVC/amorphous PCL blend. Finally, the PCL crystalline phase decreased the closer it came to the surface of the air side again. This phenomenon occurred because the forming of the amorphous phase was more thermodynamically stable than that of the crystalline phase.

The graded structure of the PVC/PCL graded blend could be schematically illustrated in Figure 3.13.

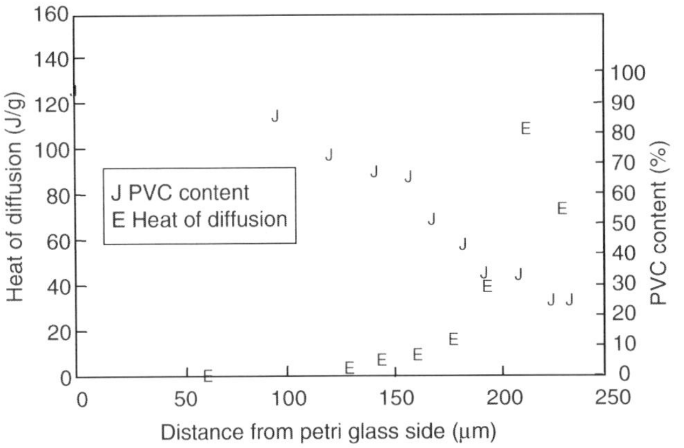

FIGURE 3.12 Graded structure of PVC/PCL graded blends (PCL crystalline).

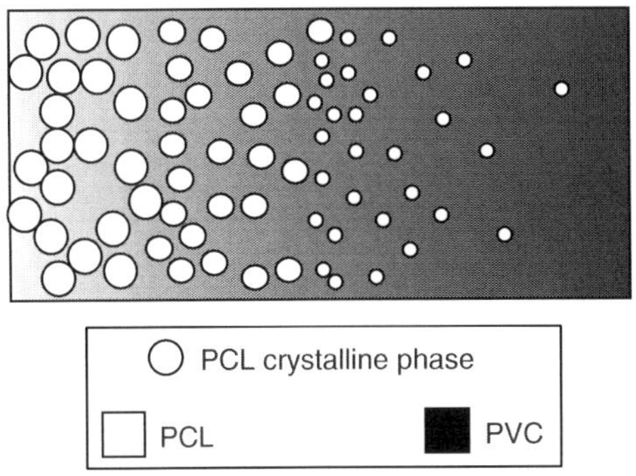

FIGURE 3.13 Schematic model of the PVC/PCL graded blend (graded structure).

3.3.3 Amorphous Polymer/Amorphous Polymer Immiscible Blend (Dissolution–Diffusion Method)

We attempted to prepare a graded PC/PS blend by the dissolution–diffusion method [24], similar to the PVC/PMMA system. In this case, the PS solution was poured onto the PC film. However, we did not obtain a graded structure, but obtained a homogeneous two-layer system, which was composed of about 50% and 0%–10% of PC, as shown in Figure 3.14. Next, the macrophase separation was observed in the former layer.

This result was thought to occur because of the following; only three factors, dissolution rate, diffusion rate, and evaporation time, could affect the process in forming a graded structure of miscible blend. However, in forming the graded structure of an immiscible blend, three additional factors also played a role: macrophase separation, surface roughness, and gravimetry, as shown in Figure 3.15. Especially, the macrophase separation may increase the size of the separated phases up to the size of the thickness in the prepared film, resulting in a possible break in the formation of a strongly graded structure, when its concentration becomes higher by the evaporation of the solvent.

Therefore, we attempted to prevent macrophase separation during graded structure formation by adding PS-*b*-PC block copolymer [45]

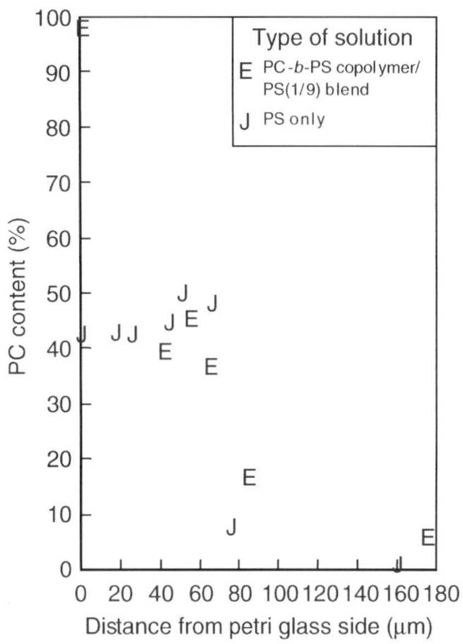

FIGURE 3.14 Graded structure of PC/PS graded blend with or without the PS-*b*-PC block copolymer.

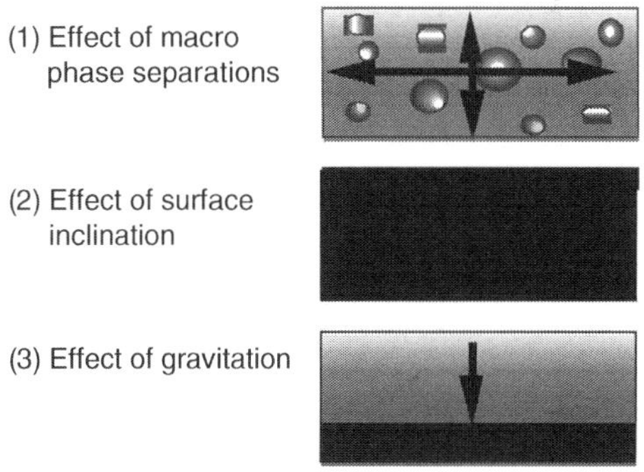

FIGURE 3.15 Other factors that have an effect on forming graded structure in an immiscible blend.

to the PS solution (PS-*b*-PC block copolymer/PS = 1/9). The copolymer may act as a compatibilizer by decreasing the interface energy of the phases. In this case, the PC segment content in the block copolymer was 46% (NMR measurement). It was found that the wide graded structure in the obtained PC/PS blend was formed in the distance range of 0–100 ∝cm from the petri glass side (Figure 3.14).

Furthermore, we attempted to prepare a graded PC/PS blend by pouring the PC solution containing the block copolymer onto the PS film. Figure 3.16 shows the change of the PC content for the direction of the film thickness. A wide graded structure was formed and confirmed not only, in the furtherest distance from, but also in the distance closest to the petri glass side. This result meant that surface roughness significantly influenced the forming of the graded structure.

Therefore, we were able to obtain the graded immiscible PC/PS blend by adding the PC-*b*-PS block copolymer. The graded structure assumed to be formed for the PC/PS graded blend could be schematically illustrated in Figure 3.17.

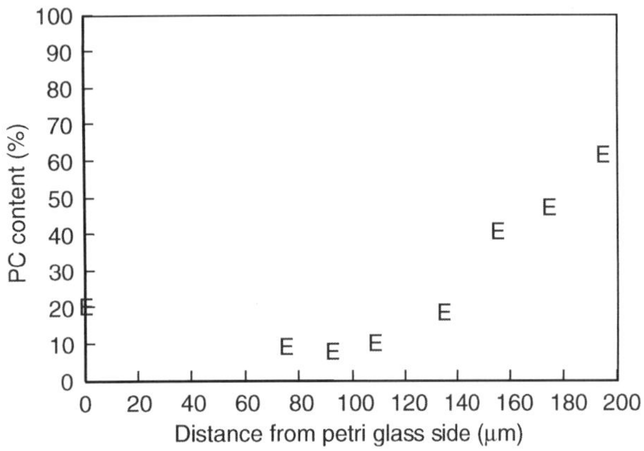

FIGURE 3.16 Graded structure of PC/PS graded blends in the case that the PC solution was poured on a PS film.

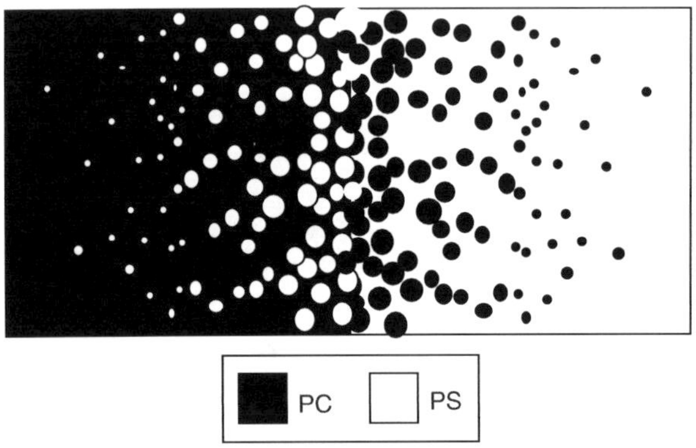

FIGURE 3.17 Schematic model of PC/PS graded blend.

3.3.4 Amorphous Polymer/Crystalline Polymer Immiscible Blend (Polymerization–Diffusion Method)

In the case of the immiscible graded blend of the PBMA/PEO system [30], we obtained the optimum conditions in the preparation by the polymerization–diffusion method. Figure 3.18 shows the PEO content of the samples for the direction of thickness, measured by the confocal Raman spectroscopy and the NMR methods. The data obtained by the NMR method were almost the same as those by the Raman method. It was found that the NMR method was available and so it was used. Figure 3.19 shows the GPC data of the layers around the thickness points (No. 1, 3, 5, 7 in Figure 3.18) as measured by NMR method. The peak corresponding to the block copolymer (EO-*b*-BMA copolymer) became larger when the number of the points was larger. Then, the molecular weight and the content of the block copolymer became larger as the number of points became larger, that is, the BMA content was larger (Table 3.2). It was considered to occur because the polymerization rate was larger with a larger amount of the BMA monomer.

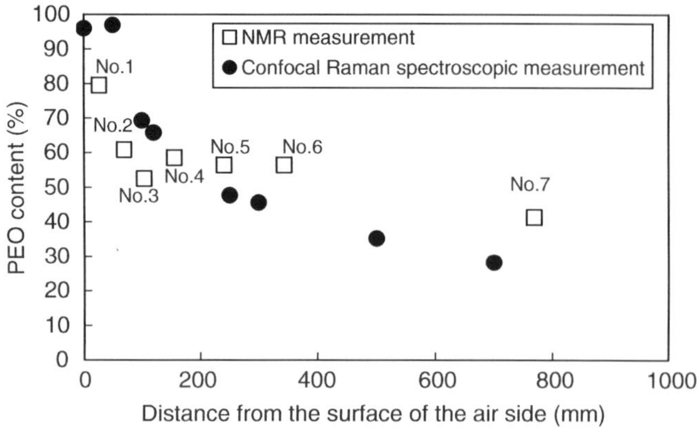

FIGURE 3.18 PEO content in the thickness direction of PEO/PBMA graded blend measured by NMR and confocal Raman spectroscopy.

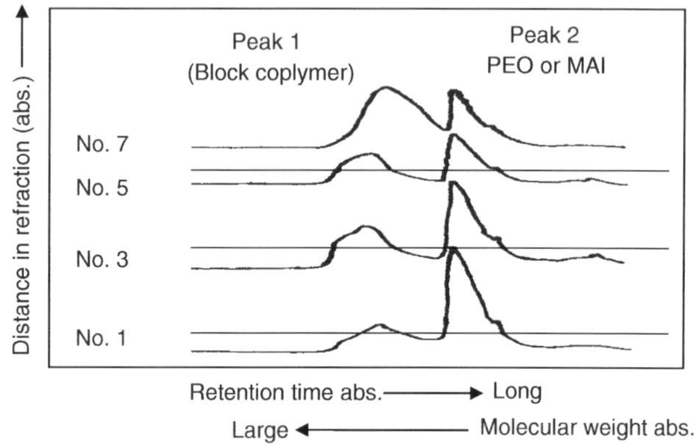

FIGURE 3.19 GPC charts in several types of the points, measured by NMR, in PEO/PBMA graded blend.

3.4 Functional and Smart Performances and the Prospect for Application

3.4.1 Functional and Smart Performances of PVC/PMA Graded Blend

Our study [6,7] found that graded polymer blends had several types of functional properties, inclusive of smart performance. Thus, the functional properties of the PVC/PMA blend containing graded structure 2 (an extremely wide graded concentration) were explained by comparing them with those of a blend containing graded structure 1 (similar to a laminate system), a perfectly miscible blend (5/5), of PVC and PMA only.

3.4.1.1 Tensile Properties

The results of the tensile properties of PVC, PMMA, the perfectly miscible blend (5/5), and the blends with graded structure 1 and 2 (blend type 1 and 2) for the vertical direction of the thickness are summarized in Table 3.3.

The tensile strength of the homogeneously miscible blend was the highest, followed by blend type 2, surpassing PVC, PMMA, and the blend type 1. This phenomenon means that the formation

TABLE 3.2 Molecular Weight and Composition of Several Types of Areas in the PEO/PBMA Functionally Graded Alloy in the Thickness Direction

	Peak 1		Peak 2		
No.	M_w	M_w/M_n	M_w	M_w/M_n	Ratio of Peak 1 to Peak 2
1	1.39×10^6	2.10	4.33×10^4	1.15	0.38
2	1.40×10^6	2.06	4.61×10^4	1.16	0.37
3	1.80×10^6	2.22	3.93×10^4	1.13	0.94
4	1.69×10^6	2.21	3.69×10^4	1.17	1.04
5	1.79×10^6	2.16	3.80×10^4	1.12	1.15
6	1.58×10^6	2.39	3.84×10^4	1.13	1.27
7	5.66×10^5	2.29	3.34×10^4	1.11	2.68

of a graded structure not only suppresses a break at the interface but also gives superior properties compared to the source materials. This was considered to occur because the blend phase with the concentration gradient had a sufficiently high tensile strength. The elongation at break of blend type 2 appeared sufficiently good. The tensile modulus of the blend type 2 was higher than that of the blend type 1. It was found that the break in the tensile stress could be suppressed by the formation of a concentration gradient.

3.4.1.2 Thermal Shock Resistance

Thermal shock resistance tests were performed by moving the specimens from one box to another (kept at 253 and 373 K) repeatedly (five times) every 30 min. The specimens were then evaluated for their thermal shock resistance by measuring a maximum angle of warp, as illustrated in Figure 3.20, and adhesive strength in shear by tension loading.

Thermal shock resistance tests of the blend type 2 with the graded structure 2 were performed and the results (maximum value of warp angle and adhesive strength in shear by tension loading) were compared with those of the blend type 1 having the graded structure 1, as shown in Table 3.3.

The film of the blend type 1 was highly warped, while that of the blend type 2 did not show any warp. The adhesive strength in shear by tension loading of the blend type 2 was higher than that of the blend type 1. The reasoning for the above properties was as follows: The differences in the expansion of the PVC (rubber state) and the PMMA (glass state) at high temperature (395 K) concentrated the warp stress at the interface and decreased the strength of the interface. However, in blend type 2, the phase containing an excellent wide concentration gradient prevented the warp stress from concentrating. Thus, the thermal shock resistance of the blend (blend type 2) with excellent wide concentration gradient was found to be superior to that of the similar blend (blend type 1) to a laminate film. The formation of an excellent wide concentration gradient was found to have improved the strength of the interface.

3.4.1.3 Smart Performance (DMA Properties)

The change in tensile storage modulus and tan δ of the PVC/PMMA blend type 2 with a wide concentration gradient around T_g was compared with the perfectly miscible blend (5/5) using the DMA measurement (temperature increasing rate: 1 K/min, frequency: 0.2 Hz). The T_g width measurements of storage modulus and half temperature of T_g width of tan δ were estimated as shown in Table 3.3.

The half width of the temperature of tan δ for the former (16 K) was significantly larger than that of the latter (10 K). Thus, the blend type 2 was confirmed to be a continuous phase, having a wide range of T_g.

TABLE 3.3 Properties of PVC/PMMA Functionally Graded Blends

		PVC/PMMA Blend				
Properties	Unit	Type 2[a]	Type 1	PMT[b]	PVC Only	PMMA Only
Tensile properties						
Tensile strength	(kgf/mm²)	6.4	4.5[c]	7.2	5.7	6.1
Elongation at break	(%)	4.5	2.8[c]	5.2	3.9	3.1
Tensile modulus of elasticity	(kgf/mm²)	200	190	220	230	230
DMA properties (tensile mode)						
T_g width of storage modulus	(K)	20	8.6, 11[c]	11	—	—
Half-temperature width T_g in tan δ	(K)	16	—	10	—	—
Thermal shock resistance						
Maximum warp angle	(degree)	9	170[d]	—	—	—
Adhesive strength in shear by tension loading	(kgf)	98	71[d]	—	—	—

[a] Blend containing graded structure 2.
[b] Perfectly miscible blend.
[c] Prepared by the hot press method.
[d] Blend containing graded structure 1.

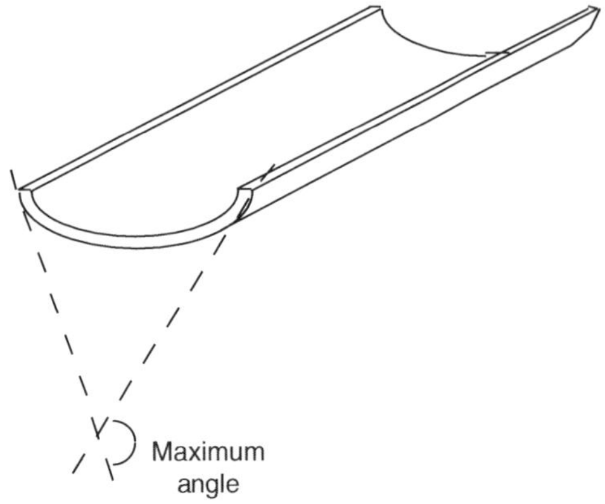

FIGURE 3.20 Measurement method of maximum angle of warp.

As a result, tan δ of the graded blends of PVC and several types of polyalkyl methacrylate (PMA) containing graded structure 2 were measured, as shown in Figure 3.21. Tan δ of the graded PVC/PHMA blend had the widest temperature range. Therefore, it was concluded that the wide temperature range was caused by the greater differences of T_g in the pair polymers of the graded PVC/PHMA blend.

Finally, we investigated the optimum conditions for preparing the graded PVC/PHMA blend with the wider temperature range of tan δ. We obtained a PVC/PHMA blend containing an excellent graded structure 2, which showed a peak of tan δ in a much wider temperature range in comparison with those of the blend containing the graded structure 1 and perfectly miscible blend (5/5), as shown in Figure 3.22.

In both systems of the PVC/PMMA and the PVC/PHMA blends, we found that the tensile storage modulus of the blend type 2 contained an excellent graded structure 2 that began to decrease at a lower temperature than that of the perfectly miscible blend (5/5) and did not have a terrace, while that of a similar blend containing the graded structure 1 in a laminate had some terraces.

Sandwich steel beams combining a polymer have been used as damping materials [46]. It is known that the damping efficiency is maximum in the temperature range at which the used polymer has a peak of tan δ. Therefore, it is expected that an excellent graded blend with a peak of tan δ in a much wider temperature range is useful as a damping material in a large temperature range. This is what the graded polymer blend is expected to be for smart materials. The following principle reflects the reasoning for this condition for smart materials.

An excellent graded blend was used for the polymer combined with the steel plates, as shown on the right of Figure 3.23. T_g of the graded blend decreases when shifting occurs from the left to the right side. At the highest temperature, that is, the same temperature as the higher T_g of the pair polymers in the blend, the area in the farthest left side shows the highest and best damping performance. As the area shifts to the right side, there is a decrease of temperature. Finally, at the lowest temperature, that is, the same temperature as the lower T_g of the pair polymers in the blend, the area on the farthest right side shows the highest

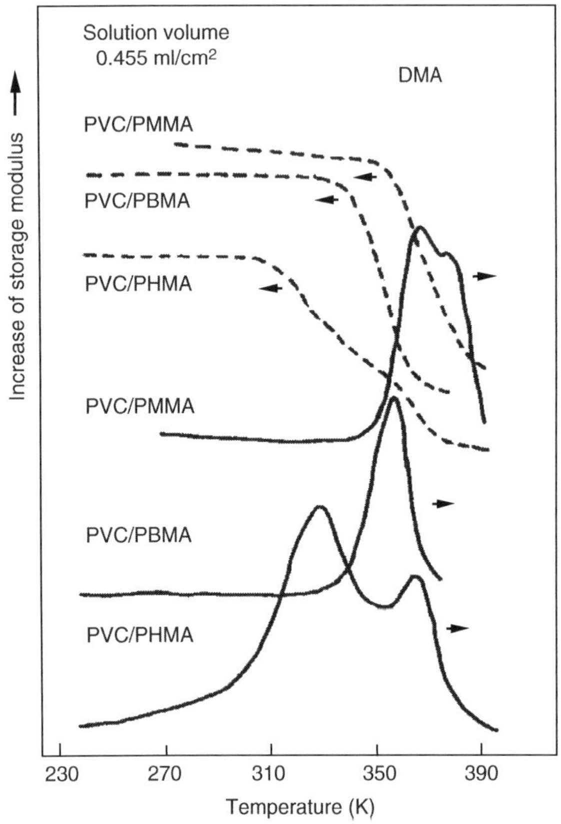

FIGURE 3.21 DMA data of PVC/PMA graded blends.

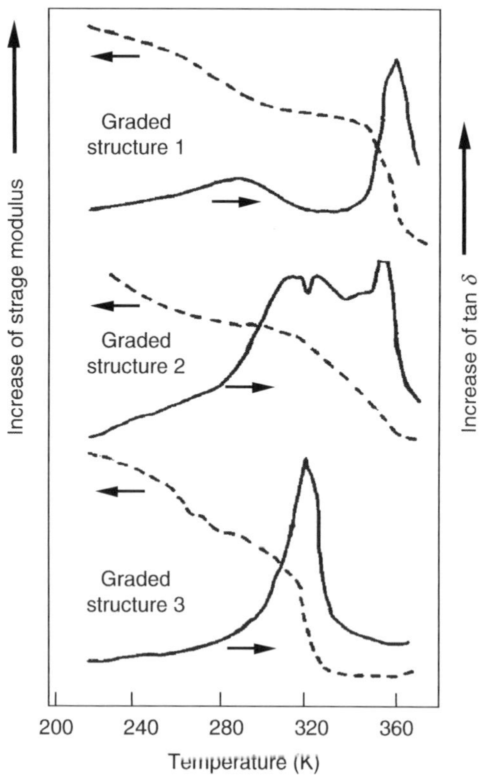

FIGURE 3.22 DMA data of PVC/PHMA graded blends.

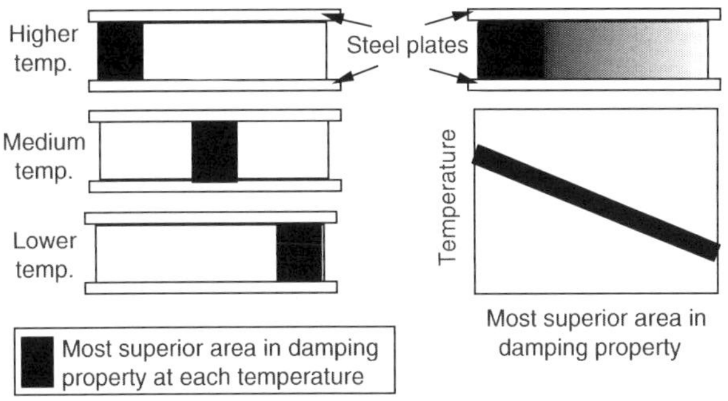

FIGURE 3.23 Schematic model of the so-called smart performance in the damping property of steel plate combined with a functionally graded blend.

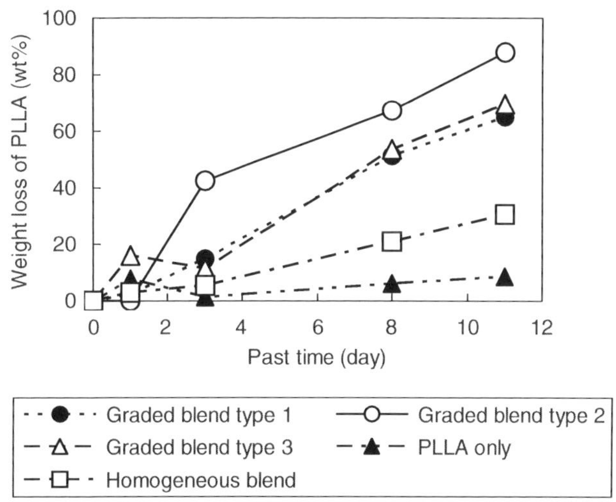

FIGURE 3.24 Changes of net weight loss of PLLA in various types of graded blends, homogeneous blend, and PLLA only, in the biodegradation test.

and best excellent damping performance. Therefore, the area showing the high damping performance shifts with the changing temperature. This performance is thus considered as one of the so-called "smart performances."

3.4.2 Functional and Smart Performances of PEO/PLLA Graded Blend

PLLA is known as biomass and biodegradable polymer and its biodegradability was enhanced by blending with PEO. Thus, we prepared the PEO/PLLA graded blend, which was expected to have both high biodegradability and tensile strength in the vertical direction of thickness [26]. The various types of graded PLLA/PEO blends, homogeneous blend, and PLLA only were subjected to degradation by a proteinase K enzyme. The biodegradability was evaluated by the net weight loss of PLLA calculated by difference of weightbefore and after testing. The net weight loss in all of the graded blends was higher than those in the homogeneous blend and the PLLA only (Figure 3.24). The net weight loss of graded blend type 2 with the best excellent graded structure was the highest. Thus, the graded structure was considered to largely increase biodegradability. It was confirmed by an SEM observation of these materials (Figure 3.25). The net weight loss of all the graded blends, where a porous structure formed before the biodegradation test, did not result in increase to a great extent. It was

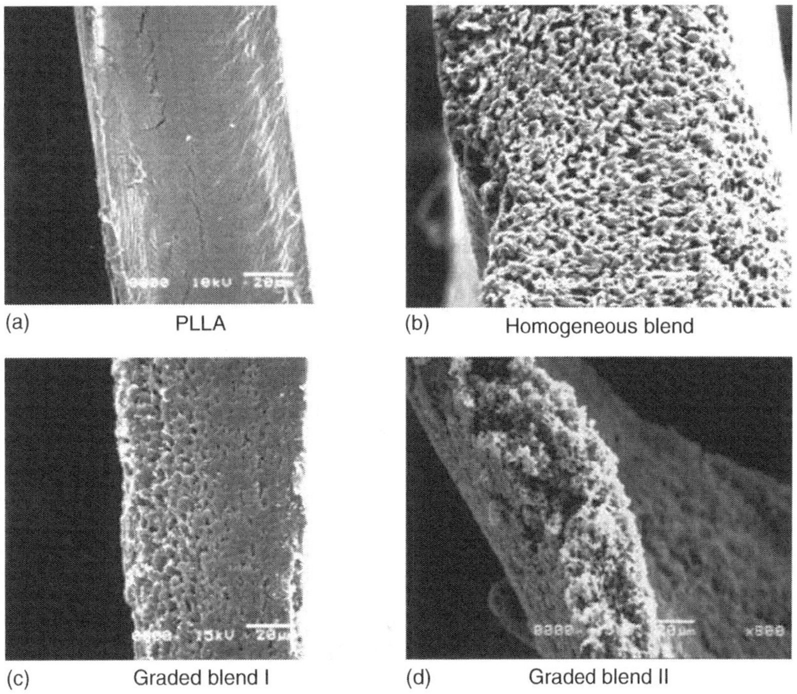

FIGURE 3.25 SEM photograph of the cross-section of the sample after biodegradation test for 11 days.

TABLE 3.4 Tensile Strength of Several Types of Materials

Materials	Tensile Strength (MPa)
Graded blend 1	2.46
Graded blend 2	3.05
Graded blend 3	3.20
Homogeneous blend	1.78
PLLA only	2.88

considered as the following: PEO, not only dissolved into water resulting in increasing the surface area attacked by the enzyme, but also, absorbed the PLLA oligomers that acted as acid catalysts of water decomposition and subsequently, it promoted the decomposition of PLLA. Furthermore, the strengths of all the graded blends were larger than that of the homogeneous blend (Table 3.4). Therefore, it was concluded that the graded structure enhanced the biodegradability while maintaining its high tensile strength.

3.4.3 Functional and Smart Performances of PEO (or PEO/LiOCl$_4$)/PBMA Graded Blend

PEO/PBMA graded blend was prepared by the polymerization–diffusion method [30]. It was found that the water vapor permeability across the graded blend films was different in the PEO content on the side of the higher pressure of water vapor (humidity: 90%) at 313 K. Thus, the permeability of the vapor from the PEO-rich side (direction 1) was always higher than that of the vapor from the PBMA-rich side (direction 2) (Figure 3.26). Therefore, it was found that the graded structure could cause anisotropy in the permeability of the water vapor.

(PEO/LiClO$_4$)/PBMA graded blend was prepared using EO-*b*-BMA block copolymer by the diffusion–dissolution method [28]. The change in the elastic modulus of the graded blend in the thickness direction was estimated as shown in Figure 3.27. The modulus of the graded blend decreased gradually around 45%, while that of the laminate began to decrease immediately (Figure 3.28). Thus, it was concluded that the graded blend was prevented from its breakaway at the interface

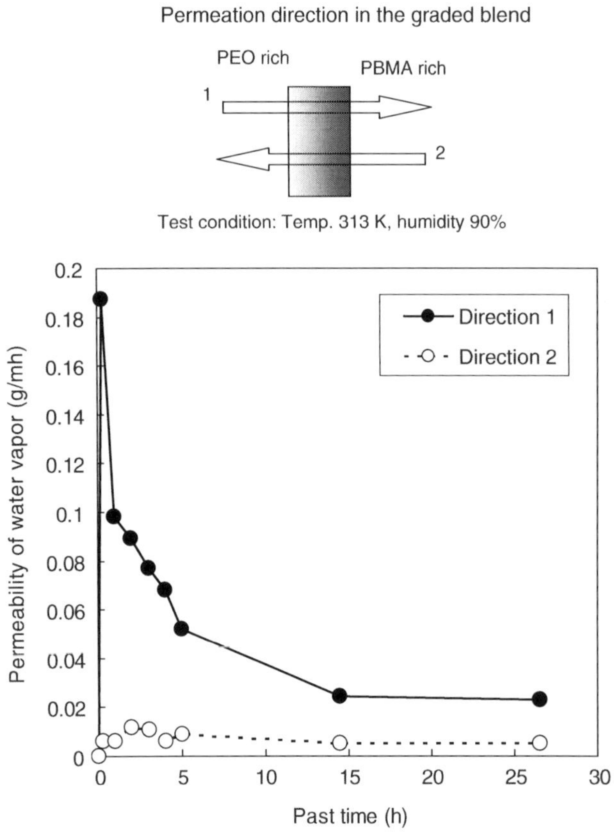

FIGURE 3.26 Permeability of water vapor of PEO/PBMA graded blend in both types of the thickness directions.

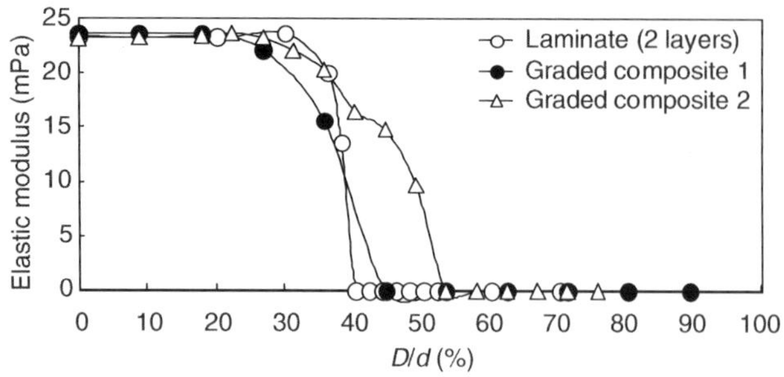

FIGURE 3.27 Tensile elastic modulus of several types of materials in the thickness direction (*D/d*: (Distance from petri glass side)/(thickness of sample)).

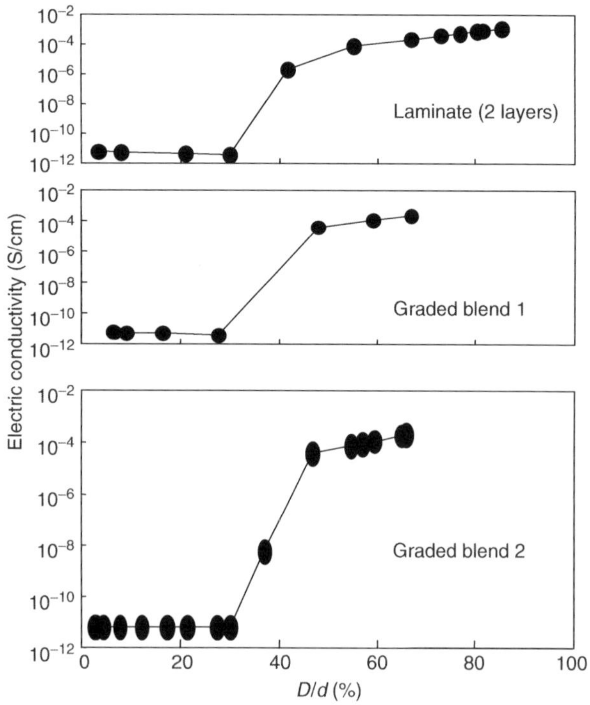

FIGURE 3.28 Electric conductivity of several types of materials in the thickness direction (*D/d*: (Distance from petri glass side)/(thickness of sample)).

TABLE 3.5 Possibility of Applications of Functionally Graded Polymer Blend

Expected Functional Property	Application
Relaxation of thermal stress	Mechanical device for antiabrasion
	Sporting goods
	Construction materials
Prevention of vibration and sound	Vibration and sound proof
Electromagnetic materials	Electromagnetic shield
	Copy machine device
Photomaterials	Optic fiber
	Lens
Medical materials	Artificial internal organs
	Artificial blood vessels and organs
Packing materials	Waterproof adhesive
Chemical	Chemical resistance materials

by its graded structure. However, all of the electric conductivities of the graded blends and the laminate began to similarly increase immediately around 40%.

3.5 Prospects for Application in the Functionally Graded Blends

Functionally graded polymer blends are expected to be used in the future as a substitute for laminates because of their superiority in their strength and thermal shock resistance. The superiority, we believe, is caused by the lack of interface, which suppresses the break at the interface and the thermal stress. Furthermore, the excellent wide compositional gradient for the graded structure results in improvement in several types of physical properties. Therefore, some new functional performances are expected, which could be caused by an improved physical property gradients, and can be applied in various types of applications, as shown in Table 3.5.

References

1. Society of Functionally Graded Material, ed., *Functionally Graded Materials*, Kogyo Chyosakai, Tokyo, 1993.
2. T. Kitano, *Kogyo Zairyo*, 43(6),112, 1994.
3. M. Takayanagi, The 23rd Colloquium of Structure and Property of Polymer, Tokyo, 1993.
4. Y. Agari, *Func Grad.. Mater.*, 16(4), 32, 1996.
5. Y. Agari, *Koubunshi Kako*, 46(6), 251, 1997.
6. Y. Agari, M. Shimada, A. Ueda, and S. Nagai, *Macromol. Chem. Phys.*, 2017, 1996.
7. Y. Agari, M. Shimada, A. Ueda, T. Anan, R. Nomura, and Y. Kawasaki, *Func. Grad. Mater.*, 1996, 761, 1997.
8. Y. Agari, M. Shimada, M.Ueda, R. Nomura, and Y. Kawasaki, *Polym. Preprints*, 47, 701, 1998.
9. Y. Agari, M. Shimada, H. Shirakawa, R. Nomura, and Y. Kawasaki, *Polym. Preprints*, 48, 698, 1999.
10. J. Z. Yu, C. Lei, and F. K. Ko, *Annu. Tech. Conf. SPE*, 52, 2352, 1994.
11. M. Omori, A. Okubo, K. Gilhwan, and T. Hirai, *Func. Grad. Mater.*, 1996, 764, 1997.
12. M. Funabashi, T. Kitano, *Seni-Gakkaishi*, 50(12), 573, 1994.
13. C. M. Thai, T. Kato, and A. Yoshizumu, *J. Thermosetting Plastics, Japan,* 16(3), 126, 1995.
14. X. M. Xie, M. Matsuoka, and K. Takemura, *Polymer*, 33(9), 1996, 1992.
15. S. Kanayama and T. Umemura, *Seikei Kako*, 2(4), 216, 1995
16. Y. Kano, S. Akiyama, H. Sano, and H. Yui, *J. Electron Microsc.*, 44(5), 344, 1995.
17. S. Akiyama and Y. Kano, *Kagaku To Kogyo*, 71, 44, 1997.
18. S. Murayama, S. Kuroda, and Z. Osawa, *Polymer*, 34(18), 3893, 1993.
19. E. Jabbari and N. A. Peppas, *Macromolecules*, 26, 2175, 1993.
20. P. F. Nealey, R. E. Cohen, and S. Argon, *Macromolecules*, 27, 4193, 1994.
21. K. C. Farinas, L. Doh, S. Venkatraman, and R. O. Potts, *Macromolecules*, 27, 5220, 1994.
22. T. E. Shearmur, A. S. Clough, D. W. Drew, M. G. D. van der Grinten, and R. A. L. Jones, *Macromolecules*, 29, 7269, 1996.
23. M. A. Parker and D. Vesely, *J. Polym. Sci., Part B Polym. Phys.*, 24, 1869, 1986.
24. Y. Agari, M. Shimada, A. Ueda, T. Koga, R. Nomura, and Y. Kawasaki, *Polym. Preprints Jpn.*, 45, 2241, 1996.
25. Y. Agari, M. Shimada, A. Ueda, T. Koga, R. Nomura, and Y. Kawasaki, *Polym. Preprints Jpn.*, 46, 657, 1997.
26. Y. Agari, Y. Kano, K. Sakai, and R. Nomura, *Polym. Preprints, Jpn.*, 51, 1065, 2002.
27. Y. Agari, Y. Anan, R. Nomura, and Y. Kawasaki, *Polymer*, 48, 1139, 2006.
28. Y. Agari, T. Morita, M. Shimada, and R. Nomura, *J. Jpn. Soc. Col. Mater.*, 75, 474, 2002.
29. Y. Agari, K.Ohishi, M. Shimada, and R. Nomura, *Polym. Preprints Jpn.*, 51, 647, 2002.
30. Y. Agari, T. Yamamoto, and R. Nomura, *Kagaku to Kogyo (Japan)*, 80, 138, 2006.
31. M. Kryszewski and G. Czeremuszkin, *Plaste u Kautchuk*, 11, 605, 1980.
32. P. Milczarek and M. Kryszewski, *Coll. Polym. Sci.*, 265, 481, 1987.
33. Y. Koike, H. Hidaka, and Y. Ohtsuka, *Appl. Opt.*, 22, 413, 1983.
34. Y. Koike, N. Tanio, E. Nihei, and Y. Ohtsuka, *Polym. Eng. Sci.*, 29(17), 1200, 1989.
35. C. F. Jasso and E. Mendizabal, *Annu. Tech. Conf. SPE*, 50, 2352, 1992.
36. S. Ashai, *Polym. Preprints, Jpn.*, 27, 18, 1978.
37. S. Ashai, The 6th Symposium Functionally Graded Materials, 61, 1993.
38. S. Ashai, *Kagaku To Kogyo*, 71, 50, 1997.
39. Y. Tsukahara, N. Nakamura, T. Hashimoto, and H. Kawai, *Polym. J.*, 12(12), 455, 1980.

40. D. Greszta, K. Matsuoka, and K. Matyaszewski, *Polym. Reprints ACS*, 37, 569, 1996.
41. M. Furukawa, T. Okazaki, and T. Yokoyama, *Polym. Preprints, Jpn.*, 45, 2239, 1996.
42. G. B. Park, M. Hirata, Y. Kagari, T. Matsunaga, J. Gong, Y. Osada, and D. C. Lee, *Polym. Preprints, Jpn.*, 45, 1836, 1996.
43. Y. Ulcer, M. Cakmak, and C. M. Hsiung, *J. Appl. Polym. Sci.*, 60(1), 125, 1996.
44. Y. Agari and A. Ueda, *J. Polym. Sci., Part B Polym. Phys.*, 32, 59, 1994.
45. M. Shimada, Y. Agari, and Y. Makimura, *Polym. Preprints, Jpn.*, 45, 1958, 1996.
46. D. J. Mead, *J. Sound Vib.*, 83, 363, 1982.

4

Structural Application of Smart Materials

R. Sreekala
Structural Engineering Research Centre, CSIR

K. Muthumani
Structural Engineering Research Centre, CSIR

4.1 Introduction

The development of durable and cost-effective high-performance construction materials and systems is important for the economic well-being of a country mainly because the cost of civil infrastructure constitutes a major portion of the national wealth. To address the problems of deteriorating civil infrastructure, research is very essential on smart materials. The research and development projects aiming to apply advanced technologies, such as new materials and new structural systems, can improve the performance of the buildings, reduce the expense of maintenance, and eventually ensure the future sustainability of the buildings. This chapter highlights the use of smart materials for the optimal performance and safe design of buildings and other infrastructures particularly those under the threat of earthquake and other natural hazards. Smart materials with embedded desired functions such as sensing and processing or with improved structural performances such as strength, ductility, usability, and low cost are the features that need to be explored and applied in structures.

4.2 Materials and Application

4.2.1 Materials

4.2.1.1 Shape Memory Alloys

The term shape memory alloys (SMA) refers to the ability of certain alloys (Ni–Ti, Cu–Al–Zn, etc.) to undergo large strains, while recovering their initial configuration at the end of the deformation process spontaneously or by heating without any residual deformation. The particular properties of SMAs are strictly associated with a solid–solid phase transformation, which can be thermal or stress-induced. Currently, SMAs are mainly applied in medical sciences, electrical, aerospace, and mechanical engineering and the recent studies indicate that they can open new applications in civil engineering specifically in seismic protection of buildings.

Properties, which enable Ni–Ti wires for civil engineering application, are as follows:

1. Repeated absorption of large amounts of strain energy under loading without permanent deformation
2. Possibility to obtain a wide range of cyclic behavior from supplemental and fully recentering to highly dissipating by simply varying the characteristics of SMA components
3. Usable strain range of 70%
4. Extraordinary fatigue resistance under large strain cycles

4.2.1.1.1 Substitute for Steel?

It is reported that the fatigue behavior of Cu–Zn–Al SMAs is comparable with steel [1]. If larger diameter rods can be manufactured, it has a potential for use in civil engineering applications. Use of fiber-reinforced plastics with SMA reinforcements requires future experimental investigations.

4.2.1.2 Piezoelectric Materials

Piezoelectricity was discovered as early as 1880 by Curie brothers [2]. When integrated into a structural member, a piezoelectric material generates an electric charge or voltage in response to mechanical forces or stresses. This phenomenon is called direct piezoelectric effect. This is useful for sensing. The converse piezoelectric effect can be used for control. Among the wide variety of sensing and actuation devices, the advantages of using piezoelectric actuators and sensors include their effectiveness over a wide frequency range, simplicity, reliability, compactness, and lightweight. Even though lot of progress has been made in the laboratory level structural testing, their usage in large-scale civil engineering structures is limited due to high-voltage requirements.

4.2.1.3 Engineered Cementitious Composites

Engineered cementitious composites (ECC) is mortar or concrete reinforced by chopped fiber. Such composite materials have been microstructurally designed using micromechanical principles. ECC exhibits strain hardening with large strain capacity and shear ductility, and good damage-tolerant mechanical behavior. The use of high-performance cementitious structural elements as energy dissipation devices and damage-tolerant elements helps to achieve a better performing and a damage-tolerant structural system. Development of concrete encased steel column elements also comes under this category.

4.2.1.3.1 Carbon Fiber Reinforced Concrete

Its ability to conduct electricity and most importantly capacity to change its conductivity with mechanical stress makes a promising material for smart structures. It is evolved as a part of densified reinforced composites (DRC) technology. The high density coupled with a choice of fibers ranging from stainless steel to chopped carbon and kelvar, applied under high pressure, gives the product outstanding qualities as per DRC technology. This technology makes it possible to produce surfaces with strength and durability superior to metals and plastics.

4.2.1.3.2 Smart Concrete

A mere addition of 0.5% specially treated carbon fibers enables the increase of electrical conductivity of concrete. Putting a load on this concrete reduces the effectiveness of the contact between each fiber and the surrounding matrix and thus slightly reduces its conductivity. On removing the load, the concrete regains its original conductivity. Because of this peculiar property, the product is called "smart concrete." The concrete could serve both as a structural material as well as a sensor.

The smart concrete could function as a traffic-sensing recorder when used as road pavements. It has got higher potential and could be exploited to make concrete reflective to radio waves and thus suitable for use in electromagnetic shielding. The smart concrete can be used to lay smart highways to guide self-steering cars, which at present follow tracks of buried magnets. The strain-sensitive concrete might even be used to detect earthquakes.

4.2.1.4 Electro- and Magnetorheological Fluids

Electro- and magnetorheological (ER/MR) fluids have essential characteristics that change from free-flowing, linear viscous fluid, to a semisolid with a controllable yield strength in milliseconds when exposed to an electric and magnetic field. These fluids are variable contenders for the development of controllable devices. It is intended to develop a structure that controls its stiffness and damping characteristic to behave adaptively against earthquake or wind forces and achieve safety and function by using ER/MR devices with lesser energy. A large-scale MR damper of 20 t capacity has been constructed and tested [3] and the experimental results indicate the applicability of MR damper in the semiactive control mode in the real-world applications.

4.2.1.5 Induced Strain Actuators

Induced strain actuators (ISA) change their own shapes according to external electric and magnetic fields and vice versa. These materials have been widely used for small and precision machines because of some advantages such as small sizes, rapid reaction, high power, and high accuracy, etc. ISA materials act as sensors because they cause change in electric or magnetic fields under deformation. ISA materials could be used to develop smart members to realize smart, comfortable, and safe structures. ISA materials can be also utilized for vibration mode control of structural members and sensor development. They are widely used in active floor vibration control.

Possible applications of ISA include

1. Long span structure
2. Axial force and friction control (for base isolator, including trigger application)
3. Active sound transparency (noise control) [4]
4. Wireless sensors of deformation

4.3 Structural Uses

4.3.1 Active Control of Structures

The concept of adaptive behavior has been an underlying theme of active control of structures, which are subjected to earthquake and other environmental type of loads. The structure adapts its dynamic characteristics to meet the performance objectives at any instant. A futuristic smart bridge system (an artist's rendition) is shown in Figure 4.1 [5].

A thermomechanical approach to develop a constitutive relation for bending of a composite beam with continuous SMA fibers embedded eccentric to neutral axis was used by Sun and Sun [6]. The authors concluded that SMAs can be successfully used for the active structural vibration control. Thompson et al. [7] also conducted an analytical investigation on the use of SMA wires to dampen the dynamic response of a cantilever beam constrained by SMA wires.

To date, active structural control has been successfully applied to over 20 commercial buildings and more than 10 bridges. One example of active control is the Kurusima bridge in Shikoku area

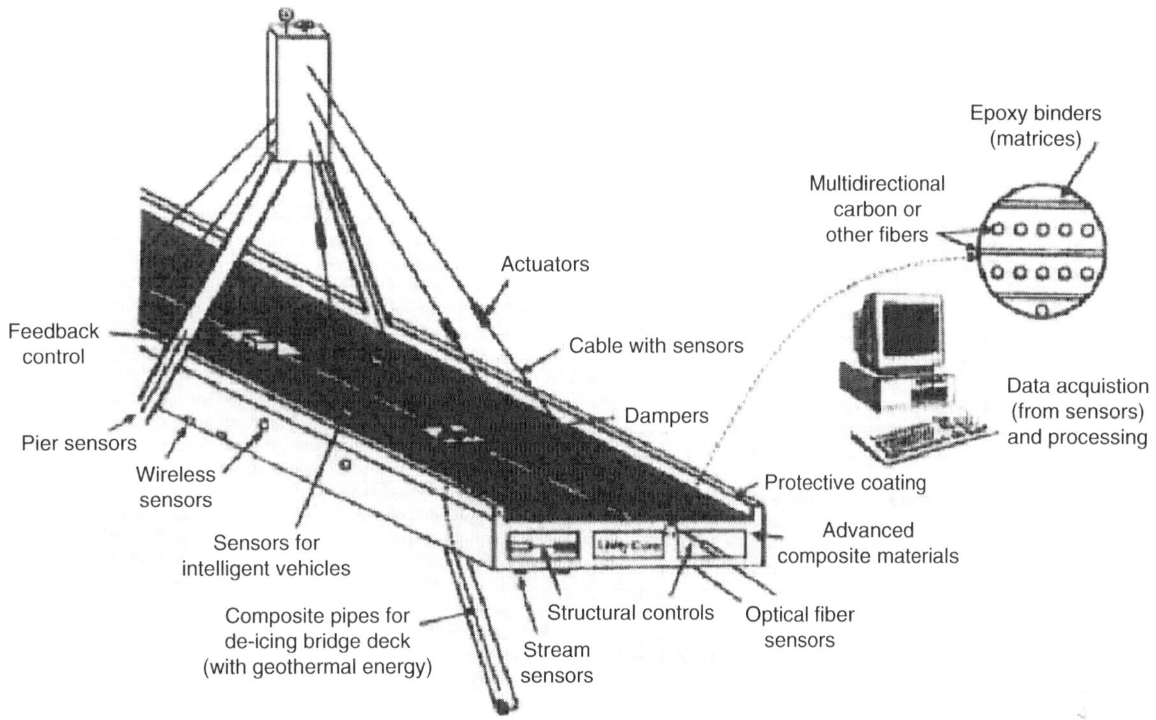

FIGURE 4.1 Futuristic smart bridge system (an artist's rendition). (*Courtesy*: *USA Today*, 3 March 1997.) J. Holnicki-szulc and J. Rodellar (eds.), *Smart Structures*, Requirements and potential applications in mechanical and civil engineering, NATO Science Series 3. High Technology, Vol. 65, Springer 1999.

in Japan. The bridge was designed with the application of active vibration control as integrated structural components. Several modes of the bridge tower, which were anticipated to be excited by wind vortex, were carefully protected by appropriate controllers during the construction phase. It therefore made it possible for the tower of this bridge to be built much lighter and more slender than one following traditional design. Active tuned mass dampers have been installed in the 11-story building, the Kyobashi Seiwa building in Tokyo (the first full-scale implementation of active control technology) and the Nanjing Communication Tower in Nanjing, China.

There are two serious challenges that remain before active control can gain general acceptance by engineering and construction professionals at large. They are (1) Reduction of capital cost and maintenance (2) Increasing system reliability and robustnessActive control systems consist a set of sensors, a controller, an active control system (actuators), and an external power supply. A schematic sketch of active control system is shown in Figure 4.2.

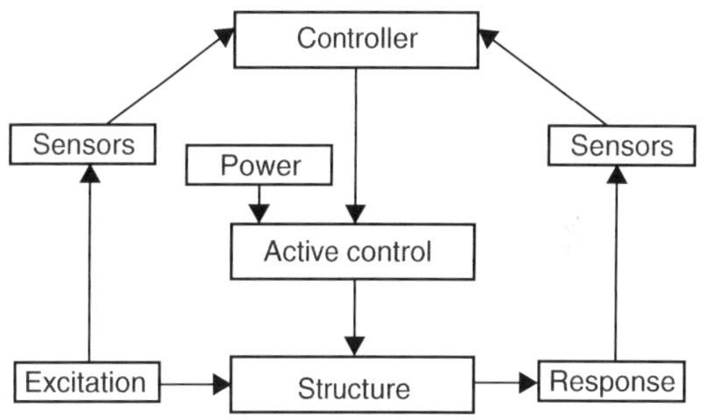

FIGURE 4.2 Schematic sketch of active control system.

Nowadays, much work on structural control is focused on intelligent structures, developments of actuating materials and piezoceramics. Due to its limited frequency bandwidth, SMA has traditionally been used for passive strategies such as dampers or other types of energy dissipation devices instead of being used in active control strategies as actuators. On the other hand, it is well known that a SMA actuator is capable of producing relatively large control forces despite its slow response time. This unique characteristic of SMA makes it very attractive for civil engineering control applications where large forces and low-frequency band width are mostly encountered. They conducted experimental studies on active control of a five-story building model with SMA actuators [8]. It was proven that despite its slow response, it is feasible to use SMA for active control of civil engineering structures. But while selecting the alloy type, utmost care should be taken in specifying the temperature range or the transition temperature or the process in alloy-making so as to suit our requirements. Dynamics of SMA should be considered in control design.

4.3.2 Passive Control of Structures

Two families of passive seismic control devices exploiting the peculiar properties of SMA kernel components have been implemented and tested within the Memory Alloys for New Seismic Isolation and Energy Dissipation Devices (MANSIDE) project. They are special braces for framed structures and isolation devices for buildings and bridges.

4.3.3 Hybrid Control

The term hybrid control generally refers to a combined passive and active control system. Since a portion of the control objective is accomplished by the passive system, less active control effort, implying less power resource, is required. Similar control resource savings can be achieved using the semiactive control scheme where the control actuators do not add mechanical energy directly to the structure, hence bounded-input/bounded-output stability is guaranteed. Semiactive control devices are often viewed as controllable passive devices. A side benefit of hybrid and semiactive control systems is that, in the case of a power failure, the passive components of the control still offer some degree of protection unlike a fully active control system. The materials described above can be used to make the hybrid control scheme workable. In the case of ER/MR dampers, hybrid control scheme is a viable option for realistic structural control.

4.3.4 Smart Material Tag

These smart material tag can be used in composite structures. These tags can be monitored externally throughout the life of the structure to relate the internal material condition. Such measurements as stress, moisture, voids, cracks, and discontinuities may be interpreted via a remote sensor [6].

4.3.5 Retrofitting

SMAs can used as self-stressing fibers and thus they can be applied for retrofitting. Self-stressing fibers are the ones in which reinforcement is placed into the composite in a nonstressed state. A prestressing force is introduced into the system without the use of large mechanical actuators by providing SMAs. These materials do not need specialized electric equipments nor do they create safety problems in the field. Treatment can be applied at any time after hardening of the matrix instead of during its curing and hardening. Long- or short-term prestressing is introduced by triggering the change in SMA shape using temperature or electricity. They make active lateral confinement of beams and columns a more practical solution. Self-stressing jackets can be manufactured for rehabilitation of existing infrastructure or for new construction.

4.3.5.1 Restoration of Cultural Heritage Structures Using SMA Devices

An innovative technique using superelastic SMA devices for the restoration of a cultural heritage structure especially masonry buildings were implemented under the framework of the European Commission-funded ISTECH Project. Masonry buildings are largely vulnerable to earthquakes because of their low resistance and ductility during earthquake ground motion. To enhance the seismic behavior of cultural heritage structures, the most common method traditionally used has been the introduction of localized reinforcements. Usually steel bars or cables served this purpose by increasing stability and ductility. But in many cases, these reinforcement techniques prove inadequate to prevent collapse.

The development of the connection technique was based on the idea of using the unique properties of Ni–Ti alloys especially its super elasticity and high resistance to corrosion [9]. The idea was to connect the external walls to the floors, the perpendicular walls or the roof with an SMA Device that should behave as follows:

1. Under low-intensity horizontal actions (wind, small intensity earthquakes), the device remains stiff, as traditional steel connections do, not allowing significant displacements.
2. Under higher intensity horizontal actions (i.e., strong earthquakes), the stiffness of the device decreases, allowing "controlled displacements," which should reduce amplification of accelerations (as compared to stiff connections) and permit the masonry to dissipate part of the transmitted energy, mainly owing to elasticity exploitation and microcracks formation in the brick walls; consequently, the structure should be able to sustain a high-intensity earthquake without collapse, though undergoing some minor damage.
3. Under extraordinary horizontal actions, the stiffness of the device increases and thus prevents instability.

4.3.5.2 SMA for Seismic Retrofit of Bridges

Unseating of supports was the major cause of bridge failures during earthquakes. Retrofit measures to reduce the likelihood of collapse due to unseating at the supports have been in place for many years. The damage to bridges in the recent Chi-Chi, Kobe, and Northridge earthquakes indicate the need to provide better methods of reducing the damaging effects of earthquakes in bridges. The use of restrainer cables and restrainer bars to limit the relative hinge displacement became popular in the United States following the collapse of several bridges due to loss of support during the 1971 San Fernando earthquake. Recent earthquakes have demonstrated that restrainers were effective in some cases. However, many bridges with restrainers sustained serious damage or collapse. Bridges that had been retrofitted with restrainer cables failed in both the 1989 Loma Prieta and 1994 Northridge earthquakes. Failure of Japanese restraining devices also occurred during the 1995 Kobe earthquake. Experimental tests of restrainer cables have shown that failure occurs in the connection elements or the through-punching shear in the concrete diaphragm. In addition, restrainers do not dissipate any significant amount of energy, since they are generally designed to remain elastic. Analytical studies of bridge and restrainer systems have demonstrated that a very large number

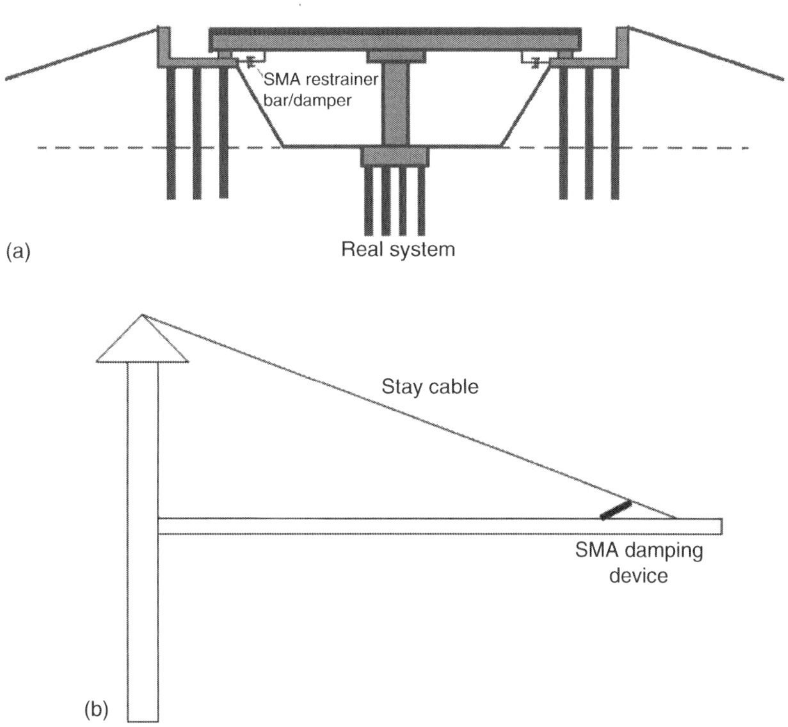

FIGURE 4.3 SMA restrainer bar used in multispan simply supported bridge abutments and intermediate piers configuration.

of restrainers are often required to limit joint movement to acceptable levels, particularly for high seismic loads. In those cases, the excessive number of restrainers would induce large forces in other components of the bridge, such as bearings and columns. The shortcomings of traditional restrainers can potentially be addressed with the use of SMA restrainers. The SMA restrainers, in the superelastic phase, act as both restrainers and dampers (Figure 4.3).

Energy dissipation and base isolation are found to be optimal candidates for structural control of structures. As far as bracing systems are concerned, until now all the applications and the research studies on this technique were focused on the energy dissipation capability. For seismic retrofitting purposes, the supplemental recentering devices (SRCD) are found to be useful as they provide forces to recover the undeformed shape of the structure at the end of the action. In existing structures, in fact, particularly when they were designed without any seismic provision, the energy dissipation can turn out to be insufficient to limit damage to structural elements. It would then be necessary to strengthen some elements to fully achieve the design objectives. Local strengthening would imply expensive work, also involving nonstructural parts. Retrofitting could turn out to be economically in convenient, and, yet some residual displacement could occur in case elastoplastic devices are used. An alternative strategy can be pursued by using SMA devices having supplemental force to recover the undeformed structural configuration, resulting in the elimination of any residual displacement, while accepting yielding in structural elements. A comparison of properties of Ni–Ti with steel is given in the Table 4.1.

TABLE 4.1 Comparison of Properties of Ni–Ti SMA with Typical Structural Steel

Property	Ni–Ti SMA	Steel
Recoverable elongation (%)	8	2
Modulus of elasticity (MPa)	8.7×10^4 (A)	2.07×10^5
	1.4×10^4 (M)	
Yield strength (MPa)	200–700 (A)	248–517
	70–140 (M)	
Ultimate tensile strength (MPa)	900 (f.a.)	448–827
	2000 (w.h.)	
Elongation at failure (%)	25–50 (f.a.)	20
	5–10 (w.h.)	
Corrosion performance	Excellent	Fair

Note: f.a. denotes fully annealed and w.h. denotes work hardened, which are two types of treatment given to the alloy. A and M denote the two phases of the alloy namely, austenite and martensite.

4.3.6 Self-Healing

Experimentally proved self-healing behavior [10], which can be applied at microlevel of a material, widens their spectrum of use. Here significant deformation beyond the first crack can be fully recovered and cracks can be fully closed.

4.3.7 Self-Stressing for Active Control

Self-stressing for active control can be used with cementitious fiber composites with some prestress, which impart self-stressing thus avoiding difficulties due to the provision of large actuators in active control, which require continuous maintenance of mechanical parts and rapid movement, which in turn created additional inertia forces.

In addition to SMAs, some other materials such as polymers can also be temporarily frozen in a prestrained state that have a potential to be used for manufacturing of self-stressing cementitious composites [1].

4.3.8 Structural Health Monitoring

Use of piezotransducers, surface bonded to the structure or embedded in the walls of the structure can be used for structural health monitoring and local damage detection. Problems of vibration and UPV testing can be avoided here. Jones et al. [11] applied neural networks to find the magnitude and location of an impact on isotropic plates and experimented using an array of piezotransducers surface bonded to the plate. Figure 4.4 shows a typical health monitoring setup making use of optic fibers.

4.3.9 Active Railway Track Support

Active control system for sleepers is adopted [5] to achieve speed improvements on existing bridges and to maintain the track in a straight and nondeformed configuration as the train passes. With the help of optimal control methodology, the train will pass the bridge with reduced track deflections and vibrations and thus velocity could be safely increased. Figure 4.5 shows various positions of the train with and without active railway track support.

4.3.10 Active Structural Control against Wind

Aerodynamic control devices to mitigate the bidirectional wind induced vibrations in tall buildings are energy efficient, since the energy in the flow is used to produce the desired control forces. Aerodynamic flap system (AFS) is an active system driven by a feedback control algorithm based on information obtained from the vibration sensors [5]. The area of flaps and angular amplitude of rotation are the principal design parameters.

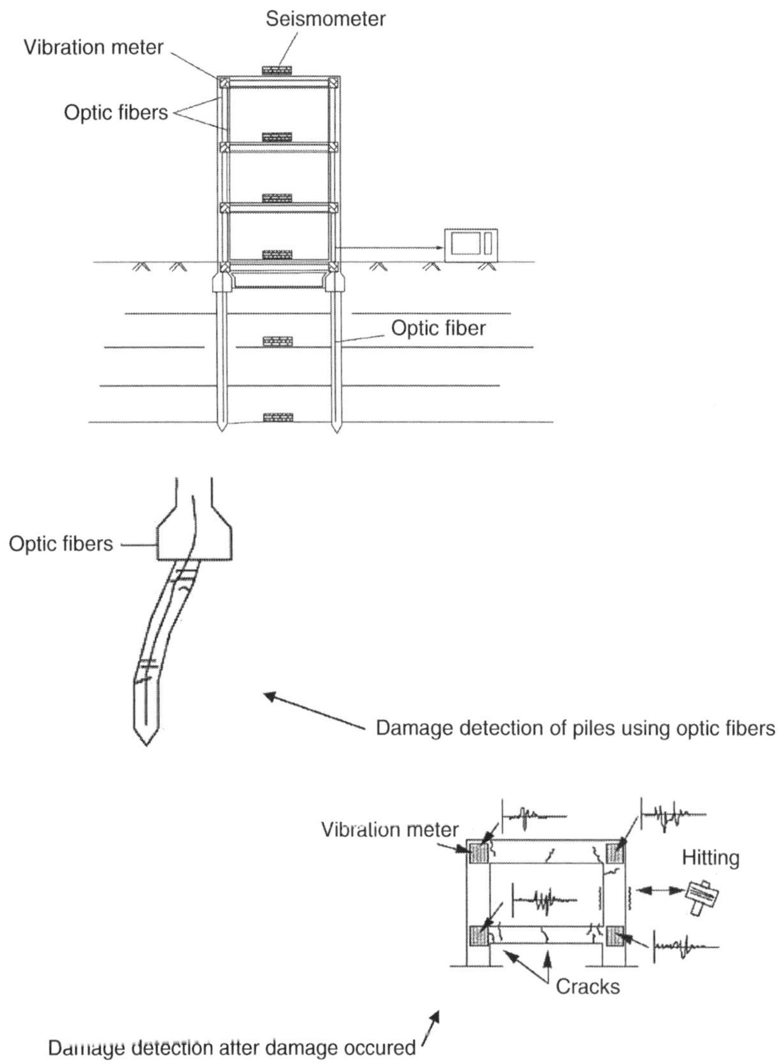

FIGURE 4.4 Health monitoring setup.

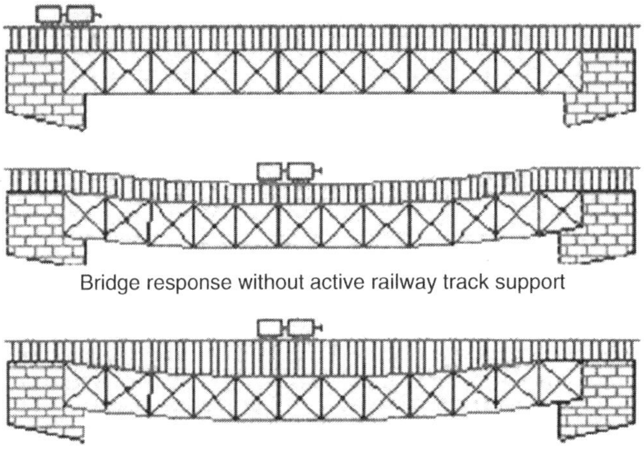

FIGURE 4.5 Active railway track support concept.

4.4 Conclusion

The technologies using smart materials are useful for both new and existing constructions. Of the many emerging technologies available, the few described in this chapter need further research to evolve the design guidelines of systems. Codes, standards, and practices are of crucial importance for the further development.

Acknowledgment

The authors gratefully acknowledge the Director, Structural Engineering Research Centre (SERC), CSIR, India for the constant encouragement and support rendered in preparation of this manuscript and also giving permission to publish it. The kind support and guidance of all the team members of Structural Dynamics Laboratory deserve acknowledgment.

References

1. N. Krstulovic-Opara and A.E. Naaman, Self Stressing Fiber Composites, *ACI Structural Journal*, 97, March–April 2000, 335–344.
2. W.G. Cady, *Piezo Electricity*, Dover, New York, 1964.
3. G. Yang, B.F. Spencer Jr., J.D. Carlson, and M.K. Sain, Large scale MR fluid dampers: Modeling and dynamic performance considerations, *Engineering Structures,* 24, 2002, 309–323.
4. A. Sampath and B. Balachandran, Active control of multiple tones in an endosure, *The Journal of the Acoustical Society of America*, 106-1, July 1999, pp. 211–225.
5. J. Holnicki-szulc and J. Rodellar (eds.), *Smart Structures*, Requirements and potential applications in mechanical and civil engineering, NATO Science Series., 3. High Technology, Vol. 65, Springer 1999.
6. G. Sun and C.T. Sun, Bending of shape memory alloy reinforced composite beam, *Journal of Materials Science*, 30(13), 1995, 5750–5754.
7. P. Thomson, G.J. Balas, and P.H. Leo, The use of shape memory Alloys for passive structural damping, *Smart Materials Structure,* 4, 1995, 36–42, IOP publishing limited.
8. J. Li, B. Samali, and C. Chapman, Experimental realisation of active control of a five storey building model using SMA actuators, *Advances in Mechanics of Structures and Materials*, Loo, Chowdhury and Fragomeni (eds.), 699–704.
9. M.G. Castellano and G. Manos, The ISTECH Project: Use of SMA in the Seismic Protection of Monuments, *Monument-98* (Workshop on Seismic Performation of Monuments), Lisbon, 1998.
10. D.J. Hannant and J.G. Keer, Autogeneous healing of Ti based sheets, *Cement and Concrete Research*, 13, 1983, 357–365.
11. R.T. Jones, J.S. Sirkis, and E.J. Friebele, Detection of impact location and magnitude for isotropic plates using neural networks, *Journal of Intelligent Material Systems and Structures*, 7, 1997, 90–99.

5

Composite Systems Modeling—Adaptive Structures: Modeling and Applications and Hybrid Composites

5.1 Hybrid Composites

S. Padma Priya

5.1.1 Introduction

Composites are materials made by the synergistic assembly of two or more constituting materials (matrix and reinforcement), engineered in such a way that they form a single component and yet can be distinguished on a macroscopic level. Matrix encases the reinforcements and acts as a binder for the fibers. Additionally, the matrix holds the reinforcements in a fixed position and the reinforcement enhances the properties of the composite system by imparting its mechanical and physical properties. Composite materials can be engineered with desired properties by choosing and varying the concentration of different types of matrices and reinforcements. The durability of composites is very high; they have very high strength to weight ratios, are resistant to environmental corrosion, and provide ease of use concept.

Composites are materials known from ancient days, when bricks were made (mud reinforced with straw). Until today, we have applications for components of aerospace materials. Mother Nature is the first maker of composites and some examples of natural composites include wood, bone, etc.

Composites are classified into several broad categories depending upon the type of matrices and reinforcements used. Depending on the types of matrices used they can be classified as

1. Metal matrix composites (MMC)
2. Ceramic matrix composites
3. Polymer matrix composites

Depending upon the reinforcement types they can be classified as

1. Fiber reinforced composites
2. Fabric reinforced composites
3. Particulate composites

In order to further enhance the properties of the composite materials, hybrid composites were developed. The behavior of hybrid composites appears to be simply a weighted sum of the individual components in which there exists a more favorable balance between the advantages and disadvantages. The hybrid composites offer advantages regarding structural integrity and sustained load under crash and impact conditions. Hybrid composites are influenced by a large number of microstructural parameters such as type of reinforcement, volume proportion of reinforcement, weave pattern of the fabric used as reinforcement, volume fraction of the reinforcement, etc. There are several types of hybrid composites and can be characterized as: (1) interply or tow-by-tow, in which tows of two or more constituent types of fiber are mixed in a regular or random manner; (2) sandwich hybrids, also known as core-shell, in which one material is sandwiched between two layers of another; (3) interply or laminated, where alternate layers of the two (or more) materials are stacked in a regular manner; (4) intimately mixed hybrids, where the constituent fibers are made to mix as randomly as possible so that no overconcentration of any one type is present in the material; (5) other kinds, such as those reinforced with ribs, pultruded wires, thin veils of fiber, or combinations of the above.

Hybridization can be done through many ways as follows:

1. Reinforcements
2. Matrices
3. Both reinforcements and matrices

5.1.1.1 Reinforcement Hybridized Composites

These are the composites, which consist of two or more types of reinforcements embedded in a single matrix. The reinforcements can be of different types as follows.

5.1.1.1.1 Synthetic (Man-Made) Fibers

This composite is composed of two types of reinforcements and both of them are synthetic. Composites of such types have been developed in order to achieve superior mechanical properties and high heat resistance capabilities. Various types of synthetic reinforcements have been developed such as glass fiber, carbon fiber, Polyetheretherketone (PEEK), rayon, nylon, Kevlar, polyester, acrylic, olefin, vinyl, aramid, etc. The following fibers can be hybridized (e.g., carbon fiber/PEEK hybrid fabric, polyethylene [PE]/glass fabric, carbon fiber/aramid fibers) in different (weight or volume) ratios according to the desired end property. For instance, if less weight however, a strong composite is required, a hybrid fabric can be considered. This would have one fabric composed of different types of fibers in it and some examples for this type of reinforcement are carbon/aramid, aramid/glass, and carbon/glass.

Of the materials mentioned above the PEEK material is very much in demand. PEEK is a linear aromatic thermoplastic based on the following repeating unit in its molecules.

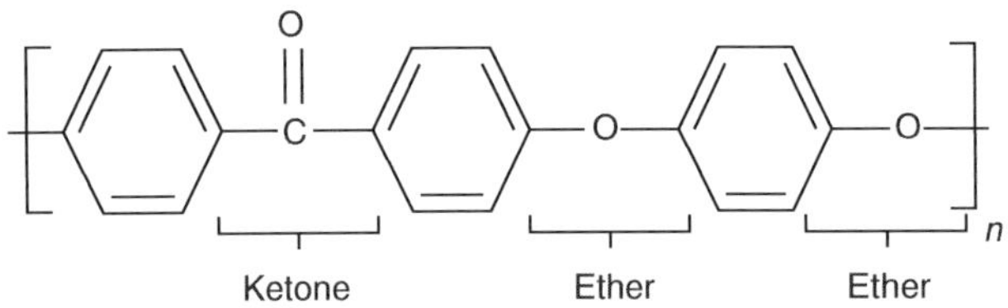

Polyether ether ketone

Continuous-carbon fiber reinforced PEEK composites are known in the industry as aromatic polymer composite (APC). PEEK is a semicrystalline polymer, and amorphous PEEK is produced when the melt is quenched.

The presence of fibers in PEEK composites tends to increase the crystallinity to a higher level because the fibers act as nucleation sites for crystal formation. Increasing crystallinity increases both the modulus and the yield strength of PEEK but reduces its strain to failure.

PEEK's maximum continuous-use temperature is 250°C. PEEK is the foremost thermoplastic matrix that may replace epoxies in many aerospace structures. Its outstanding property is its high fracture toughness, which is 50–100 times higher than that of epoxies. Another important advantage of PEEK is its low water absorption, which is less than 0.5% at 23°C compared to 4%–5% for conventional aerospace epoxies. Being semicrystalline, it does not dissolve in common solvents.

5.1.1.1.2 Natural Fibers

A composite in which its reinforcements come from natural resources is considered natural. Natural fibers, because they are naturally renewable, can be imbedded into biodegradable polymeric materials, which can produce a new class of biocomposites. Recently, society has realized that unless the environment is protected, it will be threatened due to the loss of natural resources. Conservation of forests and optimal utilization of agricultural and other renewable resources like solar energy and seawater has become almost mandatory. In this view, many attempts have been made to replace high-strength synthetic fibers with natural fibers. A single natural fiber may not impart all the properties for the final composite, thus it is hybridized with other types of natural fibers. Some of the known natural fibers are wool, cotton, silk, linen, hemp, ramie, jute, coconut, pinewood, pineapple, angora, mohair, alpaca, kapok, flax, sisal, bast fiber, kenaf, wood fiber, bamboo, banana, etc. To tailor the properties of a natural fiber reinforced composites, it is hybridized with the second type of natural fiber. Researchers have conducted studies on thermal conductivity and specific heat of jute/cotton, sisal/cotton, and ramie/cotton hybrid fabric-reinforced composites.

Small inorganic particles have also been used as fillers to prepare particulate composites. The most relevant reason to use filler in the composite is to reduce the overall cost of the material. Fillers are also used to increase flame retardancy, surface hardness, esthetic appeal, and the thermal properties of the composites. Some of the types of fillers that have been used are fly ash, silica, mica, granite powder, wood flour, aluminum oxide, carbon black, etc. Using such fillers and natural fibers or fabric, a new class of hybrid composites has been produced. One of the reinforcements is filler while the other is the fiber or the fabric, and here the filler particles are embedded between the fibrous reinforcement and the matrix. Other research developments underway have used sisal/saw dust in conventional composites and they possess an ease of processing, are environment friendly, and are economically affordable.

5.1.1.1.3 Natural and Synthetic Fibers

This type of composite consists both of natural and synthetic reinforcements in the composite system. Natural fiber reinforced composites do not always fulfill all the properties required for some technical applications. In such cases, hybridization with small amounts of synthetic fibers allows these natural fiber composites to be more suitable for technical applications. The most common synthetic fiber used to hybridize the natural fiber as a reinforcement is glass fiber. Some examples of these types of fibers are sisal/glass, silk/glass, pineapple fiber/glass, jute/glass, etc. With the addition of a small amount of glass fabric to the natural fiber reinforced composite, the mechanical strength and the chemical resistance of the composite has been shown to increase considerably (see Table 5.1 and Figure 5.1).

5.1.1.1.4 *Miscellaneous Reinforcements*

Miscellaneous reinforcement is also done to achieve required properties in the composites and some of them are synthetic fiber–particulate (glass along with fly ash) composites used to increase the abrasion resistance and compression strength of the composites, blending various levels of glass to wollastonite allows for tailored composites with high-strength properties or good dimensional stability. Fiber metal laminates (FML) offer significant improvements over currently available materials for aircraft structures due to their excellent mechanical characteristics and relatively low density and some examples for these types of composites are aluminum 2024 alloy; carbon fiber/epoxy (Ep), and aluminum 2024 alloy/glass fiber/Ep composites.

TABLE 5.1 Properties of Selected Commercial Reinforcing Fibers

Fiber	Typical Diameter (μm)[a]	Specific Gravity	Tensile Modulus (GPa)	Tensile Strength (GPa)	Strain to Failure (%)	Coefficient of Thermal Expansion (10^{-6}/°C)[b]	Poisson's Ratio
Glass							
E-glass	10	2.54	72.4	3.45	4.8	5	0.2
S-glass	10	2.49	86.9	4.30	5.0	2.9	0.22
PAN carbon							
T-300[c]	7	1.76	231	3.65	1.4	−0.6, 7–12	0.2
AS-1[d]	8	1.80	228	3.10	1.32		
AS-4[d]	7	1.80	248	4.07	1.65		
T-40[c]	5.1	1.81	290	5.65	1.8	−0.75	
IM-7[d]	5	1.78	301	5.31	1.81		
HMS-4[d]	8	1.80	345	2.48	0.7		
GY-70[e]	8.4	1.96	483	1.52	0.38		
Pitch carbon							
P-55[c]	10	2.0	380	1.90	0.5	−1.3	
P-100[c]	10	2.15	758	2.41	0.32	−1.45	
Aramid							
Kevlar 49[f]	11.9	1.45	131	3.62	2.8	−2, 59	0.35
Kevlar 149[f]	1.47	1.79	3.45	1.9			
Technora[g]		1.39	70	3.0	4.4	−6	
Extended-chain PE							
Spectra 900	38	0.97	117	2.59	3.5		
Spectra 1000	27	0.97	172	3.0	2.7		
Boron	140	2.7	393	3.1	0.79	5	0.2
SiC							
Monofilament	140	3.08	400	3.44	0.86	1.5	
Nicalon (multifilament)[h]	14.5	2.55	196	2.75	1.4		
Al_2O_3							
FiberFP[f]	20	3.95	379	1.90	0.4	8.3	
Al_2O_3–SiO_2[i]							
Fiberfrax (discontinuous)	2–12	2.73	103	1.03–1.72	−1.72		

[a] 1 μm = 0.0000393 in.
[b] 1 m/m per °C = 0.556 in./in. per °F.
[c] Amoco.
[d] Hercules.
[e] BASF.
[f] DuPont.
[g] Teijin.
[h] Nippon carbon.
[i] Carborundum.

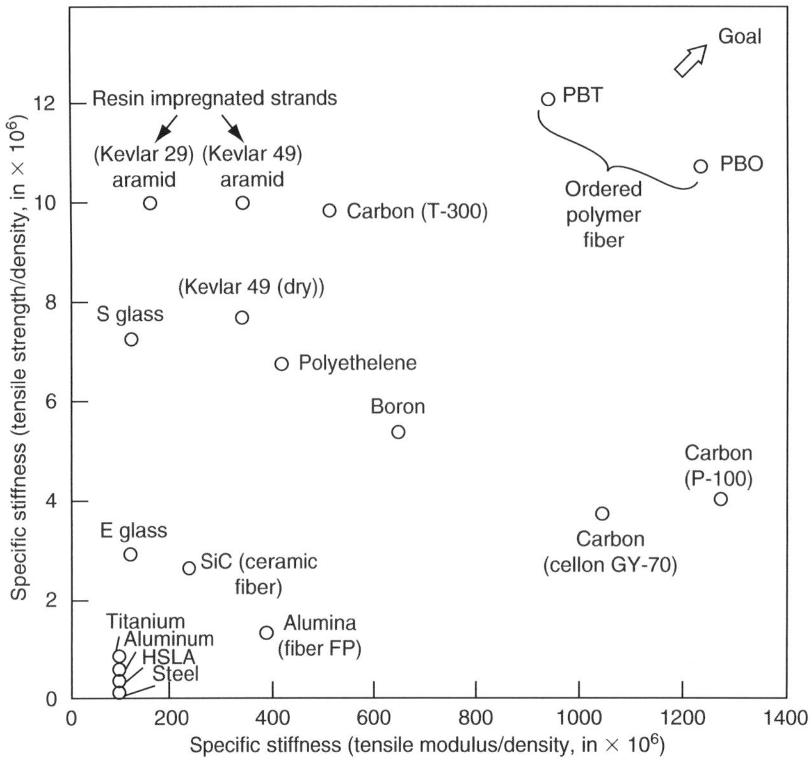

FIGURE 5.1 Specific properties of advanced fibers.

5.1.1.2 Matrix Hybridized Composites

The matrix is also one of the important constituents of the composites. Hybridizing a matrix with another matrix also means toughening of the matrix. Such matrices when used to prepare the composites also can lead to a hybrid composite. Toughening of the matrix is carried out in order to reduce the brittleness of the binder matrix and as a consequence to improve the (mechanical and chemical resistance) properties of the composites. An unmodified matrix usually consists of single-phase materials, while the addition of modifiers turns the toughened matrix into a multiphase system. When modifier domains are correctly dispersed in discrete forms throughout the matrix, the fracture energy or toughness can be greatly improved. In the case of metals, different types of alloys have been used to produce the composites. Polymer blends, by definition, are physical mixtures of structurally different homopolymers or copolymers. In polymer blends or polymer alloys, the mixing of two or more polymers provides a new material with a modified array of properties. For polymers, there are as known there are two types of matrices: thermoplastic (polymethyl methacrylate, polycarbonate [PC], polybutylene terephthalate [PBT], polystyrene [PS], etc.) and thermosets (Ep, unsaturated polyester, etc). Toughening of thermoplastics is usually carried out by blending them with different types of thermoplastic materials and elastomers in order to improve their mechanical properties and the flexibility of the matrix. Examples include blends of PC and a thermoplastic copolyether/ester/elastomers, which is shown to have a significant effect on the properties. Polyoxymethylene/elastomer/filler ternary composites have been prepared, in which a thermoplastic polyurethane and inorganic filler, $CaCO_3$, were used to achieve balanced mechanical properties for polyoxymethylene. Two other thermoplastics have also been blended together to achieve composites with high-performance properties as well as improve the mechanical properties of the composite: polyamide (PA) dispersed in a PE thermoplastic as a matrix for glass fibers. Researchers have also found that there is a significant increase in the degradation properties of thermoplastic blends with the addition of ceramic materials (PE-*co*-ethyl acrylate with a polyisobutyl methacrylate polymer reinforced with ceramic oxide powders). Some other examples of hybrid thermoplastic matrices are PC/acrylonitrile-butadiene-styrene ABS blend, PS/PC, PVA/polymethyl methacrylate, etc. Though a variety of thermoset materials have been used to prepare matrix hybrid composites, toughening of these composites by using Ep and unsaturated polyester are probably the most widely used combination.

Different kinds of modifiers have been studied to improve the toughness or ductility of cured thermoset resins. They can be classified as liquid rubbers, engineered thermoplastics, reactive diluents, and inorganic particles. Some of the following examples illustrate this type of toughening: Ep toughened with polymethylmethacrylate (PMMA), Ep toughened with PC, Ep toughened with PBT, Ep with reactive liquid rubber such as Ep phenol liquid with cashew nut shells, unsaturated PE with Ep, unsaturated PE toughened with PMMA, unsaturated PE with PBT, etc. Significant increases in some mechanical- and chemical-resistant properties were found for all of the above combinations of cured thermoset resins.

5.1.1.3 Reinforcement and Matrix Hybridized Composites

Composites of this type consist of multiphase materials (four phases) in which both the matrix and the reinforcements have been hybridized. In one way, it can be said that it is the combination of the above two types. This area of research is a new and challenging concept because as the incorporation of materials increases, there are more interphases in the composite system and if there are any differences in the interfaces of the materials, it could lead to the failure of the complete system. Care should be always taken to meet these future challenges in order to achieve a material with superior mechanical properties. It should also be noted that the viscosity of the toughened matrix resin be sufficient enough to wet the hybrid reinforcements. One of the examples that has been developed is silk/glass fabric and used as a reinforcement in thermoplastic-toughened Ep resin to achieve high mechanical properties.

The potential for the usage of hybrid composites far outweighs any negative aspects since there are a vast number of plastic/plastic, plastic/metal aramid-reinforced aluminum laminate (ARALL), metal/metal (MMC), metal/ceramic, ceramic/ceramic (space shuttle outer tile), and ceramic/plastic composite combination systems yet to be explored, investigated, developed, and in use.

5.1.1.4 Hybrid Composite Applications

Hybrid composite materials (HCM) represent the newest of the various composite materials currently under development. The hybrid composite category covers both the hybridizing of a composite material with other materials (either other composites or base unreinforced materials) and composites using multiple reinforcements. Further, this category covers the use of multiple materials (at least one of which is a composite) in structural applications and highlights the multiple uses and advantages of composite materials.

Hybrid composites can be divided into five major subcategories: (1) HCM, (2) selective reinforcements, (3) thermal management, (4) smart skins and structures, and (5) ultralightweight materials.

5.1.1.4.1 HCM

HCM are defined as a composite material system derived from the integrating of dissimilar materials, at least one of which is a basic composite material. A typical example of a HCM is a reinforced polymer composite combined with a conventional unreinforced homogenous metal. The HCM blends the desirable properties of two or more types of materials into a single material system, which displays the beneficial characteristics of the separate constituents. An existing example of a hybrid composite is

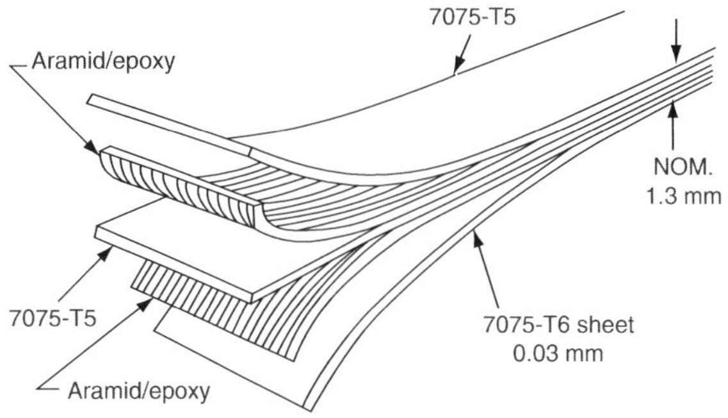

FIGURE 5.2 ARALL is an example of an HCM in production.

ARALL, which consists of high-strength aluminum alloy sheets interleafed with layers of aramid fiber-reinforced adhesive as illustrated in Figure 5.1. The ARALL hybrid composite (Figure 5.2) is baselined on several secondary structural composites of fixed-wing subsonic aircraft. A second example of a hybrid composite is a carbon–carbon composite (CCC) with a single side application of the refractory metal rhenium. This carbon–carbon–rhenium material is being developed for thermal management heat pipes on space-based radiator systems. Other examples include interpenetrating polymer networks (IPNs), which are hybrid resin matrices consisting of thermoset and thermoplastic resin combinations. Still another HCM concept involves multiple reinforcement types within a common matrix such as chopped fibers and continuous fibers within a polymer matrix.

Other HCMs include new composite materials such as nanocomposites, functionally gradient materials (FGMs), hybrid materials (Hymats), IPNs, microinfiltrated macrolaminated composites (MIMLCs), and liquid crystal polymers (LCPs), which may force development of previously uneconomical process routes if they offer the path to a technical solution for advanced system capability. These materials present opportunities for reducing the number of stages in turbine engines and in so doing may be economically beneficial even at a higher material cost because the smaller number of stages leads to greater economy of use.

HCM technology is in its infancy in comparison to that of the other types of composite materials. Whereas ARALL and IPNs have been used for the past decade, the other types of HCM are truly embryonic. From nearly all aspects of research, these HCM offer great potential for structural applications; however, their widespread use remains a decade or more in the future.

5.1.1.4.2 Selective Reinforcement

Selective reinforcement is the category of hybrid composites that provide reinforcement to a structural component in a local area or areas by means of adding a composite material. An example of this is the use of superplastic forming-diffusion bonding (SPF/DB) as a means of integrating a titanium-reinforced MMC into a base titanium structure. As part of the initial design approach, consideration must be given to the tooling required for placement of the reinforcing material within the structure. This is accomplished by building into the form tooling areas in which the MMC material is placed that permit SPF expansion of the base titanium to the MMC during processing and allowing diffusion bonding to occur, thus accomplishing integral reinforcing of the final structure.

The selective reinforcement design approach allows the aerospace designer to utilize the more costly, higher performance materials only where they are required and the less expensive materials in areas where they can perform the job. This approach leads to an optimization of both cost and performance in the most ideal case. Similarly, problems associated with use of the reinforcing composite, such as low mechanical joint strength, can be eliminated by not reinforcing the area where the mechanical joining occurs. In reality, while this approach can be very efficient, it requires the introduction of multiple materials to solve the design problem and may in turn result in increased fabrication costs and risks.

5.1.1.4.3 Thermal Management

In the field of thermal management, composite materials and HCM can be innovatively constructed to effectively limit the maximum temperature of structural hardware and rapidly transfer heat from hot areas to cool areas. This capability arises from the unmatched thermal conductivity of graphite fiber, which is higher (in the fiber direction) than that of oxygen-free high-conductivity copper (OFHC). The graphite fibers act as heat paths and, by suitable arrangement in the structure of interest, can remove heat by transmitting it along its length. In contrast, the matrix materials can act as thermal insulators so that thermal conduction through the thickness is lower by orders of magnitude than in-plane. This allows designers to develop a structure that is a thermal insulator in some directions but a thermal conductor in others.

5.1.1.4.4 Smart Skins and Smart Structures

Smart skins and smart structures are related in that each contains embedded, nonstructural elements. A smart structure contains sensors that monitor the health of the structure itself, such as fiber optics to determine temperature and structural deformations or cracks. A smart skin contains circuitry and electronic components that enable the skin to double as part of the electronic system of the parent vehicle, be it an aircraft or a missile.

Smart structure technology, like that of smart skins, is still evolving. The present technology consists of the incorporation of sensors into structural elements in the material processing stage to better control cure (or consolidation), and so on.

5.1.1.4.5 Ultralightweight Materials

The category of ultralightweight materials includes the emerging family of liquid crystal ordered polymers, which by virtue of their molecular structure, exhibit extremely high specific strength and specific stiffness. These highly directional materials are similar to composite materials in that the long, ordered molecular chains within the polymer act very much like reinforcing fibers in a composite material. Examples of these materials are poly-*p*-phenylene benzobisthiazole (PBZT) and poly-*p*-phenylene benzobisoxazole (PBO). Another example is gel-spun PE. These polymers exhibit extremely high specific strength and specific stiffness and can potentially be used as reinforcing fibers in composite laminates, as rope or cable, or as a self-reinforced thin film structure. These ordered polymers, in the form of thin films, find use in shear webs and skin applications for aircraft. These thin films can also be processed into honeycomb for lightweight structural applications. It has been estimated that a shear web made of PBZT would be one-eighth the weight of an aluminum web and one-sixth the weight of a graphite Ep web.

5.1.2 Future Directions

A type of metallic structure, which has only recently received attention is nanostructure, in which the microstructural dimensions are in the nanometer range (10^{-8} to 10^{-9} m). These nanostructures can be tailored to be either basically equiaxed or layered in nature, depending on production conditions.

These nanostructures offer novel combinations of strength, ductility, and stiffness not possible using conventional processing techniques. In the layered configuration when the thickness of the layers is below approximately 100 nm, so-called Koehler strengthening occurs in conjunction with an apparent modulus increase, opening up the possibility of varying and controlling both strength and stiffness by varying layer thickness. Practical techniques to produce nanostructures from metals include the electron beam coevaporation process developed by the Royal Academy of Engineering (RAE), which is capable of 18 kg/h deposition rates and mechanical alloying. The realms of possibilities here seem almost boundless, with the caveat of also including ceramic fibers in an overall tailored structure.

Aluminum-based alloys have been used to develop hybrid engineered materials such as MMC, mechanically alloyed dispersion strengthened aluminum (DISPAL), and ARALL. The latter two materials offer high-temperature capability and outstanding fatigue performance, respectively.

A number of production approaches have been studied, evaluated, and used to produce MMCs including powder metallurgy (P/M), casting, an in situ (XD) precipitation technique, and plasma spraying. The result is increased stiffness (by a factor of 3) and enhanced temperature capability.

In the casting approach, a particulate is dispersed uniformly in the matrix to give attractive levels of strength and stiffness, the latter of which is close to quasi-isotropic graphite-Ep. The XD technique (XD is a trademark of Martin-Marietta Corporation now Lockheed-Martin) is a process in which an in situ precipitation technique is used for the formation of the reinforcement directly in the matrix of interest. Using XD technology, it is possible to simultaneously produce a number of different reinforcements in a given matrix (Al, Ti, Cu, and intermetallics), thus allowing design of the microstructure to meet a specific need, e.g., fine particles for strength and whiskers for

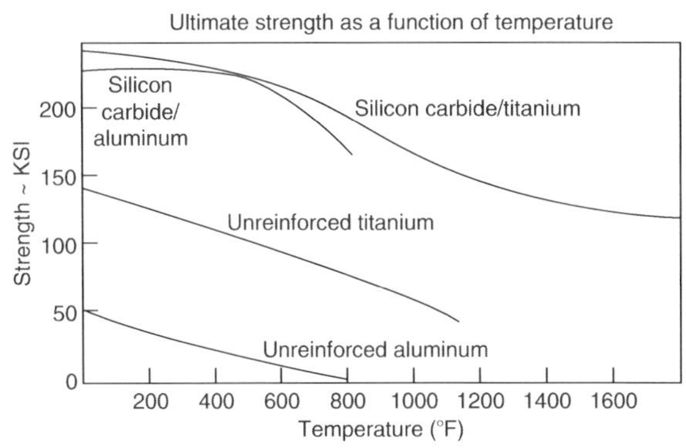

FIGURE 5.3 Ultimate tensile strength as a function of temperature for monolithic and SiC_f reinforced aluminum and titanium.

creep-rate reduction. Thus, XD technology is a very useful technique for achieving designer or engineered microstructures. Such flexibility at the microstructure level allows concurrent design, i.e., the simultaneous design of the material and component.

Titanium alloys reinforced with continuous ceramic fibers such as SiC enhanced the strength (Figure 5.3) and modulus (by a factor of 2).

It is projected that reinforced titanium aluminides may be useful up to 1000°C. The estimated strength value for unidirectional reinforced titanium aluminides with SiC fibers (SCS-6/Ti_xAl_yM) is nearly twice that of Inconel 718, over twice that of Ti-6Al-4 V at 540°C, and double that of unreinforced titanium at 760°C.

Acknowledgment

I thank my parents, brother, and dearest friend Dr. Ramakrishna for their support and encouragement to write this chapter.

5.2 Design of an Active Composite Wing Spar with Bending–Torsion Coupling

Carlos Silva, Bruno Rocha, and Afzal Suleman

5.2.1 Introduction

Adaptive materials are nowadays already reliable for industrial applications and can be easily introduced in systems newly designed, with their introduction already taken into consideration, or in older systems as a retrofit. They have the advantage of their application being easily achieved, independently of the base material, since for instance they can be bonded (surface glued or embedded) to metal, carbon, composite, etc.

Adaptive systems can improve aircraft reliability, reducing failures through vibration control or reduction. Vibration reduction means damage, wear, and failure probability reduction, which leads to larger intervals between maintenance operations. This directly leads to a high cost reduction.

Aeronautic structures are flexible, specially the lifting surfaces. As the airplane is flying, structural deformation takes place. This leads to a change in the aerodynamic loads produced by them due to the shape modification. The aerodynamics forces increment or decrement consequently results in a new structural deformation. This is a two-way interaction between the structure and the fluid [1]. The behavior can be stable, if an equilibrium condition is attained, or unstable, where catastrophic failure can occur. Nonetheless, the simple vibration phenomena will cause its damage during the life of the structures.

Vibration control, which provides vibration reduction, flutter and divergence attenuation, and delay, can be achieved using actuators to send pulses to the structure to oppose its sensed vibration based on a frequency response methodology. In addition, a multicell cross-section for wing spars can be used. These spars when built from fiber composite laminate materials, with the preferable fiber orientation misaligned with span direction, possess a flexure–torsion coupling. This means that when these spars are subjected to flexure, they will present some amount of torsion.

For certain fiber angles and multicellular cross-section configurations, certain types and intensities of that coupling can be achieved, such that it can be utilized to counteract aeroelastic deformations and phenomena, as flutter and divergence attenuation.

Recently, at the Portuguese Air Force Aeronautics Laboratory, Suleman et al. have reported several studies on aeroelastic control and flutter alleviation in adaptive flight vehicles [2–4]. These studies led to two active wing concepts: the active spar and active skin concepts and both wind-tunnel and flight tests were conducted using a remote piloted vehicle (RPV) [5]. A significant increase in the flutter speed and decrease in the amplitude of vibrations was observed. These conclusions were validated for realistic flight conditions [6,7] with the successful results, a analogous proportional control system was applied in this study.

5.2.1.1 Bending–Torsion Coupling Spars

Piezoelectric (PZT) actuators and sensors are nowadays reliable for industrial applications. They have the advantage of their application being easily achieved, independently of the base material. For instance, they can be bonded (surface glued or embedded) to metal, carbon, composite, etc. The coupling between the actuator and base structure can be easily enhanced in composite structures if one uses a misaligned layer stacking. If a flat beam is considered, the misalignment is the fiber angle deviation from the length direction. The misalignment can vary from 0° (aligned) to 90° (perpendicular alignment). A structure built with this type of internal layout when loaded vertically will

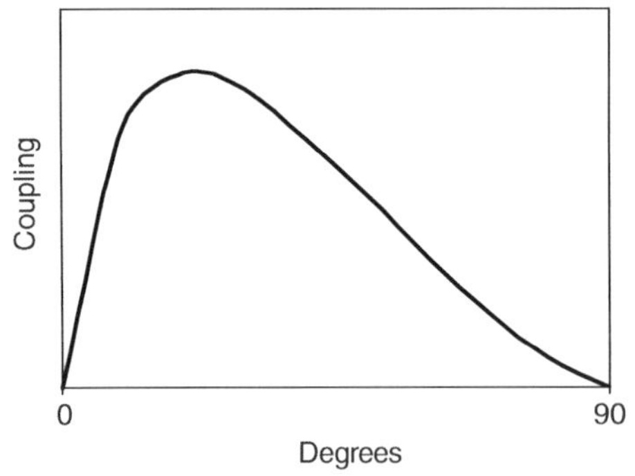

FIGURE 5.4 BTC evolution.

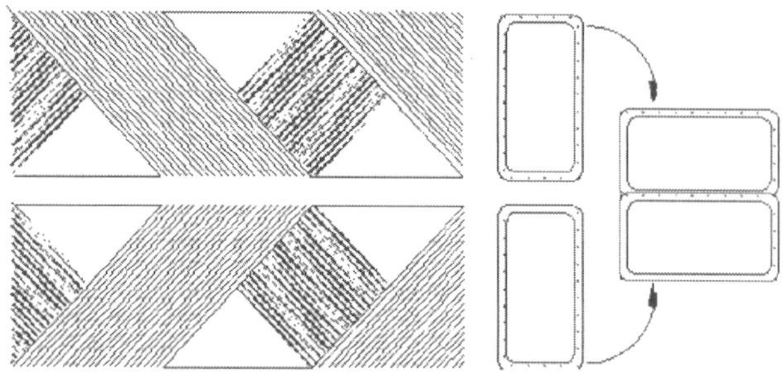

FIGURE 5.5 Semispars and fiber layout.

deform vertically and twist. The same effect can be attained if a torsion moment is applied. An inherent consequence is that the vertical displacement and torsion are no longer linear with respect to the specific load applied.

Early studies in this field were carried out by Garfinkle and Pastore [8]. They analyzed the passive and augmented active behavior of structures exhibiting bending–torsion coupling (BTC) as shown in Figure 5.4.

5.2.1.2 Actuators and Sensors

The ideal actuator would have a high-energy (mechanical stress/strain) actuation capability with low energy input (voltage/current), wide frequency range, linear and unidirectional behavior, and convenient shape. Ideal sensors would have a noise-free response, high precision, and fidelity (low mechanical energy stimulation sensitive), linear behavior, wide frequency sensing range, and convenient shape. Both would be easily integrated or bonded into the structures.

Due to this type of structure dimensions, low-frequency requirements, PZT actuators and sensors were selected. From the various types of actuators and sensors available in the market, QP40W from QuickPack ACX was selected as the actuator and PZT made of material BM500 from Sensor Technology Ltd. were chosen as the sensor.

5.2.2 Multicell Cross-Section Spar Design

One of the challenges was to design, optimize, manufacture, and test a continuous filament multicell cross-section composite spar with BTC and with embedded PZT actuators (Figure 5.5).

The resulting multicell spar will have its upper wall composite fibers oriented in the same direction as the lower wall fibers (Figure 5.5). When subjected to bending loads, this multicell spar will present bending and also torsion deformations (BTC).

Orienting the upper and lower wall fibers from trailing edge to leading edge, from root to wing tip, when the wing spar bends upwards, it twists in a manner that the angle of attack is decreased (leading edge down). With this kind of orientation, a desirable behavior is attained. This counteracts the dynamical behavior of flutter.

This is a passive characteristic of these types of spars. Using an active system, this characteristic can be augmented by applying actuators on the upper and lower multicell spar walls.

In order to determine the misalignment, a global optimization approach through metamodeling was programmed. For the type of spar presented, considering its constraints and defined objective function, the design variables set was calculated [9]. The optimized solution defined the final design to be manufactured, as follows: span, 1 m; height, 35 mm; wall thickness, 1 mm; and fiber orientation, 10°.

5.2.2.1 Actuators and Sensors Positioning

In order to determine the most advantageous positions to place both actuators and sensors, computational analyses were carried out. With respect to vibration reduction, the location of sensors and actuators was determined by finite element method (FEM) analysis of the vibration modes. This methodology revealed the location of points with larger strain values (near the wing root). The advantage provided by this approach is that a small and light actuator manages to change dramatically the vibration parameters of the wing because it is placed near the points of higher strain. Also, the higher strains permit better sensing definition by the sensors due to its larger amplitudes. This positioning is also advantageous for the frequency-based simple harmonic motion (SHM).

Despite the fact that the main present work is not on SHM, related constraints were taken into account. With respect to wave propagation methods, it is useful to position the sensors as far away as possible from the actuators. Consequently, the spar was divided into two equal zones, from now on called cells, each one defined by a pair of actuators and one of sensors. The cells have the actuators on its root and sensors on its end on the spanwise direction. The upcoming disadvantage for the vibration reduction is only related to the sensors amplitude output decrease, which can be easily overcome.

5.2.2.2 Active System

Figure 5.6 shows the active system scheme. Starting from the PZT sensor, it is assumed that the PZT signal is proportional to

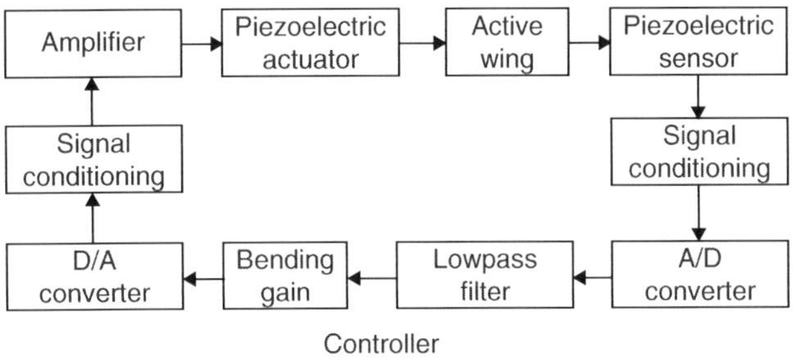

FIGURE 5.6 Proportional control scheme.

the spar's vertical displacement. This signal has to be conditioned because the control, carried out by a MicroAutoBox DSpace control module, has to be matched to the I/O requirements. Inside the control module, a Matlab Simulink model, previously uploaded and running as a flash application on the controller memory, implements the direct proportional control law (see Figure 5.7).

A low pass filter is included to retain only the frequencies below 50 Hz. The main concern is the low frequency present in the first bending vibration mode also avoids interferences coming from the power current. Afterwards, a gain is applied to the input signal and it is finally sent as an output back to the signal conditioning board. The referred gain has to be determined through experiments. Once again the I/O has to be matched to the one required by the amplifier, which by its turn, sends the final signal to the PZT actuators. Using this simple control scheme, the actuators will always counteract the spar's vertical movement.

5.2.3 Results

Several experiments were conducted during the work progress. So, spar, wing, and RPV final assembly were all tested. The experiments comprehended static and dynamic on passive and active solutions. Results obtained include sensor output analysis, fast Fourier transform (FFT), and damping calculations (see Figure 5.8).

5.2.3.1 Spar

The spar with the actuators and sensors was clamped on its root and an impact was given to the tip. Figure 5.8 shows the sensor output time evolution. Using the data attained, the first natural frequency and damping coefficient were determined for both situations.

From the results shown, it is easily seen that the improvement is significant. Up to 80% of damping increase can be achieved and up to 12%, first natural frequency augment can be attained (see Table 5.2).

5.2.3.2 Wing Design and Fabrication

After finishing the spar's experiments, a wing was built with it to be tunnel tested. This wing was intended to be used in the future as a specimen capable to allow the implementation of SHM methods. Thus, the model was designed taking into consideration two SHM methods, namely vibration based and wave

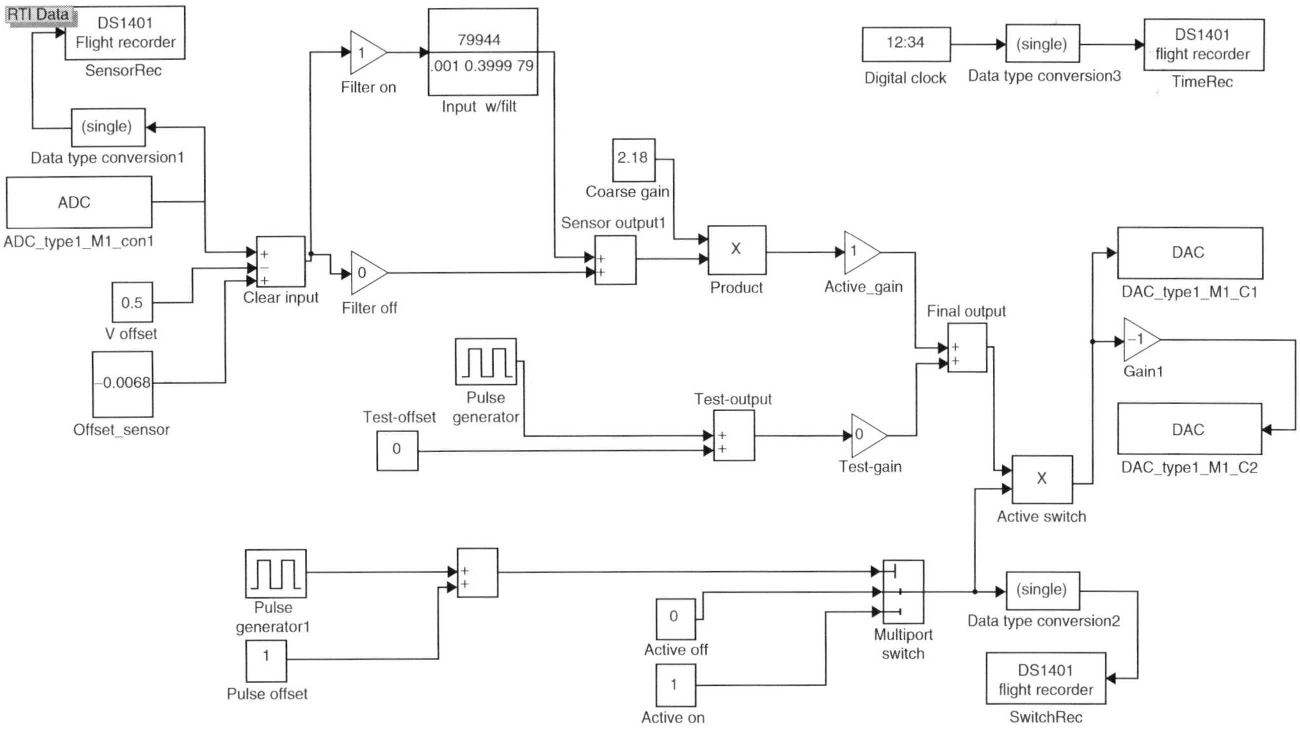

FIGURE 5.7 SIMULINK control law model.

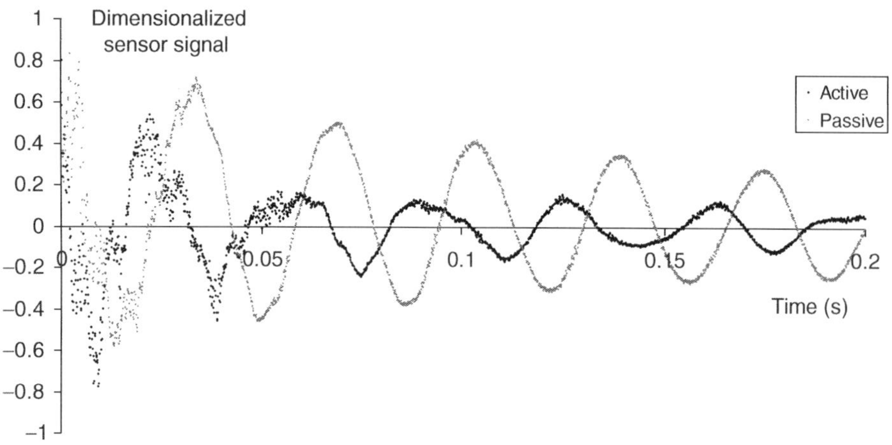

FIGURE 5.8 Passive and active responses.

TABLE 5.2 Passive and Active Damping Coefficients and First Natural Frequencies

Coefficient	Damping Frequency (Hz)	First Natural Frequency
Passive spar	0.054	28.92
Active spar	0.1	32.46
Increase	80%	12%

propagation. One main feature of the wave propagation method application is that an attached wing spar rib or any kind of stiffener limits wave propagation. Thus, in order to have space to place the actuators and sensors and to allow the wave to propagate, the general structure of this small wing model for wind tunnel testing was divided only into two cells with a continuous wing spar. However, to assure the aerodynamic shape of the wing nonstructural ribs (not attached to the main spar) were positioned spanwise, 8 cm apart from each other. Beyond these two cells, there was a structural rib at the wing root and another one 4 cm apart spanwise, connected to each other by foam (in addition to the wing spar). This was intended to form a reinforced section to guarantee the wing–fuselage connection. Also, the wing skin was not structural and could be made of a transparent film, which allowed seeing the internal wing structure, actuators/sensors positioning, systems, etc. Figure 5.9 shows the wind tunnel testing apparatus.

For each test, a sensor signal amplitude output versus time sample was retrieved and an FFT analysis was done. For instance, Figure 5.10 represents the FFT results obtained for the 20 m/s velocity for the passive wing and active wing with a gain of 50% max. A clear decrease in the vibration amplitude near for the first natural frequency can be observed.

Maximum signal output amplitudes were retrieved and Figure 5.11 shows the results. As shown, the amplitude decreased from passive to active tests. Generally, this amplitude tends to be smaller when a higher gain is applied. Figure 5.12 shows the percentage of maximum sensor signal output amplitude decrease: a maximum 8.83% decrease is reached for 30 m/s of velocity and a gain of 50% max.

As regards the damping analysis, studies were concentrated around the first natural frequency. Having the sensor signal output, FFT amplitude versus frequency graphs, damping was calculated, retrieving the frequency value for the peak amplitude and frequency values (around the previous) obtained for maximum amplitude divided by the square root of 2. With these

FIGURE 5.9 Wind tunnel test setup.

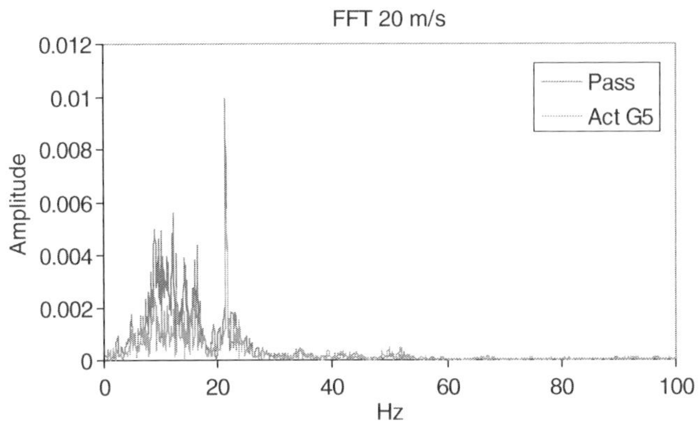

FIGURE 5.10 The 20 m/s passive and active wing FFT analysis.

three values, it was possible to determine a damping coefficient proportional value (not the damping coefficient itself) by subtracting the highest and lowest frequency values divided by the value of the central frequency.

With this algorithm, all the damping coefficients were calculated and plotted on the same graph (Figure 5.13) along with the respective quadratic polynomial trendlines. These last ones, allowed us to predict flutter speed (that corresponds to the velocity for the zero damping coefficient, i.e., when the trendlines would ideally cross the *x*/velocity axis).

It is noted that the flutter speed increases from passive to active tests, and it reaches its highest value for the 50% max gain. Also, damp improvements are observed from passive to active tests, and they are higher also for the 50% max gain. Overactuation and delays influence the active gain results for gains higher than 50%, resulting in poorer outcomes. Figure 5.14 shows the percentage damping improvements: a maximum 93% improvement is reached for 45 m/s of velocity and using 50% maximum gain.

The maximum damp improvement, now obtained, is higher than the one obtained for the spar when its mechanical tests were performed (80% max improvement). This increase can be explained by the BTC, which had no influence on the spar behavior when its mechanical tests were done, but which is essential on wind tunnel tests. On these final tests, this coupling affected the aeroelastic behavior of the wing. When the wing bends and consequently rotates along the spanwise axis due to the coupling, the aerodynamics of the wing will be influenced, i.e., aerodynamic

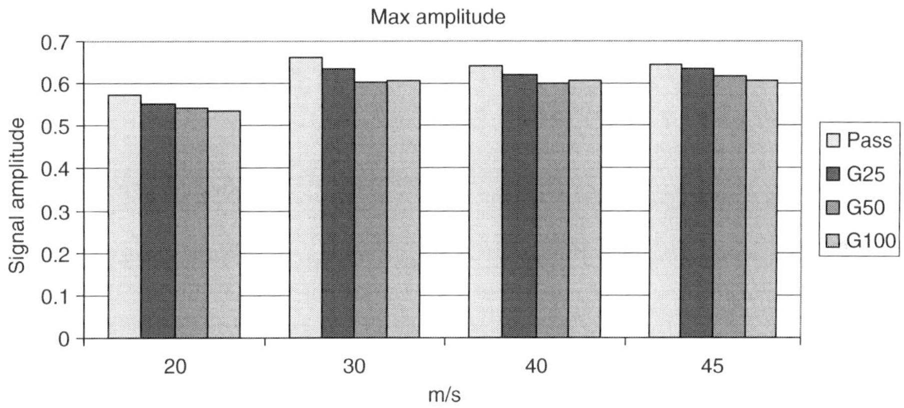

FIGURE 5.11 Wing maximum amplitude.

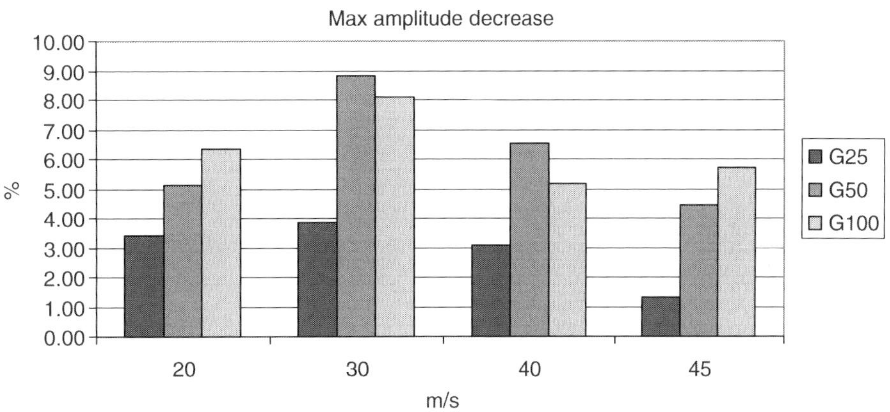

FIGURE 5.12 Wing maximum amplitude decrease.

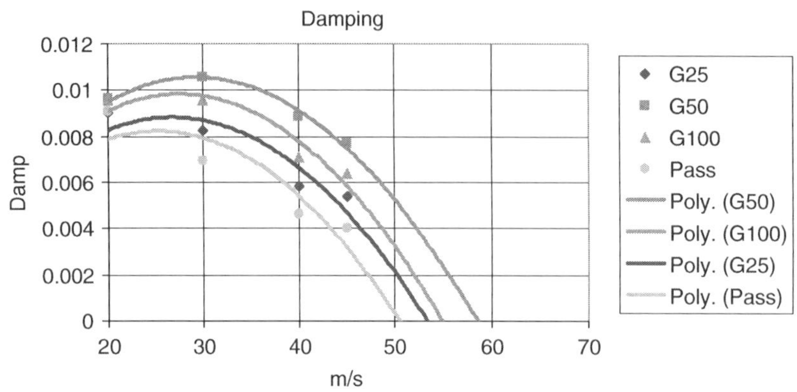

FIGURE 5.13 Damping for 25%, 50%, and 100% of maximum gain.

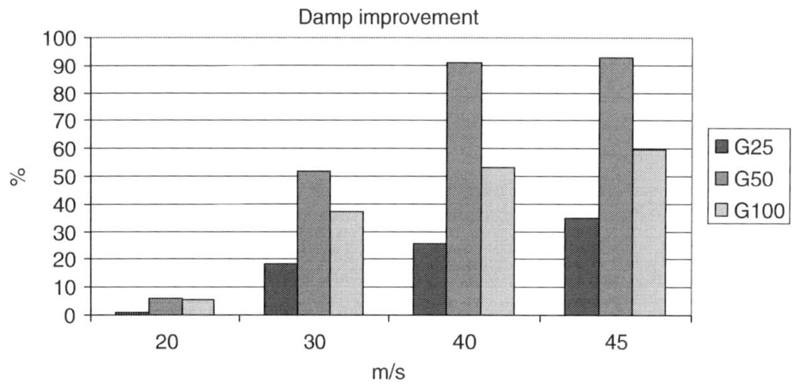

FIGURE 5.14 Damping improvement for 25%, 50%, and 100% of maximum gain.

loads applied and their distribution is altered, affecting the wing structure, changing consequently the wing structure deformations.

5.2.3.2.1 0°/10° Wing Performance Comparison

In order to truly assess the fiber misalignment behavior, a similar 0° fiber-oriented spar was manufactured and tested. Figure 5.15 presents the amplitude variation for both spars at different velocities. It can be inferred that the misalignment on spars (to a certain extent) by itself can lead to an improvement in the wing's behavior. Afterwards, it can be clearly augmented by implementing active materials.

5.2.3.3 RPV: Flight Tests

After the successful wind-tunnel testing, the wing was prepared to be installed on an RPV for flight testing. It does not compromise

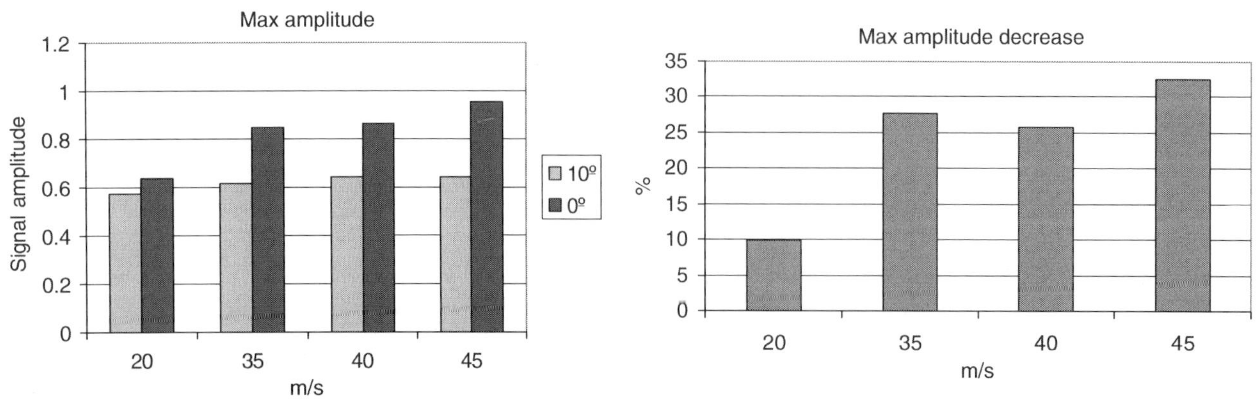

FIGURE 5.15 0° and 10° maximum amplitude and its comparison.

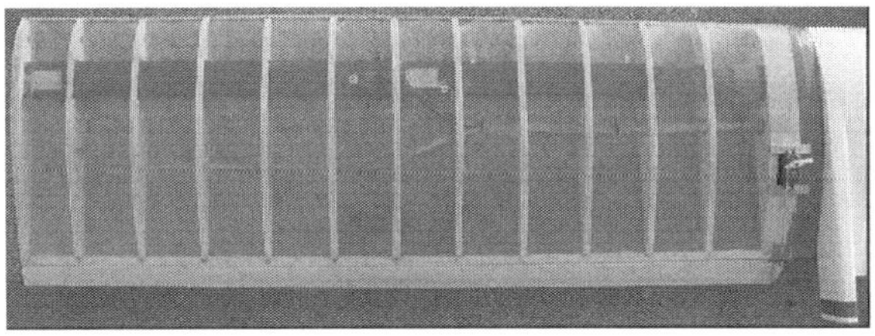

FIGURE 5.16 Flight testing modified wing.

the desired behavior but induces some mass increase. The wing must also be equipped with an aileron and respective servo action mechanism (Figure 5.16). Its root had to be reinforced in order to receive the clamping pin provided by the fuselage.

To carry out the flight tests, the amplifier also had to be changed. The one used on previous tests was adequate and was replaced by a small battery-powered one. Figures 5.17 and 5.18 show all the active equipment gear on board and in-ground. During the ground equipment testing, an interference was detected between the engine and the sensor system. Due to this condition, it was opted to do gliding flight testing.

The procedure consisted in flying the RPV as high as possible, which was limited by the pilot's observation capability. Afterwards the engine was cut off and a glide descent aimed to a speed of 20 m/s was performed. The airspeed was monitored in the PC using the Jet-Tronic software, which received the pilot telemetry signal from onboard and the testing time was being correlated with speed, and both values registered. This operation revealed that it was very difficult to conduct, even with two assistants, because of the difficulty in keeping constant conditions.

Two consecutive flights were carried out. Figure 5.19 shows the flight data recorded, including the altitude gain. Figure 5.20 is focused on the glide part of both flights. The ramp functions represent the flight time.

Flying the RPV must be done not very far from the pilot, and the aircraft performed a series of turns that influenced the results. In the turns, effective wing lift increased and the vibrations were higher. This may explain why sometimes when the control was ON (and was supposed to decrease vibrations), some very large amplitude peaks appeared. Due to this, no maximum or minimum comparison was done because they do not represent the general wing's behavior.

If one considers that a window represents a state of active or inactive, flight 1 had six windows and flight 2 had five windows. Selecting joined windows believed to represent the most equivalent flight conditions, we achieved the following FFT analysis, shown in Figures 5.21 and 5.22.

A clear improvement can be observed. Another interesting fact is that the improvement seems to be higher in the last part of both flights, the final straight approach. This can be justified by the fact that the wind conditions were more severe near the ground, thus implying more vibrations on the wing. Adding to this, since the airplane was aligned with the runway less pilots, inputs were necessary to guide the RPV to the landing.

5.2.4 Concluding Remarks

The proposed solutions have resulted in a successful implementation of a new spar concept for a conventional wing. The proposed solution consists of a multicell cross-section spar with misaligned fibers, which resulted in an improved aeroelastic performance and suppression of vibrations and consequent postponement of flutter.

The first step in the process was to design the spar itself. The conceptual design of the spar indicated that it should have more than one section, and the fiber orientation should be aligned at

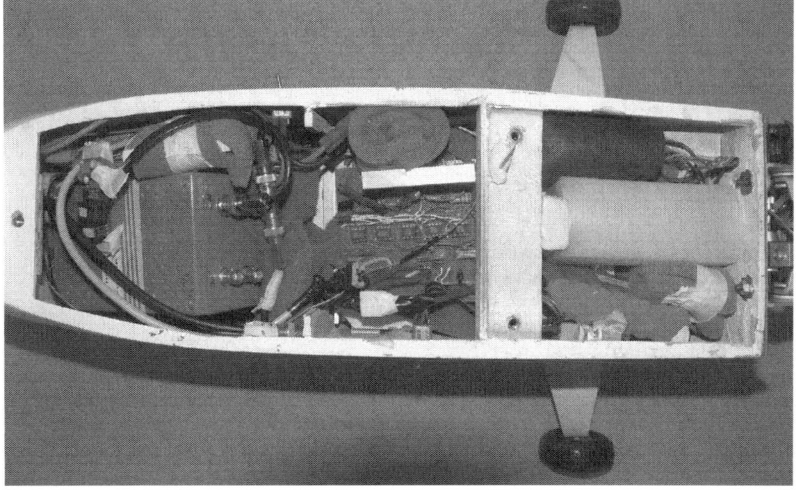

FIGURE 5.17 RPV internal equipment placing.

FIGURE 5.18 Flight test setup on runway.

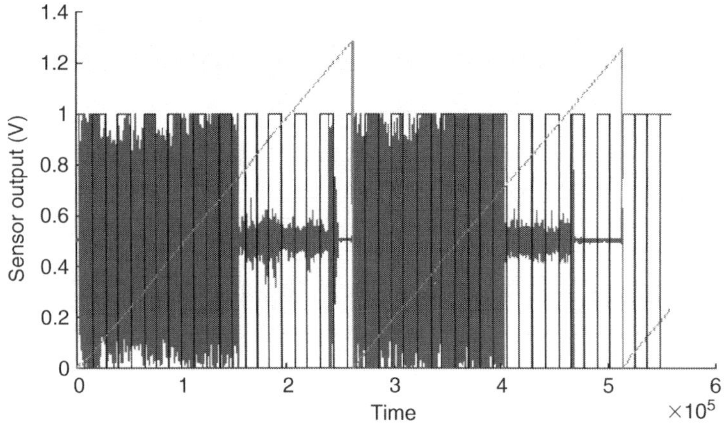

FIGURE 5.19 Flight test sensor pattern.

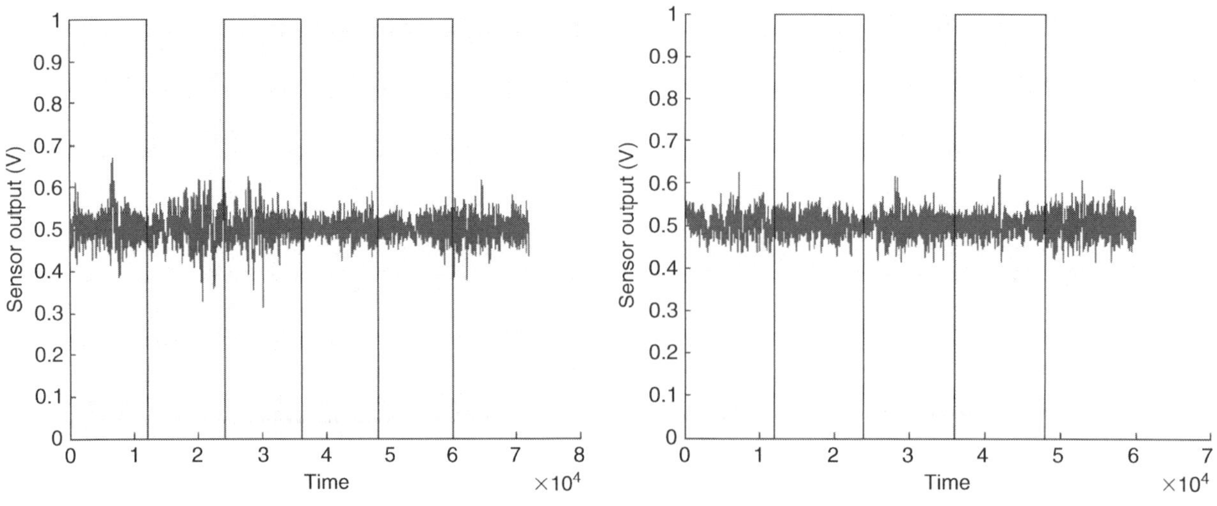

FIGURE 5.20 First and second gliding flights sensor output.

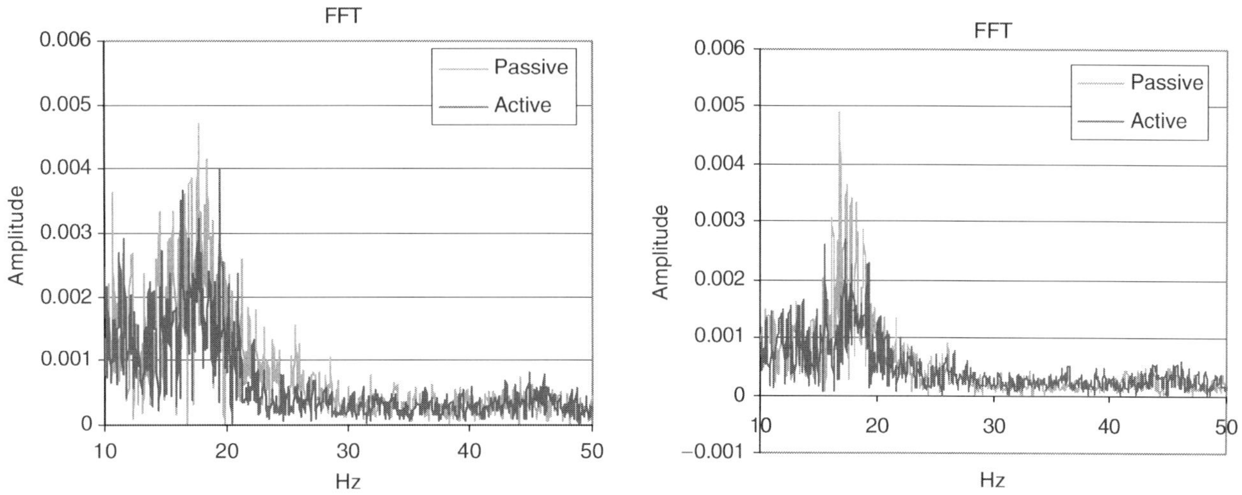

FIGURE 5.21 Second/third and fifth/sixth windows passive/active FFT analysis.

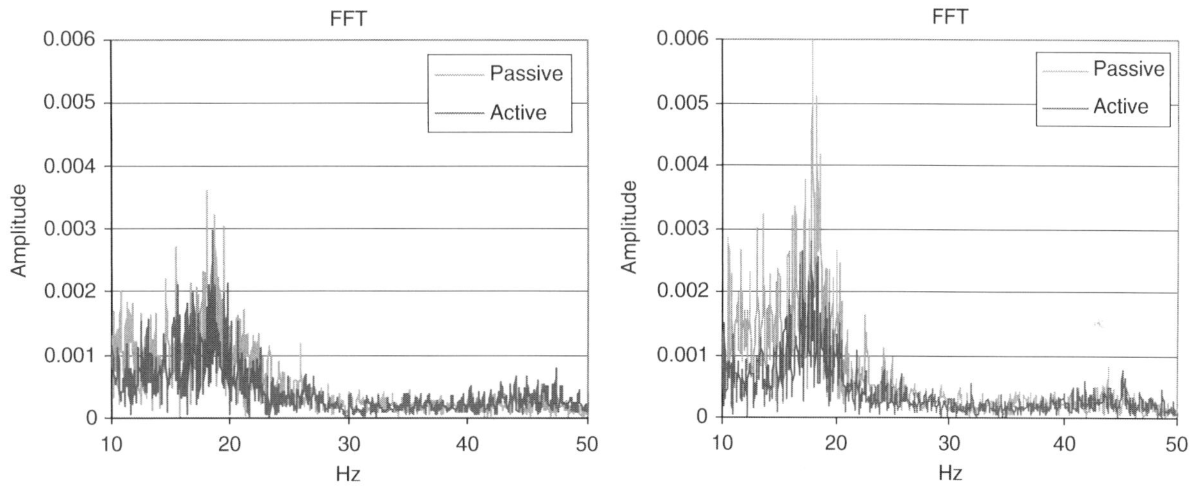

FIGURE 5.22 Second/third and fourth/fifth windows passive/active FFT analysis.

an angle with respect to the spanwise direction. On a planview, these fibers must start in the trailing edge and end in the leading edge such that the leading edge decreases its displacement when the wing experiences a lift load. For a simple constant rectangular cross-section spar with an overall constant thickness and manufactured with one material, this is the only way of achieving the desired BTC.

Next, the issue was to decide the final dimensions and the right misalignment of the fibers. On one hand, a small misalignment could result in a poor BTC but good vertical stiffness and high natural bending modes frequencies. On the other hand, a high misalignment would present a better coupling but poor stiffness and low natural bending frequencies. To determine the best solution, an optimization problem was defined and solved.

The use of active materials was tested right way at this very early stage of work. Sensors and actuators positions were carefully studied. Having in mind future objectives such as structural health monitoring, other than vibration reduction and flutter suppression, two pairs of actuators were placed, one at the root and the other at the midspan. In each case, one actuator was glued on the upper and the other on the lower part of the spar. Two pairs of sensors were also glued to the spar, but for this work, only one was necessary. Its position is in the midspan and both upper and lower positions were suitable for its task.

Experimental active tests were then run and compared to passive ones. The active methodology consisted of reading the sensors strain and feed them back to the actuators, via a proportional control system, to oppose the movement. Substantial damping and torsional natural frequency increases were attained. This presented a good motivation for continuing the path followed so far.

At this stage, the next milestone was identified. A wing was built with one of the manufactured spars. Since this spar was intended to be used for future SHM methods testing, a wing composed of two main cells was designed. One of these methods was based on wave propagation. Starting from the root, this wing had a reinforcement box for clamping in wind tunnel and the RPV fuselage. Then it had two cells. Each one was defined by two structural ribs (attached to the spar). Between these two, there were "form" ribs, which were not attached and were placed to guarantee the airfoil shape and transmit the aerodynamic loads to the edges. These were then transmitted to the spar by the attached structural ribs. This way, mechanical waves generated by the actuators could progress through the spar in each cell.

Next, wind tunnel tests were carried out. These were run at several stabilized velocities, starting from 20 m/s to a maximum near the flutter speed initiation behavior, approximately around 50 m/s. A similar wing was built at the same time, using a 0° spar to test and compare with the native behavior. The results showed that the native performance is clearly improved when applying a misalignment to the fibers. An approximately 30% reduction in the maximum vertical displacement amplitude was found for a test run at 45 m/s. It also presented a stable vertical displacement behavior as the velocity was increased. The 0° spar tended to increase this parameter as the velocity rose.

When applying active control using the piezo actuators, this aeroelastic performance of the wing was enhanced. A structural damping increase around 90% was attained for 45 m/s. The flutter was predicted to be delayed by 8 m/s, which corresponds to a 16% increase. The best results were found when using a 50% gain increase in the proportional feedback control system.

At this stage, the second milestone was completed and the main objective of the work satisfied, since the wind tunnel tests closely reproduced the structure's response to the real airflow inputs. Nevertheless, flight testing can push the limits and test other aspects and act as a determinant to assess the systems viability. So, the final milestone was to adapt this wing to the RPV and test it in flight. To do so, the control equipment had to be put onboard. The energy source was changed to batteries and an RPV, which indicated airspeed measuring equipment, was installed. The RPV was manually piloted from the ground. The control system was preprogrammed with the control law and set to save all the flight data for future postprocessing. During the flight, the active system was automatically set on and off in intervals of 12 s. Preflight tests revealed engine interferences of an undetermined source. Gliding flights option was taken and spectral frequency analyses were carried out. The results showed a significant vibration reduction despite the reduced available amplification provided by the internal amplifier.

This final work stage closed the loop of the spar development program and on the passive and active system implementation. As a final conclusion, it can be said that both the passive and active multicell cross-section spars proved to be a viable solution in active aeroelastic control solutions in adaptive flight vehicles.

References

1. Bisplinghoff, R.L., Ashley, H., and Halfman, R.L., *Aeroelasticity*, Dover, New York, 1996.
2. Suleman, A., Costa, A.P., and Moniz, P., Experimental flutter and buffeting suppression using piezoelectric actuators and sensors, *Proceedings of the Smart Structures and Materials—3674*, 1999.
3. Amprikidis, M. and Cooper, J., Adaptive internal structures for active aeroelastic control, *Proceedings of the International Forum on Aeroelasticity and Structural Dynamics—IFASD 2003*, 2003.
4. Rocha, J., Moniz, P., Costa, A., and Suleman, A., On active aeroelastic control of an adaptive wing using piezoelectric actuators, *Journal of Aircraft*, 42(1), pp 272-278, January—February, 2005.
5. Costa, A., Novel Concepts for Piezoelectric Actuated Adaptive Aeroelastic Aircraft Structures, PhD thesis, Instituto Superior Técnico, Portugal, 2002.
6. Rocha, J., Suleman, A., Costa, A., Moniz, P., and Santos, D., Research and development of an active aeroelastic adaptive flight demonstrator, CEAS/AIAA/NVvL IFASD 2003, Netherlands, 2003.
7. Rocha, J., An Experimental Study and Flight Testing of Active Aeroelastic Aircraft Wing Structures, MaSC thesis, University of Victoria, Canada, 2005.
8. Garfinkle, M. and Pastore, C., Intrinsically smart coupled box beams, Retrieved February 5, 2007, from http://www.pages.drexel.edu/~garfinkm/Spar.html.
9. Silva, C., Rocha, B., and Suleman, A., A metamodeling optimization approach to a wing spar design, 49th AIAA/ASME/ASCE/AHS/ASC Structures, Structural Dynamics and Materials, Schaumburg IL, USA, 2008.

6

Ferromagnetic Shape Memory Alloy Actuators

Yuanchang Liang
University of Washington

Minoru Taya
University of Washington

6.1 Introduction

6.1.1 Review of Shape Memory Alloy

Shape memory alloys (SMAs), such as Ni–Ti (Nitinol) alloy, have a smaller hysteresis of (austenite ⇔ martensite) phase transformation than that in steel [1]. There are numerous papers that have examined SMAs. This is because SMAs are scientifically attractive and can be used as sensors and actuators. The phase transformation in SMAs is reversible and usually can be characterized by the latent heat and the change of lattice parameters. To obtain reversible deformation, a heating or cooling device is usually incorporated. SMAs also have been used extensively as key actuator materials with applications in aerospace structures (such as chevron) and medical implant devices (such as stent), where temperature or stress is a driving force to obtain large strain recovery (shape memory effect [SME]) or large "elastic" strain (superelastic behavior).

The stress–temperature phase diagram of SMAs, shown schematically in Figure 6.1, is often used to characterize the phase transformation between stress and temperature. Usually, it is constructed by the stress–strain curves tested at different temperatures. The intersections with the temperature axis (stress = 0), T_{Ms}, T_{Mf}, T_{As}, and T_{Af} are usually taken from the results by differential scanning calorimetry (DSC), which are martensite at the start, martensite at the finish, austenite at the start, and austenite at the finish temperatures, respectively. These transformation temperatures are functions of stress where the larger the stress, the higher the transformation temperature. Figure 6.2a and b schematically show loading and unloading curves of an SMA tested at temperatures (T) higher than T_{Af} and below T_{Ms}, respectively. As the SMA is loaded at $T > T_{Af}$ from the stress-free stage, O, the stress–strain curve initially follows a straight line (OA) with the slope of E_A (Young's modulus of the austenite phase). This corresponds to the elastic behavior of the SMA with the 100% austenite phase until it reaches the stress (σ_{Ms}) of the onset stress-induced martensitic (SIM) phase transformation (austenite → martensite), point A. Upon further loading, it follows a gradual straight line until point B where the curve becomes steep again in a slope of E_M (Young's modulus of the martensite phase). Between points B and C, the SMA is under an elastic deformation with 100% martensite phase until it is yielded at the yield stress (σ_y), point C. When the stress is over σ_y, the 100% martensite phase of the SMA undergoes plastic deformation by a dislocation gliding mechanism. If under a constant temperature, $T > T_{Af}$, the loading curve continues up to some point below σ_y followed by unloading, the stress–strain path will follow the elastic unloading along the straight line (CD) until the stress reaches the onset of reverse transformation (martensite → austenite) at point D. Further unloading will render the reverse transformation by following a gradual straight line until the curve hits the elastic loading line at point E. The unloading beyond point E follows the elastic deformation path (EO) to the origin (O). This loop (OABDEO) in Figure 6.2a is called "superelastic" (SE) behavior of the SMA where the strain at point B is usually in the order of 5% for the polycrystal SMA. On the other hand, if the loading–unloading process is conducted at a constant temperature, $T < T_{MS}$ (Figure 6.2b), the curve follows the path OAB and leaves a residual strain (ε_{res}). This residual strain is recoverable when the temperature rises over T_{Af} because the martensite to austenite phase transformation occurs. The curve will follow the strain axis (BO) and return to the origin (O), i.e., zero strain. This loop (OABO) is the so-called SME. ε_{res} is usually on the order of 5% for the polycrystal SMA. Unlike the path OA of SE in Figure 6.2a, the path OA in Figure 6.2b is not linear

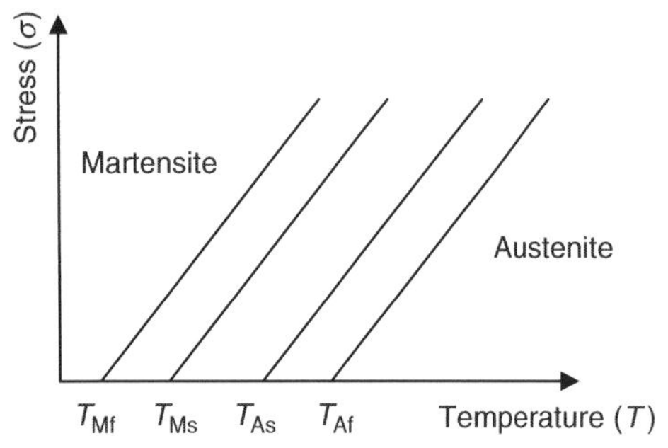

FIGURE 6.1 Stress–temperature diagram of an SMA.

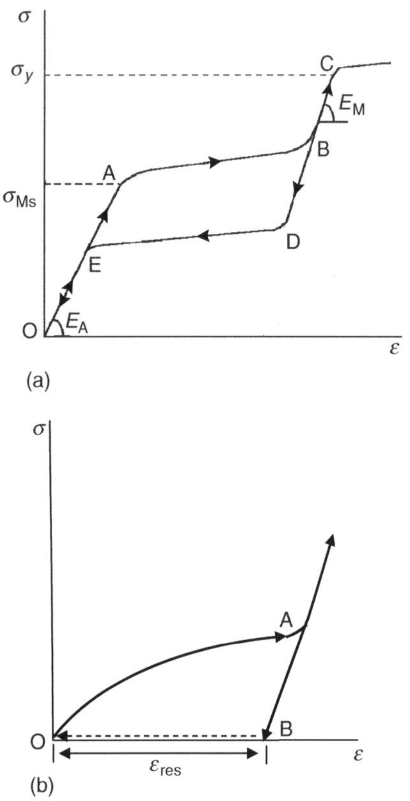

FIGURE 6.2 Stress–strain curve of an SMA with (a) SE loop at $T > T_{Af}$ and (b) SME loop at $T < T_{Ms}$.

because it is not an elastic deformation. The SMA is in the fully martensite phase at the temperature below T_{Ms} and the loading path OA is involved in the martensite variant rearrangement. However, if the SMA is loaded beyond point A in Figure 6.2b, it is under the elastic deformation with a slope of E_M similar to BC in Figure 6.2a.

6.1.2 Ferromagnetic Shape Memory Alloy

As described before, the phase transformation of SMAs can be controlled by stress and temperature. Their relationship can be described by a stress–temperature phase diagram as shown in Figure 6.1. Recently, attention has been paid to some SMAs accompanying a change in ferromagnetism at phase transformation and having low hysteresis. This is because the transformation characteristics such as transition point (martensite start temperature, T_{Ms}) and macroscopically observed strain, caused by the transformation, are possibly controlled by an applied magnetic field (H). The response of transformation is fast in this case because the characteristic time is governed by the formation and growth of martensite, which is induced by an applied magnetic field. It is fast, thus, it is plausible to produce an actuator having reversible straining with quick response to a signal imposed or detected. Such an alloy with both ferromagnetic property and SMA behaviors is called ferromagnetic shape memory alloy (FSMA) and it is considered as a strong candidate for fast responsive actuator material. The relationship of phase transformation, stress (σ), temperature (T), and magnetic field (H) can be presented as a three-dimensional phase transformation diagram as shown schematically in Figure 6.3.

Currently, only a limited numbers of FSMAs have been found as listed in Table 6.1. Many researchers have extensively studied NiMnGa, Fe–Ni–Co–Ti, Fe–Pt, and Fe–Pd FSMAs to examine SME and superelasticity. These alloys have also been considered as a possible candidate actuator materials. Among these alloys, NiMnGa [4,5] and Fe–Ni–Co–Ti [20,21] alloys have been known to exhibit good SMEs. However, the martensitic transformation start temperature (T_{Ms}) of the latter alloy is below −150°C. It is too low to be practically used. Further, Kakeshita et al. even reported magnetoelastic martensitic transformation in Fe–Ni–Co–Ti by using a strongly applied magnetic field up to 2.3×10^7 A/m. They also reported that martensitic transformation is not induced by an applied magnetic field until 4.8×10^6 A/m is reached [25]. These results show that a very strong applied magnetic field is needed to directly induce martensitic transformation. This is not suitable

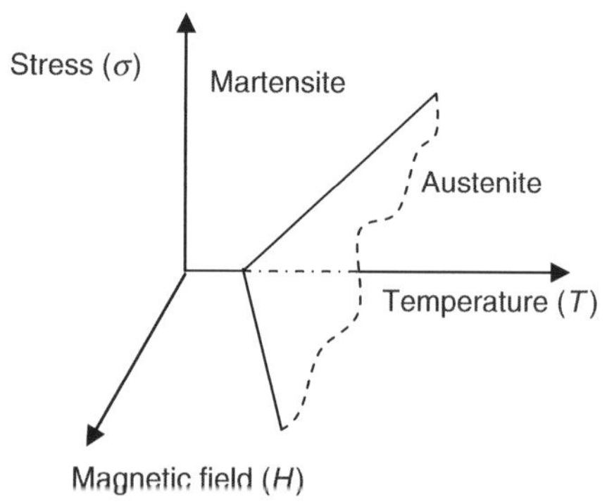

FIGURE 6.3 Schematics of three-dimensional phase transformation diagram of FSMAs.

TABLE 6.1 List of Current FSMAs

Type	Alloy	T_{Ms}	References
Heusler	Ni_2MnGa	−150°C to 150°C	[2–19]
	Fe–Ni–Co–Ti	<−150°C	[20–29]
Iron-based alloys	Fe-23–25Pt	<−48°C	[30–38]
	Fe-28–31Pd	−197°C to 80°C	[39–53]
	Fe–Pd–Ni	<RT	[54,55]
Cobalt-based alloy	NiAlCo		[56,57]

for applications as an actuator material where a compact electromagnet is needed.

However, the variant rearrangement mechanism under constant magnetic field maybe still useful for some applications. The change of the surface morphology on a single crystal specimen can be observed during the process of cooling. As shown in Figure 6.4a of a single crystal FePd whose temperature is just below T_{Mf} (= 0°C), the boundary and the contrast of the two variants are not distinct. When this specimen is further cooled at −20°C, its surface morphology exhibits a distinct boundary with a clear contrast of black and white colors (Figure 6.4b). This is because the lattice parameter change (e.g., *c/a* ratio, Figure 6.4c [43,44]) of the FePd martensite gradually decreases with decreasing temperature. The gradual change of the lattice parameter results in an increase of the angle φ formed by two martensite variants (CV1 and CV2) (Figure 6.4d). As φ increases during cooling, the contrast of the boundary between CV1 and CV2 (Figure 6.4a and b) also is enhanced. The change of the volume fraction between variants (variant rearrangement) such as CV1 and CV2 in Figure 6.4d will result in additional straining of the FePd specimen. This can be done by applying a magnetic field since the FePd is a ferromagnetic material. Such a straining, which is due to the magnetic field–induced martensite variants rearrangement, can be detected. When the FePd single crystal specimen is fully covered by the pair of CVs at −20°C, a magnetic field is applied along the length of the specimen. As shown in Figure 6.5, the strain of the magnetic field–induced martensite rearrangement is a function of the magnetic flux density, where the negative magnetic flux density means that the direction of the field is opposite. A strain of 0.4% is obtained and it is reversible and repeatable with any bias stress, although some studies have reported [20] that the application of a bias stress is required simultaneously with the applied magnetic field in order to induce such a reversible strain in FSMAs as shown in Figure 6.5.

A single crystalline FePd specimen is subjected to a compression test at −20°C and the amount of strain accompanied with the stress-induced variant rearrangement is measured. Surface morphology change of the specimen is also observed during the tests. Figure 6.6 shows the surface morphology change during the first loading and the stress–strain curves of the first two cycles of loading and unloading tests. During the first loading process (Figure 6.6a through e), the volume fraction of CV1 increases at the expense of CV2. When the stress reaches 10 MPa, the specimen is fully covered by a single variant of CV1 and the process of the variant rearrangement is completed (Figure 6.6e), while the strain reaches to 5.2%. During the first unloading process, the volume fraction of CV1 decreases and the unloading curve (white circles in Figure 6.6f) is slightly lower than that the loading curve (black circles). The specimen exhibits about 5.0%

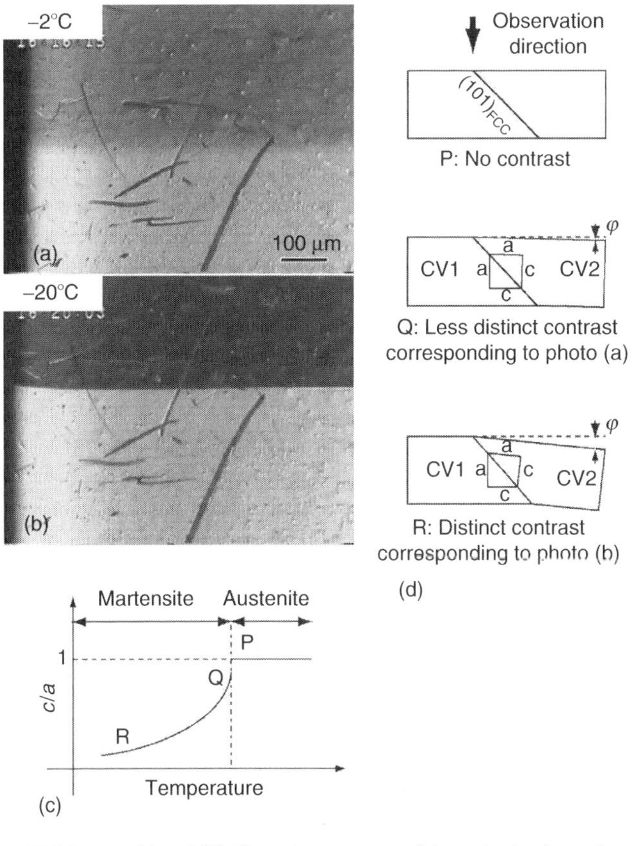

FIGURE 6.4 (a) and (b) Changing contrast of the twinning boundary after martensitic transformation. (c) Schematic illustration of the relationship between *c/a* ratio and temperature in Fe–Pd alloy. (d) Increasing angle φ with decreasing temperature.

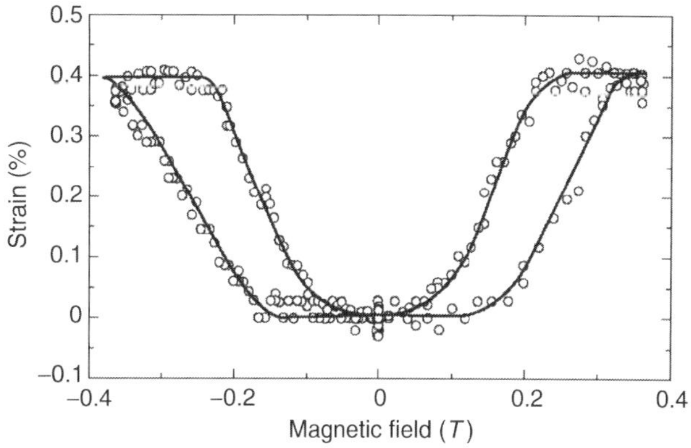

FIGURE 6.5 Magnetic field-induced strain as a function of magnetic flux density measured in Fe–Pd single crystal.

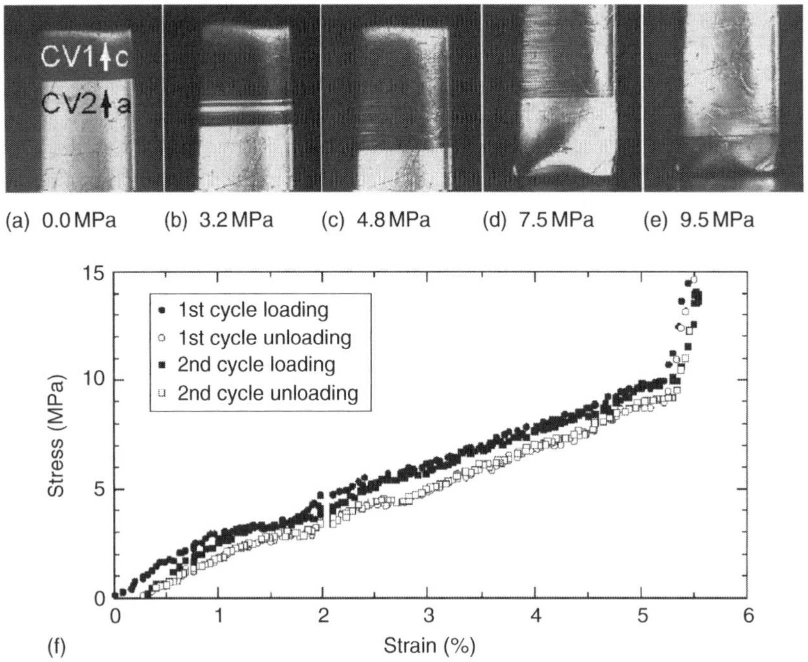

FIGURE 6.6 (a)–(e) Change of volume ratio of CVs for the first loading and (f) stress–strain curves for the first two cycles of loading and unloading.

reversible strain and 0.3% irreversible strain in the first loading unloading cycle. The stress–strain curve of the second test cycle (square dots in Figure 6.6f) almost follows the one of the first cycle. However, no reversible strain is measured in the second test and the repeatable strain is 5.0%. This repeatable strain under compressive stress could be utilized as a useful actuation method if the applied magnetic flux gradient that provides the stress is applied to the specimen so as to induce the strain by the above variant rearrangement mechanism.

Similarly, the superelastic behavior, SME, and magnetic properties have been studied in Fe–Pt alloys by some groups [30–38]. As shown in Table 6.1, Fe–Pt alloys also have the T_{Ms} much lower than room temperature [30,31]. Kakeshita et al. [38] have also reported that the martensitic transformation can be induced by a large applied magnetic field, as large as 2.0×10^7 A/m. The T_{Ms} of Fe–Pd alloys can be tailored by compositions of Pd between 80°C and −200°C [39,40,43]. However, currently most of FSMA actuator studies have focused on NiMnGa and Fe–Pd alloys.

6.1.3 Driving Mechanisms for FSMA-Based Actuators

FSMAs have both characteristics of thermoelastic SMAs and ferromagnetic properties. In principle, the phase transformation and straining also can be controlled by the magnetic field. The relationship of phase transformation, stress (σ), temperature (T), and magnetic field (H) can be presented as a three-dimensional phase transformation diagram as shown schematically in Figure 6.3. The three-dimensional surface in the diagram separates the martensite and austenite phases. This diagram can help us to determine a favorable driving mechanism for FSMAs as actuator materials, especially as controlled solely by an applied magnetic field. This control method by the magnetic field on FSMAs will be faster than the one by thermal conduction on SMAs. Several driving mechanisms (by the magnetic field) of actuators based on FSMAs have been proposed and studied. The following sections summarize the three main driving mechanisms.

6.1.3.1 Magnetic Field–Induced Phase Transformation

The phase transformations of FSMAs can be controlled or affected by a magnetic field. This is called the magnetic field–induced martensitic phase transformation [7,60]. Therefore, the superelasticity may also be controlled or produced by the magnetic field and a large reversible displacement is obtained. This property is advantageous when applied to an actuator device. The forward transformation (austenite → martensite), induced by a magnetic field, has been found in NiMnGa, while the reverse transformation (martensite → austenite), induced by a magnetic field, also occurs in FePd [60]. As shown in Figure 6.7, the slope, dH/dT, of FePd is negative and that of NiMnGa is positive. This phenomenon can be explained by the Clapeyron–Clausius relationship [59]

$$\frac{\mathrm{d}H}{\mathrm{d}T} = \frac{h}{T_{M_s}(M_M - M_A)} \tag{6.1}$$

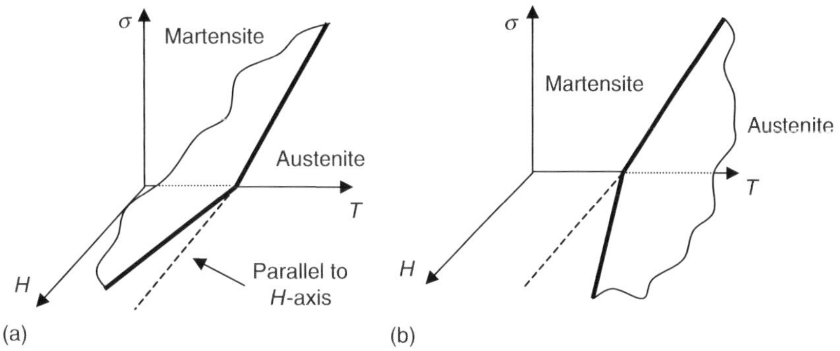

FIGURE 6.7 Schematics of the phase transformation diagram of (a) FePd and (b) NiMnGa. (From Liang, Y., Kazo, H., and Taya, M., *Mech. Mater.*, 38, 564, 2006. With permission.)

where
- h is the enthalpy
- H is the magnetic field
- M_M is the magnetization of martensite phase
- M_A is that of the austenite phase

For the case of FePd, $(M_M - M_A)$ is negative before the saturation of magnetization [60]. Hence, Equation 6.1 implies that dH/dT is negative, resulting in promoting the martensite → austenite phase transformation by applying the magnetic field. However, for the case of NiMnGa, its magnetization in the martensite phase is larger than that in the austenite phase [4,58], thus, dH/dT in Equation 6.1 is positive, resulting in promoting the austenite → martensite phase transformation. It has been reported that the magnetic field can induce phase transformation in FSMAs, but this effect is very small in both NiMnGa and FePd alloys because of a small magnetization difference $(M_M - M_A)$ between martensite and austenite phases.

6.1.3.2 Variant Rearrangement in Fully Martensite Phase by Magnetic Field

In addition to the above superelasticity, a rearrangement (change) in variants in a fully martensitic phase due to the application of stress in SMAs will occur and additional strain can be obtained as shown in Figure 6.2b. Here, stress controls variants with different transformation strain. This variant rearrangement induced by the application of the magnetic field is also possible in FSMAs. Many studies have concentrated on this mechanism [2,5,6,8–19,28,29,53]. In this case, the magnetic field controls variants with different magnetization. Variants having different magnetization may have different transformation strains. Thus, the magnetic field may change overall strain of an FSMA, when it is in a martensitic state. This is also advantageous when an FSMA is used as an actuator. If the magnetic field can really discriminate martensite variants with a different transformation strain, an FSMA can be used as a high damping material as well. First, the movement of variant or martensite boundaries accompanies energy dissipation. This is manifested by a hysteresis, even though it is small in thermoelastic transformation. Secondly, mechanical vibration interacts with the oscillating magnetic field that moves boundaries of magnetic domain or variant in martensite phase. Currently, the variant rearrangement induced by the magnetic field is effective in a single crystal FSMA, i.e., NiMnGa. However, a bias stress is needed in order to induce single variant in single crystal FSMA so a large strain can be obtained. Although 5% reversible strain has been reported by this driving mechanism, the output stress is only as high as 10 MPa. This is not suitable for actuators that require a high output force. Furthermore, NiMnGa is very brittle; therefore, it is only suitable for the applications where NiMnGa is under a compression mode.

6.1.3.3 SIM Phase Transformation by Magnetic Field Gradient

It is well-known that ferromagnetic material feels a force when it is in a nonuniform magnetic field. This force can be large enough to induce the SIM in a superelastic FSMA. This is called hybrid mechanism [59]. The hybrid mechanism consists of chain reactions: Magnetic field gradient causes magnetic force or moment, which introduces stress in an FSMA → SIM transformation → the FSMA becomes much softer and exhibits a large deformation. In reality, the hybrid mechanism can also be utilized, when an actuator is designed. That is, a large force can be applied to an actuator by a compact electromagnet with a high magnetic field gradient, resulting in a larger displacement with fast response. In this driving mechanism, FePd FSMA, which exhibits 1.6% superelastic strain, is more suitable because it has a large magnetization factor, with the same order as iron [59]. This can be easily seen by a demonstration as shown in Figure 6.8, where two identical cantilever beams made of FePd and NiMnGa are subjected to a magnetic field. As shown in Figure 6.8b, the FePd beam bends significantly because FePd has a much larger magnetization than NiMnGa. In addition, FePd is very ductile, which is suitable for any actuator application.

In summary, Table 6.2 shows the comparison among the three driving mechanisms proposed for FSMAs. Mechanism (a) is not practical because a large magnetic field is needed to change the transformation temperatures (i.e., inducing phase transformation). For example, for NiMnGa in Table 6.2, T_{Ms} can be changed

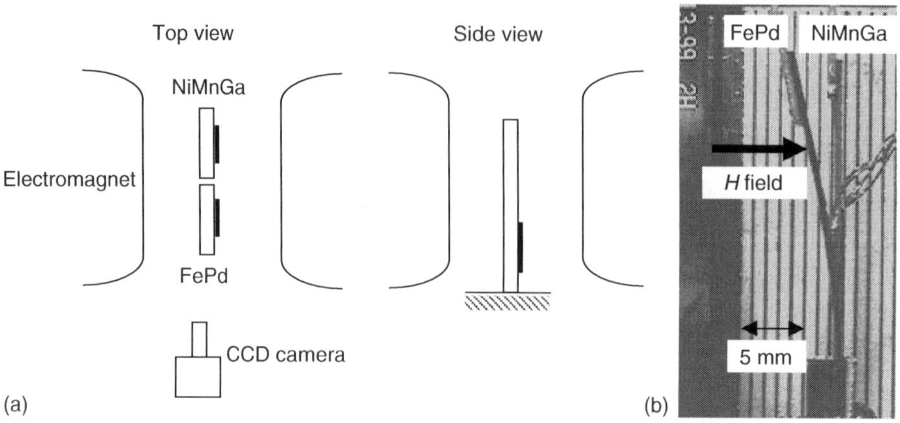

FIGURE 6.8 Bending comparison between Fe–Pd and NiMnGa cantilever beams: (a) setup and (b) after applied *H*-field.

TABLE 6.2 Summary and Comparison of Magnetic Field (*H*) Actuation Mechanisms

Mechanism	Method	Phase	Reported by	NiMnGa	Fe–Pd
Magnetic field-induced phase transformation	T_{Ms} is changed by *H*-field → forward or reverse transformation	A → M or M → A	NiMnGa Vasil'ev et al. [2] FePd Liang et al. [59]	$\Delta T_{Ms} \approx 9°C$ at $H = 4 \times 10^6$ A/m Inoue (Material Science, UW)	$\Delta T_{Ms} \approx -0.2°C$ at $H = 8 \times 10^5$ A/m
Variants rearrangement	*H*-field applied to martensite → variant change → deformation of the martensite	M	NiMnGa Chernenko et al. [62] Murray et al. [63] James et al. [4] FePd James and Wuttig [50]	The shear stress acting on martensite under $H = 3.2 \times 10^5$ A/m, (calculated) = 2.7 MPa	10 MPa
Hybrid mechanism	Force induced by *H*-field gradient applied to austenite → SIM	A → M	FePd Liang et al. [59]	Superelasticity (no *H*-field) single crystal: 1.6%, 120 MPa, compression Murray et al. [25]	Polycrystal: 1.5%, 450 MPa in tension

by a 4×10^6 A/m magnetic field, however, the electromagnet will be too huge to use for practical applications. A conventional compact solenoid cannot produce such a high magnetic field. Mechanism (b) is not achievable for polycrystal FSMA because only a very small strain is available. For example, only the order of 10^{-4} strain can be obtained for polycrystal FePd in a field up to 8×10^5 A/m [61]. Even though a large strain can be produced by a small magnetic field in single crystals of NiMnGa using mechanism (b), the output force is still as small as several MPa. Therefore, for the current available FSMA such as the FePd alloy, mechanism (c) (hybrid mechanism) shows the best performance among these mechanisms.

6.2 FSMA-Based Actuator

Previously, three driving mechanisms were summarized for the magnetic field for FSMAs to be used as actuator materials. These mechanisms are (a) magnetic field–induced phase transformation, (b) variant rearrangement in fully martensite phase by the magnetic field, and (c) hybrid mechanism (SIM transformation by the magnetic field gradient). Table 6.2 shows that the third mechanism (hybrid mechanism) has the best performance and is applicable for a compact actuator system. The hybrid mechanism is based on the force (or moment) acting on a ferromagnetic actuator material due to the magnetic field gradient. Then, this force (or moment) introduces a SIM, which will cause the material (initially in the austenite phase) becoming softer (martensite phase) and then able to produce large deformation and force. When the magnetic field is removed, the force is also removed from the FSMA material and the material will spring back because of superelasticity.

As mentioned before, polycrystal FePd has been identified as a promising FMSA for compact actuators based on the hybrid mechanism. It is because FePd can exhibit superelasticity at room temperature and has a large magnetization on the same order as iron and is so ductile, which makes it suitable for many applications. In a test, a cantilever beam made of polycrystalline FePd was simultaneously actuated with a ferromagnetic Fe beam by a

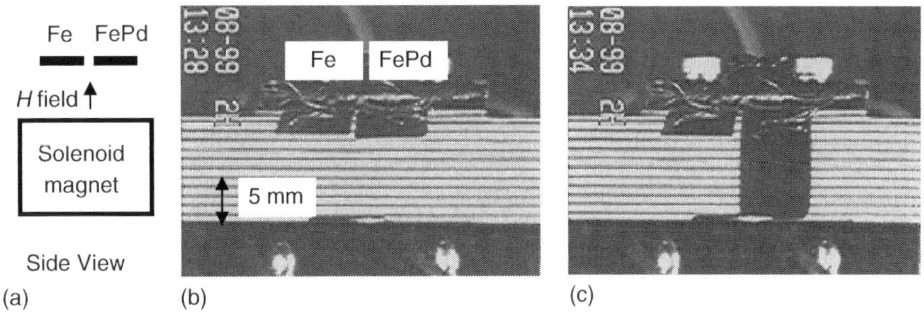

FIGURE 6.9 Bending of polycrystal FePd and Fe cantilever beams under applied magnetic field: (a) the schematic of the experimental setup (the side view from the tip of the beams) (b) before and (c) after applying the field.

portable electromagnet ($H = 8 \times 10^4$ A/m). Both beams have the same geometry and Figure 6.9a shows the experimental setup. It is clearly demonstrated in Figure 6.9c that the FePd beam has a much larger deflection than the Fe beam. The deflection is due to the hybrid mechanism. Since the magnetization of FePd in austenite is similar to that of Fe and the ratio of the Young's modulus of Fe (210 GPa) to FePd (80 GPa in the austenite phase, 30 GPa in the martensite phase) is about in the 3–7 range, the ratios of the bending deflection of FePd to Fe could be also 3–7. However, the demonstration results shows that the tip of the FePd beam exhibits a 7 mm displacement during the application of the magnetic field, while the Fe beam shows almost no deflection. This implies that the FePd beam is not only very ductile but also undergoes austenite $\rightarrow$ martensite phase transformation (SIM). As the gap between the specimen and the electromagnet becomes smaller during the deformation, the larger magnetic force leads to more phase change and deformation. This comparison shows that the FePd FSMA as an actuator material is superior to the conventional material. Both the cantilever beams spring back to the original position when the electromagnet is turned off.

Because FePd is very ductile, it is also possible to make FePd into various shapes. Figure 6.10 shows the various shapes of polycrystalline FePd specimens, such as rod, wire, and helical spring. Therefore, this material can be used without limitations of manufacturing. The helical FePd spring can be produced by winding and shape memorizing the FePd wire, which can be extruded from a solid cylinder. A series of pictures were taken to exhibit the FePd spring actuated under a compact electromagnet and shown in Figure 6.11 [64]. The spring actuation is driven by a modified electromagnet [65], which can produce a much higher magnetic field gradient than a conventional solenoid. By gradually increasing the electrical current, the FePd spring partially shrinks due to magnetic attraction. The partial shrinkage of the spring becomes part of the yoke and attracts the rest of the turns of the spring as the electrical current increases. Figure 6.11b shows the complete shrinkage of the FePd spring. Upon removing the electrical power, the spring completely returns to its initial length.

Although FePd appears as a promising actuator material based on the hybrid mechanism, it is very expensive because of

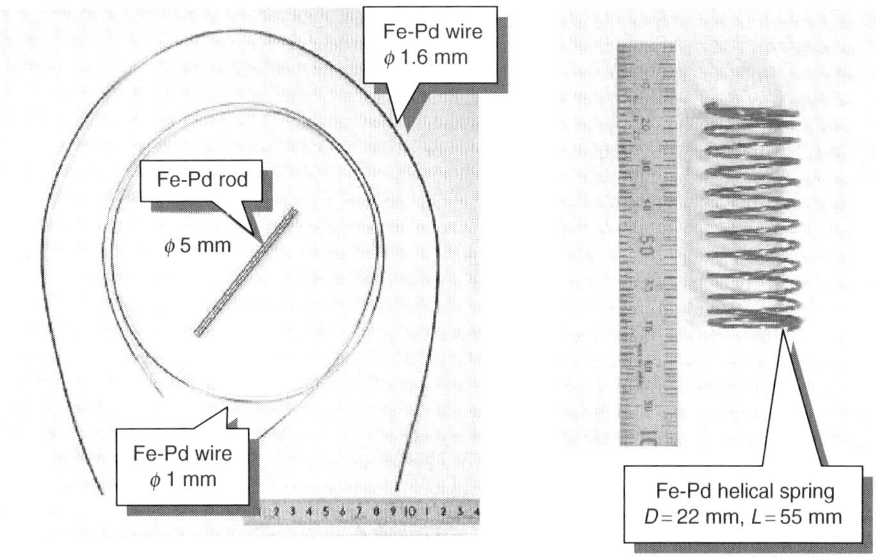

FIGURE 6.10 Different shapes of polycrystalline FePd specimens.

FIGURE 6.11 Actuation of the FePd spring actuator: (a) power off and (b) power on.

Pd. An alternative way to reduce the cost of the material is to use the FSMA composite, which is composed of the ferromagnetic material and superelastic SMA, where the ferromagnetic material is provides a large force due to the magnetic field gradient, resulting in a large deformation on superelastic SMA (i.e., NiTi) due to the SIM transformation. Several cases of FSMA composites have been studied and the laminated FSMA composite is easily made without losing its performance and is the most cost-effective [66], for example, the laminated plate (Figure 6.12a) and wire of concentric cylinders (Figure 6.12b). As shown in Figure 6.12a, the outer layer of superelastic NiTi SMA can sustain large stress and the inner core ferromagnetic material is subjected to modest stress, thus, leveraging the extra stress bearing capacity of superelastic NiTi while protecting an otherwise brittle soft ferromagnetic material.

Several FSMA composite actuators have been made. One of the examples is the torque actuator based on the FSMA composite [67] and its design concept is illustrated in Figure 6.13 where (a) and (b) denote the cases of switch on and off of the actuator system, respectively. The torque actuator consists of an inner rod, which will rotate counterclockwise upon switching on the electromagnet system, thus, attracting the FSMA plate spring to its inner wall. The rotating motion of the inner rod will provide the torque work for a dead load that is hanging on the rod by a pulley or belt. The FSMA composite is composed of a ferromagnetic material and a superelastic grade SMA, and it is subjected to a bending moment, which is not uniform over the length of the FSMA composite due to its varying curvature. In the first prototype torque actuator, the FSMA composite consists of NiTi superelastic wires and several cylindrical soft iron rods as shown in Figure 6.14a. The requirement for designing the FMSA composite based torque actuators is to induce a large stress so that the SMA plate of the FSMA composite can reach the onset of SIM transformation, while the stress in the ferromagnetic rods remains below its plastic yield stress. The FSMA composite will be attracted to the inner wall of the actuator due to strong magnetic flux gradient upon switching on the electromagnetic system. The first prototype torque actuator has produced 0.736 N

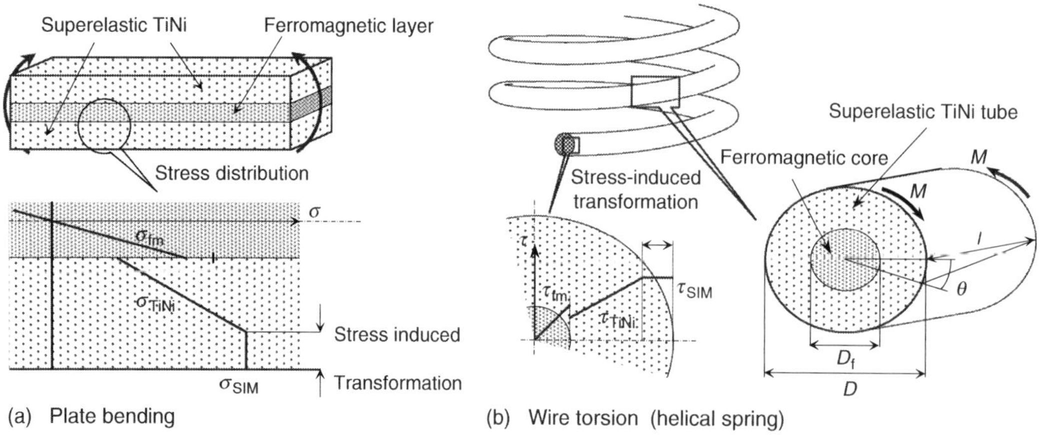

FIGURE 6.12 Two types of FSMA composites composed of soft ferromagnetic core and superelastic NiTi, (a) laminated composite plate and (b) concentric cylinder composite.

FIGURE 6.13 Photos of the torque actuator made of FSMA composite (a) switch on and (b) switch off.

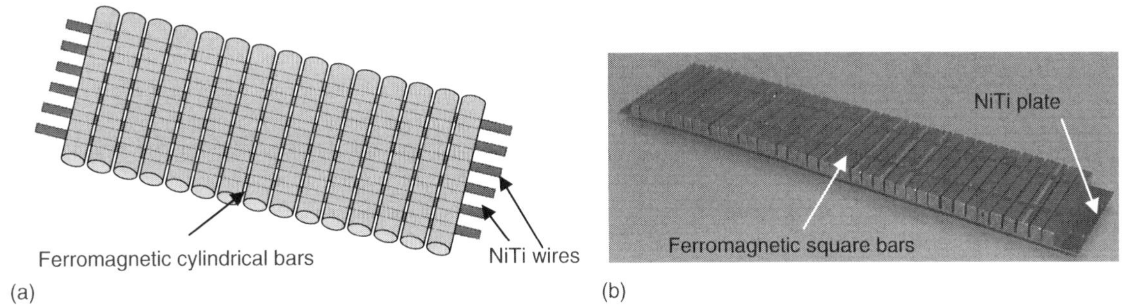

FIGURE 6.14 Two types of FSMA composites for the torque actuator (a) made of NiTi superelastic wires and cylindrical soft iron bars and (b) made of NiTi superelastic plate and square soft iron bars.

m with a 40° maximum angle. Further improvement has been done by using the FSMA composite plate spring, which is made of a superelastic Ni–Ti sheet and square soft iron bars (Figure 6.14b). Because of the higher stiffness of the structure and the stronger magnetic attracting force, it can provide a 4.8 N m torque with a 102° rotation angle.

A synthetic jet actuator based on the FSMA composite membrane has also been constructed [68]. The composite membrane was driven by the electromagnetic system and oscillated to create a synthetic jet flow through the exit hole. The FSMA composite membrane was composed of a superelastic NiTi thin sheet and a ferromagnetic soft iron pad. Figure 6.15a and b shows the parts of the membrane actuator, respectively. It has two chambers in the center divided by the composite membrane. The thin laminated yoke is used in the electromagnet unit in order to eliminate the eddy current for the high-frequency operation. When the

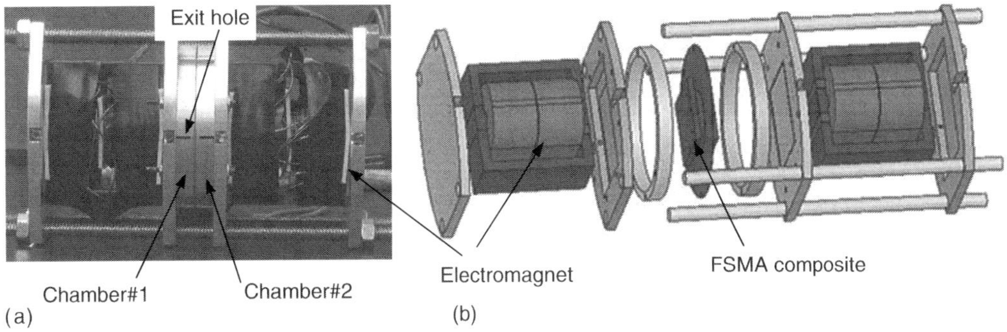

FIGURE 6.15 (a) Photo of the synthetic jet membrane actuator and (b) the schematic of actuator parts. (From Liang, Y., Kuga, Y., and Taya, M., *Sens. Actuators A*, 125, 512, 2005. With permission.)

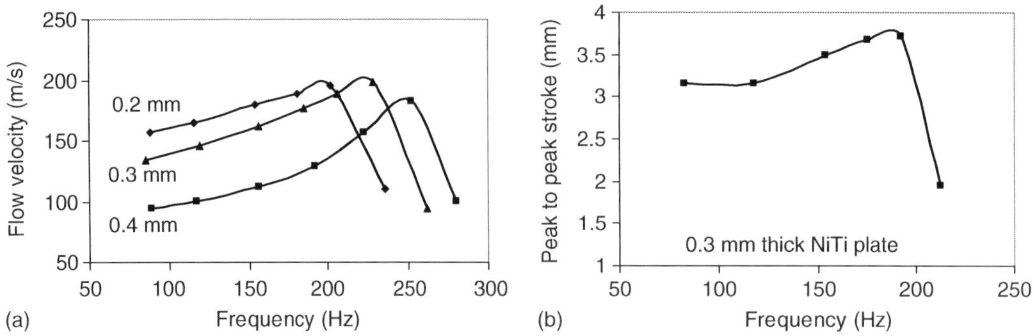

FIGURE 6.16 (a) Synthetic jet flow velocity as a function of frequency produced by 0.2, 0.3, and 0.4 mm thick NiTi sheets of the composite membrane and (b) peak to peak stroke of the composite membrane. (From Liang, Y., Kuga, Y., and Taya, M., *Sens. Actuators A*, 125, 512, 2005. With permission.)

FSMA composite membrane oscillates close to its resonance frequency, the system efficiency is optimized and the membrane exhibits a large stroke to produce a strong jet flow. Since the NiTi sheet of the composite membrane is superelastic, it can sustain large stresses without plastic deformation. The synthetic jet flow velocity from the membrane actuator was measured by the hotwire probe at the exit hole. Figure 6.16a shows the results of jet flow velocities as a function of frequency by using different thicknesses of NiTi thin sheets of the composite membrane. The maximum jet velocity, 190 m/s at 220 Hz, was obtained by using the 0.3 mm thick NiTi sheet of the composite membrane. Its displacement–frequency response is shown in Figure 6.16b where the peak to peak stroke of the membrane is 3.72 mm at 195 Hz. The energy density of the FSMA composite membrane was reported as 30 kJ/m^3 and its power density was 6000 kW/m^3 [68]. Both the energy density and the power density can be further improved if the composite membrane exhibits a much larger stroke. Also, the higher frequency response of the composite membrane can further increase the power density.

References

1. Z. Nishiyama, *Martensitic Transformation*, Academic Press, London, 1978.
2. M. Wuttig, L. Liu, K. Tsuchiya, and R. D. James, *J. Appl. Phys.*, 87, 2000, 4707.
3. R. D. James and K. F. Hane, *Acta Mater.*, 48, 2000, 197.
4. P. J. Webster, K. R. A. Ziebeck, S. L. Town, and M. S. Peak, *Phil. Mag. B*, 49, 1984, 295.
5. R. C. O'Handley, *J Appl. Phys.*, 83, 1998, 3263.
6. V. A. Chernenko, E. Cesari, V. V. Kokorin, and I. N. Vitenko, *Scripta Metal. Mater.*, 33, 1995,1239.
7. S. J. Murray, M. Farinelli, C. Kantner, J. K. Huang, S. M. Allen, and R. C. O'Handley, *J. Appl. Phys.*, 83, 1998, 7297.
8. A. N. Vasil'ev, A. D. Bozhko, V. V. Khovailo, I. E. Dikshitein, V. G. Shavrov, V. D. Buchelnikov, M. Matsumoto, S. Suzuki, T. Takagi, and J. Tani, *Phys. Rev. B*, 59, 1999, 1113.
9. A. D. Bozhko, A. N. Vasil'ev, V. V. Khovailo, I. E. Dikshitein, V. V. Koledov, S. M. Seletski, A. A. Tulaikova, A. A. Cherechukin, and V. D. Buchelnikov, *J. Exp. Theoret. Phys.*, 88, 1999, 954.
10. R. D. James, R. Tickle, and M. Wuttig, *Mater. Sci. Eng.*, A273, 1999, 320.
11. R. Tickle and R. D. James, *J. Magn. Mater.*, 195, 1999, 627.
12. K. Ullakko, J. K. Huang, V. V. Kokorin, and R. C. O'Handley, *Scripta Mat.*, 36, 1997, 1133.
13. H. Miki, T. Takagi, M. Matsumoto, J. Tani, K. Yamauchi, T. Abe, H. Nakamura, and V. Khovailo, Proc. Symp. on Dynamics of Electro-Magnetic Field, 1999, p. 326.
14. Y. Ezer, A. Sozinov, G. Kimmel, V. Etelaniemi, J. T. Glavatskaya, A. D'Anci, C. Podgursky, V. K. Lindroos, and K. Ullako, *Proc. SPIE*, 3675, 1999, 244.
15. M. Wuttig, L. Liu, K. Tsuchiya, and R. D. James, *J. Appl. Phys.*, 87, 2000, 4701.
16. Qi-Pan and R. D. James, *J. Appl. Phys.*, 87, 2000, 4702.
17. K. Tsuchiya, D. Ohtoyo, H. Nakamura, M. Umemoto, and P. G. McCormik, *Mater. Trans. JIM*, 8, 2000, 938.
18. K. Tsuchiya, H. Nakamura, M. Umemoto, and H. Ohtsuka, *Trans. MSRJ*, 25, 2000, 517.
19. K. Tsuchiya, D. Ohtoyo, M. Umemoto, and H. Ohtsuka, *Trans. MSRJ*, 25, 2000, 521.
20. S. J. Murray, M. Marioni, S. M. Allen, R. C. O'Handley, and T. A. Lograsso, *Appl. Phys. Lett.*, 77, 2000, 886.
21. R. C. O'Handley, S. J. Murray, M. Marioni, H. Nembach, and S. M. Allen, *J. Appl. Phys.*, 87, 2000, 4712.
22. S. J. Murray, M. A. Marioni, A. M. Kukla, J. Robinson, R. C. O'Handley, and S. M. Allen, *J. Appl. Phys.*, 87, 2000, 5774.
23. Yu. N. Koval, V. V. Kokorin, and L. G. Khandros, *Phys. Metal. Metall.*, 48, 1979, 162.
24. T. Maki, K. Kobayashi, M. Minato, and I. Tamura, *Scripta Metall.*, 18, 1984, 1105.
25. T. Kakeshita, K. Shimizu, T. Maki, I. Tamura, S. Kijima, and M. Date, *Scripta Metall.*, 19, 1985, 973.
26. T. Maki, S. Furutani, and I. Tamura, *ISIJ Int.*, 29, 1989, 438.
27. N. Jost and M. Huhner, *Pract. Metal.*, 26, 1989, 295.

28. E. Hornbogen and N. Jost, *J.Phys.IV, Suppl. J. Phys.III*, C4, 1991, 199.
29. T. Kikuchi and S. Kajiwara, *Mater. Trans. JIM*, 34, 1993, 907.
30. G. Kubla and E. Hornbogen, *J.Phys.IV, Suppl. J. Phys. III*, C2, 1995, 304.
31. S. J. Murray, R. Hayashi, M. Marioni, S. M. Allen, and R. C. O'Handley, *Proc. SPIE*, 3675, 1999, 204.
32. R. Hayashi, S. J. Murray, M. Marioni, S. M. Allen, and R. C. O'Handley, *Sensor and Actuator-A*, 81, 2000, 219.
33. D. P. Dunne and C. M. Wayman, *Metall.Trans.A*, 4, 1972, 137, 147.
34. M. Umemoto and C. M. Wayman, *Acta Met.*, 26, 1978, 1529.
35. M. Foos, C. Frantz, and M. Gantois, *Acta Metall.*, 29, 1981, 1091.
36. C. M. Wayman, *Scripta Metal.*, 5, 1981, 489.
37. K. Sumiyama, M. Shiga, and Y. Nakamura, *Phys. Status Solidi A*, 76, 1983, 747.
38. T. Kakeshita, T. Shimizu, S. Funada, and M. Date, *Trans. JIM*, 25, 1984, 837.
39. R. Oshima, S. Sugimoto, M. Sugiyama, and F. E. Fujita, *Trans. JIM*, 26, 1985, 523.
40. A. Cevecka, H. Otsuka, and H. K. D. H. Bhadeshia, *Mater. Sci. Tech*. Feb, 1995, 109.
41. M. Johmen, T. Fukuda, T. Takeuchi, T. Kakeshita, R. Oshima, and S. Muto, Proc. SMART2000, Sendai, 26.
42. M. Matsui, T. Shimizu, H. Yamada, and K. Adachi, *J. Mag. Mag. Mater.*, 1980, 1201, 15–18.
43. T. Sohmura, R. Oshima, and F. E. Fujita, *Scripta Metall.*, 14, 1980, 855.
44. R. Oshima, *Scripta Metall.*, 15, 1981, 829.
45. M. Matsui and K. Adachi, *J. Mag. Mag. Mater.*, 1983, 115, 31–34.
46. M. Sugiyama, R. Oshima, and F. E. Fujita, *Trans. JIM.*, 25, 1984, 585.
47. M. Sugiyama, S. Harada, and R. Oshima, *Scripta Metall.*, 19, 1985, 315.
48. M. Matsui, J. P. Kuang, T. Totani, and K. Adachi, *J. Mag. Mag. Mater.*, 1986, 911, 54–57.
49. M. Sugiyama, R. Oshima, and F. E. Fujita, *Trans. JIM*, 27, 1986, 719.
50. S. Muto, R. Oshima, and F. E. Fujita, *Acta Metall.*, 38, 1990, 685.
51. R. Oshima, S. Muto, F. E. Fujita, T. Hamada, and M. Sugiyama, *MRS. Intl. Mtg. Adv. Mat.*, 9, 1989, 475.
52. R. Oshima, M. Sugiyama, and F. E. Fujita, *Metall. Trans*. A, 19A, 1988, 803.
53. H. Seto, Y. Noda, and Y. Yamada, *J. Phys. Soc. Jpn.*, 57, 1988, 3668.
54. R. Oshima, *Bull. Jpn. Inst. Metals*, 28, 1989, 493.
55. R. Oshima, S. Muto, and F. E Fujita, *Mater. Trans. JIM*, 3, 1992, 197.
56. H. Morito, A. Fujita, K. Fukamichi, R. Kainuma, and K. Ishida, *Appl. Phys. Lett.*, 81, 2002, 1657.
57. K. Oikawa, L. Wulff, T. Iijima, F. Gejima, T. Ohmori, and A. Fujita, *Appl. Phys. Lett.*, 79, 2001, 3290.
58. K. Ullakko, J. K. Huang, C. Kantner, and R. C. O'Handley, R. C., *Appl. Phys. Lett.*, 69, 1996, 1966–1968.
59. Y. Liang, H. Kato, and M. Taya, *Mech. Mater.*, 38, 2006, 564–570.
60. Y. Liang, Y. Sutou, T. Wada, C. Lee, M. Taya, and T. Mori, *Scripta Mater.*, 48, 2003, 1415–1419.
61. T. Wada, Y. Liang, H. Kato, T. Tagawa, M. Taya, and T. Mori, *Mater. Sci. Eng. A*, 361, 2003, 75–82.
62. H. Hamada, J. H. Lee, K. Mizuuchi, M. Taya, and K. Inoue, *Metall. Trans.*, 29A, 1998, 1127–1135.
63. D. S. Ford and S. R. White, *Acta Mater.*, 44, 1996, 2295–2307.
64. T. Wada and M. Taya, *Proc. SPIE*, 4699, 2002, 294–302.
65. K. Oguri, Y. Ochiai, Y. Nishi, S. Ogino, and Y. Uchikawa, Proceedings of 9th Symposium on Intelligent Materials, March 16, 2000, Tokyo, Japan, pp. 24–25.
66. M. Kusaka and M. Taya, *J. Composite Mater*, 38, 2004, 1011–1035.
67. V. Chen, M. Taya, J. K. Lee, M. Kusaka, and T. Wada, *Proc. SPIE*, 5390, 2004, 309–316.
68. Y. Liang, Y. Kuga, and M. Taya, *Sens. Actuators A Phys*, 125/2, 2005, 512–518.

7

Aircraft Applications of Smart Structures

Johannes Schweiger
German Aerospace Society

7.1 Introduction

Probably the most famous photography in aviation (Figure 7.1) depicts not only the first successful manned powered flight on December 17, 1903 by the Wright brothers, but it also shows the first successful airplane with an active structure. The Wright brothers design must definitely be called smart with respect to many aspects and design features. They were among the first pioneers in aviation who had realized that uncoupled control about all three vehicle axes was required. They had done systematic experimental aerodynamic research to achieve maximum possible aerodynamic performance. And they had learned how to design and manufacture light weight structures in their bicycle shop. Their solution for adequate roll control of the airplane, however, was more than one century ahead of the state-of-the-art in aviation technology. Since the centennial celebrations

FIGURE 7.1 First flight of Wright Flyer I on December 17, 1903 (available as photo hard copy, postcard).

of this remarkable event, no single airplane exists yet, which makes use of smart structure concepts to control the flight of the vehicle.

Rather than fighting the low torsional stiffness of their braced biplane wing design, they used this characteristic in a positive way. By the side-way motions of the pilot, lying on a sliding cradle, the wing tips were twisted in opposite directions by the wires attached to the cradle, thus producing the desired aerodynamic loads to roll the airplane. Figure 7.2 from Orville Wright's book [1] demonstrated this principle, which is also a very good example of the importance of integrated or multidisciplinary design concepts, especially in aeronautics. Unfortunately, this knowledge was lost and forgotten over the years, mainly because of more expert knowledge in single disciplines and more formal and bureaucratic processes for the design of new airplanes. Only in recent years, some prophets in aerospace have tried to spread the news about this old idea again and develop some new ones. Weisshaar [2], for example, quoted the success of the Wright Flyer as a good example for the need for integrated design methods in 1986. The Wright Flyer example also demonstrates that smart aircraft structures do not necessarily rely only on advanced active materials.

Even earlier than that, active structure concepts were studied. Alois Wolfmüller (1864–1948), the producer of the world's first series motorcycle, had bought the second of the production glider model "Normal-Segelapparat" (normal soaring apparatus) from Otto Lilienthal in 1894 [3]. Both aviation pioneers were communicating about improvements of performance and maneuverability by controlling the air loads through flexible wing twist. Wolfmüller tried to improve the performance by introducing a flexible hinge to the wings in order to modulate aerodynamic control forces from the flexible deformations.

Today, aircraft control is achieved by means of control surfaces attached to the main aerodynamic surfaces. These devices, aileron, elevator, and rudder, create the required forces and moments to control the motion of the aircraft about all three axes in space. Depending on the size and speed of the aircraft, these surfaces are actuated manually or by hydraulic systems. If the Wright brothers had used separate ailerons to roll the airplane, the additional structural weight might have been too much for the available power from the engine.

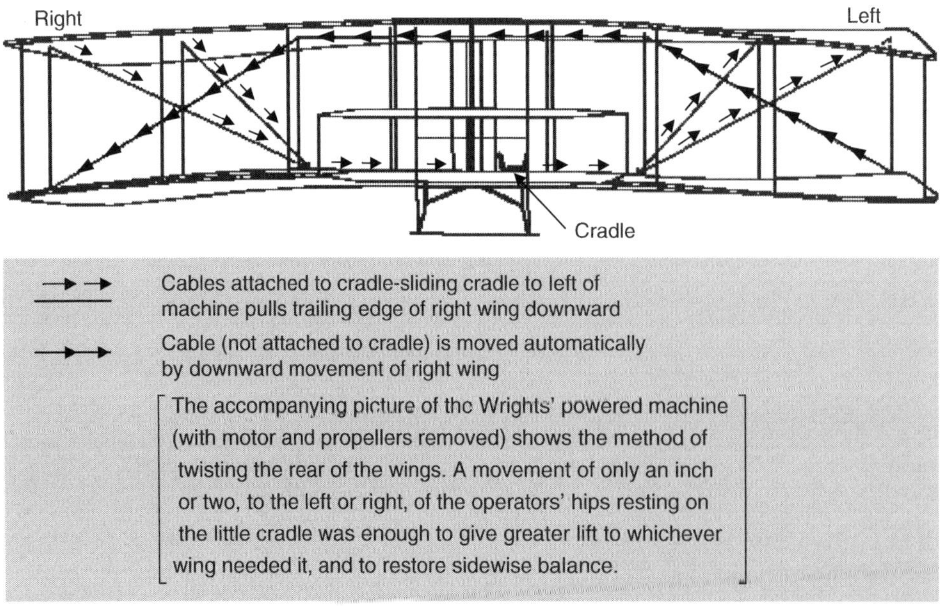

FIGURE 7.2 Active structural concept of the Wright Flyer I for roll control. (Adapted from Wright, O., *How We Invented the Airplane*, Dover Publications, Mineola, NY, 1988.)

The idea of active or smart structures for the control of air vehicles is as old as the earliest known attempts to fly with heavier-than-air machines. Early attempts by mankind to fly were usually based on the efforts to understand and copy the flight of birds. Besides the difficulty to control an unstable flying vehicle, which requires the high-performance computing power of our days or the complex neural network of animals to control their muscular system, it is even more difficult to sense and actuate the dynamic motions of continuously deforming, flexible aerodynamic surfaces.

In most of these efforts, it was the pilot himself who was supposed to actuate bird-like aerodynamic surfaces to produce the required lift force, create the forward thrust, stabilize the vehicle, and apply the necessary control inputs in time. All these early efforts were doomed to failure from the beginning for a multitude of reasons:

- Very limited knowledge of the laws of aerodynamics and flight mechanics.
- Insufficient load capacity of the available materials, only primitive manufacturing techniques, and only a rudimentary understanding of structural mechanics.
- Not enough sustained power output from the human body to produce the required thrust, or a lack of engines with sufficient power density.
- Efforts trying to copy structural design principles from nature did not take into account the scaling laws of physics; the resulting designs were too heavy, too fragile, or too flexible.
- In most cases, also the too complex efforts required to stabilize and control a flying vehicle without natural stability.

Only in recent years, after one century plus since the first successful powered flight, it was possible to design, build, and fly vehicles powered by the human body for the sole purpose of winning trophies.

For these reasons, the first successes in aviation were only possible by design concepts with almost "rigid" surfaces and natural stability of the vehicles. Nevertheless, a major contribution to the success of the Wright brothers was their "smart structures" flight control system for the roll axis. They had a combustion engine with sufficient energy density available "just in time."

7.2 Smart Structures for Flight in Nature

Although complete plants do not fly, they have to withstand aerodynamic loads by proper aeroelastic reactions, and sometimes their existence relies on their aerodynamic performance. The seeds of some plants are optimized for long distance flight in shape and mass distribution by millions of generations in a genetic optimization process. The interest in micro air vehicles in recent years has increased aerodynamic research efforts in this area [4]. Although not for free flight, the leaves of trees and their stems are built to withstand strong winds. The joints between leaves and branches have to have the right amount of flexibility in bending and torsion to reduce aerodynamic loads and at the same time avoid excessive unsteady loads from flutter.

More interesting for aircraft applications of smart structures is the flight of animals. As already mentioned above, early aviators tried to learn from birds. However, the required knowledge about the physics of flight was missing for these early attempts. The complex interactions of steady and unsteady aerodynamic forces with the active motions and passive deformations of wings and feathers are still not completely understood today. We are only now beginning to understand the functions and importance of the individual components for the efficiency of animal flight. Pendleton [5] gives some good examples from prehistoric flying saurian to modern birds with feathers.

But even more astonishing are the achievements in the art of flying for another species. Insects show by far the widest variety and most advanced structural concepts of active and passive control. An ordinary fly can land and take off on the tip of your nose or on the ceiling of a room. Dragon flies like the one shown in Figure 7.3 use a combination of passive mass balance for flutter prevention at the wing tips, and an advanced structural design with stiff chitin "spar" elements, supporting the membrane skins, and flexible hinges from Resilin to adjust the shape for all flight conditions.

The variety and large number of flying members in the family of insects has been attributed to their ability to fly, which offers the advantages to reach and conquer new territories more easily.

In the context of formal optimization methods in aeronautics, genetic algorithms became fashionable in recent years. The question in the context of technical products is: can we really wait as long as in nature to get better products? Or should we better continue to rely on gradient-based methods, which can be seen as targeted, "artificial genetic manipulations," similar to the "biological engineering" approaches today?

FIGURE 7.3 "Structural" design concept for dragon flies.

7.3 General Remarks on Aircraft Design Aspects

One reason why we have not seen more progress in smart structures applications in aeronautics may be the lack of understanding of the interactions between the different classical disciplines in aircraft design, and between these disciplines and the specialists from the smart materials area for each others needs and capabilities.

In order to assess the possibilities of smart structures applications for aircraft control, it is advisable to look at some aspects of aircraft design first. The structural engineers are usually concerned with the strength of their design for critical conditions. And they may be surprised in some cases, how relatively small deformations of the structure can have severe impacts on the aerodynamic performance or on the maneuverability. The aerodynamic and flight control engineers want the structure "as rigid as possible," but at the same time with different shapes for different flight conditions. These requirements may attract the smart structures specialists to make the design deformable and meet at the same time tight requirements for the external shape at changing flight conditions. This requires a look at the stiffness of the structure from strength needs, as well as at the additional internal loads created by the active deformations. And finally, aeroelastic aspects like flutter stability and effectiveness of deformed aerodynamic surfaces have to be considered.

For the aerodynamic engineer, an aircraft design is always a compromise between different flight conditions: a small flat wing for cruise with low drag at high speed, and a large, cambered wing for take-off and landing at low speed. This can partially be met by extendable control surfaces, attached to the fixed surface by complex kinematic systems. On the other hand, the complexity of these systems will increase the structural weight. It looks therefore attractive to replace these mechanisms and integrate their functions into one actively deformable structure. To consider such options, it is necessary to look at basic structural design requirements for aircraft wings. They have to carry the loads from between 2.5 and 9.0 times the total weight of the airplane. And they must be strong enough in torsion to keep the aerodynamic shape and transmit the loads from control surfaces. To achieve this at an acceptable structural weight, sophisticated lightweight design concepts were developed during the first half century of manned aviation. Besides improvements on the materials and manufacturing side, it is mainly the principle of lightweight design by shape within the prescribed aerodynamic shape.

Simplified, this principle leads to the placements of the material to the external shape of the airfoil and to closed cross sections of maximum area. This will provide maximum strength for a fixed amount of material. At the same time, this structure will also have maximum stiffness. This fact must be kept in mind when active deformable structures concepts are considered for airframe components.

7.4 Traditional Active or Adaptive Aircraft Control Concepts

One kind of active aircraft control concept was already developed and demonstrated in the early 1980s: artificial stabilization of the airplane's flight path by fast motions of the control surfaces, commanded by a digital flight control system. Today this system helps to reduce trim drag and increase agility of all modern fighter aircraft by avoiding negative contributions from a stabilizer surface to the total lift force in order to establish natural static stability.

Most airplanes have adaptive wings by means of additional deployable surfaces like slats at the leading edge or Fowler flaps at the trailing edge to provide additional lift for take-off and landing, or provide clean flow conditions for extreme maneuvers. The resulting effects are usually a combination of increased wing surface, increased camber, and accelerated air flow. These concepts are well described in most aerodynamic textbooks like Ref. [6] where also more sophisticated or exotic concepts like telescopic wings can be found. Variable wing sweep is rather common for combat aircraft, among them the American F-111. The mission adaptive wing (MAW) demonstrator version of this aircraft used a complex mechanical system to adjust the wing camber for changing flight conditions [5]. However, this system proved too complex and heavy for applications on production aircraft.

Active flutter stability enhancement by means of control surface deflections to create unsteady aerodynamic damping forces was also demonstrated in the 1970s [7]. There are two main reasons why we do not see them on today's airplanes:

- Flutter within the flight envelope at failure cases of the system is considered too critical.
- Performance of these systems was not very good because of the loss of static aeroelastic effectiveness of the used control surfaces with increasing airspeed; it is typically close to zero for ailerons mounted near the wing tip at the trailing edge.

The last point also applies to the effectiveness of maneuver load alleviation systems by means of symmetrical deflected outboard aileron surfaces.

7.5 Range of Active Structures and Materials Applications in Aeronautics

Besides aircraft control, including load alleviation, aerodynamic performance improvement by shape, and static aeroelastic applications for maneuverability and static stability enhancement, which will be the main points of the following sections, the following main additional applications in aeronautics are addressed by theoretical and experimental research:

Health monitoring. Because piezoelectric materials can serve as actuators and sensors as well, these materials can be used to monitor changing static or dynamic internal loading conditions, resulting from airplane maneuvers or from failures in the structure. Although these functions have been widely discussed in the literature for two decades now, no application to a production airplane is known to the author to date.

Vibration control. This application can be subdivided into two categories.

For the treatment of local or global vibrations of structural components. This was also the first flight test demonstration of piezomaterials, applied to a skin panel affected by engine noise of the Rockwell B-1B bomber [8].

For the reduction of vibration levels on equipment, mounted to the aircraft structure internally or externally. This can be done to support the integrity of the equipment, or to improve the performance of equipment acting as sensor systems. Because rather small deformations and forces have to be handled here, the most promising and near-term applications of smart materials on airplanes in this area were already predicted in the survey paper by Crowe and Sater [9].

Active flutter suppression. A lot of research effort has been dedicated to this task. The NASA wind tunnel test program "Piezoelectric Aeroelastic Response Tailoring Investigation" (PARTI) [10] demonstrated this technology. Applications to real aircraft seem to be rather unlikely because of the above-mentioned safety aspects. Panel flutter as a special case for the instability of individual skin panels at supersonic speed is also addressed in many publications. The required effort for the installation and control of the active devices seems to be inappropriate compared to a simple structural reinforcement of the critical panels.

Acoustic noise reduction. Several research projects are dealing with cabin noise reduction, especially for turbo-prop airplanes.

7.6 Aircraft Structures

7.6.1 Definitions for a Structure

To be able to design active structures systems for airplane control, it is necessary to understand the functions of the structure, the requirements that define its properties, and the reasons why existing aircraft structures are built in a specific way.

The main functions of the airframe are

- To bear and carry the loads acting on the vehicle.
- Provide space and support for engines, equipment, payload, and fuel.
- Satisfy the tight rigidity and smoothness requirements for the external shape from aerodynamics.

In other words, the external shape of the vehicle is defined by aerodynamic performance requirements, and the structure has to meet its requirements within this shape at minimum weight. This requires a careful combination of the shape and load-carrying purpose of the individual structural elements with the best available material for this purpose. Early designs used fabric to collect the aerodynamic pressure loads on a wing surface and transfer them to the fuselage through a wooden framework, supported by thin wires. As soon as it was discovered that the aerodynamic drag of these wires is 100 times higher than that of a carefully designed airfoil with the same thickness, design efforts concentrated on cantilevered wings where the loads could completely be transferred internally. Aluminum skins paved the way to modern monocoque airframes, where the main loads are carried by a monolithic shell structure, which mainly needs the internal structure only to maintain the shape. The shape of this shell provides highest resistance against bending and torsion loads at minimum weight.

7.6.2 Rigidity of Wing Structures

Considering active deformations of a wing structure, it is useful to look at the differences in the possible passive deformations under different loading conditions. The lift force is the largest component of the aerodynamic forces. It corresponds to multiples of the total weight of the airplane; 2.5 times for transport aircraft and currently up to nine times for fighters. At the same time, the external shape of a wing has the smallest dimension in its height and the largest in spanwise direction. This means that the structure needs the largest cross sections of the skins on the upper and lower surface. To actively deform a wing in bending would therefore be difficult but not impossible. However, the bending deformation of an unswept wing has no impact on the aerodynamic characteristics or loading conditions. Only swept wings are sensitive in this respect. Swept forward, the bending will increase the local aerodynamic angle of attack in streamwise direction, as indicated in Figure 7.4. This will increase the bending moment and causes structural divergence at a flight

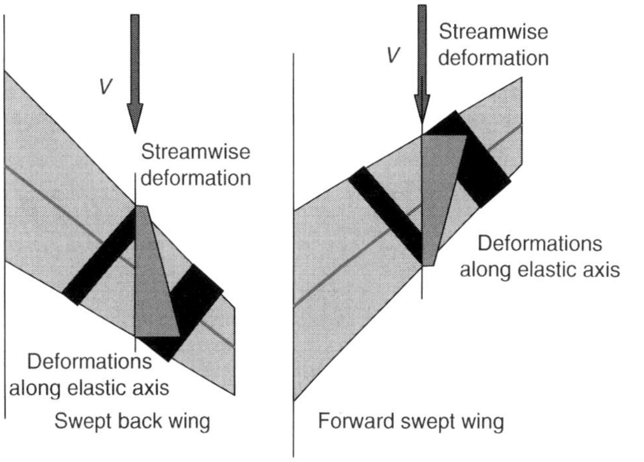

FIGURE 7.4 Bending deformation of swept wings and impacts on the aerodynamic angle-of-attack.

speed, called divergence speed, which depends on the bending stiffness and geometrical properties of the wing. The same effect causes a reduction of the bending moment on a swept back wing under load, which acts as a passive load alleviation system. But to actively control these deformations by internal forces would mean to stretch and compress the skins in spanwise direction, which is a rather difficult task.

The aerodynamic drag forces, acting in streamwise direction, are smaller than the lift forces by a factor of 10. At the same time, the shape of the airfoil creates a high static moment of resistance in this direction. For these reasons, the loads in this direction need no special attention in the structural design. An active deformation would be both very difficult and meaningless.

Torsional loads on a wing can be very high, depending on the chordwise center of pressure locations and on additional forces from deflected control surfaces. A center of pressure in front of the fictive elastic axis through the wing cross sections will cause torsional divergence at a certain flight speed, and a center of pressure too far behind the elastic axis will cause the wing to twist against the desired angle of attack or control surface deflection. The wing torsional deformation is therefore very sensitive for the loads acting on the wing, for the spanwise lift distribution, which is important for the aerodynamic drag and for the effectiveness of the control surfaces. As mentioned before, a closed torque box with maximum cross section is desirable for the structural designer. But the possibility to adjust the torsional flexibility would also allow several options for the active control of aerodynamic performance, load distribution, and control effectiveness. This active control by internal forces could theoretically be achieved by active materials in the skins, or by an internal torque device, fixed at the wing root and attached to the wing tip. Such a device, based on shape memory alloy (SMA), was already demonstrated on a wind-tunnel model [4]. For practical applications, the parameters that define the torsional stiffness of the structure with a closed cross section should be kept in mind. It is proportional to the square of the complete cross-sectional area, and linearly proportional to the average thickness of the skin. This demonstrates how difficult it would be to modify the stiffness by changes to the skins or by an internal torque tube with a smaller cross section.

The most often mentioned application of active structures for aircraft application is camber control and the integration of control surface functions into the main surface by camber control. This would mean a high chordwise bending deformation of the wing box. As mentioned above for the spanwise bending, the skins would have to be stretched and compressed considerably, but this time based on a smaller reference length and with a smaller moment arm. For this reason, we do not see chordwise bending deformations on conventional wings under load. Aeroelastic tailoring, addressed below, by means of adjusting the carbon fiber plies in thickness and direction to meet desired deformation characteristics, was also addressing camber control in the 1970s and 1980s as one specific option. Because of the above mentioned constraints, no application of this passive aeroelastic control feature on a realistic wing design is known to the author.

7.6.3 Structures and Mechanisms

To a certain extent, the main functions of structures and mechanisms are opposite: a structure must provide rigidity and a mechanism must provide large defined motions between parts. If active structures are considered, both functions must be integrated into the structure, the performance of this system should be better, and the total weight should be lower compared to a conventional design. This shows the difficulty of the task to develop active structures for aircraft control.

The intention to make the structure more flexible in order to allow deformations is therefore a contradiction. To allow deformations without producing internal forces, hinges are required. If this function is desired within the structure, the structure has to become more flexible in distinguished small regions. But this attempt will create high internal loads in regions with small cross sections. And the desired deformations for aircraft control functions will produce a high number of load cycles.

7.6.4 Passive Materials for Aircraft Structures

Lightweight aircraft structures are obtained by optimal shape and the best suitable materials for the load levels, and type of loading, Figure 7.5 shows the achievements in aircraft design by new materials. Unfortunately for active structures concepts, today's high-performance composite materials are stiffer than previous aluminum structures. It is therefore a misbelief by some active aircraft structures researchers to talk about "highly flexible" composite structures for their designs. On the other hand, today's skepticism against future applications of smart materials may be as wrong as the statement in one of the earliest textbooks on aircraft structures, where the author states that metal will never be used in aircraft structures because of its too high density compared to metal [11]. Obviously, the author was only looking at iron at that time.

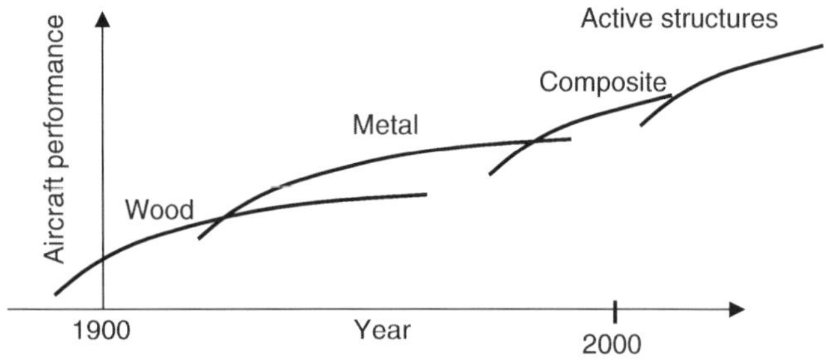

FIGURE 7.5 A/C performance improvements from materials.

The figure above also indicated the typical performance trends for new technologies. When they are introduced, they are inferior to the best available state-of-the-art technology at that time. The book "The Innovator's Dilemma" by Clayton M. Christensen [12] describes this trend for several new, disruptive technologies. First, applications are typical on low-cost, low-performance products.

7.6.5 Typical Load Requirements for Aircraft Structures

A typical fighter aircraft has to be designed to carry a load of nine times its own weight. Applied to a car with an empty weight of 1 t plus 0.5 t payload, this would mean an external load of 14.5 t. The wings for a transport aircraft have to carry 2.5 times the total weight. This shows that airplane structures have to be strong, which means that they also are rather stiff. The upper and lower skins of the torque box have a typical thickness in the order of 10 to 20 mm, compared to body of a car below 1 mm.

7.7 Smart Materials for Active Structures

Smart materials for active structure applications are mainly interesting because of their high energy density. On the other hand, their strain or stroke capacity is rather limited, as compared with other materials for aircraft structures, or with other actuators. And they are rather heavy. Probably the most complete survey paper on this topic with the title "smart aircraft structures" by Crowe and Sater [9] was presented in 1997 at the AGARD Symposium on Future Aerospace Technology in the Services of the Alliance. It tries to classify the different concepts and gives an overview on recent and ongoing research activities. There is also a prediction for future applications in real systems with different purposes and for different classes of airplanes.

The piezoelectric materials and SMA are mainly addressed for potential applications. Whereas piezoelectric materials are usually applied as patches with multiple layers, distributed over the surface, or as concentrated stack actuators, SMAs have mainly been investigated in the form of wires or torque tubes for the active deformation of aerodynamic surfaces. To date, the achieved results are not very promising for aircraft control by active structures. For this purpose, which means large static deformations in a rather short time, piezomaterials respond fast enough, but the stroke is very limited. SMAs on the other hand, could provide larger deformations and higher forces, but are not fast enough for flight control inputs, and their thermal energy supply issues are rather complex within the airframe and aircraft environment.

Because of these limitations, the Defense Advanced Research Projects Agency (DARPA) launched an ambitious research program in 1999, called compact hybrid actuators (CHAP), with the aim to multiply the stroke and force output of current actuators by a factor of 10.

7.8 Role of Aeroelasticity

7.8.1 Reputation of Aeroelasticity

Some years after the Wright brothers' success with their active wing, designers began to fear the flexibility of the structure. The famous MIT Lester B. Gardner Lecture "History of Aeroelasticity" from Raymond L. Bisplinghoff [13] quotes many of the early incidents involving aeroelastic phenomena, and the famous comment from Theodore von Kármán: "Some fear flutter because they do not understand it, and some fear it because they do." Also quoted from a review paper on Aeroelastic Tailoring by T. A. Weisshaar [2]: "As a result, aeroelasticity helped the phrase 'stiffness penalty' to enter into the design engineer's language. Aeroelasticity became, in a manner of speaking, a four-letter-word....it deserves substantial credit for the widespread belief that the only good structure is a rigid structure." The role of aeroelasticity in aviation is depicted in Figure 7.6. It shows the impact on aircraft performance over the years, mainly caused by increasing speed. But the upper dot in 1903 also indicates that aeroelasticity can also act in a positive way, if properly used and understood, also today, and on faster airplanes. Smart structures concepts will help to reverse this negative trend of aeroelastic impacts on aircraft performance. Similar to Figure 7.5, the progress in aeronautics can also be connected to the progress in aeroelasticity and related external stimuli and events, as drafted in Figure 7.7.

7.8.2 Aeroelastic Effects

Because of the difficulty to describe aeroelastic effects by proper theoretical models, involving a good description of the structure with its flexibility and structural dynamic characteristics as well as steady and unsteady aerodynamic properties, solutions were limited in the early years of aviation to selected cases with only a few degrees of freedom. More general solutions required modern computers with respect to storage space and computing time. The aeroelastic triangle (Figure 7.8), for the first time quoted by Collar [14] in 1946, describes the involved types of forces in the different aeroelastic phenomena. Looking at these forces and

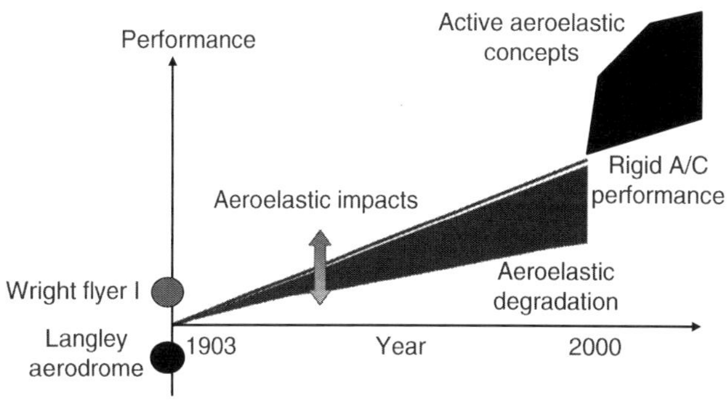

FIGURE 7.6 Impact of aeroelasticity on aircraft performance.

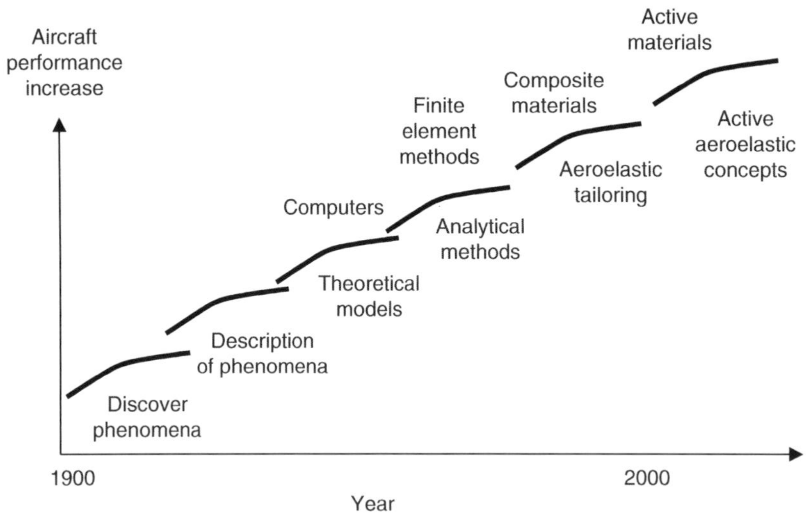

FIGURE 7.7 Relationship between aircraft performance, advances in aeroelasticity, and external stimuli.

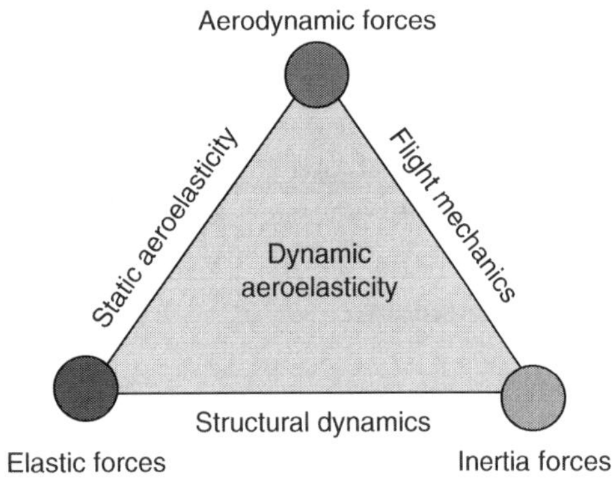

FIGURE 7.8 Aeroelastic triangle.

interactions, it becomes obvious that smart structures for aeronautical applications will have a close relationship with aeroelasticity in most cases.

7.8.2.1 Static Aeroelasticity

In static aeroelastic effects, no inertia forces are involved by definition. This is true for aileron reversal, an effect, where the rolling moment due to a control surface deflection changes the sign at a certain flight speed due to opposite deformation of the fixed surface in front of the control surface. This effect has to be avoided within the flight envelope of the aircraft in order to avoid deterioration of the pilot when he moves the stick to roll the airplane.

If the Wright brothers had used conventional ailerons on their first airplanes, they might have experienced aileron reversal because of the low torsional stiffness of their wings, even at very low speeds. On the other hand, the Wright brothers' main competitor, Samuel P. Langley, was very likely less fortunate with his Aerodrome designs because of insufficient aeroelastic stability [13] after scaling the successful smaller unmanned vehicle to larger dimensions.

For fighter airplanes, it is not sufficient to avoid aileron reversal. Even at the worst flight condition, a high roll rate must be achieved to provide high agility. This is usually done by reinforcing the wing structure because the basic static design of a fighter wing yields rolling moment effectiveness slightly above or even below zero at the worst flight condition. In the case of the U.S. aircraft F-18, the basic design had to be revised after delivery of the first batch of production aircraft. An additional weight of 200 lbs per wing side was reported to the author for the Israel Lavi lightweight fighter aircraft project to provide sufficient roll power. In addition to the loss of roll power, the adverse deformation of the control surface required larger control surface deflections, which resulted in higher hinge moments and required stronger actuators. The difficulty in predicting the most effective distribution of additional stiffness for improved roll effectiveness, especially in conjunction with the introduction of modern composite materials with highly anisotropic stiffness properties for airframe design, inspired the development of formal mathematical structural optimization methods [15].

Aileron reversal has usually the most severe static aeroelastic impact on aerodynamic forces and moments. But all other aerodynamic performance or control characteristics of an airplane are affected as well by static aeroelastic deformations and aerodynamic load redistributions to a more or less severe degree. Weisshaar [16], for example, mentions the excessive trim drag due to aeroelastic wing deformations on the delta wing of a supersonic transport aircraft.

Roll control improvement by means of active concepts was and still is the most often studied application of active concepts for static aeroelastic phenomena on aircraft. Although active structures or materials are not involved, the Active Aeroelastic Wing project [17] or Active Flexible Wing project, as it was called before, has undergone flight test trials in 2001 on a modified

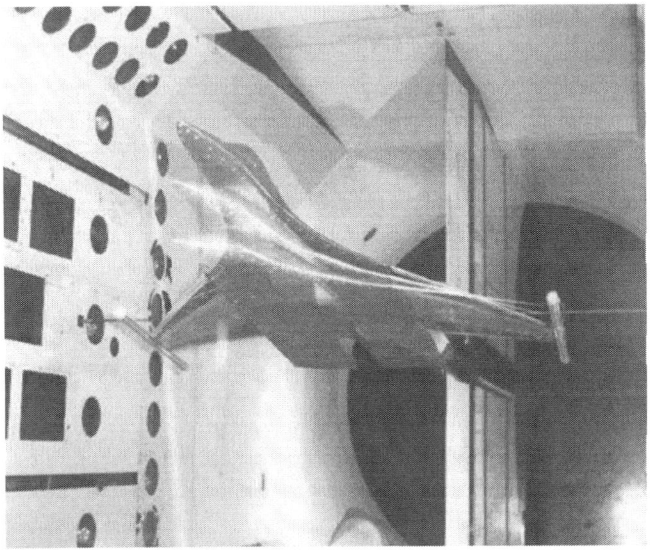

FIGURE 7.9 Active Aeroelastic Wing model mounted in the NASA transonic dynamics wind tunnel at the NASA Langley Research Center. http://www.dfrc.nasa.gov/Newsroom/.X-Press/special-editions/AAW/images/121703/windtunnel-500.jpg (July 6, 2008).

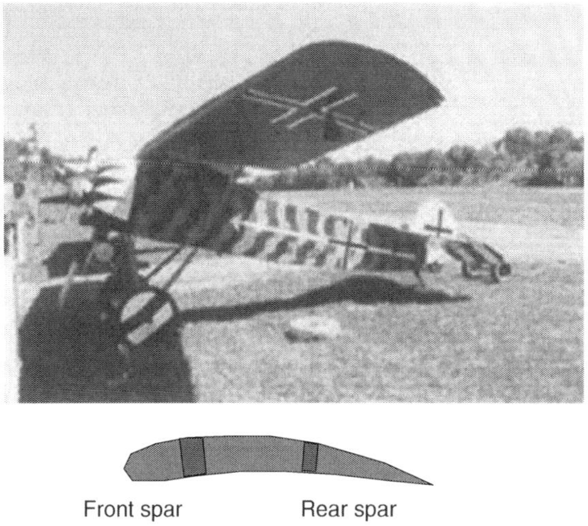

FIGURE 7.10 Fokker D-VIII monoplane, where aeroelastic divergence caused several fatal accidents after reinforcement of the rear spar. (Modified by author, photo from Internet.)

F-18. This concept originated in several theoretical studies and wind tunnel demonstrations in the 1980s. A summary of these activities is presented in a special edition of the *Journal of Aircraft* in 1995 [18]. Figure 7.9 depicts the wind tunnel model installed in the transonic dynamics tunnel at the NASA-Langley Research Center (NASA-LARC).

Static aeroelastic effectiveness losses for the lateral stability and rudder yawing moment are a well-known design driver for vertical tails. Surprisingly, almost nobody has looked so far for smart structures concepts to obtain better designs. Sensburg [19] suggested a smart passive solution, called the Diverging Tail, by means of aeroelastic tailoring the composite skins and modifying the fin root attachment to a single point aft position, to achieve higher yawing moments compared to a rigid structure.

Aeroelastic divergence was the most severe instability for early monowing airplanes. If the wing main spar is located too far behind the local aerodynamic center of pressure (at 25% chord), a lack of torsional stiffness will cause the wing structure to diverge and break at a certain speed. As Anthony Fokker describes in his book [20], sufficient strength of the design had already been demonstrated by proof load and flight tests for his D-VIII (Figure 7.10) when regulations called for a reinforced rear spar with proportional strength capacity to the front spar. This redistribution of stiffness caused torsional divergence under flight loads. This example also demonstrates the potential effects and impact from smart structures applications to an airplane structure.

The introduction of high-strength composite materials, with the possibility to create bending–torsion coupling effects from the anisotropic material properties, caused a renaissance of the forward swept wing in the late 1970s [2], which was ruled out before for higher sweep angles because of the bending–torsion divergence, as depicted in Figure 7.4.

Static aeroelasticity also includes all effects on aerodynamic load distributions, the effectiveness of active load alleviation systems by control surfaces, and flexibility effects on the aerodynamic performance. In this case, the variable inertia loads from payload or fuel on the structural deformations have to be considered simultaneously.

7.8.2.2 Dynamic Aeroelasticity

Flutter is the best known dynamic aeroelastic stability problem. It belongs to the category of self-excited oscillation systems. In this case, any small external disturbance from a control surface command or atmospheric turbulence, which excites the eigenmodes of the structure, will at the same time create additional unsteady aerodynamic forces. Depending on the mass and stiffness distribution, and on the phase angles between the involved vibration modes, the aerodynamic forces will dampen the oscillations, or enforce them in the case of flutter.

Active control for flutter stability enhancement by means of aerodynamic control surfaces was fashionable in the late 1970s [21]. In this case, the effectiveness of the system depended on the static aeroelastic effectiveness of the activated control surfaces. Mainly because of safety aspects, but also because of limited effectiveness, none of these systems has entered service so far. Active control by active structures devices was a popular research topic in recent years [10], but it is doubtful if we will ever see applications because of the same reasons.

Panel flutter is a special case, where only individual skin elements of the structure (panels) are affected. This usually happens at low supersonic speeds, and only structural elements with low static load levels, like fairings, can usually be affected. Active control by smart materials is possible, but there are no considerable impacts on aircraft effectiveness.

Buffeting is a forced vibration where turbulent flow generated by one aerodynamic surface excites this surface itself or another surface located in the region of the turbulent flow. Also here, aerodynamic control surfaces located on the affected part can be used to counteract the vibrations. Compared to flutter, the aerodynamic effectiveness of these surfaces is additionally reduced because of the turbulent flow conditions. Active structures systems are more effective in this case. For this reason, and because the required active deformations are small, the first large-scale active structures application on aircraft dealt with the buffeting problem of fighter aircraft vertical tails at extreme maneuver conditions. After several theoretical [22] and small-scale experimental studies [23], full-scale ground tests were performed in a joint Australian–Canadian–U.S. research program [24] on an F-18 and in a German program for a simplified fin structure of the Eurofighter [25]. In both cases, piezoelectric material was used.

7.8.3 Aeroelastic Tailoring and Structural Optimization

Weisshaar [26] was one of the first researchers who tried to give aeroelasticity a better reputation when modern fiber reinforced composite materials with highly anisotropic directional stiffness properties were considered for primary aircraft structures. They provided the possibility to tailor the materials' directional stiffness within the composite lay-up in order to meet desired deformation characteristics for improved aeroelastic performance. Together with formal mathematical optimization methods for the structural design, this approach allowed for the minimization of the impact from aeroelasticity.

Any improvement of a technical system is often referred to as an optimization. In structural design, this expression is today mainly used for formal analytical and numerical methods. Some years after the introduction of finite element methods (FEM) for the analysis of aircraft structures, the first attempts were made to use these tools in an automated design process. Although the structural weight is usually used as the objective function for the optimization, the major advantage of these tools is not the weight saving, but the fulfillment of aeroelastic constraints. Other than static strength requirements, which can be met by adjusting the dimensions of the individual finite elements, the sensitivities for the elements with respect to aeroelastic constraints cannot be expressed so easily. The option to tailor the composite material's properties by individual ply orientations and different layer thickness for the individual orientations required and inspired the development of numerical methods [27].

7.9 Overview of Smart Structures Concept for Aircraft Control

7.9.1 Classification of Concepts

Active structures concepts for aircraft control can be subdivided into categories by

- Purpose of the active system
- Types of devices to activate the structure

In the first case, the intended concepts are aiming at improvements of

- Control effectiveness
- Aerodynamic drag reduction by adaptive shape
- Load alleviation by adaptive deformation
- Stabilizer effectiveness for trim and static stability

As mentioned earlier, the majority of concepts are aiming at improved roll control power because it usually has the highest sensitivity from structural deformations.

A classification by actuation devices can be given by

- Activation of a passive structure by conventional or novel aerodynamic control surfaces
- Active structural elements
- Actuators or connecting elements with adaptive stiffness between structural components

An additional classification can be made by

- Concepts, where aeroelastic effects are intentionally used
- Concepts without special aeroelastic considerations

As far as aeroelasticity is addressed by the concepts, the intended improvements are aiming at the high-speed part of the flight envelope, where aeroelastic effects become more important. In the case of exploiting aeroelastic effects in a positive sense, this also means that active aeroelastic effects can usually only be used in a beneficial way at higher speeds. An exception will be shown below.

7.9.2 Fictitious Control Surface Concepts

In order to evaluate the potential benefits of smart structure concepts as well as the required energy to activate them, it is useful to start with a "virtual concept," assuming that the intended structural deformation is created by any device. Khot et al. [28] call this the "fictitious control surface" concept. They investigated the aeroelastic loss of roll control power for a conventional trailing edge control surface, and then tried to retwist the wing by supplying the same amount of strain energy that was created by the aileron deflection. The main purpose of this effort was the analytical evaluation of the required energy that is required to maintain a constant roll rate with increasing dynamic pressure. The result, which showed an increase of energy versus dynamic pressure with the same gradient as the reduction of effectiveness, however, may be misleading. The achievable rolling moment from a deformation depends on the position, where the deformation is initiated by an internal force or by a control surface deflection. In the same way as on a rigid wing, where a trailing edge surface is much more effective than a leading edge surface, there are more or less effective regions on a flexible wing, where a deflection of a control surface or a deformation of the structure by internal forces results in different rolling moments and requires different efforts to create the deflection or deformation.

7.9.3 Variable Shear Stiffness Spar Concept

In a similar way to the fictitious control surface, a study by Griffin and Hopkins [29] used a "fictitious" variable stiffness spar concept to modulate the rolling moment effectiveness of a generic F-16 wing model. They assumed a small outboard trailing edge control surface on an analytical F-16 wing model for roll control, which would operate in a conventional mode at low dynamic pressures and the negative "postreversal" effectiveness could be enhanced by turning the spar web shear stiffness off at high dynamic pressures. This concept was explained in a simplified way by "link elements," attached to the upper and lower spar caps by bolts and removable pins. The basic principle of this concept was also experimentally verified on an aeroelastic wind tunnel model for an unswept, rectangular wing with removable spars [30]. Unfortunately, no references were found, for more technical smart structures solutions if they were ever investigated for this concept.

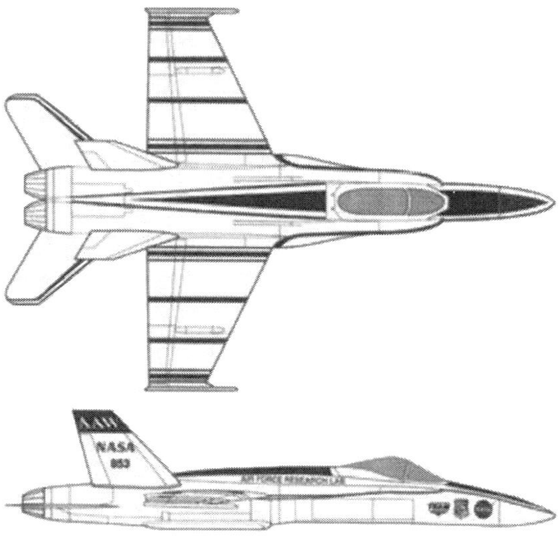

FIGURE 7.11 F-18 Active Aeroelastic Wing demonstrator aircraft. (From Sanders, B., Flick, P., and Sensburg, O. *International Forum on Aeroelasticity and Structural Dynamics*, Madrid, Spain, 2001.)

7.9.4 Innovative Control Effector Program

In the Innovative Control Effector (ICE) program from NASA, Langley [31,32], the positions and required amount of small, "fictitious control surfaces" were determined by means of a genetic optimization process for an advanced "blended wing-body" configuration. These "control effectors" were elements of the surface grid in the analytical aerodynamic model that created the "virtual" shape change. Ref. [31] gives an excellent overview on all research activities within the NASA's Morphing Program.

7.9.5 Active Flow Control Actuators

Also as a part of the NASA Morphing Program, "synthetic jet actuators were developed and tested" [31]. This device is based on a piezoelectrically driven diaphragm, which sucks and blows air through a small orifice. It was originally developed for cavity noise control. To use it for aircraft control, where much higher forces are required, the power output needs to be multiplied.

7.9.6 Innovative Aerodynamic Control Surface Concepts

Although there are no active structural components involved, these concepts can also be considered as "smart structures." In this case, the active deformation of the structure is actuated by aerodynamic control surfaces. The January/February 1995 edition of the *Journal of Aircraft* [18] was a special issue, dedicated to the U.S. Active Flexible Wing Program, which started in 1985 and turned later into the Active Aeroelastic Wing Program. This basic idea was the improvement of roll effectiveness for a fighter aircraft wing by the combination of two leading edge and two trailing edge control surfaces, which could also be operated beyond reversal speed. This concept was demonstrated on an aeroelastic wind tunnel model, with tests started in 1986.

After theoretical studies on F-16 and F-18 wings, reported by Pendleton [5,17,33], the F-18, depicted in Figure 7.11, was selected as the candidate for flight test demonstrations, which are expected to started in 2001. For this purpose, the wing structure was returned into the original stiffness version, which had shown aileron reversal in early flight tests.

Flick and Love [34] performed a study on wing geometry sensitivities with respect to the potential improvements from active aeroelastic concepts based on the combination of leading and trail edge surfaces. The results, as shown in Figure 7.12 from Ref. [5], indicates only very small advantages for low-aspect ratio wings. The theoretical studies for a generic wing model of the Eurofighter wing by the Flick, however, resulted in large improvements also for this configuration, as can be seen in Figure 7.6.

Active aeroelastic concepts research by TsAGI in Russia has already demonstrated impressive improvements in flight test. In addition to also using leading edge control surfaces to improve roll performance, a small control surface was mounted at the tip launcher. Figure 7.13 from Ref. [35] shows the achievable improvement compared to the trailing edge aileron only. Note the size of the special surface in comparison with the conventional aileron.

For high aspect ratio transport aircraft wings, especially in combination with a winglet, similar devices could be used not only for roll control, but also for adaptive induced drag reduction, or load alleviation, as indicated in Figure 7.14 for a concept, called active wing tip control (AWTC) by Schweiger and Sensburg [36]. In this case, the winglet root provides sufficient space and

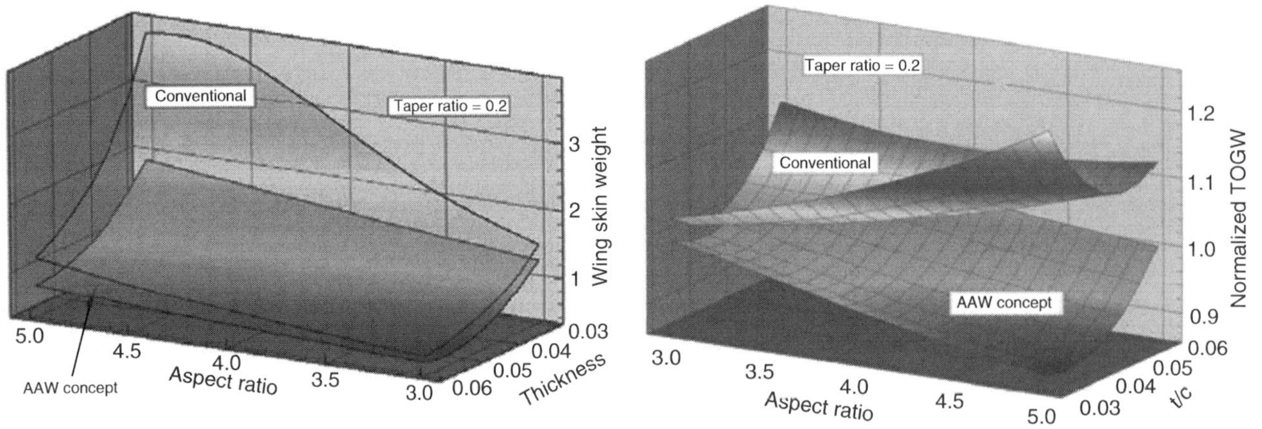

FIGURE 7.12 AAW technology advantages and wing geometry sensitivities for light weight fighter wings. (Results from Flick, P.M. and Love, M.H. *RTA Meeting on Design Issues*, Ottawa, Canada, October 1999; figures from Sanders, B., Flick, P., and Sensburg, O. *International Forum on Aeroelasticity and Structural Dynamics*, Madrid, Spain, 2001.)

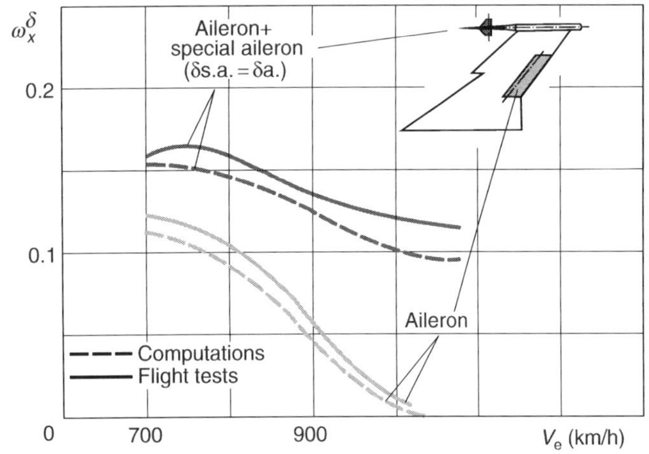

FIGURE 7.13 AAW technology in Russia: flight test results and comparison with analysis for a special wing tip aileron. (Results adapted from Kuzmina, S.I., Amiryants, G.A., Ishmuratov, F.Z., Mosunov, V.A., and Chedrik, V.V. *International Forum on Aeroelasticity and Structural Dynamics*, , Madrid, Spain, 2001.)

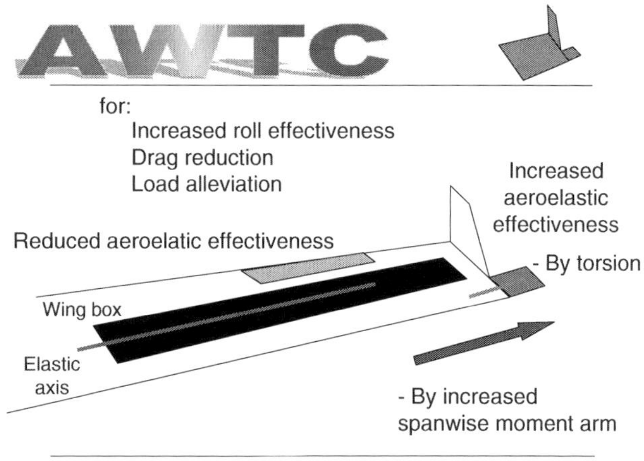

FIGURE 7.14 Advantages of an AWTC device on a transport aircraft wing.

structural rigidity to integrate the control device and its actuation system. With respect to flutter stability, the forward position of the masses increases the flutter stability, which is reduced by the aft position of the winglet.

Of course, the static aeroelastic effectiveness of a control surface is also important for dynamic applications like flutter suppression or load alleviation for buffeting of vertical tails. This fact is very often forgotten in favor of optimizing the control laws.

7.9.7 Active Structures and Materials Concepts

Dynamic applications for flutter suppression [10] or buffetting load alleviation [37,38] by means of piezoelectric material were demonstrated on wind tunnel models or in full scale ground tests. The involved mass and complexity, mainly for the electric amplifiers, precludes practical applications at the moment. For dynamic applications however, a semiactive solution with shunted piezomaterials [39] with very little energy demand is an interesting option.

The use of piezoelectric materials for static deformations is limited by the small strain capacity, as well as by the stiffness of the basic structure. Because of these facts, some researchers realized rather early that it is not advisable to integrate the active material directly into the load-carrying skins. In order to achieve large deflections, it is necessary to amplify the active material's stroke and to uncouple the to-be-deformed (soft) part of the passive structure from the (rigid) main load carrying part. Because this will usually cause a severe "strength penalty" for the main structure of conventional airplanes, practical applications are limited to unusual configurations like small Unmanned aerial vehicles (UAVs) or missiles [40]. As an example, Barrett [41] developed such a device, where the external shell of a missile fin is twisted by a PZT bender element.

Compared to piezoelectric materials, which responds very fast, SMAs are rather slow but they can produce high forces. This precludes applications which require speeds for the adaptation which are equal to or higher than the speed of flight or the speed of aircraft control device motions.

Two typical applications of SMAs were investigated in the DARPA/AFRL/NASA Smart Wing Program [5,31], an SMA torque tube to twist the wing of 16% scale wind tunnel model of a generic fighter aircraft, and SMA wires to actuate the hingeless trailing edge control surfaces. Concerning the torque tube, the ratio between the torque tube cross section and the wing torque box cross section should be kept in mind. For the replacement of conventional control surfaces, the efforts to create the deformation on a realistic structure still need to be addressed. And, what is even more important than the limitations with respect to the actuation speed, the aeroelastic aspects should be kept in mind from the beginning, in order to evaluate and optimize the effectiveness of such concepts. As depicted in Figure 7.15 from Ref. [5], the effectiveness of the conformal trailing edge control surface is better than the conventional control surface at low speed, but becomes worse with dynamic pressure. As mentioned in this reference, such concepts are not developed to replace existing systems, but to demonstrate capabilities of active materials. If this is the case, realistic applications still need to be discovered.

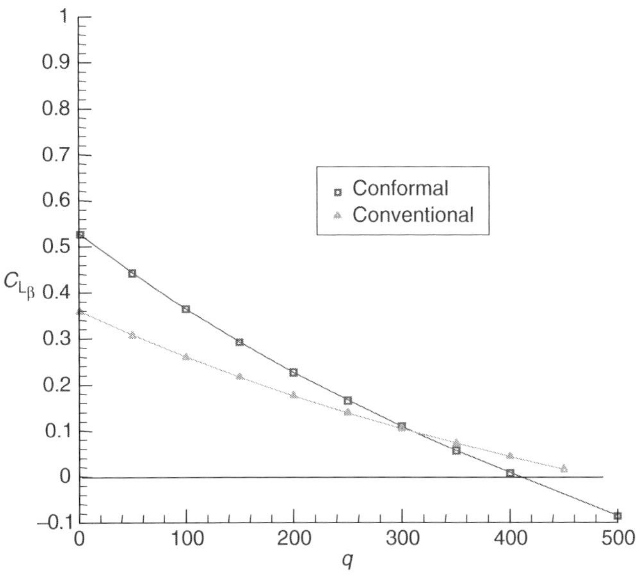

FIGURE 7.15 Comparison of rolling moment effectiveness for conventional and conformal trailing edge control surface. (From Sanders, B., Flick, P., and Sensburg, O., in: Pendleton, E. W. Ed., *International Forum on Aeroelasticity and Structural Dynamics*, Madrid, Spain, 2001.)

Smart materials applications on small remote piloted vehicles (RPVs) are currently being investigated at the Smart Materials Laboratory of the Portugese Air Force together with the Instituto Superior Técnico in Lisbon [42].

7.9.8 Other Innovative Structure Concepts

Because of the limited stroke of active materials and the inherent stiffness of a minimum weight aircraft structure, some researchers try to amplify the stroke by sophisticated kinematic systems and enable larger deformations more easily by "artificially" reducing the structural stiffness. So far, all these concepts show the following disadvantages:

- High complexity for the actuation system
- Higher energy demand compared with the actuation of conventional control surfaces
- Additional internal loads in the structure from the forced deformation
- Additional structural weight from the reduced strength capacity
- Reduced static aeroelastic effectiveness because of additional flexibility in the rear wing area
- Reduced aeroelastic stability (flutter) from the reduced stiffness

As one example, such a system was described by Monner et al. [43]. This paper summarizes the active structure research activities by the German aerospace research establishment DLR applied to an Airbus type transport aircraft wing.

An old idea, the pneumatic airplane, as depicted in Figure 7.16, may be useful, if applied to small UAVs (for storage), or, on larger airplanes to selected structural elements, like spar webs, in order to adjust the shape by a variable pneumatic stiffness to control the aeroelastic load redistribution.

7.9.9 Adaptive All-Movable Aerodynamic Surfaces

Adaptive rotational attachment or actuation stiffness for all-movable aerodynamic surfaces can be seen as a special class of active aeroelastic structures concepts. If properly designed, this concept will provide superior effectiveness compared to a rigid structure also at low speeds. Other active aeroelastic concepts show their advantages only with increasing speed, in the same way as negative aeroelastic effects are increasing.

As an example, a fixed root vertical tail can be made more effective if the structure is tailored in such a way that the elastic axis is located behind the aerodynamic center of pressure. This wash-in effect will, for example, increase the lateral stability compared to a conventional design on a swept back vertical tail, as depicted in Figure 7.17. The so-called diverging tail [19] has an improved effectiveness, but also experiences higher bending moments.

Instead of tailoring the structure, which essentially will always create a (minimized) weight increase, the tail can be designed as

FIGURE 7.16 Goodyear Inflatoplane (1950s). http://upload.wikimedia.org/wikipedia/en/C/C6/Goodyear-inflatoplane.jpg (July 6, 2008).

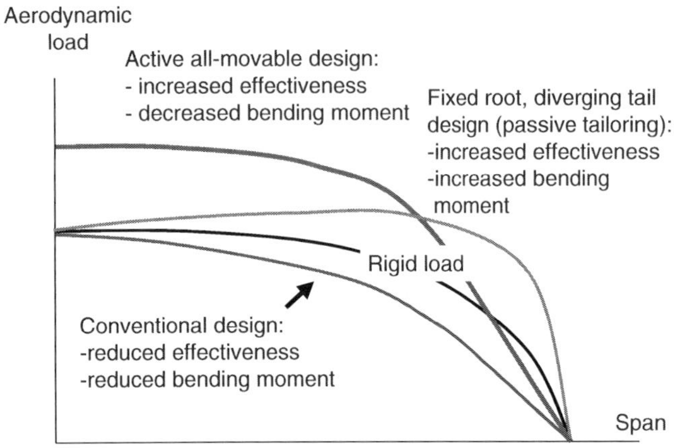

FIGURE 7.17 Aerodynamic load distribution for different design approaches on a vertical tail.

a reduced size all-movable surface, and the location of the spigot axis is used to tailor the wash-in effect, while the attachment stiffness is adjusted to the desired effectiveness. This would allow us to obtain the required effectiveness also at low speeds with a smaller tail. As described in Ref. [44], the proper shape of the surface in conjunction with the spigot axis location will also enhance the flutter stability. Figure 7.18 depicts the resultant effectiveness for different spigot axis locations at different Mach numbers (and dynamic pressure) with variable stiffness.

The crucial element of the all-movable surface with adaptive attachment stiffness is the attachment or actuation component. This can, for example, be a mechanical spring with variable stiffness and a conventional hydraulic actuator. Or as a more advanced system, a hybrid actuator with smart material elements like magnetorheological fluids. The objective of the current DARPA program "Compact Hybrid Actuators" is aimed at the development of such components with high energy density and 10 times the stroke of current systems.

Of course, such systems can also be used for horizontal stabilizers or outboard wing sections. Compared to a horizontal tailplane, where the fuselage flexibility causes aeroelastic effectiveness losses, a forward surface can exploit additional benefits from the fuselage flexibility.

7.10 Quality of the Deformations

The amount of required internal energy for the desired deformations strongly depends on the involved static aeroelastic effectiveness. As depicted in Figure 7.19, the aerodynamic loads can either deform the structure in the wrong direction and require additional efforts to compensate the deformations caused by external loads, or the internal, initial deformation is used in such a way, that the desired deformations are only triggered, and the major amount of required energy is supplied by the air at no costs. In the first case, the required deformation generated by internal forces will already create a high level of internal strain in the structure, resulting in reinforcements and extra structural weight. In the second case, the required internal actuation forces and the strain levels will be much smaller. For a favorable solution, the design process must aim at reducing the total load level of the structure in the "design case," thus reducing the required total weight for structure and actuation system compared to a conventional design, as indicated in Figure 7.20, and achieve a better performance.

With respect to specific type of deformation, the required effort and achievable results strongly depend on the typical properties of a wing structure, which in most cases can be described as a beam. First of all, lift forces will create bending deformations in the direction of the lift force. Because drag forces are much smaller (1/10) and because of the shape of the airfoil shape, in-plane bending deformations can be neglected. Depending on the chordwise location of the resulting lift force relative to the beam (torque box) shear center location, it is possible to twist the wing. This can, for example, be used to reduce the bending deformation. Because of the high stiffness of a modern wing in a chordwise section, and because of the resulting aerodynamic pressure distribution, that is usually acting in one direction, a chordwise bending deformation (camber) is very difficult to achieve by internal or external forces. This is also true for a reduced thickness rear section of a wing, for example to replace a deflected control surface.

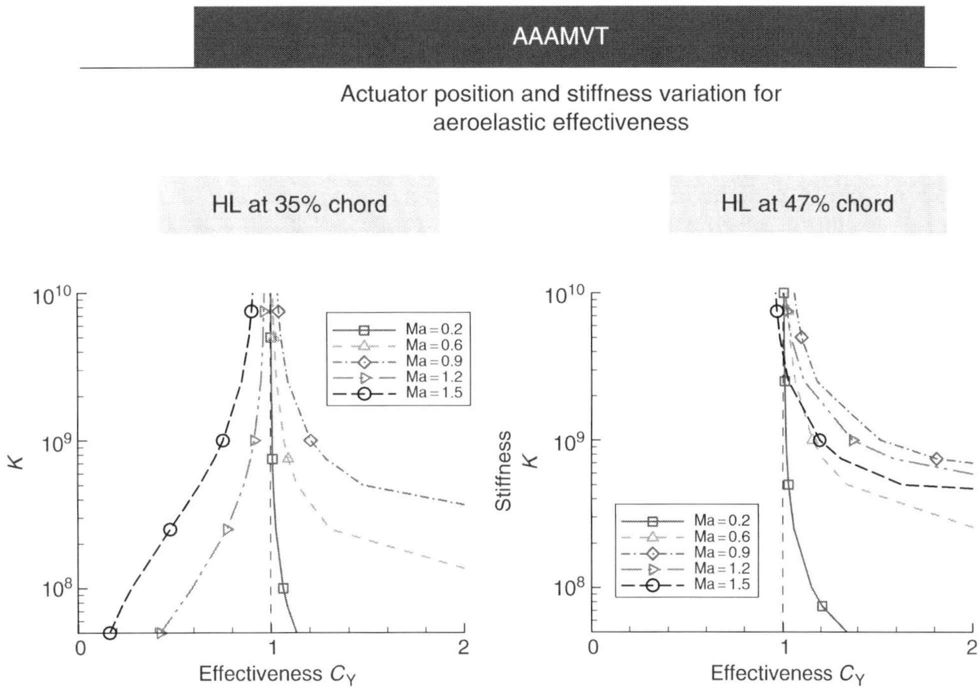

FIGURE 7.18 Achievable aeroelastic effectiveness with variable attachment stiffness for different locations of the spigot axis.

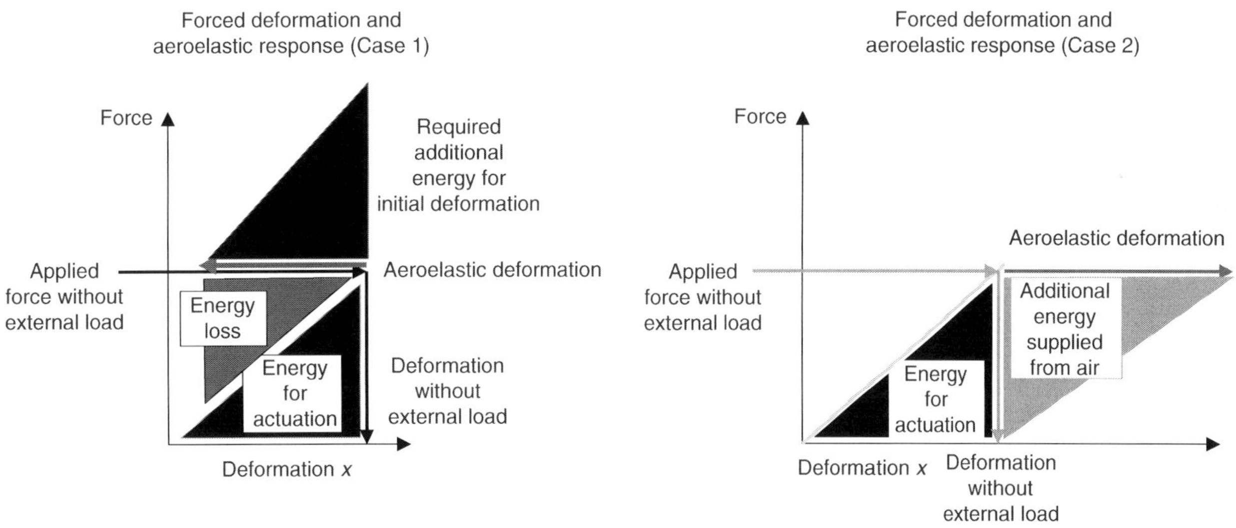

FIGURE 7.19 Forced structural deformation and aeroelastic response for different design approaches.

7.11 Achievable Amount of Deformation and Effectiveness from Different Active Aeroelastic Concepts

Classical active aeroelastic concepts rely on the adaptive use of aerodynamic control surfaces and their aeroelastic effectiveness at various flight conditions. On conventional designs, aeroelastic effect becomes more pronounced with increasing airspeed, as demonstrated in Figure 7.21 for the potential losses and gains.

Conventional active aeroelastic concepts exploit the increasing effectiveness in the upper half of this figure, as well as the recovering effectiveness of a conventional aileron beyond reversal speed. A combined operation of leading and trailing

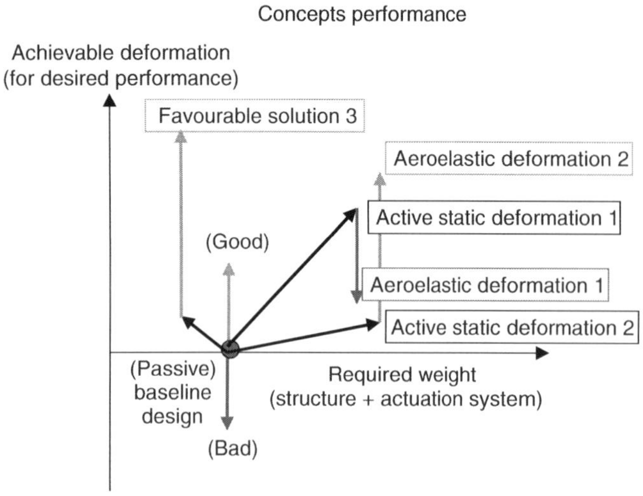

FIGURE 7.20 Performance of active structures concepts with respect to weight and performance.

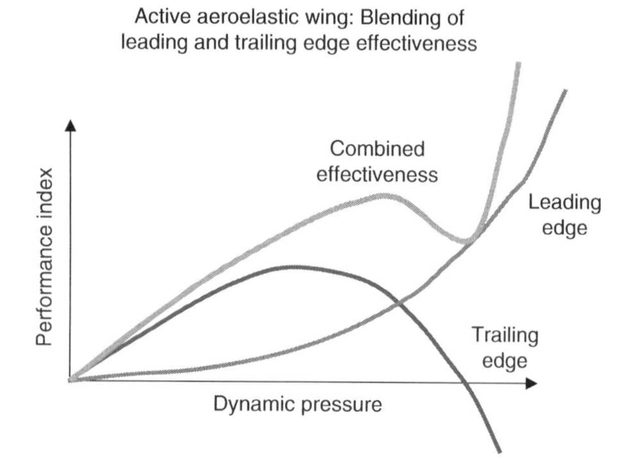

FIGURE 7.22 Achievable roll performance by combined leading and trailing edge.

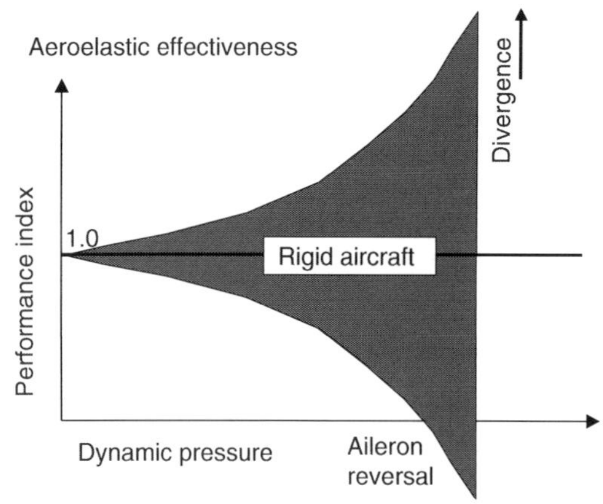

FIGURE 7.21 Typical range of aeroelastic effectiveness.

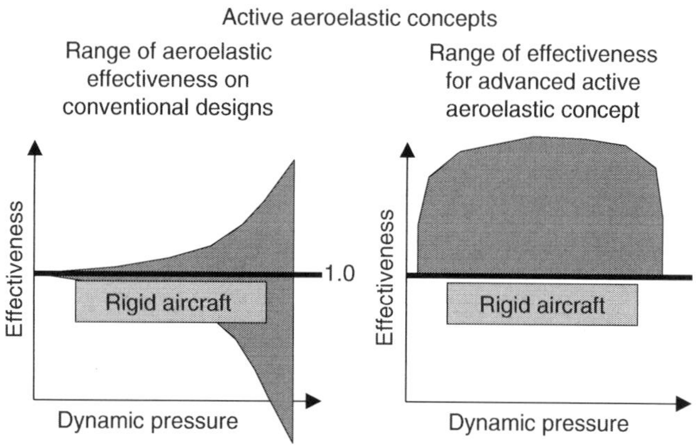

FIGURE 7.23 Aeroelastic effectiveness for conventional and adaptive-all-movable active aeroelastic concepts.

edge surface results in achievable roll rate as indicated in Figure 7.22.

In order to better exploit aeroelastic effects in a beneficial way, it is required to increase the aeroelastic sensitivity of the design in a wider range of the flight envelope. This can, for example, be achieved by an all-movable aerodynamic surface with adaptive rotational attachment stiffness. This will provide high effectiveness also at low speeds, while excessive loads from diverging components or flutter instabilities at high speeds can be avoided.

The usable aeroelastic effectiveness for conventional concepts is rather limited between take-off and cruise speed. Aileron reversal usually occurs between cruise speed and limit speed, and too high effectiveness of leading edge surfaces must be avoided at limit speed. On the other hand, adaptive all-movable concepts can provide high effectiveness at all speeds and avoid excessive loads at the high end of the speed envelope, as indicated in Figures 7.23 and 7.24. This means that a stabilizer surface can be built smaller than it would be required from a "rigid" aerodynamic low speed performance requirement.

7.12 Need for the Analysis and Analytical Design Optimization of Active Structures Concepts

Of course, active materials and structural components, together with the stimulating forces, require a correct description in the theoretical structural or multidisciplinary analysis and optimization (MDAO) models and methods. Once this is provided, the actively deforming structure needs another approach for the static aeroelastic analysis. For an aircraft with conventional control surfaces, the deflections of selected control surfaces can be predescribed for the aeroelastic analysis. For the actively deformed structure, the initial deformations without external loads first need to be determined, for example, by a static analysis.

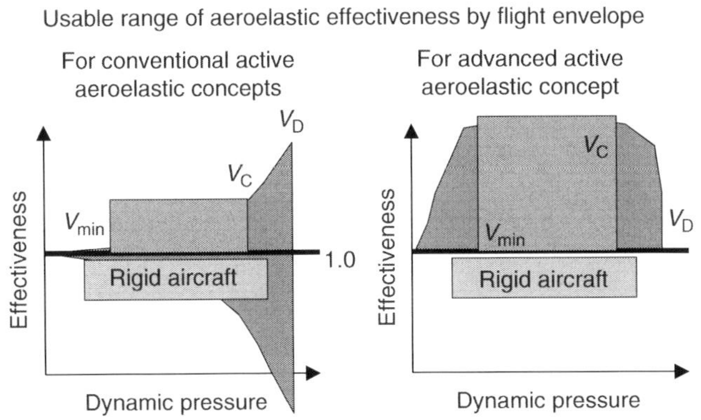

FIGURE 7.24 Usable range of aeroelastic effectiveness for conventional and advanced active aeroelastic concepts.

As described above, the achievable deformations in conjunction with the distribution of external aerodynamic loads are essential for the effectiveness of active structures concepts for aircraft control. This requires efficient tools and methods for the simultaneous, multidisciplinary analytical design. To find the best design, this task involves the optimization of

- External shape
- Arrangement of the passive structure (topology)
- Sizing of the passive structure
- Placement and sizing of the active elements
- Control concept for the active components

The aims of this approach are not only the optimum result for the objective function (minimum weight, aerodynamic performance) and fulfillment of all constraints like strength), but also optimization of additional objectives, like minimum energy. As depicted in Figure 7.25 for the optimization results of a passive structure with different constraints for the required rolling

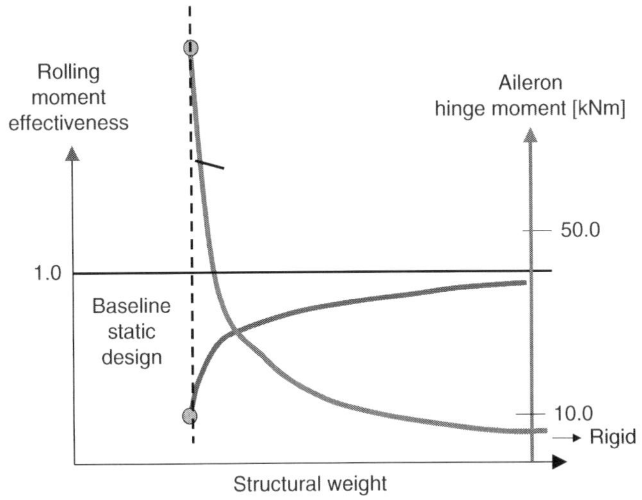

FIGURE 7.25 Optimization results for the rolling moment and hinge moment of a trailing edge aileron a low aspect ration fighter wing.

moment effectiveness, the required energy to actuate the control surface can be considerably reduced, even if the required (low) roll rate is already met.

MDO does not mean to combine single discipline analysis tools by formal computing processes. It means first a good understanding of what is going on. This is already essential for a conventional design. Only with this understanding, the creative design of an active concept can start.

It is also very important to choose the proper analysis methods for the individual disciplines. Usually, not the highest level of accuracy is suitable for the simulation of important effects for other disciplines. This also refers to refinement of the analysis models, where local details are usually not interesting for interactions. It is more important to keep the models as versatile as possible for changes of the design concepts and to allow the simulation as many variants as possible. This also means an efficient process for the generation of models, including the knowledge of the user for this process. Fully automated model generators can create terrible results if the user cannot interpret or understand the modeling process.

In the world of aerodynamics, the design of the required twist and camber distribution for a desired lift at minimum drag is also an optimization task. Assuming that minimum drag is achieved by an elliptical lift distribution along the wing span, this task can be solved by a closed formal solution and potential flow theory. More sophisticated numerical methods are required for the 2D-airfoil design or for Euler and Navier-Stokes Computational fluid dynamics (CFD) methods, which are now maturing for practical use in aircraft design.

For the conceptual aircraft design, formal optimization methods have been used for many years. Here, quantities like direct operating costs (DOC) can be expressed by rather simple equations and the structural weight can be derived from empirical data. Formal methods like optimum control theory are also available for the design of the flight control system.

So one might think that these individual optimization tasks could easily be coupled for one global aircraft optimization process. The reasons why this task is not so simple is the different nature of the design variables in individual disciplines, and their cross sensitivities with other disciplines. The expression multidisciplinary optimization (MDO) summarizes all activities in this area, which have been intensified in recent years. It must be admitted that today most existing tools and methods in this area are still single discipline optimization tasks with multidisciplinary constraints.

In order to design and analyze active aeroelastic aircraft concepts, especially when they are based on active materials or other active structural members, new quantities are required to describe their interaction with the structure, the flight control system, and the resulting aeroelastic effects.

7.13 Summary and Conclusions

In the same way as it was wrong in the past to demand that an aircraft design be as rigid as possible, it is wrong now to demand a design that is as flexible as possible. It is sometimes said that smart structures concepts could completely replace conventional control surfaces. But this looks very unrealistic, at least at the moment. The major difficulties for a successful application are that the limited deformation capacity of active materials as well as their strain allowables is usually below those of the passive structure. However, this can be resolved by a proper design of the interface between passive and active structure. But the essential difficulties are the stiffness and strain limitations of the passive structure itself. It cannot be expected that the material of the passive structure just needs to be replaced by more flexible materials without an excessive weight penalty. It is also not correct to believe that an active aeroelastic concept will become more effective if the flexibility of the structure is increased. The aeroelastic effectiveness depends on proper aeroelastic design, which needs a certain rigidity of the structure to produce the desired loads. A very flexible structure would also not be desirable from the standpoint of aerodynamic shape, stability of the flight control system, and transmission of static loads.

Because large control surface deflections are required at low speeds, where aeroelastic effects on a fixed surface are small, it is more realistic to use conventional control surfaces for this part of the flight envelope and make use of active aeroelastic deformation only at higher speeds. This would still save weight on the control surfaces and their actuation system due to the reduced loads and actuation power requirements.

To produce usable deformations of the structure also at low speeds, all-movable aerodynamic surfaces with a variable attachment stiffness are an interesting option. This concept relies on developmental efforts for active devices with a wide range of adjustable stiffness.

The reasons why we have not seen more progress to date for the successful demonstration of smart structures concepts in aeronautics may be because

- Specialists in aircraft design do not know enough about the achievements in the area of smart materials and structures.
- Smart materials and actuation system specialists who try to find and demonstrate applications in aeronautics do not know or care enough about real world conditions for airplane structures.

What we need is more awareness on both sides as well as stronger efforts to learn from each other and work together.

Although there are strong doubts about useful applications of smart structures for aircraft control, it should always be remembered how often leading experts had been wrong in the past with their predictions in many cases even on their own inventions. Norman R. Augustine quotes some of them in his famous book *Augustine's Laws* [45]:

- "The [flying] machines will eventually be fast; they will be used in sport but they should not be thought of as commercial carriers."—Octave Chanute, Aviation Pioneer, 1910.
- "The energy produced by the breaking down of the atom is a very poor kind of thing. Anyone who expects a source of power from the transformation of these atoms is talking moonshine."—Ernest Rutherford, Physicist, ca. 1910.
- "Fooling around with alternative currents is just a waste of time. Nobody will use it, ever. It's too dangerous… it could kill a man as quick as a bolt of lightning. Direct current is safe."—Thomas Edison, Inventor, ca. 1880.

Also quoted by Augustine [45], the eminent scientist Niels Bohr remarked: "Prediction is very difficult, especially about the future."

At the moment, it looks more realistic and feasible that new hybrid, concentrated active devices, positioned between a passive but properly aeroelastically tailored main aerodynamic surface and the corresponding control surfaces is the way of the future.

Appendix A: Future Directions

Whereas an aerodynamic drag was always a design driver in the development of airplanes, it has now become even more important. Minimum weight designs were always required and the amount of fuel that needs to be carried has dictated the engine development, but now the price of fuel has also become an important additional design driver. The fuel bill share in the operational costs of an airplane has risen from below 10% to above 25% and soon-to-come CO_2 surcharges may double this value.

There are strong development efforts to improve the fuel efficiency of airplanes. Besides very successful efforts in the materials and structural design and manufacturing areas, more efficient engines, and fly-by-wire flight control systems, Aerodynamic designers were trying to improve airplane efficiency by adding so-called winglets to the wingtips. Studies and R7D programs with winglets on wing tips in large-scale NASA test programs have ended up in unsuccessful applications. Reasons were the addition loads that these devices add to the wing structure, which requires more strength capacity, as well as aeroelastic stability issues, because the additional mass and aerodynamic forces of these surfaces at unfavorable locations, which requires more stiffness or additional balance masses in the order of several hundred pounds.

An interesting development happened in the mid-1990s. A small company, Aviation Partners, started to develop winglets in order to improve the range of business jets. This also included the application to the Boeing Business Jet, a derivative of the highly successful 737-family.

When rather low fuel prices started soaring again in 1999, they had a product ready for applications in commercial transport 737s. This is the first known case, where a winglet as retrofit equipment added essential aerodynamic performance improvement in the order of 3%–4% drag reduction to a commercial transport airplane, which is usually the magnitude that can be expected from a completely new aircraft design. More research is needed or is already ongoing for future smart materials and structures applications, because the addition of wingtips to an existing wing or increasing the span of a new design has severe implications on the static and dynamic structural loads as well as on the aeroelastic (flutter) stability of the vehicle. Whereas aircraft always had to be light but also cost-efficient in recent decades, much higher development and manufacturing costs for more sophisticated design are justified today.

A.7.1 Aerodynamic Drag and Structural Design Issues

Minimizing the aerodynamic drag is a classical design problem in aviation. The essential drag components are

- Term that is independent of the lift forces, which is essentially defined by friction forces and thus from the overall shape and surface area of the vehicle as well as the surface roughness.
- Lift-dependent, so-called (lift-)induced drag term, which is proportional to the squares mass of the vehicle, divided by the squared span, and a correction term for the distribution of the aerodynamic forces along the span.

This last term says that an elliptic shape of the lift forces will provide minimum drag. This shape on the other hand determines the required structural weight to carry the loads. A mass increase for higher loads will therefore have a severe impact on the drag. But this assumption for the minimum drag does not yet take into account that a higher span with a different load distribution might provide a superior solution, a fact that was already recognized by Ludwig Prandtl in the 1920 [46]. If the span is not fixed, an aerodynamic load distribution with higher loads in the inboard section and reduced loading toward the tip will give a better overall design. Aircraft designers with only aerodynamics in mind often do not consider this fact. Prandtl's assumption and similar, modified versions of it like the Refs. [47,48] are however based on simplified assumptions for the wing structure's load capacity and its mass (a simple beam model). Today and in the future, advanced analytical design and optimization methods and tools will allow much better approaches for this kind of optimization, although the links between the structural and aerodynamic worlds could and should be much better and stronger than they are today.

And, even more important, all these assumptions to date consider the same load distribution in cruise flight and at extreme, "once-in-a-lifetime" ultimate design loads. Although aeroelastic load redistributions are considered in the determination of aerodynamic conditions and in the assessment of structural design loads, they are not yet present and are not exploited in an integrated design process.

A great deal of effort has been spent developing the aerodynamic design of winglets and other wingtip devices, aiming at the reduction of the energy of the wing tip vortices that originate at the tips due to the pressure difference between upper and lower surface. Because a wing with infinite span produces no lift-induced drag, multiplanes with many slender wings or ring wings have always been present in aerodynamic research. Winglets are considered as a design option, where the horizontal span is fixed and the vertical extension helps to reduce the energy of the tip vortices. In this approach, it is also often assumed that the constant horizontal span adds less additional loads to the structure than a horizontal extension.

A.7.2 New Structural Research Efforts and Achievements

The DARPA Morphing Aircraft Program was initially mainly aiming at creating (military) flight vehicles for multipurpose missions by radical shape changes [49]. Today, as some wind tunnel demonstrators have already shown and successfully been tested, some of the newly developed active and passive materials and structures concept for this initial purpose will now certainly become very valuable to help reduce the induced drag and structural loads on new designs for commercial transport aircraft wings.

Another significant European research program Active Aeroelastic Aircraft Structure (3AS) developed in 2005 was mainly aimed at civil applications [50–53]. Some new active and passive structural design concepts, mainly based on the exploitation of beneficial aeroelastic effects, were developed there and successfully demonstrated on wing tunnel models.

Whereas aeroelastic tailoring by means of sweeping the orientations of high elastic modulus carbon fibers and composing their lay-ups was essentially meant for applications on thin, plate-like wings structures for fighter airplanes to improve their already high maneuverability at extreme flight conditions, some designers hope that such features can also be applied to rather thick, slender transport aircraft wings to achieve favorable structural and aerodynamic loading conditions. Although carbon fiber composites, which have been successfully used for almost all commercial transport airplane structures, including the fuselage, it should be mentioned here that the design space and design conditions are different from a fighter. The wing geometric properties as well as the lower dynamic pressure will show smaller aeroelastic tailoring effects in this case. Today it looks like aerodynamic drag, fuel price, and CO_2 issues have already created a vast new area for future smart materials and structures applications.

A.7.3 Example for the Interaction of Structural, Aerodynamic, and Aeroelastic Constraints for Different Wing Tip Design Concepts

An example was chosen to depict the potential benefits of a proper, multidisciplinary design approach. Different wing tip

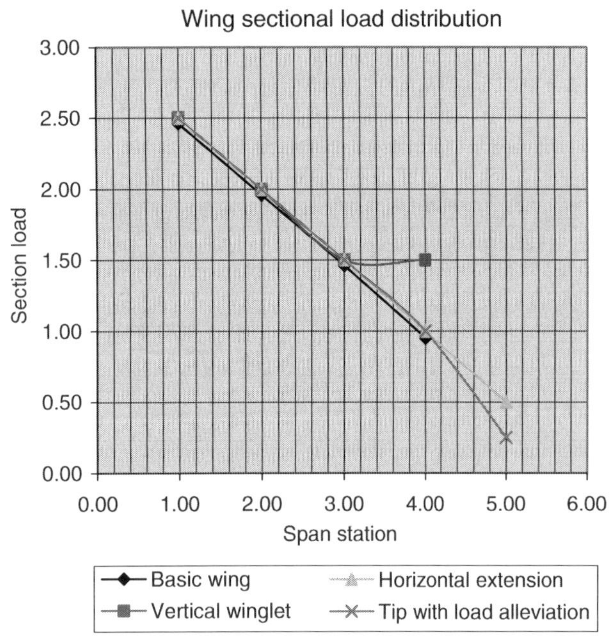

FIGURE A.7.1 Wing span load distributions for the four different design concepts.

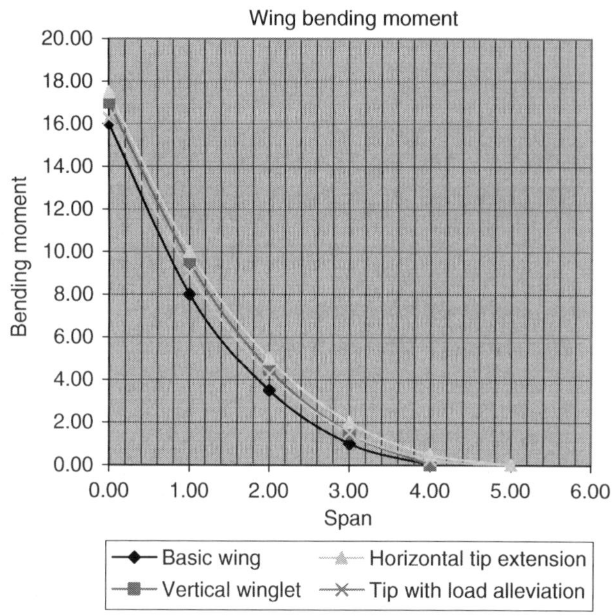

FIGURE A.7.3 Bending moment distributions along wing span for the four different design concepts.

design concepts and design approaches were compared to an overall optimum design with respect to minimum drag and fuel consumption.

Figure A.7.1 depicts the load distribution along the wing span for a baseline design, the wing with a vertical winglet added, a horizontal span extension, and a tip with a built-in load alleviation concept. They shall all have the same load distribution on the inboard sections. The resulting shear force distributions are plotted in Figure A.7.2, which show higher levels for all three modifications.

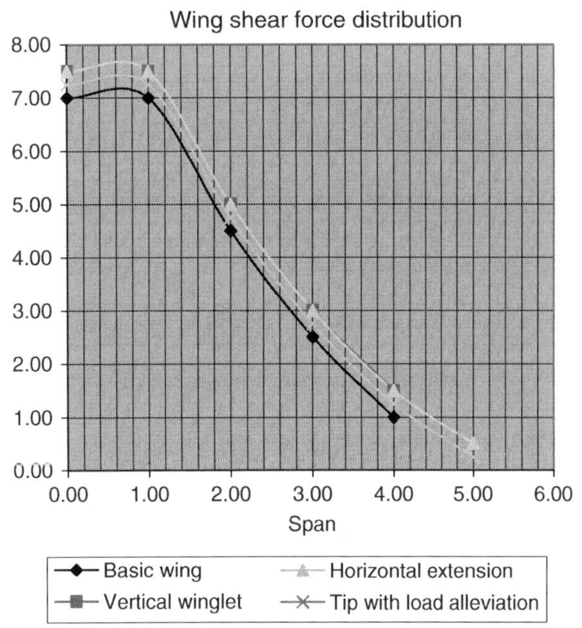

FIGURE A.7.2 Shear force distributions along wing span for the four different design concepts.

The advantage of the design with load alleviation on the tip (reduction of the tip load by 50%) becomes obvious in Figure A.7.3, where the bending moment distributions are plotted. The reduction of the small tip load results in a considerable reduction of the wing root bending moment compared to the other two tip modifications.

Rather than trying to modify the deformation characteristics of the complete wing, for example by means of aeroelastic tailoring, the composite cover skin thickness distributions and fiber orientations, it is sufficient in this case to give a "clever twist" under load to the small tip section, for example by means of a smart (passive or active design) tip attachment concept, which can be considered as a local, novel aeroelastic tailoring approach.

References

1. Wright, O.: *How We Invented the Airplane*. Dover Publications, Mineola, NY, 1988.
2. Weisshaar, T. A.: Aeroelastic Tailoring—Creative Use of Unusual Materials. AIAA-87-0976-CP.
3. Schwipps, W.: *Schwerer als Luft—die Frühzeit der Flugtechnik in Deutschland*. Bernhard & Graefe Verlag, Koblenz, Germany, 1984.
4. Woods, M. I., Henderson, J. F., and Lock, G. D.: Energy requirements for the flight of micro air vehicles. *The Aeronautical Journal*, March 2001.

5. Sanders, B., Flick, P., and Sensburg, O.: Controllable aeroe lastic lifting systems. In: Pendleton, E. W. Ed., International Forum on Aeroelasticity and Structural Dynamics 2001, Madrid, Spain.
6. Bertin, J. J. and Smith, M. L.: *Aerodynamics for Engineers*. Prentice-Hall, Englewood Cliffs, NJ, 1979.
7. Sensburg, O., Hönlinger, H., Noll, T. E., and Huttsell, L. J.: Active flutter suppression on an F-4F aircraft. *Journal of Aircraft*, 19(5), 1981.
8. Larson, C. R., Falanges, E., and Dobbs, S. K.: Piezoceramic active vibration suppression control system development for the B-1B aircraft. SPIE's 5th Annual International Symposium Smart Structures and Materials. San Diego, CA, 1998.
9. Crowe, C. R. and Sater, J. M.: Smart aircraft structures. AGARD Symposium on Future Aerospace Technology in the Services of the Alliance. CP-600, Vol. I, Paris, 1997.
10. McGowan, A. R., Heeg, J., and Lake, R. C.: Results of wind-tunnel testing from the piezoelectric aeroelastic response tailoring investigation. Proceedings of the 37th AIAA/ASME/ASCE/AHS/ASC Structures, Structural Dynamics and Materials Conference, Salt Lake City, UT, 1996.
11. Skopik, O. L.: *Wie Berechnet und Konstruiert Man Selbst Ein Flugzeug*. Vienna, 1917.
12. Christensen, C. M.: *The Innovator's Dilemma*. Harvard Business School Press, Boston, MA, 1997.
13. Flomenhoft, H. I.: *The Revolution in Structural Dynamics*. Dynaflow Press, Palm Beach Gardens, FL, 1997.
14. Collar, A. R.: The expanding domain of aeroelasticity. *Journal of the Royal Aeronautical Society*, 613–636, August 1946.
15. Schweiger, J., Simpson, J., Weiss, F., Coetzee, E., and Boller, Ch.: Needs for the analysis and integrated design optimization of active and passive structure for active aeroelastic wings. SPIE's 6th Annual International Symposium on Smart Structures and Materials. Newport Beach, CA, 1999.
16. Weisshaar, T. A.: Air vehicle design—new horizons? RTO Meeting Proceedings 36 (AC/323(AVT)TP/17) from the RTO AVT Specialists' Meeting on Structural Aspects of Flexible Aircraft Control. Ottawa, Canada, October 1999.
17. Pendleton, E. W., Bessette, D., Field, P. B., Miller, G. D., and Griffin, K. E.: Active aeroelastic wing flight research program: Technical program and model analytical development. *Journal of Aircraft* 37(4), July–August 2000.
18. *Journal of Aircraft*, 32(1), Special Section: Active Flexible Wing, 1995.
19. Sensburg, O., Schneider, G., Tischler, V., and Venkayya, V.: A unique design for a diverging flexible vertical tail. Specialists' Meeting on Structural Aspects of Flexible Aircraft Control. RTA Meeting on Design Issues. Ottawa, Canada; October 1999.
20. Fokker, A. H. G. and Gould, B.: Flying Dutchman. *The Life of Anthony Fokker*. Henry Holt and Company, New York, year unknown.
21. Noll, T., Hönlinger, H., Sensburg, O., and Schmidt, K.: Active flutter suppression design and test, a Joint U.S.–F.R.G. Program. ICAS- CP 80-5.5. Munich, Germany, 1980.
22. Lazarus, K. B., Saarmaa, E., and Agnes, G. S.: An active smart material system for buffet load alleviation. SPE Vol. 22447/179, Bellingham, WA, 1995.
23. Hauch, R. M., Jacobs, J. H., Ravindra, K., and Dima, C.: Reduction of vertical tail buffet response using active control. AIAA-CP-95-1080, Washington, 1995.
24. Moses, R. W.: Contributions to active buffeting alleviation programs by the NASA Langley Research Center. CP, 40th AIAA-SDM conference, St. Louis, MO, 1999.
25. Simpson, J. and Schweiger, J.: Industrial approach to piezoelectric damping of large fighter aircraft components. SPIE's 5th Annual International Symposium Smart Structures and Materials. San Diego, CA, 1998.
26. Weisshaar, T. A.: Divergence of forward swept composite wings. *Journal of Aircraft*, 17(6), June 1980.
27. Schweiger, J., Krammer, J., and Hörnlein, H.: Development and application of the integrated structural design tool LAGRANGE. 6th AIAA/NASA/ISSMO Symposium on Multidisciplinary Analysis and Optimization. CP-4169, Seattle, WA, 1996.
28. Khot, N., Eastep, F., and Kolonay, R.: A method for enhancement of the rolling maneuver of a flexible wing. AIAA-CP-96-1361.
29. Griffin, K. and Hopkins, M.: Aeroelastic tailoring for smart structures. 36th AIAA Structures, Structural Dynamics, and Materials Conference. New Orleans, 1995.
30. Giese, C. L., Reich, G. W., Hopkins, M. A., and Griffin, K. E.: An investigation of the aeroelastic tailoring for smart structures concept. AIAA-96-1575-CP, 1996.
31. McGowan, A. R., Horta, L. G., Harrison, J. S., and Raney, D. L.: Research activities within NASA's Morphing Program. RTO Meeting Proceedings 36 (AC/323(AVT)TP/17) from the RTO AVT Specialists' Meeting on Structural Aspects of Flexible Aircraft Control. Ottawa, Canada, October 1999.
32. Padula, S. L., Rogers, J. L., and Raney, D. L.: Multidisciplinary techniques and novel aircraft control systems. 8th AIAA/NASA/ISSMO Symposium on Multidisciplinary Analysis and Optimization. AIAA-CP-2000-4848, Long Beach, CA, 2000.
33. Pendleton, E., Griffin, K. E., Kehoe, M. W., and Perry, B.: A Flight Research Program for Active Aeroelastic Wing Technology. Conference Proceedings AIAA-96-1574-CP.
34. Flick, P. M. and Love, M. H.: The impact of active aeroelastic wing technology on conceptual aircraft design. Specialists' meeting on structural aspects of flexible aircraft control. RTA Meeting on Design Issues. Ottawa, Canada, October 1999.
35. Kuzmina, S. I., Amiryants, G. A., Ishmuratov, F. Z., Mosunov, V. A., and Chedrik, V. V.: Some applications of active aeroelasticity concepts to aircraft design. International Forum on Aeroelasticity and Structural Dynamics 2001, Madrid, Spain.
36. Schweiger, J. and Sensburg, O.: A critical review of efforts and achievements to improve aircraft performance or efficiency by active structures concepts. CP, International

Forum on Aeroelasticity and Structural Dynamics 2001, Madrid, Spain.
37. Moses, R. W.: Contributions to active buffeting alleviation programs by the NASA Langley Research Center. CP, 40th AIAA-SDM Conference, St. Louis, MO, 1999.
38. Simpson, J. and Schweiger, J.: Industrial approach to piezoelectric damping of large fighter aircraft components. SPIE's 6th Annual International Symposium Smart Structures and Materials. San Diego, CA, 1998.
39. McGowan, A. R.: A feasibility study of using shunted piezoelectrics to reduce aeroelastic response. SPIE's 5th Annual International Symposium Smart Structures and Materials. Newport Beach, CA, 1999.
40. Weisshaar, T. A.: Aeroelasticity's role in innovative UAV design and optimization—the way things ought to be. 41st Annual Israel Conference on Aerospace Sciences, Tel Aviv and Haifa, February 2001.
41. Barrett, R.: Active aeroelastic tailoring of an adaptive flexspar stabilator. *Smart Materials and Structures* No. 5, IOP Publishing Ltd., Bristol, UK, 1996.
42. Costa, P., Moniz, P. A., and Suleman, A.: Design and testing of an adaptive RPV aeroelastic demonstrator. 42nd AIAA SDM Conference, Seattle, WA, 2001, AIAA-2001-1361.
43. Monner, H. P., Breitbach, E., Bein, Th., and Hanselka, H.: Design aspects of the adaptive wing—the elastic trailing edge and the local spoiler bump. *The Aeronautical Journal*, 104(1032), 2000
44. Schweiger, J., Weiss, F., and Kullrich, T.: Active aeroelastic design of a vertical tail for a fighter aircraft. 8th AIAA/USAF/NASA/ISSMO Symposium on Multidisciplinary Analysis and Optimization, Long Beach, CA, September 2000.
45. Augustine, N. R.: *Augustine's Laws*. AIAA, Reston, VA, 1997.
46. Prandtl, L.: Induced drag of multiplanes, *NACA TN* 182, 1924.
47. Klein, G. and Viswanathan, S.: *Zeitschrift für Angewandte Mathematik und Physik*, 24(6), November, 1973.
48. Löbert, G.: Spanwise lift distribution of forward-and aft-swept wings in comparison to the optimum distribution form, AIAA-81-4187, *Journal of Aircraft*, 18(6), June 1981
49. McGowan, A. M.: Morphing technologies for future aircraft. Online transcript of speech at the Air & Space Conference and Technology Exposition. Washington, D.C., September 26, 2006.Cited at the Air Force Association URL: http://www.afa.org/media/scripts/conf2006_McGowan.asp [cited 31 March 2007].
50. Johannes, S., Afzal, S., Svetlana, K., and Vasilij, C.: MDO concepts for a European research project on active aeroelastic structures. 9th AIAA/NASA/ISSMO Symposium on Multidisciplinary Analysis and Optimization. CP, Atlanta, GA, 2002.
51. Johannes, S. and Svetlana, K.: Successful Russian–European cooperation in the active aeroelastic aircraft research. ILA Conference, Berlin, Germany, 2004.
52. Moulin, B., Feldgun, V., Karpel, M., Anguita, L., Rosich, F., and Climent, H.: Alleviation of dynamic gust loads under special control surfaces. CP, International Forum on Aeroelasticity and Structural Dynamics 2005, Munich, Germany.
53. Amiryants, G. A., Mullov, Yu. M., Shalaev, S. V., and Zichenkov, M. Ch.: Design, manufacture and wind tunnel testing of the multi-functional European research aeroelastic model (EuRAM). CP, European Conference for Aerospace Sciences, Moscow, 2004.

8

Smart Battery Materials

Arumugam Manthiram
The University of Texas at Austin

8.1 Introduction

Batteries are the major power sources for portable electronic devices as well as for automobile starting and ignition. The worldwide battery market is currently worth over \$40 billion. Batteries are also intensively pursued for a widespread hybrid electric vehicle (HEV) and plug-in hybrid electric vehicle (PHEV) applications to address the growing environmental concerns and increasing global energy demands. Moreover, batteries are critical as energy storage devices for effectively utilizing solar and wind energies, which are intermittent in nature. The performance factors of batteries are linked to the materials used as well as the engineering involved in fabricating them. This chapter, after providing a brief background to the electrochemical concepts involved in batteries, focuses on the materials aspects of high energy density (lithium ion) batteries.

8.2 Electrochemical Concepts Involved in a Battery

A battery is an electrochemical device, which converts stored chemical energy directly into electrical energy [1–5]. It consists of an anode, a cathode, and an electrolyte (Figure 8.1). During the battery operation, while the anode (A) undergoes an oxidation reaction as shown in Equation 8.1, the cathode (X) undergoes a reduction reaction as shown in Equation 8.2 by accepting the electrons released at the anode through the external circuit:

$$\mathrm{A} \rightarrow \mathrm{A}^{n+} + ne^- \tag{8.1}$$

$$\mathrm{X} + ne^- \rightarrow \mathrm{X}^{n-} \tag{8.2}$$

During this process, the electrolyte acts as a medium for the transfer of charge between the anode and cathode in the form of ions inside the cell, resulting in the formation of AX as the reaction product as shown in Equation 8.3:

$$\mathrm{A} + \mathrm{X} \rightarrow \mathrm{AX} \tag{8.3}$$

The free energy change involved in this chemical reaction is tapped out as useful electrical energy. The battery performance is characterized by several electrochemical performance factors, which are determined by both the intrinsic properties of the anode, cathode, and electrolyte materials used as well as the engineering involved in designing and fabricating the battery. Some of the performance parameters are briefly discussed below.

The capacity Q of an electrode (anode or cathode) material is given as the product of the current I flowing through the external circuit and the time t, which in turn is related to the number of electrons n involved in the reaction as shown by

$$Q = It = nFN_{\mathrm{m}} \tag{8.4}$$

where F and N_{m} are, respectively, the Faraday constant (96,487 C/mol) and number of moles of the anode A or cathode X. Specific capacity is expressed as Ah/kg, which is the capacity delivered by 1 kg of the electrode material. For example, one gram-equivalent (atomic or molecular weight in grams divided by the number of electrons n involved in the reaction) of an

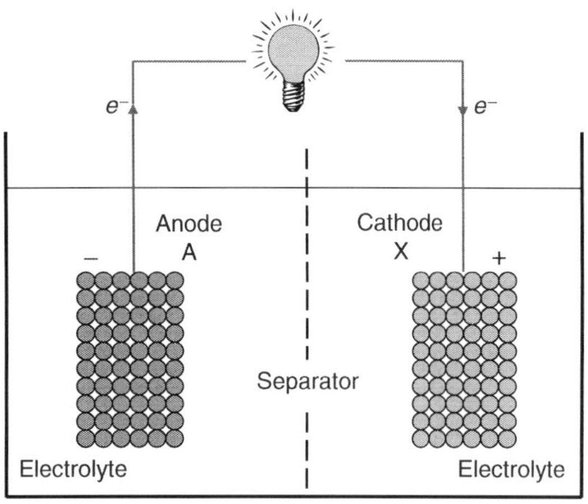

FIGURE 8.1 Operating principles of a battery, which is an electrochemical cell.

electrode material will deliver a capacity of 96,487 C or 26.8 A h. The cell voltage is the difference between the electrode potentials of the anode and cathode materials. Specific energy is a product of the specific capacity and cell voltage, which is expressed in W h/kg. Specific power is the product of the cell voltage and the current flowing, which is expressed in W/kg. The capacity, energy, and power can also be specified per unit volume. For example, energy density and power density are expressed, respectively, in W h/L and W/L.

8.3 Types of Batteries

Batteries can be classified into primary and secondary batteries. Primary batteries are nonrechargeable as the chemical reactions involved in the electrode materials are irreversible, while secondary batteries are rechargeable as the chemical reactions involved are reversible. Some examples of primary and secondary battery systems are given in Tables 8.1 and 8.2 [2]. The tables give the cell reactions, cell voltage, and cell capacity for each system. While most of the primary and secondary battery systems in Tables 8.1 and 8.2 are based on aqueous electrolytes, the primary battery systems employing lithium anode in Table 8.1 and the secondary lithium ion battery system in Table 8.2 are based on nonaqueous electrolytes. The cell voltages of the battery systems based on aqueous electrolytes are limited to ≤2.1 V due to a narrower thermodynamic, electrochemical stability window of water. In other words, it is due to a smaller energy separation E_g between the highest occupied molecular orbital (HOMO) and lowest unoccupied molecular orbital (LUMO) of water and a consequent susceptibility of water to unwanted reduction/oxidation (redox) reactions at higher cell voltages. In contrast, a wider electrochemical stability window offered by the nonaqueous electrolytes in the cases of both primary lithium batteries and secondary lithium ion batteries allows in principle the realization of cell voltages of as high as 5 V.

8.4 Lithium Ion Batteries

Since the use of nonaqueous electrolytes allows the realization of high cell voltages of around 4 V in practical cells, lithium ion cells offer much higher volumetric (W h/L) and gravimetric (W h/kg) energy densities compared to other rechargeable systems as seen in Figure 8.2 [6]. As a result, they have become attractive power sources for portable electronic devices such as laptop computers and cell phones, revolutionizing the electronics industry. Although the concept of rechargeable lithium bat-

TABLE 8.1 Major Primary Battery Systems and Their Characteristics

Battery	Anode	Cathode	Cell Reaction	Cell Voltage (V)	Capacity (mA h/g)[a]
Leclanche	Zn	MnO_2	$Zn + 2MnO_2 \rightarrow ZnO \cdot Mn_2O_3$	1.6	224
Magnesium	Mg	MnO_2	$Mg + 2MnO_2 + H_2O \rightarrow Mn_2O_3 + Mg(OH)_2$	2.8	271
Alkaline MnO_2	Zn	MnO_2	$Zn + 2MnO_2 \rightarrow ZnO + Mn_2O_3$	1.5	224
Mercury	Zn	HgO	$Zn + HgO \rightarrow ZnO + Hg$	1.34	190
Zinc–air	Zn	O_2	$Zn + 0.5O_2 \rightarrow ZnO$	1.65	658
Li–SO_2	Li	SO_2	$2Li + 2SO_2 \rightarrow Li_2S_2O_4$	3.1	379
Li–MnO_2	Li	MnO_2	$Li + MnO_2 \rightarrow LiMnO_2$	3.1	286

[a] Based on active electrode materials (cathode and anode) only.

TABLE 8.2 Major Secondary Battery Systems and Their Characteristics

Battery	Anode	Cathode	Cell Reaction	Cell Voltage (V)	Capacity (mA h/g)[a]
Lead–acid	Pb	PbO_2	$Pb + PbO_2 + 2H_2SO_4 \rightarrow 2PbSO_4 + 2H_2O$	2.1	120
Nickel–cadmium	Cd	NiOOH	$Cd + 2NiOOH + 2H_2O \rightarrow 2Ni(OH)_2 + Cd(OH)_2$	1.35	181
Nickel–hydrogen	H_2	NiOOH	$H_2 + 2NiOOH \rightarrow 2Ni(OH)_2$	1.5	289
Nickel-metal hydride	MH	NiOOH	$MH + NiOOH \rightarrow M + Ni(OH)_2$	1.35	206
Lithium ion	Li	$Li_{0.5}CoO_2$	$0.5Li + Li_{0.5}CoO_2 \rightarrow LiCoO_2$	3.7	137

[a] Based on active electrode materials (cathode and anode) only.

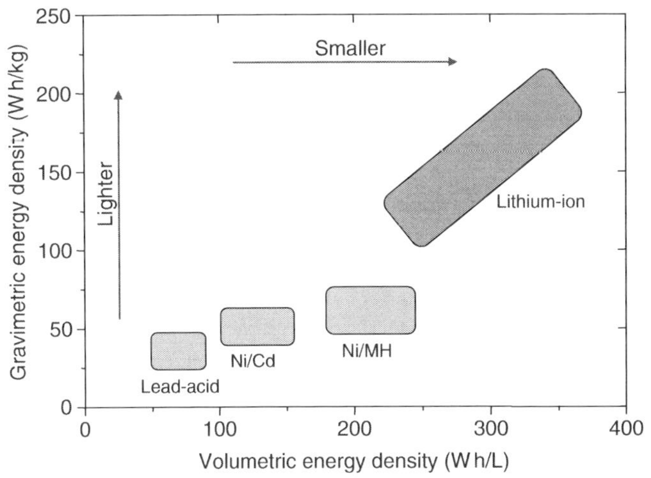

FIGURE 8.2 Comparison of the energy densities (gravimetric and volumetric) of various rechargeable battery systems.

teries was initially demonstrated with a sulfide cathode such as TiS_2 and a metallic lithium anode about three decades ago [7–9], the dendritic short-circuiting and the safety concerns associated with the metallic lithium anode remained as an impediment for several years to commercialize the rechargeable lithium battery technology. It is because of the design and development of smart materials as cathodes and anodes [7–17], the present-day lithium ion battery technology has become a commercial reality, following the first announcement by Sony in 1991.

Figure 8.3 shows the operating principles involved in a lithium ion cell. It involves the reversible extraction and insertion of lithium ions from or into two lithium insertion hosts during the charge–discharge process. For example, in commercial lithium ion cells, the lithium ions are extracted from the layered $LiCoO_2$ cathode and inserted into the layered graphite during charge, while the electrons flow from the cathode to the anode through the external circuit. During discharge, exactly the opposite happens. Since the process involves a shuttling or rocking of lithium ions between the cathode and anode during the charge–discharge process, the cell is also some times referred to as rocking chair cell. Although the concept looks simple, the anode and cathode materials should satisfy several criteria outlined below in order for them to be successful:

- Difference in the lithium chemical potential between the cathode and anode should be large to maximize the cell voltage. The cell voltage is determined by the energies involved in both the electron transfer and the Li^+ transfer. While the energy involved in electron transfer is related to the work functions of the cathode and anode, the energy involved in Li^+ transfer is determined by the crystal structure and the coordination geometry of the site into/from which Li^+ ions are inserted or extracted [18]. If we consider only the electron transfer, then the M^{n+} ion in the insertion compound $Li_xM_yO_z$ should have a high oxidation state in order for it to be employed as a cathode and a low oxidation state in order for it to be employed as an anode.
- Insertion compound $Li_xM_yO_z$ should allow an insertion or extraction of a large amount of lithium x to maximize the cell capacity. This depends on the number of available lithium ion sites and the accessibility of multiple valences for M in the insertion host.
- Insertion compound $Li_xM_yO_z$ should support both high electronic and lithium ion conductivities to offer high power density.
- Lithium insertion and extraction process should be reversible with no or minimal structural changes in the host structure over the entire range x of lithium insertion or extraction in order to provide good cycle life.
- Insertion compound $Li_xM_yO_z$ should be chemically stable without undergoing any reaction with the electrolyte over the entire range x of lithium insertion or extraction.
- Insertion compound $Li_xM_yO_z$ should be low cost and environmentally benign to be employed in practical cells and commercially feasible.

Considering some of these criteria, transition metal oxides crystallizing in layered $LiMO_2$, spinel LiM_2O_4, and olivine $LiMPO_4$ structures (M = transition metal) have become appealing as cathodes, while carbon-based materials such as graphite have become attractive anodes for lithium ion batteries. This is because while these oxides offer high voltages of 3.5–4.3 V versus metallic lithium, graphite offers a low voltage of <0.5 V versus metallic lithium, resulting in a cell voltage of 3–4 V on coupling these oxide cathodes and graphite anode. The high voltage offered by these oxide cathodes is due to a large

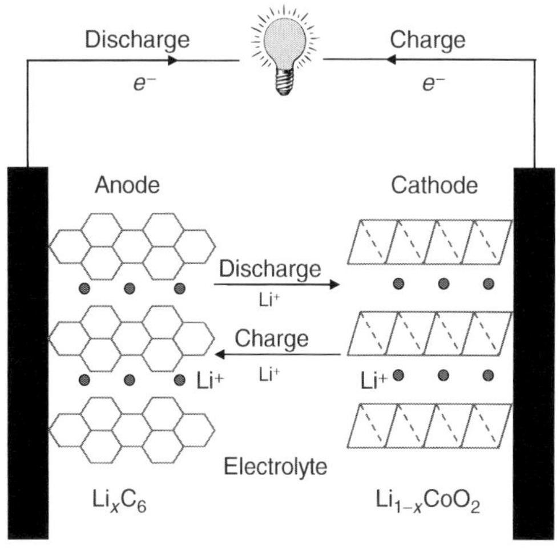

FIGURE 8.3 Operating principles of a lithium ion cell, illustrating the charge–discharge process.

Madelung energy in oxides and the consequent ability to stabilize high oxidation states such as $Co^{3+/4+}$. However, these oxide cathodes differ significantly in some of their performance factors such as energy and power densities as well as cost and environmental impact. Also, although the lithium ion battery technology has revolutionized the portable electronics market, they are yet to enter the HEV and PHEV arena. Moreover, the rapid developments in the miniaturization and features of portable electronics devices demand an increase in the energy density of the lithium ion batteries while lowering the cost of electrode materials. The sections below present the recent developments and current status of the smart cathode and anode materials mentioned above.

8.4.1 Layered Oxide Cathodes

Oxides with a general formula $LiMO_2$ (M = Co and Ni) adopt a layered structure in which the Li^+ and M^{3+} ions occupy the alternate (111) planes of a rock salt structure as shown in Figure 8.4 to give a layer sequence of –O–Li–O–M–O– along the *c*-axis. With an oxygen stacking sequence of ABCABC along the *c*-axis, this structure is designated as O_3 layer structure as the Li^+ ions occupy the octahedral sites and there are three MO_2 sheets per unit cell. While the Li^+ ions present between the strongly bonded MO_2 layers provide facile, reversible Li^+-ion diffusion, the edge-shared MO_6 octahedral arrangement with a direct M–M interaction provides good electronic conductivity. As a result, for example, layered $LiCoO_2$ offers acceptable power capability (charge–discharge rate) for portable applications such as laptop computers and cell phones. On the other hand, a large work function associated with the highly oxidized $Co^{3+/4+}$ couple provides a high cell voltage of around 4 V and the discharge voltage does not change significantly with the degree of lithium extraction/insertion x in $Li_{1-x}CoO_2$ (Figure 8.5). The high energy density with acceptable power capability and cycle life has made $LiCoO_2$ an attractive cathode, and commercial lithium ion cells presently use predominantly layered $LiCoO_2$ as the cathode.

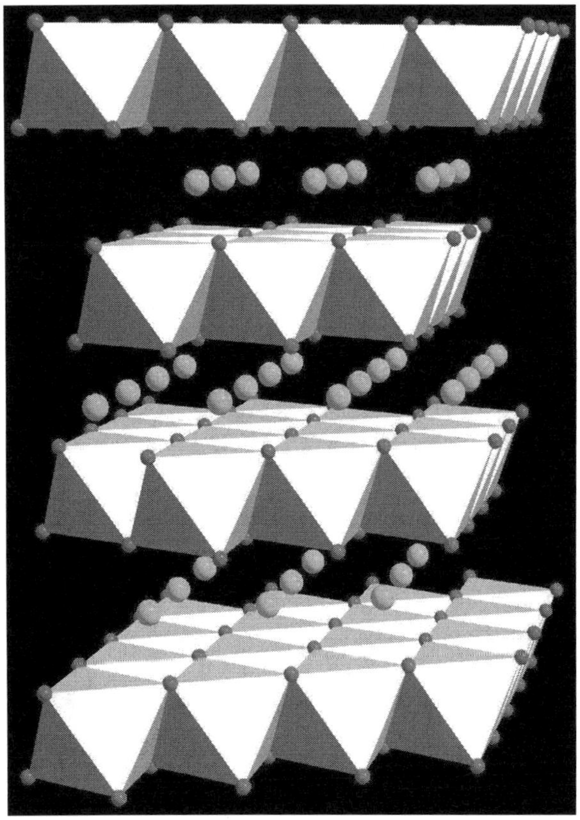

FIGURE 8.4 Crystal structure of layered $LiMO_2$ (M = Co or Ni).

However, only 50% of the theoretical capacity of $LiCoO_2$, which corresponds to a reversible extraction or insertion of 0.5 lithium per Co and a practical capacity of 140 mA h/g, can be utilized in practical cells as seen in Figure 8.5. Although this limitation was originally attributed to structural distortions around $x = 0.5$ in $Li_{1-x}CoO_2$, resulting from an ordering of Li^+ ions [19], recent characterization of chemically delithiated samples suggest that chemical instability arising from a significant overlap of $Co^{3+/4+}$:3d energy band with the top of the O^{2-}:2p band may play a critical role in controlling the reversible capacity values [20]. This conclusion is consistent with the introduction of holes (removal of electrons) into the O^{2-}:2p band rather than into the $Co^{3+/4+}$:3d band during electrochemical charge as revealed by x-ray absorption spectroscopic [21] and electron energy loss spectroscopic [22] studies. Additionally, Co is relatively expensive and toxic. Moreover, the chemical instability of $LiCoO_2$ leads to safety concerns under the conditions of overcharge, which becomes a serious issue particularly in the case of large batteries such as those necessary for HEV and PHEV applications. Also, the power capability of $LiCoO_2$ is not attractive for vehicle applications.

The above drawbacks associated with $LiCoO_2$ have created interest in the development of alternative cathode materials. In this regard, the nickel-rich $LiNi_{0.85}Co_{0.15}O_2$ that has the layered structure shown in Figure 8.4 became appealing a few years ago [23] as it offers a higher capacity of 180 mA h/g as seen in Figure 8.5 and Ni is less expensive than Co. However, more in-depth investigation has shown that $LiNi_{0.85}Co_{0.15}O_2$ suffers from impedance growth during cycling at elevated temperatures, which may partly be related to the migration of $Ni^{2+/3+}$ ions from the transition metal plane to the lithium plane at elevated temperatures [24]. Recently, layered compositions such as $LiNi_{1/3}Mn_{1/3}Co_{1/3}O_2$ [25–27] and $LiNi_{1/2}Mn_{1/2}O_2$ [28,29] have become attractive as they offer capacities close to 200 mA h/g at a lower cost compared to $LiCoO_2$. However, their power capability is lower than that of $LiCoO_2$ due to some cation disordering between the transition metal and lithium planes, and the power capability may not be adequate for vehicle applications.

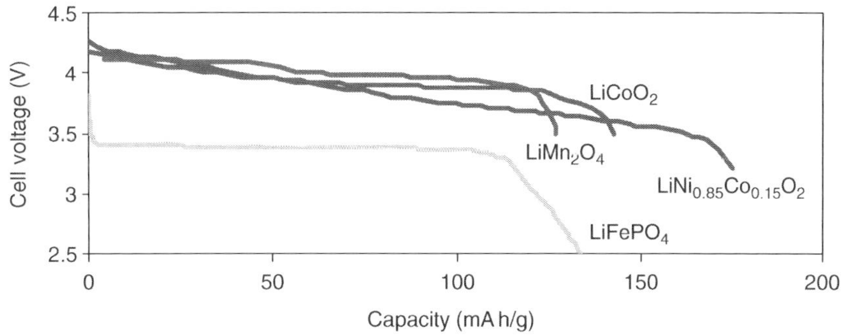

FIGURE 8.5 Comparison of the discharge curves of layered $LiCoO_2$, layered $LiNi_{0.85}Co_{0.15}O_2$, spinel $LiMn_2O_4$, and olivine $LiFePO_4$.

More recently, complex layered oxide solutions between Li[M]O_2 (M = Ni, Mn, and Co) and $Li[Li_{1/3}Mn_{2/3}]O_2$ that have the same O_3 layered structure shown in Figure 8.4 have become appealing as they exhibit high capacities of up to 280 mA h/g [29–32], which is two times higher than that of $LiCoO_2$. However, these oxides encounter an irreversible loss of oxygen gas from the lattice during first charge and a huge irreversible capacity loss of 40–100 mA h/g in the first cycle. Also, they need a high cutoff charge voltage of 4.8 V and slow charge–discharge rate (low power capability). The irreversible capacity loss could be suppressed to some extent by a surface modification of the oxides with inert oxides like Al_2O_3 [32]. The amount of oxygen loss from the lattice and the reversible capacity values have been found to increase with increasing lithium content in the transition metal layer in such $Li[M_{1-\eta}Li_\eta]O_2$ solid solutions [33,34]. Also, the irreversible oxygen loss from the lattice has been found to be sensitive to the substitution of other elements like Al^{3+} for Li^+ and F^- for O^{2-} [35]. Optimization of such complex oxide solid solutions as well as development of robust electrolytes that are stable up to 4.8 V could make these high-capacity cathodes attractive for portable electronic devices although their low power capability due to cation disorder may not be adequate for vehicle applications.

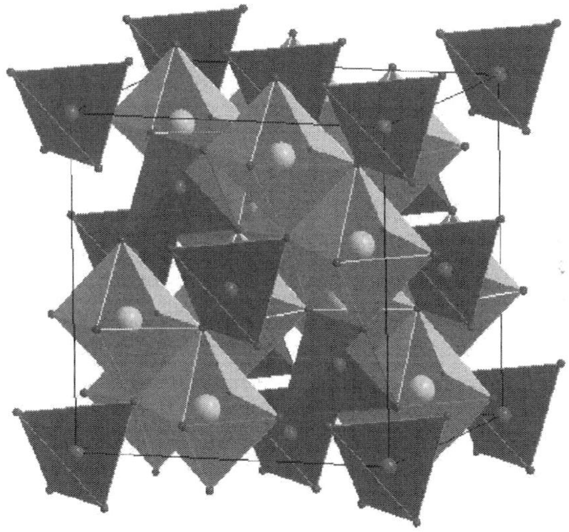

FIGURE 8.6 Crystal structure of spinel $LiMn_2O_4$, illustrating the three-dimensional framework with LiO_4 tetrahedra and MnO_6 octahedra.

8.4.2 Spinel Oxide Cathodes

Although layered oxides exhibit high capacity, some of their performance parameters such as power capability are often limited by cation disorder between the transition metal and lithium planes as well as structural and chemical instabilities occurring during cycling. The compositions with low cation disorder tend to transform from the initial O3 structure to, for example, O1 structure at deep charge [20,27,36]. The limited power capability as well as the safety and cost concerns particularly in compositions with high Co contents makes them less appealing for HEV and PHEV applications. In this regard, $LiMn_2O_4$ crystallizing in the spinel structure has appealing for vehicle applications as Mn is inexpensive and environmentally benign. Also, the three-dimensional spinel structure (Figure 8.6) with LiO_6 tetrahedra and edge-shared MnO_6 octahedra provides good structural stability during the cycling while the lying of the $Mn^{3+/4+}$:3d band well above the O^{2-}:2p band offers good chemical stability unlike, for example, layered $LiCoO_2$.

However, $LiMn_2O_4$ that has been investigated extensively over the years has been plagued by severe capacity fade particularly at elevated temperatures. Several mechanisms such as Mn dissolution from the lattice [37], structural (Jahn–Teller) distortion during over discharge [38], formation of two cubic phases during the charge–discharge process [39], and loss of crystallinity have been proposed to account for the capacity fade [40]. Several approaches such as cationic and anionic substitutions as well as surface modification have been pursued to improve the capacity retention, but they could not completely overcome the problem. More recently, from a systematic investigation of a number of singly and doubly substituted spinel compositions, initial Mn valence and the lattice parameter difference Δa between the two cubic phases formed during the charge–discharge process have been found to play a critical role in controlling the capacity retention [41–43].

However, cationic substitutions often result in an increase in the oxidation state of Mn and a consequent decrease in capacity values to unattractive levels of <100 mA h/g. This problem has been overcome by a partial substitution of F^- for O^{2-} in stabilized cation-substituted spinel oxide compositions. For example, stabilized spinel oxyfluoride compositions such as $LiMn_{1.8}Li_{0.1}Ni_{0.1}O_{3.8}F_{0.2}$ exhibit excellent capacity retention at elevated temperatures with a capacity of >100 mA h/g, high power capability, superior storage properties, and low irreversible capacity loss in the first cycle [44]. The excellent electrochemical performance is found to be due to a smaller lattice mismatch (low Δa) between the two cubic phases formed during the charge–discharge process, suppressed manganese dissolution, and an initial Mn valence of >3.6+. From a systematic investigation of several spinel compositions, the capacity fade is found to bear a clear relationship to initial Mn valence, Δa, and Mn dissolution [45]. The Mn dissolution could also be suppressed further by mixing the stabilized spinel oxyfluoride compositions with a layered oxide like $LiCoO_2$ and charging to high enough voltages (4.7 V) in the first cycle [46]. This is because the overcharged layered $Li_{1-x}CoO_2$ traps the trace amounts of protons present in the electrolyte and thereby suppresses the disproportionation of Mn^{3+} and the Mn dissolution from the spinel lattice. This composite strategy involving a spinel + layered oxide mixture could become a viable strategy to adopt the spinel cathodes for automotive applications.

In addition to the 4 V spinel cathodes described above, 5 V spinel cathodes based on $LiMn_{1.5}Ni_{0.5}O_4$ [47] are appealing particularly for high-power applications. As in the case of 4 V spinel cathodes, the electrochemical performance of 5 V spinel also bears a clear relationship to the lattice mismatch among the three cubic phases formed during the charge–discharge process [48]. Stabilized spinel compositions such as $LiMn_{1.42}Ni_{0.42}Co_{0.16}O_4$ that involve a much smaller lattice mismatch among the cubic phases exhibit better cyclability. However, development of robust electrolyte compositions that offer long-term stability at 5 V is critical to realize the full potential of these 5 V cathodes.

8.4.3 Olivine Oxide Cathodes

Another candidate that exhibits excellent structural and chemical stabilities along with superior safety characteristics is $LiFePO_4$ crystallizing in the olivine structure as shown in Figure 8.7 [17]. The excellent safety is due to the chemically more stable $Fe^{2+/3+}$ redox couple and the covalently bonded PO_4 groups. Despite the $Fe^{2+/3+}$ couple, $LiFePO_4$ exhibits a high discharge voltage of 3.6 V as seen in Figure 8.5 due to the inductive effect caused by the countercation P^{5+} and the lowering of the $Fe^{2+/3+}$ redox energy [49,50]. However, the major drawback with olivine $LiFePO_4$ is the low electronic and lithium ion conductivities and the consequent decrease in the power capability. The low electronic conductivity is due to the highly localized $Fe^{2+/3+}$ redox couple, the presence of FeO_6 octahedra, and PO_4 tetrahedra, and the coexistence of $LiFePO_4$ and $FePO_4$ as line phases during the charge–discharge process without any detectable mixed valent $Fe^{2+/3+}$.

FIGURE 8.7 Crystal structure of olivine $LiFePO_4$, consisting of FeO_6 octahedra and PO_4 tetrahedra.

However, this problem has been overcome significantly by making the $LiFePO_4$ powder at the nanoscale to reduce the lithium diffusion length as well as by an intimate mixing or coating with conductive carbon to improve the electrical conductivity [51–53]. Although conventional $LiFePO_4$ powder with a larger particle size undergoes a two-phase reaction involving $LiFePO_4$ and $FePO_4$ without any solid solubility range during the charge–discharge process, reducing the particle size to the nanoscale is believed to shift the miscibility gap to lower temperatures and provide some solid solubility range and mixed valency. With nanoscale powder and conductive carbon along with associated engineering, significant progress has been made recently by A123 company to employ $LiFePO_4$ for high-power applications. $LiFePO_4$ is currently employed in lithium ion batteries used in power tools, and it is intensively being pursued for automotive applications.

8.4.4 Carbon Anodes

With a lightweight and low electrochemical potential lying close to that of metallic lithium, graphite has become an attractive anode, and commercial lithium ion cells currently use mostly the carbon anodes [54,55]. It offers a theoretical capacity of 372 mA h/g, which corresponds to an insertion of one lithium per six carbon atoms ($x = 1$ in Li_xC_6). One of the drawbacks with the carbon anodes is the occurrence of significant amount of irreversible capacity loss during the first discharge–charge cycle due to unwanted, irreversible side reactions with the electrolyte. Although natural graphite cannot be charged with electrolytes consisting of propylene carbonate (PC) as it leads to an evolution of gas at around 1 V, this problem could be overcome with electrolytes consisting of other solvents such as ethylene carbonate

(EC) and diethyl carbonate (DEC). Additionally, hard carbons (glassy carbon) obtained by a thermal decomposition of phenolic and epoxy resins and products from petroleum pitch show higher capacities than graphite as they could accommodate extra lithium at the edges and surfaces of the graphene sheets and into nanometer size cavities. Also, hard carbons can be used with PC-based electrolytes unlike graphite. However, the hard carbons show a sloping discharge profile between 0 and 1 V unlike graphite, which shows a nearly flat discharge profile between 0 and 0.3 V.

8.5 Conclusions

Lithium ion batteries have revolutionized the portable electronics market because of the development of smart cathode, anode, and electrolyte compositions. Although only 50% of the theoretical capacity of the currently employed layered $LiCoO_2$ cathode can be used in practical cells, significant progress is being made in recent years and complex layered oxide solid solutions delivering two times higher capacity than $LiCoO_2$ have been identified. These high-energy density cathodes are attractive for portable applications, but further work is needed to improve the charge–discharge rates and electrolyte stability to higher charge voltages as well as to address the irreversible release of oxygen gas during first charge. Despite the success in portable devices, lithium ion batteries are yet to enter the HEV and PHEV arena mainly due to the high cost, safety concerns, and limited power capability of the layered $LiCoO_2$ cathode. In this regard, olivine $LiFePO_4$ and stabilized spinel $LiMn_2O_4$ (cation- and anion-substituted) cathodes are appealing as they offer high power capability and better safety while both Fe and Mn are inexpensive and environmentally benign. Accordingly, these cathodes are being pursued for HEV and PHEV applications.

Although most of the lithium ion cells are currently made with carbon anode, several new materials systems such as nanostructured tin alloys and nanocomposites based on Si are being intensively pursued. Some major issues with some of these materials are large volume changes occurring during the charge–discharge processes and high irreversible capacity loss during the first cycle. Another attractive anode system is the spinel $Li_4Ti_5O_{12}$, but the drawback is its higher discharge voltage (~1.5 V vs. lithium) and the consequent decrease in cell voltage as well as lower capacity (~175 mA h/g) compared to the carbon anode.

Overall, continued focus on the development of new, smart materials and processes has the potential to increase the energy density of the lithium ion batteries for portables as well as to make them viable for HEV and PHEV applications. The interest in HEV and PHEV is becoming more prominent and automakers around the globe are beginning to be intensively involved in this effort. Successful development of lithium ion technology for automobiles will have a profound impact in addressing one of the critical challenges facing humankind in the twenty-first century: increasing energy needs, environmental concerns, and global warming.

Acknowledgment

This work was supported by the Welch Foundation grant F-1254.

References

1. C. A. Vincent and B. Scrosati, *Modern Batteries: An Introduction to Electrochemical Power Sources*, Edward Arnold, London, 1984, p. 16.
2. D. Linden, ed., *Handbook of Batteries*, 2nd edn., McGraw-Hill, New York, 1995.
3. J. O. Besenhard, ed., *Handbook of Battery Materials*, Wiley-VCH, Weinheim, 1999.
4. P. J. Gellings and H. J. M. Bouwmeester, *The CRC Handbook of Solid State Electrochemistry*, CRC Press, Boca Raton, FL, 1997.
5. D. R. Lide, ed., *CRC Handbook of Chemistry and Physics*, 81st edn., CRC Press, Boca Raton, FL, 2000, p. 821.
6. Lithium Battery Energy Storage (LIBES) Publication, Technological Research Association, Tokyo, 1994.
7. M. S. Whittingham and A. J. Jacobson, *Intercalation Chemistry*, Academic Press, New York, 1982.
8. J. P. Gabano, *Lithium Batteries*, Academic Press, London, 1983.
9. H. V. Venkatasetty, *Lithium Battery Technology*, John Wiley, New York, 1984.
10. G. Pistoia, ed., *Lithium Batteries: New Materials, Developments and Perspectives*, Vol. 5, Elsevier, Amsterdam, 1994.
11. C. Julien and G. A. Nazri, *Solid State Batteries: Materials Design and Optimization*, Kluwer, Boston, MA, 1994.
12. G.-A. Nazri and G. Pistoia, eds., *Science and Technology of Lithium Batteries*, Kluwer Academic, Boston, MA, 2003.
13. M. Wakihara and O. Yamamoto, ed., *Lithium Ion Batteries: Fundamentals and Performance*, Wiley-VCH, Weinheim, 1998.
14. K. Mizushima, P. C. Jones, P. J. Wiseman, and J. B. Goodenough, *Mater. Res. Bull.* 15, 783, 1980.
15. J. B. Goodenough, K. Mizushima, and T. Takeda, *Jpn. J. Appl. Phys.* 19, 305, 1983.
16. M. M. Thackeray, W. I. F. David, P. G. Bruce, and J. B. Goodenough, *Mater. Res. Bull.* 18, 461, 1983.
17. A. K. Padhi, K. S. Nanjundaswamy, and J. B. Goodenough, *J. Electrochem. Soc.* 144, 1188, 1997.
18. M. K. Aydinol and G. Ceder, *J. Electrochem. Soc.* 144, 3832, 1997.
19. J. N. Reimers and J. R. Dahn, *J. Electrochem. Soc.* 139, 2091, 1992.
20. J. Choi, E. Alvarez, T. A. Arunkumar, and A. Manthiram, *Electrochem. Solid State Lett.* 9, A241, 2006.
21. L. A. Montoro, M. Abbate, and J. M. Rosolen, *Electrochem. Solid State Lett.* 3, 410, 2000.
22. A. Hightower, J. Graetz, C. C. Ahn, P. Rez, and B. Fultz, *198th Meeting of the Electrochemical Society*, Phoenix, AZ, October 22–27, 2000, Abstract No. 177.

23. W. Li and J. Curie, *J. Electrochem. Soc.* 144, 2773, 1997.
24. A. M. Kannan and A. Manthiram, *J. Electrochem. Soc.* 150, A349, 2003.
25. T. Ohzuku and Y. Makimura, *Chem. Lett.* 642, 2001.
26. K. M. Shaju, G. V. S. Rao, and B. V. R. Chowdari, *Electrochim. Acta* 48, 145, 2002.
27. J. Choi and A. Manthiram, *J. Electrochem. Soc.* 152, A1714, 2005.
28. T. Ohzuku and Y. Makimura, *Chem. Lett.* 744, 2001.
29. Z. Lu, D. D. Macneil, and J. R. Dahn, *Electrochem. Solid State Lett.* 4, A191, 2001.
30. S. H. Kang, Y. K. Sun, and K. Amine, *Electrochem. Solid-State Lett.* 6, A183, 2003.
31. Y. J. Park, Y. S. Hong, X. Wu, K. S. Ryu, and S. H. Chang, *J. Power Sources* 129, 288, 2004.
32. Y. Wu and A. Manthiram, *Electrochem. Solid State Lett.* 9, A221, 2006.
33. A. R. Armstrong, M. Holzapfel, P. Novak, C. S. Johnson, S. Kang, M. M. Thackeray, and P. G. Bruce, *J. Am. Chem. Soc.* 128, 8694, 2006.
34. T. A. Arunkumar, Y. Wu, and A. Manthiram, *Chem. Mater.* 19, 3067, 2007.
35. Y. Wu and A. Manthiram, *Electrochem. Solid State Lett.* 10, A151, 2007.
36. S. Venkatraman, J. Choi, and A. Manthiram, *Electrochem. Commun.* 6, 832, 2004.
37. R. J. Gummow, A. de Kock, and M. M. Thackeray, *Solid State Ionics* 69, 59, 1994.
38. M. M. Thackeray, Y. Shao-Horn, A. J. Kahaian, K. D. Kepler, E. Skinner, J. T. Vaugney, and S. A. Hackney, *Electrochem. Solid State Lett.* 1, 7, 1998.
39. Y. Shin and A. Manthiram, *Electrochem. Solid-State Lett.* 5, A55, 2002.
40. H. Huang, C. A. Vincent, and P. G. Bruce, *J. Electrochem. Soc.* 146, 3649, 1999.
41. Y. Shin and A. Manthiram, *Chem. Mater.* 15, 2954, 2003.
42. Y. Shin and A. Manthiram, *J. Electrochem. Soc.* 151, A204, 2004.
43. A. Manthiram, J. Choi, and W. Choi, *Solid State Ionics* 177, 2629, 2006.
44. W. Choi and A. Manthiram, *Electrochem. Solid-State Lett.* 9, A245, 2006.
45. W. Choi and A. Manthiram, *J. Electrochem. Soc.* 154, A792, 2007.
46. A. Manthiram and W. Choi, *Electrochem. Solid State Lett.* 10, A228, 2007.
47. F. G. B. Ooms, E. M. Kelder, J. Schoonman, M. Wagemaker, and F. M. Mulder, *Solid State Ionics* 152, 143, 2002.
48. T. A. Arunkumar and A. Manthiram, *Electrochem. Solid State Lett.* 8, A403-A405, 2005.
49. A. Manthiram and J. B. Goodenough, *J. Power Sources* 26, 403, 1989.
50. A. Manthiram and J. B. Goodenough, *J. Solid State Chem.* 71, 349, 1987.
51. A. Yamada, S. C. Chung, and K. Hinokuma, *J. Electrochem. Soc.* 148, A224, 2001.
52. H. Huang, S. C. Yin, and L. F. Nazar, *Electrochem. Solid State Lett.* 4, A170, 2001.
53. S. Y. Chung, J. T. Bloking, and Y. M. Chiang, *Nature Materials* 1, 123, 2002.
54. N. Imanishi, Y. Takeda, and O. Yamamoto, in *Lithium Ion Batteries: Fundamentals and Performance*, M. Wakihara and O. Yamamoto, ed., Wiley-VCH, Weinheim, 1998, p. 98.
55. M. Winter and J. O. Besenhard, in *Lithium Ion Batteries: Fundamentals and Performance*, M. Wakihara and O. Yamamoto, ed., Wiley-VCH, Weinheim, 1998, p. 127.

9

Piezoelectric and Electrostrictive Ceramics Transducers and Actuators

9.1 Smart Ferroelectric Ceramics for Transducer Applications

A.L. Kholkin, D.A. Kiselev, L.A. Kholkine, and A. Safari

9.1.1 Introduction

In the fast developing world, the use of smart materials becomes more and more important to implement sophisticated functions of the designed device starting from the materials level. In a common definition [1], the smart materials differ from the ordinary materials in that they can perform two or several functions sometimes with a useful correlation or feedback mechanism between them. In the case of piezoelectric or electrostrictive materials, it means that the same material can be used for both sensor and actuator functions. A piezoelectric or an electrostrictive sensor converts mechanical variable (displacement or force) into a measurable electrical quantity by means of a direct piezoelectric or electrostrictive effect. Alternatively, the actuator converts the electrical signal into useful displacement or force. Typically, the term transducer is used to describe actuator (transmitting) and sensor (receiving) functions. Since piezoelectrics and electrostrictors inherently possess both direct (sensor) and converse (actuator) effects, they can be considered as smart materials in the sense indicated above. The degree of smartness can vary in piezoelectric and electrostrictive materials: it is very often that the merely smart material (only sensor and actuator functions) can be engineered into a "very smart" (i.e., tunable) device or even into an "intelligent structure" where the sensor and actuator functions are mutually connected via integrated processing electronics, sometimes with learning function.

The growth in the transducer market has been rapid and is predicted to continue on its current pace in the beginning of twenty-first century. Piezoelectric and electrostrictive sensors and actuators take up a significant portion of the transducer market with a growing trend, especially due to automobile production, active vibration damping, navigation systems, and medical imaging. In this chapter, the principles of piezoelectric and electrostrictive sensors and actuators will be considered along with the properties of the most useful materials and examples of successful devices.

9.1.2 Piezoelectric and Electrostrictive Effects in Ceramic Materials

Piezoelectricity, discovered in Rochelle salt by Jacques and Pierre Curie, is the term used to describe the ability of certain materials to develop an electric charge that is directly proportional to an applied mechanical stress (Figure 9.1a). The piezoelectric materials also show the converse effect, i.e., they deform (strain) proportionally to an applied electric field (Figure 9.1b). To exhibit piezoelectricity, the crystal should belong to one of the 20 noncentrosymmetric crystallographic classes. An important subgroup of piezoelectric crystals is ferroelectrics, which possess the averaged dipole moment per unit cell (spontaneous polarization) that can be reversed by an application of the external electric field. Above the certain temperature (Curie point), most ferroelectrics lose their ferroelectric and piezoelectric properties and become paraelectrics, i.e., materials having centrosymmetric crystallographic structure with no spontaneous polarization. Electrostriction is a second order effect, which refers to the ability of all materials to deform under the application of an electric field. The phenomenological master equation describing the deformations of an insulating crystal subjected to both elastic stress and electric field is given by

$$x_{ij} = s_{ijkl}X_{kl} + d_{mij}E_m + M_{mnij}E_mE_n, \tag{9.1}$$

where

x_{ij} are the components of elastic strain
s_{ijkl} is the elastic compliance tensor
X_{kl} are the stress components
d_{mij} are the piezoelectric tensor components
M_{mnij} is the electrostrictive tensor
E_m and E_n are the components of an external electric field

In this equation, the Einstein summation rule is used for repeating indices. Typically, the electrostriction term ($\propto E_mE_n$) is more than an order of magnitude smaller than the piezoelectric term in Equation 9.1, i.e., the electrostrictive deformations are much smaller than piezoelectric strains. In this case, under zero stress, Equation 9.1 simply transforms to

$$x_{ij} \approx d_{mij}E_m. \tag{9.2}$$

Equation 9.2 describes the converse piezoelectric effect where the electric field changes dimensions of the sample (Figure 9.1b). In centrosymmetric materials, the piezoelectric effect is absent and the elastic strain is only due to the electrostriction. In single domain ferroelectrics having centrosymmetric paraelectric phase, the piezoelectric and electrostriction coefficients can be described in terms of their polarization and dielectric constant. For example, longitudinal coefficients (both electric field and deformation are along tetragonal axis, symmetry *4mm*) can be described as follows:

$$d_{33} = 2Q_{11}\varepsilon_0\varepsilon_{33}P_3, \tag{9.3a}$$

$$M_{11} = Q_{11}(\varepsilon_0\varepsilon_{33})^2, \tag{9.3b}$$

where ε_{33} and P_3 are the dielectric constant and polarization along the polar direction, $\varepsilon_0 = 8.854 \times 10^{-12}$ F/m is the permittivity of vacuum, and Q_{11} is the polarization electrostriction coefficient, which couples longitudinal strain and polarization due to general electrostriction equation:

$$S_3 = Q_{11}P_3^2. \tag{9.4}$$

The mathematical definition of the direct piezoelectric effect where applied elastic stress causes charge on the major surfaces of the piezoelectric crystal is given by

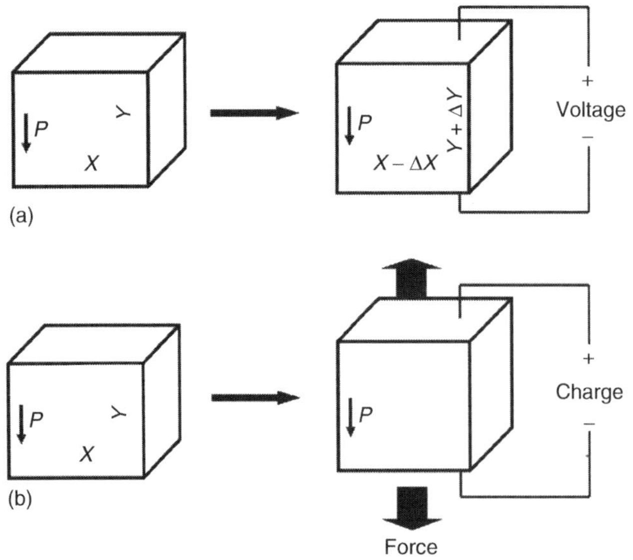

FIGURE 9.1 Schematic representations of the direct and converse piezoelectric effect: (a) an electric field applied to the material changes its shape and (b) a mechanical force on the material yields an electric field across it.

$$P_m = d_{mi} X_i, \tag{9.5}$$

where P_m is the component of electrical polarization. In the case of electrostriction (centrosymmetric crystals), no charge appears on the surface of the crystal upon stressing, and the converse electrostriction effect is simply a change of the inverse dielectric constant under mechanical stress:

$$\Delta(l/(\varepsilon\varepsilon_0)) = 2Q_{11} X_3. \tag{9.6}$$

It should be noted that the reduced matrix notation [2] for the piezoelectric and electrostriction coefficients, and stress tensor is used in Equations 9.3 through 9.6.

The piezoelectric and electrostrictive effects were described for the case of single domain crystals in which the spontaneous polarization is constant everywhere. A technologically important class of materials is piezoelectric and electrostrictive ceramics, which consist of randomly oriented grains, separated by grain boundaries. Ceramics are much less expensive in processing than single crystal and typically offer comparable piezoelectricandelectrostrictiveproperties.Apparently,innonferroelectric ceramics, piezoelectric effect of individual grains is canceled out by averaging over the entire sample and the whole structure will have a macroscopic center of symmetry and negligible piezoelectric properties. Only electrostriction can be observed in such ceramics. Sintered ferroelectric materials (single crystals or ceramics) consist of regions with different orientations of spontaneous polarization, the so-called ferroelectric domains. Domains appear when the material is cooled down through the Curie point to minimize the electrostatic and elastic energy of the system. Domain boundaries or domain walls are movable under applied electric field, so the ferroelectric can be poled, i.e., domains become oriented in the crystallographic direction closest to the direction of applied electric field (Figure 9.2). Typically, poling is performed under high electric field at elevated temperature to facilitate domain alignment. As a result, initially centrosymmetric ceramic sample loses the inversion center and becomes piezoelectric (symmetry $\infty\infty m$). There are three independent piezoelectric coefficients: d_{33}, d_{31}, and d_{15}, which relate longitudinal, transverse, and shear deformations, respectively, to the applied electric field (see Figures 9.1 and 9.3).

Another set of materials constants that is frequently used to characterize the piezoelectric properties of ceramics are the piezoelectric voltage coefficients, g_{ij} defined in a matrix notation as

$$E_i = g_{ij} X_j, \tag{9.7}$$

where E_i are the components of electric field arising due to the external stress X_j. The charge coefficients d_{ij} and voltage coefficients are related by the following equation:

$$g_{ij} = d_{ij}/(\varepsilon_0 \varepsilon_{33}). \tag{9.8}$$

An important property of piezoelectric and electrostrictive transducers is their electromechanical coupling coefficient, k, defined as

$$k^2 = \text{resulting mechanical energy/input electrical energy,}$$
$$\text{or} \tag{9.9}$$
$$k^2 = \text{resulting electrical energy/input mechanical energy.}$$

The coupling coefficient represents the efficiency of the piezoelectric in converting electrical energy into mechanical energy and vice versa. Since the energy conversion is never complete, the coupling coefficient is always less than unity.

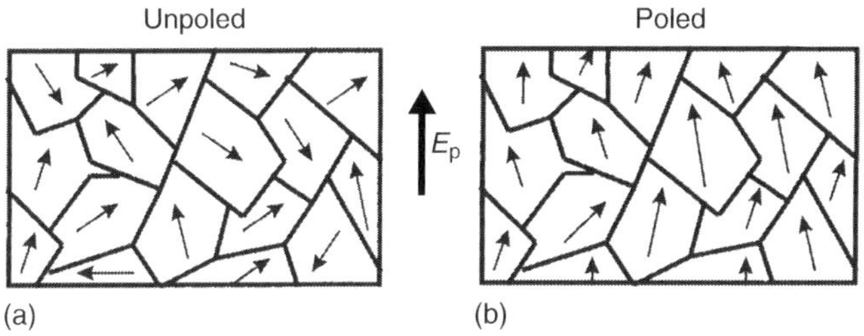

FIGURE 9.2 Schematic of the poling process in piezoelectric ceramics: (a) in the absence of an electric field, the domains have random orientation of polarization with zero piezoelectric activity; (b) the polarization within the domains and grains are aligned in the direction of the electric field.

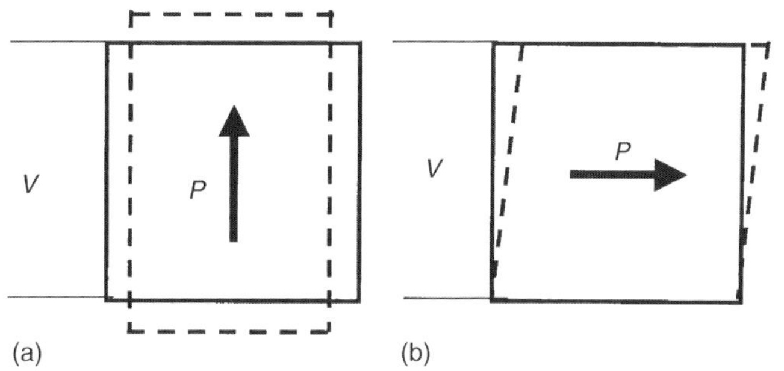

FIGURE 9.3 Schematic of the longitudinal (a), transverse (a), and shear deformations (b) of the piezoelectric ceramic material under applied electric field.

9.1.3 Measurements of Piezoelectric and Electrostrictive Effects

Different means have been proposed to characterize the piezoelectric and electrostrictive properties of ceramic materials. The resonance technique involves the measurement of the characteristic resonance frequencies when a suitably shaped specimen is driven by a sinusoidally varying electric field. To a first approximation, the behavior of a poled ceramic sample close to its fundamental resonance frequency can be represented by an equivalent circuit as shown in Figure 9.4a. The schematic behavior of the reactance of the sample as a function of frequency is shown in Figure 9.4b. By measuring the characteristic resonance frequencies of the sample, the material constants including piezoelectric coefficients can be calculated. The equations used for the calculations of the electromechanical properties are described in the IEEE Standard on Piezoelectricity [3]. The simplest example of piezoelectric measurements by resonance technique relates to a ceramic rod (typically 6 mm in diameter and 15 mm long) poled along its length. It can be shown that the coupling coefficient for this configuration (k_{33}) is expressed as a function of the series and parallel resonance frequencies, f_s and f_p, respectively:

$$k_{33} = (\pi/2)(f_s/f_p)\tan[(\pi/2)(f_p - f_s)/2]. \qquad (9.10)$$

The longitudinal piezoelectric coefficient d_{33} is then calculated using k_{33}, elastic compliance s_{33}, and low-frequency dielectric constant ε_{33}:

$$d_{33} = k_{33}(\varepsilon_{33}s_{33})^{1/2}. \qquad (9.11)$$

In a similar manner, other coupling coefficients and piezoelectric moduli can be derived using different vibration modes of the ceramic sample. The disadvantage of the resonance technique is that measurements are limited to the specific frequencies determined by the electromechanical resonance. The resonance measurements are not easy in electrostrictive samples, which require application of a strong dc bias field to induce piezoelectric effect in relaxor ferroelectrics (see Section 9.1.4).

Subresonance techniques are frequently used to evaluate piezoelectric properties of ceramic materials at frequencies much lower than the fundamental resonance frequency of the sample. They include both the measurements of the piezoelectric charge under the action of mechanical force (direct piezoelectric effect) and the measurement of electric field-induced displacement (converse piezoelectric effect). It can be shown that piezoelectric coefficients obtained by direct and converse piezoelectric effects are thermodynamically equivalent. Electrostrictive properties of ceramics are most easily determined by measurements of displacements as a function of the electric field or polarization. Thus M and Q electrostriction coefficients can be evaluated according to Equations 9.1 and 9.4, respectively. Alternatively, Equations 9.3b and 9.6 can be also used for electrostriction measurements.

A direct technique is widely used to evaluate the sensor capabilities of piezoelectric and electrostrictive materials at sufficiently low frequencies. The mechanical deformations can be applied in different directions to obtain different components of the piezoelectric and electrostriction tensors. In the simplest case, the metal electrodes are placed onto the major surfaces of the piezoelectric transducer normal to its poling direction (Figure 9.1b). Thus the charge is produced on the electrode plates upon mechanical loading, which is proportional to the longitudinal piezoelectric coefficient d_{33} and the force F exerted on the ceramic sample: $Q = d_{33}F$. The charge can be measured by means of charge amplifier using an etalon capacitor in the feedback loop. Details of the direct piezoelectric measurements can be found in a number of textbooks (see, e.g., Ref. [4]).

Measurements of electric field-induced displacements can be performed by a number of techniques including strain gauges, linear variable differential transformers (LVDTs), capacitance method, fiber optic sensor, and laser interferometry. Metal wire strain gauges are the most popular sensors used to measure strain with the resolution of about 10^{-6}. The strain gauge is glued to the ceramic sample and the resistance of gauge changes according to its deformation. The resistance variation is measured by a precise potentiometer up to the frequency of several MHz. Several gauges need to be used in order to get the complete set of the piezoelectric and electrostrictive coefficients of the sample.

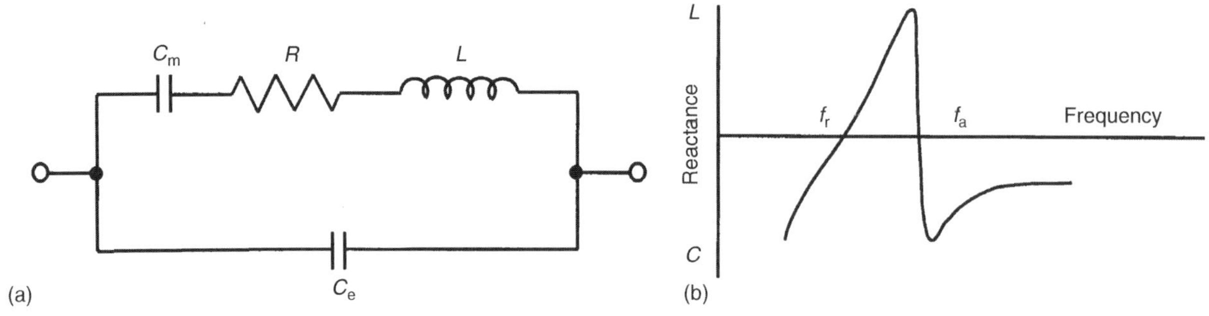

FIGURE 9.4 (a) Equivalent circuit of the piezoelectric sample near its fundamental electromechanical resonance (top branch represents the mechanical part and bottom branch represents the electrical part of the circuit) and (b) electrical reactance of the sample as a function of frequency.

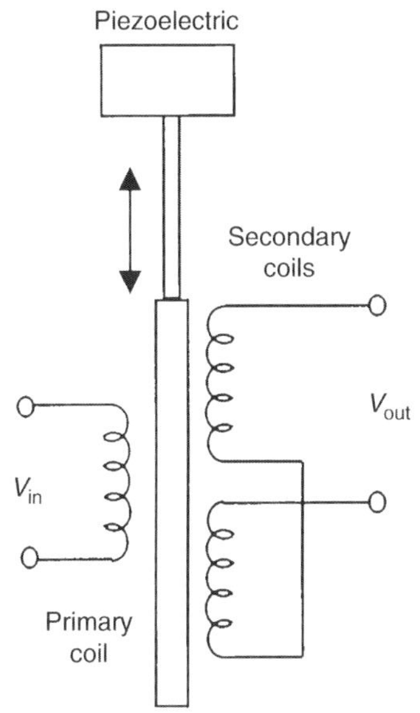

FIGURE 9.5 Principle of LVDT used for the measurement of the electric field-induced deformations in a piezolectric sample.

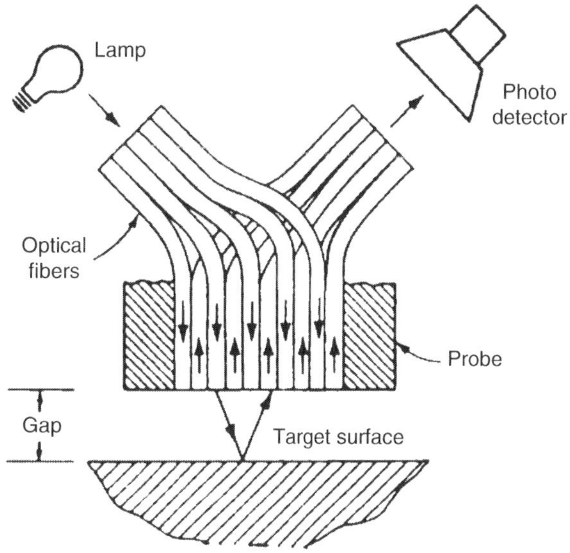

FIGURE 9.6 Schematic of the fiber-optic photonic sensor used for nondestructive evaluation of electric field-induced strains.

The design of LVDT is illustrated in Figure 9.5. The moving surface of the sample is attached to the magnetic core inserted in the center of primary and secondary electromagnetic coils. The change of the core position causes the change of the mutual inductance of the coils. The primary coil is supplied by an ac current, and the signal in the secondary coils is proportional to the displacement of the core. The response speed is limited by the frequency of ac signal and the mechanical resonance of the coil, and typically does not exceed 100 Hz. The resolution is sufficiently high (10–100 nm) and depends on the number of turns.

Capacitive technique for strain measurements is based on a change of the capacitance of the parallel-plate capacitor with the air gap between two opposite plates. One of the plates is rigidly connected to the moving surface of the sample and another plate is fixed by the holder. The capacitance change due to vibration of the sample can be measured precisely by a zero-point potentiometer and a lock-in amplifier. Therefore, a high resolution (in the Angstrom range) can be achieved by this technique. The measurement frequency must be much lower than the frequency of an ac input signal and typically does not exceed 100 Hz.

All the previously described techniques require a mechanical contact between the sample and the measurement unit. This, of course, limits the resolution and maximum frequency, and prevents the accurate measurements of the phase angle between the driving voltage and displacement (so-called piezoelectric loss). Significant force is exerted on the moving surface, which in the case of ceramic thin films, may damage the sample. Therefore, the noncontact measurements are preferred for the accurate determination of electric field-induced displacements of piezoelectric and electrostrictive materials. Photonic fiber-optic sensor offers noncontact measurements of the displacement of a flat reflecting surface. The principle of operation is illustrated in Figure 9.6 [5]. The sensor head consists of a bundle of transmitting and receiving optical fibers located in the immediate vicinity of the vibrating surface. The intensity of the reflected light depends on the gap width between the moving object and the probe. This intensity versus width dependence allows for the accurate determination of the displacement in both dc and ac modes. Using a lock-in amplifier for the measurements of the output signal, which is proportional to the light intensity, the resolution of the order of 1 Å can be achieved [5]. The frequency response is determined by the frequency band of the photodiode and amplifier (typically of the order of several hundreds of kHz). A modification of this techniques based on the cantilever gently touching the sample has been recently introduced to avoid the influence of the surface roughness and to decrease the electrode size [6]. This allowed to measure piezoelectric and electrostrictive displacements in thick films.

Another technique allowing noncontact accurate measurement of the electric field-induced displacements is the optical interferometry. Interferometric methods of measuring small displacements include homodyne [7], heterodyne [8], and Fabri-Perot [9] techniques. The common technique is a homodyne interferometer with an active stabilization of the working point to prevent drift due to thermal expansion. When two laser beams of the same wavelength λ interfere, the light intensity changes periodically ($\lambda/2$ period) with the change of the optical path length difference between two beams. When one of the beams is reflected by the moving object, the intensity of the output light changes depending on the electric field-induced displacement of the sample. In the early measurements, simple Michelson interferometer was used [10] which, in principle, enables very high resolution (~10^{-5} Å). However, the measurements are limited to a narrow frequency range since the sample is attached to a rigid substrate and displacement of only the front surface of the sample is monitored [10]. The error due to bending effect of the sample can be very high especially in the case of ferroelectric thin films. In response to that, a double beam (Mach–Zender) interferometer was suggested in which the difference of the displacements of both major surfaces of the sample is taken into account [11]. The modified version of the double-beam interferometer specially adapted for the thin film measurement demonstrated the resolution as high as 10^{-4} Å in the frequency range (10–10^5 Hz) and a long-term stability (<1%) [12].

9.1.4 Common Piezoelectric and Electrostrictive Materials

9.1.4.1 Single Crystals

A number of single crystals (ferroelectric and nonferroelectric) have demonstrated piezoelectricity. Though nonferroelectric piezoelectrics exhibit much lower piezoelectric coefficients than the ferroelectric materials, the former are still extensively used in some applications requiring either high-temperature stability or low loss. The most important material is quartz (crystalline SiO_2), which possess small but very stable piezoelectric properties (e.g., $d_{11} = 2.3$ pC/N, x-cut [2]). $LiNbO_3$ and $LiTaO_3$ are ferroelectrics with high Curie temperatures (1210°C and 660°C, respectively) that are mostly used in surface acoustic wave (SAW) devices. Recent investigations [13] have shown that rhombohedral single crystals in the $Pb(Zn_{1/3}Nb_{1/3})$-$PbTiO_3$ system exhibit exceptionally large piezoelectric coefficients ($d_{33} = 2500$ pm/V) and coupling coefficients ($k = 0.94$). Ultrahigh strain of 1.7% was observed in these materials under high electric field. These single crystals are now being intensively investigated and show a significant promise for next generation of smart materials.

9.1.4.2 Piezoelectric and Electrostrictive Ceramics

As indicated earlier, the randomness of the grains in as-prepared polycrystalline ferroelectric ceramics yields a nonpiezoelectric centrosymmetric material. Piezoelectric behavior is induced by "poling" the ceramic (Figure 9.2). Although all of the domains in a ceramic can never be fully aligned along the poling axis due to symmetry limitations, the ceramic ends up with a net polarization along the poling axis and sufficiently high piezoelectric properties.

The largest class of piezoelectric ceramics is made up of mixed oxides containing corner-sharing octahedra of O^{2-} ions. The most technologically important materials in this class are perovskites with a general formula ABO_3 where A = Na, K, Rb, Ca, Sr, Ba, Pb, etc. and B = Ti, Sn, Zr, Nb, Ta, or W. Piezoelectric ceramics having this structure include barium titanate ($BaTiO_3$), lead titanate ($PbTiO_3$), lead zirconate titanate ($PbZr_xTi_{1-x}O_3$ or PZT), lead lanthanum zirconate titanate {$Pb_{1-x}La_x(Zr_yTi_{1-y})_{1-x/4}O_3$, or PLZT}, and lead magnesium niobate {$PbMg_{1/3}Nb_{2/3}O_3$ or PMN}. Several of these ceramics are discussed below.

The piezoelectric effect in $BaTiO_3$ was discovered in the 1940s [14] and it became the first piezoelectric ceramic developed. The Curie point of $BaTiO_3$ is about 120°C–130°C. Above 130°C, a

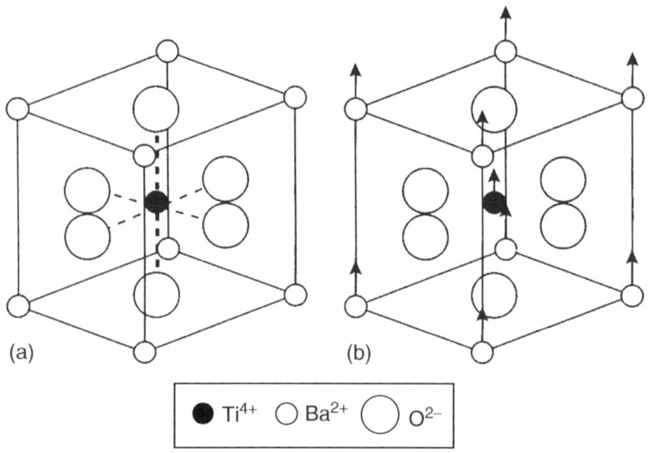

FIGURE 9.7 Crystal structure of $BaTiO_3$: (a) above the Curie point, the cell is cubic; (b) below the Curie point, the cell is tetragonal with Ba^{2+} and Ti^{4+} ions displaced relative to O^{2-} ions.

nonpiezoelectric cubic phase is stable, where the center of positive charge (Ba^{2+} and Ti^{4+}) coincides with the center of the negative charge (O^{2-}) (Figure 9.7a). When cooled below the Curie point, a tetragonal structure (shown in Figure 9.7b) develops where the center of positive charge is displaced relative to the O^{2-} ions, leading to the formation of electric dipoles. The piezoelectric coefficients of $BaTiO_3$ are relatively high: $d_{15} = 270$ and $d_{33} = 190$ pC/N [14]. The coupling coefficient for $BaTiO_3$ is approximately 0.5. Due to its high dielectric constant, $BaTiO_3$ is widely used as a capacitor material.

Lead titanate, $PbTiO_3$ first reported to be ferroelectric in 1950 [14], has a similar structure to BaTiO but with a significantly higher Curie point ($T_C = 490$°C). Pure lead titanate is difficult to fabricate in a bulk form. When cooled through the Curie temperature, the grains go through a cubic to tetragonal phase change, leading to large strain and ceramic fracturing. This spontaneous strain has been decreased by additions of dopants such as Ca, Sr, Ba, Sn, and W. Calcium-doped $PbTiO_3$ [15] has a dielectric constant of about 200, and longitudinal piezoelectric coefficient (d_{33}) of 65 pC/N. As a result of sufficiently high piezoelectric coefficient and low dielectric constant, the voltage piezoelectric coefficients are exceptionally high, which led to the applications of lead titanate as hydrophones and sonobuoys [16].

PZT is a binary solid solution of $PbZrO_3$ (an antiferroelectric orthorhombic structure) and $PbTiO_3$ (a ferroelectric tetragonal perovskite structure) [14]. It has a perovskite structure, with the Zr^{4+} and Ti^{4+} ions occupying the B site of the ABO_3 structure at random (Figure 9.8). PZT shows a temperature-independent morphotropic phase boundary (MPB) between tetragonal and rhombohedral phases at the Zr:Ti ratio of 52:48 (Figure 9.9). Recently, a narrow region of the monoclinic phase was observed in poled PZT ceramics [17]. The existence of this phase may explain easy rotation of the polarization between tetragonal and rhombohedral phases. As a result of the large number of polarization orientations, this material shows efficient

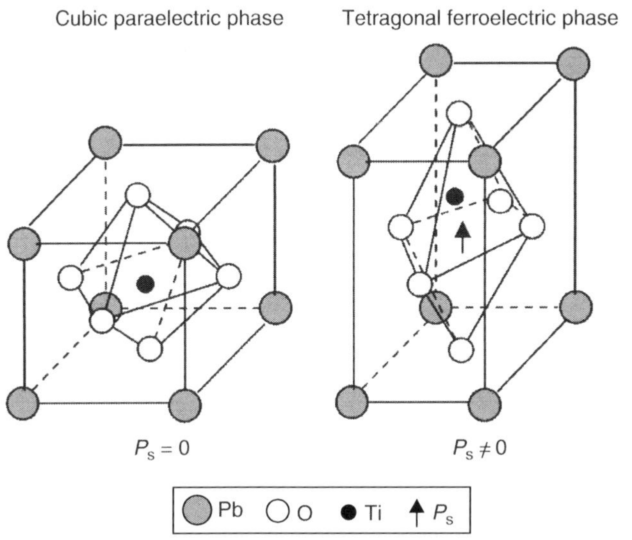

FIGURE 9.8 Schematic of the crystal structure of $Pb(Zr,Ti)O_3$ ceramics: above the Curie point, the cell is cubic (left); below the Curie point, the cell is tetragonal/rhombohedral.

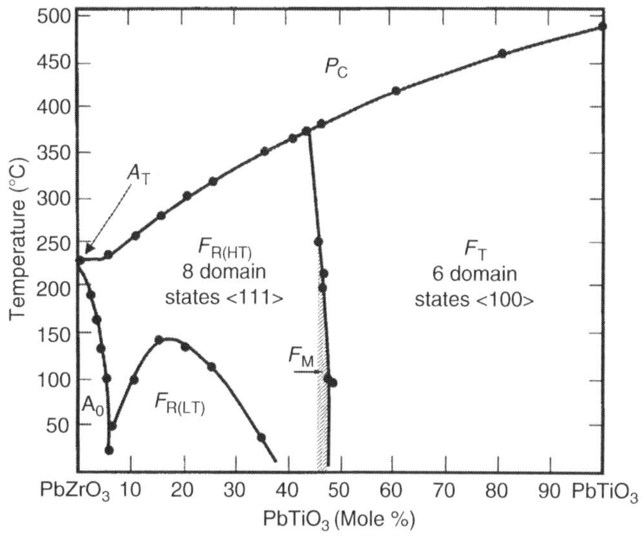

FIGURE 9.9 Phase diagram of PZT piezoelectric ceramics as a function of mole % of $PbTiO_3$.

poling and large enhancement in the piezoelectric properties at this composition due to domain wall motion. Piezoelectric PZT at the MPB is usually doped by a variety of ions to form what are known as "hard" and "soft" PZTs. Hard PZT (abbreviated typically as PZT 4) is doped with acceptor ions, such as K^+ or Na^+ at the A site, or Fe^{3+}, Al^{3+} or Mn^{3+} at the B site. This doping lowers the piezoelectric properties and makes the PZT more difficult to pole or depole. Soft PZT (PZT 5H) is doped with donor ions, such as La^{3+} at the A site, or Nb^{5+} or Sb^{5+} at the B site. It has very high piezoelectric properties and is easy to pole or depole. The comparison of piezoelectric properties of several major piezoelectric ceramics is given in Table 9.1.

Another group of perovskite ceramics is represented by PMN ($PbMg_{1/3}Nb_{2/3}O_3$), which belongs to a family of so-called relaxor ferroelectrics or ferroelectrics with a diffuse phase transition. Unlike normal ferroelectrics, which have well-defined Curie points to separate ferroelectric and paraelectric phases, relaxor ferroelectrics exhibit a broad transition peak in the weak-field dielectric constant [18]. This kind of transition is often referred to as diffuse phase transition. A distinct feature of relaxor ferroelectrics is a strong frequency dispersion of the dielectric constant and a shift of its maximum with frequency. It is suggested that relaxor properties are due to local inhomogeneity of B-ions (e.g., Mg^{2+} and Nb^{5+}) in the perovskite lattice. Relaxors do not possess a piezoelectric effect without a dc bias field, which breaks nonpiezoelectric cubic phase into the rhombohedral ferroelelectric phase. The advantage of using relaxors as the actuators is due to their large induced polarization and electrostrictive strain of the order of 10^{-3} combined with negligible hysteresis. A comparison of electric field-induced strain in typical piezoelectric (PZT) and electrostrictive ceramics (PMN) is shown in Figure 9.10.

9.1.4.3 Processing of Piezoelectric Ceramics

The electromechanical properties of piezoelectric ceramics are largely influenced by their processing conditions. Each step of the process must be carefully controlled to yield the best product. Figure 9.11 is a flowchart of a typical oxide manufacturing process for piezoelectric ceramics. The high-purity raw materials are accurately weighed according to their desired ratio, and mechanically or chemically mixed. During the calcination step, the solid phases react to yield the piezoelectric phase. After

TABLE 9.1 Comparison of Piezoelectric and Related Properties of Several Important Piezoelectric Materials

Parameter	Quartz	$BaTiO_3$	PbTiO3:Sm	PZT 4	PZT 5H	LF4T
d_{33} (pC/N)	2.3	190	65	289	593	410
g_{33} (10^{-3} V m/N)	57.8	12.6	42	26.1	19.7	20.2
k_t	0.09	0.38	0.50	0.51	0.50	0.45
k_p		0.33	0.03	0.58	0.65	0.6
$\varepsilon_{33}^X/\varepsilon_0$	5	1700	175	1300	3400	2300
Q_m	$>10^5$		900	500	65	—
T_c (°C)		120	355	328	193	253

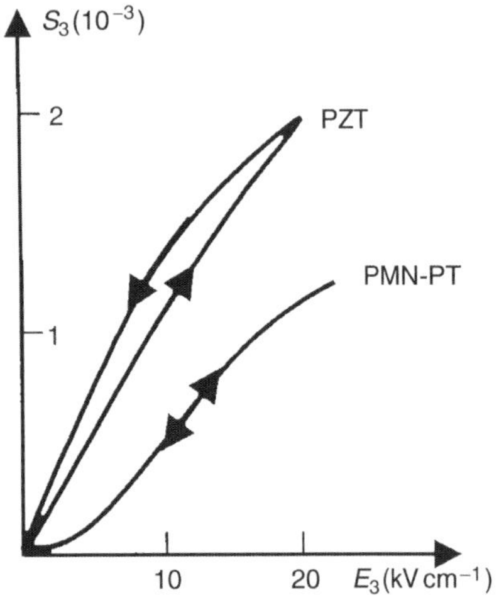

FIGURE 9.10 Comparison of the electric field-induced strain in typical piezoelectric (PZT) and relaxor (0.9PMN-0.1PT) ceramics.

calcining, the solid mixture is ground into fine particles by milling. Shaping is done by a variety of ceramic processing techniques including powder compaction, tape casting, slip casting, or extrusion. During the shaping operation, organic materials are typically added to the ceramic powder to improve its flow and binding characteristics. These organics are removed in a low-temperature (500°C–600°C) binder burnout step. After binder burnout, the ceramic structure is sintered to an optimum density at an elevated temperature. For the lead-containing piezoelectric and electrostrictive ceramics ($PbTiO_3$, PZT, PMN), sintering is performed in sealed crucibles with an optimized PbO atmosphere. This is because lead loss occurs in these ceramics above 800°C.

9.1.5 Piezoelectric Composites

Single-phase piezoelectric and electrostrictive materials are not ideally suited for hydrostatic and ultrasonic applications where the ceramic element radiates and receives the acoustic waves. Though the d_{33} and d_{31} piezoelectric coefficients are exceptionally high in PZT ceramics, their hydrostatic voltage response is relatively low due to the high dielectric constant and low hydrostatic charge coefficient $d_h = d_{33} + 2d_{31}$. Since $d_{31} \approx -0.4d_{33}$ in PZT ceramics [14], the hydrostatic sensor capabilities of single-phase ceramics are rather low. In addition, the high density of ceramics results in high acoustic impedance mismatch between the transducer and the medium in which the acoustic waves are propagating. On the other hand, piezoelectric polymers have low density (low impedance) and low dielectric constant and piezoelectric coefficients. In the past three decades, researchers have been focusing on methods to combine the best characteristics of ceramics and polymers to overcome the aforementioned deficiencies. Combination of a piezoelectric ceramic with a polymer allows one to tailor piezoelectric properties of the composites. The mechanical and electrical properties of a composite strongly depend upon the characteristics of each phase and the manner in which they are connected. In a diphasic composite, there are 10 different ways in which the materials can be oriented in three-dimensional (3D) space [19]. The possible connectivity patterns are 0–0, 1–0, 2–0, 3–0, 1–1, 2–1, 3–1, 2–2, 3–2, and 3–3. As a matter of convention, the first and second numbers in the connectivity denote the continuity of the piezoelectric and polymer phases, respectively. Figure 9.12 shows some of the composites made in the past 30 years [20]. The important connectivity patterns are 0–3, 1–3, 3–3, and 2–2. The 0-3 composites are made with a homogeneous distribution of piezoelectric ceramic particles within polymer matrix. The primary advantage of these composites is their ability to be formed into shapes while remaining piezoelectrically active. The disadvantage is that the 0–3 composites cannot be sufficiently poled because the ceramic phase is not self-connected in the poling direction. On the other hand, 3–0 composites, which are simply the ceramic matrix with low concentration of polymer inclusions or voids, can be effectively poled and exhibit hydrostatic properties superior to that of single phase PZT [20]. In composites with 3–3 connectivity, the piezoceramic and polymer phases are continuous in three dimensions in the form of two interlocking skeletons having intimate contact with one another. The first composite with 3–3 connectivity was formed by the replamine process using coral skeleton [21]. Another effective method of making 3–3 composites is called BURPS (acronym for BURnedout Plastic Spheres) [22] with properties similar to the replamine composites. A mixture of PZT powder and burnable plastic spheres are used to fabricate the PZT–polymer composites. Other techniques, such as relic processing [23] and distorted reticulated ceramics [24], have been developed to fabricate 3–3 composites. Recently, fused deposition modeling (FDM) and fused deposition of ceramics (FDC) have been used to make ladder and 3D honeycomb composites [25]. In the FDM technique, a 3D plastic mold is

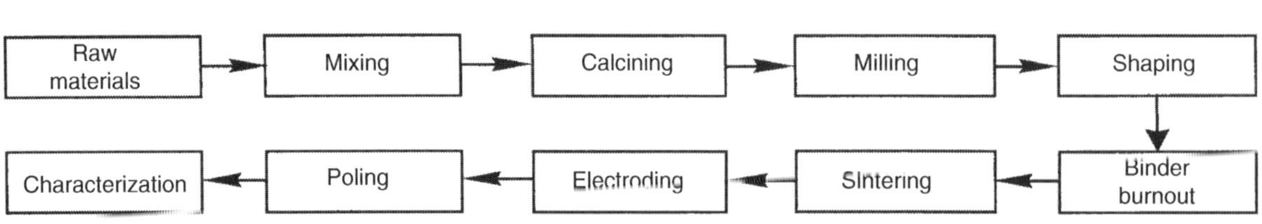

FIGURE 9.11 Flowchart for the processing of piezoelectric ceramics.

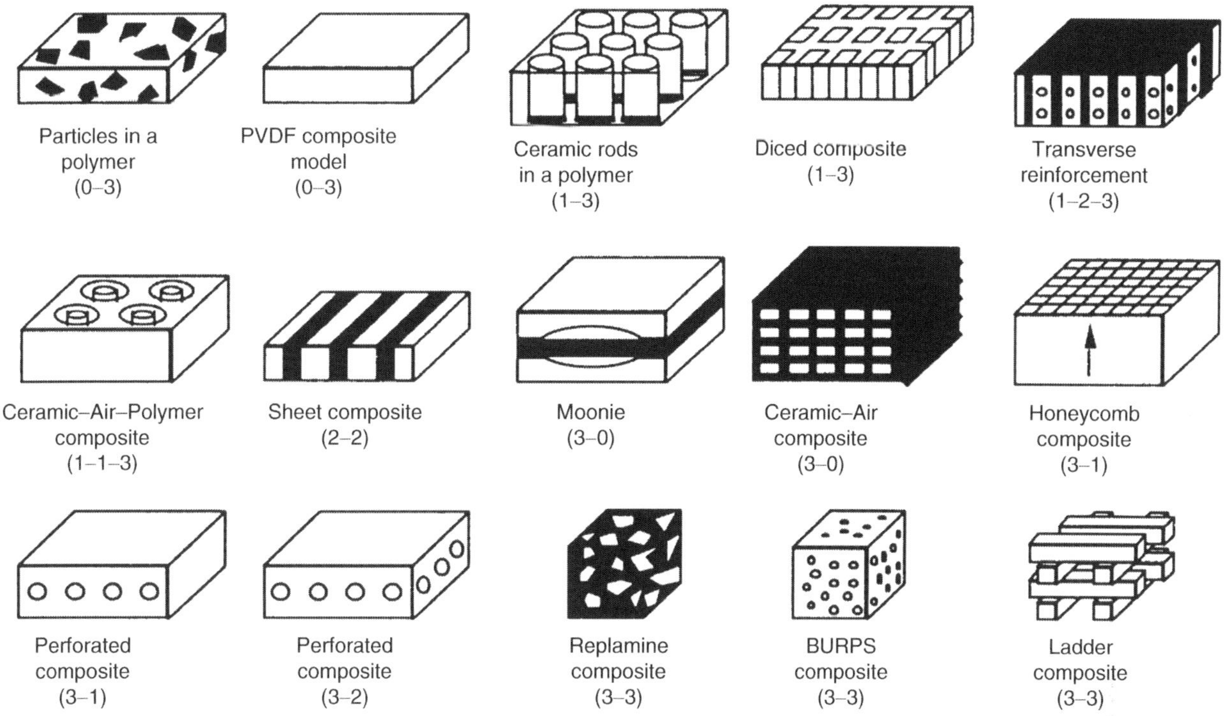

FIGURE 9.12 Schematic of various piezoelectric composites of different connectivities.

prepared and filled with a PZT slurry. The FDC process directly deposits a mixture of PZT and polymer in the form of a 3D ladder structure. Either structure is heat-treated to burn the organic, sintered, and embedded in epoxy polymer.

The most extensively studied and used in transducer applications are the composites with 1–3 connectivities. They consist of individual PZT rods, or fibers, aligned in the direction parallel to poling and embedded in a polymer matrix. A decoupling of d and d_{33} and d_{31} responses of the PZT elements significantly enhances dh. The rod diameter, spacing between them, composite thickness, volume fraction of PZT, and polymer compliance influences the composite performance. The most common methods of forming 1–3 composites are dice and fill technique [26] and injection molding [27]. In the former method, the composite is fabricated by dicing deep grooves in perpendicular directions into a solid sintered block of poled PZT. The grooves are backfilled with polymer and the base is removed via either grinding or cutting. In the latter method, a thermoplastic mixture of ceramic powder and organic binder is injected into a cooled mold. The process can be used to form composites with a variety of rod sizes, shapes, and spacings. This technique has recently been used for the mass production of SmartPanels™ by Materials Systems, Inc. [28].

9.1.6 Applications of Piezoelectric and Electrostrictive Ceramics

Piezoelectric and electrostrictive effects provide a direct coupling between mechanical and electrical quantities. Therefore, they have been extensively used in a variety of electromechanical devices for both sensor and actuator applications. The direct piezoelectric effect is apparently used for the generation of charge (voltage) in applications such gas igniters, acoustic pressure sensors, vibration sensors, accelerometers, hydrophones, etc. [29]. Using a converse effect, mechanical displacements are produced under the applied electric field. The best known examples of actuators are the piezoelectric motors, piezoelectrically driven relays, ink-jet heads for printers, noise cancellation systems, VCR head trackers, precise positioners, deformable mirrors for correction of optical images, etc. [30]. Acoustic and ultrasonic vibrations can be generated by an ac field at resonance conditions and detected by a piezoelectric receiver. Very often acoustic sender and receiver are combined in the same device, the piezoelectric transducer. Ultrasonic transducers are used for medical imaging, nondestructive testing, fish finders, etc. [31]. Another important application of piezoelectrics at high frequencies include frequency control, bulk and SAW resonators, filters, and delay lines.

The ultrasonic transducers operate in a so-called pulse-echo mode. The transducer sends an acoustic wave, which is reflected from the interfaces and is received by the same transducer. These echoes vary in intensity according to the type of interface (e.g., tissue and bone). Therefore, an ultrasonic image is created, which represents the mechanical properties of the human bodies. Thus anatomic structures of different organs can be recognized in a real time. Clearly, when a low intensity of ultrasonic waves is used and a transducer is sensitive enough, ultrasonic medical imaging is one of the safest diagnostic techniques. The transducer is composed of matching and backing layers and piezoelectric material itself. The backing and matching layers are added to the transducer to reduce the acoustic impedance mismatch between the imaged object and a transducer and to damp acoustic backwave. Composite materials instead of single-phase materials are frequently used in order to increase performance of the transducers [20].

Actuator applications require significant displacement and generative force to be produced under moderate electric fields. For precise positioning, a reproducibility of the displacements is also required. In this case, electrostrictive materials (PMN or its solid solution with PT) are preferred because of their small hysteresis and small remanence. Figure 9.13 shows several possible designs of the piezoelectric and electrostrictive actuators. In simple structures, as shown in Figure 9.13a through d, the actuator displacements are solely due to d_{33}, d_{31}, or d_{15} effects of the ceramic rod, plate, or tube. Since strain is limited to 10^{-3}, the typical displacement of a 1 cm long actuator is of the order of 10 μm. The multilayer actuator (Figure 9.13b) utilizes the parallel connection of the ceramic plates cemented together. In this case, the displacement of many individual sheets of a piezoelectric ceramics is summed up. The advantage of the multilayer actuators is the reduced operating voltage, fast speed, and large generative force. The useful design is a piezoelectric tube (Figure 9.13c), which is poled and driven by the voltage applied in radial direction (through the wall width). The axial response is due to d_{31} coefficient of the material and is proportional to the length/width ratio. The radial response can be tuned to almost zero by manipulating the geometry of the tube

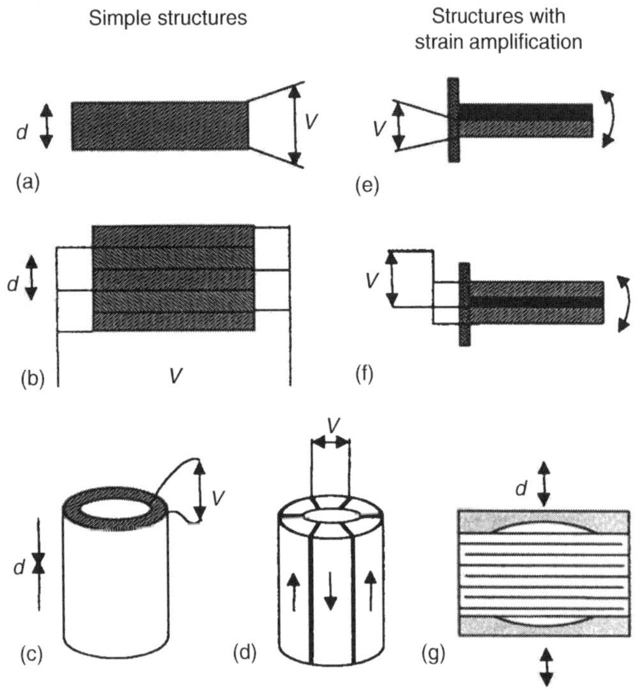

FIGURE 9.13 Typical actuator designs: simple structures [(a)–(d)] and structures with strain amplification [(e)–(g)].

[32]. This configuration is beneficial in suppressing unwanted lateral displacements. Another important design is a shear actuator (Figure 9.13d) directly transforming voltage applied normal to the polarization vector into the pure rotation due to d_{15}-coefficient [33].

As has been previously indicated, all the simple structures are based on pure d_{33}, d_{31}, or d_{15} displacements, so the displacements are limited to tens of microns. The amplification of the strain at the expense of generative force can be achieved by using monomorph and bimorph structures (Figure 9.13e and f). These types of actuators produce large displacements (up to several millimeters) but low generative force and slow response. Another way of strain amplification is represented by flextensional transducers. One of the designs called MOONIE is shown in Figure 9.13g [34]. This type of actuator utilizes the bending effect of the moon-shaped metallic cap attached to both sides of a multilayer actuator. The d_{31} motion of the actuator is amplified by bending of the metallic cap. Another examples of flextensional actuators are RAINBOWs and CYMBALs (not shown in the figure) [35,36]. Flextensional actuators have intermediate characteristics between multilayers and bimorphs and are now intensively used in various actuator applications. An example of the smart structure using flextensional actuator (MOONIE) is shown on Figure 9.14. The actuator portion of the device consists of the standard MOONIE and a small piezoelectric ceramic embedded in the upper cap that serves as a sensor. The sensor detects vibrations normal to the actuator surface and, via a feedback loop, sends a signal of appropriate amplitude and phase to the actuator so that it effectively cancels the external vibration. Potential applications of the smart structure shown in Figure 9.14 include active optical systems, rotor suspension systems, and other noise cancellation devices.

Recent trends to miniaturization have resulted in the extensive use of piezoelectric and electrostrictive materials in microelectromechanical systems (MEMS). Since miniaturization of bulk ceramics is limited, these materials are used in a thin or thick

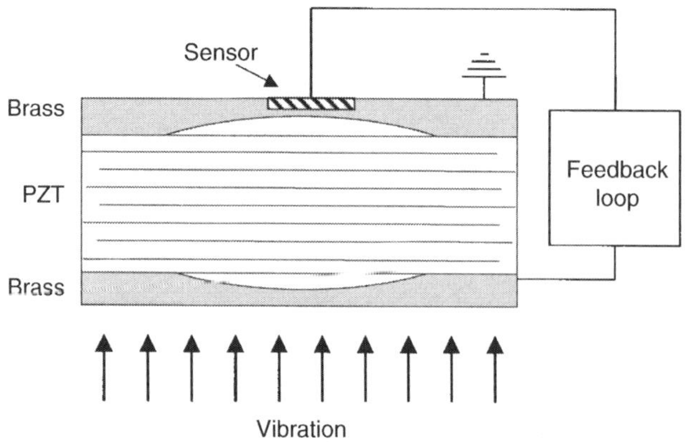

FIGURE 9.14 Example of the smart system using a PZT sensor incorporated in MOONIE actuator. (Adapted from Dogan, A., Uchino, K., and Newnham, R.E., *IEEE Trans. Ultrasonics Ferroelectrics Frequency Control*, 44, 597, 1997.)

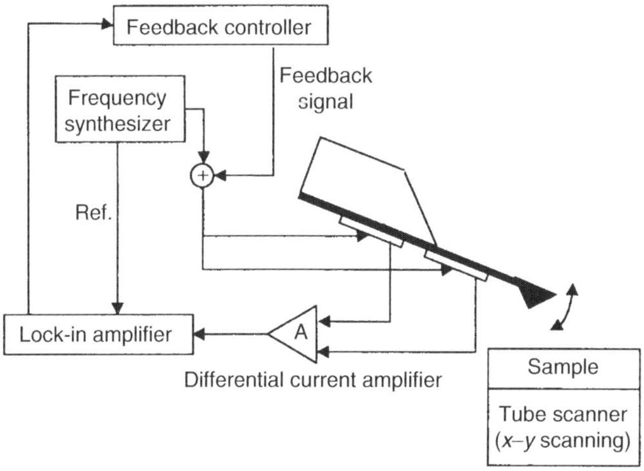

FIGURE 9.15 Schematic of the AFM cantilever sensor and actuator based on the PZT thin film. (Adapted from Fujii, T. and Watanabe, S., *Appl. Phys. Lett.*, 68, 487, 1996.)

film form. Thin film actuators based on piezoelectric effect in PZT materials have been demonstrated by several groups. They include micromotors [37], acoustic imaging devices [38], components for atomic force microscope (AFM) [39], and micropumps [40]. The design of the AFM utilizing PZT film for both sensing and actuating functions is shown in Figure 9.15. The excitation ac voltage signal superimposed on the actuation dc voltage is applied to PZT film deposited on the Si cantilever. The vibrational amplitude, which is sensitive to the atomic force between the tip and investigated surface, is detected by measuring the difference between the cantilever current and reference current. The feedback system maintains this current constant while scanning in x–y plane. This system does not require the optical registration of the vibration that makes PZT-based AFM compact and allows the multiprobe systems to be achieved. Since PZT film is very sensitive to vibrations, the vertical resolution of such AFM approaches that of conventional systems. This electromechanically driven AFM gives an excellent example of using piezoelectric ceramic thin film as smart material. The examples of other promising applications can be found in a recent review [41].

9.1.7 Current Research and Future Trends

Piezoelectric and electrostrictive ceramics currently mostly rely on lead-oxide based ABO3 perovskites due to their excellent properties as sensors and actuators. However, with the increased public awareness on health problems associated with lead and environmental protection policies, research will be focused on finding lead-free compounds having similar piezoelectric properties to PZT. The newly developed lead-free piezoelectric ceramics with comparable piezoelectric performances to those of PZT materials [42] consist of a new composition based on (K,Na) NbO_3 alkaline niobate system with textured microstructure. As in the case of PZT, the highest properties are observed near the MPB between orthorhombic and tetragonal crystal phases of the alkaline niobate system that generated high piezoelectric constant ($d_{33} \approx 400$ pC/N, see Table 9.1). Relaxor single-crystal materials with giant piezoelectric effect already found a wide range of applications in composite transducers for medical imaging to MEMS. The current research has focused on the preparation of these materials in the ceramic form and understanding the nature of their unusually high piezoelectric performance [43]. The current trend to miniaturization will continue giving rise to complex sensors and actuators integrated directly on a Si chip [41]. Using modern patterning techniques, the scale of piezoelectric and electrostrictive ceramics can be reduced to submicron dimensions, giving rise to nanoscale materials (see, e.g., Ref. [44] and references therein) that can be used in nanoelectromechanical systems (NEMS). The research on piezoelectric and electrostrictive materials operating under severe external conditions (temperature, pressure, harsh chemical environments) will be obviously continued for applications in space and deep ocean exploration, noise cancellation in airplanes and helicopters, etc.

References

1. R. B. Newnham and G. R. Ruschau, *J. Intell. Mater. Syst. Struct.* 4, 289, 1993.
2. J. F. Nye, *Physical Properties of Crystals*, Oxford: Oxford University Press, 1985.
3. IEEE Standards on Piezoelectricity, IEEE Std. 176, 1978.
4. V. F. Janas, A. Safari, A. Bandyopadhyay, and A. Kholkin, in *Measurements, Instrumentation and Sensor Handbook*, J. G. Webster ed., Boca Raton, FL: CRC Press.
5. MTI-2000 Fotonic™ Sensor, Instruction Manual, MTI Instruments Division, Latham NY.
6. N. Vyshatko, P. Brioso, P. Vilarinho, and A. L. Kholkin, *Rev. Sci. Instr.* 76, 085101, 2005.
7. S. Sizgoric and A. A. Gundjian, *Proc. IEEE* 57, 1312, 1969.
8. R. M. De La Rue, R. F. Humphryes, I. M. Mason, and E. A. Ash, *Proc. IEEE* 119, 117, 1972.
9. V. E. Bottom, *Rev. Sci. Instrum.* 35, 374, 1964.
10. Q. M. Zhang, W. Y. Pan, and L. E. Cross, *J. Appl. Phys.* 65, 2807, 1988.
11. W. Y. Pan and L. E. Cross, *Rev. Sci. Instrum.* 60, 2701, 1989.
12. A. L. Kholkin, Ch. Wuethrich, D. V. Taylor, and N. Setter, *Rev. Sci. Instrum.* 67, 1935, 1996.
13. S. -E. Park and T. R. Shrout, *J. Appl. Phys.* 82, 1804, 1997.
14. B. Jaffe, W. R. Cook, Jr., and H. Jaffe, *Piezoelectric Ceramics*, Marietta, OH: R.A.N. 1971.
15. Piezoelectric Products, Sensor Technology Limited, Collingwood, Ontario, Canada 1991.
16. K. M. Rittenmyer and R. Y. Ting, *Ferroelectrics* 110, 171, 1990.
17. R. Guo, L. E. Cross, S. -E. Park, B. Noheda, D. E. Cox, and G. Shirane, *Phys. Rev. Lett.* 84, 5423, 2000.
18. L. E. Cross, in *Ferroelectric Ceramics: Tutorials, Reviews, Theory, Processing and Applications*, N. Setter and E. Colla eds., Basel: Birkhouser, 1993.

19. R. E. Newnham, D. P. Skinner, and L. E. Cross, *Mat. Res. Soc. Bull.* 13, 525, 1978.
20. T. R. Gururaja, A. Safari, R. E. Newnham, and L. B. Cross, in *Electronic Ceramics*, L. M. Levinson ed., New York: Marcel Dekker, p. 92.
21. D. P. Skinner, R. E. Newnham, and L. E. Cross, *Mat. Res. Soc. Bull.* 13, 599, 1978.
22. T. R. Shrout, W. A. Schulze, and J. V. Biggers, *Mat. Res. Soc. Bull.* 14, 1553, 1979.
23. S. M. Ting, V. F. Janas, and A. Safari, *J. Am. Ceram. Soc.* 79, 1689, 1996.
24. M. J. Creedon and W. A. Schuize, *Ferroelectrics* 153, 333, 1994.
25. A. Bandyopadhayay, R. K. Panda, V. F. Janas, M. K. Agarwala, S. C. Danforth, and A. Safari, *J. Am. Ceram. Soc.* 80, 1366, 1997.
26. H. P. Savakus, K. A. Klicker, and R. E. Newnham, *Mat. Res. Soc. Bull.* 16, 677, 1981.
27. R. L. Gentilman, D. F. Fiore, H. T. Pham, K. W. French, and L. W. Bowen, *Ceram. Trans.* 43, 239, 1994.
28. D. Fiore, R. Toni, and R. Gentilman, *Proc. of the 8th US–Japan Seminar on Dielectric and Piezoelectric Ceramics*, Plymouth, MA, 1997, p. 135.
29. A. J. Moulson and J. M. Herbert, *Electroceramics: Materials, Properties, Applications*, London: Chapman & Hall, 1990.
30. K. Uchino, *Piezoelectric Actuators and Ultrasonic Motors*, Boston: Kluwer, 1997.
31. J. M. Herbert, *Ferroelectric Transducers and Sensors*, New York: Gordon & Breach, 1982.
32. Q. M. Zhang, H. Wang, and L. E. Cross, *Proc. SPIE, Smart Structures and Materials* 1916, 244, 1993.
33. A. E. Glazounov, Q. M. Zhang, and C. Kim, *Appl. Phys. Lett.* 72, 2526, 1998.
34. A. Dogan, Q. C. Xu, K. Onitsuka, S. Yoshikawa, K. Uchino, and R. E. Newnham, *Ferroelectrics* 156, 1, 1994.
35. G. Haertling, *Bull. Am. Ceramic Soc.* 73, 93, 1994.
36. A. Dogan, K. Uchino, and R. E. Newnham, *IEEE Trans. Ultrasonics Ferroelectrics Frequency Control* 44, 597, 1997.
37. P. Muralt, M. Kohli, T. Maeder, A. Kholkin, K. Brooks, N. Setter, and R. Luthier, *Sensors Actuators* A48, 157, 1995.
38. R. E. Newnham, J. F. Fernandez, K. A. Markowski, J. T. Fielding, A. Dogan, and J. Wallis, *Mat. Res. Soc. Proc.* 360, 33, 1995.
39. T. Fujii and S. Watanabe, *Appl. Phys. Lett.* 68, 487, 1996.
40. J. Bernstein, K. Houston, L. Niles, K. Li, H. Chen, L. E. Cross, and K. Udayakumar, *Proc. of the 8th. Symp. on Integrated Ferroelectrics*, Tempe, AR, 1996, p. 68.
41. N. Setter, D. Damjanovic, L. Eng, G. Fox, S. Gevorgian, S. Hong, A. Kingon, H. Kohlsedt, N. Y. Park, G. B. Stephenson, I. Stolichnov, A. Tagantsev, D. Taylor, T. Yamada, and S. Streiffer. *J. Appl. Phys.* 100, 151606, 2006.
42. Y. Saito, H. Takao, T. Tani, T. Nonoyama, K. Takatori, T. Homma, T. Nagaya, and M. Nakamura, *Nature* 432, 84, 2004.
43. Z. Kutnjak, J. Petzelt, and R. Blinc, *Nature* 441, 956, 2006.
44. A. Gruverman and A. L. Kholkin, *Rep. Prog. Phys.* 69, 2443, 2006.

9.2 Smart Ceramics: Transducers, Sensors, and Actuators

Kenji Uchino and Yukio Ito

9.2.1 Introduction

9.2.1.1 Smart Material

Let us start with the smartness of the material. Table 9.2 lists the various effects that relate the input (electric field, magnetic field, stress, heat, and light) to the output (charge/current, magnetization, strain, temperature, and light). Conducting and elastic materials that generate current and strain outputs, respectively, for the input, voltage, or stress (well-known phenomena) are sometimes called trivial materials. Conversely, pyroelectric and piezoelectric materials that generate an electric field from the input of heat or stress (unexpected phenomena) are called "smart" materials. These off-diagonal couplings have a corresponding converse effect such as electrocaloric and converse-piezoelectric effects, and both sensing and actuating functions can be realized in the same materials. Intelligent materials should possess a drive/control or processing function, which is adaptive to the change in environmental conditions in addition to the actuation and sensing functions. Note that the ferroelectric materials exhibit most of these effects, except magnetic-related phenomena. Thus, the ferroelectrics are said to be very smart materials.

The actuator in a narrow meaning stands for materials or devices that generate mechanical strain (or stress) output. As indicated by the solid columnar border in Table 9.2, solid state actuators use converse piezoelectric, magnetostriction, elasticity, thermal expansion, or photostriction phenomena. A shape-memory alloy is a kind of thermally expanding material. On the other hand, a sensor requires charge and current output in most cases. Thus, conducting and semiconducting, magnetoelectric, piezoelectric, pyroelectric, and photovoltaic materials are used for detecting electric fields, magnetic fields, stress, heat, and light, respectively (see the dashed columnar border in Table 9.2).

In this sense, piezoelectric materials are most popularly used in smart structures and systems, because the same material is applicable for both sensors and actuators, in principle. We treat the piezoelectric transducers, sensors, and actuators mainly in this chapter. Even though transducers, in general, are devices that convert input energy to a different energy type of output, the piezoelectric "transducer" is often used to denote a device that possesses both sensing and actuating functions, exemplified by underwater sonar.

9.2.1.2 Piezoelectric Effect

Certain materials produce electric charges on their surfaces as a consequence of applying mechanical stress. When the induced charge is proportional to the mechanical stress, it is called direct piezoelectric effect and was discovered by J. and P. Curie in 1880.

TABLE 9.2 Various Effects in Ferroelectric and Ferromagnetic Materials

Input → Material device → Output

Input \ Output	Charge Current	Magnetization	Strain	Temperature	Light
Electric field	Permittivity Conductivity	Electro magnetic effect	Converse piezoeffect	Electric caloric effect	Electric-optic effect
Magnetic field	Magnetic electric effect	Permeability	Magneto-striction	Magnetic caloric effect	Magnetic optic effect
Stress	Piezoelectric effect	Piezomagnetic effect	Elastic constant	—	Photoelastic effect
Heat	Pyroelectric effect	—	Thermal expansion	Specific heat	—
Light	Photovoltaic effect	—	Photostriction	—	Refractive index

Diagonal coupling (shaded box) = Smart material
Off-diagonal coupling (dashed box) = Sensor; (solid box) = Actuator

Materials that show this phenomenon also conversely have a geometric strain generated that is proportional to an applied electric field. This is the converse piezoelectric effect. The root of the word "piezo" is the Greek word for "pressure;" hence the original meaning of the word piezoelectricity implied "pressure electricity" [1,2].

Piezoelectric materials couple electrical and mechanical parameters. The material used earliest for its piezoelectric properties was single-crystal quartz. Quartz crystal resonators for frequency control appear today at the heart of clocks and are also used in TVs and computers. Ferroelectric polycrystalline ceramics, such as barium titanate and lead zirconate titanate (PZT), exhibit piezoelectricity when electrically poled. Because these ceramics possess significant and stable piezoelectric effects, that is, high electromechanical coupling, they can produce large strains or forces and hence are extensively used as transducers. Piezoelectric polymers, notably polyvinylidene difluoride (PVDF) and its copolymers with trifluoroethylene and piezoelectric composites that combine a piezoelectric ceramic and a passive polymer have been developed and offer high potential. Recently, thin films of piezoelectric materials are being researched due to their potential use in microdevices (sensors and microelectromechanical systems [MEMS]). Piezoelectricity is being extensively used in fabricating various devices such as transducers, sensors, actuators, surface acoustic wave (SAW) devices, and frequency controls.

We describe the fundamentals of piezoelectric effect first, then present a brief history of piezoelectricity, followed by present day piezoelectric materials that are used, and finally various potential applications of piezoelectric materials are presented.

9.2.2 Piezoelectricity

9.2.2.1 Relationship between Crystal Symmetry and Properties

All crystals can be classified into 32 point groups according to their crystallographic symmetry. These point groups are divided into two classes; one has a center of symmetry and the other lacks it. There are 21 noncentrosymmetric point groups. Crystals that belong to 20 of these point groups exhibit piezoelectricity. Although cubic class 432 lacks a center of symmetry, it does not permit piezoelectricity. Of these 20 point groups, 10 polar crystal classes contain a unique axis, along which an electric dipole moment is oriented in the unstrained condition.

The pyroelectric effect appears in any material that possesses a polar symmetry axis. The material in this category develops electric charge on the surface owing to the change in dipole moment as temperature changes. The pyroelectric crystals whose spontaneous polarization are reorientable by applying an electric field of sufficient magnitude (not exceeding the breakdown limit of the crystal) are called ferroelectrics [3,4]. Table 9.3 shows the crystallographic classification of the point groups.

9.2.2.2 Piezoelectric Coefficients

Materials are deformed by stresses and the resulting deformations are represented by strains ($\Delta L/L$). When the stress X (force per unit area) causes a proportional strain x,

$$x = sX, \tag{9.12}$$

where all quantities are tensors; x and X are second rank and s is fourth rank. Piezoelectricity creates additional strains by an applied electric field E. The piezoelectric equation is given by

TABLE 9.3 Crystallographic Classification According to Crystal Centrosymmetry and Polarity

Polarity	Symmetry	Crystal System										
		Cubic		Hexagonal		Tetragonal		Rhombohedral		Orthorhombic	Monoclinic	Triclinic
Nonpolar (22)	Centro (11)	m3m	m3	6/mmm	6/m	4/mmm	4/m	$\bar{3}$m	$\bar{3}$	mmm	2/m	
Nonpolar (22)	Non-centro (21)	432, $\bar{4}$3m	23	622, $\bar{6}$m2	$\bar{6}$	422, $\bar{4}$2m	$\bar{4}$	32		222		
Polar (Pyroelectric) (10)	Non-centro (21)			6mm	6	4mm	4	3m	3	mm2	2, m	1

Note: ——— Inside the bold line are piezoelectrics.

$$x_{ij} = s_{ijkl}X_{kl} + d_{ijk}E_k, \quad (9.13)$$

where

E is the electric field

d is the piezoelectric constant, which is a third rank tensor

This equation can be also expressed in a matrix form such as given for a poled ceramic:

$$\begin{bmatrix} x_1 \\ x_2 \\ x_3 \\ x_4 \\ x_5 \\ x_6 \end{bmatrix} = \begin{bmatrix} 0 & 0 & d_{31} \\ 0 & 0 & d_{31} \\ 0 & 0 & d_{33} \\ 0 & d_{15} & 0 \\ d_{15} & 0 & 0 \\ 0 & 0 & 0 \end{bmatrix} \begin{bmatrix} E_1 \\ E_2 \\ E_3 \end{bmatrix}. \quad (9.14)$$

Another frequently used piezoelectric constant is g, which gives the electric field produced when a stress is applied ($E = gT$). The g constant is related to the d constant through the relative permittivity ε:

$$g = \frac{d}{\varepsilon\varepsilon_0}, \quad (9.15)$$

where ε_0 is the permittivity of free space. A measure of the effectiveness of the electromechanical energy conversion is the electromechanical coupling factor k, which measures the fraction of the electrical energy converted to mechanical energy when an electric field is applied or vice versa when a material is stressed [5]:

$$k^2 = \text{(Stored mechanical energy/input electrical energy)} \quad (9.16)$$

or

$$= \text{(Stored electrical energy/input mechanical energy)}. \quad (9.17)$$

Let us calculate k^2 in Equation 9.16, when an electric field E is applied to a piezoelectric material. Because the input electrical energy is $(1/2)\cdot\varepsilon_0\cdot\varepsilon E^2$ per unit volume and the stored mechanical energy per unit volume under zero external stress is given by $(1/2)\, x^2/s = (1/2)\,(d\,E)^2/s$, k^2 can be calculated as

$$k^2 = [(1/2)\,(d\,E)^2/s]/[(1/2)\,\varepsilon_0\varepsilon\,E^2] = d^2/\varepsilon_0\varepsilon s. \quad (9.18)$$

Note that k is always less than 1. Typical values of k are 0.10 for quartz, 0.4 for $BaTiO_3$ ceramic, 0.5–0.7 for PZT ceramic, and 0.1–0.3 for PVDF polymer. Another important material parameter is the mechanical quality factor Q_m, which determines the frequency characteristics. Q_m is given by an inverse value of mechanical loss:

$$Q_m = \frac{1}{\tan\delta}. \quad (9.19)$$

A low Q_m is preferred for sensor applications in general to cover a wide frequency range. Conversely, a high Q_m is most desired for ultrasonic actuations (e.g., ultrasonic motors [USM]) to suppress heat generation through the loss.

9.2.2.3 History of Piezoelectricity

As stated already, Pierre and Jacques Curie discovered piezoelectricity in quartz in 1880. The discovery of ferroelectricity accelerated the creation of useful piezoelectric materials. Rochelle salt was the first ferroelectric discovered in 1921. Until 1940, only two types of ferroelectrics were known, Rochelle salt and potassium dihydrogen phosphate and its isomorph. From 1940 to 1943, unusual dielectric properties such as an abnormally high dielectric constant were found in barium titanate ($BaTiO_3$) independently by Wainer and Salmon, Ogawa, and Wul and Golman. After the discovery, compositional modifications for $BaTiO_3$ led to improvement in the temperature stability and high-voltage output. Piezoelectric transducers based on $BaTiO_3$ ceramics became well-established in a number of devices.

In 1950s, Jaffe and coworkers established the PZT system induces strong piezoelectric effects. The maximum piezoelectric response was found for PZT compositions near the morphotropic phase boundary between the rhombohedral and tetragonal phases. Since then, the PZT system containing various additives has become the dominant piezoelectric ceramic for a variety of applications. The development of PZT-based ternary solid solution systems was a major success of the piezoelectric industry for these applications.

In 1969, Kawai et al. discovered that certain polymers, notably PVDF, are piezoelectric when stretched during fabrication. Such piezoelectric polymers are also useful for some transducer applications. In 1978, Newnham et al. improved composite piezoelectric materials by combining a piezoelectric ceramic with a passive polymer whose properties can be tailored to the requirements of various piezoelectric devices.

Another class of ceramic material has recently become important: relaxor-type electrostrictors, such as lead magnesium niobate (PMN), typically doped with 10% lead titanate (PT), which have potential applications in the piezoelectric actuator field. A recent breakthrough in the growth of high-quality, large, single-crystal relaxor piezoelectric compositions has created interest in these materials for applications ranging from high-strain actuators to high-frequency transducers for medical ultrasound devices due to their superior electromechanical characteristics. More recently, thin films of piezoelectric materials such as zinc oxide (ZnO) and PZT have been extensively investigated and developed for use in MEMS devices.

9.2.3 Piezoelectric Materials

This section summarizes the current status of piezoelectric materials: single crystal materials, piezoceramics, piezopolymers, piezocomposites, and piezofilms. Table 9.4 shows the material parameters of some representative piezoelectric materials described here [6,7].

9.2.3.1 Single Crystals

Piezoelectric ceramics are widely used at present for a large number of applications. However, single-crystal materials retain their utility; they are essential for applications such as frequency stabilized oscillators and surface acoustic devices. The most popular single-crystal piezoelectric materials are quartz, lithium niobate ($LiNbO_3$), and lithium tantalate ($LiTaO_3$). The single crystals are anisotropic in general and have different properties depending on the cut of the materials and the direction of bulk or surface wave propagation.

Quartz is a well-known piezoelectric material. α-Quartz belongs to the triclinic crystal system with point group 32 and has a phase transition at 537°C to the β-type that is not piezoelectric. Quartz has a cut with a zero temperature coefficient of the resonance frequency change. For instance, quartz oscillators using the thickness shear mode of the AT-cut are extensively used as clock sources in computers and as frequency stabilized oscillators in TVs and video cassette recorder (VCRs). On the other hand, an ST-cut quartz substrate that has X-propagation has a zero temperature coefficient for SAWs and so is used for SAW devices that have highly stabilized frequencies. Another distinguishing characteristic of quartz is its extremely high mechanical quality factor $Q_m > 10^5$.

Lithium niobate and lithium tantalate belong to an isomorphous crystal system and are composed of oxygen octahedron. The Curie temperatures of $LiNbO_3$ and $LiTaO_3$ are 1210°C and 660°C, respectively. The crystal symmetry of the ferroelectric phase of these single crystals is 3m and the polarization direction is along the *c*-axis. These materials have high electromechanical coupling coefficients for SAWs. In addition, large single crystals can easily be obtained from their melts by using the conventional Czochralski technique. Thus, both materials are very important in SAW device applications.

9.2.3.2 Perovskite Ceramics

Most of the piezoelectric ceramics have the perovskite structure ABO_3, shown in Figure 9.16. This ideal structure consists of a simple cubic unit cell that has a large cation A at the corner, a smaller cation B in the body center, and oxygens O in the centers of the faces. The structure is a network of corner-linked oxygen octahedra surrounding B cations. The piezoelectric properties of perovskite-structured materials can be easily tailored for applications by incorporating various cations in the perovskite structure.

9.2.3.2.1 Barium Titanate

Barium titanate ($BaTiO_3$) is one of the most thoroughly studied and most widely used piezoelectric materials. Figure 9.17 shows the temperature dependence of dielectric constants in $BaTiO_3$

TABLE 9.4 Material Parameters of Representative Piezoelectric Materials

Parameter	Quartz	$BaTiO_3$	PZT 4	PZT 5H	$(Pb, Sm)TiO_3$	PVDF-TrFE
d_{33} (pC/N)	2.3	190	289	593	65	33
g_{33} (10^{-3} V m/N)	57.8	12.6	26.1	19.7	42	380
k_t	0.09	0.38	0.51	0.50	0.50	0.30
k_p		0.33	0.58	0.65	0.03	
$\varepsilon_3^T/\varepsilon_0$	5	1700	1300	3400	175	6
Q_m	$>10^5$		500	65	900	3–10
T_C (°C)		120	328	193	355	

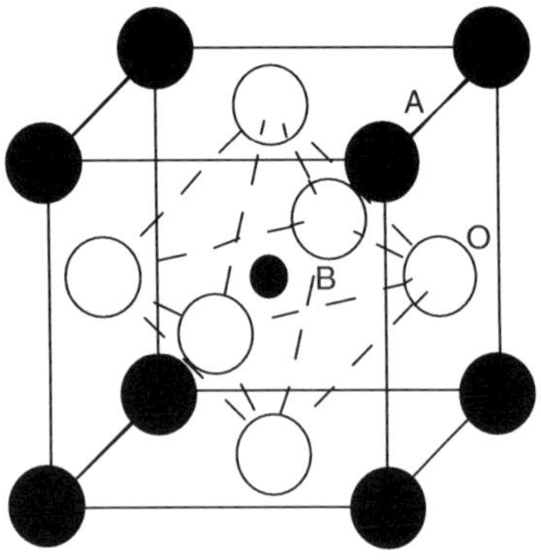

FIGURE 9.16 Perovskite structure ABO_3. (Modified from *Encyclopedia of Smart Materials.*)

that demonstrate the phase transitions in $BaTiO_3$ single crystals. Three anomalies can be observed. The discontinuity at the Curie point (130°C) is due to a transition from a ferroelectric to a paraelectric phase. The other two discontinuities are accompanied by transitions from one ferroelectric phase to another. Above the Curie point, the crystal structure is cubic and has no spontaneous dipole moments. At the Curie point, the crystal becomes polar and the structure changes from a cubic to a tetragonal phase. The dipole moment and the spontaneous polarization are parallel to the tetragonal axis. Just below the Curie temperature, the vector of the spontaneous polarization points in the [001] direction (tetragonal phase) and below 5°C, it reorients in the [011] (orthrhombic phase), and below −90°C in the [111] (rhombohedral phase). The dielectric and piezoelectric properties of ferroelectric ceramic $BaTiO_3$ can be affected by its own stoichiometry, microstructure, and by dopants entering into the A or B site solid solution. Modified $BaTiO_3$ ceramics that contain dopants such as Pb or Ca ions have been used as commercial piezoelectric materials.

9.2.3.2.2 Lead Zirconate–Lead Titanate

Piezoelectric Pb(Ti, Zr)O_3 solid solutions (PZT) ceramics are widely used because of their superior piezoelectric properties. The phase diagram of the PZT system (Pb(Zr_xTi_{1-x})O_3) is shown in Figure 9.18. The crystalline symmetry of this solid solution is determined by the Zr content. PT also has a tetragonal ferroelectric phase of the perovskite structure. As the Zr content x increases, the tetragonal distortion decreases, and when $x > 0.52$, the structure changes from the tetragonal 4 mm phase to another ferroelectric phase of rhombohedral 3 m symmetry. Figure 9.19 shows the dependence of several d constants on the composition near the morphotropic phase boundary between the tetragonal and rhombohedral phases. The d constants have their highest values near the morphotropic phase boundary. This enhancement in the piezoelectric effect is attributed to the increased ease of reorientation of the polarization in an electric field. Doping the PZT material with donors or acceptor changes the properties dramatically. Donor doping with ions such as Nb^{5+} or Ta^{5+} provides soft PZTs like PZT-5, because of the facility of domain motion due to the charge compensation of the Pb vacancy, which is generated during sintering. On the other hand, acceptor doping with Fe^{3+} or Sc^{3+} leads to hard PZTs such as PZT-8 because oxygen vacancies pin the domain wall motion.

9.2.3.2.3 Lead Titanate

$PbTiO_3$ has a tetragonal structure at room temperature and has large tetragonality $c/a = 1.063$. The Curie temperature is 490°C. Densely sintered $PbTiO_3$ ceramics cannot be obtained easily because they break up into powders when cooled through the Curie temperature. This is partly due to the large spontaneous strain that occurs at the transition. PT ceramics modified by small amounts of additives exhibit high piezoelectric anisotropy. Either (Pb, Sm)TiO_3 [8] or (Pb, Ca)TiO_3 [9] has extremely low planar coupling, that is, a large k_t/k_p ratio. Here, k_t and k_p are thickness-extensional and planar electromechanical coupling

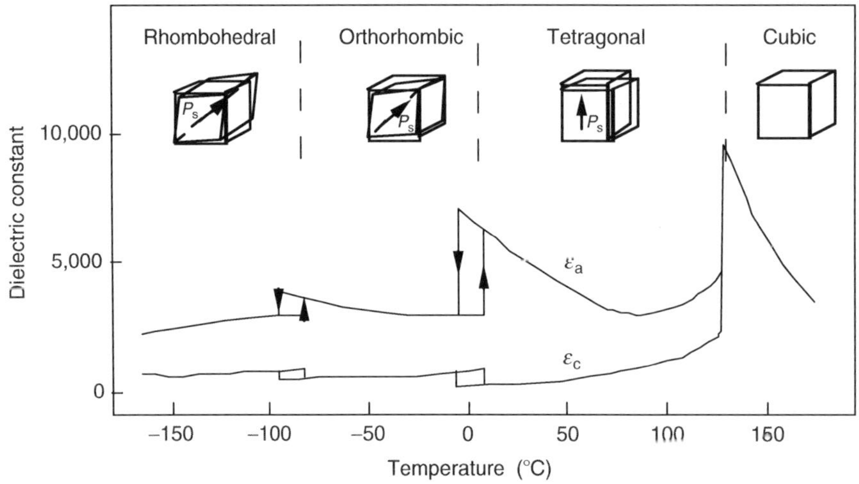

FIGURE 9.17 Dielectric constants of $BaTiO_3$ as a function of temperature. (Modified from *Encyclopedia of Smart Materials.*)

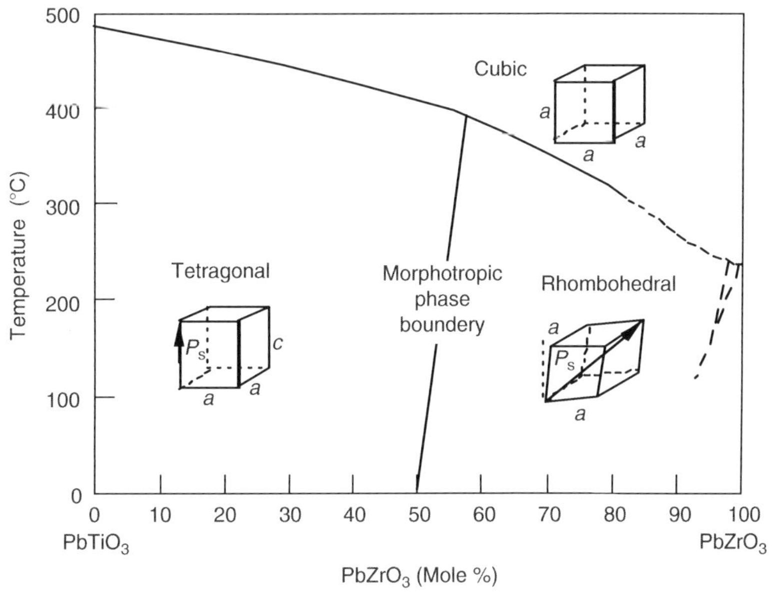

FIGURE 9.18 Phase diagram of the PZT system. (Modified from Ceramics, transducers, in *Encyclopedia of Smart Materials.*)

factors, respectively. (Pb, Nd)(Ti, Mn, In)O_3 ceramics that have a zero temperature coefficient of SAW delay have been developed as superior substrate materials for SAW devices [10].

9.2.3.2.4 *Relaxor Ferroelectrics*

Relaxor ferroelectrics differ from normal ferroelectrics; they have broad phase transitions from the paraelectric to the ferroelectric state, strong frequency dependence of the dielectric constant (i.e., dielectric relaxation) and weak remanent polarization at temperatures close to the dielectric maximum. Lead-based relaxor materials have complex disordered perovskite structures of the general formula $Pb(B_1, B_2)O_3$ ($B_1 = Mg^{2+}, Zn^{2+}, Sc^{3+}$, $B_2 = Nb^{5+}, Ta^{5+}, W^{6+}$). The B site cations are distributed randomly in the crystal. The characteristic of a relaxor is a broad and frequency dispersive dielectric maximum. In addition, relaxor-type materials such as lead magnesium niobate $Pb(Mg_{1/3}Nb_{2/3})O_3$–lead titanate $PbTiO_3$ solid solution [PMN-PT] exhibit electrostrictive phenomena that are suitable for actuator applications. Figure 9.20 shows an electric-field-induced strain curve that was observed for 0.9PMN-0.1PT and reported by Cross et al. in 1980 [11]. Note that a strain of 0.1% can be induced by an electric field as small as 1 kV/mm and that hysteresis is negligibly small for this electrostriction.

Because electrostriction is the secondary electromechanical coupling observed in cubic structures, in principle, the charge is

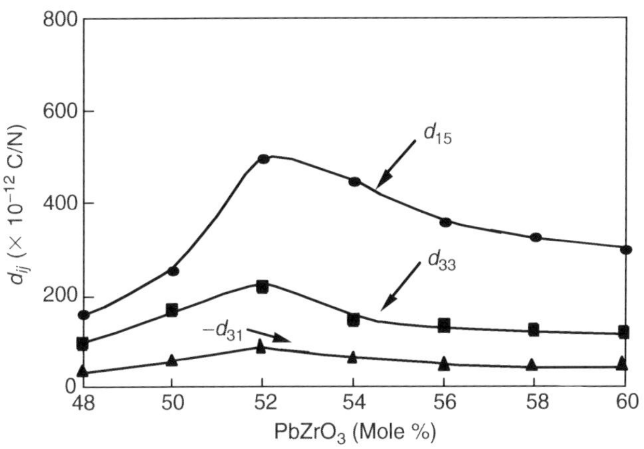

FIGURE 9.19 Piezoelectric *d* strain coefficients versus composition for the PZT system. (Modified from Ceramics, transducers, in *Encyclopedia of Smart Materials.*)

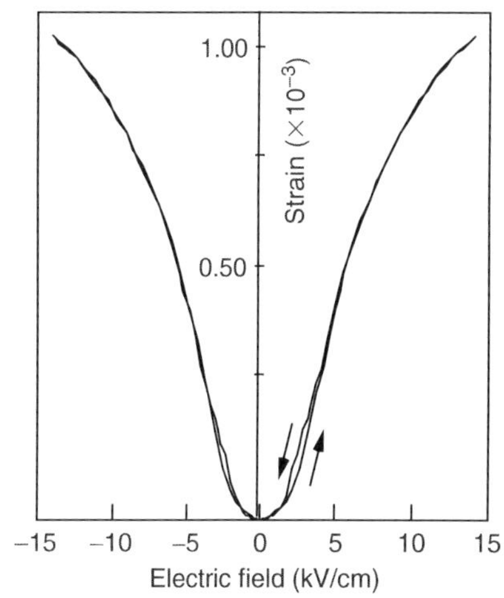

FIGURE 9.20 Field-induced electrostrictive strain in 0.9PMN-0.1PT. (Modified from Ceramics, transducers, in *Encyclopedia of Smart Materials.*)

not induced under applied stress. The converse electrostrictive effect, which can be used for sensor applications, means that the permittivity (first derivative of polarization with respect to an electric field) is changed by stress.

In relaxor ferroelectrics, the piezoelectric effect can be induced under a bias field, that is, the electromechanical coupling factor k_t varies as the applied DC bias field changes. As the DC bias field increases, the coupling increases and saturates. These materials can be used for ultrasonic transducers that are tunable by a bias field [12].

The recent development of single-crystal piezoelectrics started in 1981, when Kuwata, Uchino, and Nomura first reported an enormously large electromechanical coupling factor k_{33}=92%–95% and a piezoelectric constant d_{33} = 1500 pC/N in solid-solution single crystals between relaxor and normal ferroelectrics, $Pb(Zn_{1/3}Nb_{2/3})O_3$-$PbTiO_3$ [13]. After about 10 years, Y. Yamashita et al. (Toshiba) and T. R. Shrout et al. (Penn State) independently reconfirmed that these values are true, and much more improved data were obtained in these several years, aimed at medical acoustic applications [14]. Important data have been accumulated for $Pb(Mg_{1/3}Nb_{2/3})O_3$ [PMN], $Pb(Zn_{1/3}Nb_{2/3})O_3$ [PZN], and binary systems of these materials combined with $PbTiO_3$ (PMN-PT and PZN-PT) for actuator applications. Strains as large as 1.7% can be induced practically for a morphotropic phase boundary composition of the PZN-PT solid-solution single crystals. Figure 9.21 shows the field induced strain curve for [001] oriented 0.92PZN-0.08PT [14]. It is notable that the highest values are observed for a rhombohedral composition only when the single crystal is poled along the perovskite [001] axis, not along the [111] spontaneous polarization axis.

9.2.3.3 Polymers

PVDF or PVF_2 is a piezoelectric when stretched during fabrication. Thin sheets of the cast polymer are drawn and stretched in the plane of the sheet in at least one direction and frequently also

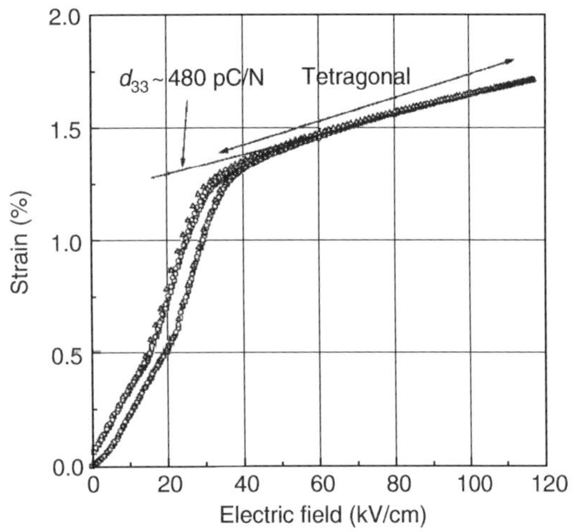

FIGURE 9.21 Field-induced strain curve for [001] oriented 0.92PZN-0.08PT. (Modified from *Encyclopedia of Smart Materials.*)

in the perpendicular direction to convert the material into its microscopically polar phase. Crystallization from a melt forms the nonpolar α-phase, which can be converted into another polar β-phase by uniaxial or biaxial drawing; these dipoles are then reoriented by electric poling. Large sheets can be manufactured and thermally formed into complex shapes. Copolymerization of vinilydene difluoride with trifluoroethylene (TrFE) results in a random copolymer (PVDF-TrFE) that has a stable, polar β phase. This polymer does not need to be stretched; it can be poled directly as formed. The thickness-mode coupling coefficient of 0.30 has been reported. Such piezoelectric polymers are used for directional microphones and ultrasonic hydrophones.

9.2.3.4 Composites

Piezocomposites comprised of piezoelectric ceramics and polymers are promising materials because of excellent tailored properties. The geometry of two-phase composites can be classified according to the connectivity of each phase (0, 1, 2, or 3 dimensionality) into 10 structures; 0-0, 0-1, 0-2, 0-3, 1-1, 1-2, 1-3, 2-2, 2-3, and 3-3 [15]. A 1-3 piezocomposite, or PZT–rod/polymer–matrix composite is a most promising candidate. The advantages of this composite are high coupling factors, low acoustic impedance (square root of the product of its density and elastic stiffness), a good match to water or human tissue, mechanical flexibility, a broad bandwidth in combination with a low mechanical quality factor, and the possibility of making undiced arrays by structuring only the electrodes. The thickness-mode electromechanical coupling of the composite can exceed the k_t (0.40–0.50) of the constituent ceramic and almost approaches the value of the rod-mode electromechanical coupling, k_{33} (0.70–0.80), of that ceramic [16]. The acoustic match to tissue or water (1.5 Mrayls) of typical piezoceramics (20–30 Mrayls) is significantly improved by forming a composite structure, that is, by replacing a heavy, stiff ceramic by a light, soft polymer. Piezoelectric composite materials are especially useful for underwater sonar and medical diagnostic ultrasonic transducers.

9.2.3.5 Thin Films

Both zinc oxide (ZnO) and aluminum nitride (AlN) are simple binary compounds that have Wurtzite-type structures, which can be sputter-deposited in a *c*-axis-oriented thin film on a variety of substrates. ZnO has reasonable piezoelectric coupling and its thin films are widely used in bulk acoustic and SAW devices. The fabrication of highly *c*-axis oriented ZnO films has been extensively studied and developed. The performance of ZnO devices is, however, limited due to their small piezoelectric coupling (20%–30%). PZT thin films are expected to exhibit higher piezoelectric properties. At present, the growth of PZT thin film is being carried out for use in microtransducers and microactuators. A series of theoretical calculations on perovskite-type ferroelectric crystals suggests that large *d* and *k* values of magnitudes similar to those of PZN-PT can also be expected in PZT. Crystal orientation dependence of piezoelectric properties was

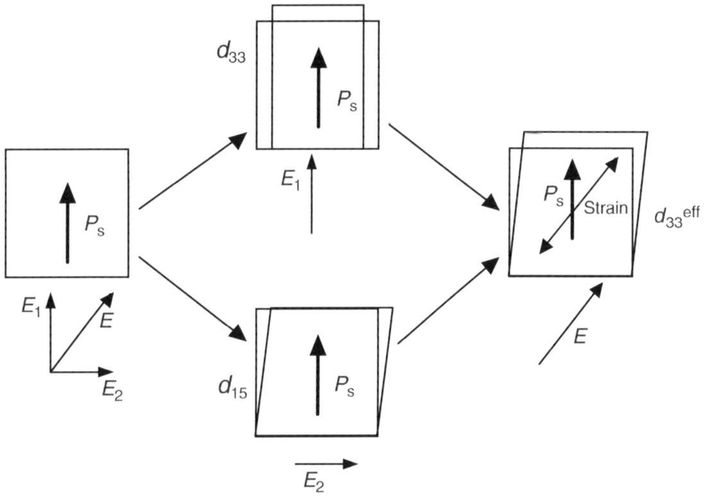

FIGURE 9.22 Principle of the enhancement in electromechanical couplings in a perovskite piezoelectric. (Modified from *Encyclopedia of Smart Materials.*)

phenomenologically calculated for compositions around the morphotropic phase boundary of PZT [17]. The maximum longitudinal piezoelectric constant d_{33} (four to five times the enhancement) and the electromechanical coupling factor k_{33} (more than 90%) in the rhombohedral composition were found at angles of 57° and 51°, respectively, canted from the spontaneous polarization direction [111], which correspond roughly to the perovskite [100] axis.

Figure 9.22 shows the principle of the enhancement in electromechanical couplings. Because the shear coupling d_{15} is the highest in perovskite piezoelectric crystals, the applied field should be canted from the spontaneous polarization direction to obtain the maximum strain. Epitaxially grown, [001] oriented thin/thick films using a rhomboherial PZT composition reportedly enhance the effective piezoelectric constant by—four to five times.

9.2.4 Applications of Piezoelectricity

Piezoelectric materials can provide coupling between electrical and mechanical energy and thus have been extensively used in a variety of electromechanical devices. The direct piezoelectric effect is most obviously used to generate charge or high voltage in applications such as the spark ignition of gas in space heaters, cooking stoves, and cigarette lighters. Using the converse effect, small mechanical displacements and vibrations can be produced in actuators by applying an electric field. Acoustic and ultrasonic vibrations can be generated by an alternating field tuned at the mechanical resonance frequency of a piezoelectric device, and can be detected by amplifying the field generated by vibration incident on the material, which is usually used for ultrasonic transducers. Another important application of piezoelectricity is frequency control. The application of piezoelectric materials ranges over many technology fields, including ultrasonic transducers, actuators and USMs; electronic components such as resonators, wave filters, delay lines, SAW devices, and transformers and high-voltage applications; gas ignitors, ultrasonic cleaning, and machining. Piezoelectric-based sensors, for instance, accelerometers, automobile knock sensors, vibration sensors, strain gages, and flow meters have been developed because pressure and vibration can be directly sensed as electric signals through the piezoelectric effect. Examples of these applications are given in the following sections.

9.2.4.1 Pressure Sensor/Accelerometer/Gyroscope

The gas igniter is one of the basic applications of piezoelectric ceramics. Very high voltage generated in a piezoelectric ceramic under applied mechanical stress can cause sparking and ignite the gas.

Piezoelectric ceramics can be employed as stress sensors and acceleration sensors, because of their "direct piezoelectric effect." Kistler (Switzerland) is manufacturing a 3-D stress sensor. By combining an appropriate number of quartz crystal plates (extensional and shear types), the multilayer device can detect three-dimensional stresses [18].

Figure 9.23 shows a cylindrical gyroscope commercialized by NEC-Tokin (Japan) [19]. The cylinder has six divided electrodes; one pair is used to excite the fundamental bending vibration mode and the other two pairs are used to detect the acceleration. When rotation acceleration is applied around the axis of this gyro, the voltage generated on the electrodes is modulated by the Coriolis force. By subtracting the signals between the two sensor electrode pairs, a voltage directly proportional to the acceleration can be obtained. This type of gyroscope has been widely installed in handheld video cameras to monitor the inevitable hand vibration during operation and to compensate for it electronically on a display by using the sensed signal.

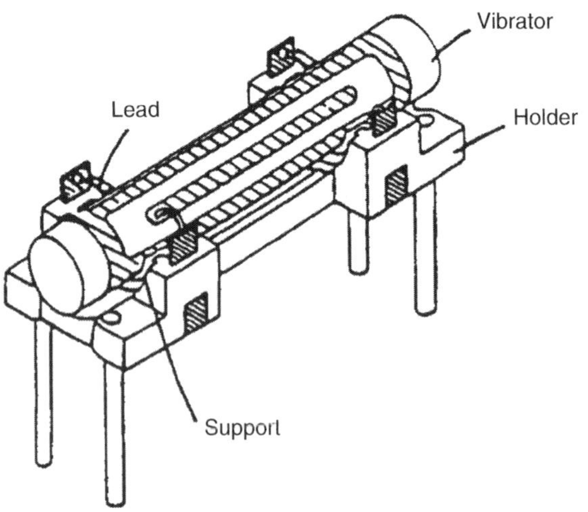

FIGURE 9.23 Cylindrical gyroscope commercialized by NEC-Tokin (Japan). (Modified from *Encyclopedia of Smart Materials.*)

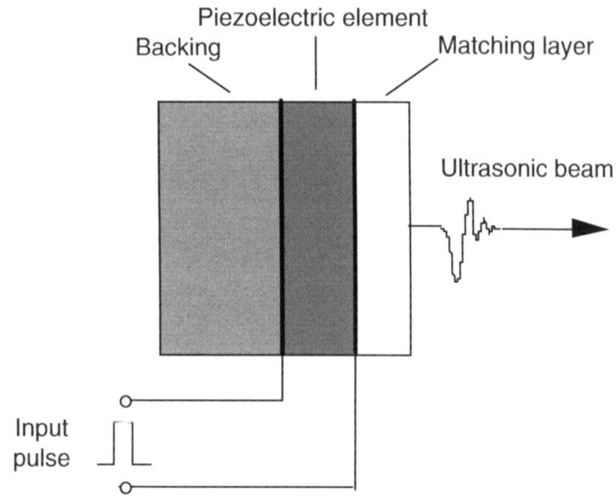

FIGURE 9.24 Geometry of the fundamental transducer for acoustic imaging. (Modified from *Encyclopedia of Smart Materials.*)

9.2.4.2 Ultrasonic Transducer

One of the most important applications of piezoelectric materials is based on ultrasonic echo field [20,21]. Ultrasonic transducers convert electrical energy into a mechanical form when generating an acoustic pulse and convert mechanical energy into an electrical signal when detecting its echo. Nowadays, piezoelectric transducers are being used in medical ultrasound for clinical applications that range from diagnosis to therapy and surgery. They are also used for underwater detection, such as sonars and fish finders, and nondestructive testing.

The ultrasonic transducers often operate in a pulse-echo mode. The transducer converts electrical input into an acoustic wave output. The transmitted waves propagate into the body, and echoes are generated that travel back to be received by the same transducer. These echoes vary in intensity according to the type of tissue or body structure, and thereby create images. An ultrasonic image represents the mechanical properties of the tissue, such as density and elasticity. We can recognize anatomical structures in an ultrasonic image because the organ boundaries and fluid-to-tissue interfaces are easily discerned. The ultrasonic imaging can also be done in real time. This means that we can follow rapidly moving structures such as heart without motional distortion. In addition, ultrasound is one of the safest diagnostic imaging techniques. It does not use ionizing radiation like x-rays and thus is routinely used for fetal and obstetrical imaging. Useful areas for ultrasonic imaging include cardiac structures, the vascular system, the fetus, and abdominal organs such as liver and kidney. In brief, it is possible to see inside the human body by using a beam of ultrasound without breaking the skin.

There are various types of transducers used in ultrasonic imaging. Mechanical sector transducers consist of single, relatively large resonators that provide images by mechanical scanning such as wobbling. Multiple element array transducers permit the imaging systems to access discrete elements individually and enable electronic focusing in the scanning plane at various adjustable penetration depths by using phase delays. The two basic types of array transducers are linear and phased (or sector). Linear array transducers are used for radiological and obstetrical examinations, and phased array transducers are useful for cardiological applications where positioning between the ribs is necessary.

Figure 9.24 shows the geometry of the basic ultrasonic transducer. The transducer is composed mainly of matching, piezoelectric material, and backing layers [22]. One or more matching layers are used to increase sound transmissions into tissues. The backing is attached to the transducer rear to damp the acoustic return wave and to reduce the pulse duration. Piezoelectric materials are used to generate and detect ultrasound. In general, broadband transducers should be used for medical ultrasonic imaging. The broad bandwidth response corresponds to a short pulse length that results in better axial resolution. Three factors are important in designing broad bandwidth transducers. The first is acoustic impedance matching, that is, effectively coupling the acoustic energy to the body. The second is high electromechanical coupling coefficient of the transducer. The third is electrical impedance matching, that is, effectively coupling electrical energy from the driving electronics to the transducer across the frequency range of interest. The operator of pulse-echo transducers is based on the thickness mode resonance of the piezoelectric thin plate. The thickness mode coupling coefficient, k_t, is related to the efficiency of converting electric energy into acoustic and vice versa. Further, a low planar mode coupling coefficient, k_p, is beneficial for limiting energies from being expended in a nonproductive lateral mode. A large dielectric constant is necessary to enable a good electrical impedance match to the system, especially in tiny piezoelectric sizes.

Table 9.5 compares the properties of ultrasonic transducer materials [7,23] Ferroelectric ceramics, such as PZT and modified PT, are almost universally used as ultrasonic transducers. The success of ceramics is due to their very high electromechanical

TABLE 9.5 Comparison of the Properties of Ultrasonic Transducer Materials

	PZT Ceramic	PVDF Polymer	PZT–Polymer Composite	ZnO Film
k_t	0.45–0.55	0.20–0.30	0.60–0.75	0.20–0.30
Z (Mrayls)	20–30	1.5–4	4–20	35
$\varepsilon_{33}^T/\varepsilon_0$	200–5000	10	50–2500	10
tan δ (%)	<1	1.5–5	<1	<1
Q_m	10–1000	5–10	2–50	10
ρ (g/cm^3)	5.5–8	1–2	2–5	3–6

coupling coefficients. In particular, soft PZT ceramics such as PZT-5A and 5H type compositions are most widely used because of their exceedingly high coupling properties and because they can be relatively easily tailored, for instance, in the wide dielectric constant range. On the other hand, modified PTs such as samarium -doped materials have high piezoelectric anisotropy: the planar coupling factor k_p is much less than the thickness coupling factor k_t. Because the absence of lateral coupling leads to reduced interference from spurious lateral resonances in longitudinal oscillators, this is very useful in high-frequency array transducer applications. One disadvantage to PZT and other lead-based ceramics is their large acoustic impedance (approximately 30 kg m^{-2} s^{-1} (Mrayls) compared to body tissue (1.5 Mrayls). Single or multiple matching layers of intermediate impedances need to be used in PZT to improve acoustic matching.

On the other hand, piezoelectric polymers, such as PVDF-trifluoroethylene, have much lower acoustic impedance (4–5 Mrayls) than ceramics and thus match soft tissues better. However, piezopolymers are less sensitive than the ceramics and they have relatively low dielectric constants that require large drive voltage and giving poor noise performance due to mismatching of electrical impedance.

Piezoelectric ceramic/polymer composites are alternatives to ceramics and polymers. Piezocomposites that have 2-2 or 1-3 connectivity are commonly used in ultrasonic medical applications. They combine the low acoustic impedance advantage of polymers with the high sensitivity and low electrical impedance advantages of ceramics.

The design frequency of a transducer depends on the penetration depth required by the application. Resolution is improved as frequency increases. Although a high-frequency transducer can produce a high-resolution image, higher frequency acoustic energy is more readily attenuated by the body. A lower frequency transducer is used as a compromise when imaging deeper structures. Most of medical ultrasound imaging systems operate in the frequency range from 2 to 10 MHz and can resolve objects approximately 0.2–1 mm in size. At 3.5 MHz, imaging to a depth of 10–20 cm is possible, and at 50 MHz, increased losses limit the depth to less than 1 cm. Higher-frequency transducers (10–50 MHz) are used for endoscopic imaging and for catheter-based intravascular imaging. Ultrasonic microscopy is being done at frequencies higher than 100 MHz. The operating frequency of the transducer is directly related to the thickness and velocity of sound in piezoelectric materials employed. As frequency increases, resonator thickness decreases. For a 3.5 MHz transducer, PZT ceramic thickness must be roughly 0.4 mm. Conventional ceramic transducers, such as PZT, are limited to frequencies below 80 MHz because of the difficulty of fabricating thinner devices [24]. Piezoelectric thin-film transducers such as ZnO have to be used for microscopic applications (at frequencies higher than 100 MHz, corresponding to a thickness of less than 20 μm) [25].

9.2.4.3 Resonator and Filter

When a piezoelectric body vibrates at its resonant frequency, it absorbs considerably more energy than at other frequencies, resulting in a fall of the impedance. This phenomenon enables using piezoelectric materials as wave filters. A filter is required to pass a certain selected frequency band or to stop a given band. The band width of a filter fabricated from a piezoelectric material is determined by the square of the coupling coefficient k. Quartz crystals that have very low k value of about 0.1 can pass very narrow frequency bands of approximately 1% of the center resonance frequency. On the other hand, PZT ceramics whose planar coupling coefficient of about 0.5 can easily pass a band of 10% of the center resonance frequency. The sharpness of the passband depends on the mechanical quality factor Q_m of the materials. Quartz also has a very high Q_m of about 10^6, which results in a sharp cutoff of the passband and well-defined frequency of the oscillator.

A simple resonator is a thin disk electroded on its plane faces and vibrating radially for applications in filters whose center frequency ranges from 200 kHz to 1 MHz and whose bandwidth is several percent of the center frequency. The disk diameter must be about 5.6 mm for a frequency of 455 kHz. However, if the required frequency is higher than 10 MHz, other modes of vibration such as the thickness extensional mode are exploited, because of its smaller size disk. Trapped-energy type filters made from PZT ceramics have been widely used in the intermediate frequency (IF) range, for example, 10.7 MHz for FM radio receivers and transmitters. By employing the trapped-energy phenomenon, the overtone frequencies are suppressed. The plate is partly covered with electrodes of a specific area and thickness. The fundamental frequency of the thickness mode beneath the electrode is less than that of the unelectroded portion because of the extra inertia of the electrode mass. The longer wave characteristic of the electrode region cannot propagate in the unelectroded region. The higher-frequency overtones can propagate into the unelectroded region. This is called the trapped-energy principle. Figure 9.25 shows a schematic drawing of a trapped-energy filter. In this structure, the top electrode is split so that coupling between the two parts is efficient only at resonance. More stable filters suitable for telecommunication systems have been made from single crystals such as quartz or $LiTaO_3$.

9.2.4.4 Piezoelectric Transformer

The transfer of vibration energy from one set of electrodes to another on a piezoelectric ceramic body can be used to transform voltage. The device is called a piezoelectric transformer. Recently,

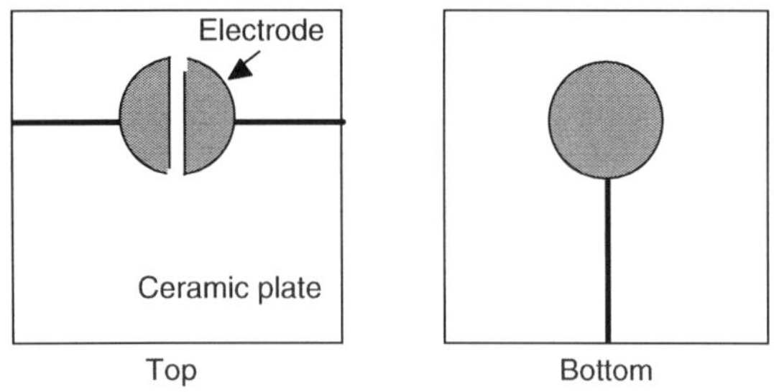

FIGURE 9.25 Trapped-energy filter. (Modified from *Encyclopedia of Smart Materials.*)

office automation equipment that has a liquid crystal display such as notebook-type personal computers and car navigation systems has been successfully commercialized. This equipment that uses a liquid crystal display requires a very thin transformer without electromagnetic noise to start the glow of a fluorescent backlamp. This application has recently accelerated the development of the piezoelectric transformers. Figure 9.26 shows the basic structure, where two differently poled parts coexist in one piezoelectric plate. The plate has electrodes on half of its major faces and on an edge. The plate is then poled in its thickness direction at one end and parallel to the long axis over most of its length. A low-voltage AC supply is applied to the large-area electrodes at a frequency that excites a length extensional mode resonance. Then, a high-voltage output can be taken from the small electrode and from one of the larger electrodes. Following the proposal by Rosen mentioned before, piezoelectric transformers of several different structures have been reported [26]. A multilayer type transformers are proposed to increase the voltage step-up ratio [27]. The input part has a multilayer structure and has internal electrodes, and the output electrodes are formed at the side surface of the half of the rectangular plate. This transformer uses the piezoelectric longitudinal mode for the input and output parts.

9.2.4.5 Saw Device

SAW, also called a Rayleigh wave, is composed of a coupling between longitudinal and shear waves in which the SAW energy is confined near the surface. An associated electrostatic wave exists for a SAW on a piezoelectric substrate that allows electroacoustic coupling via a transducer. The advantages of SAW technology are that a wave can be electroacoustically accessed and trapped at the substrate surface and its velocity is approximately 10^4 times slower than an electromagnetic wave. The SAW wavelength is of the same order of magnitude as line dimensions that can be photolithographically produced, and the lengths for both short and long delays are achievable on reasonable size substrates [28,29].

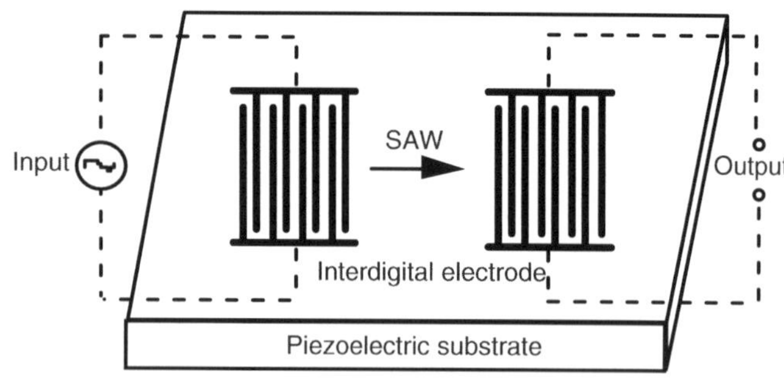

FIGURE 9.27 Typical SAW bidirectional filter that consists of two interdigital transducers. (Modified from *Encyclopedia of Smart Materials.*)

There is a very broad range of commercial system applications, including front-end and IF filters, community antenna television (CATV), and VCR components, synthesizers, analyzers, and navigators. In SAW transducers, finger electrodes provide the ability to sample or tap the wave, and the electrode gap gives the relative delay. A SAW filter is composed of a minimum of two transducers. A schematic of a simple SAW bidirectional filter is shown in Figure 9.27. A bidirectional transducer radiates energy equally from each side of the transducer. Energy not received is absorbed to eliminate spurious reflection.

Various materials are currently being used for SAW devices. The most popular single-crystal SAW materials are lithium niobate and lithium tantalate. The materials have different properties depending on their cuts and the direction of propagation. The fundamental parameters are the SAW velocity, the temperature coefficients of delay (TCD), the electromechanical coupling factor, and the propagation loss. SAWs can be generated and detected by spatially periodic, interdigital electrodes on the plane surface of a piezoelectric plate. A periodic electric field is produced when an RF source is connected to the electrode, thus permitting piezoelectric coupling to a traveling surface wave. If an RF source of a frequency f is applied to an electrode whose periodicity is p, energy conversion from an electrical to mechanical form will be maximum when

$$f = f_0 = \frac{V_s}{p}, \tag{9.20}$$

where

V_s is the SAW velocity

f_0 is the center frequency of the device

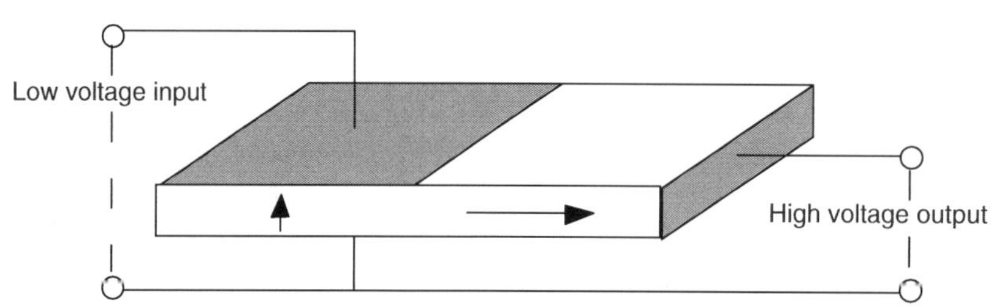

FIGURE 9.26 Rosen-type piezoelectric transformer. (Modified from *Encyclopedia of Smart Materials.*)

SAW velocity is an important parameter that determines the center frequency. Another important parameter for many applications is the temperature sensitivity. For example, the temperature stability of the center frequency of SAW bandpass filters is a direct function of temperature coefficient for the velocity and delay time of the material used. The first-order TCD time is given by

$$\left(\frac{1}{t}\right)\left(\frac{dt}{dT}\right)=\left(\frac{1}{L}\right)\left(\frac{dL}{dT}\right)-\left(\frac{1}{V_s}\right)\left(\frac{dV_s}{dT}\right), \quad (9.21)$$

where
$t = L/V_s$ is the delay time
L is the SAW propagation length

The surface wave coupling factor, k_s^2, is defined in terms of the change in SAW velocity that occurs when the wave passes across a surface coated by a thin massless conductor, so that the piezoelectric field associated with the wave is effectively shorted-circuited. The coupling factor, k_s^2, is expressed by

$$k_s^2 = 2\frac{(V_f - V_m)}{V_f}, \quad (9.22)$$

where
V_f is the free surface wave velocity
V_m is the velocity on the metallized surface

In SAW applications, the value of k_s^2 relates to the maximum bandwidth obtainable and the amount of signal loss between input and output that determines the fractional bandwidth versus minimum insertion loss for a given material and a filter. Propagation loss, one of the major factors that determines the insertion loss of a device, is caused by wave scattering by crystalline defects and surface irregularities. Materials that have high electromechanical coupling factors combined with small TCD time are likely to be required. The free surface velocity, V_0, of the material is a function of cut angle and propagative direction. The TCD is an indication of the frequency shift expected from a transducer due to a temperature change and is also a function of the cut angle and the propagation direction. The substrate is chosen on the basis of the device's design specifications for operating temperature, fractional bandwidth, and insertion loss.

Table 9.6 shows some important material parameters of representative SAW materials. Piezoelectric single crystals such as 128°*Y*–*X* (128°-rotated-*Y*-cut and *X*-propagation)—$LiNbO_3$ and *X*-112°Y (*X*-cut and 112°-rotated-*Y*-propagation)—$LiTaO_3$ have been extensively employed as SAW substrates for VIF filters. ZnO thin films, *c*-axis oriented and deposited on a fused quartz, glass, or sapphire substrate, have also been commercialized for SAW devices.

9.2.4.6 Actuators

Currently another important application of piezoelectric materials exists in the actuator field [30]. Using the converse piezoelectric effect, a small displacement can be produced by applying an electric field to a piezoelectric material. Vibrations can be generated by applying an alternating electric field. There is a demand in advanced precision engineering for a variety of types of actuators that can adjust position precisely (micropositioning devices), suppress noise vibrations (dampers), and drive objects dynamically (USMs). These devices are used in areas, including optics, astronomy, fluid control, and precision machinery. Piezoelectric strain and electrostriction induced by an electric field are used for actuator applications.

Figure 9.28 shows the design classification of ceramic actuators. Simple devices composed of a disk or a multilayer type use the strain

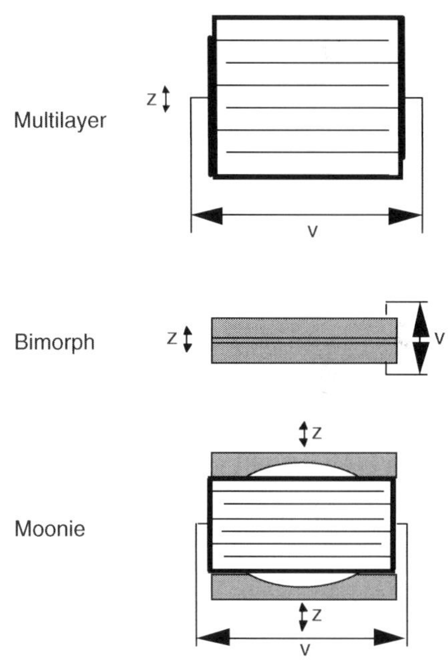

FIGURE 9.28 Structures of ceramic actuators. (Modified from *Encyclopedia of Smart Materials.*)

TABLE 9.6 Material Parameters for Representative SAW Materials

	Material	Cut–Propagation Direction	k^2 (%)	TCD (ppm/C)	V_0 (m/s)	ε_r
Single crystal	Quartz	ST—X	0.16	0	3158	4.5
	$LiNbO_3$	128°Y—X	5.5	−74	3960	35
	$LiTaO_3$	X112°—Y	0.75	−18	3290	42
	$Li_2B_4O_7$	(110) —<001>	0.8	0	3467	9.5
Ceramic	PZT-In($Li_{3/5}W_{2/5}$)O_3		1.0	10	2270	690
	(Pb,Nd)(Ti, Mn, In)O_3		2.6	<1	2554	225
Thin film	ZnO/glass		0.64	−15	3150	8.5
	ZnO/Sapphire		1.0	−30	5000	8.5

induced in a ceramic by the applied electric field directly. Complex devices do not use the induced strain directly but use the amplified displacement through a special magnification mechanism such as unimorph, bimorph, and moonie. The most popularly used multilayer and bimorph types have the following characteristics: The multilayer type does not have a large displacement (10 μm), but has advantages in generation force (1 kN), response speed (10 μs), lifetime (10^{11} cycles), and the electromechanical coupling factor k_{33} (0.70). The bimorph type has a large displacement (300 μm), but has disadvantages in generation force (1 N), response speed (1 ms), lifetime (10^8 cycles), and the electromechanical coupling factor k_{eff} (0.10). For instance, in a 0.65 PMN–0.35 PT multilayer actuator with 99 layers of 100 μm thick sheets ($2 \times 3 \times 10\,mm^3$), a 8.7 μm displacement is generated by a 100 V voltage, accompanied by a slight hysteresis. The transmitted response of the induced displacement after the application of a rectangular voltage is as quick as 10 μs. The multilayer has a field-induced strain of 0.1% along the length [30].

Unimorph and bimorph devices are defined by the number of piezoelectric ceramic plates: only one ceramic plate is bonded onto an elastic shim, or two ceramic plates are bonded together. The bimorph causes bending deformation because each piezoelectric plate bonded together produces extension or contraction in an electric field. In general, there are two types of piezoelectric bimorphs: the antiparallel polarization type and the parallel polarization type, as shown in Figure 9.29. Two poled piezoelectric plates $t/2$ thick and L long are bonded so that their polarization directions opposite or parallel to each other. In the cantilever bimorph configuration where one end is clamped, the tip displacement δ_z under an applied voltage V is

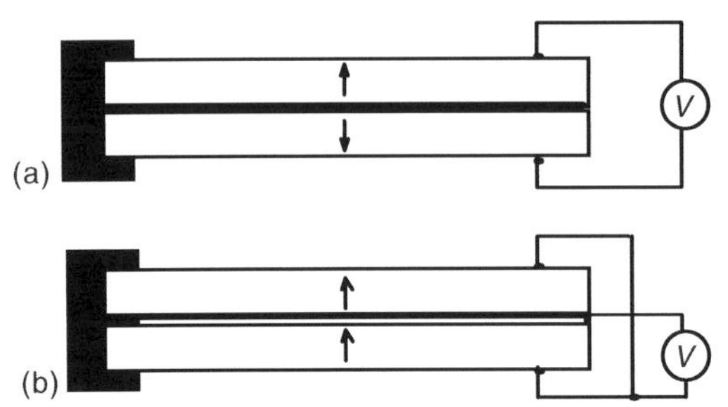

FIGURE 9.29 Two types of piezoelectric bimorphs: (a) antiparallel polarization type and (b) parallel polarization type. (Modified from *Encyclopedia of Smart Materials*.)

$$\delta_z = (3/2)d_{31}(L^2/t^2)\ V \text{ (antiparallel type)} \tag{9.23}$$

$$\delta_z = 3d_{31}\ (L^2/t^2)\ V \text{ (parallel type).} \tag{9.24}$$

The resonance frequency f_r for both types is given by

$$f_r = 0.16\ t/L^2(\rho s_{11}^{E})^{-1/2}, \tag{9.25}$$

where

ρ is the density

s_{11}^{E} is the elastic compliance

A metallic sheet (called a shim) is occasionally sandwiched between the two piezoelectric plates to increase the reliability; the structure can be maintained even if the ceramics fracture. Using the bimorph structure, a large magnification of the displacement is easily obtainable. However, the disadvantages include a low response speed (1 kHz) and low generative force [30].

A composite actuator structure called "moonie" has been developed to amplify the small displacements induced in piezoelectric ceramics [31]. The moonie consists of a thin single or multilayer element and two metal plates that have narrow moon-shaped cavities bonded together. This device has characteristics intermediate between the conventional multilayer and bimorph actuators; it has an order of magnitude larger displacement (100 μm) than the multilayer, and much larger generative force (100 N) and quicker response (100 μs) than the bimorph.

Some examples of applications of piezoelectric and electrostrictive actuators are described here. The piezoelectric impact dot-matrix printer is the first mass-produced device that uses multilayer ceramic actuators (Figure 9.30) [32]. The advantage of a piezoelectric printer head compared to conventional magnetic type are low energy consumption, low heat generation, and fast printing speed. Longitudinal multilayer actuators do not have a

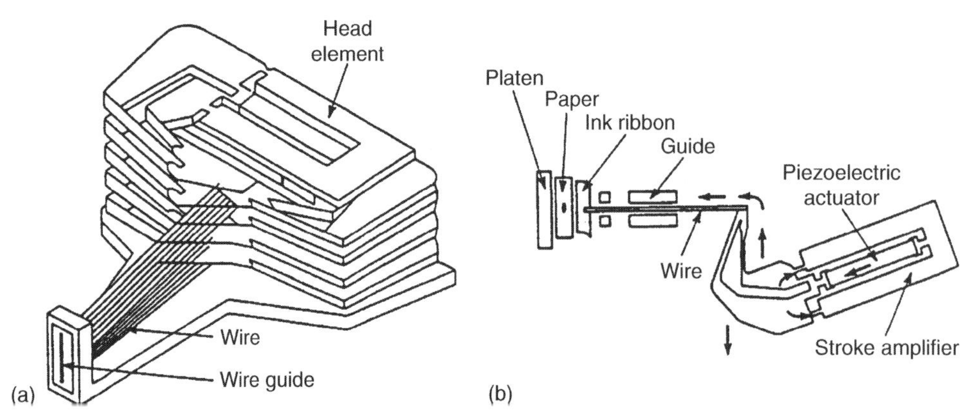

FIGURE 9.30 Impact dot matrix printer head commercialized by NEC (Japan). (Modified from *Encyclopedia of Smart Materials*.)

large displacement and thus a suitable displacement magnification mechanism is necessary. The displacement induced in a multilayer actuator pushes up the force point and its displacement magnification is carried out through hinge levers to generate a large wire stroke. When the displacement in the piezoactuator is 8 μm, the wire stroke of 240 μm can be obtained; the magnification rate is 30 times.

Bimorph structures are commonly used for VCR head tracking actuators because of their large displacements. An auto tracking scan system uses the piezoelectric actuators so that the head follows the recording track even driven in both still and quick modes [33]. As can be anticipated, the bimorph drive is inevitably accompanied by a rotational motion. A special mechanism has to be employed to obtain perfectly linear translational motion. Piezoelectric pumps for gas or liquid that use the alternating bending motion of the bimorph have been developed for intravenous drip injection in hospitals and for medication dispensing for chemotherapy, chronic pain, and diabetes [34]. Piezoelectric fans for cooling electronic circuits are made from a pair of bimorphs that are driven out of phase so as to blow effectively [35]. Furthermore, piezobimorph type camera shutters have been widely commercialized by Minolta [36].

Lenses and mirrors in optical control systems require micropositioning and even the shapes of mirrors are adjusted to correct image distortions. For instance, a space-qualified active mirror called articulating fold mirror utilizes six PMN electrostrictive multilayer actuators to position and tilt a mirror tip precisely to correct the focusing aberration of the Hubble space telescope [37].

Piezoelectric actuators are also useful for vibration suppression systems of an automobile. An electronic controlled shock absorber was developed by Toyota [38]. Piezoelectric sensors that detect road roughness are composed of five layers of 0.5 mm thick PZT disks. The actuator is made of 88 layers of 0.5 mm thick disks. Under 500 V, it generates about 50 μm displacement, which is magnified by 40 times by a piston and plunger pin combination. This stroke pushes the change valve of the damping force down and then opens the bypass oil route, leading to decrease in flow resistance. This electronically controlled shock absorber has both controllability and provides comfort simultaneously.

The U.S. Army is interested in developing a rotor control system for helicopters, because a slight change in the blade angle drastically enhances controllability. Figure 9.31 shows a bearingless rotor flexbeam that has piezoelectric strips attached [39]. Various types of PZT-sandwiched beam structures have been investigated for such a flexbeam application and for active vibration control.

In order to increase the diesel engine efficiency, high-pressure fuel and quick injection control are required. In one engine cycle (typically 60 Hz), the multiple injections should be realized in a very sharp shape for the diesel engine. For this purpose, piezoelectric actuators, specifically multilayer types, are adopted. The highest reliability of these devices at an elevated temperature (150°C) for a long period (10 years) has been achieved. The so-called common-rail type injection valves have been widely commercialized by Siemens, Bosch, and Denso Corp. (Figure 9.32) [40].

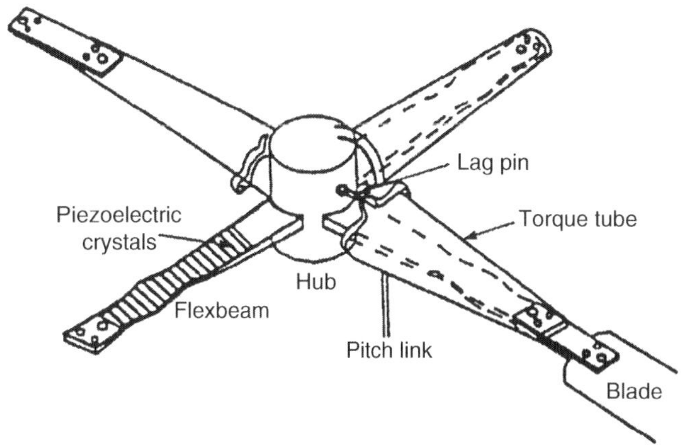

FIGURE 9.31 Bearingless rotor flexbeam with attached piezoelectric strips. (Modified from *Encyclopedia of Smart Materials.*)

9.2.4.7 USMs

An USM is an example of piezoelectric actuator that uses resonant vibration. Linear motion in USMs is obtained by frictional force from the elliptical vibration. The motor consists of a high-frequency power supply, a vibrator, and a slider. The vibrator is composed of a piezoelectric driving component and an elastic vibratory part, and the slider is composed of an elastic moving part and a friction coat. The characteristics of USMs are low speed and high torque compared to the high speed and low torque of conventional electromagnetic motors [41,42].

USMs are classified into two types: the standing-wave type and the propagating-wave type. The displacement of a standing wave is expressed by

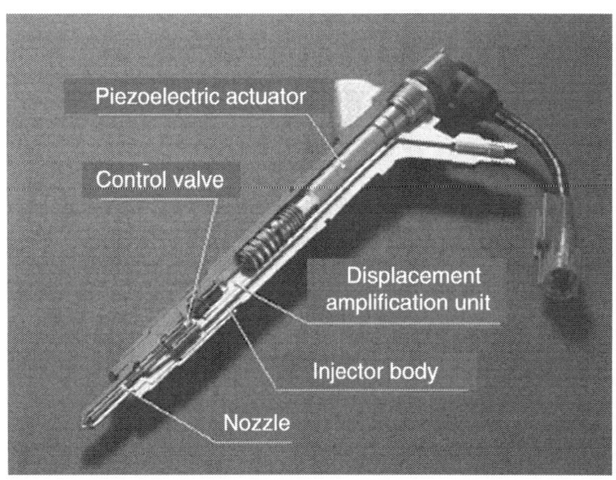

FIGURE 9.32 Common rail type diesel injection valve with a piezoelectric multilayer actuator. (Courtesy of Denso Corp.) (Modified from *Encyclopedia of Smart Materials.*)

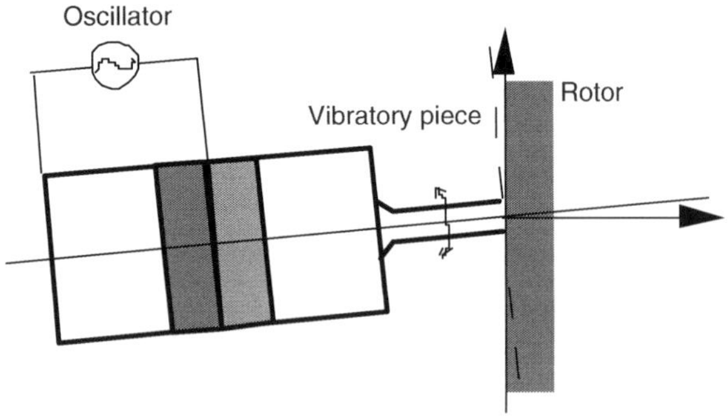

FIGURE 9.33 Vibratory coupler type USM. (Modified from *Encyclopedia of Smart Materials.*)

$$x_s(x,t) = A\cos(kx)\cos(\omega t), \tag{9.26}$$

for a propagative wave displacement is given by

$$\begin{aligned} x_p(x,t) &= A\cos(kx - \omega t) \\ &= A\cos(kx)\cos(\omega t) \\ &\quad + A\cos(kx - \pi/2)\cos(\omega x - \pi/2). \end{aligned} \tag{9.27}$$

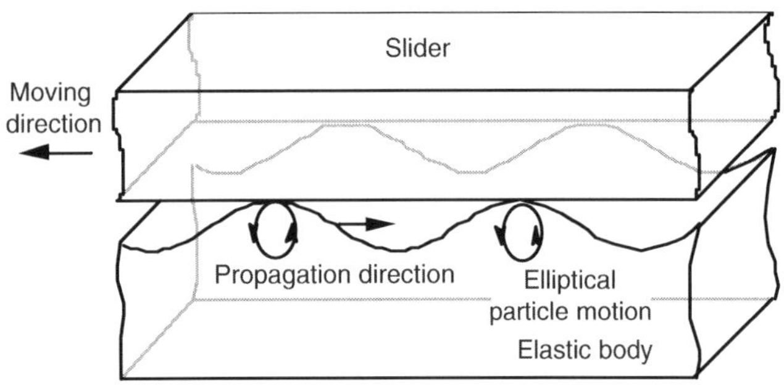

FIGURE 9.34 Principle of the propagating wave type USM. (Modified from *Encyclopedia of Smart Materials.*)

A propagating wave can be generated by superimposing two standing waves whose phases differ from each other by 90° in both time and space. The standing-wave type is sometimes called a vibratory coupler or a "woodpecker" type; a vibratory piece is connected to a piezoelectric driver, and the tip portion generates flat-elliptic movement (Figure 9.33). The vibratory piece is attached to a rotor or a slider at a slight cant angle. The standing-wave type has high efficiency and up to 98% of the theoretical value. However, a problem of this type is lack of control in both clockwise and counterclockwise directions. The principle of the propagation type is shown in Figure 9.34. In the propagating-wave type, also called the "surfing-type," a surface particle of the elastic body draws an elliptical locus due to coupling of the longitudinal and transverse waves. This type generally requires two vibrational sources to generate one propagating wave; this leads to low efficiency (not more than 50%), but is controllable in both the rotational directions. An ultrasonic rotary motor is successfully used in autofocusing camera to produce precise rotational displacements. The advantages of this motor over the conventional electromagnetic motor are silent drive (inaudible), thin motor design, and energy savings.

Tiny conventional electromagnetic motors, smaller than 1 cm, that have sufficient energy efficiency are rather difficult to produce. Therefore, the USM is gaining widespread attention. USMs whose efficiency is independent of size are superior in the minimotor area. The Penn State University and Samsung Electromechanics developed a zoom mechanism for a cellular phone camera with two microrotary motors [43–45]. A micro motor called "metal tube type" consisting of a metal hollow cylinder and two PZT rectangular plates is used as basic microactuators (see Figure 9.35a). When we drive one of the PZT plates, Plate X, a bending vibration is excited basically along x-axis. However, because of an asymmetrical mass (Plate Y), another hybridized bending mode is excited with some phase lag along y-axis, leading to an elliptical locus in a clockwise direction, like a "Hula-Hoop" motion. The rotor of this motor is a cylindrical rod with a pair of stainless ferrule pressed with a spring. The assembly is shown in Figure 9.35b. The metal

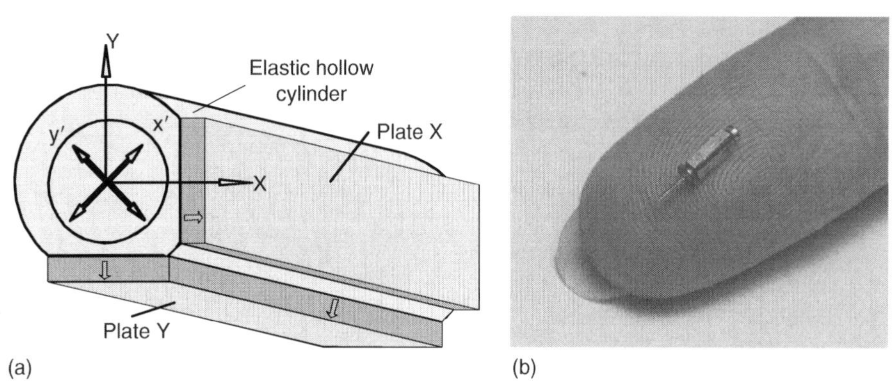

FIGURE 9.35 "Metal tube" motor that uses a metal tube and two rectangular PZT plates. (a) Schematic structure and (b) photo of a world smallest motor (1.5 mmϕ). (Modified from *Encyclopedia of Smart Materials.*)

cylinder motor 2.4 mm in diameter and 12 mm in length was driven at 62.1 kHz in both rotation directions. A no-load speed of 1800 rpm and an output torque up to 1.8 mN m are obtained for rotation in both directions under an applied voltage of 80 V. Rather high maximum efficiency of about 28% for this small motor is a noteworthy feature. The smallest camera in the world adopted two micro-USMs with 2.4 mm diameter and 14 mm length to control zooming and focusing lenses independently in conjunction with screw mechanisms.

References

1. B. Jaffe, W. Cook, and H. Jaffe, *Piezoelectric Ceramics*, London: Academic Press, 1971.
2. W. G. Cady, *Piezoelectricity*, New York: McGraw-Hill, revised edition by Dover, 1964.
3. F. Jona and G. Shirane, *Ferroelectric Crystals*, London: Pergamon Press, 1962.
4. M. E. Lines and A. M. Glass, *Principles and Applications of Ferroelectric Materials*, Oxford: Clarendon Press, 1977.
5. IEEE Standard on Piezoelectricity, New York: IEEE, Inc., 1978.
6. Landold and Boernstein, Numerical data and functional relationships. *Science and Technology: Crystal and Solid State Physics*, Vol.11, Berlin: Springer-Verlag, 1979.
7. W. A. Smith, *Proc. SPIE—The International Society for Optical Engineering*, 1733, 1992.
8. H. Takeuchi, S. Jyomura, E. Yamamoto, and Y. Ito, *J. Acoust. Soc. Am.*, 74, 1114, 1982.
9. Y. Yamashita, K. Yokoyama, H. Honda, and T. Takahashi, *Jpn. J. Appl. Phys.*, 20 (Suppl. 20-4), 183, 1981.
10. Y. Ito, H. Takeuchi, S. Jyomura, K. Nagatsuma, and S. Ashida, *Appl. Phys. Lett.*, 35, 595, 1979.
11. L. E. Cross, S. J. Jang, R. E. Newnham, S. Nomura, and K. Uchino, *Ferroelectrics*, 23, 187, 1980.
12. H. Takeuchi, H. Masuzawa, C. Nakaya, and Y. Ito, Proc. IEEE 1990 Ultrasonics Symposium, 697, 1990.
13. J. Kuwata, K. Uchino, and S. Nomura, *Jpn. J. Appl. Phys.*, 21, 1298, 1982.
14. T. R. Shrout, Z. P. Chang, N. Kim, and S. Markgraf, *Ferroelec. Lett.*, 12, 63, 1990.
15. R. E. Newnham, D. P. Skinner, and L. E. Cross, *Mater. Res. Bull.*, 13, 525, 1978.
16. W. A. Smith, Proc. 1989 IEEE Ultrasonic Symposium, 755, 1989.
17. X. H. Du, J. Zheng, U. Belegundu, and K. Uchino, *J. Appl. Phys. Lett.*, 72, 2421, 1998.
18. Kistler, Stress Sensor, Production Catalog, Switzerland.
19. Tokin, Gyroscope, Production Catalog, Japan.
20. B. A. Auld, *Acoustic Fields and Waves in Solids*, 2nd edn., Melbourne: Robert E. Krieger, 1990.
21. G. S. Kino, *Acoustic Waves: Device Imaging and Analog Signal Processing*, Englewood Cliffs, NJ: Prentice-Hall, 1987.
22. C. S. Desilets, J. D. Fraser, and G. S. Kino, *IEEE Trans. Sonics Ultrason.*, SU-25, 115, 1978.
23. T. R. Gururaja, *Am. Ceram. Soc. Bull.*, 73, 50, 1994.
24. F. S. Foster, L. K. Ryan, and D. H. Turnbull, *IEEE Trans. Ultrason. Ferroelec. Freq. Cont.*, 38, 446, 1991.
25. Y. Ito, K. Kushida, K. Sugawara, and H. Takeuchi, *IEEE Trans. Ultrason. Ferroelec. Freq. Cont.*, 42, 316, 1995.
26. C. A. Rosen, Proc. Electronic Component Symp., p. 205, 1957.
27. S. Kawashima, O. Ohnishi, H. Hakamata, S. Tagami, A. Fukuoka, T. Inoue, and S. Hirose, Proc. IEEE Int'l Ustrasonic Symp. '94, France, Nov., 1994.
28. C. Campbell, *Surface Acoustic Wave Devices and Their Signal Processing Applications*, San Diego, CA: Academic Press, 1989.
29. H. Matthews, *Surface Wave Filters*, New York: Wiley Interscience, 1977.
30. K. Uchino and J. R. Giniewicz, *Micromechatronics*, New York: Marcel Dekker, 2003.
31. Y. Sugawara, K. Onitsuka, S. Yoshikawa, Q. C. Xu, R. E. Newnham, and K. Uchino, *J. Am. Ceram. Soc.*, 75, 996, 1992.
32. K. Yano, T. Inoue, S. Takahashi, and I. Fukui, *Proc. Jpn. Electr. Commun. Soc.*, pp. 1–159, Spring, 1984.
33. A. Ohgoshi and S. Nishigaki, *Ceramic Data Book '81*, Tokyo: Institute of Industrial Manufacturing Techology, p. 35, 1981.
34. T. Narasaki, Japan Patent Disclosure, S. 57-137671, 1978.
35. M. Yorinaga, D. Makino, K. Kawaguchi, and M. Naito, *Jpn. J. Appl. Phys.*, 24, Suppl. 24-3, 203, 1985.
36. Minolta Camera, Product Catalog "Mac Dual I, II".
37. J. L. Fanson and M. A. Ealey, Active and Adaptive Optical Components and Systems II, SPIE 1920, Albuquerque, 1993.
38. Y. Yokoya, *Electron. Ceramics*, 22(111), 55, 1991.
39. F. K. Straub, *Smart Mater. Struct.*, 5, 1, 1996.
40. A. Fujii, Proc. JTTAS Meeting on Dec. 2, Tokyo, 2005.
41. S. Ueha and Y. Tomikawa, *Ultrasonic Motors*, Oxford: Clarendon Press, 1993.
42. K. Uchino, *Piezoelectric Actuators and Ultrasonic Motors*, Boston, MA: Kluwer Academic, 1996.
43. B. Koc, S. Cagatay, and K. Uchino, *IEEE Ultrasonic Ferroelec. Freq. Control Trans.*, 49(4), 495–500, 2002.
44. S. Cagatay, B. Koc, and K. Uchino, IEEE Trans. UFFC, 50(7), 782–786, 2003.
45. K. Uchino, Proc. New Actuator 2004, Bremen, Germany. June 14–16, p. 127, 2004.

9.3 Noncontact Ultrasonic Testing and Analysis of Materials

Mahesh C. Bhardwaj

9.3.1 Introduction

Noncontact (air–gas coupled) uses of sound waves are perhaps buried in the antiquities of our civilization. In modern times, presumably one of the first applications is related to Antarctic's ice thickness measurement by sending high-intensity sound waves from an airplane in the 1920s. But close to reality, noncontact ultrasound (NCU) is highly desirable for testing and analyzing early stage materials formation (green ceramics, powder metals, elastomers, composite prepregs, etc.), and for contact and liquid-sensitive materials (porous materials, food, pharmaceutical and hygroscopic materials, burn and wounded victims, or where contact with the test media is simply a nuisance). Materials such as these cannot be reliably tested by conventional ultrasound, which utilizes liquid coupling of the transducer to the test material. Liquid coupling is necessary in order to efficiently transmit ultrasound in the material as explained by Brunk [1] in this book.

However, there are realities that defy NCU, which are: extremely high absorption of ultrasound by air or other gases, phenomenally high acoustic impedance (*Z*) mismatch between air and the test media, and inefficient ultrasound transduction from a piezoelectric material into the rarified media like air or other gases (Figure 9.36). The first two are natural phenomena about which nothing can be done. Thus we are left to do "something" to the piezoelectric material. To this effect, a handful of researchers have been busy developing transducer designs that are based upon acoustic impedance, *Z* matching to air by utilizing a variety of polymers as the final matching layer on the piezoelectric material, Fox et al. [2], Bhardwaj [3], Yano et al. [4], Haller and Khouri-Yakub [5]. These transducers exhibit reasonable efficiency, but only for limited NCU applications, and that too at relatively low frequencies such as <500 kHz. Further, in order to investigate attenuative materials such as concrete, very thick multilayered composites and sandwich structures, rocks, wood, lumber, etc. polymer *Z*-matched transducers have to be excited with abnormally high powers, thus risking transducer failure besides causing other problems.

In 1997, Bhardwaj after decades of attempts finally succeeded in producing a transducer that is characterized by phenomenally high transduction in air [6] from 50 kHz to more than 5 MHz frequencies (Figure 9.37). This transducer is *Z*-matched with compressed fiber material the acoustic impedance of which is extremely close (arguably the closest) to that of air, which is primarily responsible for extraordinary transduction in air and other gases. Simply known as NCU transducers, their efficiency relative to polymer *Z*-matched transducers is order of magnitude more. Further, in ambient air, the sensitivity of the NCU transducers is 14 to 30 dB lesser (depending on the frequency) than their equivalent contact and immersion counterparts, when tested with appropriate solids or with water.

Considering the magnitude of NCU applications made possible by these transducers, we believe a new field of significance in materials and biomedical diagnostics has now become a reality. Details of NCU transducers, acoustic measurements, and applications have been provided by Bhardwaj in the *Encyclopedia of Smart Materials* [7]. In this chapter, our focus is on providing the reader with introduction to NCU transducers, systems, and applications for a wide range of materials by utilizing the well-known ultrasonic techniques.

9.3.2 NCU Transducers

It should be pointed out that early NCU transducers that are based upon conventional solid and polymer matrix piezoelectric

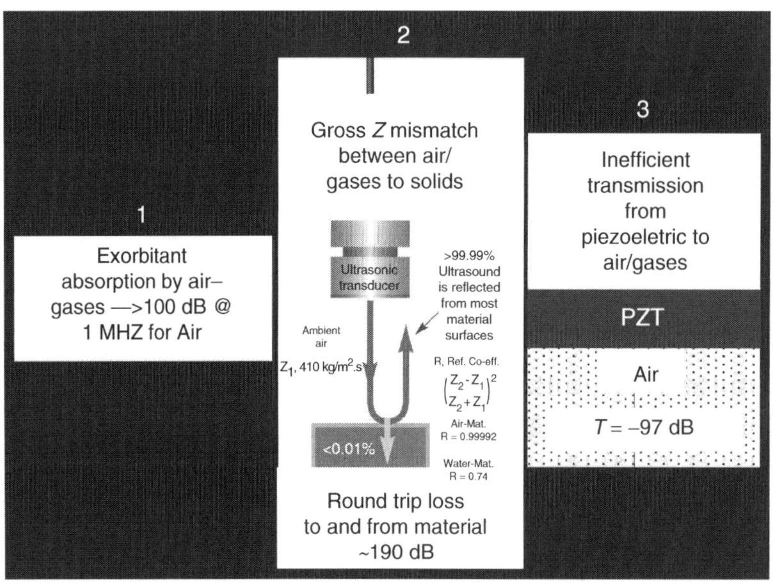

FIGURE 9.36 Realities that defy NCU. Ultrasound absorption by air–gases and massive acoustic impedance mismatch between air and test media are natural phenomena about which nothing can be done. Therefore, piezoelectric material must be treated in a manner that it exudes enormous acoustic pressure in air.

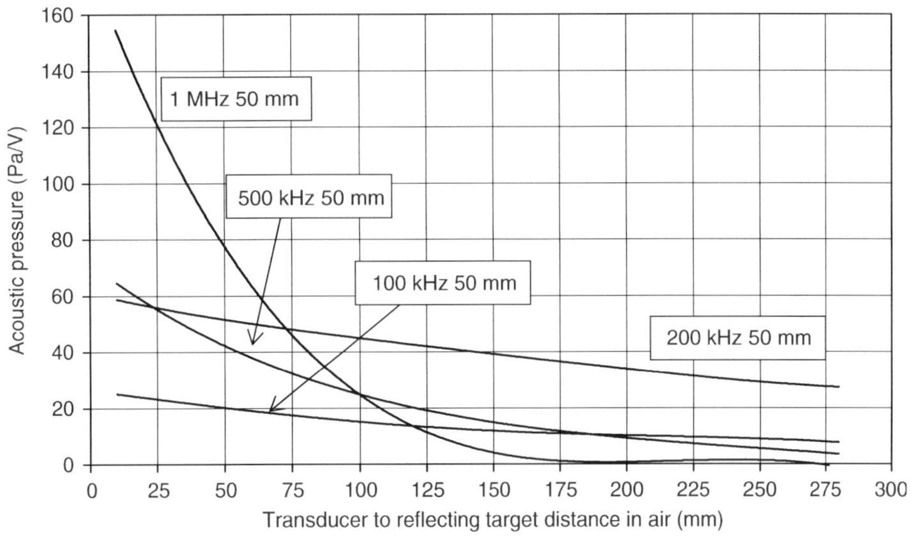

FIGURE 9.37 Acoustic pressure (Pa/V) of several NCU transducers as a function of reflecting target distance from the transducer in ambient air.

materials allowed relatively easy production from 1 to 5 MHz frequencies. Dissatisfied with the performance and signal-to-noise ratio (SNR) of NCU transducers between 50 and 500 kHz based on conventional piezoelectric materials, forced us to think even more unconventionally. In 2002, this resulted in the development of a new piezoelectric composite. Known as gas matrix piezoelectric (GMP) composite [8], it is characterized by extremely unusual features. For example, the thickness coupling constant of this material is nearly equal to longitudinal coupling constant. GMP is also characterized by zero acoustic cross-talk, very large displacement in the coupling direction, negative Poisson's ratio, very low mechanical Q, low dielectric constant, low density, and many other useful properties. Further, the manufacturability of GMP allows the production of extremely large transducers, hitherto considered arduous or impossible. GMP is particularly suitable between <50 and >500 kHz. This material has not only enhanced the performance of NCU transducers, but has also elevated those contact and immersion applications that require low ultrasound frequencies.

NCU transducers have been successfully produced from <50 kHz to >5 MHz and in the dimensional range from <1 mm to >250 mm in planar, point and cylindrical focused configurations. For high-throughput applications besides the large transmitters, multielement arrays have also be produced and deployed in the industry. Table 9.7 provides the salient features of NCU transducers and Figure 9.38 shows a number of these devices varying in frequency and dimensions.

TABLE 9.7 Salient Characteristics and Features of NCU Transducers

Characteristics/Features	Description
Frequency range	<50 KHz to >5 MHz
Dimensional range	<1 mm to >250 mm (extremely large dimensions also possible)
Shape	Circular, square, or rectangular
Field geometry	Planar, point, and cylindrical focused
Sensitivity	Extremely high in ambient air—typically 14–30 dB below conventional contact and immersion transducers
Bandwidth	Typically between 30% and 50% at bandwidth center frequency
Housing	Two-part aluminum protected with plastic cover
Coaxial connector	Standard BNC
Mechanical construction	Robust, factor suitable
Environment limitations	Temperature: −20 to 60°C. RH: Up to 80%, higher with special provisions

FIGURE 9.38 Wide range of NCU transducers including large transmitters and multielement arrays varying in frequency from 50 kHz to 5 MHz.

9.3.3 NCU System and Signal Processing

NCU transducers can be used with conventional pulsers and receivers (designed for contact and immersion ultrasonic testing), but only for limited applications. However, for wide range of materials and testing objectives, it is necessary that suitable transducer excitation and receiver amplification be used. One such system is Ultran's 2nd Wave M510, which is composed of computer controlled burst pulser, single and multichannel receiver amplifier, and high-speed signal processing (Figure 9.39). This system is sufficient to generate NCU images of materials in various formats and measurements, which can also be postprocessed to convert acoustic data into characteristics and features relevant to the test material.

FIGURE 9.39 NCU Ultran's 2nd Wave M510 system shown here with an *X–Y* raster scanning stage for imaging applications.

9.3.4 NCU Techniques and Applications

Techniques such as through transmission, T-R pitch-catch, and single transducer pulse-echo that are well known in conventional ultrasound can also be applied to NCU. In this section, we provide brief introduction to these techniques and exhibit NCU analysis of a wide range of materials. Details of materials analyzed are given in the figure captions.

9.3.4.1 Direct or through Transmission NCU

This technique, which requires access to both sides of the test material (Figure 9.40) is the easiest to use in NCU mode. By utilizing suitable transducers here, we show imaging and analysis of a number of materials:

- Uncured (prepregs), cured Carbon Fiber Reinforced Plastic (CFRP), C-C, and sandwich composites (Figures 9.41 through 9.45).
- Rubber, plastics, and foams (Figures 9.46 through 9.48).
- Green, sintered, and fired ceramics and refractories (Figures 9.49 through 9.53)
- Concrete (Figure 9.54)
- Lumber and wood (Figures 9.55 and 9.56)
- Food (Figure 9.57)

9.3.4.2 T-R Pitch-Catch Same Side Reflection NCU

By suitably angulating transmitting and receiving transducers on the same side of the test material where access is only from one side, it is also possible to launch bulk waves in materials (Figure 9.58). By doing so, either the far side reflection or reflections from within the material can be investigated to decipher the internal features or condition of the

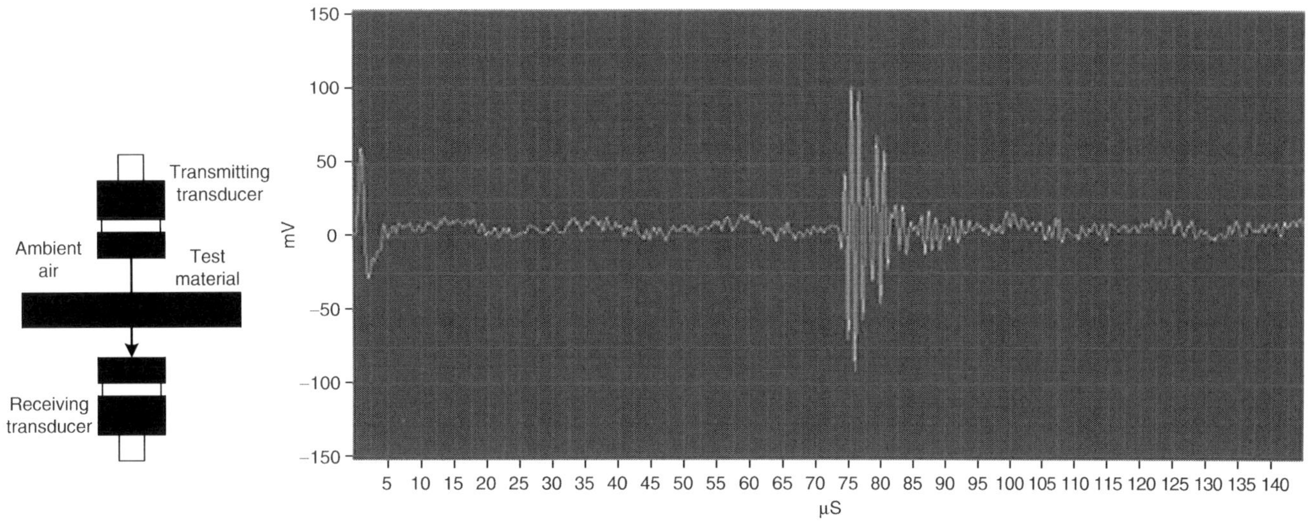

FIGURE 9.40 NCU direct transmission technique showing an example of ultrasound transmission signal through a material. In this case from 8 mm CFRP composite with 1 MHz transducers.

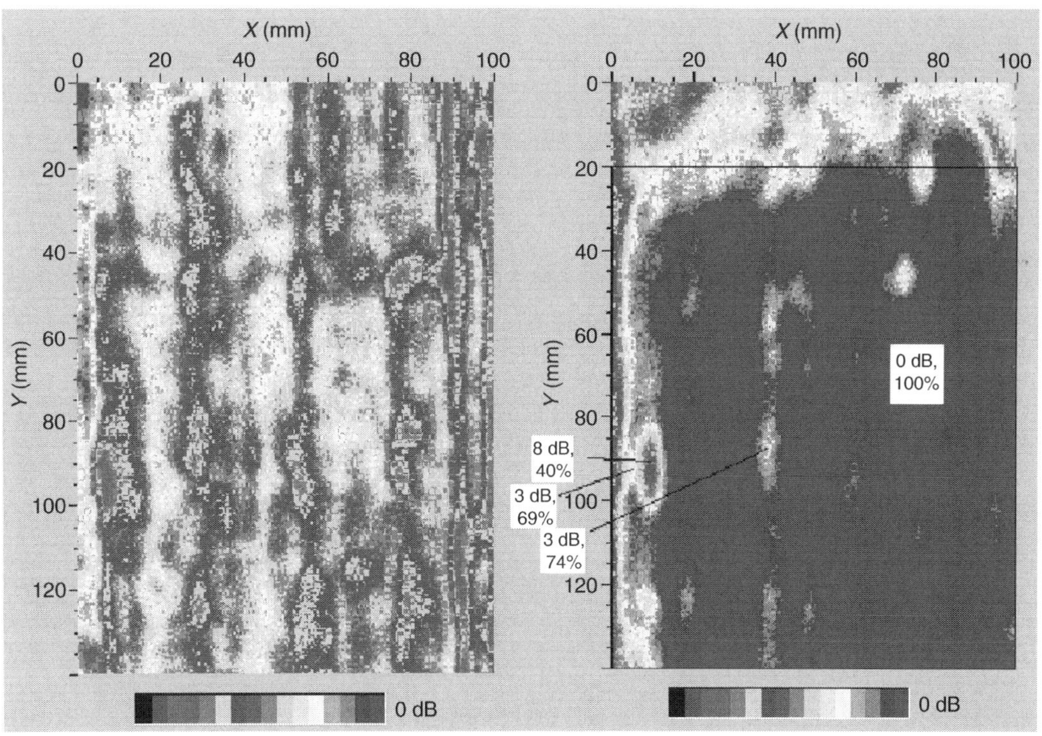

FIGURE 9.41 Uncured (left) and cured (right) CFRP stacks transmission NCU images. Darker regions in the uncured material image are indicative of poor adhesion. During the curing process, several such areas are bonded, while some are not such as the ones indicated by the lighter areas in the cured material. These observations were also confirmed by optical imaging of sample cross-sections. Transducers: 500 kHz, 12.5 mm active diameter. (Reproduced from HEXCEL Composites, U.K.)

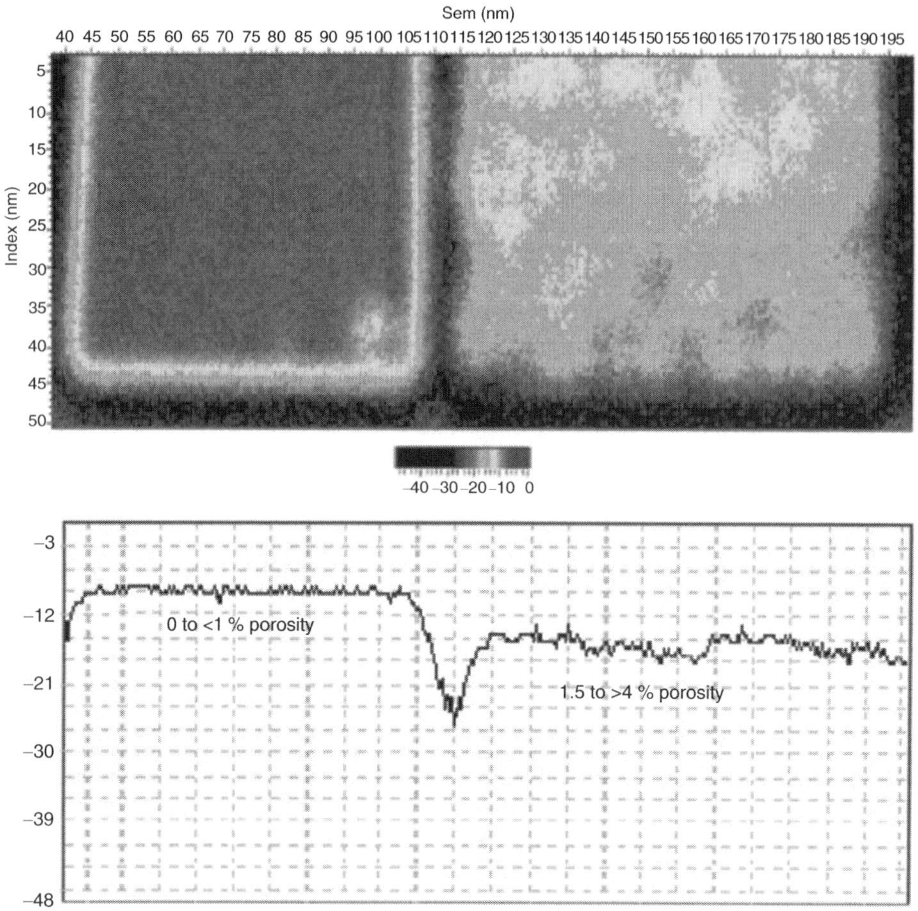

FIGURE 9.42 Varying porosity (10 mm) cured CFRP composites NCU imaging and analysis, left 0% to 0.1% porosity and right 1.5% to >4% porosity. Porosity variation across both samples is also shown by cross-sectional profile, bottom, where *y*-axis scale is in dB. Transducers: 500 kHz, 12.5 mm active diameter.

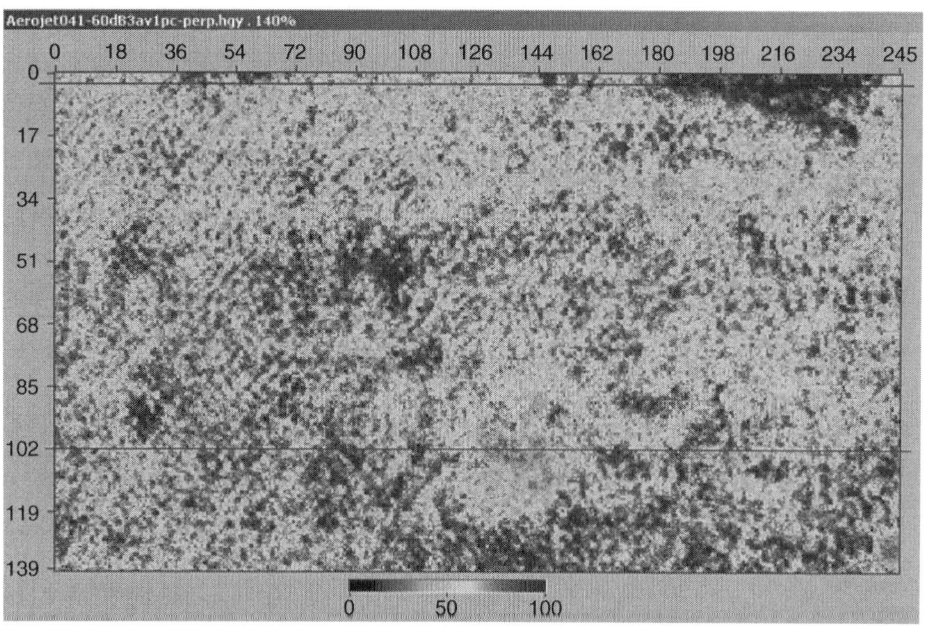

FIGURE 9.43 Graphitized woven carbon composite NCU transmission image exhibiting gross porosity and apparent defects. Transducers: Focused 500 kHz 12.5 × 12.5 mm active area.

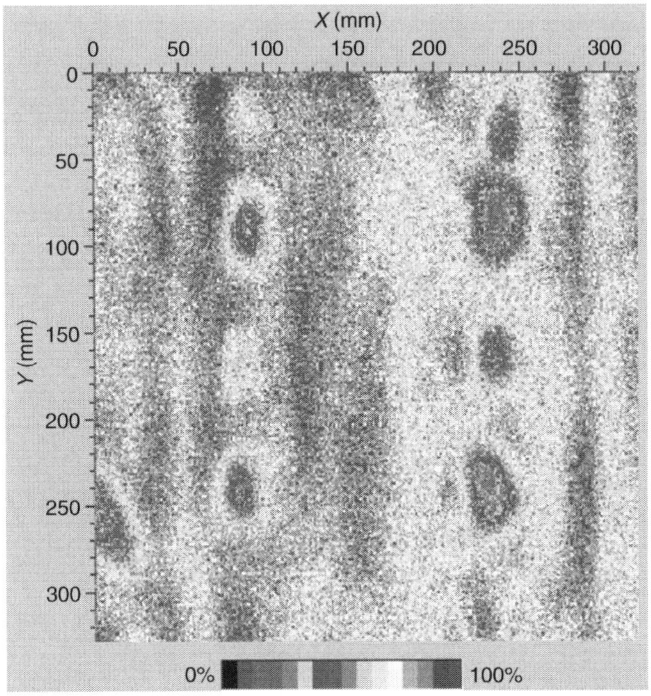

FIGURE 9.44 Nomex honeycomb (50 mm thick) CFRP skins sandwich NCU transmission imaging showing artificially created defects at skin honeycomb interfaces. Transducers: 140 kHz, 25 mm active diameter.

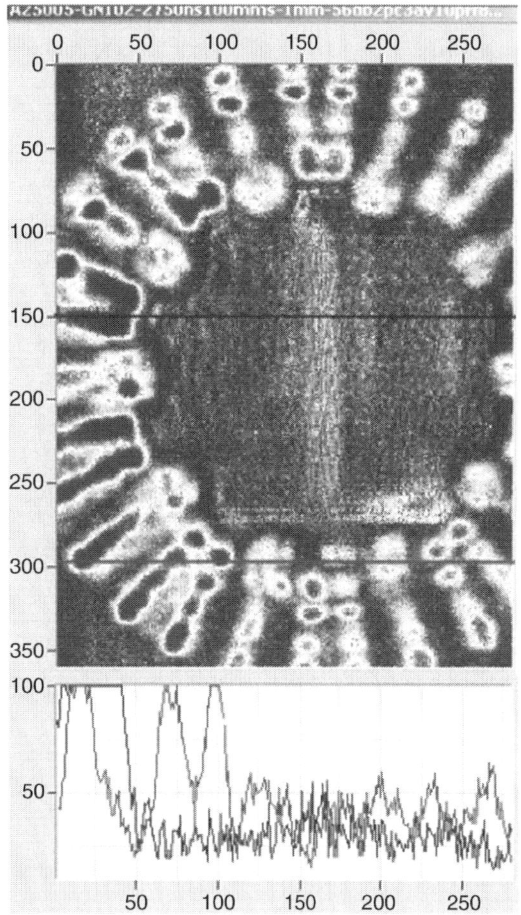

FIGURE 9.45 2D carbon–carbon disc brake NCU transmission imaging showing regions of very high and very low ultrasound transmission, presumably indicative of varying material density. Transducers: Focused 200 kHz, 25 mm active diameter.

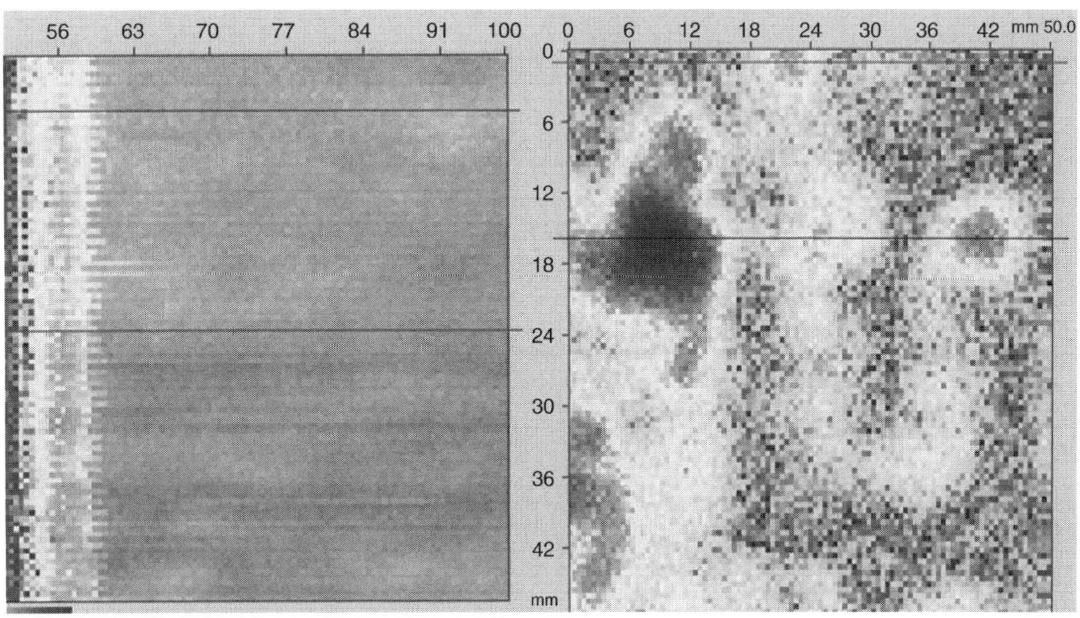

FIGURE 9.46 Tire rubber mixes NCU imaging. Left: Homogeneous mix. Right: Heterogeneous mix. Transducers: 500 kHz, 12.5 mm active diameter.

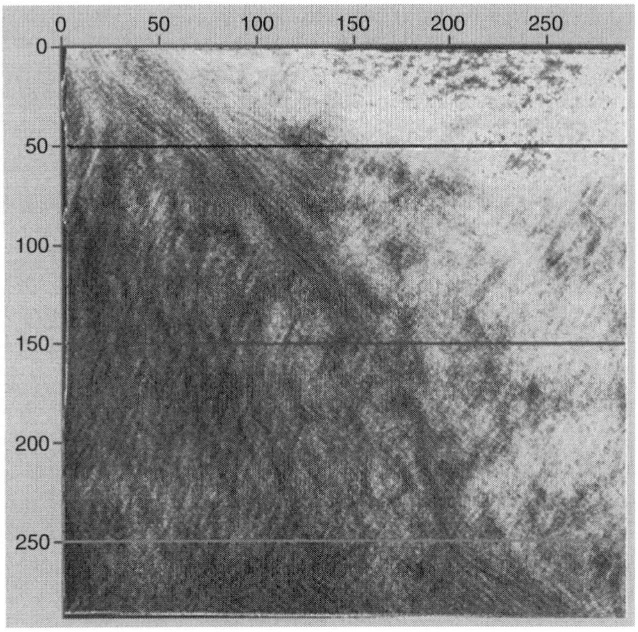

FIGURE 9.47 Sheet molded plastic (SMP) NCU imaging showing unwanted fiber orientation. Transducers: Focused 500 kHz, 12.5 × 12.5 mm.

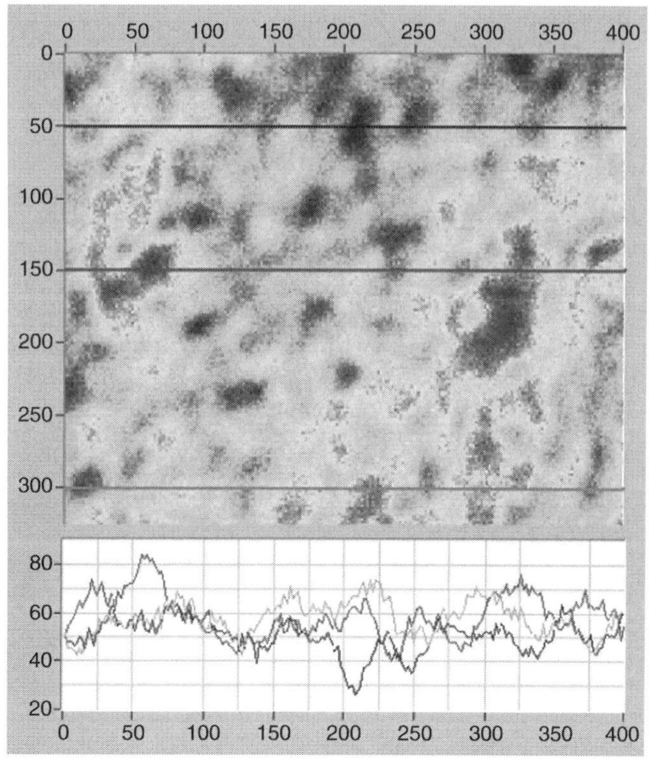

FIGURE 9.48 A 100 mm polyurethane foam NCU imaging, presumably showing porosity and density variations. Transducers: 200 kHz, 25 mm active diameter.

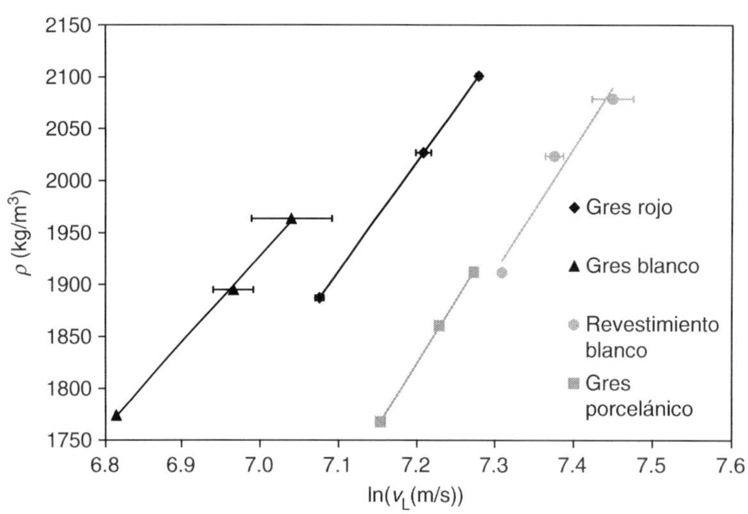

FIGURE 9.49 Green ceramic floor tiles NCU velocity–density relationships for a variety of compositions. Transducers: 500 kHz, 19 mm active diameter.

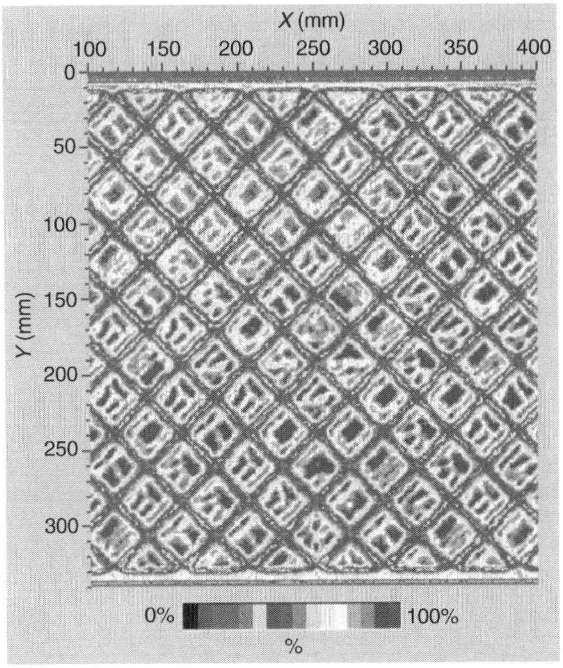

FIGURE 9.50 Green ceramic floor tile high-resolution NCU image. Transducers: Focused 500 kHz, 19 mm active diameter.

material. An example of this is shown in Figure 9.59. Here we see the artificially embedded defects in a 4 mm thick panel of CFRP composite.

9.3.4.3 Single Transducer Pulse-Echo NCU

Under ambient conditions, it is very easy to obtain high-quality signals corresponding to the reflection from the material surface (Figure 9.60). By monitoring the amplitude and time of flight of surface reflections, materials can be characterized for surface roughness, texture, etc. Figures 9.61 and 9.62 provide examples to this effect.

While highly desired, pulse-echo NCU technique to obtain the far side materials reflection under ambient conditions is at best extremely arduous, if not impossible at the time of this writing. However, if the test is conducted under high air–gas pressure, then it is relatively easy to observe far side reflections in pulse-echo NCU mode. Figure 9.63 shows an example of multiple thickness reflections from 10 mm steel at 5 bar air pressure by using a special broadband 3 MHz transducer.

9.3.5 Perusal of NCU

In this chapter, we have provided an introduction to an extraordinarily high air–gas transduction piezoelectric transducer. Examples of materials testing by NCU with well-known ultrasonic techniques have also been given. In our laboratory, work continues to enhance NCU by considering the geometrical acoustic in air and ultrasound interaction with materials. This work exhibits the unique nature of NCU with respect to very high resolution and detectability of discontinuities in materials, besides other applications of significance.

Underscoring the significance and novelty of NCU, in recent years, several ultrasound and materials research institutes and organizations have undertaken the task of pursuing this field. It suffices to say that their work also concerns testing and analysis of industrial materials, food, biomedical,

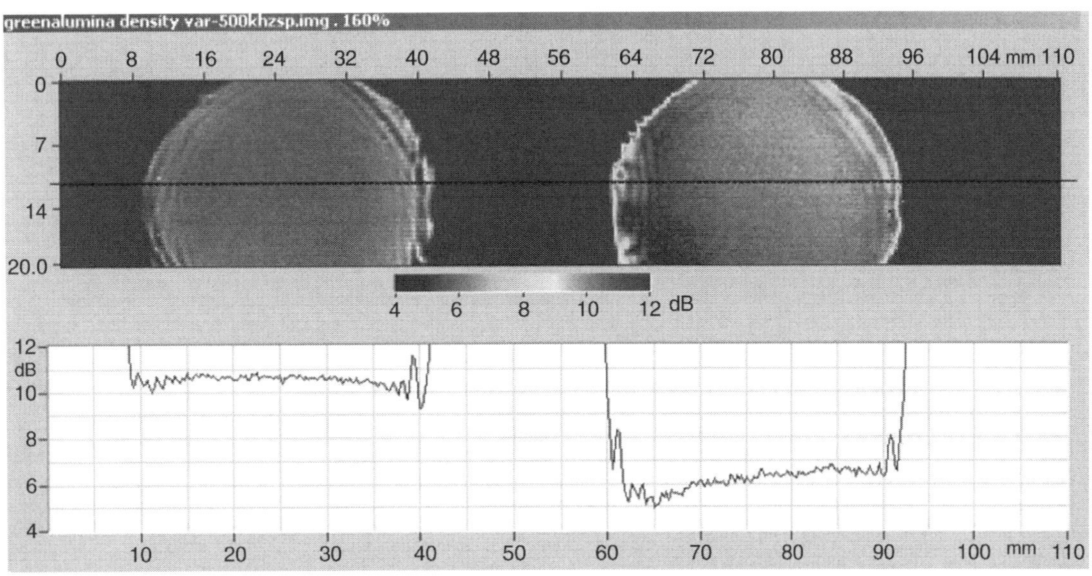

FIGURE 9.51 Green alumina varying in density NCU image analysis. Left: Pressed at 9000 psi, right: pressed at 15,000 psi. Note as the density increases (higher pressing pressure), in NCU mode ultrasound transmission decreases. Transducers: 500 kHz, 12.5 mm active diameter.

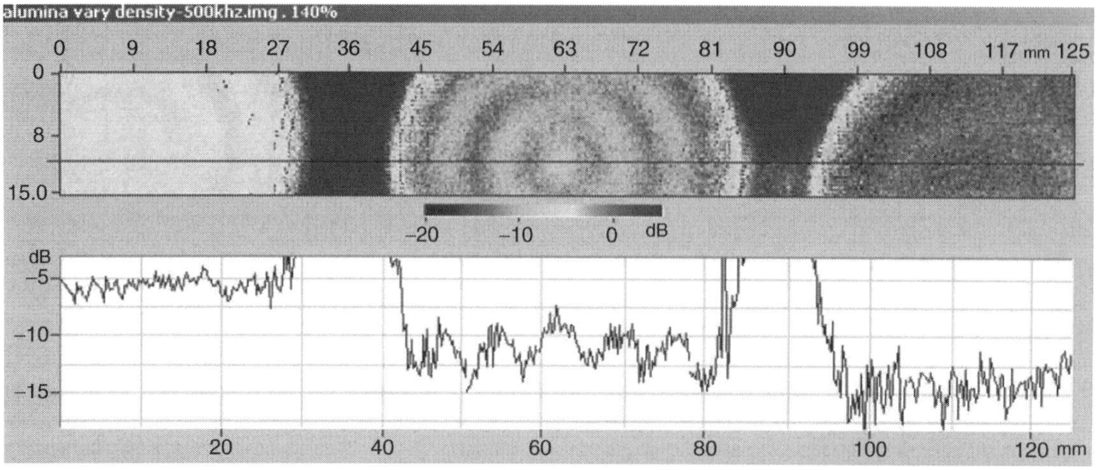

FIGURE 9.52 Sintered alumina NCU transmission imaging and analysis of samples varying in density. Left: 65% (2.58 g/cc), middle: 70% (2.8 g/cc), right: 77% (3.08 g/cc). Transducers: 500 kHz, 12.5 mm active diameter.

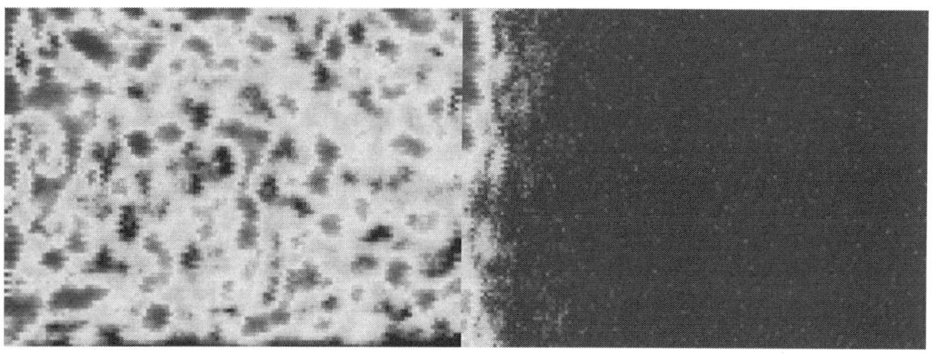

FIGURE 9.53 Fired silica refractory NCU imaging. Left: Defect-free. Right: Cracked. Transducers: 500 kHz, 12.5 mm active diameter.

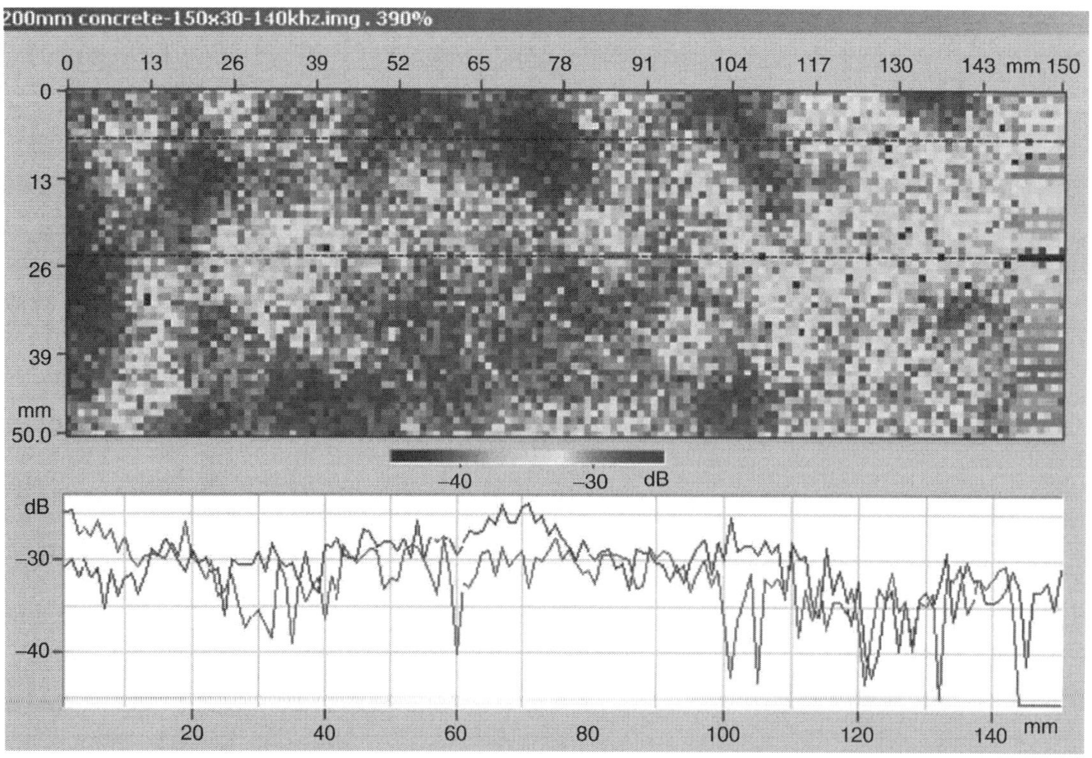

FIGURE 9.54 Concrete NCU (200 mm) imaging showing internal defects. Transducers: Mixed 140 kHz.

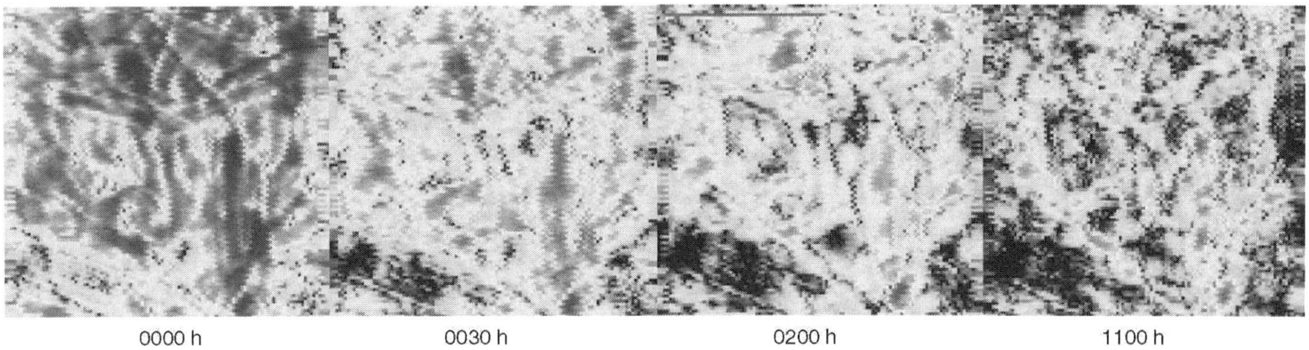

FIGURE 9.55 Oriented strand board (OSB) NCU imaging as a function of moisture content. Left to right: Dry, 30 min soak, 2 h soak, and 11 h soak. Transducers: 200 kHz. Note as the moisture content increases the NCU transmission decreases, which is more evident in the colored images.

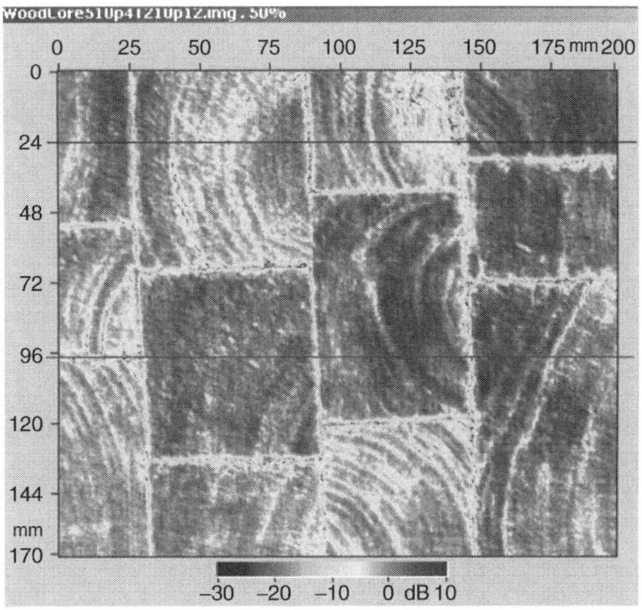

FIGURE 9.56 Porous wood core NCU high-resolution imaging. Transducers: Mixed 1 MHz.

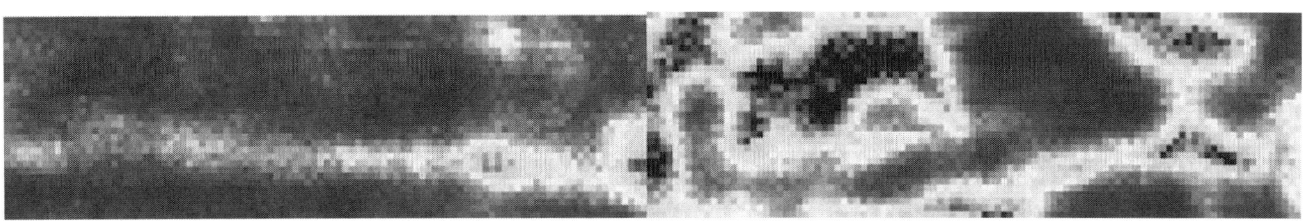

FIGURE 9.57 Cheddar cheese NCU imaging. Left: Reduced fat. Right: Extra sharp. It appears as the fat content in cheese increases so does the ultrasound attenuation. Transducers: 1 MHz, 12.5 mm active area diameter.

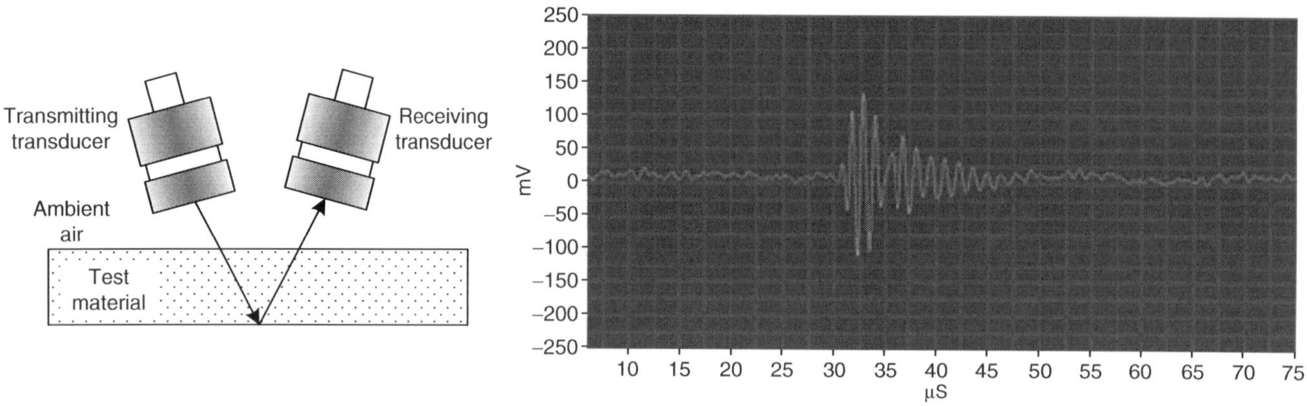

FIGURE 9.58 NCU T-R pitch-catch reflection technique showing far side (bottom surface) reflection from a material. In this case the bottom surface of 12.5 mm thick aluminum with 1 MHz transducers.

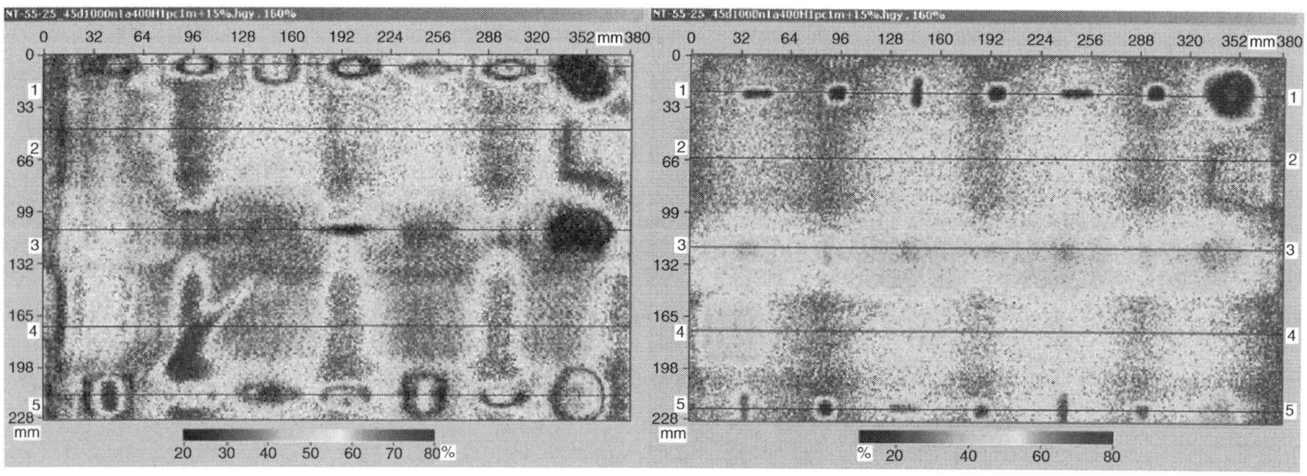

FIGURE 9.59 Same side NCU image of a CFRP panel with embedded defects (left). As a comparison, the right hand image shows similar image, but acquired by direct transmission mode. Transducers: 500 kHz, 12.5 mm active diameter.

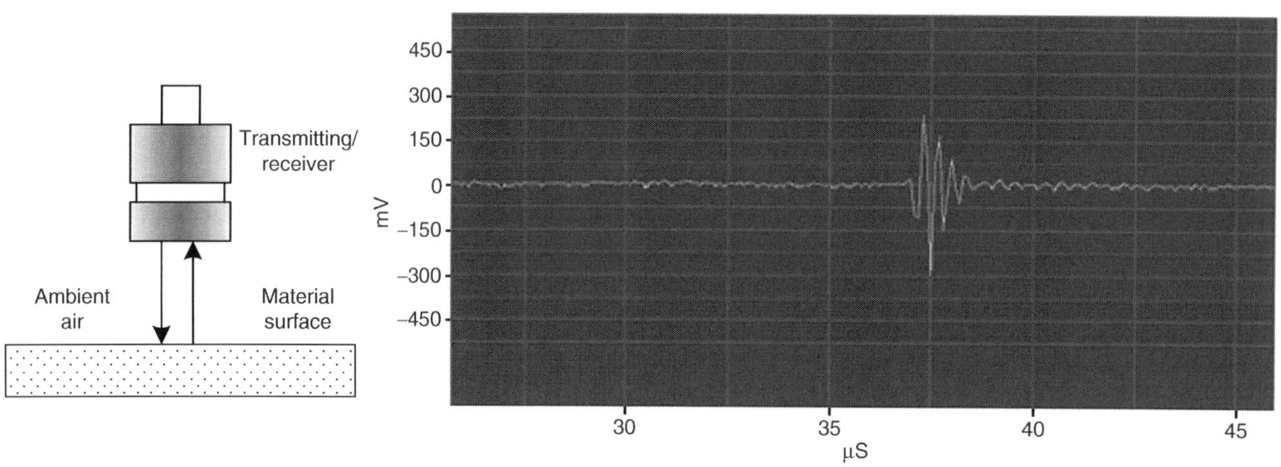

FIGURE 9.60 NCU single transducer, pulse-echo technique showing surface reflection signal from a material. In this case the surface of dense alumina with 3 MHz transducer.

FIGURE 9.61 Very high-resolution NCU surface images of familiar objects generated by 3 MHz sharply focused transducer.

and other applications. A partial list of these organizations is as following:

- Iowa State University, United States
- University of Bordeaux, France
- University of Irvine at California, United States
- University of Ancona, Italy
- Institute of Technical Ceramics, Spain
- Yamanashi University, Japan
- California State University at Carson, United States
- Katholieke Hogeschool (KATHO), Belgium
- Katholieke University at Leuven, Belgium
- Stanford University at Stanford, United States
- University of Windsor, Canada

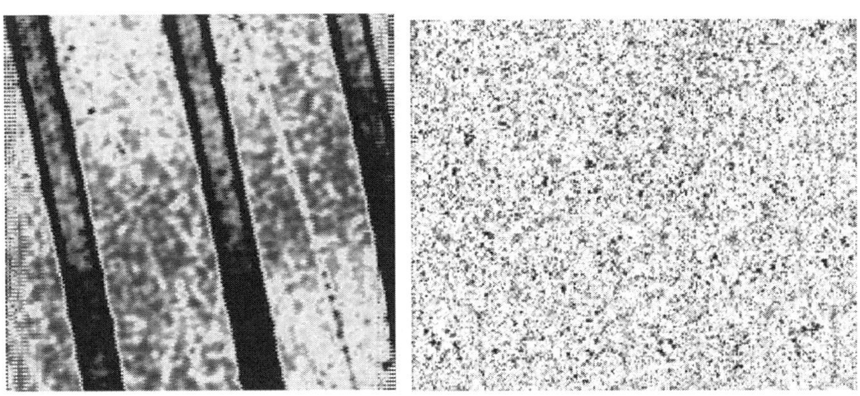

FIGURE 9.62 Surface imaging of other materials: Left: Semiconductor polishing pad showing surface roughness. Right: Fine celled polymer foam showing its texture, the details of which are more evident in color image.

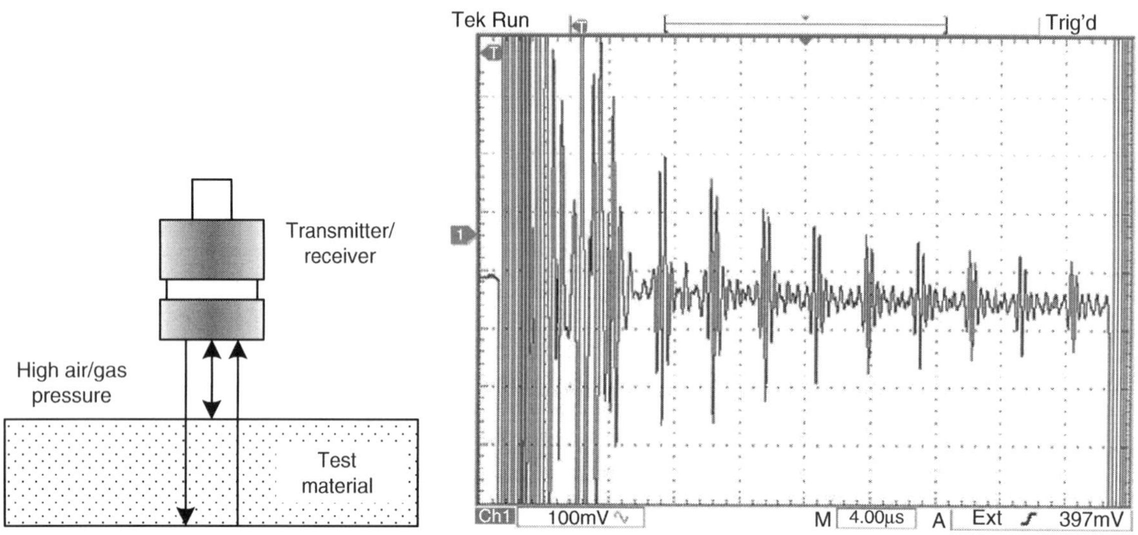

FIGURE 9.63 NCU single transducer, pulse-echo at high gas pressure. In this case, far side (bottom surface) reflections from a 10 mm steel plate at 5 bar air pressure by using a broadband 3 MHz NCU transducer. (Reproduced from Gas Technology Institute, U.S.A.)

- Fraunhofer Institute at Saarbrucken, Germany
- Penn State University at University Park, United States
- Queens University at Kingston, Canada
- Warwick University, United Kindom
- National Physical Laboratory, United Kindom
- Los Alamos National Laboratory, United States
- Gas Technology Institute, United States
- Southwest Research Institute, United States
- Nagoya Institute of Technology, Japan
- Johns Hopkins University, United States

It is our hope that the materials and biomedical industries, research institutes, and government agencies take a serious look at the status of modern ultrasound, particularly NCU transducers and associated testing know-how and its necessity in solving important process and quality-related material testing needs. The advantages of NCU are too numerous and far beyond conventional ultrasound to be ignored. We believe sooner it is applied for specific tasks along with increment in practical educational foundation, better it would be for our complex socio-technical world.

References

1. Brunk, J.A., Ultrasonic nondestructive testing and materials characterization, this volume.
2. Fox, J.D., Khuri-Yakub, B.T., and Kino, G.S., High frequency wave measurements in air, 1983 IEEE Ultrasonics Symposium, pp. 581–592, 1983.
3. Bhardwaj, M.C., Modern ultrasonic transducers, Commercial Catalog, Ultran Laboratories, Inc., 1986.
4. Yano, T., Tone, M., and Fukumoto, A., Range finding and surface characterization using high frequency air transducer, *IEEE Trans. UFFC*, 34(2), 222–236, 1987.
5. Haller, M.I. and Khuri-Yakub, B.T., 1–3 composites for ultrasonic air transducer, IEEE Ultrasonics Symposium, pp. 937–939, 1992.
6. Bhardwaj, M.C., Ultrasonic transducer for high transduction in gases and method for non-contact transmission in solids, U.S. Patent 6,311,573, November 6, 2001. Japan Patent 3225050, August 24, 2001. European Cooperation Treaty, Pending. WIPO, PCT/US98/12537.
7. Bhardwaj, M.C., Non-destructive evaluation: introduction of non-contact ultrasound, *Encyclopedia of Smart Materials*, M. Schwartz ed., John Wiley & Sons, New York, pp. 690–714, 2002.
8. Bhardwaj, M.C., Gas matrix piezoelectric composite, U.S. and international patents pending.

10

Chitosan-Based Gels and Hydrogels

10.1 Chitosan-Based Gels

Kang De Yao, Fang Lian Yao, Jun Jie Li, and Yu Ji Yin

10.1.1 Introduction

Chitosan is a natural biopolymer obtained by deacetylation of chitin, which is produced from marine shellfish, such as crabs, shrimp, fungal cell walls, and other biological sources. Chemically it is a linear cationic poly(β-(1–4)-2-amino-2-deoxy-D-glucan) derived from chitin, a poly(β-(1–4)-2-amino-2-acetamido-2-deoxy-D-glucan) by deacetylation. Chitosan is described in terms of the degree of deacetylation (DDA) and average molecular weight. And next to cellulose, it is the second most plentiful biomass and is already known as a biocompatible and biodegradable material. Many researchers have examined tissue response to various chitosan-based implants. Results indicate that these materials evoke minimal foreign body reactions [1].

Gels consist of 3D polymer networks swollen in swellers, whereas hydrogels swell in water. Smart (intelligent) gels (or hydrogels) can swell or contract in response to stimulus changes, e.g., temperature, pH, ionic strength, chemical, and electrical fields.

Chitosan has been used for developing smart gels via networks or complex formations. Due to their various properties that depend on environmental variables such as pH, ionic strength, and electric field, these materials have potential applications such as biomaterials, separation membranes, and field-responsive materials.

10.1.2 Supramolecular Interactions and Gel Formation

We know that chitosan-based gels are chemically or physically cross-linked networks. To make them smart, special supramolecular interactions are often introduced into the network to provide reversible responsiveness to environmental stimuli. Therefore, different interpolymer complexes are formed within the chitosan-based networks. These complexes can form or dissociate in response to changes in their environment. They can be divided into the following major categories based on the dominant type of interaction.

1. Hydrogen bonding complexes
2. Polyelectrolyte complexes (PEC)
3. Grafted and block network
4. Self-assembly
5. Coordination complexes

The presence of amino groups in chitosan ($pK_a = 6.5$) results in pH-sensitive, temperature-responsive, and electric field-sensitive behavior. Research efforts have been aimed at tailoring the properties of chitosan through cross-linking (include covalence cross-link and physical cross-link), functionalization, copolymerization, and blending to fit smart requirements.

10.1.2.1 Hydrogen Bond Complexes

The preparation of a hydrogel containing a covalently cross-linked chitosan minimally requires chitosan and a cross-linker in an appropriate solvent, usually water. The amino groups of glucosamine residue within chitosan chains can serve as cross-linking sites, for example, reacting with glutaraldehyde (GA) [2,3], glyoxa [4], proanthocyanidin [5], 1-ethyl-3-(3-dimethylaminopropyl) carbodiimide (EDC), or genipin [6] to form imine cross-linking between linear chitosan chains that lead to gel formation.

Other components can be added, such as additional polymers to form a hybrid polymer network (HPN) or a semi- or full-IPN (interpenetrating polymer network). Auxiliary molecules can also be used to initiate reactions during the preparation of the network. An IPN is defined as a combination of two cross-linked polymers; at least one of them synthesizes or cross-links in the presence of the other. If one of the polymers is linear (without being cross-linked), a semi-IPN results. The IPN technique can be used to improve the properties of chitosan-based gels.

Lee and Chen [7] examined GA-cross-linked chitosan–isopropylacrylamide(chitosan/PNIPAAm) semi-IPN and IPN hydrogels. They found that the swelling ratios of IPN gels are lower than those of semi-IPN gels and the reason is that the addition of cross-linked chitosan in the PNIPAAm hydrogel makes the network structure of the gel denser and more hydrophobic. The swelling ratios of PNIPAAm/chitosan gels slightly decrease with the addition of chitosan, but the swelling ratios are not significantly affected by the amount of chitosan added to the PNIPAAm gel.

Because of the existence of many $-NH_2$ groups on chitosan chains, hydrogels composed of chitosan are also used as pH-sensitive materials [8]. A novel pH-sensitive hydrogel composed of N,O-carboxymethyl chitosan and alginate cross-linked by genipin (NOCC/alginate) and the swelling ratio of this hydrogel at pH 1.2 was about 2.5 within 30 min subsequent to swelling, while it was approximately 6.5 at pH 7.4 (as shown in Figure 10.1). This finding may contribute to the fact that, at low pH, the swelling ratio was limited due to formation of hydrogen bonds between NOCC (–COOH and –OH) and alginate (–COOH and –OH). However, at higher pH 7.4, the carboxylic groups on the genipin-cross-linked NOCC/alginate hydrogel became progressively ionized (–COO–). In this case, the hydrogel swelled more significantly due to a large swelling force created by the electrostatic repulsion between the ionized acid groups. The swelling ratio of the hydrogel with alginate was significantly lower than that without alginate due to the formation of hydrogen bonds between NOCC ($-NH_2$ and –OH) and alginate (–OH) [9].

Yao et al. [10] examined GA cross-linked chitosan–poly-(propylene glycol) (PPG), semi-IPN as a pH-sensitive hydrogel. Chitosan–PPG semi-IPN shows pH-dependent swelling; the highest swelling degree (SD) is at pH 4.0 and the lowest at pH 7.0 (Figure 10.2). The structure features of chitosan–PPG are reversible (e.g., $-NH_3$ to $-NH_2$) when the gels are transferred from a pH 1.0 to a pH 7.8 buffer [11], because it is a hydrophobic and water-insoluble polymer and the incorporated PPG is expected to decrease the equilibrium SD in a low pH medium. In addition, linear PPG in the network can enhance the flexibility of a semi-IPN. For the same reason, at a higher pH it may not fully ionize all the $-NH_2$, the pH-sensitive, biodegradable, semi-IPN chitosan–polyvinyl alcohol (PVA) hydrogel cross-linked with GA and shows low-bound water concentrations (C_{BW}, mg/mg dry hydrogel). On the other hand, if the pH goes too low, the high concentration of H^+ will actually impair the swelling, so the maximum C_{BW}, mg/mg dry hydrogel would appear at pH 3 medium (Figure 10.3) [12,13].

The SD of hydrogels based on [*N*-(2-carboxybenzyl) chitosan, CBCS] (CBCSG) was significantly increased with the raise of the pH in the range of pH 5.0–9.0 and the lowest swelling occurred almost always at pH 5.0 among CBCSGs, as shown in Figure 10.4 [14]. This maybe depends mainly on the osmotic pressure difference between the inside of the gel and the surroundings caused by the redistribution of mobile ions [15], and along with decreasing pH. The amount of –COO is gradually reduced inside the hydrogels, which leads to a decrease in osmotic pressure and makes the SD of the hydrogels smaller. When the pH was reduced to 1, the –COO was completely inhibited, and the dominant ionic groups were $-NH_3^+$, but the number is small because the majority of the $-NH_2$ groups were linked, therefore the CBCSGs swelled sluggishly as pH diminished in the range of pH 1.0–5.0.

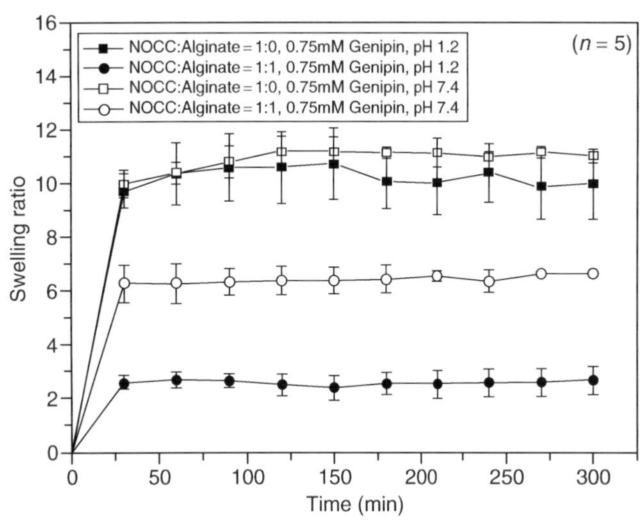

FIGURE 10.1 Swelling characteristics of the NOCC and NOCC/alginate hydrogels cross-linked with 0.75 mM genipin at pH 1.2 or 7.4 at 37°C.

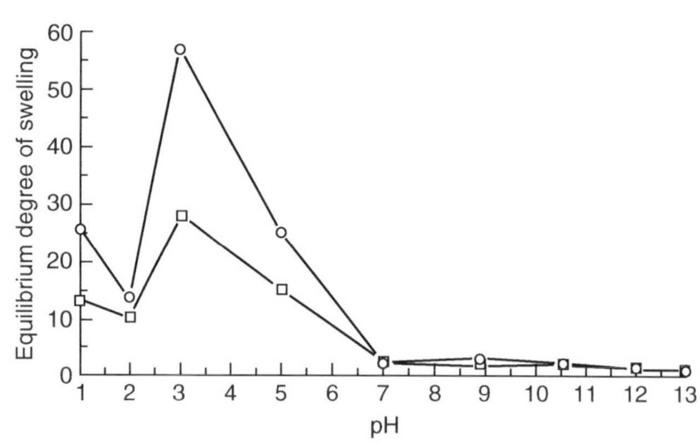

FIGURE 10.2 Equilibrium degree of swelling vs. pH for the semi-IPN, synthesized from different amounts of cross-linking agent. Molar ratios of $-CHO:-NH_2$ are 2.21° and 4.83°, respectively.

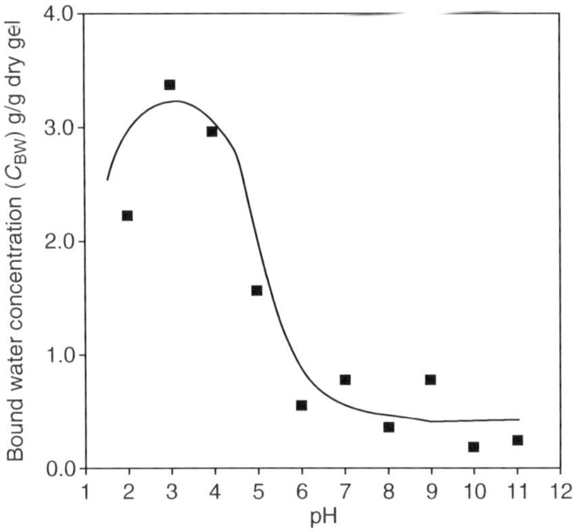

FIGURE 10.3 Bound water concentration in the chitosan–PVA hydrogel at different pH values (chitosan:PVA molar ratio 1:10, GA concentration 33.3 M).

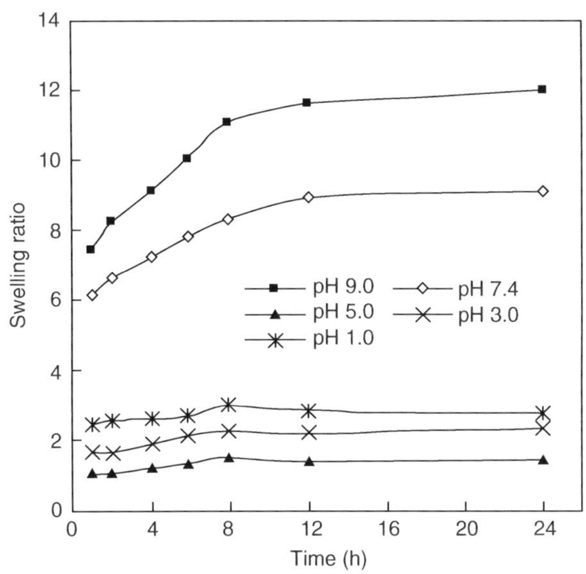

FIGURE 10.4 Swelling behavior of CBCSG in different buffer solutions.

Collagen is known to be the most promising of biomaterials and has been found diverse applications, which can form a complex with chitosan [16]. The collagen–chitosan complex can be used to mimic the components of the native extracellular matrix (ECM). It has also been reported that the hybrids of collagen–chitosan manufactured by cross-linking [17]. Gelatin is a denatured form of collagen, composed of glycine, praline, hydroxyproline, arginine, and other amino acids. The amphiphilic protein (isoelectric point = 4.96) can provide amino groups for cocross-linking with chitosan via GA thus preparing a chitosan–gelatin (CG) HPN. The pH-sensitive swelling behavior of HPN gel is displayed in Figure 10.5. The data show that the degree of swelling declines

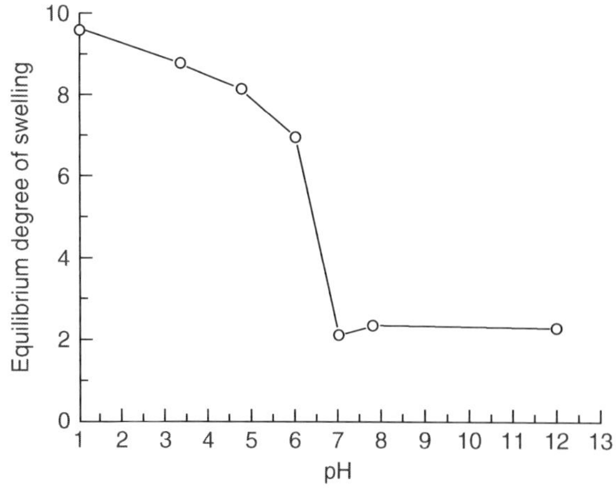

FIGURE 10.5 Swelling behavior of CG hybrid network specimen with a –CHO:–NH_2 molar ratio of 10 in solutions of different pH with ionic strength I = 0.1 at 37°C.

sharply at pH 7.0 and this can be explained by the fact that the hydrogen bonds within the chirosan–gelatin HPN dissociate in an acidic medium [18].

The properties of many chitosan-based hydrogen bond complexes depends on not only pH but also temperature. Carmen et al. investigated the swelling behavior of different crosslink chitosan–poly (*N*-isopropylacrylamide) interpenetrated networks (CS/PNIPA). The SD of all the CS/PNIPA IPN decrease with the enhancement of temperature, and the SD at low pH 3 is higher than that at high pH 8 for any temperature (as shown in Figure 10.6) [19]. Temperature/pH-sensitive poly(2-ethyl-2-oxazoline)/chitosan(PEtOz/CS) IPN hydrogels exhibited considerable shrinkage over the temperature range of 35°–45°, and thermoresponsive behavior has decreased with increasing chitosan content [20].

The electrical field is another important factor, which influences the properties of chitosan-based hydrogels. The response of electrosensitive hydrogels is generally exhibited in the form of either swelling and shrinking or bending behaviors. Figure 10.7 shows the degree of bending of the chitosan–poly(hydroxyethyl methacrylate) semi-IPN hydrogel as a function of the various applied voltages in an aqueous 1.0% by weight NaCl solution. The degree and the increase in the speed of bending as well as an increase in the voltage applied across the hydrogel indicates that the bending is induced by the electric current, however, the bending did not occur in pure water [21]. The chitosan–polyallylamine IPN hydrogel exhibited the same characteristic [22].

10.1.2.2 PEC

PEC are generally obtained either by the reaction of polycations and polyanions or by polymerizing monomers that have suitable functional groups onto polymeric templates of known structures. PECs have numerous applications such as membranes, medical prosthetic materials, environmental sensors, and chemical detectors. Chitosan, a cationic polysaccharide, has been complexed with anionic polymers. The properties of PEC are mainly

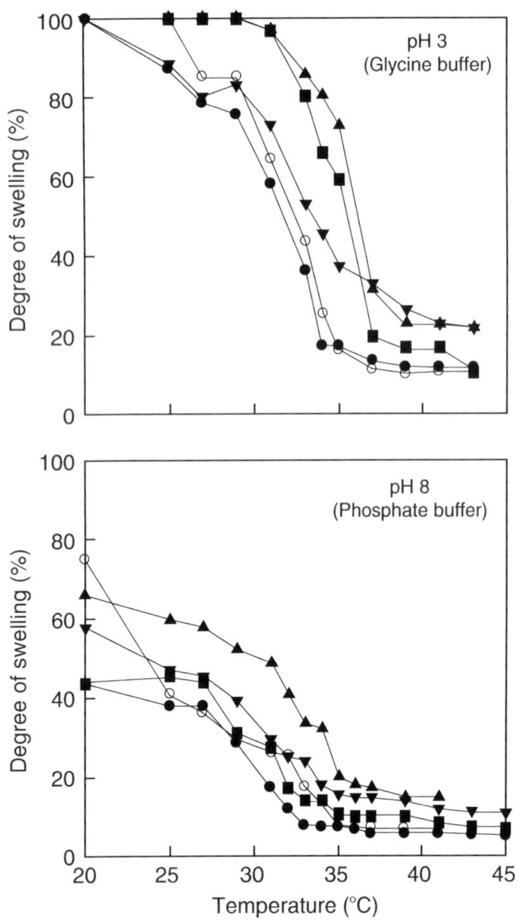

FIGURE 10.6 Dependence of the degree of swelling (%) of PNIPA/CS IPNs on the pH and temperature of the aqueous medium; IPN-0 (O), IPN-1 (●), IPN-3 (■), IPN-5 (▼), and IPN-7 (▼). PNIPA/chitosan-I IPNs exhibited a similar pattern. (IPN-X, where X denotes the concentration (Vol.%) of GA.

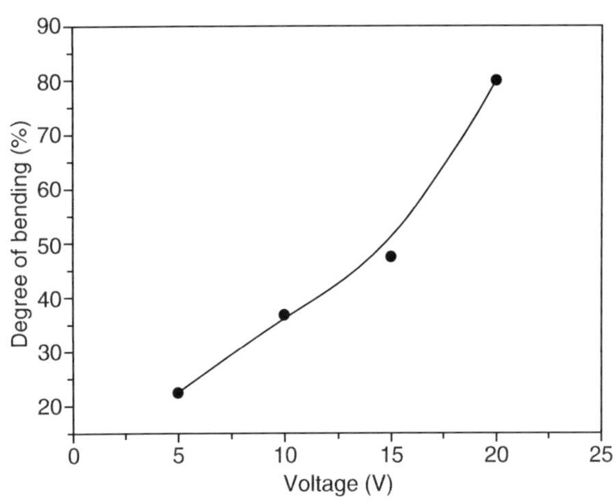

FIGURE 10.7 Degree of bending of chitosan/PHEMA semi IPN in response to various voltages in a 1.0 wt % NaCl solution at $T = 35°C$.

determined by the degree of interaction between the individual polymers. This latter condition depends essentially on their global charge densities and determines their relative composition in the PEC. The lower the charge density of the polymer, the higher is the polymer proportion in the PEC, since more polymeric chains are required to react with the other polymers. A PEC can be formed via two ionizable polymers with opposite charges individually. This means that a PEC-forming reaction can only occur at pH values in the vicinity of the pK_a interval of the two polyelectrolytes.

CS is a pH-sensitive polymer (pK_a is ca. 6.5) whose amino groups can be protonated depending on the pH of the environment. When CS is mixed with ampholytic gelatin (Gel, its pH_{iso} is approximately 4.7), which in the case of pH of medium above the isoelectric point of Gel, and where a net charge is negative one, there occurs electrostatic interaction between the ammonium ions of CS salts and carboxylate groups of Gel. Yao et al. [23] found that there was strong interaction between Gel and CS in the aqueous medium that was enough to form PEC in situ, and the CS/Gel PEC only yielded at pH value larger than 4.7 and above pH 6.2 CS could precipitate from the solution.

Pectin is an acidic polysaccharide that has a repeating unit of α-(1,4)-L-rhamnose units. A PEC has been formatted from anionic pectin and cationic chitosan. The PEC swells obviously at pH < 3 and pH > 8, and does not swell in the range of 3 < pH < 8. Moreover, its degree of swelling in an acidic medium is by far higher than that in an alkaline medium. The swelling of the PEC correlates with its composition and is also affected by the degree of deacetylation (DDA) and the methoxy level of pectin [24].

Alginate comprises a linear chain of (1,4)-linked β-D-mannuronate and α-L-glucuronate residue arranged blockwise. A gel-like chitosan/alginate PEC was at one time formed with coexisting polyions of opposite charges ($-NH_3^+$ and $-COO^-$). Polysaccharides (chitosan and alginate) have bulky pyranose rings and a highly stereoregular configuration in their rigid linear backbone chains. At a given pH, the composition of the PEC shifted to a lower alginate content as the degree of N-acetylating of chitosan increased and the higher the pH, the lower the alginate content of the PEC for a given chitosan sample [25]. W. Argüelles-Monal et al. [26] found that the degree of polymerization of chitosan and the chemical composition of alginic acid have shown to have an influence on the PEC formation.

Yoshitsune et al. [27,28] found that the chitosan/polyalkylene-oxide–maleic acid copolymer (CS/PAOMA) PEC films swelled at a low pH, shrunk at pH between 4.8 and 6.5 and their swelling enhanced after pH 6.5. This swelling behavior can be attributed to the electrostatic interaction between the protonation of amino groups in CS and carboxyl groups in the PAOMA. The pK_a of CS is 6.5 and the pK_a of PAOMA is 4.8. In different pH medium corresponding groups (e.g., NH3+ and COO−) play different roles between intermolecules.

10.1.2.3 Grafted and Block Network

Many polymers, such as phosphatidylethanolamine, poly(ethylene glycol) (PEG), poly(vinyl acetate), poly(vinyl alcohol), poly(3-hydroxyalkanoate), linoleic acid, poly(L-lactic acid), and

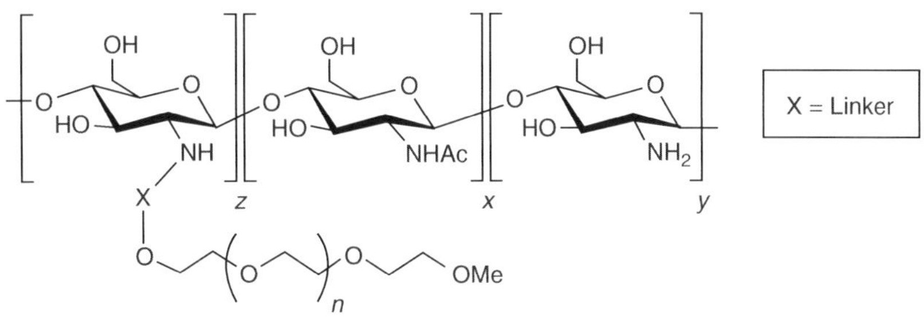

FIGURE 10.8 Structure of chitosan-*g*-PEG.

so on, have been grafted onto the chitosan in order to obtain some type of special performance.

PEG is a water-soluble polymer that has low toxicity and immunogenicity. Hydrophilic PEG can form a diffuse layer containing water molecules, which has protein resistance when PEG is adsorbed on a surface. Many researchers have turned their attention to graft copolymers of chitosan and PEG, as shown in Figure 10.8, because of their high biocompatibility [29–31]. Hydrogen bonds exist between the hydrogen of the amino groups of chitosan and the oxygen of the polyether in alkaline regions, whereas the hydrogen bonds dissociate at acidic pH. Additionally, the SD is very sharp at a high pH 6 [32], and the polymers were soluble in water over a wide pH range within a short time and the viscosities of the solution were very low. The degradation rate of the copolymer declined with the enhancing degree of substitution of PEG to chitosan, which was probably due to the relative stability of PEG to lysozyme [33].

Poly(α-hydroxy acids) generates acidic degradation products at the implanted site, which evokes undesirable tissue reaction [34]. The acid by-product may lead to local disturbances due to poor vascularization in the surrounding tissue. Chitosan may be combined with acid-producing biodegradable polymers, so that local toxicity due to the acid by-products can be alleviated. Yao et al. obtained the cytocompatible poly(chitosan-*g*-L-lactic acid) through grafting oligo(L-lactic acid) onto the amino groups in chitosan without a catalyst. These graft copolymers had a higher strength than chitosan and their swelling behavior was influenced by pH (as shown in Figure 10.9). When a pH < 2 and with an increase in the buffer pH, the concentration of the charged ionic groups in the films also increased. The swelling of the samples were shown to increase due to the enhancement of the osmotic pressure and charge repulsion. While at a higher pH, the degree of ionization was reduced due to the deprotonation of the amino units of chitosan and the swelling of the films decreased. In addition, the hydrophobic side-chain aggregation and hydrogen bonds in the copolymers were much stronger, which also lead to lower swellability of the samples [35,36].

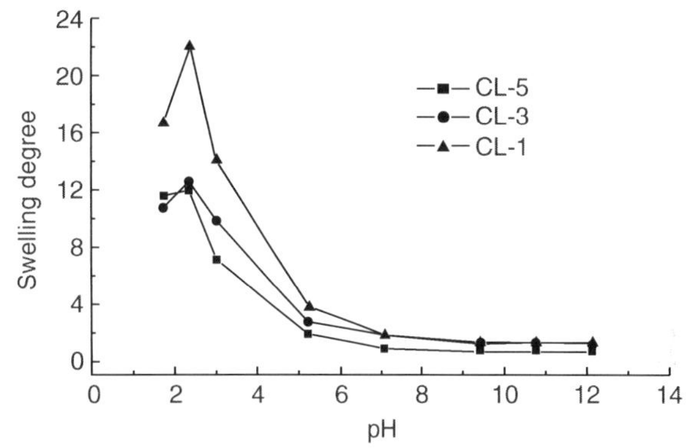

FIGURE 10.9 Effect of the LA/CS feed ratio and pH value on the equilibrium water uptake of the CL films (CL-1, CL-3, and CL-5 denote that LA/CS = 0.5, 2.0, and 4.0(w/w), respectively).

10.1.2.4 Self-Assembly

The technique of self-assembly layer by layer (LbL) PEC multilayer films opens up many new opportunities for us to achieve the ideal model surface whose properties are controllable [37]. The self-assembly polyelectrolyte systems are sensitive to the fabrication orders in terms of the orders of static electric interactions. It has long been known that one of the challenges of the materials construction is the controllable properties of the materials, including shape, size, morphology, thickness, roughness, chemical and transport properties, degree of interpenetration [38], and surface evidence associated with the mechanisms of the absorption and formation of polyelectrolytes on charged surfaces [39].

Poly(acrylic acid) (PAA) is a synthetic anionic electrolyte that has been synthesized for use in transmucosal drug delivery and self-curing membranes with CS. CS-PAA with a nanofibrous structure have been formed by self-assembling. The nanofibers are of different lengths ranging from 1 μm to several microns, but the widths of the nanofibers are on a nanoscale from 50 to 150 nm. In the CS/PAA system, ionic interactions occur between the two oppositely charged macromolecules forming a polycomplex that precipitates as an insoluble aggregate because the charged groups, which are responsible for solubility (carboxylic groups of PAA in the form of COO^- and amino groups of CS in the form of NH_3^+), are involved in the complex [40].

The formation of nanoparticles by self-assembly is an energy-saving process. Self-assembling nanoparticles can be formed spontaneously under mild conditions. Figure 10.10 shows the

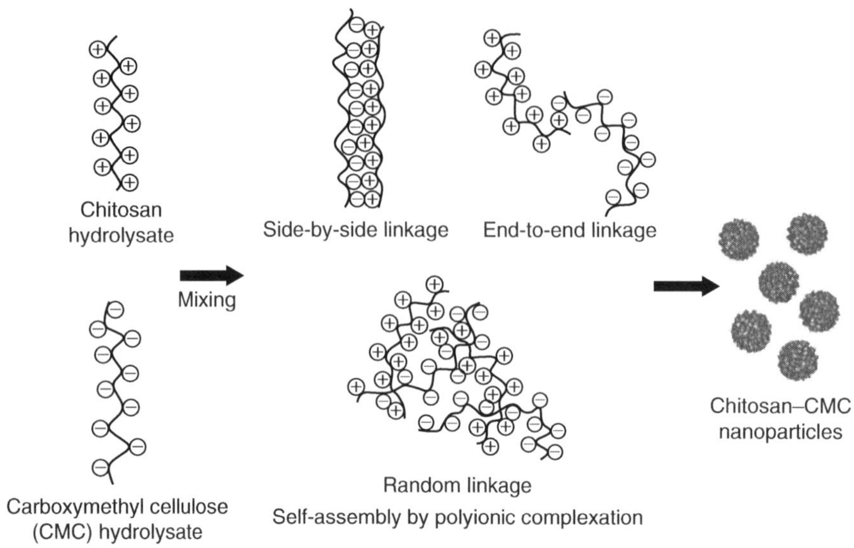

FIGURE 10.10 Schematic illustration of the formation of chitosan–CMC nanoparticles.

formation process for chitosan/carboxymethyl cellulose (CS–CMC) nanoparticles. The size of distribution was relatively narrow and the mean particle size decreased from 226 to 165 nm with the decreasing molecular weight of chitosan hydrolysate from 9.5 to 6.8 kDa. Moreover, with a decreasing pH, the average size was enhanced however; the size was not affected by low temperature [41].

10.1.2.5 Coordination Complexes

Chitosan has one amino group in addition to hydroxyl groups in the repeating unit. So chitosan can coordinate strongly with transition-metal ions [42] where the precious metal and rare metal ions form coordination complexes [42]. They may be used as absorbents for metal ions and separation membranes. The order for the stability contents of these chitosan–transition-metal complexes is Mn <Fe<C<Ni<Cu<Zn, which is known as the Irving-Williams order. The difference in the stability of complexes may cause a variance in the degree of swelling of a complex membrane whose metal/glucosamine ratio ranges from 1:8 to 1:64 [43]. Domard et al. [44] confirms that in a weak alkaline medium, the free amine function is the most favorable site for the coordination with one copper II ion, leading to a stable pending complex.

10.1.3 Applications

10.1.3.1 Controlled Release Matrixes

The aim of the controlled release of drugs is either to modulate tissue drug levels or spatially or temporarily place a drug in some region of the body to maintain efficacy and depress side reactions. Therefore, a matrix used for controlled release should offer modulating ability for drug delivery in addition to maintaining drug stability with the matrix. Chitosan has structure characteristics similar to those glycosaminoglycans (GAGs); its smart material properties have been explored for use in controlled release matrixes. The use of chitosan in the development of oral sustained release stemmed from the intragastric "floating tables" of chitosan. Chitosan has gel-forming properties in a low pH range in addition to its antacid, antiulcer characteristics. Thus, it has promising potential for use in oral sustained delivery systems. Gastrointestinal, respiratory, ophthalmologic, cervical, and vaginal mucins are hydrophilic saline gels that are thickened by nature in anionic glycoproteins. In searching for drug delivery systems for such substrates, suitable cationic polymers are needed. Chitosan is known to be bioadhesive to mucosal surfaces, such as in the eye, nose, and vagina.

The DDA of chitosan and molecular weight are the main factors that affect the role of chitosan in therapeutic and intelligent drug delivery systems [45]. The DDA also controls the degree of crystallinity and hydrophobicity in chitosan due to variations in hydrophobic interactions. These hydrophobic interactions ultimately control the loading and release characteristics of chitosan matrices. Gupta and Jabrail [46] revealed that microspheres prepared using chitosan with 62% (w/w) of DDA were suitable for sustained release of centchroman in comparison to microspheres prepared using chitosan with low (48% w/w) and high (75% w/w) DDA.

The cross-link of chitosan microspheres has an influence on the release behavior for drug, chitosan microspheres, which when cross-linked by genipen may delay and extend drug delivery [47]. The pH-sensitive CG HPN microsphere has been prepared via the inverse suspension cross-linking with GA. The cimetidine-loaded microspheres can release the drug alternately pulsed and change the pH of the medium. The HPN microspheres appear to be a promising carrier for drug delivery [48].

Chitosan–starch fibers are excellent biomaterials for drug release. Their release behavior can be controlled by changing the

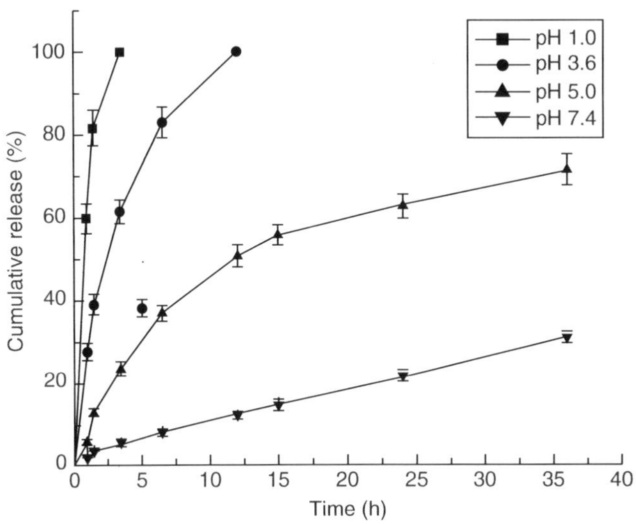

FIGURE 10.11 pH influence of the release medium on the drug controlled release process from chitosan/starch fibers (30/70, w/w).

composition ratio, the drug-loaded amount, pH, and ionic strength of the release system. For example, the release was accelerated with the decrease of the pH (Figure 10.11), because the electrostatic interaction between anions and chitosan was greatly influenced by the pH of the solutions; the decrease of pH also weakened the salt bonds and therefore, facilitated fiber swelling. As a result, the drug release was accelerated. On the other hand, the pH also has a slight effect on the solubility of the drug. A higher pH leads to a better solubility of the drug, which results in a higher drug release rate. However, compared to the strong influence of pH on the fiber matrix, pH effects on drug could be neglected [49,50].

Macroporous chitosan scaffolds reinforced by β-tricalcium phosphate (β-TCP) and calcium phosphate invert glasses also were used as a drug carrier for controlled drug release. This scaffold effectively reduced the initial burst release of antibiotic gentamicin sulfate (GS) in PBS. In comparison with pure chitosan scaffolds, the composite scaffolds showed a lower initial burst release followed by a slow sustained release of GS up to 3 weeks after immersing them in PBS. A possible mechanism for controlled GS release from the composite scaffolds might be the formation of strong cross-linking between the amine groups of chitosan and phosphate ions in PBS solution. The composite scaffolds would form more cross-linked bonds than the origin chitosan scaffold due to the presence of a higher concentration of phosphates in PBS solution [51].

Protein and enzymes represent a growing and promising field of therapeutics and are currently administrated by injection. As a result of development of biotechnology, many peptides and proteins have been explored as a new generation of drugs. Maintaining their bioactivity in the drug delivery is one of the key problems for these drugs. Oral peptide drug delivery is the easiest method of administration; however, proteins are quickly denatured and degraded in the stomach. Moreover, the intestinal transport of peptide drugs is generally very poor. Chitosan can enhance the intestinal transport of peptide drugs by increasing the paracellular permeability of the intestinal epithelium. For example, chitosan nanocapsules will increase the intestinal absorption of salmon calcitonin [52,53], and the extra PEG affect with the release behavior of the resulting nanocapsules for chitosan–PEG nanocapsules. Additionally, the increase in the pegylation degree of chitosan can lead to a significant decrease in the percentage of salmon calcitonin released. This could be related to both the greater amount of peptide associated with the chitosan–PEG nanocapsules and to the different composition and organization of the coating [30].

The progress in biotechnology, coupled with an increased understanding of the molecular mechanism of a variety of diseases at the gene level, has effected dramatic changes in therapeutic modalities. Recombinant DNA itself has been used like a "drug" in gene therapy, where genes are applied to produce therapeutic proteins in the patient. Oligonucleotides, relatively small synthetic DNA designed to hybridize specific mRNA sequences, have been used to block gene expression. To achieve these goals, the gene drugs must be administered via an appropriate route and be delivered into the intercellular site of the target cells where gene expression occurs.

Gene drugs have substantial problems such as polyanionic, nucleic, and low permeability. Therefore, a suitable carrier system is the key to successful in vivo gene therapy. Considering that viral vectors have a number of potential limitations involving safety, a cationic lipid polymer developed as NDA carriers can improve in vivo lipid transfection efficiency. Chitosan has been considered to be a good candidate for gene delivery systems. It has also been found to interact with negatively charged DNA in a similar fashion [54].

Chitosan has been shown to condense DNA effectively and protect DNA from nuclease degradation [55,56]. However, poor water solubility and low transfection efficiency of chitosan are major drawbacks for its use as a gene delivery carrier [57]. The water solubility of chitosan can be improved by various approaches such as quaternization of the amino group, N-carboxymethylation, and PEGylation. An N-dodecylated chitosan (CS-12) was synthesized and assembled with DNA to form a DNA–CS-12 complex. Its AFM image (Figure 10.12) revealed that the DNA was aggregated and the complex appeared globular in structure. The diameter varied from 90 to 230 nm, and the average height was approximately 18 nm. A single globule consisted of 40–115 DNA molecules [58,59]. Yang et al. [31] synthesized the folate-PEG-grafted chitosan (FA–PEG–CS) for targeted plasmid DNA delivery to tumor cells, and the results showed that the water solubility of FA-PEG-CS and PEG-CS were excellent and increased as a function of the PEG grafting degree. Moreover, there was strong binding between the modified chitosan and the DNA [60]. The PEG–CS DNA complexes injected via the bile duct showed an improved gene expression in a rat liver compared with unmodified chitosan [61].

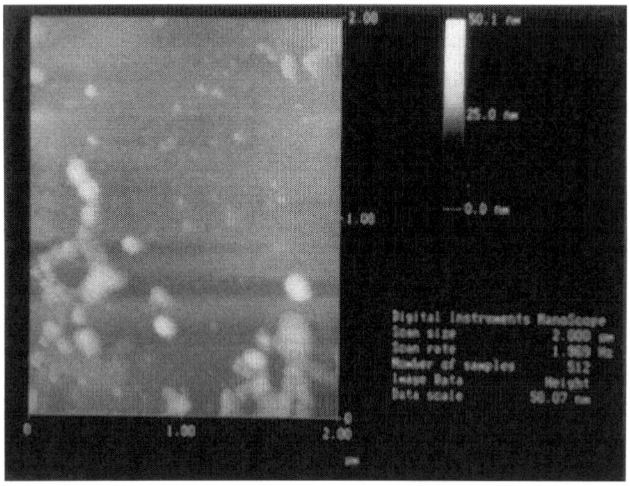

FIGURE 10.12 AFM images of DNA–CS-12 complex. The chitosan–DNA charge ratio was 2/1 and the measurement was carried out in a tapping mode.

10.1.3.2 Separation Membranes

Chemically modified chitosan membranes have been used in various fields. For example, metal-ion separation, gas separation, reverse osmosis, prevaporation separation of alcohol–water mixtures, ultrafiltration of biological macromolecular products, and the affinity precipitation of protein isolates.

Prevaporation is a very useful membrane separation technique for separating organic liquid mixtures, such as azetropic mixtures and mixtures of materials that have boiling points. In prevaporation, the characteristics of permeation and separation are significantly governed by the solubility and diffusion of the permeate.

The separation mechanism of prevaporation is based on the solution-diffusion theory, in that the adsorption–diffusion–desorption process of the component in the feed comes across the membrane from one side to the other. Therefore, prevaporation properties can be improved by enhancing the adsorption of one component in the feed to the membrane or accelerating the diffusion of one component in the feed through the membrane.

Chitosan has been used to from prevaporation membranes for separating water–alcohol mixtures and has shown good performance in dehydrating alcohol solutions. CS/PVA blend membranes have been formed by aqueous solution-casting, followed by cross-linking with formaldehyde or GA; these membranes showed improved strength and flexibility under both wet and dry conditions as compared to those in membranes of the individual component. However, cross-linking of the membranes with GA reduced the hydrophilicity of the blend membranes [62,63]. CS/PVA blend membranes exhibited higher flux than a CS membrane. With an increasing amount of PVA in the blend, polymer chains became more flexible, thereby increasing the free-volume space of the blend matrix and increasing the flux [64]. With an increasing membrane thickness, selectivity has improved, but the flux has decreased [65].

Smart polymers are the basis for a new protein isolation technique–affinity precipitation. The precipitation applies a ligand coupled to a water-soluble polymer known as an affinity macroligand, which forms a complex with the target. The phase separation of the complex triggered by small changes in environment, for example, pH, temperature, ionic strength, or addition of reagent, makes the polymer backbone insoluble; afterward, the target protein can be recovered via elution or dissolution. Chitosan itself has been successfully used as a macroligand for affinity precipitation of isolated wheat germ agglutinin and formed its extract and glycosides from cellulose preparation by changes in the pH of media [66].

Long-term stability of operation is critical for membranes when used in commercial practice. Chitosan–silica hybrid material (CSHM) based membranes, which were obtained through blending chitosan and γ-(glycidyloxypropyl)trimethoxysilane (GPTMS) in acetic acid aqueous solution, had a significant improvement in their long-term stability. CSHM-5 (GPTMS with 5% (wt %)) had an initial flux of 1730 g/(m^2 h) and showed the best long-term stability on prevaporative dehydration of 70 wt %. isopropanol–water (IPA–H_2O) solution (as shown in Table 10.1). The dehydration performance of CSHM-5 was maintained after 140 days where CSHM-10 failed in 4 days in the test owing to its brittleness [67].

Yao et al. prepared CG network films cross-linked by GA and synthesized nanohydroxyapatite (nHA) on it in situ. The results showed that the exit chemical interactions between nHA and CG templates, the COO^-, C=O, or $-NH_2$ groups may be the especially active sites for the coordination of Ca^{2+} to form ion complexes, which are the active sites of nucleation and the formation nuclei of nHA controlled by these special groups. In order to modulate the size of nHA crystalline, one can change the charged density of a CG network by changing the content CS and Gel in the CG network film [68]. Rusua et al. also prepared the size-controlled hydroxyapatite nanoparticles in a chitosan matrix [69].

Chitosan is also a cationic polyelectrolyte, and chitosan membranes have been used for active chloride ion conductivity in aqueous solutions. In the same way, if these membranes were set in a closed circuit, this kind of active ionic conductivity should be present. In its natural state (dry state), a chitosan membrane

TABLE 10.1 Long-Term Stability Test on Various Membranes for Pervaporation Dehydration of 70 wt % Isopropanol–Water (IPA–H_2O) Solution, where *X* (wt % Based on Chitosan) Indicates the Amount of GPTMS Added into the Preparation Solution

Membrane	Thickness (μm)	Initial Flux (g/(m^2 h))	Membrane Life (days)
Pure CS	11.8 ± 1.4	4470	1
CSHM-5	12.8 ± 1.0	2000	47
CSNM-2.5	12.0 ± 1.2	1730	141
CSHM-7.5	13.8 ± 0.5	1560	70
CSHM-10	14.2 ± 0.4	1240	4

has very low electrical conductivity. However, when it is swollen in water, its amino groups can be protonated and thus may contribute to ionic conduction in the membrane. A chitosan membrane made from unmodified chitosan with a high molecular weight and an appropriate DDA has an ionic conductivity of around 10^{-4} S cm^{-1} after hydration for 1 h at 25°C. However, this is still relatively low for practical applications [70]. Compared with the uncross-linked chitosan membrane, the cross-linked chitosan membranes with lower DDA values and higher molecular weights have a relatively high conductivity [71] and the ionic conductivity of hydrated modified chitosan membranes has been greatly improved over that of unmodified chitosan membrane. Since ionic conductance will occur only after membranes are hydrated, it can be concluded that the hydrated properties of membranes will critically affect the ion permeability through the membranes, and the membranes with a relatively high swelling index may allow ions to go through more easily in the swollen state of the membranes [72]. Lithium (Li^+) and proton (H^+)-conducting chitosan-based electrolytes have been prepared using the solution cast technique and have shown excellent conductivity [73]. The chitosan-based polymer electrolyte has the potential for use in practical electrochemical devices.

10.1.3.3 Immobilization Supports

Immobilization is an effective measure for using enzymes or microbial cells as recoverable, stable, and specific industrial biocatalysts. Immobilization must be carried out under mild conditions. Immobilized enzymes can be easily separated from soluble reaction products and unreacted substrate, thus simplifying the work-up and preventing protein contamination of the final product [74].

An ideal support for enzyme immobilization should be chosen to achieve essential properties such as chemical stability, hydrophilicity, rigidity, mechanical stability, large surface area, and resistance to microbial attack. Several methods have been used to modify the supports to improve stability and mechanical strength, and to modify different functional groups that may be superior for enzyme immobilization. One of the ways to improve these properties is the grafting of a monomer onto a matrix. Chitosan has been used as a support for enzyme immobilization, since it offers considerable advantages such as form versatility (powder, gel beads, flocks, fibers, capsules, and membranes), biodegradability, low toxicity, high affinity toward proteins and good biocompatibility [75]. Chitosan beads were used for immobilizing pepsin and the pepsin immobilized onto chitosan beads exhibits an improved resistance against thermal denaturation, thermal stability, as well as storage stability of immobilized pepsin which was greater than that of free pepsin [76]. Hydrophobically modifying chitosan can increase the enzyme activity, which is a more optimal matrix for enzyme immobilization [77].

Pancreatic lipase, fungal laccases, cellulose, trypsin, glutamater dehydrogenase, penicillin acylase, and glucosidase have been immobilized via chitosan gels. The result is that immobilization improves the stability of enzymes.

10.1.3.4 ECM for Tissue Engineering

Various synthetic and naturally derived hydrogels have recently been used as artificial ECMs for cell immobilization, cell transplantation, and tissue engineering. Native ECMs are complex chemically and physically cross-linked networks of proteins and GAGs. Artificial ECMs replace many functions of the native ECM, such as organizing cells into a 3-D architecture, providing mechanical integrity to new tissue, and providing a hydrated space for the diffusion of nutrients and metabolites to and from cells. Chitosan is similar to GAGs. Therefore, it is promising for application as a biomaterial in addition to its use as a controlled delivery matrix.

Chitosan is a basic polysaccharide, so it is possible to evaluate the percentage of amino functions, which remain charged at the pH of cell cultivation (7.2–7.4). These cationic charged materials have a definite influence on cell attachment by their possible interaction with negative charges located at the cell surface. Nomizu and coworkers developed biomaterials using integrin-binding peptides covalently bound to chitosan membranes and demonstrate that peptide–chitosan membranes can regulate specific integrin-mediated cell responses and are useful constructs as ECM mimetics [78].

In order to improve the cytocompatibility, Yao et al. prepared the CG–hyaluronic acid scaffold. The scaffold showed a higher water uptake and retention ability and fibroblast and keratinocytes were cocultured in the scaffold. The cells grew and proliferated well and they exhibited a strong viability. Keratinocytes were cocultured with fibroblasts in CG–hyaluronic acid scaffolds and this construct resulted in an artificial bilayer skin in vitro. The artificial skin obtained was flexible and had good mechanical properties, which suggested that CG–hyaluronic acid scaffolds are suitable for preparing a bilayer skin substitute [79,80].

Mesenchymal stem cells (MSC), a heterogeneous stem cell population within bone marrow, are reliant in the extracellular environment and not only can they survive but also can differentiate with a similar structure to GAGs and collagen. The PEC derived from gelatin and chitosan has also been used as supporting scaffolds, mimicking the natural ECM, for tissue engineering bone and cartilage with promising results. Zhao et al. examined the effects of the incorporation of hydroxyapatite in the chitosan and gelatin polymer network on human MSC microenvironment formation and 3D tissue development in vitro. The human MSCs maintained the higher propagation and multilineage differentiation potentials with an enhanced osteogenic differentiation upon induction in 3-D scaffolds [81].

10.1.3.5 Field-Responsive Materials

Chitosan-based gels consist of a positive charged network and a fluid (e.g., water) that fills the interstitial space of the network. The gels exhibit a variety of unique field-responsive behaviors, such as electromechanical (EMC) phenomena.

EMC behavior deals with the contraction of polymers in an electric field. The effect of an electric field on polyelectrolyte hydrogels relates to the proton action of its alkaline amino groups

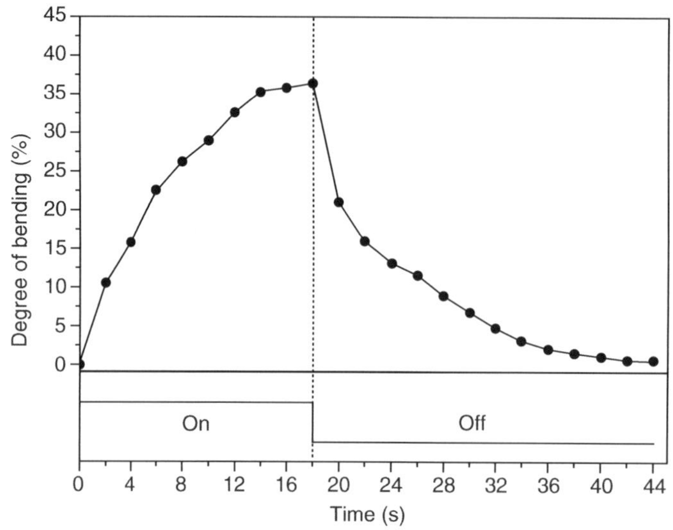

FIGURE 10.13 Bending kinetics of the chitosan/PHEMA semi-IPN, in a 1.0 wt % NaCl solution, voltage (10 V), at $T = 35°C$.

and the redistribution of mobile counter ions. The EMC behavior of chitosan and poly(hydroxyethyl methacrylate) in a 1.0 wt % NaCl solution in a 10 V electric field is shown in Figure 10.13 [21]. The figure also shows that the reversible bending behavior of chitosan–PAN semi-IPN hydrogel depends on the application of the electric field [82].

References

1. J.-K.F. Sub and H.W.K. Matthew, Application of chitosan-based polysaccharide biomaterials in cartilage tissue engineering: A review, *Biomaterials* **21**, 2589–2598, 2000.
2. G.A.F. Robert and K.E. Taylor, The formation of gels by reaction of chitosan with glutaraldehyde. *Makromol. Chem.* **190**, 995–960, 1989.
3. M. Ruiza, A.M. Sastreb, and E. Guibal, Palladium sorption on glutaraldehyde-crosslinked chitosan, *React. Funct. Polym.* **45**, 155–173, 2000.
4. M.N. Khalid, L. Ho, J.L. Agnely, J.L. Grossiord, and G. Couarraze, Swelling properties and mechanical characterization of a semiinterpenetrating chitosan/polyethylene oxide network; Comparison with a chitosan reference gel, *STP Pharm. Sci.* **9**, 359–364, 1999.
5. S. Kim, M.E. Nimni, Z. Yang, and B. Han, Chitosan/gelatin-based films crosslinked by proanthocyanidin. *J. Biomed. Mater. Res. Part B: Appl. Biomater.* **75B**, 442–450, 2005.
6. Y.-C. Kuo and C.-Y. Lin, Effect of genipin-crosslinked chitin–chitosan scaffolds with hydroxyapatite modifications on the cultivation of bovine knee chondrocytes, *Biotechnol. Bio Eng.* **95**(1), 132–144, 2006.
7. W.F. Lee and Y.J. Chen, Studies on preparation and swelling properties of the nisopropylacrylamide/chitosan semi-IPN and IPN hydrogels, *J. Appl. Polym. Sci.* **82**, 2487–2496, 2001.
8. X. Qu, A. Wirsen, and A.C. Albertsson, Novel pH-sensitive chitosan hydrogels: Swelling behavior and states of water, *Polymer* **41**, 4589–4598, 2000.
9. S.C. Chen, Y.C. Wu, F.L. Mi, Y.H. Lin, L.C. Yu, and H.W. Sung, A novel pH-sensitive hydrogel composed of *N,O*-carboxymethyl chitosan and alginate cross-linked by genipin for protein drug delivery, *J. Control. Release* **96**, 285–300, 2004.
10. K.D. Yao, T. Peng, M.F.A. Goosen, J.M. Min, and Y.Y. He, pH-sensitivity of hydrogels based on complex forming chitosan–polyether interpenetrating polymer network, *J. Appl. Polym. Sci.* **48**(2), 354–353, 1993.
11. T. Peng, K.D. Yao, C. Yuan, and M.F.A. Goosen, Structural changes of pH-sensitive chitosan/polyether hydrogels in different pH solution, *J. Polym. Sci., Part A.* **32**(3), 591–593, 1994.
12. T. Wang and S. Gunasekaran, State of water in chitosan–PVA hydrogel, *J. Appl. Polym. Sci.* **101**, 3227–3232, 2006.
13. T. Wang, M. Turhan, and S. Gunasekaran, Selected properties of pH-sensitive, biodegradable chitosan–poly(vinyl alcohol) hydrogel, *Polym. Int.* **53**, 911–918, 2004.
14. Y. Lin, Q. Chen, and H. Luo, Preparation and characterization of *N*-(2-carboxybenzyl)chitosan as a potential pH-sensitive hydrogel for drug delivery, *Carbohyd. Res.* **342**, 87–95, 2007.
15. E.K. Sarkyt and B.S. Vladimir, Swelling, shrinking, deformation, and oscillation of polyampholyte gels based on vinyl 2-aminoethyl ether and sodium acrylate, *Langmuir* **15**, 4230–4235, 1999.
16. A. Sionkowska, M. Wisniewski, J. Skopinska, C.J. Kennedy, and T.J. Wess, Molecular interactions in collagen and chitosan blends, *Biomaterials* **25**, 795–801, 2004.

17. Z. Chen, X. Mo, and F. Qing, Electrospinning of collagen–chitosan complex, *Mater. Lett.* **61**, 3490–3494, 2007.
18. K.D. Yao, Y.J. Yin, M.X. Xu, and Y.F. Wang, Investigation of pH-sensitive drug delivery system of chitosan/gelatin hybrid polymer network, *Polym. Int.* **38**(1), 77–82, 1995.
19. G. Ivarez-Lorenzo, A. Concheiro, A.S. Dubovik, N.V. Grinberg, T.V. Burova, and V.Ya. Grinberg, Temperature-sensitive chitosan-poly(*N*-isopropylacrylamide) interpenetrated networks with enhanced loading capacity and controlled release properties, *J. Control. Release* **102**, 629–641, 2005.
20. S.J. Kim, K.J. Lee, I.Y. Kim, D.I. Shin, and S. Kim, Temperature and pH-response swelling behavior of poly(2-ethyl-2-oxazoline)/chitosan interpenetrating polymer network hydrogels, *J. Appl. Polym. Sci.* **99**, 1100–1103, 2006.
21. S.J. Kim, S.R. Shin, S.M. Lee, I.Y. Kim, and S.I. Kim, Electromechanical properties of hydrogels based on chitosan and poly(hydroxyethyl methacrylate) in NaCl solution, *Smart. Mater. Struct.* **13**, 1036–1039, 2004.
22. S.J. Kim, S.J. Park, M.-S. Shin, and S.I. Kim, Characteristics of electrical responsive chitosan/polyallylamine interpenetrating polymer network hydrogel, *J. Appl..Polym. Sci.* **86**, 2290–2295, 2002.
23. Y. Yin, Z. Li, Y. Sun, and K. Yao, A preliminary study on chitosan/gelatin polyelectrolyte complex formation, *J. Mater. Sci.* **40**, 4649–4652, 2005.
24. K.D. Yao, H.L. Tu, F. Chang, J.W. Zhang, and J. Liu, pH sensitivity of the swelling of a chitosan–pectin polyelectrolyte complex, *Angew. Makromol. Chem.* **245**, 63–72, 1997.
25. K.Y. Lee, W.H. Park, and W.S. Ha, Polyelectrolyte complexes of sodium alginate with chitosan or its derivatives for microcapsules, *J. Appl. Polym. Sci.* **63**(4), 425–432, 1997.
26. L. Becherán-Marón, C. Peniche, and W. Argüelles-Monal, Study of the interpolyelectrolyte reaction between chitosan and alginate: Influence of alginate composition and chitosan molecular weight, *Int. J. Biol. Macromol.* **34**, 127–133, 2004.
27. T. Yoshizawa, Y. Shin-ya, K.-J. Hong, and T. Kajiuchi, pH- and temperature-sensitive permeation through polyelectrolyte complex films composed of chitosan and polyalkylene-oxide–maleic acid copolymer, *J. Membrane Sci.* **241**, 347–354, 2004.
28. T. Yoshizawa, Y. Shin-ya, K.-J. Hong, and T. Kajiuchi, pH- and temperature-sensitive release behaviors from polyelectrolyte complex films composed of chitosan and PAOMA copolymer, *Eur. J. Pharm. Biopharm.* **59**, 307–313, 2005.
29. E. Fernandez-Megia, R. Novoa-Carballal, E. Quiñoá, and R. Riguera, Conjugation of bioactive ligands to PEG-grafted chitosan at the distal end of PEG, *Biomacromolecules* **8**, 833–842, 2007.
30. C. Prego, D. Torres, E. Fernandez-Megia, R. Novoa-Carballal, E. Quiñoá, and M.J. Alonso, Chitosan–PEG nanocapsules as new carriers for oral peptide delivery effect of chitosan pegylation degree, *J. Control. Release,* **111**, 299–308, 2006.
31. P. Chan, M. Kurisawa, J.E. Chung, and Y.-Y. Yang, Synthesis and characterization of chitosan-*g*-poly(ethylene glycol)-folate as a non-viral carrier for tumor-targeted gene delivery, *Biomaterials* **28**, 540–549, 2007.
32. M.S. Beena, T. Chandy, and C.P. Sharma, Heparin immobilized chitosan–poly(ethylene glycol) interpenetrating network: Antithrombogenicity. *Art. Cells Blood Subst. Immobil. Biotech.* **23**(2), 175–192, 1995.
33. Y. Hu, H. Jiang, C. Xu, Y. Wang, and K. Zhu, Preparation and characterization of poly(ethylene glycol)-*g*-chitosan with water and organosolubility, *Carbohyd. Polym.* **61**, 472–479, 2005.
34. C.A. Landes and S. Kriener, Resorbable plate osteosynthesis of sagittal split osteotomies with major bone movement, *Plast. Reconstr. Surg.* **111**, 1828–1840, 2003.
35. F. Yao, W. Chen, H. Wang, H. Liu, K. Yao, P. Sun, and H. Lin, A study on cytocompatible poly(chitosan-*g*-L-lactic acid), *Polymer* **44**, 6435–6441, 2003.
36. Y. Fanglian, L. Chang, C. Wei, B. Yun, T. Zhiyuan, and Y. Kangde. Synthesis and characterization of chitosan grafted oligo(L-lactic acid), *Macromol. Biosci.*, **3**, 653–656, 2003.
37. A.R. Esker, C. Mengel, G. Wegner, Ultrathin films of a polyelectrolyte with layered architecture, *Science* **280**, 892–895, 1998.
38. P.T. Hammond, Recent explorations in electrostatic multilayer thin film assembly, *Curr. Opin. Colloid Interf. Sci.* **4**, 430–442, 2000.
39. Q. Feng, G. Zeng, P. Yanga, C. Wang, and J. Caia, Self-assembly and characterization of polyelectrolyte complex films of hyaluronic acid/chitosan, *Colloid. Surface A* **257–258**, 85–88, 2005.
40. C.-Y. Chen, J.-W. Wang, and M.-H. Hon, Polyion complex nanofibrous structure formed by self-assembly of chitosan and poly(acrylic acid), *Macromol. Mater. Eng.* **291**, 123–127, 2006.
41. S. Ichikawa, S. Iwamoto, and J. Watanabe, Formation of biocompatible nanoparticles by self-assembly of enzymatic hydrolysates of chitosan and carboxymethyl cellulose, *Biosci. Biotechnol. Biochem.*, **69**(9), 1637–1642, 2005.
42. G. Huai-min and C. Xian-su, Study of cobalt(II)–chitosan coordination polymer and its catalytic activity and selectivity for vinyl monomer polymerization, *Polym. Adv. Technol.* **15**, 89–92, 2004.
43. N. Kobota, Permeability properties of chitosan–transition metal complex membrane, *J. Appl. Polym. Sci.* **64**, 819–822, 1997.
44. R. Terreux, M. Domard, C. Viton, and A. Domard, Interactions study between the copper II ion and constitutive elements of chitosan structure by DFT calculation, *Biomacromolecules* **7**, 31–37, 2006.
45. D. Thacharodi and K.P. Ruo, Propranolol hydrochloride release behavior of cross-linked chitosan membranes, *J. Chem. Technol. Biotechnol.* **58**, 177–181, 1993.
46. K.C. Gupta and F.H. Jabrail, Effects of degree of deacetylation and cross-linking on physical characteristics, swelling and release behavior of chitosan microspheres, *Carbohyd. Polym.* **66**, 43–54, 2006.
47. Y. Yuan, B.M. Chesnutt, G. Utturkar, W.O. Haggard, Y. Yang, J.L. Ong, and J.D. Bumgardner, The effect of cross-linking of

chitosan microspheres with genipin on protein release, *Carbohyd. Polym.* **68**, 561–567, 2007.
48. K.D. Yao, M.X. Xu, Y.J. Yin, J.Y. Zhao, and X.L. Chen, pH-sensitive chitosan/gelatin hybrid polymer network microspheres for delivery of cimetidine, *Polym. Int.* **39**, 333–337, 1996.
49. Q. Wang, N. Zhang, X. Hu, J. Yang, and Y. Du, Chitosan/starch fibers and their properties for drug controlled release. *Eur. J. Pharm. Biopharm.* **66**, 39–404, 2007.
50. Q. Wang, Y.M. Du, and L.H. Fan, Properties of chitosan/poly (vinyl alcohol) films for drug controlled release, *J. Appl. Polym. Sci.* **96**(3), 808–813, 2005.
51. Y. Zhang and M. Zhang, Calcium phosphate/chitosan composite scaffolds controlled in vitro antibiotic drug release, *J. Biomed. Mater. Res.* **62**, 378–386, 2002.
52. C. Prego, M. García, D. Torres, and M.J. Alonso, Transmucosal macromolecular drug delivery, *J. Control. Release*, **101**, 151–162, 2005.
53. C. Prego, M. Fabre, D. Torres, and M.J. Alonso, Efficacy and mechanism of action of chitosan nanocapsules for oral peptide delivery, *Pharm. Res.* **23**(3), 549–556, 2006.
54. F.C. MacLaughlin, R.J. Mumper, J. Wang, J.M. Tagliaferri, I. Gill, M. Hinchcliffe, et al., Chitosan and depolymerized chitosan oligomers as condensing carriers for in vivo plasmid delivery, *J. Control. Release* **56**, 259–272, 1998.
55. S. Mao, X. Shuai, F. Unger, M. Simon, D. Bi, and T. Kissel, The depolymerization of chitosan: Effects on physicochemical and biological properties, *Int. J. Pharm.* **281**, 45–54, 2004.
56. W.G. Liu and K.D. Yao, Chitosan and its derivatives—a promising nonviral vector for gene transfection, *J. Control. Release* **83**, 1–11, 2002.
57. M. Prabaharan and J.F. Mano, Chitosan-based particles as controlled drug delivery systems, *Drug Deliv.* **12**(1), 41–57, 2005.
58. W.G. Liu, K.D. Yao, and Q.G. Liu, Formation of a DNA/N-dodecylated chitosan complex and salt-induced gene delivery, *J. Appl. Polym. Sci.* **82**, 3391–3395, 2001.
59. W.G. Liu and K.D. Yao, Chitosan and its derivatives—a promising non-viral vector for gene transfection, *J. Control. Release* **83**, 1–11, 2002.
60. P. Chan, M. Kurisawa, J.E. Chung, and Y.-Y. Yang, Synthesis and characterization of chitosan-*g*-poly(ethylene glycol)-folate as a non-viral carrier for tumor-targeted gene delivery, *Biomaterials,* **28**, 540–549, 2007.
61. X. Jiang, H. Dai, K.W. Leong, H.Q. Hao, and Y.Y. Yang, PEG-*g*-chitosan complexes for liver targets gene delivery, Abstract for International Conference on Materials For Advanced Technologies, P12, Singapore, 2003.
62. Y.J. Ming, S.W. Su, L.T. Lang, and Y.M. Chen, Evaluation of chitosan/PVA blended hydrogel membranes, *J. Membr. Sci.* **236**, 39–51, 2004.
63. S.B. Bahrami, S.S. Kordestani, H. Mirzadeh, and P. Mansoon, Poly(vinyl alcohol)–chitosan blends: Preparation, mechanical and physical properties, *Iranian Polym. J.* **12**, 139–146, 2003.
64. K.S.V. Krishna Rao, M.C.S. Subha, M. Sairam, N.N. Mallikarjuna, and T.M. Aminabhavi, Blend membranes of chitosan and poly(vinyl alcohol) in pervaporation dehydration of isopropanol and tetrahydrofuran, *J. Appl. Polym. Sci.* **103**, 1918–1926, 2007.
65. D. Anjali Devi, B. Smitha, S. Sridhar, T.M. Aminabhavi, Dehydration of 1,4-dioxane through blend membranes of poly(vinyl alcohol) and chitosan by pervaporation, *J. Membrane Sci.* **280**, 138–147, 2006.
66. I.Y. Galaev, M.N. Gupta, and B. Mattiasson, Use smart polymers for bioseparations, *Chemtech*, **26**(12), 19–25, 1996.
67. Y.-L. Liu, Y.-H. Su, K.-R. Lee, and J.-Y. Lai, Crosslinked organic–inorganic hybrid chitosan membranes for pervaporation dehydration of isopropanol–water mixtures with a long-term stability, *J. Membrane. Sci.* **251**, 233–238, 2005.
68. J. Li, Y.P. Chen, Y. Yin, F. Yao, and K. Yao, Modulation of nano-hydroxyapatite size via formation on chitosan-gelatin network film in situ, *Biomaterials*, **28**(5), 781–790, 2007.
69. V.M. Rusua, C.-H. Ng, M. Wilke, B. Tiersch, P. Fratzl, and M.G. Pete, Size-controlled hydroxyapatite nanoparticles as self-organized organic–inorganic composite materials, *Biomaterials*, **26**, 5414–5421, 2005.
70. Y. Wan, K.A.M. Creber, B. Peppley, and V.T. Bui, Ionic conductivity of chitosan membranes, *Polymer* **44**, 1057–1065, 2003.
71. Y. Wan, K.A.M. Creber, B. Peppley, and V. Tam Bui, Ionic conductivity and related properties of crosslinked chitosan membranes, *J. Appl. Polym. Sci.* **89**, 306–317, 2003.
72. Y. Wan, K.A.M. Creber, B. Peppley, and V. Tam Bui, Ionic conductivity and tensile properties of hydroxyethyl and hydroxypropyl chitosan membranes, *J. Polym. Sci. Part B: Polym. Phys.* **42**, 1379–1397, 2004.
73. T. Winie, S.R. Majid, A.S.A. Khiar, and A.K. Arof, Ionic conductivity of chitosan membranes and application for electrochemical devices, *Polym. Adv. Technol.* 17, 523–527, 2006.
74. J. Lalonde and A. Margolin, Immobilization of enzymes. In: Drauz, K. and Waldmann, H., editors. *Enzyme Catalysis in Organic Synthesis.* Weinheim: Wiley–VCH, pp. 163–184, 2002.
75. N. Ha and S.P. Meyers, Preparation and characterization of chitin and chitosan, a review, *J. Aquat. Food. Prod. Tech.* **4**, 27–52, 1995.
76. G.D. Altun and S.A. Cetinus, Immobilization of pepsin on chitosan beads, *Food Chem.* **100**, 964–971, 2007.
77. T. Klotzbach, M. Watt, Y. Ansari, and S.D. Minteer, Effects of hydrophobic modification of chitosan and Nafion on transport properties, ion-exchange capacities, and enzyme immobilization, *J. Membrane. Sci.* **282**, 276–283, 2006.
78. M. Mochizuki, N. Yamagata, D. Philp, K. Hozumi, T. Watanabe, Y. Kikkawa, Y. Kadoya, H.K. Kleinman, and M. Nomizu, Integrin-dependent cell behavior on ECM peptide-conjugated chitosan membranes, *Biopolymers (Pept. Sci.)*, **88**(2), 122–130, 2007.
79. J.S. Mao, H.F. Liu, Y.J. Yin, and K.D. Yao, The properties of chitosan–gelatin membranes and scaffolds modified with hyaluronic acid by different methods, *Biomaterials*, **24**, 1621–1629, 2003.
80. H. Liu, J. Mao, K. Yao, G. Yang, L. Cui, and Y. Cao, A study on a chitosan–gelatin–hyaluronic acid scaffold as artificial skin

in vitro and its tissue engineering applications, *J. Biomater. Sci. Polymer Edn.* **15**(1), 25–40, 2004.

81. F. Zhao, W.L. Graysona, T. Maa, B. Bunnellb, W.W. Lu, Effects of hydroxyapatite in 3-D chitosan–gelatin polymer network on human mesenchymal stem cell construct development, *Biomaterials*, **27**, 1859–1867, 2006.
82. S.J. Kim, S.R. Shin, J.H. Lee, S.H. Lee, and S.I. Kim, Electrical response characterization of chitosan/polyacrylonitrile hydrogel in NaCl solutions, *J. Appl. Polym. Sci.* **90**, 91–96, 2003.

10.2 Chitosan-Based Hydrogels in Biomedical and Pharmaceutical Sciences

Claire Jarry and Matthew S. Shive

10.2.1 Introduction

In the past few decades, hydrogels based on natural biopolymers such as chitosan, collagen [1], alginate [2], and hyaluronic acid [3] have been widely studied because of their biological properties and potential uses in biomedical and pharmaceutical fields. The polysaccharide chitosan, in particular, has demonstrated unique properties, which it brings to chitosan-based hydrogels, including biocompatibility and if desired, biodegradability, as well as meeting acceptable levels of efficacy and safety. Indeed, chitosan-based hydrogels have been investigated for specific applications in areas as varied as drug delivery systems for controlled release of drugs [4,5] and proteins [6], to scaffolds for tissue engineering [7]. In addition to hydrogels, chitosan in the form of films, beads, microspheres, and solutions has been well-studied [8], and owing to chitosan's biological and physicochemical properties, have led to the development of numerous innovations not only in the pharmaceutical [8–10] and biomedical [11] industries, but also cosmetics [12], paper improvement [13], waste water treatment [14], food and agriculture [15], and textile [16] industries. Numerous properties and applications of chitosan are presented in Table 10.2.

10.2.2 Chitosan and Chitosan Derivatives

Chitosan is a natural copolymer of D-glucosamine and *N*-acetyl-D-glucosamine units belonging to the family of β(1–4)-linked polysaccharides. Chitosan is derived from chitin, which is

TABLE 10.2 Applications of Chitosan and Chitosan Derivatives

Property	Industry	Applications	References
Absorption enhancement	Pharmaceutical	Increase permeability of tight junctions for hydrophilic macromolecules for oral drug and nasal vaccine deliveries	[6,86–88]
Angiogenic	Pharmaceutical and biomedical	Cartilage regeneration and tissue engineering	[4,76]
Antimicrobial	Food and agriculture	Seed and food preservation	[15,83,89–91]
	Cosmetics	Body lotion, toothpaste	[12]
Cationic	Pharmaceutical and biomedical	Gene delivery through anionic complexation, Drug delivery from microspheres	[46,47,92,93]
Cell substrate	Biomedical	Bone regeneration	[66,94]
	Pharmaceutical	Fat and cholesterol binding	[95,96]
Chelation	Waste water treatment	Removal of metal ions	[14]
	Chemistry	Separation of inorganic anions	[97]
	Textile	Dye sorption	[98]
Coating	Paper	Paper mechanical properties improvement	[13]
	Biomedical	Improvement of osseointegration of orthopedic devices	[99]
	Cosmetics	Hair conditioning	[100]
Cytocompatibility	Biomedical	Intervertebral disk regeneration bone and cartilage regeneration	[41,73,79,81]
Hemostatic	Biomedical	Hemostatic dressings	[101–103]
Gelling	Pharmaceutical	Local drug delivery	[5,59]
Mucoadhesive	Biomedical	Postsurgical adhesion prevention and tissue adhesive	[31,32]
		Cartilage regeneration	[75–77]
Osteoconductive	Biomedical	Bone regeneration	[61,68,104]
Wound healing	Biomedical	Wound occlusion	
		Wound dressings	[85,105,106]

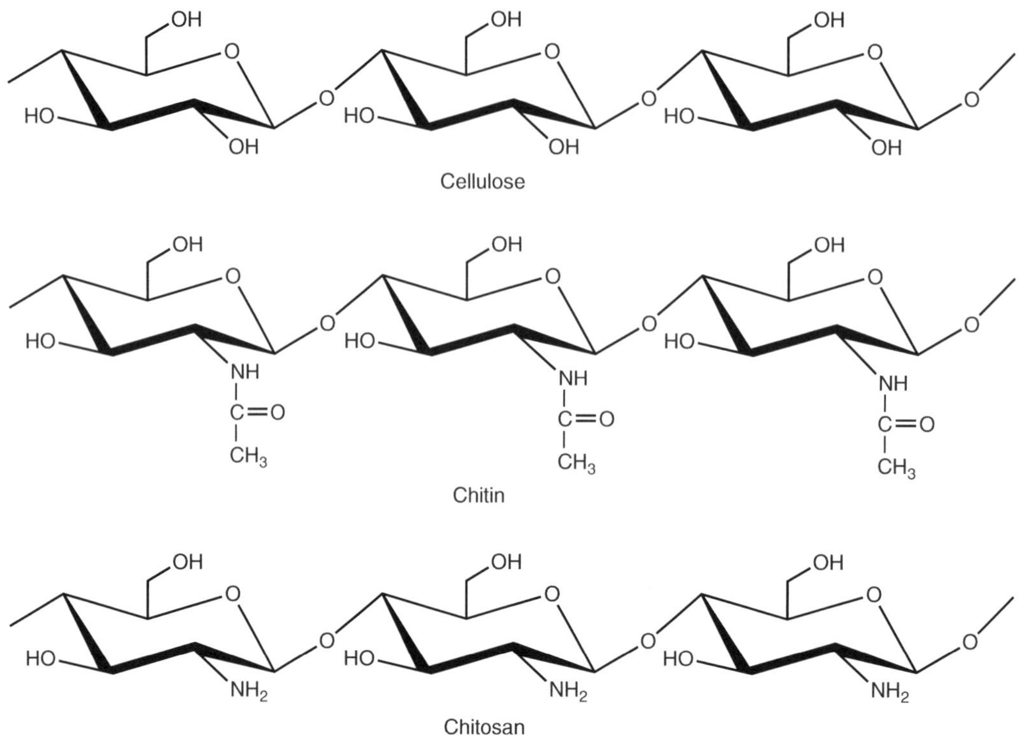

FIGURE 10.14 Structures of cellulose, chitin, and chitosan.

plentiful in nature, and is found in exoskeletons of crustaceans, as well as some insects, yeast, and fungi. Very similar to cellulose in structure, chitin has N-acetyl groups on its C_2 units as opposed to the hydroxyl groups of cellulose (Figure 10.14). The partial substitution of chitin's N-acetyl groups for NH_2 groups (deacetylation) under alkaline conditions yields chitosan and the degree of substitution is called DDA. Chitosan with a DDA between 70% and 95% has shown utility in the pharmaceutical and biomedical fields. At or below its pK_a of ~6.3 [17], chitosan is positively charged in dilute acids since the free NH_2 groups are protonated, thus overcoming the associative forces between chains and permitting solubilization. On the other hand, if the solvent pH is above the pK_a, the fraction of protonated NH_2 groups on the chitosan is insufficient and the polymer remains insoluble. DDA is one parameter that plays an important role in the biodegradability of chitosan, among other things. Indeed, the degradation of chitosan in vivo may occur primarily through the endogenous enzyme lysozyme, and is inversely proportional to its DDA, i.e., a lower DDA permits degradation to a greater extent than a higher DDA, and in cases of very high DDA (>90%), enzymatic degradation slows such that these chitosans are considered nonbiodegradable [18–21].

Further chemical modification of the chitosan chain other than deacetylation has been studied as a way of bringing new functionalities to this polymer. Derivatization of chitosan through chemical substitution of the primary NH_2 groups can result in desired functional groups randomly distributed along the backbone. The modified chitosan, or chitosan derivative, and the substituted functional groups can bring improved physicochemical properties to the polymer. For example, the mucoadhesivity of chitosan for a drug delivery application was improved using thiol substitution. The thiolated chitosan enhanced the pH-dependent formation of both inter- and intramolecular disulfide bonds with mucus glycoproteins, and prolonged the controlled release of therapeutic agents by providing a strong cohesion and stability to the delivery system [22].

10.2.3 Hydrogels

By definition, a gel is a colloidal dispersion (dispersed solid in a liquid solvent) that spontaneously forms a homogenous and fine network (coagulation). Both the liquid and solid are in a continuous phase bringing attributes of both to the gel. By convention, when the solvent is water, the term hydrogel is used. A xerogel describes the removal of the liquid phase leaving only the gel network structure. In the pharmaceutical field, hydrogels are commonly applied as drug delivery systems because they can be designed to respond to physiological environments. For instance, dramatic behavioral changes in swelling properties and mechanical strength can be induced as consequence of exposure to different stimuli such as pH, temperature, local protein concentrations, and ionic strengths [23]. Hydrogels can be classified into two groups, chemical or physical, depending on the nature of the bonds between the network chains.

10.2.3.1 Chemical Hydrogels

Chemical hydrogels are so called because of covalent bonding between hydrophilic macromolecules crosslinking a three-dimensional network. The stabilizing cross-links give these irreversible hydrogels the ability to swell in water while maintaining that network. Chitosan chemical hydrogels most commonly achieve this cross-linking using glutaraldehyde [24–26], but sclearaldehyde [27], glyoxal [28,29], genipin [30], and ultraviolet irradiation [31,32] have also been employed. Use of such chemical cross-linkers, however, has raised safety concerns regarding human use due to the potential toxicity of free unreacted molecules and has motivated further development of alternative types of chitosan hydrogels.

10.2.3.2 Physical Hydrogels

Physical hydrogels are formed through the physical gelation of chitosan as a consequence of modifying the polymers hydrophilic–hydrophobic balance, which permits the formation of both hydrophobic interactions as well as hydrogen bonding. These hydrogels are often reversible because, as opposed to the chemical hydrogels, they are maintained via much weaker intermolecular forces, such as electrostatic or hydrophobic interactions as well as hydrogen bonding. Several parameters or conditions can be regulated to control the properties of these gels, including pH, media, temperature, chitosan DDA, and charge density. Intelligent use of chitosan derivatives, in particular, can facilitate the manipulation of these parameters, thus enhancing specific chitosan interactions and further improving the hydrogel characteristics, in general. This strategy was used with a polyethylene glycol (PEG) derivative that increased the solubility of chitosan in neutral pH [33], grafted N-isobutyryl groups, which imparted thermosensitivity to a hydrogel [34], alkyl chains that favored the formation of hydrophobic domains thus improving gelation [35,36], a polyampholyte hydrogel, i.e., a hydrogel that contains both positive and negative charges and thus, being pH-dependant, that was achieved via carboxymethyl groups [37], and an ethylenediaminetetraacetic acid (EDTA) derivative that improved the antimicrobial nature of topical gels [38].

Physical chitosan hydrogels can also be achieved through the complexation of polyelectrolytes of opposite charges, called ionotropic gelation [39]. Chitosan, by its cationic nature, will ionically interact with polyanion polymers, such as alginate [40,41], pectin [42,43], xanthan [44], collagen [45], and DNA [46,47].

A special group of physical hydrogels have been studied called injectable thermosensitive hydrogels, which have been adapted for in vivo use, where they solidify in situ upon injection, and are attractive in many fields due to less invasive delivery. Although hydrogel systems, which gel with a decrease in temperature, may be more familiar (e.g., poly(vinyl alcohol), gelatin, or agar), those that are liquid at room temperature or below and require an increase in temperature to solidify may be more amenable to pharmaceutical and biomedical applications. Poloxamer, a copolymer of poly(ethylene oxide) and poly(propylene oxide), and poly(*N*-isopropyl acrylamide) are the most studied systems of this kind [48]. Few thermosensitive systems, however, have contained chitosan due to its physiologically incompatible low pK_a (6.3), and the systematic formation of unusable hydrated gel-like precipitates at pH values above it. Attempts to develop chitosan-based systems that are both injectable and liquid at physiological pH, as well as thermosensitive have proven to be challenging. One such attempt grafted PEG onto chitosan to increase its solubility to a pH near 7. By controlling both the degree of substitution of PEG on the chitosan chain as well as the concentrations of both polymers, an injectable thermosensitive [30] and thermoreversible [49] hydrogel was obtained, which gelled with an increase in temperature above 25°C. While not fully understood, the mechanism of gelation seems to include hydrophobic interactions between chitosan polymer chains, hydrogen bonding between hydroxyl and amino groups of chitosan, and intermolecular bonding between PEG chains.

Another interesting system utilizing chitosan and polyol salts was successfully developed [50,51], yielding an injectable and thermosensitive hydrogel at a pH near 7 called chitosan-GP. Glycerophosphate (GP), a polyol-phosphate salt, was added to an aqueous chitosan solution where it acted as a buffering agent and permitted chitosan solutions with pH values of 7 without precipitation or gelation. Increasing temperatures increased gelation involving a combination of electrostatic repulsions between like-charged chitosan chains, electrostatic attraction between the oppositely charged chitosan and GP, and hydrophobic interchain attractions. The temperature of gelation in this chitosan–GP system can be controlled, as shown in Figure 10.15, primarily through the regulation of the solution pH. If desired, gelation at human body temperature can be achieved, such as a solution with a 2% chitosan concentration and a pH of 7.2, which gels at 37°C, while slight acidification to pH 6.85 increases the gelation temperature to nearly 50°C.

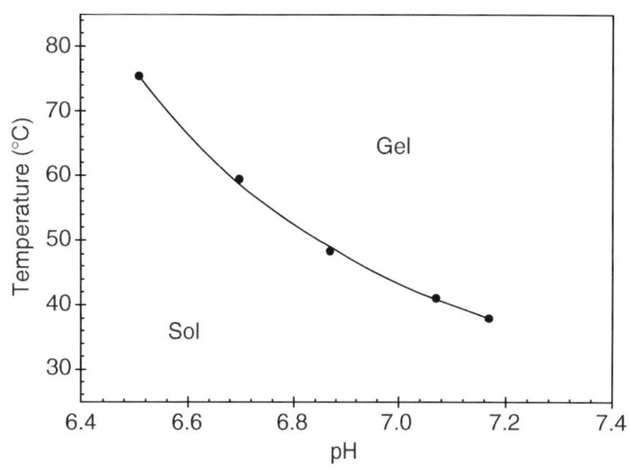

FIGURE 10.15 pH-Dependent gelation temperature of chitosan–GP solutions. (Reprinted from Chenite, A., Buschmann, M., Wang, D., Chaput, C., and Kandani, N., *Carbohydr. Polym.*, 46, 39, 2001. With permission.)

10.2.4 Chitosan Hydrogel Applications

Chitosan-based hydrogels have been widely studied in the development of controlled drug delivery systems, tissue engineering, and other biomedical applications.

10.2.4.1 Drug Delivery Systems

Drug delivery systems are designed to act as depots, safely releasing physically entrapped bioactive molecules at a desired rate. In most cases, drug delivery systems based on chitosan or chitosan derivative hydrogels are initially in a dry state, such as films, tablets, or microspheres. The dry hydrogel swells following contact with biological fluids, thus permitting dissolution of the drug and subsequently its diffusion and release through the polymeric matrix of the hydrogel. Furthermore, the utility of the chitosan matrix in protecting sensitive drugs, proteins, and peptides from harsh environmental conditions at the delivery site is particularly interesting in applications targeting the gastrointestinal tract, where these proteins and peptides have short half-lives and are poorly absorbed when administrated orally, due to their hydrophilic nature and large molecular size. The susceptibility of such molecules to degradation by proteolytic enzymes and harsh conditions can be decreased by use of chitosan, improving their absorption from the gastrointestinal tract when used in the form of gel beads [52,53], microspheres [54,55], or noncovalently cross-linked chitosan hydrogels [56,57]. Chitosan microspheres encapsulated insulin and prevented its proteolytic enzymatic degradation, and improved its bioavailability by controlling the release rate of the drug from the microspheres [54]. pH-sensitive microspheres of chitosan/sodium tripolyphosphate/dextran sulfate resisted hydrolysis in strong acid and biodegradation in enzymatic surroundings and achieved successful intestinal delivery of the hydrophobic drug, ibuprofen [58].

Typically, the systemic administration of repeated doses of a therapeutic agent leads to large plasma drug concentration variations. Sustained release of macromolecules was achieved over a period of several hours to a few days, using the chitosan-GP injectable formulation [5]. The same system loaded with paclitaxel for local delivery demonstrated the ability to inhibit tumor growth in a subcutaneous tumor mouse model at the same level as four intravenous injections of Taxol® (registered trademark of Bristol-Myers Squibb Co., New York), but with less toxicity [59]. Because the release is usually monitored by the diffusion of the drug through the matrix, a burst release could still occur in the first hours after implantation. To manage this burst effect and to maintain therapeutic plasma levels of the drug, additional cross-links can be introduced into a physical hydrogel to further reduce the mobility of polymer chains and thus slow drug release. Improved sustained release can be achieved using the addition of another interactive polymer, such as chondroitin sulfate, into the chitosan hydrogel beads to prolong the release of drug [52], or by increasing gel hydrophobicity and thereby limiting both drug dissolution and chain relaxation in nonconvalently cross-linked hydrogels [56]. Other techniques can be employed to prolong protein release, such as cross-linking PEG-grafted chitosan with genipin, transforming a physical hydrogel into an insoluble chemical hydrogel [30]. Also, the rate of bioerosion and biodegradation of these hydrogels can be manipulated by using chitosans with lower DDAs, which degrade more rapidly and release drug faster than those containing more permanent chitosans with higher DDAs. In the latter case, the delivery rate will be more dependent on drug diffusion from the hydrogel matrix than degradation.

10.2.4.2 Tissue Engineering

Tissue engineering is the biomedical field approaching the repair of damaged tissues through the creation of new tissue or the modification or repair of existing tissues. Hydrogels, in general, are an attractive material for tissue engineering because they are structurally similar to the ECM of many tissues, can often be processed under relatively mild conditions, and may be applied in vivo in a minimally invasive manner. Because of regenerative and healing properties associated with chitosan and chitosan derivatives [11], chitosan hydrogels have been used as scaffold for applications in tissue engineering of bone, cartilage and intervertebral disk, as well as in wound repair.

10.2.4.2.1 Bone Repair

Chitosan exhibits osteoconductivity [60–62] i.e., it supports the attachment of new bone cells (osteoblasts) and osteoprogenitor cells, and facilitate the formation of new bone (osteogenesis) [63]. Consequently, it has been evaluated as a potential vehicle for bioceramics particles as an alternative to the gold standard of autografting (same donor/receptor) and its associated difficulties of procurement morbidity, increased operative time, and limited availability especially when treating large osseous defects. Due to a chemical composition similar to that of natural bone and excellent osteoconductivity [64,65], calcium phosphate bioceramics have been mixed with biodegradable polymers, which act as carriers to facilitate implantation and to favor bone regeneration as both the polymer and the calcium phosphate degrade over time. And while biphasic calcium phosphate (BCP), tricalcium phosphate, and hydroxyapatite have been mixed with chitosan to produce sponges [66], pastes or composites [67–69], and cements [70], use of chitosan hydrogels has been quite limited [41]. The chitosan–GP hydrogel was combined with BCP particles, resulting in a fully injectable solution capable of forming a solid bone graft substitute at body temperature [71]. BCP is known to stimulate and promote bone regeneration [65,72], and combined with the osteoconductivity of chitosan, this injectable system could be applied in both orthopedic and dental applications. Studies using environmental scan electronic microscopy (ESEM) showed that BCP particles are uniformly embedded into the hydrogel matrix (Figure 10.16) where it is hypothesized to act synergistically with chitosan–GP to stimulate the formation of new bone.

FIGURE 10.16 Granules of BCP embedded in chitosan-GP as viewed under ESEM. Addition of BCP to chitosan-GP results in a fully injectable solution capable of forming a solid bone graft substitute at body temperature.

10.2.4.2.2 Cartilage Repair

Cartilage damage represents a major clinical challenge for orthopedic surgeons because cartilage does not have intrinsic properties, primarily due to the lack of vascularity and a cell population incapable of mounting a repair response. These cells, called chondrocytes, are embedded in the highly hydrated cartilage tissue composed primarily of type II collagen and GAGs. Having a similar macromolecular structure to cartilage, hydrogels of different types have been investigated to mimic the cartilage ECM to which cells can attach and proliferate. This tissue engineering approach requires the isolation of articular chondrocytes or their precursor cells, followed by the cell proliferation in vitro and the seeding of cultured cells into a hydrogel matrix that will eventually be implanted into the damaged cartilage area [73,74]. Unfortunately, processes involving in vitro cell culture for clinical use are prohibitive due to their complexity and extreme cost. Another form of chitosan-GP has been investigated in the treatment of articular cartilage damage of the knee. The chitosan–GP matrix, when mixed with autologous whole blood and surgically applied, demonstrated in animals the ability to regenerate reproducible high-quality cartilage [75–77] through a mechanism involving vascularization of the underlying bone and an increase in stem cell migration. Called BST-CarGel® (registered trademark of Bio syntech Canada Inc., Laval, Canada), this material has advanced to the clinical testing stage as a medical device [78].

10.2.4.2.3 Intervertebral Disk Replacement

Current treatment of intervertebral disk degeneration involves surgical excision of the damaged tissue, insertion of a metal cage or prosthetic device to restore the intervertebral space, and fusion of the vertebral bone. These treatments provide symptomatic relief, but may compromise spine biomechanics and also exacerbate the degenerative process in adjacent spinal segments. Another approach would be to use a delivery system to implant, i.e., cells which can be transformed into a desired intervertebral cell type. Ideally, the delivery system would be injectable, the delivered cells will eventually regenerate the tissue in the intervertebral space, and the disk should obtain sufficient mechanical strength to resist long-term and cyclical biomechanical compressions. Hydroxybutyl chitosan [79], chitosan–genipin [80], and chitosan–GP [81] injectable thermosensitive hydrogels have been evaluated for such a delivery system. These gels have shown in early stage research to have the ability to maintain intervertebral disk and mesenchymal stem cell viability, although attaining sufficient biomechanical properties in the disk is still problematic.

10.2.4.2.4 Wound Healing

Wound healing encompasses such clinical problems as chronic ulcers and burns. An ideal wound dressing should, in addition of being biocompatible, protect the wound from bacterial infection and provide a moist healing environment. A hydrogel dressing made from chitosan, with its wound healing [82] and antimicrobial properties [83], could therefore be a perfect marriage of these properties. Indeed, photocross-linkable chitosan hydrogels applied on full thickness skin incisions on the back of mice significantly induced wound contraction and accelerate wound closure [84]. And the negative effects of oxygen free radicals on wound healing were eliminated by the antioxidant taurine incorporated into a chitosan hydrogel [85]. Finally, the chitosan–GP hydrogel has been applied to diabetic ulcers on the foot. Called BST-DermOn™ (trademark of Bio syntech Canada Inc., Laval, Canada), this material has advanced to the clinical testing stage as a medical device [78].

10.2.5 Conclusion

Chitosan-based hydrogels draw their versatility from the biological and physicochemical properties associated with chitosan and chitosan derivatives. As evidenced by their varied uses from controlled drug delivery to tissue engineering, the physical forms of chitosan hydrogels appear to hold the most promise due mainly to their improved safety. In particular, the development of injectable and thermosensitive chitosan hydrogels has opened the door for innovative treatments using less invasive delivery methods.

References

1. Purna, S.K. and Babu, M., Collagen based dressings—a review. *Burns,* 26, 54, 2000.
2. Tonnesen, H.H. and Karlsen, J., Alginate in drug delivery systems. *Drug. Dev. Ind. Pharm.,* 28, 621, 2002.
3. Price, R.D., et al., The role of hyaluronic acid in wound healing: assessment of clinical evidence. *Am. J. Clin. Dermatol.,* 6, 393, 2005.

4. Ishihara, M., et al., Chitosan hydrogel as a drug delivery carrier to control angiogenesis. *J. Artif. Organs,* 9, 8, 2006.
5. Ruel-Gariepy, E., et al., Characterization of thermosensitive chitosan gels for the sustained delivery of drugs. *Int. J. Pharm.,* 203, 89, 2000.
6. Bernkop-Schnurch, A., Chitosan and its derivatives: potential excipients for peroral peptide delivery systems. *Int. J. Pharm.,* 194, 1, 2000.
7. Gutowska, A., Jeong, B., and Jasionowski, M., Injectable gels for tissue engineering. *Anat. Rec.,* 263, 342, 2001.
8. Felt, O., Buri, P., and Gurny, R., Chitosan: a unique polysaccharide for drug delivery. *Drug Dev. Ind. Pharm.,* 24, 979, 1998.
9. Singla, A.K. and Chawla, M., Chitosan: Some pharmaceutical and biological aspects–an update. *J. Pharm. Pharmacol.,* 53, 1047, 2001.
10. Paul, W. and Sharma, C.P., Chitosan, a drug carrier for the 21st century: A review. *S.T.P. Pharma. Sci.,* 10, 5, 2000.
11. Mattioli-Belmonte, M., Muzzarelli, B., and Muzzarelli, R.A.A., Chitin and chitosan in wound healing and other biomedical applications. *Carbohydr. Eur.,* 19, 30, 1997.
12. Muzzarelli, R., et al., Characteristic properties of *N*-carboxybutyl chitosan. *Carbohydr. Polym.,* 11, 307, 1989.
13. Mucha, M. and Miskiewicz, D., Reactive chitosan blends for paper improvement, in *Chitosan in Pharmacy and Chemistry*. Muzzarelli, R.A.A. and Muzzarelli, C., Eds., Atec, Italy, 2002, p. 251.
14. Lalov, I., et al., Treatment of waste water from distilleries wtih chitosan. *Wat. Res.,* 34, 1503, 2000.
15. El Gaouth, A., Arul, J., and Asselin, A., Potential use of chitosan in postharvest preservation of fruits and vegetables, in *Advances in Chitin and Chitosan*. Brine, C.J., Sandford, P.A., and Zikakis, J.P., Eds., Elsevier Applied Science, New York, 1992, p. 440.
16. Lim, S.H. and Husdon, M., Review of chitosan and its derivatives as antimicrobial agents and their uses as textile chemicals. *J. Macromol. Sci., Part C Polym. Rev.,* C43, 223, 2003.
17. Muzzarelli, R.A.A., Muzzarelli, C., and Terbojevich, M., Chitin chemistry, upgrading a renewable resource. *Carbohydr. Eur.,* 19, 10, 1997.
18. Gori, S., Gélification et dégradation d'un biopolymère par voie enzymatique, in Masters of Applied Sciences in Biomedical Engineering, Montréal, École Polytechnique de Montréal, 2002.
19. Vårum, K.M., et al., In vitro degradation rates of partially N-acetylated chitosans in human serum. *Carbohydr. Res.,* 299, 99, 1997.
20. Nordtveit, R.J., Vårum, K.M., and Smidsrød, O., Degradation of fully water-soluble, partially N-acetylated chitosans with lysozyme. *Carbohyd. Polym.,* 23, 253, 1994.
21. Sashiwa, H., et al., Lysozyme susceptibility of partially deacetylated chitin. *Int. J. Biol. Macromol.,* 12, 295, 1990.
22. Bernkop-Schnurch, A., Hornof, M., and Guggi, D., Thiolated chitosans. *Eur. J. Pharm. Biopharm.,* 57, 9, 2004.
23. Gupta, P., Vermani, K., and Garg, S., Hydrogels: from controlled release to pH-responsive drug delivery. *Drug Discov. Today,* 7, 569, 2002.
24. Argüelles-Monal, W., et al., Rheological study of the chitosan/glutaraldehyde chemical gel system. *Polym. Gels Networks,* 6, 429, 1998.
25. Goycoolea, F.M., et al., Effect of chemical crosslinking on the swelling and shrinking properties of thermal and pH-responsive chitosan hydrogels. *Macromol. Biosci.,* 3, 612, 2003.
26. Roberts, G.A.F. and Taylor, K.E., Chitosan gels, 3. The formation of gels by reaction of chitosan with glutaraldehyde. *Makromol. Chem.,* 190, 951, 1989.
27. Guo, B., Elgsaeter, A., and Stokke, B.T., Gelation kinetics of sclealdehyde–chitosan co-gels. *Polym. Gels Networks,* 6, 113, 1998.
28. Park, H., Park, K., and Kim, D., Preparation and swelling behavior of chitosan-based superporous hydrogels for gastric retention application. *J. Biomed. Mater. Res. A,* 76, 144, 2006.
29. Park, H. and Kim, D., Swelling and mechanical properties of glycol chitosan/poly(vinyl alcohol) IPN-type superporous hydrogels. *J. Biomed. Mater. Res. A,* 78A, 662, 2006.
30. Bhattarai, N., et al., PEG-grafted chitosan as an injectable thermosensitive hydrogel for sustained protein release. *J. Control. Release,* 103, 609, 2005.
31. Yeo, Y., et al., Peritoneal application of chitosan and UV-cross-linkable chitosan. *J. Biomed. Mater. Res. A,* 78A, 668, 2006.
32. Ono, K., et al., Photocrosslinkable chitosan as a biological adhesive. *J. Biomed. Mater. Res.,* 49, 289, 2000.
33. Ouchi, T., Nishizawa, H., and Ohya, Y., Aggregation phenomenon of PEG-grafted chitosan in aqueous solution. *Polymer,* 39, 5171, 1998.
34. Felix, L., et al., Kinetics of gelation and thermal sensitivity of *N*-isobutyryl chitosan hydrogels. *Biomacromolecules,* 6, 2408, 2005.
35. Rinaudo, M., et al., Specific interactions in modified chitosan systems. *Biomacromolecules,* 6, 2396, 2005.
36. Holme, K.R. and Hall, L.D., Chitosan derivatives bearing C_{10}-alkyl glycoside branches: A temperature-induced gelling polysaccharide. *Macromolecules,* 24, 3828, 1991.
37. Chen, L., Tian, Z., and Du, Y., Synthesis and pH sensitivity of carboxymethyl chitosan-based polyampholyte hydrogels for protein carrier matrices. *Biomaterials,* 25, 3725, 2004.
38. Valenta, C., Christen, B., and Bernkop-Schnurch, A., Chitosan–EDTA conjugate: a novel polymer for topical gels. *J. Pharm. Pharmacol.,* 50, 445, 1998.
39. Berger, J., et al., Structure and interactions in chitosan hydrogels formed by complexation or aggregation for biomedical applications. *Eur. J. Pharm. Biopharm.,* 57, 35, 2004.
40. Lin, Y.H., et al., Physically crosslinked alginate/*N,O*-carboxymethyl chitosan hydrogels with calcium for oral delivery of protein drugs. *Biomaterials,* 26, 2105, 2005.
41. Park, D.J., et al., Injectable bone using chitosan–alginate gel/mesenchymal stem cells/BMP-2 composites. *J. Craniomaxillofac. Surg.,* 33, 50, 2005.

42. Marudova, M., Macdougall, A.J., and Ring, S.G., Pectin–chitosan interactions and gel formation. *Carbohydr. Res.*, 339, 1933, 2004.
43. Nordby, M.H., et al., Thermoreversible gelation of aqueous mixtures of pectin and chitosan. Rheology. *Biomacromolecules*, 4, 337, 2003.
44. Chellat, F., et al., In vitro and in vivo biocompatibility of chitosan–xanthan polyionic complex. *J. Biomed. Mater. Res.*, 51, 107, 2000.
45. Tan, W., Krishnaraj, R., and Desai, T.A., Evaluation of nanostructured composite collagen–chitosan matrices for tissue engineering. *Tissue Eng.*, 7, 203, 2001.
46. Guang Liu, W. and De Yao, K., Chitosan and its derivatives—a promising non-viral vector for gene transfection. *J. Control. Release*, 83, 1, 2002.
47. Borchard, G., Chitosans for gene delivery. *Adv. Drug Deliv. Rev.*, 52, 145, 2001.
48. Ruel-Gariepy, E. and Leroux, J.C., In situ-forming hydrogels—review of temperature-sensitive systems. *Eur. J. Pharm. Biopharm.*, 58, 409, 2004.
49. Bhattarai, N., Matsen, F.A., and Zhang, M., PEG-grafted chitosan as an injectable thermoreversible hydrogel. *Macromol. Biosci.*, 5, 107, 2005.
50. Chenite, A., et al., Rheological characterisation of thermogelling chitosan/glycerol-phosphate solutions. *Carbohydr. Polym.*, 46, 39, 2001.
51. Chenite, A., et al., Novel injectable neutral solutions of chitosan form biodegradable gels in situ. *Biomaterials*, 21, 2155, 2000.
52. Kofuji, K., et al., The controlled release of a drug from biodegradable chitosan gel beads. *Chem. Pharm. Bull.* (*Tokyo*), 48, 579, 2000.
53. Shu, X.Z. and Zhu, K.J., Controlled drug release properties of ionically cross-linked chitosan beads: The influence of anion structure. *Int. J. Pharm.*, 233, 217, 2002.
54. Wang, L.Y., et al., Preparation and characterization of uniform-sized chitosan microspheres containing insulin by membrane emulsification and a two-step solidification process. *Colloids Surf. B Biointerfaces*, 50, 126, 2006.
55. Giunchedi, P., et al., Formulation and in vivo evaluation of chlorhexidine buccal tablets prepared using drug-loaded chitosan microspheres. *Eur. J. Pharm. Biopharm.*, 53, 233, 2002.
56. Martin, L., et al., The release of model macromolecules may be controlled by the hydrophobicity of palmitoyl glycol chitosan hydrogels. *J. Control. Release*, 80, 87, 2002.
57. Knapczyk, J., Chitosan hydrogel as a base for semisolid drug forms. *Int. J. Pharm.*, 93, 233, 1993.
58. Lin, W.C., Yu, D.G., and Yang, M.C., pH-sensitive polyelectrolyte complex gel microspheres composed of chitosan/sodium tripolyphosphate/dextran sulfate: Swelling kinetics and drug delivery properties. *Colloids Surf. B Biointerfaces*, 44, 143, 2005.
59. Ruel-Gariepy, E., et al., A thermosensitive chitosan-based hydrogel for the local delivery of paclitaxel. *Eur. J. Pharm. Biopharm.*, 57, 53, 2004.
60. Muzzarelli, R.A., et al., Stimulatory effect on bone formation exerted by a modified chitosan. *Biomaterials*, 15, 1075, 1994.
61. Muzzarelli, R.A., et al., Osteoconductive properties of methylpyrrolidinone chitosan in an animal model. *Biomaterials*, 14, 925, 1993.
62. Muzzarelli, R.A., et al., Osteoconduction exerted by methylpyrrolidinone chitosan used in dental surgery. *Biomaterials*, 14, 39, 1993.
63. Klokkevold, P.R., et al., Osteogenesis enhanced by chitosan (poly-*N*-acetyl glucosaminoglycan) in vitro. *J. Periodontol.*, 67, 1170, 1996.
64. Legeros, R.Z., et al., Biphasic calcium phosphate bioceramics: preparation, properties and applications. *J. Mater. Sci. Mater. Med.*, 14, 201, 2003.
65. Daculsi, G., Biphasic calcium phosphate concept applied to artificial bone, implant coating and injectable bone substitute. *Biomaterials*, 19, 1473, 1998.
66. Seol, Y.J., et al., Chitosan sponges as tissue engineering scaffolds for bone formation. *Biotechnol. Lett.*, 26, 1037, 2004.
67. Carey, L.E., et al., Premixed rapid-setting calcium phosphate composites for bone repair. *Biomaterials*, 26, 5002, 2005.
68. Liu, H., et al., Novel injectable calcium phosphate/chitosan composites for bone substitute materials. *Acta Biomater.*, 2, 557, 2006.
69. Mukherjee, D.P., et al., An animal evaluation of a paste of chitosan glutamate and hydroxyapatite as a synthetic bone graft material. *J. Biomed. Mater. Res. B Appl. Biomater.*, 67, 603, 2003.
70. Leroux, L., et al., Effects of various adjuvants (lactic acid, glycerol, and chitosan) on the injectability of a calcium phosphate cement. *Bone*, 25, 31S, 1999.
71. Jarry, C., Shive, M., and Chenite, A., *An injectable biomaterial for bone repair, material science forum*, 539-543, 535, 2007.
72. Nery, E.B., et al., Tissue response to biphasic calcium phosphate ceramic with different ratios of HA/beta TCP in periodontal osseous defects. *J. Periodontol.*, 63, 729, 1992.
73. Suh, J.K. and Matthew, H.W., Application of chitosan-based polysaccharide biomaterials in cartilage tissue engineering: A review. *Biomaterials*, 21, 2589, 2000.
74. Hoemann, C.D., et al., Tissue engineering of cartilage using an injectable and adhesive chitosan-based cell-delivery vehicle. *Osteoarthritis Cartilage*, 13, 318, 2005.
75. Hoemann, C., et al., Chitosan-glycerol phosphate/blood implants elicit hyaline cartilage repair integrated with porous subchondral bone in microdrilled rabbit defects. *Osteoarthritis Cartilage*, 15, 78, 2007.
76. Chevrier, A., et al., Chitosan-glycerol phosphate/blood implants increase cell recruitment, transient vascularization and subchondral bone remodeling in drilled cartilage defects. *Osteoarthritis Cartilage*, 15, 316, 2007.
77. Hoemann, C.D., et al., Chitosan-glycerol phosphate/blood implants improve hyaline cartilage repair in ovine microfracture defects. *J. Bone Joint Surg. Am.*, 87, 2671, 2005.

78. www.clinicaltrials.gov
79. Dang, J.M., et al., Temperature-responsive hydroxybutyl chitosan for the culture of mesenchymal stem cells and intervertebral disk cells. *Biomaterials,* 27, 406, 2006.
80. Mwale, F., et al., Biological evaluation of chitosan salts cross-linked to genipin as a cell scaffold for disk tissue engineering. *Tissue Eng.,* 11, 130, 2005.
81. Roughley, P., et al., The potential of chitosan-based gels containing intervertebral disc cells for nucleus pulposus supplementation. *Biomaterials,* 27, 388, 2006.
82. Muzzarelli, R.A.A., Human enzymatic activities related to the therapeutic administration of chitin derivatives. *Cell. Mol. Life Sci.,* 53, 131, 1997.
83. Rabea, E.I., et al., Chitosan as antimicrobial agent: Applications and mode of action. *Biomacromolecules,* 4, 1457, 2003.
84. Ishihara, M., et al., Photocrosslinkable chitosan as a dressing for wound occlusion and accelerator in healing process. *Biomaterials,* 23, 833, 2002.
85. Degim, Z., et al., An investigation on skin wound healing in mice with a taurine–chitosan gel formulation. *Amino Acids,* 22, 187, 2002.
86. Thanou, M., Verhoef, J.C., and Junginger, H.E., Chitosan and its derivatives as intestinal absorption enhancers. *Adv. Drug Deliv. Rev.,* 50 Suppl 1, S91, 2001.
87. Thanou, M., Verhoef, J.C., and Junginger, H.E., Oral drug absorption enhancement by chitosan and its derivatives. *Adv. Drug Deliv. Rev.,* 52, 117, 2001.
88. Illum, L., et al., Chitosan as a novel nasal delivery system for vaccines. *Adv. Drug Deliv. Rev.,* 51, 81, 2001.
89. Hirano, S., Applications of chitin and chitosan in the ecological and environmental fields, in *Applications of Chitin and Chitosan.* Goosen, M., Eds., Lancaster, PA, 1997, p. 31.
90. Struszczyk, H. and Pospieszny, H., New applications of chitin and its derivatives in plant protection, in *Applications of Chitin and Chitosan.* Goosen, M., Eds., Lancaster, PA, 1997, p. 171.
91. Freepons, D., Enhancing food production with chitosan seed-coating technology, in *Applications of Chitin and Chitosan.* Goosen, M., Eds., Lancaster, PA, 1997, p. 129.
92. Lavertu, M., et al., High efficiency gene transfer using chitosan/DNA nanoparticles with specific combinations of molecular weight and degree of deacetylation. *Biomaterials,* 27, 4815, 2006.
93. Sinha, V.R., et al., Chitosan microspheres as a potential carrier for drugs. *Int. J. Pharm.,* 274, 1, 2004.
94. Lahiji, A., et al., Chitosan supports the expression of extracellular matrix proteins in human osteoblasts and chondrocytes. *J. Biomed. Mater. Res.,* 51, 586, 2000.
95. Bokura, H. and Kobayashi, S., Chitosan decreases total cholesterol in women: A randomized, double-blind, placebo-controlled trial. *Eur. J. Clin. Nutr.,* 57, 721, 2003.
96. Muzzarelli, R., Clinical and biochemical evaluation of chitosan for hypercholesterolemia and overweight control. *EXS,* 87, 293, 1999.
97. Takayanagi, T. and Motomizu, S., Chitosan as cationic polyelectrolyte for the modification of electroosmotic flow and its utilization for the separation of inorganic anions by capillary zone electrophoresis. *Anal. Sci.,* 22, 1241, 2006.
98. Jocic, D., et al., Dye sorption by wool treated with low temperature plasma and chitosan, in *Chitosan in Pharmacy and Chemistry.* Muzzarelli, R.A.A. and Muzzarelli, C., Eds., Atec, Italy, 2002, p. 317.
99. Bumgardner, J.D., et al., Chitosan: potential use as a bioactive coating for orthopaedic and craniofacial/dental implants. *J. Biomater. Sci. Polym. Ed.,* 14, 423, 2003.
100. Lang, G., Wendel, H., and Konrad, E., Quaternary hydroxyethyl-substituted chitosan derivatives, cosmetic compositions based thereon and processes for the production thereof, Patent, US4772690, 1988.
101. Wedmore, I., et al., A special report on the chitosan-based hemostatic dressing: Experience in current combat operations. *J. Trauma.,* 60, 655, 2006.
102. Lee, K.Y., Ha, W.S., and Park, W.H., Blood compatibility and biodegradability of partially N-acylated chitosan derivatives. *Biomaterials,* 16, 1211, 1995.
103. Rao, S.B. and Sharma, C.P., Use of chitosan as a biomaterial: studies on its safety and hemostatic potential. *J. Biomed. Mater. Res.,* 34, 21, 1997.
104. Kawakami, T., et al., Experimental study on osteoconductive properties of a chitosan-bonded hydroxyapatite self-hardening paste. *Biomaterials,* 13, 759, 1992.
105. Ueno, H., Mori, T., and Fujinaga, T., Topical formulations and wound healing applications of chitosan. *Adv. Drug Deliv. Rev.,* 52, 105, 2001.
106. Muzzarelli, R., et al., Biochemistry, histology and clinical uses of chitins and chitosans in wound healing. *EXS,* 87, 251, 1999.

11

Films, Coatings, Adhesives, Polymers, and Thermoelectric Materials

11.1 Smart Adhesives

James A. Harvey and Susan Williams

Smart materials are materials that respond to their environments in a timely manner. Smart materials can receive, transmit, or process a stimulus and respond by producing a useful effect that may include a signal that the materials are acting upon it. Some of the stimuli that may act upon these materials are strain, stress, temperature, chemicals (including pH stimuli), electric field, magnetic field, hydrostatic pressure, different types of radiation, and other forms of stimuli. Another important criterion for a smart material is its ability to receive stimuli and responding to the stimuli to produce a useful effect that is and it must be reversible. Another feature that is an important factor in determining if a material is smart pertains to its asymmetrical nature. From the purist point of view, materials are smart if at some point within their performance history, they act reversible to a stimulus. Materials that formally had the label of being smart include piezoelectric materials, electrostrictive materials, electrorheological materials, magnetorheological materials, thermoresponsive materials, pH-sensitive materials, UV-sensitive materials, smart polymers, smart gels (hydrogels), smart catalysts, and shape memory alloys [1]. With the progress experienced by smart materials and the successful introduction of nanotechnology, a new class of smart materials has been developed. This class consists of the smart adhesives.

Smart adhesives do not have the full range of capabilities or actions as the typical smart materials. However, smart adhesives do have the abilities to do many smart-like actions. These include, but not limited to the following: mixing, curing monitoring the presence of hazardous gases and changes in temperatures via color changes, and the release of adhesion upon demand.

The development of a truly smart adhesive hinges around the ability of the adhesive to behave as if it could have a brain, nerves, and muscles. This type of behavior has not yet been developed. This indeed will be accomplished through the efforts of adhesive, smart adhesives, and smart materials practitioners.

TABLE 11.1 Examples of Smart Adhesive Properties and Typical Applications

Smart Adhesive Properties	Applications
Aesthetics—color surface finish	Color matching
	High contrast for visible security
	Cross-pigmentation for two-part security
	Opaque or translucent
Disassembly	Disbond on demand for materials recycling
Materials mimicry	Ability to be machined, drilled, or cut
	Magnetic
Thermal conductivity	Thermal management in electronics and electric motors
	Heat sink attachment
Electrical conductivity	Organic solder replacement
Electrical resistance/insulation	Electrical terminal potting
Energy and vibration adsorption	Thread locking
	Gas seal in threading fittings
	Mirror bonding in transportation
	Joint disassembly
Heat resistance, fire resistance, fire extinguishing	Mass transport and aerospace interior trim
	Hot joints in electrical devices or high-temperature environments
Auto-surface coupling	Minimizing surface treatment on aluminum and composites
Self-indicating—presence of cured adhesive	UV fluorescence for automated quality
	Color change on cure for hazardous gas fittings
Security	Leak sealing and repair
Ultralow density	Weight saving in transportation—particularly for aerospace duties
Vibration adsorption	Noise abatement, vibration damping
Regulatory compliance	Intrinsically safe in use: medical devices, drug-delivery patches and food contact (direct or indirect).

Source: From Fakley, M., *Chem. Ind.*, 21, 691, 2001. With permission.

One of the initial mentions of the term smart adhesives in the open literature was a review article with that title. The foci of the review were how to work smartly and the factors to consider when choosing an adhesive. The article did contain a table (Table 11.1) that is considered to be the forerunner of many concepts and applications of smart adhesives [2].

Another review discussed the uses of smart adhesives in the information technological fields. Such usages described the use of surface mount adhesives to fix and hold soldering and other operations. Single-component heat-cured epoxies were used in this application. Other technology-related smart adhesives like conformal coatings provide thin dielectric coatings that extend circuit board lifetimes by protecting components and traces from corrosion, shorts, and mechanical damage. Solid systems such as UV curing acrylics and silicones were used in this function [3].

An interesting and comprehensive review of medical adhesives for use in products that are applied directly to the body or inside the body by surgical procedures has been reported. The article focused on pressure-sensitive adhesives and tissue adhesives. These two classes of adhesives were further divided into synthetic and biological adhesives. The report cited the progress gained in the understanding of wound management and the development and formulation of polymers for pressure-sensitive adhesives that facilitated the wound healing process. Adhesives with high-moisture vapor permeation allow wound dressings to control exudates more effectively. Trauma management has seen advances made in smart pressure-sensitive adhesives that changed adhesive properties with external stimuli [4].

Researchers reported an example of smart adhesion at a polymer/metal (oxide) interface that responded reversibly to changes in temperature. The temperature dependence in this system arose from the rubber elasticity of the polymer, 1,4-polybutadiene, and mirrored the interfacial behavior of the same polymer against water. These systems offer unique opportunities for designing responsive materials whose properties can be actively controlled [5].

Interesting uses of the waste from the chromium compound treatment of the collagen polymers resulting from the treatment of leather from the tanning process were reported in the literature. Covalent intramolecular and intermolecular cross-links were formed between modified lysine side chains within the collagen fibrils. High molecular weight versions expanded the potential commercial applications to the veterinary, medical, and cosmetic fields. It should be pointed out that the biopolymers isolated from chrome-tanned collagenic wastes were applied directly to the following industries: paper, wood, ink, textile and leather, as binders, biodegradable biopolymers, and protein-based self-feeding biodegradable flower pots, fillers, retanning agents, and casein substitute in photographic adhesives. A second possibility would be to produce low formaldehyde content ecoadhesives and collagen-based smart fabric formed by chemical reaction with these biopolymers [6].

In the realm of smart materials, smart adhesives, and other smart types, there exists a unique class of materials, which utilize a smart material in conjunction with a nonsmart material. These loosely belong to the classification of smart structures. There are smart structures that monitor the behavior of adhesives or enhance the manufacture of adhesives.

The first example of these types of smartness is the use of piezoelectrics to monitor the behavior of adhesives. The action of piezoelectrics is discussed in Chapter 9. Metalized poly(vinylidene fluoride) films were etched to provide multipoint sensors, which were imbedded within adhesive joints to measure the peel stresses. Peel stresses were thought to be the most critical stresses responsible for failure of a variety of adhesive joints. The technique successfully demonstrated the peel stress trends expected in single and double lap joints and butt joints. Adhesive cure monitoring and void/porosity detection were also performed with ultrasound using the film [7–9].

Other literature citations describing modifications or improvements to the use of piezoelectrics to monitor adhesive joints have been recently published. These along with the previously mentioned efforts indicate the potential of having wires embedded with a bond to monitor the behavior, which in turn can lead to the prediction of the lifetime (or out of service time) of a structure [10–13]. This cited literature pertains to the through-life, nondestructive monitoring of adhesively bonded structures. It presented a discussion of the concept of microelectromechanical systems (MEMS) smart sensors. The MEMS smart sensors were permanently installed in such large structures as ships, aircrafts, or land vehicles [14].

Another approach of incorporating smart materials with adhesives involves the addition of smart particles in the adhesive formulation to increase the reactivity of the adhesive. Nanoparticles with crystal structure having properties of ferromagnetic, ferrimagnetic, superparamagnetic, or piezoelectric materials were heated by electric, magnetic, or electromagnetic alternating fields aided in the hardening of adhesives [15,16].

As expected, the reverse has been accomplished. By heating nanoparticles with crystal structure having properties of ferromagnetic, ferrimagnetic, superparamagnetic, or piezoelectric materials by electric, magnetic, or electromagnetic alternating fields, thermoplastic adhesives can be loosened [17].

Another concept that allows one to work smarter with adhesives consists of depositing adhesives or adhesion control agents in a manufacturing process using ink jet printing heads. The authors are advocates of this technique of applying minute amounts to confined or limited spaces. We would like to caution users of this application of chemistries that the consistence of the material is the same after deposition and is the same as formulated or initially manufactured [18].

Packaging of microelectronic devices has become more and more important. Applications for interconnection technologies range from highly specialized processors to be assembled on flex for mobile phones, highly defined packages as for hearing aids to low-cost transponders as for smart cards and smart labels. Each of these interconnection types has different requirements in technologies and costs. The use of isotropic conductive adhesive (ICA) for bumping and assembly for smart cards results in a number of advantages such as a simple versatile process, lower temperatures, and environment friendliness. Using an anisotropic conductive adhesive (ACA) or a nonconductive adhesive (NCA) for smart label production allows a high throughput from reel to reel [19].

One of the applications of special adhesives has been in the fabrication of smart cards. UV-curable ACAs were used for these mobile electronic products. The adhesives were subjected to different mechanical or environmental attacks like warps, stretching, changing humidity and temperature, which could affect the performance and reliability of the electronic products [20].

A simple-design Cu coil plated on polyethylene terephthalate was used as the microstrip antenna of the smart card. The entire fabrication process of the UV-curable ACA was divided into three parts: high-power UV curing, chip-on-flex (COF) bonding, and postcuring. By varying the UV curing and postcuring parameters, a number of contactless smart cards were made [21].

A class of smart materials that has paralled the growth of and is closely related to the smart adhesives are the smart coatings. The advances of the smart coatings can be illustrated in the topics covered at a smart coating conference held in February 2005.

Examples of smart coatings included the formation of an enigmatic polymer with the unique property of being hydrophilic (water loving) when dry and hydrophobic (water hating) when wet. The researchers produced an antimicrobial coating by adding hydantoin into fluorine-containing polymer chains. The researchers claimed that water-induced hydrophobic surfaces would result in new medicine devices, switching devices, and drag-reducing coatings [22].

A new fluoropolymer additive for coating has been developed by Ausimont. This new coating additive assists in the removal of graffiti from outside surfaces without affecting the coating and weathering properties of the surface [23].

Millenium Chemicals commercialized a polysiloxane silicone coating, Ecopaint. This unique coating in conjunction with either titanium dioxide or calcium carbonate or a combination of the two inorganics renders respirator-causing problems, nitrogen oxides (NO_x gases), harmless. Buildings with this type of system will be more environmentally friendly [24].

In conclusion, one of the authors remembers an article by a group of French adhesive scientists in which they commented on the status of adhesion theories. They indicated that there was still a great deal of information needed to explain why things stick together. The authors must take these remarks into account and apply similar methodologies to the development of smart adhesives. The growth of smart adhesives is in its early development stages. Most examples of the smart adhesives found in the open literature do not truly fill all the criteria of being smart that is having a "brain," "nerves," and "muscles." With time and research and development efforts, this will change.

References

1. Harvey, J.A. Smart materials, in *Mechanical Engineers' Handbook*, 3rd ed., Kutz, M. (ed.), John Wiley & Sons, New York, 2005.
2. Fakley, M. *Chemistry & Industry*, 21, 691–695, 2001.
3. Birkett, D. *Chemistry in Britain*, 38(11), 29–31, 2002.
4. Webster, I. and West, P.J. in *Polymeric Biomaterials*, 2nd ed., Dumitriu, S. (ed.), Marcel Dekker, New York, pp. 703–737, 2002.
5. Khongtongad, S. and Ferguson, G.S. *Journal of the American Chemical Society*, 124(25), 7254–7255, 2002.
6. Cot, J. *Journal of the American Leather Chemists Association*, 99(8), 322–350, 2004.
7. Dillard, D.A., Anderson, G.L., and Davis, D.D. Jr., *Journal of Adhesion*, 29(1–4), 245–255, 1989.
8. Anderson, G.L. and Dillard, D.A. NTIS. Report (ARO-23759.3-EG-F; Order No. AD-A233 880 1991).
9. Anderson, G.L., Mommaerts, J., Tang, S.L., Duke, J.C., and Dillard, D.A. Proceedings of a Conference on Recent Advances in Adaptive Sensing Materials and Their Applications, Rogers, C.A and Rogers, R.C. (eds.), pp. 266–273, 1992.
10. Shankar, K., Tahtali, M., Chester, R., and Torr, G., *Proceedings of SPIE-The International Society for Optical Engineering*, 4317 (Experimental Mechanics), 363–368, 2001.
11. Kwon, J.W., Chin, W.S., and Lee, D.G. *Journal of Adhesion Science and Technology*, 17(6), 777–796, 2003.
12. Hodges, C.A., Mossi, K.M., and Scott, L.A. *Proceedings of SPIE-The International Society for Optical Engineering*, 5053 (Active Materials: Behavior and Mechanics), 467–474, 2003.
13. Hwang, H.Y. and Lee, D.G. *Journal of Adhesion Science and Technology*, 19(12), 1053–1080, 2005.
14. Wilson, A., Christina, O.-J., and Muscat, R. *Materials Australia*, 34(3), 15–17, 2002.
15. Kirsten, C.N., Henke, G., Claudia, M.-J., Unger, L., and Meier, F. PCT Int. Appl. WO 2002018499 A1 20020307, 2002.
16. Kirsten, C.N. PCT Int. Appl. WO 2002012409 A1 20020214, 2002.
17. Kirsten, C.N., Chrisophliemk, P., Nonninger, R., Schirra, H., and Schmidt, H. Ger. Offen., DE 19924138 A1 20001130, 2000.
18. Saksa, T.A. and Lee, S. U.S. Pat. Appl. Publ., US 2003070740 A1 20030417, 2003.
19. Seidowski, T., Kriebel, F., and Galties, J. Proceedings—International Conference on Adhesive Joining and Coating Technology in Electronics Manufacturing, presented at Adhesive in Electronics 2000, 4th, Espoo, Finland, pp. 52–54, June 18–21, 2000.
20. Lee, K., Ng, K.T., Tan, C.W., Chan, Y.C., and Cheng, L.M. Proceedings of International Conference on the Business of Electronic Product Reliability and Liability, Shanghai, China, pp. 134–139 Apr. 27–30, 2004.
21. Tan, C.W., Siu, Y.M., Lee, K.K., Chan, Y.C., and Cheng, L.M. Proceedings of International Conference on the Business of Electronic Product Reliability and Liability, Shanghai, China, pp. 140–144, Apr. 27–30, 2004.
22. NASA Tech Brief Briefs, Insider, May 5, 2005.
23. Locaspi, A. and Marchetti, R. Paint & Coatings Industry, March, 2000.
24. www.SpecialChem.com, Intelligent Adhesives, Sept. 14, 2005.

11.2 Oxides as Potential Thermoelectric Materials

S. Hébert and A. Maignan

11.2.1 Introduction

The search for new sources of energy is an active topic. Among them, the thermoelectric generation is a powerful technique, which implies no mechanical movements and no fluid, and is thus a reliable technique to recover waste heat and generate electricity. Due to the small conversion efficiencies, thermoelectric generation is for the moment restricted to specific applications such as spatial applications (with radioisotopic thermoelectric generator). However, recovering industrial waste heat could be an efficient way to generate power [1], and thermoelectric materials operating at high temperatures (T up to 1400°C for steel industry [1]) are needed. At such temperatures, the classical thermoelectric materials as Bi_2Te_3 suffer from thermal decomposition or detrimental oxidations. Recently, this has created a rush for new thermoelectric materials such as skutterudites, clathrates, among others. The principle of a thermoelectric device is shown in Figure 11.1. It is based on pairs of n- and p-type legs electrically and thermally assembled in serial and parallel modes, respectively. The intrinsic performances of a thermoelectric material, n- or p-type, are defined by its figure of merit, $Z = S^2/\rho\kappa$, or its power factor PF $= S^2/\rho$, where S is the Seebeck coefficient, ρ the electrical resistivity, and κ the thermal conductivity. The dimensionless ratio ZT should be at least equal to 1 for applications. The discovery of a large thermoelectric power in the metallic oxide Na_xCoO_2 has shown that oxides are promising materials for thermoelectric applications [2]. Due to their good thermal stability in high temperatures and in air, they can be used for recovery of waste heat and thermoelectric generation at high temperatures. Following this report, several studies have been dedicated to the search for new good thermoelectric oxides. We will present here the results obtained in our group. Among them, the misfit oxides look promising. This family of oxides was first discovered in our laboratory [3], and their layered structure consists in the stacking of CoO_2 layers of the CdI_2 type as in Na_xCoO_2, separated by NaCl-like layers. Their transport properties are very close to the ones of Na_xCoO_2, with a large thermopower associated to a metallic behavior above 100 K [4]. A ZT value close to 1.2 at 973 K has been estimated by Funahashi

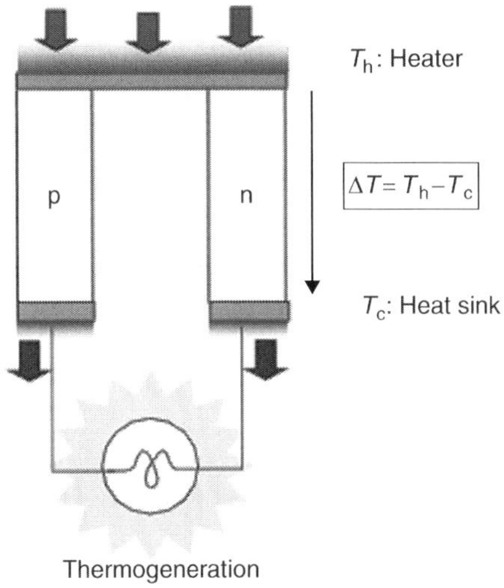

FIGURE 11.1 Schematic description of a thermogeneration device showing one pair made of n and p legs.

et al. [5]. The optimization of the thermoelectric properties of this family of oxides has been investigated and this work has then been extended to the search for new thermoelectric oxides. In the following, the chapter will be divided in two parts. First, the results concerning semiconducting oxides and the possibility to create n- and p-type thermoelements will be presented. Then, the results obtained in the case of metallic oxides will be described. Most of the results will be analyzed using the Heikes formula [6], which states that, at high temperature, the Seebeck coefficient is a direct measure of the entropy per carrier, and thus mainly depends on the carrier concentration. We will show how the spin and orbital degeneracy terms, as proposed by Doumerc [7] and Koshibae et al. [8], can enhance the Seebeck coefficient in these oxides.

11.2.2 Perovskite $LaCoO_3$: A n- or p-Type Oxide

The Heikes formula has been proposed to calculate the Seebeck coefficient of localized carriers [6]. This formula is valid in the limit of infinite temperature and can be written as:

$$S = \frac{-k_B}{|e|} \ln\left(\frac{1-x}{x}\right) \tag{11.1}$$

where x is the carrier density. For small doping, it is thus theoretically possible to create n- or p-type thermoelements, with large Seebeck coefficient.

Following this formula, small doping has been used in $LaCoO_3$ to introduce mixed cation valency of Co^{2+}/Co^{3+} (n-type) or Co^{3+}/Co^{4+} (p-type). Figure 11.2 presents the Seebeck coefficient of $LaCoO_3$, $La_{0.98}Sr_{0.02}{}^{2+}Co^{3+/4+}O_3$, and $La_{0.99}Ce_{0.01}{}^{4+}Co^{2+/3+}O_3$. As expected, in the stoichiometric compound, the Seebeck coefficient is very large, and positive (+600 μV/K at 300 K), suggesting a small concentration of holes in the parent compound ($x \sim 9 \times 10^{-4}$ Co^{4+}/Co following the Heikes formula). S is reduced by the introduction of Sr^{2+}, and the sign of S changes when Ce^{4+} is introduced. Using the Heikes formula, a good agreement is obtained between the nominal and actual doping concentration in the case of p-type materials, with $x \sim 2.22 \times 10^{-2}$ Co^{4+}/Co to be compared to the nominal 2×10^{-2} Co^{4+}/Co calculated from the chemical formula. The agreement is not so good in the n-type case with $x \sim 3.6 \times 10^{-2}$ Co^{2+}/Co, to be compared to 10^{-2},

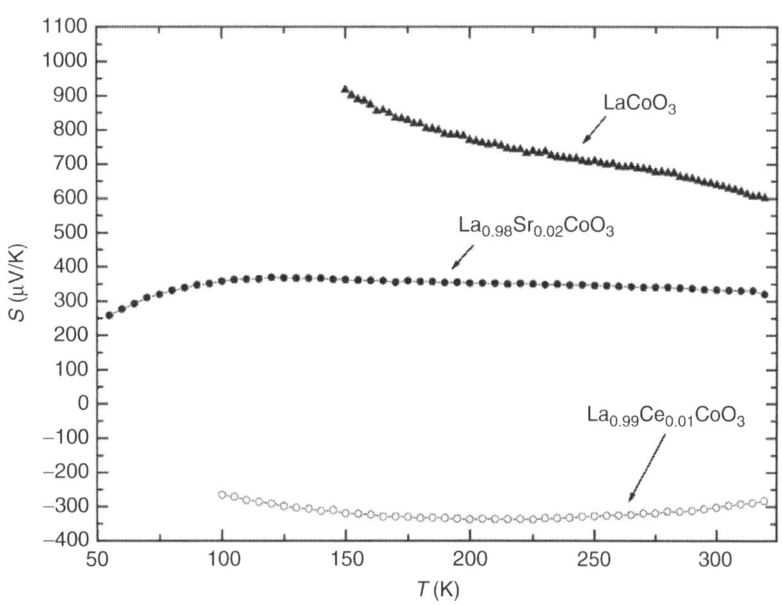

FIGURE 11.2 $S(T)$ of $LaCoO_3$ and p ($La_{0.98}Sr_{0.02}{}^{2+}Co^{3+/4+}O_3$) and n ($La_{0.99}Ce_{0.01}{}^{4+}Co^{2+/3+}O_3$) type doped $LaCoO_3$.

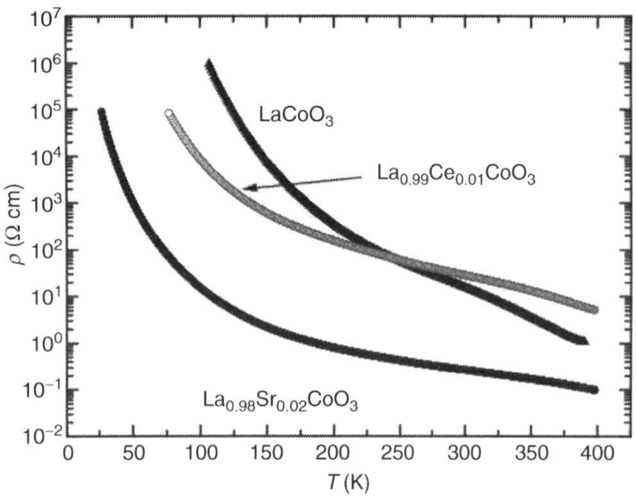

FIGURE 11.3 $\rho(T)$ of $LaCoO_3$, $La_{0.98}Sr_{0.02}{}^{2+}Co^{3+/4+}O_3$, and $La_{0.99}Ce_{0.01}{}^{4+}Co^{2+/3+}O_3$.

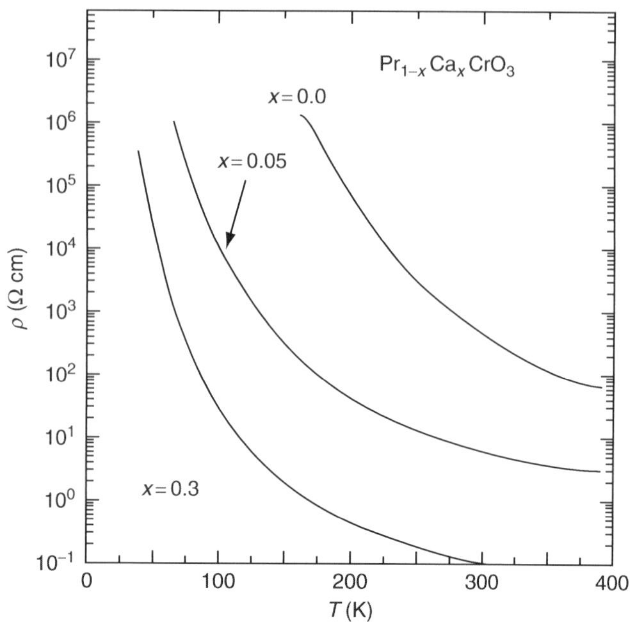

FIGURE 11.4 $\rho(T)$ of the $Pr_{1-x}Ca_xCrO_3$ series.

but this difference could be linked to a problem of homogeneity due to the very small dopant concentration. Also, other terms linked to the spin and orbital degeneracies should be added to the Heikes formula, as shown by Taskin et al. [9]. The most important result is that n- or p-type materials can be created by doping this oxide with holes or electrons [10].

Concerning the thermoelectric performances, a strong asymmetry is observed for the resistivities, with large ρ in $La_{0.99}Ce_{0.01}{}^{4+}Co^{2+/3+}O_3$, which will lead systematically to a smaller power factor for n-type $LaCoO_3$ (Figure 11.3). As shown in Ref. [11], carrier transport in doped $LaCoO_3$ is easier in the case of holes than in the case of electrons. This phenomenon, called the "spin blockade," is due to the spin states of cobalt and to the different filling of the orbitals: a hole can hop in the t_{2g} orbitals, while hopping is more difficult in the case of electron doping, due to the formation of a (Co^{3+}+ electron), which is not energetically favorable.

The performances of the n-type $LaCoO_3$ are thus always smaller. Nevertheless, the possibility to create n- and p-type thermoelements starting from a stoichiometric compound is very appealing. This result can be extended to other compounds, as shown in the 1D $Ca_3Co_2O_6$ [12].

11.2.3 Orthochromites $Pr_{1-x}Ca_xCrO_3$: Role of the Spin and Orbital Degeneracies

Following the first papers reporting on the Heikes formula, many calculations have been performed to generalize it, taking into account the possible spin [7] and orbital degeneracies [8]. Among them, a detailed analysis has been performed by Marsh and Parris to calculate the thermopower of systems with t_{2g} orbitals ($LaCrO_3$ [13]), or with e_g orbitals (i.e., $LaMnO_3$ [14]). Taking into account the polaronic nature of transport in these materials, and the spatial and spin degeneracies associated to the orbitals, they calculate the Seebeck coefficients of these materials. Following these results, we have measured the thermopower of the $Pr_{1-x}Ca_xCrO_3$ system for $x \le 0.3$ [15].

Figure 11.4 presents the resistivity curves. These materials are semiconductors, and for $x = 0$, $\rho \sim 100\ \Omega$ cm at 400 K. As x increases, the carrier concentration increases and the semiconducting-like resistivity decreases.

The thermopower data are presented in Figure 11.5. The Seebeck coefficient of $x = 0$ is positive, and very large, with S~900 μV/K at 300 K. The introduction of carriers induces a strong decrease of S as x increases, but S remains positive as expected for hole doping (introduction of Cr^{4+} in the Cr^{3+} matrix through the substitution of Ca^{2+}). S is almost constant from 200 to 300 K, and the high-temperature measurements show that S is independent of temperature at least up to 650 K (Figure 11.6).

The constant value of S is consistent with polaronic transport. The evolution of S measured at 300 K, as a function of x, is presented in Figure 11.7. The solid line represents the theoretical evolution of S from the Heikes formula (Equation 11.1). The dotted line is the evolution of S versus x calculated from the Marsh and Parris formula [13], which takes into account the spin and orbital degeneracies:

$$S = \frac{-k_B}{|e|}\ln\left(\frac{1-x}{x}\right) + \Delta S_d, \text{ with } \Delta S_d = \frac{k_B}{|e|}\ln(\Gamma_{orb}\Gamma_{spin}) \quad (11.2)$$

where Γ_{orb} and Γ_{spin} are the orbital and spin degeneracies. For Cr^{3+}/Cr^{4+}, in the case of weak magnetic coupling, $\Delta S_d = +69.9$ μV/K [13]. Figure 11.7 shows that the experimental results fit perfectly well

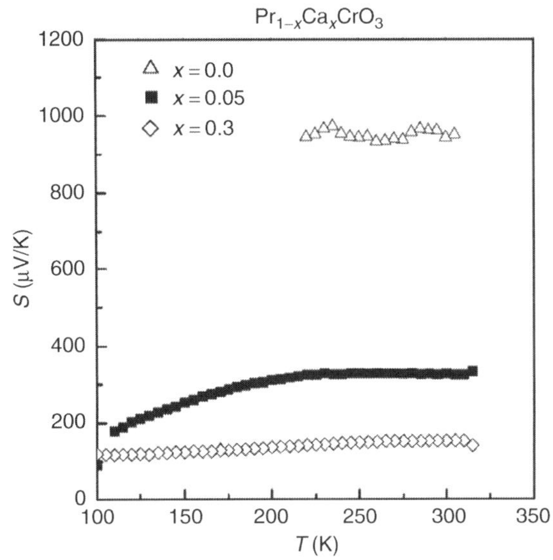

FIGURE 11.5 $S(T)$ of the $Pr_{1-x}Ca_xCrO_3$ series.

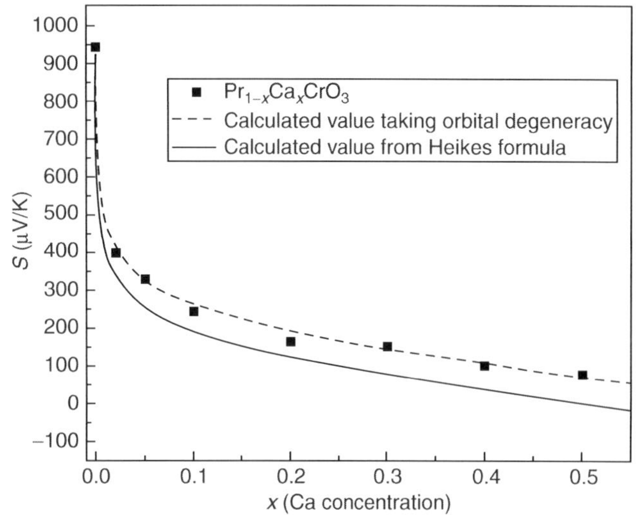

FIGURE 11.7 S as a function of the Ca^{2+} concentration. The solid line corresponds to the Heikes formula (Equation 11.1), and the dotted line to the Marsh and Parris model, which takes into account the spin and orbital degeneracies (Equation 11.2).

with the theory. The spin and orbital degeneracies can strongly enhance the thermopower, and maximizing this term is crucial to optimize the Seebeck coefficient.

11.2.4 Large Thermopower in Metallic Oxides: The Misfit Layer Oxides

Na_xCoO_2 is at the origin of the search for new thermoelectric oxides. Na_xCoO_2 is a layered material with CoO_2 layers of tilted edge-shared octahedra separated by randomly filled Na layers. Two models have been proposed to explain the large S value. The first one takes into account the spin and orbital degeneracies associated to a mixture of low spin Co^{3+} and low spin Co^{4+}, which is responsible for a large entropy and therefore a large S [8]. This model should be applied only for localized systems and cannot explain the metallicity of Na_xCoO_2. The second one has been proposed by Singh who calculated the band structure of Na_xCoO_2 [16]. Due to the rhombohedral symmetry of the CoO_2 layers, the t_{2g} orbitals are split in two sublevels. There would therefore be two types of carriers, light ones in the e_g' band responsible for metallicity, and heavy ones in the a_{1g} band responsible for a large thermopower.

The misfit layered oxides have been discovered in our laboratory a decade ago [3]. Their structure has the same CoO_2 layers of tilted edge-shared octahedra, separated in this case by NaCl-like layers (Figure 11.8). There can be three or four separating layers. Recently, new misfits with two separating layers have been reported [17,18]. The structure can be described with two monoclinic sublattices, with common a, c, and β parameters but

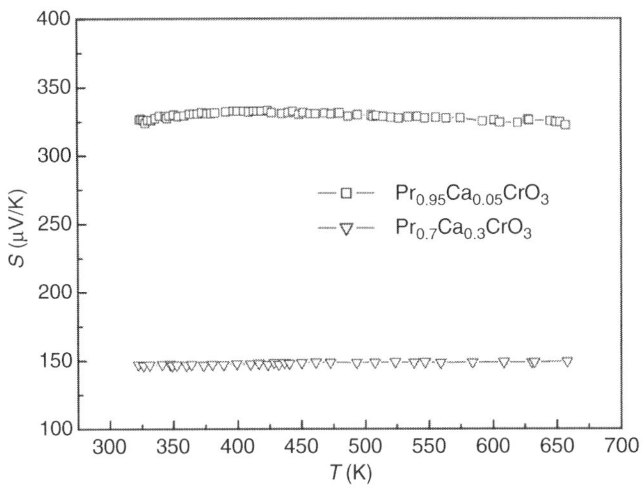

FIGURE 11.6 $S(T)$ measured up to 700 K of $Pr_{1-x}Ca_xCrO_3$ (x = 0.05 and 0.3).

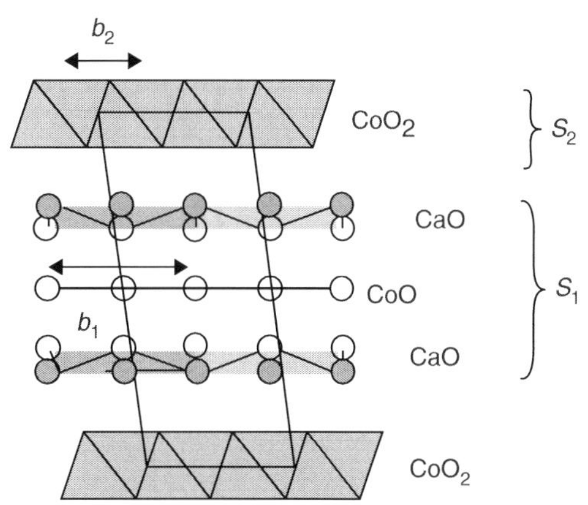

FIGURE 11.8 Schematic description of the misfit structure of $[Ca_2CoO_3][CoO_2]_{1.62}$, with the CoO_2 layers and the NaCl-like layers.

incommensurate b parameters, with the ratio $b_1/b_2 \sim 1.6$–2 (b_1 for NaCl-like layer and b_2 for CoO_2 layer).

The misfit oxides possess common features with Na_xCoO_2: they are metallic, with a large thermopower at room temperature [4]. The aim of our work was to dope the CoO_2 layers and understand the evolution of S (and ρ) with doping.

11.2.4.1 Influence of Doping

Substitutions take place most of the times in the NaCl-like layers, except for the case of Rh, as detailed in the following discussion. The doping of the CoO_2 layers is therefore induced by the charge balance between the two sublattices: the NaCl-like layers are positively charged (α, which is unknown), while the CoO_2 layers is negatively charged, the two sublattices being electrostatically bounded. Taking into account this charge balance, the following equations can be written for the Co valency:

$$v_{Co} = x + 3 = -\frac{\alpha}{b_1/b_2} + 4$$

where x is the Co^{4+} concentration.

x depends both on α and on the b_1/b_2 ratio, where b_1 and b_2 are the b parameters of the NaCl-like layers and CoO_2 layers, respectively. Another important parameter is the oxygen stoichiometry, which directly affects α.

The investigation of the different families of misfits has shown a general trend. First, the thermopower does not change much, from +90 μV/K for $[Tl_{0.81}Co_{0.2}Sr_{1.99}O_3]^{RS}[CoO_2]_{1.79}$ [19] to 170 μV/K for $[Ca_2Co_{0.6}Ti_{0.4}O_3][CoO_2]_{1.62}$ [20]. Second, two different families of behaviors are observed. Some misfits show a metallic behavior down to low T (2 K), and a small positive magnetoresistance (MR), as for $[Tl_{0.81}Co_{0.2}Sr_{1.99}O_3]^{RS}[CoO_2]_{1.79}$ and $[Bi_2Ba_{1.8}Co_{0.2}O_4][CoO_2]_2$ [21]. On the other hand, most of them show a strong increase of resistivity at low T, with a strong negative MR. The thermopower is always smaller in the first family.

Due to the large number of unknown parameters (oxygen stoichiometry in the two sublattices, cobalt valency in the NaCl-like layer, among others), it is very difficult to precisely know the Co valency in the CoO_2 layer and check the validity of the Koshibae's formula [8]. We have nevertheless found that, for a given family of misfit, S always increases when b_1/b_2 decreases (Figure 11.9).

The same trend is observed in the $[Pb_{0.7}Sr_{2-x}Ca_xCo_{0.3}O_3][CoO_2]b_1/b_2$ system (Figure 11.10) where S increases from S = 120 μV/K for $x = 0$ ($b_1/b_2 = 1.79$) to S = 165 μV/K for $x = 2$ ($b_1/b_2 = 1.62$) [22]. Also, this evolution is observed from 110 μV/K for $[Sr_2CoO_3][CoO_2]_{1.80}$ (b_1/b_2 = 1.80) to 120 μV/K for $[Ca_2CoO_3][CoO_2]_{1.62}$ ($b_1/b_2 = 1.62$) [23].

Following the v_{Co} equation, $v_{Co} = x + 3 = -(\alpha/(b_1/b_2)) + 4$, x theoretically decreases when b_1/b_2 decreases. From the Koshibae's formula, S increases when x decreases. The results presented here can therefore qualitatively be understood in this framework: S should increase when b_1/b_2 decreases.

For a constant b_1/b_2, it is also possible to modify α, and thus v_{Co}. By doping $[Ca_2CoO_3][CoO_2]_{1.62}$ with Ti^{4+} [20], or Pb^{3+}, b_1/b_2 is unchanged (1.62), but S is equal to 165 μVK with Ti^{4+} or 160 μV/K for Pb^{3+}, instead of 120 μV/K. If the oxygen content is unchanged, α is increased by the substitution, and v_{Co} decreases, consistently with an increase of S through the Koshibae's formula.

Recent experiments of annealing and titration by Karppinen et al. have shown that doping indeed modifies the S value and that the smaller content of Co^{4+} corresponds to the larger S [24].

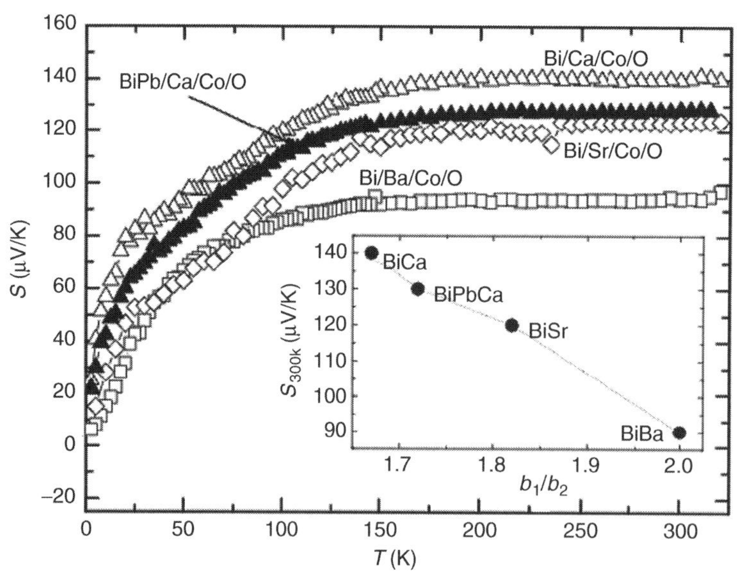

FIGURE 11.9 $S(T)$ of the Bi-based family of misfit. Inset: S at 300 K as a function of b_1/b_2.

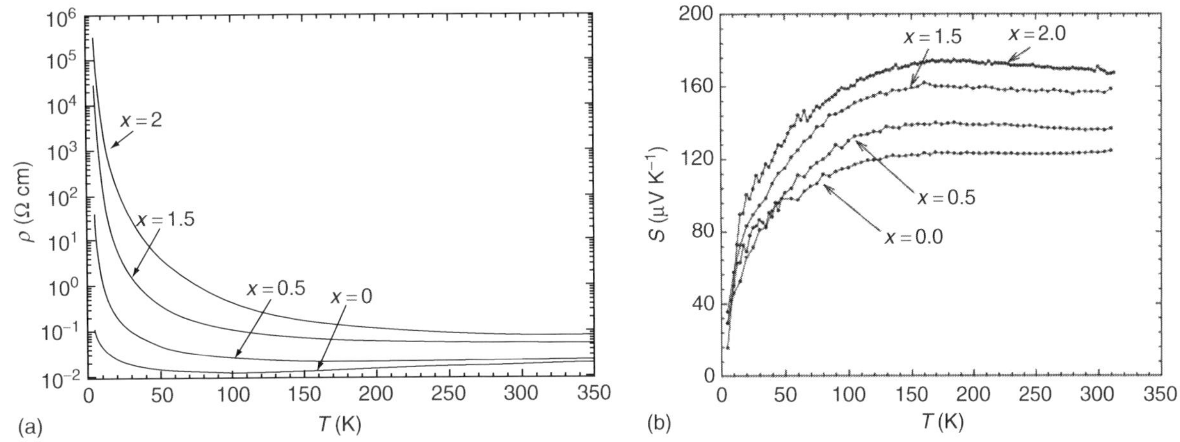

FIGURE 11.10 (a) $\rho(T)$ of the Pb/Sr/Ca/Co/O misfits; (b) $S(T)$ of the Pb/Sr/Ca/Co/O misfits.

The quantitative analysis of S as a function of doping can unfortunately not be made here due to the number of unknown quantities. Other techniques are now needed to evaluate the Co^{4+} concentration and make a more quantitative analysis.

11.2.4.2 Importance of Low Spin State

Following the generalized Heikes formula, the large entropy associated to the low spin states of Co^{3+} and Co^{4+} in the CoO_2 layers would be at the origin of the large thermopower. Interestingly, rhodium can partially [25] or totally be substituted to cobalt in the CoO_2 layer [26]. Rhodium is isoelectronic to cobalt, and always low spin, with Rh^{3+} (t_{2g}^{6}) and Rh^{4+} (t_{2g}^{5}) electronic configuration. We have thus investigated the magnetotransport properties of $[Bi_{1.95}Ba_{1.95}Rh_{0.1}O_4][RhO_2]_{1.8}$, knowing that le Rh species are in the low spin states, to compare them to the cobalt misfits [27].

The material is metallic and shows a large S of +90 μV/K at 300 K (Figures 11.11 and 11.12). These properties are very close to the ones measured in BiBaCoO [21], and this similarity is indirect proof of the low spin states proposed in the case of cobalt misfits. Also, a small positive MR of +5% is observed at low temperature in a magnetic field of 7 T. This is, to our knowledge, a unique case of MR in the rhodium oxides, and this emphasizes the peculiarity of the transport properties observed in these CdI_2-like layers.

11.2.4.3 Electrical Resistivity and Thermal Conductivity

Two different kinds of behaviors are observed in the misfits. At low temperature, only two misfits measured so far are metallic down to 2 K [19], or with a small increase of resistivity at low temperature [21], associated with a small positive MR. All the others present a transition to a more resistive behavior below 100 K, associated to a strong negative MR (−90% at 2.5 K in 7 T for BiCaCo [28]). In most of them, an increase of Seebeck coefficient is associated to an increase of

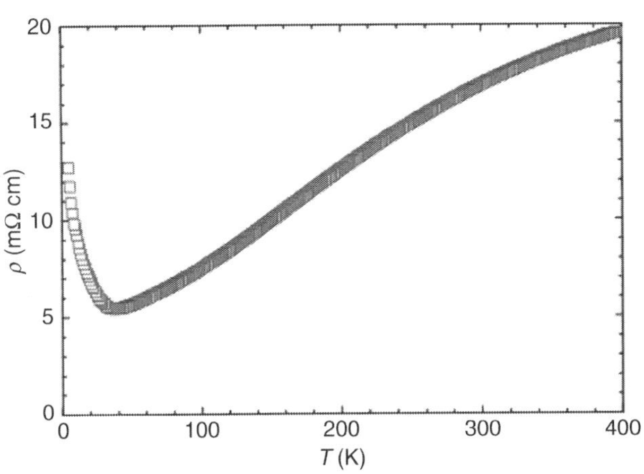

FIGURE 11.11 $\rho(T)$ of the Bi/Ba/Rh/O samples.

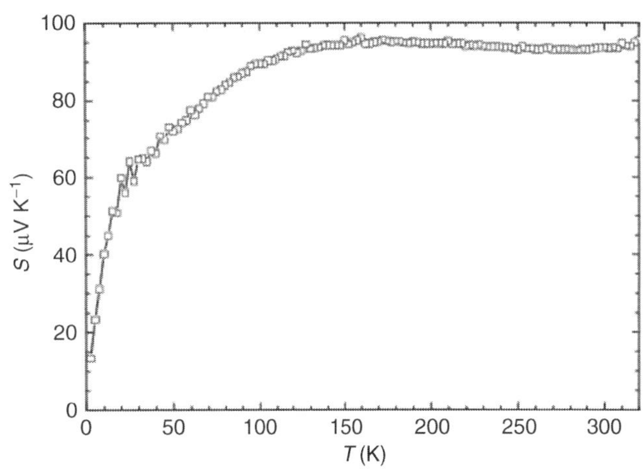

FIGURE 11.12 $S(T)$ of Bi/Ba/Rh/O.

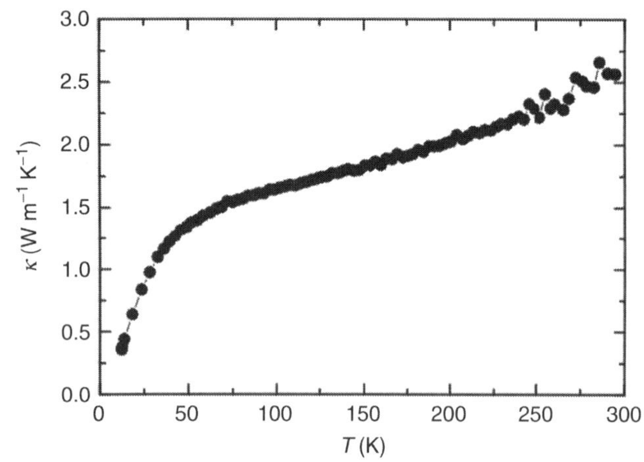

FIGURE 11.13 $\kappa(T)$ of $[Ca_2CoO_3][CoO_2]_{1.62}$.

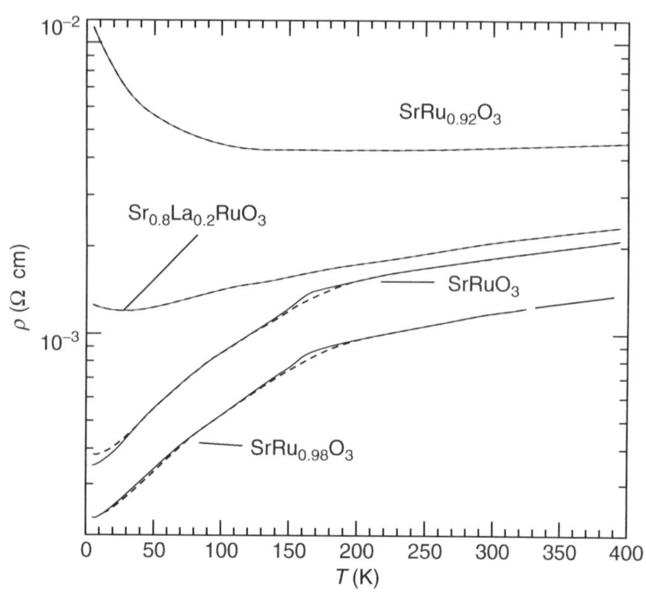

FIGURE 11.14 $\rho(T)$ of $SrRuO_3$ and related compounds, measured under 0T for the solid line and under 7T for the dashed line.

the resistivity and a transition to a more localized behavior, in the whole temperature range, as shown in Figure 11.10a for the PbSrCaCoO series. Nevertheless, in all these materials, the resistivity remains in the same range at room temperature with $\rho \sim$ 10–100 mΩ cm.

The thermal conductivity measured in polycrystals is always small with $\kappa \sim$ 2–3 W/m/K at 300 K (Figure 11.13). Following the Wiedemann-Franz law, it can be shown that the electronic part of the thermal conductivity is very small, with $\kappa_{el} \sim 0.03$ W/m/K at 300 K. These small values are close to the ones reported in disordered materials [29] and show that these materials could be good example of phonon glasses and electron crystals (PGEC) [30].

Combining all these values, maximum power factors close to 2×10^{-4} W/m/K² can be obtained at 300 K. By optimizing the texturation process in these materials, ZT values close to 1 have been reported [5].

11.2.5 $SrRuO_3$: A Metallic Perovskite with a Thermoelectric Power Driven by the Spin Degeneracy Term

The coexistence of a large thermopower in the metallic sodium cobaltites or misfits raises many questions. In particular, the importance of the spin and orbital degeneracies is very interesting and maximizing these terms should be crucial to get large Seebeck coefficients. Following this idea, we have decided to investigate other metallic oxides. Among them, the ruthenium oxides have been intensively investigated due to the ferromagnetism coexisting with good metallicity in $SrRuO_3$ [31], and to the superconductivity observed in Sr_2RuO_4 [32]. Also electronic correlations are expected to be less important in these 4d materials than in the 3d [33].

Only few reports can be found in the literature about the thermoelectric properties of $SrRuO_3$ and related compounds. The thermopower of $SrRuO_3$ is close to +21 μV/K at 300 K [34]. The thermoelectric properties of $SrRuO_3$ and related compounds have been investigated, as a function of doping. Four different samples have been investigated to modify the formal ruthenium oxidation state from 3.8 (for $Sr_{0.8}La_{0.2}RuO_3$), 4 (for $SrRuO_3$ and $SrRu_{0.97}O_{2.94}$), to 4.35 (for $SrRu_{0.92}O_3$).

Figure 11.14 shows the resistivity of the four different samples investigated here. $SrRuO_3$ is metallic with a clear change of slope around $T_c \sim 160$ K. A similar behavior is observed for $SrRu_{0.97}O_{2.94}$, with an even smaller resistivity. This counterintuitive result might be only related to extrinsic effects such as a reduction of the grain boundaries scattering. When the formal oxidation state is different from 4, the resistivity increases. ρ is still metallic in $Sr_{0.8}La_{0.2}RuO_3$, but larger, and ρ shows a more localized behavior in the case of $SrRu_{0.92}O_3$, with $d\rho/dT < 0$ at low T.

The thermopower is presented in Figure 11.15. At low temperatures, all the curves are very close, with $S \sim T$ up to 100–150 K. For $T >$ 100–150 K, S saturates in the case of $SrRuO_3$ and $SrRu_{0.97}O_{2.94}$ to $S_{300K} \sim 33$ μV/K. For $Sr_{0.8}La_{0.2}RuO_3$, S is only slightly affected with $S_{300K} \sim 33$ μV/K, still increasing at 300 K, and the accident observed at $T_c = 160$ K in $SrRuO_3$ [35] has disappeared. On the other hand, in $SrRu_{0.92}O_3$, $S_{300K} \sim 21$ μV/K and $dS/dT < 0$. For the four different samples, S is modified mostly at $T > 100$ K, but the differences are not very large.

The spin degeneracy associated to the mixed valency Ru^{3+}/Ru^{4+} or Ru^{4+}/Ru^{5+} can be calculated following Doumerc [7]: $S_{spin} = (-k_B/|e|)\ln[(2S_n + 1)/(2S_{n+1} + 1)]$, and with $S = 1/2$ for Ru^{3+}, $S = 1$ for Ru^{4+}, and $S = 3/2$ for Ru^{5+}, we obtain the two limits: $S = +35$ μV/K for Ru^{3+}/Ru^{4+} and $S = +25$ μV/K for Ru^{4+}/Ru^{5+}. Surprisingly, these two values correspond to the experimental limits found in our experiment. The spin degeneracy term directly gives a good estimate of the Seebeck coefficient. Again, this result is not in agreement with the metallic behavior of most of the samples. All metallic oxides do not behave like $SrRuO_3$, as

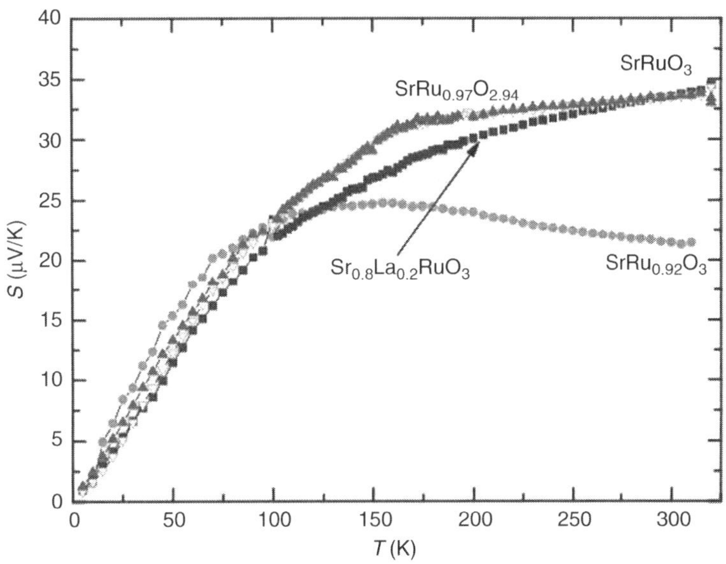

FIGURE 11.15 $S(T)$ of $SrRuO_3$ and related compounds.

for example $La_{1-x}Sr_xTiO_3$, which can be described by a single band model, with S linear in T [36]. Nevertheless, misfits and ruthenates possess similar features (S almost constant from 100 K to high T, with the S value depending on the spin degeneracy), which are also observed in other oxides. For example, in $BaMoO_3$, S is constant from 200 to 1000 K, close to −30 μV/K. The spin degeneracy term gives −25 μV/K for Mo^{3+}/Mo^{4+} ($S = 3/2/S = 1$) or −35 μV/K for Mo^{4+}/Mo^{5+} ($S = 1/S = 1/2$) [37]. This shows that this spin degeneracy term is important to maximize the Seebeck coefficient.

11.2.6 Conclusion

The present results obtained for oxides show the richness of their thermoelectric properties. It must be outlined that for several examples among those given here, these oxides synthesized in air are chemically stable up to very large temperatures (~1000°C). Interestingly, for such high temperatures, the oxides characterized by semiconducting-like electrical resistivities exhibit low-resistivity values. In this respect, it is remarkable that either metallic or semiconducting-like oxides can have large Seebeck coefficient values, which is counterintuitive when compared to classical thermoelectric materials. The origin of the large thermoelectric power in oxides reflects the strong contribution of the spin or orbital degeneracies to the Heikes formula. Again, this experimental conclusion is not predicted in the case of a metal but the example of the metallic ruthenates with constant Seebeck values over a wide range of temperatures demonstrates that theoretical models have still to be developed in the future to explain these properties. With regard to possible applications of thermoelectric oxides, the third important physical parameter is their thermal conductivity. The electronic part is very small due to small charge carrier mean free path, and is mainly phononic and small (~ 1 to 10 W/K/m). At present, though several devices for waste-heat conversion into electricity based on all oxides materials have been made, the best results for temperature gradients of about 500°C remain lower than 1 W/cm^2. This tells us that new oxides with enhanced thermoelectric properties are still needed to increase the efficiency of such devices. In particular, oxides containing transition metal cations with large spin and orbital entropies appear to be the most promising candidates.

Acknowledgments

The authors would like to thank their collaborators in the Crismat laboratory: Denis Pelloquin, Christine Martin, Maryvonne Hervieu, Bernard Raveau, Richard Retoux, Delphine Flahaut, Yannick Klein, Cédric Yaicle, Sudipta Pal, Vincent Hardy, and Christophe Goupil. Also fruitful collaborations with Jiri Hejtmanek (Institute of Physics, Prague) and Bogdan Dabrowski (Argonne National Laboratory) are acknowledged.

References

1. T. Kajikawa, *Thermoelectric Handbook, Macro to Nano*, M. Rowe (ed.), CRC Press, Boca Raton, FL, 2006.
2. I. Terasaki, Y. Sasago, and K. Uchinokura, *Phys. Rev. B* 56, R12685, 1997.
3. P. Boullay, B. Domengès, M. Hervieu, D. Groult, and B. Raveau, *Chem. Mater.* 8, 1482, 1996.
4. A. C. Masset, C. Michel, A. Maignan, C. Martin, and B. Raveau, *Phys. Rev. B* 62, 166, 2000.
5. R. Funahashi, I. Matsubara, H. Ikuta, T. Takeuchi, U. Mizutani, and S. Sodeoka, *Jpn. J. Appl. Phys.* 39, L1127, 2000.
6. P. M. Chaikin and G. Beni, *Phys. Rev. B* 13, 647, 1976.
7. J. P. Doumerc, *J. Solid Stat. Chem.* 110, 419, 1994.

8. W. Koshibae, K. Tsutsui, and S. Maekawa, *Phys. Rev. B* 62, 6869, 2000.
9. A. A. Taskin, A. N. Lavrov, and Y. Ando, *Phys. Rev. B* 73, 121101, 2006.
10. A. Maignan, D. Flahaut, and S. Hébert, *Eur. Phys. J. B* 39, 145, 2004.
11. A. Maignan, V. Caignaert, B. Raveau, D. Khomskii, and G. Sawatzky, *Phys. Rev. Lett.* 93, 026401, 2004.
12. S. Hébert, D. Flahaut, C. Martin, S. Lemonnier, J. Noudem, C. Goupil, and A. Maignan, *Progress in Solid State Chemistry*, 35, 457, 2007.
13. D. B. Marsh and P. E. Parris, *Phys. Rev. B* 54, 7720, 1996.
14. D. B. Marsh and P. E. Parris, *Phys. Rev. B* 54, 16602, 1996.
15. S. Pal, S. Hébert, C. Yaicle, C. Martin, and A. Maignan, *Eur. Phys. J. B* 53, 5, 2006.
16. D. J. Singh, *Phys. Rev. B* 61, 13397, 2000.
17. H. Yamauchi, K. Sakai, T. Nagai, Y. Matsui, and M. Karppinen, *Chem. Mater.* 18, 155, 2006.
18. M. Isobe, M. Onoda, M. Shizuya, M. Tanaka, E. Takayama-Muromachi, *Journal of the American Chemical Society*, 129, 14585, 2007.
19. S. Hébert, S. Lambert, D. Pelloquin, and A. Maignan, *Phys. Rev. B* 64, 172101, 2001.
20. A. Maignan, S. Hébert, L. Pi, D. Pelloquin, C. Martin, C. Michel, M. Hervieu, and B. Raveau, *Cryst. Eng.* 5, 299, 2002.
21. M. Hervieu, A. Maignan, C. Michel, V. Hardy, N. Créon, and B. Raveau, *Phys. Rev. B* 67, 045112, 2003.
22. A. Maignan, S. Hébert, D. Pelloquin, C. Michel, and J. Hejtmanek, *J. Appl. Phys.* 92, 1964, 2002.
23. D. Pelloquin, S. Hébert, A. Maignan, and B. Raveau, *Solid Stat. Sci.* 6, 167, 2004.
24. M. Karppinen, H. Fjellvag, T. Konno, Y. Morita, T. Motohashi, and H. Yamauchi, *Chem. Mater.* 16, 2790, 2004.
25. D. Pelloquin, S. Hébert, A. Maignan, and B. Raveau, *J. Solid Stat. Chem.* 178, 769, 2005.
26. S. Okada and I. Terasaki, *Jpn. J. Appl. Phys.* 44, 1834, 2005.
27. Y. Klein, S. Hébert, D. Pelloquin, V. Hardy, and A. Maignan, *Phys. Rev. B* 73, 165121, 2006.
28. A. Maignan, S. Hébert, M. Hervieu, C. Michel, D. Pelloquin, and D. Khomskii, *J. Phys. Cond. Matt.* 15, 7055, 2003.
29. D. G. Cahill, S. K. Watson, and R. O. Pohl, *Phys. Rev. B* 46, 6131, 1992.
30. G. A. Slack, in *CRC Handbook of Thermoelectrics*, D. M. Rowe (ed.), CRC Press, Boca Raton, FL, 1995, p. 407.
31. P. B. Allen, H. Berger, O. Chauvet, L. Forro, T. Jarlborg, A. Junod, B. Revaz, and G. Santi, *Phys. Rev. B* 53, 4393, 1996.
32. Y. Maeno, H. Hashimoto, K. Yoshida, S. Nashizaki, T. Fujita, J. G. Bednorz, and F. Lichtenberg, *Nature* 372, 532, 1994.
33. I. I. Mazin, D. and J. Singh, *Phys. Rev. B* 56, 2556, 1997.
34. K. Yamaura, D. P. Young, and E. Takayama-Muromachi, *Phys. Rev. B* 69, 024410, 2004.
35. Y. Klein, S. Hébert, A. Maignan, S. Kolesnik, T. Maxwell, and B. Dabrowski, *Phys. Rev. B* 73, 052412, 2006.
36. T. Okuda, K. Nakanishi, S. Mizasaka, and Y. Tokura, *Phys. Rev. B* 63, 113104, 2001.
37. K. Kurosaki, T. Oyama, H. Muta, M. Uno, and S. Yamanaka, *J. Alloys Comp.* 372, 65, 2004.

11.3 Electrically Conductive Adhesives

Yi Li, Myung Jin Yim, Kyoung-sik Moon, and C.P. Wong

11.3.1 Introduction

Electrically conductive adhesives (ECAs) are composites of polymeric matrices and electrically conductive fillers. The polymeric resin, such as an epoxy, a silicone, or a polyimide, provides physical and mechanical properties such as adhesion, mechanical strength, impact strength, and the metal filler (such as, silver, gold, nickel, or copper) conducts electricity. Metal-filled thermoset polymers were first patented as ECAs in the 1950s [1–3]. Recently, ECA materials have been identified as one of the major alternatives for lead-containing solders for microelectronics packaging applications. ECAs offer numerous advantages over conventional solder technology, such as environmental friendliness, mild processing conditions (enabling the use of heat-sensitive and low-cost components and substrates), fewer processing steps (reducing processing cost), low stress on the substrates, and fine pitch interconnect capability (enabling the miniaturization of electronic devices) [4–7]. Therefore, conductive adhesives have been used in flat panel displays such as liquid crystal display (LCD), and smart card applications as an interconnect material and in flip-chip assembly, chip-scale package (CSP), and ball grid array (BGA) applications as a replacement for solder. However, no currently commercialized ECAs can replace tin-lead metal solders in all applications due to some challenging issues such as lower electrical conductivity, conductivity fatigue (decreased conductivity at elevated temperature and humidity aging or normal use condition) in reliability testing, limited current-carrying capability, and poor impact strength. Table 11.2 gives a general comparison between tin-lead solder and generic commercialized ECAs [8].

Depending on the conductive filler loading level, ECAs are divided into ICAs, ACAs, and NCAs. For ICAs, the electrical conductivity in all x-, y-, and z-directions is provided due to high filler content exceeding the percolation threshold. For ACAs or NCAs, the electrical conductivity is provided

TABLE 11.2 Conductive Adhesives Compared with Solder

Characteristic	Sn/Pb Solder	ECA
Volume resistivity	0.000015 Ω cm	0.00035 Ω cm
Typical junction R	10–15 mW	<25 mW
Thermal conductivity	30 W/m °K	3.5 W/m °K
Shear strength	2200 psi	2000 psi
Finest pitch	300 μm	<150–200 μm
Minimum processing temperature	215°C	150°C–170°C
Environmental impact	Negative	Very minor
Thermal fatigue	Yes	Minimal

only in z-direction between the electrodes of the assembly. Figure 11.16 shows the schematics of the interconnect structures and typical cross-sectional images of flip-chip joints by ICA, ACA, and NCA materials illustrating the bonding mechanism for all three adhesives.

11.3.2 Isotropically Conductive Adhesives

ICAs, also called as "polymer solder," are composites of polymer resin and conductive fillers. The adhesive matrix is used to form a mechanical bond for the interconnects. Both thermosetting and thermoplastic materials are used as the polymer matrix. Epoxy, cyanate ester, silicone, polyurethane, etc. are widely used thermosets, and phenolic epoxy, maleimide acrylic preimidized polyimide, etc. are the commonly used thermoplastics. An attractive advantage of thermoplastic ICAs is that they are reworkable, i.e., can easily be repaired. A major drawback of thermoplastic ICAs, however, is the degradation of adhesion at high temperature. Another drawback of polyimide-based ICAs is that they generally contain solvents. During heating, voids are formed when the solvent evaporates. Most of commercial ICAs are based on thermosetting resins. Thermoset epoxies are by far the most common binders due to the superior balanced properties, such as excellent adhesive strength, good chemical and corrosion resistances, and low cost, while thermoplastics are usually added to allow softening and rework under moderate heat. The conductive fillers provide the composite with electrical conductivity through contact between the conductive particles. The possible conductive fillers include silver (Ag), gold (Au), nickel (Ni), copper (Cu), and carbon in various forms (graphites, carbon nanotubes, etc.), sizes, and shapes. Among different metal particles, silver flakes are the most commonly used conductive fillers for current commercial ICAs because of the high conductivity, simple process, and the maximum contact with flakes. In addition, silver is unique among all the cost-effective metals by nature of its conductive oxide. Oxides of most common metals are good electrical insulators, and copper powder, for example, becomes a poor conductor after aging. Nickel- and copper-based conductive adhesives generally do not have good resistance stability, because both nickel and copper are easily oxidized. ICAs have been used for die attach adhesives [9,10]. Recently, ICAs have also been considered as an alternative to tin/lead solders in surface mount technology (SMT) [11,12], flip-chip [13], and other applications and a large amount of effort has been conducted to improve the properties of ICAs in the past few years.

11.3.2.1 Improvement of Electrical Conductivity of ICAs

Polymer-based ECAs typically have lower electrical conductivity than Sn/Pb solders. To enhance the electrical conductivity of ICAs, various methods have been used and significant improvement of conductivity of ICA has been achieved.

11.3.2.2 Increase of Polymer Matrix Shrinkage

In general, ICA pastes exhibit insulative property before cure, but the conductivity increases dramatically after cure. ICAs achieve electrical conductivity during the polymer curing process caused by the shrinkage of polymer binder. Accordingly, ICAs with high

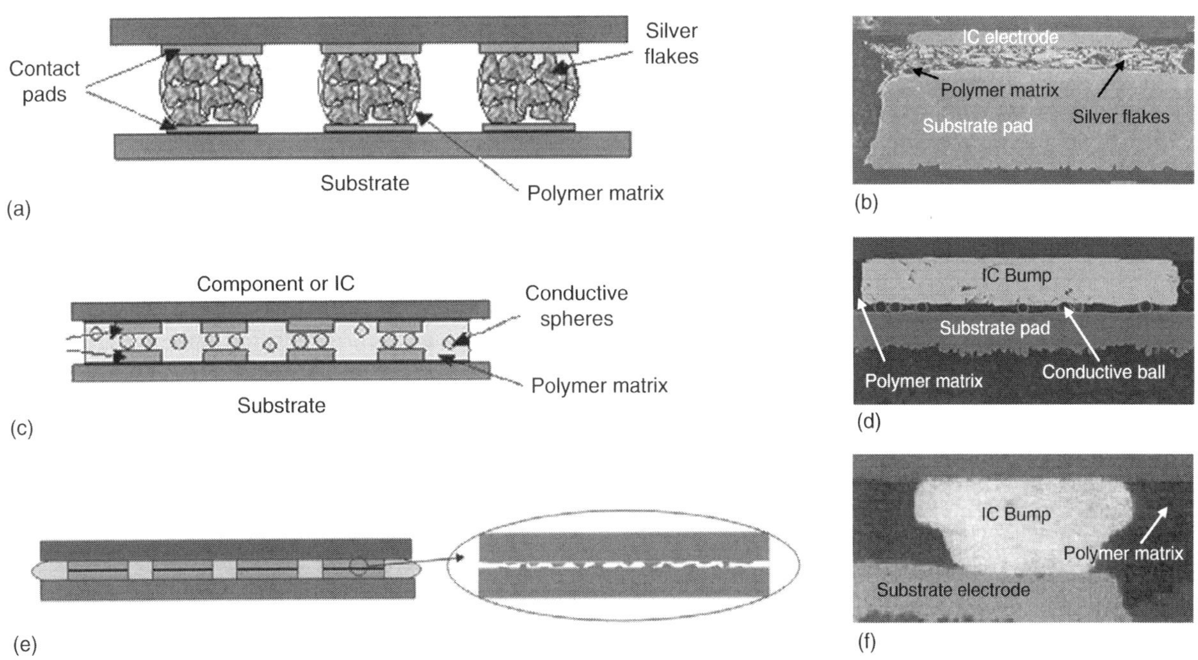

FIGURE 11.16 Schematic illustrations and cross-sectional views of (a,b) ICA, (c,d) ACA, and (e,f) NCA flip-chip bondings.

cure shrinkage generally exhibit higher conductivity [14]. With increasing cross-linking density of ICAs, the shrinkage of the polymer matrix increases, and subsequently, the resistivity of ICAs decreases. For epoxy-based ICAs, a small amount of a multifunctional epoxy resin can be added into an ICA formulation to increase cross-linking density, shrinkage, and thus increase electrical conductivity.

11.3.2.3 In Situ Replacement of Lubricants on Ag flakes

An ICA is generally composed of a polymer binder and Ag flakes. There is a thin layer of organic lubricant on the Ag flake surface. This lubricant layer plays an important role for the performance of ICAs, including the dispersion of the Ag flakes in the adhesives and the rheology of the adhesive formulations [15,16]. This organic lubricant layer, typically a fatty acid such as stearic acid, forms a silver salt complex between the Ag surface and the lubricant. However, this lubricant layer affects conductivity of an ICA because it is electrically insulating. To improve conductivity, the organic lubricant layer should be partially or fully removed or replaced during the curing of ICA. Short chain dicarboxylic acid is one of the suitable lubricant removers because of the strong affinity and more acidic of carboxylic functional group (–COOH) to silver. With the addition of only small amount of short chain dicarboxylic acid, the conductivity of ICA can be improved significantly due to the easier electronic tunneling/transportation between Ag flakes and subsequently the intimate flake-flake contact [17].

11.3.2.4 Incorporation of Reducing Agent in Conductive Adhesives

Silver flakes are by far the most used fillers for conductive adhesives due to the unique properties of the high conductivity of silver oxide compared to other metal oxides, most of which are insulative. However, the conductivity of silver oxide is still inferior to metal itself. Therefore, incorporation of reducing agents would further improve the electrical conductivity of ICAs. Aldehydes are one of the reducing agents that can be introduced. Obviously improved conductivity was reported in ICAs due to the reaction between aldehydes and silver oxides that exist on the surface of metal fillers during the curing process [18]:

$$\text{R-CHO} + \text{Ag}_2\text{O} \rightarrow \text{R} - \text{COOH} + 2\text{Ag} \tag{11.3}$$

The oxidation product of aldehydes, carboxylic acids, which are stronger acids and have shorter molecular length than stearic acid, can also partially replace or remove the stearic acid on Ag flakes and contribute to the improved electrical conductivity.

11.3.2.5 Low-Temperature Transient Liquid Phase Fillers

Another approach for improving electrical conductivity is to incorporate transient liquid-phase metallic fillers in ICA formulations. The filler used is a mixture of a high-melting-point metal powder (such as Cu) and a low-melting-point alloy powder (such as Sn-Pb or Sn-In). The low-melting-alloy filler melts when its melting point is achieved during the cure of the polymer matrix. The liquid phase dissolves and wets the high melting point particles. The liquid exists only for a short period of time and then forms an alloy and solidifies. The electrical conduction is established through a plurality of metallurgical connections in situ formed from these two powders in a polymer binder (Figure 11.17). The polymer binder with acid functional ingredient fluxes both the metal powders and the metals to be joined and facilitates the transient liquid bonding of the powders to form a stable metallurgical network for electrical conduction, and also forms an interpenetrating polymer network providing adhesion. High electrical conductivity can be achieved using this method [19,20]. One critical limitation of this technology is that the numbers of combinations of low and high melting fillers are limited. Only certain combinations of two metallic fillers, which are mutually soluble, exist to form this type of metallurgical interconnection.

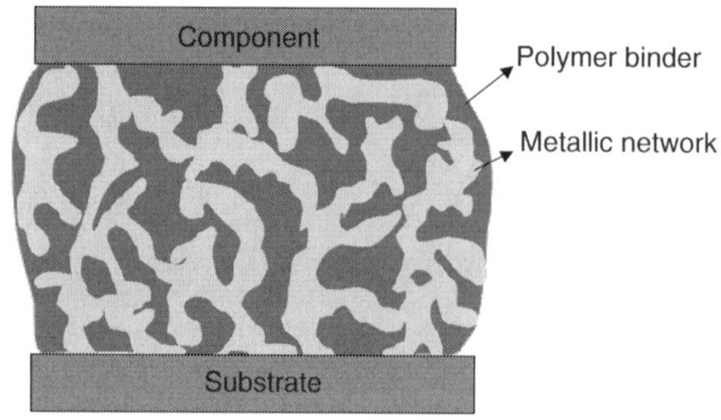

FIGURE 11.17 Diagram of transient liquid phase sintering conductive adhesives.

11.3.2.6 Low-Temperature Sintering of Nanosilver Fillers

Recently, nanosized conductive particles were proposed to be used as conductive fillers in ICAs for high-performance and fine-pitch interconnects. Although the nanosilver fillers in ICAs can reduce the percolation threshold, there has been concern that incorporation of nanosized fillers may introduce more contact spots due to high surface area and consequently induce higher resistivity compared to micron-sized fillers. A recent study showed that nanosilver particles could exhibit sintering behavior at curing temperature of ICAs [21]. Typically, applicavtion of nanofillers increases the contact resistance and reduces the electrical performance of the ICAs. The number of contacts between the small particles is larger than that between the large particles as shown in Figure 11.18a and b. The overall resistance of the ICA formulation is the sum of the resistance of fillers, the resistance between fillers, and the resistance between filler and pads (Equation 11.4). In order to decrease the overall contact resistance, the reduction of the number of contact points between the particles may be effective. If nanoparticles are sintered

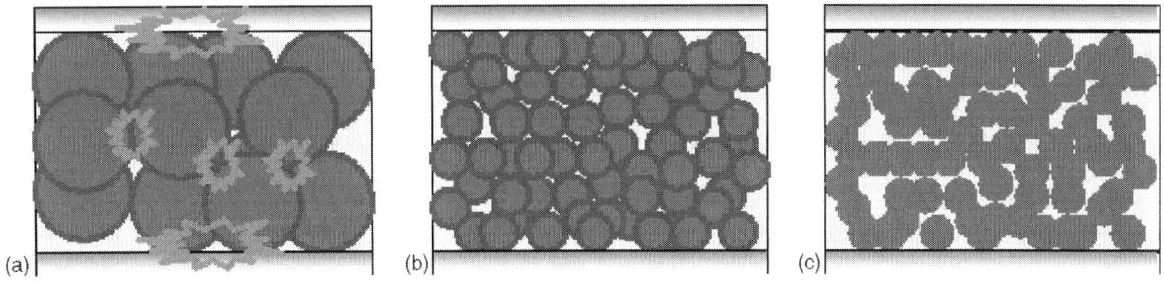

FIGURE 11.18 Schematic illustration of particles between the metal pads. (From Jiang, H.J., Moon, K., Lu, J., and Wong, C.P., *J. Electron. Mater.*, 34, 1432, 2005.)

together, then the contact between fillers will be fewer. This will lead to smaller contact resistance (Figure 11.18c). By using effective surfactants for the dispersion and effective capping those nanosized silver fillers in ECAs, obvious sintering behavior of the nanofillers can be achieved. The sintering of nanosilver fillers improved the interfacial properties of conductive fillers and polymer matrices and reduced the contact resistance between fillers. Therefore, an improved electrical conductivity of nanosilver-filled ICAs can be achieved.

$$R_{\text{total}} = R_{\text{btw fillers}} + R_{\text{filler to bond pad}} + R_{\text{fillers}} \tag{11.4}$$

11.3.2.2 Reliability Enhancement of ICA Interconnects

One of the critical reliability issues of ICAs is that contact resistance between an ICA and nonnoble metal finished components increases dramatically during an elevated temperature and humidity aging. The National Center of Manufacturing and Science (NCMS) defined the stability criterion for solder replacement conductive adhesives as a contact resistance shift (increase) of less than 20% after aging at 85°C/85%RH for 500 h [22]. However, most currently commercialized ICAs cannot pass the reliability test on nonnoble metal surfaces such as Sn/Pb, Sn, Cu, Ni, etc. However, some of the improved ICAs have been developed recently to satisfy the reliability requirements.

11.3.2.2.1 Mechanism Underlying Unstable Contact Resistance

Simple oxidation and galvanic corrosion of the nonnoble metal surfaces are the two possible mechanisms for unstable contact resistance of ICAs. Studies have shown that galvanic corrosion at the interface between an ICA and nonnoble metal is the dominant mechanism for the unstable contact resistance [23,24] (Figure 11.19). Under the aging conditions (for example, 85°C/85%RH), the nonnoble metal acts as an anode, and is oxidized by losing electrons, and then turns into metal ion ($M - ne^- = M^{n+}$). The noble metal acts as a cathode and its reaction generally is $2H_2O + O_2 + 4e^- = 4OH^-$. Then M^{n+} combines with OH^- to form a metal hydroxide then further oxidizes to metal oxide. After corrosion, a layer of metal hydroxide or metal oxide is formed at the interface. Because this layer is electrically insulating, the contact resistance increases dramatically. A galvanic corrosion process has several characteristics: (1) occurs only under wet conditions, (2) an electrolyte must be present, and (3) oxygen generally accelerates the process, and (4) dissimilar metals should be present. Based on the mechanism of unstable contact resistance of ICAs, several methods can be applied to stabilize the contact resistance and improve the reliability.

11.3.2.2.2 Low-Moisture Absorption

Moisture in polymer composites has been known to have an adverse effect on both mechanical and electrical properties of epoxy laminates. Effects of moisture absorption on conductive adhesive joints include degrading bulk mechanical strength,

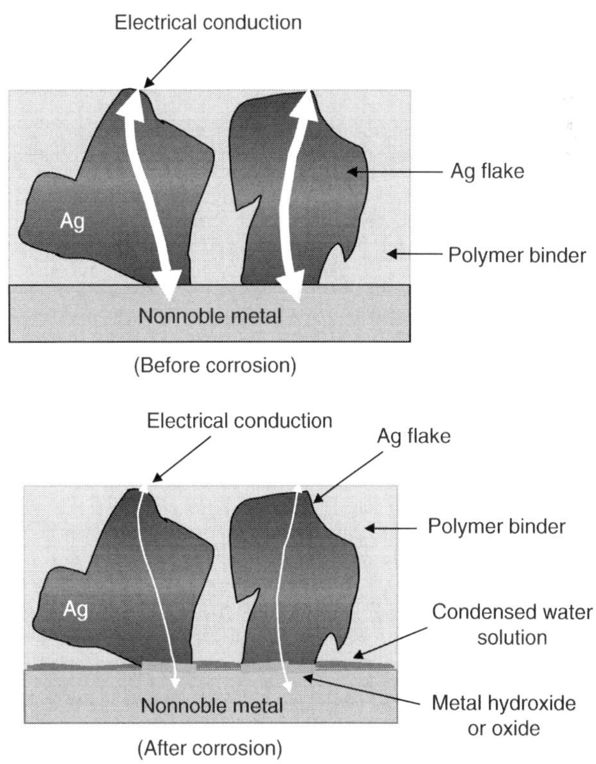

FIGURE 11.19 Metal hydroxide or oxide formation after galvanic corrosion.

decreasing interfacial adhesion strength, and causing delamination, promoting the growth of voids present in the joints, giving rise to swelling stress in the joint interfaces and inducing the formation of metal oxide layers resulted from corrosion. The water condensed from the absorbed moisture at the interface between an ECA and metal surface forms the electrolyte solution required for galvanic corrosion. Therefore, one way to prevent galvanic corrosion at the interface between an ICA and the nonnoble metal surface and achieve high reliability is to select ICAs with lower moisture absorption. Typically, ECAs formulated with high purity, low moisture absorption, and less polar resins have lower moisture absorption and contact resistance [25].

11.3.2.2.3 Oxygen Scavengers

Since oxygen accelerates galvanic corrosion, oxygen scavengers can be used in ICAs to slow down the corrosion rate [23]. When ambient oxygen molecules diffuse through the polymer binder, they react with the oxygen scavenger and are consumed. The main mechanism for oxygen scavengers to inhibit the corrosion is the cathodic mechanism, which is based on the lowering of oxygen concentrations. In the system consisting of metal, water, oxygen, and oxygen scavenger, the cathodic reaction of O_2 competes with its chemical reduction. If the overall corrosion process is cathodically controlled, the reduction in corrosion rate depends on the relative magnitudes of the rates of the electrochemical cathodic process R_{el} and the chemical process R_{ch}. The overall rate of oxygen reduction R_{ox} is given by the sum of the rates of the two processes indicated in the following equation:

$$R_{corr} = R_{ox} - R_{ch} \tag{11.5}$$

Since the rate of corrosion, R_{corr}, is equal to the rate of the electrochemical cathode process R_{el}, the corrosion rate is given by

$$R_{corr} = R_{ox} - R_{ch} \tag{11.6}$$

It can be seen from Equation 11.6 that the smaller the corrosion rate, the greater the rate of the chemical reduction of oxygen. Therefore, the reactivity of an oxygen scavenger with oxygen is an important consideration of its properties. Some commonly used oxygen scavengers include sulfites (such as sodium sulfates (Na_2SO_4), hydrazine ($H_2N\text{-}NH_2$), carbohydrazide ($H_2N\text{-}NH\text{-}CO\text{-}NH\text{-}NH_2$), diethylhydroxylamine ($(C_2H_5)_2N\text{-}OH$), and hydroquinone ($HO\text{-}C_6H_4\text{-}OH$).

11.3.2.2.4 Corrosion Inhibitors

Another method of preventing galvanic corrosion and stabilizing contact resistance is the use of corrosion inhibitors in ICA formulations [25,26]. In general, organic corrosion inhibitors are chemicals that adsorb on metal surfaces and act as a passivation barrier layer between the metal and the environment by forming an inert film over the metal surfaces [27–31]. Thus, the metal finishes can be protected. Some chelating compounds are especially effective in preventing metal corrosion. Appropriate selection of corrosion inhibitors can be very effective in protecting the metal finishes from corrosion. However, the effectiveness of the corrosion inhibitors is highly dependent on the types of contact surfaces, which is governed by the affinity of the passivating materials toward the coated metals. Effective corrosion inhibitors have been discovered for Sn/Pb, Cu, Al, and Sn surfaces, respectively [23,26,31].

11.3.2.2.5 Sacrificial Anode

To improve the contact resistance stability, applying a sacrificial anode is another efficient method. For ECAs during aging, the larger the difference in electrochemical potential, the faster the corrosion develops. Table 11.3 shows the electrode potential values of some common metals. Generally, metals with a low potential tend to corrode faster and show increased contact resistance than those with a high potential value. Therefore, when applying sacrificial materials with lower electrochemical potential than those of electrode-metal pads into ICAs, the sacrificial materials are preferably corroded first and, thus, can protect the metal finishes. This corrosion control is very important in reliability issues of the conductive adhesive joints. The addition of low corrosion potential individual metals, metal mixtures, or metal alloys reduces the electrode potential of ICAs, or in other words, narrows down the potential gap between the ICA and the metal finishes. Thus, these sacrificial anode materials act as an anode in this configuration and they are corroded first instead of the metal finishes, resulting in protecting the surfaces at the cathode [32–34].

11.3.2.2.6 Oxide-Penetrating Particles

Another approach of improving contact resistance stability during aging is to incorporate some electrically conductive particles, which have sharp edges, into the ICA formulations. The particle is called oxide-penetrating filler. Force must be provided to drive the oxide-penetrating particles through oxide layer and hold them against the adherent materials. This can be accomplished by employing polymer binders that show high shrinkage when cured [35]. (Figure 11.20) This concept is used in polymer-solder, which has good contact resistance stability

TABLE 11.3 Electrode Potentials of Selected Metal–Ion Pairs

Electrode Reaction	Standard Potential (eV)
$Au - 3e^- = Au^{3+}$	1.50
$Pt - 2e^- = Pt^{2+}$	1.20
$Ag - e^- = Ag^-$	0.80
$2H_2O + O_2 + 4e^- = 4OH^-$	0.40
$Cu - 1e^- = Cu^+$	0.34
$Cu - 2e^- = Cu^{2+}$	0.52
$2H^+ + 2e^- = H_2$	0.00
$Pb - 2e^- = Pb^{2+}$	−0.13
$Sn - 2e^- = Sn^{2+}$	−0.14
$Ni - 2e^- = Ni^{2+}$	−0.25

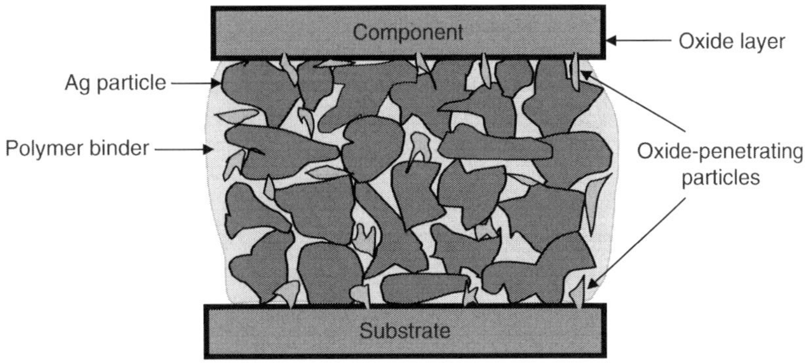

FIGURE 11.20 Joint connected with an ICA containing oxide-penetrating particles and silver powders.

with standard surface-mounted devices (SMDs) on both solder-coated and bare circuit boards.

11.3.2.3 Enhancement of Adhesion Strength

Another issue of conductive adhesive is the lower adhesion strength. High adhesion strength is a critical parameter in fine pitch interconnection that is fragile to shocks encountered during assembly, handling, and lifetime usage.

There are two types of adhesion mechanisms, physical bonding and chemical bonding, which contribute to the overall adhesion strength of polymer on a surface [36]. Chemical bonding involves the formation of covalent or ionic bonds to link between the polymer and the substrate. Physical bonding involves mechanical interlocking or physical adsorption such as van der Waals force between the polymer and the surface of the substrate. In cases where the molecules of the polymer are highly compatible with the molecules of the substrate, they interact to form an interdiffusion layer. In mechanical interlocking, polymer and substrate interact on a more macroscopic level, where the polymer flows into the crevices and the pores of substrate surface to establish adhesion. Therefore, a polymer is expected to have higher adhesion on a rougher surface because there is more surface area and "anchors" to allow for interlocking between the polymer and the substrate.

11.3.2.3.1 Plasma Cleaning and Vacuum Process

Plasma cleaning of surfaces has been considered as one of the effective approaches to enhance the adhesion strength of conductive adhesives [37]. During the plasma etching process, the plasma charged particles, such as radicals react with the contaminants and those long chains organic molecules can be broken down into small gaseous particles (mostly gaseous water and carbon-oxide conjunctions). These particles can be evacuated from the system by the vacuum pump. After the surface is cleaned of oxide, a layer of atoms available in the chamber will deposit on the clean surface, protecting the metal from new oxidation. Furthermore, the plasma can also etch the surface to enhance the mechanical interlocking mechanism for adhesion improvement. With plasma treatment, the adhesion strength of conductive adhesives could be improved. However, the improvement also requires a high vacuum process due to the possible corrosion on the highly cleaned surface caused by the moisture in ECAs. After vacuum process, the surface was changed by removal of air and water and the contraction of adhesives.

11.3.2.3.2 Application of Coupling Agents in ECA

Another approach to improve adhesion is by using coupling agents [38]. Coupling agents are organofunctional compounds based on silicon, titanium, or zirconium. For example, R-X-$(O\text{-}R')_3$, where, X = Si, Ti or Zr, R = organic chain that interacts with the polymer, and R′ = organic chain that interacts with the substrate. A coupling agent consists of two parts and acts as intermediary to "couple" the inorganic substrate and polymer.

Silane-coupling agents have been commonly used to improve the adhesion performance. In conductive adhesives, application of specific silane-coupling agents with appropriate concentration can increase the adhesion strength on different metal surfaces. Although silane-coupling agents are mostly used, some other coupling agents with various functional groups, such as thiol, carboxylate-coupling agents, are also used for specific surfaces [39].

11.3.2.3.3 Optimization of Elastic Modulus

In order to enhance the adhesion, another approach is to lower the elastic modulus of adhesive resins. By using low elastic modulus resins, the thermal stress at the adhesion interface can be reduced and, thus, the improved the adhesion strength [40,41]. In general, the adhesion strength increases with lowering the elastic modulus. However, too low modulus value deteriorates the cohesive force and thus, decreases the adhesion strength. Therefore, the elastic modulus needs to be optimized to improve the adhesion properties. In addition to the methods listed above, some other factors such as curing conditions, surface roughness and structures of IC packaging may also affect the adhesion strength of conductive adhesives.

11.3.2.4 Improvement of Impact Performance

Impact performance is a critical property of solder replacement ICAs. Effort has been continuing in developing ICAs that have higher impact strength and can pass the drop test, a standard test used to evaluate the impact strength of ICAs.

Nanosized metal particles were used in ICAs to improve the electrical conduction and mechanical strength. Using nanosized particles, agglomerates are formed due to surface tension effect, and high-power ultrasonication is needed to disperse these agglomerates. Another approach is to simply decrease the filler loading to improve the impact strength [42]. However, such a process reduces the electrical properties of the conductive adhesives. Recent development was reported where conductive adhesives were developed using resins of low modulus so that this class of conductive adhesives could absorb the impact energy developed during the drop [43]. Conformal coating with silicone or polyurethanes of the SMDs was used to improve mechanical strength. It was demonstrated that conformal coating could improve the impact strength of conductive adhesives joints [44]. Furthermore, elastomer-modified epoxy resins with enhanced loss modulus (high tan θ values) were also used to enhance the damping properties and the impact performance of ICAs [45].

11.3.2.5 Improvement of Reliability in Thermomechanical Cycling Test

The poor thermal cycling (TC) performance of the ECA joints has been another reliability issue for board level interconnects. It has been found rather difficult to control the TC failures of the ECA joints. Generally, the failure of the electrical interconnection during the TC test could be caused by many factors such as coefficient of the thermal expansion (CTE) mismatch between the IC component chip, the interconnection materials, and the substrates, elastic moduli difference of these components, adhesion strength of the interconnect materials on the IC chip and the substrate, the mechanical properties of the IC chips, the glass transition or the softening point of the ECA materials, moisture uptake in the interface and the bulk ECAs, the surface or interface property change and so forth. Especially, the thermal stress in the ECA joints generated by a huge temperature difference during the TC and the interfacial delamination due to the adhesion degradation could be the critical reasons. In this aspect, a feasible solution to the TC failure problem is to introduce flexible molecules into the epoxy resin matrix. By releasing the thermal stress with the flexible molecules, the thermomechanical stresses could be dramatically reduced and the ECA/component joint interfaces could keep intact through the TC test [46]. In addition, application of low CTE adhesives by using high loading of fillers has been proved to induce lower shear strain induced by CTE mismatch between chip and board under temperature cycling [47–49].

11.3.2.6 Electrochemical Migration Control

Although silver is the most widely used conductive filler in ICAs, silver migration has long been a reliability concern in the electronic industry. Metal migration is an electrochemical process, whereby metal (e.g., silver), in contact with an insulating material, in a humid environment and under an applied electric field, leaves its initial location in ionic form and deposits at another location [50]. It is considered that a threshold voltage exists above which the migration starts. Such migration may lead to a reduction in electrical spacing or cause a short circuit between interconnections. The migration process begins when a thin continuous film of water forms on an insulating material between oppositely charged electrodes. When a potential is applied across the electrodes, a chemical reaction takes place at the positively biased electrode where positive metal ions are formed. These ions, through ionic conduction, migrate toward the negatively charged cathode and over time, they accumulate to form metallic dendrites. As the dendrite growth increases, a reduction of electrical spacing occurs. Eventually, the dendrite silver growth reaches the anode and creates a metal bridge between the electrodes, resulting in an electrical short circuit [51].

Although other metals may also migrate under specific environment, silver is more susceptible to migration, mainly due to the high solubility of silver ion, low activation energy for silver migration, high tendency to form dendrite shape and low possibility to form stable passivation oxide layer [52–54]. The rate of silver migration is increased by (1) an increase in the applied potential; (2) an increase in the time of the applied potentials; (3) an increase in the level of relative humidity; (4) an increase in the presence of ionic and hydroscopic contaminants on the surface of the substrate; and (5) a decrease in the distance between electrodes of the opposite polarity.

In order to reduce silver migration and improve the reliability, several methods have been reported. The methods include: (1) alloying the silver with an anodically stable metal such as palladium [51] or platinum [55]; (2) using hydrophobic coating over the PWB to shield its surface from humidity and ionic contamination [56], since water and contaminates can act as a transport medium and increase the rate of migration; (3) plating of silver with metals such as tin, nickel, or gold to protect the silver fillers and reduce migration; (4) coating the substrate with polymer [57]; (5) applying benzotriazole (BTA) and its derivatives in the environment [58]; (6) employing siloxane epoxy polymers as diffusion barriers due to the excellent adhesion of siloxane epoxy polymers to conductive metals [59]; and (7) chelating silver fillers in ECAs with molecular monolayers [60].

11.3.3 Anisotropic Conductive Adhesives and Films

ACAs or anisotropic conductive films (ACFs) provide unidirectional electrical conductivity in the vertical or *Z*-axis. This directional conductivity is achieved by using a relatively low volume loading of conductive fillers (5 to 20 vol.%). The low-volume loading is insufficient for interparticle contact and prevents conductivity in the *X*–*Y* plane of the adhesives. The *Z*-axis adhesive, in film or paste form, is interposed between the surfaces to be connected. Heat and pressure are simultaneously applied to this stack-up until the particles bridge the two conductor surfaces.

The ACF bonding method is a thermocompression bonding process as shown in Figure 11.21. In case of tape-carrier packages

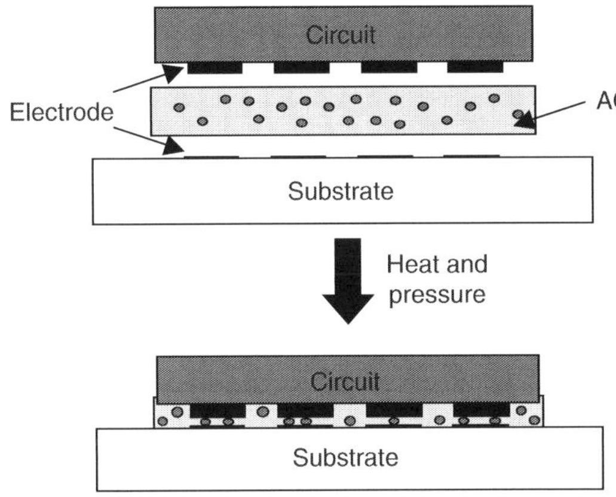

FIGURE 11.21 Thermocompression bonding using ACF.

(TCP) bonding, the ACF material is attached on a glass substrate and preattached. Final bonding is established by the thermal cure of the ACF resin, typically at 150°C–220°C, 3–20 s, and 100–200 MPa, and the conductive particle deformation between the electrodes of TCP and the glass substrate by applied bonding pressure.

Interconnection technologies using ACFs are major packaging methods for flat panel display modules and have high resolution, are light weight, have thin profile, and low consumption power [61], and already successfully implemented in the forms of out lead bonding (OLB), flex to PCB bonding (PCB), reliable direct chip attach such as chip-on-glass (COG), chip-on-film (COF) for flat panel display modules [62–65], including LCD, plasma display panel (PDP), and organic light emitting diode display (OLED). As for the small and fine pitched bump of driver ICs to be packaged, fine pitch capability of ACF interconnection is much more desired for COG, COF, and even OLB assemblies. There have been advances in development works for improved material system and design rule for ACF materials to meet fine pitch capability and better adhesion characteristics of ACF interconnection for flat panel displays.

In addition to the LCD industry, ACA/ACF is now finding applications in flex circuits and SMT for CSP, application-specific integrated circuit (ASIC), and flip-chip attachment for cell phones, radios, personal digital assistants (PDAs), sensor chip in digital cameras, and memory chip in laptop computers. In spite of the wide applications of ACA/ACF, there are some key issues that hinder their implementations as high power devices. The ACA/ACF joints generally have lower electrical conductivity, poor current carrying capability, and electrical failure during TC.

11.3.3.1 Improvement of Electrical Properties of ACA/ACF

As the conductivity of ACA joints is directly determined by the mechanical contact between the terminals of chips and the electrodes on chip carriers, the bonding force plays a critical role in the electric performance. High bonding pressure is favorable to an intimate contact, and thus to a low contact resistance. In addition, because the bonding of all bumps of a chip is performed simultaneously, uniform conductivity of all joints in the chip requires the bonding situation of every joint to be completely the same.

11.3.3.1.1 Organic Self-Assembly Monolayer Enhancement

In order to enhance the electrical performance of ACA materials, organic monolayers have been introduced into the interface between metal fillers and the metal-finished bond pad of ACAs [66,67]. These organic molecules adhere to the metal surface and form physiochemical bonds, which allow electrons to flow, as such, it reduces electrical resistance and enables a high current flow. The unique electrical properties are due to their tuning of metal work functions by those organic monolayers. The metal surfaces can be chemically modified by the organic monolayers and the reduced work functions can be achieved by using suitable organic monolayer coatings. An important consideration when examining the advantages of organic monolayers pertains to the affinity and thermal stability of organic compounds to specific metal surfaces. Table 11.4 gives the examples of molecules preferred for maximum interactions with specific metal finishes.

11.3.3.1.2 Low-Temperature Sintering of Nano Ag Filled ACA

One of the concerns for ACA/ACF is the higher joint resistance since interconnection using ACA/ACF relies on mechanical contact, unlike the metal bonding of soldering. An approach to minimize the joint resistance of ACA/ACF is to make the conductive fillers fuse each other and form metallic joints such as metal solder joints. However, to fuse metal fillers in polymers does not appear feasible, since a typical organic printed circuit board ($T_g \sim 125°C$), on which the metal filled polymer is applied, cannot withstand such a high temperature; the melting temperature (T_m) of Ag, for example, is around 960°C. Research showed that T_m and sintering temperatures of materials could be dramatically reduced by decreasing the size of the materials [68–70]. It has been reported that the surface premelting and sintering processes are a primary mechanism of the T_m depression of the fine nanoparticles (<50 nm). For nanosized particles, sintering

TABLE 11.4 Potential Organic Monolayer Interfacial Modifiers for Different Metal Finishes

Formula	Compound	Metal Finish
R-S-H	Thiols	Au, Ag, Cu, Pt, Zn
R-COOH	Carboxylates	Fe/Fe_xO_y, Ti/TiO_2, Ni, Al, Ag
R-C≡N	Cyanides	Au, Ag, Pt
R-N=C=O	Isocyanates	Pt, Pd, Rh, Ru
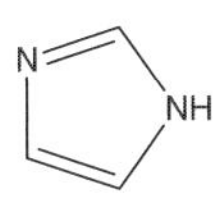	Imidazole and derivatives	Cu
R-SiOH	Organosilicone derivatives	SiO_2, Al_2O_3, quartz, glass, mica, ZnSe, GeO_2, Au

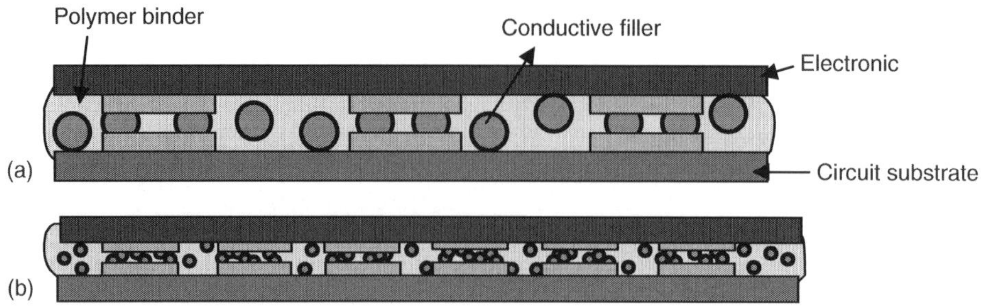

FIGURE 11.22 Schematic illustrations of ACA/ACF with (a) microsized conductive fillers and (b) nanosized conductive fillers.

behavior could occur at much lower temperatures, as such, the use of the fine metal particles in ACAs would be promising for fabricating high electrical performance ACA joints through eliminating the interface between metal fillers. The application of nanosized particles can also increase the number of conductive fillers on each bond pad and result in more contact area between fillers and bond pads as illustrated in Figure 11.22. For the sintering reaction in a certain material system, temperature and duration are the most important parameters, in particular, the sintering temperature.

11.3.3.2 Improvement of Thermal Properties of ACA/ACF

The thermal performance of an adhesively assembled chip is of vast interest as power dissipation in the high-performance, high-frequency devices. The current high-performance microprocessor operates in excess of 5 GHz, as a consequence, enormous heat has been generated which needs to be dissipated for reliability concern. Power dissipations have been simulated by Sihlbom et al. for both ICA and ACA flip-chip joints [71]. They concluded that the ACA flip-chip joint is more effective in transferring heat to the substrate from the powered chip than the ICA joint, because the adhesive thickness is so much thinner than the ICA joint.

For the ACA interconnect joints to deliver high current, not only a low electrical resistance, but also a high thermal conductivity of the interconnect materials is required. The higher thermal conductivity can help dissipate heat more efficiently from adhesive joints generated at high current. Therefore, a higher thermal conductivity could contribute to an improved current-carrying capability. With the introduction of suitable organic monolayer treatment, an improved interface property between metal fillers and polymer resins could be achieved, which enhanced the thermal conductivity of ACA joints. Studies show that organic monolayer enhanced ACA joints with best electrical properties also had a higher thermal conductivity. It is also reported that with the addition of high thermal conductivity fillers (e.g., SiC, AlN) into the ACA formulation, the higher thermal conductivity could be achieved [72,73], which also rendered a high current-carrying capability. Furthermore, carbon nanotubes, with high thermal conductivity (>3000 W/k from theoretical calculations) can also be aligned within the composite matrix to enhance the thermal conductivity.

11.3.3.3 ACAs for High-Frequency Interconnections

Recently, high-frequency modeling and the characterization for ACA flip-chip interconnects have been performed to understand the high-frequency characteristics of flip-chip interconnects using ACFs, and thereby design higher performance ACF materials and bump system [74–76].

The effect of low dielectric filler additions on high-frequency behavior of ACF was investigated. The extracted impedance model parameters of a 100 μm100 μm bonding pad were presented for conventional ACF with the conductive ball only, and ACF with the conductive ball and SiO_2 filler [77]. In ACF flip-chip interconnects at high frequency, interconnection capacitance formed is relatively high due to the high dielectric constant of the ACF resin and the large area, the small gap of the parallel metal plate structure, compared with the solder ball flip-chip structure. Therefore, the resonance frequency of the ACF flip-chip interconnect is lower than the solder ball flip-chip interconnects. In addition, the bump metallurgy can also affect the high-frequency behaviors of ACF interconnect.

High-frequency performance of several flip-chip interconnects using ACFs at RF and high-frequency range were demonstrated and the ACF flip-chip assembly was proved as a simple and cost effective method for high frequency devices.

11.3.3.4 Nanowire ACF for Ultrafine Pitch Flip-Chip Interconnection

In order to satisfy the reduced I/O pitch and avoid electrical shorts, a possible solution is to use high-aspect-ratio metal post. Nanowires exhibit high possibilities due to the small size and extremely high aspect ratio. In the literature, nanowires have been shown to be applied to field effective transistor (FET) sensor for gas detection, magnetic hard-disk, nanoelectrodes for electrochemical sensor, thermal-electric device for thermal dissipation, temperature control, among others [78–80]. To prepare nanowires, it is important to define nanostructures on photoresist. High-cost lithographic methods such as e-beam, X-ray, or scanning probe lithography have been used but the length of nanowires cannot be achieved in micrometers. Another less expensive alternative is electrodeposition of metal into nanoporous template such as anodic aluminum oxide (AAO) [81] or a block-copolymer self-assembly template [82]. The disadvantages of a block-copolymer template include thin thickness (that means short nanowires), nonuniform distribution, and poor parallelism of nanopores. However, AAO has benefits of higher thickness (>10 μm), uniform pore size and density, larger size and very parallel pores.

11.3.3.5 Reliability of ACA/ACF

Reliability is always one of the most important considerations to determine the value of materials. Failure behaviors of the interconnection are characterized by performing accelerated aging tests such as high and low temperature storages, temperature cycling, and humidity storage. Some main degradation mechanisms have been found, including oxide growth on less noble metallization causing increased contact resistance, stresses due to the thermal load or influence of humidity inducing cracks within or delamination of the adhesive layer.

11.3.3.5.1 Effects of Bonding Conditions

Bonding conditions of ACF, such as bonding temperature, temperature ramping rate, and pressure affect the curing of ACF materials and the bonding quality of ACF joints. It has been reported that higher bonding temperatures resulted in higher cross-linking density and stronger adhesion strength of ACF [83,84]. The increased adhesion strength with increasing bonding temperature is considered to be due to the interdiffusion or reaction between ACF polymeric material and the flex. In addition, it has been found that ACF joints with higher degrees of curing showed smaller increase in contact resistance after aging. At lower bonding temperatures and subsequently lower degree of cure, the conductive particles can migrate to the polymer layer. Therefore, the conductive particles could be squeezed out of the joints due to the applied pressure through the soft adhesive. As a result, in some joints, higher and unstable contact resistance and even some open joints were found. At higher bonding temperature and higher degree of cure, the conductive particles remain throughout the ACF joints.

The bonding pressure also affects the reliability of ACF joints significantly. The lower the used pressure, the lower stress accumulated at the interface and consequently the higher reliability after TC test. However, pressure has to be large enough to ensure electrical conductivity and strong bonding.

11.3.3.5.2 Effect of Mechanical Properties on the Reliability of ACF Joints

Since thermal stress or influence of humidity-induced cracks is the main degradation mechanism of ACF joints, the thermomechanical properties of ACF materials play a significant role in the joint reliability during the TC test [85,86]. Generally, materials with low CTE should show improved reliability in temperature cycling [87]. This is because matched CTEs between the matching electrode pairs of the IC and the substrate (two contact partners) is necessary to minimize thermomechanical stresses on conductive joints within the polymeric layer (Figure 11.23).

To enhance the reliability of ACA flip-chip assemblies, nonconductive fillers, such as silica fillers were added in the ACA materials with different filler loadings [88]. It has been reported that the modulus of ACA materials increased and CTE below the T_g decreased as the content of silica increased. Consequently, the reliability of ACA flip-chip assemblies using ACA with higher content of nonconductive fillers was significantly improved than that of flip-chip assembly using conventional ACAs.

11.3.4 Nonconductive Adhesives and Films

ECA joints can be formed without any conductive fillers. The electrical connection of NCAs is achieved by sealing the two contact partners under pressure and heat. Thus, the small gap contact is created, approaching the two surfaces to the distance of the surface asperities, as illustrated in Figure 11.16e. The formation of contact spots depends on the surface roughness of the contact partners. Approaching the two surfaces enables a small number of contact spots to form, which allows the electric current to flow. When the parts are pressed together during the sealing process, the number and area of the single contact spots are increased according to the macroscopic elasticity or flexibility of the parts and the microhardness and plasticity of the surfaces, respectively.

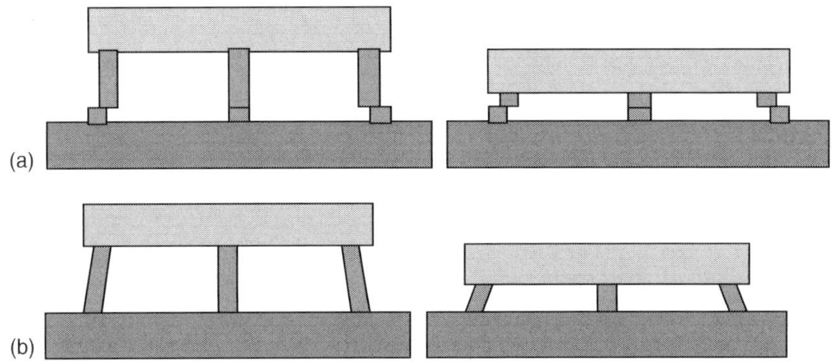

FIGURE 11.23 (a) Deformation mechanism of the ACA joint due to thermally induced stress and (b) schematic diagram showing that thermal mismatch strain reduces, as solder bump height increases.

11.3.4.1 Electrical Properties of NCA

Conductive joints with NCA/NCF provide a number of advantages compared to other adhesive bonding techniques. NCA/NCF joints avoid short-circuiting and are not limited, in terms of particle size or percolation phenomena, to a reduction of connector pitches. The size reduction of electronic devices can be realized by the shrinkage of the package and the chip. Further advantages include cost-effectiveness, ease of processing regarding the possibility of nonstructured adhesive application, good compatibility with a wide range of contact materials, and low-temperature cure. In fact, the pitch size of the NCF joints can be limited only by the pitch pattern of the bond pad, rather than the adhesive materials.

In spite of the advantages of NCF joints, there are still some challenging properties of NCF. The electrical conduction of NCF between the two surfaces of the IC bump and the substrate bond pad is achieved by bonding the two contact parts with pressure and heat. Thus, the direct mechanical contact is formed by the very fine up-and-hill structure of the bottom and top pad surfaces with the roughness during bonding and a permanent joint is then formed by NCF resin curing or solidification. Since the electrical conductivity of NCF is achieved through physical and mechanical contact and no metallurgical joints are formed, it has limited electrical conductance and current-carrying capability. Low contact resistance and high current-carrying capability of NCF joints are demanding properties for lead-free solder alternatives and high-current density application. To ensure low contact resistance and high current density, interface between electrodes plays important role. The interfaces between electrodes for NCF joints should be defect free and occupy the stable contact area even under high electrical current and harsh environments. This interface control in NCF joints contributes to performance and reliability of their joints in electronic packaging. In most NCF joints, electrical conductance between contacts is dependent upon the constriction resistance and tunneling resistance due to the presence of ultrathin insulating film between contacts. The control of the tunneling resistance is important in reducing the contact resistance for NCF joints [89–91]. Self-assembled π-conjugated molecular wires were recently discovered to functionalize the NCF joints and tailor their physical and chemical properties. Conjugated molecules tend to have a smaller band gaps between the highest occupied molecular orbital (HOMO) and the lowest unoccupied molecular orbital (LUMO) and possess delocalized conjugated π-electrons that can contribute to conduction. In semiconductors, self-assembled molecular wires have seen importance in tuning the metal work function (ϕ) and electrical conduction of metal–molecule contact [92–96].

11.3.4.2 NCAs with Improved Reliability for Flip-Chip Assembly

For the full implementation of flip-chip using NCAs, it is necessary to provide good reliability data to prove the availability of NCAs flip-chip technology. The most commonly observed flip-chip failure has occurred during the TC test, which is due to the thermal expansion mismatch between chips and substrates. Therefore, the problem of CTE mismatch between chips and substrates becomes serious with the NCAs flip-chip assembly because of high CTE of NCAs materials without any filler. Unlike ACAs with some conductive fillers, even small delamination or deformation of the NCA joint can easily lose the electrical joint. For this reason, novel NCAs that have low CTE for underfill-like function has been developed. Figure 11.24 shows the schematic illustration of flip-chip CSP using NCF as first level interconnection and its cross-section view of Au stud bump joint [97].

The addition of nonconducting silica filler of the NCA composite materials has control on the curing behavior, thermomechanical properties, and reliability for the NCA flip-chip assembly on an organic substrate. The content of nonconducting filler was optimized for the desirable thermomechanical properties of NCA composite materials such as proper curing profile, high

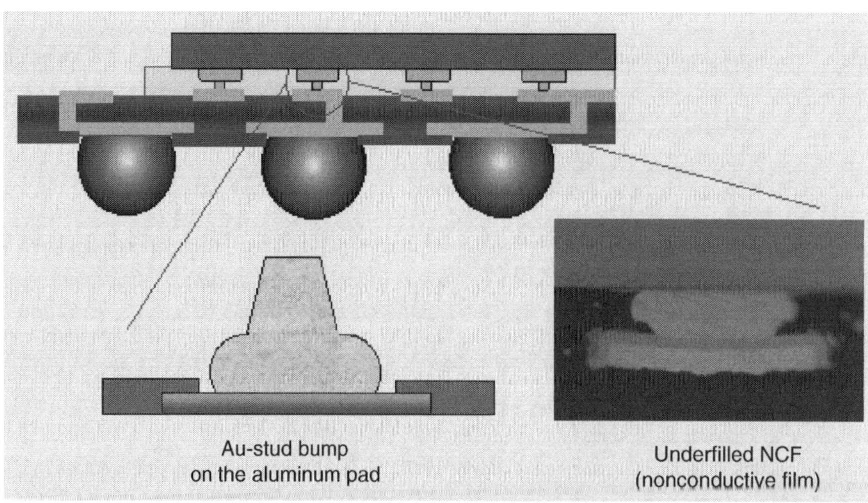

FIGURE 11.24 Schematics of flip-chip CSP using NCF and cross-section of NCF interconnection.

T_g, low CTE and modulus, and strong adhesion. These effects of nonconducting filler addition on the NCA material properties were verified by reliability tests. The reliability of NCA flip-chip assembly by using a modified NCA with nonconducting fillers is significantly higher than that of a commercial ACF. Therefore, the incorporation of nonconductive fillers in the NCA composite material significantly improves the reliability of flip-chip CSP using NCAs materials. NCA materials continue to increase their applications with low cost, finer pitch interconnection, high reliability, and processability. But the technical concerns for process and performance of NCAs such as high bonding pressure and electrical instability at high temperature should be resolved [98]. Recently, nanosized metallic fillers were incorporated into NCA formulation for improved electrical and thermal conductivity and reduced bonding pressure [99].

11.3.5 Summary

Polymer–metal-based ECAs have been evolved to meet the high electrical/mechanical/thermal performance, fine pitch capability, low-temperature process, and strong adhesion and reliability requirements for electronic packaging module and assemblies. As one of the most promising lead-free alternatives in electronic and optoelectronic device packaging interconnect materials, ECAs have shown remarkable advantages and attracted many research interests. Significant improvements in the electrical, mechanical, and thermal properties of different types of conductive adhesives have been achieved. ECAs with high performance and acceptable reliability have been developed for applications in die attach, flip-chip and surface-mount interconnect applications. These lead-free materials and processing methods have great potential to replace the lead-containing solders in the electronic industry.

References

1. H. Wolfson and G. Elliot, Electrically Conducting Cements Containing Epoxy Resins and Silver, U.S. Patent, 2,774, 747, 1956.
2. K.R. Matz, Electrically Conductive Cement and Brush Shunt Containing the Same, U.S. Patent, 2,849,631, 1958.
3. D.P. Beck, Printed Electrical Resistors, U.S. Patent, 2,866,057, 1958.
4. Y. Li, K. Moon, and C.P. Wong, *Science*, 308, 2005, 1419–1420.
5. J.S. Hwang, (Ed.), *Environment-Friendly Electronics: Lead-Free Technology*, Electrochemical Publications Ltd., Port Erin, UK, 2001, Chapter 1, pp. 4–10.
6. Y. Li and C.P. Wong, *Mater. Sci. Eng.*, *R* 51, 2006, 1–35.
7. J. Lau, C.P. Wong, N.C. Lee, and S.W.R. Lee, In *Electronics Manufacturing: with Lead-Free, Halogen-Free, and Conductive-Adhesive Materials*, McGraw Hill, New York, 2002.
8. K. Gilleo, in: J.S. Hwang (Ed.), *Environment-Friendly Electronics: Lead-free Technology*, Electrochemical Publications Ltd., Port Erin, UK, 2001, Chapter 24.
9. R.L. Dietz et al., *Soldering Surface Mount Tech.*, 9, 1997, 55.
10. J. Miragliotta, R.C. Benson, T.E. Phillips, and J.A. Emerson, in *Proc. Mater. Res. Soc. Symp.*, 515, 1998, 245.
11. D. Cavasin, K. Brice-Heams, and A. Arab, in Proc. 53rd Electronic Components and Technology Conf., 2003, pp. 1404–1407.
12. R. Kisiel, *J. Electron. Packag.*, 124, 2002, 367.
13. H. de Vries, J. van Delft, and K. Slob, *IEEE Trans. Compon. Packag. Technol.*, 28, 2005, 499.
14. D. Lu, Q.K. Tong, and C.P. Wong, *IEEE Trans. Compon. Packag. Manuf. Technol. Part C*, 22, 1999, 223–227.
15. E.M. Jost and K. McNeilly, Proceedings of International Society for Hybrids and Microelectronics Society, 1987, pp. 548–553.
16. D. Lu, Q.K. Tong, and C.P. Wong, *IEEE Trans. Compon. Packag. Manuf. Technol. Part A*, 22–3, 1999, 365–371.
17. Y. Li, K. Moon, and C.P. Wong, *IEEE Trans. Compon. Packag. Technol.*, 29–1, 2006, 173–178.
18. Y. Li, K. Moon, A. Whitman, and C.P. Wong, *IEEE Trans. Compon. Packag. Technol.*, 29–4, 2006, 758–763.
19. C. Gallagher, G. Matijasevic, and J.F. Maguire, Proceedings of 47th IEEE Electronic Components and Technology Conference, May 18–21, 1997, pp. 554–560.
20. J.W. Roman and T.W. Eagar, Proceedings of the International Society for Hybrids and Microelectronics Society (San Francisco, CA), Oct. 1992, p. 52.
21. H.J. Jiang, K. Moon, J. Lu, and C.P. Wong, *J. Electron. Mater.*, 34, 2005, 1432–1439.
22. M. Zwolinski, J. Hickman, H. Rubon, and Y. Zaks, Proceedings of the 2nd IEEE International Conference on Adhesive Joining & Coating Technology in Electronics Manufacturing, (Stockholm, Sweden), June 1996, p. 333.
23. D. Lu, Q.K. Tong, and C.P. Wong, *IEEE Trans. Compon. Packag. Manuf. Technol. Part C*, 22(3), 1999, 228–232.
24. K. Persson, A. Nylund, J. Liu, and I. Olefjord, Proceedings of the 7th European Conference on Applications of Surface and Interface Analysis, Gothenburg, June 1997.
25. D. Lu and C.P. Wong, *J. Appl. Polym. Sci.*, 74, 1999, 399–406.
26. Y. Li, K. Moon, and C.P. Wong, *J. Adhesion Sci. Technol.*, 19–16, 2005, 1427–1444.
27. C. Cheng, G. Fredrickson, Y. Xiao, Q.K. Tong, and D. Lu, U.S. Patent No. 6,344,157, 2002.
28. G. Trabanelli and V. Carassiti, in: G. Fontana and R.W. Staehle (Eds.), *Advanced Corrosion Science and Technology*, Plenum Press, New York, 1970, Vol. 1, pp. 147–229.
29. G. Trabanelli, in: F. Mansfeld (Ed.), *Corrosion Mechanisms*, Marcel Dekker, New York, 1987, pp. 119–164.
30. O.L. Riggs, Jr., in: C.C. Nathan (Ed.), *Theoretical Aspects of Corrosion Inhibitors and Inhibition*, The National Association of Corrosion Engineers (NACE), Houston, TX, 1973, pp. 2–27.
31. L.J. Matienzo, F.D. Egitto, and P.E. Logan, *J. Mater. Sci.*, 38, 2003, 4831–4842.
32. H. Li, K. Moon, and C.P. Wong, *J. Electron. Mater.*, 33, 2004, 106–113.

33. H. Takezawa, T. Mitani, T. Kitae, H. Sogo, S. Kobayashi, and Y. Bessho, Proceedings of 8th IEEE International Symposium on Advanced Packaging Materials: Processes, Properties and Interfaces, Atlanta, GA, March 3–6, 2002, pp. 39–143.
34. K. Moon, S. Liong, H. Li, and C.P. Wong, *J. Electron. Mater.*, 33–2, 2004, 1381–1388.
35. D. Durand, D. Vieau, A.L. Chu, and T.S. Weiu, U.S. Patent 5,180,523, November 1989.
36. A.N. Gent and G.R. Hamed, in: J.I. Kroschwitz, H.F. Mark, N.M. Bikales, C.G. Overberger, and G. Menges, (Eds.), *Encyclopedia of Polymer Science and Technology*, Wiley, New York, Vol. 1, 1985.
37. J.E. Morris and S. Probsthain, Proceedings. 4th IEEE International Conference on Adhesive Joining and Coating Technology in Electronics Manufacturing, June 8–21, 2000, pp. 41–45.
38. S. Liong, C.P. Wong, and W.F. Burgoyne, *IEEE Trans. Compon. Packag. Technol.*, 28–2, 2005, 327–336.
39. K. Moon, C. Rocket, and C.P. Wong, *J. Adhesion Sci. Technol.*, 18–2, 2004, 153–167.
40. A. Nagai, K. Takemura, K. Isaka, O. Watanabe, K. Kojima, K. Matsuda, and I. Watanabe, 2nd IEMT/IMC Symposium, 15–17, April 1998, pp. 353–357.
41. I. Watanabe, T. Fujinawa, M. Arifuku, M. Fujii, and Y. Gotoh, Proceedings of 9th IEEE International Symposium on Advanced Packaging Materials: Processes, Properties and Interfaces, Atlanta, GA, March 24–26, 2004, pp. 11–16.
42. S. Macathy, Proceedings of Surface Mount International, San Jose, CA, August 27–31, 1995, pp. 562–567.
43. S.A. Vona and Q.K. Tong, Proceedings of 4th International Symposium and Exhibition on Advanced Packaging Materials, Processes, Properties and Interfaces, Braselton, GA, March 14–16, 1998, pp. 261–267.
44. J. Liu and B. Weman, Proceedings of the 2nd International Symposium on Electronics Packaging Technology, December 9–12, 1996, pp. 313–319.
45. D. Lu and C.P. Wong, Proceedings—IEEE International Symposium on Advanced Packaging Materials: Processes, Properties and Interfaces, Braselton, GA, March 6–8, 2000, pp. 24–31.
46. H. Li, K. Moon, Y. Li, L. Fan, J. Xu, and C.P. Wong, Proceedings of 54th IEEE Electronic Components and Technology Conference, Las Vegas, Nevada, June 1–4, 2004, pp. 165–169.
47. J.H. Zhang, and Y.C. Chan, *J. Electron. Mater.*, 32, 2003, 228–234.
48. W.-S. Kwon, M.-J. Yim, K.-W. Paik, S.-J. Ham, and S.-B. Lee, *J. Electron. Packag.*, 127, 2005, 86–90.
49. M.J. Rizvi, Y.C. Chan, C. Bailey, H. Lu, and A. Sharif, *Soldering Surface Mount Technol.*, 17, 2005, 40–48.
50. G. Davies and J. Sandstrom, Circuits Manufacturing, October 1976, pp. 56–62.
51. G. Harsanyi and G. Ripka, *Electrocomp. Sci. Tech.*, 11, 1985, 281–290.
52. G.D. Giacomo, in: J. McHardy and F. Ludwig (Eds.), *Electrochemistry of Semiconductors and Electronics: Processes and Devices*, Noyes Publications, Park Ridge, NJ, 1992, Chapter 6, pp. 255–295.
53. R. Manepalli, F. Stepniak, S.A. Bidstrup-Allen, and P.A. Kohl, *IEEE Trans. Adv. Packag.*, 22, 1999, 4–8.
54. G.D. Giacomo, *Reliability of Electronic Packages and Semiconductor Devices*, McGraw-Hill, New York, 1997, Chapter 9.
55. R. Wassink, Hybrid Circ. no. 13, May 1987, pp. 9–13.
56. A. Der Marderosian, Ratheon Co. Equipment Division, Equipment Development Laboratories, pp. 134–141.
57. H. Schonhorn and L.H. Sharpe, Prevention of Surface Mass Migration by a Polymeric Surface Coating. U.S. Patent 4,377,619, 1983.
58. V. Brusic, G.S. Frankel, J. Roldan, and R. Saraf, *J. Electrochem. Soc.*, 142, 1995, 2591–2594.
59. P.-I. Wang, T.-M. Lu, S.P. Murarka, and R. Ghoshal, U.S. Patent No. 2,005,023,6711, 2005.
60. Y. Li and C.P. Wong, *Appl. Phys. Lett.*, 89, 2006, 112112.
61. I. Watanabe et al., In *Proc. Asia Display/IDW*, pp. 553–556, 2001.
62. H. Nishida, K. Sakamoto, and H. Ogawa, *IBM J. Res. Dev.*, 42, 1998, 517.
63. D.J. Williams et al., *Soldering Surface Mount Tech.*, 1993, 4.
64. J. Liu, A. Tolvgard, J. Malmodin, and Z. Lai, *IEEE Trans. Compon. Packag., Manuf. Technol.*, 22, 1999, 186.
65. P. Clot, J.F. Zeberli, J.M. Chenuz, F. Ferrando, and D. Styblo, in Proc. Electronics Manufac. Technol. Symp., 24th IEEE/CPMT, 36, 1999.
66. Y. Li, K. Moon, and C.P. Wong, *J. Electron. Mater.*, 34–3, 2005, 266–271.
67. Y. Li, K. Moon, and C.P. Wong, *J. Electron. Mater.*, 34–12, 2005, 1573–1578.
68. K. Moon, H. Dong, R. Maric, S. Pothukuchi, A. Hunt, Y. Li, and C.P. Wong, *J. Electron. Mater.*, 34, 2005, 132–139.
69. Y. Matsuba, *Erekutoronikusu Jisso Gakkaishi*, 6(2), 2003, 130–135.
70. Y. Li, K. Moon, and C.P. Wong, *J. Appl. Polym. Sci.*, 99, 2006, 1665.
71. A. Sihlbom, R. Sihlbom, and J. Liu, Proceedings of IEEE 2nd Electronic Packaging Technology Conference, December 8–10, 1998, Singapore, pp. 251–257.
72. M.J. Yim, H.-J. Kim, and K.-W. Paik, *J. Electron. Mater.*, 34, 2005, 1165–1171.
73. L. Ekstrand, H. Krstiansen, and J. Liu, Proceedings of the IEEE 28th Int. Spring Seminar on Electronics Technology, Wiener Neustadt, Austria, May 19–20, 2005, pp. 35–39.
74. M.J. Yim, W.H. Ryu, Y.D. Jeon, J.H. Lee, S.Y. Ahn, J.H. Kim, and K.W. Paik, *IEEE Trans. Compon. Packag., Manuf. Technol.*, 22, 1999, 575.
75. R. Sihlbom, et al., *IEEE Trans. Compon. Packag. Manuf. Technol.*, 21, 1998, 469.
76. G. Dou, Y.C. Chan, J.E. Morris, and D.C. Whalley, *Soldering Surf. Mount Technol.*, 18, 2006, 3.
77. M.J. Yim, I.H. Jeong, H.K. Choi, J.S. Hwang, J.Y. Ahn, W.S. Kwon, and K.W. Paik, *IEEE Trans. Comp. Packag. Technol.*, 28, 2005, 789.

78. C.M. Lieber, *Science*, 293, 2001, 1289–1292.
79. G.A. Prinz, *Science*, 282, 1998, 1660.
80. C.R. Martin and V.P. Menon, *Anal. Chem.*, 67, 1995, 1920–1928.
81. J.M. Xu, *Appl. Phys. Lett.*, 79, 2001, 1039–1041.
82. T.P. Russell, *Science*, 290, 2000, 2126–2129.
83. Y.W. Chiu, Y.C. Chan, and S.M. Lui, *Microelectron. Reliab.*, 42, 2002, 1945–1951.
84. L. Cao, Z. Lai, and J. Liu, Proceedings of the IEEE 6th CPMT Conference on High Density Microsystem Design and Packaging and Component Failure Analysis (HDP'04), Shanghai, China, June 30–July 3, 2004, pp. 254–258.
85. L. Cao, Z. Lai, and J. Liu, *J. Electron. Packag.*, 127, 2005, 43–46.
86. R.A. Islam, Y.C. Chan, and B. Ralph, *J. Mater. Res.*, 19, 2004, 1662–1668.
87. A. Schubert, Proceedings of IEEE 4th International Symposium on Advanced Packaging Materials, Braselton, GA, March 15–18, 1998, pp. 153–160.
88. M.-J. Yim and K.-W. Paik, *IEEE Trans. Compon. Packag. Technol.*, 24-1, 2001, 24–32.
89. R. Holm, In *Electrical Contacts*, Springer, New York, 1967.
90. M. Chin, K.A. Iyer, and S.J. Hu, *IEEE Trans. Compon. Packag. Technol.*, 27(2), 2004, 317–326.
91. L. Kogut and K. Komvopoulos, *J. Appl. Phys.*, 95(2), 2004, 576–585.
92. B. de Boer, A. Hadipour, M.M. Mandoc, T. van Woudenbergh, and P.W.M. Blom, *Adv. Mater.*, 17, 2005, 621–625.
93. T. Dadosh, Y. Gordin, R. Krahne, I. Khivrich, D. Mahalu, V. Frydman, J. Sperling, A. Yacoby, and I. Bar-Joseph, *Nature*, 436, 2005, 677–680.
94. Q. Sun, A. Selloni, and G. Scoles, *J. Phys. Chem. B*, 110, 2006, 3493–3498.
95. G. Heimel, L. Romaner, J.-L. Brédas, and E. Zojer, *Phys. Rev. Lett.*, 96, 2006, 196806.
96. Y. Li, M.J. Yim, and C.P. Wong, *J. Electron. Mater.*, 36-5, 2007, 549–554.
97. M.J. Yim, J.S. Hwang, W.S. Kwon, K.W. Jang, and K.W. Paik, *IEEE Trans. Electr. Packg. Manuf.*, 26, 2003, 150.
98. H.C. Cheng, C.L. Ho, K.N. Chiang, and S.M. Chang, *IEEE Trans. Compon. Packag. Tech.*, 27, 2004, 398.
99. Y. Li, M.J. Yim, K. Moon, and C.P. Wong, Proceedings of 57th IEEE Electronic Components and Technology Conference, Reno, Nevada, May 29–June 1, 2007, pp. 1350–1355.

11.4 Electrochromic Sol–Gel Coatings

L.C. Klein

11.4.1 Definition of Electrochromism

Certain inorganic compounds, especially oxides of polyvalent metals, exhibit coloration that depends on the oxidation state of their cations. This property leads to electrochromism, which is a reversible and visible change in transmittance or reflectance. The oxidation–reduction reactions are electrochemically induced, using low voltages, on the order of ±1 V dc.

An electrochromic (EC) device is a multilayer construction for which one of the layers shows EC properties (Figure 11.25) [1–5]. An EC device operates on the principle of a galvanic cell. The best known EC material is tungsten trioxide (WO_3), which forms deep blue alkali and hydrogen tungsten bronzes (M_xWO_3) on reduction. The reaction is expressed by the following equation:

$$\underset{\text{(Clear)}}{\overset{6+}{W}O_3 + xM^+ + xe'} = M_xWO_3 = \underset{\text{(Blue)}}{M_x\overset{5+}{W_x}\;\overset{6+}{W_{1-x}}O_3} \tag{11.7}$$

where M is hydrogen or alkali, e′ is an electron, and $0 < x < 1$. Typically, a low-voltage EC device with an EC cathode is colored in the charged state and bleached in the discharged state.

A major application of EC devices is so-called "smart windows." These electronically controlled windows, which can lighten or darken depending on charge insertion/extraction, are designed to adjust to the amount of sunlight, the time of the day or the season of the year. Control of transmittance through windows is a factor in the energy usage in a building, as well as the creation of a pleasant environment in interior spaces. Using them on buildings reduces the heating lost through architectural windows, if they are colored on bright, summer days and bleached on cloudy, winter days.

EC windows that fulfill these expectations have been demonstrated using thin film technologies other than sol–gel processing [6,7]. Substitution of a sol–gel layer for any one of the layers—the EC electrode, the counter electrode, or the electrolyte—is a breakthrough in making affordable EC windows [8–10].

11.4.2 Principle of the Galvanic Cell

Recall for a moment, the basic galvanic cell. The cell consists of an anode, a cathode, and an electrolyte. In an EC device, the layers are referred to as electrode, counterelectrode, and electrolyte. The performance of an EC device is shown schematically in Figure 11.26. Assuming that the EC device is operating with a lithium ion ($M = Li^+$), on charging the Li is oxidized at the cathode (electrode)

$$Li \rightarrow Li^+ + e' \tag{11.8}$$

and reduced at the anode (counterelectrode)

$$Li^+ + e' \rightarrow Li \tag{11.9}$$

In the case of a reversible cell, both the anode and cathode are intercalation compounds with a layered or framework structure. In addition to tungsten oxides, other common electrode materials

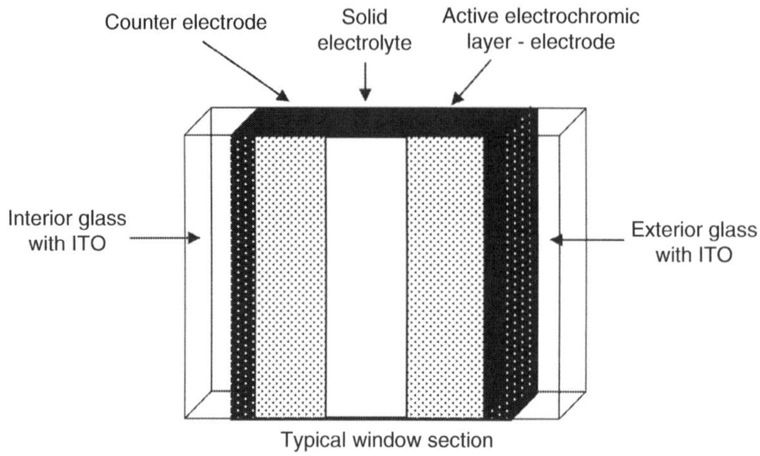

FIGURE 11.25 Schematic of an EC device, substrate (ITO-coated commercial float glass), electrode (tungsten oxide), solid electrolyte (lithium containing inorganic gel), counterelectrode (vanadium oxide), and substrate.

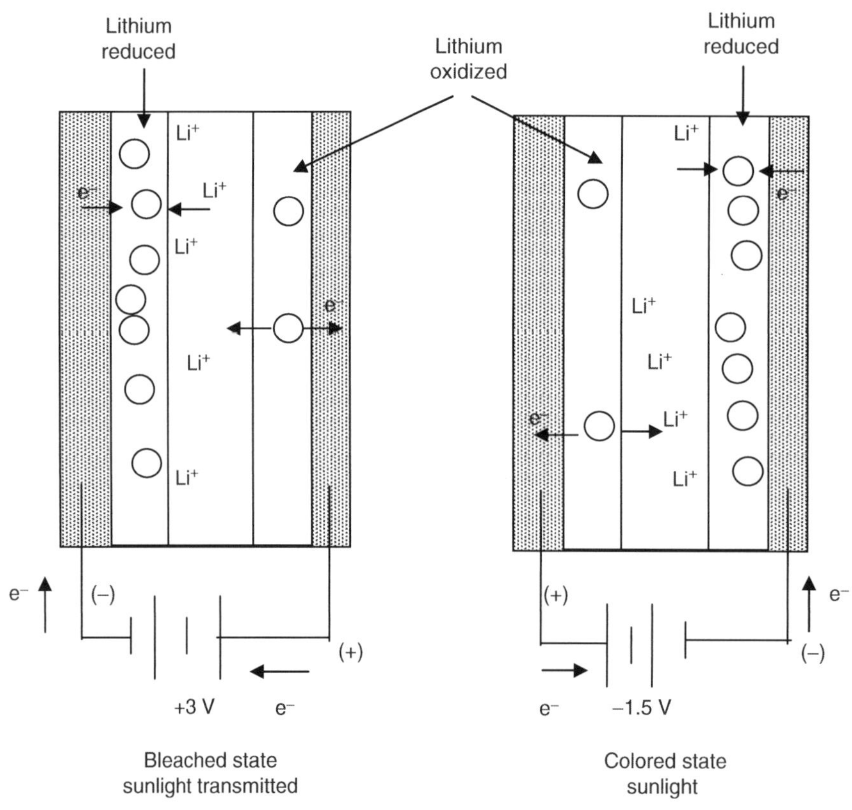

FIGURE 11.26 Schematic operation of the EC device, comparing the colored state (charged) and bleached state (discharged). The open circle indicates a lithium, where the oxidation/reduction reaction takes place, according to Equations 11.8 and 11.9.

are vanadium oxides, niobium oxides, and titania-doped cerium oxides [11–13]. The requirements for the electrodes are:

- High coloration efficiency
- Fast switching between colored and bleached states
- High diffusivity for lithium
- High capacity for lithium
- Compatibility with electrolyte

A critical feature in a galvanic cell is the electrolyte, which has to transport ions back and forth between the cathode and anode. The electrolyte has to have

- Wide electrochemical potential window
- Stability during charge/overcharge and discharge/over-discharge

Most commercial electrolytes are polymers and "gel" electrolytes [14]. Some problems with commercial electrolytes are that aqueous electrolytes may generate H_2 and nonaqueous electrolytes have conductivities that are too low.

Solid electrolytes are attractive because there are effectively no moving parts, meaning no mobile protons. While a solid electrolyte is functioning as a separator and membrane, it is mechanically rigid, holding the components of the EC device fixed. The problem is that most solid electrolytes do not have high enough ionic conductivity at ambient temperature [15]. One way around this problem is thin film technology [16], which is, in part, the motivation for using sol–gel processing.

11.4.3 Sol–Gel Processing

The definition of a gel is a material having solid-like properties, meaning a gel has fixed shape. On the one hand, the organic "gels," which are frequently used in batteries, are rigid, but only in the sense that they "set" when shear stress is absent. Inorganic gels, on the other hand, that result from the sol–gel process are irreversible gels (see Ref. [17] for general background). These gels are based on an oxide network. Not only are these rigid structures, they are inert in contact with organics and solvents.

A natural advantage of the sol–gel process is the fact that a film of oxide can be placed on a substrate directly from solution. Other methods of depositing films of oxides are sputtering or chemical vapor deposition (CVD). These methods may succeed in some cases, but the sol–gel process is simpler than either one.

Dipping and spinning are common processes for preparing sol–gel thin films. Among the available techniques, dip-coating is more widely used. Its advantages include the following:

- Deposits uniform thin films on large areas
- Allows the coating of any size or shape
- Allows control of film microstructure
- Delivers multicomponent films without changing the deposition equipment
- Provides flexibility in manufacturing

Another important consequence of the sol–gel process is the fact that the film is less than 1 μm thick. While the film is rigid, it is thin enough to resist cracking, so that dimensional changes in the electrodes can be accommodated even when the electrolyte is in place.

Sol–gel thin films used in EC devices have exceptional requirements:

- Exhibit significant spectral switching over the visible and infrared regions of the solar spectrum.
- Have acceptable durability, both chemical and physical, over a long time of exposure (~30 year lifetime of architectural windows [18]).

There are three basic approaches to the preparation of transition metal oxides [19–21]. One of the routes is to prepare a solution using alkoxides, metal organic compounds with alkoxy groups. The available alkoxides are only slightly soluble in common organic solvents and tend to be very reactive with water. The second route is to use salts such as oxychlorides as the network former oxide precursor. The third route is to use colloidal sols.

A sol, by definition, is a colloidal suspension of small particles (1–500 nm) in a liquid. Oxide sols are obtained after hydrolysis–condensation reactions of the precursors. Hydrolysis occurs when metal cations (M) are solvated by water molecules. Condensation occurs as soon as hydroxyl groups are present in the coordination sphere of M. These reactions lead to M-O-M bonds. When the number of links is high enough, the sol becomes a gel. The gel point occurs when the solid phase becomes continuous. After this point, the gel continues to age, primarily causing changes in structure. Aging is followed by drying, when evaporation of the liquid can cause further shrinkage of the gel structure.

11.4.3.1 Tungsten Oxide Gels for the Active Electrochromic Layer

While there are several ways to obtain WO_3 thin films [11], the least expensive precursor is tungstic acid. The reasons for choosing tungstic acid sols are:

- Their low cost
- Their aqueous nature
- Their ability to be recycled
- The presence of W–O–W bonds in the sols
- Their amorphous nature and nondirectional morphology

A drawback of tungstic acid sols is that they gel in 30 min.

A solution of aqueous sodium tungstate, $Na_2WO_4 \cdot 2H_2O$, is made by mixing tungstic acid powder, sodium hydroxide, and water. The solution is poured through a proton exchange resin, which is conditioned before use by washing with 2 N HCl. Excess acid is removed from the column by washing with distilled water.

The sol is collected in a beaker containing a defined amount of solvent. The sol, which is clear yellow, contains polymeric species. The mechanism of polymerization is approximately:

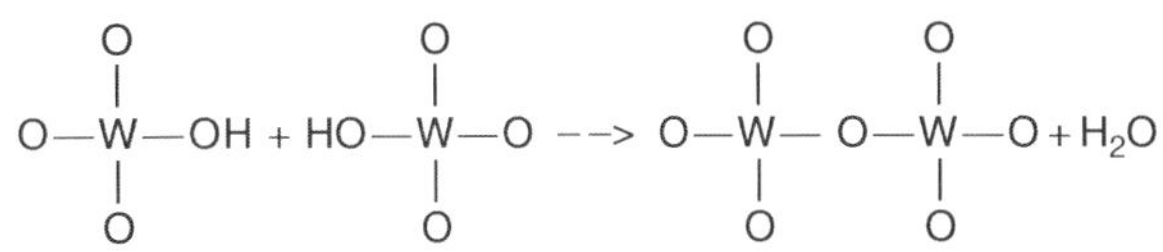

(11.10)

As-prepared WO_3 films, which are poorly crystallized, have a higher coloring efficiency than crystallized WO_3. Crystallization of the WO_3 thin films is expected to occur after a heating at 400°C, so films should not crystallize when heat treated for 1 h only at 90°C.

Comparisons between uncoated glass and coated glass showed that the transmittance in the visible range is lower by about 10%–15% for coated samples. Nevertheless, the transmittance of the coated glass is acceptable for most applications. The WO_3 sol is deposited on commercial float glass, which is coated with indium tin oxide (ITO).

11.4.3.2 Vanadium Oxide Gels for the Counter Electrode

Again, there are several ways to obtain V_2O_5 thin films [22–24]. These include the nitric acid dissolution of vanadate salts, proton exchange, hydrolysis of vanadium alkoxides, and hydration of amorphous vanadia. The least expensive precursor is vanadia powder that is melted at 1000°C (T_{melt} = 690°C) and quenched into water to produce amorphous vanadia. This glassy material is soluble in water and yields a golden brown, colloidal sol that is suitable for dip coating. The behavior of the vanadia sols parallels, in many respects, the tungstic acid sols, so that thin, transparent coatings can be produced.

11.4.3.3 Solid Electrolyte for Lithium Ion Conduction

Sol–gel processing has been used for more than 15 years to prepare solutions containing oxide components desirable for fast ion conductors (FIC) [15]. Fast ion-conducting gels and glasses are desirable because they

- Are isotropic.
- Have few restrictions on composition.
- Have more channels for fast ion conduction than crystalline compounds.

Ionic transport in solids is an activated process. In most FIC, the activation energy is indicative of the energy of motion. The desired features in a FIC are

- Large number of mobile ions
- Large number of sites for the ions
- Relatively low activation energy
- Wide compositional range
- Ease of fabrication in complex shapes and thin films

Classically, high ionic conductivity requires high alkali concentration. The system investigated originally for the solid electrolyte was lithium silicates. In the Li_2O-SiO_2 gels, the second component was introduced typically as lithium nitrate. The nitrate salt dissolves in water and then crystallizes in the dried gel as nitrate nanocrystals [25]. Alternatively, lithium hydroxide was used to create very high pH conditions, increasing the chance of a reaction between LiOH and the silica network to give Si-O-Li bonds [26]. The alkoxide-based gels, or so-called polymerized gels, showed values of ionic conductivity that are stable and in the range needed for operation of a lithium battery at around 250°C. The lithium nitrate gels had a characteristic behavior described by a composite mechanism [27].

Thermodynamic calculations were performed in order to find out which oxides, such as alumina, zirconia, or titania, could improve the stability of the gel. Using the standard heats of formation, it is easy to see that no oxide addition makes lithium silicates stable compared to lithium metal. At best, the addition improves the kinetic stability at low temperatures. Alumina is one candidate for increasing the kinetic stability of the lithium silicate gels.

Ionic conductivity measurements were carried out by the ac complex impedance method using a Solartron 1250 frequency response analyzer and 1186 electrochemical interface, which were programmed by a Hewlett-Packard 9816 desktop computer for data collection and analysis. Experiments were performed on pressed pellets with both sides coated with a layer of platinum paste to provide contacts with platinum electrodes.

Ionic conductivity measurements were performed on gels heated from 100°C to around 360°C at a heating rate of 10°C/min in the frequency range of 10 Hz to 65 kHz. Measurements were repeated and reproduced to eliminate any concern that there was a contribution to the measurements from protons. In bulk materials, the conductivities are too low at room temperature for use in an EC device.

In the case of measurements on thin films, the surface of a microelectrode (Microsensor Systems SAW 302), consisting of an interdigitated array of gold electrodes deposited on a quartz crystal substrate, was coated with the solution [28]. A sample containing 10 mol% Al_2O_3 was selected because measurements on the bulk showed good ionic conductivity for this composition. The values of ionic conductivity measured on thin films matched those for bulk ionic conductivity.

While stable and reproducible conductivities have been demonstrated in thin films, the values of conductivity are still too low. Further advances in EC devices are dependent on finding a solution to the electrolyte problem. Some recent studies on organically modified electrolytes are leading in the right direction [29,30].

11.4.4 Conclusions

In further development of all sol–gel EC devices, physical properties need to be considered, in addition to the ionic conductivity. Of course, the electrolyte has to have

- High electronic resistivity
- High ionic conductivity
- No defects

In addition, all of the layers require

- Film thickness less than about 1000 nm (generally less than 500 nm)
- High transparency over the visible range
- Good adhesion between adjacent layers
- Thermodynamic stability with adjacent layers over a wide electrochemical window

Overall, an EC device should operate under the approximate conditions of +2.5 V (100 s)/–1.5 V (100 s) for more than 10,000 cycles at 15 mC/cm^2.

For practical EC windows, there has been significant recent progress using a variety of technologies [31,32]. Sol–gel processing, of course, is only one route under investigation. There continues to be interest in sol–gel processing because precursors for all of the layers are readily available. The resulting gels are inorganic and rigid. In particular, the sol–gel electrolyte operates while playing the role of a mechanical divider between the electrodes. Finally, sol–gel processing is a simple process for large area coatings.

Acknowledgments

This work in the past was supported by the Assistant Secretary for Conservation and Renewable Energy, Office of Transportation Technologies, Electric and Hybrid Propulsion Division of the US Department of Energy under Contract No. DE-AC03-76SF00098, Subcontract No. 4593610 with the Lawrence Berkeley Laboratory. My thanks go to G. Amatucci, L. Laby, G. Lous, A. Wojcik, M. Greenblatt, A. Salkind, and J. Van Dine for many useful discussions.

References

1. C. G. Granqvist, *Journal of the European Ceramic Society*, 25, 2005, 2907.
2. F. G. K. Baucke, *Materials Science and Engineering*, B10, 1991, 285.
3. C. B. Greenberg, *Thin Solid Films*, 251, 1994, 81.
4. C. B. Greenberg, *Journal of Electrochemical Society*, 140, 1993, 11.
5. I. D. Watkins, G. E. Tulloch, T. Maine, L. Spiccia, D. R. McFarlane, and J. L. Woolfrey, in *Sol–Gel Processing of Advanced Materials*, L. C. Klein, E. J. A. Pope, S. Sakka, and J. L. Woolfrey (eds.), Ceramic Transactions, Vol. 81: American Ceramic Society, Westerville, OH, 1998, p. 223.
6. J. -G. Zhang, C. E. Tracy, D. K. Benson, and S. K. Deb, *Journal of Materials Research* 8, 1993, 2649.
7. J. P. Cronin, D. J. Tarico, J. C. L. Tonazzi, A. Agrawal, and S. R. Kennedy, *Solar Energy Materials and Solar Cells*, 29, 1993, 371.
8. J. E. Van Dine, V. D. Parkhe, L. C. Klein, and F. A. Trumbore, U.S. Patent 5,404,244, April 4, 1995.
9. J. E. Van Dine, V. D. Parkhe, L. C. Klein, and F. A. Trumbore, U.S. Patent 5,659,417, August 19, 1997.
10. J. E. Van Dine, V. D. Parkhe, L. C. Klein, and F. A. Trumbore, U.S. Patent 5,699,192, December 16, 1997.
11. J. Livage, in *Solid State Ionics III*, G. -A. Nazri, J. M. Tarascon, and M. Armand (eds.), MRS Symposium Proceedings, Vol. 293, Pittsburgh, PA, 1993, p. 261.
12. A. Pawlicka, C. Avellaneda, M. Schmitt, S. Heusing, and M. A. Aegerter, in *Sol–Gel Processing of Advanced Materials*, L. C. Klein, E. J. A. Pope, S. Sakka, J. L. Woolfrey (eds.), Ceramic Transactions, Vol. 81: American Ceramic Society, Westerville, OH, 1998, p. 229.
13. B. Munro, P. Conrad, S. Kramer, H. Schmidt, and P. Zapp, *Solar Energy Materials and Solar Cells*, 54, 1998, 131.
14. C. Marcel and J. -M. Tarascon, *Solid State Ionics*, 143, 2001, 89.
15. J. P. Boilot and Ph. Colomban, in *Sol–Gel Technology for Thin Films, Fibers, Preforms, Electronics and Specialty Shapes*, L. C. Klein (ed.), Noyes Publication, NJ, 1988, p. 303.
16. B. Wang, M. Greenblatt, J. Yan, and Y. Wu, *Journal of Sol–Gel Science and Technology*, Park Ridge, NJ, 2, 1994, 323.
17. C. J. Brinker and G.W. Scherer, *Sol–Gel Science: The Physics and Chemistry of Sol–Gel Processing*, Academic Press, Boston, MA, 1990.
18. C. E. Tracy, J.-G. Zhang, D. K. Benson, A. W. Czanderna, and S. K. Deb, *Electrochimica Acta* 44, 1999, 3195.
19. A. Chemseddine, M. Henry, and J. Livage, Revue de Chimie Minerale 21, 1984, 487.
20. B. Munro, S. Kramer, P. Zapp, and H. Krug, *J. Sol-Gel Science and Technology*, 13, 1998, 673.
21. B. Pecquenard, H. Lecacheux, S. Castro-Garcia, and J. Livage, *J. Sol–Gel Science and Technology* 13, 1998, 923.
22. J. Livage and J. Lemerle, *Ann. Rev. Mat. Sci.*, 12, 1982, 103.
23. S. Mege, M. Verelst. P. Lecante, E. Perez, F. Amsart, and J. M. Savariault, *Journal of Noncrystalline Solids*, 238, 1998, 37.
24. R. D. Cussler, M. G. Kulkarni, and E. L. Cussler, *Journal of Membrane Science*, 127, 1997, 153.
25. S. -P. Szu, L. C. Klein, and M. Greenblatt, *Journal of Noncrystalline Solids*, 121, 1990, 119.
26. S. -P. Szu, M. Greenblatt, and L. C. Klein, *Journal of Noncrystalline Solids*, 124, 1990, 91.
27. S. -P. Szu, M. Greenblatt, and L. C. Klein, *Solid State Ionics*, 46, 1991, 291.
28. L. Laby, L. C. Klein, J. Yan, and M. Greenblatt, *Solid State Ionics*, 81, 1995, 217.
29. K. Dahmouche, M. Atik, N. C. Mello, T. J. Bonagamba, H. Panepucci, M. A. Aegerter, and P. Judeinstein, *Journal of Sol–Gel Science and Technology*, 8, 1997, 711.
30. P. Judeinstein, J. Titman, M. Stamm, and H. Schmidt, *Chemical Materials*, 6, 1994, 127.
31. C. G. Granqvist, *Electrochimica Acta*, 44, 1999, 3005.
32. C. G. Granqvist, A. Azens, A. Hjelm, L. Kullman, G. A. Niklasson, D. Ronnow, M. Stromme Mattsson, M. Veszelei, and G. Vaivars, *Solar Energy*, 63, 1998, 199.

12

Cure and Health Monitoring

Tatsuro Kosaka
Osaka City University

12.1 Introduction

Because safety is the most important keyword, companies have made enormous efforts to assure that the quality of their products when shipped and when in operation perform frequent and accurate inspections. The inspection methods have been improved to become more accurate, quicker, and more labor saving therefore reducing cost and time. However, even the most efficient inspection method requires downtime of the machines and cannot insure the safety when an emergent accident occurs in service. In order to avoid these problems, successive inspection methods have been desired.

Cure and health monitoring is a new technology to monitor materials and structures from birth to death successively. During the manufacturing process of polymer products, the cure monitoring is applied to observe processes of the cure reaction, temperature, pressure, residual stress, homogeneity, and integrity over the materials. After the shipment, built-in sensors measure mechanical and thermal responses such as strain, stress, and temperature in real time and report the information to remote operating centers. The main difference between the traditional inspection method and the cure and health monitoring method is the use of built-in sensors and remote monitoring technologies. Recent developments of sensor technologies has enabled us to install miniature sensors into materials and structures for observing the internal conditions in real time, and with the evolution of the Internet has made it practical to collect data from remote sensing systems at high speed. This paper introduces the cure and health monitoring method from the viewpoint of built-in sensors.

12.2 Cure Monitoring

The curing process is one of the most important steps in the manufacturing of thermosetting polymers because it significantly controls the properties of the materials. Monitoring of the cure process is essential in order to find the best conditions for temperature and pressure. During the curing process, the material changes from liquid to solid state by a chemical reaction, and therefore, any property such as density, elasticity, viscosity, glass temperature, crystallinity, and dielectricity dramatically changes. Many cure monitoring methods have been proposed but the most popular methods are thermal analyses by differential scanning calorimeters (DSC), viscoelastic analyses by thermomechanical analyzers (TMA), and chemical analyses of the molecular structures by infrared spectrometers. The degree of cure (DOC), which is defined by a ratio between reaction heats and total exothermic heats, is the most important index to evaluate during the curing process, and thus the first purpose of cure monitoring is to obtain DOC.

Although these traditional methods are useful and sophisticated, they are available only in the laboratory due to the requirement of small samples. On the other hand, the cure monitoring of practical products has become important with the increasing demand of large and complex-shaped polymer matrix composites (PMC) because of the local concentrations of the incomplete cure, the residual stress and voids, which sometimes occur. In order to monitor the internal state of the practical products, miniature and embeddable sensors are necessary. Table 12.1 lists the sensors available for cure monitoring with measurable values and usability for the cure monitoring.

Optical fiber sensors have advantages due to their small size, flexibility, locality, and high-temperature resistance for cure monitoring. In addition, many kinds of physical values such as temperature, pressure, strain, refractive index, and molecular structure can be monitored. The optical fibers used in the spectrometers and reflectometers require exposed surfaces of their core in order to transmit light through materials. Three types of sensor structures are the transmission type, reflection type, and evanescent type and have been proposed for use as illustrated in Figure 12.1. The evanescent type sensor utilizes evanescent light leaked from the core of the stripped optical fiber. Fourier transform

TABLE 12.1 Embeddable Sensors for Cure Monitoring

	Measurable Values	Use in Operation
Optical fiber sensors		
Spectrometers	Molecular structures	Poor
Reflectometers	Refractive index	Poor
Strain/temperature sensors	Strain, temperature	Good
Dielectric sensors	Permittivity, loss factor	Poor
Piezoelectric sensors		
Impedance measurement	Impedance	Poor
Ultrasonic measurement	Sound velocity	Good

infrared spectroscopy (FTIR) and UV fluorimetry are available as fiber optic spectrometers. The FTIR measures the absorption spectrum of the infrared light whose bands identify the specific molecular structures. Since the specific molecular groups appear or disappear by the advance of the cure reaction, the DOC can be obtained from changes in the magnitude of the specific absorption band. On the other hand, the UV fluorimetry is a technique based on the photoluminescent phenomenon where the incident light irradiates the fluorescent material. The emission light from the fluorescent curing agent can be used for the cure monitoring of polymers. It has been reported that the peak of the excitatory spectra of epoxy resin shifts during cure. Although the spectrometers are very powerful in monitoring the curing process, the typical measuring speed is slow.

The fiber optic reflectometers measure Fresnel's reflection rate at the boundary between the optical fiber and polymer. The reflection rate varies monotonically when the refractive index of the polymer increases during the curing process. It should be noted that temperature monitoring near the fiber tip is necessary because the temperature effect on the refractive index cannot be negligible. The main advantage of this method is a simple, low-cost, and fast measurement system. The strain and temperature sensors are mainly used for health monitoring and are mentioned later. Although the strain is not convenient in evaluating the DOC, the measurements of the residual strain during the curing process yield benefits especially if the sensors are installed before the process is initiated. Any type of fiber optic strain sensors can be utilized for the purpose.

A dielectric sensor consists of two parallel electrodes whose pattern is printed on a thin nonconductive base as shown in Figure 12.2. Since the electrodes filled with the polymer forms an equivalent RC circuit, the sensor is able to measure the complex dielectric constant of the polymer. For the cure monitoring, the reciprocal of a product of the alternating current frequency and the loss factor, which is called ion viscosity, is more convenient than the permittivity because it varies on a logarithmic scale during the curing process. In the literature, it has been shown that the log ion viscosity of epoxy resin behaves similar to the DOC during the cure process.

Piezoelectric materials show the reversible piezoelectric effect, in which external applied voltage changes the material shapes. Therefore, piezoelectric materials can be used as actuators as well as sensors. For cure monitoring, piezoelectric sensors have been employed to monitor the impedance of polymers or to measure the ultrasonic velocity. For the impedance measurement, the piezoelectric wafer embedded in the viscous polymer can be modeled by a mass-spring-damper system and represented as an electromechanical circuit. Therefore the electrical response for the resonance frequency is sensitive to changes in shear for the modulus and viscosity of the curing polymer around the sensor. On the other hand, the piezoelectric sensors/actuators can be used to transmit and receive the ultrasonic signals propagating through the polymer and then the sound velocity, which is proportional to the square root of the stiffness, can be obtained. It has been reported that the behavior of the sound velocity of the polymer can be monitored during the curing process.

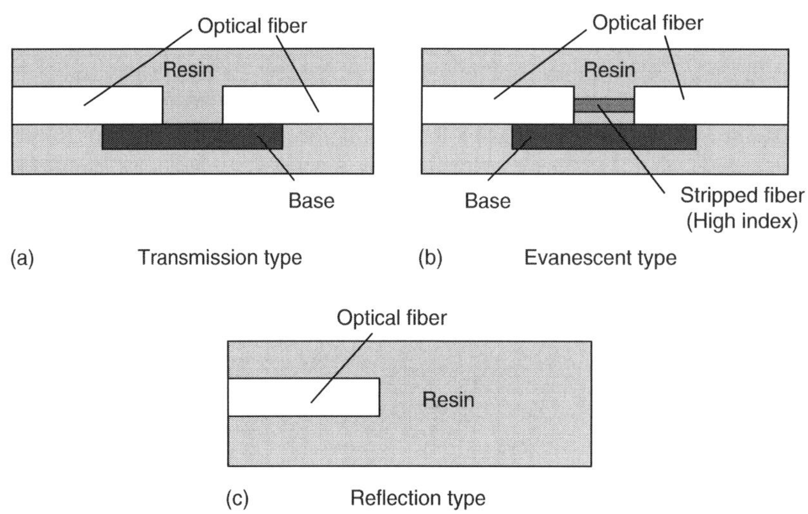

FIGURE 12.1 Illustrations of the sensing parts of fiber optic spectrometers and reflectometers.

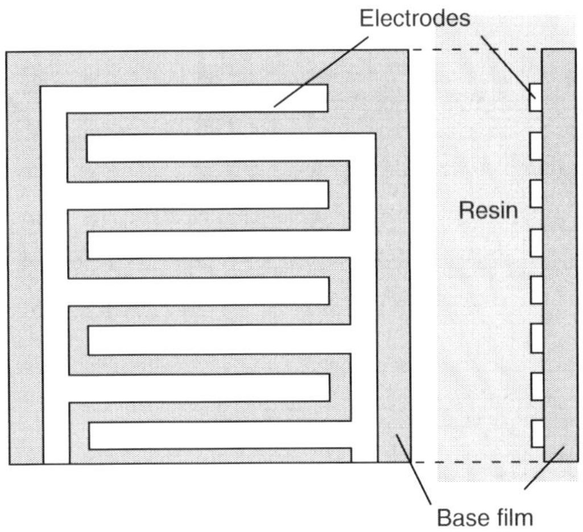

FIGURE 12.2 Interdigital dielectric sensors where the electrodes were printed on the base film.

12.3 Health Monitoring

Like a human body, structures deteriorate in long-term use by mechanical and thermal fatigue, overload, impact load, and chemical corrosion, etc. The deterioration of structures, which often causes catastrophic accidents, is the most significant problem, and therefore, periodic and accurate inspections are essential in order to ensure safe operation. Although the most common inspection technique is a visual inspection, damages in materials that are internal are not detectable. Thus, nondestructive evaluation (NDE) techniques have been developed. Although the NDE techniques such as an ultrasonic scan, X-radiography, and thermography are widely used, they can be used only during the downtime of the machines due to the large and heavy apparatuses.

With health monitoring the use of built-in sensors is an attractive approach to monitor structures successfully. Especially in aeronautical and civil engineering, successive health monitoring has been expected to dramatically improve safety because shocking disasters such as the crash of aircraft and collapse of buildings have occurred recently. In health monitoring, temperature distribution, strain responses, vibration responses, and damage initiation as well as development are the targets to be monitored since they lead to degradation and destruction of structures. Table 12.2 lists the major sensors and sensing methods available for health monitoring with their measurable values and their area of coverage.

Optical fiber sensors are the most promising for internal health monitoring in the long term due to their small size, light weight, high mechanical performance, and high durability. There are many types of optical fiber sensors that have been developed, which are more accurate, lower in cost, and have a higher functionality, as listed in Table 12.3. In the table, optical fiber sensors have been categorized by their optical systems, which strongly affect costs, sensing speed, and sensing area.

Optical loss sensors are based on the loss occurring in the sensing parts. The earliest idea of health monitoring by the sensors using the fiber breaks to denote the damage progression of materials can be seen in literature. Today, many kinds of optical loss sensors have been proposed as shown in Figure 12.3. The microbend sensor uses optical loss changes shown by damage development or the strain of a material. The tapered sensors have their sensing parts constructed to induce a leak of light and then the optical loss appears as a function of strain. The vibration sensor can measure transverse vibration when applied to the sensor from the transmission loss caused by the optical axial slippage between the input and output fibers. The advantage of the optical loss sensor is the simplest and cheapest optical system consisting of a laser-emitting diode (LED), a photodetector (PD), and a communication optical fiber. In addition to the cost advantage, the sensor system naturally has a function of high-speed measurement due to their high optical frequency.

The Michelson interferometric, polarimetric, and laser Doppler velocitymeter (LDV) sensors are fundamental sensing methods which use optical fibers. For cure monitoring, the interference between two polarized lights is often measured by the polarimetric sensing device. The optical system is very simple, but it is difficult to maintain stability for these sensors when the two different optical fibers for measuring and reference are employed. As for LDV sensors, some ideas to amplify Doppler's effect in the sensing part have been proposed recently in order to improve their sensitivity and stability.

Although the extrinsic Fabry-Perot interferometer (EFPI) sensor is one of the optical interferometers, the difference from the others is that the input and output lights propagate in the same fiber. Therefore, the sensor shows very good stability when

TABLE 12.2 Sensors and Sensing Techniques for Health Monitoring

	Measurable Values	Measurable Area
Optical fiber sensors	Temperature, strain, vibration, velocity, damages	Around fiber
Piezoelectric sensors	Vibration, impact, acoustic emission, damages	Multi points
Electric resistance measurement	Strain, damages	Whole

TABLE 12.3 Optical Fiber Sensors for Health Monitoring

	Measurable Values	Multiplexing	Optical System
Optical loss	Damage, strain, vibration	Switching	Power meter
Michelson	Strain, vibration	Switching	Interferometer
Polarimeter	Strain, vibration	Switching	Interferometer
LDV	Velocity	Switching	Interferometer
EFPI	Strain	Switching	Interferometer
FBG, LPG	Strain, temperature	Frequency domain	Spectrometer
Distributed	Strain, temperature	Distributed	OTDR

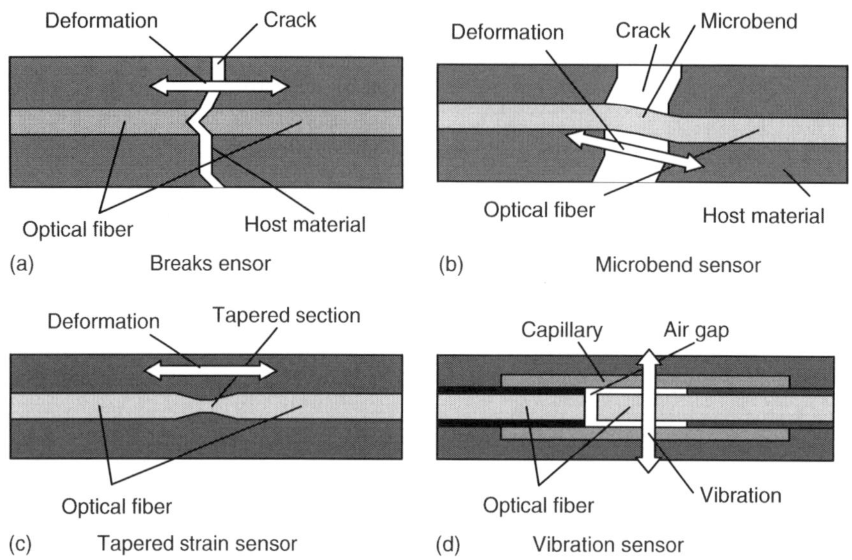

FIGURE 12.3 Optical loss sensors for health monitoring.

it is embedded in materials. The sensor is constructed from an input/output fiber, a reflector, a glass capillary, and the two flat-end optical fibers, which compose a Fabry-Perot interferometer as illustrated in Figure 12.4. The typical gauge length and strain resolution are 1 mm and 10^{-6}, respectively. The EFPI sensor has been mainly applied to monitor strain while specialized types have been used to measure the pressure of existing temperature.

The fiber Bragg grating (FBG) sensor is the most popular sensor for health monitoring due to the high sensitivity, durability, and its multiplexing capability. The FBG sensor has a sensing part whose typical length is 10 mm in which the refractive index of the core varies periodically to form a Bragg grating. When a broadband light is incident into the sensor part, the narrow-band diffracted light reflects and the center wavelength changes proportionally to strain and temperature. The typical resolutions are 10^{-6} strains and 0.1°C. Although the FBG sensor is not suited for measuring nonuniform strain distribution such as strain concentration around a crack tip, however, this weak point becomes an advantage for the purpose of monitoring damage development by placing the sensor near neighboring weak parts of structures. The long-period grating (LPG) sensor is similar to the FBG sensor but the period of the grating is much longer. Since the strain and temperature sensitivities of the several loss peaks of the transmitted light are different from each other, the LPG sensor can measure strain and temperature simultaneously by the single sensor. The main advantage of the FBG sensors for health monitoring is their ease and quick multiplexed measurements in the frequency domain.

The distributed fiber optic sensors can measure strain and temperature distribution along the optical fibers by measuring the reflected light at any part. An optical time domain reflectometer (OTDR) can be employed to scan the intensity, spectrum, and

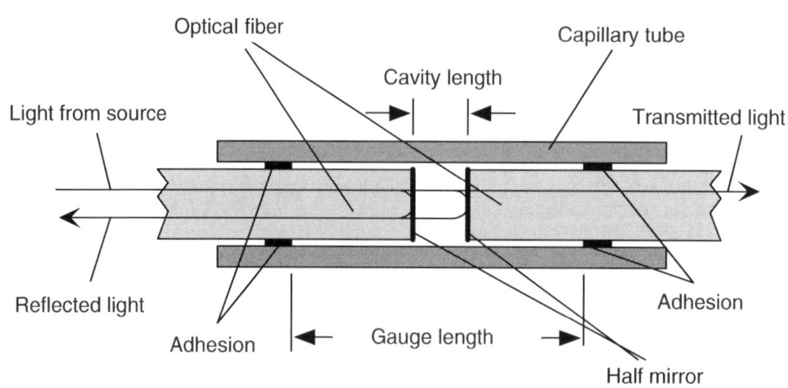

FIGURE 12.4 Structure of the EFPI sensor.

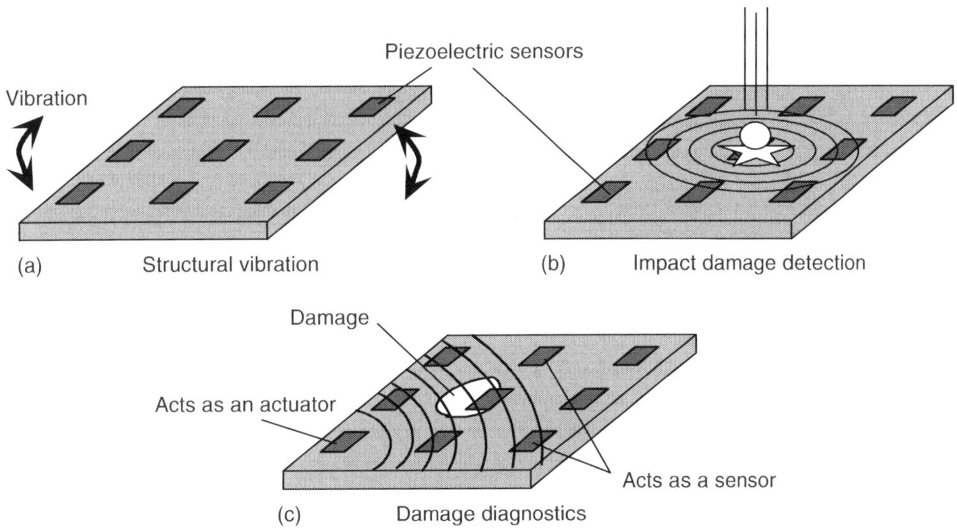

FIGURE 12.5 Health monitoring by piezoelectric sensors/actuators.

position of the reflected light such as Raman scattering and Brillouin scattering. The distributed sensors are practical for the point of wide area measurement of large structures but the typical spatial resolution is almost 1 m and the scanning speed is slow.

Lead zirconate titanate (PZT) ceramic wafers and polyvinylidene fluoride (PVDF) films are popular piezoelectric sensors for health monitoring. PZT is more powerful than PVDF and operates very efficiently as an actuator as well as a sensor while PVDF can be formed into any desired shape due to its high flexibility. Typical health monitoring methods using piezoelectric sensors are structural vibration monitoring, impact monitoring, and damage diagnostics as illustrated in Figure 12.5. In structural vibration monitoring, the dynamic strain responses are measured by the sensors and then modal shapes and frequencies are analyzed, which change when the structures deteriorate. The impact monitoring is also important for the structure because the impact damages often cause destruction of structures. The sensors measure the impact signals and the impact energy, location, and generated damages can be estimated. The passive damage diagnostics use acoustic emission (AE) signals, which occur through damage initiation and progress. The active method has been achieved by allowing the piezoelectric sensors/actuators to operate and thus transmit and receive the diagnostics signals. Although the active damage diagnostics require a power source, they are the most powerful because they act at any time without structural vibration, impact or damage development.

The electric resistance measurement of carbon reinforced composites and carbon/steel reinforced concretes is attractive to monitor strain and damages. The main advantage for this method is the unnecessity of embedding of sensors because the materials work as sensors. The resistance changes by deforming the shape of the materials and the damage initiation. The mechanism of the resistivity variation is very complicated and particularly for the reinforcing structures of the materials because the complex current paths are constructed mainly by the conductive reinforcements as illustrated in Figure12.6. This method has an interesting function in which the resistance can record the degree of damages.

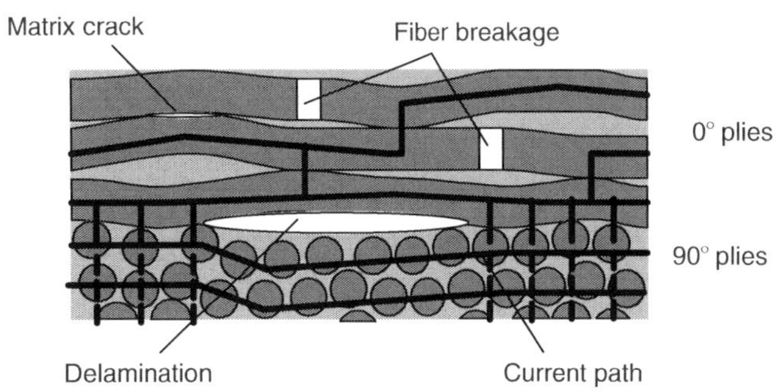

FIGURE 12.6 Current paths in CFRP laminates with damages.

Bibliography

Abry, J.C., Bochard, S., Chateauminois, A., Salvia, M., and Giraud, G. 1999. In situ detection of damage in CFRP laminates by electrical resistance measurement. *Composite Science and Technology*, 59:925–935.

Arregui, F.J., Matias, I.R., and Amo, M.L. 2000. Optical fiber strain gauge based on a tapered single-mode fiber. *Sensors and Actuators*, 79:90–96.

Doyle, C. and Fernando, G. 1998. Detecting impact damage in a composite material with an optical fiber vibration sensor system. *Smart Materials and Structures*, 7:543–549.

Glossop, N.D.W., Dubois, S., Tsaw, W., LeBlanc, M., Lymer, J., Measures, R.M., and Tennyson, R.C. 1990. Optical fiber damage detection for an aircraft composite edge. *Composites*, 21:71–80.

Islam, A.S. and Craig, K.C. 1994. Damage detection in composite structures using piezoelectric materials. *Smart Materials and Structures*, 3:318–328.

Kageyama, K., Murayama, H., Ohsawa, I., Kanai, M., Nagata, K., Machijima, Y., and Matsumura, F. 2005. Acoustic emission monitoring of a reinforced concrete structure by applying new fiber-optic sensors. *Smart Materials and Structures*, 14:52–59.

Kalamkarov, A.L., Fitzgerald, S.B., and MacDonald, D.O. 1999. The use of Fabry Perot fiber optic sensors to monitor residual strains during pultrusion of FRP composites. *Composites: Part B*, 30:167–175.

Kim, J.S. and Lee, D.G. 1996. Measurement of the degree of cure of carbon fiber epoxy composite materials. *Journal of Composite Materials*, 30:1436–1457.

Leung, C.K.Y., Elvin, N., Olson, N., Morse, T.F., and He, Y.F. 2000. A novel distributed optical crack sensor for concrete structures. *Engineering Fracture Mechanics*, 65:133–148.

Liu, Y.M., Ganesh, C., Steele, J.P.H., and Jones, J.E. 1997. Fiber optic sensor development for real-time in-situ epoxy cure monitoring. *Journal of Composite Materials*, 31:87–102.

Murukeshan, V.M., Chan, P.Y., Ong, L.S., and Seah, L.K. 2000. Cure monitoring of smart composites using fiber Bragg grating based embedded sensors. *Sensors and Actuators: A*, 79:153–161.

Okabe, Y., Yashiro, S., Kosaka, T., and Takeda, N. 2000. Detection of transverse cracks in CFRP composites using embedded fiber Bragg grating sensors. *Smart Materials and Structures*, 9:832–838.

Paik, H.J. and Sung, N.H. 1994. Fiberoptic intrinsic fluorescence for in-situ cure monitoring of amine cured epoxy and composites. *Polymer Engineering and Science*, 34:1025–1032.

Powerll, G.R., Crosby, P.A., Waters, D.N., France, C.M., Spooncer, R.C., and Fernando, G.F. 1998. In-situ cure monitoring using optical fibre sensors—a comparative study. *Smart Materials and Structures*, 7:557–568.

Rath, M., Döring, J., Stark, W., and Hinrichsen, G. 2000. Process monitoring of moulding compounds by ultrasonic measurements in a compression mould. *NDT & E International*, 33:123–130.

Smith, J., Brown, A.W., DeMerchant, M.D., and Bao, X. 1999. Simultaneous strain and temperature measurement using a Brillouin scattering based distributed sensor. *Proc. SPIE*, 3670:366–373.

Todoroki, A., 1998. Health monitoring of internal delamination cracks for graphite/epoxy composites by electric potential method. Proc. 4th ESSM and 2nd MIMR Conference, pp. 429–434.

Tracy, M., Roh, Y.S., and Chang, F.K., 1996. Impact damage diagnostics for composite structures using built-in sensors and actuators. Proc. of Third ICIM/ECSSM '96, pp. 118–123.

Vries, M., Bhatia, V., D'Alberto, T., Arya, V., and Claus, R.O. 1998. Photoinduced granting-based optical fiber sensors for structural analysis and control. *Engineering Structures*, 20:205–210.

Wang, X., Ehlers, C., Kissinger, C., Neitzel, M., Ye, L., and Mai, Y.W. 1998. Sensitivity of piezoelectric wafers to the curing of thermoset resins and thermoset composites. *Smart Materials and Structures*, 7:113–120.

13

Drug Delivery Systems

13.1 Smart Drug Delivery Systems

Il Keun Kwon, Sung Won Kim, Somali Chaterji, Kumar Vedantham, and Kinam Park

13.1.1 Introduction

Polymers have been indispensable in preparing various controlled release formulations. Of the many routes of drug administration, oral delivery is most attractive due to high patient convenience and compliance, but it is often very challenging. For example, oral delivery of drugs that are poorly soluble, or are not stable in the gastrointestinal tract, or have high molecular weights has been difficult. Delivery of high molecular weight drugs, such as peptide and protein drugs and oligonucleotides, by parenteral administration has also been difficult due to their short circulation half-lives as well as lack of suitable delivery systems for long-term delivery. A number of approaches have been explored to increase the bioavailability of hydrophilic/hydrophobic drugs, peptides, and proteins by conjugating drugs with natural or synthetic macromolecules or loading in polymeric drug delivery systems. Polymers with functional properties have been used to overcome hurdles in delivery of diverse drugs.

Stimuli–responsive polymers are polymers which respond with physical or chemical transformations in response to changes in environmental conditions, such as temperature, pH, mechanical stress, electric/magnetic response, light, and ionic strength. Figure 13.1 schematically presents the concept of a stimuli–responsive drug delivery system using glucose-dependent self-regulated insulin release as an example. The sensor that recognizes the stimulus, glucose in this case, should have high specificity and sensitivity. The information process should handle the signal as fast as possible, and so the kinetics of information processing is highly important. The actuator should have the ability to reverse the action, i.e., transition between on and off states, with sufficient magnitude to be useful. Stimuli–responsive polymers can be in the form of a drug reservoir, which can release its contents in response to environmental signals. This release can be via changes in their shapes, surface properties, solubility parameters, or by molecular self-assembly, and sol–gel transition [1–7].

Nanotechnology, in combination with diverse targeting strategies, can bring advantages to drug delivery, overcoming the limitations of conventional therapeutics [8]. The advantages include delivery of much higher therapeutic payloads per target biorecognition event; the ability to carry multiple, potentially different targeting agents for enhanced selectivity; the ability to integrate means to bypass biological barriers; and the colocalized delivery of multiple agents, resulting in targeted combination therapy. This chapter summarizes various stimuli-modulated drug delivery systems as well as the applications of nanoparticles in drug delivery.

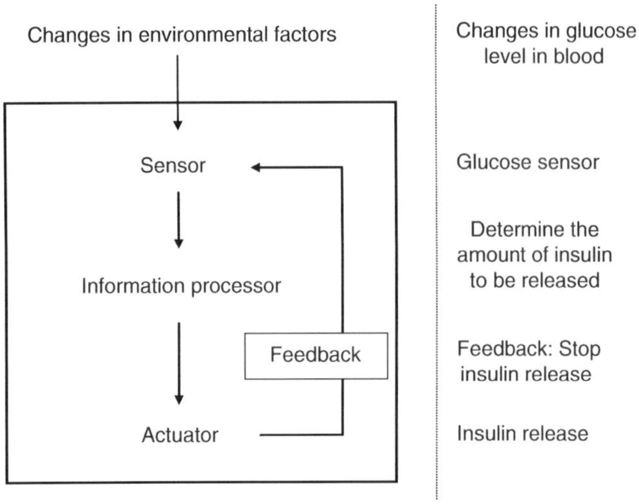

FIGURE 13.1 Schematic illustration of the on–off drug release in a stimuli-modulated drug delivery system. Glucose-dependent modulated insulin release is shown on the right as an example.

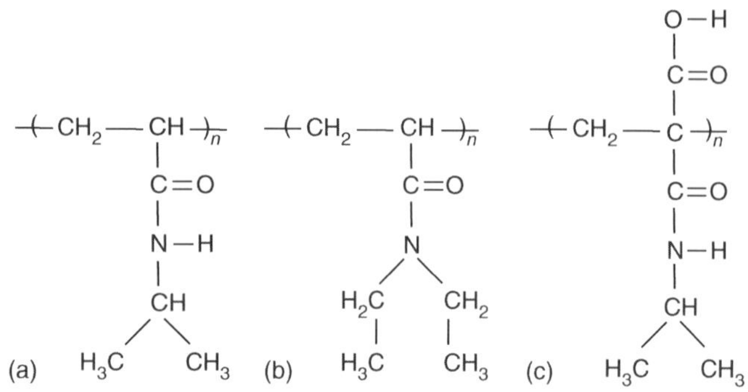

FIGURE 13.2 Chemical structures of poly(*N*-isopropylacrylamide) (a), poly(*N*,*N*′-diethylacrylamide) (b), and poly(2-carboxyisopropylacrylamide) (c).

13.1.2 Stimuli-Modulated Release Systems

13.1.2.1 Thermoresponsive Polymers

The signals and physiological conditions that can trigger environment-sensitive polymers are temperature, pH, mechanical stimulus, biomolecule concentration, electric field strength, and salt concentration. Of these, temperature is the most commonly studied triggering signal for modulated drug delivery [7,9,10]. Temperature is a useful stimulus for controlled drug delivery, because human body temperature deviates from 37°C in the presence of pyrogens [8,9]. The common characteristic of temperature-sensitive polymers is the presence of a hydrophobic group, such as methyl, ethyl, and propyl groups. The properties of thermoresponsive polymers are governed by lower critical solution temperature (LCST), which is defined as the temperature at which the polymer solution undergoes a phase transition from a soluble to an insoluble state [11]. This phase transition is mainly determined by the hydrophilic–hydrophobic balance of the polymer [12]. In general, the solubility of most of the polymers increases with increase in temperature. In the case of polymers that exhibit LCST, increase in temperature decreases the water solubility due to hydrophobic associations of polymer molecules and reduction in hydrogen bonding between polymer and water molecules [10].

The study of thermoresponsive gels started in 1978 when Tanaka reported the thermodynamics behind the collapse of a polyacrylamide gel [13]. Most commonly studied synthetic thermoresponsive polymers are poly(*N*-isopropylacrylamide) (PNIPAAm), poly(*N*,*N*′-diethylacrylamide) (PDEAAm), and poly(2-carboxyisopropylacrylamide) (PCIPAAm) (Figure 13.2). PNIPAAm is one of the most useful thermoresponsive polymers in drug delivery, because its LCST in water is around 32°C, close to the body temperature [14]. The LCST can be altered by adjusting the ratio of the hydrophilic and the hydrophobic segments of the polymer [15–17].

An interesting finding regarding temperature-modulated release was that PNIPAAm hydrogels with more hydrophobic comonomers, such as *n*-butylmethacrylate, formed a dense skin during the deswelling process when the temperature increased above its LCST [18–20]. This dense skin formation (surface regulation) is able to block the release of loaded drugs from the matrix or diffusion through the membrane, resulting in the reversible off-control for solute release. P(NIPAAm-*co*-methacrylic acid) was applied for the pulsatile release of a thrombolytic agent, streptokinase, by simultaneous pH and temperature changes, which could occur in blood pH and temperature at the site of blood clotting [21]. The pulsatile release pattern of loaded streptokinase was obtained by fluctuating temperature and pH, simultaneously. The opposite on–off drug release pattern (i.e., release, or "on-control," at high temperature) was obtained by using the interpenetrating polymer network (IPN) structures of poly(acrylamide-*co*-butylmethacrylate) and poly(acrylic acid) (PAAc) [22] or poly(*N*,*N*-dimethylacrylamide [DMA]) and PAAc [23]. The polymer networks are formed by hydrogen bonds at a lower temperature. At a higher temperature, the hydrogen bonds are dissociated, resulting in release of the loaded drug.

The sensitivity of sol–gel transition can be accelerated by structural molecular design, e.g., making comb-type PNIPAAm [24,25] and by changing the synthesis condition, such as a weak alkaline condition, for P(NIPAAm-*co*-AAc) [26]. Dissociation of the carboxyl groups (–COOH) of AAC to carboxylate anions ($-COO^-$) results in electrostatic repulsion between the carboxylate anions, leading to expanded conformation of the polymer chains. The expanded conformation induces fast temperature sensitivity and improves the oscillating swelling–deswelling property.

A large number of poly(ethylene oxide) (PEO) and poly(propylene oxide) (PPO) block copolymers possess an inverse temperature-sensitive micellization and gelation property. Many are commercially available under the names of Pluronics (or Poloxamers) and Tetronics [27]. Gelation of PEO-PPO-PEO can be induced by the three-dimensional packing of micelles formed by the hydrophilic–hydrophobic balance at over the critical temperature [28]. The PPO hydrophobic block can be replaced with other hydrophobic groups, such as poly(1,2-butylene oxide) (PBO) [29,30] and polyesters, such as poly(L-lactic acid) (PLLA), and (DL-lactic acid-*co*-glycolic acid) (PLGA) [31–33]. The PEO-PLLA-PEO triblock copolymer, which combined thermogelation and biodegradability with longer gel duration, showed the sol state at 45°C and transformed to a gel at a lower temperature (e.g., 37°C) [32]. This transition helps loading and delivering of a drug at a sustained rate in the body. A series of low molecular weight PLGA-PEG-PLGA copolymers also showed similar thermoreversible sol–gel transitions to that of PEO-PLLA-PEO block copolymer [34]. The lower sol–gel transition is related to the increase in aggregation numbers and size of micelles consisting of a polyester core and a PEO shell at higher temperatures, and the upper gel–sol transition induced from the shrinking of micelles or shape changes caused by dehydration of the PEO block [32]. The sol–gel transition temperature is a function of both concentration and composition of the block copolymers. The ABA-type triblock copolymers have potential as an injectable drug delivery system with their sharp phase transition property.

Some natural polymers including proteins [35], polysaccharides [36,37], and their complexations [38–40] possess thermoresponsive properties and have been used to form hydrogels by chemical or physical cross-linking. For example, gelatin combined with chitosan [38] or polyol salt [39] has been used for injectable drug delivery. Other naturally available thermoresponsive polymers are cellulose derivatives. Methylcellulose (MC) and hydroxypropylmethylcellulose (HPMC) are typical examples. The transition temperatures of MC and HPMC are known to be 40–50°C and 75–90°C, respectively [41]. Kawasaki *et al.* studied xyloglucan, a naturally occurring thermoresponsive polysaccharide, which is derived from tamarind seeds as a vehicle for oral delivery. Degradation of xyloglucan by β-galactosidase results in the formation of a product that undergoes sol–gel transition [42]. Recently, recombinant artificial elastin-like polypeptides (ELPs), which are biopolymers with repeating pentapeptides (Val-Pro-Gly-*X*-Gly, where the "guest residue" *X* can be any of the natural amino acids except proline), were reported to undergo a thermally reversible phase transition [43,44]. With current advances in biotechnology, coupling of stimuli-sensitive peptide motifs to a synthetic polymer or other polypeptide can create novel stimuli-sensitive hybrid systems [45].

13.1.2.2 pH-Sensitive Polymers

The pH variation in the GIT is pronounced. The stomach has an acidic environment, pH of 1–2 in a fasting condition, and pH 4 during digestion. The pH in the duodenum is 5.5 due to the mixing of acidic chyme with bicarbonate from pancreatic juices. The extracellular and intracellular pH values in most cancers are more acidic (about pH 6.7 and pH < 7.0, respectively) than in normal tissues or cells [46–48]. Such differences in pH values, even though small, can be exploited for cancer targeted drug delivery using pH-responsive polymeric drug delivery system.

The pH-sensitive polymers include ionizable groups such as carboxylic, sulfonic acid, or basic amino groups that can accept or release protons in response to the environmental pH [10]. Figure 13.3 shows chemical structures of representative

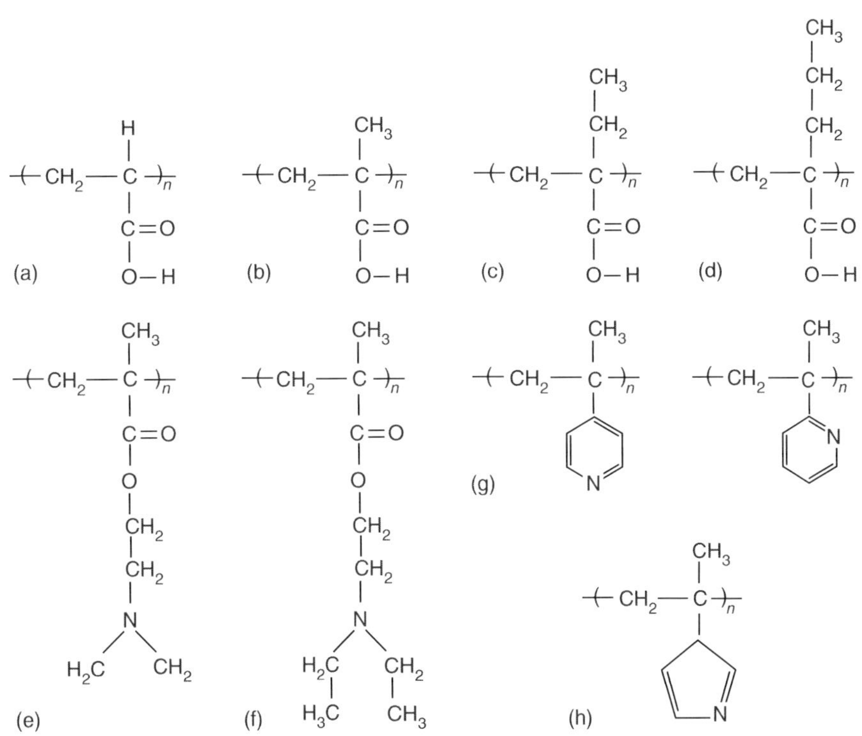

FIGURE 13.3 Chemical structures of representative pH-responsive polymers: (a) poly(acrylic acid), (b) poly(methacrylic acid), (c) poly(3-ethyl acrylic acid), (d) poly(2-propyl acrylic acid), (e) poly(*N,N'*-dimethylaminoethyl methacrylate), (f) poly(*N,N*-diethylaminoethyl methacrylate), (g) poly(4 or 2-vinypyridine), and (h) poly(vinyl imidazole).

pH-responsive polyacids or polybases. Depending on pH, such polyelectrolytes undergo a transition between protonation and deprotonation due to their pK_a or pK_b. In general, weak acids are water-soluble by deprotonation while weak bases become less soluble. Thus, in an aqueous environment, the ionization leads to swelling of polymer hydrogels as well as disintegration of physically assembled polymer particles [48]. In addition to pH, however, pH sensitivity also is influenced by other parameters, such as ionic strength or hydrogen bonding. For example, PAAc, with its pK_a around 4–4.5 [49], has extended chain structure at pH above the pK_a value, because of the polymer chain hydration and electrostatic repulsion of charged polymer chains. However, at a low pH, PAAc can make a complex with any proton-accepting polymers, such as PEG or polyvinylpyrrolidone (PVP) [49,50]. Complex formation through hydrogen bonding between these polymers leads to squeezing out of water molecules from the complex. On the other hand, high ionic strength in polymer solution usually masks the charge of a polymer enough to reduce the electrostatic repulsion between polymer chains. Thus, the pH-sensitive hydrogel showed higher swelling capacity at low ionic strength than at high ionic strength [51].

The pH-sensitive polymers, which can bypass drug resistance and deliver sufficient amounts of the anticancer drug, were evaluated both *in vitro* and *in vivo* experiments [52–54]. A new pH-sensitive polymer based on sulfonamide derivatives has been used for targeting to a receptor that was overexpressed by acidic extracellular matrix (ECM) of cancers [55–57]. Other pH-sensitive polymers, block copolymers of poly(L-histidine), which is a polybase, and PEG, were also synthesized to make polymeric micelles that can respond to the local pH changes in the body [58,59]. The local pH change around cancer cells has also been used to deliver DNA from nanoparticle complexes formed by electrostatic interactions between DNA, cationic polymer (PEI), and amphiphilic copolymer. The complex formed at physiological pH dissociated to release DNA at pH 6.6 [60]. Chitosan, a polysaccharide, is also known to be complexed by citrate in the pH range of 4.3–7.6 [61]. Thus, chitosan, being a pH-sensitive natural polymer, can be used to deliver drugs in a controlled manner by exploiting the pH-sensitive formation and disintegration of chitosan-containing polyelectrolyte complexes.

Exploring the wide diversity in the behavior of polymer gels, Luo *et al.* stumbled upon an interesting property in a class of gels-silicone based polysilamines [62]. The chemistry of this novel class of hydrogels comprises of alternating diamine and organosilyl units. Surprisingly, these gels hardened on swelling based on a reversible rod–globule transition. In addition, in aqueous media, polysilamine hydrogels exhibit reproducible swelling/syneresis behavior in response to the protonation degree of the network. The gel network chains expand and rigidify their conformation on protonation and anion binding, which induce the gel swelling consistently. Since a drawback in the use of most conventional hydrogels as biomedical devices is the decrease in mechanical properties on swelling, the unique property of this class of silicone-based hydrogels makes them attractive for use in controlled drug delivery, especially as pump systems in pulsatile drug delivery.

13.1.2.3 Biomacromolecule-Sensitive Polymers

13.1.2.3.1 Glucose

In the treatment of diabetes caused by the malfunction of pancreas, glucose-responsive polymers, which undergo swelling changes in response to glucose, can construct self-regulated insulin-delivery systems, in which an appropriate amount of insulin can be released in response to the blood glucose concentration.

Glucose-sensitive polymers have been divided into boronic acid derivatives, glucose oxidase (GOx) conjugates, and concanavalin A (ConA) conjugates. The boronic acid moiety can accept one hydroxyl ion to be negatively charged when over a certain pH (pK_a of boronic acid). The ionized form of boronic acid can bind to a *trans*-diol such as glucose with high affinity [63,64]. pH-sensitive polymers in coupling with the GOx can change the hydrophilicity depending on the existence of glucose, since it can degrade glucoses to yield proton ions, which lead the local environments to be acidic. PAAc was used for forming a gate to regulate solute permeability in a membrane. If, at high glucose concentration, GOx generates a large amount of gluconic acid, local pH decreases and protonated AAc causes low permeability [65,66]. Likewise, sulfonamide-based hydrogels also decreased its swelling capacity at high glucose concentration [67]. Since the ConA can hold four glucose units per molecule, ConA blended with glucose-bearing polymers forms cross-linked glucose-sensitive hydrogels. As the concentration of glucose increases, the polymer network becomes loose enough to speed up the loaded drug (e.g., insulin) release [68–70]. Glucose-responsive hydrogels have been manipulated with ConA and glycopolymers that contains lectin-interacting saccharide moieties [71]. You *et al.* designed PEO-*b*-poly(2-glucosyloxyethyl acrylate) designed for a glucose-responsive micellar structure, which might release the loaded insulin only when the glucose concentration in blood was increased [72]. It is noted, however, that ConA may leak out into the environment during swelling in an aqueous glucose solution, since it is not covalently bound in the hydrogel networks [73].

13.1.2.3.2 Glutathione

Glutathione is an important biomolecule in maintaining the homeostasis of cellular redox state. Kakizawa *et al.* reported glutathione-sensitive micelles that were ready to be cleaved inside cell rather than circulating blood [74]. Usually, the micelles selectively deliver therapeutic oligodeoxynucleotides (ODNs) or small interfering RNAs (siRNAs) into cells without degradation during blood circulation, which have therapeutic efficacy only inside a cell. A conjugate between PEG and ODN inhibiting luciferase expression via disulfide bond successfully prevented luciferase gene expression in HuH-7 cells, because of the reaction between disulfide bonds and intracellular glutathiones [75]. PEG-*b*-thiolated poly(L-lysine) formed micelle with a cross-linked core due to disulfide bridges. In this case, intracellular glutathiones made the micelle so unstable that the loaded ODN or siRNA were liberated from the ionic complex [74,76]. El-Sayed *et al.* recently proposed a pH- and glutathione-sensitive polymer to deliver therapeutic ODNs [77]. This dual stimuli-sensitive polymer provided tunable range of pH to maximize the therapeutic efficiency by varying the composition of pH-sensitive and hydrophobic moiety (butyl acrylate) or glutathione-responsive moiety (pyridyl disulfide acrylate).

13.1.2.4 Other Stimuli-Responsive Polymers

13.1.2.4.1 Mechanical Stress

Lee *et al.* prepared a degradable and injectable hydrogel as an ECM, which was designed to respond to mechanical stimuli in the body. The basic structure was based on polyguluronate (PG) prepared from the alginate. After oxidization, aldehyde groups on the PG were treated with dihydrazide to form a cross-linked hydrogel, which released a protein drug in a pulsatile manner [78]. Pressure-sensitive polymers (PSPs) or adhesives (PSAs) have been widely studied for the transdermal delivery of drugs [79]. PSAs should have lower surface energy than (or, at least equal to) that of skin where they are to be attached. The adhesiveness can be increased with increase of pressure, which makes the elastic polymer temporally viscous. Osada and Matsuda studied a shape-memory polymer. A poly(*N,N′*-methylenebisacrylamide) hydrogel was thermally deformed, and then recovered its original shape by cooling to room temperature [80]. The shape memory properties of thermoplastic polyurethane allow design of a fully polymeric self-expandable drug eluting stent [81].

13.1.2.4.2 Electrical Signal

In a pioneering work by Okahata *et al.* [82], a thin, but porous nylon-2,12 capsule was prepared from ethylenediamine and 1,10-bis(chlorocarbonyl)decane. The pore was filled with bilayer-forming amphiphiles such as dialkyldimethylammonium bromide (cationic), sodium didodecylphosphate (anionic), and 1,3-dihexadecyl-*rac*-glycero-2-phosphocholine (zwitter ionic), which resulted in reversible gating of intracapsular permeants (e.g., glucose, ions) by the external electric field (0–80 V). A model protein drug, zinc insulin, was loaded into a complex of polyethyloxazoline (PEOx) and poly(methacrylic acid) (PMAA) [83]. Because the complex was based on the hydrogen bonding between two polymers, the complex could form a stable matrix that was easily erodible at pH over 5.4 by deionization of acid moieties of PMAA. Electrical current might hydrolyze water molecules, leading to many OH^- ions. Thus, the matrix could turn on the insulin–release by application of the electrical current (5 mA). Another pulsatile drug (insulin) delivery with an electroresponsive hydrogel was studied [84]. The gel led to plasma glucose pulsatility when stimulated with a constant current of 1.0 mA (0.36 mA/cm^2) *in vivo*. The mechanism of insulin release from the cross-linked polymer, which has a high pK_a (10.4), is associated with electrokinetic flow of solvated insulin with water; that is, transportation process of counterions (electrophoresis) and water molecules (electroosmosis) in the cross-linked polyelectrolyte gel network.

13.1.2.4.3 Ultrasound

With the development of biodegradable polymers, ultrasound has been used to externally regulate the release profile of a drug loaded in a carrier [85,86]. For example, Pluronic micelles containing doxorubicin showed a fast drug release rate as a function of ultrasound frequency as well as power density, which effectively enhanced intracellular drug uptake in a breast cancer cell line with multidrug resistance (MDR) [87]. Ultrasound can be utilized not only to facilitate the degradation rate of biodegradable polymers, but also to modulate the mass transport property of nondegradable polymers. Lavon *et al.* investigated the effect of the ultrasound on model drug release from a nonerodible polymer hydrogel, poly(hydroxyethyl methacrylate-*co-N,N′*-dimethylaminoethyl methacrylate) (PHEMA-*N,N′*-DMAEMA) [88]. The ultrasound could enhance the drug release rate from a matrix where the mass transport is limited.

13.1.2.4.4 Magnetic Field

Polymeric coating onto the magnetic or composite particles has been used for the magnetic resonance imaging (MRI) agents [89,90]. Hsieh *et al.* prepared a polymer matrix (ethylene–vinyl acetate copolymer) containing magnetic beads and a model protein drug, bovine serum albumin (BSA) [91]. When exposed to an oscillating magnetic field, the polymer matrix released BSA in a sustained manner, with the small magnetic beads embedded in the matrix might enlarge either pores or channels where the BSA moved through.

13.1.2.4.5 Redox State

At their early stage of development, redox-sensitive polymers have been used to modulate the permeability of vesicle membranes [92]. For example, a recent study using ABA-type triblock copolymer of PEG-*b*-poly(propylene sulfide) (PPS)-*b*-PEG reported disintegration of the vesicle membranes responding to the oxidation signal [93]. The hydrophobic PPS block is oxidized to be more hydrophilic, which finally leads to vesicle disruption.

13.1.2.4.6 Light

In a study by Shimoboji et al., biological activity of a protein conjugated to a photoresponsive polymer was regulated by light irradiation [94]. Two kinds of photosensitive polymers prepared in the study were *N,N*-DMA-*co*-4-phenylazophenyl acrylate and DMA-*co*-*N*-4-phenylazophenyl acrylamide. The former became soluble in water by irradiating ultraviolet (UV), but became insoluble by irradiating visible light, while the latter possessed an opposite solubility profiles. By conjugation of each polymer to a genetically engineered streptavidin, on–off switching of biotin affinity induced by UV or visible light was confirmed. Tao *et al.* engineered a hollow sphere with controllable optical property as well as membrane permeability in response to the light [95].

13.1.2.4.7 Enzymes

Recent studies on the stimuli-sensitive polymers have quickly moved toward the enzyme-sensitive polymers. Because a disease usually accompanies at least one physiologically or biochemically impaired function, detection and utilization of such a small change can be greatly useful for targeted drug delivery. It was already reported that doxorubicin conjugation to PEG [96] or *N*-(2-hydroxypropyl)methacrylamide copolymer [97] via peptide linkers could be cleaved by lysosomal enzymes, leading to a great potential for cancer treatment. Recently, enzyme-sensitive hydrogels also have been intensively developed for tissue engineering [98].

13.1.2.4.8 Ionic Strength

Most of ionic strength-sensitive polymers have also demonstrated pH-sensitivity, because these polymers are basically polyelectrolytes. Ion–polymer interaction results in reducing electrostatic repulsion between ionic polymer backbones, which makes it difficult for solutes to pass through. One recent study reported a specific ion selectivity of a polymer microcapsule to control the solute release [99]. Poly(NIPAAm-*co*-benzo-*18*-crown[6]-acrylamide) showed Ba^{2+}-selective closure of the membrane pores, while Na^+ did not have much influence on the membrane permeability.

13.1.3 Nanoparticles in Drug Delivery

Recent advances in nanotechnology utilizing various polymers and hydrogels enhanced the ability to design better drug delivery systems. Among them, injectable nanoparticles are of great interest for application in cancer treatments. The size of nanoparticles can be small enough to have intracapillary or transcapillary passage. When the nanoparticles have long circulating stealth property achieved by PEGylation, nonspecific opsonization and interaction with cells and release of the loaded drug during circulation can be minimized [8]. Uptake of nanoparticles by cells or tissues was much higher as compared to larger-size microparticles [100]. In a size-dependent cytotoxicity study, nanoparticles less than 150 nm in diameter induced significant apoptosis in the endothelial cells. In 2005, the U.S. Food and Drug Administration (FDA) approved intravenously administered paclitaxel-loaded 130 nm albumin nanogels (Abraxane™) for cancer therapy [101].

MDR is a major problem in chemotherapy, and it occurs when tumor cells acquire the ability to pump toxic anticancer agents out through their cell membrane before they have the ability to kill the cell. To overcome this problem, "smart drug delivery systems," such as pH-dependent drug delivery, drug targeting, and molecular imaging of drug-containing nanoparticles, have been investigated. The targeting nanoparticle can be obtained by conjugation of site-recognition moieties, such as synthetic ligands [102], growth factors [103], antibodies [104], and folic acid (folate) [105,106], to the surface of the nanoparticles.

Lee *et al.* have developed targeted polymeric micelles using a pH-sensitive polymer, a copolymer of poly(L-histidine) and PEG, linked to folate of which receptor is overexpressed by many types of tumors. The polymers can deliver sufficient amounts of the anticancer drug doxorubicin to kill otherwise resistant tumor cells [59]. Broz *et al.* have constructed a new class of nanoparticles using an easily constructed biocompatible polymer, biotin-functionalized (poly(2-methyloxazoline)-*b*-poly(dimethylsiloxane)-*b*-poly(2-methyloxazoline), which is capable of delivering therapeutic agents to a wide variety of cells involved in cancer and other diseases [107]. These nanoparticles, which were called nanocontainers, automatically incorporated payload molecules as they self-assembled in water. The fluorescent probe-labeled nanocontainers, ranging from 100 to 250 nm, were targeted against the scavenger receptor A1 from macrophages, an important cell in human disease. Soppimath et al. constructed pH-triggered thermally responsive nanoparticles from poly(NIPAAm-*co*-*N,N'*-DMA-*co*-10-undecenoic acid) [108]. At pH 7.4, the nanoparticles released an initial burst of doxorubicin accounting for 18% of the loaded drug. In contrast, at pH 6.6 and lower the nanoparticles released approximately 70% of the loaded drug over the course of 48 h. Xu *et al.* generated dendritic core-shell nanocarriers based on hyperbranched polyethyleneimine cores and different shells containing aliphatic chains and PEG chains, respectively [109]. Fast pH-sensitive cleavage of the imine bond occurred at pH 5–7 while relatively high stability of the imine bond was observed at a pH of 8. This pH shift corresponds to the pH shift observed in malignant tissues from that of normal tissues and hence is a valuable trigger for releasing encapsulated drugs at the target site. Photodynamic therapy (PDT), which uses a light-sensitive chemical known as a photosensitizer (such as porphyrins, which is a component of hemoglobin) to produce cell-killing "reactive oxygen species," has become an important option for the treatment of esophageal cancer and nonsmall cell lung cancer [110].

While there is no doubt that nanotechnology for drug delivery has brought a new dimension to drug delivery technologies, further development and improvements are necessary for the nanoparticulate formulations to be useful in clinical applications.

13.1.4 Conclusions

The ideal drug delivery systems should keep a drug at a desired therapeutic level at the target site while avoiding frequent administration. With the rapid advances in polymer fabrication keeping pace with the strides in nanotechnology, an ever-increasing number of smart polymers with unique characteristics are being engineered. With the rapid development of such stimuli-responsive polymers, a whole line of novel drug delivery platforms with smart functionalities and enhanced patient compliance are bound to emerge.

Acknowledgment

This study was supported in part by grants from NIH through 511-1336-1761 and from Purdue Cancer Center.

References

1. Bignotti, F., Penco, M., Sartore, L., Peroni, I., Mendichi, R., Casolaro, M., and D'Amore, A., Synthesis, characterisation and solution behaviour of thermo- and pH-responsive polymers bearing L-leucine residues in the side chains, *Polymer*, 41, 8247, 2000.
2. Gan, L.H., Gan, Y.Y., and Deen, G.R., Poly(*N*-acryloyl-*N*′-propylpiperazine): A new stimuli-responsive polymer, *Macromolecules*, 33, 7893, 2000.
3. Jeong, B. and Gutowska, A., Lessons from nature: Stimuli-responsive polymers and their biomedical applications, *Trends Biotech.*, 20, 305, 2002.
4. Kaneko, Y. and Okano, T., Temperature-responsive hydrogels as intelligent materials, in *Biorelated Polymers and Gels*, Okano, T., Ed., Academic Press, London, 1998, Chapter 2.
5. Peng, T. and Cheng, Y.L., PNIPAAm and PMAA co-grafted porous PE membranes: Living radical co-grafting mechanism and multi-stimuli responsive permeability, *Polymer*, 42, 2091, 2000.
6. Pinkrah, V.T., Snowden, M.J., Mitchell, J.C., Seidel, J., Chowdhry, B.Z., and Fern, G.R., Physicochemical properties of poly(*N*-isopropylacrylamide-*co*-4-vinylpyridine) cationic polyelectrolyte colloidal microgels, *Langmuir*, 19, 585, 2003.
7. Kikuchi, A. and Okano, T., Hydrogels: stimuli-sensitive hydrogels, in *Polymeric Drug Delivery Systems*, Kwon, G.S., Ed., Taylor & Francis Group, Boca Raton, FL, 2005, Chapter 7.
8. Kwon, I.K., Jeong, S.H., Kang, E., and Park, K., Nanoparticulate drug delivery for cancer therapy, in *Polymeric Drug Delivery Systems*, Nalwa, H.S., Webster, T., Eds., American Scientific Publishers, New York, 2006, Chapter 3.
9. Bae, Y.H., Smart polymers in drug delivery, *Pharm. News*, 9, 417, 2002.
10. Qiu, Y. and Park, K., Environment-sensitive hydrogels for drug delivery, *Adv. Drug. Deliv. Rev.*, 53, 321, 2001.
11. Kwon, Y.M. and Kim, S.W., Thermosensitive biodegradable hydrogels for the delivery of therapeutic agents, in *Polymeric Drug Delivery Systems*, Kwon, G.S., Ed., Taylor & Francis Group, Boca Raton, FL, 2005, Chapter 6.
12. Song, S.-C., Lee, S.B., Jin, J.-I., and Sohn, Y.S., A new class of biodegradable thermosensitive polymers. I. Synthesis and characterization of poly(organophosphazenes) with methoxy-poly(ethylene glycol) and amino acid esters as side groups, *Macromolecules*, 32, 2188, 1999.
13. Tanaka, T., Dynamics of critical concentration fluctuations in gels, *Phys. Rev.*, 17, 763, 1978.
14. Schild, H.G., Poly(*N*-isopropylacrylamide): Experiment, theory and application, *Prog. Polym. Sci.*, 17, 163, 1992.
15. Feil, H., Bae, Y.H., Feijen, J., and Kim, S.W., Mutual influence of pH and temperature on the swelling of ionizable and thermosensitive hydrogels, *Macromolecules*, 25, 5528, 1992.
16. Hirotsu, S., Coexistence of phases and the nature of first-order phase transition in poly(*N*-isopropylacrylamide) gels, *Adv. Polym. Sci.*, 10, 1, 1993.
17. Irie, M., Stimuli-responsive poly(*N*-isopropylacrylamide). Photo- and chemical-induced phase transitions, *Adv. Polym. Sci.*, 110, 49, 1993.
18. Bae, Y.H., Okano, T., and Kim, S.W., "On–off" thermocontrol of solute transport. I. Temperature dependence of swelling of *N*-isopropylacrylamide networks modified with hydrophobic components in water, *Pharm. Res.*, 8, 531, 1991.
19. Bae, Y.H., Okano, T., and Kim, S.W., "On–off" thermocontrol of solute transport. II. Solute release from thermosensitive hydrogels, *Pharm. Res.*, 8, 624, 1991.
20. Yoshida, R., Sakai, K., Okano, T., and Sakurai, Y., Modulating the phase transition temperature and thermosensitivity in *N*-isopropylacrylamide copolymer gels, *J. Biomat. Sci. Polym. Ed.*, 585, 1994.
21. Brazel, C.S. and Peppas, N.A., Pulsatile local delivery of thrombolytic and antithrombotic agents using poly(*N*-isopropylacrylamide-*co*-methacrylic acid) hydrogels, *J. Control. Rel.*, 39, 57, 1996.
22. Katono, H., Maruyama, A., Sanui, K., Ogata, N., Okano, T., and Sakurai, Y., Thermoresponsive swelling and drug release switching of interpenetrating polymer networks composed of poly(acrylamide-*co*-butyl methacrylate) and poly(acrylic acid), *J. Control. Rel.*, 16, 215, 1991.
23. Aoki, T., Kawashima, M., Katono, H., Sanui, K., Ogata, N., Okano, T., and Sakurai, Y., Temperature-responsive interpenetrating polymer networks constructed with poly(acrylic acid) and poly(*N,N*-dimethylacrylamide), *Macromolecules*, 27, 947, 1994.
24. Annaka, M., Tanaka, C., Nakahira, T., Sugiyama, M., Aoyagi, T., and Okano, T., Fluorescence study on the swelling behavior of comb-type grafted poly(*N*-isopropylacrylamide) hydrogels, *Macromolecules*, 35, 8173, 2002.
25. Yoshida, R., Uchida, K., Kaneko, Y., Sakai, K., Kikuchi, A., Sakurai, Y., and Okano, T., Comb-type grafted hydrogels with rapid de-swelling response to temperature changes, *Nature*, 374, 240, 1995.
26. Zhang, X.-Z., Yang, Y.-Y., Wang, F.-J., and Chung, T.-S., Thermosensitive poly(*N*-isopropylacrylamide-*co*-acrylic acid) hydrogels with expanded network structures and improved oscillating swelling–deswelling properties, *Langmuir*, 18, 2013, 2002.
27. Bromberg, L.E. and Ron, E.S., Temperature-responsive gels and thermogelling polymer matrixes for protein and peptide delivery, *Adv. Drug. Deliv. Rev.*, 31, 197, 1998.
28. Zhang, K. and Khan, A., Phase behavior of poly(ethylene oxide)–poly(propylene oxide)–poly(ethylene oxide) triblock copolymers in water, *Macromolecules*, 28, 3807, 1995.
29. Calvo, P., Remunan-Lopez, C., Vila-Jato, J.L., and Alonso, M.J., Chitosan and chitosan/ethylene oxide-propylene oxide

block copolymer nanoparticles as novel carriers for proteins and vaccines, *Pharm. Res.*, 14, 1431, 1997.
30. Shinkai, M., Yanase, M., Honda, H., Wakabayashi, T., Yoshida, J., and Kobayashi, T., Intracellular hyperthermia for cancer using magnetite cationic liposomes: *In vitro* study, *Jpn. J. Cancer Res.*, 87, 1179, 1996.
31. Jeong, B., Bae, Y.H., and Kim, S.W., Thermoreversible gelation of PEG-PLGA-PEG triblock copolymer aqueous solutions, *Macromolecules*, 32, 7064, 1999.
32. Jeong, B., Bae, Y.H., Lee, D.S., and Kim, S.W., Biodegradable block copolymers as injectable drug-delivery systems, *Nature*, 388, 860, 1997.
33. Jeong, B., Lee, D.S., Shon, J.-I., Bae, Y.H., and Kim, S.W., Thermoreversible gelation of poly(ethylene oxide) biodegradable polyester block copolymers, *J. Polym. Sci. Polym. Chem.*, 37, 751, 1999.
34. Lee, D.S., Shim, M.S., Kim, S.W., Lee, H., Park, I., and Chang, T., Novel thermo-reversible gelation of biodegradable PLGA-*block*-PEO-*block*-PLGA triblock copolymers in aqueous solution, *Macromol. Rapid. Commun.*, 22, 587, 2001.
35. Kuijpers, A.J., Engbers, G.H.M., Feijen, J., De Smedt, S.C., Meyvis, T.K.L., Demeester, J., Krijgsveld, J., Zaat, S.A.J., and Dankert, J., Characterization of the network structure of carbodiimide crosslinked gelatin gels, *Macromolecules*, 32, 3325, 1999.
36. Dentini, M., Desideri, P., Crescenzi, V., Yuguchi, Y., Urakawa, H., and Kajiwara, K., Solution and gelling properties of gellan benzyl esters, *Macromolecules*, 32, 7109, 1999.
37. Ramzi, M., Rochas, C., and Guenet, J.-M., Structure–properties relation for agarose thermoreversible gels in binary solvents, *Macromolecules*, 31, 6106, 1998.
38. Chen, T., Embree Heather, D., Wu, L.-Q., and Payne Gregory, F., *In vitro* protein–polysaccharide conjugation: Tyrosinase-catalyzed conjugation of gelatin and chitosan, *Biopolymers*, 64, 292, 2002.
39. Chenite, A., Chaput, C., Wang, D., Combes, C., Buschmann, M.D., Hoemann, C.D., Leroux, J.C., Atkinson, B.L., Binette, F., and Selmani, A., Novel injectable neutral solutions of chitosan form biodegradable gels in situ, *Biomaterials*, 21, 2155, 2000.
40. Nordby Marianne, H., Kjoniksen, A.-L., Nystrom, B., and Roots, J., Thermoreversible gelation of aqueous mixtures of pectin and chitosan. Rheology, *Biomacromolecules*, 4, 337, 2003.
41. Ruel-Gariepy, E. and Leroux, J.-C., In situ-forming hydrogels-review of temperature-sensitive systems, *Eur. J. Pharm. Biopharm.*, 58, 409, 2004.
42. Kawasaki, N., Ohkura, R., Miyazaki, S., Uno, Y., Sugimoto, S., and Attwood, D., Thermally reversible xyloglucan gels as vehicles for oral drug delivery, *Int. J. Pharm.*, 181, 227, 1999.
43. Furgeson Darin, Y., Dreher Matthew, R., and Chilkoti, A., Structural optimization of a "smart" doxorubicin–polypeptide conjugate for thermally targeted delivery to solid tumors, *J. Control. Rel.*, 110, 362, 2006.
44. Meyer, D.E. and Chilkoti, A., Purification of recombinant proteins by fusion with thermally-responsive polypeptides, *Nat. Biotechnol.*, 17, 1112, 1999.
45. Wang, C., Stewart, R.J., and Kopecek, J., Hybrid hydrogels assembled from synthetic polymers and coiled-coil protein domains, *Nature*, 397, 417, 1999.
46. Simon, S., Roy, D., and Schindler, M., Intracellular pH and the control of multidrug resistance, *Proc. Natl. Acad. Sci. USA*, 91, 1128, 1994.
47. Tannock, I.F. and Rotin, D., Acid pH in tumors and its potential for therapeutic exploitation, *Cancer Res.*, 49, 4373, 1989.
48. Na, K. and Bae, Y.H., pH-sensitive polymers for drug delivery, in *Polymeric Drug Delivery Systems*, Kwon, G.S., Ed., Taylor & Francis Group, Boca Raton, FL, 2005, Chapter 3.
49. Pradip, Maltesh, C., Somasundaran, P., Kulkarni, R.A., and Gundiah, S., Polymer–polymer complexation in dilute aqueous solutions: Poly(acrylic acid)–poly(ethylene oxide) and poly(acrylic acid)–poly(vinylpyrrolidone), *Langmuir*, 7, 2108, 1991.
50. Fujiwara, M., Grubbs, R.H., and Baldeschwieler, J.D., Characterization of pH-dependent poly(acrylic acid) complexation with phospholipid vesicles, *J. Colloid Interf. Sci.*, 85, 210, 1997.
51. Gemeinhart, R.A., Chen, J., Park, H., and Park, K., pH-sensitivity of fast responsive superporous hydrogels, *J. Biomter. Sci. Polym. ed.*, 11, 1371, 2000.
52. Taillefer, J., Brasseur, N., Van Lier, J.E., Lenaerts, V., Le Garrec, D., and Leroux, J.C., In-vitro and in-vivo evaluation of pH-responsive polymeric micelles in a photodynamic cancer therapy model, *J. Pharm. Pharmacol.*, 53, 155, 2001.
53. Gaur, U., Sahoo, S.K., De, T.K., Ghosh, P.C., Maitra, A., and Ghosh, P.K., Biodistribution of fluoresceinated dextran using novel nanoparticles evading reticuloendothelial system, *Int. J. Pharm.*, 202, 1, 2000.
54. Potineni, A., Lynn, D.M., Langer, R., and Amiji, M.M., Poly(ethylene oxide)-modified poly(*b*-amino ester) nanoparticles as a pH-sensitive biodegradable system for paclitaxel delivery, *J. Control. Release*, 86, 223, 2003.
55. Chu, L.-Y., Yamaguchi, T., and Nakao, S., A molecular-recognition microcapsule for environmental stimuli-responsive controlled release, *Adv. Mater*, 14, 386, 2002.
56. Kang, S.I. and Bae, Y.H., pH-induced volume-phase transition of hydrogels containing sulfonamide side group by reversible crystal formation, *Macromolecules*, 34, 8173, 2001.
57. Kang, S.I. and Bae, Y.H., pH-Induced solubility transition of sulfonamide-based polymers, *J. Control. Rel.*, 80, 145, 2002.
58. Kim, G.M., Bae, Y.H., and Jo, W.H., pH-induced micelle formation of poly(histidine-*co*-phenylalanine)-*block*-poly(ethylene glycol) in aqueous media, *Macromol. Biosci.*, 5, 1118, 2005.
59. Lee, E.S., Na, K., and Bae, Y.H., Polymeric micelle for tumor pH and folate-mediated targeting, *J. Control. Rel.*, 91, 103, 2003.

60. Sethuraman, V.A., Na, K., and Bae, Y.H., pH-responsive sulfonamide/PEI system for tumor specific gene delivery: An *in vitro* study, *Biomacromolecules*, 7, 64, 2006.
61. Shu, X.Z., Zhu, K.J., and Song, W., Novel pH-sensitive citrate cross-linked chitosan film for drug controlled release, *Int. J. Pharm.*, 212, 19, 2001.
62. Luo, L., Kato, M., Tsuruta, T., Kataoka, K., and Nagasaki, Y., Stimuli-sensitive polymer gels that stiffen upon swelling, *Macromolecules*, 33, 4992, 2000.
63. Hisamitsu, I., Kataoka, K., Okano, T., and Sakurai, Y., Glucose-responsive gel from phenylborate polymer and poly(vinyl alcohol): Prompt response at physiological pH through the interaction of borate with amino group in the gel, *Pharm. Res.*, 14, 289, 1997.
64. Matsumoto, A., Yoshida, R., and Kataoka, K., Glucose-responsive polymer gel bearing phenylborate derivative as a glucose-sensing moiety operating at the physiological pH, *Biomacromolecules*, 5, 1038, 2004.
65. Chu, L.Y., Li, Y., Zhu, J.H., Wang, H.D., and Liang, Y.J., Control of pore size and permeability of a glucose-responsive gating membrane for insulin delivery, *J. Control. Rel.*, 97, 43, 2004.
66. Chu, L.Y., Liang, Y.J., Chen, W.M., Ju, X.J., and Wang, H.D., Preparation of glucose-sensitive microcapsules with a porous membrane and functional gates, *Colloids Surf. B Biointerf.*, 37, 9, 2004.
67. Kang, S.I. and Bae, Y.H., A sulfonamide based glucose-responsive hydrogel with covalently immobilized glucose oxidase and catalase, *J. Control. Rel.*, 86, 115, 2003.
68. Obaidat, A.A. and Park, K., Characterization of protein release through glucose-sensitive hydrogel membranes, *Biomaterials*, 18, 801, 1997.
69. Sato, K., Kodama, D., and Anzai, J., Sugar-sensitive thin films composed of concanavalin A and sugar-bearing polymers, *Anal. Sci.*, 21, 1375, 2005.
70. Tanna, S., Sahota, T.S., Sawicka, K., and Taylor, M.J., The effect of degree of acrylic derivatisation on dextran and concanavalin A glucose-responsive materials for closed-loop insulin delivery, *Biomaterials*, 27, 4498, 2006.
71. Miyata, T., Jikihara, A., and Nakamae, K., Preparation of poly(2-glucosyloxyethyl methacrylate)–concanavalin A complex hydrogel and its glucose-sensitivity, *Macromol. Chem. Phys.*, 197, 1135, 1996.
72. You, L.-C., Lu, F.-Z., Li, Z.-C., Zhang, W., and Li, F.-M., Glucose-sensitive aggregates formed by poly(ethylene oxide)-*block*-poly(2-glucosyloxyethyl acrylate) with concanavalin A in dilute aqueous medium, *Macromolecules*, 36, 1, 2003.
73. Miyata, T., Jikihara, A., Nakamae, K., and Hoffman, A.S., Preparation of reversibly glucose-responsive hydrogels by covalent immobilization of lectin in polymer networks having pendant glucose, *J. Biomater. Sci. Polym. Ed.*, 15, 1085, 2004.
74. Kakizawa, Y., Harada, A., and Kataoka, K., Glutathione-sensitive stabilization of block copolymer micelles composed of antisense DNA and thiolated poly(ethylene glycol)-*block*-poly(L-lysine): A potential carrier for systemic delivery of antisense DNA, *Biomacromolecules*, 2, 491, 2001.
75. Oishi, M., Hayama, T., Akiyama, Y., Takae, S., Harada, A., Yamasaki, Y., Nagatsugi, F., Sasaki, S., Nagasaki, Y., and Kataoka, K., Supramolecular assemblies for the cytoplasmic delivery of antisense oligodeoxynucleotide: polyion complex (PIC) micelles based on poly(ethylene glycol)-*SS*-oligodeoxynucleotide conjugate, *Biomacromolecules*, 6, 2449, 2005.
76. Kakizawa, Y., Furukawa, S., and Kataoka, K., Block copolymer-coated calcium phosphate nanoparticles sensing intracellular environment for oligodeoxynucleotide and siRNA delivery, *J. Control. Rel.*, 97, 345, 2004.
77. El-Sayed, M.E., Hoffman, A.S., and Stayton, P.S., Rational design of composition and activity correlations for pH-responsive and glutathione-reactive polymer therapeutics, *J. Control. Rel.*, 104, 417, 2005.
78. Lee, K.Y., Alsberg, E., and Mooney, D.J., Degradable and injectable poly(aldehyde guluronate) hydrogels for bone tissue engineering, *J. Biomed. Mater. Res.*, 56, 228, 2001.
79. Venkatraman, S. and Gale, R., Skin adhesives and skin adhesion. 1. Transdermal drug delivery systems, *Biomaterials*, 19, 1119, 1998.
80. Osada, Y. and Matsuda, A., Shape memory in hydrogels, *Nature*, 376, 219, 1995.
81. Wache, H.M., Tartakowska, D.J., Hentrich, A., and Wagner, M.H., Development of a polymer stent with shape memory effect as a drug delivery system, *J. Mater. Sci. Mater. Med.*, 14, 109, 2003.
82. Okahata, Y., Hachiya, S., Ariga, K., and Seki, T., The electrical breakdown and permeability control of a bilayer-corked capsule membrane in an external electric field, *J. Am. Chem. Soc.*, 108, 2863, 1986.
83. Kwon, I.C., Bae, Y.H., and Kim, S.W., Electrically erodible polymer gel for controlled release of drugs, *Nature*, 354, 291, 1991.
84. Kagatani, S., Shinoda, T., Konno, Y., Fukui, M., Ohmura, T., and Osada, Y., Electroresponsive pulsatile depot delivery of insulin from poly(dimethylaminopropylacrylamide) gel in rats, *J. Pharm. Sci.*, 86, 1273, 1997.
85. Agrawal, C.M., Kennedy, M.E., and Micallef, D.M., The effects of ultrasound irradiation on a biodegradable 50%–50% copolymer of polylactic and polyglycolic acids, *J. Biomed. Mater. Res.*, 28, 851, 1994.
86. Kost, J., Leong, K., and Langer, R., Ultrasound-enhanced polymer degradation and release of incorporated substances, *Proc. Natl. Acad. Sci. USA*, 86, 7663, 1989.
87. Marin, A., Sun, H., Husseini, G.A., Pitt, W.G., Christensen, D.A., and Rapoport, N.Y., Drug delivery in pluronic micelles: Effect of high-frequency ultrasound on drug release from micelles and intracellular uptake, *J. Control. Rel.*, 84, 39, 2002.
88. Lavon, I. and Kost, J., Mass transport enhancement by ultrasound in non-degradable polymeric controlled release systems, *J. Control. Rel.*, 54, 1, 1998.

89. Gupta, A.K. and Curtis, A.S., Surface modified superparamagnetic nanoparticles for drug delivery: Interaction studies with human fibroblasts in culture, *J. Mater. Sci. Mater. Med.*, 15, 493, 2004.
90. Diwan, M., Elamanchili, P., Lane, H., Gainer, A., and Samuel, J., Biodegradable nanoparticle mediated antigen delivery to human cord blood derived dendritic cells for induction of primary T cell responses, *J. Drug Target.*, 11, 495, 2003.
91. Hsieh, D.S., Langer, R., and Folkman, J., Magnetic modulation of release of macromolecules from polymers, *Proc. Natl. Acad. Sci. USA*, 78, 1863, 1981.
92. Okahata, Y., Ariga, K., and Seki, T., Redox-sensitive permeation from a capsule membrane grafted with viologen-containing polymers, *J. Chem. Soc. Chem. Commun.*, 73, 1988.
93. Napoli, A., Valentini, M., Tirelli, N., Muller, M., and Hubbell, J.A., Oxidation-responsive polymeric vesicles, *Nat. Mater.*, 3, 183, 2004.
94. Shimoboji, T., Ding, Z.L., Stayton, P.S., and Hoffman, A.S., Photoswitching of ligand association with a photoresponsive polymer–protein conjugate, *Bioconjug. Chem.*, 13, 915, 2002.
95. Tao, X., Li, J., and Mohwald H., Self-assembly, optical behavior, and permeability of a novel capsule based on an azo dye and polyelectrolytes, *Chemistry,* 10, 3397, 2004.
96. Veronese, F.M., Schiavon, O., Pasut, G., Mendichi, R., Andersson, L., Tsirk, A., Ford, J., Wu, G., Kneller, S., Davies, J., and Duncan, R., PEG-doxorubicin conjugates: Influence of polymer structure on drug release, *in vitro* cytotoxicity, biodistribution, and antitumor activity, *Bioconjug. Chem.*, 16, 775, 2005.
97. Kovar, M., Kovar, L., Subr, V., Etrych, T., Ulbrich, K., Mrkvan, T., Loucka, J., and Rihova, B., HPMA copolymers containing doxorubicin bound by a proteolytically or hydrolytically cleavable bond: comparison of biological properties *in vitro*, *J. Control. Rel.*, 99, 301, 2004.
98. Lutolf, M.P. and Hubbell, J.A., Synthetic biomaterials as instructive extracellular microenvironments for morphogenesis in tissue engineering, *Nat. Biotechnol.*, 23, 47, 2005.
99. Chu, L.-Y., Yamaguchi, T., and Nakao, S., A molecular-recognition microcapsule for environmental stimuli-responsive controlled release, *Adv. Mater.*, 14, 386, 2002.
100. Desai, M.P., Labhasetwar, V., Amidon, G.L., and Levy, R.J., Gastrointestinal uptake of biodegradable microparticles: effect of particle size, *Pharm. Res.*, 13, 1838, 1996.
101. Gupta, R.B., *Fundamentals of Drug Nanoparticles*, Taylor & Francis Group, New York, 2006.
102. Nasongkla, N., Shuai, X., Ai, H., Weinberg, B.D., Pink, J., Boothman, D.A., and Gao, J., Bioorganic chemistry: cRGD-functionalized polymer micelles for targeted doxorubicin delivery, *Angew Chem. Int. Ed.*, 43, 6323, 2004.
103. Kurihara, A., Deguchi, Y., and Pardridge, W.M., Epidermal growth factor radiopharmaceuticals: In chelation, conjugation to a blood–brain barrier delivery vector via a biotin–polyethylene linker, pharacokinetics, and *in vivo* imaging of experimental brain tumors, *Bioconjugate Chem.*, 10, 502, 1999.
104. Kao, G.Y., Chang, L.-J., and Allen, T.M., Use of targeted cationic liposomes in enhanced DNA delivery to cancer cells, *Cancer Gene Ther.*, 3, 250, 1996.
105. Ping, P., Wang, W., Chen, X., and Jing, X., Poly(epsilon-caprolactone) polyurethane and its shape-memory property, *Biomacromolecules*, 6, 587, 2005.
106. Salmaso, S., Semenzato, A., Caliceti, P., Hoebeke, J., Sonvico, F., Dubernet, C., and Couvreur, P., Specific antitumor targetable b-cyclodextrin-poly(ethylene glycol)-folic acid drug delivery bioconjugate, *Bioconjug. Chem.*, 15, 997, 2004.
107. Broz, P., Benito, S.M., Saw, C., Burger, P., Heider, H., Pfisterer, M., Marsch, S., Meier, W., and Hunziker, P., Cell targeting by a generic receptor-targeted polymer nanocontainer platform, *J. Control. Rel.*, 102, 475, 2005.
108. Soppimath, K.S., Tan, D.C.-W., and Yang, Y.-Y., pH-Triggered thermally responsive polymer core-shell nanoparticles for drug delivery, *Adv. Mat.*, 17, 318, 2005.
109. Xu, S., Kraemer, M., and Haag, R., pH-responsive dendritic core-shell architectures as amphiphilic nanocarriers for polar drugs, *J. Drug Target*, 14, 367, 2006.
110. Sharman, W.M., van Lier, J.E., and Allen, C.M., Targeted photodynamic therapy via receptor mediated delivery systems, *Adv. Drug. Deliv. Rev.*, 56, 53, 2004.

13.2 Drug Delivery Systems: Smart Polymers

Joseph Kost

13.2.1 Introduction

The rapid advancement of biomedical research has led to many creative applications for biocompatible polymers. As modern medicine discerns more mechanisms, both of physiology and of pathophysiology, the approach to healing is to mimic, or if possible, to recreate the physiology of healthy functioning. Thus, the area of smart polymers for responsive drug delivery has evolved. The developments fall under two categories: externally regulated or pulsatile systems (also known as "open-loop" systems) and self-regulated systems (also known as "closed-loop"). The following chapter outlines the fundamentals of this research area.

13.2.2 Development of Controlled Drug Delivery Systems

13.2.2.1 Control of Drug Concentration Levels Over Time

While newer and more powerful drugs continue to be developed, increasing attention is being given to the methods of administering these active substances. In conventional drug delivery, the drug concentration in the blood rises when the drug is taken, then peaks, and declines. Maintaining drug in the desired therapeutic range with just a single dose, or targeting the drug to a specific area (lowering the systemic drug level), are goals that have been successfully attained with commercially available controlled release devices [1,2]. However, there are many clinical situations where the approach of a constant drug delivery rate is insufficient, such as the delivery of insulin for patients with diabetes mellitus, antiarrhythmics for patients with heart rhythm disorders, gastric acid inhibitors for ulcer control, nitrates for patients with angina pectoris, as well as selective β-blockade, birth control, general hormone replacement, immunization, and cancer chemotherapy. Furthermore, studies in the field of chronopharmacology indicate that the onset of certain diseases exhibit strong circadian temporal dependence. Thus, treatment of these diseases could be optimized through the use of responsive (smart) delivery systems [3,4], which are, in essence, man-made imitations of healthy functioning.

13.2.2.2 Classification of "Smart" Polymers

Smart controlled drug release devices can be classified as open or closed loop systems. Open-loop control systems are those where information about the controlled variable is not automatically used to adjust the system inputs to compensate for the change in the process variables. In the controlled drug delivery field, open-loop systems are known as pulsatile or externally regulated. The externally controlled devices apply external triggers for pulsatile delivery such as ultrasonic, thermal, electric, light, and chemical or biochemical agents.

Closed-loop control systems, on the other hand, are defined as systems where the controlled variable is detected, and as a result, the system output is adjusted accordingly. In the controlled drug delivery field, closed-loop systems are known as self-regulated, in which the release rate is controlled by feedback information, without any external intervention. The self-regulated systems utilize several approaches for the rate control mechanisms such as: pH-sensitive polymers, enzyme–substrate reactions, pH-sensitive drug solubility, competitive binding, antibody interactions, and metal concentration-dependent hydrolysis [5–7].

Many approaches for mimicking the physiological healthy state are undergoing research. The focus of this chapter is on smart polymers; therefore, other important areas, such as using pumps for controlled drug delivery, liposomes, microencapsulation of living cells or gene therapy, are not covered. An additional area of significant research that is not covered in this chapter is site-directed or targeted drug delivery, where the release is constant (as in chemotherapy for cancer treatment). Here, the fundamental principles and recent advances of responsive drug delivery for both pulsatile and self-regulated systems are reviewed.

13.2.3 Pulsatile Systems

13.2.3.1 Ultrasonically Stimulated Systems

Release rates of substances can be repeatedly modulated at will from a position external to the delivery system by ultrasonic irradiation [8]. When bioerodible polymers were used as the drug carrier matrices, both the polymer erosion and drug release rates were enhanced when samples were exposed to ultrasound. The system's response to the ultrasonic triggering was rapid and reversible. The releasing agents, *p*-nitroaniline, *p*-aminohippurate, BSA, and insulin, were tested for integrity following exposure to ultrasonic energy, and were found to be intact. Enhanced drug release rate was also observed when nonerodible polymeric systems (where drug release rate is diffusion dependent) were exposed to ultrasound [8].

When diabetic rats, receiving nonerodible hydrophilic polymer implants containing insulin, were exposed to ultrasound, a sharp drop in blood glucose levels was observed after the irradiation. This indication of a rapid release of insulin in the implanted site has suggested the feasibility of ultrasound mediated drug delivery [8,9]. Recently ultrasound was also applied, in the clinic, for the enhancement of drugs transport through the skin [10].

13.2.3.2 Electrically Stimulated Systems

Electrically controlled systems provide drug release by the application of electric field on a biodegradable polymer [11], rate-limiting membrane/hydrogel [12], and/or directly on the solute [13]. The electrophoretic migration of a charged macrosolute within a hydrated membrane results from the combined response to the electrical forces on the solute and its associated counterions in the adjacent electrolyte solution [13]. Approaches involving microelectromechanical systems (MEMS) [6] or microchips are being developed. For example, carbon nanotubes [14] or polymeric biodegradable microchips [15], containing nanosize drug reservoirs, might be able to deliver pharmaceuticals for long time periods in a controlled manner by electrically opening the caps of these reservoirs. Electrically controlled membrane permeability has been also of interest in the field of electrically controlled or enhanced transdermal drug delivery (e.g., iontophoresis, electroporation) [16].

13.2.3.3 Photostimulated Systems

Photoinduced phase transition of gels [17] and polymers degradation [18] have been also proposed for temporal drug delivery. Photoresponsive gels reversibly change their physical or chemical properties upon photoradiation, thus affecting the release rate of drugs incorporated in the polymer. A photoresponsive polymer consists of a photoreceptor, usually a photochromic chromophore, and a functional part. The optical signal is captured by the photochromic molecules and then the isomerization of the chromophores in the photoreceptor converts it to a chemical signal.

13.2.3.4 Calcium-Stimulated Systems

An example for these is a calcium-responsive biodegradable drug delivery system [19], where a nonactive enzyme (alpha amylase) is incorporated into a matrix composed of its substrate (starch). The system responds to a trigger molecule (calcium) that causes the nonactive enzyme to become active. The matrix then starts to degrade, resulting in release of the drug incorporated in the matrix.

13.2.4 Self-Regulated Systems (Environmentally Responsive Systems)

Polymers that alter their characteristics in response to changes in their environment have been of great recent interest. Several research groups have been developing drug delivery systems, based on these responsive polymers, that more closely resemble the normal physiological process [3,4]. In these devices, drug delivery is regulated by means of an interaction with the surrounding environment (feedback information) without any external intervention. The most commonly studied polymers, having environmental sensitivity, are either pH- or temperature-sensitive. There are also inflammation-sensitive systems and systems utilizing enzymes and specific binding interactions, which will be discussed as well.

13.2.4.1 Temperature-Sensitive Systems

Temperature-sensitive polymers are based on polymer–water interactions, especially specific hydrophobic–hydrophilic balancing effects, and the configuration of side groups. When polymer networks swell in a solvent, there is usually a negligible or small positive enthalpy of mixing or dilution. Although a positive enthalpy change opposes the process, the large gain in the entropy drives it. In aqueous polymer solutions, the opposite is often observed. This unusual behavior is associated with a phenomenon of polymer phase separation as the temperature is raised to a critical value, known as the LCST. Polymers characterized by LCST usually shrink, as the temperature is increased through the LCST. Lowering the temperature below LCST results in the swelling of the polymer. Bioactive agents such as drugs, enzymes, antibodies, and genes may be immobilized on or within the temperature sensitive polymers. Responsive drug/genes release patterns regulated by environmental temperature changes have been recently demonstrated by several groups [4,20,21].

13.2.4.2 pH-Sensitive Systems

The pH range of fluids in various segments of the GIT may provide environmental stimuli for responsive drug release. Studies by several research groups [3,20,21] have been performed on polymers containing weakly acidic or basic groups in the polymeric backbone. The charge density of the polymers depends on pH and ionic composition of the outer solution (the solution into which the polymer is exposed). Altering the pH of the solution will cause swelling or deswelling of the polymer. Polyacidic polymers will be unswollen at low pH, since the acidic groups will be protonated and hence unionized. With increasing pH, polyacidic polymers will swell. The opposite holds for polybasic polymers, since the ionization of the basic groups will increase with decreasing pH. Thus, drug release from devices made from these polymers will display pH-dependent release rates.

Lowman and Peppas [22] proposed pH-responsive drug delivery based on a complexation–decomplexation mechanism. Copolymer networks of PMAA grafted with poly(ethylene glycol) exhibit pH-dependent swelling behavior due to the reversible formation–dissociation of interpolymer complexes. Drug delivery systems based on this copolymer showed pH-dependent drug delivery rates.

Heller and Trescony [23] were the first to propose the use of pH-sensitive bioerodible polymers. In their approach, described in the section on systems utilizing enzymes, an enzyme–substrate reaction produces a pH change that is used to modulate the erosion of pH-sensitive polymer containing a dispersed therapeutic agent.

13.2.4.3 Inflammation-Responsive Systems

Inflammation-responsive drug delivery systems are based on biodegradable hydrogels that are specifically degraded by hydroxyl radicals or enzymes produces at inflammatory sites [24]. The radicals or specific enzymes cause degradation of the polymer and thus release of the bioactive agent, making this delivery system potentially useful for treatment of osteoarthritis.

13.2.4.4 Systems Utilizing Antibody Interactions

Pitt [25] proposed utilizing morphine (hapten)–antibody interactions to suppress enzymatic degradation and permeability of polymeric reservoirs or matrix drug delivery systems. The delivery device consists of the drug naltrexone contained in a polymeric reservoir or dispersed in a polymeric matrix configuration. The device is coated by covalently grafting morphine to the surface. Exposure of the grafted surface to antibodies to morphine results in coating of the surface by the antibodies, a process that can be reversed by exposure to exogenous morphine. The presence of the antibodies on the surface, or in the pores of the delivery device, will block or impede the permeability of naltrexone in a reservoir configuration or enzyme-catalyzed surface degradation and concomitant release of the drug from a matrix device. The reversible binding of antigen to antibody serve also as the basis for hydrogel swelling that could lead to release of a bioactive agent [26,27].

13.2.4.5 Systems Utilizing Enzymes

13.2.4.5.1 Urea-Responsive Delivery

Heller and Trescony [23] were the first to attempt using immobilized enzymes to alter local pH and thus cause changes in polymer erosion rates. The proposed system is based on the conversion of urea to NH_4HCO_3 and NH_4OH by the action of urease. As this reaction causes a pH increase, a polymer that is subjected to increased erosion at high pH is required.

13.2.4.5.2 Morphine Triggered Naltrexone Delivery System

Heller [28] also proposed a biodegradable drug delivery system that would be passive until the appearance of morphine external to the device. Activation of release is based on the reversible inactivation of enzymes achieved by the covalent attachment of hapten, close to the active site of the enzyme–hapten conjugate, with the hapten antibody. Because the antibodies are large molecules, access of the substrate to the enzyme's active site is sterically inhibited, thus effectively rendering the enzyme inactive.

Triggering of drug (naltrexone) release is initiated by the appearance of morphine (hapten) in the tissue, causing dissociation of the enzyme–heptan–antibody complex, rendering the enzyme active and therefore release of naltrexone. Many of the glucose-responsive systems discussed in the next section also utilize enzymes.

13.2.4.6 Glucose-Responsive Insulin Delivery

The present modes of treatment, including insulin pumps, not fully mimic the physiology of insulin secretion. Therefore the development of a smart insulin-delivery system could significantly help patients with diabetes in controlling their blood glucose level, thus avoiding the various severe diabetes complications including eye disease, gangrene of the extremities, cardiovascular disease, and renal failure. The development of glucose-sensitive insulin-delivery systems has made use of several approaches, including polymer–complex system, immobilized GOx in pH-sensitive polymers, and competitive binding.

13.2.4.6.1 Polymer Complex System

Kitano *et al.* [29] proposed a glucose sensitive insulin release system based on a sol–gel transition. A phenylboronic acid (PBA) moiety was incorporated into poly(*N*-vinyl-2-pyrrolidone) (poly(NVP-*co*-PBA)). Insulin was incorporated into a polymer gel formed by a complex of poly(vinyl alcohol) with poly(NVP-*co*-PBA). PBA can form reversible covalent complexes with molecules having diol units, such as glucose or PVA. With the addition of glucose, PVA in the PVA–boronate complex is replaced by glucose. This leads to a transformation of the system from gel to sol state, facilitating the release of insulin from the polymeric complex.

13.2.4.6.2 Competitive Binding

The basic principle of competitive binding and its application to controlled drug delivery was first presented by Brownlee and Cerami [30]. They suggested the preparation of biologically active glycosylated insulins, which are complementary to the major combining site of carbohydrate binding proteins such as Concavalin A (Con A). Con A is immobilized on sepharose beads. The glycosylated insulin is displaced from the Con A by glucose in response to, and proportional to, the amount of glucose present, which competes for the same binding sites. Kim *et al.* [31] found that the release rate of glycosylated insulin also depends on the binding affinity of the insulin derivative to the Con A and can be influenced by the choice of saccharide group in the glycosylated insulin. By encapsulating the glycosylated insulin-bound Con A with a suitable polymer that is permeable to glucose and insulin, the glucose influx and insulin efflux would be controlled by the encapsulation membrane.

Similarly, hydrogels or membranes formed by mixing Con A with copolymers containing allyl glucose will undergo a reversible sol–gel phase transition in the presence of glucose due to the competitive binding between the free glucose and Con A (Con A acts as a cross-linker). The free glucose would replace the Con A and therefore disrupt the cross-links and make the polymer more permeable to insulin [32].

13.2.4.6.3 Immobilized GOx in pH Sensitive Polymers

Responsive drug delivery systems based on pH-sensitive polymers have been developed along three different approaches: pH-dependent swelling, degradation, and solubility.

13.2.4.6.3.1 pH-Dependent Solubility Glucose-dependent insulin release was proposed by Langer and coworkers [33,34] based on the fact that insulin solubility is pH-dependent. Insulin was incorporated into ethylene vinyl acetate (EVAc) copolymer matrices in solid form. GOx was immobilized to sepharose beads, which were incorporated along with insulin into the EVAc matrices. When glucose entered the matrix, the produced gluconic acid caused a rise in insulin solubility and consequently enhanced release.

13.2.4.6.3.2 pH-Dependent Degradation Heller [28] suggested a system in which insulin is immobilized in a pH-sensitive bioerodible polymer, which is surrounded by a hydrogel containing immobilized GOx. When glucose diffuses into the hydrogel and is oxidized to gluconic acid, the resultant lowered pH triggers enhanced polymer degradation and release of insulin from the polymer in proportion to the concentration of glucose.

13.2.4.6.3.3 pH-Dependent Swelling Systems based on pH-sensitive polymers consist of immobilized GOx in a pH-responsive hydrogel, enclosing a saturated insulin solution or incorporated with insulin were first proposed by Horbett *et al.* [35] and Ishira *et al.* [36]. As glucose diffuses into the hydrogel, GOx catalyzes its conversion to gluconic acid, thereby lowering the pH in the microenvironment of the hydrogel and causing swelling. Since insulin should permeate the swelled hydrogel more rapidly, faster delivery of insulin in the presence of glucose is anticipated. As the glucose concentration decreases in response to the released insulin, the hydrogel should contract and decrease the rate of insulin delivery. Up to now, several pH-responsive hydrogels have been devised [37]. However, none of these systems could fully mimic the physiology of insulin secretion as yet. For glucose-responsive self-regulated insulin release systems, biocompatibility, bioavailability, long-term stability and system response kinetics are essential to ensure safety and efficacy [38–43].

13.2.5 Concluding Remarks

The approaches discussed represent attempts conducted over the past three decades to achieve responsive release. It should be pointed out that these drug delivery systems are still in the developmental stage and much research will have to be conducted for such systems to become practical clinical alternatives. Critical considerations are the biocompatibility and toxicology of these multicomponent polymer-based systems, the response times of these systems to stimuli, the ability to provide practical levels of the desired drug, and addressing necessary formulation issues in dosage or design (e.g., shelf life, sterilization, reproducibility). A key issue in the practical utilization of the pulsatile, externally triggered systems (i.e., ultrasound, electrically regulated, and photoresponsive) will be the design of small portable trigger units that the patient can easily use. Ideally, such systems could be worn by the patient, such as a wristwatch-like system, and it could be either preprogrammed to go on and off at specific times or the patient could turn it on when needed. A critical issue in the development of responsive, self-regulated systems, such as those containing enzymes or antibodies, is the stability and/or potential leakage and possible immunogenicity of these bioactive agents. The successful development of responsive polymer delivery systems will be a significant challenge. Nevertheless, the considerable pharmacological benefit these systems could potentially provide, particularly given ongoing research in biotechnology, pharmacology and medicine, which may provide new insights on the desirability and requirements for pulsatile release, should make this an important and fruitful area for future research.

References

1. Langer, R., Where a pill won't reach, *Sci. Am.*, 288, 50, 2003.
2. Langer, R., Drug delivery and targeting, *Nature*, 392, 5, 1998.
3. Kost, J. and Langer, R., Responsive polymeric delivery systems, *Adv. Drug Deliver. Rev.*, 46, 125, 2001.
4. Kopecek, J., Smart and genetically engineered biomaterials and drug delivery systems, *Eur. J. Pharm. Sci.*, 20, 1, 2003.
5. Goldbart, R., Traitel, T., Lapidot, S., and Kost, J., Enzymatically controlled responsive drug delivery systems, *Polym. Advan. Technol.*, 13, 1006, 2002.
6. Galaev, I.Y. and Mattiasson, B., "Smart" polymers and what they could do in Biotechnology and Medicine, *Trends Biotechnol.*, 17, 335, 1999.
7. Kaetsu, I. *et al.*, Intelligent type controlled release systems by radiation techniques, *Radiat. Phys. Chem.*, 55, 193, 1999.
8. Kost, J., Leong, K., and Langer, R., Ultrasound-enhanced polymer degradation and release of incorporated substances, *Proc. Natl. Acad. Sci.*, 86, 7663, 1989.
9. Miyazaki, S., Yokouchi, C., and Takada, M., External control of drug release: controlled release of insulin from a hydrophilic polymer implants by ultrasound irradiation in diabetic rats, *J. Pharm. Pharmacol.*, 40, 716, 1988.
10. Mitragotri, S. and Kost, J., Low-Frequency Sonophoresis: A Review, *Adv. Drug Deliver. Rev.*, 56, 589, 2004.
11. Kwon, I.C., Bae, Y.H., and Kim, S.W., Electrically erodible polymer gel for controlled release of drugs, *Nature*, 354, 291, 1991.
12. Madou, M.J. and He, K.Q., Exploitation of a novel artificial muscle for controlled drug delivery, *Polym. Mater. Sci. Eng.*, 83, 405, 2000.
13. Grodzinsky, A.J. and Grimshaw, P. E., Electrically and chemically controlled hydrogels for drug delivery, in *Pulsed and self regulated drug delivery*, Kost, J., Ed., CRC Press, Boca Raton, 1999, 47.
14. Bianco, A., Kostarelos, K., and Prato, M., Applications of carbon nanotubes in drug delivery, *Curr. Opin. Chem. Biol.*, 9, 674, 2005.
15. Grayson, A. *et al.*, Multiple-pulse drug delivery from a resorbable polymeric microchip device, *Nat. mater.*, 2, 767, 2003.
16. Prausnitz, M.R., Mitragotri, S., and Langer, R., Current status and future potential of transdermal drug delivery, *Nat. Rev. Drug Discov.*, 3, 115, 2004.
17. Mamada, A. *et al.*, Photoinduced phase transition of gels, *Macromolecules*, 23, 1517, 1990.
18. Yui, N., Okano, T., and Sakurai, Y., Photo-responsive degradation of heterogeneous hydrogels comprising crosslinked hyaluronic acid and lipid microspheres for temporal drug delivery, *J. Control. Release*, 26, 141, 1993.
19. Goldbart, R. and Kost, J., Calcium responsive bioerodible drug delivery systems, *Pharm. Res.*, 16, 1483, 1999.
20. Alexander, C., Temperature- and pH-responsive smart polymers for gene delivery, *Expert Opin. Drug Deliver.*, 3, 573, 2006.
21. de Las Heras Alarcam, C., Pennadam, S., and Alexander, C., Stimuli responsive polymers for biomedical applications, *Chem. Soc. Rev.*, 3, 276, 2005.
22. Lowman, A.M. and Peppas, N.A., Pulsatile drug delivery based on a complexation-decomplexation mechanism, in *Intelligent Materials for Controlled Release*, Dinh, S., DeNuzzio, J., and Comfort, A., Eds., ACS Symposium Series 728, American Chemical Society, 30, 1999.
23. Heller, J. and Trescony, P.V., Controlled drug release by polymer dissolution II. An enzyme mediated delivery system, *J. Pharm. Sci.*, 68, 919, 1979.
24. Yui, N., Okano, T., and Sakurai, Y., Inflammation responsive degradation of crosslinked hyaluronic acid gel, *J. Control. Release*, 22, 105, 1992.
25. Pitt, C.G., Self-regulated and triggered drug delivery systems, *Pharm. Int.*, 88, 91, 1986.
26. Miyata, T., Asami, N., and Uragami, T., A reversibly antigen-responsive hydrogel, *Nature*, 24, 766, 1999.
27. Nakayama, G.R., Roskos, K., Fritzinger, B., and Heller, J., A study of reversibly inactivated lipases for use in morphine-triggered naltrexone delivery system, *J. Biomed. Mater. Res.*, 29, 1389, 1995.
28. Heller, J., Feedback-controlled drug delivery, in *Controlled drug delivery: challenges and strategies*, Park, K., Ed., American Chemical Society, Washington, DC, 1997,127.
29. Kitano, S., Kitano, S., Koyama, Y., Kataoka, K., Okanu, T., and Sakurai, Y., A novel drug delivery system utilizing a glucose responsive polymer complex between poly(vinyl alcohol) and poly(*N*-vinyl-2-pyrrolidone) with phenylboronic acid moiety, *J. Control. Release*, 19, 161, 1992.
30. Brownlee, M. and Cerami, A., A glucose-controlled insulin-delivery system: semisynthetic insulin bound to lectin, *Science*, 26, 1190, 1979.
31. Kim, S.W., Pai, C.M., Makino, L.A., Seminoff, L.A., Holmberg, D.L., Gleeson, J.M., Wilson, D.E., and Mack, E.J., Self regulated glycosylated insulin delivery, *J. Control. Release*, 11, 193, 1990.

32. Kim, J.J. and Park, K., Modulated insulin delivery from glucose sensitive hydrogel dosage forms, *J. Control. Release*, 77, 29, 2001.
33. Brown, L., Edelman, E., Fischel-Ghodsian, F., and Langer, R., Characterization of glucose-mediated insulin release from implantable polymers, *J. Pharm. Sci.*, 85, 1341, 1996.
34. Fischel-Ghodsian, F., Brown, L., Mahiowitz, E., Branderburg, D., and Langer, R., Enzymatically controlled drug delivery, *Proc. Natl. Acad. Sci.*, 85, 2403, 1988.
35. Horbett, T., Kost, J., and Ratner, B., Swelling behavior of glucose sensitive membranes, *Am. Chem. Soc., Div. Polym. Chem.*, 24, 34, 1983.
36. Ishihara, K., Kobayashi, M., and Shonohara, I., Control of insulin permeation through a polymer membrane with responsive function for glucose, *Macromol. Chem., Rapid Commun.*, 4, 327, 1983.
37. Kost, J., Intelligent Drug Delivery Systems in *The Encyclopedia of Controlled Drug Delivery*, Mathiowitz, E. Ed., John Wiley & Sons, NY, 2000, 445.
38. Podual, K., Doyle, F.J., and Peppas, N.A., Glucose-sensitivity of glucose oxidase-containing cationic copolymer hydrogels having poly(ethylene glycol) grafts, *J. Control. Release*, 67, 9, 2000.
39. Chu, L.Y. *et al.*, Control of pore size and permeability of a glucose-responsive gating membrane for insulin delivery, *J. Control. Release*, 97, 43, 2004.
40. Misra, G. and Siegel, R., Long term autonomous rhythmic hormone release across hydrogel membrane, *J. Control. Release*, 81, 1, 2002.
41. Traitel, T., Cohen, Y., and Kost, J., Characterization of glucose-sensitive insulin release systems in simulated *in vivo* conditions, *Biomaterials*, 21, 1679, 2000.
42. Matsumoto, A., Yoshida, R., and Kataoka, K., Glucose-responsive polymer gel bearing phenylborate derivative as a glucose-sensing moiety operating at the physiological pH, *Biomacromolecules*, 3, 1038, 2004.
43. Peppas, N.A., Is there a future in glucose-sensitive, responsive insulin delivery systems? STP, *Pharma. Sci.*, 14, 247, 2004.

14

Fiber Optic Systems: Optical Fiber Sensor Technology and Windows

14.1 Introduction and Application of Fiber Optic Sensors

Nezih Mrad and Henry C.H. Li

14.1.1 Introduction

Recent advances in sensors, microelectronics, and adaptive signal processing technologies have significantly shaped the fundamental approaches to dealing with traditional problems. Imagine a world where mechanical systems achieve optimum functionality demonstrated by biological systems such as the human brain and body. These systems that are known as smart, adaptive, and intelligent could mimic human muscular, sensory, and nervous systems by employing embedded sensors, actuators, advanced signal processing and control capabilities. Micro and nanosensors act as the system's nerve endings sending signals to the processor (brain), which in turn sends signals to the actuator that take on the role of responsive or adaptive muscles, leading to smart and adaptive response. Furthermore, driven by increased safety and reliability at reduced cost and enhanced performance, key technologies such as electrorheological fluids, shape memory alloys, piezoelectric materials, magnetostrictive and electrostrictive materials, and micro and nanoelectromechanical devices (MEMS/NEMS) are identified to play a major role in the development of such adaptive systems. One main experienced limitation is the integration of sensory nodes and sensor networks within a complex system. Among the few sensors suited for incorporation into materials and structures and for the realization of such adaptive and smart systems, optical fibers are considered to lead the pack in sensor technology. If successfully implemented, this technology is posed to revolutionize the current approaches to systems' health monitoring and management, performance assessment, and implementation of adaptive systems and structures.

Compared to more traditional measurement techniques, fiber optic sensors offer unique capabilities such as monitoring the manufacturing process of composite and metallic parts, performing nondestructive testing once fabrication is complete, enabling structural and component health monitoring for prognostics health management, and structural control for component life extension. Because of their very low weight, small size, high bandwidth, and immunity to electromagnetic and radio frequency interferences, fiber optic sensors have significant performance advantages over traditional sensors. In fact, in recent years, optical fibers are steadily becoming more cost-effective due to recent advances in telecommunications, optoelectronic, and micro and nano industries. In contrast to classical sensors that are largely based on measurement of electrical parameters such as resistance or capacitance, fiber optic sensors make use of a variety of novel phenomena inherent in the structure of the fiber itself. Some of these phenomena are briefly presented in this document along with an introduction to the technology and its application in several industrial sectors.

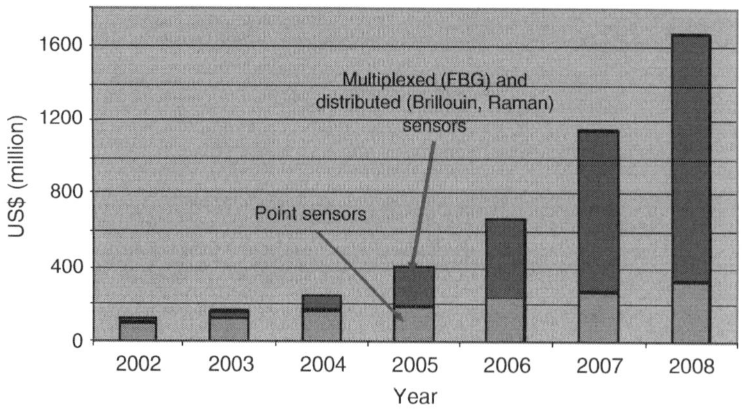

FIGURE 14.1 Estimation of worldwide fiber optic sensor market.

According to the Business Communications Company, Inc. (BCC) [1], which is one of the market research companies, fiber optic sensors are classified as by-product of optical fiber research and development for the communication industry. This industry, regardless of its sluggish recent market, remains the largest application market for this sensor technology, in addition to medicine, defense, and aerospace applications. The world market for fiber optic sensors was worth $175 million in 1998 and expected to reach $600 million by 2011. However other estimates expect the market to be worth $1600 million by 2008 (Figure 14.1) [2]. Figure 14.1 further illustrates the potential for distributed versus discrete sensor type. According to BCC, the total worldwide revenues for fiber optic sensors are projected to increase from a current estimate of $288.1 million to $304.3 million in 2006 with an increase at a modest average annual growth rate (AAGR) of 4.1% to $371.8 million by 2011. BCC further concludes that while fiber optic sensors technology has tremendous potential, this potential has only been realized within the telecommunications market. Figure 14.2 [1] illustrates an example of the market potential for only extrinsic and intrinsic sensor type. Extrinsic sensors are forecast to increase at an AAGR of 4.1% until 2011, to $274.4 million whereas intrinsic sensors will rise from $79.9 million in 2006 to $97.4 million in 2011 with an AAGR of 4%. Intrinsic sensors market revenues lag that of extrinsic sensors; however, this category of sensors is where the greatest opportunity for new market entrants is expected. Currently, the largest application segment for such intrinsic sensors is military technologies that tend to be very specialized and project focused, seldom leading to mainstream market consumption resulting in market success. Growth in intrinsic sensors is expected to be rather slow in some application markets due to ongoing challenges with the technology.

14.1.2 Fiber Optic Sensors

Fiber optic sensor development capitalized on the successful discovery of communication optical fibers in the 1950s and has been underway since the early 1970s. In recent years, accelerated progress was experienced due to the significant development of new, low-cost materials, and devices and the emergence of micro and nanotechnologies for the telecommunications industry. The shape and form of optical fibers are similar to those reinforcing fibers used in fiber-reinforced composite materials. However, the diameter of optical fibers is much larger, usually in the order of 40–250 μm, compared to glass and carbon fibers used in composites that are typically 10 μm or smaller. Optical fibers consist of a light waveguide inner silica-based core surrounded by an annular-doped silica cladding that is protected by a polymer coating (Figure 14.3). This optical fiber can also be made using other materials such as plastic [3]. The fiber core refractive index is relatively large compared to that of the cladding index. The change in refractive indices, between the core and the cladding, provides the required mechanics for light propagation within the fiber core. Depending on the wavelength of the light input, waveguide geometry and distribution of its

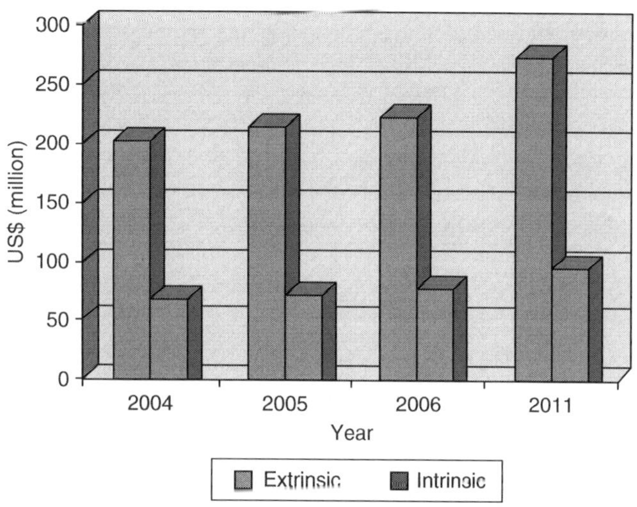

FIGURE 14.2 Worldwide fiber optic sensor revenues by category.

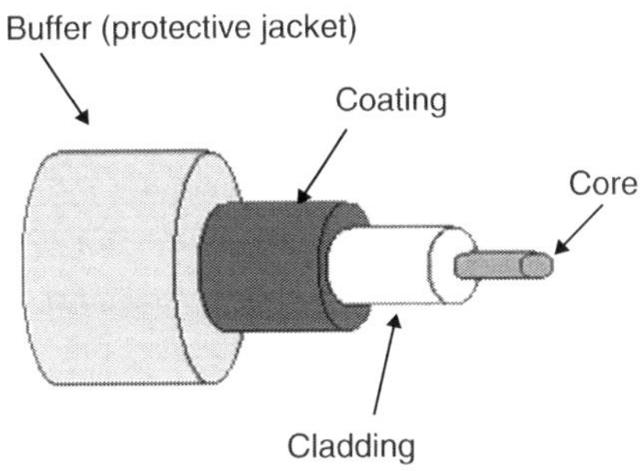

Sensor parameter	Single mode fiber (μm)	Multimode fiber (μm)	Material
Core	2–10	50–150	Silica-based
Cladding	80–120	100–250	Doped silica
Coating	250	250	Polymer
Buffer	900	900	Plastic

FIGURE 14.3 Schematic of the basic structure of an optical fiber.

refractive indices, several modes can propagate through the fiber, resulting in the so-called, single and multimode optical fibers. Both fiber types are used in the construction of fiber optic sensors. However, single-mode fibers are more sensitive to strain variation and are thus the preferred choice. Table 14.1 [4] provides a summary of the typical properties of various optical fiber types (e.g., single and multimode).

From a practical point of view, optical fiber sensors can be divided into two broad classes, namely short and long gauge length optical fiber sensors, also known as discrete and distributed. Short gauge length sensors are comparable to those conventional sensors such as resistive foil strain gauges and thermocouples. They typically provide a measure of physical parameters over distances of several millimeters (≤20 mm). Long gauge length sensors have gauge lengths ranging from several centimeters to hundreds of meters and kilometers. They typically measure physical parameters over distances of few centimeters to few meters (40 cm to 5 m). The latter type is generally known as distributed, spatially distributed, distributed-effect, integrating, and/or averaging sensors. Figure 14.4 illustrates the classification and configuration of these sensors. The two classes of sensors can also be divided into two types, intrinsic and extrinsic. This refers to the sensing region of the fiber sensor being either outside or inside the fiber, respectively. Both intrinsic and extrinsic sensors configurations that also employ single mode and multimode fibers are illustrated in Figure 14.5 [5]. In the extrinsic type, the optical fiber acts only as a light-transmitting and communication medium. Received light intensity is affected by the variation in the physical parameters in the transducer (sensitive) element. The transducer element does not necessarily consist of optical fibers and the optical effect occurs in a different medium. Such sensors usually employ multimode optical fibers. In the intrinsic type, the optical fiber itself acts as the sensing element, which includes both transducer and communication media. When the conditions of the sensed medium change, the light propagation properties of the optical fiber also change, providing a measurement of the change in condition. Sensors of this type are immune to dirty environments due to their closed optical path and are generally found in fiber optic gyroscopes and fiber optic hydrophones. These are expected to have the greatest developmental opportunity regardless of their slow market growth.

TABLE 14.1 Typical Properties of Various Optical Fiber Types

Type[a]	Loss (dB/km)	Numerical Aperture (NA)	Core Size (OD Microns)	Core to Cladding Ratio	Bandwidth (MHz km)
Multimode step index					
Glass–clad–glass (bundle)	400–600	0.4–0.6	50–70	0.9–0.95	20
Plastic–clad–silica	3–10	0.3–0.4	200–600	0.7	20
Glass–clad–glass	2–6	0.2–0.3	50–200	0.4–0.8	20
Single mode (low loss)					
Glass–clad–glass	2–6	0.15	5–8	0.04	1000

[a] A variety of material combinations can be used for both core and cladding; however, it is generally application dependent. Some combination examples include: cladding and core of quartz glass; plastic core and cladding; normal glass core and cladding; plastic cladding and quartz glass core (known as plastic-cladded fibers [PCFs]).

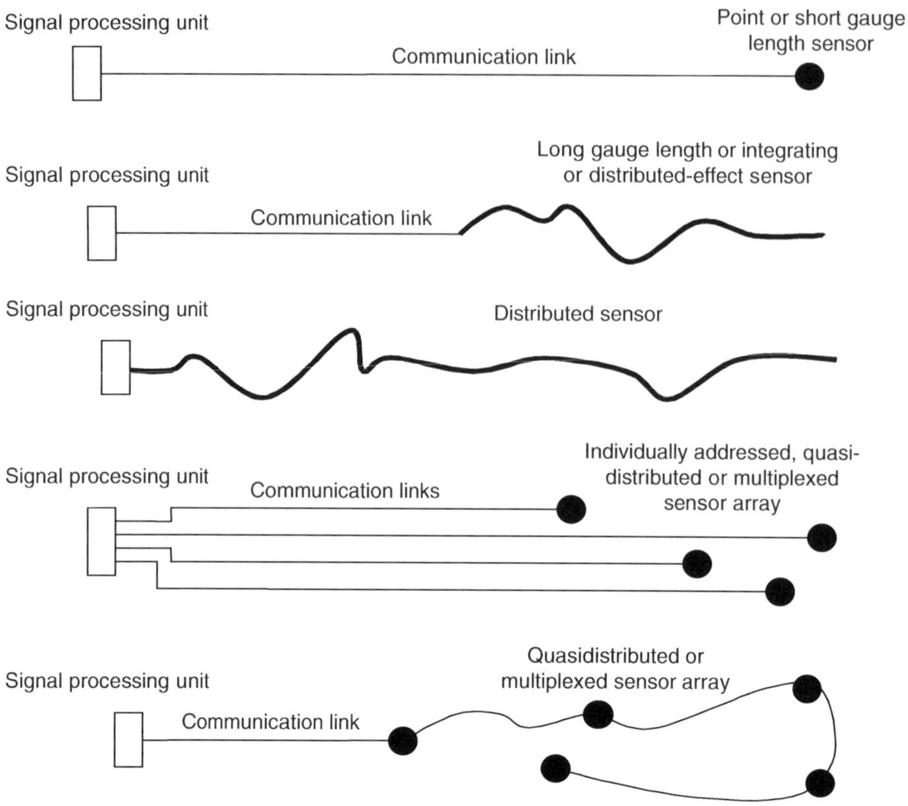

FIGURE 14.4 Classification of optical fiber sensors.

Traditionally, optical fiber sensors are classified by their transducer or modulation mechanisms. The transducer mechanism affects the properties of the transmitted light in the fiber, such as intensity, polarization, phase, and modal content (wavelength). Sensors with these transducer mechanisms are known as intensity, polarimetric, phase, or modal content sensors, respectively. Fiber optic strain, temperature, pressure, and chemical sensors have been developed, which employ many different optical modulation mechanisms including intensity (distributed microbend), phase (interferometric), wavelength (Bragg gratings), and state of polarization (polarimetric). Some of the technology benefits, concerns, and constraints are provided in Table 14.2.

14.1.3 Optical Fiber Modulation Mechanisms

14.1.3.1 Intensity-Modulated Sensors

Intensity-modulated sensors rely on the reduction of transmitted light intensity through an optical fiber as a function of the perturbing environmental parameters [6,7]. The light intensity attenuation or loss can be associated with transmission, reflection,

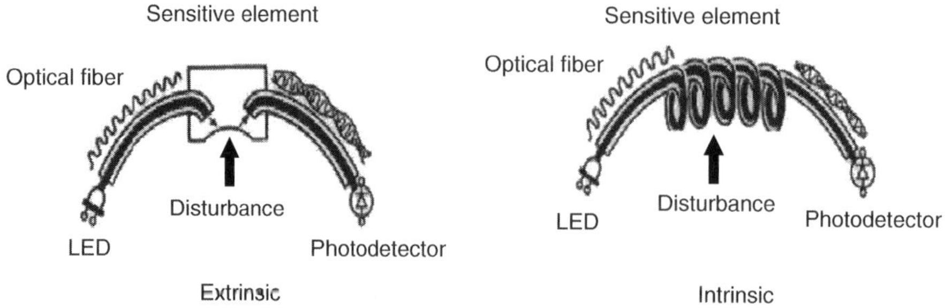

FIGURE 14.5 Extrinsic and intrinsic optical fiber sensors configurations.

TABLE 14.2 Partial List of Applications, Benefits, and Concerns of Optical Fiber Sensors

Applications	Benefits	Concerns
Industrial processes	Compact size and low weight	Moderate to high cost
Industrial control	Geometric versatility and flexibility	Lack of standardization
Biotechnology	EMI/RFI immunity	Impractical modulation units
Energy industry (generation, distribution, exploration, and extraction)	High speed and high bandwidth capacity	Calibration difficulties due to drift
Off-shore oil rigs	Remote sensing capability	Selectivity
Undersea pipelines	Tolerance to extreme environments	Long-term stability, durability, and reliability
Navel vessels	Intrinsic safety	Lack of suitable temperature compensation
Military/aerospace industry	Low maintenance	Accuracy/linearity
Spacecraft and space structures	High sensitivity to multiple quantities	Low demand
Robotic systems	Distributed sensing potential	Complex signal processing
Structural monitoring	Passive effect on measured environment	Potential damage when handling
Aerospace guidance and control	Increased networking and multiplexing capabilities	Isolation from unwanted parameters
Passive/active damping	Potentially inexpensive	Availability of optical sources
Damage localization in civil, mechanical, aerospace structures	Greater sensitivity	Low general awareness
Automotive monitoring industry	Electrical passiveness	Availability of suitable instrumentation
Intelligent transportation systems	Low maintenance and reduced maintenance costs	Existence of competing technologies
Utility infrastructure	Wide dynamic range	Embedding protocol
Environmental monitoring	Insensitivity to corrosion	Bonding robustness
Intelligent building systems	Minimum impact on mechanical properties of host structure	Multiplexing capability
Security systems	Ideally suited to be embedded into composite structures	Technological (e.g., temperature compensation, signal handling, multiplexing, optical signal processing)
Load cells	Potentially high temperature capability	Operational (e.g., in-service applications)
Pressure vessel monitoring	Both point and distributed configuration	Demand (e.g., military continue to be the primary market)

microbending, absorption, and scattering (Figure 14.6) [4,8]. These sensors are simple to construct, are low cost, and do not require complex signal interpretation techniques. However, like all intensity-modulated sensors, the signal can be affected by the light source's fluctuations affecting the accuracy of the measurement. In addition, they are difficult to calibrate and require a fiber optic network to determine the location of the desired measurement (e.g., damage). A classical application of these single use sensors is large area damage detection in structural components [6].

14.1.3.2 Polarimetric Sensors

Polarimetric sensors rely on coherent interference between two light beams from a common source traveling along two different (orthogonal) polarization axes of a common optical fiber. This sensor uses a highly birefringent (HiBi) fiber that divides linearly incident polarized light into the two orthogonal modes traveling along the fiber at different phase velocities. This phase velocity difference prevents the transfer of optical energy between states and thus maintains the polarization state of

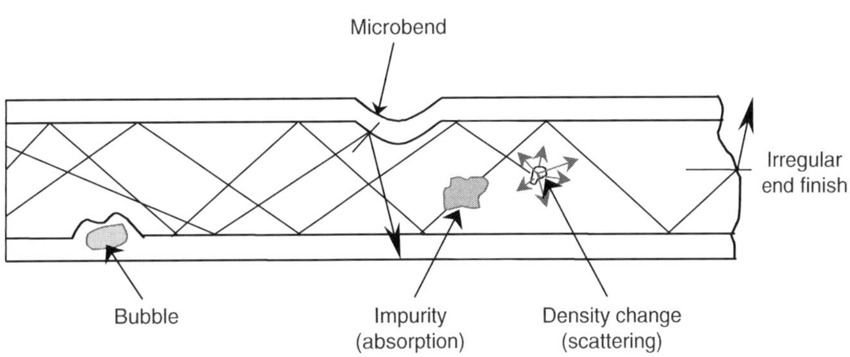

FIGURE 14.6 Schematic of the causes of light intensity attenuation in an optical fiber.

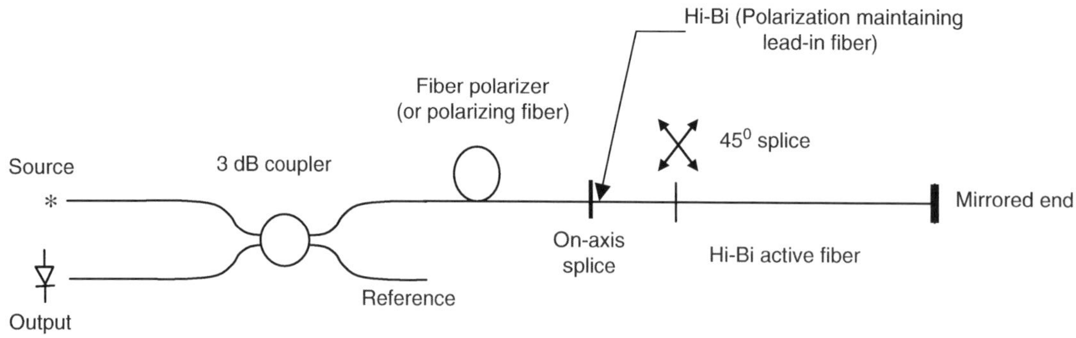

FIGURE 14.7 Schematic configuration of a polarimetric sensor configuration.

light transmitted through the fiber. Figure 14.7 illustrates a schematic configuration of a polarimetric sensor [8]. A transverse loading of the fiber will produce a change in the birefringence that manifests itself as a change in the polarization state at the fiber output. Various experimental arrangements have been developed to convert this polarization state change to a change in optical output power [9–11]. The major disadvantage of this type of sensor is the three-dimensional nature of strain sensitivity in these fibers, which can be difficult to resolve. Various approaches have recently been investigated to address this limitation such as writing gratings or overlaying two gratings with significantly different wavelengths into the birefringent fiber [12].

14.1.3.3 Interferometric Sensors

Interferometric sensors rely on the phase change between two interfering paths of light in response to external stimuli, such as strain, pressure, temperature, and/or chemical reaction. The two paths of light can either come from two different fibers, such as the case for Mach–Zehnder and Michelson sensors, or from a single optical fiber, such as the case in Fabry-Perot and Bragg grating sensors [3]. In a Mach–Zehnder interferometric sensor, the sensing fiber is either embedded or bonded to a host structure while the reference fiber remains external to it, as shown in Figure 14.8 [8]. The necessity of such reference fiber makes the use of such a system impractical for large-scale applications due to the requirement of preserving the state of the reference arm. In a Michelson interferometric sensor, both the sensing and reference fibers are embedded or bonded to the host structure. Both fibers have mirrored ends and the sensing region is formed by the differential path length between the two parallel fibers, as shown in Figure 14.9 [8]. Although this type of sensor possesses high sensitivity and is simple to construct, it suffers from phase preservation problems at the fiber–host interface, greater intrusion to the host structure (due to the fact that the sensing and reference fibers must be closely spaced), and higher susceptibility to noise [3].

Unlike Mach–Zehnder and Michelson sensors, Fabry–Perot interferometric (FPI) sensor only requires a single optical fiber and can be constructed in two configurations: intrinsic (IFPI) and extrinsic (EFPI). As shown in Figure 14.10 [8], the sensing region of the IFPI sensor is formed by two fusion-spliced reflective mirrors in the optical fiber whereas the sensing region of the EFPI sensor is formed using two optical fibers fused to a hollow glass tube by either fusion-splicing or adhesive bonding, forming the sensing cavity (air gap). The mirrored ends of two optical fibers are perpendicular to the fiber axis and separated by an air

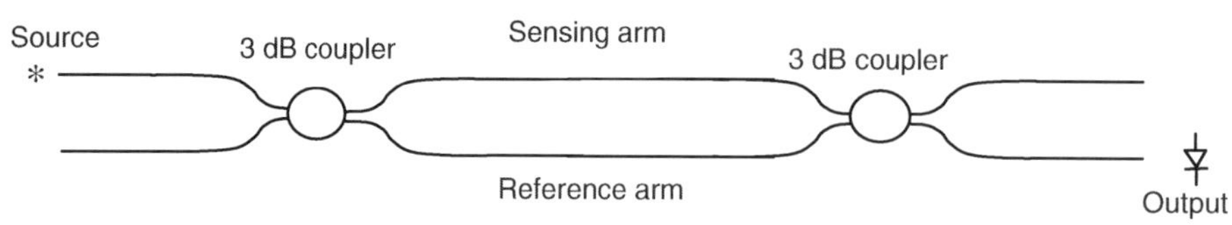

FIGURE 14.8 Schematic configuration of a Mach–Zehnder interferometer.

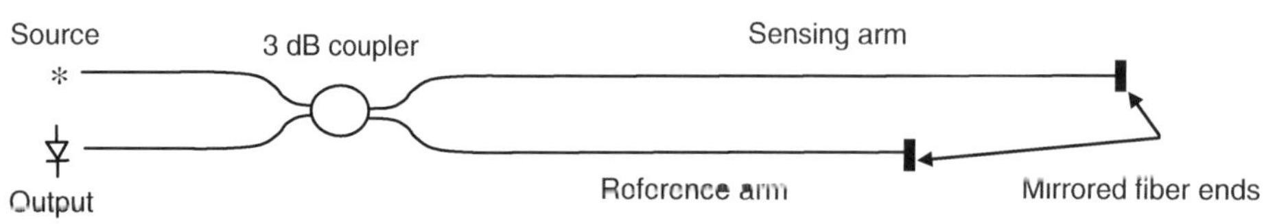

FIGURE 14.9 Schematic configuration of a localized Michelson interferometer configuration.

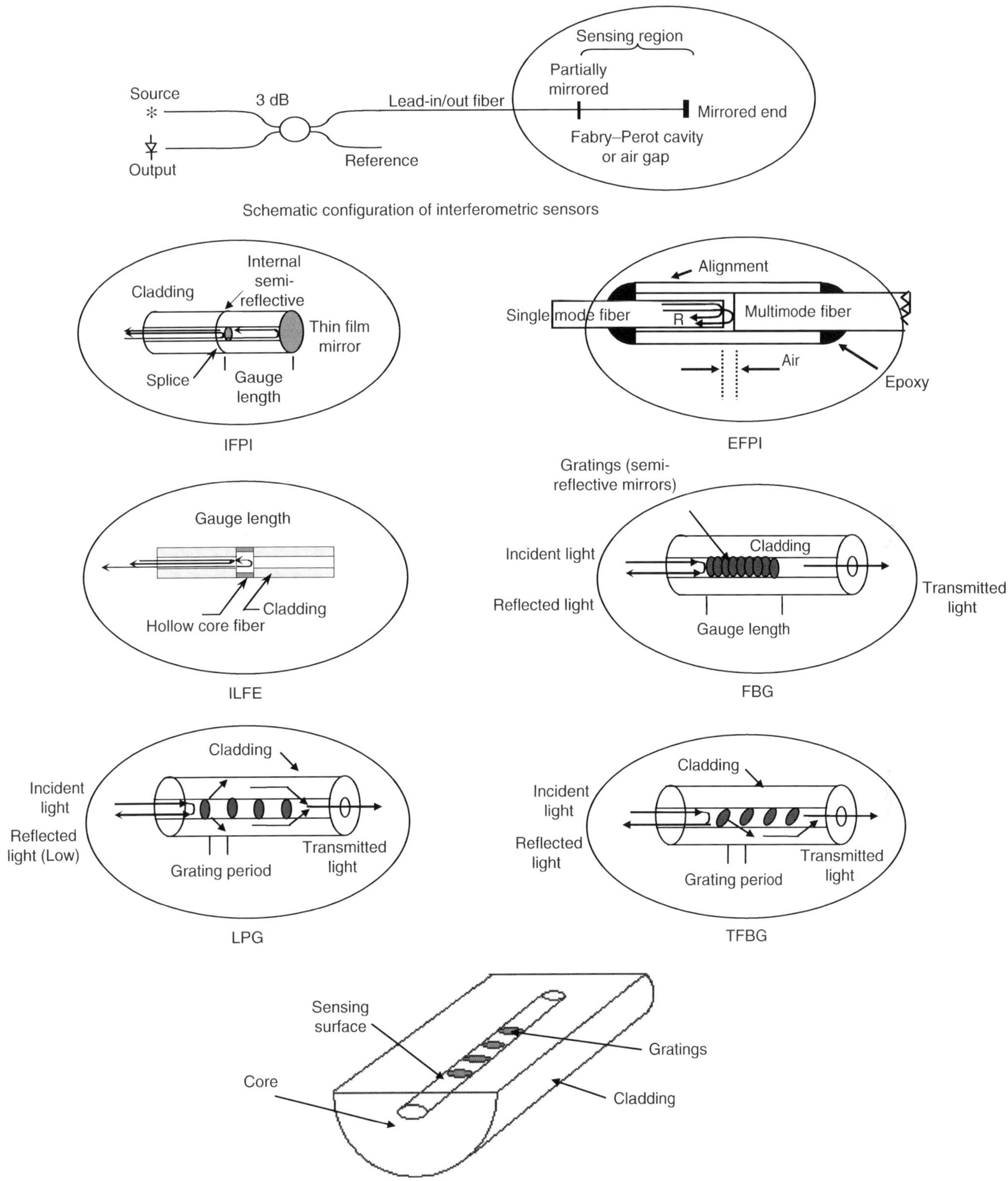

FIGURE 14.10 Schematic configuration of IFPI, EFPI, ILFE, FBG, LPG, TFBG, and D-shaped optical fiber.

gap, forming the EFPI. Any changes in the air gap length of the EFPI or gauge length of the IFPI affect the phase change between the lead-in/lead-out and the reflection fibers, resulting in the appropriate measurement.

It is well known that fusion splicing or adhesive bonding of the optical fibers during the construction of these sensors inadvertently introduce potential weak points at the physical connections for both IFPI and EFPI. In particular, severe stress concentrations exist in EFPI sensors at the alignment tube fusion points. It is also established that the relatively large size of the EFPI may compromise the structural strength and integrity if embedded into host materials. These sensors are known to have reduced multiplexing and networking capability with the need for sophisticated interrogation equipment [3]. To reduce the stress concentration effect, an inline fiber etalon (ILFE) was proposed [13]. This sensor, illustrated in Figure 14.10 [8], provides the same properties as EFPI with a well-defined microsize gauge length (~20 μm), reduced temperature sensitivity, and high transverse strain insensitivity. These added advantages over EFPI come at the expense of increased fabrication difficulties, limited multiplexing capabilities, and reduced technology exposure.

Unlike Fabry Perot and ILFE sensors, the transducer element of a fiber Bragg grating (FBG) sensor is an integral part of the optical fiber, providing this type of sensor with increased networking, multiplexing, and reduced intrusiveness capabilities. The light in the sensing fiber of the FBG is also split into two beams at the beginning of the grating. One light beam is reflected back at the gratings (the semireflective mirrors), while the other light beam traverses the grating through the sensing region only to enter the new gratings as illustrated in Figure 14.11. It is noted that the reflection wavelength is a function of the average refractive index and the grating pitch. Straining the fiber causes a change in the grating pitch and hence the reflected wavelength. This single-mode sensor converts phase changes into spectral information more suitable for measurement. The gratings are formed through periodic modulation of core refractive index using ultraviolet laser or phase mask technology [14,15]. These gratings only reflect a narrow band of wavelengths propagating in the fiber. FBGs show excellent linearity between measured strain and applied stress [16,17]. With appropriate calibration and thermal compensation, accurate strain measurement can be obtained. The increased multiplexing capability using this single-fiber provides distributed measurements over small and large surface areas, through wavelength division multiplexing (WDM) [17,18] or optical frequency domain reflectometry (OFDR) [16,19], making them more suitable for large-scale applications where the weight penalties associated with a large number of single-ended sensors can easily be mitigated. Drawbacks accompanying this type of sensor include the high cost of interrogation system, sensitivity to temperature variations, and limited dynamic response (restricted by the interrogation hardware) amongst others. Nevertheless, the overall benefits of Bragg grating sensors make them the ideal choice amongst the family of fiber optic sensors for use in health monitoring systems for both composite and metallic structures.

14.1.4 Emerging Optical Fiber Concepts

In recent years, several fiber optic sensor concepts have emerged as a result of the increasing interest in online in-situ structural health monitoring and in response to addressing some of the limitations presented by current technologies. These concepts are generally based on the above presented modulation mechanisms with novel approaches for transducer fabrication and interrogation. Among the several concepts that can be found on the open literature and briefly listed in Table 14.3, only three are briefly presented: long period gratings (LPG), tilted fiber Bragg gratings (TFBG), and D-shaped fiber Bragg gratings (DFBG). These sensors are mainly introduced to either enhance the exciting fiber optic sensor capabilities, address current limitation such as temperature compensation, dynamic and range, high multiplexing and distributed sensing, or to increase sensing system sensitivity to different parameters.

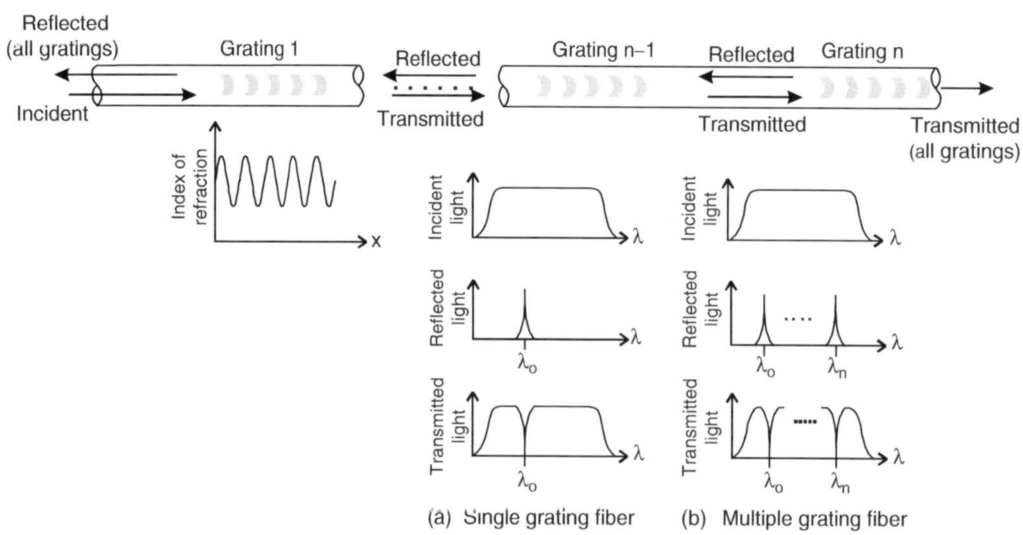

FIGURE 14.11 Schematic configuration of a FBG sensor.

TABLE 14.3 Some Concepts on Optical Fiber Sensor Technology and Application

Concept	Parameter Measurement	Applications
Sagnac interferometer	Rotation	Global positioning systems, electric current sensing
Fluorescent-based fiber sensors	In vivo chemical agent measurements	Chemical/biological applications
White light interferometry	Dispersion of optical materials, displacement, surface profiles	Photonics, material engineering
Magnetic field sensors	AC/DC magnetic fields	Banking (credit card detection) and computer engineering (CD reading)
Semiconductor absorption sensors	Temperature	In harsh conditions (such as radiation, high voltage, and EMI prone areas)
Velocity sensors	In vivo flow	Medical/clinical environments
Ultrasonic/laser NDE sensors	Structural elements and faults/defects	Concrete testing—civil engineering
Long period Bragg gratings (LPGs)	Strain, external refractive index, temperature, bend radius	SHM, civil, chemical engineering, photonics
D-type fiber Bragg gratings (DFBG)	Bending, temp, strain, etc.	Structural health monitoring
Tilted fiber Bragg gratings (TFBG)	Simultaneous temperature and strain, wavelength interrogation, refractive index	Aerospace, civil, manufacturing, photonics
Optical time domain reflectometry (OTDR)	Fiber breaks, loss spectra, strain temperature	SHM and photonics
Polarimetric fiber optic sensors	External perturbations	Civil, mechanical, aerospace, etc.

14.1.4.1 Long Period Bragg Grating Sensors

LPG sensors do not require the fabrication complexity observed with regular FBGs [20] and have very low back reflection, low insertion loss, and exhibit polarization independence [21]. The LPG periodicity is typically above 100 μm, compared to the submicron pitch of regular Bragg gratings. These sensors work in transmission relying on the coupling between the fundamental core mode and the propagating cladding modes (Figure 14.10) [22,23]. This produces a series of interference fringes within the attenuation bands in the transmission spectrum centered at specific wavelengths of the LPG fiber [24]. Changes in the center wavelengths of these bands depend on the grating period/pitch, the length of the sensor and external factors such as temperature, strain, external refractive index, and bend radius. The latter external perturbations inherently modify the physical characteristics of the fiber in terms of its grating pitch and its refractive index of the core and cladding modes [22] and are of significant interest in the field of smart structures, chemical/biological sensing, optical communication, and health monitoring applications. Their exceedingly high sensitivity to transverse as well as longitudinal strain and their acute sensitivity to temperature advocate their potential utilization in applications, requiring multiaxis strain and temperature sensing [22,25]. LPGs can also be customized to largely remove the sensitivity to all but one selected parameter [21,26]. These sensors have the difficulty of allowing localized measurements and embedding into composite structures. Interrogation of the output of an LPG and multiplexing capabilities are also deemed to be complex [27]. Methods have been proposed based on derivative spectroscopy interrogation techniques employed with two pairs of FBGs and LPGs [24].

14.1.4.2 Tilted Fiber Bragg Grating Sensors

As the name suggests, TFBG, unlike traditional gratings, are offset vertically (angular orientation) inside the fiber core. They are produced by rotating the incidence of the interfering laser light used to write the grating from the usual perpendicular orientation. This results in gratings that are inclined or tilted to the fiber axis. The periodicity, and hence the Bragg wavelength of the tilted grating, is greater than that in the untilted configuration. Tilting of the grating is a simple method of increasing the Bragg wavelength without needing extra writing equipment, although it has been shown that the grating reflectivity decreases steadily to zero for tilt angles from 0° to 45° [28]. A tilted Bragg grating can be combined with a regular grating to separate the cross-sensitivity of temperature and strain effects [29,30]. As shown in Figure 14.10, the coupling of light from the fiber core is directed to toward the cladding, which suggests that the narrow attenuation bands in the transmission spectrum are sensitive to the external refractive index [31].

14.1.4.3 DFBG Sensors

These are Bragg gratings, usually of long period, written into D-shaped optical fiber (Figure 14.10). These sensors show strong fiber orientation dependence in their spectral response when subjected to bending [32,33]. DFBGs can form the basis for new vector sensors with the ability to measure curvature with direction recognition. They can also be used as temperature sensors [34]. The prime reason for the usage of the D-shaped fiber sensors in the industry as opposed to standard circular fibers is the ability to interact effectively with the propagating optical field in the area external to the fiber. This is because the core of the DFBG is closer to the outer regions beyond the flattened cladding [35]. The optical field basically allows the interaction between the lightwave and any external perturbations, in other words defining the sensitivity of the sensor to environmental disturbances [36]. A proposed configuration is a hybrid FBG-LPG sensor inscribed in a D-shaped fiber that allows for both temperature

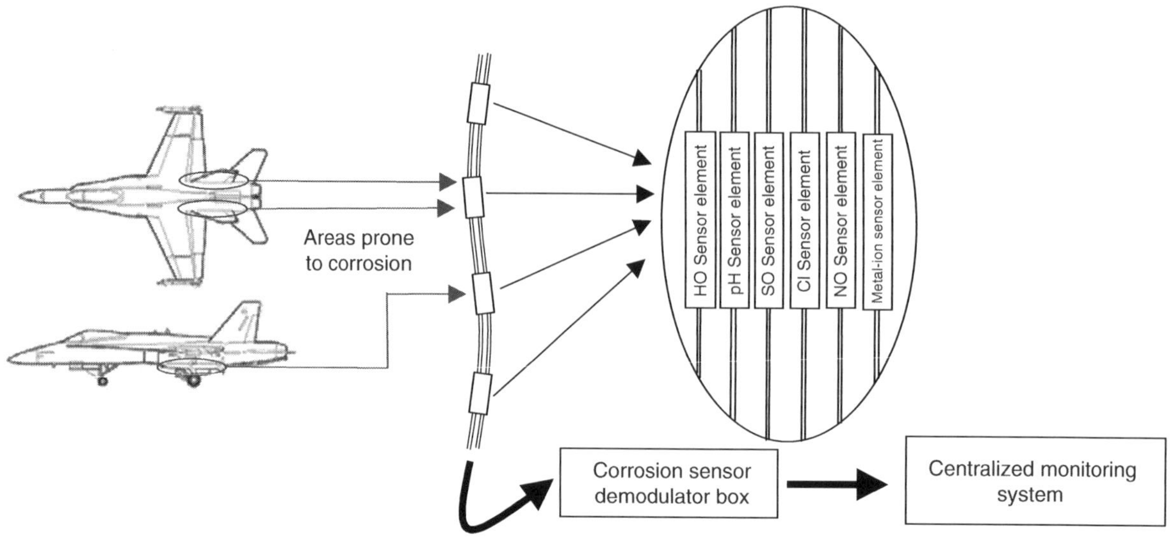

FIGURE 14.12 Illustration of a fighter jet health monitoring system.

and external refractive index sensitivity [37]. One drawback when dealing with D-shaped sensors is that the sensitivity to curvature of these FBGs is as pronounced as their sensitivity to other physical parameters such as axial strain and temperature. One proposition is to join two D-shaped FBGs at their horizontal level planes, thus making it less susceptible to the effects of temperature and axial strain changes [27].

14.1.5 Selected Applications of Fiber Optic Sensors

Optical fiber sensors provide the ability to measure various physical parameters, such as temperature, pressure, strain, displacement, rotation, magnetic/electric field, and corrosion (Table 14.3) [3,7,38–41]. To date, it is known that more than 60 different parameters can be measured using fiber optic sensors. In addition, these sensors can be used in applications where no electrical analog is suitable. When embedded directly into materials, optical fiber sensors can be used to provide in-situ process monitoring [42,43]. In conjunction with nondestructive evaluation schemes, they can also be used to monitor and evaluate not only the integrity of these structures once they have been manufactured, but also to check for flaws that may have resulted from processing and handling. These sensors can further be connected into a health-monitoring network to determine maintenance requirements and operational readiness, thus increasing worthiness, reducing frequent maintenance dependency costs, and providing engineers with an opportunity to improve designs and safety of their respective products. Figure 14.12 illustrates a concept of a potential health monitoring system for a typical fighter jet.

Such concept is increasingly emerging in all sectors of the industry, particularly space and aerospace. In aerospace, sensors are used widely to monitor vital components of aircraft, fighter jets, and rocket engines. Fiber optic sensors are used in the Arnold Engineering Development Center to supervise the air engine and fuel pressure in the combustor of a dual combustor ramjet (DCR) (Figure 14.13 [44]). In this case, a pressure transducer based on reflected light and Fabry Perot interferometry principles are used (Figure 14.14 [44]). A light-emitting diode (LED) light source incident on the fiber is directed toward a diaphragm, which deflects due to pressure fluctuation. The light is reflected back toward a photodiode that measures the intensity of the reflected light, which is determined to be proportional to the deflection experienced by the transducer diaphragm [44]. It is determined that these types of sensors are more difficult to manufacture due to their significant precision requirements [44].

NASA employed fiber optic Bragg grating temperature and strain sensors for the structural health monitoring in the X-38 spacecraft (Figure 14.15). These sensors are either attached to the structure or strain isolated for temperature compensation application depending on the situation of interest [45]. Upon reentry, the aft or rear section of the vehicle experiences a significant amount of structural load (due to drag) and thermal gradient. Eight strain and four temperature sensors experienced temperatures and strains ranging from −40°C to 190°C and −1000 με to 3000 με, respectively [45]. These Bragg grating sensors are determined to be the best alternative to using conventional techniques with strain gauges and thermocouples requiring additional electronic equipment and controls. Previous work on the X33 reusable space vehicle focused on the use of fiber optic sensors for structural health monitoring of the composite fuel tank. Figure 14.16 [46] illustrates a postinstallation check of the instrumentation of a fiber optic sensing system on the composite fuel tank prototype of the X33 reusable space vehicle.

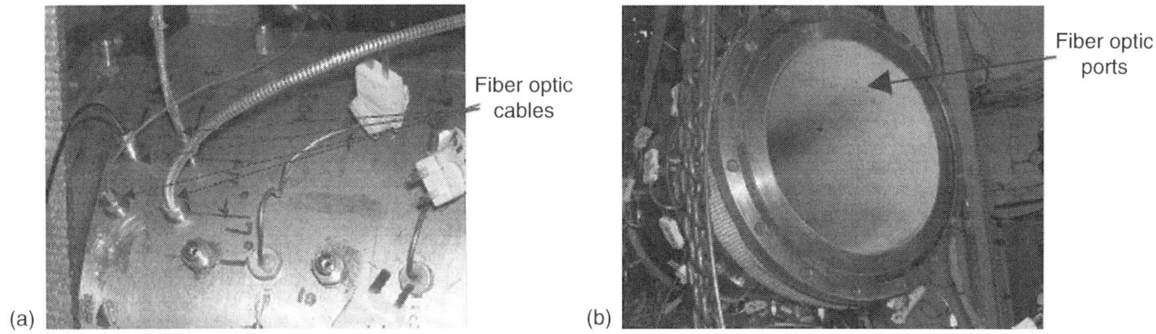

FIGURE 14.13 Installation of fiber optic sensor in a combustor of the DCR.

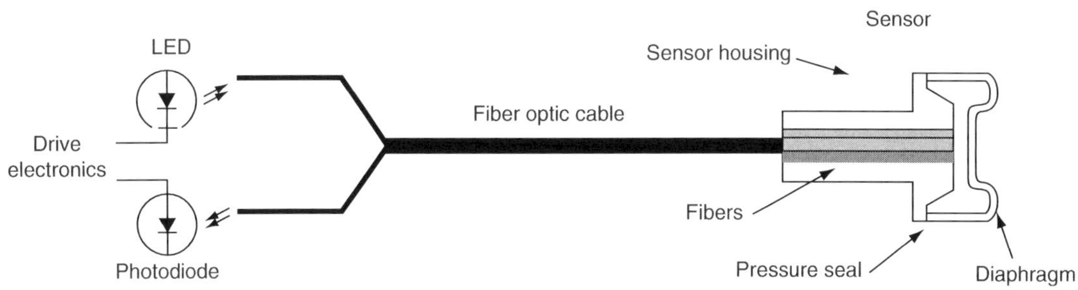

FIGURE 14.14 Configuration of pressure transducer sensor used in the compressor of DCR.

FIGURE 14.15 Illustration of the X-38 reentry vehicle and the sensor targets.

The European Space Agency (ESA) has also vigorously pursued the use fiber optic sensors in various aerospace projects undertaken or in progress [47], such as their use of embedded sensors on their next generation launchers. These rocket launchers are generally exposed to extreme mechanical and thermal stresses because of their operational harsh environments. As part of their reusable launch vehicles (RLV), ESA has conducted assessments on the use of palladium-coated D-shaped FBG sensors in their

FIGURE 14.16 Fiber optic sensors for pressure monitoring in the composite fuel tank of the X-33 reusable space vehicle.

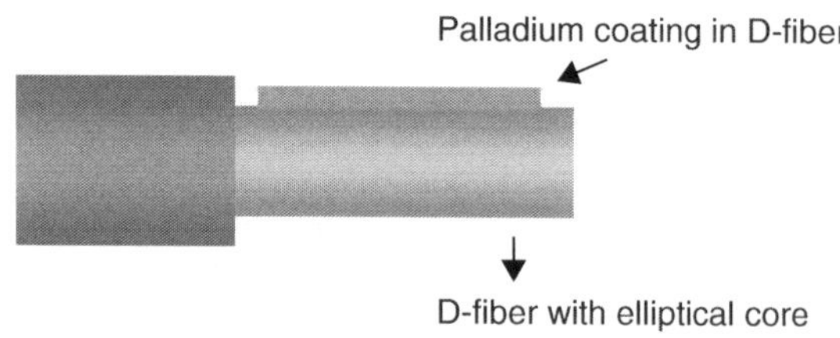

FIGURE 14.17 Palladium-coated fiber Bragg sensor.

cryogenic fuel tanks to monitor the changes in strain, temperature, and any hydrogen leakage that might occur [45,47]. Figure 14.17 illustrates such coated sensor that has strain response ranging from 1000 to 3000 $\mu\varepsilon$ for temperatures as low as −253°C. It is observed that the palladium coating decreased the response time of the sensor at low temperatures (≤30°C).

Recent advent of composite repair technology for aging metallic military aircraft structures has seen the development of "smart patches" with embedded fiber optic strain sensors for the assessment of patch integrity, which detect damage through a measurable loss in load transfer in the region of repair [48–50]. Significant efforts by several defense organizations focused on the exploitation of these sensors for bonded composites patch repair integrity assessment, monitoring, and repair certification. Arrayed fiber optic Bragg grating sensors were bonded to and embedded into a simulated repair to monitor any patch disbond or delamination. As shown in Figure 14.18 [51], the multiplexing flexibility of these sensors provided a suitable less intrusive approach to repairs monitoring. The same figure further illustrated the effectiveness of these sensors in disbond and delamination monitoring for specific repair configuration. The

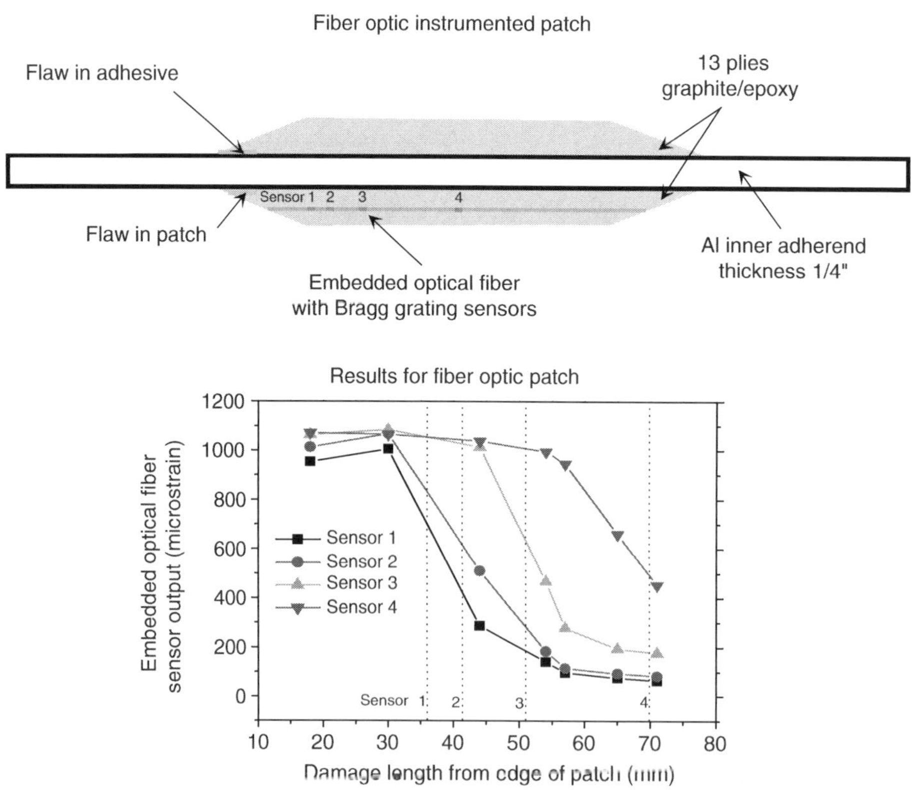

FIGURE 14.18 FBG in bonded repair health monitoring.

FIGURE 14.19 Fiber optic application to dams, power lines, and smart buildings.

environmental issues impacting sensors performance in actual aircraft operating environment remain unaddressed, leading to reduced product commercialization and market penetration. More recently, the health monitoring of composite marine structures using fiber optic sensors is also beginning to attract noticeable research attention [52–54], in addition to the significant research that has expended on the implementation of fiber optic strain sensors for delamination detection and load monitoring in aerospace carbon/epoxy composite structures as well as traditional metallic airframes [16,19,55–58].

In industrial engineering applications, infrastructural health monitoring has significant and detrimental impact on power supply, general health, and medical services. As witnessed by the 1998 ice storm of eastern Canada, 4%–5% of the population, or approximately 140,000 people, were forced to spend at least one night in one of the 454 emergency shelters, which the disaster organization had established due to the highly uninstrumented infrastructure. Distributed or multiplexed fiber optic sensors suitable for large-scale monitoring can be found in mines and tunnels, where pressure, temperature, and strains are monitored for the purpose of structural integrity monitoring and safety enhancement. New generation building structures, "smart buildings," employ distributed fiber optic sensor networks for earthquake monitoring, fire detection, and noise monitoring (Figure 14.19). These sensors are usually interfaced with a control system for structural adaptive control for the purpose of enhancing structural safety and reducing earthquake-induced damage. Distributed, multiplexed, or highly networked fiber optic sensors are seen to advance the highway system and contribute to the intelligent transportation systems (ITS). Sensors are embedded into pavement for the dual purpose of pavement deterioration monitoring and weigh-in-motion (WIM). Figure 14.20 [59] illustrates a schematic of a Bragg grating–based WIN and SHM and control system.

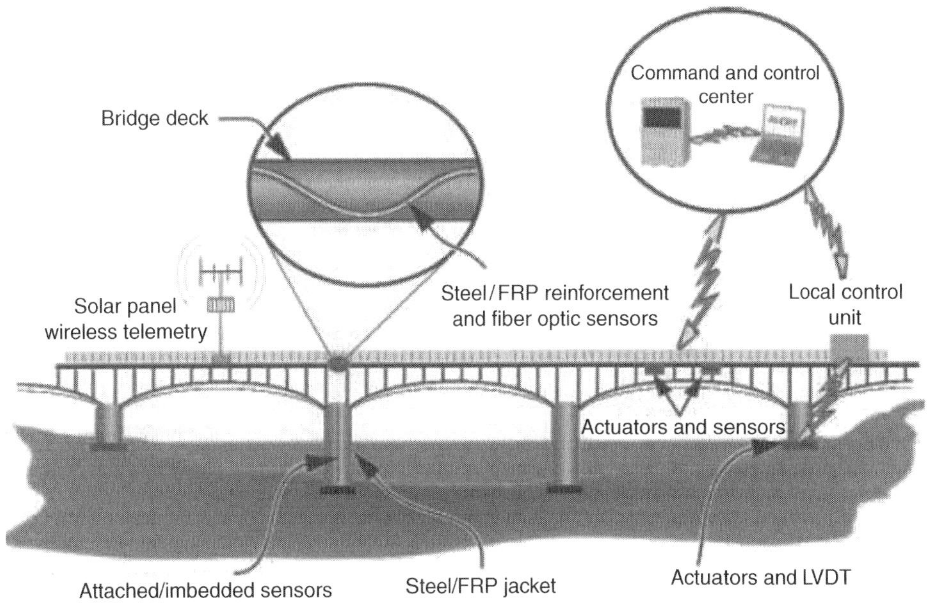

FIGURE 14.20 Fiber optic sensor network for WIM and SHM and control.

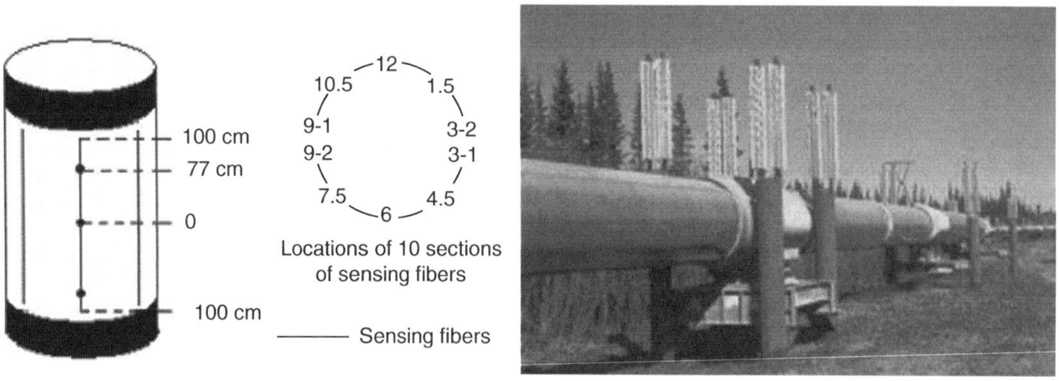

FIGURE 14.21 Location of the sensors around the pipe section.

In the oil and gas industry, there is a growing concern with the maintenance costs associated with repair or remediation of underground pipelines that are susceptible to various faults including dents, leakages, buckling, and corrosion. These faults can significantly compromise the integrity of the pipeline and cause major economic and health disruption. The use of distributed Brillouin fiber optic sensors (DBFS) for environmental and pipeline strain monitoring has been underway for sometime now (see Figure 14.21) [60]. These temperature-compensated strain or pressure sensors have been demonstrated to have high resolution to temperature and strain and provide both measurements in harsh environments.

14.1.6 Conclusions

With the emergence of areas, such as diagnostics, prognostics, health monitoring and management, and smart structures comes the need for advanced sensory systems capabilities. Despite the existence of an extensive list of sensors, advanced features, such as reduced multiplexing electronics, and multisensor data fusion, calls for new and more advanced sensor concepts such as fiber optic sensors. This chapter focused on providing a brief overview of fiber optic sensor technologies and its applications to selected industries. Several long- and short-gauge fiber optic sensors were introduced. These sensors included Mach–Zehnder and Michelson interferometers, polarimetric, interferometric, long and short period Bragg grating sensors, IFPI and EFPI sensors, ILFE sensors, among others. The benefits, applications, and concerns associated with several of these sensors were also highlighted.

Long gauge fiber optic sensors are mainly employed in civil infrastructure applications, such as bridges, pipelines, power lines, dams, skyscrapers, mines, and underground excavation. Although these interferometric and Brillouin scattering-based sensors exhibit the potential for use in large areas, structural applications for simultaneous strain/pressure and temperature measurements, they lack the spatial resolution required from some applications like aerospace. These sensors also suffer from the elevated cost associated with their manufacturing, operation, and signal demodulation.

Short gauge length sensors provide potential solutions to challenges in the aerospace field and continue to provide technological advances for the development of smart structures and their applications. Bragg grating and EFPI optical fiber sensors were found to be the most promising and relatively mature technology for integration into smart structures and health monitoring and management. These sensors provide high resolution, improved accuracy, high sensitivity and millimeter-size gauge lengths suitable for confined environments. Bragg grating sensors continue to be the undisputed champion because of their superior multiplexing capabilities; however, they still suffer from temperature and strain cross-sensitivity.

Regardless of the advances in fiber optic sensor technology, illustrated by the applications presented in this document and others, issues related to advances in optical signal processing, development of specialty optical fiber, system integration, and system economics need to be addressed as these form the nervous system of structures from the stages of its inception to its retirement. Current efforts are underway in introducing new designs, such as tilted Bragg grating, dual laid LPG, D-shaped Bragg grating, and integrating different sensing mechanisms to render the sensing systems more efficient and cost-effective.

References

1. Business Communications Company, Inc. http://www.bccresearch.com/ias/IAS002C.asp
2. Krohn, D.A., Light wave venture LLC, in *Proceedings of SPIE*, 2004, 5589, 34.
3. Zhou, G. and Sim, L.M., Damage detection and assessment in fibre-reinforced composite structures with embedded fibre optic sensors—review, *Smart Materials and Structures*, 11, 925, 2002.

4. Krohn, D.A., Fiber optic sensors: Fundamentals and applications, *Instrument Society of America*, 138, 1988.
5. Hauptmann, P., *Sensors: Principles and Applications*, Prentice Hall, Englewood Cliffs, NJ, 1993.
6. Rippert, L., Wevers, M., and Van Huffel, S., Optical and acoustic damage detection in laminated CFRP composite materials, *Composites Science and Technology*, 60(14), 2713, 2000.
7. Badcock, R.A. and Fernando, G.F., An intensity-based optical fibre sensor for fatigue damage detection in advanced fibre-reinforced composites, *Smart Materials and Structures*, 4(4), 223, 1995.
8. Mrad, N., Optical fiber sensor technology: Introduction and evaluation and application, in *Encyclopedia of Smart Materials*, Schwartz, M., (ed.), John Wiley and Sons, Inc., New York, 2, 715, 2002.
9. Murukeshan, V.M., Chan, P.Y., Ong, L.S., and Asundi, A., On-line health monitoring of smart composite structures using fiber polarimetric sensor, *Smart Materials and Structures*, 8, 544, 1999.
10. Grossmann, B.G. and Huang, L.-T., Fiber optic sensor array for multi-dimensional strain measurement, *Smart Materials and Structures*, 7(2), 159, 1998.
11. Lo, Y.L., Sirkis, J.S., and Ritchie, K.T., A study of the optomechanical response of a diametrically loaded high-birefringent optical fiber, *Smart Materials and Structures*, 4(4), 327, 1995.
12. Udd, E., Nelson, D.V., Lawrence, C.M., and Ferguson, B.A., Three axis strain plus temperature fiber grating sensor, in *Proceedings of SPIE*, 2718, 1996.
13. Sirkis, J.S., Fiber optic sensors for smart structures: Basics and applications, in *Proceedings of SPIE's Fifth Annual International Symposium on Smart Structures and Materials*, San Diego, CA, 1–5 March 1998.
14. Wosinski, L., Dainese, M., Swillo, M., UV-photosensitivity and refractive index engineering in silica-based planar lightwave circuits, in *Proceedings of 2005 Seventh International Conference on Transparent Optical Networks*, Barcelona, Spain, 3–7 July 2005, 1, 343.
15. Rizvi, N.H., Gower, M.C., Goodall, F.C., Arthur, G., and Herman, P., Excimer laser writing of submicrometre period fibre Bragg gratings using phase-shifting mask projection, *Electronics Letters*, 37(11), 901, 1995.
16. Brown, T., Wood, K., Childers, B., Cano, R., Jensen, B., and Rogowski, R., Fiber optic sensors for health monitoring of morphing aircraft, in *Proceedings of SPIE Sixth Annual International Symposium on Smart Structures and Materials*, Newport Beach, USA, 1999, 3674, 60.
17. Simpson, J.O., Wise, S.A., Bryant, R.G., Cano, R.J., Gates, T.S., Hinkley, J.A., Rogowski, R.S., and Whitley, K.S., Innovative materials for aircraft morphing, in *Proceedings of SPIE'S Fifth Annual International Symposium on Smart Structures and Materials*, San Diego, CA, 1998, 3326, 240.
18. Anastasi, R.F. and Lopatin, C., Application of a fiber optic distributed strain sensor system to woven E-Glass composite, NASA TM-2001-211051, 2001.
19. Prosser, W.H., Allison, S.G., Woodard, S.E., Wincheski, R.A., Cooper, E.G., Price, D.C., Hedley, M., Prokopenko, M., Scott, D.A., Tessler, A., and Spangler, J.L., Structural health management for future aerospace vehicles, in *Proceedings of the Second Australasian Workshop on Structural Health Monitoring*, Monash University, Melbourne, Australia, 16–17 December, 2004, p. 1.
20. Kolokoltsev, O.V., Svyryd V.A., Llamas I.F., and Romero C.L.O., Synthesis of nonuniform long-period optical gratings with a continuous refractive index profile, Investigation, Revista Mexicana De F'Isica, 51(6), 610, December 2005.
21. Wang, Y.P., Rao, Y.J., Ran, Z.L., Zhu, T., and Zeng, X.K., Bend-insensitive long-period fiber grating sensors, *Optics and Lasers in Engineering*, 41, 233, 2004.
22. James, S.W. and Tatam, R.P., Optical fibre long-period grating sensors: Characteristics and application, *Measurement Science and Technology*, 14, R49, 2003.
23. Laffont, G. and Ferdinand, P., Tilted short-period fibre-Bragg-grating induced coupling to cladding modes for accurate refractometry, *Measurement Science and Technology*, 12, 765, 2001.
24. Allsop, T., Earthrowl, T., Reeves, R., Webb, D.J., and Bennion, I., The interrogation and multiplexing of long period grating curvature sensors using a Bragg grating based, derivative spectroscopy technique, *Measurement Science and Technology*, 15, 44, 2004.
25. Liu, Y., Zhang, L., and Bennion I., Fibre optic load sensors with high transverse strain sensitivity based on long-period gratings in B/Ge co-doped fibre, *Electronics Letters*, 35(8), 661, 1999.
26. Chen, S., Tong, Z., Zhao, Q., Liu, Z., and Dong, X., A smart bending sensor with a novel temperature and strain-insensitive long-period grating, *Sensors and Actuators A*, 116, 103, 2004.
27. Araujo, F.M., Ferreira, L.A., Santos, J.L., and Farahi, F., Temperature and strain insensitive bending measurements with d-type fibre Bragg gratings, *Measurement Science and Technology*, 12, 829, 2001.
28. Erdogan, T. and Sipe, J.E., Tilted fiber phase gratings, *Journal of Optical Society of America A*, 13, 296, 1996.
29. Kang, S.C., Kim, S.Y., Lee, S.B., Kwon, S.W., Choi, S.S., and Lee, B., Temperature-independent strain sensor system using a tilted fiber Bragg grating demodulator, *IEEE Photonics Technology Letters*, 10(10), 1461, 1998.
30. Caucheteur, C., Lhomme, F., Maakaroun, F., Chah, K., Blondel, M., and Megret, P., Simultaneous strain and temperature sensor using superimposed tilted Bragg gratings, in *Proceedings of IEEE/LEOS Symposium*, Benelex Chapter, Ghent, Belgium, 2004, p. 219.

31. Caucheteur, C. and Mégret, P., Demodulation technique for weakly tilted fiber Bragg grating refractometer, *IEEE Photonics Technology Letters*, 17(12), 2703, 2005.
32. Allsop, T., Dobb, H., Mezentsev, V., Earthgrowl, T., Gillooly, A., Webb, D.J., and Bennion, I., The spectral sensitivity of long period gratings fabricated in elliptical core D-shaped optical fibre, *Optics Communications*, 259, 537, 2006.
33. Zhao, D., Zhou, K., Chen, X., Zhang, L., Bennion, I., Flockhart, G., MacPherson, W.N., Barton, J.S., and Jones, J.D.C., Implementation of vectorial bend sensors using long-period gratings UV-inscribed in special shape fibres, *Measurement Science and Technology,* 15, 1647, 2004.
34. Smith, K.H., Ipson, B.L., Lowder, T.L., Hawkins, A.R., Selfridge, R.H., and Schultz, S.M., Surface-relief fiber Bragg gratings for sensing applications, *Applied Optics*, 45, 1669, 2006.
35. Allsop, T., Gillooly, A., Mezentsev, V., Earthgrowl-Gould, T., Neal, R., Webb, D.J., and Bennion, I., Bending and orientational characteristics of long period gratings written in D-shaped optical fiber, *IEEE Transactions on Instrumentation and Measurement*, 53(1), 130, 2004.
36. Ghandehari, M., Materials health monitoring in smart cable structures, CANSMART 2006, in *Proceedings of the International Workshop on Smart Materials and Structures*, Toronto, Ontario, Canada, 2006, p. 213.
37. Chen, X., Zhou, K., Zhang, L., and Bennion, I., Simultaneous measurement of temperature and external refractive index by use of a hybrid grating in D fiber With enhanced sensitivity by HF etching, *Applied Optics*, 44(2), 178, 2005.
38. Morey, W.W., Meltz, G., and Weiss, J.M., Evaluation of a fiber Bragg grating hydrostatic pressure sensor, in *Proceedings of the 8th International Conference on Optical Fiber Sensors*, Monterey Marriott, Monterey, CA, January 29–31, 1992.
39. Kanellopoulos, S.E., Handerek, V.A., and Rogers A.J., Simultaneous strain and temperature sensing with photogenerated in-fiber gratings, *Optics Letters*, 20(3), 333, 1995.
40. Cavaleiro, P.M., Araujo, F.M., and Lobo Ribeiro, A.B., Metal-coated fibre Bragg grating sensors for electric current metering, *Electronics Letters*, 23(11), 1133, 1998.
41. Luo, F., Liu, J., and Chen, S., Fiber optic distributed sensing scheme for monitoring structural strain and deformation, *Optical Engineering*, 36(5), 1548, 1997.
42. Lawrence, C.M., Nelson, D.V., Spingarn, J.R., and Bennett, T.E., Measurement of process-induced strains in composite materials using embedded fiber optic sensors, in *Proceedings of the SPIE International Society for Optical Engineering, Fiber Optic Smart Structures and Skins*, 1996, 2718, 60.
43. Levy, R.L. and Schwab, S.D., Monitoring the composite curing process with a fluorescence-based fiber-optic sensor, *Polymer Composites*, 12(2), 96, 1991.
44. Land, H.B., III and Eddins, C.L., Optical pressure measurement: Using fiber optic transducers in hypersonic flight vehicles, *IEEE Instrumentation & Measurement Magazine*, 7(3), 38, 2004.
45. Ecke, W., Latka, I., Willsch, R., Reutlinger, A., and Graue, R., Fibre optic sensor network for spacecraft health monitoring, *Measurement Science and Technology*, 12, 974–980, 2001.
46. Troy, C.T., Fiber optic smart structures: A technology ahead of its time is finally winning acceptance, *Photonics Spectra*, 31(5), 112, 1997.
47. Mckenzie, I. and Karafolas, N., Fiber optic sensing in space structures: The experience of the European space agency, in *Proceedings of SPIE 17th International Conference on Optic Fibre Sensors*, Voet, M., Willsch, R., Ecke, W., Jones, J., and Culshaw, B. (eds.), 2005, 5855, p. 262.
48. Davis, C., Baker, W., Moss, S., Galea, S.C., and Jones, R., In situ health monitoring of bonded composite repairs using a novel fiber Bragg grating sensing arrangement, in *Proceedings of SPIE Smart Materials II*, 2002, 4934, p. 140.
49. Budtsev, Y., Gorbatov, N., Tur, M., Green, A.K., Kressel, I., Ben-Simon, U., Ghilai, G., Sharfir, E., Berkovic, G., and Gali, S., Smart bonded composite repairs for aging aircraft, in *Proceedings of the Second European Workshop on Structural Health and Monitoring, Munich*, Germany, 7–9 July, 2004, p. 341.
50. Li, H.C.H., Dupouy, O., Herszberg, I., Stoddart, P.R., Davis, C.E., and Mouritz, A.P., Health monitoring of bonded composite repairs using fibre optic sensors, in *Proceedings of SPIE, Smart Structural Materials and NDE for Health Monitoring and Diagnosis*, 2006, 6174, p. 367.
51. Mrad, N., Progress towards fibre optic smart structures, The Department of National Defence, Defence R&D Canada, Technical Memorandum, Publication No. DRDC Atlantic TM 2003-200, 2004.
52. Li, H.C.H., Davis, C., Herszberg, I., Mouritz, A.P., Galea, S.C., and Thomson, R.S., Application of fibre optic strain sensors for the health monitoring of adhesively bonded composite ship joints, in *Proceedings of the Fourth International Workshop on Structural Health and Monitoring*, Stanford University, CA, 2003, 1327.
53. Wang, G., Pran, K., Sagvolden, G., Havsgard, G. B., Jensen, A.E., Johnson, G.A., and Vohra, S.T., Ship hull structure monitoring using fibre optic sensors, *Smart Material and Structure*, 10, 472, 2001.
54. Davies, H.L., Everall, L.A., and Gallon, A.M., Application of smart optical fiber sensors for structural load monitoring, in *Proceedings of SPIE Smart Structure and Materials, Industrial and Commercial Applications of Smart Structures Technologies*, 2001, 4332, p. 114.
55. Udd, E., Schulz, W.L., Seim, J.M., Trego, A., Haugse, E., and Johnson, P.E., Use of transversely loaded fiber grating strain sensors for aerospace applications, in *Proceedings of SPIE*, 2000, 3994, 96.

56. Okabe, Y., Mizutani, T., Yashiro, S., and Takeda, N., Detection of microscopic damage in composite laminates with embedded small-diameter fiber Bragg grating sensors, *Composite Science and Technology*, 62, 951, 2002.
57. Takeda, N., Okabe, Y., Kuwahara, J., Kojima, S., and Ogisu, T., Development of smart composite structures with small-diameter fiber Bragg grating sensors for damage detection: Qualitative evaluation of delamination length in CFRP laminates using lamb wave sensing, *Composites Science and Technology*, 65, 2575, 2005.
58. Guemes, A., Pintado, J.M., Frovel, M., and Menendez, J.M., FBG based SHM systems in monolithic composite structures for aerospace, in *Proceedings of the Second European Workshop on Structural Health and Monitoring*, Munich, Germany, 7–9 July, 2004, 870.
59. Akhras, G., Smart materials and smart systems for the future, *Can. Military J.*, 25, 2000.
60. Zou, L., Bao, X., Ravet, R., and Chen, L., Distributed Brillouin fiber sensor for detecting pipeline buckling in an energy pipe under internal pressure, *Applied Optics*, 45(14), 3372, 2006.

14.2 Smart Windows

John Bell

14.2.1 Outline

Smart windows encompass a wide range of technologies. The common feature of all smart window technologies is the ability to switch the optical transmittance and reflectance of the window. While this characteristic is of interest for a range of applications, including optical computing, display devices, and architectural glazing, the principal application considered here for smart windows is architectural glazing. This application places a number of constraints on the technology under consideration, including two particularly important constraints: large area (up to 2 × 2 m); and long lifetime (in excess of 10^4 switching cycles for up to 30 years).

This chapter only considers technologies that target this application, although many of the technologies discussed also have applications in other areas, as outlined in the major recent reviews by Granqvist et al. [1]. The chapter also only considers technologies that are actively under development for architectural glazing applications. As a result, some technologies, such as photochromic materials, are not discussed despite significant research in the 1980s.

An introduction to smart windows is given in Section 14.2.2, followed by a discussion of the physics of building windows in Section 14.2.3, which is necessary to understand the differences between the types of smart windows and smart window technologies, which are discussed in Section 14.2.4. This chapter does not attempt to encapsulate all the information regarding electrochromic materials, but focuses on the key principles for smart window applications, and some of the key areas of critical importance for these applications. Detailed recent reviews of electrochromic materials can be found elsewhere [1–3].

Inorganic electrochromic smart windows have been studied for longer than most other systems, and still appear to be the most likely candidates for commercial development in the near future. They have many advantages over other technologies, either in the energy benefits, which can result from their use, the ease of control of the devices, and the ability to independently control the devices for a range of different environmental conditions. They are also specularly transmitting in all states. Consequently, Section 14.2.5 is devoted to a more detailed discussion of the materials, device structures, and control methods for electrochromic smart windows. Section 14.2.6 gives a brief consideration of future trends in smart window systems.

14.2.2 Introduction

The term “smart window” was introduced in the mid-1980s by Professor Claes Granqvist to describe optically switchable electrochromic glazings. These devices exemplify the fundamental characteristic of all “smart windows”: controllable variation in the optical transmittance of the window. In the case of electrochromic smart windows, the variation in optical transmittance occurs through the double injection of electrons and ions (usually H^+ or Li^+ ions) into an electrochromic material such as WO_3. In WO_3, this leads to the development of either a broad absorption band or a reflection edge (depending on the details of the material preparation), leading to a low transmittance state. The process is (ideally) reversible, with the extraction of the ions and electrons returning the material to a transparent state. The control of the process is accomplished by the application of a small voltage (or the passing of a small current), which controls the ion injection and extraction process. Since the initial discovery of electrochromism in thin film WO_3 by Deb et al. [4] in 1969, there has been an enormous amount of research directed toward the goal of large area switchable windows for architectural applications. This initially focused on electrochromic systems, covering a wide range of materials and device structures. However during the 1980s and subsequently, several alternate optical switching systems have been developed, which fall into the general category of smart windows. These include other electrically activated systems such as suspended particle devices [5] and phase-dispersed liquid crystals [6], temperature-controlled switchable devices such as thermochromic [7] and thermotropic devices [8], and recently developed gasochromic devices, which are controlled by the use of reducing or oxidizing gas mixtures in a window unit [8]. Recent research in the past 5 years has explored a vast range of new materials for electrochromism, including many organic materials, used in both devices [9–11], and exploring basic electrochromic properties of new polymer systems (see, e.g., Refs. [3,12–14], and further references in Ref. [1]).

14.2.3 Architectural Glazing Applications for Smart Windows

The potential reductions in heating, cooling, and lighting energy use, which can result from the use of switchable glazings in buildings, has provided the impetus for the majority of this research. There have been over 700 U.S. patents lodged in this field since 1976. It is now well established that reductions in energy consumption of up to 50% are possible with the use of electrochromic glazings in commercial buildings, with savings of 20%–30% obtainable in most climatic conditions. Savings are highest in hot climates such as the Middle East (up to 49% energy reductions [15]), Australasia, and the tropics [16,17], and southern Europe [18]. However the energy performance is very dependent on climate and building type, as well as the properties of the electrochromic glazing. Two recent studies conclude that use of electrochromic smart windows is not justifiable on economic grounds in Hong Kong [19] as only 6.6% energy reduction could be achieved in a high-rise apartment building.

Windows are also the primary application for thermochromic and thermotropic windows, which change transmittance as a function of temperature, and can therefore reduce transmittance of infrared radiation (i.e., heat) when the temperature is high. Studies of thermotropic smart windows also show that significant energy savings can be obtained in appropriate climates [20]. However, some of the other "smart" windows, such as the polymer-dispersed liquid crystal (PDLC) devices, do not provide significant energy savings, but are used as privacy screens. In order to understand the role of switching in a smart window, it is necessary to understand the nature of glazings and how the optical properties of a window affect its performance and utility.

14.2.3.1 Physics of Windows

Common to all applications of glazing materials is the need for transmission of light, and in most applications, the reflectivity is also very important. In order to quantify the light transmission properties, the transmittance and reflectance spectra, $T(\lambda)$ and $R(\lambda)$ can be measured and used to define several different average transmittance and reflectance quantities. The most important of these are the solar and visible transmittance and reflectance:

$$T_{\text{sol}} = \frac{\int_0^\infty T(\lambda)\varphi_{\text{sol}}(\lambda)\,d\lambda}{\int_0^\infty \varphi_{\text{sol}}(\lambda)\,d\lambda}; \quad R_{\text{sol}} = \frac{\int_0^\infty R(\lambda)\varphi_{\text{sol}}(\lambda)\,d\lambda}{\int_0^\infty \varphi_{\text{sol}}(\lambda)\,d\lambda} \tag{14.1}$$

$$T_{\text{vis}} = \frac{\int_{370}^{770} T(\lambda)\varphi_{\text{vis}}(\lambda)\,d\lambda}{\int_{370}^{770} \varphi_{\text{vis}}(\lambda)\,d\lambda}; \quad R_{\text{vis}} = \frac{\int_{370}^{770} R(\lambda)\varphi_{\text{vis}}(\lambda)\,d\lambda}{\int_{370}^{770} \varphi_{\text{vis}}(\lambda)\,d\lambda} \tag{14.2}$$

where φ_{sol} and φ_{vis} are the solar spectrum and visible response of the human eye, respectively. Usually the air mass 1.5 solar spectrum is used to define the solar averages, although it is not necessarily the most appropriate at all locations, and there can be significant differences between different solar spectra [21]. The difference in the spectral ranges for φ_{sol}, φ_{vis}, and φ_{th},

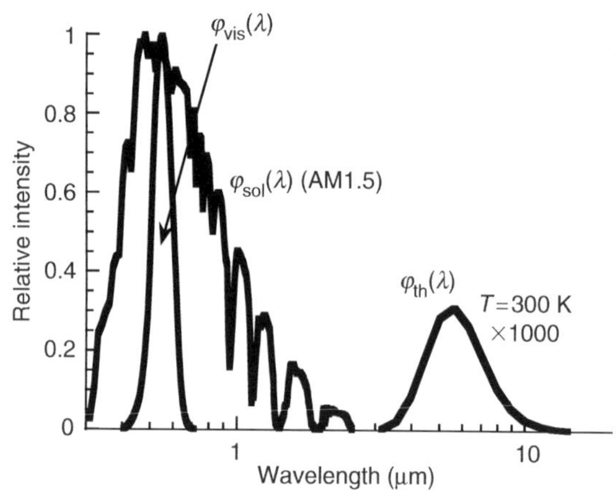

FIGURE 14.22 Three components related to the ambient radiation environment, which need to be considered for design of windows. The three distinct spectral regions correspond to the wavelengths $0.3 < \lambda < 2.5\,\mu m$ (solar radiation $\varphi_{\text{sol}}(\lambda)$), $\lambda > 3\,\mu m$ (thermal IR), shown here as $\varphi_{\text{th}}(\lambda)$ for $T = 300\,K$, and $0.37 < \lambda < 0.77\,\mu m$ corresponding to the visible response of the human eye, $\varphi_{\text{vis}}(\lambda)$.

illustrated in Figure 14.22, immediately gives rise to the concept of spectral selectivity, which is central to many advanced window glazing systems. This refers to the ability of a window to transmit, for example, visible radiation (high T_{vis}), but to reflect heat (low T_{sol} and high R_{sol}). This is fully explained in Ref. [22]. The selectivity $T_{\text{vis}}/T_{\text{sol}}$, which has a maximum value of approximately 2, can therefore be used as a figure-of-merit for glazings. Another important optical quantity is the solar absorption, $A_{\text{sol}} = 1 - T_{\text{sol}} - R_{\text{sol}}$. A window with high absorption of solar radiation will heat up considerably, and this heat can be transmitted into the building, negating the reduction in insolation, which results from the low solar transmittance [23].

While these are the basic quantities of interest, there are several derived quantities, such as the total solar energy transmittance (TSET) [21], which are commonly used to classify window glazings. In the case of switchable glazings, the key parameters of interest are the change in the visible and solar transmittance between the two states of the window. The values usually quoted are the normal–normal transmittance and reflectance, i.e., the fraction of an incident beam normally incident on the window, which is specularly transmitted. The spectral response typical of different types of smart windows is illustrated in Figure 14.23. While most windows are designed to have low diffuse scattering, many of the smart windows available or in development switch between a specularly transmitting and reflecting state, and a diffusely transmitting and reflecting state. In this case the normal–hemispherical or sometimes the hemispherical–hemispherical transmittance and reflectance are quoted.

Even for the purposes of minimizing energy use in buildings, there is no single figure-of-merit for smart windows, which can be used to identify the optimum smart window. In general, the important characteristics, and those quoted in this work where

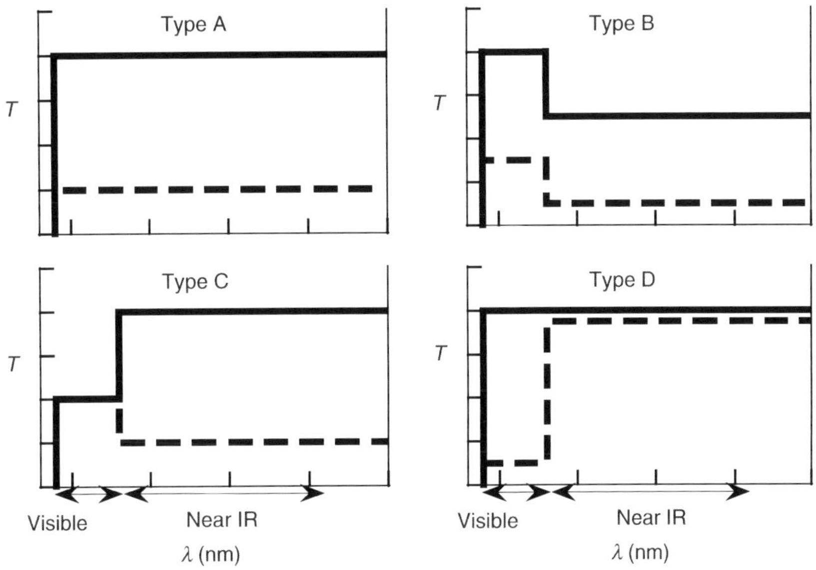

FIGURE 14.23 Schematic illustration of the different types of switching of transmitted radiation, which are possible with different switchable window systems.

they are available, are the selectivity, defined above, and the contrast ratio, or amount of switching, in both the visible and solar spectral ranges: $T_{vis}^{clear}/T_{vis}^{colored}$ and $T_{sol}^{clear}/T_{sol}^{colored}$ respectively. The contrast ratio is frequently quoted as the change in optical density, $\Delta OD = \log_e T_{clear}/T_{colored}$. A complete description of the window behavior involves the variation of the selectivity or contrast a function of the switching parameter, as is schematically illustrated for several different devices in Figure 14.24. Some authors prefer to use the *g*-factor or TSET rather than the solar transmittance, as this include the effects of thermal conduction through the window and secondary heat gains which arise from absorption in the glazing [21]. A good single number which is frequently used to rate a smart window is the ratio of the maximum visible transmittance to the minimum value of TSET ($T_{vis}^{clear}/T_{vis}^{colored}$). Another approach to defining the performance of smart windows in buildings is the energy cost for a building when different types of windows are used. Selkowitz et al. [24] introduced this concept through a graph, which compares the cooling and lighting costs (the costs most affected by advanced glazing in commercial buildings) for different types of switchable glazings.

There are other important aspects of windows and the materials used to make windows, such as durability [25], cost [26], color [27] and haze, or the amount of radiation that is diffusely transmitted [28].

14.2.4 Survey of Smart Windows

The principal smart window systems are summarized in Table 14.4. This table shows the switching mechanism, the method by which this mechanism is operated, whether the device exhibits memory and the type of spectral response (selectivity and contrast) that occurs. The table also includes a control parameter, which in most cases is the same as the operational switching parameter. However, in the context used here, the control parameter is one which can be measured and provides a degree of control over the optical state of the device. This is extremely important for practical

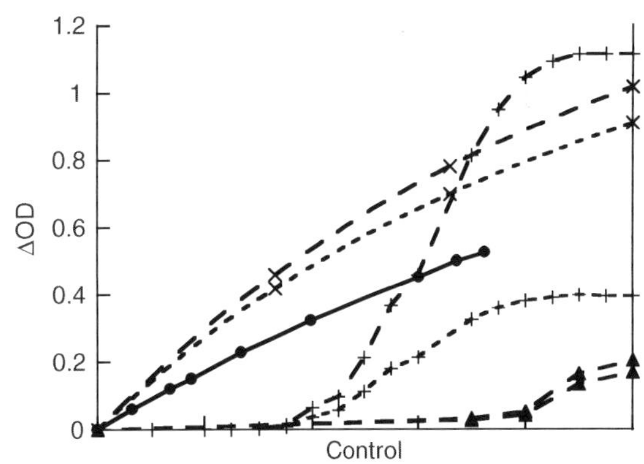

FIGURE 14.24 Variation in optical density for four differently electrically activated smart window devices. (●) A sol–gel deposited device manufactured by Sustainable Technologies Australia Ltd, and based on WO_3; (×) a sputtered device manufactured by Asahi Glass Co, and based on a WO_3-NiO complementary device (see Section 14.2.4.2); (+) an organic electrochromic from Gentex Corporation, based on viologens; (▲) a PDLC device manufactured by 3M Corporation. The different curves represent different spectra used in calculating the optical density: —— a broadband transmittance; - - - the visible transmittance T_{vis}; and — — — the solar transmittance T_{sol}. The divergence in the optical density for visible and solar transmittance for the Gentex viologen-based device is a reflection of the type of spectral change (Type D in Figure 14.23).

TABLE 14.4 Characteristics of Smart Window Systems

System	Mechanism of Switching	Operational Switching Parameter	Control Parameter	Switching Behavior	Spectral Response (See Figure 14.23)	Memory	Selectivity	Contrast	References
							Colored/Clear	Visible/Solar	
Electrochromic (organic)	Reduction of organic species (e.g., viologen)	Electrical control–voltage/current (approximately 2–3 V DC; 50–1000 μA/cm²)	Voltage	Threshold voltage, then approximately linear response to saturation	Type D	No	0.2/1.3 specular	13/2.5	[29]
Electrochromic (inorganic)	Reduction of electrochromic (e.g., WO_3) by electrochemical ion insertion		Charge injected into electrochromic layer (mC/cm²)	Approximately linear change in optical density with charge	Type A and B	Yes	1.4–1.9/1.3 specular	3–10/3–11 depends on system	[29]
							2.3/1.55 specular	17/26	[30] (classic)
							2.8/2.5 specular	17/19	(Sea green)
Thermochromic	Metal–insulator transition	Temperature (T_{switch} depends on material)	Glazing temperature (°C)	Rapid change at T_c (transition point)	Type C	N/A	1.0–1.2/<1 specular	~1/1.1–1.6	[7,43]
Thermotropic	Phase separation leading to multiple scattering	Temperature (T_{switch} depends on material)	Glazing temperature (°C)	Gradual transition (in hydrogel, from 30°C to 65°C)	Type A	N/A	1.5–1.8/1.3–1.7 (hydrogel)	4.6/5 (hydrogel)	[8,47]
							1.1/1.2 (cloud gel) scattering/specular	15/16 (cloud gel)	
PDLC	Alignment of liquid crystal particles in electric field	Electrical control (~100 V AC—voltage depends on system)	Applied Voltage	Highly nonlinear response	Type D	No	0.9/1.0 scattering/specular (with haze)	1.6/1.5	[6,29]
Suspended particle devices	Alignment of suspended particles in electric field	Electrical control (50–200 V AC)	Applied voltage	Linear response 0–50 V, saturation by ~100 V	Type D	No	N/A	14/-	[5]
		Power 2–10 W/m²					Specular		
Gasochromic	Reduction of gasochromic (e.g., WO_3) by catalytic ion insertion	Gas control concentration of H_2 in window cavity	Concentration of H_2 in window cavity	Nonlinear response to gas concentration	Type A and B	Yes	1.2–1.7/1.0–1.5 specular	4.2/4.9	[8,51,52]

applications of smart windows where it is essential to be able to control the state of a device, as well as to be able to switch the device. As can be seen in Figure 14.24, of the electrically activated devices, the inorganic electrochromic devices provide the most linear response with respect to the control parameter, which in this case is injected charge density.

14.2.4.1 Electrochromic Smart Windows

Electrochromic materials undergo a change in their optical properties when an electric field is applied. There is a wide range of both organic and inorganic materials that exhibit electrochromism. Inorganic electrochromic materials, most commonly transition metal oxides, change from a clear to a colored state when ions and electrons are inserted into their lattice, under the application of a small DC, leading to optical absorption. Extraction of these ions and electrons is effected by a reversal of the electric field and the material returns to the clear state. Organic electrochromic materials involve stoichiometric redox couples, where the oxidized and reduced forms have different colors. The application of an electrical potential sufficient to change the oxidation state of the material therefore causes a color change.

14.2.4.1.1 Inorganic Electrochromic Smart Windows

Many inorganic electrochromic materials maybe produced in thin films on transparent electrically conducting substrates, thereby allowing their use in window fabrication. Tungsten oxide (WO_3) is the most commonly used inorganic electrochromic material and is the compound in which electrochromism was first observed. Tungsten oxide can be colored by the injection of electrons and small ions including H^+, Na^+, K^+, and Li^+, as schematically illustrated in the reaction scheme in Equation 14.3a. Intercalation of lithium and electrons into tungsten oxide forms the tungsten bronze Li_xWO_3 (reversible for $0 < x < 0.4$ [31]), which is deep blue in color. In order to fabricate an electrochromic device, a source of ions and electrons is required. The most common device structure being used for inorganic electrochromic devices is illustrated in Figure 14.25. There are a number of variants of this device structure with different components, and this is discussed in Section 14.2.4. The most widely used counterelectrode materials are cerium-titanium oxide (CeO_2-TiO_2) [32] vanadium oxide (V_2O_5) [33] and derivatives. In a WO_3-V_2O_5 device, the complementary reaction in the counterelectrode is illustrated in Equation 14.3b, with Li^+ as the mobile ionic species [34]:

$$\underset{\text{transparent}}{WO_3 + xLi^+ + xe^-} \underset{+V}{\overset{-V}{\rightleftarrows}} \underset{\text{deep blue}}{Li_xWO_3} \quad \text{Working electrode} \qquad (14.3a)$$

$$Li_yV_2O_5 \underset{-V}{\overset{+V}{\rightleftarrows}} Li_{y-x}V_2O_5 + xLi^+ + xe^- \quad \text{Counterelectrode} \qquad (14.3b)$$

The change in optical density is usually proportional to the charge density injected into the electrode [35,36], and the coloration efficiency, $CE = \Delta OD/Q_{in}$, is frequently used to characterize the performance of a film or a device. For WO_3-based electrochromics, a typical coloration efficiency is between 55 and 70 mC/cm^2 at $\lambda = 700$ nm [37].

The electrochromic material changes its optical constants, usually through the development of an absorption band [38], or though an increase in the effective free electron density in the material [39]. These give rise to optical changes known as "absorption modulation" and "reflectance modulation" [22], respectively, according to whether the change in transmittance results from an increase in optical absorption or reflectance. Reflectance modulation is the preferable optical change for glazing materials, as these materials do not absorb radiation significantly in the colored state, minimizing secondary heat gain, and reducing the TSET value of the window. The specific behavior depends on the microstructure of the material, with crystalline WO_3 films exhibiting reflectance modulation, and amorphous WO_3 films showing absorption modulation. The coloration efficiency spectra in Figure 14.26 illustrate the change in the absorption characteristics of sputtered WO_3 as the deposition temperature is increased. Despite the better performance that can be obtained with reflectance modulation, the majority of prototype devices

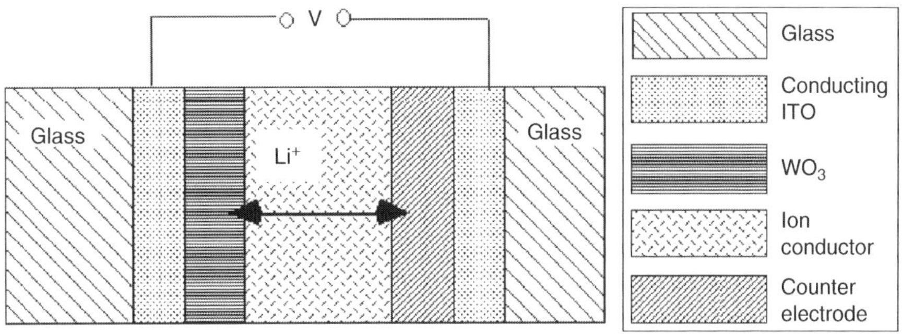

FIGURE 14.25 Spectral coloration efficiency, which is directly proportional to change in optical density in sputtered WO_3. This shows the development of an absorption band in WO_3 during ion injection. Note the shift in the absorption band peak position as the substrate temperature increases and the film becomes more crystalline, representing the change from absorption to reflection modulation.

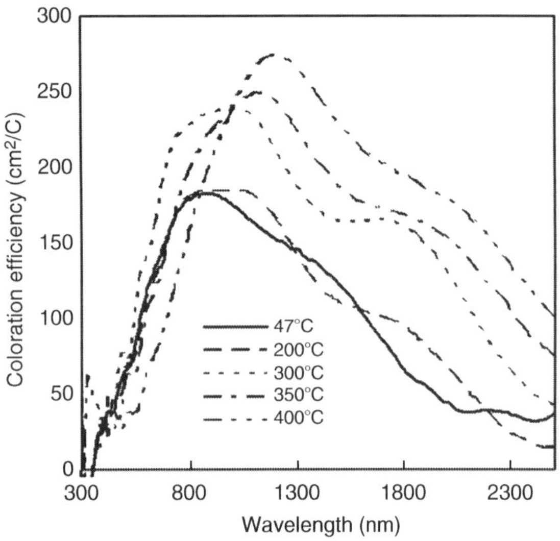

FIGURE 14.26 Schematic illustration of the structure of a typical inorganic electrochromic device, showing the five-layer structure.

produced to date use absorption modulation, as it is hard to produce crystalline WO_3 in conjunction with the other layers in the device.

Devices based on inorganic electrochromics are usually spectrally selective to some degree (type B behavior in Figure 14.23), and it is possible to strongly modulate the solar transmittance of the window, while still maintaining a moderate visible transmittance, thus allowing significant daylighting in a building. This makes WO_3 an excellent candidate for smart windows where a low solar transmittance is a major requirement and daylighting is a direct advantage.

14.2.4.1.2 *Organic Electrochromic Smart Windows*

The absorption coefficients of organic electrochromic materials are generally extremely high when compared to inorganic materials, so it is possible to produce devices that undergo drastic optical changes involving very intense colors. The majority of organic electrochromic materials are conducting polymers. Some of these organic materials, such as polyaniline, have different colors for the various oxidation states, so it is possible to achieve multiple coloration by a stepwise oxidation or reduction [40]. The transmittance of organic electrochromic materials depends on the specific compound; however, most of these materials switch primarily in the visible region, with little modulation of solar transmittance when compared to inorganic systems (type D spectral behavior). This makes organic electrochromics less suitable for smart window applications than inorganic electrochromics.

The most widely studied family of organic electrochromic materials is the viologens, which are currently used in a commercial, variable reflectance mirror for automobile rear view mirrors (Gentex Corporation). In this application, it is free from many of the problems associated with solar/UV degradation, which will occur in architectural glazing applications of smart window. Polyaniline is a promising candidate for organic electrochromic smart windows due to its high stability and ability to undergo multiple color changes [40,41].

14.2.4.2 Thermochromic Devices

Thermochromic materials change their optical properties when heated, and return to the original state when cooled to the initial temperature [42]. The most widely studied inorganic thermochromic material, and the best candidate for large area thermochromic smart window applications, is VO_2 and various doped forms of VO_2 [43]. This material undergoes a metal–insulator phase transition at 68°C, exhibiting metallic properties at high temperature and associated infrared reflectivity, as shown in Figure 14.27a. The visible part of the spectrum is largely unaffected by the transition (type C spectral behavior), so the principal change that occurs at switching is in the solar transmittance. The visible transmittance decreases slightly, but there is a moderately strong absorption band in the visible spectrum in the low-temperature insulating state, which means that the visible transmittance is relatively low in both states.

The major area of research for development of large area smart windows, which are useful for energy control in buildings, is the development of materials that have a higher visible transmittance, and, most importantly, with a transition temperature close to room temperature. This means that the windows will switch to a low transmittance state to prevent overheating in the building. Both of these areas have been the subject of considerable research with numerous efforts to dope VO_2 with various elements to achieve these goals. Fluorination of VO_2 [7] increases the visible transmittance from 27% to 35% (in the low-temperature state), but also reduces the effect of switching, with a change in solar transmittance from 44% to 40% at the transition. However, the efforts to reduce the transition temperature have been much more successful, with a significant reduction in the thermochromic transition temperature from 68°C to approximately 10°C for $W_{0.032}V_{0.968}O_2$ [43,44]. This is illustrated in Figure 14.27b.

Apart from the relatively poor spectral selectivity and contrast ratio of thermochromic devices, the inability to independently control the state of thermochromic devices makes it difficult to envisage wide ranging applications as smart windows.

14.2.4.3 Thermotropic Devices

Thermotropic devices, originally classed with thermochromic devices in the literature [45,46], also change their optical properties with increasing temperature. However, the mechanism of the change is sufficiently different from typical inorganic thermochromic materials that they are now recognized as a different class of material. Thermotropic materials change between a specularly transmitting transparent state to a white, diffusely reflecting state as their temperature rises. The change is reversible. This change occurs through a phase separation, which occurs in the materials at elevated temperatures. This transforms the material from a homogeneous optical material to a heterogeneous, two-phase material at high temperature. The size of the

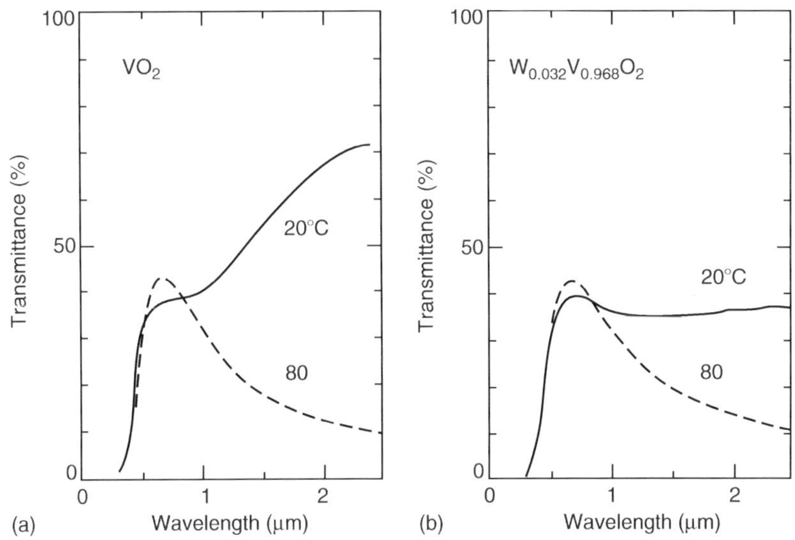

FIGURE 14.27 Transmittance of (a) VO_2 and (b) W-doped VO_2 at temperatures above (80°C) and below (20°C) the thermochromic transition. The spectra are of type C in both cases (see Figure 14.23). (Reprinted from Sobhan, M.A., Kivaisi, R.T., Stjerna, B., and Granqvist, C.-G., *Solar Energy Mater. Solar Cells*, 44, 451–455, 1996. With permission.)

particles of the two phases is such as to lead principally to backscattering of incident radiation [47].

There are two main types of thermotropic materials: hydrogels and polymer blends. Both systems have a similar device structure, as illustrated in Figure 14.28. In both cases, the materials form homogeneous transparent mixtures at low temperatures, in which the two components are bound together at a molecular or molecular cluster (macromolecular) level. The bonding is usually hydrogen bonding and these bonds can be broken as the temperature increases from room temperature. Once the bonds break, the materials separate into distinct domains with different refractive indices and cause scattering. If the particle size is appropriate (close to the wavelength of light over the solar spectral range), then strong scattering occurs.

14.2.4.3.1 Hydrogels

This is one class of thermotropic materials and the thermotropic material CloudGel, commercially available from Suntek, is of this type. As the name suggests, a hydrogel is a water–polymer mixture. Another example of these materials is the TALD hydrogel, which is a mixture of colloidal particles of a polyether–water mixture, embedded in a carboxyvinyl polymer and water gel [47]. The colloidal particles are soluble in the gel at room temperature and below, but as the temperature increases, water is expelled from the colloidal particles and the polyether particles become more dense and increase their refractive index.

The switching temperature of the hydrogel materials can be varied between 5°C and 60°C by changing the proportions of the components. The materials switch over a reasonably wide temperature range, as illustrated in Figure 14.29. The size of the particles in the phase-separated state, and the thickness of the hydrogel layer (typically 1 mm) ensure that the majority of the radiation is backscattered by multiple scattering in the high-temperature state (see Figure 14.29). This means that these thermotropic materials have good potential as energy-efficient windows.

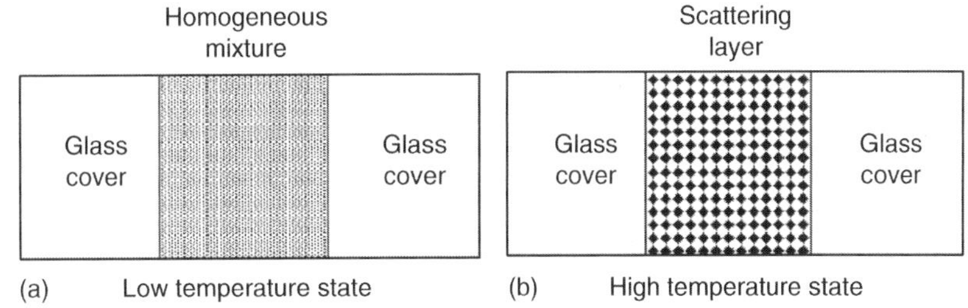

FIGURE 14.28 Structure of a thermotropic laminate. In the low temperature state (a), the device is fully transparent, (b) but in the high-temperature state the thermotropic state separates into discreet particles and the layer becomes scattering.

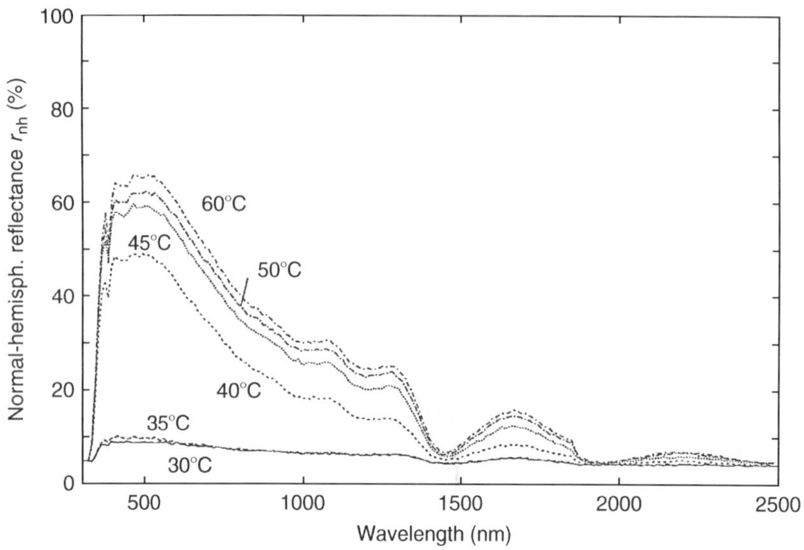

FIGURE 14.29 Normal-hemispherical reflectance of a thermotropic hydrogel laminate at a range of temperatures below (30°C) and above (>35°C) the thermotropic transition temperature. The spectra are of type B, demonstrating a high degree of selectivity.

14.2.4.3.2 Polymer Blends

Polymer blend thermotropic materials have the advantage of not containing liquid water. An example of a polymer blend hydrogel is a high refractive index (1.585) cross-linked matrix of polystyrol and polyhydroxy ethyl methacrylate, with a second, low refractive index (1.45) phase of polypropylene oxide interpenetrating the matrix. At low temperatures, the lower refractive index component is hydrogen bonded to the matrix. At higher temperatures, these bonds break and the polypropylene oxide component forms small inclusions in the matrix, leading to scattering. The size of the scattering particles and the transition temperature depend on the proportions of the two components and the degree of cross-linking of the matrix.

14.2.4.3.3 Applications

The principal factors constraining the application of thermotropic devices are the inability to independently control the device transmittance and the scattering state at high temperatures. The latter means that thermotropic devices are unlikely to be used in view windows, but they have applications in skylights and for overheating protection on brick facades. These applications have been extensively investigated [47], and large area windows can be routinely produced. Figure 14.30 shows a number of thermotropic windows under partial shading, illustrating the sensitivity to temperature. These windows were part of a study to investigate energy savings and acceptance of the windows by building occupants. In this work, significant energy savings were reported (up to 17%), and over 75% of building occupants reported improved comfort conditions [20].

14.2.4.4 PDLC Devices

There are many classes of liquid crystals. Most liquid crystals consist of thin, needle-like, or rod-shaped organic molecules, which exhibit significant ordering properties. There are several classes of electro-optic devices, which can be formed using liquid crystals, but only two classes, guest–host systems and PDLCs, have been extensively investigated for use in smart windows. Lampert [48] provides an excellent introduction to the physics of liquid crystals, and Basturk and Grupp [49] and Montgomery [6] provide good reviews of the basics of both types of liquid crystal devices.

FIGURE 14.30 Thermotropic windows (in various different configurations of inner and outer glass in each separate pane), illustrating the transition between the transparent (low temperature, and in this case, unshaded state) and the opaque state. (From Inoue, T., *Energy Build.* 35, 463, 2003. With permission.)

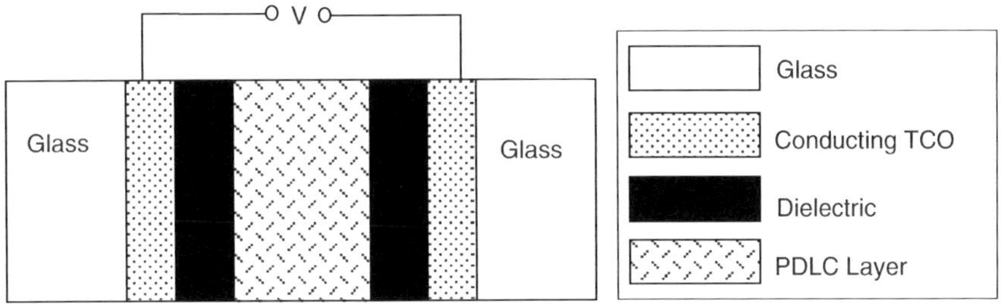

FIGURE 14.31 Structure of a PDLC device. Unlike the electrochromic device, it is not a conducting device, with the switching dependent on the electric field across the PDLC layer.

The principal liquid crystal devices currently available are the PDLC devices, which are commercially available from 3M under the name "Privacy Film." This device consists of a PDLC film sandwiched between indium-doped tin oxide (ITO)-coated sheets of glass as illustrated in Figure 14.31. The PDLC film consists of spherical liquid crystal droplets embedded in a polymer matrix. The typical droplet diameter is about 1 μm or less, i.e., of the same order of magnitude as light wavelengths. On application of a voltage to the ITO electrodes, the liquid crystal molecules are orientated in the field direction, resulting in index matching between the droplets and the matrix and thus high transmittance for light propagating parallel to the applied field. When the field is switched off, the liquid crystal molecules reorient randomly, the (unmatched) value of the extraordinary refractive index also becomes effective, and causes scattering at the droplet surfaces; the film appears milky.

While PDLC devices act very effectively as a privacy screen, with essentially zero specular transmittance in the zero voltage state, and good transmittance but with some diffuse scattering in the on state, the devices are not particularly effective for energy conservation as the majority of the scattered radiation is forward scattered. This is illustrated in Figure 14.32, which shows the total transmittance (including the diffuse transmittance). While the device is totally opaque in the "off" state (0 V), the total transmittance is still substantial. Another drawback with the PDLC devices, and all liquid crystal systems, is the lack of memory in the devices. This is intrinsic to liquid crystal systems, as the alignment of the liquid crystal molecules is caused by dipolar polarization in the applied field. The curves in Figure 14.32 also show the highly nonlinear behavior of the switching with the applied voltage, which is the control parameter in Figure 14.23 for the PDLC device.

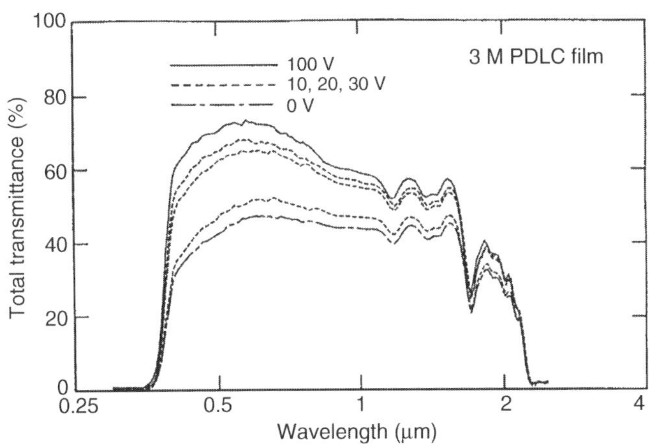

FIGURE 14.32 Normal-hemispherical transmittance spectra for the PDLC device for applied voltages of 0, 10, 20, 30, and 100 V. The device is opaque (in the sense of being unable to perceive images through the device in the off state [0 V]), but still transmits significant energy. (Reproduced from Dr. Arne Roos. With permission.)

14.2.4.5 Suspended Particle Devices

Suspended particle devices, also termed light valves, consist of a suspension of fine particles, which in the "off" state (no applied voltage) are strongly absorbing, but which can be aligned by the application of an electric field, leading to a transparent "on" state. The device requires the suspension to be constrained between conducting electrodes, and in the large area devices currently being commercialized by Hankuk Glass, these are deposited on a PET film, which is then laminated to glass [5]. The suspended particles are usually needle-shaped particles of polyiodides or parapathite approximately 1 μm long, suspended in an organic liquid [45]. In current devices, methods have been developed to encapsulate microdroplets containing the suspended particles in a polymer matrix. The device structure is essentially identical to the PDLC device (Figure 14.30), with the suspended particle layer replacing the PDLC layer.

In the absence of an applied electric field, the particles are randomly arranged and absorb the incident radiation. The absorption in all states (from zero applied voltage to the maximum applied voltage) depends on the concentration of particles in the film, and the color can also be altered by the choice of particles. Increasing the applied voltage increases the transmittance approximately linearly to about 50 V, and the transmittance then starts to saturate, reaching saturation at about 100 V. The major drawback with the devices for use as smart windows is that they do not provide significant spectral selectivity (figures for solar transmittance are not available) or significant reductions in solar transmittance. As with PDLC devices, they are principally privacy and visual comfort devices, however, these devices have a high viewing angle unlike liquid crystal devices.

14.2.4.6 Gasochromic Devices

Gasochromic smart windows are devices that color by a complex process involving the adsorption and dissociation of a gas onto a catalyst, followed by a "spill over" [50] of the dissociated gas molecules into the gasochromic layer causing coloration [8,51–53]. Gasochromic devices are similar to electrochromic devices in that they both color when another species is inserted into the atomic network of the host material. Tungsten oxide is a good material for both electrochromic and gasochromic devices. The major differences between electrochromic and gasochromic devices lie in the different mechanisms of coloration and the relative structural simplicity of the gasochromic devices.

Simple gasochromic films have a laminar structure involving a substrate (usually glass), the gasochromic layer, and a catalyst layer. These films are fabricated into gasochromic smart windows by adding another pane of glass a short distance from the catalyst surface, so that a small cavity is formed inside the device, as shown in Figure 14.33. Various gas mixtures can then be introduced into this interior cavity to color and bleach the window. Gasochromic tungsten oxide colors according to the following simplified redox reaction, when hydrogen gas is used to induce coloration:

$$\underset{\text{transparent}}{x\text{H} + \text{WO}_3} \rightleftarrows \underset{\text{blue}}{\text{H}_x\text{WO}_3} \tag{14.4}$$

This reaction is essentially identical to the reaction involved in the coloration of electrochromic WO_3. The principal difference is that the catalyst layer allows the dissociation of H_2 gas into an activated species, which can diffuse into and react with WO_3. In an electrochromic system, the positive and negative charges are introduced via an electrical circuit and the polymer electrolyte and counterelectrode are used to separate the charge and store cations.

Pure hydrogen is not required to color gasochromic films and very dilute, nonexplosive mixtures can be used. Bleaching of gasochromic films is usually accomplished by filling the window cavity with a dilute oxygen mixture; however, bleaching can also be achieved using argon or a vacuum [52]. Partial coloration or bleaching of gasochromic devices can be achieved by replacement of the gas in the window cavity with an inert gas such as argon once a suitable level of coloration has been attained.

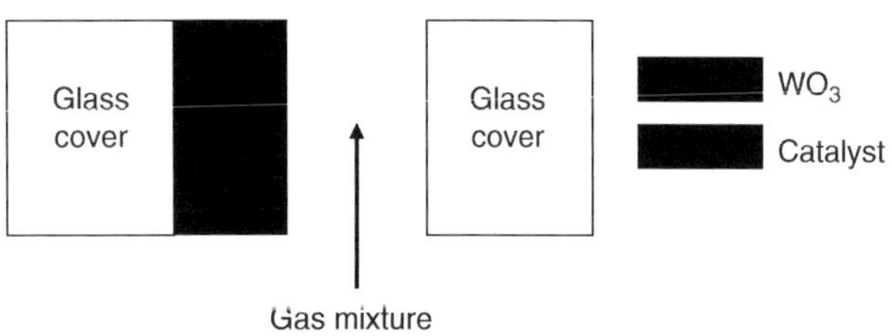

FIGURE 14.33 Typical structure of a gasochromic device.

14.2.4.6.1 Gasochromic Technology

Gasochromic technology has stemmed largely from research into electrochromism, and tungsten oxide is the most commonly reported gasochromic material [8,51,52,53]. Platinum is frequently used as the catalyst layer due to its ability to dissociate hydrogen and oxygen gas molecules. Some new systems with a gasochromic layer based on alloys of lanthanides have been recently reported [54–56]. Catalyst-coated lanthanum hydrides (LnH_x) undergo extreme color changes when x is varied from 2 (film reflective in blue region/transmitting in red) to 3 (yellow/transparent) [54]. Exposure of catalyst-covered LnH_2 films to hydrogen gas quickly results in the formation of the transparent LnH_3 species. When exposed to air or a vacuum, the film reverts back to the LnH_2 form. The function of the catalyst is to allow the dissociation of gas molecules, but also prevents the underlying film from rapid oxidation in air [56]. The addition of magnesium reduces the absorption in the transparent state and the transmission in the opaque state, while increasing the opaque state reflectivity, and also has the advantage of producing more color-neutral windows [54,56]. A particularly impressive system involves a gadolinium-magnesium alloy $Gd_{40}Mg_{60}$ that is reported to switch from a silver white mirror with almost no transmission and high reflection, to a fully transparent color-neutral state [54].

14.2.4.6.2 Applications

The function of the glass pane used opposite the gasochromic layer is to contain the gas used for switching, and ideally has a high visible transmittance. Ordinary float glass is suitable, however, glass with low-emittance coatings can be used to improve the thermal performance. Owing to the small number of layers required for gasochromic smart windows, the dynamic transmittance range is very high when compared to electrochromics and other chromogenic glazings. Gasochromic smart windows have been reported with visible transmittances ranging from 76% (bleached state) to 18% (colored state). The advantages associated with the relative structural simplicity of gasochromic smart windows are somewhat offset by the complexity of the infrastructure needed to hold and transfer the gas mixtures required for their operation. The gas system must be capable of storing gas, preparing and filtering gas mixtures, and delivering these to a bank of multiple windows. Reference [8] offers some innovative solutions to the problems of gas support.

As for all smart windows, lifetime is a critical issue for gasochromic devices. The principal degradation mechanism for gasochromics is catalyst poisoning and varying water content of the films [53]. Recent results are promising however, and one gasochromic smart window 1.1 × 0.6 m in size has been cycled 10,000 times with no significant reduction in its performance [52].

14.2.5 Electrochromic Smart Windows

As noted in Section 14.2.2, one of the most promising candidates for a future commercial smart windows is the electrochromic glazing based on inorganic electrochromic materials. Inorganic electrochromic smart windows are also the closest to commercial production for large area architectural applications with substantial potential to reduce building energy use and hence the discussion below is based on those windows with inorganic materials, and in particular tungsten oxide.

The method of window control is also very important as, in order to optimize the energy performance, it is essential to be able to set specific optical states independent of temperature and irradiance incident on the window, and the lifetime of the electrochromic films and entire devices can be reduced by inappropriate control methods [57].

Various different types of electrochromic material and window testing and control [58] have been used: constant voltage control, trapezoidal voltage waveform [25], cyclic voltammetry (triangular voltage waveform) [59], and constant current control [60]. There have also been many theoretical analyses of the charge injection and extraction processes. It is widely recognized that the coloration process is based on double injection of ions and electrons forming the tungsten bronze M_xWO_3 [34]. There are various theories regarding the mechanisms limiting the coloration current, including the presence of a barrier at the ion-injecting interface (electrolyte/WO_3 interface) [34,61]; diffusion-limited motion of ions within the electrochromic films [62,63] and the series resistance of the cell [64,65]. Impedance spectroscopy has been used by a number of authors recently to study ion injection kinetics in electrochromics [66,67], and this holds some promise as a useful tool for the analysis of degradation of smart windows in operation.

FIGURE 14.34 Large scale prototype electrochromic window. (From Zini, M., *Build. Environ.* 41, 1262, 2006. With permission.)

14.2.5.1 Electrochromic Smart Window Structures

The basic multilayer electrochromic device structure is shown in Figure 14.25. An example of a smart window is shown in Figure 14.34. It is possible to classify various electrochromic devices according to the structure and materials used to achieve the ion injection and extraction processes in a device. There are three principal generic device types, which are essentially all the same as that shown in Figure 14.25, but with different approaches to the ion transport and ion storage layers. The materials used in these devices are discussed in more detail in Section 14.2.5.

14.2.5.1.1 Type 1—Ion Conducting Layer and Passive Counterelectrode

The most widely used device structure has an ion transport layer (also called the electrolyte layer) formed between the working electrode and an ion storage electrode (also called the counterelectrode). The fundamental requirement of the ion storage layer is charge capacity, i.e., the amount of charge (number of ions) that can reversibly be stored in the layer. The charge capacity is normalized to electrode area and measured in mC/cm^2. The charge capacity of the ion storage electrode should be in excess of charge required for sufficient coloration of the working electrode. Typical values are between 10 and 30 mC/cm^2. It is also essential to achieve high transparency of the counterelectrode, which is often in conflict with the requirement for high charge capacity. Another requirement for the ion storage layer in this structure is a low coloration efficiency. A good ion storage electrode should have a very large charge capacity and be optically passive. An example of a device of this type is the near commercial window produced by Sustainable Technologies Australia. This device uses a modified WO_3 electrochromic electrode and a V_2O_5-based counterelectrode, with a lithium-doped cross-linked polymer electrolyte. Spectra of the colored and bleached states of this device are shown in Figure 14.35.

14.2.5.1.2 Type 2—Combined Ion Conducting Layer and Counterelectrode

In the second approach to device design, the means of charge delivery and charge storage are realized in one layer. This is the ion-conducting layer, which contains dispersed redox couples, thus in the deintercalated state ions are stored in the electrolyte. An example of this type of device is based on WO_3 as the working electrode, with a solid polymer electrolyte containing lithium polyorganodisulfide redox salts [68]. These devices have the advantage of simplicity of manufacture, requiring deposition of one less layer than Type 1 devices. In addition, the redox salts are generally optically passive, so charge capacity of the counterelectrode is not an issue. However, the redox reaction only occurs at the surface of the layer and transport of the electrons across the polymer layer from the transperent electrical conductor (TEC) on the counterelectrode side of the device can limit the device.

14.2.5.1.3 Type 3—Ion Transport Layer and Complementary Counterelectrode

The third device structure is very similar to the first type as there are separate ion transport and counterelectrode layers.

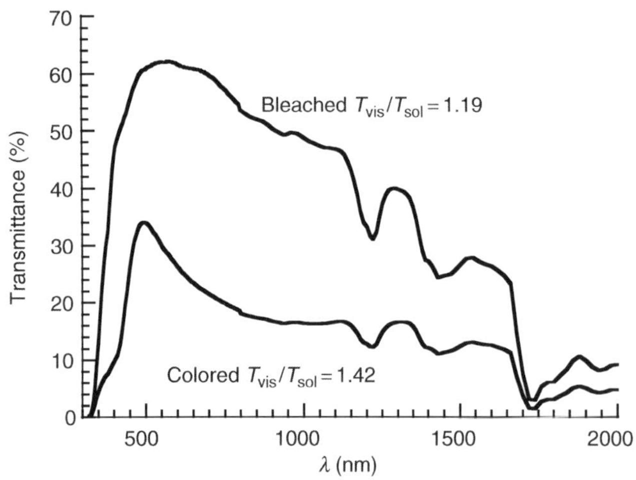

FIGURE 14.35 Transmittance spectra for colored and bleached states of an STA electrochromic device. The selectivity in both states are shown on the graph, indicating the increase in selectivity in the colored state. The dynamic range for the device is T_{sol} = 51.5% to T_{sol} = 20.6%, with injected charge of 10 mC/cm^2.

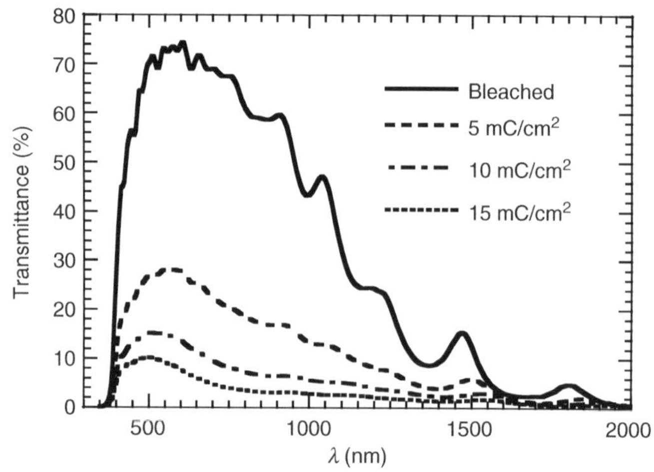

FIGURE 14.36 Transmittance spectra of an Asahi electrochromic device in the bleached state and for three colored states, with 5, 10, and 15 mC/cm^2 injected into the WO_3 layer at a constant potential of 1.5 V.

However, the counterelectrode material is a electrochromic material, which is complementary to the working electrode, i.e., it is an anodically coloring electrochromic material, which darkens upon extraction of ions from the counterelectrode and bleaches upon injection of ions [69]. In this way, both electrodes contribute to the coloration of the electrochromic device, enhancing the overall performance of the device. This type of device has the advantage of requiring less overall charge transport to achieve the same level of coloration as a device of type 1. Hence the charge capacity requirements of the electrodes are lower than for type 1 devices. A device of this type has been developed by the Asahi Glass Company (Japan). This device consists of the following structure: Glass/TEC (200 nm)/NiO_x (500 nm)/Ta_2O_5 (500 nm)/WO_3 (500 nm)/TEC (500 nm)/adhesive film (250 μm)/glass. The spectral response of this device is shown in Figure 14.36 for four different levels of charge injection. Note that the change in optical density per unit charge is significantly larger for this device than for the STA device of type 1 (see Figure 14.35) owing to the coloration of both the working electrode (WO_3) and the counterelectrode (NiO_x).

14.2.5.2 Materials Used in Electrochromic Devices

14.2.5.2.1 Electrochromic Materials

Tungsten oxide is currently the most commonly used inorganic electrochromic material. Coloration during ion insertion (H^+, Li^+, Na^+, and K^+) and bleaching on ion extraction is termed cathodic coloration. In an electrochromic smart window, thin film electrodes must be in contact with the working and counterelectrodes and an electrolyte is required for ion transport. Use of electrochromic materials in window fabrication necessitates the transparency of all components and it is these constraints that limit the number of materials suitable for electrochromic device fabrication.

Numerous transition metal oxides are suitable to be used as the electrochromic layer in smart windows, and a detailed discussion of electrochromism in these materials is given in Ref. [70]. The most widely studied electrochromic material is WO_3 and doped WO_3, where the dopant is another transition metal oxide such as TiO_2 or MoO_3. Tungsten oxide films are commonly prepared by thermal evaporation, sputtering, chemical vapor deposition (CVD), spray deposition, anodization, and sol–gel deposition. A detailed discussion of these deposition techniques and their effects on the properties of the deposited materials is given in Refs. [70,71] and references therein. On the basis of the amount of charge that can be inserted, the coloration efficiency and the transmittance change of the film, the electrochromic performance of WO_3 deposited by most techniques is very similar, with the most important effect being increased crystallinity in the films sputtered onto heated substrates (above approximately 350°C), or for sol–gel deposited films heat treated at temperatures of 350–400°C after deposition. This increase in crystallinity is usually accompanied by an increase in reflectance modulation in the films.

Other inorganic electrochromic systems, which have received significant attention, are NiO, TiO_2, IrO_2, Nb_2O_5, and more recently, SnO_2 and Pr_2O_3. A variety of techniques have been used for the deposition of all of these coatings, but none of these materials show the electrochromic efficiency of WO_3.

14.2.5.2.2 Counterelectrode Materials

NiO and V_2O_5 have been extensively studied as potential counterelectrode materials for electrochromic devices [70], while CeO_2-TiO_2, and more recently CeO_2-ZrO_2, have been extensively studied in the past few years [32]. NiO is particularly attractive as it colors anodically (i.e., on the extraction of ions), and hence complements both the coloration and bleaching of the cathodically coloring WO_3 layer [69] (see above). V_2O_5 has excellent charge capacity and can be produced relatively easily using sol–gel deposition [33], but it has the disadvantage of an absorption edge encroaching into the visible region of the spectrum, resulting in a yellowish coloration in the films [72]. Initially (up to the early 1990s) sputtering and evaporation dominated the deposition of electrochromics, but sol–gel deposition is becoming very widespread except for NiO, although there has been one report of sol–gel-deposited NiO [73].

14.2.5.2.3 *Ion Transport Layer*

The function of the electrolyte layer is to allow ions to travel between the working and counterelectrodes, hence it is ionically conducting. Low electrical conductivity of the electrolyte layer is advantageous because it reduces the internal leakage current in a device and allows devices to stay colored for long periods of time without the need for an external power source.

Electrolytes used in electrochromic smart windows can be classified as either polymeric or superionic conductors. Polymeric electrolytes are prepared as a liquid with some dissolved lithium salt, and are then used to laminate the working and counterelectrode layers together. The polymer therefore allows ionic conduction while giving the device its mechanical strength. Superionic conductors such as M-β-alumina (where M is a monovalent metal such as Li, Na, or K) allow ions to move through the host lattice of alumina. These electrolytes are advantageous because they are not prone to damage by factors such as UV irradiation and can be deposited directly onto the electrochromic layer, however they may have a nonzero electrical conductivity, which leads to a reduced memory in the colored state. The Asahi device, which uses Ta_2O_5 as an ion conducting layer, shows poor memory compared to devices that use polymer-based electrolytes (e.g., the STA device discussed above) [29].

14.2.5.2.4 *Transparent Electronic Conductors*

The transparent electrical conductors in most electrochromic devices are thin films of metal oxides such as indium-doped tin oxide (SnO_2:In or ITO) and fluorine-doped tin oxide (SnO_2:F or FTO). These materials are commercially available precoated onto glass substrates by spray pyrolysis methods [74]. Some electrochromic devices utilize conducting polymeric layers such as ITO-PET (ITO coated onto polyethylene terephthalate) as the transparent electrical conductors, which allows for the fabrication of polymeric solid-state electrochromic devices [75].

14.2.5.3 Control of Electrochromic Smart Windows

The application of electrochromic smart windows in architectural glazings requires that the devices can perform tens of thousands of switching cycles over a period of 20–30 years, with minimal degradation in optical performance. The lifetime of a smart window is measured in number of coloration/bleaching cycles when the transmittance in each cycle is modulated between specified maximum and minimum values. However, the transmittance of the window, particularly of an installed window, is not easily accessible. As illustrated in Figure 14.24, the optical density of an electrochromic window is in a one-to-one (and very nearly linear) relationship with injected charge density. This relationship is also almost independent of temperature, so a good switching algorithm for electrochromic smart windows can be based on charge control. When considered as an electrochemical system, an electrochromic device is very similar to a rechargeable battery. The energy in the battery is stored and released by means of reversible redox reactions. In electrochromic devices, these reactions result in optical changes. The durability of batteries has been investigated extensively over the last 50 years and several empirical relationships between the lifetime and operational parameters have been found [76]. In particular, battery lifetime is found to be inversely proportional to the operational current, so the switching current must be small enough to provide a long lifetime.

In the design of a control algorithm for electrochromic windows, two fundamental limits that restrict fast switching of the window must be considered:

- Thermodynamic limit (overvoltage): the voltage must be maintained below the potential at which destructive side reaction(s) occur.
- Kinetic limit (excessive charging current): large currents cause irreversible cation trapping on the surface of the film [60] owing to charge "pile-up" at the surface. This will usually be caused by slow cation diffusion into the electrochromic layer.

Of course, if the voltage is too small, the desirable electrochromic reaction can never occur, and if the charging current is too small, the switching time of the window is unacceptably long. Therefore, both current and voltage need to be maintained within fixed limits to achieve optimum device lifetime, with charge used as an independent parameter to control the optical state of the device.

Both thermodynamic (redox potentials) and kinetic (resistance, diffusion coefficients) properties of electrochromic window are temperature-dependent, thus they generally require adjustments in switching parameters for changing environmental conditions. For example, a constant applied voltage will not provide constant contrast ratio at different temperatures owing to the temperature dependence of the film potential on temperature. At high temperatures, a voltage suitable for low temperature could exceed the potential of side reactions (thermodynamic limit) and will lead to higher currents and possible kinetic device failure.

Typical cycles of electrochromic window performed at 50°C and 18°C are presented in Figure 14.37. The charge intercalated into working electrode is the same in both cycles, therefore the changes in optical transmittance are almost identical. The charging current density is also identical at the two temperatures, however, at 50°C the window required significantly lower coloring voltage since diffusion coefficients and electrolyte conductance increase with temperature. The control algorithm implemented to produce these results includes constant current charging until the injected charge exceeds a given value Q_{in}. If voltage exceeds a predetermined safe value V_{max}, the system automatically switches to a constant voltage mode. In the bleaching part

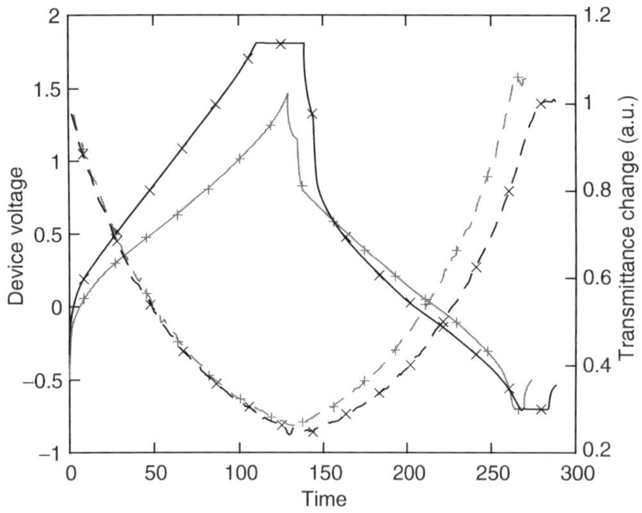

FIGURE 14.37 Electrochromic device performance at 50°C (×) and 18°C (+). The solid lines represent the device voltage and the dashed lines the relative transmittance of the device. In both cycles the operational parameters are: current density = 0.1 mA/cm², Q_{in} = 13 mC/cm²; Q_{out} = −13 mC/cm², preset switching voltages are: V_{max} = 1.8 V, V_{min} = −0.7 V. Note that at 50°C voltage does not reach V_{max}. The high-temperature cycle period is shorter and lower voltages are required, however, transmittance change in the two cycles is almost identical.

of a cycle extracted charge shall not exceed Q_{out}, and voltage shall be lower in magnitude than a predetermined maximum bleaching voltage V_{min}. Methods for the control of electrochromic devices using these techniques are described in patents [77,78] and the theoretical background is outlined in Refs. [57,60].

14.2.6 Future Directions

Smart window prototypes of sufficient scale for full-scale testing are now available, as evidenced by the examples illustrated in Figures 14.30 and 14.34 above. There is a growing body of evidence that these windows will provide significant energy savings for commercial buildings in appropriate climates (not ably hot, arid climates) provided appropriate window control strategies are employed. Smart windows based on sputtered tungsten oxide electrochromics are now [79] commercially available, although previous commercial products released by Pilkington (see Pilkington Company History [80]) appear to have been withdrawn from the market. There are still significant questions about the durability of such windows, and the major research efforts in the immediate future will be to improve window durability during cycling. The advantage of the electrochromic windows in terms of independent control of the optical state (using injected charge density) is actually a disadvantage in this context as it allows overdriving of the device, while the gasochromic and thermally switched devices are intrinsically self-limiting in terms of switching the windows. However, other durability issues exist with these devices.

In order to improve the performance of smart windows in architectural glazing applications, two issues need to be addressed: the development of truly reflective smart window devices (like crystalline electrochromic materials, thermotropic glazings, and some of the recently discovered gasochromic materials), which are fully specular and (relatively) color neutral; and controlling the thermal emittance of glazings. The latter will broaden the range of climates in which smart windows can be usefully used, and will also open up a new range of nonarchitectural applications.

The future of smart windows in architectural applications is very bright; the growing trend toward sustainable buildings and energy conservation, coupled with some recent reviews of the total life cycle performance [81] and user acceptance [82] of smart windows, which show that these systems offer a genuine technology for reducing greenhouse emissions and are acceptable to users, suggest that widespread uptake of these technologies is likely within the next 10 years.

References

1. C.G. Granqvist, E. Avendaño, and A. Azens, Electrochromic coatings and devices: Survey of some recent advances, *Thin Solid Films* 442, 201–211, 2003.
2. E. Avendaño, L. Berggren, G.A. Niklasson, C.G. Granqvist, and A. Azens, Electrochromic materials and devices: Brief survey and new data on optical absorption in tungsten oxide and nickel oxide films, *Thin Solid Films* 496, 30–36, 2006.
3. P.R. Somani and S. Radhakrishnan, Electrochromic materials and devices: Present and future, *Materials Chemistry and Physics* 77, 117–133, 2002.
4. S.K. Deb, A Novel Electrophotographic system. *Appl. Oppt. Supp.* 3, 192–195, 1969.
5. B.-S. Yu, E.-S.Kim, and Y.-W. Lee, Developments in suspended particle devices (SPD), *SPIE* 3138, 217–225, 1997.
6. G.P. Montgomery, Polymer-dispersed and encapsulated liquid crystal films, in C.M. Lampert and C.G. Granqvist, eds., *Large Area Chromogenics: Materials and Devices for Transmittance Control*, SPIE, Washington, DC, 1998, p. 577.
7. K.A. Khan and C.G. Granqvist, Thermochromism of sputter deposited vanadium oxyfluoride coatings, in C.M. Lampert and C.G. Granqvist, eds., *Large Area Chromogenics: Materials and Devices for Transmittance Control*, SPIE, Washington, DC, 1998, p. 160.
8. A. Georg, W. Graf, D. Schweiger, V. Wittwer, P. Nitz, and H. Wilson, Switchable glazing with a large dynamic range in total solar energy transmittance (TSET), *Solar Energy* 62(3), 215–228, 1998.
9. A. Bessière, C. Duhamel, J.-C. Badot, V. Lucas, and M.-C. Certiat, Study and optimization of a flexible electrochromic device based on polyaniline, *Electrochim. Acta* 49, 2051–2055, 2004.
10. L.-M. Huang, C.-H. Chen, and T.-C. Wen, Development and characterization of flexible electrochromic devices based

on polyaniline and poly(3,4-ethylenedioxythiophene)-poly(styrene sulfonic acid), *Electrochim. Acta* 51, 5858–5863, 2006.
11. S.-W. Huang and K.-C. Ho, An all-thiophene electrochromic device fabricated with poly(3-methylthiophene) and poly(3,4-ethylenedioxythiophene), *Sol. Energy Mater. Sol. Cells* 90, 491–505, 2006.
12. P. Manisankar, C. Vedhi, G. Selvanathan, and H.G. Prabu, Electrochemical synthesis and characterization of novel electrochromic poly(3,4-ethylenedioxythiophene-co-diclofenac) with surfactants, *Electrochim. Acta* 51, 2964–2970, 2006.
13. G.-S. Liou, N.-K. Huang, and Y.-L. Yang, New soluble triphenylamine-based amorphous aromatic polymides for high performance blue-emitting hole-transporting and anodically electrochromic materials, *Polymer* 47, 7013–7020, 2006.
14. G.-S. Liou, S.-H. Hsiao, and T.-H. Su, Synthesis, luminescence and electrochromism of aromatic poly(amine-amide) s with pendent triphenylamine moieties, *J. Mater. Chem.* 15, 1812–1820, 2005.
15. A.B.S. Bahaj et al., Potential of emerging glazing technologies for highly glazed buildings in hot arid climates, *Energy Buildings* (2007), doi:10.1016/j.enbuild.2007.05.006
16. J.M. Bell, G.B. Smith, A.T. Baker, L.A. Evans, I.L. Skryabin, J. Barczynska, S.G. Bosi, J.J. Bright, D.R. Macfarlane, B.O. West, L. Spiccia, G. Voelkel, A. Koplik, B.Greaves, G.E. Tulloch, S.M. Tulloch, S.M. Jenkins, I.D. Watkins, M. Williams, D. Frost, G. Vogelman, and G. Evans, Intelligent Glazing Materials for Energy Efficient Heating, Lighting and Cooling of Buildings, end of project report, Energy Research and Development Corporation, Canberra, May 1996.
17. G.L. Yoong and G.E. Tulloch, Smart windows-major energy savings for the built environment in the tropics. Proceedings of the Eurosun, Copenhagen, Denmark, June 19–22, 2000.
18. F. Gugliermetti and F. Bisegna, Visual and energy management of electrochromic windows in Mediterranean climate, *Building Environment* 38, 479–492, 2003.
19. F. Yika and M. Bojic, Application of switchable glazing to high-rise residential buildings in Hong Kong, *Energy and Buildings* 38, 463–471, 2006.
20. T. Inoue, Solar shading and daylighting by means of autonomous responsive dimming glass: Practical application, *Energy Buildings* 35, 463–471, 2003.
21. W.J. Platzer, Determination of key glazing parameters: Part 1—Total solar energy transmittance. Proceedings of Windows Innovations Conference, WIC '95, pp. 22–31, Toronto, Canada, 1995.
22. C.G. Granqvist. Introduction to materials science for solar energy conversion systems, in A.A.M. Sayigh, ed., *Materials Science for Solar Energy Conversion Systems*, Pergamon Press, New York, 1991, p. 1.
23. ASHRAE Handbook of Fundamentals, American Society of Heating, Refrigeration and Air Conditioning Engineers, Inc., Atlanta, 1981, Chapter 27.
24. S.E. Selkowitz and C.M. Lampert. Applications of large area chromogenics to architectural glazing, in C.M. Lampert and C.G. Granqvist, eds., *Large Area Chromogenics: Materials and Devices for Transmittance Control*, SPIE, Washington, DC, 1998, p. 22.
25. A.W. Czanderna, J.-G. Zhang, C.E. Tracy, D.K. Benson, and S.K. Deb, Accelerated life testing of large area electrochromic devices for window application. *Proc. SPIE*, 3138, 68–77, 1997.
26. R. Sullivan, S. Selkowitz, P. Lyons, P.C. Thomas, I. Heimonen, O. Andresen, H. Aschehoug, P. Simmler, Eggimann, and T. Frank, Energy simulation studies in IEA/SHC Task 18 advanced glazings and associated materials for solar and building applications. Proceedings of Windows Innovations Conference, WIC '95, pp. 635–662, Toronto, Canada, 1995.
27. P.K. Nair, M.T.S. Nair, A. Fernandez, and M. Ocampo, Prospects of chemically deposited chalcogenide thin films for solar control applications. *J. Phys. D*. 22(6), 829–836, 1989.
28. J. Ferber and W. Platzer, Correct optical measurement of scattering samples. *Proc. SPIE* 2255, 708–717, 1994.
29. M.D. Rubin, IEA Task 18, Project B3: Chromogenic glazing, Final report, 1997.
30. Sage Electrochromics, http://www.sage-ec.com/pages/technol.html (accessed 15 October 2007).
31. J.-G. Zhang, D.K. Benson, C.E. Tracy, and S.K. Deb, The influence of microstructure on the electrochromic properties of Li_xWO_3 films: Part II. Limiting mechanisms in colouring and bleaching processes. *J. Mater. Res.* 8(10), 2657–2667, 1993.
32. C.G. Granqvist, A. Azens, A. Hjelm, L. Kullman, G.A. Niklasson, D. Ronnow, M.S. Mattson, M. Veszelei, and G. Vaivars, Recent advances in electrochromics for smart windows applications. *Solar Energy* 64(4), 199–216, 1998.
33. G.E. Tulloch, I.L. Skryabin, G. Evans, and J.M. Bell, Operation of electrochromic devices prepared by sol–gel methods. *Proc. SPIE* 3136, 426–432, 1997.
34. B.W. Faughnan and R.S. Crandall, Electrochromic display based on WO_3 in J.I. Pankove, ed., *Topics in Applied Physics V40, Display Devices*, Spring-Verlag, New York, 1980, Chapter 5.
35. T. Kase, T. Miyamoto, T. Yoshimoto, Y. Ohsawa, H. Inaba, and K. Nakase, Performance of tungsten oxide/prussian blue device, in C.M. Lampert and C.G. Granqvist, eds., *Large Area Chromogenics: Materials and Devices for Transmittance Control*, SPIE, Washington, DC, 1998, p. 504.
36. J.M. Bell, J. Barczynska, L.A. Evans, K.A. MacDonald, J.Wang, D.C. Green, and G.B. Smith, Electrochromism in sol-gel deposited TiO_2 films, *Proc. SPIE*. 2255, 324–331, 1994.
37. A. Agrawal, J.P. Cronin, and R. Zhang, Review of solid state electrochromic coatings produced using sol–gel techniques. *Solar Energy Mater. Solar Cells* 31(1), 9–22, 1993.
38. C. Wang and J.M. Bell, Influence of deposition temperature on electrochromic properties of sputtered WO_3 thin films. *Solar Energy Materials and Solar Cells* 43(4), 377–391, 1996.

39. J.S.E.M. Svensson and C.G. Granqvist, Modulated transmittance and reflectance in crystalline electrochromic WO_3 films: Theoretical limits. *Appl. Phys. Lett.* 45(8), 828–830, 1984.
40. S.C. Yang, Conducting polymer as electrochromic material: Polyaniline, in C.M. Lampert and C.G. Granqvist, eds., *Large Area Chromogenics: Materials and Devices for Transmittance Control*, SPIE, Washington, DC, 1998, p. 335.
41. R. Jiang and S. Dong, Chromatic reaction of polyaniline film and its characterization, *J. Chem. Soc., Faraday Trans.* 85(7), 1585, 1989.
42. C.M. Lampert and C.G. Granqvist. Introduction to chromogenics, in C.M. Lampert and C.G. Granqvist, eds., *Large Area Chromogenics: Materials and Devices for Transmittance Control*, SPIE, Washington, DC, 1998, p. 2.
43. M.A. Sobhan, R.T. Kivaisi, B. Stjerna, and C.-G. Granqvist, Thermochromism of sputter-deposited $W_xV_{1-x}O_2$ films. *Solar Energy Materials and Solar Cells* 44(4), 451–455, 1996.
44. G.V. Jorgenson and J.C. Lee. Thermochromic materials and devices: Inorganic systems, in C.M. Lampert and C.G. Granqvist, eds., *Large Area Chromogenics: Materials and Devices for Transmittance Control*, SPIE, Washington, DC, 1998, p. 142.
45. C.M. Lampert, *Science and Technology of Electrochromics: Lecture Notes*, SPIE Short Course, 1993.
46. A. Beck, W. Körner, H. Scheller, J. Fricke, W.J. Platzer, and V. Wittwer, Control of solar insolation via thermochromic light switching gels. *Solar Energy Materials and Solar Cells* 36, 339–347, 1995.
47. H.R. Wilson, Potential of thermotropic layers to prevent overheating—a review, *Proc. SPIE* 2255, 214–225, 1994.
48. C.M. Lampert. Introduction to liquid crystals, in C.M. Lampert and C.G. Granqvist, eds., *Large Area Chromogenics: Materials and Devices for Transmittance Control*, SPIE, Washington, DC, 1998, p. 550.
49. N. Basturk and J. Grupp. Liquid crystal guest host devices and their use as light switches, in C.M. Lampert and C.G. Granqvist, eds., *Large Area Chromogenics: Materials and Devices for Transmittance Control*, SPIE, Washington, DC, 1998, p. 557.
50. S. Khoobiar, Particle to particle migration of hydrogen atoms on platinum–alumina catalysts from particle to neighbouring particles. *J. Phys. Chem.* 68(2), 411–412, 1964.
51. D. Schweiger, A. Georg, W. Graf, and V. Wittwer, Examination of the kinetics and performance of a catalytically switching (gasochromic) device. *Solar Energy Materials and Solar Cells* 54, 99–108, 1998.
52. A. Georg, W. Graf, R. Neumann, and V. Wittwer, Stability of gasochromic WO_3 films. To be published in Proceedings of Eurosun 98, Portoroz, Slovenia, 1998.
53. A. Georg, W. Graf, R. Neumann, and V. Wittwer, Mechanism of the gasochromic colouration in WO_3 films. To be published in Proceedings of the Third International Meeting on Electrochromics (IME-3), London, England, 1998.
54. P. van der Sluis, New Electrochromic materials based on metal hydrides. To be published in Proceedings of the Third International Meeting on Electrochromics (IME-3), London, England, 1998.
55. K. von Rottkay, M. Rubin, F. Michalak, R. Armitage, T. Richardson, and J. Slack, Effect of hydrogen insertion on the optical properties of Pd-coated magnesium lanthanides. To be published in Proceedings of the Third International Meeting on Electrochromics (IME-3), London, England, 1998.
56. P. van der Sluis, M. Ouwerkerk, and P.A. Duine, Optical switches based on magnesium lanthanide alloy hydrides. *Appl. Phys. Lett.* 70(25), 3356–3358, 1998.
57. J.M. Bell, I.L. Skryabin, and G. Vogelmann, Towards a model for large area electrochromic device, Proceedings of the Third Symposium on Electrochromic Materials, The Electrochemical Society, 96-24, pp. 396–403, 1997.
58. C.M. Lampert and C.G. Granqvist, eds., *Large Area Chromogenics: Materials and Devices for Transmittance Control*, SPIE, Washington, DC, 1998, Parts 5, 6, 7 and 9.
59. J. Göttsche, A. Hinsch, and V. Wittwer, Electrochromic mixed WO_3-TiO_2 thin films produced by sputtering and the sol-gel technique: A comparison. *Solar Energy Materials and Solar Cells* 31(3), 415–428, 1993.
60. J. Wang, J.M. Bell, and I.L. Skryabin, Simulation model for electrochromic process at WO_3 electrode by a current step and its application to the smart window system. *Proc. SPIE* 3138, 20–30, 1997.
61. S.K. Mohapatra, Electrochromism in Li_xWO_3. *J. Electrochem. Soc.* 125(2), 284–288, 1978.
62. B. Reichman and A.J. Bard, A digital simulation model for electrochromic processes at WO_3 electrodes. *J. Electrochem. Soc.* 127(3), 647–654, 1980.
63. C. Ho, I.D. Raistrick, and R.A. Huggins, Application of A-C techniques to the study of lithium diffusion in tungsten trioxide thin films. *J. Electrochem. Soc.* 127(2), 343–350, 1980.
64. B. Vuillemin and O. Bohnke, Kinetics study and modelling of the electrochromic phenomenon in amorphous tungsten trioxide thin films in acid and lithium electrolytes. *Solid State Ionics* 68, 257–267, 1994.
65. O. Bohnke, M. Rezrazi, B. Vuillemin, C. Bohnke, P.A. Gillet, and C. Rousselot, "In situ" optical and electrochemical characterization of electrochromic phenomena into tungsten trioxide thin films. *Solar Energy Materials and Solar Cells* 25, 361–374, 1992.
66. M. Strømme, J. Isidorsson, G.A. Niklasson, and C.G. Granqvist, Impedance studies on Li insertion electrodes of Sn oxide and oxyfluoride. *J. Appl. Phys.* 80(1), 233–241, 1996.
67. C. Wang, J.M. Bell, and I.L. Skryabin, Degradation mechanisms in sol–gel deposited electrochromic devices. Proceedings of the 2nd International Meeting on Electrochromism— Durability Forum, San Diego, CA, 1996. (To appear in Solar Energy Materials and Solar Cells).
68. C.M. Lampert, S.J. Visco, M.M. Doeff, Y.P. Ma, Y. He, and J.-C. Giron, Characteristics for laminated electrochromic devices using polyorganodisulphide electrodes. *Proc. SPIE* 2017, 143–154, 1993.

69. X. Hu, X. Chen, and Z. Li, All-solid state electrochromic device with NiO/WO_3 complementary structure. *Proc. SPIE* 3138, 58–62, 1997.
70. C.G. Granqvist, *Handbook of Inorganic Electrochromic Materials*, Elsevier, Amsterdam, 1995.
71. J.M. Bell and J.P. Matthews, Glazing Materials. *Materials Forum* 22, 1–24, 1998.
72. S.F. Cogan, Electrochromic Vanadium Pentoxide, in C.M. Lampert and C.G. Granqvist, eds., *Large Area Chromogenics: Materials and Devices for Transmittance Control*, SPIE, Washington, DC, 1998, p. 313.
73. M. Takeski, K. Yoshimura, and S. Tanemura, Preparation and characterization of electrochromic nickel oxide thin film by sol–gel method. Proceedings of Windows Innovations Conference, WIC '95, pp. 365–371, Toronto, Canada, 1995.
74. P.F. Gerhardinger and R.J. McCurdy, Float-line deposited transparent conductors—implications for the PV industry. *Mater. Res. Soc. Symp. Proc.* 426, 399–410, 1996.
75. M.A. De Paoli, G. Casalbore-Miceli, E.M. Girotto, and W. A. Gazotti Jr., All polymeric solid-state electrochromic devices. To be published in Proceedings of the Third International Meeting on Electrochromics (IME-3), London, England, 1998.
76. V.S. Bagotzky and A.M. Skundin, *Chemical Power Sources*, Academic Press, London, 1980.
77. J.M. Bell and I.L. Skryabin, Method of control of an electrochromic device, Australian Patent PO3010, 1996.
78. R.A. Batchelor, Electrochromic Devices, PCT/GB97/00199, 2/02/96.
79. Sage Electrochromics Inc, http://www.sage-ec.com (accessed 15 October 2007).
80. Pilkington Company History, 1950–1999, http://www.pilkington.com/about+pilkington/company+briefing/company+history/1950+-+1999.htm (accessed 15 October 2007).
81. E. Syrrakou, S. Papaefthimiou, and P. Yianoulis, Eco-efficiency evaluation of a smart window prototype, *Science of the Total Environment* 359, 267–282, 2006.
82. M. Zinzi, Office worker preferences of electrochromic windows: a pilot study, *Building and Environment* 41, 1262–1273, 2006.

15

Flip-Chip Underfill: Materials, Process, and Reliability

Zhuqing Zhang
Hewlett-Packard Company

C.P. Wong
Georgia Institute of Technology

15.1 Introduction

The brain of modern electronics is the integrated circuit (IC) on the semiconductor chip. In order for the brain to control the system, interconnects need to be established between the IC chip and other electronic parts, power and ground, and inputs and outputs. The first-level interconnect usually connects the chip to a package made of either plastics and or ceramics, which in turn is assembled onto a printed circuit board (PCB). Three main interconnect techniques are used: wire-bonding, tape automated bonding (TAB), and flip-chip. In a wire-bonded package, the chip is adhered to a carrier substrate using die-attach adhesive with the active IC facing up. A gold or aluminum wire is bonded between each pad on the chip and the corresponding bonding surface on the carrier as shown in Figure 15.1. The chip and the wire interconnections are usually protected by encapsulation. TAB, on the other hand, uses a prefabricated lead frame carrier with copper leads adapted to the IC pads. The copper is usually gold-plated to provide a finish for bonding to the IC chip pads. The chip is attached onto the carrier and either thermosonic/thermocompression bonding or Au/Sn bonding is used to establish the interconnect. Both wire-bonding and TAB interconnects are limited to peripheral arrangement and therefore low input/output (I/O) counts. Flip-chip, however, can utilize the entire semiconductor area for interconnects. In a flip-chip package, the active side of an IC chip is faced down toward and mounted onto a substrate [1]. Interconnects, in the form of solder bumps, stud bumps, or adhesive bumps, are built on the active surface of the chip, and are joined to the substrate pads, in either a melting operation, adhesive joining, thermosonic, or thermocompression process. Figure 15.2 shows an example of solder-bumped chip surface for flip-chip interconnect. Since flip-chip was first developed about 40 years ago, many variations of the flip-chip design have been developed, among which, the controlled collapse chip connection (also known as C4) invented by IBM in 1960s is the most important form of flip-chip [2]. Compared with conventional packaging using wire-bonding technology, flip-chip offers many advantages such as high I/O density, short interconnects, self-alignment, better heat dissipation through the back of the die, smaller footprint, lower profile, and high throughput, etc. The outstanding merits of flip-chip have made it one of the most attracting techniques in modern electronic packaging, including MCM modules, high-frequency communications, high-performance computers, portable electronics, and fiber optical assemblies.

Until the late 1980s, flip-chips were mounted onto silicon or ceramic substrates. Low-cost organic substrates could not be used due to the concern of the thermal–mechanical fatigue life of the C4 solder joints. This thermal–mechanical issue mainly arises from the coefficient of thermal expansion (CTE) mismatch between the semiconductor chip (typically Si, 2.5 ppm/°C) and the substrate (4–10 ppm/°C for ceramics and 18–24 ppm/°C for organic FR4 board). As the distance from the neutral point (DNP) increases, the shear stress at the solder joints increases accordingly. So with the increase in the chip size, the thermal–mechanical

FIGURE 15.1 First level interconnect using wire-bonding.

FIGURE 15.2 Area array solder bumps for flip-chip interconnect.

reliability becomes a critical issue. Organic substrates have advantages over ceramic substrates because of their low cost and low dielectric constant. But the high CTE differences between the organic substrates and the silicon chip exert great thermal stress on the solder joint during temperature cycling.

In 1987, Hitachi first demonstrated the improvement of solder fatigue life with the use of filled resin to match solder CTE [3]. This filled resin, later called "underfill" was one of the most innovative developments to enable the use of low-cost organic substrate in flip-chip packages. Underfill is a liquid encapsulate, usually epoxy resins heavily filled with fused silica (SiO_2) particles, which is applied between the chip and the substrate after flip-chip interconnection. Upon curing, the hardened underfill exhibits high modulus, low CTE matching that of the solder joint, low moisture absorption, and good adhesion toward the chip and the substrate. Thermal stresses on the solder joints are redistributed among the chip, underfill, substrate, and all the solder joints, instead of concentrating on the peripheral solder joints. It has been demonstrated that the application of underfill can reduce the all-important solder strain level to 0.10–0.25 of the strain in joints, which are not encapsulated [4,5]. Therefore, underfill can increase the solder joint fatigue life by 10 to 100 times. In addition, it provides an environmental protection to the IC chip and solder joints. Underfill becomes the practical solution to extending the application of flip-chip technology from ceramics to organic substrates, and from the high-end to cost-sensitive products. Today, flip-chip is being extensively studied and used by almost all major electronic companies around world including Intel, AMD, Hitachi, IBM, Delphi, Motorola, Casio, etc.

15.2 Conventional Underfill Materials and Process

The generic schematic of a flip chip package is shown in Figure 15.3. Conventional underfill is applied after the flip-chip interconnects are formed. The resin flows into the gap between the chip and the substrate by a capillary force. Therefore, it is also called "capillary underfill." A typical capillary underfill is a mixture of liquid organic resin binder and inorganic fillers. The organic binders are often epoxy resin mix, although cyanate ester or other resin has been used for underfill application as well. Figure 15.4 shows the chemical structure of some commonly used

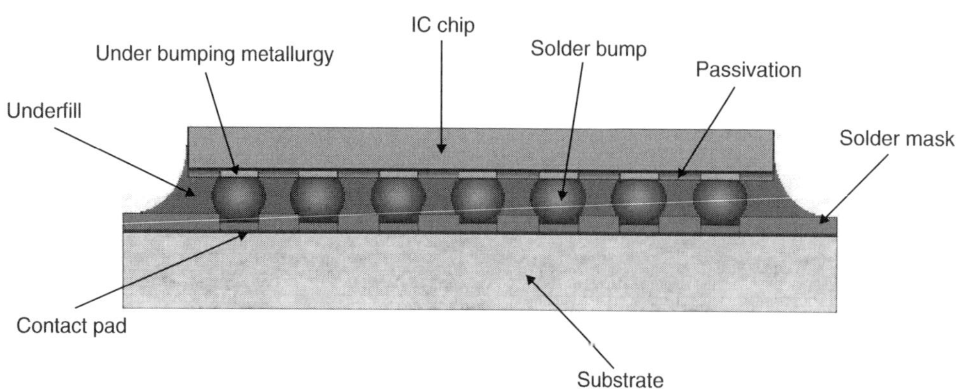

FIGURE 15.3 Generic configuration of C4 with underfill.

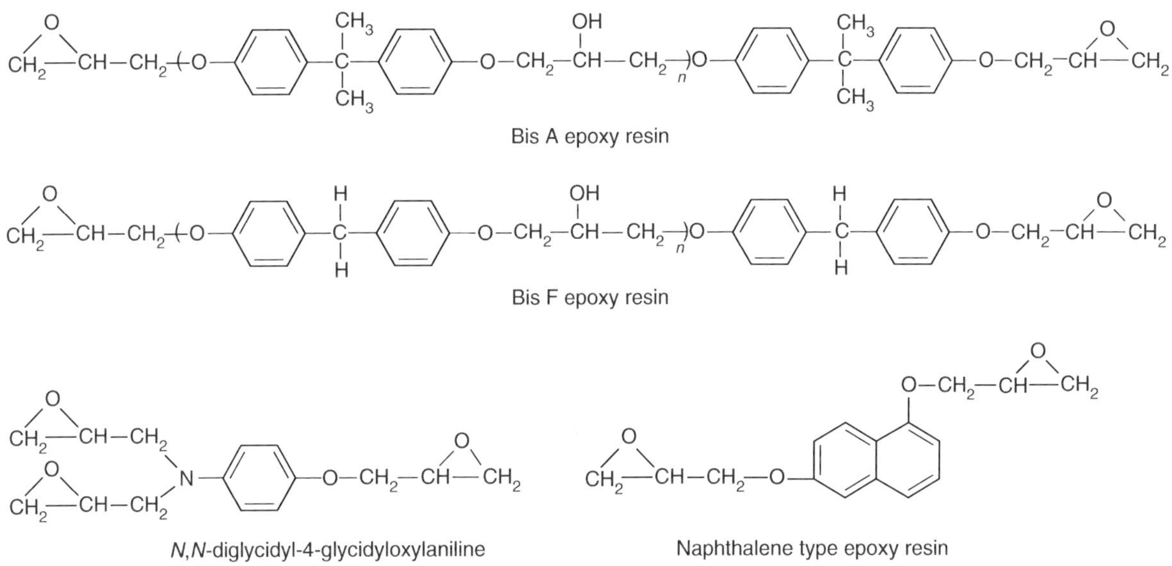

FIGURE 15.4 Typical epoxy resin structures used in underfill.

epoxy resins. In addition to epoxy resin, a hardener is often used to form a cross-linking structure upon curing. Sometimes a latent catalyst is incorporated to achieve long pot life and fast curing. Inorganic fillers typically used in underfill formulation are micron-sized silica. The silica fillers are incorporated into the resin binder to enhance the material properties of cured underfill such as low CTE, high modulus, and low moisture uptake, etc. Other agents that can be found in an underfill formulation include adhesion promoters, toughening agents, and dispersing agents, etc. These chemicals are incorporated to help the resin mixing and enhance the cured underfill properties.

Figure 15.5 shows the process steps of flip-chip with conventional underfill. Separate flux dispensing and cleaning steps are required before and after the assembling of the chip, respectively. After the chip is assembled onto the substrate, the underfill is usually needle-dispensed and is dragged into the gap between the chip and the substrate by a capillary force. Then a heating step is needed to cure the underfill resin to form a permanent composite.

The flow of the capillary underfill has been extensively studied since it is considered to be one of the bottlenecks for the flip-chip process. The capillary flow is usually slow and can be incomplete, resulting in voids in the packages and also nonhomogeneity in the resin/filler system. The filling problem becomes even more serious as the chip size increases. The flow modeling of flip chip underfill is often approximated as the viscous flow of the underfill adhesive between two parallel plates. One can use the Hele-Shaw model to simulate the underfill flow with the above approximation. The time required to fill a chip of length L can be calculated as [6]

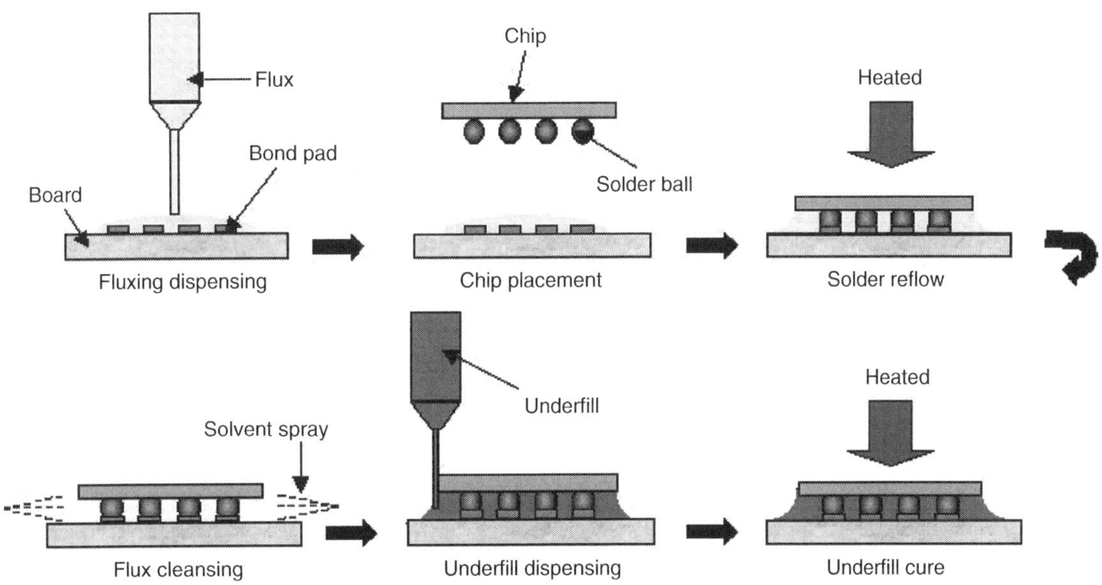

FIGURE 15.5 Flip-chip process using conventional underfill.

$$f(x)=\left(\frac{\beta}{x}\right)\left(\frac{x}{\theta}\right)^{\beta}\exp\left(-\left(\frac{x}{\theta}\right)^{\beta}\right) \quad (15.1)$$

where
- η is the underfill viscosity
- σ is the coefficient of the surface tension
- θ is the contact angle
- h is the gap distance

It is easily seen that a larger chip with a smaller gap distance would require longer time to fill.

The above approximation does not take the existence of solder bumps into account. It is shown that the approximation breaks down when the spacing between bumps is comparable to the gap height [7]. Therefore, this model cannot apply to high-density area array flip-chip applications. Using transparent quartz dies assembled onto different substrates, Nguyen et al. observed the flow of commercial underfills and used a 3D PLICE-CAD to model the underfill flow front [8]. A comparison between the peripheral and area array chips showed that the bumps enhanced the flatness of the flow front by providing periodic wetting sites. A racing effect along the edges was observed. Voids can be formed at the merging of flow fronts. The merging of the fronts also produced streaks, which are zones of no- or slow-moving fluids, leading to higher potential for filler settling.

Recent development in underfill flow models has also considered the effect of the contact angle on solder and bump geometry. A study by Young and Yang used a modified Hele-Shaw model considering the flow resistance in both the thickness direction between the chip and substrate, and the plane direction between solder bumps [9]. It was found that the capillary force parameter would approach a constant value at very large pitch for the same gap height. As the bump pitch is reduced, the capillary force will increase to a maximum as a result of underfill wetting on the solder given the contact angle on the solder is small, and then quickly drops to zero as the pitch approaches the bump diameter. Their study also showed that a hexagonal bump arrangement is more efficient to enhance the capillary force at critical bump pitch.

15.3 Reliability of Flip-Chip Underfill Packages

The reliability of a flip-chip package can be evaluated in a number of different methods, including thermal cycling, thermal shock, pressure-cook test, etc. The lifetime of a solder joint interconnect during temperature cycling can often be described by statistical models such as Weibull distribution. The probability density function (PDF) of the Weibull distribution is given by

$$f(x)=\left(\frac{\beta}{x}\right)\left(\frac{x}{\theta}\right)^{\beta}\exp\left(-\left(\frac{x}{\theta}\right)^{\beta}\right) \quad (15.2)$$

where
- x is the thermal cycling life as a random variable
- θ is the characteristic life
- β is the shape parameter

The mean time to failure (MTTF), which is the expectation of the time to failure, for the Weibull distribution is

$$\text{MTTF} = \theta\Gamma\left(1+\frac{1}{\beta}\right) \quad (15.3)$$

where Γ is the gamma function. It is generally believed that fatigue of the solder joints is a major reason for structure and electrical failures. The solder fatigue life can be described as a function of inelastic shear strain in the Coffin-Manson equation [10,11]:

$$N_f = \frac{1}{2}\left(\frac{\Delta\gamma}{2\varepsilon_f'}\right)^{1/c} \quad (15.4)$$

where
- N_f is the number of cycles to fatigue failure
- $\Delta\gamma$ is the inelastic shear strain
- ε_f' is the fatigue ductility coefficient
- c is the fatigue ductility exponent

Other strain-based fatigue equations have been proposed, among which Solomon's model is often used [12]:

$$N_f = \left(\frac{\theta}{\Delta\gamma_p}\right)^{1/\alpha} \quad (15.5)$$

where $\Delta\gamma_p$ is the percentage inelastic shear strain, θ and α are constants.

It has been shown that the use of the underfill can increase the lifetime of the solder joints by at least an order of magnitude during thermal cycling [13]. It was found that in an underfilled flip-chip package, the fatigue life is highly dependent on the material properties of the underfill. The analytic model by Nysaether et al. [14] showed that while an underfill without filler increased the lifetime by a factor of 5–10, a filled underfill with a lower CTE gave a 20–24-fold increase in lifetime. For both filled and nonfilled samples, the lifetime is nearly constant regardless of DNP, indicating that the underfill effectively couples the stress among all the solder joints.

Many numerical models have been developed to study the solder fatigue life of a flip-chip package with or without underfill. The polymeric nature of the underfill material requires careful characterization for correct material property input to the numerical models. The modulus of a polymeric material is not only a function of temperature, but also a function of time, i.e., it is a viscoelastic material. Thermal mechanical analyzer (TMA) and dynamical mechanical analyzer (DMA) are typically used to characterize the viscoelastic properties of the underfill material. Dudek et al. characterized four commercial electronic polymers and used finite element (FE) analyses to study the effect of die size and underfill material properties on the thermomechanical reliability of the flip-chip on board (FCOB) package [15]. They found that although the use of underfill can effective reduce the shear strain, it can also cause bump creep strain in the transverse board direction due to stretching and compressing of the bump during thermal cycling. This load is due to the CTE mismatch between the solder and underfill/solder-mask layer. Underfill with CTE that matched the solder material

(22–26 ppm/°C) gives the best thermal cycling life based on the creep strain criterion.

The function of the underfill in a flip-chip package is stress redistribution, not stress reduction. A rigid underfill material mechanically couples the device and the substrate, changing partially the shear stress experienced by the solder joints into bending stress on the whole structure. Shrinkage of the underfill during cure and the CTE mismatch during cooling after cure can generate large stress on the Si chip, resulting in die crack in some cases. Palaniappan et al. performed in-situ stress measurements in the flip-chip assemblies using a test chip with piezoresistive stress-sensing devices [16]. The study concluded that the underfill cure process generates large compressive stress on the active die surface, indicating a complex convex bending state in the flip chip. The level of stress measured can lead to Si fracture. The residual die stress was found to be strongly dependent on underfill CTE, modulus, and T_g. A FE analysis by Mercado et al. on the die edge cracking in flip-chip PBGA packages also concluded that the energy release rate for horizontal Si facture increases with underfill modulus and CTE [17].

In addition to temperature-related thermomechanical failure, moisture-induced failures such as delamination and corrosion are common for a flip-chip underfill package. Highly accelerated stress test (HAST) is often used to determine the temperature and moisture sensitivity of the package. The test uses hash environment conditions such as high temperature, high humidity, and high pressure. A typical test condition can be 121°C, 100% relative humidity (RH), and 2 atm pressure. It has also been known as the autoclave or pressure cooker test (PCT). The moisture absorbed by the polymeric materials can hydrolyze the interfacial bonds between underfill and the die, resulting in delamination starting from the corner of the die, which further promotes the moisture diffusion along the interface. The moisture at the interface can cause corrosion of the solder joints and metal traces on the substrate. Delamination decouples the underfill with the Si die and can cause stress concentration on the surrounding solder joints, leading to early fatigue failure of those joints. The absorbed moisture also causes hygroscopic swelling. Lahoti et al. studied the combined effect of moisture and temperature on the reliability of flip-chip ball grid array (FCBGA) packages using FE analysis. The simulation results revealed the significance of hygroscopic-induced tensile stress on the under bump metallurgy (UBM) and interdielectric layer (ILD) [18].

Interfacial delamination of underfill to various materials such as die passivation, solder material, and solder mask on the substrate is a leading cause for failure in flip-chip underfill packages. One way to improve the reliability under temperature humid aging is to incorporate adhesion promoters, or coupling agents, into underfill to increase adhesion of underfill to the surrounding materials. Luo et al. studied six different coupling agents and their effect on underfill. The authors found that the incorporation of the coupling agents clearly affected the curing profile and bulk property of the underfill, such as T_g and modulus. The effect of coupling agents on adhesion and adhesion retention after temperature moisture aging was highly dependent on coupling agent type and interacting surfaces. The addition of titanate and zirconate coupling agents can improve adhesion of epoxy underfill with BCB-passivated silicon. However, with the addition of the same coupling agents, the adhesion strength of underfill with polyimide passivation decreased after aging at 85C/85% RH [19].

TABLE 15.1 Desirable Underfill Properties for Flip-Chip Packages

Curing Temperature	<150°C
Curing time	<30 min
T_g	>125°C
Working life (viscosity double @ 25°C)	>16 h
CTE (α_1)	22–27 ppm/°C
Modulus	8–10 GPa
Fracture toughness	>1.3 MPa*m
Moisture absorption (8 h boiling water)	<0.25%
Filler contents	<70 wt 1%

In summary, many studies have concluded that the underfill material property is one of the key factors determining the reliability of the package. The general guideline on the material properties of underfill for flip-chip packages can be summarized in Table 15.1. However, one has to keep in mind that different failure modes coexist in a reliability test, which sometimes present conflicting requirements on underfill. For instance, to effectively couple the stress on the solder joints, high-modulus underfill is desired. On the other hand, high underfill modulus can lead to a high residual stress and therefore die crack. Another example is the filler loading. Low CTE requirement on underfill indicates high filler loading. However, an underfill with higher filler loading typically has a higher viscosity, causing difficulty in underfill dispensing. The result might be underfill voids and nonuniformity, which would cause reliability issues. Therefore, the choice of underfill highly depends on the application, e.g., die size, passivation material, substrate material, type of solder, and environment conditions the package will be subjected to during application, etc.

15.4 New Challenges to Underfill

As silicon technology moves to sub-0.1 μm feature size, the demands for packaging also are involved as the bump pitch gets tighter, bump size smaller, die size larger for future flip-chips. As a result, the capillary underfill process faces tremendous challenges. As it was discussed previously, the underfill flow problem is aggravated as the size of the chip becomes larger and the gap between the chip and substrate gets smaller. Among the emerging development of flip chip, lead-free solder and low-*K* (dielectric constant) ILD/Cu present new challenges to underfill [20].

High-lead and lead-tin eutectic solders have been widely used for chip-package interconnections. Recent environmental legislations toward toxic materials and consumers' demand for green electronics have pushed the drive towards lead-free solders. Alternatives have been proposed using multiple combinations of elements like tin, silver, copper, bismuth, indium, and zinc, most

TABLE 15.2 Possible Lead-Free Alloys

Alloy	Melting Point (°C)
Sn96.5/Ag3.5	221
Sn99.3/Cu0.7	227
Sn/Ag/Cu	217 (ternary eutectic)
Sn/Ag/Cu/X(Sb, In)	Ranging according to compositions, usually above 210
Sn/Ag/Bi	Ranging according to compositions, usually above 200
Sn95/Sb5	232–240
Sn91/Zn9	199
Bi58/Sn42	138

of which require increased reflow temperature profiles during the soldering process relative to the well-known tin-lead alloys. Table 15.2 shows some of the common lead-free solders.

Among the several lead-free candidate solders, the near ternary eutectic Sn–Ag–Cu (SAC) alloy compositions, with melting temperatures around 217°C, are becoming consensus candidates. The optimal composition 95.4 Sn/3.1Ag/1.5Cu has provided a combination of good strength, fatigue resistance, and plasticity [21]. In addition, the alloy has sufficient supply and adequate wetting characteristics.

The use of the Sn/Ag/Cu solder presents two major challenges on the flip-chip assembly process. First, since the melting point of the alloy is more than 30°C higher than that of the eutectic Sn/Pb alloy, the process temperature is raised by 30°C–40°C. The high process temperature has a great impact on the substrate since the conventional FR-4 material has a T_g at around 125°C and also subjects the attached components to a higher thermal stress. Higher warpage is introduced when the board is subjected to higher temperature reflow. There have been considerable research in the high T_g substrate for the lead-free process. The second challenge comes from the flux chemistry. Since the current fluxes in use are usually designed for eutectic Sn/Pb solder, they either do not have high enough activity or do not possess sufficient thermal stability at high temperature. So generally, the wetting behavior of the lead-free solders is not as good as that of the eutectic Sn/Pb solder [22,23].

With the trends of a lead-free solder interconnect, the underfill for flip-chip in package application faces new challenges of compatibility with higher reflow temperature. High-temperature reflow causes component damage due to the enhanced level of material degradation, moisture ingress, and mechanical expansion. Therefore, the thermal stability, adhesion to various interfaces, strength, and fracture toughness of the underfill need to be improved. The SAC alloy does not plastically deform as much as the eutectic PbSn solder. The creep deformation is less at a lower stress level and more at higher stress level compared to the PbSn solder, which indicates that the choice of the underfill would depend on the application needs. A temperature cycle with a large temperature difference and lower dwelling times could induce more creep and therefore requires more protection from the underfill [24]. An evaluation of underfill materials for lead-free application conducted by Intel Corporation [25] showed that the majority of the failures occurred during the moisture sensitive level (MSL) 3 followed by 260°C reflow. Delamination seemed to be the common failure mechanism after the high-temperature reflow. This failure was also correlated to materials with low filler content and low coupling agent content. In general, materials with high filler content (and therefore low CTE, high modulus, and low moisture uptake) and good adhesion were compatible with the lead-free process.

As the IC fabrication moves toward small feature and high density, the interconnect delay becomes dominant. This calls out for new interconnect and ILD materials. The Cu metallurgy and low-*K* ILD has been successfully implemented to increase device speed and reduce power consumption. These low-*K* materials tend to be porous and brittle, having high CTE and low mechanical strength, compared to the traditional ILD materials such as SiO_2. The CTE mismatch between the low-*K* ILD and the silicon die creates a high thermomechanical stress at the interface. Therefore, the choice of underfill becomes critical since it not only protects the solder joints by stress redistribution, but also need to protect the low-*K* IDL and its interface with the silicon. Critical material properties of underfill to achieve reliability requirement for the low-*K* ILD package include the T_g, CTE, and the modulus. However, the optimal combination of these properties is still controversial.

Five underfills were evaluated for the low-*K* flip chip package by Tsao et al. [26]. Both modeling and experimental evaluation indicated that the low T_g and low stress-coupling index underfills yielded better reliability in the low-*K* flip-chip package. Two moderately low T_g underfills (T_g between 70°C and 120°C) showed good potential in protecting both the solder joints and low-*K* interface. An underfill with a very low T_g (lower than 70°C), on the other hand, failed to protect the solder joints during the thermal cycling test. A study conducted by LSI Logic Corp and Henkel Loctite (now called Henkel) Corp [27], on the other hand, indicated that underfills with high T_g and low modulus is advantageous for the low-*K* flip chip. The low modulus of the underfill exerts lower stress on the package and therefore reducing the stress on the low-*K* layer, preventing underfill delamination and die cracking. The high T_g prevents solder bump fatigue by maintaining a low CTE over the temperature cycling. The high T_g, low modulus underfill developed by Henkel exhibited good manufacturability and reliability in the package qualification testing including JEDEC preconditioning, thermal cycling, biased humidity testing, and high-temperature storage.

With all the new challenges to the flip-chip and underfill technology, capillary underfill is still the main packaging technology for flip-chip devices. However, the continuing shrinking of pitch distance and gap height will eventually post limitations on the capillary flow. The industry has started to look for alternatives to capillary underfill. The following sections describe the new developments in underfill material and processes.

15.5 No-Flow Underfill

The idea of integrated flux and underfill was patented by Pennisi and Papageorge in Motorola back in 1992 [28]. It triggered the research and development of the no-flow underfill process. The first no-flow underfill process was published by Wong and Baldwin in 1996 [29]. The schematic process steps are illustrated in Figure 15.6. Instead of underfill dispensing after the chip assembly in the conventional process, in a no-flow underfill process, the underfill is dispensed onto the substrate before the placement of the chip. Then the chip is aligned and placed onto the substrate and the whole assembly goes through solder reflow where the interconnection through solder balls is established while the solder melts. This novel no-flow process eliminates the separate flux dispensing and flux cleaning steps, avoids the capillary flow of underfill, and finally combines the solder bump reflow and underfill cured into a single step, hence, improves the production efficiency of the underfill process. It is a step forward for the flip-chip to be compatible with surface mount technology (SMT).

The key to the success of a no-flow underfill process lies in the underfill material. The first patent on the no-flow underfill material was by Wong and Shi at Georgia Institute of Technology [30]. The two critical properties of no-flow underfill to enable this new process to have a latent curing ability and the built-in fluxing capability. The nature of the no-flow underfill process requires that the underfill has enough reaction latency to maintain its low viscosity until the solder joints are formed. Otherwise, gelled underfill would prevent the melting solders from collapsing onto the contact pads, resulting in a low yield of the solder joint. On the other hand, elimination of the postcure is desired since postcure takes additional off-line process time, adding to the cost of this process. Many latent catalysts for epoxy resins have been explored for the application of no-flow underfill. In the material system that Wong and Shi designed, Co(II) acetylacetonate was used as the latent catalyst [31,32], which gave enough curing latency for no-flow underfill. The advantage of metal chelates lies not only in its latent acceleration, but also in the wide curing range they offer. By exploring different metal ions and chelates, the curing behavior of different epoxy resins could be tailored to the application of no-flow underfill for lead-free solder bumped flip-chip [33]. Since lead-free solders usually have a higher melting point than eutectic SnPb solder, no-flow underfill for lead-free bumped flip-chip requires higher curing latency to ensure the wetting of the lead-free solder on the contact pad. Zhang et al. explored 43 different metal chelates and developed no-flow underfill compatible with lead-free solder reflow [33]. Successful lead-free bumped flip-chip on a board package using the no-flow underfill process has been demonstrated [34].

Despite the importance of the curing process of no-flow underfill, there is little study on the curing kinetics and its relation to the reflow profile. In an attempt to develop systematic methodology to characterize the curing process of no-flow underfill, Zhang and Wong used an autocatalytic curing kinetic model with temperature-dependent parameters to predict the evolution of degree of cure (DOC) during the solder reflow process [35]. Figure 15.7 shows the result of DOC calculation of a no-flow underfill in eutectic SnPb and lead-free solder reflow process. If the DOC of the underfill at the solder melting temperature is lower than the gel point, the molten solder would be allowed to wet the substrate and make the interconnection. Another approach is the in situ measurement of viscosity of no-flow underfill using microdielectrometry by Morganelli and Wheelock [36]. Since the viscosity is related to the ionic conductivity, the dielectric properties of the underfill can be used for the in-situ analysis of the no-flow underfill in the reflow process, which can be used to predict the solder wetting behavior.

The other key property for no-flow underfill is the fluxing capability. In a conventional flip-chip process, flux is used to reduce and eliminate the metal oxide on the solder and to prevent it from being reoxidized under high temperature. Instead of applying flux, no-flow underfill is dispensed before the chip placement. Hence, the self-fluxing capability is required to facilitate solder wetting. To achieve this goal, research has been done to develop reflow-curable polymer fluxes [37]. A comprehensive study on the fluxing agent of no-flow underfill material was carried out by Shi et al. [38–40], which included the relationship between the surface composite on the Cu pad and the fluxing capability of no-flow underfill, and also the effect of the addition of the fluxing agent on the curing and material properties of no-flow underfill.

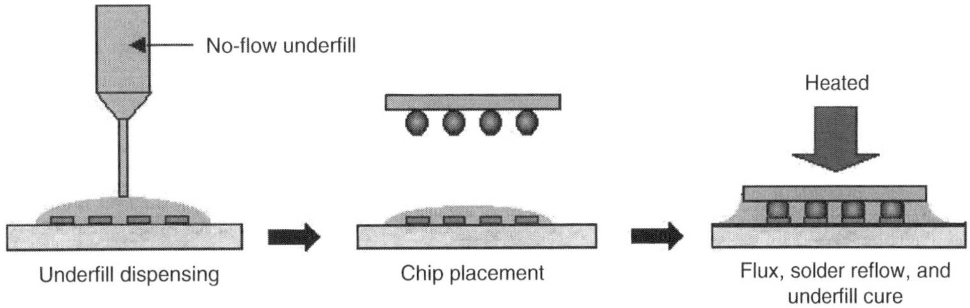

FIGURE 15.6 Flip-chip process using no-flow underfill.

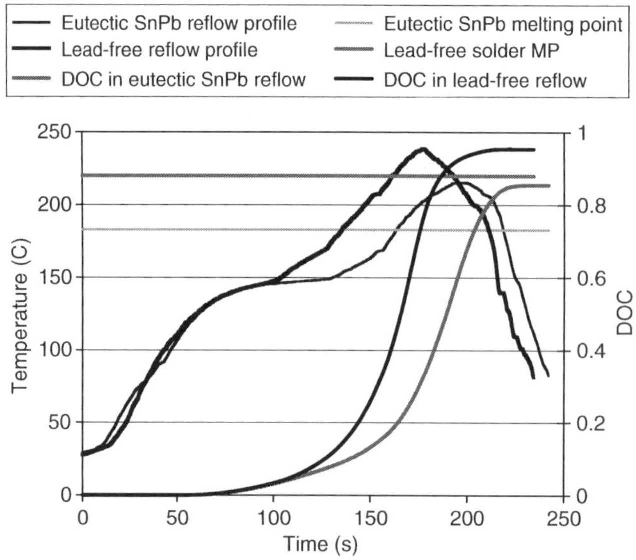

FIGURE 15.7 DOC evolution of a no-flow underfill in eutectic SnPb and lead-free reflow process.

The process of no-flow underfill has always attracted much attention in the assembly industry. Voids formation is often observed in many flip-chip no-flow underfill packages. The origin of the voids could be the out-gassing of the underfill, moisture in the board, and trapped voids during assembly, etc. They are usually tacked to a solder bump or in between two bumps [41,42]. Figure 15.8 shows an example of underfill voids in a no-flow underfill package observed with scanning acoustic microscope. Voids in the underfill, especially voids near the solder bumps lead to early failure through a number of ways including stress concentration, underfill delaminate, and solder extrusion. Studies by Wang et al. have showed that solder bridging might result from the solder bump extrusion through the microvoids trapped between adjacent bumps [43]. The material and process factors influencing the voiding behavior are complicated and interacting. It has been shown that the outgassing of anhydride could cause severe voiding, if the curing latency is high and also the reflow temperature is high; hence, the voiding becomes more prominent in a lead-free reflow process [44]. The important process parameters that affects underfill voiding in a no-flow process include the underfill dispensing pattern, the solder mask design, the placement force and speed, and the reflow profile, etc. [45,46]. Before assembly, the PWB substrate needs to be baked to dry out any moisture to prevent voiding from the board [42]. It has been shown that in some cases a fast gelation of underfill is desired to minimize the voiding while in other cases, extending duration at high temperature can "push" out the voids [42,47]. In short, with the right material and process parameters, voiding in no-flow underfill can be minimized. However, the process window is usually very narrow. An important point was raised by Zhao and Wong [34] that for a small circuit board where the temperature distribution is more homogenous, it is relatively easy to develop a "good" reflow profile while for complex SMT assemblies involving multiple components and significant thermal mass difference across the board, the optimization of the reflow process presents a great challenge.

The reliability of a flip-chip no-flow underfill package has been evaluated by many researchers. Discrepancies exist among these reports because the process and reliability of the no-flow

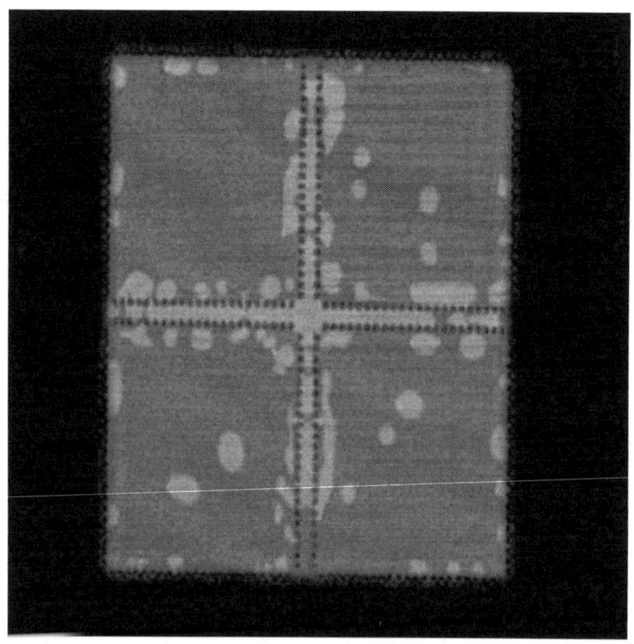

FIGURE 15.8 Example of voids in a no-flow underfill package.

underfill package depends largely on the package designs including the size of the chip, the pitch, the surface finish of the pad, etc. Among the earliest reporters on no-flow underfill, Gamota and Melton compared the reliability and typical failure mode of a conventional underfill package and no-flow underfill packages [48]. They found that in a conventional underfill assembly, the failure of the assembly mainly resulted from the interfacial delamination between the underfill and the chip passivation. However, with unfilled materials in no-flow underfill, good interfacial integrity was observed, and the assembly failed mainly due to the fracture through the solder interconnects near the PCB. Since the no flow underfill was unfilled, the CTE was high. They argued that the relative localized CTE mismatch between the chip, the underfill, and the PCB resulted in a high local stress field which initiated fracture in the solder interconnects. No-flow underfill without silica fillers or very low filler loadings is not only high in CTE, but also low in fracture toughness [49]. Combined with high CTE mismatch, the low fracture toughness leads to early underfill cracking both inside the bulk and in underfill fillet. Fillet cracking causes delamination between the underfill and the die passivation and/or between the underfill and the board, while bulk cracking can initiate solder joint cracking and solder bridging [50]. These all become the common failure modes for the flip-chip no-flow underfill package. Efforts have been made to enhance the toughness of the no-flow underfill materials through the incorporation of toughening agents [51]. The effect of the glass transition temperature (T_g) of the no-flow underfill on the reliability of the package has been controversial. It is usually believed that the T_g of the underfill should exceed the upper limit of the temperature cycling (125°C or 150°C) to ensure consistent material behavior during the reliability test. However, some tests have shown that low T_g (~70°C) underfill material performed better in liquid-to-liquid thermal shock (LLTS) [52]. The research by Zhang et al. on the development of nonanhydride based no-flow underfill [53] also showed that high T_g is not critical to reliability. Although the CTE of the underfill above T_g is much higher than that below T_g, the modulus of the underfill decreases dramatically; so the overall stress in the underfill does not necessarily increase when the environment temperature exceeds its T_g. But high T_g might result in a higher residue stress inside the underfill after the material cools down after curing, which leads to an early crack in the underfill.

15.5.1 Approaches of Incorporating Silica Fillers into No-Flow Underfill

The previous research has shown that the correlation between the material properties and package reliability in the case of flip-chip underfill is very complicated. It is difficult to separate the effect of each factor since the material properties are often correlated with each other. However, it is generally agreed that low CTE and high modulus are favorable for high interconnect reliability [54]. Hence, the inclusion of silica fillers into the underfill is critical to enhance the reliability. However, since the underfill is predeposited on the board before the chip assembly in a no-flow process, the fillers are easily trapped in between the solder bump and contact pad and hinder the interconnection [55]. Thermocompression reflow (TCR) has been used to exclude the silica filler from the solder joint [56]. The process step is illustrated in Figure 15.9. In a TCR process, the underfill is dispensed onto a preheated substrate. The chip is then picked and bonded to the substrate and held at an elevated temperature under force for a certain period of time for solder joint formation. The assembly is postcured afterwards. It was found that the bonding force and temperature were important factors influencing yield. A detailed study was carried out by Kawamoto et al. at NAMICS Corporation to determine the effect of filler on the solder joint connection in a TCR-like no-flow underfill process [57]. The study used two different sizes of silica fillers at different loading levels. It was found that a good solder connection can be made with underfill with up to 60 wt% filler loading without filler surface treatment. Smaller filler tended to increase the viscosity of the underfill and more fillers remained at the solder interface due to the larger number of fillers at the same weight percentage loading. The study also found that proper surface treatment of the fillers can lower the underfill viscosity and increase yield at high filler loading.

Other approaches have been explored to incorporate silica fillers into no-flow underfill. In a novel patented process, Zhang et al. used a double-layer no-flow underfill [58], in which two layers of no-flow underfill are applied. The bottom layer underfill is relatively high in viscosity and is not filled with silica fillers. It is applied onto the substrate first; then the upper layer underfill, which is filled with silica fillers, is dispensed. The chip is then placed onto the substrate and reflowed, during which the solder joints are formed and the underfill is cured or partially cured.

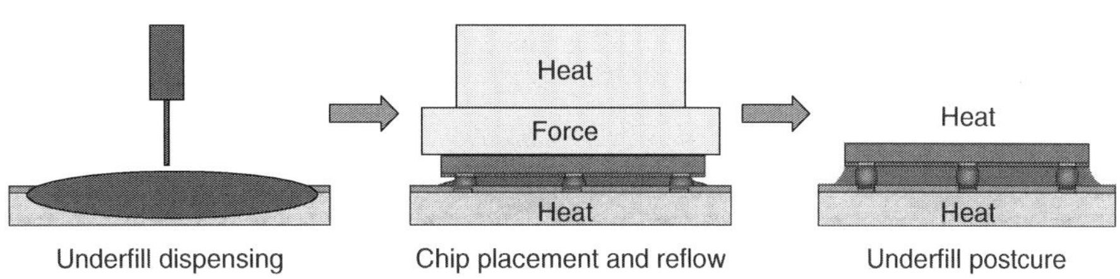

FIGURE 15.9 TCR for flip-chip.

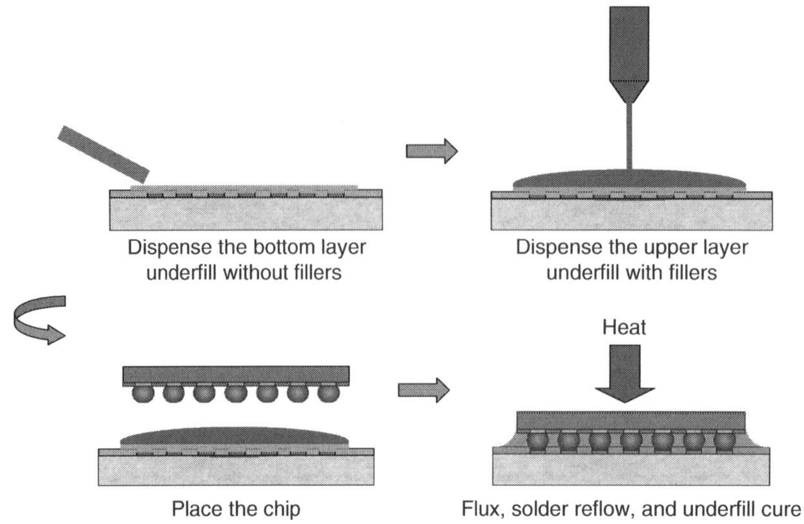

FIGURE 15.10 Double-layer no-flow underfill process.

The process flow chart is illustrated in Figure 15.10. It was demonstrated that high yield was achieved using an upper layer underfill of 65 wt% silica filled [59]. Further investigations on the process indicated that factors affecting the interconnection yield of the double-layer no-flow underfill are complicated and interacting with each other [60]. The process window is narrow and the thickness and the viscosity of the bottom layer underfill are essential to the wetting of the solder bumps. And of course, it adds on another step in the flip-chip process and has a higher process cost.

The recent advances in nanoscience and nanotechnology have enabled innovative research in materials for electronic packaging. It was found that nanosized silica fillers with surface modification can be mixed with thermosetting resins to provide a uniform dispersion of nonagglomerated particles. Used as no-flow underfill, the nanocomposite materials allowed 50 wt% filler loading with a good interconnect yield [61]. This high-performance no-flow underfill developed by 3M used 123 nm silica filler. With filler loading of 50 wt%, the CTE of the material was 42 ppm/°C and the good interconnect yield was achieved using a PB10 die (5°5 mm, 64 peripheral bumps). A joint research study was conducted by 3M and Georgia Institute of Technology in the process and a reliability evaluation of the nanosilica-incorporated no-flow underfill [62]. Figure 15.11 shows a SEM picture of the solder joint in the presence of no-flow underfill with nanosilica fillers. A 1.5× increase of characteristics life was observed in the air-to-air thermal cycling (AATC) reliability test with the nanosilica fillers. Although the nanocomposite no-flow underfill material shows good potential for a highly reliable flip-chip package using a SMT-friendly no-flow underfill process, the fundamental mechanism of the nanosilica interaction with the solder joints and the underfill is still not well understood. Since nanosize particles have a large surface area and tend to form irregular agglomerations, which increase their difficulty to be incorporated into a binder, surface treatment of nanosilica is of great importance in formulating an underfill. A fundamental study on the surface modification of nanosize silica for underfill application was carried out by Sun et al. [63]. They found that the type of the surface treatment was the primary factor affecting the property of the formulation. Using an epoxy silane, the authors showed that the viscosity of the composite underfill was greatly reduced.

In summary, the invention of no-flow underfill greatly simplifies the flip-chip underfill process and draws flip-chip toward SMT. A successful no-flow underfill process requires careful investigation into the materials and process parameters. A great deal of the research effort has been devoted to the materials, process, and reliability of flip-chip no-flow underfill assembly. Since the underfill does not contain silica filler and hence behaves differently from the conventional underfill, the

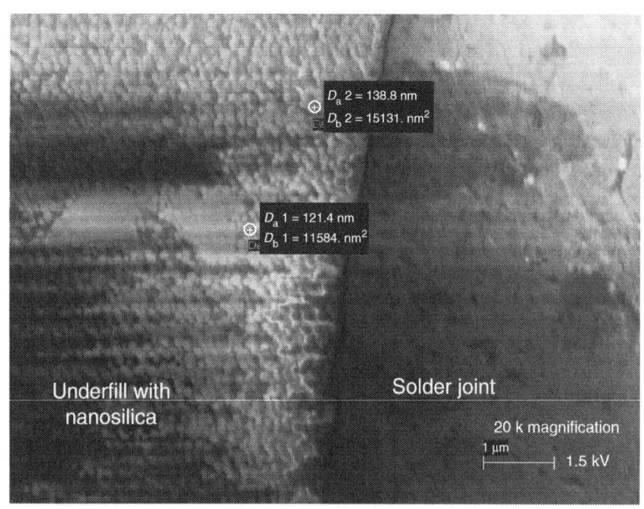

FIGURE 15.11 SEM picture of a solder joint with nano-silica incorporated no-flow underfill.

failure modes and reliability concerns are sometimes also different from the conventional flip-chip underfill assembly. There are several ways to enhance the reliability of a flip-chip no-flow underfill package. One way is to enhance the fracture toughness of the underfill without degrading other material properties to prevent underfill cracking in the thermal cycling. Or, low T_g and low modulus materials have been used to decrease the stress in the underfill. However, this approach diminishes the role of the underfill as a stress redistribution layer, and although it does decrease the stress in underfill, it cannot prevent solder joint fatigue failure from the thermomechanical stress, especially in the case of the large chip, high I/O counts, and small pitch size applications. Another way is to add silica fillers into the underfill and match the properties of a conventional underfill. In order to overcome the difficulty of filler entrapment, different approaches have been explored. However, these approaches are less SMT transparent and diminish the low-cost purpose of a no-flow underfill process. Nanosilica-incorporated no-flow underfill has shown the potential of a highly reliable flip-chip package with a SMT-friendly no-flow underfill process. However, fundamental understanding of the nanosilica and its interaction with underfill and solder is still lacking and further development is needed to optimize the materials and processes.

15.6 Molded Underfill

Epoxy molding compounds (EMCs) have been practiced in component packaging for a long time. The novel idea of combining overmolding and the underfill together results in a molded underfill [64,65]. Molded underfill is applied to a flip-chip in a package via a transfer molding process, during which the molding compound not only fills the gap between the chip and the substrate but also encapsulates the whole chip [66]. It offers the advantages of combining the underfilling and transfer molding into one step for reduced process time and improved mechanical stability [67]. It also utilizes EMCs, which have long been proven to provide superior package reliability. Compared with the conventional underfill, which is usually filled with silica at around 50–70 wt%, molded underfill can afford a much higher filler content up to 80 wt%, which offers a low CTE closely matched with the solder joint and the board. Also, compared with the conventional molding compound, molded underfill requires fillers in smaller sizes, which also can contribute to lowering the CTE of the material [68]. Molded underfill is especially suitable for flip-chip in package to improve production efficiency. It was reported that a fourfold production rate increase can be expected using molded underfill versus a conventional underfill process [69].

Molded underfill resembles the pressurized underfill encapsulation [70] in the mold design and process except that the materials in use are not liquid encapsulants that only fill up the gap between the chip and the substrate, but rather molding compounds that overmold the entire components. Figure 15.12 shows a design of the mold for FCBGA components using molded underfill.

The design of the mold faces the challenge that the flip-chip geometry has a higher resistance to the mold flow so that air might be trapped under the chip. In fact, voids have been observed in the molded underfill packages using an acoustic microscope [71]. Several molding processes can be used to minimize this geometry effect [72]. One way is to use mold vents as shown in Figure 15.12 and to use also geometrical optimization to create similar flow resistance over and under the chip. One can also use vacuum-assisted molding to prevent air entrapment. Another approach is to design a cavity in the substrate as shown in Figure 15.12. Though it requires a special design on the substrate, this method has proved to be a robust process and is commonly adopted.

Important process parameters in a molded underfill process include the molding temperature, clamp force, and injection pressure [56]. High-temperature molding is favored for lower viscosity of the molding compound and hence better flow properties and less stress on the solder joint. However, the upper limit of the molding temperature is the melting point (T_m) of the solder material. Temperature near T_m combined with high injection pressure might cause the solder to melt and even the die to be "swept" away from the site. Also a low-T_g substrate is likely to be damaged at high molding temperature and high clamp force. Flash is affected by both the clamp force and the injection pressure. The overflow of the molding compound might contaminate other contact pads or testing pads on the substrate. Bump cracking and die cracking are likely to occur as a result of high injection pressure. In short, a successful molded underfill process requires a combined effort in material selection, mold design, and process optimization. But the potential cost reduction and reliability enhancement of molded underfill is attracting great efforts in the industry.

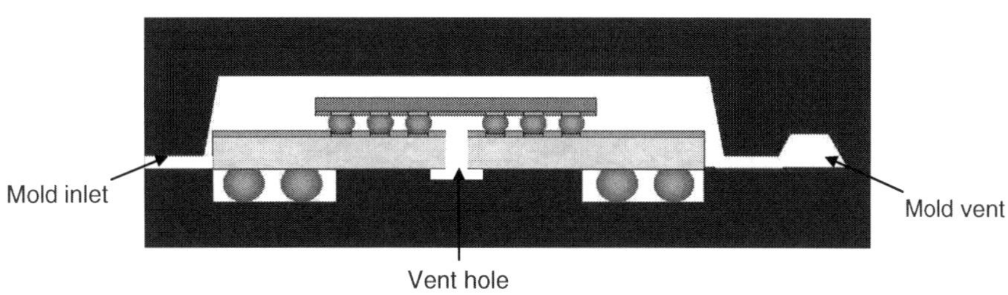

FIGURE 15.12 Design of flip-chip BGA with molded underfill.

15.7 Wafer Level Underfill

The invention of no-flow underfill eliminates the capillary flow and combines fluxing, solder reflow, and underfill curing into one step, which greatly simplifies the underfill process. However, as pointed out previously in the chapter, no-flow underfill has some inherent disadvantages including the unavailability of a heavily filled material, which is a big concern for high-reliability packages. Also, the no-flow process still needs an individual underfill dispensing step and therefore is not totally transparent to standard SMT facilities. An improved concept, wafer level underfill was proposed as a SMT-compatible flip-chip process to achieve low cost and high reliability [73–76]. The schematic process steps are illustrated in Figure 15.13. In this process, the underfill is applied either onto a bumped wafer or a wafer without solder bumps, using a proper method, such as printing or coating. Then the underfill is B-staged and wafer is diced into single chips. In the case of unbumped wafer, the wafer is bumped before dicing when the underfill can be used as a mask. The individual chips are then placed onto the substrate by standard SMT assembly equipment.

It is noted that in some types of WLCSP, a polymeric layer is also used on the wafer scale to redistribute the I/O or to enhance the reliability. However, this polymeric layer usually does not glue with the substrate and cannot be considered as underfill. The wafer level underfill discussed here is an adhesive to glue chip and substrate together and functions as a stress-redistribution layer rather than a stress-buffering layer. The attraction of the wafer level underfill lies in the potential low-cost potential (since it does not require a significant change in the wafer backend of the line process) and high reliability of the assembly enhanced with the underfill. However, the wafer level underfill faces critical material and process challenges including uniform underfill film deposition on the wafer, B-stage process for the underfill, dicing, and storage of B-staged underfill, fluxing capability, shelf-life, solder wetting in the presence of underfill, desire for no postcure and reworkability, etc. Since the wafer level underfill process suggests a convergence of front-end and back-end of the line in package manufacturing, close cooperation between chip manufacturers, package companies, and material suppliers are required. Several research programs have been carried out the cooperatively in this area [77–79]. Innovative ways of addressing the above issues and examples of wafer level processes are presented in this chapter.

In most wafer level underfill processes, the applied underfill must be B-staged before the singulation of the wafer. The B-stage process usually involves partial curing, solvent evaporation, or both, of the underfill. In order to facilitate dicing, storage, and handling, the B-staged underfill must appear solid-like and possess enough mechanical integrity and stability after B-stage. However, in the final assembly, the underfill is required to possess "reflowability," i.e., the ability to melt and flow to allow the solder bumps to wet the contacting pads and form solder joints. Therefore, the control of the curing process and the B-stage properties of the underfill is essential for a successful wafer level underfill process. A study conducted at Georgia Institute of Technology utilized the curing kinetics model to calculate the DOC evolution of different underfills during solder reflow process [80]. Combined with the gelation behavior of the underfills, the solder wetting capability during reflow was predicted and confirmed experimentally. Based on the B-stage process window and the material properties of the B-staged underfill, a successful wafer level underfill material and process were developed. Full area array at 200 μm pitch flip-chip assembly with the developed wafer level underfill was also demonstrated [81] as shown in Figure 15.14.

The above study shows that the control of the B-stage process of the wafer level underfill is critical to achieve good dicing and storage properties and the solder interconnect on the board-level assembly. One way to avoid dicing in the presence of nonfully cured underfill is presented in Figure 15.15, a wafer scale applied

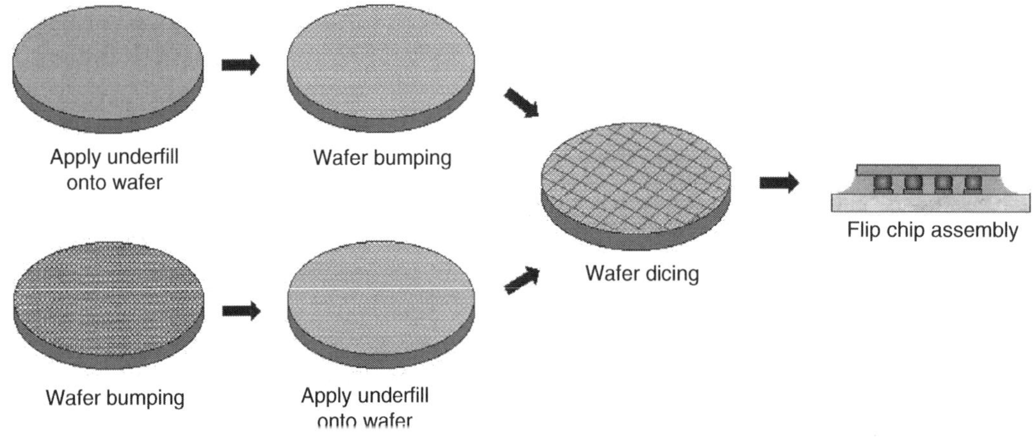

FIGURE 15.13 Process steps of wafer level underfill.

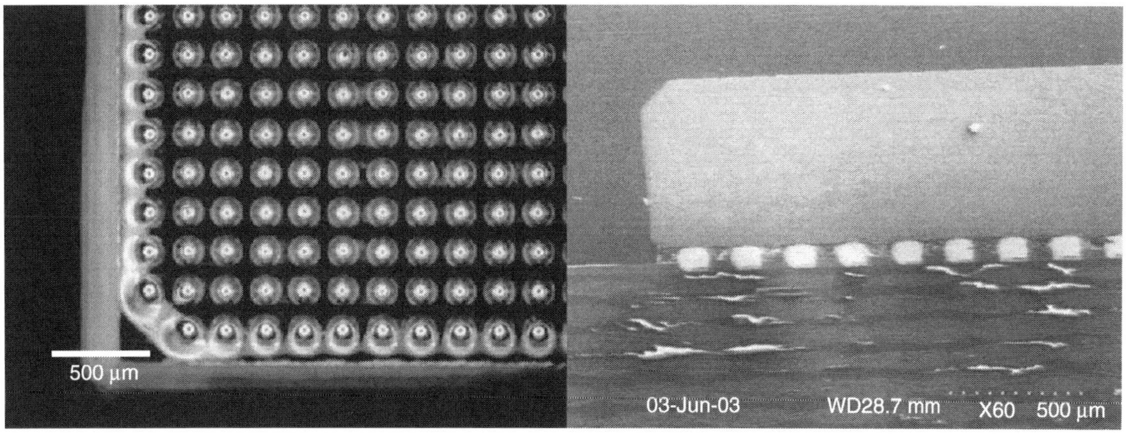

FIGURE 15.14 A 200 μm pitch wafer level underfill process: underfill coated wafer (left) and assembly on the substrate (right).

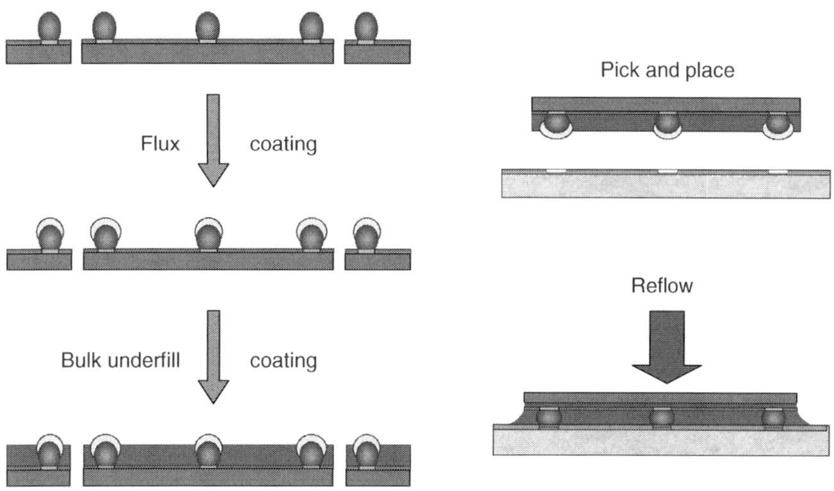

FIGURE 15.15 Wafer scale applied reworkable fluxing underfill process.

reworkable fluxing underfill process developed by Motorola, Loctite, and Auburn University [77]. Since uncured underfill materials are likely to absorb moisture that leads to potential voiding in the assembly, in this process, the wafer is diced prior to underfill coating. Two dissimilar materials are applied; the flux layer coating by screen or stencil printing and the bulk underfill coating by a modified screen printing to keep the saw street clean. The separation of the flux from the bulk underfill material preserves the shelf life of the bulk underfill as well as prevents the deposition of fillers on the top of the solder bump so as to ensure the solder joint interconnection in the flip-chip assembly. In this process, no additional flux dispensing on the board is needed and hence the underfill needs to be tacky in the flip-chip bonding process to ensure the attachment of the chip to the board, as discussed previously.

Underfill deposition on the wafer using liquid material via coating or printing requires subsequent B-staging, which is often tricky and problematic. The process developed by 3M and Delphi-Delco circumvents the B-stage step using film lamination [82]. The process steps are shown in Figure 15.16, in which the solid film comprised of thermoset/thermoplastic composite is laminated onto the bumped wafer in vacuum. Heat is applied under a vacuum to ensure the complete wetting of the film over the whole wafer and to exclude any voids. Then a proprietary process is carried out to expose the solder bump without altering the original solder shape. The subsequent flip-chip assembly is carried out in a no-flow underfill-like process in which a curable flux adhesive is applied on the board and then the assembly is reflowed.

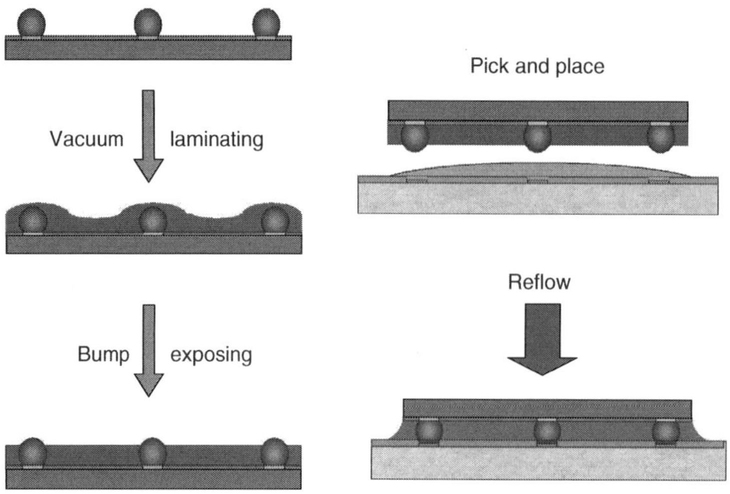

FIGURE 15.16 Wafer-applied underfill film laminating process.

Wafer level underfill can also be applied before the bumping process. Figure 15.17 shows a multilayer wafer-scale underfill process developed by Aguila Technologies, Inc. [83]. The highly filled wafer level underfill is screen printed onto an unbumped wafer and then cured. Then this material is laser-ablated to form microvias that expose the bond pads. The vias are filled with solder paste and reflowed. Bumps are formed on the top of the filled vias. The flip-chip assembly is similar to the no-flow underfill process again with a polymer flux dispensed onto the board before chip placement.

One similarity among all these three processes is the separation of flux material from the bulk underfill. Wafer level underfill processes provide the convenience of separating different functionalities by using dissimilar materials so that "the one magic material that solves everything" is not required. However, it is likely to create inhomogeneity inside the underfill layer, the impact of which on the reliability is not fully understood.

A novel photodefinable material, which acts both as a photoresist and an underfill layer applied on the wafer level, was reported by the Georgia Institute of Technology [84]. In the proposed process as shown in Figure 15.18, the wafer level underfill is applied on the unbumped wafer, and then is exposed to the UV light with a mask for cross-linking. After development, the unexposed material is removed and the bump pads on the wafer are exposed for solder bumping. The fully cured film is left on the wafer and acts as the underfill during the subsequent SMT assembly after device singulation. A polymeric flux is needed during the assembly for holding the device in place on the board and providing a fluxing capability, a process similar to the dry

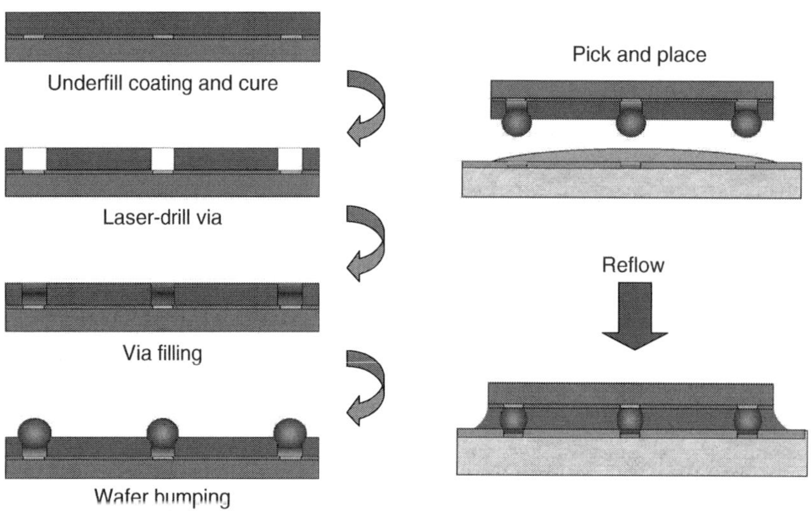

FIGURE 15.17 Multilayer wafer-scale underfill process.

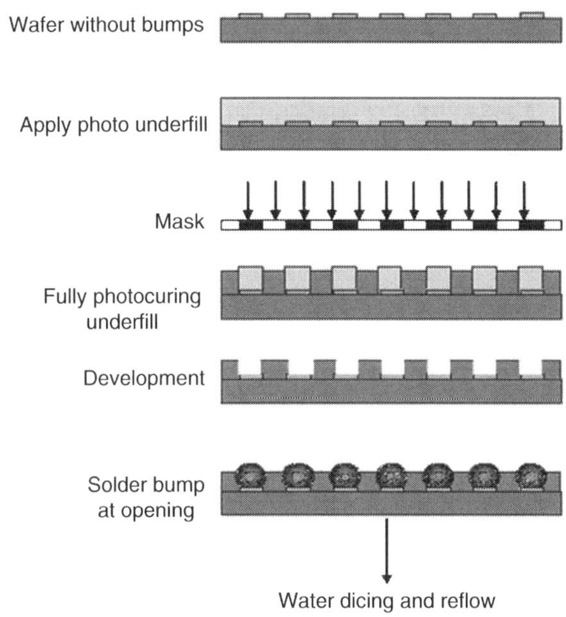

FIGURE 15.18 Photodefinable wafer level underfill process.

film laminated wafer level underfill. In order to enhance the material property, the addition of silica fillers is necessary. In this case, nanosized silica was used to avoid UV light scattering, which hinders the photocross-linking process. It also resulted in an optical transparent film on the wafer to facilitate the vision recognition during dicing and assembly process. The photodefinable nanocomposite wafer level underfill presents a cost-effective way of applying wafer level underfill and has potentially a fine-pitch capability.

Since wafer level underfill is a relatively new concept and most research is still in the process and material development stage, there are few reports on the reliability of a flip-chip package using wafer level underfill. Although there is no standard process for wafer level underfill yet, the final decision might depend on the wafer and chip size, bump pitch, and package type, etc. Like wafer level CSP, multiple solutions can coexist for wafer level underfill process.

15.8 Summary

Flip-chip offers many advantages over other interconnection technologies and is practiced in many applications. Underfill is necessary for a reliable flip-chip on an organic package but is process-unfriendly and becomes the bottleneck of a high-production flip-chip assembly. As silicon technology moves to nanometer nodes with feature size less than 65 nm, a shrinking pitch and gap distance, as well as new developments in lead-free solder and low-*K* ILD/Cu interconnect present new challenges to underfill materials and processes. Many variations of the conventional underfill have been invented that address the problem, among which, the newly developed no-flow underfill, molded underfill, and wafer level underfill have attracted much attention. The no-flow underfill process simplifies the conventional flip-chip underfill process by integrating flux into the underfill, eliminating capillary flow, and combining solder reflow and underfill cure into one step. However, the predeposited underfill cannot contain high levels of silica filler due to the interference of filler with solder joint formation. The resulting high CTE of the underfill limits the reliability of the package. Various ways have been explored to enhance the reliability through improved fracture toughness of the underfill material, low T_g, and low modulus underfill, and the incorporation of fillers using other process approaches. Molded underfill combines underfill and overmold together and is especially suitable for flip-chip in packaging to improve the capillary underfill flow and the production efficiency. Careful material selection, mold design, and process optimization are required to achieve a robust molded underfill process. Wafer level underfill presents a convergence of front-end and back-end in packaging manufacturing and may provide a solution for low-cost and a highly reliable flip-chip process. Various material and process issues including underfill deposition, wafer dicing with underfill, shelf-life, vision recognition, chip placement, and solder wetting with underfill, etc., have been addressed through novel material developments and different process approaches. Although the research is still in its early stages, there is no standard in the process yet developed. Additionally, there have been considerable successes in demonstrating the process, which looks promising for future packaging manufacturing. All of these three approaches require close cooperation between the material suppliers, package designers, assembly companies, and maybe chip manufacturers. A good understanding in both the materials and the processes and their interrelationship is essential to achieve successful packages for the future success of these developments.

References

1. C.P. Wong, S. Lou, and Z. Zhang, Flip the chip, *Science*, 290, 2269, 2000.
2. E. Davis, W. Harding, R. Schwartz, and J. Coring, Solid logic technology: Verdatile high performance microelectronics, *IBM Journal of Research & Development*, 8, 102, 1964.
3. F. Nakano, T. Soga, and S. Amagi, Resin-insertion effect on thermal cycle resistivity of flip-chip mounted LSI devices, the Proceedings of the International Society of Hybrid Microelectronics Conference, p. 536, 1987.
4. Y. Tsukada, Surface laminar circuit and flip-chip attach packaging, Proceedings of the 42nd Electronic Components and Technology Conference, p. 22, 1992.
5. B. Han and Y. Guo, Thermal deformation analysis of various electronic packaging products by Moire and microscope Moire interferometry, *Journal Electronic Packaging*, 117, 185, 1995.
6. S. Han and K.K. Wang, Analysis of the flow of encapsulant during underfill encapsulation of flip-chips, *IEEE Transactions on Components, Packaging, and Manufacturing Technology, Part B*, 20(4), 424–433, 1997.

7. S. Han, K.K. Wang, and S.Y. Cho, Experimental and analytical study on the flow of encapsulant during underfill encapsulation of flip-chips, Proceedings of the 46th Electronic Components and Technology Conference, pp. 327–334, 1996.
8. L. Nguyen, C. Quentin, P. Fine, B. Cobb, S. Bayyuk, H. Yang, and S.A. Bidstrup-Allen, Underfill of flip chip on laminates: Simulation and validation, *IEEE Transactions on Components and Packaging Technology*, 22(2), 168–176, 1999.
9. W.B. Young and W.L. Yang, Underfill of flip-chip: The effect of contact angle and solder bump arrangement, *IEEE Transactions on Advanced Packaging*, 29(3), 647–653, 2006.
10. Coffin, A study of the effect of cyclic thermal stresses on a ductile metal, *ASME Transactions*, 76, 931–950, 1954.
11. Manson, Fatigue: A complex subject-some simple approximations, *Experimental Mechanics*, 5, 193–226, 1965.
12. H.D. Solomon, Fatigue of 60/40 solder, *IEEE Transactions on components, Hybrids and Manufacturing Technology*, CHMT-9(4), 423–432, 1986.
13. H. Bressers, P. Beris, J. Caers, and J. Wondergerm, Influence of chemistry and processing of flip chip underfills on reliability, 2nd International Conference on Adhesive Joining and Coating Technology in Electronics Manufacturing, Stockholm, Sweden, 1996.
14. J.B. Nysaether, P. Lundstrom, and J. Liu, Measurements of solder bump lifetime as a function of underfill material properties, *IEEE Transactions on Components, Packaging and Manufacturing Technology, Part A*, 21(2), 281–287, 1998.
15. R. Dudek, A. Schubert, and B. Michel, Analyses of flip chip attach reliability, Proceedings of 4th International Conference on Adhesive Joining and Coating Technology in Electronics Manufacturing, pp. 77–85, 2000.
16. P. Palaniappan, P. Selman, D. Baldwin, J. Wu, and C.P. Wong, Correlation of flip chip underfill process parameters and material properties with in-process stress generation, Proceedings of the 48th Electronic Components and Technology Conference, pp. 838–847, 1998.
17. L. Mercado and V. Sarihan, Evaluation of die edge cracking in flip-chip PBGA packages, *IEEE Transactions on Components and Packaging Technologies*, 26(4), 719–723, 2003.
18. S.P. Lahoti, S.C. Kallolimath, and J. Zhou, Finite element analysis of thermo-hygro-mechanical failure of a flip chip package, Proceedings of IEEE 6th International Conference on Electronic Packaging Technology, 2005.
19. S. Luo and C.P. Wong, Effect of coupling agents on underfill material in flip chip packaging, Proceedings of 2000 International Symposium on Advanced Packaging Materials, pp. 183–188, 2000.
20. T. Chen, J. Wang, and D. Lu, Emerging challenges of underfill for flip chip application, Proceedings of the 54th Electronic Components and Technology Conference, pp. 175–179, 2004.
21. J.S. Hwang, Lead-free solder: the Sn/Ag/Cu system, *Surface Mount Technology*, 18, July 2000.
22. B. Huang and N.C. Lee, Prospect of lead free alternatives for reflow soldering, *Proceedings of SPIE—The International Society for Optical Engineering*, 3906, 771, 1999.
23. A. Butterfield, V. Visintainer, and V. Goudarzi, Lead-free solder paste flux evaluation and implementation in personal communication devices, Proceedings of the 50th Electronic Components and Technology Conference, p. 1420, 2000.
24. S. Mahalingam, K. Goray, and A. Joshi, Design of underfill materials for lead free flip chip application, Proceedings of 2004 IEEE International Society Conference on Thermal Phenomena, pp. 473–479, 2004.
25. C.K. Chee, Y.T. Chin, T. Sterrett, Y. He, H.P. Sow, R. Manepali, and D. Chandran, Lead-free compatible underfill materials for flip chip application, Proceedings of the 52nd Electronic Components and Technology Conference, pp. 417–424, 2002.
26. P. Tsao, C. Huang, M. Li, B. Su, and N. Tsai, Underfill characterization for low-*k* dielectric/Cu interconnect IC flip-chip package reliability, Proceedings of the 54th IEEE Electronic Components and Technology Conference, pp. 767–769, 2004.
27. S. Rajagopalan, K. Desai, M. Todd, and G. Carson, Underfill for low-K silicon technology, Proceedings of 2004 IEEE/SEMI International Electronics Manufacturing Technology Symposium, 2004.
28. R. Pennisi and M. Papageorge, Adhesive and Encapsulant Material with Fluxing Properties, U.S. Patent 5,128,746, (July 7, 1992).
29. C.P. Wong and D. Baldwin, No-Flow Underfill for Flip-Chip Packages, U.S. Patent Disclosure, April 1996.
30. C.P. Wong and S.H. Shi, No-Flow Underfill of Epoxy Resin, Anhydride, Fluxing Agent and Surfactant, U.S. Patent 6,180,696, (Jan. 30, 2001).
31. C.P. Wong, S.H. Shi, and G. Jefferson, High performance no flow underfills for low-cost flip-chip applications, Proceedings of the 47th Electronic Components and Technology Conference, p. 850, 1997.
32. C.P. Wong, S.H. Shi, and G. Jefferson, High performance no-flow underfills for flip-chip applications: Material characterization, *IEEE Transactions on Components, Packaging, and Manufacturing Technology, Part A: Packaging Technologies*, 21(3), 450, 1998.
33. Z. Zhang, S.H. Shi, and C.P. Wong, Development of no-flow underfill materials for lead-free bumped flip-chip applications, *IEEE Trans. on Components, and Packaging Technologies*, 24(1), 59–66, 2000.
34. Z. Zhang and C.P. Wong, Development of no-flow underfill for lead-free bumped flip-chip assemblies, Proceedings of Electronics Packaging Technology Conference, pp. 234–240, Singapore, 2000.
35. Z. Zhang and C.P. Wong, Study and modeling of the curing behavior of no-flow underfill, Proceedings of the 8th International Symposium and Exhibition on Advanced Packaging Materials Processes, Properties and Interfaces, pp. 194–200, Stone Mountain, Georgia, 2002.

36. P. Morganelli and B. Wheelock, Viscosity of a no-flow underfill during reflow and its relationship to solder wetting, Proceedings of the 51st Electronic Components and Technology Conference, pp. 163–166, 2001.
37. R.W. Johnson, M.A. Capote, S. Chu, L. Zhou, and B. Gao, Reflow-curable polymer fluxes for flip chip encapsulation, Proceedings of International Conference on Multichip Modules and High Density Packaging, pp. 41–46, 1998.
38. S.H. Shi and C.P. Wong, Study of the fluxing agent effects on the properties of no-flow underfill materials for flip-chip applications, Proceedings of the 48th Electronic Components and Technology Conference, p. 117, 1998.
39. S.H. Shi, D. Lu, and C.P. Wong, Study on the relationship between the surface composition of copper pads and no-flow underfill fluxing capability, Proceedings of the 5th International Symposium on Advanced Packaging Materials: Processes, Properties and Interfaces, p. 325, 1999.
40. S.H. Shi and C.P. Wong, Study of the fluxing agent effects on the properties of no-flow underfill materials for flip-chip applications, *IEEE Transactions on Components and Packaging Technologies, Part A: Packaging Technologies*, 22(2), 141, June 1999.
41. P. Palm, K. Puhakka, J. Maattanen, T. Heimonen, and A. Tuominen, Applicability of no-flow fluxing encapsulants and flip chip technology in volume production, Proceedings of the 4th International Conference on Adhesive Joining and Coating Technology in Electronics Manufacturing, pp. 163–167, 2000.
42. K. Puhakka and J.K. Kivilahti, High density flip chip interconnections produced with in-situ underfills and compatible solder coatings, Proceedings of the 3rd International Conference on Adhesives Joining and Coating Technology in Electronics Manufacturing, pp. 96–100, 1998.
43. T. Wang, T.H. Chew, Y.X. Chew, and L. Foo, Reliability studies of flip chip package with reflowable underfill, Proceedings of the Pan Pacific Microelectronic Symposium, Kauai, Hawaii, February 2001, pp. 65–70.
44. Z. Zhang and C.P. Wong, Assembly of lead-free bumped flip-chip with no-flow underfills, *IEEE Transactions on Electronics Packaging Manufacturing*, 25(2), 113–119, 2002.
45. D. Miller and D.F. Baldwin, Effects of substrate design on underfill voiding using the low cost, high throughput flip chip assembly process, Proceedings of the 7th International Symposium on Advanced Packaging Materials: Processes, Properties and Interfaces, pp. 51–56, 2001.
46 R. Zhao, R.W. Johnson, G. Jones, E. Yaeger, M. Konarski, P. Krug, and L. Crane, Processing of fluxing underfills for flip chip-on-laminate assembly, Presented at IPC SMEMA Council APEX 2002, Proceedings of APEX, San Diego, CA, pp. S18-1-1–S18-1-7, 2002.
47. T. Wang, C. Lum, J. Kee, T.H. Chew, P. Miao, L. Foo, and C. Lin, Studies on a reflowable underfill for flip chip application, Proceedings of the 50th Electronic Components and Technology Conference, pp. 323–329, 2000.
48. D. Gamota and C.M. Melton, The development of reflowable materials systems to integrate the reflow and underfill dispensing processes for DCA/FCOB assembly, *IEEE Transactions on Components and Packaging Technologies, Part C*, 20(3), 183, July 1997.
49. X. Dai, M.V. Brillhart, M. Roesch, and P.S. Ho, Adhesion and toughening mechanisms at underfill interfaces for flip-chip-on-organic-substrate packaging, *IEEE Transactions on Components and Packaging Technologies*, 23(1), 117–127, 2000.
50. B.S. Smith, R. Thorpe, and D.F. Baldwin, A reliability and failure mode analysis of no flow underfill materials for low cost flip chip assembly, Proceedings of 50th Electronic Components & Technology Conference, pp. 1719 1730, 2000.
51. K.S. Moon, L. Fan, and C.P. Wong, Study on the effect of toughening of no-flow underfill on fillet cracking, Proceedings of the 51st Electronic Components and Technology Conference, pp. 167–173, 2001.
52. H. Wang and T. Tomaso, Novel single pass reflow encapsulant for flip chip application, Proceedings of the 6th International Symposium on Advanced Packaging Materials: Process, Properties, and Interfaces, pp. 97–101, 2000.
53. Z. Zhang, L. Fan, and C.P. Wong, Development of environmental friendly non-anhydride no-flow underfills, *IEEE Transactions on Components and Packaging Technologies*, 25(1), 140–147, 2002.
54. S.H. Shi, Q. Yao, J. Qu, and C.P. Wong, Study on the correlation of flip-chip reliability with mechanical properties of no-flow underfill materials, Proceedings of the 6th International Symposium on Advanced Packaging Materials: Processes, Properties and Interfaces, pp. 271–277, 2000.
55. S.H. Shi and C.P. Wong, Recent advances in the development of no-flow underfill encapsulants—a practical approach towards the actual manufacturing application, Proceedings of the 49th Electronic Components and Technology Conference, p. 770, 1999.
56. P. Miao, Y. Chew, T. Wang, and L. Foo, Flip-chip assembly development via modified reflowable underfill process, Proceedings of the 51st Electronic Components and Technology Conference, pp. 174–180, 2001.
57. S. Kawamoto, O. Suzuki, and Y. Abe, The effect of filler on the solder connection for no-flow underfill, Proceedings of the 56th Electronic Components and Technology Conference, pp. 479–484, 2006.
58. Z. Zhang, J. Lu, and C.P. Wong, Provisional Patent 60/288, 246: A Novel Process Approach to Incorporate Silica Filler into No-Flow Underfill, 5–2–2001.
59. Z. Zhang, J. Lu, and C.P. Wong, A novel approach for incorporating silica fillers into no-flow underfill, Proceedings of the 51st Electronic Components and Technology Conference, pp. 310–316, 2001.
60. Z. Zhang and C.P. Wong, Novel filled no-flow underfill materials and process, Proceedings of the 8th International Symposium and Exhibition on Advanced Packaging Materials Processes, Properties and Interfaces, pp. 201–209, 2002.

61. K.M. Gross, S. Hackett, D.G. Larkey, M.J. Scheultz, and W. Thompson, New materials for high performance no-flow underfill, Symposium Proceedings of IMAPS 2002, Denver, September 2002.
62. K. Gross, S. Hackett, W. Schultz, W. Thompson, Z. Zhang, L. Fan, and C.P. Wong, Nanocomposite underfills for flip chip application, Proceedings of the 53rd Electronic Components and Technology Conference, pp. 951–956, 2003.
63. Y. Sun, Z. Zhang, and C.P. Wong, Fundamental research on surface modification of nano-size silica for underfill applications, Proceedings of the 54th Electronic Components and Technology Conference, pp. 754–760, 2004.
64. P.O. Weber, Chip Package with Molded Underfill, U.S. Patent 6,038,136, March 14, 2000.
65. P.O. Weber, Chip Package with Transfer Mold Underfill, U.S. Patent 6, 157,086, December 5, 2000.
66. K. Gilleo, B. Cotterman, and T. Chen, Molded underfill for flip chip in package, High Density Interconnection, p. 28, June 2000.
67. T. Braun, K.F. Becker, M. Koch, V. Bader, R. Aschenbrenner, and H. Reichl, Flip chip molding—recent progress in flip chip encapsulation, Proceedings of 8th International Advanced Packaging Materials Symposium, pp. 151–159, March 2002.
68. F. Liu, Y.P. Wang, K. Chai, and T.D. Her, Characterization of molded underfill material for flip chip ball grid array packages, Proceedings of the 51st Electronic Components and Technology Conference, pp. 288–292, 2001.
69. L.P. Rector, S. Gong, T.R. Miles, and K. Gaffney, Transfer molding encapsulation of flip chip array packages, IMAPS Proceedings, pp. 760–766, 2000.
70. S. Han and K.K. Wang, Study on the pressurized underfill encapsulation of flip chips, *IEEE Transactions on Components, Packaging, and Manufacturing Technology, Part B: Advanced Packaging*, 20(4), 434–442, 1999.
71. L.P. Rector, S. Gong, and K. Gaffney, On the performance of epoxy molding compounds for flip chip transfer molding encapsulation, Proceedings of the 51st Electronic Components and Technology Conference, pp. 293–297, 2001.
72. K.F. Becker, T. Braun, M. Koch, F. Ansorge, R. Aschenbrenner, and H. Reichl, Advanced flip chip encapsulation: transfer molding process for simultaneous underfilling and postencapsulation, Proceedings of the 1st International IEEE Conference on Polymers and Adhesives in Microelectronics and Photonics, pp. 130–139, 2001.
73. S.H. Shi, T. Yamashita, and C.P. Wong, Development of the wafer-level compressive-flow underfill process and its required materials, Proceedings of the 49th Electronic Components and Technology Conference, p. 961, 1999.
74. S.H. Shi, T. Yamashita, and C.P. Wong, Development of the wafer-level compressive-flow underfill encapsulant, Proceedings of the 5th International Symposium on Advanced Packaging Materials: Processes, Properties and Interfaces, p. 337, 1999.
75. K. Gilleo and D. Blumel, Transforming Flip chip into csp with reworkable wafer-level underfill, Proceedings of the Pan Pacific Microelectronics Symposium, p. 159, 1999.
76. K. Gilleo, Flip Chip with Integrated Flux, Mask and Underfill, W.O. Patent 99/56312, November 4, 1999.
77. J. Qi, P. Kulkarni, N. Yala, J. Danvir, M. Chason, R.W. Johnson, R. Zhao, L. Crane, M. Konarski, E. Yaeger, A. Torres, R. Tishkoff, and P. Krug, Assembly of flip chips utilizing wafer applied underfill, Presented at IPC SMEMA Council APEX 2002, Proceedings of APEX, San Diego, CA, pp. S18-3-1–S18-3-7, 2002.
78. Q. Tong, B. Ma, E. Zhang, A. Savoca, L. Nguyen, C. Quentin, S. Lou, H. Li, L. Fan, and C.P. Wong, Recent advances on a wafer-level flip chip packaging process, Proceedings of the 50th Electronic Components and Technology Conference, pp. 101–106, 2000.
79. S. Charles, M. Kropp, R. Kinney, S. Hackett, R. Zenner, F.B. Li, R. Mader, P. Hogerton, A. Chaudhuri, F. Stepniak, and M. Walsh, Pre-applied underfill adhesives for flip chip attachment, IMAPS Proceedings, International Symposium on Microelectronics, Baltimore, MD, pp. 178–183, 2001.
80. Z. Zhang, Y. Sun, L. Fan, and C.P. Wong, Study on B-stage properties of wafer level underfill, *Journal of Adhesion Science and Technology*, 18(3), 361–380, 2004.
81. Z. Zhang, Y. Sun, L. Fan, R. Doraiswami, and C.P. Wong, Development of wafer level underfill material and process, Proceedings of 5th Electronic Packaging Technology Conference, Singapore, pp. 194–198, December 2003.
82. R.L.D. Zenner and B. S. Carpenter, Wafer-applied underfill film laminating, Proceedings of the 8th International Symposium on Advanced Packaging Materials, pp. 317–325, 2002.
83. R.V. Burress, M.A. Capote, Y.-J. Lee, H.A. Lenos, and J.F. Zamora, A practical, flip-chip multi-layer pre-encapsulation technology for wafer-scale underfill, Proceedings of the 51st Electronic Components and Technology Conference, pp. 777–781, 2001.
84. Y. Sun, Z. Zhang, and C.P. Wong, Photo-definable nanocomposite for wafer level packaging, Proceedings of the 55th Electronic Components and Technology Conference, p. 179, 2005.

16

Dielectric Cure Monitoring of Polymers, Composites, and Coatings: Synthesis, Cure, Fabrication, and Aging

David Kranbuehl
College of William and Mary

16.1 Introduction

Frequency-dependent electromagnetic sensors (FDEMS) dielectric measurements using in situ microsensors is a particularly useful technique for monitoring the changing state of a polymer during synthesis and cure and as a composite or as a coating during fabrication and aging in the application environment [1–5]. Measurements can be made in the laboratory and in the manufacturing environment to monitor the polymerization process and to monitor durability and aging in an environmental chamber or in the degradative environment. Dielectric in situ microsensor monitoring techniques can be used to monitor cure in production ovens and autoclaves on the plant floor as well as in the field environment, for example, monitoring cure of coatings on the surface of a ship in dry dock. Durability and aging can be monitored while the object is in use. Examples are a marine coating on a ship, the protective coating on the liner of an acid containing tank, a rocket propellant, an adhesive in a bond joint, or the polymer in a composite structure.

Dielectric (FDEMS) microsensing techniques ought to be more widely used, particularly in monitoring cure and aging. One reason the planar microsensor should be used more extensively is that it is ideally suited to monitoring changes in the chemical state at a particular position or ply layer in a complex composite structure. This condition is difficult to duplicate in most rheological and calorimetric measurements. Furthermore, unlike some optical and calorimetric techniques, the sensitivity does not decrease when one tries to detect changes during the late stages of cure. Remote, planar, microelectric field sensors provide simple, in situ, accurate measurements of changes in state throughout the fabrication process and with a high degree of sensitivity.

A probable reason why FDEMS is not used more widely is that there is a limited understanding by practitioners in industry of the physical principles upon which the instrumental technique is based. Often there is a desire to cut short the importance of understanding the science behind the frequency dependence of electric measurements and to simply interpret changes in the state of the instrument's signal. This simplified approach leads to misinterpretation and error.

The key to successfully using electric field sensor measurements to monitor changes in the state of a resin system during cure and aging in the use environment, is to understand the frequency dependence of the complex permittivity ε^*, a force–displacement parameter similar to the complex modulus G^* or compliance J^* in rheological measurements.

One might ask why doe one need in situ microsensors in addition to accurate composite models for fabrication. Modeling and individual thinking that lead to a procedure-driven cure cycle are beset with operating difficulties. Most notably, as has been described by Ciriscioli and Springer [6], modeling requires extensive material data characterization of resin properties, as well as fabric and tooling properties that are time consuming to measure and, most importantly, that will vary from day-to-day, batch, and lay up to lay up.

Here the use of FDEMS microsensors to monitor the changes in processing properties in a variety of manufacturing situations is described. In one particularly complex example, the buildup uses properties of an epoxy resin as a function of ply depth in the autoclave mold during cure of a 1 in. thick graphite composite as well as use of this data to verify a process model is described. Monitoring the fabrication process during resin infiltration and cure of an advanced fiber architecture composite and to verify model predictions of that fabrication process is discussed. Use of the sensor to monitor rapid cure using UV or e-beam radiation will be described. Use of the sensor output to intelligently control the fabrication process of a composite in conjunction with model predications will also be described.

Then the application of FDEMS sensor to monitor ingress of degradative components, for example, into a coating, an adhesive bond line, and a composite structure during use for the purpose of life monitoring and evaluating the time for replacement will be briefly discussed.

16.2 Instrumentation

Frequency-dependent complex dielectric measurements are made by using an impedance analyzer controlled by a microcomputer [1,2]. In the work discussed here, measurements at frequencies from 5 to 5×10^6 Hz are taken continuously throughout the entire cure process at regular intervals and converted to the complex permittivity, $\varepsilon^* = \varepsilon' - i\varepsilon''$. The measurements are made with a geometry-independent planar interdigitated microsensor. This system is used with either a Hewlett Packard or a Schlumberger impedance bridge. The system permits multiplexed measurement of several sensors. The sensor itself is planar, 1 by 0.5 in. in area and 3 mm thick. This single sensor-bridge microcomputer assembly is able to make continuous uninterrupted measurements of both ε' and ε'' over 10 decades in magnitude at all frequencies. The sensor is inert and has been used at temperatures exceeding 400°C and pressures over 1000 psi.

16.3 Theory

Frequency-dependent measurements of the dielectric impedance of a material as characterized by its equivalent capacitance, C, and conductance, G, are used to calculate the complex permittivity, $\varepsilon^* = \varepsilon' - i\varepsilon''$, where i is the imaginary number, $\omega = 2\pi f$, f is the measurement frequency, and C_0 is the equivalent air replacement capacitance of the sensor.

$$\varepsilon'(\omega) = \frac{C(\omega)\text{material}}{C_0}$$
$$\varepsilon''(\omega) = \frac{G(\omega)\text{material}}{C_0} \tag{16.1}$$

This calculation is possible when using the sensor whose geometry is invariant over all measurement conditions. Both the real and the imaginary parts of ε^* can have dipolar (d) and ionic (i)-charge polarization components.

$$\varepsilon' = \varepsilon'_{\text{d}} + \varepsilon'_{\text{i}}$$
$$\varepsilon'' = \varepsilon''_{\text{d}} + \varepsilon''_{\text{i}} \tag{16.2}$$

Plots of the product of frequency (ω) multiplied by the imaginary component of the complex permittivity $\varepsilon''(\omega)$ make it relatively easy to visually determine the $\varepsilon''(\omega)$ permittivity when the low-frequency magnitude of ε'' is dominated by the mobility of ions in the resin and when at higher frequencies, the rotational mobility of bound charge dominates ε''. Generally, the magnitude of the low-frequency overlapping values of $\omega\varepsilon''(\omega)$ can be used to measure the change with time of the ionic mobility through the parameter σ where

$$\sigma(\text{ohm}^{-1}\text{cm}^{-1}) = \omega_0\,\omega\varepsilon''_{\text{i}}(\omega)$$
$$\varepsilon_0 = 8.854 \times 10^{-14}\,\text{C}^2\,\text{J}^{-1}\,\text{cm}^{-1} \tag{16.3}$$

The changing value of the ionic mobility is a molecular probe that can be used to quantitatively monitor the viscosity of the resin during cure. The dipolar component of the loss at higher frequencies can then be determined by subtracting the ionic component:

$$\varepsilon''(\omega)(\text{dipolar}) = \varepsilon''(\omega) - \frac{\sigma}{\omega\varepsilon_0} \tag{16.4}$$

The peaks in ε'' (dipolar) (which are usually close to the peaks in ε'') can be used to determine the time or point in the cure process when the mean dipolar relaxation time, τ, has attained a specific value $\tau = 1/\omega$, where $\omega = 2\pi f$ is the frequency of measurement. The dipolar mobility as measured by the mean relaxation time τ can be used as a molecular probe of the buildup in T_g. The time occurrence of a given dipolar relaxation time as measured by a peak in a particular high-frequency value of $\varepsilon''(\omega)$ can be quantitatively related to the attainment of a specific value of the resin's T_g. Finally, the tail of the dipolar relaxation peak as monitored by the changing value of $(d\varepsilon''/dt)/\varepsilon''$ can be used for in situ monitoring during processing the buildup in degree of cure and related end-use properties such as modulus, hardness, etc., during the final stages of cure or postcure.

16.4 Isothermal Cure

The variation in the magnitude of ε'' with frequency and with time for the diglycidylether bisphenol-A amine epoxy held at

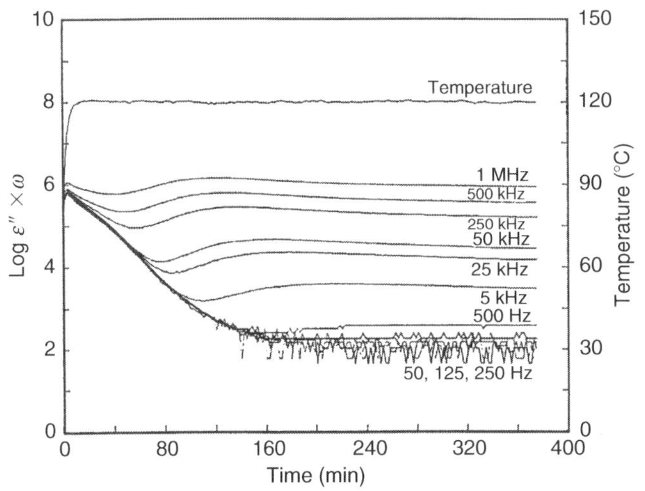

FIGURE 16.1 Dielectric loss multiplied by frequency versus time over a range of frequencies during cure of an epoxide at 120°C.

121°C is shown in Figure 16.1. The magnitude of ε'' changes four orders of magnitude during the course of the polymerization reaction. A plot of $\omega\varepsilon''$ is a particularly informative representation of the polarization process because as discussed from Equations 16.1 through 16.4, overlap of $\omega\varepsilon''(\omega)$ frequency for differing frequencies indicates when and at what time translational diffusion of charge, which can be used to monitor viscosity and reaction advancement, is the dominant physical process affecting the loss. Similarly, the peaks in $\omega\varepsilon''(\omega)$ for individual frequencies indicate the dipolar rotational diffusion relaxation time, a quantity which can be used to monitor T_g.

The frequency dependence of loss ε'' is used to find σ by determining from a computer analysis, or a log–log plot of ε'' versus frequency, the frequency region where $\omega\varepsilon''(\omega)$ is a constant. In this frequency region, the value of σ is determined from Equation 16.3.

Figure 16.2 is a plot of log(σ) versus log (viscosity [η]) constructed from dielectric data and measurements on a dynamic rheometer. The figure shows that at a viscosity less than 1 Pa s, (10 P), σ is proportioned to $1/\eta$ because the slope of log(σ) versus log(η) is approximately −1. The gel point of the polymerization reaction occurs at 90 min based on the crossover of G' and G'' measured at 40 rad s^{-1}. This result is very close to the time at which η achieves 100 Pa s, which is also often associated with gel. The region of gel marks the onset of a much more rapid change in viscosity than with σ. This change is undoubtedly because as gel occurs the viscoelastic properties of the resin involve the cooperative motion of many chains while the translational diffusion of the ions continues to involve motions over much smaller molecular dimensions.

The time of occurrence of a characteristic dipolar relaxation time can be determined by noting the time at which $\varepsilon''_{\text{dipolar}}$ achieves a maximum for each of the frequencies measured where $\tau = 1/2\ \pi f$ at the time at which $\varepsilon''_{\text{dipolar}}$ achieved a maximum for frequency f. Values of τ can be measured over a range of frequencies and temperatures. The time of occurrence of a given peak in ε'' and the corresponding τ is directly related to the time of achievement of a specific degree of reaction and the value of T_g at that point in time. This relationship between the peak in ε'' at each frequency and degree of cure and T_g is made in the laboratory in conjunction usually with differential scanning calorimetry (DSC) measurements versus time at the same temperature.

Figure 16.3 is a plot showing the correlation of σ measured with the electric field sensors versus degree of cure α as determined by DSC measurements at identical times during a 120°C cure. Figure 16.4 demonstrates how the change in dε''/dt can detect the final buildup in full cure as α approaches 1.0 with considerably more precision than a DSC measurement.

In summary, these in situ sensing measurements of ε'', σ dε''/dt, and peaks in $\varepsilon''(\omega)$ can be used to monitor changes in essential cure processing properties, viscosity, degree of cure, and buildup in T_g, during fabrication at any point in the production mold.

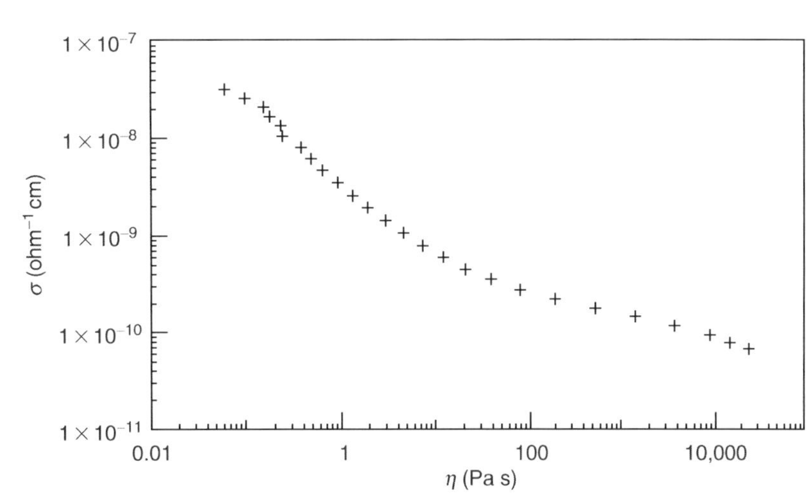

FIGURE 16.2 Correlation of changes in the specific conductivity with the viscosity during cure of an epoxide.

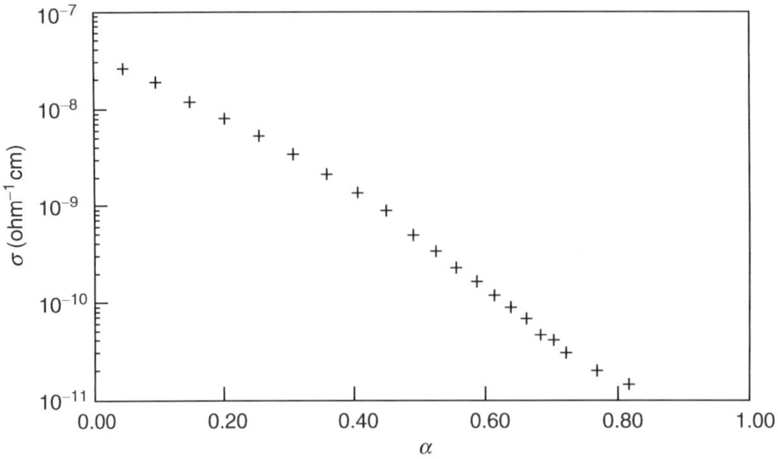

FIGURE 16.3 Correlation of changes in the specific conductivity with the reaction advancement during cure of an epoxide.

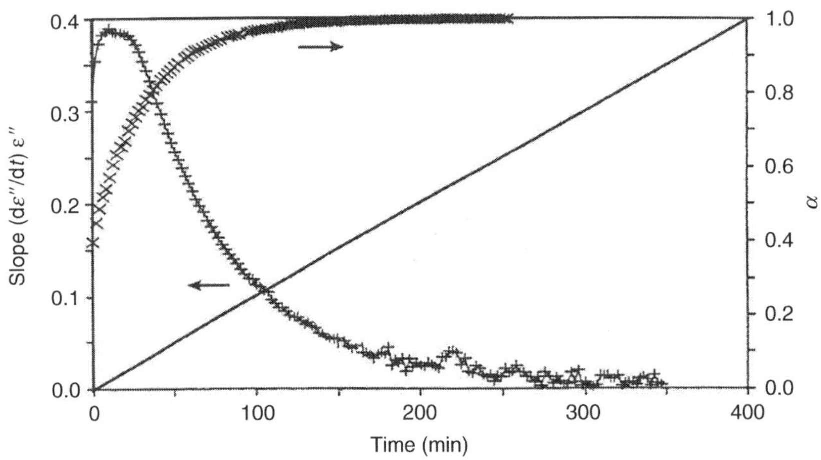

FIGURE 16.4 Correlation of changes in the rate of change of the dielectric loss with changes in the reaction advancement near the end of cure of an epoxide.

16.5 Monitoring the Cure Process in the Mold

16.5.1 Thick Laminates

Figure 16.5 is a plot of the sensors output during a typical epoxy cure cycle in the mold [7,8]. Here, the temperature both lags behind and overshoots the expected hold temperature due to the mass and heat transfer characteristics of a particular oven, autoclave, and mold. Yet, using the correlation relations developed in Figures 16.2 through 16.4 for a particular resin, the sensor output in Figure 16.5 can be used to monitor the actual state, changing viscosity, degree of cure, and T_g at a particular position in the composite laminate.

Figures 16.6 and 16.7 display the positions of sensors in a one inch thick 192 ply epoxy-photo graphite laminate.

Figure 16.8 shows the sensor monitored viscosity. From this figure, it is evident how the viscosity in the center lags behind the viscosity at the surface until the ramp approaching the final hold temperature. At about 130 min, the viscosity at the center ply catches up with the outside ply and a minimum similar viscosity is achieved at the same time uniformly throughout all 192 plys. This is a key final time for pressure consolidation of the 192 ply laminate. The fact that all plys achieve the same viscosity at the same time is undoubtedly one reason why this time–temperature cure cycle developed by trial and error is successful in creating a well-formed final product.

Figure 16.9 shows how this sensor output can be used to verify model predictions such as from the Loos–Springer process model during fabrication on the plant floor [6–8].

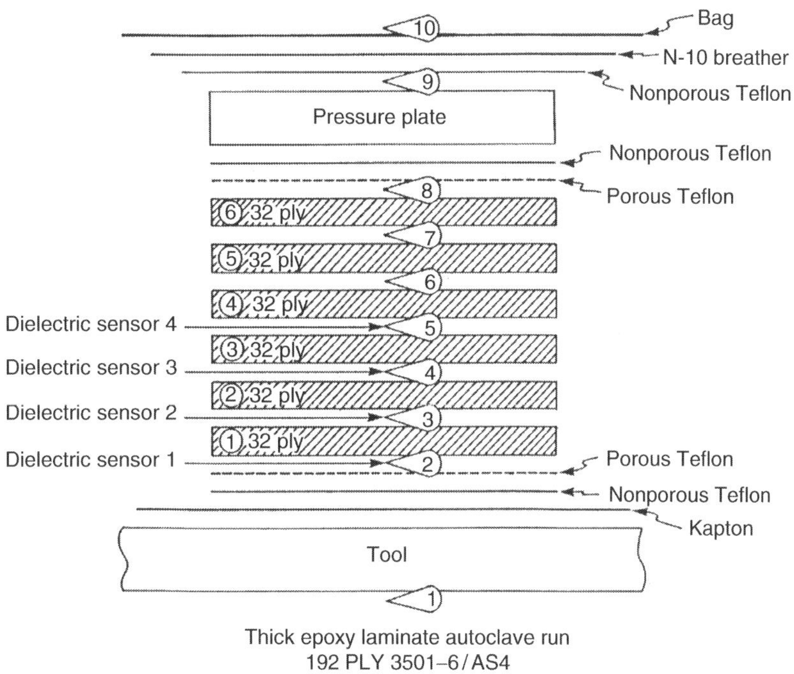

FIGURE 16.5 Positions of the sensors in a thick 192 ply epoxy laminate.

FIGURE 16.6 Photograph of the thick 192 ply epoxy laminate with the embedded sensors after cure.

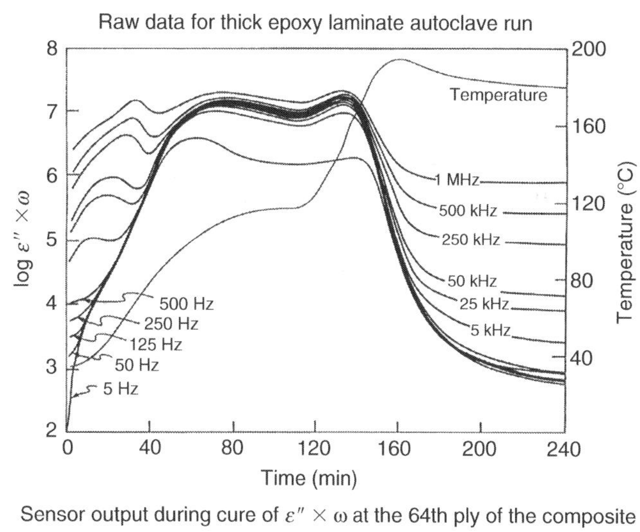

FIGURE 16.7 Dielectric loss multiplied by frequency versus time over a range of frequencies during cure of a 192 ply thick epoxy laminate at the 64th ply in an autoclave.

16.5.2 Resin Infusion of Three-Dimensionally Advanced Fiber Architecture Preforms

This is the second application where we have used electric field sensors to verify model predictions and monitor the processing properties as a function of position. Figure 16.10 is a diagram of a blade stiffened panel, which is a prototype for a wing. Here resin is placed below the panel. Then temperature is used to reduce the viscosity of the epoxy and a vacuum to draw the resin up into the panel. Sensors 2 through 9 were placed on top of the panel to monitor the time at which the viscosity reached that position and the resulting changes in viscosity. The table below reports the wetout times at each sensor. Note sensor 1 was located at the bottom of the panel and its wetout time is the time when the resin began to flow. Sensors 3, 6, and 9 were at the top of the back stiffener and wetout last.

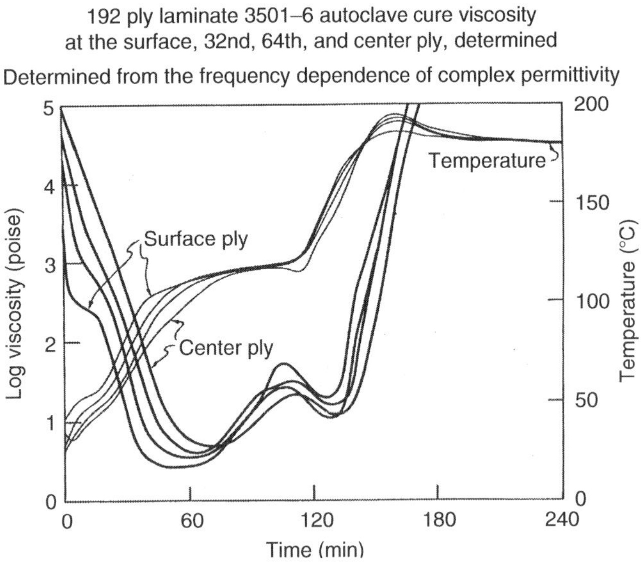

FIGURE 16.8 Sensor measured viscosity at each sensor ply location during cure in the autoclave.

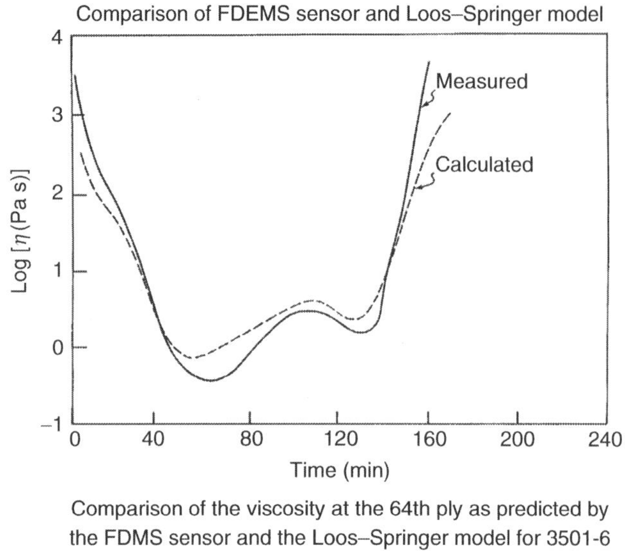

FIGURE 16.9 Comparison of the sensor measured viscosity at the 64th ply versus the predictions of the Springer–Loos model.

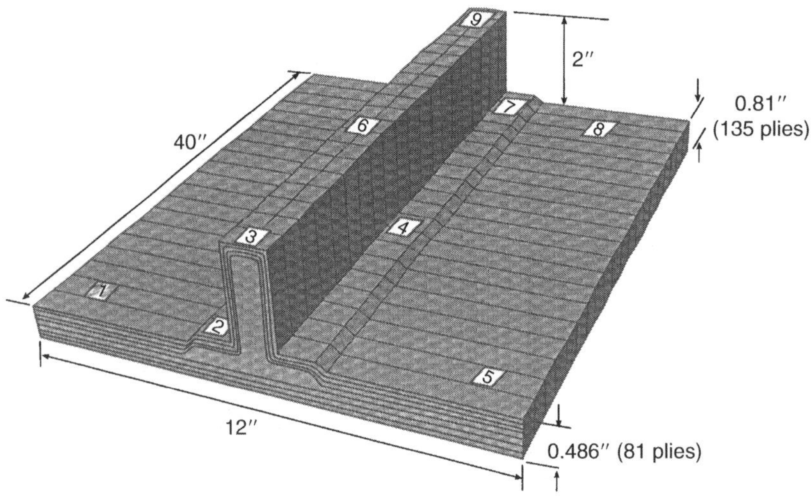

FIGURE 16.10 Drawing of the positions of the sensors in the advanced fiber architecture blade stiffened panel. The panel is impregnated with epoxy using vacuum assisted resin infusion from the bottom of the panel.

Sensor location	1	2	3	4	5	6	7	8	9
Wetout time (min)	63	104	168	114	97	68	110	100	163

Figure 16.11 reports the viscosity of the resin underneath the panel at sensor 1 and the viscosity of position 3 after resin reaches this position.

Further details of model predictions and their verification during resin infusion of three-dimensionally stitched complex fiber preforms have been previously reported [9,10].

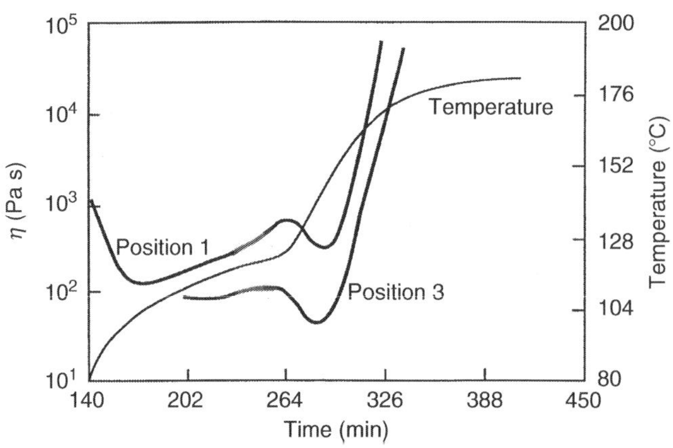

FIGURE 16.11 Sensor measured viscosity of the epoxide at the bottom of the panel, sensor 1, and at position 3 after wetout.

16.6 Automated Intelligent Closed Loop Control

Using models to predict how the cure processing properties should ideally change at each stage of fabrication, sensors can be

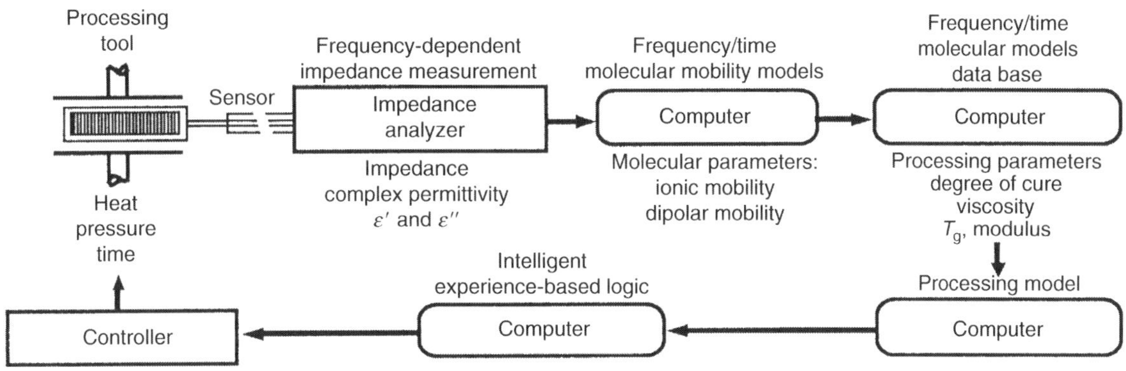

FIGURE 16.12 Schematic of the automated closed loop intelligent sensor controlled system.

used to monitor the fabrication process, verify when the appropriate stage has been successfully completed such as complete resin infusion of the graphite perform and then to adjust the temperature and pressure for the next stage, not based on time but rather the actual state of the resin–fiber system [11–14].

Figure 16.12 is a schematic of such an intelligent closed loop system. Figure 16.13 displays the temperature and viscosity during an intelligent automated cure of the 1 in. thick epoxy graphite panel described earlier. Here fabrication rules were developed based on model predictions and user expertise in which the air temperature increased and then would be held at 180°C until the viscosity at the center ply achieved the viscosity at the outer ply. Then the heat was turned off and the exotherm at the center ply increased the reaction advancement and viscosity fairly uniformly throughout the 192 plys. Twenty minutes later, after moving through the viscosity minimum under pressure, the heat was turned on to bring the panel back to 180°C. The heat was turned off after full cure was achieved as indicated by the sensors.

Based on ultrasonic C-scans a 1 in. thick panel free of voids was produced in 180 min versus 240 min using the conventional cure cycle.

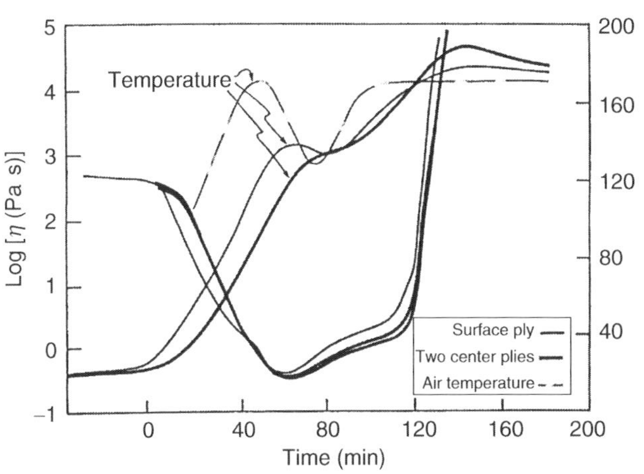

FIGURE 16.13 Viscosity at various ply positions of a thick laminate when cured using the closed loop intelligent rule based sensor controlled system.

16.7 Rapid Cure by UV and Electron Beam Radiation

Over the past decade, considerable attention has been directed to radiation-induced cure as it can be very rapid, more precisely controlled, and can reduce many problems associated with thermal cure. Applications include rapid cure of coatings, films, and points for uses such as paint on antennas, rapid cure of repair patches on a composite, and improved control and speed in filament winding [15–18].

Electric field sensors can be placed at any location in a part and are ideal for monitoring cure of a coating when placed on the substrates surface as they monitor changes in state at the bottom of the coating with one side of the resin exposed to air and the other facing the substrates surface.

Figure 16.14 displays sensor output during irradiation of a one second UV pulse and the subsequent rate of cure after the irradiation at the 8th second. Approximately nine measurements were made per second. Figure 16.15a and b displays the sensor's ability to demonstrate that one epoxy formulation reaches full cure in 60 s while the second while appearing to reach full cure in about 1 min, actually takes 8 min to achieve full cure.

16.7.1 Monitoring Cure of Coatings

In more traditional coating-cure situations, FDEMS output similar to Figure 16.16 can be used to monitor the variation in cure rate with temperature, humidity, airflow, pigment loading, catalyst concentration, thickness, age, batch, etc. [15–18].

For example, it was thought epoxy–polyamide coatings would cure at temperatures as low as 10°C (50°F), even though the time required to reach a given stage is at least double that required at 22°C [4]. To examine the effects of decreased temperature, the polymerization process was monitored in an environmental chamber, held at 11°C and 30% RH for 3 weeks. Figure 16.17 shows a plot of $\log(\varepsilon'' * 2\pi f)$ measured with the planar FDEMS sensor. The time to set-to-touch was 420 min. Dry-to-hard occurred at 695 min. The value of $\log(\varepsilon^{\circ} * 2\pi f)$ at set-to-touch was 4.1 and at dry-to-hard was 2.9 compared to 4.2 and 2.8 in the

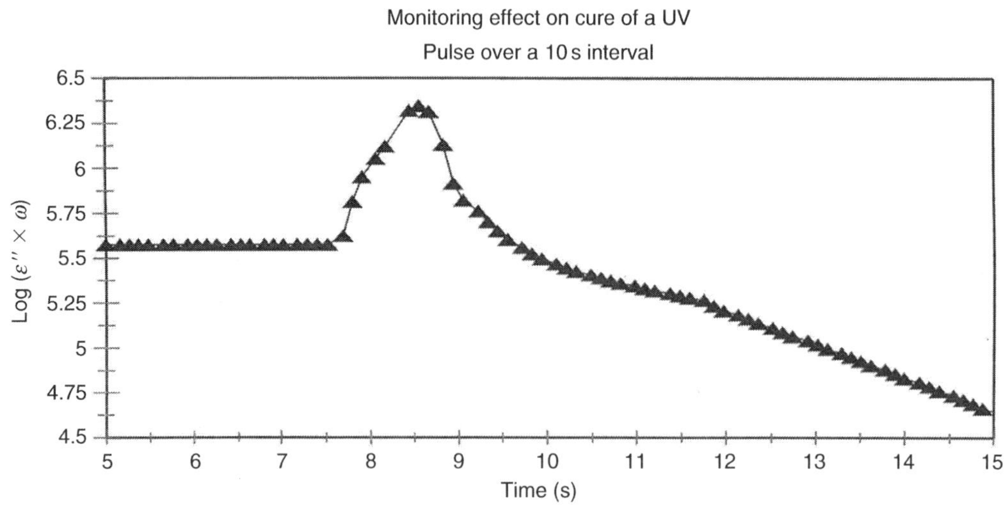

FIGURE 16.14 Dielectric loss multiplied by frequency during rapid cure of a coating after a 1s exposure of UV radiation.

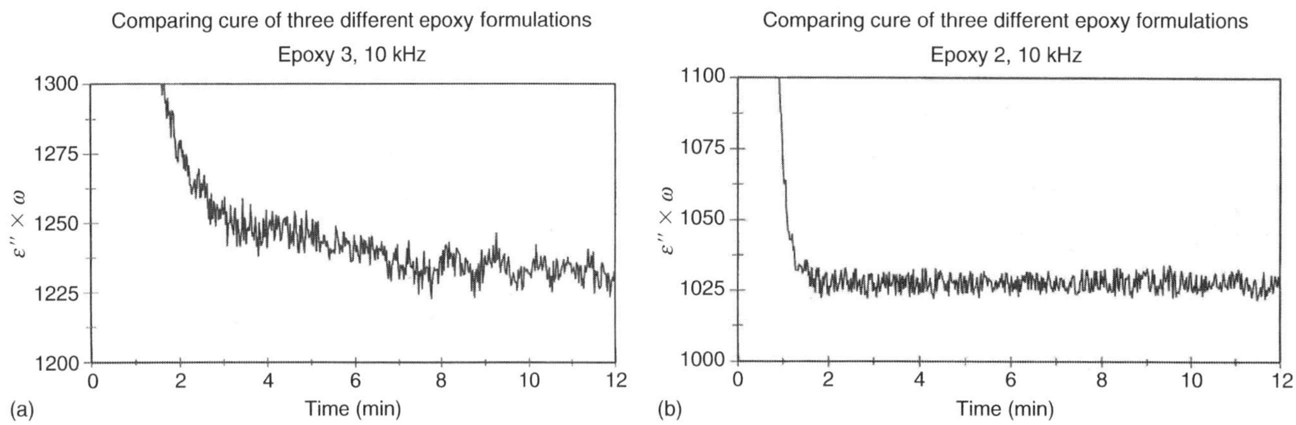

FIGURE 16.15 Dielectric loss multiplied by frequency of two different coating formulations to determine which achieves full cure most rapidly after a one second UV dose.

24°C measurement. The drop in $\varepsilon'' * 2\pi f$ is much slower than at 24°C. The sample continued to cure as seen by the continual drop in ε'' even during the third week since application. After 500 h at 11°C, the 50 kHz line still has not reached the degree of polymerization that the same coating does in 48 h at 24°C and the rate of cure as monitored by the change in ε'' is relatively flat, indicating a very slow cure rate at 11°C.

These FDEMS results clearly indicate that below 10°C, the curing is greatly retarded and full cure is not achieved. More important this epoxy-polyamide will not reach full cure until the temperature rises. Although the partly cured film may feel dry, poor resistance to abrasion, moisture, and chemicals result.

It is interesting to demonstrate that the sensor can monitor the extent to which the first coating of the Seaguard epoxy polyamide is softened by the second for varying elapsed times since the initial coat's application. Figure 16.18 displays the softening of the first coating by the second when the second coating has been applied 144 min after the first for the epoxy-polyamide system shown in Figure 16.16. The initial coating's values of $\log(\varepsilon''\omega)$ have only dropped from 5.5 to 4.5. Upon application of the second coating, the values rise to 6.0 after 60 min as the second coat resoftens the first layer.

Next, the FDEMS sensor is used to examine the extent of softening of the first coat if 26 h elapse before the second coating is applied. Figure 16.19 shows that after 26 h, the first coat has decreased its low-frequency values of $\log(\varepsilon''\omega)$ to around 2. The application of the second coat on the next day resoftens the first coat and $\log(\varepsilon''\omega)$ returns back to 5.5. Hence, 24 h has little effect on the resoftening ability of the first coat by a second coat for this paint system.

The ability of the FDEMS sensor to monitor cure in a proprietary latex coating is shown in Figure 16.20. Initially, there is a rather rapid decay in the $\log(\varepsilon''\omega)$ overlapping ionic lines from 10 to 8.5 over 30 min. This is followed by a rapid drop of over two

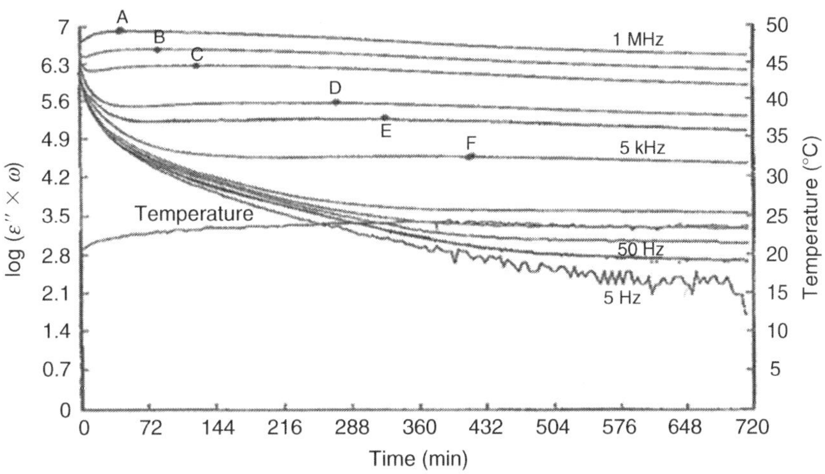

FIGURE 16.16 Decrease in the log of the product of ε'' times the frequency for a marine coating during cure at room temperature. Measurements of ε'' were made at over a range of frequencies from 5 Hz to 1 MHz. The series of peaks in ε'' at the various frequencies are each associated with the attainment of a particular value of T_g. The overlapping values of ε'' times frequency values can be used to calculate σ.

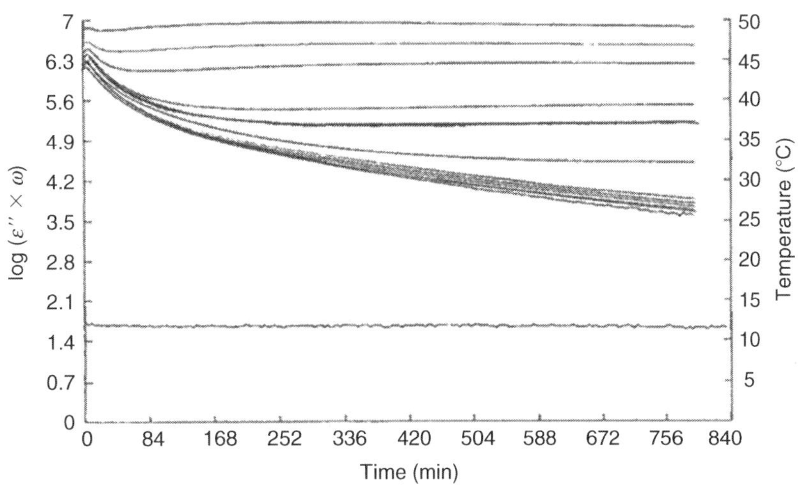

FIGURE 16.17 Cure of the same coating at 11°C.

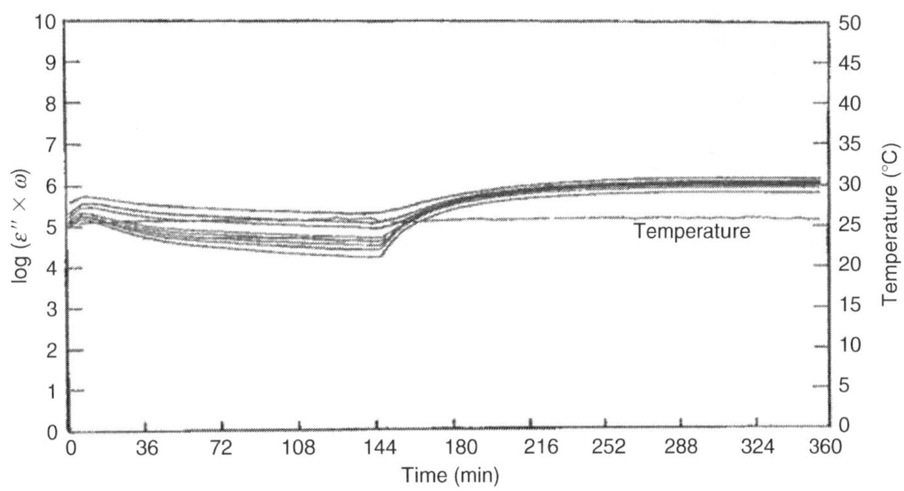

FIGURE 16.18 Similar measurement of coating cure where a second coat is applied at 144 min into cure.

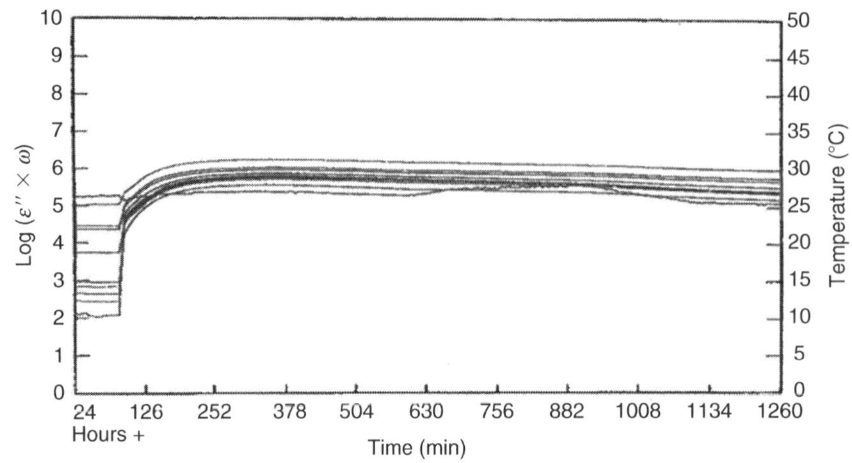

FIGURE 16.19 Similar measurements of coating cure where second coat is applied after 26 h.

decades from 8.5 to 6.2 in several more minutes. One can ascribe the initial 30 min to phase I packing of the latex spheres. There is a loss of volatiles to the point where the latex spheres touch. After this point, water is no longer the continuous medium, the latex spheres rather than the suspending fluid are the conducting medium. This transition point generated the large drop in $\varepsilon''\omega$. The rapid drop is then followed by a gradual decrease in $\log(\varepsilon''\omega)$ which is ascribed to phase two deformation of the latex spheres. There appears to be another shift in the slope around 72 min. This shift in the rate of the drop of $\log(\varepsilon''\omega)$ at 72 min is probably due to the transitions from what is described as a predominately phase II deformation and packing cure process to a phase III autohesion and diffusion cure process.

16.7.2 Monitoring Polyester Synthesis

The polyester synthesized for this application is used as a precursor for a number of coating formulations. It is a condensation reaction between two diols, neopentyl glycol and trimethylopropane, with two diacids, adipic acid and isopthalic acid. A drying agent was added to remove residual water. The reaction was carried out in the laboratory using a 3 L, fluted, round bottom flask. The flask was outfitted with a packed column condenser, an air-driven stir bar, a FDEMS sensor and a glass stopper on the flask's flute for addition of ingredients. The same FDEMS sensor was later attached to a steel pipe for insertion into the pilot plant's reactor for monitoring production batches based on the calibration of the sensor output with extent of reaction from the laboratory experiments.

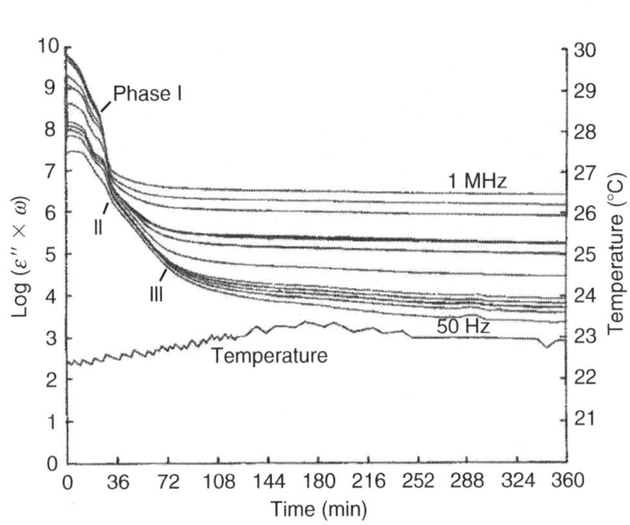

FIGURE 16.20 Use of $\log(\varepsilon'' \times \omega)$ to monitor cure of a latex coating at room temperature.

There are two issues of industrial significance which are usually monitored in a polyester synthesis: acid number and the refractive index of the distillate. The acid numbers of the samples taken during the synthesis are germane because they establish the criteria for the termination of the reaction. In an industrial setting, when a sample has an acid number of a certain value, then the next stage of the synthesis would be undertaken. The problem with this technique is in the timing. Considering the reaction is continuous and the acid number titration technique takes about 1 h to perform, any acid number data would be available at least an hour behind the actual reaction. These timing problems are considerable. There is a need to establish an in situ technique for the determination of end point criteria.

To accomplish this, samples were removed from the reactor and their acid numbers determined. Then, this acid number was correlated with the FDEMS sensor data at the time of sampling. The correlation of the absolute, online measurement of the ionic mobility, $\log(\varepsilon'' * \omega)$, against acid number is shown in Figure 16.21. Figure 16.21 shows that there is a good correlation of the sensor output with acid number. The data were normalized by taking the highest $\log(\varepsilon'' * \omega)$ value of the synthesis, called the peak, and dividing the changing $\log(\varepsilon'' * \omega)$ values by the value at the peak.

16.8 Life Monitoring

Aging, degradation, and reduction in performance properties during use in the field is another application [19]. Here sensors

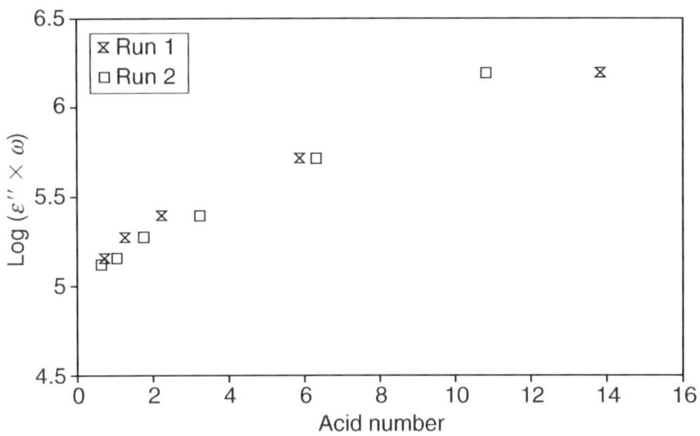

FIGURE 16.21 Monitoring the polymerization process of a polyester in the reaction vessel by the correlation of change in $\varepsilon'' \times \omega$ with the change in the acid number.

can be embedded in witness coupons exposed along side the actual composite structure to the same fluctuations in the environment during use. Sensors can also be embedded and left in the structure itself during fabrication [20–22]. Figure 16.22 displays a commercially available sensor system used to monitor aging of the polymer liner in a flexible pipe used to transport oil-gas from the ocean floor to a floating platform.

Degradation generally involves chemical processes which produce changes at the molecular level and which in one sense are the reverse of what happens during cure. Thus, electric field sensors embedded in the material can also detect aging and changes in performance properties in situ during use. As in fabrication, the sensor output needs to be calibrated with changes in mechanical properties or material properties such as degradation due to UV radiation, high temperature, atomic oxygen, or penetration of fuel oil, acid, or water. During aging, these changes occur much more slowly and are generally much smaller than during fabrication. Thus, monitoring aging requires more precision and sensitivity than in monitoring the state of a part during fabrication.

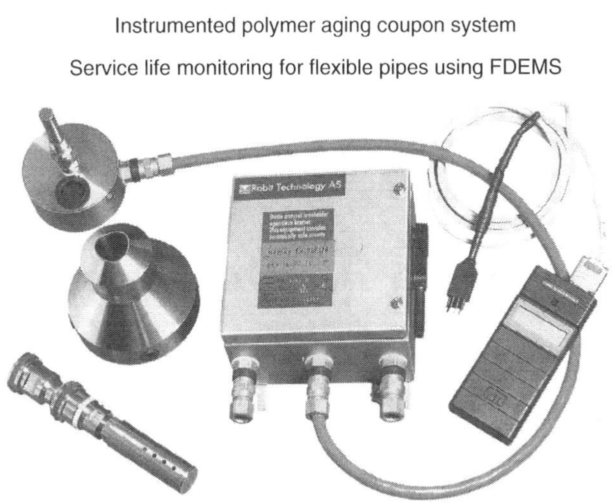

FIGURE 16.22 Commercially available electric field sensor system for monitoring aging of performance properties subsea of the polymer lining in pipes used to transport crude oil to floating platforms.

Acknowledgments

The author thanks numerous undergraduate and graduate students and Prof. Alfred C. Loos. Support from NASA Langley, National Science Administration, the Air Force Edwards California Laboratory, Boeing-McDonnell Douglas Aircraft Company, and ICI's Coatings Companies is gratefully appreciated.

References

1. D. Kranbuehl, Dielectric monitoring of polymerization and cure, invited chapter in *Dielectric Spectroscopy of Polymeric Materials*, J. Runt and J. Fitzgerald, eds. American Chemical Society, Washington, DC, pp. 303–328, 1997.
2. D. Kranbuehl, In-situ online measurement of composite cure with frequency dependent electromagnetic sensors, *Plastics, Rubber, and Composite Processing*, 16, 213–219, 1991.
3. S. Poncet, G. Boiteaux, H. Sautereau, G. Seytre, J. Rogozinski, and D. Kranbuehl, Monitoring phase separation and reaction advancement in situ in thermoplastic/epoxy blends, *Polymer*, 40, 6811–6820, 1999.
4. J.M. Brown, S. Srinivasan, T. Ward, J. McGrath, A.C. Loos, D. Hood, and D. Kranbuehl, Production of controlled networks and morphologies in toughened thermosetting resins using real time in situ curve monitoring, *Polymer*, 37, 1691–1696, 1996.
5. D. Kranbuehl, In situ frequency dependent dielectric sensing of cure, *Processing of Composites*, Hansa, Munich, 137–157, 2000.
6. P. Ciriscioli and G. Springer, *Smart Autoclave Cure of Composite*, Technomic Publishing Co., 1990.
7. D. Kranbuehl, Frequency dependent dielectric measurement sensing: Monitoring in situ the changing state of polymeric materials during processing and use in the field, invited chapter in *Encyclopedia of Smart Materials*, John Wiley, New York, pp. 456–470, 2002.
8. D. Kranbuehl, In situ frequency dependent dielectric sensing of cure, *Processing of Composites*, Hansa, Munich, 137–157, 2000.
9. D. Kranbuehl and A. Loos, Monitoring resin position, reaction advancement and processing properties, in *Resin Transfer Molding for Aerospace Applications*, Chapman & Hall, London, pp. 412–432, 1999.
10. G. Hasko, H. Dexter, A. Loos, and D. Kranbuehl, Application of science based RTM for fabricating primary aircraft structural elements, *Journal of Advanced Materials*, 26, 9–15, 1994.
11. S. Hart, D. Kranbuehl, D. Hood, A. Loos, J. Koury, and J. Harvey, FDEMS sensor-loss model QPALS intelligent cure processing of polyimides, *International SAMPE Symposium Proceedings*, 39(1), 1641–1651, 1994.

12. D. Kranbuehl, D. Hood, J. Rogozinski, W. Limburg, A. Loos, J. MacRae, and V. Hammond, In situ FDEMS sensing for intelligent automated cure in resin transfer molding, *International SAMPE Symposium Proceedings*, 40(1), 1995.
13. D. Kranbuehl, D. Hood, J. Rogozinski, R. Barksdale, A. Loos, and D. MacRae, FDEMS sensing for automated intelligent processing of polyimides, *Proceedings of ASME International Mechanical Engineering Congress H1041A*, 2, 1017–1030, 1995.
14. B.J. Lepene, T.E. Long, A. Meyer, and D. Kranbuehl, Moisture-curing kinetics of isocyanate prepolymer adhesives, *Journal of Adhesion*, 78(4), 297–312, 2002.
15. D. Kranbuehl and A. Domanski, In situ monitoring of coating polymerization, cure and aging using frequency dependent dielectric monitoring *Journal of Coatings Technology*, 48–55, June 2004.
16. D. Kranbuehl, D. Hood, J. Rogozinski, A. Meyer, and M. Neag, Monitoring the changing state of a polymeric resin coating during synthesis, cure and use, in *Progress in Organic Coatings*, Elsevier, Amsterdam, pp. 881–888, 1999.
17. D. Kranbuehl, D. Hood, C. Kellam, and J. Yang, In situ sensing for monitoring molecular and physical property changes during film formation, in *Film Formation in Waterborne Coatings*, ed., T. Prouder, American Chemical Society Symposium Series, Washington, DC, pp. 648, 96–117, 1996.
18. D. Kranbuehl, J. Rogozinski, A. Meyer, L. Hoipkemeier, and N. Nikolic, On-line in situ sensor monitoring of rapidly curing UV films in film formation in coatings, T. Provder and M. Urban, eds., American Chemical Society, Washington, DC, pp. 141–156, 2001.
19. D. Kranbuehl, In situ monitoring of coating polymerization, *Cure and Aging Coatings Technology*, 1(6), June 2004.
20. D. Kranbuehl, W. Newby-Mahler, D. Hood, S. Cuse, K. Reifsnider, and A. Loos, Use of in situ dielectric sensing for intelligent processing and health monitoring, in *Structural Health Monitoring*, F.K. Chang, ed., Technomic, Penn, pp. 33–43, 1997.
21. D. Kranbuehl, D. Hood, L. McCullough, M. Eriksen, and E. Hov, Frequency dependent electromagnetic sensing for quality control and life monitoring of polymers, Proceedings of ICC Conference on Inspection of Structural Composites, 1994.
22. D. Kranbuehl, D. Hood, J. Rogozinski, A. Meyer, E. Powell, C. Higgins, C. Davis, L. Hoipkemeier, C. Ambler, C. Elko, and N. Olukeu, Monitoring and modeling of durability of polymers used for composite offshore pipe, *Recent Developments in Durability Analysis of Composites*, 413–420, 2000.

17 Magnetorheological Fluids

J. David Carlson
Lord Corporation

17.1 Introduction to Magnetorheological Fluids

Magnetorheological (MR) fluids are suspensions of magnetically responsive particles in a liquid carrier. The essential feature of MR fluids is their ability to change from a freely flowing liquid into a semisolid when exposed to an external magnetic field as shown in Figure 17.1. Most commonly, MR fluids consist of micron-sized, nearly pure elemental iron particles suspended in a hydrocarbon-based oil. When an external magnetic field is applied to MR fluid, the fluid develops a yield strength that is more or less proportional to the strength of the applied magnetic field. This yield strength arises from the internal structure that is formed when the iron particles polarize and chain together along the magnetic field lines. This magnetic field-induced yield strength adds linearly to the off-state shear stress due to the fluid viscosity such that the slope of the stress versus shear rate curve, i.e., plastic viscosity, remains largely unchanged.

A simple, Bingham plastic model is effective at describing the basic field-dependent characteristics of MR fluid [1]. In this model, the total yield stress τ_{total} is given by

$$\tau_{total} = \tau(H) + \eta_p \dot{\gamma} \quad (17.1)$$

where

$\tau(H)$ is the yield strength caused by the applied magnetic field H

$\dot{\gamma}$ is the shear rate

η_p is the magnetic field-independent plastic viscosity defined as the slope of the shear stress versus shear strain rate relationship

The behavior of MR fluid is summarized by the graph of τ_{total} versus $\dot{\gamma}$ shown in Figure 17.2.

MR fluids belong to a class of “smart” fluids that also include electrorheological (ER) fluids. Unlike ER fluids that respond to strong electric fields, MR fluids do not require the application high voltage. As a result, they are able to achieve very high yield strengths without encountering the problem of dielectric breakdown. As a consequence, MR fluids are able to routinely reach magnetic field-induced yield strengths of 50 kPa or more with a magnetic field that can easily be generated by an electromagnet operated at low voltage and modest current. MR fluids should not be confused with colloidal ferrofluids in which the particles are about 1000 times smaller than those found in typical MR fluids. While ferrofluids experience a body force in a magnetic field gradient, they are incapable of developing the large yield strength found with MR fluids.

17.2 MR Fluid History

MR and ER fluids both have histories that date from the 1940s. Jacob Rabinow at the U.S. National Bureau of Standards (now the National Institute of Standards and Technology) was responsible for the initial discovery and early development of MR fluids [2,3]. He applied for his first MR fluid patent in 1947 [4]. MR fluids made by Rabinow for use in clutches were not dissimilar to many of those made today and exhibited comparable yield strength. Rabinow's pioneering work led to a brief period of interest in MR fluids in the 1950s and early 1960s. However, by the late 1960s, interest in MR fluids had waned and a growing interest in ER fluids began to emerge. By the late 1980s, a

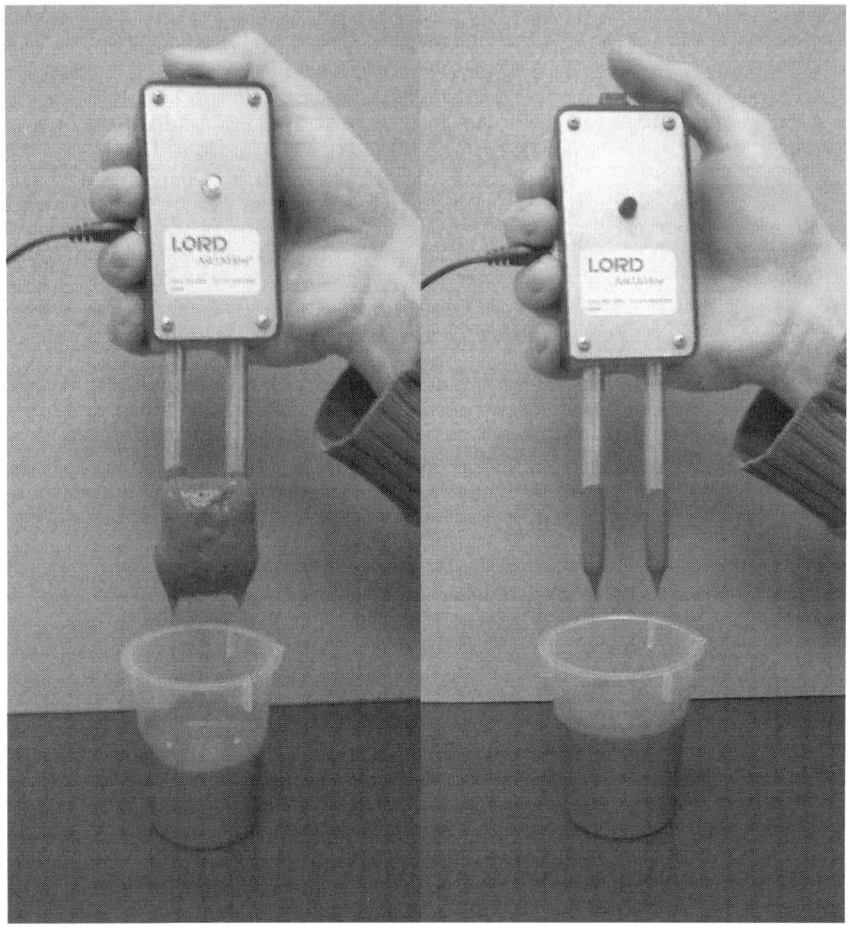

FIGURE 17.1 Demonstration of the operation of an MR fluid. On the left, the magnetic field is off. On the right, an electromagnet is energized to generate a magnetic field between the two steel prongs of the MR probe.

large number of universities and companies were actively pursuing research on ER fluids and devices [5]. Much of this work was motivated by potential automotive applications such as real-time controllable "smart" shock absorbers for improved ride and handling and controllable torque transfer devices [6]. In spite of great improvements in ER fluid formulations, ER fluid failed to emerge in any practical way. Maximum yield strengths of ER fluids remain too low for most applications. Further, they require power supplies capable of many kilovolts as well as expensive wires and connectors rated for such high voltages. The inherent temperature dependence of their electrical polarization mechanisms and a strong sensitivity to moisture and contamination makes their use outside of the laboratory difficult and costly.

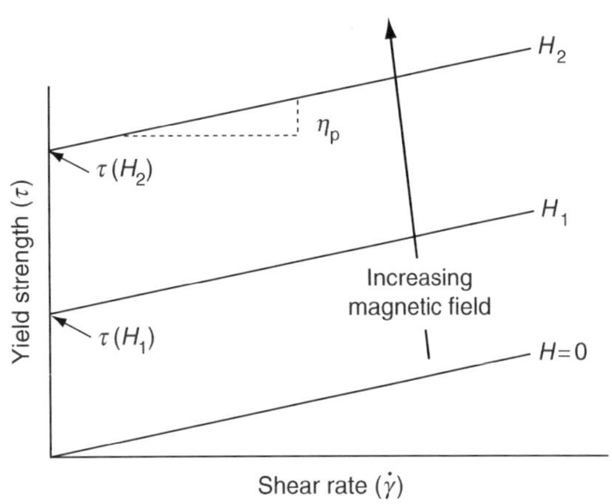

FIGURE 17.2 Behavior of an idealized, Bingham model, MR fluid in the presence of an applied magnetic field H as a function of shear rate.

Beginning in the early 1990s, a resurgence of interest in MR fluids and applications emerged in response to a number of practical limitations encountered with ER fluids [7,8]. MR fluids offered substantially higher yield strength plus the ability to operate at higher and lower temperatures. Most importantly, high voltages were not required to provide the necessary magnetic fields required to activate MR fluids. Common, low-voltage power supplies and batteries could directly power the electromagnets in MR fluid devices. This was coupled with fluid developments that led to high-durability MR fluids and the advent of modern, microprocessor-based control systems. The ultimate dream of a mass-produced "smart" MR fluid automotive shock absorber system was finally realized in early 2002 with the introduction of a real-time controlled suspension system as standard equipment on a Cadillac automobile [9].

17.3 MR Fluid Composition

The composition of a typical MR fluid can be broken into three parts: magnetically polarizable particles, a liquid vehicle, and an assortment of additives. While it is possible to create a suspension that will display a MR response with only a liquid and magnetically particles, such suspensions are highly unstable and do not perform satisfactorily. A well-formulated additive package is generally necessary to achieve a stable, long-lived MR fluid. Much of the MR fluid research activity in recent years has focused on developing proprietary additive packages that have enabled the simultaneous realization of high on-state force, low off-state forces and very long life in highly dynamic applications [10].

17.3.1 Polarizable Particles

The magnetically polarizable particles in an MR fluid are almost always nearly pure elemental iron although any magnetically polarizable particle could be used to make a functioning, albeit usually weaker, MR fluid. Powdered iron, water- or gas-atomized iron, nickel alloys, iron/cobalt alloys, magnetic stainless steels, and ferrites are all possible. The maximum yield strength of MR fluid scales approximately as the square of the saturation magnetization J_s of the particles [11]. In this regard, the best particles one could use would be alloys of iron and cobalt such as Permendur with a saturation magnetization of about 2.4 T [12,13]. Unfortunately, the very high cost of cobalt, which accounts for about 50% of these alloys, makes them prohibitively expensive for most applications. By far, the most cost-effective MR fluid particle is pure elemental iron with a saturation magnetization of 2.1 T. Virtually all other metals, alloys, and oxides have saturation magnetizations significantly lower than that of iron, resulting in substantially weaker MR fluids.

The most widely used material for MR fluid particles is carbonyl iron. Carbonyl iron is the common name given to iron particles formed from the thermal decomposition of iron pentacarbonyl. Key physical properties of carbonyl iron powder are the very spherical shape of the particles and the fine particle size in the 1–10 μm range. Other forms of elemental iron powder, such as water atomized or electrolytic, are also possible. Typical iron particle volume fractions range between 20% and 45%. Choosing the ideal volume fraction for a particular application is always a trade-off. While the maximum on-state yield strength scales more or less proportionate to the iron particle volume fraction, the off-state viscosity also increases at an even faster rate with increasing particle content.

17.3.2 Liquid Vehicle

The most common liquid vehicles for MR fluids are hydrocarbon oils. These can be mineral oils, synthetic oils, or mixtures of both. In general, hydrocarbon oils are the vehicle of choice because of their good lubricity, durability, and availability of a wide range of well-behaved and well understood additives. Synthetic poly(alpha-olefins) (PAOs) are often chosen because of their well-controlled properties over a broad range of temperatures. Other liquid vehicles for special-purpose MR fluids are silicone oils and water. While generally inferior to hydrocarbon oils in terms of lubricity, durability, and amenability to many additives, silicone oils do offer a somewhat broader temperature range with less variation in viscosity than synthetic hydrocarbons. Silicone oils are often chosen in situations where a hydrocarbon vehicle may not be compatible with other materials exposed to the MR fluid such as rubber seals and diaphragms. Care must be taken when using silicone oils to avoid conditions that might promote cross-linking of the oil, leading to viscosity increase and gum formation.

Water-based MR fluids offer the highest on-state yield strength and lowest off-state viscosity for a given particulate loading of any MR fluid. However, the high vapor pressure of water means that significant evaporative liquid loss must be considered. Water is only used as a vehicle for MR fluid in situations where evaporation is not a concern. This generally means that water-based MR fluid is only used in systems that are totally and absolutely sealed. Water-based fluids are usually not appropriate for systems containing a dynamic, sliding seal such as a shock absorber since the film of water invariably left on the surface of the exposed shaft will evaporate and lead to a progressive loss of fluid.

17.3.3 Additives

A wide variety of proprietary additives similar to those found in commercial lubricants are formulated into MR fluids. Such additives are used to discourage sedimentation prevent agglomeration, enhance lubricity, prevent oxidation, modify viscosity, and inhibit wear. Unlike colloidal ferrofluids, the relatively large particle size and large difference between the specific gravity of the magnetic particles and the carrier liquid can cause rapid settling in a MR fluid [14]. Sedimentation is typically controlled by the use of organic or inorganic thixotropic agents and surfactants. The thixotropic networks that form impart a small, zero-field yield strength to the MR fluid that inhibits flow at ultralow shear rates. Such thixotropic networks are very weak so that as the fluid is purposely sheared at higher rates, the network collapses, allowing the fluid to shear thin to a desirable low off-state dynamic viscosity.

Except for very special cases, such as dampers designed for seismic damage mitigation in civil engineering structures, lack of complete suspension stability is not a necessity. MR fluid devices such as dampers and shock absorbers that are used in very dynamic applications are efficient mixing devices. As long as the MR fluid does not settle into a hard sediment, normal motion in these devices is usually adequate to remix any stratified MR fluid back to a homogeneous condition [10]. In a well-designed MR fluid, damper remixing occurs within one or two strokes. While the MR fluid in these dampers will stratify over time, the particle-rich sediment remains soft and easily remixed.

17.4 Properties of Typical MR Fluids

While there are a number of "standard" MR fluids commercially available today, MR fluids for significant applications are generally tailored to meet the specific requirements of the application. The liquid type and viscosity, particle details, and volume fraction and additive package will all be chosen to optimize the MR fluid for the specific conditions of the application. More than 99% of all commercial MR fluids are based on synthetic hydrocarbon oils.

The most important property is the magnetic field dependent yield strength. This yield strength adds to the viscous-dependent stress at any given shear rate or speed. Maximum yield strengths are typically in the range of 25–100 kPa and depend most strongly on the volume fraction of iron particles in the fluid. For magnetic fields below about 100 kA/m, the field the developed yield strength is nearly linearly proportional to magnetic field. Above 100 kA/m, the increase in yield strength progressively rolls-off and eventually completely saturates at a field strength of about 400–500 kA/m.

The B–H curve for MR fluids throughout their useful range is nonlinear and of such a magnitude as to place them into a very unique category of materials intermediate between low-susceptibility materials like aluminum and ferrous materials like steel. MR fluids have initial (very low field) relative magnetic permeability in the range of 4–8. As the MR fluid yield strength begins to saturate at higher magnetic fields so also the slope of the B–H curves roll-off and approaches a slope of 1 at fields in the range of 400–500 kA/m. The maximum magnetic field strength that is useful in a practical sense is usually about 250 kA/m.

The off-state property most critical for dynamic mechanical applications is the field-independent plastic viscosity η_p. This viscosity creates shear rate or velocity-dependent forces and torques that are always present in a MR device regardless of whether a magnetic field is applied or not. This viscosity is also responsible for most of the temperature dependence observed in the force output of a device. The magnitude of η_p is controlled by the viscosity of the liquid vehicle and the particle volume fraction. Plastic viscosities at room temperature typically range from 50 to 200 mPa s although much higher values are easily possible with higher viscosity carrier liquids. The viscous off-state force in MR device can become quite large at the shear rates (10^4 to $> 10^5\,s^{-1}$) routinely encountered in MR shock absorbers, clutches, and brakes. Graphs of the on-state yield strength versus magnetic field and B–H curves for typical oil-based MR fluids having a variety of iron particle loadings are given in Figures 17.3 and 17.4. Table 17.1 lists the key physical properties for several commercial, MR fluids available from Lord Corporation [15–17]. More details about MR fluid properties are discussed in the reference by Black and Carlson [14].

The following pair of empirical equations is useful for describing the salient properties of most MR fluids [18]. The first equation gives the magnitude of the induced yield strength $\tau(H)$ as a function of the applied magnetic field H while the second gives the relationship between magnetic flux density B and the applied magnetic field intensity H in the MR:

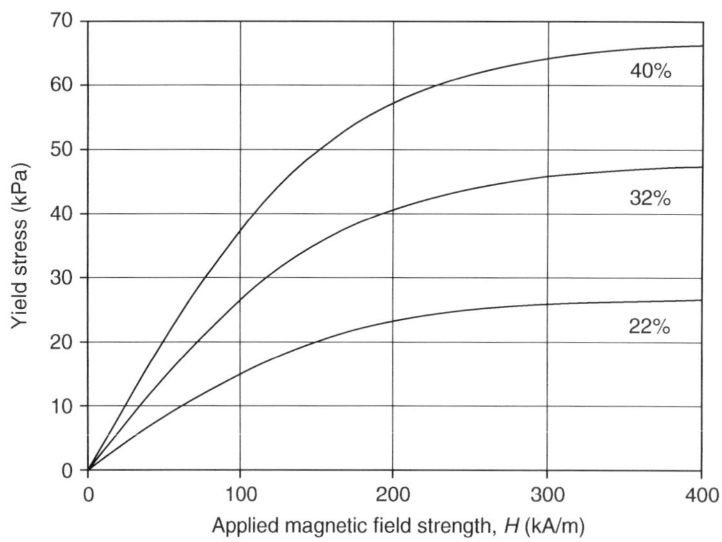

FIGURE 17.3 Typical yield stress versus magnetic field curves for several oil-based MR fluids having iron particle volume fractions ranging from 22% to 40% measured in parallel plate mode at a maximum shear rate of 262 s^{-1}.

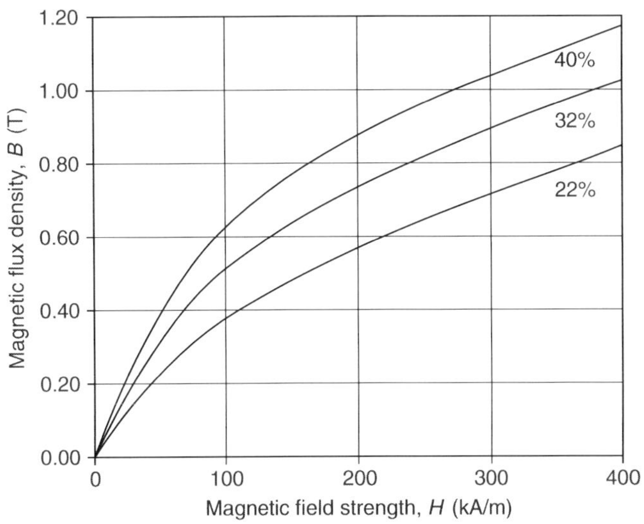

FIGURE 17.4 Typical *B–H* curves for several oil-based MR fluids having iron particle volume fractions ranging from 22% to 40%.

$$\tau(H) = C \times 271{,}700 \times \phi^{1.5239} \times \tanh(6.33 \times 10^{-6} H) \quad (17.2)$$

$$B = 1.91 \times \phi^{1.133}[1 - \exp(-10.97\mu_0 H)] + \mu_0 H \quad (17.3)$$

where
- ϕ is the volume fraction of iron particles
- τ is in Pa
- B is in T
- H is in A/m
- μ_0 is the magnetic constant equal to $4\pi \times 10^{-7}$ Henry/m
- constant C equals 1.0, 1.16, or 0.95, depending on whether the carrier fluid is hydrocarbon oil, water, or silicone oil

These equations have been developed to provide a convenient and practical description of the rheological and magnetic properties of virtually any MR fluid.

17.5 Basic Operating Modes

MR fluids are typically used in either a valve mode or a direct shear mode as illustrated in Figure 17.5. Examples of valve mode devices include servo-valves, dampers, shock absorbers, and actuators. Examples of direct shear mode devices include clutches, brakes, chucking and locking devices, some dampers, and structural composites. A less well-known squeeze-film mode has been explored only minimally and will not be discussed here.

17.5.1 Valve Mode

In the valve mode, MR fluid flows through a flow channel and a magnetic field is applied transverse to the flow direction. The yield strength that develops in the fluid establishes a pressure threshold for any fluid flow. Varying the magnetic field allows one to vary this pressure threshold, thus creating a controllable and proportionate valve mechanism without the need for moving mechanical parts. A well-designed magnetic circuit with electromagnet and ancillary high magnetic permeability steel flux conduits and pole pieces defining the flow channel are all that is required. Valve mode operation is most commonly found in MR fluid dampers and shock absorbers. Typically, in these devices, a

TABLE 17.1 Properties of Representative, Commercial Oil-Based MR Fluids

Property	Normal Range	MRF-122-2ED	MRF-132AD	MRF-140CG
Carrier liquid	—	Hydrocarbon	Hydrocarbon	Hydrocarbon
Particle volume fraction, ϕ	0.20–0.45	0.22	0.32	0.40
Particle weight fraction	0.70–0.90	0.72	0.81	0.854
Density (g/cm^3)	2–4	2.38	3.09	3.64
Yield strength (kPa) at 100 kA/m	10–55	13	23	38
Yield strength (kPa) at 200 kA/m	20–80	23	42	57
Yield strength (kPa) at saturation	25–100	~29	~45	~66
Plastic viscosity (mPa s) at 40°C, $\dot{\gamma} > 500\,s^{-1}$	50–200	61	92	280
Operating temperature (°C)	—	−40 to 130	−40 to 130	−40 to 130
Magnetic permeability, relative at low field	3.5–10	~4	~6	~8
Response time (s)	<0.001	<0.001	<0.001	<0.001
Thermal conductivity[a] (W/m °C), at 25°C	—	0.21–0.81	0.25–1.06	0.28–1.28
Specific heat (J/g °C), at 25°C	0.6–1.0	0.94	0.80	0.71
Coefficient of thermal expansion[b] 0–50°C, $(\Delta V/V)$/°C	$2–7 \times 10^{-4}$	6.5×10^{-4}	5.5×10^{-4}	5.0×10^{-4}

Sources: Lord Corporation, MRF-122-2ED magneto-rheological fluid, Lord Technical Data Sheet, DS7014, 2006; Lord Corporation, MRF-132DG magneto-rheological fluid, Lord Technical Data Sheet, DS7015, 2006; Lord Corporation, MRF-140CG magneto-rheological fluid, Lord Technical Data Sheet, DS7012, 2006.

[a] Values were calculated with and without magnetic field applied.

[b] Calculated values.

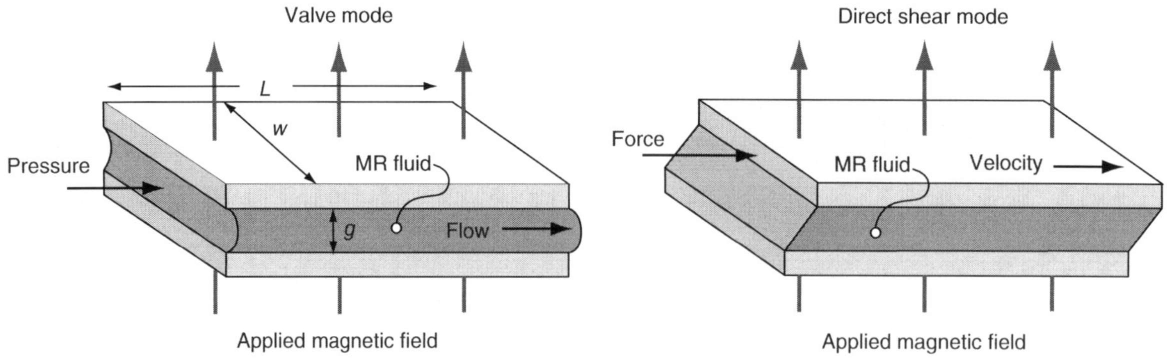

FIGURE 17.5 Two most common operation modes for MR fluids—valve mode and direct shear mode.

piston moves back and forth in a tubular housing that is filled with MR fluid. The MR fluid is forced to pass through an orifice in or around the piston.

The pressure drop developed by a valve-mode device can be divided into two components, the pressure ΔP_η due to the fluid viscosity and ΔP_{MR} due the magnetic field-induced yield. These pressures may be approximated by [6,8]

$$\Delta P_\eta = \frac{12\eta_p QL}{g^3 w} \tag{17.4}$$

$$\Delta P_{MR} = \frac{c\tau(H)L}{g} \tag{17.5}$$

Q is the volumetric flow rate. The parameter c has a value that ranges from a minimum value of 2 for $\Delta P_{MR}/\Delta P_\eta$ less than ~1 to a maximum value of 3 for $\Delta P_{MR}/\Delta P_\eta$ greater than ~100. The total pressure drop in a valve-mode device is approximately equal to the sum of viscous and magnetic field-induced contributions, ΔP_η and ΔP_{MR}. The total damping force developed by a valve-mode damper will thus be the total pressure drop multiplied by the piston area.

17.5.2 Direct Shear Mode

In a direct shear mode device, a layer of MR fluid is constrained between the poles and a force is applied that causes one pole to move laterally relative to the other. The amount of force needed to cause the fluid to shear then depends on the applied magnetic field strength and the developed shear strength in the MR fluid. Again, a well-designed magnetic circuit with electromagnet and ancillary high magnetic permeability steel flux conduits and pole pieces are used to apply the magnetic field to the MR fluid. Direct shear devices are most commonly found in clutches and brakes. In these cases, the device is configured such that a layer of MR fluid resists shear when one member is rotated relative to another. Common arrangements are disc and drum types of clutches and brakes that produce a controllable amount of torque coupling.

The force developed by a direct-shear device can be divided into F_η the force due to the viscous drag of the fluid and F_{MR} the force due to magnetic field-induced shear stress [6,8]:

$$F_\eta = \frac{\eta_p SLw}{g} \tag{17.6}$$

$$F_{MR} = \tau(H)Lw \tag{17.7}$$

where S is the relative velocity. The total force developed by the direct-shear device is the sum of F_η and F_{MR}.

17.6 MR Fluid Applications

MR fluid technology has progressed to the point where it is routinely used on a commercial scale to provide semiactive control in a variety of automotive and industrial applications. MR fluids have been used commercially since the mid-1990s. Today, the greatest driving force behind MR fluid technology is primary automotive suspension systems.

In January 2002, General Motors Corporation introduced the Cadillac Seville STS automobile with a MagneRide™ suspension system having real-time controllable MR fluid shock absorbers and struts as standard equipment [9]. The MR fluid shock absorbers were made by Delphi Corporation and the MR fluid was manufactured by Lord Corporation. MR fluid-based suspension systems are now available on many models of automobile including Cadillac (SRX, XLR, XLR-V, STS, and DTS), Chevrolet (Corvette), Buick (Lucerne), Ferrari (599 GTB Fiorano), Audi (TT and R8), Holden (HSV Commodore), and Honda Acura (MDX) with many more planned.

All commercial MR fluid controlled primary suspension systems are based on a monotube shock absorber that utilizes a single-stage, axisymmetric MR valve contained in the piston. Such a MR fluid piston/valve design efficiently provides high on-state and low off-state force in a very compact and robust package. The controllers for these systems process inputs from relative position sensors at each wheel. In addition,

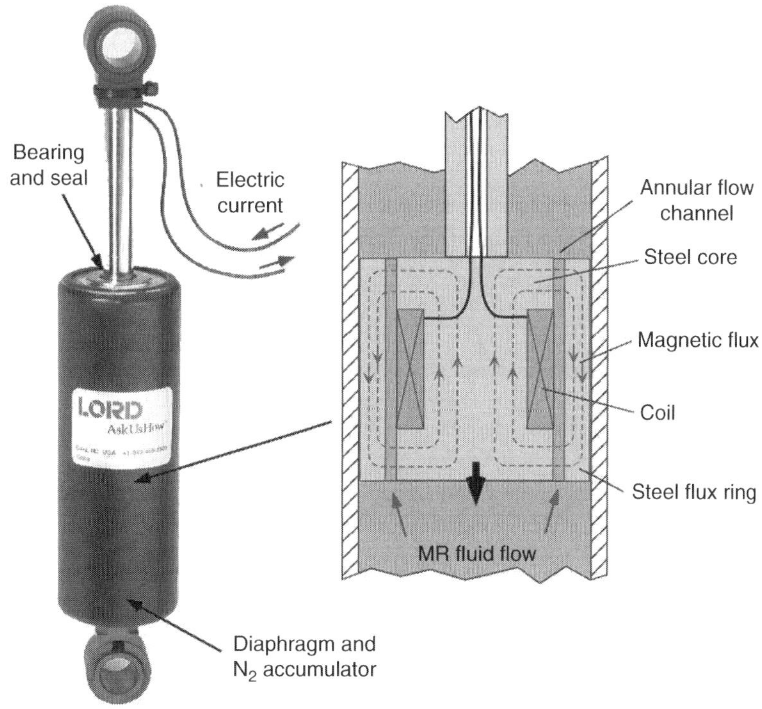

FIGURE 17.6 Basic MR fluid damper with axisymmetric valve geometry.

inputs from a lateral accelerometer yaw rate sensor, steering angle sensor, and speed sensor all feed by way of a CAN bus [19] into the controller. The control algorithms simultaneously optimize a wide range of factors that influence overall handling and ride comfort. System response time is less than 10 ms.

Common to virtually all commercial MR fluid shock absorbers is single-coil, axisymmetric MR fluid valve having an annular flow path as shown in Figure 17.6. This simple design has proven to be very reliable and is amenable to efficient, mass production. The straight flow path without any bends or changes in flow direction enables very low off-state forces without compromising the on-state. Automotive MR fluid shock absorbers have high durability with an operational life in excess of 150,000 km in a vehicle. Commercial MR fluid dampers are available in a variety of configurations including: coil-over shock, air-lift shock, or Macpherson strut.

Other, commercially significant, application areas include steer-by-wire (haptic) systems and secondary vehicle suspension systems in heavy-duty vehicles. Additional developing areas that are poised to become commercially significant include engine mounts, fan clutches, all-wheel drive clutches, position locking and detent devices, crash protection systems, civil engineering infrastructure dampers, military vehicle suspensions, and prosthetic devices. Current production of MR fluid is on the order of 100,000 L per year. In 2007, it is estimated that more than half a million MR fluid devices are in use worldwide [20].

References

1. Phillips, R.W., *Engineering Applications of Fluids with a Variable Yield Stress*, PhD thesis, University of California, Berkeley, CA, 1969.
2. Rabinow, J., The magnetic fluid clutch. *AIEE Trans.*, **67**, 1308, 1948.
3. Magnetic fluid clutch, *National Bureau of Standards Technical News Bulletin*, **32(4)**, 54, 1948.
4. Rabinow, J., Magnetic Fluid Torque and Force Transmitting Device, U.S. Patent 2, 575, 360, 1951.
5. Carlson, J.D., Sprecher, A.F., and Conrad, H., *Electrorheological Fluids*, Technomic, Lancaster, 1990.
6. Duclos, T.G., Design of devices using electro-rheological fluids, Society of Automotive Engineers, Technical Paper Series, 881134, Warrendale, 1988.
7. Carlson, J.D., The promise of controllable fluids, in *Proceedings of Actuator 94*, Borgmann, H. and Lenz, K., Eds., AXON Technologie, Bremen, 1994, p. 266.
8. Carlson, J.D., Catanzarite, D.M., and St. Clair, K.A., Commercial magneto-rheological fluid devices, *Int. J. Mod. Phys. B*, **10**, 2857, 1996.
9. Carlson, J.D., MR fluid technology—commercial status in 2006, in *Proceedings of 10th International Conference on ER Fluids and MR Suspensions*, Gordaninejad, F., Ed., World Scientific, Singapore, 389, 2007.
10. Carlson, J.D., What makes a good MR fluid, *J. Intelligent Mater. Syst. Struct.*, **13**, 431, 2003.

11. Ginder, J.M., Davis, L.C., and Elie, L.D., Rheology of magnetorheological fluids: Models and measurements, *Int. J. Mod. Phys. B*, **10**, 3293, 1996.
12. Margida, A.J., Weiss, K.D., and Carlson, J.D., Magnetorheological materials based on iron alloy particles, *Int. J. Mod. Phys. B*, **10**, 3335, 1996.
13. Carlson, J.D. and Weiss, K.D., Magnetorheological Materials Based on Alloy Particles. U.S. Patent 5, 382, 373, 1995.
14. Black, T. and Carlson, J.D., Magnetizable fluids, in *Synthetics, Mineral Oils, and Bio-Based Lubricants*, Rudnick, L.R., Ed., Taylor and Francis, Boca Raton, FL, 2006, Chapter 35.
15. Lord Corporation, MRF-122–2ED magneto-rheological fluid, Lord Technical Data Sheet, DS7014, 2006.
16. Lord Corporation, MRF-132DG magneto-rheological fluid, Lord Technical Data Sheet, DS7015, 2006.
17. Lord Corporation, MRF-140CG magneto-rheological fluid, Lord Technical Data Sheet, DS7012, 2006.
18. Carlson, J.D., MR fluids and devices in the real world, *Int. J. Mod. Physics B*, **19**, 1463, 2005.
19. International Organization for Standardization, ISO-11898, 2003.
20. Toscano, J.R., Lord Corporation, personal communication, 2007.

18

Intelligent Processing of Materials

J.A. Güemes
Universita Politécnica Madrid

J.M. Menéndez
AIRBUS

18.1 Concept of Intelligent Materials Processing

Intelligent materials processing (IPM) is achieved by the integration of online sensors, process models, and adaptive control strategies. Intelligent processing means a control system that uses prior knowledge and process models to manipulate the processing conditions in response to available online sensor data. In brief, intelligent processing is process control by objectives, rather than control following prescribed parameters. The central elements of IPM are (1) process understanding expressed in terms of a process model, (2) real-time information on processing parameters and material condition obtained with online process sensors, and (3) a model-based sensing and control strategy to achieve the desired characteristics in the finished product.

The traditional approach to material processing establishes a "process window," namely a range of acceptable values for the main control parameters such as initial composition, temperature, and pressure. These parameters were established from previous experience, or by trial and error, and they may be readjusted to accord the results of the actual production. Metal welding is one such example of this off-line control procedure. Depending on the thickness, composition, and other characteristics, the process parameters are selected from standard processing charts. The quality of the execution relies on the operator's skill. Final quality control through nondestructive procedures (NDE), like x-rays, maybe needed on certain structures to guarantee the compliance of the metal weld to the required properties.

To make the metal welding process "intelligent," one would need not only to have a control on the electrodes voltage and the welding speed, as it is done in conventional processing, but also to have sensors that collect data from the material in the molted area, in the same way as the operator's eye gets information about its shape and color. A computer-based model would be used to predict the final microstructure based on the values of the monitored variables and the setting of control parameters. The system then would predict how to modify these control parameters during the subsequent steps of the process in order to achieve the desired final microstructure. Although our technology has not yet been able to produce such an "intelligent welding machine" to replace the skilled operator, there is an ongoing project on "Weld process Sensing, Modeling and Control" sponsored by NIST; details can be found in Ref. [1].

The simple welding example help to identify the two main issues that need to be addressed in order to implement "intelligent processing" on any material:

1. Availability of reliable sensors to monitor the process variables. Such sensors must be able to work at the processing environment without disturbing the process itself. For instance, while thermal analysis is a powerful way to understand the material state and its transformations, so it is very useful for off-line control, this sensing principle cannot be applied for online control.
2. Full understanding of the process, the interdependence among the process variables that can be sensed and the process parameters that can be controlled, in order to achieve predictive capability on the material evolution. The sensing and processing times must be short compared to the physical times. The faster computers are, the higher will be the number of processes that can be real-time controlled.

It must be emphasized that "intelligent processing" is not the same as "automated processing," but a step ahead. Automated processing has the capability of keeping the process parameters inside the process window. It may even include some sensors to keep records of the process variables at intermediate steps or continuously. But automated processing cannot react to major disturbances, other than triggering an alarm, and it cannot perform continuous adjustments of the control parameters to achieve a predefined final state. Differences are summarized in Table 18.1.

The main benefit from intelligent processing is that it reduces the time spent in moving a design from project desks into production. Nowadays designers use software tools to evaluate the process constraints and the feasibility of different concepts, known as "virtual manufacturing." However, several manufacturing tests are still required for experimental adjustment of the control parameters before production can begin. With IPM, these parameters would adjust by themselves, producing a high-quality product at the first attempt. Significant impact on costs would be achieved by reducing the time elapsed between the conceptual design and the final product. Structural components would not have to be mass-produced to be economical. To exemplify this assessment, let us think about the resin transfer molding (RTM) process for composite materials. The software simulating the resin flow is a valid tool for virtual manufacturing, but it relies on assumed data for perform permeability; slight differences in the compaction of the perform induce strong changes in the flow, the predictions are only qualitative. Only when combining the numerical model with local measurements, as will be shown later in this chapter, the predictions may be quantitative.

A more ambitious benefit is claimed in Ref. [2]:

> The focus of materials process research is to establish theories and scientific methods for in situ self improvement of the design and control of material processing using self-directed computer aided systems. The specific objectives are: [1] generate new methods and knowledge relative to the design process, [2] integrate, synthesize and generalize new knowledge in the form of axioms, and [3] automatically incorporate this new knowledge for use in improving materials research and process control metrics (i.e., quality, time and cost). Long term objective is to develop a language, principled in theories, for the development of virtual material processing systems and integrated, intelligent manufacturing systems across a Materials and Processing Information Highway.

These general points apply to any engineering material, and industrial examples may be found in every technological area, from polymers to metals, ceramics, and composites. The different levels of "intelligence" correspond to the accuracy and predictive capability of the mathematical models of the process, the availability of in-situ real-time sensors and the efficiency of the control algorithms to find the optimal control strategy.

The idea of IPM started around 1990 with the advent of high-speed, low-cost computers. Web sites of laboratories with specific activity on IPM are given in Refs. [3–7]. The link to the University of Virginia deals with metal processing, while the next two are more specific for composites. A link to the International Conferences on "Intelligent Processing and Manufacturing of Materials" is available in Ref. [6]. Information on active or completed projects is available in Ref. [7].

18.2 Processing of Composite Materials

Although initially restricted to military aircrafts, composite parts have been intensively used during the last decade in commercial airplanes to reduce weight and improve performances, the upcoming generation of aircrafts gives a clear indication of this continuing trend. The new Airbus A380, the latest challenge of the European aerospace industry, will include more weight in composite materials than any other aircraft before. This development was conditioned by the need of substantial reductions of the operative costs compared to previous generations of airliners.

The attempts for cost reduction are advancing in two directions:

- *Design for manufacturing*: Design concepts must take into account the limitations of the process in order to make designs better adapted to the manufacturing processes.

TABLE 18.1 Differences between Manual, Automated, and Intelligent Processing

Manual Processing	Automated Processing	Intelligent Processing
The process parameters are selected from processing charts	A "process window" is preestablished for the raw material worst-case condition	Model-based control algorithms calculate and modify in real-time the control parameters to achieve the goal product
Process execution relies on operator's ability	Process parameters are kept inside the "process window" by online controllers	Real-time sensors monitor the process evolution
Off-line quality control by NDE	The process would not react to batch variation in the material composition, part thickness, environmental changes	A detailed math model correlates the required final properties with the measured variables and the control parameters
Example: Manual soldering of metal sheets	Example: Common autoclave cure of composite materials	Example: Smart cure of composites

- *New manufacturing processes*: The development of innovative technologies includes automated procedures and smart manufacturing.

Current composite technologies have reached a good level of automation. Cutting, hot drape forming, laser projection, tape lay-up, tow placement, and finishing operations are fully automated. Several automated production equipment and machines were developed during the 1990s and applied to a variety of components. Highly automated composites manufacturing facilities for primary composite aircraft structures are now in service.

Most of the composites used for aerospace structures are made by stacking layers of prepreg material at prescribed directions onto a tool, then compacting and curing the stack to transform it into a monolithic solid or laminate. The prepreg layers are done as a fabric or UD tape of high-strength/stiffness fibers, embedded into a polymeric resin. Composite material processing in general, and autoclave curing in particular, is a complex process, with significant batch to batch variability, where some of the critical quality variables, as void content, cannot be measured before the end of the cure, when they cannot be corrected. This lack of access to the quality variables during the cure is what is forcing to the autoclave cure process to be so conservative, otherwise the rejection rate of cured composite parts would be unacceptably high.

Smart technologies have to include the following improvements:

- Modeling techniques of material behavior and process parameters
- Development and validation of real-time process monitoring techniques, such as dielectrometry, optical fibers, pressure sensors, and others
- Development of intelligent regulation systems, with implementation of modeling and process monitoring techniques

The results are expected to be a decrease of development and manufacturing costs and lead times, enhancement of quality and reliability of composite components, reduction of scraps, and reduction of qualification procedures due to better understanding and higher reliability of process as well as fast adaptation to the change in raw materials.

The process monitoring techniques range from simple pressure and temperature data logging to a distributed sensor network on the process variables at the manufacturing facility, the tooling, and the part itself. The introduction of monitoring techniques is the basis for a subsequent real-time control of the process, whereby the process parameters are modified to optimize the product.

18.2.1 Cure of Thermosetting Resins

Even there is an increasing interest in thermoplastic resins as matrices for composites, still most of the high-performance composites materials are done with thermosetting resins, as epoxies, bismaleimides (BMIs), and cyanate ester. These resins undergo a complex physicochemical change during the cure process. From the chemical point of view, the small molecules composing the resin (molecular weights ranging from 200 to 2000 AMU) start reacting each other when temperature is high enough, and, at the end of a process called "cure cycle," a network of primary bonds is formed, transforming the resin into a non-melting rigid material. This evolution is measured by the degree of cure, defined as the percentage of reaction accomplished. As the chemical reaction is exothermic, the degree of cure is experimentally obtained as the ratio of the heat released up to a given time, to the total heat liberated at the full cross-linking.

Under a physical point of view, heating liquefies the uncured resin, so its behavior may be described by a viscous flow. When the resin is reacting, its molecules increase its length and the viscosity raises. Well before the end of the cure, a gel state is reached and the resin cannot longer flow. The reacting mixture is not yet as stiff as it will become when the cure is completed and it behaves as an amorphous elastomeric solid. This important event is called gel transition, and it is the limit for the resin ability to adapt to the shape of the mold and for the voids to collapse. Cure needs to be fully completed to attain the best properties of the resin, particularly its final glass transition temperature, which is related to the maximum in-service temperature.

The externally controlled process parameters are the autoclave temperature and pressure, and the vacuum level, acting all of them as the boundary condition of the uncured laminate. Currently, in the automated procedure, they follow a preestablished time pattern, called "cure cycle."

The internal variables of the material, such as temperature, pressure, resin viscosity, glass transition temperature, degree of cure, void size, void percentage and gas pressure inside the voids, are functions depending on the spatial location and time. The required final properties of the material are adequate strength and stiffness values. Both are correlated with the resin and void contents, and the internal residual stresses that appear during the cure cycle. The maximum service temperature is directly proportional to the final T_g.

Since 1982, different authors have proposed mathematical models with several levels of complexity, most of them based on phenomenological laws, trying to justify experimental findings [8–10]. Essentially, all of them include five submodels:

1. *Thermochemical model*: Predicts the degree of cure as a function of the temperature and time.
2. *Thermal transfer model*: Calculates the internal temperature distribution as a balance between conductivities, thermal capacitances, and the thermal overshoot due to the internal exothermic reactions.
3. *Rheological model*: Predicts the viscosity of the resin as a function of temperature and degree of cure. It also predicts the T_g and the gel time. It is usually connected to the resin flow model, along the fiber and through the thickness. This model calculates the internal pressure, the compaction of the laminate, and the final resin contents.

4. *Void model*: Establishes the worst case scenario for voids that come from air initially trapped between adjacent plies, or later generated at high temperatures either from volatile substances dissolved in the uncured resin or from previous water condensation. Resin pressure should make these voids collapse before gel time is reached.
5. *Residual stresses*: The chemical shrinkage of the matrix and the anisotropic expansion of each lamina create an internal stress field, which may produce warping or other undesired geometrical distortions; these internal stresses may be particularly high at the free edges, decreasing the laminate failure loads.

The details for these mathematical models are beyond the scope of this book. Interested readers may find a detailed description in Ref. [10]; even these models rely on many empirical parameters, and in some cases, like in the void model, on heuristic assumptions, the models are powerful tools for understanding how a change in the autoclave process parameters, or in the tooling and bagging material, may affect the final part quality.

18.2.2 Sensors and Sensing Techniques for Composites Manufacturing

Conventional sensors are used currently to monitor the autoclave temperature and pressure. However, there is much research [11–17] being done to develop different kinds of sensors for in-situ monitoring of variables that so far remain unknown (ply compaction pressure) can be roughly estimated (cure residual stresses), or can only be measured by laboratory techniques (T_g, degree of cure, etc.). There is another chapter in this book dealing with sensors for cure monitoring, written by T. Fukuda and T. Kosaka; to avoid redundancies, only a brief summary is done, together with a more detailed description on a fiber optic cure sensor [11] on which we pioneered the developments.

Dielectric sensors [12] are being used to measure the electrical capacitance and conductivity of the resin during the curing process. A drop in the conductivity can be correlated with the initiation of a three-dimensional structure and hence the gelation and vitrification processes. Figure 18.1 shows a conductivity plot for the 8552 resin from which the minimum viscosity, gel point, and vitrification point can be identified.

Ultrasonic sensors [13] use acoustic pulses in either the through-transmission or the pulse-echo mode. As the resin cures and gets harder, the speed of the sound through it increases and the attenuation decreases, approaching the limit values characteristic of the completely cured composite. In the pulse-echo mode, the sensor sends a pulse and then receives the resulting echo from the interface. As the resin cures, the return time shortens compared to the original time when uncured.

Fiber optic-based techniques (evanescent wave interaction, transmission spectrum analysis, Fresnel reflection, fluorescence, Raman spectrum measurement, etc.) are good candidates for the validation of cure cycles in autoclave. This is because of the small size of the optical fiber and because the optical fiber can be easily brought inside the autoclave. A comprehensive review of more than 200 references on process monitoring of fiber reinforced composites using optical fiber sensors was recently done by Fernando and Degamber [16]. Changes in the infrared spectrum of the resin are a good indicator of the cure evolution. Fourier transform infrared spectroscopy (FTIR) provides an attractive approach, since the information is spectrally encoded, and does not change with variation in the intensity of the light source. Standard optical fibers are not transparent to mid-IR, but still sensitive changes can be detected at the near-IR band (4000 to 9000 cm^{-1}) for epoxy and BMI resins. There are commercial probes available for IR spectroscopy, based on diffuse reflection, but because of their large size, they cannot be embedded in composites. Measurements of evanescent wave interactions is a promising development [16].

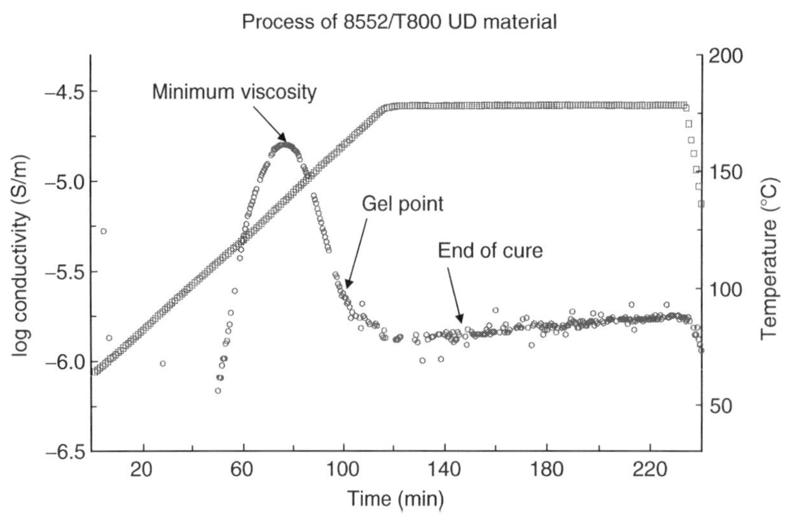

FIGURE 18.1 Typical conductivity plot for 8552 resin.

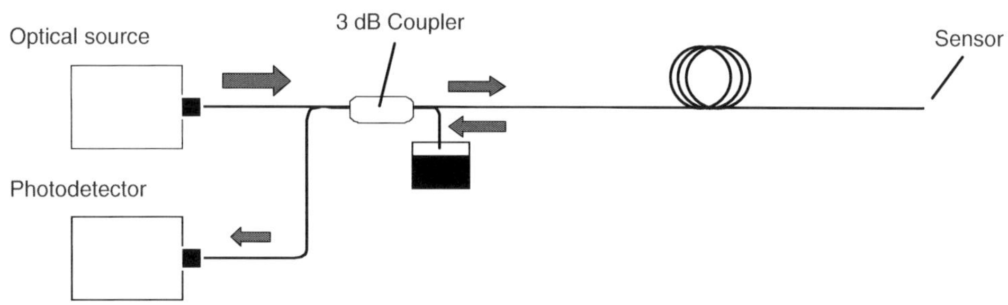

FIGURE 18.2 Schematic of a Fresnel-based reflectometer.

Changes in the refractive index of the resin during the cure cycle are a simple but effective approach to monitor the cure cycle. Devices developed for this purpose are known as fiber optic Fresnel reflectometers. The cleaved tip of an optical fiber operates as a sensor because the percentage of the back-reflected light is proportional to the refractive index of the material outside the fiber. Basic scheme is shown at Figure 18.2. Nevertheless, two main issues have to be solved before a practical implementation of this concept:

- Since these optical devices are intensity sensors, they are also sensitive to drifts in the optical power of the source and perturbations in the optical path (e.g., micro and macrobending of the optical fiber).
- Refractive index of the resin depends on both the degree of cure and the temperature. A temperature sensor has to be used together with the refractive index sensor.

To overcome these issues, we proposed and successfully demonstrated [11] a new scheme (Figure 18.3), where the photodetector was substituted by an optical spectrum analyzer and which includes two fiber Bragg gratings (FBGs) as internal references.

FBGs have proven their versatility and feasibility in many applications as fiber-optic-based smart structures. In this application, they are used as very accurate references to remove the undesirable perturbations on the signal reflected back by the fiber core–resin interface. Bragg gratings operate in this technique as narrowband mirrors, reflecting a constant percentage of the optical power that passes through them. Therefore, the gratings are used to detect the amount of the light power that reaches a certain point of the optical path. One device, located close to the optoelectronic system, operates as a main reference. The other FBG at the tip of the probe compensates the possible perturbations in the optical path affording also a measurement of the temperature. Figure 18.4 shows the results of a series of tests of isothermal cure cycles obtained by this method.

18.2.3 Fiber Optic Sensors Applied to RTM Process Monitoring

RTM is an adequate process for the production of medium to high series, limiting the human interaction to supervising and control tasks. A dry preform of fibers is placed inside a closed mold, vacuum is applied at the venting ports, the resin gets into the mold through the inlet ports, and must spread out into the preform before reaching the exit or being gelled. To achieve a structural quality similar to that obtained with autoclave curing, a high fiber volume is required, meaning that the preform has to be closely packed, which makes the resin's flow difficult. Should dry spots occur and the resin gel before the end of the injection, the part would be rejected.

The setup of a RTM process is complex, involving a large amount of variables in the tool design, the material, the injection equipment, and the process itself. Long development periods and a high number of initial scraps units are usual. Process modeling techniques are used during the development of successful molding strategies, reducing the number of trial and error tests. But modeling techniques are not a panacea; in practice, it is very difficult to incorporate all the process coefficients. Some coefficients introduced in the model are only rough approximations. Variations in the raw materials, equipment performance, or environmental conditions can result in erratic results for injections

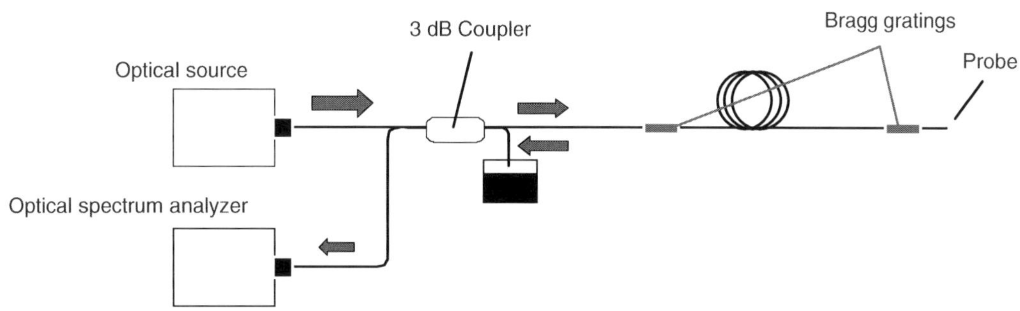

FIGURE 18.3 Schematic of a Bragg grating referenced cure monitoring system.

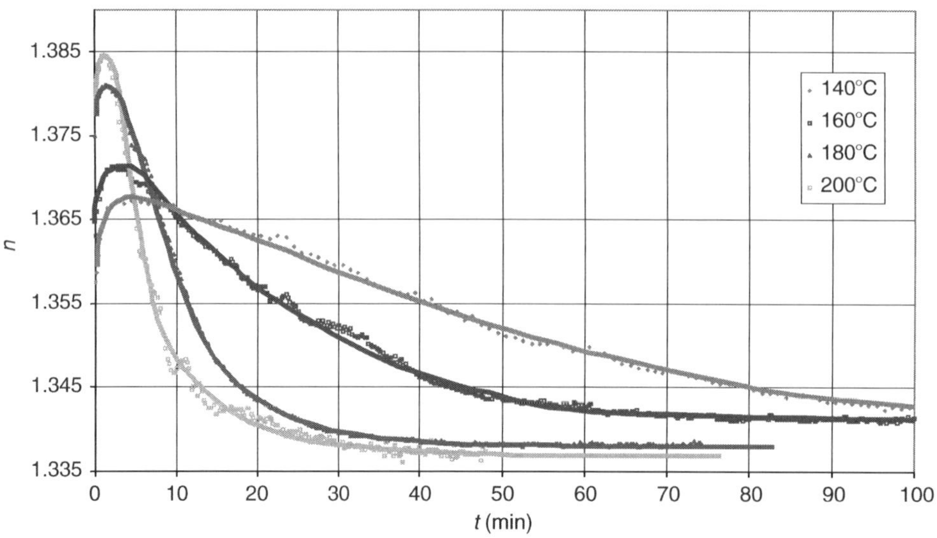

FIGURE 18.4 Isothermal variation of the refractive index of the resin Hexcel 8552 obtained using an optical probe embedded in an 8552/T800 prepreg laminate at 140°C, 160°C, 180°C, and 200°C.

apparently included in the process windows. Knowledge of the resin flow through the dry preform is the most critical process variable.

Some techniques are available to monitor the advance of the resin flow front inside the mold during RTM processes. Fiber optic sensors can easily be integrated in composite materials and so it makes this approach attractive. The arrival of resin can be detected by measuring the change in the refractive index at the interface. Fresnel reflection-based sensors were chosen as the most adequate technique available due to their inherent advantages:

- High signal-to-noise ratio due to the high difference between the refractive indexes of the air and the resin (a refractive index of 1.3 to 1.5, typical of epoxy resins, promotes a change in the signal of two orders of magnitude)
- Direct interpretability of the results
- Small size of the sensor. A cleaved optical fiber (0.125 mm of diameter), easily embeddable in the preform, reaching remote areas of the composite part under analysis with very low mechanical interference
- Simplicity of the optoelectronic equipment (a light source, a photodetector, an optical coupler and an optical switch), data acquisition, and data processing hardware and software
- Versatility of the configuration of the sensing network that can be embedded into the preform or integrated in the mold

Figure 18.5 provides a schematic of the system: light coming out from an optical source is launched into an optical fiber through a coupler. The tip of the optical fiber, which is cleaved

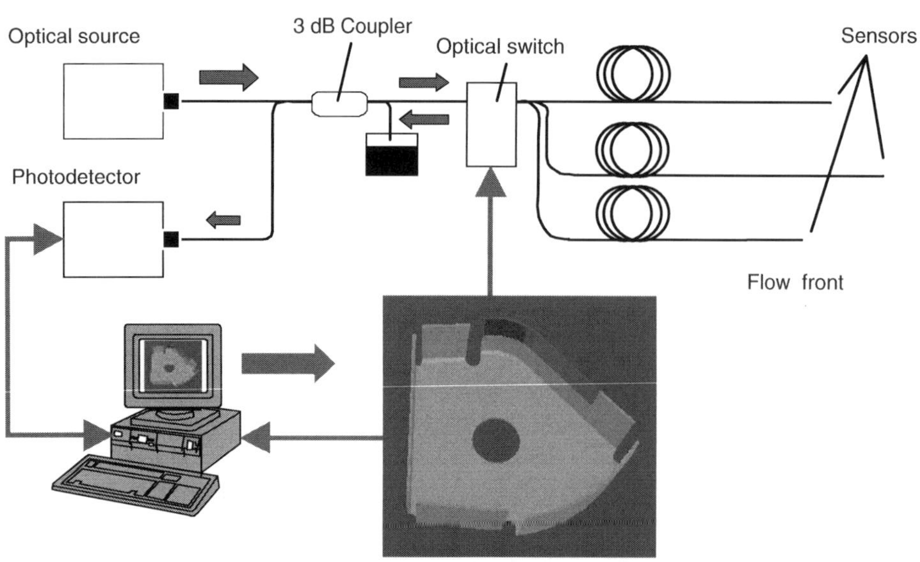

FIGURE 18.5 Schematic of a Fresnel-based RTM flow monitoring system.

FIGURE 18.6 Picture of the RTM injection system.

very precisely, is immersed in the medium under analysis, in this case, resin. The fiber tip interface acts as a sensor: most of the light that reaches this surface passes through it, but a small percentage of it (about 2%) is reflected back according to Fresnel law; this percentage will drop to 0.1% when resin wets the interface. The reflected light travels back through the coupler and is detected by a photodetector. An optical switch multiplexes several sensors, allowing for the interrogation of them at different preselected positions. Commercially available optical fiber switches, from telecommunication applications, allow to interrogating up to 32 optical fibers, with a commutation time of some milliseconds, much better than required, having in mind the characteristic time of the RTM process.

Figure 18.6 gives an example on a flow monitoring system implemented in a conventional RTM injection equipment during the manufacturing of carbon fiber reinforced plastic (CFRP) ribs of the elevator leading edge of the A-340 600 aircraft (Figure 18.7). Nine optical fibers were brought inside the mold through the vacuum port using three Poly Tetrafluroethylene (PTFE) tubes. The procedure to introduce the tubes into the mold without causing vacuum leakage is straightforward, using a T fitting in the vacuum port (Figure 18.8a). The sensors are not attached to the mold, but located inside the preform, bonded with a small drop of cyanoacrylate, in different positions in the same interface, as shown in Figure 18.8b. Injection of the rib was carried out in special conditions to slow the flow of the resin in the mold (vacuum was applied).

Every optical fiber sensor survived the assembly of the mold and the injection process, showing a high signal-to-noise ratio. Six over eight sensors detected the presence of resin. The time of transit of the resin through the mold was registered as indicated in Figure 18.9. Here, the complex geometry of the part justifies the use of computational models to design the mould and optimize the location of input and output ports. Several injection strategies (considering different material permeability at different positions in the mold) were previously tested to correlate the simulated flow to the results obtained in real injections. Experimental data of permeability were used as input in these models to obtain a map of times of resin transit. These maps were compared to the data given by the fiber optic monitoring system. Results presented in Figure 18.10 (numbers indicate the order of the activation of the sensors)

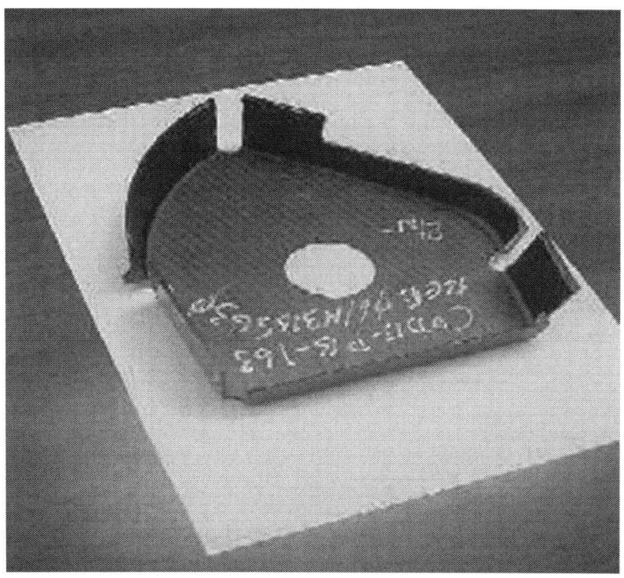

FIGURE 18.7 Composite part produced by RTM.

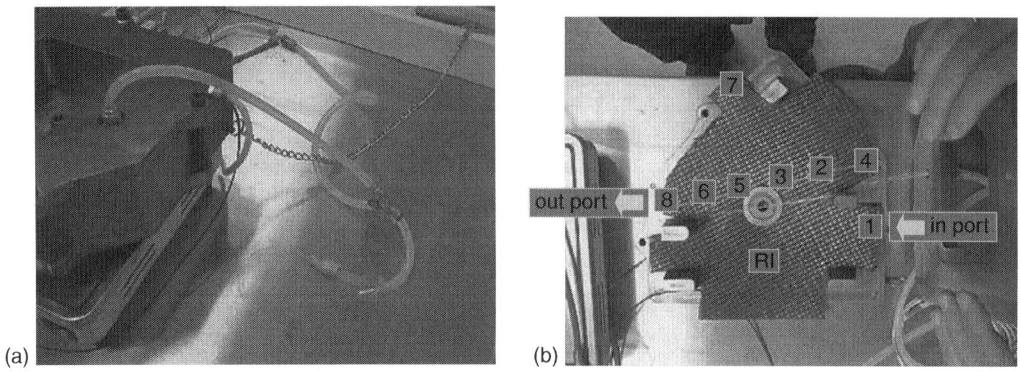

FIGURE 18.8 (a) Detail of the T fitting with the vacuum port and the optical fiber ingress. (b) Picture of the position of the sensors attached to the preform.

show that the real evolution of the resin qualitatively corresponds to the results obtained through computational methods. However, besides important quantitative differences, the detection of dry spots in two sensing points justify the use of monitoring systems in RTM processes during the development stage: Minor variations on the preform condition bring about large changes in the results. This development supports the use of intelligent control systems to optimize the filling of molds with complex geometry.

An advantage of using optical fibers for RTM flow detection is that they can later be used to monitor the cure and to obtain the residual stresses. This is simply done by adding a Bragg grating to the flow monitoring sensor, as explained in previous paragraphs. Thus, the same device is used to monitor the arrival of the resin with one of the systems and, after this moment, the evolution of the refractive index with the other system. No negative effects over the sensors have been detected during the different steps of the process: manufacture of the preform, integration of the sensors, assembly of the mold, heating of the mold, injection of the resin, and cure of the resin. Figure 18.11 shows the evolution of the refractive index during the curing process and the temperature in the coolest and the hottest point in the mold during the process. While the Bragg grating is very sensitive to the temperature, it is found that changes in the refractive index are due both to changes in the degree of cure and to changes in the temperature, which pose difficulties to the analysis of the experimental results. It was found a low resolution of the technique after the gel time.

It can be concluded that a fiber optic flow monitoring system can detect the resin front progression in real molding with very few limitations, and it has demonstrated as a valuable tool in the development stages of RTM process. Fiber optic cure monitoring systems are in an early stage of development, improvements in their sensitivity, particularly at high degrees of cure, are required.

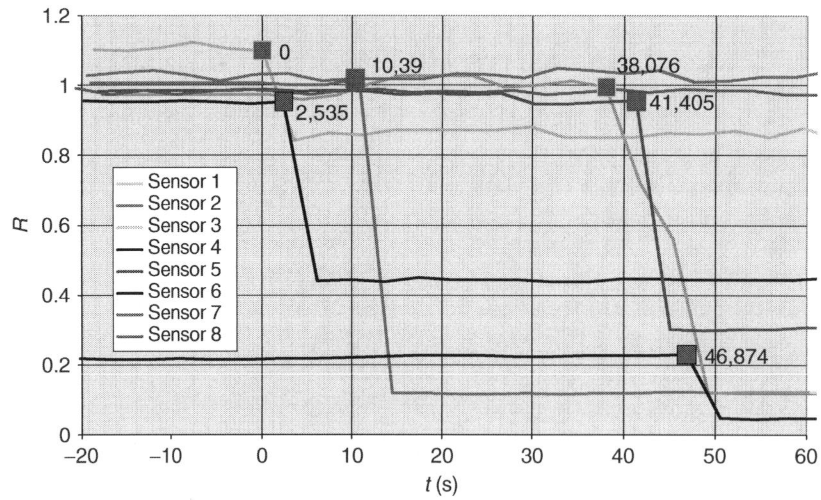

FIGURE 18.9 A340-600 elevator rib: time of transit of the resin. Experimental values.

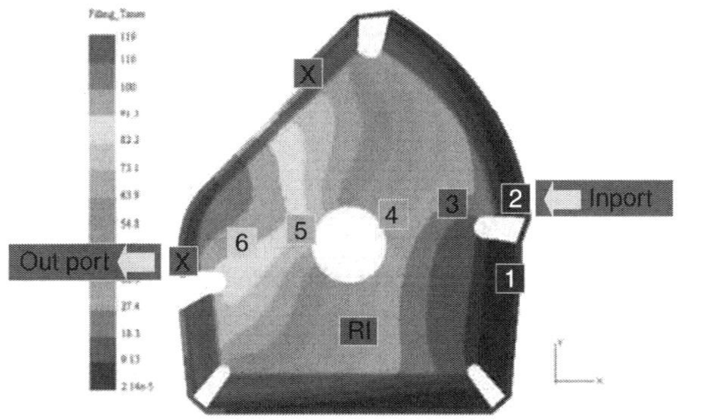

FIGURE 18.10 A340-600 elevator rib: time of transit of the resin calculations.

18.3 Intelligent Processing of Metallic Materials

References [19–22] are representative of the work done in the field of intelligent processing of metals and metal matrix composites (MMC). Most of the papers deal with intelligent hot isostatic pressing (HIP) as an advanced process to fabricate near-net shape parts with low ductility, difficult to melt intermetallic alloys of aerospace interest, as titanium aluminures. The consolidation process includes solid state reactions, adding a difussional flow to the classical plastic yielding and creep.

The processing parameters during the HIP process are temperature, pressure, and time. A practical way to understand the densification response for a powder material is through the use of HIP maps, firstly introduced by Ashby [23], for constant temperatures and pressures. When dealing with MMC, an additional problem appears; high temperatures and pressures help the compaction, but may degrade the composite by fiber–matrix reaction or fiber breakage. The final objective is to achieve the maximum densification at minimum time without causing fiber degradation.

The laws for plastic deformation and creep flow are given in Ref. [22], together with kinetic equations for diffusion, fiber–matrix reactivity and probability of fiber fracture. This model allows to simulate the outcome to a dynamic hipping law and coupled to a predictive control algorithm to find the best process cycle for a given set of initial conditions, materials properties, machine dynamics, and desired product goal state.

Eddy current probes were done using boron nitride preforms and platinum windings. These probes measure changes in the thickness of composite samples to a precision better than 20 ∝m during HIP cycles up to temperatures in excess of 1000°C.

The authors conclude that the extension of IPM as a feedback control method would allow to material engineers to design processes to attain a goal-state microstructure without long experimental programs. Materials could be built optimally adapted to its applications. The main limitation comes from the availability of noninvasive sensor techniques, based on eddy currents, laser ultrasonics, etc., that may inform on evolving microstructure.

18.4 Conclusion

In order to maintain the technological lead in the area of advanced materials, it is essential to build the quantitative knowledge of the relationships between material properties and processes. The development of advanced models is essential for the development of manufacturing systems must be done parallel to advances in sensor and control technology. Whereas a model provides an understanding on the physics and chemistry involved, and explains the relationship between dependent and

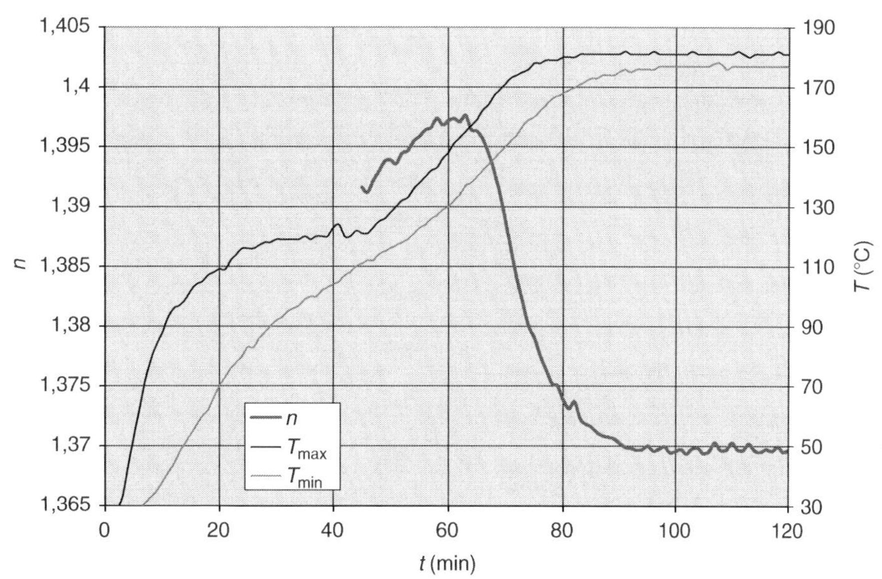

FIGURE 18.11 Mold temperatures and refractive index of the resin during curing.

control variables, sensors provide information related to the state of the material and process variables. The control function maintains quality through the use of data provided by sensors as inputs to the models, and manages the process to achieve the desired product.

Pioneering experiences have demonstrated the feasibility of the IPM concepts, both for advanced composites and metallic materials. Among the benefits for composite materials are the reduction of development costs and lead times, the enhancement of quality and reliability, and a simplification of the qualification procedures. For metallic materials, most of the experiences are concentrated on consolidation of intermetallic powders, but other processes as forging and soldering are also investigated. For both cases, the availability of affordable sensors is the critical issue for its industrial implementation.

References

1. www.msel.nist.gov.
2. LeClair-S-R and Jackson-A, Work Unit Directive (WUD54) Amendment: Materials Process Design. NASA no. 19980069695. 1996.
3. www.ipm.virginia.edu.
4. www.ipc.northwestern.edu.
5. www.ccm.udel.edu.
6. www.ipmm.mining.ubc.ca.
7. www.ims.org.
8. A.C. Loos and Springer G., Curing of epoxy matrix composites, *Journal of Composite Materials*, 21:243–261, 1983.
9. S.G. Advani and P. Simacek, Modelling and simulation of flow, heat transfer and cure (in resin transfer molding). In *Resin Transfer Moulding for Aerospace Sciences*. Dordrecht, Netherlands and Norwell, MA: Kluwer Academic, 1998, pp. 225–281.
10. R.S. Davé and A. Loos (Eds.), *Processing of Composites*, Munich: Hanser Verlag, 2000.
11. F. Rodríguez-Lence, P. Muñoz-Esquer, J.M. Menéndez, C. Pardo de Vera, S. Díaz, and J.A. Güemes. Smart sensors for resin flow and composite cure monitoring. Proceedings of the 12th International Conference on Composite Materials, ICCM-12, Paris, France, 1999.
12. C.W. Lee and B.P. Rice, resin transfer process monitoring and control, *SAMPE Journal*, 34(6), 48–55, 1998.
13. D.D. Shepard and K.R. Smith, A new ultrasonic measurement system for the cure monitoring of thermosetting resins and composites, *Journal of Thermal Analysis*, 49(1), 95–100, 1997.
14. G. Maistros, D. Bofilios, G. Cracknell, and A. Milburn, Intelligent curing of composites: Integration of dielectric cure monitoring with a knowledge-based system for efficient component recovery, Proceedings of the 19th International SAMPE Europe Conference of the Society for the Advancement of Materials and Process Engineering, pp. 733–743, Paris, 1998.
15. S. Cossins, M. Connell, and B. Cross et al., In situ near-IR cure monitoring of a model epoxy matrix composite, *Applied Spectroscopy*, 50(7), 900–905, 1996.
16. G.F. Fernando and B. Degamber, Process monitoring of fibre reinforced composites using optical fibre sensors, *Inttnal Materials Reviews*, 51(2), 65–106, 2006.
17. Y.M. Liu, C. Ganesh, J.P.H. Steele, and J.E. Jones, Fiber optic sensor development for real-time in-situ epoxy cure monitoring, *Journal of Composite Materials*, 31(1), 87–102, 1997.
18. J.L. Kardos, M.P. Dudukovic, and R. Dave. *Advances in Polymer Science*, Vol. 80 Epoxy resin and composites. In: Karel Dusek, ed. Berlin: Springer Verlag, 1986, pp. 101–124.
19. D.G. Backman, E.S. Rusell, D.Y. Wey, and Y. Pang, Intelligent prodessing for metal matrix composites, Proceedings of the Symposium on Intelligent Processing of Materials, Minerals, Metals and Materials Society, Warrendale, PA, 1989.
20. T.F. Zahrah, F. Charron, and N.M. Wereley, Model-based consolidation for near-net shape, AIAA/ASME Structures, Structural Dynamics and Materials Conference, April 1994, pp. 2315–2323.
21. H.N.G. Wadley, Intelligent processing of smart materials, AGARD LS-205 on Smart Structures and Materials: Implications for Military Aircraft of New Generation, 1996.
22. H.N.G. Wadley and R. Vancheeswaran, The intelligent processing of materials: An overview and case study, *Journal of Materials*, 19–30, January 1998.
23. W.B. Li, M.F. Ashby, and K.E. Easterling, On densification and shape-change during hot isostatic pressing, *Acta Metallurgica*, 35, 2831, 1987.

19

Magnets, Magnetic, and Magnetostrictive Materials

19.1 Magnets, Organic/Polymer

Joel S. Miller and Arthur J. Epstein

Magnetism has enabled the development and exploitation of fundamental science ranging from quantum mechanics, to condensed matter chemistry and physics, to materials science. Control of the magnetic behavior of materials has enabled the wide spread availability of low-cost electricity and electric motors, to the development of telecommunication devices (microphones, televisions, telephones, etc.) as well as the magnetic storage for computers. Magnets are suitable components for sensors and actuators and are crucial for smart materials and systems. Magnetic materials known from time immemorial are comprised of either transition or rare earth metals or their ions with spins residing in d- or f-orbitals, respectively, e.g., Fe, Gd, CrO_2, $SmCo_5$, $Co_{17}Sm_2$, and $Nd_2Fe_{14}B$. These materials are prepared by high-temperature metallurgical methods and generally are brittle, and chemically reactive. The latter half of the twentieth century has witnessed the replacement of many metal and ceramic materials with lightweight polymeric materials. While this has been primarily occurred for structural materials, examples in increasing numbers have also occurred for electrically conducting and optical materials. More recently, examples new of magnetic materials [1] have been reported and undoubtedly the twenty-first century will see the commercialization of organic/polymeric magnets [2].

Magnetism is a direct consequence of the spin or exchange coupling of unpaired electron spins. Noncoupled, independent electron spins (Figure 19.1a) lead to paramagnetic behavior, which can be quantitatively modeled by the Curie law and the Brillouin expression. Weak coupling (opposed or antiferromagnetic, or in alignment or ferromagnetic) leads to a modification of behaviors and is modeling with the Curie-Weiss law. Strong coupling can lead to magnetic ordering and alignment of spins (Figure 19.1b) and consequently a substantial magnetic moment and a ferromagnet, while that of opposed spins (Figure 19.1c) to an antiferromagnet as the net moments cancel. In contrast, the incomplete cancellation of spins can lead to a net magnetic moment and a ferrimagnet (Figure 19.1d) or a canted antiferromagnet (or weak ferromagnet) (Figure 19.1e).

The discovery of organic-based magnets with spins residing in p-orbitals have expanded the following properties already associated with magnets to include, e.g., solubility, modulation of the properties via organic chemistry synthetic methods, and low-temperature (nonmetallugical) processing, enhancing their technological importance and value as part of "smart" materials/systems.

A few organic nitroxides order ferromagnetically below a T_c of 1.5 K. These include one polymorph of 4-nitrophenyl nitronyl nitroxide reported by Kinoshita and coworkers with a T_c of 0.6 K (Figure 19.2a) [1c], and a polymorph of dinitroxide reported by Rassat and coworkers with a T_c of 1.48 K (Figure 19.2b) [3]. In addition, several canted antiferromagnets, e.g., one polymorph of

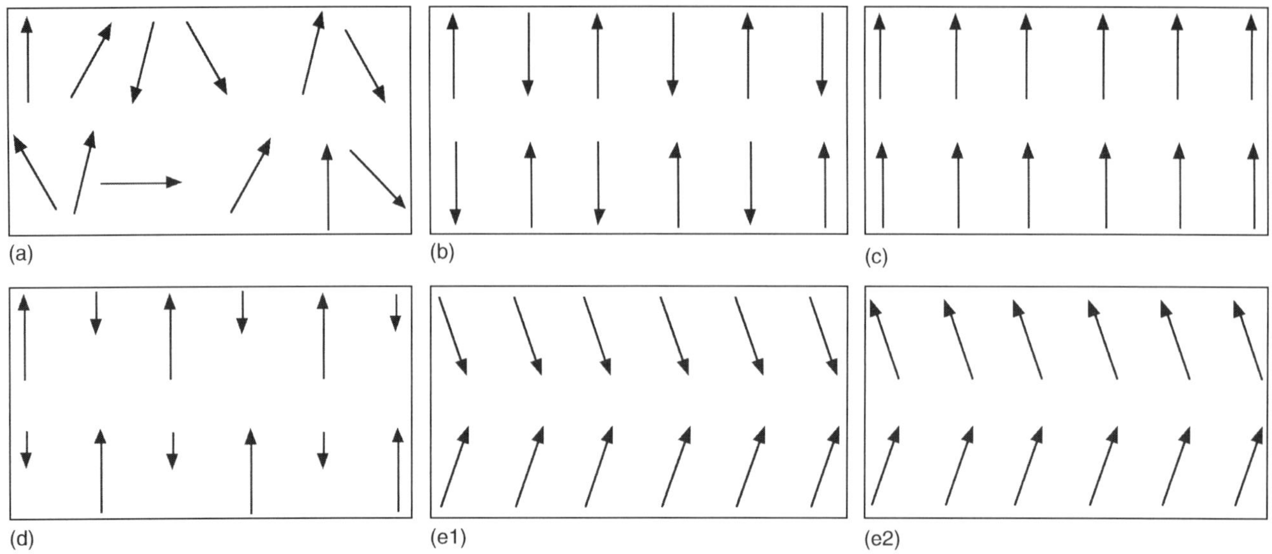

FIGURE 19.1 Two-dimensional spin alignment for (a) paramagnet, (b) antiferromagnet, (c) ferromagnet, (d) ferrimagnet, and (e) canted antiferromagnet behavior.

4-NCC$_6$F$_4$CNSSN (T_c = 36 K [4] that is increased to 65 K under 16 kbar pressure [4b]) (Figure 19.2c), and the electron transfer salt $\{[(H_3C)_2N]_2C_2[N(CH_3)_2)_2]\}^+$ $[C_{60}]$ (T_c = 16 K) [5], have been reported. These examples of organic-based magnets, albeit with very low T_c values, are crystalline solids, and not polymers. Reports of magnetic ordering in C_{60}-based polymers, however, exist [6].

Organic-based magnets possessing unpaired electron spins in both p- and d-orbitals also have been reported [1a–1c]. This includes ionic decamethylferrocenium tetracyanothanide, $\{Fe[C_5(CH_3)_5]_2\}[TCNE]$ (T_c = 4.8 K; TCNE = tetracyanoethylene), Figure 19.3), exhibiting the first evidence for magnetic hysteretic behavior in an organic-based magnet [7,8]. This magnet has an alternating $D^+ A^- D^+ A^-$ {D = $\{Fe[C_5(CH_3)_5]_2\}^+$; A = $[TCNE]^-$} solid state structure (Figure 19.4). The observed 16,300 emu Oe/mol saturation magnetization (M_s) is in excellent agreement with the calculated value of 16,700 emu Oe/mol for single crystals aligned parallel to the chain (Fe···Fe) axis. This value is 37% greater than that observed for iron metal on either a Fe or mole basis. Hysteresis loops with a coercive field of 1 kOe are observed at 2 K (Figure 19.5) [1b,7,8].

Each D and A has a single spin (S = 1/2). Above 16 K, the magnetic susceptibility behaves as expected for a one-dimensional (1D) ferromagnetically coupled Heisenberg chain with $J/k_B = 27$ K [8]. Below this temperature, the susceptibility diverges as $(T - T_c)^{-\gamma}$ as anticipated for a 1D Heisenberg-like system approaching a 3D magnetically ordered state. Spontaneous magnetization below the 4.8 K ordering temperature follows $(T_c - T)\beta$ with $\beta \sim 0.5$. Hysteresis loops are well defined with coercive field (H_{cr}) of 1 kOe at 2 K (Figure 19.5), indicating substantial domain walls pinning.

Replacement of Fe by Mn [9] and Cr [10] and substitution of TCNE with TCNQ [11–14] [7,7,8,8-tetracyano-*p*-quinodimethane (Figure 19.3c)], leads to ferromagnets with T_c reduced for TCNQ substitution and enhanced when Mn is utilized (Table 19.1 and Figure 19.6). Partial substitution of spinless (S = 0) $\{Co[C_5(CH_3)_5]_2\}^+$ for (S = 1/2) $\{Fe[C_5(CH_3)_5]_2\}^+$ in $\{Fe[C_5(CH_3)_5]_2\}^+$ $[TCNE]^-$ leads to a rapid decrease in T_c as a function of the fraction of spinless sites (1 − x) occurs, e.g., 2.5% substitution of $\{Fe[C_5(CH_3)_5]_2\}^+$ sites by $\{Co[C_5(CH_3)_5]_2\}^+$ decreases T_c by 43% [15]. Details of the magnetic properties can be found in reviews [16].

Gatteschi, Rey, and coworkers [1d] demonstrated and subsequently Iwamura and coworkers [17] have shown that covalent polymers comprised of bis(hexfluoroacetylacetonate)-manganese(II) (Figure 19.7) and nitroxides bound to the Mn(II) sites order as ferrimagnets. Using more complex nitroxides, Iwamura and coworkers have prepared related systems with T_c values ~46 K.

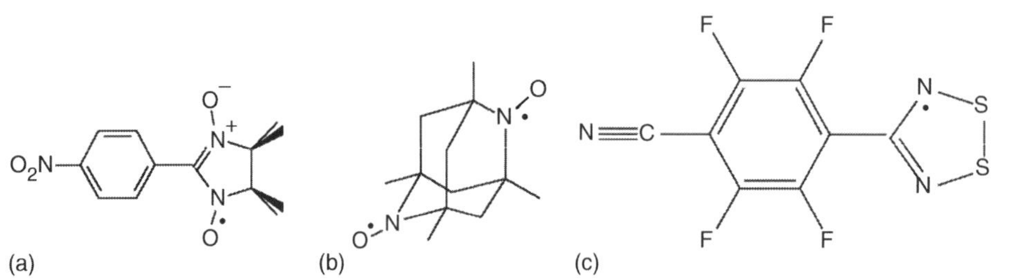

FIGURE 19.2 Structures of 4-nitrophenyl nitronyl nitroxide, which orders as a ferromagnet at 0.6 K (a), a dinitroxide, that orders as a ferromagnet at 1.48 K (b), and 4-NCC$_6$F$_4$CNSSN, which orders as a canted antiferromagnet at 36 K (c). The unpaired electron spins are designated by a dot (·).

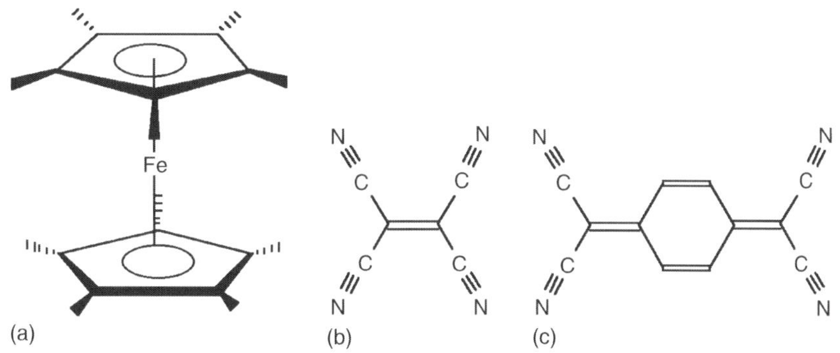

FIGURE 19.3 Molecular structures of $Fe[C_5(CH_3)_5]_2$ (a), TCNE (b), and TCNQ (c).

Using TCNE electron-transfer salts of Mn^{II}(porphyrin)'s, e.g., [MnTPP][TCNE] (H_2TPP = *meso*-tetraphenylporphyrin, Figure 19.8 [18], magnets with $T_c < 28\,K$) [19] based upon metallomacrocycles can also be prepared. Both the $[TCNE]^-$ and the nitroxides each have one spin; however, the Mn(II) in the former 1D polymeric chain has five spins ($S = 5/2$), while the Mn(III) in the latter polymeric chain has four spins ($S = 2$). In both cases, the Mn and organic spins couple antiferromagnetically, leading to ferrimagnetic ordering. The solid state motif is distinctly different than that for $\{M[C_5(CH_3)_5]_2\}^+$ [TCNE] (Figure 19.4) as $[TCNE]^-$ does not coordinate to the M in the latter system. Thus, the bonding of $[TCNE]^-$ to Mn is a model for the bonding of $[TCNE]^-$ to V in the $V[TCNE]_x \cdot z$(solvent) room temperature magnet.

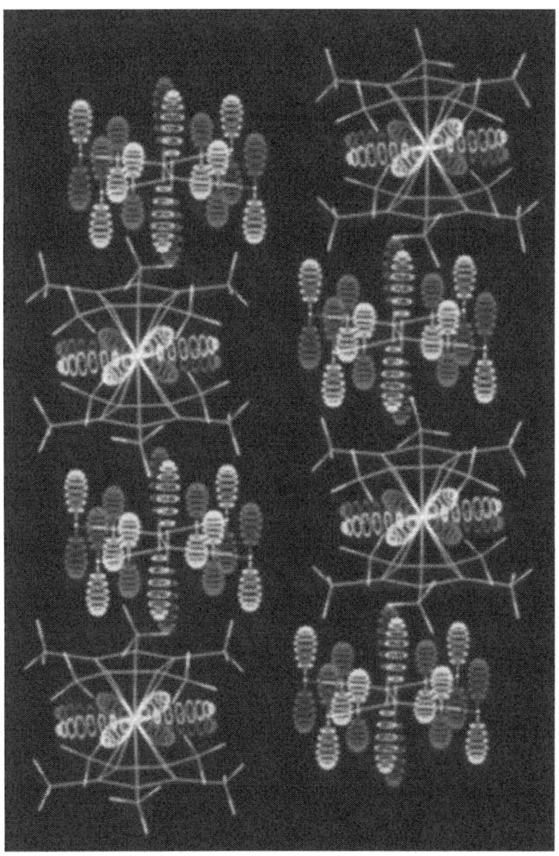

FIGURE 19.4 Crystal structure depicting segments of two parallel out-of-registry chains of $\{Fe[C_5(CH_3)_5]_2\}$[TCNE] showing the orbitals possessing the largest density of unpaired electrons.

As a consequence of the alternating $S = 2$ and $S = 1/2$ chain structure, the $[MnTPP]^+$ $[TCNE]^-$ 1D chain system, which forms a large family of magnets, is a model system for studying a number of unusual magnetic phenomena. This includes the magnetic behavior of mixed quantum/classical spin systems [20a] and the effects of disorder [20b], and the role of classical dipolar interation (in contrast to quantum mechanical exchange), and fractal interactions [21] in achieving magnetic ordering [20c]. Due to the single ion anisotropy for $[MnTPP]^+$ and the large difference in orbital overlaps along and between chains, this class of materials often undergo lattice- and spin-dimensionality cross-overs as a function of temperature [20d]. Metamagnetic-like behavior is noted for many members of this

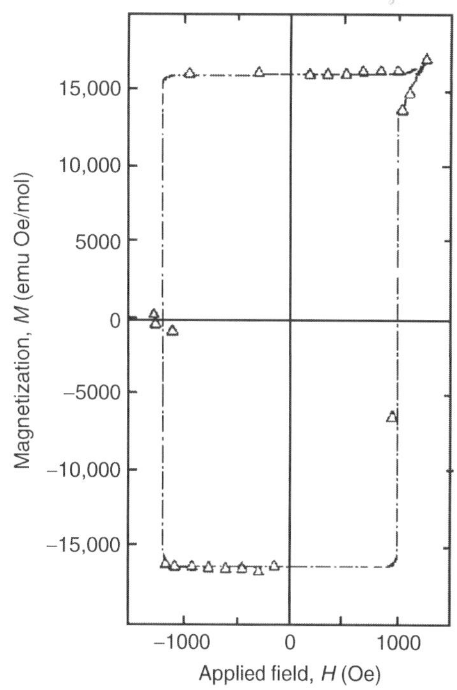

FIGURE 19.5 2 K hysteresis loops with a coercive field (H_{cr}) of 1 kOe for $\{Fe[C_5(CH_3)_5]_2\}$[TCNE].

TABLE 19.1 Summary of the Critical Temperatures (T_c) and Coercive Fields (H_{cr}) for $\{M[C_5(CH_3)_5]_2\}[A]$ (M = Cr, Mn, Fe; A = TCNE, TCNQ)

	[TCNE]•⁻			[TCNQ]•⁻		
M	FeC	Mn	Cr	Fe	Mn	Cr
SD⁺	1/2	1	3/2	1/2	1	3/2
SA⁻	1/2	1/2	1/2	1/2	1/2	1/2
T_c (K)	4.8	8.8	3.65	3.0	6.3	3.3
θ (K)	+16.9	+22.6	+22.2	+3.8	+10.5	+11.6
H_{cr} (kOe) (K)	1.0 (2)	1.2 (4.2)	[a]	[b]	3.6 (3)	[a]
References	[7,8]	[9]	[10]	[11]	[12]	[13]

[a] Not observed.
[b] Not reported.

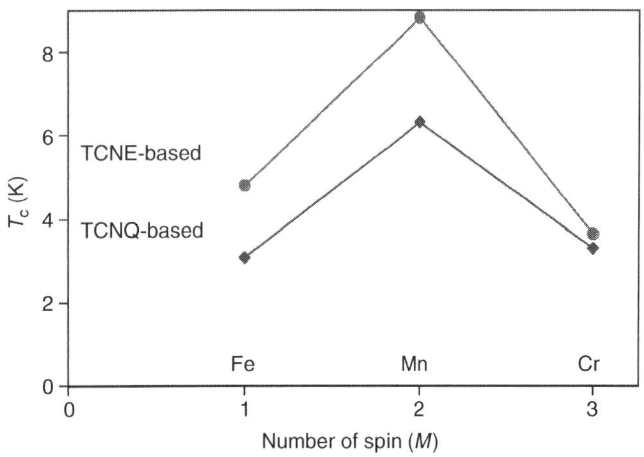

FIGURE 19.6 Critical temperature (T_c) for $\{M[C_5(CH_3)_5]_2\}[A]$ (M = Fe, Mn, Cr; A = TCNE, TCNQ).

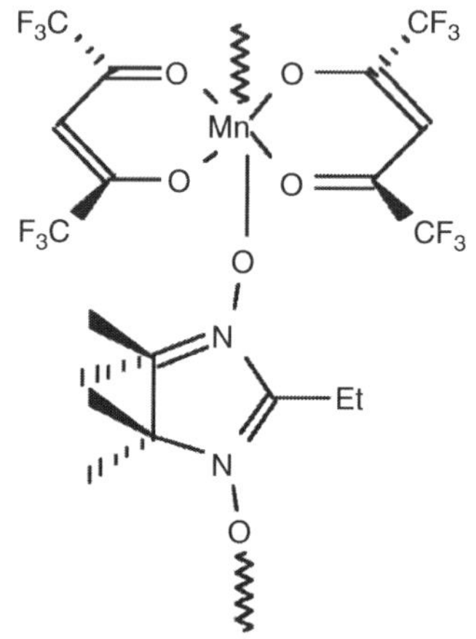

FIGURE 19.7 Structure of $Mn^{II}(hfac)_2NITEt$ (hfac = hexafluoro-acetylacetonate; NITEt = ethyl nitronyl nitroxide).

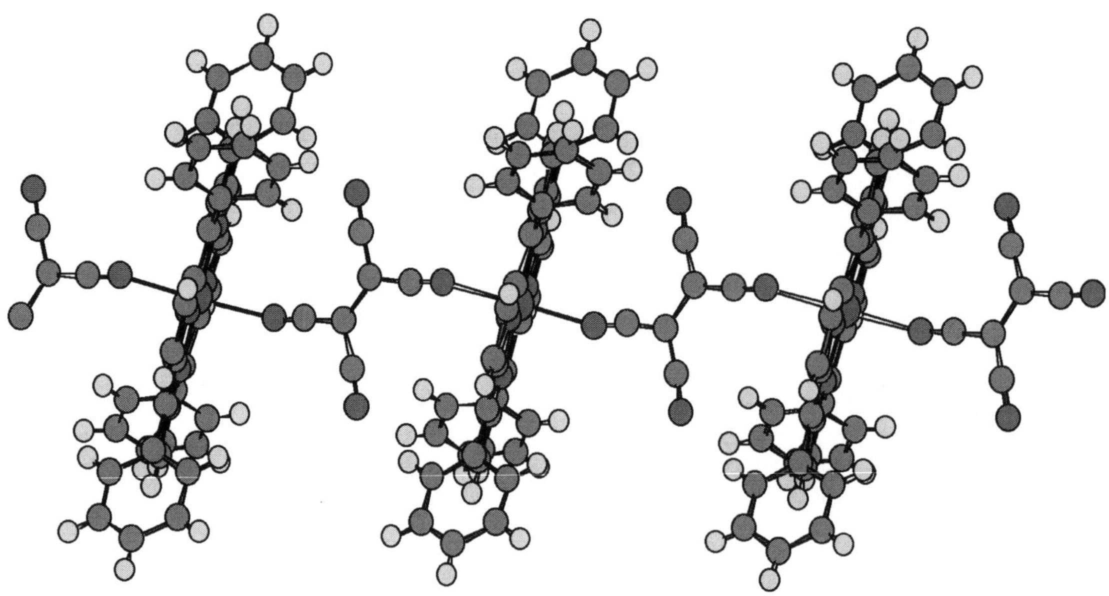

FIGURE 19.8 Segment of an 1D ···D⁺ A⁻D⁺ A⁻··· chain of $[MnTPP][TCNE] \cdot 2C_6H_5CH_3$ showing $[MnTPP]^+$ *trans*-μ-*N*-σ-bonding to $[TCNE]^-$.

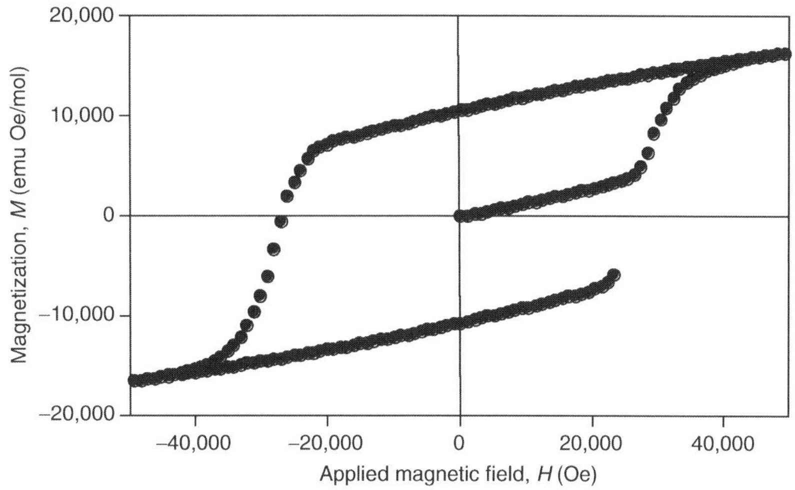

FIGURE 19.9 Metamagnetic and hysteresis with 27,000 Oe critical, H_c, and coercive, H_{cR}, fields at 2 K for *meso*-tetrakis(4-bromophenyl) porphinatomanganese(III) tetracyanoethanide. (From Rittenberg, D.K., Sugiura, K-i., Sakata, Y., Mikami, S., Epstein, A.J., and Miller, J.S., *Adv. Mater.*, 12, 126, 2000.)

family at low temperature with unusually high critical fields (H_c) and unusually high coercive fields (H_{cr}) as large as ~27,000 Oe and may have substantial remanent magnetizations (Figure 19.9) [20e].

19.1.1 V(TCNE)$_x$ · z(Solvent) Room Temperature Magnets

Reaction of $V(C_6H_6)_2$ [22a] or $V(CO)_6$ [22b] and TCNE in a variety of solvents [22c], e.g., dichloromethane, leads to loss of the benzene ligands and immediate formation of V(TCNE)$_x$·zCH$_2$Cl$_2$ ($x \sim 2$; $y \sim 1/2$). Due to its extreme air and water sensitivities as well as insolubility, compositional inhomogenieties within and between preparations occur, and the structure remains elusive. This material, however, is the first ambient temperature organic- or polymer-based magnet ($T_c \sim 400$ K). The proposed structure has each V being octahedrally coordinated with up to six ligands (N ligands from different TCNE ligands) and each TCNE is reduced and is either planar or twisted and bound to up to four V atoms. Recently, solvent-free thin magnetic films of V(TCNE)$_x$ composition on a variety of substrates, e.g., silicon, salt, glass, aluminum, have been prepared [23]. Enhanced stability can be achieved via overcoating the V(TCNE)$_x$ films with Parylene [24].

The inhomogeneities in the structure is expected to lead to variations in the magnitude, not sign, of the exchange between the V^{II} ($S = 3/2$) and $[TCNE]^-$ ($S = 1/2$), and the disorder should result in a small anisotropy [22].

19.1.2 M(TCNE)$_2$·z(CH$_2$Cl$_2$) (M = Mn, Fe, Co, Ni) High Room Temperature Magnets

Additional members of the family of high-T_c organic-based magnets M(TCNE)$_2$·z(CH$_2$Cl$_2$), (M = Mn, Fe, Co, Ni; TCNE = tetracyanoethylene) have been prepared [25]. These materials have T_c' values of 97, 75, 44, and 44 K, respectively. Field cooled/zero field cooled magnetization studies suggest that while the Mn [26] system is a reentrant spin glass, the Fe [27] system is a random anisotropy system. Both systems exhibit complex behavior below T_c. For example, hysteresis curves for the Fe compound, taken at 5 K, are constricted, with a spin-flop shape, indicating ferrimagnetic behavior, and field- and zero-field-cooled magnetization studies reveal magnetic irreversibilities below T_c for both compounds. Static and dynamic scaling analyses of the dc magnetization and ac susceptibility data for the Mn compound show that this system undergoes a transition to a 3D ferrimagnetic state at T_c, followed by a reentrant transition, to a spin glass state at $T_g = 2.5$ K. For M = Fe, the results of static scaling analyses are consistent with a high-T transition to a correlated sperimagnet, while below ~20 K, there is a cross-over to sperimagnetic behavior.

Solid solutions of V$_x$M$_{2-x}$[TCNE]$_2$ (M = Fe, Co, Ni) composition have been prepared that enables control to the coercive field and T_c [28]. In addition, photoinduced magnetic behavior, which is wavelength dependent, is observed for Mn(TCNE)$_2$ · zCH$_2$Cl$_2$) [29].

In addition to V(TCNE)$_2$-based magnets, substitution of TCNE with α-α′-dicyanoperfluorostilbene (DCNPFS), tetracyanopyrazine (TCNP), tetracyanobenzene (TCNB), and TCNQ form V(A)$_2$ (A = DCNPFS [30], TCNP [31], TCNB [32]) with $T_c \sim 200$ K, and V(TCNQ)$_2$ with $T_c \sim 50$ K [33]. The reaction of TCNB and $V(CO)_6$ also forms a magnet whose T_c exceeds room temperature ($T_c \sim 325$ K) [32].

19.1.3 Hexacyanometallate Room Temperature Magnets

Although technically not organic-based magnets, several materials termed molecule-based magnets are prepared by

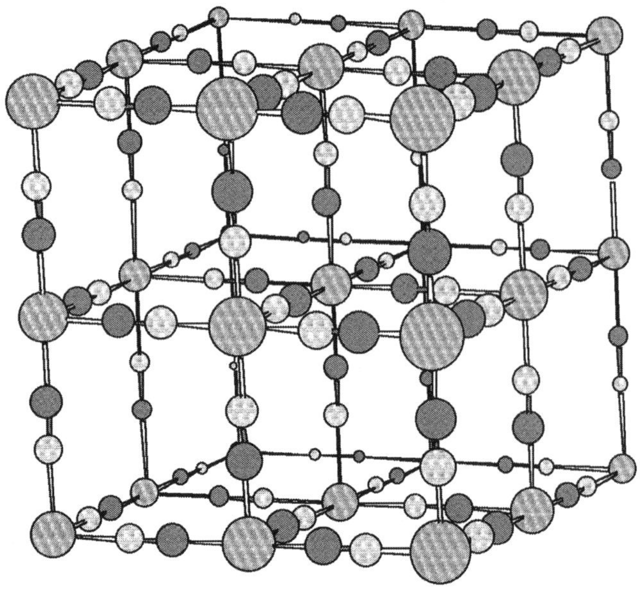

FIGURE 19.10 Idealized structure of $Fe^{III}_4[Fe^{II}(CN)_6]_3 \cdot zH_2O$, Prussian blue with ···$Fe^{III}$-N≡C-$Fe^{II}$-C≡N-$Fe^{III}$··· linkages along each of the three unit cell axes.

the same organic chemistry methodologies and have been reported to magnetically order in many cases, and above room temperature in a few cases. Prussian blue, $Fe^{III}_4[Fe^{II}(CN)_6]_3 \cdot zH_2O$, possesses a 3D network structure, with ···Fe^{III}–N≡C–Fe^{II}–C≡N–Fe^{III}··· linkages along each of the three identical unit cell axes of the cubic unit cell (Figure 19.10) and is a prototype structure that stabilizes both ferro- and ferrimagnetic order. Replacement of iron with other spin-bearing metal-ions leads to strong magnetic coupling and magnetic ordering with high T_c values. Examples include ferromagnetic $CsNi^{II}[Cr^{III}(CN)_6] \cdot 2H_2O$ (T_c = 90 K) [34], and ferrimagnetic $CsMn^{II}[Cr^{III}(CN)_6] \cdot H_2O$ (T_c = 90 K) [35], as well as thin films (≤2 μm), either oxidized or reduced, of $Cr^{III}[Cr^{III}(CN)_6]_{0.93}[Cr^{II}(CN)_6]_{0.05}$ (T_c = 260 K) [36] that exhibit negative magnetization (Figure 19.11). Verdaguer and coworkers reported that ferrimagnetic $V^{II}_{0.42}V^{III}_{10.58}[Cr^{III}(CN)_6]_{0.86} \cdot 2.8\ H_2O$ magnetically ordered above room temperature (T_c = 315 K) [37]. Further study of this class of materials has lead to an enhancement of the T_c to ~100°C (373 K) (Figure 19.12) [38]. Specific details on the properties of this class of magnets can be found in reviews [39].

19.1.4 Uses of Organic/Polymeric Magnets

The magnetic especially in conjunction with other physical properties may lead to their use in smart materials in the future [2]. This includes the next generation of electronic, magnetic, or photonic devices ranging from magnetic imaging to data storage and to static and low-frequency magnetic shielding and magnetic induction. The latter application in static and low-frequency magnetic shielding and magnetic induction is particularly promising [2b], as relatively high initial permeabilities have been reported for the $V(TCNE)_x \cdot z(CH_2Cl_2)$ materials. This, combined with their low density (~1 g/cm³), relatively low resistivity (~10^3 ohm-cm), and low power loss (as low as ~2 erg/(cm³ cycle)) suggest that future generations of these and related may be practical, especially for applications requiring low weight. Feasibility for $V(TCNE)_x \cdot z(CH_2Cl_2)$ in magnetic shielding applications is illustrated in Figure 19.13. Other potential applications include photoinduced magnetic effects [29,40], and spintronic applications [41].

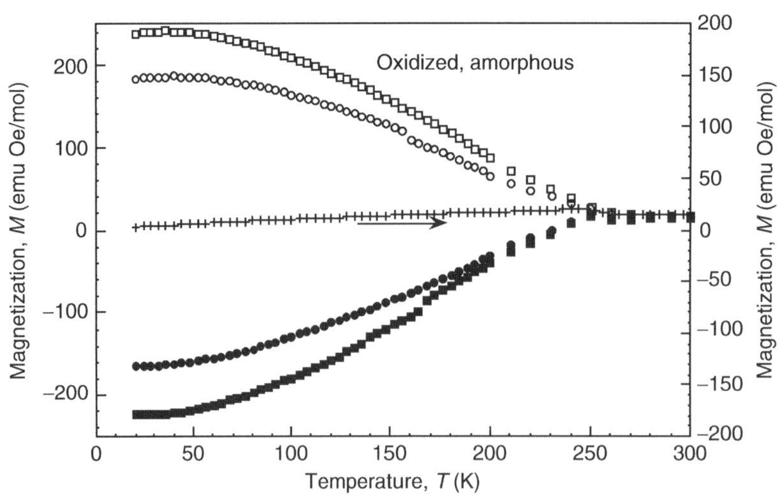

FIGURE 19.11 Temperature-dependent magnetization of the amorphous film of $Cr^{III}[Cr^{III}(CN)_6]_{0.98}-[Cr^{II}(CN)_6]_{0.02}$, zero field cooled (+) and field cooled in 5 Oe (○), −5 Oe (●), 10 Oe (□), −10 Oe (■) after 2 min oxidation at −0.2 V upon warming in a 5 Oe field. (From Buschmann, W.E., Paulson, S.C., Wynn, C.M., Girtu, M., Epstein, A.J., White, H.S., and Miller, J.S., *Adv. Mater.*, 9, 645, 1997; *Chem. Mater.*, 10, 1386, 1998.)

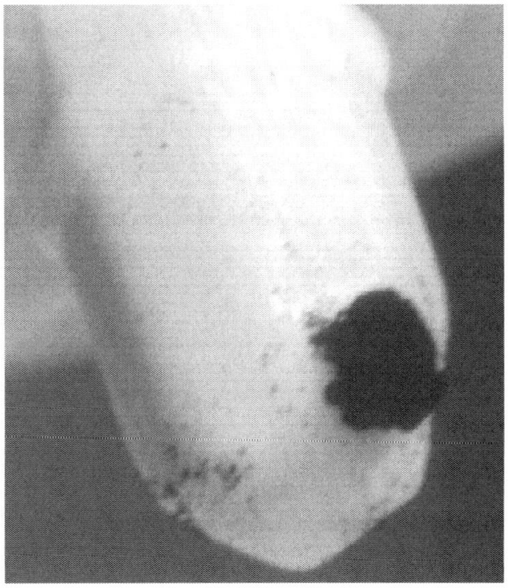

FIGURE 19.12 Sample of a $V[Cr(CN)_6]_y \cdot zH_2O$ magnet (T_c = 372 K) being attracted to a Teflon coated magnet at room temperature in the air. (Hatlevik, O., Buschmann, W.E., Zhang, J., Manson, J.L., and Miller, J.S., *Adv. Mater.*, 11, 914, 1999.)

FIGURE 19.13 Folded soft iron rods (staples) are shown attracted to a $SmCo_5$ permanent magnet (left). With a pellet of $V(TCNE)_x \cdot z(CH_2Cl_2)$ at room temperature shielding the magnetic field, the soft iron rods hang freely.

Acknowledgments

The authors gratefully acknowledge the extensive contributions of their collaborators, students, and postdoctoral associates in the studies discussed herein. The authors also gratefully acknowledge the continued partial support by the Department of Energy Division of Materials Science (Grant Nos. DE-FG03-93ER45504, DE-FG02-86ER45271, and DE-FG02-01ER45931), the U.S. Air Force Office of Scientific Research (Grant No. AFOSR FA9550-06-01-0175), and National Science Foundation (Grant No. 0553573).

References

1. (a) Blundell, S. J., Pratt F. L. *J. Phys.: Condens. Matter*, 2004, *16*, R771. (b) Miller, J. S., Epstein, A. J., *Angew. Chem. Int. Ed.* 1994, *33*, 385. Miller, J. S., Epstein, A. J., *Chem. Eng. News*, 1995, *73*#40, 30. (c) Kinoshita, M., *Jap. J. Appl. Phys.* 1994, *33*, 5718. (d) Gatteschi, D. *Adv. Mat.* 1994, *6*, 635. Coronado, E., Gatteschi, D., *J. Mater. Chem.* 2006, *16, 2513.*
2. (a) Landee, C. P., Melville, D., Miller J. S., in *NATO ARW Molecular Magnetic Materials*, Kahn, O., Gatteschi, D., Miller, J. S., Palacio, F., Eds. 1991, *E198*, 395. (b) Morin, B. G., Hahm, C., Epstein, A. J., Miller, J. S., *J. Appl. Phys.* 1994, *75*, 5782. (c) Miller, J. S., Epstein, A. J., *Chemtech* 1991, *21*, 168. (d) Miller, J. S., *Adv. Mater.* 1994, *6*, 322.
3. Chiarelli, R., Rassat, A., Dromzee, Y., Jeannin, Y., Novak, M. A., Tholence, J. L., *Phys. Scrip.* 1993, *T49*, 706.
4. (a) Palacio, F., Antorrena, G., Castro, M., Burrile, R., Rawson, J., Smith, N. B., Bricklebank, N., Novoa, J., Ritter, C., *Phys. Rev. Lett.* 1997, *79*, 2336. (b) Mito, M., Kawae, T., Takagi, S., Matsushita, D. H., Rawson, J. M., Palacio, F., *Polyhed.* 2001, *20*, 1509.
5. Omerzu, A., Arcon, D., Blinc, R., Mihailovic, D., in *Magnetism—Molecules to Materials*, J. S. Miller and M. Drillon, Eds., Wiley-VCH, Mannheim, Vol. 2, p. 123 (2001).
6. (a)Han, K.-H., Spemann, D., Höhne, R., Setzer, A., Makarova, T., Esquinazi, P., Butz, T., Carbon, 2003, *41*, 785. (b) Makarova, T.L., Sundqvist, B., Höhne, R., Equinazi, P., Kapelvich, Y., Scharff, P., Davydov, V., Kashevarova, L.S., Rakhmania, A.V., *Nature*, 2006, 440, 707.
7. (a) Miller, J. S., Calabrese, J. C., Epstein, A. J., Bigelow, R. W., Zhang, J. H., Reiff, W. M., *J. Chem. Soc. Chem. Commun.* 1986, 1026. (b) Miller, J. S., Calabrese, J. C., Rommelmann, H., Chittipeddi, S. R., Zhang, J. H., Reiff, W. M., Epstein, A. J., *J. Am. Chem. Soc.* 1987, *109*, 769.
8. Chittipeddi, S., Cromack, K. R., Miller, J. S., Epstein, A. J., *Phys. Rev. Lett.* 1987, *58*, 2695.
9. Miller, J. S., McLean, R. S., Vazquez, C., Calabrese, J. C., Zuo, F., Epstein, A. J., *J. Mater. Chem.* 1993, *3*, 215.
10. Yee, G. T., Manriquez, J. M., Dixon, D. A., McLean, R. S., Groski, D. M., Flippen, R. B., Narayan, K. S., Epstein, A. J., Miller, J. S., *Adv. Mater.* 1991, *3*, 309.
11. Broderick, W. E., Eichorn, D. M., Lu, X., Toscano, P. J., Owens, S. M., Hoffman, B. M., *J. Am. Chem. Soc.* 1995, *117*, 3641.
12. Broderick, W. E., Thompson, J. A., Day, E. P., Hoffman, B. M., *Science* 1990, *249*, 410.
13. Broderick, W. E., Hoffman, B. M., *J. Am. Chem. Soc.* 1991, *113*, 6334.
14. Taliaferro, M. L., Palacio, F., Miller, J. S., *J. Mater. Chem.* 2006, *16*, 2677.
15. Narayan, K. S., Kai, K. M., Epstein, A. J., Miller, J. S., *J. Appl. Phys.* 1991, *69*, 5953; Narayan, K. S., Morin, B. G., Miller, J. S., Epstein, A. J., *Phys. Rev. B* 1992, *46*, 6195.
16. Yee, G. T., Miller, J. S., in *Magnetism—Molecules to Materials*, J. S. Miller and M. Drillon, Eds., Wiley-VCH, Mannheim, Vol. 5, p. 223 (2005). Coronado, E., Galàn-Mascarós, J. R., Miller, J. S., in *Comprehensive Organometallic Chemistry*, Vol. 3 (2006).

17. Inoue, K., Hayamizu, T., Iwamura, H., *Mol. Cryst. Liq. Cryst.* 1995, *273*, 67. Izoka, A., Murata, S., Sugawara, T., Iwamura, H., *J. Am. Chem. Soc.* 1985, *107*, 1786; *J. Am. Chem. Soc.* 1987, *109*, 2631.
18. Miller, J. S., Epstein, A. J., *J. Chem. Soc., Chem. Commun.* 1998, 1319.
19. Brandon, E. J., Rittenberg, D. K., Arif, A. M., Miller, J. S., *Inorg. Chem.* 1998, *37*, 3376.
20. (a) Miller, J. S., Calabrese, J. C., McLean, R. S., Epstein, A. J., *Adv. Mater.* 1992, *4*, 498. (b) Brinckerhoff, W. B., Morin, B. G., Brandon, E. J., Miller, J. S., Epstein, A. J., *J. Appl. Phys.* 1996, *79*, 6147. (c) Wynn, C. M., Girtu, M. A., Brinckerhoff, W. B., Sugiura, K-I., Miller, J. S., Epstein, A. J., *Chem. Mater.* 1997, *9*, 2156. (d) Wynn, C. M., Girtu, M. A., Miller, J. S., Epstein, A. J., *Phys. Rev. B* 1997, *56*, 315. (e) Rittenberg, D. K., Sugiura, K-i., Sakata, Y., Mikami, S., Epstein, A. J., Miller, J. S., *Adv. Mater.* 2000, *12*, 126.
21. Etzkorn, S. J., Hibbs, W., Miller, J. S., Epstein, A. J., *Phys. Rev Lett.* 2002, *89*, 207201.
22. (a) Manriquez, J. M., Yee, G. T., McLean, R. S., Epstein, A. J., Miller, J. S., *Science* 1991, *252*, 1415. Miller, J. S., Yee, G. T., Manriquez, J. M., Epstein, A. J., in the Proceedings of Nobel Symposium #NS-81. *Conjugated Polymers and Related Materials: The Interconnection of Chemical and Electronic Structure*, Oxford University Press, Oxford, 1993, p. 461; *La Chim. La Ind.* 1992, *74*, 845. (b) Zhang, J., Zhou, P., Brinckerhoff, W. B., Epstein, A. J., Vazquez, C., McLean, R. S., Miller, J. S., *ACS Sym. Ser.* 1996, *644*, 311. (c) Thorum, M. S., Pokhodnya, K. I., Miller, J. S. *Polyhedron* 2006, *25*, 1927.
23. Pokhodnya, K. I., Epstein, A. J., Miller, J. S., *Adv. Mater.* 2000, *12*, 410.
24. Pokhodnya, K. I., Bonner, M., Miller, J. S., *Chem. Mater.* 2004, *16*, 5114.
25. Zhang, J., Ensling, J., Ksenofontov, V., Gütlich, P., Epstein, A. J., Miller, J. S., *Angew. Chem. Int. Ed.* 1998, *37*, 657.
26. Wynn, C. M., Girtu, M. A., Zhang, J., Miller, J. S., Epstein, A. J., *Phys. Rev. B* 1998, *58*, 8508.
27. Girtu, M. A., Wynn, C. M., Zhang, J., Miller, J. S., Epstein, A. J., *Phys. Rev. B* 2000, *61*, 492.
28. (a) Vickers, E. B., Senesi, A., Miller, J. S., *Inorg. Chim. Acta* 2004, *357*, 3889. (b) Pokhodnya, K. I., Vickers, E. B., Bonner, M., Epstein, A. J., Miller, J. S., *Chem. Mater.* 2004, *16*, 3218. (c) Pokhodnya, K. I., Burtman, V., Epstein, A. J., Raebiger, J. W., Miller, J. S., *Adv. Mater.* 2003, *15*, 1211.
29. (a) Pejakovic, D., Kitamura, C., Miller, J. S., Epstein, A. J., *Phys. Rev. Lett.* 2002, *88*, 57202. (b) Pejakovic, D., Epstein, A. J., Kitamura, C., Miller, J. S., *J. Appl. Phys.* 2002, *91*, 7176.
30. Fitzgerald, J. P., Kaul, B. B., Yee, G. T., *ChemComm*, 2000, 49.
31. Vickers, E. B., Selby, T. D., Miller, J. S., *J. Am. Chem. Soc.* 2004, *126*, 3716.
32. Taliaferro, M. L., Thorum, M. S., Miller, J. S., *Angew. Chem.* 2006, *45*, 5326.
33. Vickers, E. B., Selby, T. D., Thorum, M. S., Taliaferro, M. L., Miller, J. S., *Inorg. Chem.* 2004, *43*, 6414.
34. Gadet, V., Mallah, T., Castro, I., Verdaguer, M., *J. Am. Chem. Soc.* 1992, *114*, 9213.
35. Greibler, W. D., Babel, D., *Z. Naturforsch* 1982, *87b*, 832.
36. Buschmann, W. E., Paulson, S. C., Wynn, C. M., Girtu, M., Epstein, A. J., White, H. S., Miller, J. S., *Adv. Mater.* 1997, *9*, 645; *Chem. Mater.* 1998, *10*, 1386.
37. Ferlay, S., Mallah, T., Ouahes, R., Veillet, P., Verdaguer, M., *Nature* 1995, *378*, 701.
38. (a) Hatlevik, O., Buschmann, W. E., Zhang, J., Manson, J. L., Miller, J. S., *Adv. Mater.* 1999, *11*, 914. (b) Holmes, S. D., Girolami, G., *J. Am. Chem. Soc.* 1999, *121*, 5593.
39. (a) Verdaguer, M., Girolami, G. S., In *Magnetism—Molecules to Materials*, Miller, J. S., Drillon, M., Eds., Wiley-VCH: Weinheim, 2005, Vol. 5, p. 283. (b) Verdaguer, M., Bleuzen, A., Marvaud, V., Vaissermann, J., Seuleiman, M., Desplanches, C., Scuiller, A., Train, C., Garde, R., Gelly, G., Lomenech, C., Rosenman, I., Veillet, P., Cartier, C., Villain, F., *Coord. Chem. Rev.* 1999, *190–192*, 1023. (c) Verdaguer, M., Bleuzen, A., Train, C., Garde, R., Fabrizi de Biani, F., Desplanches, C., *Phil. Trans. R. Soc. Lond. A* 1999, *357*, 2959. (d) Hashimoto, K., Ohkoshi, S., *Phil. Trans. R. Soc. Lond. A* 1999, *357*, 2977.
40. (a) Sato, O., Iyoda, T., Fujishima, A., Hashimoto, K., *Science* 1996, *272*, 704. (b) Pejakovic, D. A., Manson, J. L., Miller, J. S., Epstein, A. J., *J. Appl. Phys.* 2000, *87*, 6028.
41. (a) Prigodin, V. N., Raju, N. P., Pokhodnya, K. I., Miller, J. S., Epstein, A. J., *Adv. Mater.* 2002, *14*, 1230. (b) Raju, N. P., Savrin, T., Prigodin, V. N., Pokhodnya, K. I., Miller, J. S., Epstein, A. J., *J. Appl. Phys.* 2003, *93*, 6799.

19.2 Powder Metallurgy Used for a Giant Magnetostrictive Material Actuator Sensor

Hiroshi Eda and Hirotaka Ojima

19.2.1 Introduction

Magnetostrictive materials for actuator devices have used ferrite and nickel materials in the past, however, the amount of displacement of those materials has been only 30 ppm while piezoelectric materials whose amount of displacement is 800 ppm has been used for actuator devices.

Currently, the giant magnetostrictive material (GMM) has at least 1500 ppm in the amount of displacement and GMM has generated stress three times larger than that of piezoelectric materials and is as good or better than that of piezoelectric materials [1]. Therefore, the GMM is expected to be used for actuator devices in lieu of piezoelectric materials. By changing the process of manufacture for GMM from the conventional Bridgman method to a powder-metallurgical method, we have developed the GMMs at a low cost and with a high amount of displacement.

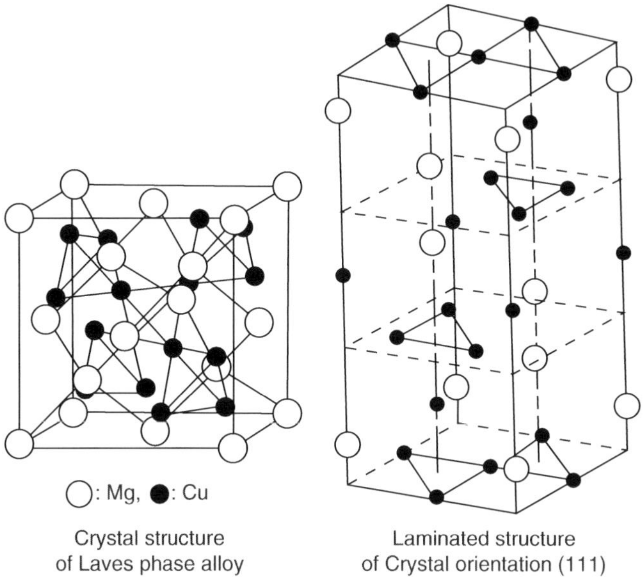

FIGURE 19.14 Crystal structure of Laves phase alloy.

At present, GMM are binary cubic compounds "RT_2" of a lanthanoid element [2] (R), which has a large magnetic moment and magnetoelastic energy (Tb, Sm) and an iron group element (T), which covers Fe, Co, Ni, Mn, etc. or pseudobinary cubic compounds "$RR'T_2$" where R′ are Dy (dysprosium), Ho (holmium) etc. [3]. However, the Curie temperature of RR′ is very low and a high magnetic field is necessary to drive the GMM from RT_2. Thus, pseudobinary cubic compounds $Tb_{0.3}Dy_{0.7}T_2$ have been commercialized and $SmR'T_2$ has been marketed. RT_2 has a Laves phase C_{15} structure shown in Figure 19.14.

19.2.2 Features of GMM

Magnetostriction is caused by the interaction of elasticity and magnetoelasticity. Magnetostrictive materials are both positive and negative magnetostrictive materials while the former has a dimension increase with a driving magnetic field ($Tb_{0.3}Dy_{0.7}T_2$) and the latter has a dimension decrease with a driving magnetic field (SmT_2). When an orthogonal driving magnetic field is applied to these materials, these materials become kinked. This phenomenon is caused by magnetizing the materials with the driving magnetic field, and when the dimensions of the materials are changed by external stress, the magnetization intensity changes.

That is, if the amount of change of the magnetization intensity is detected by a pickup coil, the GMM is potentially able to be used as a sensor for stress and torsion. Moreover, since the Young's modulus of the material changes by changes in the dimension of the material, for example, then the material whose ΛE effect is above 20% can be considered as a surface acoustic wave device [4]. These features are summarized in Figure 19.15.

19.2.3 Characteristics and Physical Properties of GMM

Currently, thin-film studies and developments of bulk GMM have been actively researched in the United States and Europe, however, much of the work on bulk GMM has been applied to practical applications. GMM actuators were mainly developed into the device, which is made with unidirectional solidification or anisotropic crystal forming in a magnetic field [5]. Important attributes consist of: magnetostriction (λ), which is large and hysteresis characteristics, which are low for a driving magnetic field (*H*), while frequency response and temperature characteristics reflect a large-bandwidth. Figure 19.16 shows an H-λ characteristic of $Tb_{0.3}Dy_{0.7}(Fe_{0.9}Co_{0.05})_{1.92}$ made by the Bridgman method (PS-2) and a powder metallurgical method (PMS-1). From this figure, the magnetostriction of PMS-1 is equivalent to that of PS-2, however, the hysteresis characteristics of PMS-1 are about 10% lower than that of PS-2. Table 19.2 shows the characteristics and physical properties of GMM made by the powder metallurgical method.

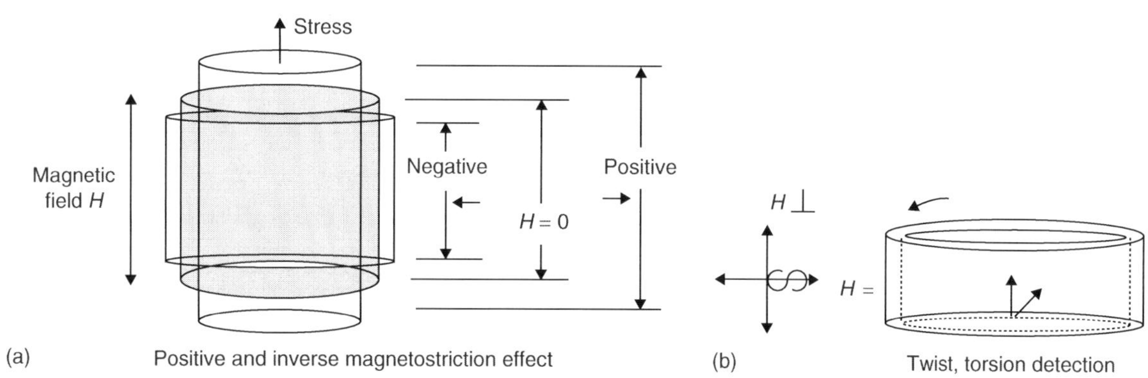

FIGURE 19.15 Positive and inverse magnetostriction effect of GMM.

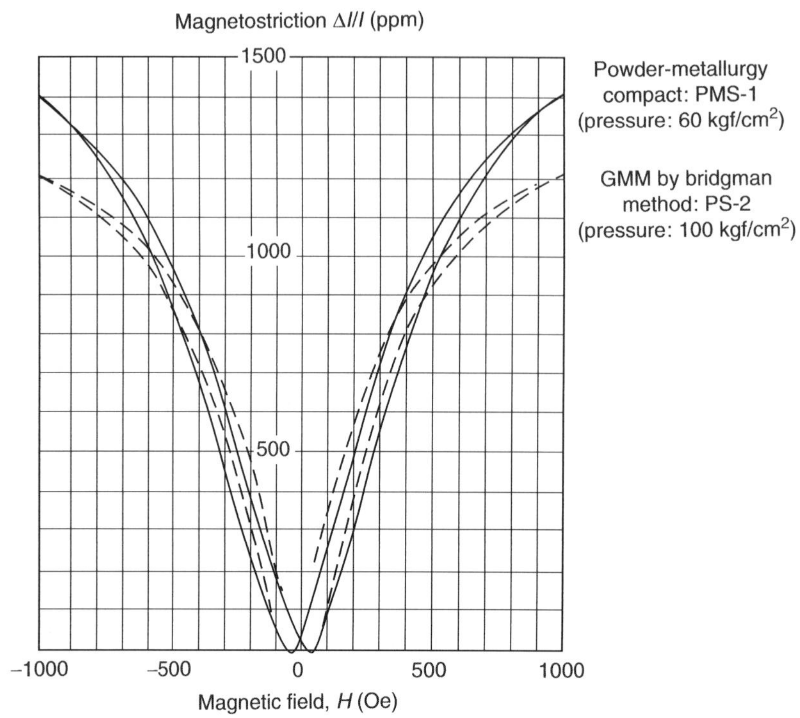

FIGURE 19.16 Magnetostriction of GMM by Bridgman method (PS-2) and powder-metallurgy method (PMS-1).

TABLE 19.2 Powder Metallurgical Process Characteristics and Properties

Material Name		PMS-1	PMH-1	PMT-1	PIS-1	PMS-2
Element Geometry		Board, Prism, Cylinder, etc.				
Displacement (ppm) (preload: 60 kg/cm²)	Magnetic field $H = 0.4$ kOe	600 ppm	600 ppm	600 ppm	350 ppm	750 ppm
	$= 1.0$ kOe	1100 ppm	1100 ppm	1100 ppm	650 ppm	1250 ppm
	$= 2.0$ kOe	1300 ppm	1250 ppm	1300 ppm	—	1400 ppm
	$= 3.0$ kOe	1450 ppm	1400 ppm	1450 ppm	—	1550 ppm
Generated stress (kgf/cm²)	Magnetic field $H = 0.4$ kOe	80(–)	80(–)	80(–)	—	100(–)
	$= 1.0$ kOe	200(–)	200(–)	200(–)	—	250(–)
	$= 2.0$ kOe	280(–)	280(–)	280(–)	—	395(–)
	$= 3.1$ kOe	390(–)	390(–)	390(–)	—	470(–)
Mechanical property	Density (g/mL)	7.95(–)	7.95(–)	7.95(–)	7.95(–)	8.50(–)
	Young's modulus (N/m²)	2.0 × 10 + E10	2.0 × 10 + E10	2.0 × 10 + E10	—	2.4 × 10 + E10
	Compressive strength (kgf/cm²)	600 × 10 + E6	600 × 10 + E6	600 × 10 + E6	←	720 × 10 + E6
	Tensile strength (kgf/cm²)	20 × 10 + E6	20 × 10 + E7	20 × 10 + E8	←	24 × 10 + E6
Thermal property	Coefficient of thermal expansion (CTE) (ppm/°C)	12(–)	←	←	←	←
	Specific heat (kJ/kg·°C)	0.35(–)	←	←	←	←
	Thermal conductivity (W/m·°C)	10/11/2007	←	←	←	←

TABLE 19.2 (continued) Powder Metallurgical Process Characteristics and Properties

Material Name		PMS-1	PMH-1	PMT-1	PIS-1	PMS-2
Element Geometry		Board, Prism, Cylinder, etc.				
Electromagnetic property	Curie temperature (°C)	380(−)	←	←	←	←
	Electrical resistance (W · m)	600 × 10 − E8	←	←	←	←
	Energy density(kJ/m^3)	10(−)	10(−)	10(−)	—	16(−)
	Relative magnetic permeability	8(−)	←	≈6(−)	≈6(−)	≈8(−)
	Coercitivity (Hc; Oe)	50	20	70	50	50
Added characteristics		Anisotropy and positive magnetostrictive standard material	Anisotropy and positive magnetostrictive low-hysteresis material	Anisotropy and positive magnetostrictive large-bandwidth temperature material	Isotropy and positive magnetostrictive powder metallurgy material	Anisotropy and positive magnetostrictive high generation stress material

19.2.4 Features of Powder Metallurgical GMM

Anisotropic GMM made by the powder metallurgical method has been processed similar to a process used to produce rare-earth magnets, and has been formed into cylindrical and C-forms by a near-net shape technique. However, the density of the device that was produced by the method above is low compared with the device made by the Bridgman method and the generated stress for the device volume is low. When compared with the Bridgman method which utilizes a solid-state sintering or molten iron coagulation method, it is possible to add materials for the improvement of the magnetism of the various elements and to realize the anisotropic GMM without Tb.

Temperature characteristics, frequency response, and hysteresis characteristics of GMM are shown in Figures 19.17 and 19.18 and Table 19.3, respectively. Magnetostrictive characteristics are directly influenced by forming the anisotropic material in a magnetic field with a powder metallurgical method, and the anisotropic orientation characteristic is important. Table 19.4 shows the magnetostriction value of the GMM when made with metallic molding and rubber molding.

19.2.5 Applications of GMM

Applications of GMM have been developed, and the bulk material of GMM has been applied to power instruments. Table 19.5 shows some examples of the applications of GMM. The features of this material are a low driving voltage, noncontact driving and low impedance. Moreover, the advantage of GMM compared with the piezoelectric material has been to be able to drive

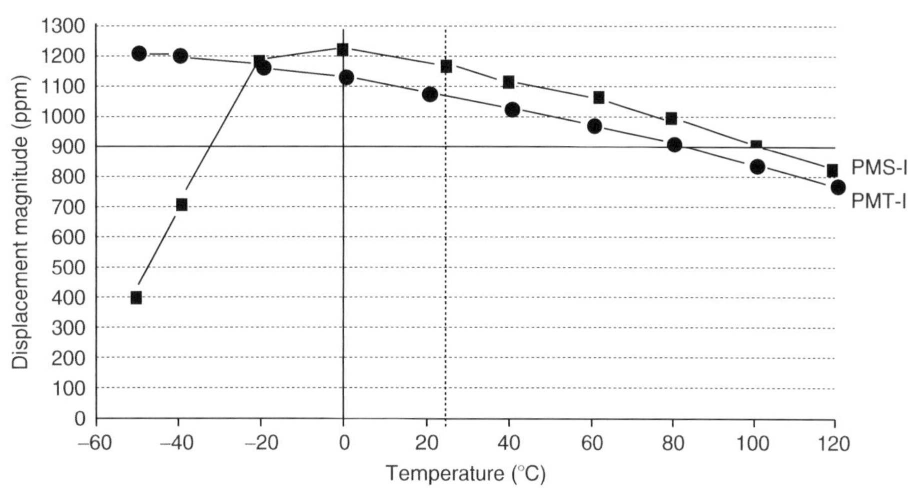

FIGURE 19.17 Temperature characteristic of GMM in magnetic field $H = 1$ kOe.

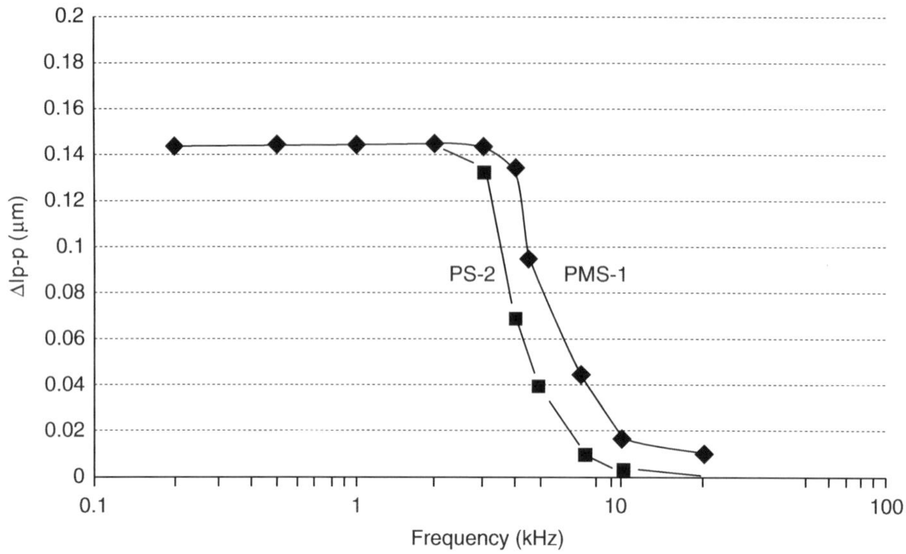

FIGURE 19.18 Frequency characteristic of PS-2 and PMS-1.

the power device in a high-temperature range and have high fatigue endurance.

However, even if the electrical resistance is low and the response is fast, these characteristics are effected by eddy-current loss. Thus, in a high-frequency drive, the shape of the GMM is necessary to be a thin plate or have an acicular shape while the laminated shape is needed for a large driving force. However, in order to drive an actuator in a high-frequency area, the design of the giant magnetostrictive actuator (GMA) is important. One must keep in mind the time constant of the coil of the GMA since the demagnetizing field by the eddy current is reduced. Figure 19.19 shows the relationship between the size of the thin plate and the frequency characteristic.

The Curie temperature of GMM based on Tb and Sm is from 380°C to 420°C. Since the temperature characteristic of GMM is superior to that of piezoelectric materials, which is about 180°C, GMA has been considered as an automotive actuator especially since the coefficient of thermal expansion of GMM is near that

TABLE 19.3 Temperature Characteristics, Frequency Response, and Hysteresis Characteristics of GMM

Characteristic (Elemental Substitution)	Coercitivity H_c (Oe)	Magnetostrictive Hysteresis $(\Delta\lambda)_{max}$	Magnetostrictive Value $\Delta l/l$ (ppm)	
			$H = 0.4\,kOe$	$H = 1.0\,kOe$
$Tb_{0.3}Dy_{0.7}Fe_{1.9}$	80	150	580	1080
Co/Fe = 3% substitution	100	170	680	1150
Co/Fe = 6% substitution	65	120	620	1110
Ni/Fe = 6% substitution	52	110	660	1150

TABLE 19.4 Magnetostriction Values of GMM

Magnetostrictive Characteristic (Mold and Methods)		Magnetostrictive Value: $\Delta l/l$ (ppm)	
		$H = 0.4\,kOe$	$H = 1.0\,kOe$
Metallic molding	Molding of longitudinal magnetic field	450	900
	Molding of transverse field	680	1200
Rubber molding	Molding of longitudinal magnetic field	600	1100
	Molding of transverse field	740	1300

TABLE 19.5 Examples of GMM Applications

	Actuator			Oscillator		Sensor	
Demands	Large Displacement	Positioning	Twist	Large Amplitude	Small Input	Vibration and Load	SAW, etc.
Typical application	High speed on–off valve, printer head, shutter, fuel injection valve, high speed brake, driller, clutch, etc.	Valve opening control, table top control, clutch, servo valve, rudder face control, focus adjustment, etc.	Stepping motor, angle control, etc.	Pump, vibration exciter, sonar, traveling wave motor, ultrasonic disk cutter, ultrasound bath, active damper, etc.	Acouophone, various speakers, ultrasonic vibrator, linear motor, etc.	Vibration sensor, ultrasonic displacement sensor, contact sensor, location sensor, magnetic field sensor, magnetostrictive applied generation of electricity, weighing sensor, etc.	SAW, AE sensor, water level indicator, etc.
Standard material PMS-1	○	≥	○	○	×	≥	≥
Wideband temperature PMT-1	○	×	≥	○	×	×	×
Low-hysteresis material PMH-1	≥	○	○	≥	○	○	○

of general metallic components. Thus, since the dimensions of GMA changes with increased temperature by the Joule heat of the coil, it is necessary to consider the design of the structure of GMA in order to reduce the change in dimensions by temperature. Many designs have been suggested. The magnetomechanical coupling factor of GMM is from 0.75 to 0.8 [6] and equivalent to that of a piezoelectric material. The devices with GMM are discussed in the next section.

19.2.5.1 All-Around Vibration Sensor/Actuator

This vibration sensor makes use of the Villari effect of GMM [7]. The sensor has a low impedance, the transmission loss is low and it is detectable in a wide temperature area. Figure 19.20 shows the sensor structure and Figure 19.21 shows the sensor's characteristics. On the other hand, if current is input into the sensing terminal, the sensor is then able to be used as the oscillator, which is applicable to controlling the table top. However, the oscillator is applicable and can be used as a traveling-wave motor in the ultrasonic range [8].

19.2.5.2 Giant Magnetostrictive Torsional Actuator/Torque Sensor

When we produced the isotopic cylindrical device of GMM and applied a cylindrical axial magnetic field as well as a circumferential

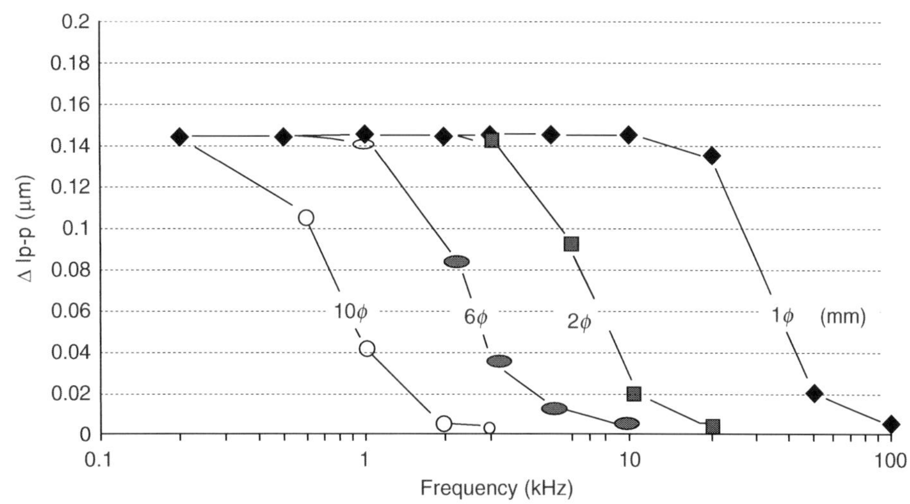

FIGURE 19.19 Frequency characteristic of GMM for material shape.

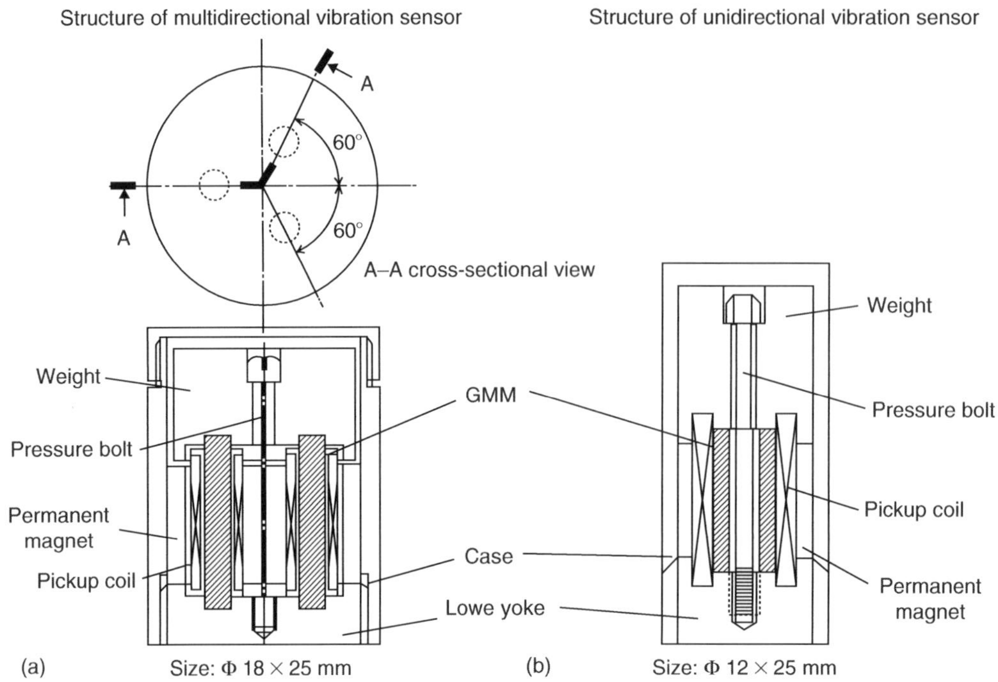

FIGURE 19.20 Structure of GMM vibration sensor.

magnetic field to the device, the device was magnetized in the synthetic direction of those magnetic fields. Therefore, if the circumferential magnetic field is fixed and the axial magnetic field has an alternating-current magnetic field, the synthetic magnetization direction is above 90° and a great amount of kinking occurs to the cylindrical device. On the other hand, if the circumferential magnetic field is fixed and a pick up coil is equipped in the axial direction and the device is twisted, the voltage is induced in the pick up coil with a magnetic flux change. Figure 19.22 shows the structure of the torsional actuator device and Figure 19.23 shows the device's characteristics.

Other applications, which are used, include industrial instruments, microrobots, and active dampers, which have been studied under the auspices of the Japan Society of Mechanical Engineers and the Japan Society for Precision Engineering in Japan.

19.2.5.3 GMA for Valve Ring Indentation

From the model of Figure 19.24, the following equation was derived.

$$\frac{m\mathrm{d}^2x}{\mathrm{d}t^2} = F_{\mathrm{B}} + F_0 \sin\omega x - \mu_{\mathrm{d}} F_{\mathrm{N}} + S\sigma_y \left[\frac{m\mathrm{d}^2x}{\mathrm{d}t^2} = F_{\mathrm{B}} + F_0 \sin\omega x - f_{\mathrm{r}} x - F_{\mathrm{p}} \right] \quad (19.1)$$

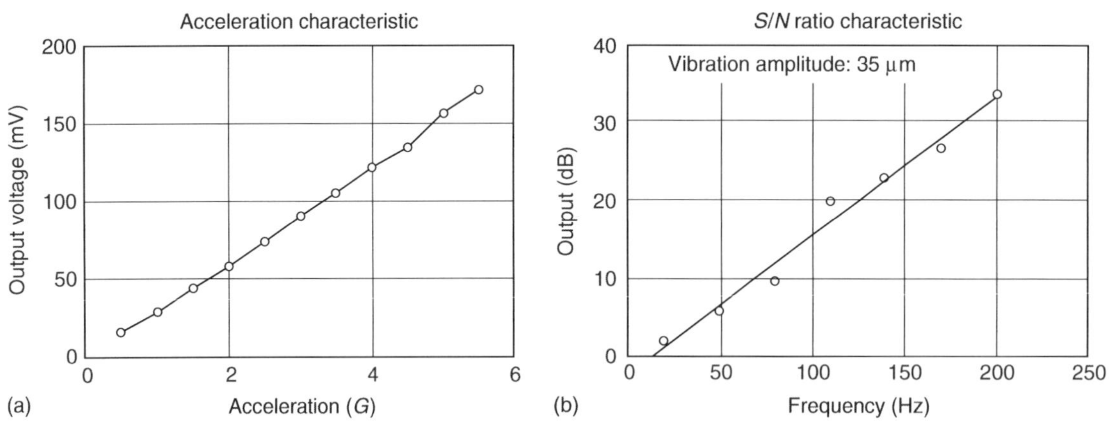

FIGURE 19.21 Characteristics of GMM vibration sensor.

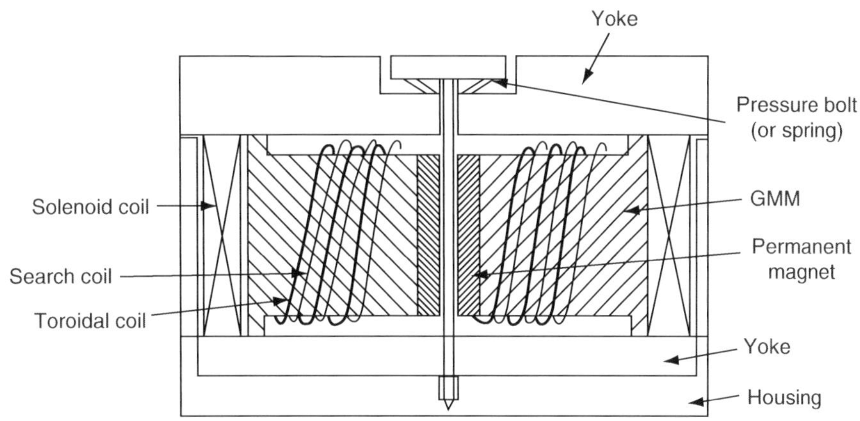

FIGURE 19.22 Structure of giant magnetostrictive torsional actuator.

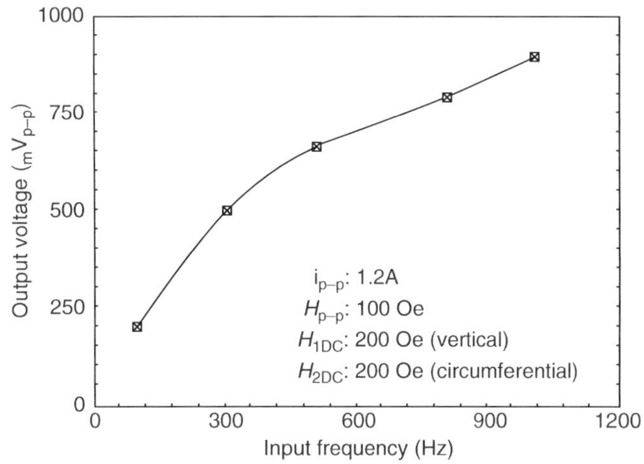

FIGURE 19.23 Characteristic of giant magnetostrictive torsional actuator.

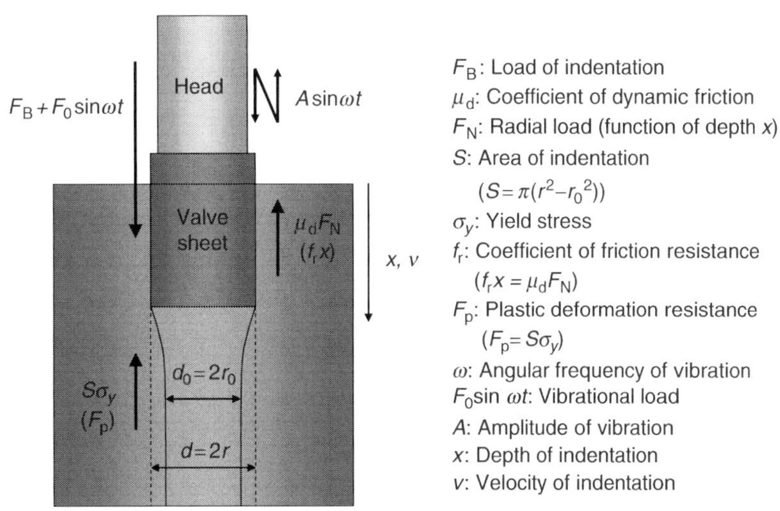

FIGURE 19.24 Model of valve ring indentation.

Assuming that the equation of motion (19.1) has equality, since the right-hand side of Equation 19.1 is positive, the following equation was derived:

$$F_B + F_0 \sin \omega x > \mu_d F_N + S\sigma, \tag{19.2}$$

When Equation 19.2 is true, the valve rings are able to be pressed into place.

19.2.6 Conclusion

It has been about 30 years since GMM were discovered at the Naval Surface Warfare Center (NSWC). During this time, scientists and researchers have commercialized GMM and the quality and cost of GMM has been improved, nevertheless the cost of GMM is still high for civil applications. Bad hysteresis characteristics for users of GMM have not improved sufficiently enough to warrant its full usage. We are keenly aware of our responsibility to resolve these current problems to enable the use GMM, and this gives impetus for ourselves as well as other users to continue to pursue developments and improvements.

References

1. J. L. Butler, Application Manual for the Design of ETREMA Terfenol-D Magnetostrictive Transducers, Image Acoustics, Inc., N. Mansfield; T. Mori; I. Uni, Giant Magnetostriction Materials and Their Applications, PhD dissertation, 1998.
2. A. E. Clark, *Ferromag. Mater.*, 1, 531, 1980; A.E. Clark and H. Eda, Giant magnetostrictive materials, *Nikkan-kogyo-shinbun* (in Japanese), 3–80, 1995.
3. J. L. Butler, Application Manual for the Design of ETREMA TERFENOL-D™, Edge Technologies, p. 25, 1988.
4. D. C. Webb, K. L. Pavis, N. C. Koon, and A. E. Gangaly, *Appl. Phys. Lett.*, 31(4), 1245, 1977.
5. G. F. Clark et al., *Philosophical Mag. B*, 1982, 146(4), 331–343.
6. Terfenol-D Notes, Edge Technologies Inc., Etrema Producys Division. March, 1990.
7. Y. Yamamoto, H. Eda, T. Mori, and R. Amen, The 8th Symposium on Electromagnetics and Dynamics, p. 79, 1996.
8. T. Mori and K. Ushida, The 8th Symposium on Electromagnetics and Dynamics, p. 85, 1996; T. Mori and H. Eda, A study on torque induction in the GMM, *J. Alloys Compounds*, 258, 93, 1997.

20

Shape-Memory Alloys and Effects: Types, Functions, Modeling, and Applications

20.1 Magnetically Controlled Shape Memory Alloys

Ilkka Aaltio, Oleg Heczko, Outi Söderberg, and Simo-Pekka Hannula

20.1.1 Activation of the Shape Memory Alloys with Magnetic Field

Shape memory alloys (SMAs) are metallic compounds. They exhibit a cubic high-temperature crystal structure, which transforms by martensitic reaction into another crystal structure (tetragonal, orthorhombic, hexagonal, etc.) when temperature decreases. Martensitic transformation enables large shape changes and the shape memory effect (SME), but heating and cooling are needed to complete the phase transformation. SME can produce large relative strains and high stress output, but usually small actuation frequency due to the slow cooling process [1–4]. The SMAs have typically twinned martensitic structure with mobile twin boundaries. The multivariant twin structure can easily be deformed to the single-variant state to yield the maximum SME shape change.

To trigger the SMAs with a magnetic field, alloys must possess magnetic ordering, i.e., to be ferromagnetic or ferri/antiferromagnetic. These alloys may exhibit (1) magnetic field-induced strain (MFIS) by magnetic shape memory effect (MSME) or by magnetic field-induced martensite transformation (MF-SIM), (2) magnetic field-assisted superelasticity (MFAS), (3) conventional magnetostriction (MS), and (4) giant magnetocaloric effect (GMCE). The latter two are not discussed here in detail, but the other phenomena are illustrated in Figure 20.1. A list of magnetically activated magnetic SMAs is presented in Table 20.1 [5–7].

20.1.1.1 Magnetic Shape Memory Effect

In MSME, the martensitic twin variant structure changes, resulting in giant strain changes of up to 10%. The origin of the effect is fundamentally different from MS, where the deformation is caused by magnetization rotation relative to crystal lattice (maximum 0.25% strain). The giant MFIS of magnetic shape memory alloys (MSMAs) is based on the rearrangement of the martensite twin variants by the twin boundary motion [8–13], resulting in structure reorientation and changes in direction of magnetization.

The MSMA(or ferromagnetic SMA, i.e., FSMA) should have a thermoelastic twinned martensite phase with nonsymmetric lattice structure (e.g., tetragonal or orthorhombic); high magnetocrystalline anisotropy and highly mobile twin boundaries [14–16]. Furthermore, a single crystal with a single twin variant structure is necessary for the maximum shape change.

High magnetic anisotropy and mobile twin boundaries ensure that the magnetic field creates a magnetic stress σ_{mag} that is larger than the twinning stress σ_{tw}, i.e., the stress needed for reorientation of the martensite structure [17–21]. The magnetic stress has its maximum in the magnetic saturation, given by $\sigma_{mag} = (1-c/a)^{-1}K_u$ where K_u is the magnetic anisotropy constant and c and a are lattice constants of the martensite structure. The detailed mechanism can be explained with the 10M (also called 5M) Ni–Mn–Ga alloy. In its layered structure, the short c-axis (in the cubic parent phase coordinates) is simultaneously the easy axis of magnetization [22,23]. When the magnetic field is applied along the crystallographic axis and, thus, perpendicularly to magnetization direction, the magnetization vector rotates toward the field (initial linear increase of the magnetization curve, Figure 20.3a) and the magnetic energy increases. When this energy exceeds the energy level needed for initiating the twin boundary movement (detwinning), the single variant structure (A) begins to convert to another variant structure (B) having its easy magnetization axis (the short c-axis) along the applied field and the stored energy decreases. This induced reorientation is observable on the sample surface as moving twin boundaries (Figure 20.2) and it is reflected in the magnetization curve as a change of initial slope (Figure 20.3a) [16,21,24].

Figure 20.2 shows a schematic representation of the MSME in actuation and for clarity, the length change is exaggerated. The c-axis of the variants A and B is shown by short arrows. The magnetic field-driven actuation sequence (H as the dashed line arrows) together with the returning prestress (σ_{return}) is shown from left to right. If no load or magnetic field is present (the far left and right sides), the twin structure of the MSM element remains in its single variant state, long or short.

With a low σ_{tw} MSMA (< 1 MPa), the required triggering field is also low. Increasing the field increases further the σ_{mag}, and the amount of the reoriented twin variants until the saturation point is reached and the whole twin structure has changed. Further increase of magnetic field does not result in a larger MFIS, i.e., maximum contraction of the sample is obtained when exchanging between two single martensite variant states. The maximum obtainable strain ε_0 is limited by the lattice distortion given by $\varepsilon_0 = 1-c/a$. In 10M martensite, ε_0 is about 0.06 (6%) and

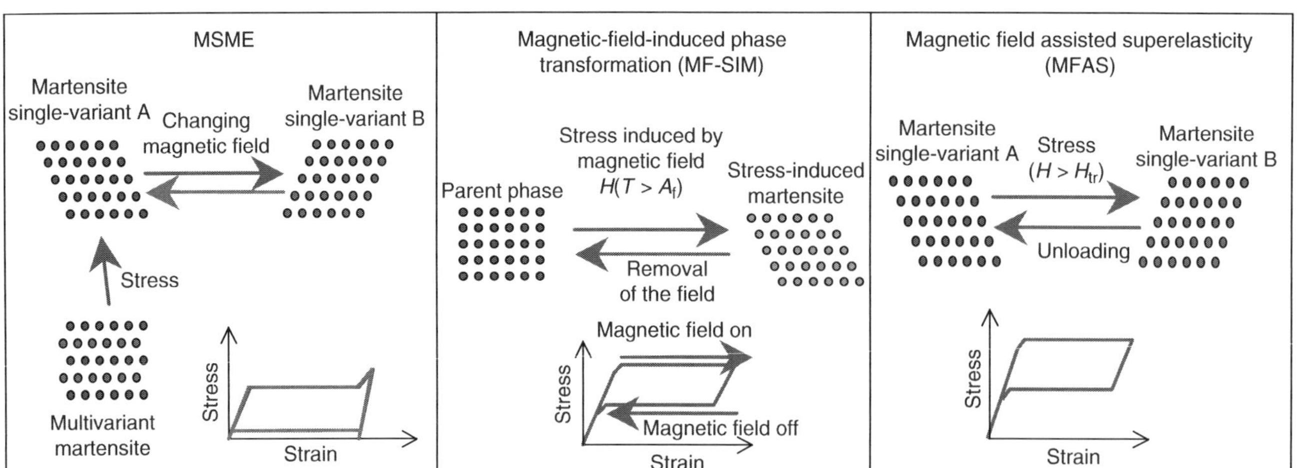

FIGURE 20.1 Three different actuation principles of magnetically controlled SMAs.

TABLE 20.1 Magnetic Field Activated SMAs and Their Properties

Alloy	Maximum MFIS and Actuation Operation Temperatures Reported	Parent Phase /Martensite Structures/ Approximate Martensitic Transformation Temperatures	Actuation Methods	Ductility
Ni–Mn–Ga	6% or 10% in SC depending on the martensite crystal structure, 6% up to 333 K	$L2_1$/10M or 14M 10M up to approximately 333 K, 14M up to approximately 368 K $L2_1$/2M usually above 373 K	MS, MSM, MF-SIM, MFAS, GMCE	Depends on the martensite crystal structure and SC/PC
Ni–Mn–Al	0.17% in SC, 0.01% in PC at 253 K	B2 (L21)/10M 12M 14M/around RT or below B2 (L21)/ 14M or 2M/above or well above RT	MSM	Improved by γ phase
Ni–Fe–Ga Ni–Fe–Ga–Co	0.02% in SC at ~100 K 0.7% in SC at 300 K	L21/10M 6M 14M/above and below RT	MS	Improved by γ phase
Co–Ni–Al	0.06% in SC at 165 K 0.013% in PC at 293 K	B2/L10 (=2M)/below, above and well above RT	MF-SIM	Improved by γ phase
Co–Ni–Ga	0.011% in MS ribbons 0.003% in PC at RT	B2/L10 (=2M)/below, above and well above RT	MF-SIM	Improved by γ phase
Fe_3Pt	~0.6% in SC at 4.2 K	Ordered fcc/fct/~100 K	MSM	Intrinsically better
Fe–Pd	3.1% in SC at 77 K 0.01%–0.05% in PC 0.06–0.17 in MS ribbons	Disordered fcc/fct/around RT or below	MSM	Intrinsically better

Sources: From Pons, J., Cesan, E., Segui, C., Masdeu, F., Santamarta, R., Ferromagnetic SMAs: Alternatives to Ni-Mn-Ga, Mat. Sci. Eng A 481, 57, 2008 presentation in ESOMAT2006, Bochum, Germany, September 16, 2006; Chernenko, V.A., Kokorin, V.V., and Vitenko, I.N., *Scripta Metall. Mater.*, 33, 1239, 1995; Lanska, N., et al., *Appl. Phys.* 95, 8074, 2004.

Note: Single crystal is denoted by SC and polycrystal by PC.

in 14M (7M) about 10% [16,25–28]. In reality, some parts of the MSMA do not change and, therefore, the measured maximum strains are somewhat lower. When the magnetic field is removed, the shape change remains (Figures 20.2 and 20.3b) unless the original twin structure is brought back by applying a magnetic field perpendicular to original one or by an external load such as a prestress spring load applied perpendicularly to the field direction (Figure 20.2) [13,16,17,28–31].

The applied external stress reduces the obtainable MFIS and increases the field needed to start MSM as the field has to work

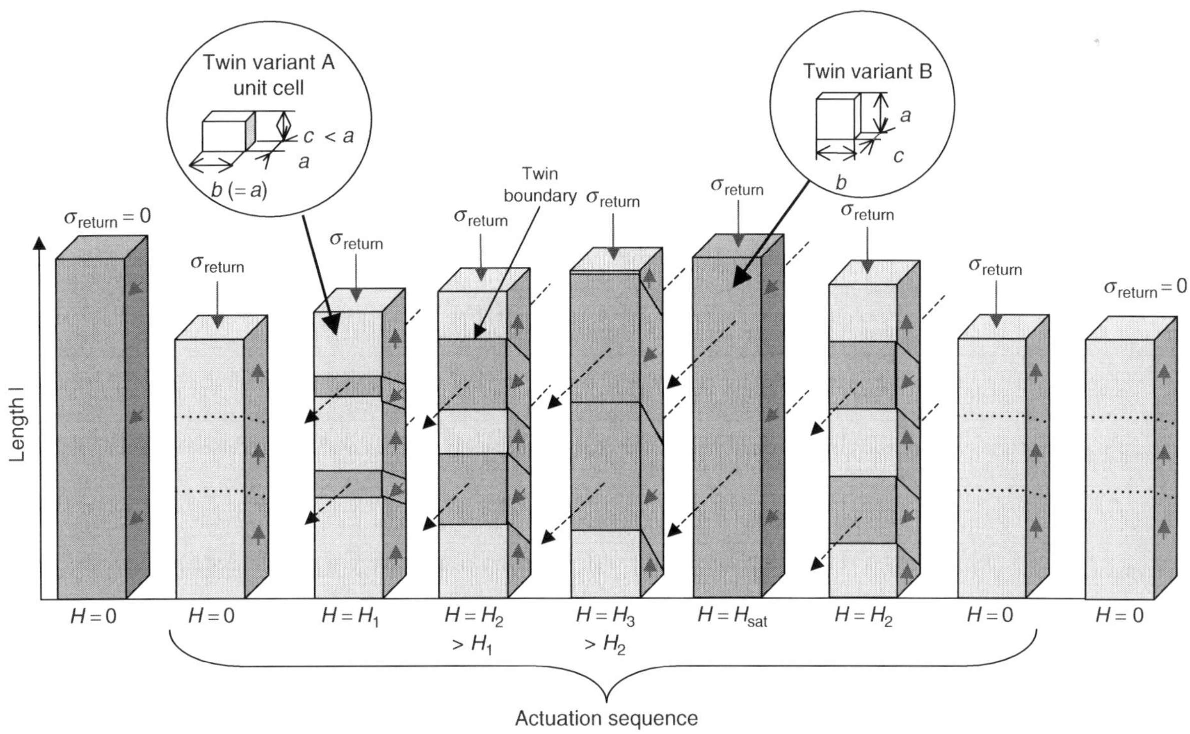

FIGURE 20.2 Schematic representation.

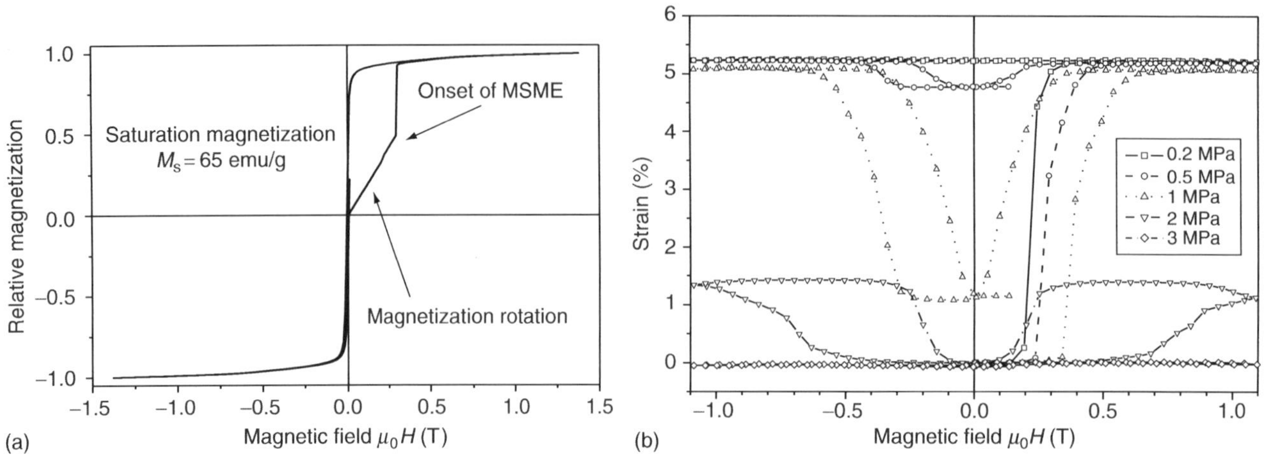

FIGURE 20.3 (a) Magnetization curve showing MSME and (b) magnetic field induced strain as a function of magnetic field H on different compressive stresses ($Ni_{49.7}Mn_{29.1}Ga_{21.2}$).

against the combination of internal σ_{tw} and external mechanical stress σ_{return}, i.e., in static state $\sigma_{mag} \geq \sigma_{tw} + \sigma_{return}$ [31,32]. In actuation, the optimum applied stress σ_{return} returns the MSM element to its initial state after the removal of the magnetic field. Too small σ_{return} causes the element to return only partially to the initial state, and too high a stress reduces the MFIS (Figure 20.3b). If the transversal stress exceeds magnetic stress σ_{mag}, the MSME is totally suppressed [16,24]. This stress limit in a linear MSME mode is called blocking stress [33,34] and above it, only the conventional MS is observed [35]. It should be noted that the MFIS change can be also other than linear, such as bending [36].

MSM-fatigue and its mechanisms are presently not very well known—the desired actuation strain strongly affects them. Sometimes there is no significant change, sometimes MFIS increases, and in the worst case the sample cracks [37,38]. Successful actuation of 200×10^6 and 100×10^6 cycles has been reported [29,39]. On nonoptimal stress modes, secondary twin variants may appear or stepwise appearing lattice defects may reduce the MFIS or produce microcracks [39].

The service temperature of MSME is limited by the crystallographic and magnetic transitions. MSME does not occur in the paramagnetic state (above the Curie point) or in cubic austenite. It disappears also in phase transformation to another type of martensite (intermartensitic reaction). Furthermore, this effect is dependent on the changes of lattice parameters, elastic constants, the twinning stress, magnetic anisotropy, and magnetization with temperature [39–42]. Decrease of temperature increases σ_{tw}, which eventually exceeds σ_{mag} and MSME is blocked, for example, in 10M Ni–Mn–Ga somewhere between 153 and 253 K [41,43].

20.1.1.2 Magnetic Field-Assisted Superelasticity

The magnetically assisted superelasticity (MFAS) of alloys occurs in the same temperature region as MSME as this effect is complementary to MSME [44]. By deforming single-variant martensite along the long crystallographic axis, the twin structure converts to the variant with the short axis in the stress direction. Application of a constant magnetic field perpendicular to the stress direction tends to convert the stress-induced variant back to the initial state (short axis along the field), when the external stress is removed. If the applied field is strong and twin mobility high enough, the variant reorientation is fully reversible. In a constant field ≥0.6 T with the 10M Ni–Mn–Ga a strain of 6% can be obtained with less than 8 MPa external stress [31,44], while with 14M structure, 10% is possible [19]. This behavior can be used for determining the material parameters needed in the MSM device design by measuring stress–strain behavior in a magnetic field [21,31,45–49].

20.1.1.3 Magnetically Activated Phase Transformation

When the stress-induced martensite formation (i.e., superelasticity) close to the austenite–martensite transformation temperature is induced by the magnetic field, it is called magnetically activated phase transformation (MFSIM). The effect originated from the magnetic energy difference between austenite and martensite phases in a magnetic field, which creates the stress triggering the SIM formation [1,48,50–55]. In Ni–Mn–Ga, this effect requires usually a large magnetic field in the order of tens of teslas, but very close to the phase transformation temperatures where the needed field may be smaller. Also, in some other alloys, such as Ni–Mn–In, 3–5 T is enough. Actuation is obtained by the repeating phase transformations when applying and removing the magnetic field. SIM-type behavior is typical for Co–Ni-based alloys (Co–Ni [52], Co–Ni–Ga [56], Co–Ni–Al [57], etc.).

20.1.1.4 Giant Magnetocaloric Effect

In addition to the dilatational changes triggered by the magnetic field, some of the ferromagnetic SMAs also show a GMCE [58–60]. This phenomenon can be used for high-temperature refrigeration [61]. To gain this behavior, the alloy should have the structural and magnetic transitions very close to each other. In GMCE material produces or absorbs heat while being heated or cooled through the transition temperature region in a magnetic field. The magnetocaloric effect can be estimated from the isothermal magnetization curves at different temperatures [58,60,62–64].

20.1.2 Production of MSMAs

MSMAs have to meet specific characteristics described previously. There are not many alloys satisfying these conditions. Microstructure and chemical composition must be correct as well as the single crystalline structure. Furthermore, the application specifications require tailoring of the MSMAs. The bulk MSM materials are produced by the single crystal growth techniques, such as Bridgman [43,65,66] or Czochralski [67]. Optimum crystal orientation of the ingots increases the yield. After casting, homogenizing and ordering heat treatments are needed. Then the annealed samples are cut, ground, and polished to the desired dimensions of the MSM element. In addition to the bulk samples, MSMAs have been produced in research scale in forms of thin films [68–72], ribbons [73,74], and composites [75].

20.1.3 Applications

MSMAs can be used for actuating applications, i.e., to produce movement, for sensor or power generation applications, and for vibration damping. Presently, the most commonly used alloys are 10M Ni–Mn–Ga; however, the 14M Ni–Mn–Ga alloys have been studied to broaden MSM region to the higher temperatures.

20.1.3.1 Actuating

The main advantages of MSM actuators are their short response time, large relative strain and stroke, and proportional controllability. Furthermore, the MSM element itself does not have to be wired. So far, the disadvantages are the limited operating temperature, relatively low force output and hysteresis.

Basic parts of a typical linear MSM actuator are the piece of MSMA, i.e., the MSM element, the magnetic circuit and the body. The direction of magnetic field is orthogonal to the direction of the induced strain (Figure 20.2). The variable field, either a rotating magnetic field (no returning force needed) or a transversal field (assisted by the mechanical returning force, Figure 20.4), is created by an electromagnetic circuit. The magnetic circuit concentrates the field to the space of the MSM element. For an actuator using 10M Ni–Mn–Ga, the necessary maximum H is approximately 500 kA/m [29]. The produced force of the actuator depends basically on the cross-sectional area and the stroke on the length of the MSM element, both depending on the MSMA properties [76]. In a design similar to Figure 20.4, the spring applies usually a prestress of 0.5–1 MPa, which the MSM element has to work against.

MSM actuators are best driven with current, because there is a delay in magnetization of voltage-controlled coils [43]. By proper electromagnetic design, unipolar current can be used. With a bias magnetic field obtained with permanent magnets, smaller peak

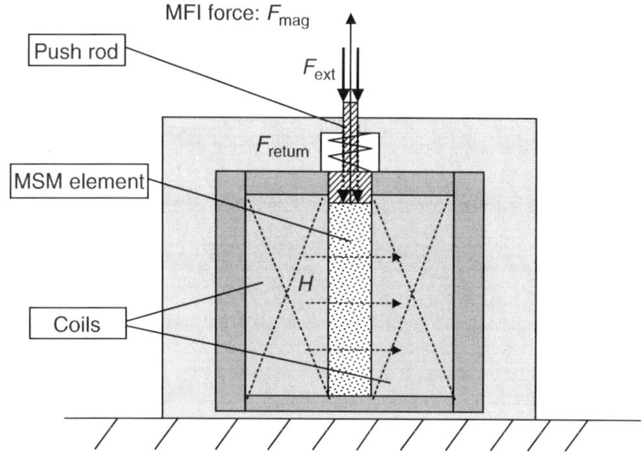

FIGURE 20.4 Cross-sectional scheme of an MSM actuator, showing the basic parts and the directions of the magnetic field H, MFI force F_{mag}, returning spring force F_{return}, and external load F_{ext}.

power and smaller coils can be used. When the actuation frequency increases, more current is needed, due to, for example, skin effect and the generated eddy currents limiting the highest frequency performance. Overheating can lead to the reverse transformation, but the actuating element recovers when it cools back to martensite range; however, overheating is not recommended [77].

The actuators exhibit a hysteresis due to the twin boundary motion, which causes a lag when reversing the actuation direction. Because σ_{tw} is usually constant in actuator use, the hysteresis can be compensated by a suitable control system. Hysteresis prevents also the overshoot and drift of the actuator [24,29]. The efficiency of MSM actuators depend strongly on the properties of the MSMA such as σ_{mag} and σ_{tw} [23,78] and on the electromagnetic design. MSM actuators can be used, for example, to produce proportional linear back-and-forth motion, inchworm type linear motors, proportional valves, and pumps [36]. In small-sized MSM actuators, response times are short. A 0.2 ms rise time in small actuators can be reached [28]. Increase of the operating temperature of MSM alloys would extend the field of applications, such as in automotive engines, where the fast response and relatively large strains provide advantages [79,80].

20.1.3.2 Sensing and Power Generation

MSMA sensors may be constructed using the inverse MSME: When the material shape changes, the effective magnetization curve of the material is altered [16,44]. This effect can be used to monitor the shape of the material by measuring its magnetization [81,82]. In addition to sensing, power generation by the inverse MSME has been demonstrated [83]. When an element of $Ni_{50.1}Mn_{28.4}Ga_{21.5}$ MSMA was compressed at a velocity of 1.5 m/s, a voltage of over 60 V was generated using a simple electromagnetic circuit.

20.1.3.3 Damping

Vibration damping mechanisms that may operate in magnetically controlled SMAs are based on the hysteretic motion of the martensite twin boundaries, motion of magnetic domain walls, or stress-induced martensite formation. The main damping mechanism is the dissipative motion of the twin boundaries. Due to the phase transformations, the internal friction and the elastic modulus of Ni–Mn–Ga is temperature dependent [84–86]. The damping elements can be constructed as composite overcoming intrinsic brittleness of the MSMA. Enhanced damping was observed in composite using MSMA powder [87]. However, damping of MSMA is presently a subject of research and no commercial products use this technology.

References

1. Otsuka, K. and Wayman, C.M., *Shape Memory Materials*. Cambridge, UK: Cambridge University Press, 1998, 284pp.
2. Duerig, T.W., *Engineering Aspects of Shape Memory Alloys*. London: Butterworth-Heinemann, 1990, 499pp.
3. Friend, C., Shape memory alloys, in *Encyclopedia of Materials: Science and Technology*, Amsterdam: Elsevier Science, 2001, pp. 8445–8451.
4. Quandt, E., Actuator materials for small scale devices, in *Encyclopedia of Materials: Science and Technology*, Amsterdam: Elsevier Science, 2001, pp. 35–38.
5. Pons, J., et al., Ferromagnetic shape memory alloys: Alternatives to Ni-Mn-Ga, Mat. Sci. Eng A 481, 57, 2008.
6. Chernenko, V.A., Kokorin, V.V., and Vitenko, I.N., The development of new ferromagnetic shape memory alloys in Ni-Mn-Ga system, *Scripta Metall. Mater.* 33, 1239, 1995.
7. Lanska, N., et al., Composition and temperature dependence of the crystal structure in Ni-Mn-Ga alloys, *Appl. Phys.* 95, 8074, 2004.
8. Rhyne, J.J., et al., Rare earth metal single crystals. I. High-field properties of Dy, Er, Ho, Tb, and Gd, *J. Appl. Phys.*, 39, 892, 1968.
9. Liebermann, H.H. and Graham C.D., Plastic and magnetoplastic deformation of Dy single crystals, *Acta Met.*, 25, 715, 1977.
10. Ullakko, K., Magnetically controlled shape memory alloys: A new class of actuator materials, *J. Mat. Eng. Perform.*, 5, 405, 1996.
11. Ullakko, K., et al., Magnetically controlled shape memory effect in Ni_2MnGa intermetallics, *Scr. Mater.*, 36, 1133, 1997.
12. O'Handley, R.C. and Allen, S.M., Shape memory alloys, magnetically activated ferromagnetic shape-memory materials, in *Encyclopedia of Smart Materials*, New York: John Wiley and Sons, 2002, pp. 936–951.
13. Söderberg, O., et al., Giant magnetostrictive materials, in *Handbook of Magnetic Materials*, Vol. 16, Buschow, K.H.J., ed., Amsterdam: Elsevier, 2006, pp. 1–39.
14. Heczko, O., Sozinov, A., and Ullakko, K., Giant field-induced reversible strain in magnetic shape memory NiMnGa alloy, *IEEE Trans. Magn.*, 36, 3266, 2000.
15. Likhachev, A.A. and Ullakko, K., Magnetic-field-controlled twin boundaries motion and giant magneto-mechanical effects in Ni-Mn-Ga shape memory alloy, *Phys. Lett. A*, 275, 142, 2000.
16. Likhachev, A.A. and Ullakko, K., The model of magnetic-field-controlled shape memory effect in NiMnGa, *J. Phys. IV Proc.*, 11, Pr8/293, 2001.
17. Sozinov, A., et al., Giant magnetic-field-induced strain in NiMnGa seven-layered martensitic phase, *Appl. Phys. Lett.*, 80, 1746, 2002.
18. Likhachev, A.A., Sozinov, A., and Ullakko, K., Magneto-mechanical cycling and modeling the external stress effect on the magnetic-field-controlled strain response in Ni-Mn-Ga, *J. Phys. IV Proc.*, 115, 95, 2004.
19. Heczko, O., Straka, L., and Ullakko, K., Relation between structure, magnetization process and magnetic shape memory effect of various martensites occurring in Ni-Mn-Ga alloys, *J. Phys. IV Proc.*, 112, 959, 2003.
20. Heczko, O., et al., Temperature dependence of magnetic anisotropy in Ni-Mn-Ga alloy exhibiting giant field-induced strain, *J. Appl. Phys.*, 91, 8228, 2002.
21. Ge, Y., et al., Crystal structure of three Ni-Mn-Ga alloys in powder and bulk materials, *J. Phys. IV Proc.*, 112, 921, 2003.
22. Straka, L. and Heczko, O., Magnetic anisotropy in Ni-Mn-Ga martensites, *J. Appl. Phys.* 93, 8636, 2003.
23. Heczko, O., Magnetic shape memory effect and magnetization reversal, *J. Magn. Magn. Mater*, 290–291, 787, 2005.
24. Suorsa, I., Performance and modeling of magnetic shape memory actuators and sensors, TKK Dissertations 4, Helsinki University of Technology, Espoo, Finland, 2005, 70pp.
25. James, R.D., Tickle, R., and Wuttig, M., Large field-induced strains in ferromagnetic shape memory materials, *Mat. Sci. Eng. A*, 273–275, 320, 1999.
26. O'Handley, R.C., et al., Phenomenology of giant magnetic-field-induced strain in ferromagnetic shape-memory materials (invited), *J. Appl. Phys.*, 87, 4712, 2000.
27. Likhachev, A.A., Sozinov, A., and Ullakko, K., Modeling the strain response, magneto-mechanical cycling under the external stress, work output and energy losses in Ni-Mn-Ga, *Mech. Mater.*, 38, 551, 2006.
28. Aaltio, I. and Ullakko, K., Magnetic shape memory (MSM) actuators, Borgmann, H., ed., Proc. ACTUATOR 2000, Messe Bremen, Germany, 2000, pp. 527–530.
29. Tellinen, J., et al., Basic properties of magnetic shape memory actuators, Borgmann, H., ed., Proc. of ACTUATOR 2002, Messe Bremen, Germany, pp. 566–569.
30. Murray, S.J., et al., 6% Magnetic-field-induced strain by twin-boundary motion in ferromagnetic Ni-Mn-Ga, *Appl. Phys. Lett.*, 77, 886, 2000.
31. Straka, L. and Heczko, O., Reversible 6% strain of Ni-Mn-Ga martensite using opposing external stress in static and variable magnetic fields, *J. Magn. Magn. Mater.*, 290–291, 829, 2005.
32. Likhachev, A.A., Sozinov, A., and Ullakko, K., Optimizing work output in Ni-Mn-Ga and other ferromagnetic shape-memory alloys, Lynch, C.S., ed., Proc. SPIE, 4699, 553, 2002.
33. Murray, S.J., et al., Giant magnetic-field-induced strain in Ni-Mn-Ga crystals: Experimental results and modeling, *J. Magn. Magn. Mater.*, 226–230, 945, 2001.

34. Malla, A., et al., Effect of composition on the magnetic and elastic properties of shape-memory Ni-Mn-Ga, Lagoudas, D.C., ed., Proc. SPIE, 5053, 147, 2003.
35. Heczko, O., Determination of ordinary magnetostriction in Ni-Mn-Ga magnetic shape memory alloy, *J. Magn. Magn. Mater.*, 290–291, 846, 2005.
36. Suorsa, I., et al., Applications of magnetic shape memory actuators, Borgmann, H., ed., Proc. ACTUATOR 2002, Bremen Messe, Germany, pp. 158–161.
37. Müllner, P., Chernenko, V.A., and Kostorz, G., Large cyclic magnetic-field-induced deformation in orthorhombic (14M) Ni-Mn-Ga martensite, *J. Appl. Phys.*, 95, 1531, 2004.
38. Heczko, O., et al., Magnetic shape memory fatigue, Armstrong, W.D., ed., Proc. SPIE 5761, 513, 2005.
39. Müllner, P., et al., Nanomechanics and magnetic structure of orthorhombic Ni-Mn-Ga martensite, Mat. Sci. Eng A 481, 66, 2008.
40. Glavatska, N., et al., Temperature dependence of martensite structure and its effect on magnetic-field-induced strain in Ni_2MnGa magnetic shape memory alloys, *J. Phys. IV Proc.*, 112, 963, 2003.
41. Heczko, O. and Straka, L., Temperature dependence and temperature limits of magnetic shape memory effect, *J. Appl. Phys.*, 94, 7139, 2003.
42. Khovailo, V.V., et al., Magnetic properties and magnetostructural phase transitions in $Ni^{2+x}Mn^{1-x}Ga$ shape memory alloys, *J. Phys.: Cond. Matter*, 1–7, arXiv:cond-mat/0405134, 2004.
43. Pagounis, E. and Quandt, E., Recent advances and challenges in magnetic shape memory materials, Borgmann, H., ed., Proc. ACTUATOR 2006, Messe Bremen, Germany, pp. 394–400.
44. Straka, L. and Heczko, O., Superelastic response of Ni-Mn-Ga martensite in magnetic fields and a simple model, *IEEE Trans. Magn.*, 39, 3402, 2003.
45. Müllner, P., Chernenko, V.A., and Kostorz, G., A microscopic approach to the magnetic-field-induced deformation of martensite (magnetoplasticity), *J. Magn. Magn. Mater.*, 267, 325, 2003.
46. Chernenko, V.A., et al., Magnetic-field-induced superelasticity of ferromagnetic thermoelastic martensites: Experiment and modeling, *Phys. Rev. B*, 69, 134410, 2004.
47. Suorsa, I. and Pagounis, E., Magnetic field-induced stress in the Ni-Mn-Ga magnetic shape memory alloy, *J. Appl. Phys.*, 95, 4958, 2004.
48. Sozinov, A., et al., Stress- and magnetic-field-induced variant rearrangement in Ni-Mn-Ga single crystals with seven-layered martensitic structure, *Mater. Sci. Eng. A*, 378, 399, 2004.
49. Likhachev, A.A., Sozinov, A., and Ullakko, K., Different modeling concepts of magnetic shape memory and their comparison with some experimental results obtained in Ni-Mn-Ga, *Mater. Sci. Eng. A*, 378, 513, 2004.
50. Kakeshita, T., Saburi, T., and Shimizu, K., Effects of hydrostatic pressure and magnetic field on martensitic transformations, *Mater. Sci. Eng. A*, 273–275, 21, 1999.
51. Hamilton, R.F., et al., Thermal and stress-induced martensitic transformations in NiFeGa single crystals under tension and compression, *Scripta Mat.*, 54, 465, 2006.
52. Liu, Y., et al., Magneto-shape-memory effect in Co-Ni single crystals, *Appl. Phys. Lett.*, 78, 3660, 2001.
53. Li, Y.X., et al., Magnetic field-controlled two-way shape memory in CoNiGa single crystals, *Appl. Phys. Lett.*, 84, 3594, 2004.
54. Chernenko, V.A., et al., Thermal and magnetic properties of stress-induced martensites in Ni-Mn-Ga alloys, *J. Phys. IV*, 5, C2/95, 1995.
55. Vasil'ev, A., et al., Structural and magnetic phase transitions in shape memory alloys $Ni_{2+x}Mn_{1-x}Ga$, *J. Magn. Mater.*, 196–197, 837, 1999.
56. Liu, J., et al., Microstructure evolution in CoNiGa shape memory alloys, *J. Alloys Comp.*, 420, 145, 2006.
57. Hamilton, R.F., et al., Pseudoelasticity in Co-Ni-Al single and polycrystals, *Acta Mat.*, 54, 587, 2006.
58. Pakhomov, A.B., et al., Magnetization and magnetocaloric effect in magnetic shape memory alloys Ni-Mn-Ga, *IEEE Trans. Magn.*, 37, 2718, 2001.
59. Marcos, J., et al., Magnetic field induced entropy change and magnetoelasticity in Ni-Mn-Ga alloys, *Phys. Rev. B*, 66, 224413, 2002.
60. Hu, F.-X., et al., Magnetic entropy change in $Ni_{50.1}Mn_{20.7}Ga_{29.6}$ single crystal, *J. Appl. Phys.* 90, 5216, 2001.
61. Zhou, X., et al., A criterion for enhancing the giant magnetocaloric effect: (Ni-Mn-Ga)—a promising new system for magnetic refrigeration, *J. Phys. Cond. Matter*, 16, L39, 2004.
62. Zhou, X., et al., Influence of the nature of the magnetic phase transition on the associated magnetocaloric effect in the Ni-Mn-Ga system, *J. Magn. Magn. Mater.*, 293, 854, 2005.
63. Marcos, J., et al., Multiscale origin of the magnetocaloric effect in Ni-Mn-Ga shape-memory alloys, *Phys. Rev. B*, 68, 094401, 2003.
64. Marcos, J., et al., Magnetic field induced entropy change and magnetoelasticity in Ni-Mn-Ga alloys, *J. Magn. Mater.*, 272–276, E1595, 2004.
65. Schlagel, D.L., et al., Chemical segregation during bulk single crystal preparation of Ni-Mn-Ga ferromagnetic shape memory alloys, *J. Alloys Compounds*, 312, 77, 2000.
66. Söderberg, O., et al., Recent breakthrough development of the magnetic shape memory effect in Ni-Mn-Ga alloys, *Smart Mater. Struct.*, 14, 223, 2005.
67. Liu, G., et al., Characterization of preferential orientation of martensitic variants in a single crystal of NiMnGa, *Solid State Commun.*, 130, 687, 2004.
68. Suzuki, M., et al., Fabrication and characterization of sputtered Ni_2MnGa thin films, *Mater. Trans. JIM*, 40, 1174, 1999.
69. Ahn, J., et al., Magnetic properties, structure and shape-memory transitions in Ni-Mn-Ga thin films grown by ion-beam sputtering, *IEEE Trans. Magn.*, 37, 2141, 2001.
70. Hakola, A., et al., Ni-Mn-Ga films on Si, GaAs and Ni-Mn-Ga single crystals by pulsed laser deposition, *Appl. Surf. Sci.*, 238, 155, 2004.

71. Kohl, M., et al., A novel actuation mechanism on the basis of ferromagnetic SMA thin films, *Sensors Actuators A Phys.*, A114, 445, 2004.
72. Dong, J.W., et al., Shape memory and ferromagnetic shape memory effects in single-crystal Ni_2MnGa thin films, *J. Appl. Phys.*, 95, 2593, 2004.
73. Heczko, O., et al., Magnetic properties of Ni-Mn-Ga ribbon prepared by rapid solidification, *IEEE Trans. Magn.*, 38, 2841, 2002.
74. Algarabel, P.A., et al., Magnetic-field-induced strain in Ni_2MnGa melt-spun ribbons, *J. Magn. Mater.*, 272–276, 2047, 2004.
75. Hosoda, H., et al., Material design and shape memory properties of smart composites composed of polymer and ferromagnetic shape memory alloy particles, *Sci. Techn. Adv. Mat.*, 5, 503, 2004.
76. Suorsa, I., et al., *Design of active element for MSM actuator*, Borgmann H., ed., Proc. ACTUATOR 2004, Bremen Messe, Germany, pp. 573–530.
77. Xiong, F., et al., Thermally induced fracture of single crystal Ni-Mn-Ga ferromagnetic shape memory alloy, *J. Alloys Comp.*, 415, 188, 2006.
78. Tellinen, J., *Efficiency of MSM material, technical notes*, Adaptamat Ltd., unpublished.
79. Pagounis, E., Magnetic shape memory actuators for motion control applications, *Eur. Design Eng.*, 96, 2004.
80. Jokinen, T., Ullakko, K., and Suorsa, I., Magnetic shape memory materials—new possibilities to create force and movement by magnetic fields, Proc. ICEMS 2001, 1, 20, 2001.
81. Müllner, P., Chernenko, V.A., and Kostorz, G., Stress-induced twin rearrangement resulting in change of magnetization in a Ni-Mn-Ga ferromagnetic martensite, *Scr. Mater.*, 49, 129, 2003.
82. Heczko, O. and Straka, L., Magnetization changes in Ni-Mn-Ga magnetic shape memory single crystal during compressive stress reorientation, *Scr. Mater.*, 54, 1549, 2006.
83. Suorsa, I., Tellinen, J., Ullakko, K., and Pagounis, E., Voltage generation induced by mechanical straining in magnetic shape memory materials, *J. Appl. Phys.*, 95, 8054, 2004.
84. Cesari, E., et al., Internal friction associated with the structural phase transformations in Ni-Mn-Ga alloys, *Acta Mater.*, 45, 999, 1997.
85. Chernenko, V.A., et al., Sequence of martensitic transformations in Ni-Mn-Ga alloys, *Phys. Rev. B*, 57, 2659, 1998.
86. Gans, E., Henry, G., and Carman, G.P., High energy absorption in bulk ferromagnetic shape memory alloys ($Ni_{50}Mn_{29}Ga_{21}$), Lagoudas, D.C., ed., Proc. SPIE, 5387, 2004.
87. Feuchtwanger, J., et al., Energy absorption in Ni-Mn-Ga-polymer composites, *J. Appl. Phys.*, 93, 8528, 2003.

20.2 SMAs in Micro- and Nanoscale Engineering Applications

Yves Bellouard

20.2.1 Introduction

Microsystems (also referred to as microelectromechanical systems [MEMS]) have flourished during the past decades to rapidly penetrate a broad range of applications like automotives, consumer products, and biomedical devices. They perform sophisticated tasks in a miniaturized volume. Common operations such as shaping or analyzing light signals, mixing, processing, or analyzing ultrasmall volumes of chemicals, sensing mechanical signals, probing gas, sequencing biomolecules are realized by these tiny machines. There are several advantages of the using microsystems, which include the reduction of consumables (for instance, the use of less chemicals in lab-on-a-chip), a faster response time (like for airbag sensors), an enhanced portability (RF-MEMS), higher printing resolution (ink-jet head), higher efficiency for microchemical reactor, among others.

For all these applications, new actuating principles that are adapted to the microscale devices are needed. In this context, SMAs have been considered early on. Over the last decade, the fabrication of microdevices such as microgrippers [1–4], microvalves and micropumps [5–8], spacers [5], actuated microendoscope [9], nerves clamp [11], tactile displays [10], to name a few, have been reported.

This chapter provides a critical overview on the use of SMA materials for microsystems: A brief introduction on SMA is first provided followed by a review of SMA microsystems designing methods. Finally, the chapter concludes with a discussion on future prospects. Specific issues related to material processing and in particular thin-film processing are not addressed. Details on these particular aspects can be found, for instance, in Refs. [2,12–14].

20.2.2 SMAs: A Brief Introduction from a Microsystem Design Prospective

SMA are intermetallics* that have two (or sometimes several) crystallographic phases† for which reversible transformations from one to the other occur through diffusionless‡ transformations (the so-called reversible martensitic transformations [15]).

Figure 20.5 briefly illustrates multiple crystallographic phases in the case of a binary alloy like Ni–Ti. At low tempera-

* Intermetallics are alloys made of two or more metallic components forming a new compound that have different properties that its constituents metals taken individually.

† Each phase is thermodynamically stable for a certain domain of temperature and pressure conditions.

‡ During the transformation, atoms collectively moves along subatomic distances so that atomic liaisons remain unchanged although the crystal lattice is distorted.

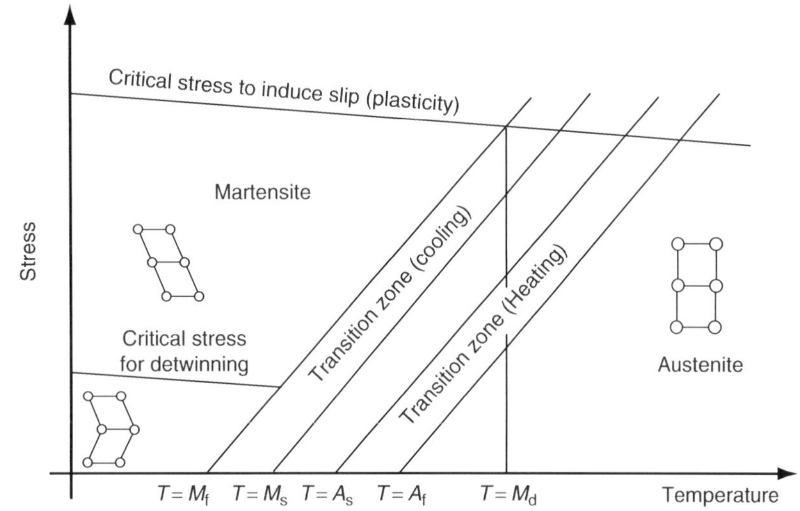

FIGURE 20.5 Two crystallographic phases of a Shape Memory Alloy represented in a stress/temperature phase diagram.

ture and low pressure, the alloy has its low-crystallographic phase called martensite. At higher temperature, the alloy has a different crystallographic structure of higher symmetry. Two remarkable effects are related with the phase change: the SME and superelasticity. As these effects have been extensively documented elsewhere (see for instance Refs. [15–17] as well as the paragraph 20.4 of this Chapter), we just briefly summarize their main features from a microsystems design prospective.

Superelasticity occurs when the martensitic phase transformation is stress-induced at a constant temperature. The transformation is characterized by a plateau and a hysteresis upon unloading. A typical loading–unloading curve is shown on Figure 20.6. The magnitude of reversible pseudoelastic strain can be as high as 8% or even more for single crystals.

The SME refers to the ability of the material, initially deformed in its low-temperature phase (called martensite), to recover its original shape upon heating to its high-temperature phase (called austenite or parent phase). The SME is a macroscopic effect of thermally induced crystallographic phase changes. A very important aspect is that it is a one-way occurrence. This phenomenon is illustrated in Figure 20.7.

Let us consider an SMA material initially in its parent phase (β) (step 3 in Figure 20.7). On cooling, the material transforms to the martensitic phase (step 2). From a crystallographic point of view, this phase is characterized by a lower symmetry than the parent phase and therefore has different possible crystallographic orientations (commonly called as variants). Multiple martensitic variants with different crystallographic orientations (for instance A, B, C, and D in Figure 20.7) nucleate so that the deformation strain

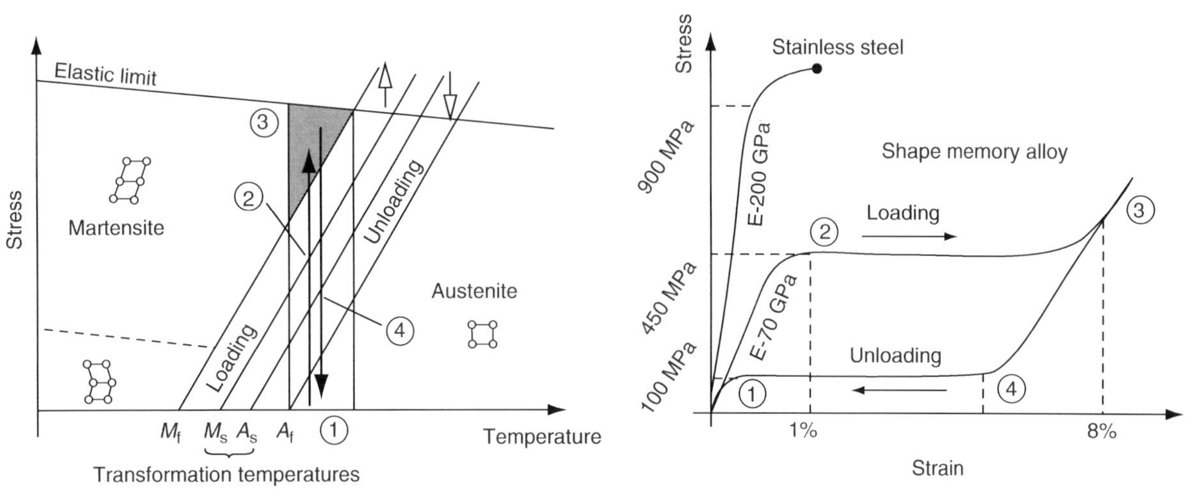

FIGURE 20.6 Left: Superelastic loading represented in the stress/temperature phase diagram-note that the loading and unloading occur at constant stress during transformation, a characteristic feature not captured by the stress/temperature phase diagram. Right: Typical tensile test curve of a superelastic material compared to the one of a stainless steel.

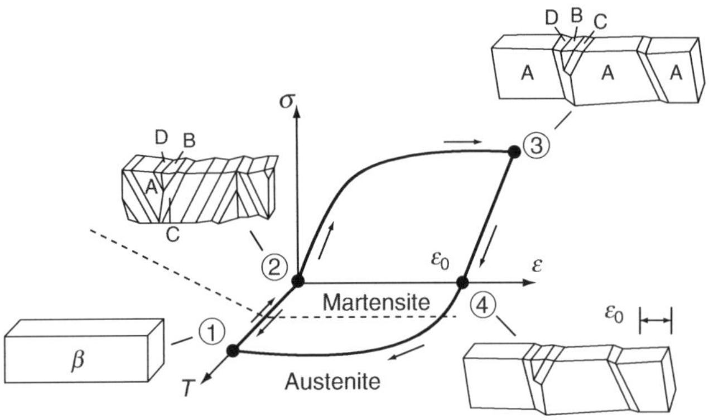

FIGURE 20.7 Illustration of the SME (one-way) (the small sketches illustrative of the material structure one adapted from [18].

energy is minimized (step 1 to step 2). These particular variants are called self-accommodating and form twin pairs. From a macroscopic point of view, at this stage, there is no change of shape as the volume globally remains the same although the crystallographic structure of the material has changed. If a stress is applied on the material (step 3 in Figure 20.7), variants that have a favorable orientation related to the stress grow at the expense of less favorably oriented ones. This effect is called detwinning as self-accommodating variants (that were somehow defining mating pairs) disappear. Once the stress is released (step 4), as the newly formed martensite variants are thermodynamically stable, the material retains the applied deformation (ε_0 in Figure 20.7). Upon heating above the martensitic phase transformation temperatures, the parent phase, which has a higher symmetry from a crystallographic point of view, nucleates and progressively replaces the martensite, causing the disappearance of the apparent deformation (ε_0). This ability to recover its original shape (i.e., the shape corresponding to the parent phase) is called the SME. Cooling again with no applied stress, self-accommodating variants will form (step 2) and no macroscopic change of volume will be observed. From a macroscopic shape change point of view, the SME is therefore not intrinsically reversible: A change of shape is only observed if nonself-accommodating variants have nucleated.

The design objective of an SMA actuator is to achieve a reversible macroscopic shape change: From a microstructure point of view, it is equivalent to have nonself-accommodating martensitic variants nucleating on cooling. In the next section, we review various approaches that have been used to fulfill this design objective.

20.2.3 SMA Microactuators: Design Principles

To provide the necessary reversible SME, two approaches have been explored: intrinsic and extrinsic methods.

Intrinsic methods consist of modifying the material microstructure so that certain martensitic variant orientations will preferably nucleate upon cooling. Extrinsic methods refer to the addition of an external element coupled to the SMA material that provides the required stress to induce stress-oriented variants.

A third method is based on monolithic design. This method is a combination of intrinsic and extrinsic methods.

20.2.4 Intrinsic Methods: Tailoring the Microstructure

20.2.4.1 Two-Way SME

The martensitic variants nucleation is influenced by oriented precipitates and defects like dislocations present in the material lattice structures. Oriented defects lead to what is commonly called the "two-way shape memory effect" (TWSME). Different thermomechanical processes inducing a TWSME have been identified [18,19] like severe deformation of the martensitic phase, thermomechanical cycling under constraints, and isothermal mechanical cycling in the austenite phase. These thermomechanical processes that led to the TWSME are sometimes called training processes to convey the idea that the material progressively memorizes a new shape.

This fascinating behavior is illustrated in Figure 20.8: A TWSME microgripper is shown [4,21]. The millimeter-size device consists of a single piece of metal laser-cut from a 180 μm-thick Ni–Ti–Cu cold-rolled sheet. It is used by a medical instrument manufacturer to assemble submillimeter cylindrical lenses. The gripper design consists of a finger that can move in the plane to press an object against a surface. The working principle is illustrated in Figure 20.8 (operating mode): Upon cooling, the gripper jaw opens and closes on heating. The heat is provided by a simple electrically resistive layer deposited on a ceramic substrate on which the gripper is glued to. The gripper parent shape (the memorized shape) is the as-cut shape. At this point, the device does not spontaneously (i.e., without any applied stress) undergo change of shape. To achieve the reversible finger motion, the gripper is deformed and constrained so that it cannot recover its original shape (step 2 in Figure 20.6). About 100 thermal cycles are applied between the austenite and martensite phases. During thermal cycling, stress builds up in the finger as oriented defects are created. Once this sequence is complete, the gripper is operational and a spontaneous change of shape is observed.

The gripper applies forces of about 15 mN, which is huge compared to the weight of the device. Time response for the jaw to open and close depends on the heating and cooling rates. Typically, it varies from a few hundreds of milliseconds to a second. The microgripper shows excellent fatigue properties (>200,000 cycles) [21] and stability with respect to mechanical perturbations [22].

Remarkable effects are associated with the stability of the TWSME [22,23]. After the training process, nonself-accommodating variants nucleate upon cooling and lead to a macroscopic shape change (the second memorized shape). Yet, the material can still be deformed isothermally in martensite so that another metastable configuration of variants is introduced. This martensite microstructure does not appear spontaneously on cooling but by mechanically deforming the material from its martensitic

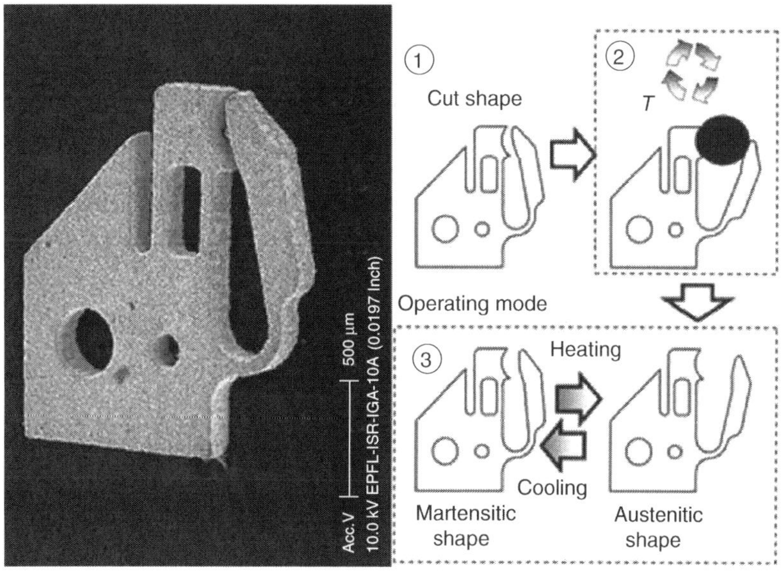

FIGURE 20.8 (Left) Scanning electron microscope picture of the TWSME microgripper for submillimeter lens micromanipulation. (From Bellouard, Y., PhD dissertation, Lausanne, 2000.) (Right) The desired shape is laser-cut (1) and thermomechanically cycled under constraint (training process) (2). The working principle is also shown (3) (scale bar is 500 μm).

shape. It is therefore a prepared state, artificially introduced that puts the material in a sort of out-of-equilibrium state, i.e., outside the sequence of spontaneous shape changes introduce by the TWSME.

The material response to this perturbation is quite spectacular and is shown for a cantilever beam in Figure 20.9. The cantilever beam is made out of a ternary alloy (Ni–Ti–Cu (5% at.) annealed at 515°C for 30 min) [23]. The figure shows the relative displacement of the cantilever. First, the cantilever is deformed in martensite (prepared state). Upon heating, the deformation slowly disappears until the phase transformation temperatures are reached. (We call this first sequence approach-to-equilibrium.) At this point, the cantilever follows the motion path introduced by the TWSME

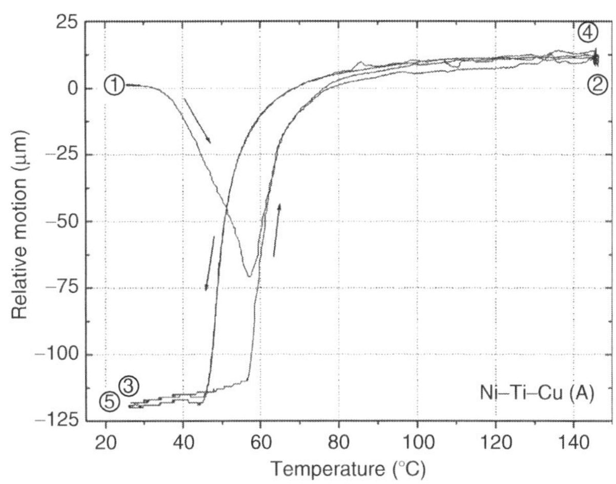

FIGURE 20.9 Out-of-equilibrium behavior in an Ni–Ti–Cu alloy.

(sequence 2–3–4–5). The prepared state is not stable as it disappears after one complete heating–cooling cycle. Noticeably, the introduced prepared state or perturbation is cancelled.

A proposed interpretation [22,23] of this phenomenon is a two-step mechanism: First, upon heating, variants introduced by the perturbation are unstable and are reoriented within the stress field introduced by the training process. Second, when the martensite to austenite transformation temperature is reached, some variants start to transform into austenite. This phenomenon was also observed in a binary alloy of different composition. Specimens, although with different chemical composition, display a consistent similar behavior in response to mechanical perturbations.

Although further studies are needed to better understand the mechanism related to this effect, the observations currently show that TWSME is a robust effect. From an application standpoint, in the case of the microgripper shown in Figure 20.6, it demonstrates that the actuator can recover from unwanted deformation introduced in the martensite.

At an even smaller scale, the use of the TWSME has recently been reported for temperature-controlled surface protrusions [24]. In that particular case, the TWSME is induced by severe plastic deformation: Spherical microindents are made on the surface of a Ni–Ti thin film in its martensite phase followed by a planarization step to restore a flat surface. Protrusions spontaneously appear upon heating and disappear on cooling.

20.2.4.2 Oriented Precipitates (All-Around Effect)

Oriented precipitates were reported to induce a reversible shape change [25,26]. The principle is explained in Figure 20.10.

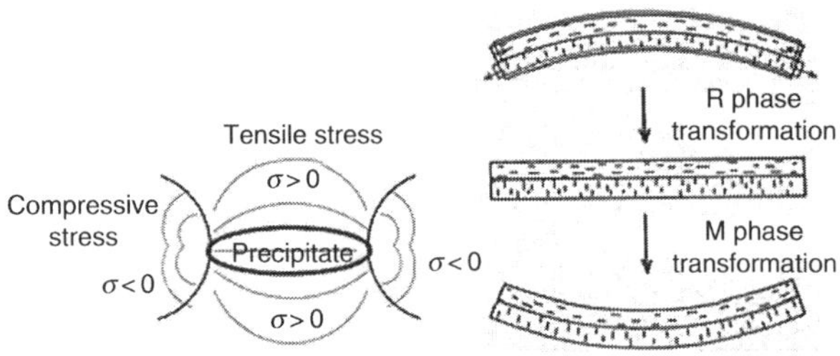

FIGURE 20.10 (Left) Stress distribution around precipitates. (Right) Schematic of the all-around effect. The specimen is aged under constraint in a curved shape. (Adapted from Kainuma, R., Matsumoto, M., and Honma, T. in *Proc. Int. Conf. Martensitic Transform.*, The Japan Institute of Metals, Nara, Japan, 1986, 717–722.)

Precipitates (Ti_3Ni_4) with ellipsoid shape form during constrained-aging of Ni-rich Ti–Ni alloys. In compression, these precipitates are roughly oriented perpendicular to the direction of applied stress while in tension, they are oriented parallel. For a beam loaded in bending, about half of the precipitates are in tensile stress while the rest of the precipitates experience a compressive stress. If the material initially flat is aged under constraint so that it remains bent during the heat treatment, precipitates with two opposing directions form across the material thickness. During the phase transformation, the material undergoes shape changes spontaneously: from curved to flat and then to curved again but in the opposite direction (as illustrated in Figure 20.10).

Numerous studies have been conducted to analyze the effects of coherent Ti_3Ni_4 precipitates on the phase transformation. Multiple-step transformations were reported and analyzed in Ni-rich Ni–Ti [27,28]. It was observed that these precipitates produce an internal stress field [27,29], which modifies the thermodynamic equilibrium locally. The R-phase and martensite morphologies are both affected by the stress field, which induces the growth of specific variants [27]. Furthermore, rather low stresses [30] on the order of a few MPa are reported to modify the precipitate distributions.

Kuribayashi and coworkers reported the use of this process for an SMA active joint for miniature robotarm [31,32]: A 5 μm-thick sputtered-deposited thin film is deposited on NaCl substrate. The film is removed from the substrate and annealed at 1073 K between two crystal plates for 10 min. Finally, the film is aged in a quartz tube of 3.5 mm in diameter at 673 K for 6 h. A drawback of the precipitation process is that it results in an increase of Ni (due to the precipitation process during aging), which results in a decrease of reverse transformation temperatures [16].

20.2.5 Extrinsic Methods

These methods consist in using an additional mechanical element to provide the necessary force to promote the nucleation of stress-induced martensitic variants. This mechanical element can be either a bias spring, another SMA (antagonistic design), a deadweight, etc. The most commonly used element is the bias spring as it is rather easy to implement.

A typical mechanical construction sketch is shown in Figure 20.11. An SMA coupled to a bias spring is preloaded so that the system consisting of the SMA actuator and the spring is under stress: Martensite variants are then reoriented to minimize the strain energy (detwinned, see stage 3 in Figure 20.7) inducing a net macroscopic deformation. Upon heating, the material transforms back to the parent phase and loads the spring as it tries to recover its original shape. The output of the mechanism is taken between the SMA and the bias spring.

A variety of small devices utilize this principle (e.g., microvalves [5–7], gripper [1,3], endoscope [8], and tactile displays [10]). As an illustration, a microvalve is shown [5] in Figure 20.12. It consists of an actuator die with a poppet controlled by a micropatterned Ni–Ti layer. At low temperature, the bias spring (a Cu–Be microfabricated layer) pushes the poppet toward the orifice and deforms the Ni–Ti ribbon (detwinning process). Resistive heating causes the ribbon to transform back to austenite (the parent phase) and to lift the bias spring up and open the valve.

Another example of construction utilizing a bias spring is shown in Figure 20.13. This microdevice is a millimeter-sized microendoscope with five degrees-of-freedom actuated by shape memory elements that are locally heated on certain portions. Sensors based on strain gauges are also incorporated on the structure. In another version, tactile sensors have been added

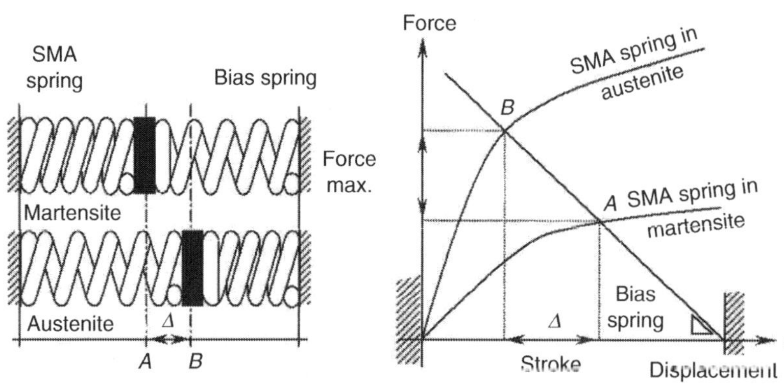

FIGURE 20.11 General principle of a spring-based design.

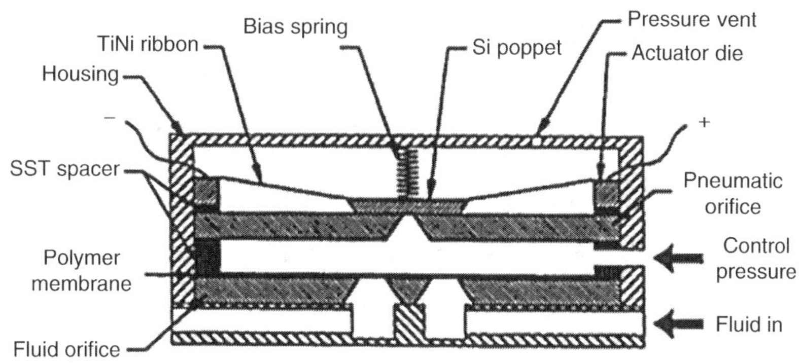

FIGURE 20.12 Microvalve with SMA actuator (Ti–Ni ribbon). The valve is shown in its close position.

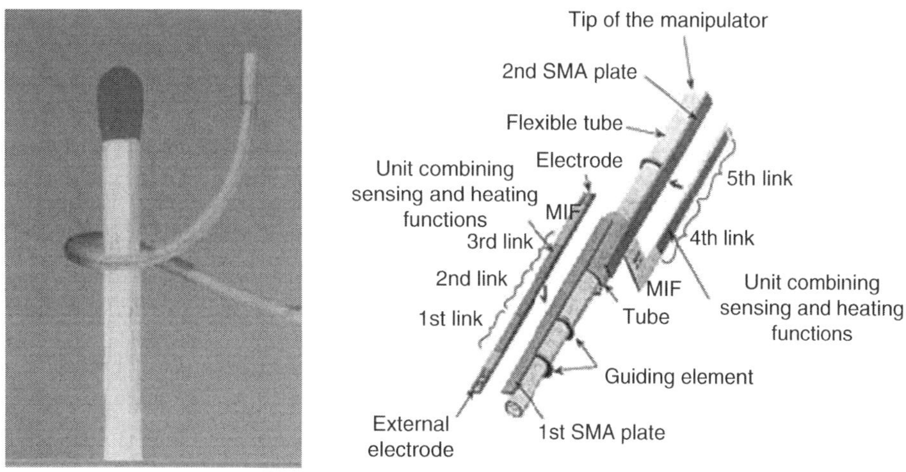

FIGURE 20.13 Olympus Co. microendoscope. (Left) The endoscope winding around a match. (Right) Exploded view of the inside structure. (From Benard, W.L., Kahn, H., Heuer, A.H., and Huff, M.A., *J. MEMS*, 7(2), 245, 1998; Kaneko, S., Aramaki, S., Arai, K., Takahashi, Y., Adachi, H., and Yanagisawa, K., *J. Intell. Mater. Syst. Struct.*, 7, 331, 1996.)

that give reflex functions to the catheter so that when the tube touches something, it automatically bends in the opposite direction. While possible for miniature devices, the preload step becomes particularly challenging to implement for submillimeter devices as it requires micromanipulation or hybrid process integration at a smaller scale.

For thin film-based devices deposited on a substrate, an elegant method to solve this issue is to take advantage of the residual stress that developed during the thin film deposition and annealing. The stress build-up mechanism during processing is schematically illustrated in Figure 20.14: A Ni–Ti film is deposited on a Si substrate. At the deposition temperature considered, the film is amorphous (no crystallographic structure is found). To be functional, an annealing step is required, which is realized by heating the film to typically 700–900 K.

Upon heating, due to the difference between the coefficient of thermal expansion (CTE) between the film and the substrate, a compressive stress builds up. Upon crystallization at a temperature T_a, the stress is relaxed and the film is under tensile stress on cooling. In the device operating mode, this biasing stress can be efficiently used to deflect for instance, a Si-cantilever beam or a membrane.

This effect has been demonstrated in a microgripper [2]. This device is shown in Figure 20.15. A 5 μm Ni–Ti–Cu thin film is deposited on a silicon substrate. The jaws are fabricated by precision sawing and bulk-micromachining of Si.

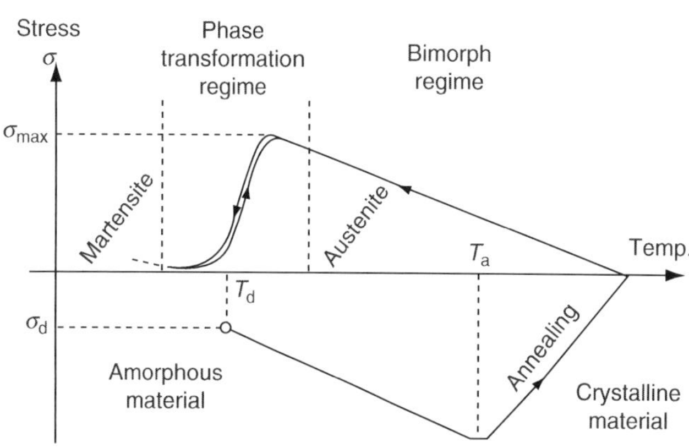

FIGURE 20.14 Mechanism of stress generation for Ni–Ti deposited on Si-substrate.

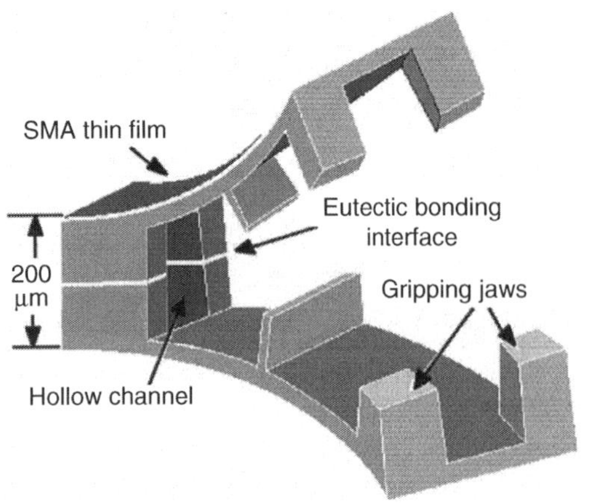

FIGURE 20.15 SMA/Si bimorph microgripper. (Adapted from Krulevitch, P., Lee, A.P., Ramsey, P.B., Trevino, J.C., Hamilton, J., and Northup, M.A., *J. MEMS* 5, 270, 1996.)

Polyimide thin film has also been proposed as a biasing mechanism to create bimorph-like device construction [33]. Along the same line, SMA composites consisting of Ti(Ni,Cu)/Mo have also been reported [14]. For this particular device, it was noted that the work output is about 10 times larger than the work output of the corresponding bimetallic effect [14,27]. Although the residual stress generated during processing of bimorph structures is an efficient mechanism to create a biasing force effect, it limits the design space to out-of-plane motion.

20.2.6 Monolithic Design: Laser Annealing of SMA (LASMA)

To bypass the need for a bias spring and to enlarge the design space, monolithic integration has been suggested [21,34]. The key idea is to tailor spatially the microstructures across the SMA element. Figure 20.16 illustrates the idea. Let us consider a microactuator consisting of two springs opposing each other. The actuator is monolithic and symmetrical but yet has different mechanical characteristics. This can be achieved for instance by local annealing [35] instead of annealing the complete material.

For thin films, unless the substrate is heated sufficiently, as-deposited, the material is amorphous. Similarly, for cold-rolled sheets, a final annealing step is necessary to restore the crystallographic structure severely deformed during the cold-rolling process.

Figure 20.17 shows, for instance, a tensile test experiment performed on a binary Ni–Ti test specimen before and after annealing. Before annealing, the transformation is suppressed and no superelasticity is observed due to the large amount of defects that has been introduced during the forming process. After annealing, the material recrystallizes and its functional properties are restored.

Local annealing can be performed by direct joule heating [34,35]: An electrical current flow between two contact points dissipates heat. This method assumes that the electrical path matches the desired mechanical element that needs to be annealed. Another approach is to use laser annealing [35]. This approach consists of scanning a laser beam over the specimen to locally heat it up to temperatures where annealing takes place. Near-infrared laser (in the 600–1000 nm range) is typically used. In this wavelength range, Ni–Ti absorbs about 30%–35% of the incident radiation (for a thin film, in near normal incidence). Typical irradiance level, i.e., the power density, required to anneal SMA microdevices is on the order of a few W/mm^2. These levels depend on the material thickness, geometry, and substrate. Process feedback can be achieved using infrared sensors or by monitoring the change of electrical resistance during annealing. The latter was demonstrated on thin film-based laser-annealed microswitch where a net decrease of resistivity was observed during annealing. In the annealing temperature ranges, Ni–Ti oxidizes rapidly in the presence of oxygen. However, due to the short high-temperature exposure time, these effects remain negligible for most cases. On thin

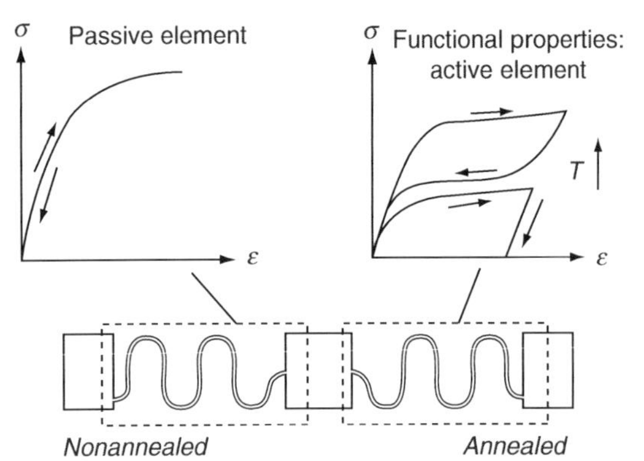

FIGURE 20.16 Schematic of monolithic integration. By locally annealing the material, mechanical properties are distributed across the material.

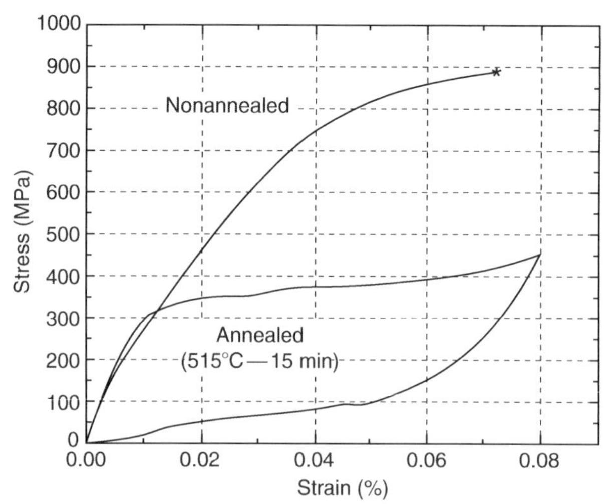

FIGURE 20.17 Tensile characteristics on a binary Ni–Ti annealed and nonannealed.

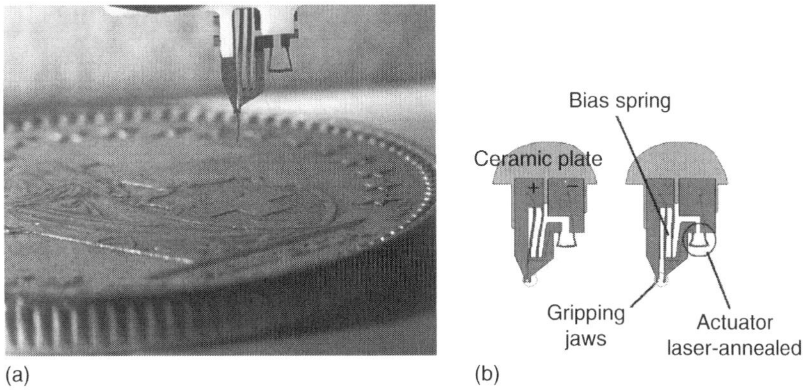

FIGURE 20.18 Laser-annealed monolithic microgripper. This example illustrates the integration of multiple functionalities in a same substrate. (From Zhang, H., Bellouard, Y., Burdet, E., Clavel, R., Poo, A.-N., and Hutmacher, D.W., *Proceedings. ICRA '04. 2004 IEEE International Conference on Robotics and Automation*, Vol. 5, 26 April–1 May 2004, pp. 4918–4924 [39].)

films, the oxide thickness of laser-annealed specimens in normal air was measured for various laser irradiances and scanning speed [36]. It was found that the oxide thickness does not exceed 250 nm for the laser power density of interest.

Wang et al. [37] performed a systematic analysis of the microstructure found for various scanning parameters and laser power. It was found that the microstructure is homogeneous in the laser spot region.

In the previous section (20.2.5), it was shown that for film deposited on a Si substrate, stress generated during deposition and annealing can be used to achieve a reversible motion. Similarly, in the case of laser annealing of freestanding microdevice, stress induced during annealing resulting from differential thermal expansion can also be exploited to introduce a two-way effect without the need to introduce a prestrain. For instance, during local laser exposure, the device can locally be put under compressive stress due to asymmetric thermal expansion; this stress is relaxed during annealing and transforms into a tensile stress on cooling [38].

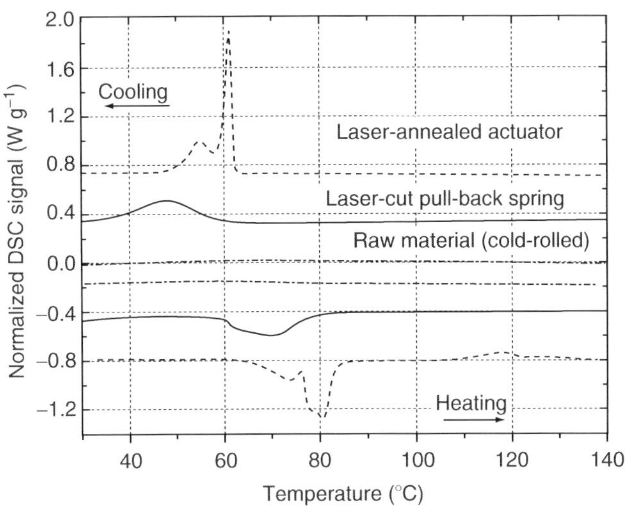

FIGURE 20.19 DSC of the microgripper main parts namely the pull-back spring and the actuator element (laser annealed). The signal for the raw material, as received, is shown for comparison. For each of the three DSC, the mass specimens are similar.

A laser-annealed microgripper is shown in Figure 20.18 [38]. This device was designed for the manipulation of scaffold parts used in bone reconstruction therapy. The raw material is a cold-rolled Ni–Ti–Cu (5% at.) sheet. The device is laser-cut using a Nd–YAG slab laser. Due to the high level of cold-rolling, as received, the raw material does not exhibit a phase transformation. This can be seen on the differential scanning calorimetry (DSC) results presented in Figure 20.19 (raw material).

The laser-cut itself induced a partial annealing effect localized in the laser-affected zone. This annealing effect is limited to the so-called heat-affected zone (typically 8–10 μm). Note that this annealing process resulting from the machining process is not desirable: it can be significantly reduced by choosing a less heat-producing micromanufacturing method (like femtosecond lasers or combined laser-machining with water jet).

The effect of laser annealing on the DSC signal is clearly visible. A sharp, well-defined, transformation peak is observed. (Smaller secondary peaks are related to the DSC specimen preparation itself that required additional laser cutting.)

From a design point of view, this device is similar to the bias-spring/actuator configuration. As the raw material is a cold-rolled sheet, prestrain is achieved by introducing an offset between the contact pads during assembly. As mentioned earlier, this prestrain may not be necessary for laser-annealed thin film-based microdevices.

20.2.7 Summary of SMA Actuator Design Methods

Figure 20.20 summarizes the various principles available to induce a reversible macroscopic shape change. The most commonly found techniques were briefly outlined in the previous

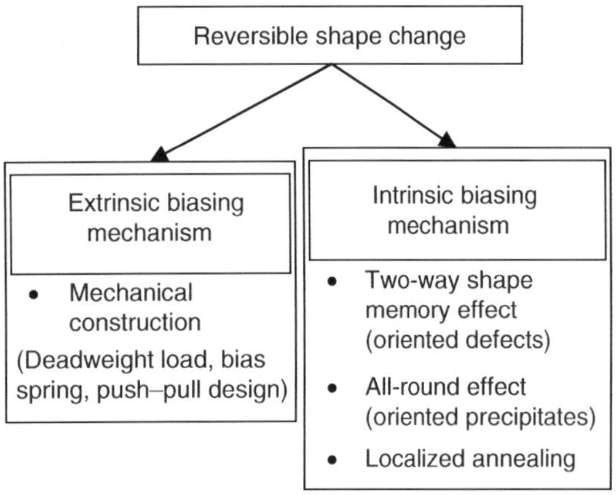

FIGURE 20.20 Outlook of methods that induce reversible shape change in SMA.

sections. In addition to these techniques, one should add ion bombardment as a means to induce localized amorphization or inhomogeneities, which, combined with unaffected zones, leads to an intrinsic two-way effect [40]. Special fabrication processes can also lead to intrinsic two-way effects: This is the case for melt-spinning (where molten metal is rapidly solidified on a high-speed spinning wheel) and thin film fabricated by stacking sequence of thin layers of Ni and Ti [41]. These two-way effects essentially result from the particular microstructure found across the material thickness.

The high output force is almost systematically cited to support the use of SMA in microsystems. While it is an important argument, it has to be balanced with known limitations of SMA that have a poor bandwidth (<100 Hz) and the lack of intrinsic reversibility.

The latter can be overcome by properly designing the SMA microsystems. Furthermore, local annealing technique also offers a way to bypass process compatibilities issues resulting from the need of high-temperature processing.

20.2.8 SMA Flexures

Flexures are used in precision engineering where highly accurate, wear-free, smooth, and repeatable motion is desired. Flexures are based on deformation of material to achieve a motion between elastically joined parts. They are used in a variety of precision mechanisms such as high-resolution balances or high-accuracy optical positioning stages.

SMA are an attractive option in designing flexures. Various examples of SMA flexures were reported [42–47]: Superelastic flexures can withstand larger deformations for the same weight as a conventional flexure. In addition, the damping properties of SMA, controllable through the phase transformation, offer new design opportunities for adaptive compliant mechanisms [47]. The martensitic phase transformation can also be used to shift the natural frequency of flexures adding useful functionalities such as vibration rejection [47].

20.2.9 Future Directions for SMA Applications in Microtechnologies

20.2.9.1 SMA on a Flexible Substrate

Electronics on flexible substrate are developing at a fast pace. The so-called polymer electronics or plastic MEMS are rapidly emerging as a powerful platform for multifunctional integration in a variety of packages and substrates. This is particularly attractive for biomedical applications where fluidics, optics, and electronics can be merged on a same, flexible substrate (smart bandage, for instance). Ni–Ti alloys being biocompatible materials are quite attractive in that context. Although early demonstrations of Ni–Ti/polyimide actuator [43] were made, more work is needed to expand the process capabilities.

20.2.9.2 SME at the Nanoscale: Toward Nano-SMA Actuator

Observing SME effect at the nanoscale has been of great interest recently both from an application and theoretical point of view. Nanoindentations experiments [48,49] have demonstrated nanoscale SME. These studies provide the direction for future nanoactuators based on SMA. However, shape memory transformation in nanopatterned SMA actuator is yet to be demonstrated.

References

1. K. Ikuta, Proc. IEEE Int. Conf. Robotics Autom., Cincinatti, 1990, pp. 215–216.
2. P. Krulevitch, A.P. Lee, P.B. Ramsey, J.C. Trevino, J. Hamilton, and M.A. Northup, *J. MEMS* 5(4), 1996, 270–282.
3. J. Hesselbach, R. Pittschellis, E. Hornbogen, and M. Mertmann, Proc. Shape Memory and Superelastic Technol., Pacific Grove, CA, 1997, pp. 251–256.
4. Y. Bellouard, R. Clavel, J.-E. Bidaux, R. Gotthardt, and T. Sidler, Proc. Shape Memory Superelastic Technol., Pacific Grove, CA, 1997, pp. 245–250.
5. A.D. Johnson and V.V. Martynov, Proc. 2nd Int. Conf. Shape Memory Superelastic Technol., Pacific Grove, CA, 1997, pp. 149–154.
6. J.D. Busch and A.D. Johnson, Shape-memory alloys microactuator, U.S. Patent 5,061–914, June 27, 1989.
7. M. Kohl, D. Dittmann, E. Quandt, B. Winzek, S. Miyazaki, and D.M. Allen, *Mater. Sci. Eng. A*, 230, 1999, 273–275.
8. W.L. Benard, H. Kahn, A.H. Heuer, and M.A. Huff, *J. MEMS*, 7(2), 1998, 245–251.
9. S. Kaneko, S. Aramaki, K. Arai, Y. Takahashi, H. Adachi, and K. Yanagisawa, *J. Intell. Mater. Syst. Struct.*, 7, 1996, 331–335.
10. H. Fischer, R. Trapp, L. Schüle, and B. Hoffmann, *J. Phys. IV France* 7, 1997, C5–609–614.
11. S. Takeuchi and I. Shimoyama, *J. MEMS*, 19(1), 2000, 24–31.
12. S. Miyazaki and A. Ishida, *Mater. Sci. Eng. A*, 273–275, 1999, 106–133.
13. Y. Fu, H. Du, W. Huang, S. Zhang, and M. Hu, *Sensors Actuators A*, 112, 2004, 395–408.

14. B. Winzek, S. Schmitz, H. Rumpf, T. Serzl, R. Hassdorf, S. Tienhaus, J. Feydt, M. Moske, and E. Quandt, *Mater. Sci. Eng. A*, 378, 2004, 40–46.
15. L. Delaey, Phase transformations in materials. In *Material Science and Technology*, vol. 5, R.W. Cahn, P. Haasen, E.J. Kramer, eds. VCH, Weinheim, 1991.
16. K. Otsuka and X. Ren, *Prog. Mater. Sci.*, 50, 2005, 511–678.
17. K. Otsuka and C.M. Wayman, eds., *Shape Memory Materials*. Cambridge University Press, Cambridge, U.K., 1998.
18. J. Perkins and D. Hogson, in *Engineering Aspects of Shape Memory Alloys*, T.W. Duerig, K.N. Melton, D. Stöckel, and C.M. Wayman, eds. Butterworth-Heinemann, London, 1990.
19. H. Scherngell and A.C. Kneissl, *Acta Mater.*, 50(2), 2002, 327–341.
20. H. Scherngell and A.C. Kneissl, *Script. Mater.*, 39(2), 205–212.
21. Y. Bellouard, PhD dissertation, Ecole Poly technique Fe'de' rale de Lausanne, 2000.
22. Y. Bellouard, R. Clavel, and R. Gotthardt, *J. Physique IV* JP 11(8), 2001, Pr8583–Pr8588.
23. Y. Bellouard, R. Clavel, R. Gotthardt, and J. van Humbeeck, *J. Physique IV* JP 112(II), 2003, 765–768.
24. Y. Zhang, Y.-T. Cheng, and D.S. Grummon, *Appl. Phys. Lett.*, 89, 2006, 41912.
25. M. Nishida and T. Honma, *Script. Metallurg.*, 18, 1984, 1293–1298.
26. R. Kainuma, M. Matsumoto, and T. Honma, Proc. Int. Conf. Martensitic Transform., The Japan Institute of Metals, Nara, Japan, 1986, pp. 717–722.
27. L. Bataillard, J.-E. Bidaux, and R. Gotthardt, *Phil. Mag.*, A 78, 1998, 327–344.
28. J. Khalil-Allafi, A. Dlouhy, and G. Eggeler, *Acta Mater.*, 50 2002 4255–4274.
29. W. Tirry and D. Schryvers, *Acta Mater.*, 53 2005, 1041–1049.
30. J. Michutta, Ch. Somsen, A. Yawny, A. Dlouhy, and G. Eggeler, *Acta Mater.*, 54, 2006, 3525–3542.
31. K. Kuribayashi, Proc. MEMS, 1990, pp. 217–221. DOI 10.1109/MEMSYS.1990.110279.
32. K. Kuribayashi, S. Shimizu, T. Nishinohara, T. Taniguchi, M. Yoshitake, and S. Ogawa, Proc. IROS 1993, pp. 1697–1702. DOI 10.1109/IROS.1993.583865.
33. K. Kuribayashi and T. Fujii, Proc. Int. Symp. Micromechatronics Human Science (MHS), 1988, pp. 165–170.
34. Y. Bellouard, R. Clavel, R. Gotthardt, J.-E. Bidaux, and T. Sidler, Actuators, Int. Conf. on New Actuators, H. Borgmann, ed., Bremen, Germany, June 17–19, 1998, pp. 502–505.
35. Y. Bellouard, T. Lehnert, J.-E. Bidaux, T. Sidler, R. Clavel, and R. Gotthardt, *Mater. Sci. Eng.*, A273–275, 1999, 733–737.
36. F. Khelfaoui, Y. Bellouard, T. Gessmann, X. Wang, J. Vlassak, and M. Hafez, Proceedings of the International Conference on Shape Memory and Superelastic Technology (SMST-2004), Germany, ASM International, 2004.
37. X. Wang, Y. Bellouard, and J.J. Vlassak, *Acta Mater.*, 2005, 4955–4961.
38. Y. Bellouard, T. Lehnert, R. Clavel, and R. Gotthardt, *J. Phys. IV*, JP 11(8), 2001, Pr8571–Pr8576.
39. H. Zhang, Y. Bellouard, E. Burdet, R. Clavel, A.-N. Poo, and D.W. Hutmacher, Robotics and automation, 2004. Proceedings. ICRA '04. 2004 IEEE International Conference on Robotics and Automation, Vol. 5, 26 April–1 May 2004, pp. 4918–4924. DOI 10.1109/ROBOT.2004.1302497.
40. D.S. Grummon and R. Gotthardt, *Acta Mater.*, 48, 2000, 635–646.
41. T. Lehnert, S. Tixier, P. Böni, and R. Gotthardt, *Mater. Sci. Eng.*, A273–275, 1999, 713–716.
42. K. Kuribayashi and T. Fujii, Proc. Int. Symp. Micromechatronics Hum. Sci., Piscataway, NJ, 1998, pp. 165–170.
43. J. Peirs, D. Reynaerts, and H. Van Brussel, *Sensors Actuators A Phys* 70(1/2), 1998, 135.
44. M. Mertmann, E. Hornbogen, J. Hesselbach, and R. Pittschellis, Proceedings of SMST, A. Pelton, D. Hodgson, and S. Russell, T. Duerig, eds., Pacific Grove, Santa Clara, March 2–6, 1997, p. 549.
45. J. Hesselbach and A. Raatz, *Mater. Sci. Forum* 394(3), 2001, 79.
46. Y. Bellouard, R. Clavel, R. Gotthardt, J.E. Bidaux, and T. Sidler, Proceedings of Actuators, H. Borgmann, ed., Bremen, 1998, p. 505.
47. Y. Bellouard and R. Clavel, Shape memory flexures, *Mater. Sci. Eng. A*, 378(25), 2004, 210–215.
48. G.A. Shaw, D.S. Stone, A.D. Johnson, A.B. Ellis, and W.C. Crone, *Appl. Phys. Lett.*, 83, 2003, 257.
49. X.-G. Ma and K. Komvopoulos, *Appl. Phys. Lett.*, 83, 2003, 3773.

20.3 Mathematical Models for Shape Memory Materials

Davide Bernardini and Thomas J. Pence

20.3.1 Introduction

The purpose of this chapter is to give a basic primer on the mathematical modeling of shape memory material behavior for the purpose of understanding the behavior of structures and mechanical devices. It is assumed that the reader is familiar with the elementary aspects of shape memory materials (SMM) like pseudoelastic behavior, SME, and their underlying solid phase transformations [27,33]. Attention is restricted to behavior that is observable at a macroscopic scale under thermal and mechanical loads involving a single component of stress, e.g., tension versus compression, as is generally sufficient for describing wires. More advanced modeling issues, many associated with detailed metallurgical aspects, are outside of the scope of the present exposition, including the behavior under multidimensional stress states; low-cycle fatigue behavior induced by plastic deformation of the phases; effects associated with plasticity, viscoelasticity, or strain gradients; and the two-way shape memory effect. The phenomenology that will be object of modeling is briefly recalled in the following section, in order to set the terminology and point out the issues relevant for the subsequent sections.

20.3.2 Experimental Observations

20.3.2.1 Thermal Transformation

In the absence of loads, the material is in the austenitic phase A at temperatures $T > A_f$, and is in the martensitic phase M at temperatures $T < M_f$. Between these temperatures, the crystallographic phase may involve a mixture of martensite and austenite, in which case it is convenient to introduce the fraction of martensite ξ. This fraction is dependent on the recent temperature history whereupon it is useful to introduce the additional material parameters M_s and A_s. Here M_s is the start temperature for the cooling conversion of austenite to martensite, the forward martensitic transformation (FwT). This kind of martensite is characterized by the absence of macroscopic transformation strain relative to austenite. This is due to the fact that at a microscopic scale several variants occur in a way that the overall transformation strain vanishes (simple twinning is a standard case). Such thermal martensite is said to be self-accommodated.

Similarly, A_s is the start temperature for the heating conversion of this martensite back to austenite, the reverse martensitic transformation (RvT). These temperatures are determined by alloy content and processing history. The standard observed ordering of these temperatures is $M_f < M_s < A_s < A_f$. In fact, the central inequality is not required by any thermodynamic principle, and the description naturally extends to the less restrictive conditions $M_f \leq M_s \leq A_f$ and $M_f \leq A_s \leq A_f$. As is conventional for cooling transformations, the forward transformation is exothermic. Conversely, the RvT is endothermic. The thermodynamically reversible latent heat released and absorbed in these transformations is denoted by H.

20.3.2.2 Isothermal Pseudoelastic Loading

The imposition of stress also leads to FwT whereas its removal facilitates RvT. If cycles of mechanical loading and unloading are applied at a sufficiently slow rate at temperatures $T > A_f$, then the typical pseudoelastic outer loop is observed. Specifically, so long as plasticity is not present, loading in tension first gives a linear elastic branch with modulus E_A and $\xi = 0$, followed by a conversion plateau associated with FwT, followed by a second linear elastic branch with modulus E_M and $\xi = 1$ (Figure 20.21). Here the conversion during FwT is no longer self-accommodated, and instead results in a crystallographic biased martensite M+ that provides transformation strain γ_+ to the material.

Since the FwT is exothermic, it would naturally lead to an increase in the SMM temperature as determined by the SMM specific heat, denoted by c. However, if the loading rate is sufficiently low, then heat transfer permits the temperature of the SMM to remain close to its original value as determined by the overall thermal environment. In this isothermal setting, the FwT plateau begins and finishes at transformation stresses $\sigma_{Ms}(T)$ and $\sigma_{Mf}(T)$. These transformation stresses are temperature dependent, as the same test carried out at a different temperature T' yields $\sigma_{Ms}(T')$ and $\sigma_{Mf}(T')$ different from the former ones. A representation of the transformation stresses versus temperature is called either a phase diagram or a state diagram (Figure 20.21). The dependence of $\sigma_{Ms}(T)$ and $\sigma_{Mf}(T)$ upon temperature for $T > A_f$ is essentially linear and the common slope for these two curves, denoted by m, is also a material parameter. The Clausius–Clapeyron condition of conventional thermodynamics gives that the slope m is related to the latent heat, the transformation temperature, and the transformation strain via a standard relation. In view of the multiple transformation temperatures associated with SMMs, this relation can be taken for modeling purposes as:

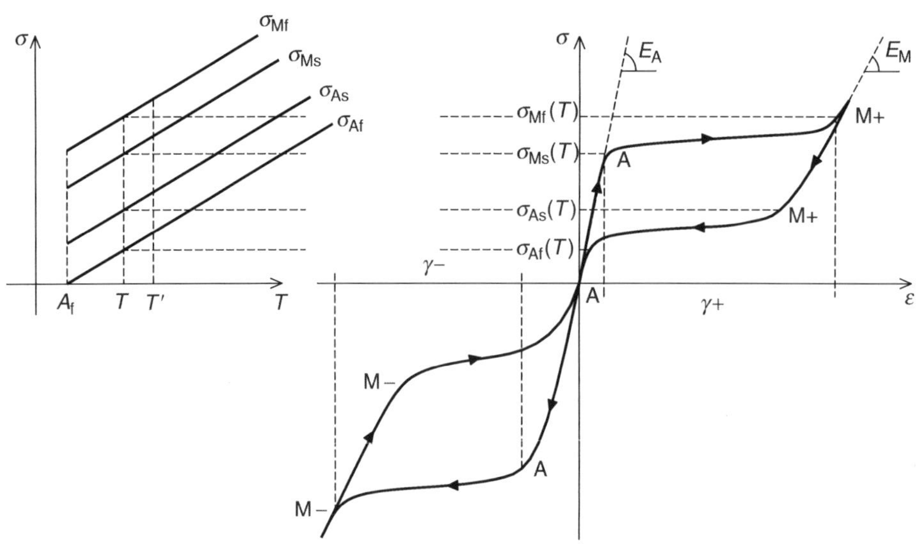

FIGURE 20.21 At a given temperature, uniaxial loading, and unloading for complete FwT and complete RvT generates the outer loop of the stress–strain diagram. Changing the temperature alters the transformation stresses.

$$m = \frac{4H}{(M_f + M_s + A_s + A_f)\gamma_+} \tag{20.1}$$

Since the high damping capacity associated with SMMs may complicate calorimetric measurement of H, Equation 20.1 is also useful for the determination of H from stress–strain data.

20.3.2.3 Isothermal Pseudoelastic Tensile Unloading

Unloading at $T > A_f$ reverses the procedure discussed above with one essential difference, namely the plateau associated with RvT is below that of the FwT (Figure 20.21). Hence, even though the transformation strain γ_+ is completely recovered, there is a stress–strain hysteresis loop. The RvT is endothermic, but sufficiently slow unloading permits the temperature to remain at the ambient, whereupon the start and finish of the lower plateau is at transformation stresses $\sigma_{As}(T)$ and $\sigma_{Af}(T)$. For $T > A_f$, these curves on the state diagram are also linear with slope m.

20.3.2.4 Isothermal Pseudoelastic Compressive Loading and Unloading

The application of the opposite sign load of compression induces a different assembly of martensite variants. The overall assembly is denoted by M− and characterized by an overall transformation strain $\gamma_- < 0$. Forward and reverse transformations in compression are thus associated with the respective conversions A → M− and M− → A (Figure 20.21). Functions analogous to $\sigma_{Ms}(T)$, $\sigma_{Mf}(T)$, $\sigma_{As}(T)$, and $\sigma_{Af}(T)$ can be constructed in the compressive regime of the state diagram, and for $T > A_f$, the slope relation (Equation 20.1) continues to hold provided that γ_+ is replaced by γ_-.

20.3.2.5 Internal Subloops

At $T > A_f$, if a tensile loading is arrested before the completion of FwT, then the material is a mixture of A and M+, say with martensite fraction value $\xi = \xi_a < 1$. Unloading then activates RvT at a stress $\sigma_{As}(T,\xi_a)$ that, in general, may be different from the one of the outer loop. If a subsequent reloading occurs before the completion of the RvT, say when $\xi = \xi_b$ (so that $0 < \xi_b < \xi_a$), then this reloading activates FwT at a stress $\sigma_{Ms}(T,\xi_b)$, giving rise to an internal pseudoelastic subloop (Figure 20.22a).

Unloading or reloading at different values of ξ yields different values of the RvT activation stresses (Figure 20.22b). The loci of potential activation of the transformations of internal loops are called subloop activation curves and may be described by functions $\sigma_{As}(T,\xi_0)$ and $\sigma_{Ms}(T,\xi_0)$ where ξ_0 denotes the fraction of martensite at the end of the last transformation process. This expresses the fact that an internal resistance must be overcome in order to activate the transformation, and this resistance depends on the actual microstructure from which the transformation evolves. On the other hand, if internal transformations are brought to completion, then the completion stress is often the same as that for the outer loop, irrespective of the particular initial microstructure. SMMs having a completion stress that is independent of the microstructure at transformation activation are said to obey the Ivshin condition for hysteresis completion (Figure 20.22b and c).

20.3.2.6 Nonisothermal Pseudoelasticity

Isothermal pseudoelasticity only takes place when the loading rate is sufficiently slow so as to permit the heat released and absorbed during transformation to be completely exchanged with the environment hence allowing the temperature of the material to remain unchanged. Conversely, if the loading rate exceeds the rate of heat exchange with the environment, then the latent heat of transformation gives rise to a change of temperature within the SMM.

Since the transformation stresses depend on the temperature, it follows that the shape of the pseudoelastic loop may be significantly different with respect to the isothermal case (Figure 20.23). Beyond a certain loading rate, this temperature change process may be treated as adiabatic (no heat exchange at all). Of course, the adiabatic treatment does not pertain during any time intervals between load change wherein the now heated or cooled SMM exchanges heat with the environment at constant load.

20.3.2.7 Reorientation

The M+ type of martensite obtained by pseudoelastic FwT at $T > A_f$ can also be obtained at temperatures $T < A_f$. For example,

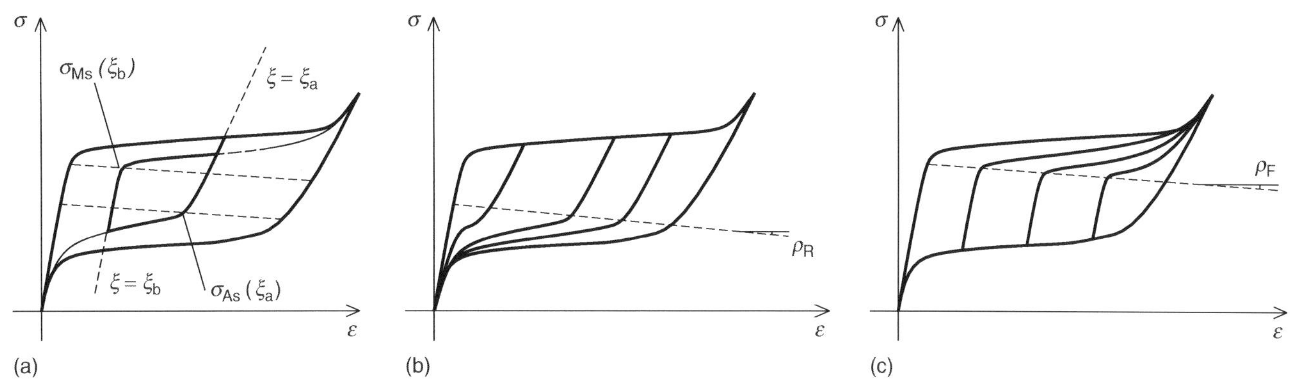

FIGURE 20.22 (a) Sublooping occurs when transformations do not go to completion. (b) Elastic unloading from arrested FwT followed by the activation of RvT at the dashed line. (c) Elastic loading from arrested RvT followed by activation of FwT at the dashed line. Both (b) and (c) show subloop completion at a common point.

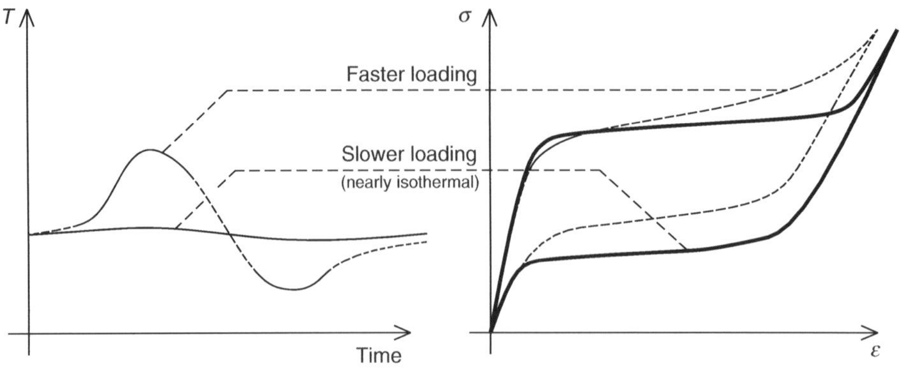

FIGURE 20.23 Sufficiently slow loading gives nearly isothermal behavior. For faster loading, the latent heat of transformation gives SMM temperature change that in turn alters the stress–strain curve.

at $T < M_f$, this is accomplished by applying stress to the self-accommodated M martensite obtained by cooling (i.e., purely thermal FwT). The applied stress then gives an M →M+ conversion that is called forward reorientation (FwR). This FwR also results in a transformation strain γ_+ within the SMM. The associated reorientation plateau begins and finishes at transformation stresses σ_{Rs+} and σ_{Rf+} (Figure 20.24). Unlike the activation stresses for A →M+ transformation, the reorientation stresses only display mild temperature dependence. Unloading does not activate the reverse transformation, so the transformation strain γ_+ remains at zero stress. Compressive loading can then initiate a corresponding M+ → M− reverse reorientation (RvR). In addition, compressive loading applied to a self-accommodated martensite M gives M → M−, which is also classified as RvR. The associated reorientation plateau in compression begins and finishes at transformation stresses σ_{Rs-} and σ_{Rf-}. There is no latent heat associated with reorientation, which is consistent with the mild to vanishing effect of temperature on the threshold stresses σ_{Rs+}, σ_{Rf+}, σ_{Rs-}, and σ_{Rf-}.

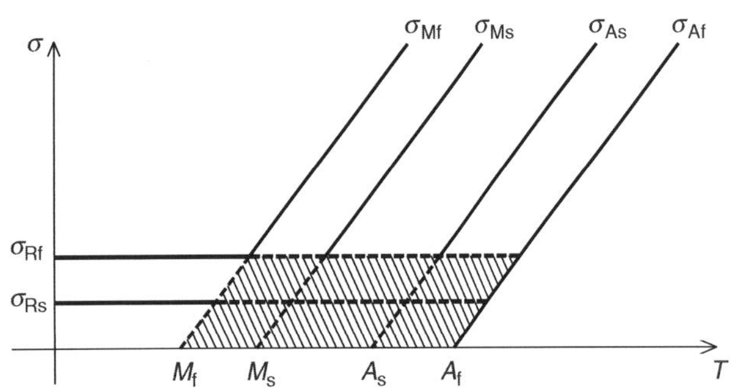

FIGURE 20.24 Extended state diagram showing outer loop activation and completion lines for FwT and RvT and similar lines for FwR. Combined transformation and reorientation can occur on the hatched region for suitably oriented (σ,T)-paths. Although not indicated here, reorientation may cause the FwT and RvT lines to become curved in the region of combined transformation and reorientation.

Pure reorientation refers to reorientation that occurs at constant phase fraction, for example at $T < M_f$ with $\xi = 1$ as discussed above. Conversely, at temperatures in the interval $M_f < T < A_f$, it is not unusual for reorientation to take place in conjunction with phase transformation. For example, cooling austenite to a temperature below M_s but above M_f gives a mixture of austenite and self-accommodated martensite. Isothermal tensile loading then converts both the A and the self-accommodated M to the oriented martensite M+, the former conversion by phase transformation and the latter conversion by reorientation. An extended state diagram for which the temperature may range from below M_f to above A_f must therefore show multiple lines associated with the start and finish of both phase transformation and reorientation (Figure 20.24).

20.3.2.8 Shape Memory Effect

The SME is a consequence of the fact that isothermal shape change due to pure reorientation can be undone by thermal RvT. In particular, starting with austenite, consider a three stage cycle composed as:

(sm1) thermal FwT (i.e., A→M by cooling in the absence of stress), followed by
(sm2) isothermal reorientation M→M+ due to loading which then persists upon unloading, followed by
(sm3) thermal RvT (i.e., M+→A by heating in the absence of stress).

Stage (sm1) gives no shape change, whereas stage (sm2) provides a seemingly plastic deformation. However, stage (sm3) reverses the shape change by purely thermal means, giving rise to the SME. At the conclusion of (sm3), the material has returned to the shape associated with the original austenite and remains at that shape during any temperature change (even a cooling associated with FwT) so long as there is no further loading (Figure 20.25).

The two-way memory effect refers to certain processes wherein multiple cycles of this kind are performed in such a manner so as

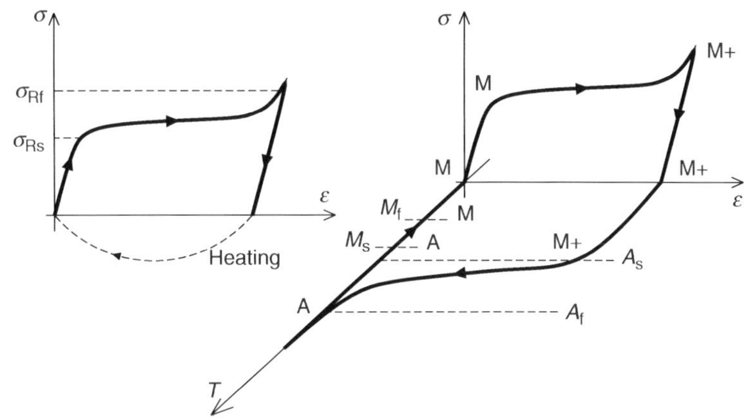

FIGURE 20.25 For $T < M_f$, loading and unloading reorients an originally self-accommodated martensite, resulting in an apparent plastic strain. Heating converts the reoriented martensite to austenite, returning the material to its original shape (the SME).

to create a defect structure (e.g., a dislocation pattern) after reorientation that serves to stabilize the deformation after stage (sm2). Such "training" can eventually lead to a situation wherein the shape change associated with reorientation can be obtained by cooling in the absence of load, at which point a stress-free cooling and heating cycle gives a reversible shape change. Models for this two-way effect generally require some notion of internal defect stress, which is a concept that we do not treat here.

20.3.3 Overview of the Literature

The modeling of the thermomechanical behavior of SMM has received increasing attention in the scientific literature for the last two decades. Presently, the literature is vast and some review papers are available (e.g., Refs. [3,21]). In the following, a quick survey will be given, more details can be found in the abovementioned reviews. Roughly speaking, most of the existing models can be classified into two broad categories.

20.3.3.1 Micromechanical Models

Such theories analyze the material at a scale in which the multiphase nature of the material is explicit. When the material is modeled as a continuum, each point represents a material element made of a single solid phase rather than a mixture of phases. Some kind of averaging procedure may then serve to recover the macroscopic behavior. In this category, there are several approaches, of which only two are mentioned here. One line of research was initiated by Patoor et al. [28] and then pursued by the same group in several papers (e.g., Ref. [35]) and also developed by Brinson and coworkers (e.g., Ref. [15]). This approach describes the macroscopic behavior via the homogenization of a representative volume element of austenite and various variants of martensites. For single crystals, this produces a detailed analysis of the microstructure evolution. It can be extended also to polycrystals taking into account the grain structure although requiring a large computational effort.

Another approach initiated by Ball and James [2] and then pursued by many authors (e.g., Refs. [7,32]) aims to determine the actual microstructure formed under given thermomechanical conditions via the minimization of a suitable energy function. This requires advanced methods of calculus of variations.

20.3.3.2 Macroscopic Models

The detailed study of the microstructure may become unpractical both for the computational effort and for the lack of detailed knowledge about the microstructure of actual materials.

This motivates approaches that directly model the macroscopic behavior without such detailed microstructural considerations. In this case, when the material is modeled as a continuum, each point represents a material element made of a mixture of phases, so that the features of the mixing has to be specified by suitable descriptors.

Such models then differ on how the effect of the microstructure is included.

- *Transition models without internal variables.* Such models describe hysteresis by considering minimization of macroscopic energy functions with multiple energy wells, each corresponding to a different microstructure, in conjunction with rules that delay the transition between wells. Such rules of delay, based in some sense on the idea that it is necessary to overdrive the system in order to transition between wells, have the effect of temporarily overruling the associated equation of state. Falk [14] initiated the development of such models in the context of SMM, while several advances were obtained, among the others, by Muller and Seelecke [25], and Puglisi and Truskinovsky [29]. This type of modeling is not considered here in detail, although it is worthwhile to note that there are deep and fundamental connections between these models and models containing internal variables that are discussed below.
- *Phenomenological models of hysteresis.* Such models of hysteresis have been widely used in fields like, for example,

that of magnetic materials, to model experimentally observed curves involving high nonlinearity and complicated looping. This may, but often does not, involve explicit focus on their link with the physical phenomena of interest. Reliability and the robustness of the model are of paramount consideration. By favorably matching the algorithm to experiment, these algorithms permit the treatment of arbitrarily complex driving input [18,23]. Preisach models for hysteresis can also be viewed as an example of this type of model [26].

- *Models for the evolution of internal variables.* Perhaps the most common macroscopic approach is based on the introduction of internal variables that describe the internal microstructure of the material. Such variables are included within the list of the independent variables of the standard constitutive functions. The evolution of the phase transformations is then modeled by suitable equations of evolution, the so-called transformation kinetics [3–6,9–13,17,19,34].

Connections can be forged between all of these approaches. In the following, our attention will be focused on models for the evolution of internal variables.

20.3.4 Formulation of SMM Models in the Framework of Thermomechanics with Internal Variables

In this setting, the state of the material element is described not only by a pair of mechanical variables (stress σ and strain ε) and a pair of thermal variables (temperature T and entropy η), but also by a list of suitable internal variables α. For one-dimensional models, the mechanical variables are scalars; for multidimensional models, many of these variables become second-order tensors. Additional control variables, for example, electric or magnetic field may be added, but this case is not considered here. Moreover, the thermomechanical formulation requires an energetic variable (like the internal energy e or Gibb's free energy $\psi = e - T\eta - \sigma\varepsilon$), and the rate of energy dissipation Γ.

All such variables are constrained by balance equations that may be stated in various slightly different ways. Following, for example, Green and Naghdi [16], there are balance of internal energy and balance of entropy

$$\dot{e} = \sigma\dot{\varepsilon} + \mathrm{r}, \qquad \dot{\eta} = \frac{1}{T}(\Gamma + r) \tag{20.2}$$

where r is the external heat supply, as well as the Clausius–Duhem inequality $\Gamma > 0$ associated with the second law of thermodynamics. Since the constitutive theory is developed in the context of homogeneous processes there is no heat flow contribution to Equation 20.2. The resulting constitutive models, however, would typically be implemented in the context of equations describing heat transfer effects and either static equilibrium or structural dynamics [20].

The main conceptual steps needed to develop a constitutive theory for SMM are usually

(i) choice of the set $\mathbf{c}$ of thermal and mechanical variables that may be controlled
(ii) choice of the set $\boldsymbol{\alpha}$ of internal variables that reasonably describe the internal microstructure
(iii) specification of constitutive equations for the remaining thermomechanical variables
(iv) specification of a constitutive equation for the energetic variable consistent with (i)
(v) specification of constitutive equations for the internal variables

For example, taking $\mathbf{c} = \{\sigma, T\}$, then steps (iii) and (iv) require the specification of functions $\{\varepsilon(\mathbf{c},\boldsymbol{\alpha}), \eta(\mathbf{c},\boldsymbol{\alpha})\}$ and for the Gibbs free energy, $\psi(\mathbf{c},\boldsymbol{\alpha})$. Step (v) is generally accomplished by specifying a transformation kinetic, typically a differential equation in the form $\dot{\alpha} = f(\mathbf{c},\alpha,\dot{\mathbf{c}})$ that prescribes the evolution of the internal variables.

Thermodynamic consistency with the above balance equations requires the following equations of state for stress and temperature:

$$\varepsilon = -\frac{\partial \psi}{\partial \sigma}, \qquad \eta = -\frac{\partial \psi}{\partial T} \tag{20.3}$$

hence making step (iii) follow from step (iv). Alternatively, step (v) may be carried out by specifying a constitutive function for the rate of energy dissipation Γ. Consistency with the balance Equation 20.2 then provides the appropriate transformation kinetic, which in this case, follows as a logical consequence, rather than being separately posed at the outset [4,30]. Accordingly, each model is thus characterized by the choice of the internal variables $\boldsymbol{\alpha}$, the Gibbs free energy function ψ, and the dissipation function Γ.

In the following, the above concepts will be discussed in some detail and exemplified along the lines of Ref. [6]. To aid in the present exposition, we consider a number of simplifying assumptions on the material behavior. Specifically, we consider a SMM that has been processed so that its thermomechanical behavior permits reasonable modeling under the following specializations:

(s1) Tension–compression symmetry so that $\gamma_+ = -\gamma_- = \gamma$, $\sigma_{Rs+} = -\sigma_{Rs-} = \sigma_{Rs}$, and $\sigma_{Rf+} = -\sigma_{Rf-} = \sigma_{Rf}$
(s2) Similar elastic properties so that $E_A = E_M = E$
(s3) Aforementioned Clausius–Clapeyron relation (Equation 20.1)
(s4) Similar FwT and RvT temperature intervals $M_s - M_f \cong A_f - A_s$
(s5) Ivshin's condition for hysteresis completion
(s6) Negligible thermal expansion of the conventional kind
(s7) No phases other than austenite and martensite that contribute to the SMM effect
(s8) No slip plasticity or viscoelasticity

None of these simplifications is central to model development and the corresponding modeling generalizations associated with dropping any of these simplifications involve varying degree of additional mathematical detail. Such generalizations range from quite straightforward (e.g., for inclusion of conventional thermal expansion) to more complicated (e.g., additional phases, such as the R-phase, that contribute to the SMM effect). By first addressing models in such specialized settings, it is also easier to understand the distinction between modeling effects that are energetic (i.e., stemming from ψ) and dissipative (stemming from Γ) so as to aid in the proper construction of more generalized models. In addition to standard treatments permitting dropping any or all of (s1)–(s8), there is also a clear path to a fully three-dimensional generalization for treating multiaxial stress and strain. The interested reader may consult Refs. [4,6] for details.

20.3.4.1 Choice of the Internal Variables (Step ii)

The role of the internal variables is central to the description of hysteresis, which is a basic feature of the thermomechanical response. This is because if ψ were only dependent on σ and T, then the equations of state (Equation 20.3) would require that the same strain would be sustained by the same stress and temperature irrespective of whether this was obtained on loading or unloading. Stress–strain hysteresis during loading and unloading is due to the presence of different microstructures, and this effect is captured by the internal variables $\boldsymbol{\alpha}$. Useful choices for $\boldsymbol{\alpha}$ depend on whether the model of interest is one- or multi-dimensional and also by the kind of behaviors to be modeled.

20.3.4.1.1 *Single Internal Variable*

If only pseudoelasticity is of concern, then a single internal variable is sufficient for the basic modeling. The martensite volume fraction ξ, which necessarily obeys $0 \le \xi \le 1$, is therefore the standard choice in almost all such models. Increasing, decreasing, and unchanging ξ are then identified respectively with FwT, RvT, and no transformation (nT). Classical models are those developed by Tanaka et al. [36] and Liang and Rogers [22], which are distinguished by the particular form of the kinetic rule for the evolution of ξ. In this context, Ivshin and Pence [19] were among the first to model nonisothermal behavior associated with heat transfer and the adiabatic limit.

Several models have been proposed in this setting. Among the others, we mention those developed by Raniecki et al. [31], Auricchio et al. [1] who pay special attention to issues related to numerical implementation, and by Matsuzaki and Kamita [24]. All such groups developed various generalizations of the model presented in the referenced papers, which therefore represent just a pointer to the literature. Once again, however, there are always exceptions, as internal variable models can be formulated with some type of proxy variable that accomplishes the same effect, for example, overall martensite transformation strain as used in the model by Bondaryev and Wayman [8].

20.3.4.1.2 *Two Internal Variables*

Models in which ξ is the only internal variable cannot distinguish between different kinds of martensite (thermal vs. stress-induced, twinned vs. untwinned) and so this restricts their range of application to the case of simple pseudoelasticity. If it is necessary to model either the SME or other effects associated with reorientation, then there is the need to distinguish between different martensite microstructures during the operation of the shape memory device, from which it follows that a more refined set of internal variables is required.

Among the earliest proposals for such a refinement was to distinguish between thermally induced martensite (which has no net transformation strain) and pure stress-induced martensite (with transformation strain) by splitting ξ into ξ_{thermal} and ξ_{sim} [10]. In particular, by allowing for the conversion sequence: thermal martensite to stress induced martensite (by loading and unloading), to austenite (by heating), to thermal martensite (by cooling), such a framework provides the simplest reasonable model of the SME.

In such models, the microstructure associated with ξ_{sim} restricts consideration to either tension or compression, but not both. In order to treat both tension and compression, further refinement is necessary. In particular, by identifying the most compressive and most tensile microstructures, indicated above by M− and M+, it is natural to introduce volume fractions ξ_- and ξ_+. Further, by identifying the particular combination of M− and M+ for which there is no net transformation, it is possible to dispense with ξ_{thermal} [37]. Thus the internal variable pair $\boldsymbol{\alpha} = \{\xi_-, \xi_+\}$ provides more generality than the pair $\boldsymbol{\alpha} = \{\xi_{\text{thermal}}, \xi_{\text{sim}}\}$ in that it continues to allow for the modeling of shape memory across the full temperature range, while also permitting compression–tension modeling. Properly formulated, such $\boldsymbol{\alpha} = \{\xi_-, \xi_+\}$ models allow for tension–compression asymmetry [3].

It is, however, difficult to easily extend $\boldsymbol{\alpha} = \{\xi_-, \xi_+\}$ models to two or three dimensions. One way to make this dimensional leap is to introduce an extended variety of microstructures, e.g., $\boldsymbol{\alpha} = \{\xi_1, \xi_2, \ldots\}$. One may then seek to determine the specific fraction of each microstructure on the basis of the stress–temperature history [28,35]. Alternatively, the type of martensite present can be described directly. In a fully three-dimensional setting one may introduce a tensor internal variable $\boldsymbol{M}$ describing the deformation between austenite and a homogenization of the portion of the microstructure in the martensite phase. Modeling can then proceed with $\boldsymbol{\alpha} = \{\xi, \boldsymbol{M}\}$ in either a finite strain setting or in a setting in which the elastic strains are treated as in the conventional linear theory of elasticity [5].

Specialized to one dimension, the type of martensite is a scalar that can be identified with transformation strain, such $\boldsymbol{\alpha} = \{\xi, \boldsymbol{M}\}$ models also give rise to a broadly useful one-dimensional theory that is able to treat pseudoelasticity, shape memory, and martensite reorientation. In this case, the scalar martensite internal variable M under simplifying assumption (s1) is required to obey $-\gamma \le M \le \gamma$. Increasing, decreasing, and unchanging M are then identified respectively with FwR, RvR, and nR.

20.3.4.2 Choice of the Free Energy Function (Steps iii and iv)

For the purposes of providing an accurate and tractable model under (s1)–(s8), the case of two internal variables $\boldsymbol{\alpha} = \{\xi, M\}$ is considered. As discussed in more generality in Refs. [4,6], an appropriate form for the free energy (step iv) in this setting is

$$\psi = -\frac{1}{2E}\sigma^2 - \xi M\sigma + c\left(T - T_o - T\ln\frac{T}{T_o}\right) - \eta_{oA}T + \frac{H}{T_{eq}}\left(T - T_{eq}\right)\xi + e_{oA} - \Omega\xi(1-\xi) + \frac{1}{2}\beta\xi M^2 \tag{20.4}$$

with material constants as described below. This form may be generalized in various ways often by addition of terms. For example, addition of a term $a_{CTE}\sigma(T - T_o)$ provides the standard treatment of conventional thermal expansion. We may also mention the possibility to adding terms that depend on the strain gradient as in Ref. [34] so as to allow for the modeling of localization phenomena.

The constitutive equations for ε, η (step iii) now follow from the Gibb's free energy $\psi(\sigma, T, \xi, M)$ via Equation 20.3 as

$$\varepsilon = \frac{1}{E}\sigma + \xi M, \quad \eta = c\ln\frac{T}{T_o} - \frac{H}{T_{eq}}\xi + \eta_{oA} \tag{20.5}$$

As indicated earlier E is the elastic modulus, c the specific heat, and H the latent heat of transformation. Morevoer T_o is any chosen reference temperature, whereas T_{eq} is an "equilibrium temperature" at which the pure phases of austenite and martensite have the same free energy. The term ξM gives the transformation strain due to the presence of martensite. In particular, for thermal martensite, it is the case that $M = 0$ whereupon there is no transformation strain irrespective of the martensite fraction ξ. The constants e_{oA} and η_{oA} give the SMM internal energy and the SMM entropy of the austenite phase at T_o when the stress vanishes. The values of e_{oA} and η_{oA} are not of consequence, since the entropy and the internal energy only enter the treatment in a natural way via their changes. From a practical point of view, both of them can be ignored (e.g., set them to zero in any calculation). There are thus two remaining material constants in ψ, namely Ω and β. An indication of their role follows upon consideration of the hypothetical state of an energy minimizing equilibrium as discussed next.

20.3.4.2.1 *Energy Minimization*

Because of hysteresis, there may be many equilibrium states consistent with a given stress and temperature, most of which are not energy minimizing. The actual equilibrium state achieved in the operation of an SMM device depends on the history of the controllable variables and so will typically not coincide with an energy minimal state. Energy minimal states, however, lie between the hysteresis extremes and so provide a natural reference state. Furthermore, the energy minimal states provide the basic form of the (σ,T) state diagram.

The hypothetical energy minimizing equilibrium state for a given (σ, T) is determined by the pair (ξ, M) that minimizes $\psi(\sigma, T, \xi, M)$ under the simultaneous constraints $0 \le \xi \le 1$ and $-\gamma \le M \le \gamma$. Requiring the vanishing of the appropriate derivatives and solving for the internal variables then gives that

$$\begin{aligned} M &= \hat{M}(\sigma) = \frac{\sigma}{\beta} \\ \xi &= \hat{\xi}(\sigma, T, M) \\ &= \frac{1}{2} - \frac{1}{2\Omega}\left(\frac{H\left(T - T_{eq}\right)}{T_{eq}} - M\sigma + \frac{\beta}{2}M^2\right) \end{aligned} \tag{20.6}$$

The first equation shows that the orientation of the martensite, as given by M, depends solely on the stress, and is determined by proportionality factor $1/\beta$. Hence β may be regarded as a reorientation modulus. On the other hand, it follows from the second equation in Equation 20.6 that the amount of martensite ξ is proportional to temperature and, for positive H and Ω, that heating yields more austenite whereas cooling promotes more martensite. In particular, Ω is a coefficient of martensite formation. Substitution of $M = \hat{M}(\sigma)$ into $\hat{\xi}\left(\sigma, T, \hat{M}(\sigma)\right)$ gives a stress dependence in the form $\sigma^2/4\Omega\beta$ so that stress of either sign promotes martensite formation.

Equation 20.6 does not acknowledge the constraints $0 \le \xi \le 1$ and $-\gamma \le M \le \gamma$. Enforcing the constraints now gives the precise description of the energy minimal values of ξ and M. Specifically

$$M = \begin{cases} -\gamma, & \text{if} \quad \hat{M}(\sigma) < -\gamma \\ \hat{M}(\sigma), & \text{if} \quad -\gamma \le \hat{M}(\sigma) \le \gamma \\ \gamma, & \text{if} \quad \hat{M}(\sigma) > \gamma \end{cases} \tag{20.7}$$

Then, using this M, the energy minimal value of ξ is

$$\xi = \begin{cases} 0, & \text{if} \quad \hat{\xi}(\sigma, T, M) < 0 \\ \hat{\xi}(\sigma, T, M), & \text{if} \quad 0 \le \hat{\xi}(\sigma, T, M) \le 1 \\ 1, & \text{if} \quad \hat{\xi}(\sigma, T, M) > 1 \end{cases} \tag{20.8}$$

In this energy minimal treatment, thermal martensite (ξ when $\sigma = 0$) is formed on the temperature interval $T_{eq} - \Omega T_{eq}/H \le T \le T_{eq} + \Omega T_{eq}$ H. Note also that a pure martensite microstructure holds at sufficiently low temperature ($\xi = 1$ if $T \le T_{eq} - \Omega T_{eq}/H$ for any σ). However, at high temperatures, it is always possible to convert austenite to martensite with sufficiently large stress magnitude (for both positive and negative σ). If $\sigma \ge \beta\gamma$, so that $M = \gamma$, it then follows that a constant value of ξ is maintained on lines in the (σ,T)-plane with slope $dT/d\sigma = \gamma T_{eq}/H$. This is the Clausius–Clapeyron equation, a result ensured by the overall thermodynamic consistency of the modeling framework. The more general Clausius–Clapeyron relation is also true: $dT/d\sigma = MT_{eq}/H$.

20.3.4.3 Choice of the Dissipation Function (Step v) and Derivation of the Transformation Kinetics

The energy minimal framework provides all the correct trends; it therefore remains to augment the framework with an appropriate description of hysteresis. Hysteresis causes the actual microstructure to lag behind that which is associated with energy minimization. Alternatively, to achieve particular values of the internal variables $\boldsymbol{\alpha} = \{\xi, M\}$, it is generally necessary to drive (σ, T) beyond that associated with energy minimization, because of the internal friction, which opposes microstructural change. This amounts to a requirement that the thermodynamic forces driving the transformations must attain a nonzero threshold in order to activate a change in the microstructure. In particular, the derivatives of the free energy with respect to the internal variables are precisely these driving forces, with $\Pi_\xi = -\partial\psi/\partial\xi$ driving the martensitic transformation and $\Pi_M = -\partial\psi/\partial M$ driving reorientation. With the expression (Equation 20.4) for ψ, it follows that

$$\begin{aligned}\Pi_\xi(\sigma,T,\xi,M) &= M\sigma - \frac{H}{T_{\text{eq}}}(T - T_{\text{eq}}) + \Omega(1-2\xi) - \frac{1}{2}\beta M^2 \\ \Pi_M(\sigma,\xi,M) &= (\sigma - \beta M)\xi\end{aligned} \tag{20.9}$$

It will be shown below that, whereas the energy minimal microstructures are formed under zero driving forces, in the presence of energy dissipation the martensite fraction ξ changes if and only if Π_ξ both attains and sustains a threshold level denoted by Λ_ξ. This threshold represents the internal friction associated with the movement of the phase boundaries at the microscale. Similarly, martensite orientation M changes if and only if Π_M both attains and sustains a corresponding threshold Λ_M.

This can be modeled by prescribing a constitutive function for the rate of energy dissipation Γ. As shown in Ref. [4], SMM models can be developed by a prescription in the form:

$$\Gamma = \Lambda_\xi \dot{\xi} + \Lambda_M \dot{M}, \tag{20.10}$$

where Λ_ξ and Λ_M are constitutive functions that, as shown below, express the abovementioned thresholds for the driving forces. In the most general case, such functions may depend on the current values of σ, T, ξ, M as well as the previous history of microstructural states via past values of the internal variables. Any specification of such functions, combined with the free energy function ψ, yields a specific model. A simple choice that yields a model of pseudoelasticity, shape memory, and reorientation effects under the assumptions listed above, is the following:

$$\begin{aligned}\Lambda_\xi &= \begin{cases} \Lambda_{\xi\uparrow} = \kappa_\xi \left(\dfrac{1 - 2\xi_0 + \xi_0\xi}{1 - \xi_0} \right) \geq 0, & \text{if} \quad \dot{\xi} > 0 \\ \Lambda_{\xi\downarrow} = -\kappa_\xi \left(\dfrac{\xi_0\xi + \xi_0 - \xi}{\xi_0} \right) \leq 0, & \text{if} \quad \dot{\xi} < 0 \end{cases} \\ \Lambda_M &= \begin{cases} \Lambda_{M\uparrow} = \kappa_M \xi \geq 0, & \text{if} \quad \dot{M} > 0 \\ \Lambda_{M\downarrow} = -\kappa_M \xi \leq 0, & \text{if} \quad \dot{M} < 0 \end{cases}\end{aligned} \tag{20.11}$$

where, as in the earlier sublooping discussion, ξ_0 denotes the fraction of martensite at the end of the last transformation process. Here κ_ξ and κ_M are material constants, which must be nonnegative so as to provide consistency with the Clausius–Duhem second law inequality. The quotient form of Λ_ξ is a consequence of the Ivshin hysteresis condition (s5), as meeting this condition requires that $\Lambda_{\xi\uparrow}$ is independent of ξ_0 when $\xi = 1$, and that $\Lambda_{\xi\downarrow}$ is independent of ξ_0 when $\xi = 0$.

There are nine combined kinetic direction possibilities for $\boldsymbol{\alpha} = \{\xi, M\}$ corresponding to the product of the three transformation possibilities: FwT, RvT, nT, with the three reorientation possibilities: FwR, RvR, nR. Eliminating the heat supply r between the balance equations for energy and entropy (Equation 20.2) and using Equation 20.3, Equation 20.10 leads to

$$(\Lambda_\xi - \Pi_\xi)\dot{\xi} + (\Lambda_M - \Pi_M)\dot{M} = 0 \tag{20.12}$$

One immediate consequence is that an assumption of zero dissipation ($\Lambda_\xi = 0$ and $\Lambda_M = 0$) leads to the theory of energy minimal microstructure discussed above. More generally, the sign conditions on Λ_ξ and Λ_M ensure that hysteresis results from the need to overdrive the system so as to overcome frictional resistance. On this basis, one may now extract the conditions distinguishing between FwT, RvT, or nT and the conditions distinguishing between FwR, RvR, or nR.

For FwT the conditions that must be satisfied are:

$$\xi < 1, \quad \frac{\partial \Pi_\xi}{\partial \sigma}\dot{\sigma} + \frac{\partial \Pi_\xi}{\partial T}\dot{T} > 0, \quad \text{and} \quad \Pi_\xi = \Lambda_{\xi\uparrow} \tag{20.13}$$

These correspond respectively to requirements that there is (a) room for additional change to ξ, (b) a control variable path direction on the state diagram that is conducive to change, and (c) sufficient driving force to overcome the frictional dissipation. In view of the expressions for Λ_ξ and Π_ξ, one thus finds that the conditions for FwT are $\xi < 1$, $M\dot{\sigma} - H\dot{T}/T_{\text{eq}} > 0$, along with

$$\begin{aligned}& M\sigma - \frac{H}{T_{\text{eq}}}(T - T_{\text{eq}}) + \Omega(1-2\xi) - \frac{1}{2}\beta M^2 \\ &= \kappa_\xi \left(\frac{1 - 2\xi_0 + \xi_0\xi}{1 - \xi_0} \right)\end{aligned} \tag{20.14}$$

The conditions for RvT are: $\xi > 0$, $M\dot{\sigma} - H\dot{T}/T_{eq} < 0$, and

$$\begin{aligned}& M\sigma - \frac{H}{T_{\text{eq}}}(T - T_{\text{eq}}) + \Omega(1-2\xi) - \frac{1}{2}\beta M^2 \\ &= -\kappa_\xi \left(\frac{\xi_0\xi + \xi_0 - \xi}{\xi_0} \right)\end{aligned} \tag{20.15}$$

If there is neither FwT nor RvT, then there is nT.

The condition for FwR is $M < \gamma$, $\dot{\sigma} > 0$, and

$$\sigma - \beta M = \kappa_M \tag{20.16}$$

The condition for RvR is $M > -\gamma$, $\dot{\sigma} < 0$, and

$$\sigma - \beta M = -\kappa_M \tag{20.17}$$

If there is neither FwR nor RvR, then there is nR.

20.3.4.3.1 *Pseudoelasticity*

In this framework, pure pseudoelasticity corresponds to either (FwT, nR) or (RvT, nR). The standard example involves loading at temperatures above A_f with either $M=\gamma$ or $M=-\gamma$. If the inequality conditions for FwT are met, then the equality condition (Equation 20.14) determines the control variable pair (σ, T) associated with activating FwT. Meeting Equation 20.14 serves to define the function $\sigma_{Ms}(T, \xi_0)$ that describes activation triggering via $\sigma = \sigma_{Ms}(T, \xi_0)$. Prior to activation $\xi = \xi_0$, but after activation, ξ will increase while ξ_0 remains fixed. Similar considerations govern RvT with Equation 20.15 replacing Equation 20.14. Triggering of either transformation requires a value Π_ξ that is proportional to the amount of transforming (diminishing) phase, i.e., $\Pi_\xi = \kappa_\xi(1-\xi_0)$ for activation of FwT, and $\Pi_\xi = -\kappa_\xi \xi_0$ for activation of RvT. Continued transformation requires increasing $|\Pi_\xi|$. Transformation completion then requires Π_ξ to attain the value κ_ξ for FwT and $-\kappa_\xi$ for RvT.

During FwT, the increase of ξ occurs so as to maintain Equation 20.14. For $\xi_0 = 0$, this gives the outer loop of the stress–strain curve, while a subloop is obtained for $0 < \xi_0 < 1$. The FwT evolution of ξ can be obtained by solving Equation 20.14 for ξ algebraically. In numerical routines, it would be typical to compute this evolution via the rate form of Equation 20.14, that is

$$\dot{\xi} = \left(2\Omega + \left(\frac{\xi_0}{1-\xi_0}\right)\kappa_\xi\right)^{-1}\left(-\frac{H}{T_{eq}}\dot{T} + M\dot{\sigma}\right) \tag{20.18}$$

Reversing the path of the control variables, e.g., by unloading, then activates RvT if the driving force Π_ξ attains the corresponding threshold $\Lambda_{\xi\downarrow}$. It then follows from Equation 20.15 that (RvT, nR) occurs according to the kinetics

$$\dot{\xi} = \left(2\Omega + \left(\frac{1-\xi_0}{\xi_0}\right)\kappa_\xi\right)^{-1}\left(-\frac{H}{T_{eq}}\dot{T} + M\dot{\sigma}\right) \tag{20.19}$$

20.3.4.3.2 *Reorientation*

Pure reorientation corresponds to (nT, FwR) or (nT, RvR). The most important physical case of pure reorientation is that which occurs when the material is fully martensitic ($\xi = 1$), for example a conversion of thermal martensite $M=0$ to the maximal stress-induced martensite $M=\gamma$ at temperatures below M_f. Observing, however, that the conditions governing FwR and RvR are independent of temperature, it follows that the variable M can evolve at any temperature. When this evolution takes place with $\xi=0$, the change is not expressed physically (this is because each appearance of M in ψ involves a multiplication by ξ), however such "virtual reorientation" is central to the mathematical model. For example at $T > A_f$, this virtual reorientation sets $M = \gamma$ for tensile loading and unloading, and sets $M = -\gamma$ for compressive loading and unloading. This setting and resetting of M occurs during the elastic loading and unloading in the austenite phase, and so does not have an effect before the onset of FwT.

In the temperature range $M_f < T < A_f$, one may encounter the four cases (FwT, FwR), (FwT, RvR), (RvT, FwR), and (RvT, RvR). In these cases, one of the equality conditions, Equation 20.14 or Equation 20.15, occurs in tandem with one of the equality conditions, Equation 20.16 or Equation 20.17, so as to give two equations for ξ and M in terms of σ, T, and ξ_0. Once again, the rate form may be practical for numerical computation. For example, for (FwT, FwR), this kinetic rate form yields

$$\dot{\xi} = \left(2\Omega + \left(\frac{\xi_0}{1-\xi_0}\right)\kappa_\xi\right)^{-1}\left(-\frac{H}{T_{eq}}\dot{T} + \frac{\sigma}{\beta}\dot{\sigma}\right), \quad \dot{M} = \frac{1}{\beta}\dot{\sigma} \tag{20.20}$$

20.3.4.4 Material Parameter Identification

The model parameters E, γ, c, and H are typically available from measurement, whereas the model parameter T_0 is chosen for convenience on the basis of a standard operating temperature, and model parameters η_{0A} and e_{0A} play no role. This leaves the specification of model parameters: T_{eq}, Ω, β, κ_M, and κ_ξ. Under the simplifying conditions (s1)–(s8) listed earlier, these may be taken as follows:

$$\begin{aligned} T_{eq} &= \frac{1}{4}(A_f + A_s + M_s + M_f) \\ \Omega &= \left(\frac{A_f - A_s + M_s - M_f}{A_f + A_s + M_s + M_f}\right)H \\ \beta &= \frac{1}{\gamma}(\sigma_{Rf} - \sigma_{Rs}),\ \kappa_M = \sigma_{Rs} \\ \kappa_\xi &= \left(\frac{A_f + A_s - M_s - M_f}{A_f + A_s + M_s + M_f}\right)H \end{aligned} \tag{20.21}$$

Observe that Ω and κ_ξ are each a multiple of $H>0$ whereupon the ordering $M_f < M_s < A_s < A_f$ ensures that $\Omega > 0$ and $\kappa_\xi > 0$.

It is instructive to consider the more general thermodynamically admissible conditions: $M_f \le M_s \le A_f$ and $M_f \le A_s \le A_f$, which still yield $\Omega \ge 0$ and $\kappa_\xi \ge 0$. Here we temporarily allow the possibility of equality in the various relations so as to remark on some special cases. The extremely special case $M_f = M_s = A_s = A_f = T_{eq}$ then corresponds to abrupt thermal transformations A $\rightarrow$ M and M $\rightarrow$ A without any hysteresis. In this case, both $\Omega = 0$ and $\kappa_\xi = 0$. The ξ versus T graph of an abrupt, hysteresis-free transformation is a step function at the common transformation temperature. This step can be modified, or "unfolded," in two canonical ways: (u1) by making it a ramp, and (u2) by splitting it into two separate vertical steps one to be traveled for A $\rightarrow$ M and the other to be traveled for M $\rightarrow$ A. The modification (u1) corresponds to a gradual but still hysteresis-free transformation and can be regarded as a pure coexistence unfolding. The modification (u2) corresponds to two abrupt transformations separated by a hysteresis interval and can be regarded as a pure hysteresis unfolding. The pure hysteresis unfolding (u2) corresponds to taking $M_f = M_s < A_s = A_f$ in which case $\Omega = 0$ and $\kappa_\xi > 0$. The pure coexistence unfolding (u1) correspond to taking $M_f = A_s < M_s = A_f$ in which case $\Omega > 0$

and $\kappa_\xi = 0$. In this sense, Ω is a direct indicator of phase coexistence and κ_ξ is a direct indicator of hysteresis. In view of the observed ordering, $M_f < M_s < A_s < A_f$, it then follows that $\kappa_\xi > \Omega > 0$, indicating the central role of hysteresis in SMM behavior.

20.3.5 Summary and Concluding Remarks

In this contribution we have attempted to provide an overview of the basic modeling issues, and a necessarily simple review of the many modeling approaches that describe SMMs. We have also provided a tractable model for both phase transformation and reorientation appropriate to uniaxial deformation. As shown by Equation 20.21, this model provides a simple correspondence between physical parameters and modeling parameters. It is to be emphasized that this model naturally generates a state diagram, whereupon the issue arises as to how to insure that such a diagram matches experiment. In this regard, the model construction was aided by the simplifying assumptions (s1)–(s8) although, as emphasized throughout, more general situations can be treated by a suitable extension of the basic formulae provided here. The corresponding generalizations require varying levels of additional mathematical framework within the model. For example, replacing Equation 20.11 by a more involved expression allows one to fit arbitrary transformation temperatures that are no longer bound by the requirement (s4). Such replacements also allow for the treatment of more subtle issues, such as the specification of the slopes ρ_r and ρ_f appearing in Figure 20.22. By such means, one can construct suitably faithful state diagrams, replicate complicated sublooping procedures, predict dissipation and damping, and more generally design shape memory devices and structures that perform as intended when incorporated into larger systems.

References

1. Auricchio, F., Taylor, R.L., and Lubliner, J. 1997. Shape-memory alloys: Macromodelling and numerical simulations of the superelastic behavior, *Comp. Meth. Appl. Mech. Eng.*, 146, 281–312.
2. Ball, J.G. and James, R.D. 1987. Fine phase mixtures as minimizers of energy, *Arch. Rat. Mech. Anal.*, 100, 13–52.
3. Bernardini, D. and Pence, T.J. 2002. Shape memory alloys: modeling, In: *Encyclopedia of Smart Materials*, vol. 2, pp. 964–980, Wiley, New York.
4. Bernardini, D. and Pence, T.J. 2002. Models for one-variant shape memory materials based on dissipation functions, *Int. J. Nonlinear Mech.*, 37, 1299–1317.
5. Bernardini, D. and Pence, T.J. 2003. A multifield theory for the modeling of the macroscopic behavior of shape memory materials, In: *Advances in Multifield Theories of Continua with Substructure*, Capriz G. and Mariano P.M. (Eds.), pp. 199–242, Birkhauser, Boston, MA.
6. Bernardini, D. and Pence, T.J. 2005. Uniaxial modeling of multivariant shape-memory materials with internal sublooping using dissipation functions, *Meccanica*, 40, 339–364.
7. Bhattacharya, K. and Kohn, R.V. 1997. Elastic energy minimization and the recoverable strains of polycrystalline shape-memory materials, *Arch. Rat. Mech. Anal.*, 139, 99–180.
8. Bondaryev, E.N. and Wayman, C.M. 1988. Some stress–strain–temperature relationships for shape memory alloys, *Metallur. Trans. A*, 19A, 2407–2413.
9. Boyd, J.G. and Lagoudas, D.C. 1996. A thermodynamical constitutive model for shape memory materials. Part I. The monolithic shape memory alloy, *Int. J. Plasticity*, 12, 805–842.
10. Brinson, L.C. 1993. One-dimensional constitutive behavior of shape memory alloys: Thermomechanical derivation with non-constant material functions and redefined martensite internal variable, *J. Intell. Mater. Syst. Struct.*, 4, 229–242.
11. Chang, B.C., Shaw, J.A., and Iadicola, M.A. 2006. Thermodynamics of shape memory alloy wire: Modeling, experiments and applications, *Cont. Mech. Therm.*, 18, 83–118.
12. De La Flor, S., Urbina, C., and Ferrando, F. 2006. Constitutive model of shape memory alloys; theoretical formulation and experimental validation, *Mat. Sci. Eng A.*, 427, 112–122.
13. Elahina, M.H. and Ahmadian, M. 2005. An enhanced SMA phenomenological model: I. The shortcomings of the existing models, *Smart Mater. Struct.*, 14, 1297–1308.
14. Falk, F. 1980. Model, free energy, mechanics and thermodynamics of shape memory alloys, *Acta Metallur.*, 28, 1773–1780.
15. Gao, X., Huang, M., and Brinson, L.C. 2000. A multivariant micromechanical model for SMAs. Part 1. Crystallographic issues for single crystal model, *Int. J. Plasticity*, 16, 1345–1369.
16. Green, A.E. and Naghdi, P.M. 1977. On thermodynamics and the nature of the second law, *Proc. Roy. Soc. Lond. A*, 357, 253–270.
17. Ikeda, T., Nae, F.A., Naito, H., and Matsuzaki, Y. 2004. Constitutive model for of shape memory alloys for unidirectional loading considering inner hysteresis loops, *Smart Mater. Struct.*, 13, 916–925.
18. Ivshin, Y. and Pence, T.J. 1994a. A constitutive model for hysteretic phase transitions, *Int. J. Eng. Sci.*, 32, 681–704.
19. Ivshin, Y. and Pence, T.J. 1994b. A thermomechanical model for a one-variant shape memory material, *J. Intell. Mat. Syst. Struct.*, 5, 455–473.
20. Lacarbonara, W., Bernardini, D., and Vestroni, F. 2004. Nonlinear thermomechanical oscillations of shape memory devices, *Int. J. Solids Struct.*, 41, 1209–1234.
21. Lagoudas, D.C., Entchev, P.B., Popov, P., Patoor, E., Brinson, L.C., and Gao, X. 2006. Shape memory alloys, Part II: Modeling of polycrystals, *Mech. Materials*, 38, 430–462.
22. Liang, C. and Rogers, C.A. 1990. One-dimensional thermomechanical constitutive relations for shape memory materials, *J. Intell. Mat. Syst. Struct.*, 1, 207–234.
23. Likhacev, A.A. and Koval, Y.N. 1992. On the differential equation describing the hysteretic behavior of shape memory alloys, *Scrip. Metallur. Mater.*, 27, 223–227.
24. Matsuzaki, Y. and Kamita, T. 1998. One-dimensional pseudoelastic theory of shape memory alloys, *Smart Mater. Struct.*, 489–495.
25. Muller, I. and Seelecke, S. 2001. Thermodynamic aspects of shape memory alloys, *Math. Computer Modelling*, 34, 1307–1355.
26. Ortin, J. 1991. Preisach modeling of hysteresis for a pseudoelastic Cu-Zn-Al single crystal, *J. Appl. Phys.*, 71, 1454–1461.

27. Otsuka, K. and Shimizu, K. 1986. Pseudoelasticity and shape memory effects in alloys, *Int. Metal Rev.*, 31, 93–114.
28. Patoor, E., Eberhardt, A., and Berveiller, M. 1988. Thermomechanical behavior of shape memory alloys, *Arch. Mech.*, 40, 755–794.
29. Puglisi, G. and Truskinovsky, L. 2005. Thermodynamics of rate-independent plasticity, *J. Mech. Phys. Solids*, 53, 655–679.
30. Rajagopal, K.R. and Srinivasa, A.R. 1999. On the thermomechanics of shape memory wires, *Z. Angew. Math. Phys.*, 50, 459–496.
31. Raniecki, B., Lexcellent, C., and Tanaka, K. 1992. Thermodynamic models for pseudoelastic behavior of shape memory alloys, *Arch. Mech.*, 44, 261–288.
32. Roubicek, T. 2004. Models of microstructure evolution in shape memory alloys, In: *Nonlinear Homogenization and Its Applications to Composites, Polycrystals and Smart Materials*, Ponte Casteneda P., Telega J.J., and Gamblin B. (Eds.), NATO Science Series, Math. Phys. Chem. 170, pp. 65–106, Kluwer.
33. Shaw, J.A. and Kyriakides, S. 1995. Thermomechanical Aspects of NiTi, *J. Mech. Phys. Solids*, 43, 1243–1281.
34. Shaw, J.A. 2002. A thermomechanical model for a 1-D shape memory alloy wire with propagating instabilities, *Int. J. Solids Struct*, 39, 1275–1305.
35. Siredey, N., Patoor, E., Berveiller, M., and Eberhardt, A. 1999. Constitutive equations for polycrystalline thermoelastic shape memory alloys. Part I. Intragranular interactions and behavior of the grain, *Int. J. Solids Struct.*, 36, 4289–4315.
36. Tanaka, K., Kobayashi, S., and Sato, Y. 1986. Thermomechanics of transformation pseudoelasticity and shape memory effect in alloys, *Int. J. Plasticity*, 2, 59–72.
37. Wu, X. and Pence, T.J. 1998. Two variant modeling of shape memory materials: Unfolding a phase diagram triple point, *J. Intell. Mat. Syst. Struct.*, 9, 335–354.

20.4 Shape Memory Alloys

Jan Van Humbeeck

SMAs are functional or adaptive materials with the following properties: superelastic behavior, SME (reversible strains of several percent during heating or cooling over a limited temperature range), generation of high recovery stresses, a work output with a high power and weight ratio, and a high damping capacity.

All these properties are more extensively described in Part 3.

20.4.1 SMAs and the Martensitic Transformation [1–4,36]

The distinctive functional properties of SMAs are closely linked to a solid–solid phase transformation occurring in a metastable solid state of some specific alloys. It occurs diffusionless and is therefore called martensitic. The solid phase obtained during cooling is called martensite. The parent phase in which the transformation occurs is called the beta (β)-phase. The temperature at which the transformation occurs can be chosen in a temperature range between −150°C and 200°C, depending on the composition and the microstructural constitution, the latter determined mainly by the thermomechanical processing.

The temperature-induced transformation is characterized by four temperatures: M_s and M_f during cooling and A_s and A_f during heating. M_s and M_f indicate the temperatures at which the transformation from the parent phase into martensite respectively starts and finishes, as illustrated in Figure 20.26.

A_s and A_f indicate respectively the temperatures at which the reverse transformation (martensite to beta) starts and finishes. The overall transformation describes a hysteresis of the order of 10–50°C. At a temperature just above the A_f temperature (limited to a few tens of °C), the martensitic phase can also be induced by straining the sample, as illustrated in Figure 20.27.

After reaching a critical stress, called $\sigma^{p\text{-}m}$, the sample will start to transform to martensite. The value of $\sigma^{p\text{-}m}$ increases linearly with temperature (averagely 5 MPa K^{-1} for polycrystalline material), starting from zero at the M_s temperature. The relation between the stress at which martensite is induced and the temperature is given by the Clausius–Clapeyron equation. During further straining, the stress at which the transformation occurs is almost constant until the material is fully transformed. Further straining will now lead to elastic loading of the martensite, followed by plastic deformation. When the strain is limited to the start of the elastic loading of the martensite and the applied stress is released again, the reverse transformation will occur at a lower stress level as during loading, leading to the reverse movement of the first induced strain. The material is said to be superelastic. Reversible strains up to 8% of the initial length can be obtained, compared to the 0.2% elastic strain of normal metallic material. Since this superelastic effect is due to a phase transformation, this effect is also

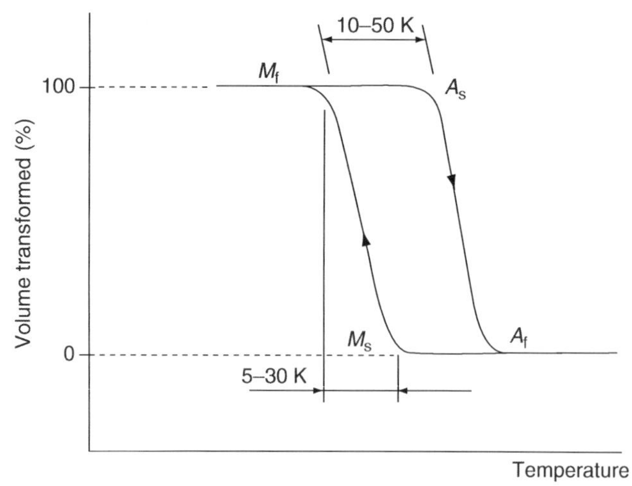

FIGURE 20.26 Schematic representation of the volume transformed as a function of temperature.

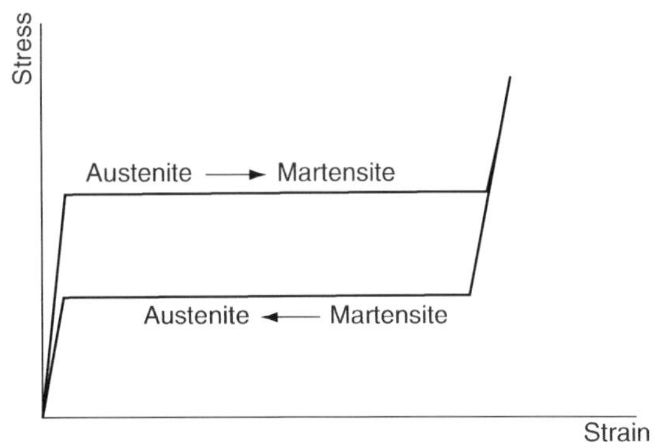

FIGURE 20.27 Superelastic behavior at constant temperature due to stress-induced transformation.

called many times such as pseudoelasticity, pseudoelastic effect, or superelasticity. The reverse deformation occurs at a lower stress plateau as during loading and shows thus an hysteresis, analogous to the temperature-induced transformation, but with a size of 50–300 MPa.

It follows that the volume transformed fraction due to temperature- or stress-induced transformation cannot be characterized by a single value of temperature or stress. It is also important to notice that many characteristics such as Young's modulus and electrical resistivity can change drastically during transformation.

The origin of the functional properties lies in the specific microstructure of the martensite. A single crystal of the parent phase (i.e., a grain) transforms to a collection of martensite crystals, called variants, which are separated by interfaces. They grow thermoelastically in a self-accommodating way. When this microstructure is deformed, reorientation of the variants occurs by displacement of the interfaces, in such a way that the variants, which lattice shear accommodates, the applied strain will be favored at the expense of the other less favored oriented variants. This occurs (principally) without plastic deformation.

The strain obtained during reorientation is thus limited. In the ultimate case, a single crystal (variant) of martensite is obtained. Further straining will lead to plastic deformation, hampering the functional properties. When a reoriented material is heated above its transformation temperatures, it will transform to the original parent phase orientation, giving rise to the so-called SME.

During pseudoelastic deformation, the variants will be introduced, which accommodate the applied strain until the whole sample has been transformed into martensite, after which classic (elastic + plastic) deformation of the sample occurs.

20.4.2 SMA Systems [5,6]

Many systems exhibit a martensitic transformation. Generally they are subdivided in ferrous and nonferrous martensites. They are summarized in Tables 20.2 and 20.3. From all systems mentioned in both tables, in fact only one major system became successful in industrial application: NiTi(X,Y) in which X,Y are elements replaced by Ni or Ti. Besides the NiTi system, a lot of attention has been given in earlier times to Cu-based alloys and to Fe–Mn–Si-based alloys. Furthermore, during recent years, special attention has been given to high-temperature SMAs (HTSMA).

TABLE 20.2 Classification of the Nonferrous Martensites

Group	Alloy System
Terminal solid solutions based on an element having allotropic phases	Cobalt and its alloys Rare earth metals and their alloys Titanium, zirconium and their alloys Alkali metals and their alloys and thallium Others as Pu, Ur, Hg, and alloys
Intermetallic solid solutions with a bcc-parent phase	β-Hume–Rothery phases of the Cu-, Ag-, Au-based alloys β-Ni–Al alloys Ni-Ti-X alloys
Alloys showing cubic to tetragonal trans. (incl. quasi-martensite)	Indium-based alloys Manganese based alloys (paramagnetic ↔ antiferromagnetic) A15-compounds Others: Ru-Ta, Ru-Nb, Y-Cu, LaCd, $LaAg_x$–In_{1-x}

Source: From Delaey, L., Chandrasekaran, M., Andrade, M., and Van Humbeeck, J., *Proceedings of the International Conference on Solid-Solid Phase Transformation*, Met. Society of AIME, 1982, p. 1429.

20.4.2.1 Fe-Based Alloys [6–10,37]

In ferrous alloys, the austenite (FCC-γ phase) can be transformed to three kinds of martensites depending on composition or stress: γ-α′ (BCC), γ → ε (HCP), and γ → FCT martensite. Although a SME has been observed for all three

TABLE 20.3 Ferrous Alloys Which Exhibit a Complete or Nearly Complete SME

Alloy	Composition	Crystal Structure of Martensite	Nature of Transformation[a]
Fe–Pt	≈ 25 at.% Pt	bct (α′)	TE
	≈ 25 at.% Pt	fct	TE
Fe–Pd	≈ 30 at.% Pd	fct	TE
Fe–Ni–Co–Ti	23%Ni–10%Co–4%Ti	bct (α′)	—
	33%Ni–10%Co–4%Ti	bct (α′)	TE
Fe–Ni–C	31%Ni–0.4%C	bct (α′)	Non-TE
Fe–Mn–Si	30%Mn–1%Si	hcp (ε)	Non-TE
	28%–33%Mn–4%–6%Si	hcp (ε)	Non-TE

Source: From Maki, T. and Tamura, T., *Proc. ICOMAT-86*, Japanese Institute of Metals, 1986, p. 963.

[a] TE, thermoelastic martensite; non-TE, nonthermoelastic martensite.

types of transformation, most attention to develop a commercial alloy has been given to the alloys with a $\gamma \rightarrow \varepsilon$ transformation. These alloys have a low stacking fault energy of austenite (Fe–Cr–Ni, Fe-high Mn alloys). The austenite to ε-martensite transformation proceeds by the $a/6$ [112] Schockley partial dislocations trailing a stacking fault ribbon on every {111} austenite plane and changing the crystal structure to martensite. The SME, which is of the one-way type, mainly results from a reverse motion of the Schockley partial dislocations during heating.

Any factor impeding the reversibility of the motion of partial dislocations will lead to an incomplete recovery and in turn to a poor SME. On an average, a SME of about 3% can be obtained with recovery stresses of a few 100 MPa.

The internal factors controlling the recovery are alloy composition, Néel temperature, transformation temperature, lattice defects, among others. External factors are applied stress and strain, deformation, recovery annealing temperature, and thermomechanical treatment.

20.4.2.2 Cu-Based Alloys [5,11–15]

Copper-based SMAs are mainly derived from Cu–Zn and Cu–Al binary systems. Both show a martensitic transformation. The composition range of these alloys corresponds to that of the well-known β Hume–Rothery phase. In most SMAs, this phase has a disordered bcc structure at high temperatures but orders to a B2, $D0_3$ or $L2_1$ form at lower temperatures. The shear elastic constant of the β-phase exhibits an anomalous behavior with decreasing temperature, i.e., it is lowered till the lattice instability with respect to {110}<110> shears at some temperature transforms β-phase to martensite. The temperature of transformation to martensite, M_s, varies with the alloy composition. The elastic anisotropy of the β-phase is much higher compared to normal metals and alloys and increases further as the martensitic transformation is approached.

Cu–Zn and Cu–Al martensites are of three types: α', β', or γ' with subscript 1, 2, or 3 added to indicate the ordered lattices in β, viz. B2 (2) or $D0_3$ (1) or L2 (3). Some conversion from one martensite structure to another, e.g., $\beta' \rightarrow \gamma'$ may also take place. The net result is a coalescence of plates within a self-accommodating group and even coalescence of groups. Heating this deformed martensite microstructure transforms it to the β-phase with the SME accompanying the structural change.

Table 20.4 summarizes three systems that were once considered as potential SMA, but so far only with limited success. The main problems encountered are the limited ductility due to a rather large grain size and large crystal anisotropy, stabilization of martensite as function of time and temperature, poor corrosion resistance, underperformance for functional properties when compared to NiTi.

20.4.2.3 Ni–Ti Alloys [4,7,16]

$Ni_{50}Ti_{50}$ and the near equiatomic Ni–Ti alloys are the best explored system of all shape memory alloys and occupies almost the whole market of SMA. $Ni_{50}Ti_{50}$ is an intermetallic phase that has some solubility at higher temperature.

The science and technology of Ni–Ti is overwhelmingly documented. The influence of composition and thermomechanical processing on the functional properties is well understood and described in literature. The basic concept of processing Ni–Ti alloys is that in order to avoid plastic deformation during shape memory or pseudoelastic loading, the martensitic and the β-phase have to be strengthened. This occurs by classic methods: strain hardening during cold deformation, solution hardening, and precipitation hardening. Ni–Ti alloys have the significant advantage that these techniques can be easily applied due to an excellent ductility and a very interesting but complicated precipitation process (Table 20.5).

The compositions of the Ni–Ti–SMA are approximately between 48 and 52 at.% Ni and the transformation temperatures of the B2 structure to the martensitic phase with a monoclinic B19′ structure are very sensitive to the nickel content (a decrease of about 150° for an increase of 1 at.% Ni). The transformation temperatures can be chosen between −40°C and +100°C.

Ni–Ti alloys show the best shape memory behavior of all SMAs. Even in polycrystalline state, 8% shape recovery is possible and 8% pseudoelastic strain is completely reversible above A_f, while the recovery stress is of the order of 800 MPa.

In some cases, the martensitic transformation is preceded by the so-called R-phase transition. The R-transition is a B2 $\leftrightarrow$ rhombohedral transformation that has also second-order characteristics.

The most specific characteristics of this R-phase transition is that it shows a clear one- and two-way memory effect of the order of 1% recoverable strain and that the hysteresis of the transformation is very small, only to the extent of a few degrees, which creates possibilities for accurately regulating devices.

TABLE 20.4 Actual (A) Industrial Cu-Based Alloys

Base Alloy	Composition Wt%	Ms (°C)	Hyst. (°C)	Other Alloying Elements in Solution (%)	Current Grain Refining Elements Producing Precipitates
Cu–Zn–Al	5–30 Zn 4–8 Al	−190 to +100	10 (β')	Ni (−5%) Mn (−12%)	Co (CoAl); B (AlB_2); Zr (?); B, Cr (Cr_xB_y)
Cu–Al–Ni	11–14.5 Al 3–5 Ni	−140 to +200	10 (β') 40 (γ')	Mn (−5%)	Ti ($(Cu, Ni)_2$ TiAl); B (AlB_{12}); Zr (?)
Cu–Al–Be	9–12 Al 0.4–1 Be	−80 to +80	6 (β')	Ni (−5%)	B (AlB_2 or AlB_{12}) Ti (Cu_2TiAl)

TABLE 20.5 Property Values of Ni-Ti and Cu-Based SMAs

Property	Unit	Ni-Ti	Cu-Based
Physical			
Melting point	°C	1300	1000
Density	10^3 kg m^{-3}	6.5	7.8
Thermal conductivity of austenite or martensite	W m^{-1} K^{-1}	18 8.6	120 —
Coefficient of thermal expansion of austenite or martensite	10^{-6} K^{-1}	11 6.6	— 16
Specific heat	J kg^{-1} K^{-1}	500	400
Transformation enthalpy	J kg^{-1}	20,000	10,000
Corrosion performance		Similar to 300 series stainless steel	Fair
Wear resistance		Good	—
Electromagnetic			
Resistivity of austenite or martensite	10^{-6} m	1.0 0.8	0.1 0.1
Magnetic permeability	<1.002	<1.002	—
Magnetic susceptibility	emu g^{-1}	3×10^{-6}	—
Mechanical			
Young's modulus (austenite)	GPa	80	80
G (austenite)	GPa	27	—
Ultimate tensile strength	MPa	700–1100 (annealed) 1300–2000 (not annealed)	800
Elongation at failure	%%	30–50	10–15
Fatigue strength $N=10^6$	MPa	350	270
Shape memory			
Transformation temperatures	°C	−100 to 100	−200 to +200
Hysteresis	°C	5–50	5–20
One-way memory strain	%%	3–8	4–6
Maximum temperature (short time)	°C	300	200
Damping capacity	%%SDC	15	80
Superelastic strain	%%	6–8	4
Superelastic energy storage	J g^{-1}	6.5	1.8
Maximum recovery stress	MPa	600–900	600
Stress rate	MPa K^{-1}	5–15	2–5
Work output	J g^{-1}	4	1

It should be noted that further cooling transforms the R-phase into B19′ martensite. During heating, generally only the reverse martensitic transformation will be observed. It has been shown that the appearance of the R-phase depends on composition, alloying elements, and thermomechanical processing. In fact, the major common point is that all effects depressing the martensitic forward transformation below room temperature will favor the appearance of the R-phase transition, which is quite stable near 30°C.

20.4.2.4 Ternary Ni–Ti Alloy Systems [17–21]

The addition of a third element opens up even more possibilities for adapting binary Ni–Ti alloys toward more specific needs of applications. Adding a third element implies a relative replacement of Ni or Ti. Therefore it must be always very well indicated which atom Ni or Ti, or both, is replaced by the third element.

Alloying third elements will influence not only the transformation temperatures but will also have an effect on hysteresis, strength, ductility, shape memory characteristics, and also on the B2→(R)→B19′ sequence.

In more application-oriented cases, one can distinguish four purposes to add third elements:

1. To decrease (Cu) or increase (Nb) the hysteresis
2. To lower the transformation temperatures (Fe, Cr, Co, Al)
3. To increase the transformation temperatures (Hf, Zr, Pd, Pt, Au)
4. To strengthen the matrix (Mo, W, O, C)

Some of the ternary alloys have been developed for large-scale applications. We will only summarize the two most well developed: Ni–Ti–Cu and Ni–Ti–Nb.

20.4.2.4.1 Ti–Ni–Cu [22,23,38]

Ternary Ti–Ni–Cu alloys in which Ni is substituted by Cu are certainly as important as binary TiNi. Increasing the Cu content decreases the deformation stress for the martensite state and decreases also the pseudoelastic hysteresis without affecting significantly the M_s temperature. However, more than 10% Cu addition embrittles the alloys, hampering the formability.

It should also be remembered that while Ti–Ni transforms from a B2 into a monoclinic phase, Ti–Ni–Cu exceeding 15 at.% Cu transforms from a B2 into an orthorhombic phase. Ti–Ni–Cu with less than 15 at.% Cu transforms in two stages.

20.4.2.4.2 Ti–Ni–Nb [24,25]

The inherent transformation hysteresis of Ni–Ti–Nb is larger than for binary Ni–Ti alloys. By the presence of a large dispersed volume fraction of deformable β-Nb particles, hysteresis can be further widened by an overdeformation of the stress-induced martensite, generally between M_s and M_d. Originally Ni–Ti–Nb (more specifically $Ni_{47}Ti_{94}Nb_9$) was developed by Raychem Corp. for clamping devices. The large shift of the reverse transformation temperatures, from below to above room temperature by deformation, allows room storage for opened couplings.

Recently, also pseudoelastic Ni–Ti–Nb alloys have been developed with three significant differences relative to binary alloys

- Stress rate is much lower.
- The $\sigma^{p\text{-}m}$ stresses are much higher.
- The superelastic window is much larger.

20.4.2.5 High-Temperature SMAs [26,27]

Actual SMAs are limited to maximal A_f temperatures of 120°C, M_s generally being below 100°C. However, since market demands on SMA have expanded greatly, the need for SMA transforming at higher temperatures than presently available is increasing. The main application areas of interest are actuators in automobile and oil industry and in safety devices. There is also an interest in robotics since SMAs with high transformation temperatures allow faster cooling, which would significantly increase the bandwidth in which the robot should operate.

In spite of the fact that many alloy systems show high transformation temperatures, no real large-scale applications have been developed. No major breakthrough has been reported yet mainly due to the following problems: (much) lower performance than the successful Ni–Ti alloys, stabilization of the martensite, decomposition of the martensite or parent phase, brittleness due to high elastic anisotropy or due to the presence of brittle phases or precipitates.

Another condition for a good SME is that the stress to induce martensite or the stress to reorient martensite is (much) lower than the critical stress for normal slip. Since this critical stress for slip is generally decreasing with increasing temperature, the above condition is quite difficult to fulfil, especially at high temperatures. A potential HTSMA should thus be designed at such a composition or thermomechanical treatment that strengthening mechanisms are incorporated to increase the critical stress for slip. Ni–Ti–Pd, Ni–Ti–Hf, Ni–Ti–Zr, Cu–Al–Ni, and Cu–Zr alloys are in decreasing importance the most promising alloy systems.

20.4.2.6 Other Types of SMAs [28,29,39]

20.4.2.6.1 β-Ti Alloys

In spite of the good biocompatibility of Ni–Ti alloys, doubts remain on the long-term stability or on the danger of bad surface treatment, leading to Ni leaching. Since Ni is known for his high allergic reaction, Ni-less SMAs could be attractive. Such alloys might be developed based on the allotropic transformation in Ti, a highly biocompatible material. Pure titanium shows an allotropic transformation from β (bcc) to α (hexagonal) at 1265 K. Transition elements (TM) stabilize the β-phase. Thus the temperature of the $(\alpha + \beta)/\beta$ transition decreases with increasing concentration of the alloying element. β-Phase Ti alloys can be martensitically transformed if they are quenched from the stable β-phase. Two types of martensite, respectively α' and α'' can be formed, depending on the composition and the solution treatment conditions.

The α'-martensite is hexagonal, while α'' has an orthorhombic structure. It is the α''-martensite that shows the SME. Ti–Mo and Ti–Nb are the basic systems that presently receive most attention.

Generally, a shape recovery in the order of 3% can be obtained based on strain-induced martensite and recovery stresses up to 170 MPa have been reported. The disadvantage is that those alloys are very prone to stabilization and decomposition due to the fact that the β-phase is retained after quenching in its metastable state and competes with ω-phase during quenching. Also spinodal decomposition of α''-martensite in Ti–Mo and Ti–Nb has been observed. The sensitivity to decomposition at moderate temperatures is less, if not, important at room temperature. Therefore, pseudoelastic β-Ti alloys could offer an interesting alternative to Ni–Ti alloys, for example, for orthodontic wires.

20.4.2.6.2 Magnetic Shape Memory Alloys [30]

Magnetic shape memory (MSM) alloys are a new class of adaptive materials that change their shape when exposed to a moderate magnetic field. The advantage of MSM compared to "classic" SMAs is their ability to generate up to 10% strains at frequencies well above 100 Hz. The MSM mechanism is based on the martensite twin boundary motion driven by the external magnetic field when the material is in complete martensite state.

The martensite crystal, after transformation from cubic austenite, consists of a mixture of tetragonal martensite variants, having different *c*-axis orientation, separated by twin boundaries. When the MSM material is exposed to an external magnetic field, the twins in a favorable orientation relative to the field direction grow at the expense of other twins.

Magnetic field increases the amount of twin variants of "preferable" orientation relative to the field, i.e., the twin boundaries move in the material. As a consequence, MSM exhibit giant MFIS but with a clear lower work output volume ratio as "classic" shape memory alloys. For the time being, the MSM effect was observed in Ni_2MnGa, Fe–Pd and Co–Ni–Al alloys. Among them, the near stoichiometric single crystalline Ni_2MnGa alloy has shown by far the best stroke performance.

Because of their large energy output, MSM materials show enormous potential in high-stroke dynamic actuator applications. On the other hand, when the material is subjected to mechanical straining, the twin variants reorient, too, which alters their magnetization and the surrounding magnetic field. This phenomenon can be used for sensor applications and for voltage generation.

One shortcoming of MSM materials in practical applications is their temperature range, which is limited by a maximum operating temperature. This temperature is determined by structural and magnetic characteristics of the alloy.

20.4.3 Functional Properties of SMAs [31,32]

The temperature-induced transformation is the basis of the one- and two-way memory effect, of the generation of recovery stresses, and of work production. The stress-induced transformation is the basis of the superelastic effect.

When the material is martensite, it is said to be in its cold shape, when the material is in the beta phase, it is said to be in its hot shape. The initial shape that the material obtains after processing is always the hot shape. This means that after cooling

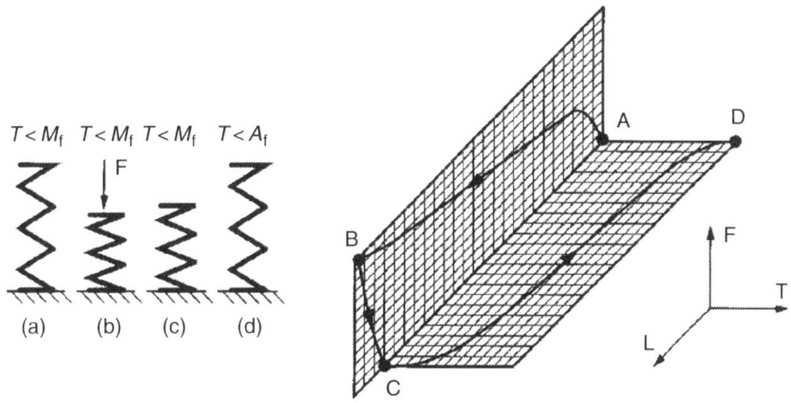

FIGURE 20.28 One-way memory effect. The sample is deformed (A→B) and unloaded (B→C) at a temperature below M_f. The apparent plastic deformation is restored during heating to a temperature above A_f (C→D). Length change, load and temperature are indicated by L, F, and T, respectively.

into martensite the defined cold shape is equal to the initial hot shape. This shape can now be further deformed to the final cold shape (limited to less than 10% strain deformation of the initial shape) to induce the specific properties as described now.

(i) *One-way SME:* The specific functional properties of SMAs are sixfold and are illustrated in Figures 20.28 through Figure 20.32 by the example of a shape memory spring. In the martensitic state, the shape memory element can be easily deformed from a fixed so-called "hot shape," retained after cooling to almost any "cold shape" (Figure 20.28). The only restriction is that the deformations may not exceed a certain strain limit (up to 8%) and should be covered solely by the reorientation of the martensite variants, which leads to a remnant deformation without inducing plasticity. These apparent plastic deformations can be recovered completely during heating, resulting in the original hot shape. This effect is called the one-way memory effect, since only the hot shape is memorized. The transition temperatures between the "cold shape condition" and "hot shape condition," A_s and A_f are determined by the alloying and processing parameters.

(ii) *Two-way memory effect* refers to the memorization of two shapes (Figure 20.29). A cold shape is spontaneously obtained during cooling. Different from the one-way memory effect, no external forces are required for obtaining the memorized cold shape. During subsequent heating, the original hot shape is restored. The two-way memory effect is only obtained after a specific thermomechanical treatment, called training.

(iii) When the shape recovery from the cold shape to the hot shape is impeded, high recovery stresses are gradually generated during heating, as illustrated in Figure 20.30. Stresses up to 800 MPa can be obtained.

(iv) Shape recovery during heating can be biased by an external force (Figure 20.31). The resulting motion against a bias force corresponds to work production, up to 5 J g^{-1}.

(v) SMEs described above require temperature changes. In contrast, the superelastic effect (Figure 20.32) is isothermal in nature and involves the storage of potential energy. Isothermal loading of the shape memory element in the hot shape condition results in large reversible deformations (up to 8%) at nearly constant stress levels (see also Figures 20.27 and 20.32). The deformations

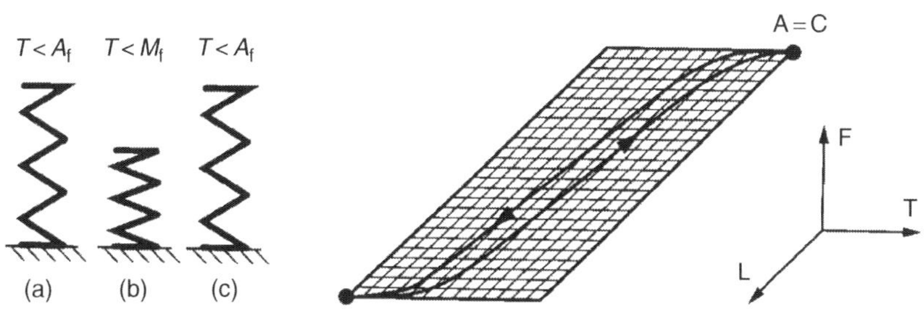

FIGURE 20.29 Two-way memory effect. A spontaneous shape change occurs during cooling to a temperature below M_f (A→B). This shape change is recovered during subsequent heating to a temperature above A_f (B→C).

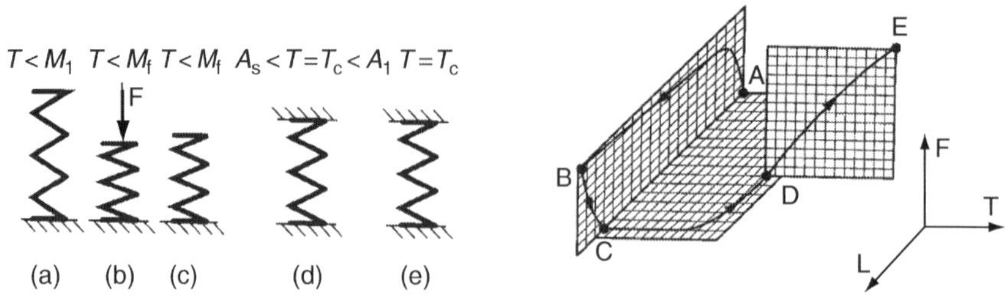

FIGURE 20.30 Generation of shape recovery stresses. The sample is deformed (A→B) and unloaded (B→C) at a temperature below M_f. Recovery stresses are generated during heating (D→E) starting from the contact temperature T_c (D), situated between A_s and A_f.

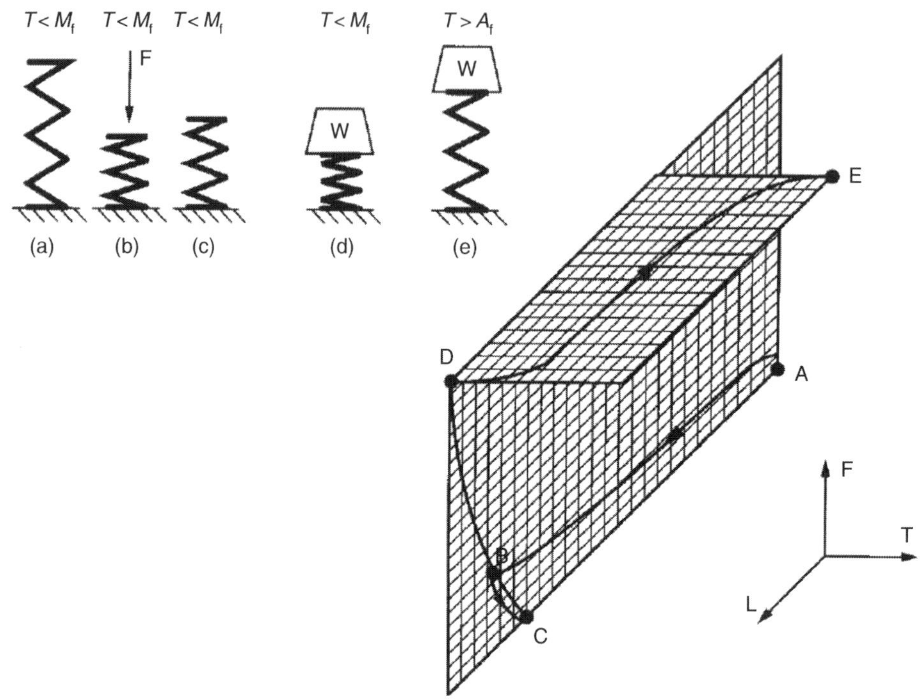

FIGURE 20.31 Work output. The sample is deformed at a temperature below M_f (A→B), followed by unloading (B→C), and repeated loading with a bias weight W (C→D). Shape recovery occurs at an opposing force W during heating to a temperature above A_f (D→E), so work is done.

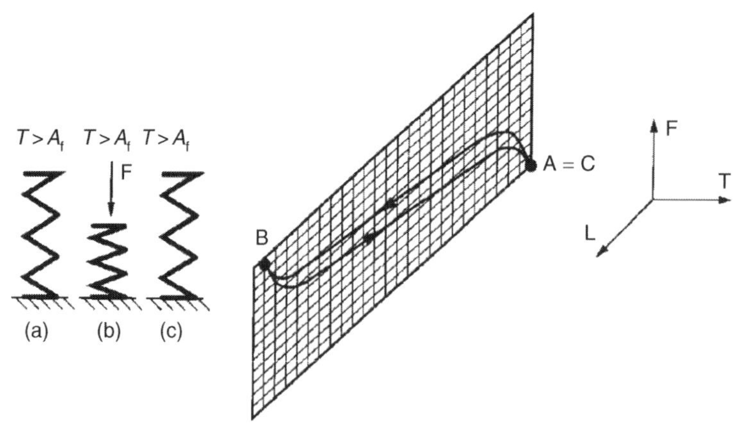

FIGURE 20.32 Superelastic effect. The sample is strongly deformed (A→B) at a temperature above A_f. During subsequent unloading a complete shape recovery occurs (B→C). The area enclosed between the loading and unloading curves is a measure of the energy dissipated in one superelastic cycle.

are completely recovered at a lower stress level during unloading. These stress levels are alloy- and temperature-dependent. In general, the stress levels increase linearly (2–15 MPa K^{-1}) with temperature.

(vi) SMAs also have a high damping capacity in the martensitic state and two-phase condition. These alloys show in the cold shape condition a strong amplitude-dependent internal friction. For impact loads, the specific damping capacity can be as high as 90%. In the hot shape condition, energy is dissipated during superelastic cycling as a result of the stress hysteresis between superelastic loading and unloading (Figure 20.27).

The functions and applications linked with those properties can be divided into five categories.

(i) One- and two-way memory effect can be used for free recovery applications. This refers to applications in which the single function of the SMA-element is to cause motions without any biasing stress.
(ii) Generation of recovery stresses can be used for diverse clamping and fixation devices, such as SMA-couplings and SMA-connectors, ranging from very small diameters (<1 mm) to very large diameters (>1 m).
(iii) Diverse actuation applications have been developed based on the work production capacity of SMAs.
(iv) Pseudoelasticity is used in many biomedical and other superelastic applications where high reversible strains have to be combined with high-stress plateaus.
(v) Damping applications can be developed based on the high damping capacity of SMAs. The high damping capacity of the martensitic phase is interesting for passive damping. The superelastic hysteresis might be interesting for earthquake damping and isolation purposes.

In the past several years, SMAs have found their specific niches in many domains of industrial activities. A steadily growing amount of different SMA applications are now produced at large volumes. Detailed descriptions of many successful or potentially successful applications can be found in recent conference proceedings from the well-known conferences ICOMAT, ESOMAT, SMST, and some locally organized conferences as well as on many of the very instructive and informative Web sites. The bibliography at the end of the reference list provides a list of books related to martensite and SMAs.

References

1. Funakubo H., *Shape Memory Alloys*, 1984, Gordon & Breach, London.
2. Delaey L., Diffusionless transformations, in *Materials Science and Technology, Vol. 5: Phase Transformations in Materials* (Haasen, P., ed.), 1991, pp. 339–404, VCH Verlagsgesellschaft mbH, Weinheim, Germany.
3. Wollants P., Roos J.R., and Delaey L., Thermally and stress induced thermoelastic martensitic transformations in the reference frame of equilibrium thermodynamics, *Prog. Mater. Sci.*, 37(3), 1993, 2427–2488.
4. Otsuka K. and Wayman C.M., *Shape Memory Materials*, 1998, 284 pp., Cambridge University Press, Cambridge.
5. Delaey L., Chandrasekaran M., Andrade M., and Van Humbeeck J., Non-ferrous martensites, classification, crystal structure, morphology, microstructure, Proceedings Int. Conf. Solid-Solid Phase Transformation, Met. Society of AIME, 1982, pp. 1429–1453.
6. Maki T. and Tamura T., Shape memory effect in ferrous alloys, Proc. ICOMAT-86, Japanese Institute of Metals, 1986, pp. 963–970.
7. Sato A., Sama K., Chishima E., and Mori T., Shape memory effect and mechanical behaviour of an Fe-30Mn-1Si alloy single crystal, *J. Physique*, 12 C4, 1982, 797–802.
8. Sato A., Chishima E., Soma K., and Mori T., Shape memory effect in $\gamma \leftrightarrow \varepsilon$ transformation in Fe-30Mn-1Si alloy single crystals, *Acta Metal.*, 30 1982, 1177–1183.
9. Gu Q., Van Humbeeck J., and Delaey L., A review of the martensitic transformation and shape memory effect in Fe-Mn-Si alloys, *J. Physique IV*, C3 4, 1994, C3-135–140.
10. Maki T., Ferrous shape memory alloys, Chapter 5 in *Shape Memory Materials* (Otsuka K. and Wayman C.M., eds.), Chapter 5, 1998, 284 pp., Cambridge University Press.
11. Van Humbeeck J. and Delaey L., A comparative review of the (potential) shape memory alloys, *The Martensitic Transformation in Science and Technology* (Hornbogen E. and Jost N., eds.), 1989, pp. 15–25, DGM, Verlag, Oberursel, Germany.
12. Van Humbeeck J., Chandrasekaran M., and Stalmans R., Copper based shape memory alloys and the martensitic transformation, Proc.-ICOMAT-92, 1993, pp. 1015–1025, Monterey Institute of Advanced Studies, Monterey, CA.
13. Tadaki T., Cu-based shape memory alloys, *Shape Memory Materials* (Otsuka K. and Wayman C.M., eds.), 1998, 284 pp., Cambridge University Press, Cambridge (Chapter 3).
14. Ahlers M., Martensite and equilibrium phases in Cu-Zn and Cu-Zn-Al alloys, *Progress in Materials Science*, vol. 30 (Christian J.W., Haasen P., and Massalski T.B., eds.), 1986, pp. 135–186.
15. Wu S.K. and Ming H., Cu-based shape memory alloys, *Engineering Aspects of Shape Memory Alloys* (Duerig T. et al., eds.), 1990, pp. 69–88, Butterworths Scientific, London.
16. Otsuka K. and Ren X., Physical metallurgy of Ti-Ni based shape memory alloys, *Prog. Mater. Sci.*, 50, 2005, 511–678.
17. Eckelmeyer K.H., The effect of alloying on the shape memory phenomenon in Nitinol, *Scripta Met.* 10, 1976, 667–672.
18. Honma T., Matsumoto M., Shugo Y., and Yamazaki I., Effects of addition of 3d transition elements on the phase transformation in TiNi compound, Proceedings ICOMAT-79, 1979, pp. 259–264.
19. Huisman-Kleinherenbrink Patricia, On the Martensitic Transformation Temperatures of NiTi and Their Dependence on Alloying Elements, PhD Thesis, University of Twente, the Netherlands, 1991.

20. Kachin V.N., Martensitic transformation and shape memory effect in B2 intermetallic compounds of Ti, *Rev. Phys. Appl.* 24 1989, 733–739.
21. Kachin V.N., Voronin V.P., Sivokha V.P., and Pushin V.G., Martensitic transformation and shape memory effect in polycomponent TiNi-based alloys, ICOMAT-95 proceedings, Part I and II, Lausanne 1995 (Gotthardt R. and Van Humbeeck J., eds.), Editions de Physique, *J. Physique IV*, 5(8), 1995, 707–172.
22. Saburi T., Takagaki T., Nenno S., and Koshino K., Mechanical behaviour of shape memory Ti-Ni-Cu alloys, *MRS Intl Mtg. Adv. Mater.*, 9, 1989, 147–152.
23. Horikawa H. and Ueki T., Superelastic characteristics of Ni-Ti-Cu-X alloys, *Trans. Mat. Res. Soc. Jpn.*, 18B, 1993, 1113–1116.
24. Melton K.N., Proft J.L., and Duerig T.W., Wide hysteresis shape memory alloys based on the Ni-Ti-Nb system, *MRS Intl Mtg. Adv. Mater.* 9, 1989, 165–170.
25. Yang J.H. and Simpson J.W., Stress-induced transformation and superelasticity in Ni-Ti-Nb alloys, ICOMAT-95 Proceedings, Part I and II, Lausanne 1995, (Gotthardt R. and Van Humbeeck J., eds.), Editions de Physique, *J. Physique, IV*, 5(8), 1995, 771–776.
26. Van Humbeeck J., High temperature shape memory alloys, *Trans. ASME*, 121, 1999, 98–101.
27. Firstov G.S., Van Humbeeck J., and Koval J.N., High temperature shape memory alloys. Some recent developments, *Mater. Sci. Eng. A*, 378 2004, 2–10.
28. Maeshima T. and Nishida M., Shape memory properties of biomedical Fi-Mo-Ag and T-Mo-Sn alloys, *Mater. Trans.*, 45(4), 2004, 1096–1100.
29. Kim H.Y., Ikehara, Y., Kim J.I., Hosoda H., and Miyazaki, S., Martensitic transformation, shape memory effect and superelasticity of Ti-Nb binary alloys, *Acta Mater.*, 54, 2006, 2419–2429.
30. O. Söderberg, O., Ge, Y., Sozinov, A., Hannula, S.-P., and Lindroos V.K., Recent breakthrough development of the magnetic shape memory effect in Ni-Mn-Ga alloys, *Smart. Mater. Struct.* 14, 2005, 223–235.
31. Van Humbeeck J., Shape memory alloys: a material and a technology. *Adv. Eng. Mater.*, 3(11), 2001, 837–850.
32. Van Humbeeck J. and Kustov S., Active and passive damping of noise and vibration through shape memory alloys: application and mechanisms, *Smart Mat. Struct.*, 14, 2005, 171–185.
33. Nishida M., Wayman C.M., and Honma T., Precipitation processes in near-equiatomic TiNi shape memory alloys, *Met. Trans.* 17A, 1986, 1505–1515.
34. Todoriki T. and Tamura H., Effect of heat treatment after cold working on the phase transformation in TiNi alloy, *Trans. Jpn. Inst. Metals*, 28, 1987, 83–94.
35. Wayman C.M., Phase transformations in NiTi type shape memory alloys, Proc. Int. Conf. On Mart. Transf. 1986, The Japanese Institute of Metals, pp. 645–652.
36. De 89.
37. Sat 89.
38. Sa 89b.
39. an H 97.

Bibliography

1. Z. Nishiyama, *Martensitic Transformation*, Academic Press, London, 1978.
2. H. Funakubo, *Shape Memory Alloys*, Gordon & Breach, London, 1986.
3. G.B. Olson and W.S. Owen, *Martensite*, ASM-international, 1992.
4. K..Otuska and C.M. Wayman, *Shape Memory Materials*, Cambridge University Press, Cambridge, 1998.
5. V.E. Gunther, *Delay Law and New Class of Materials and Implants in Medicine*, SST-Publishing, Tomsk, Northampton,MA, 2000.
6. L' Hocine Yahia, *Shape Memory Implants*, Springer, Berlin, 2000.
7. M. Kohl, *Shape Memory Actuators*, Springer, Berlin, 2004.

20.5 Smart Materials Modeling

Manuel Laso

20.5.1 Introduction

The most salient feature of modeling work in the area of smart materials is its great diversity. Materials considered as smart span a staggeringly wide range. Smart materials run the gamut from the inorganic, monolithic crystalline materials, to the organic, polymeric, semicrystalline ones. Composites, polycrystalline materials, hydrated gels, magnetostrictive/ferromagnetic tagged composites, electrochromic materials, etc. to mention but a few, further expand the range of smart materials to be modeled. The complexity that arises from this great variety of material types is compounded with the wide range of interesting properties they display. Finally, the question of the time and the length scales at which the modeling is to be implemented adds an extra level of complexity to the field: Even when applied to the very same material and the very same property, it frequently happens that different smart material modelers (i) look at the material at vastly different spatial or temporal scales, (ii) use completely unrelated modeling techniques, and (iii) even come to conclusions and modeling results, which can be unrelated for all practical purposes.

Polyvinylidene fluoride (PVDF) is often considered a smart material and provides a good illustration of this situation.

(a) At the quantum chemical (QC) level, appreciable effort has been devoted to the determination of the electronic structure of PVDF via QC methods with a view to estimate the

TABLE 20.6 Representative Modeling Levels and Goals for Major Types of Smart Materials

Type of Material	Application	Modeling Level/Technique	Modeling Goal
Piezoelectric ceramic functional gradient	Actuators	Micromechanical CEs, composites (c), FE (d)	Electromechanical coupling Mechanical response
Piezoelectric ceramic-adaptive composites	Actuators Smart structures	Micromechanical CEs, composites (c) FE (d)	Electromechanical coupling Mechanical response
	Smart rotors		
	Structural damping		
	Aerodynamic control		
Lithium insertion compounds	Smart batteries	QC (a)	Structure prediction
			Structure stability
			Energy diagrams Jahn–Teller distortion
Shape memory alloys	Mechanical Biomedical Microrobotics Microdevices Actuators	Micromechanical CEs, plasticity, composites (c) FE (d)	Shape memory effect Superelasticity Hysteresis
Smart polymers	Biomedical sensing		
	Drug delivery		
	Immobilization		
Electrostrictive ceramic	Ultrasonic transducers	FE (d)	Electromechanical coupling
	Actuators		Mechanical response
	MEMS		
Piezoelectric ceramic	Pressure sensors	FE (d)	Electromechanical coupling
	Accelerometers		Mechanical response
	Gyroscope		
	Resonators		
	Filters		
Smart coatings	Stress visualization	QC (a)	Mechanism of mechanoluminescence, band calculation
Magnetostrictive/ferromagnetic tagged composites	Composite cure and health monitoring	Micromechanical CEs (c) FE (d)	Nondestructive materials testing
Electrorheological fluids	Actuators	Colloid dynamics (c),	Structure–property relationship
	Brakes	Viscoelastic flow calculations (d)	Rheological CE
	Mechanical couplers		Device design
Magnetorheological fluids	Shock absorbers	Colloid dynamics (c),	Structure–property relationship
	Dampers	Viscoelastic flow calculations (d)	Rheological CE
	Brakes		Device design
Fiber sensors	Distributed sensors	FE, analytical (d)	Optothermomechanical behavior
	Bragg gratings		
	Fabry–Perot sensor		
	Optical signal processing		
	Intelligent transportation		
	Optical multiplexing		
Organic gelators	Molecular recognition	QC (a)	Crystal structure
		Atomistic MD and MC (b)	Molecular complementary
Gels/hydrogels	Drug delivery	Mesoscopic (c)	Gel structure, transport behavior
		Coarse-grained polymer MC (b)	
Magnetostrictive materials	Transducers	Micromechanical CEs (c)	Magnetomechanical design
	Sensors	FE (d)	
	Actuators		
	Motors		
	Magnetometers		

(continued)

TABLE 20.6 (continued) Representative Modeling Levels and Goals for Major Types of Smart Materials

Type of Material	Application	Modeling Level/Technique	Modeling Goal
Giant magnetostrictive materials	Sensors	Micromechanical CEs (c)	Magnetomechanical design
	Actuators	FE (d)	
	Positioning devices		
	Dampers		
Smart ceramics	Intelligent synthesis	Chemical thermodynamics, chemical kinetics (d)	Reaction kinetics
			Reactor design and control
Smart paints	Vibrational sensor	Micromechanical CEs (c)	Electromechanical coupling
	Atmospheric sensor	FE (d)	Mechanical response
	Human skin mimicry		
Piezoelectric polymers	Structural active elements	Atomistic MD and MC (b)	Crystal structure prediction
	Strain sensors	Micromechanical CEs (c)	Electromechanical coupling
	Transducers	FE (d)	Polarization response
	Active and passive vibration control		
Functionally graded polymer blends	Damping	Micromechanical CEs (c)	Electromechanical coupling
	Artificial tissue	FE (d)	Polarization response
	Medical applications		
Smart perovskites	Nonvolatile memories electromechanical conversion fuel cells	QC (a) atomistic MD and MC (b)	Crystal structure Piezoelectricity Pyroelectricity Ionic conductivity
Smart skins	Sound control	Micromechanical CEs (c)	Electromechanical coupling
		FE (d)	
Thermoresponsive inorganic materials	Temperature-sensing responsive devices	QC (a) Atomistic MD and MC (b)	Crystal structure Phase transitions
Electrochromic materials	Smart windows	QC (a)	Band structure
	Architectural glazing		Transmittance
	Thermochromic devices		Ion conduction
	Thermotropic devices		

monomeric dipole. Very frequently, QC methods are applied to quite small fragments of the polymeric chains, sometimes even *in vacuo*. This applies especially to those most advanced and sophisticated QC methods available today.

(b) Somewhat similar work has also been performed at the atomistic level although the assignment of electronic distribution to individual atoms is not directly based on a fundamental QC approach. At this level, the description of PVDF is based on a classical picture of atoms as sites interacting via an empirical force field. Typical modeling goals at this level are the correct calculation of crystalline polymorphs, and an estimation of piezoelectric, dielectric, and elastic properties via molecular dynamics (MD) and Monte Carlo (MC) methods. Prediction of electric properties is frequently limited by the use of partial charges to describe the spatial electron distribution.

(c) At the mesoscopic level, questions such as the stacking of folded PVDF lamellae, the prediction of semicrystalline morphology, and the prediction of piezoelectric moduli based on homogenization are addressed.

(d) At the macroscopic level, modeling PVDF often refers to the simulation and design of a particular geometry or device based on a purely continuum-mechanical description of the material via partial differential equations. All smart material information is condensed in a few macroscopic parameters, like piezoelectric moduli, elastic compliances, etc. At this level, modeling goes well beyond the material itself and is intimately linked to the design of a specific device for a specific function.

In spite of the general propensity to consider one's own specialty field as slightly more central and momentous than others fields, or to consider, say, quantum mechanics more fundamental than finite elements (FE), scientists and engineers working at any of the four levels just described are justly and equally entitled to claiming their work as smart materials modeling. For some specific materials, it also happens that modeling efforts at the different levels develop more or less independently of each other, with little if any communication among them. This state of affairs is easily detected by the corresponding sets of cited literature being often disjointed or having minimal overlap.

As a natural consequence, the description of smart materials and their equilibrium and nonequilibrium properties makes it necessary to resort to most of the modeling techniques, which are considered basic and fundamental in materials science, and, in addition, to some rather specialized ones. Therefore, this chapter cannot be more than a tabular overview of techniques and a basic description of the general modeling setting where these techniques are applied. The very modest goal of this chapter is thus to provide a roadmap of the modeling landscape in the context of smart materials and a set of pointers to the fundamental literature.

Specialists in a given area will find this necessarily cursory treatment of their field very superficial and lacking in detail. Specialists and generalists both may, however, find more useful the general directions pertaining to other fields and the summary overview of other modeling smart materials techniques in use.

20.5.2 Conservation Laws, Equations of State, and Constitutive Equations

Although at first seemingly removed from the description of specific modeling techniques, it is quite useful to get started in the area by describing the general setting in which most modeling efforts are embedded. The three blocks depicted in Figure 20.33 correspond to the three main types of equations required to formulate a calculation, design, or modeling task in the most general sense.

1. Macrofield (conservation) equations (Block 1) express very fundamental laws, like the constancy of mass, momentum, energy, electric charge, mole number in the absence of chemical reaction, etc. In the present context, it is largely irrelevant whether we think of them as axiomatic or as originating from deeper symmetry principles. Conservation equations are of very general validity and are typically formulated as partial differential equations. They describe the relationship among macroscopic fields via their time and space derivatives. They hence contain basic dynamic information for the evolution of macroscopic fields of variables like internal energy, entropy, stress, velocity, temperature, electric charge, electric field, density, composition, etc. They are often based on the assumption of local equilibrium, i.e., deviations from local equilibrium are vanishingly small, and a few, macroscopically smooth fields suffice to describe the state of the material. These equations do not form a closed set, i.e., they cannot be solved on their own. Most importantly, they contain no information whatsoever about the medium (material) in which these fields exist, so they are not material specific.
2. Equations of state (EOS) (Block 2) describe the equilibrium behavior (properties) of a specific material. EOS are not universal laws, but statements of relationships between some of the fields in Block 1 that appear to be a valid description of the static (equilibrium) thermodynamic properties for a given material. These relationships are very frequently formulated as algebraic equations. One of the simplest examples of a Block 2 EOS is the assumption of constant mass density for a particular material. While some EOS can be derived from more basic principles, like equilibrium statistical mechanics, many practical EOS are empirical correlations of adequate flexibility and containing a sufficient number of parameters to be adjusted to the material under consideration.
3. Constitutive equations (CEs (Block 3) describe the nonequilibrium behavior of a specific material. Again, they are not universal laws, but more or less plausible statements of relationships between some of the fields in Block 1 that appear to describe the dynamic behavior of a given material when driven out of equilibrium by an external influence. Examples of well-known CEs are the assumption that the local electric current density in a material is proportional to the local electric field, the proportionality constant being a (scalar or tensorial) electrical conductivity (microscopic Ohm's law),* or Hooke's relationship between stress and strain. CEs often appear in the form of local

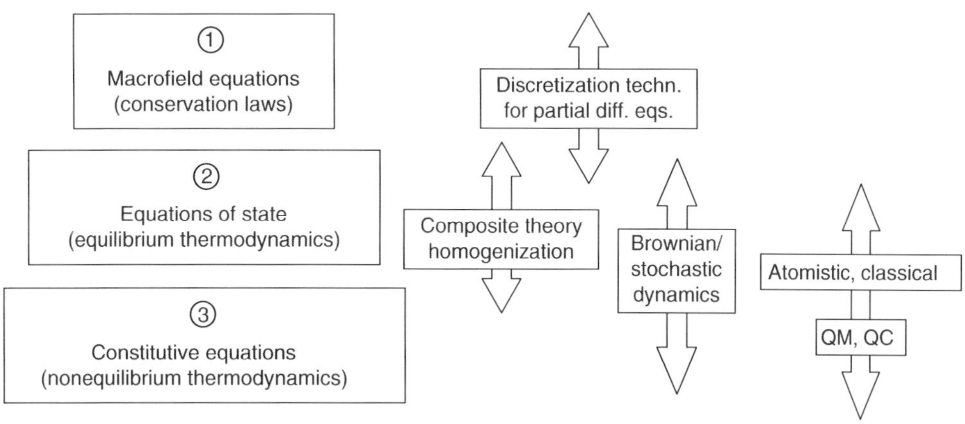

FIGURE 20.33 General formulation of modeling in smart materials; levels of description.

relationships between generalized fluxes and generalized forces (spatial gradients, driving forces) in the spirit of linear irreversible thermodynamics. In other cases, CEs are formulated as differential, integral, or integro differential equations. Material properties appear as characteristic magnitudes relating several fields, e.g., a Newtonian viscosity relates the rate of strain tensor and the stress tensor.

In the general case, only the complete set of equations that appear in the three blocks is solvable. Its solution consists of the macroscopic fields and their time dependence, possibly including an asymptotic approach to a steady state.

In the PVDF example, a set of stress, strain, electric field, and electric displacement fields (i.e., their values at every point in space and their variation in time) can be obtained by writing down and solving a set of equations in Blocks 1 to 3. Since the previous discussion is based on a purely macroscopic level of description (level (d) in the list in Section 20.5.1), it would be perfectly adequate and self-contained if the modeling goal were to achieve a quantitative description of the electric polarization of a PVDF piezoelectric of a given equilibrium shape when subjected to a given stress. In order to perform modeling at this macroscopic level, elastic compliances, dielectric permittivities, and piezoelectric moduli must be available, coming either from an experiment or possibly from a lower level modeling technique.

Modeling at levels (a), (b), and (c) are however quite different: now the goal is not to compute the solution of a complex (heterogeneous) problem for a particular material geometry. It is rather the prediction under very simple (e.g., homogeneous) conditions and starting from some basic principles (the meaning of basic being now level-dependent) of some or all of the physical, chemical, and thermodynamic properties that appear in the EOS and in the CEs, so that the modeler at level (d) can take them as input. In this sense, one of the goals of levels (a), (b), and (c) is to focus on Blocks 2 and, above all, 3 and pass on the information, which is required at level (d) to deal with the complete set of Blocks 1, 2, and 3.

It is however also true that information passing takes place across these three first levels (a), (b), and (c):

- Some of the parameters appearing in the force field for classical MD modeling (level b) of PVDF may come from semiempirical or ab initio QC calculations (level a).
- Atomistically computed crystal and amorphous matrix compliances via Parrinello–Rahman MD (level b) can be used to predict overall PVDF elastic compliance via homogenization methods borrowed from the theory of composites (level c).

Thus, what is an output at a given level represents a necessary input at the next higher level. In this sense, it is now possible to assign modelers (a), (b), and (c) to the EOS and CE blocks. Modeler (d) must include all three blocks in the modeling work, see Table 20.6.

The reader looking for an entry point in the field of modeling smart materials is urged to first devote some effort to carefully consider at what level or levels his or her modeling should take place. A good deal of frustration can be avoided by a judicious choice of the correct description level.

It should also be mentioned that multiscale modeling methods exist and are gaining increasing popularity. Such hierarchical modeling techniques are undergoing rapid development and are bound to have a significant impact on the modeling of smart materials in the near future. The key idea behind multiscale modeling is to link in a single modeling approach two or more techniques residing at differential time/length scales ((a) through (d) above). As a prototypical example, the possibility of linking QC (density functional theory [DFT], actually) with classical MD and FE modeling techniques in a single calculation [1] has already been demonstrated for crack propagation in Si.

20.5.3 Informal Classification of Modeling Techniques for Smart Materials

is a fairly comprehensive list of materials considered as smart together with their applications and suitable modeling techniques. As expected, modeling goals range from the very microscopic, low-level ones (band calculations, crystal structure prediction, etc.) to the completely phenomenological and macroscopic (design of transducers, actuators, etc.). Furthermore, at the largest spatial scales (level d above) the modeling of the material itself, i.e., its equations of state and CEs, blends into the overall problem formulation, i.e., the design of a smart structure. Since the goal of the present chapter is the modeling of smart materials, it would be desirable to draw as clear a dividing line as possible in order to separate the modeling of the material itself from that of the structure.

This separation is closely linked to the idea of homogeneity: Except for materials whose smart function depends on being intrinsically graded, the modeling of the material itself is best addressed in a homogeneous setting. It is also possible that the goal of the modeling is precisely the determination of average (homogenized) properties. Unfortunately, the line that separates modeling a material from modeling a structure is often blurred.

We can group the modeling methodologies that appear in Table 20.6 in a few rather well-defined families. The following list refers to some of the most basic literature in each area. The list and the references at the end of the chapter are by no means exhaustive, but they represent useful entry points:

- Quantum chemistry methods [2]
  - *Ab initio* QC, including Car–Parrinello MD [3]
  - Semiempirical [4]
- Classical atomistic methods

* As a side comment, EOS and CEs can be detected by their carrying a surname: On the one hand, conservation laws are universal and one would be hard pressed to associate a person's name with their discovery. However, many EOS and all CEs are only plausible postulates but by no means universal laws. The inventor of such a plausible postulate seems to be naturally entitled to append his/her name to the corresponding equation, hence Fourier's, Fick's, Newton's, and Ohm's laws for heat, mass, momentum, and electric charge transport, respectively.

- MD [5]
- MC and hybrid MC [6]
- Mesoscopic methods
  - Brownian/stochastic dynamics [7]
  - Lattice Boltzmann [8]
- Coarse-graining methods:
  - Micromechanical modeling [9]
  - Composite theory [10]
- Discrete techniques for partial differential equations:
  - Galerkin and its variants (Petrov–Galerkin, generalized Galerkin) [11]
  - Collocation [12]
  - Meshless methods [13]
- Nonequilibrium thermodynamics
  - Linear irreversible thermodynamics [14]
  - GENERIC [15]

As a general introduction to the field of smart materials, and not only their modeling, Ref. [16] is highly recommended.

20.5.4 Conclusion

The great diversity and the range of smart materials make it very difficult to draw general conclusions about their modeling. Only a few general guidelines can be suggested:

- Strive for a clear separation between modeling a smart material, and modeling a smart device or structure. Be aware that in some cases, this separation is not possible.
- Prior to any modeling work, carefully consider the general modeling situation depicted. Decide at which levels the smart material is to be modeled.
- Clearly distinguish equilibrium and nonequilibrium modeling goals.
- Clearly distinguish between homogeneous and nonhomogeneous situations.
- If multilevel modeling is involved, pay extra attention to consistent coarse-graining and thermodynamic admissibility [15].

References

1. G. Lu, E.B. Tadmor, and E. Kaxiras, *Physical Review B* **73**, 024108, 2006.
2. J.P. Lowe and Peterson, K.L., *Quantum Chemistry*, Academic Press, New York, 2005.
3. D. Marx and Hutter, J., in *Modern Methods and Algorithms of Quantum Chemistry*, ed. J. Grotendorst, John von Neumann Institute for Computing, Julich, NIC Series, Jülich, 2000, Vol. 1, p. 301.
4. M.C. Zerner, *Semiempirical Molecular Orbital Methods*, VCH Publishers, Weinheim, 1991.
5. D. Frenkel and Smit, B., *Understanding Molecular Simulation*, Academic Press, London, 2001.
6. A. Leach, *Molecular Modeling: Principles and Applications*, Prentice Hall, Englewood Cliffs, NJ, 2001.
7. H.C. Öttinger, *Stochastic Processes in Polymeric Fluids: Tools and Examples for Developing Simulation Algorithms*, Springer, Berlin, 1996.
8. S. Succi, *The Lattice Boltzmann Equation: For Fluid Dynamics and Beyond*, Clarendon Press, Oxford, 2001.
9. J. Aboudi, *Mechanics of Composite Materials: A Unified Micromechanical Approach*, Elsevier, Amsterdam, 1991.
10. T.W.C. Derek Hull, D.R. Clarke, and S. Suresh *An Introduction to Composite Materials*, Cambridge University Press, Cambridge, UK, 1996.
11. A. Quarteroni and Valli A., *Numerical Approximation of Partial Differential Equations*, Springer-Verlag, Berlin and Heidelberg, 1994.
12. B.L. Nicola Bellomo, Revelli, R., and Ridolfi, L., *Generalized Collocation Methods*, Birkhauser Verlag AG, 2007.
13. Y.C. James Lee and Azim Eskandarian, *Meshless Methods in Solid Mechanics*, Springer-Verlag, New York, 2006.
14. S.R.D.G.A.P. Mazur, *Non-equilibrium Thermodynamics*, Dover, 1984.
15. H.C. Ottinger, *Beyond Equilibrium Thermodynamics*, John Wiley, Chichester, 2005.
16. M. Schwartz, *Encyclopedia of Smart Materials*, John Wiley, Chichester, 2002.

20.6 On the Microstructural Mechanisms of SMEs

Monica Barney and Michelle Bartning

SMAs have the unique ability to recover their exact original shape even after large macroscopic displacements. Typical elastic–plastic materials begin to permanently deform at strains less than 1%, while in SMAs, this can be as high as 8% [1]. The "magic" behind this property is rooted in the martensitic phase transformation that is driven by either a change in temperature or a change in stress [2,3]. While the existence of this solid-state transformation is necessary for SME to occur, it is not a sufficient condition (martensitic transformations were first seen in steels, where SMEs are not observed [4]). The property of thermoelasticity is critical in determining whether a material that undergoes this type of phase transformation will exhibit shape memory. It is necessary to discuss the microscopic details of this reaction to develop a systematic understanding of the macroscopic effects of thermal shape memory (commonly called shape memory) and mechanical shape memory (commonly called superelasticity) that are of interest to engineers developing practical applications. The more common descriptions of SMAs discuss these two effects separately, giving the impression that they are only arbitrarily related and just happen to occur in the same material. In the second half of this chapter, we will show that these two macroscopic effects lie on a continuum that is governed by test temperature [5,6]. Within this temperature spectrum, the material ranges from stable, nontransforming austenite to stable, nontransforming martensite. Although much of the following discussion applies to all SMAs, some of the details are specific to the most widely used alloy: Ni-rich NiTi (Nitinol) that has been processed with solution treatment followed by aging.

20.6.1 Microscopic Mechanisms

The ability of SMAs to not only recover large amounts of deformation but also return to the exact same shape lies in the details of the first-order martensitic phase transformation. It is important to understand this reaction to be able to fully exploit the resulting macroscopic properties, so we will spend some time describing it here. The easiest way to organize the details of this reaction in the proper context is to review the thermal SME first. We first start with the parent phase, called austenite, which has the unusual combination of being a high-temperature and high-symmetry phase. As the specimen is cooled to a characteristic temperature, the material suddenly changes to the lower-symmetry phase, called martensite. This change happens suddenly, as it is displacive (i.e., diffusionless) in nature, with the microstructure changing from a high-symmetry crystal structure (specifically, simple cubic (B2) in NiTi [7]) to a lower symmetry one (monoclinic (B19') in NiTi [7]), with only very small shifts required of the atoms. The crystallographic arrangements of the two structures are shown in Figure 20.34a, with the black circles representing the nickel atoms [8]. As martensite nucleates within the austenite, these two configurations must find a way to fit together in order to form an interface free from macroscopic stresses and dislocation formation if the material is to exhibit reversibility of this phase change. Also, the new phase must fit into the space (volume and shape) that the austenite formerly occupied. This ability is called shape accommodation [4], and can be achieved by either twinning or slip [9,10]. The ability to accomplish this by twinning is one of the most critical ingredients of materials that can undergo a reversible martensitic reaction, which gives rise to shape memory behavior. Though twin boundaries are highly mobile, twinning can accommodate only small volume changes as compared to slip [9]. This limitation places strict constraints on the differences in lattice parameters of the two phases, such that the austenite is required to have a cubic crystal structure [4]. In this way, the interface is coherent and the volume change between the two is very small. If this criterion is not met, plastic deformation accompanies the phase transformation, destroying reversibility. To maintain this required compatibility, the martensite forms a twinned structure, where the different orientation of each twin is only a symmetry-related crystallographic difference, called a variant [3,4]. The twin boundaries are low energy, since they can move by nucleating new, though energetically equivalent, variants. To see this

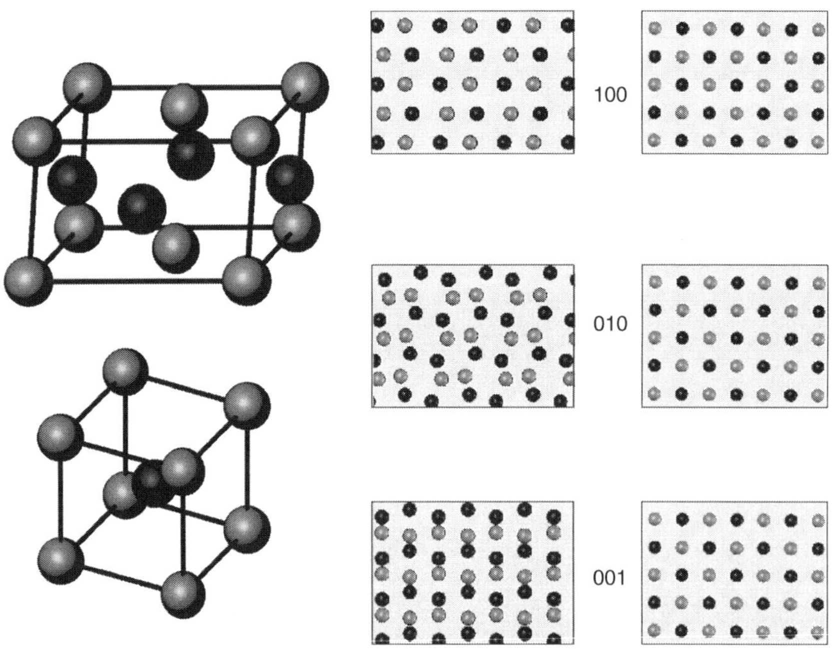

FIGURE 20.34 (a) Schematics of the two basic crystal structures in Nitinol. The upper structure is martensite (monoclinic, B19'), while the simple cubic (B2) austenite is drawn in the lower left hand corner and 100 family of planes (b) The three members of the (100) family of planes are drawn for the martensite (left-hand side) and austenite (right-hand side). Note the degeneracy of the cubic symmetry causing only one variant to be possible. All black circles represent the nickel atoms. (From Bronfenbrenner, D., 2006, with permission.)

more clearly, the three members of the {100} family of planes are drawn for each of the phases in Figure 20.34b. Because of the cubic symmetry, all of the {100} planes of austenite appear exactly the same (i.e., it is univariant [3]). For the low-symmetry martensite, they look different, but are all of the same energy. As a result of this difference, when the martensite transforms back to the austenite upon heating (the last step of the thermal SME), all of the martensite variants transform to the only austenite variant has, causing it to remember the exact original shape.

20.6.2 Macroscopic Effects

We are now in a better position to answer the seemingly straightforward question, "If you pull on NiTi, what happens?" There is not, however, a single answer to this question, as it depends on the relative relationship between the ambient temperature while pulling and the transition temperatures of the material. Figure 20.35 defines four of the important transition temperatures: martensite start (M_s), martensite finish (M_f), austenite start (A_s), and austenite finish (A_f). These temperatures can be determined by DSC, a technique that measures the heat flow into (endothermic reaction) or out of (exothermic reaction) the sample while varying the temperature [11,12]. The sample starts off in the austenite phase at high temperature (upper right-hand corner of Figure 20.35); upon cooling to M_s, martensite starts to form. Between M_s and M_f, the thermally induced martensite is only partially stable and some austenite is still present. Once the M_f temperature is reached, the phase transformation is complete and further decreasing the temperature will only cause thermal contraction of the martensite. Upon increasing the temperature to A_s, the material

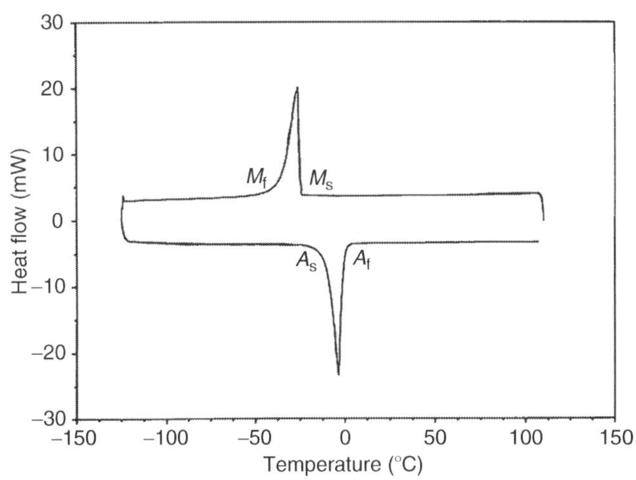

FIGURE 20.35 Plot of the thermal transitions typical in Nitinol as measured by DSC, with the characteristic temperatures labeled. Note that the transitions upon cooling do not occur at the same temperature as heating (i.e., $M_s \neq A_f$, $M_f \neq A_s$), a sign of hysteresis.

transforms back to the austenitic phase, completing when the A_f is reached. After this point, more heat supplied only causes thermal expansion of the austenite. This method determines the thermal transitions in the absence of mechanical driving force, making it easier to understand which type of driving force is behind the observed macroscopic (stress–strain) mechanical response. These transition temperatures are strongly affected by the heat treatment performed. Solution treating and aging Ni-rich NiTi gives a fine array of precipitates needed to stabilize the martensite and raise the A_f into a range suitable for medical devices. These defects also raise the stress needed for slip [13], an important consideration for defining the window over which superelasticity can occur. This will be discussed further in the next section.

Usually, the mechanical properties of NiTi are described by two limiting cases of the stress–strain relationships, called superelasticity and shape memory. However, to capture the full range of the mechanical performance possible for SMAs, stress, strain, and temperature must all be displayed together in some fashion, either by a three-dimensional plot of all these parameters (see Figure 20.36) or in a series of stress–strain curves, as displayed in Figure 20.37a-h. In this figure, the test temperature decreases from a-h [14], and the left-hand column shows unloading behavior while the right-hand side shows strain-to-failure curves. The plot at the highest temperature (Figure 20.37a) shows the response typical of an elastic–plastic material; in this case, it is stable, nontransforming austenite. Because high temperature destabilizes the martensite [15], it is possible to increase temperature to a point where stress-inducing martensite is no longer possible; this temperature is called the martensite deformation temperature M_d [16]. It represents the point at which the critical stress for forming martensite is greater than that required to move a dislocation [17]. Therefore, the top curve (Figure 20.37a) shows the expected behavior at temperatures above M_d.

Below M_d, it becomes possible to stress-induce martensite. This effect is shown in Figure 20.37b-c and 20.37d-e, where upon reaching the elastic limit of the austenite, a constant-stress plateau takes place before elastic–plastic deformation of martensite occurs. In this constant-stress region, stress-induced martensite (SIM) is created and this new phase forms in the detwinned state (unlike when it is thermally induced) since the loading direction biases martensite formation toward one variant [9]. It is increasingly easier to stress induce martensite as the temperature is lowered, since less mechanical driving force for transformation is required upon approaching the region where martensite becomes thermally stable (starting at M_s, becoming fully stable at M_f). This is evidenced by the drop in the SIM plateau with decreasing temperature, as seen in Figure 20.38 (from Ref. [9], see also Figure 2 in Ref. [6]).

Upon unloading, SIM is no longer stable and transforms back to austenite along the lower plateau. Full recovery upon unloading only happens if the temperature is above the A_f. This

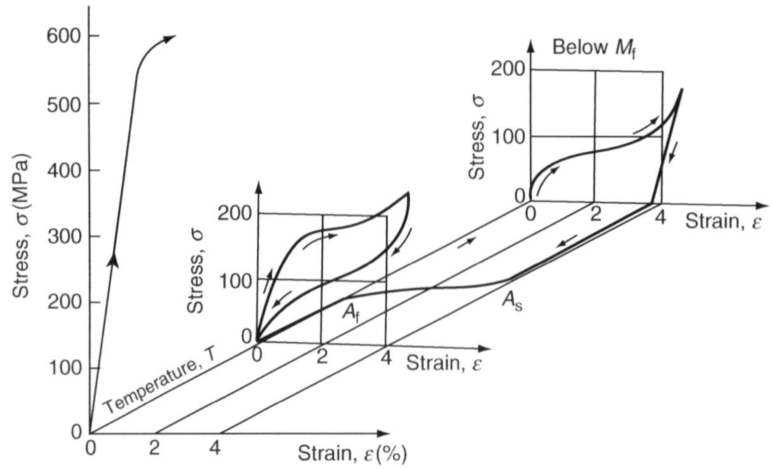

FIGURE 20.36 Three-dimensional plot of stress, strain, and temperature, with temperature increasing into the page (from Wayman, C.M. and Duerig T.W., *Engineering Aspects of SMAs*, Butterworth-Heinemann, London, 1990, p. 3). The curve closest to the reader shows elastic–plastic deformation of austenite, which occurs if the temperature is above M_d. The middle curve displays the behavior that takes place below M_d, but above A_f, commonly referred to as superelasticity. The final part of the schematic displays the entire path taken for the limiting case of shape memory. Following the arrows in the diagram, the sample is cooled to M_f, deformed, unloaded, and heated through A_s to a temperature above A_f for strain full recovery.

is the limiting case of superelasticity (see Figure 20.39 [18]), which occurs in the temperature range of $A_f < T < M_d$. If the SIM is formed below A_f, not all of the strain is recovered upon unloading, leaving what is commonly called permanent set [17]. This term can be misleading, since it is used to describe all unrecovered strain left in the material after removing the mechanical driving force. However, heating above A_f supplies thermal driving force to recover some types of permanent set. A better description is to refer to the permanent set as residual strain that can be made up of thermally recoverable strain, plastic strain, or a combination of both.

If the operating temperature is lowered below M_f, the martensite is fully stable and deforms by first detwinning (as evidenced by a constant stress plateau, see Figure 20.37f and g,

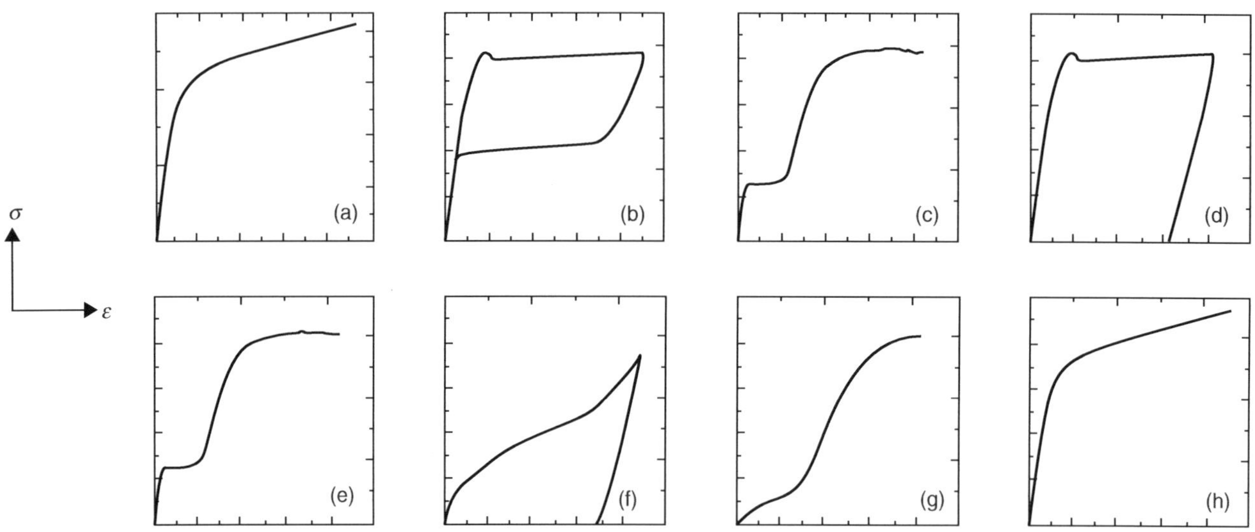

FIGURE 20.37 Alternate way of displaying the thermoelastic interdependence on stress, strain, and temperature. Temperature decreases from a–h in the figure, while strain is on the horizontal axis and stress on the vertical. The plots represent characteristic mechanical behavior under the following temperature conditions: (a) $T > M_d$, stable, nontransforming austenite elastically then plastically deforms; (b) $A_f < T < M_d$, superelastic austenite, where (c) shows the elastic–plastic deformation of the stress induced martensite (SIM); (d) $M_s < T < A_s$, thermally recoverable SIM, where (e) shows the elastic–plastic deformation of the SIM; and (f) $T < M_f$, where martensite with mobile twin boundaries is present. Here the stress plateau occurs upon martensite detwinning, and (g) shows the mechanical response once that mechanism in exhausted; and (h) $T \ll M_f$, where martensite with immobile boundaries exists. The plot shows no detwinning; only elastic–plastic deformation of the martensite is allowed. (Adapted from McKelvey, A.L., PhD dissertation, University of California, Berkeley, 1999. With permission.)

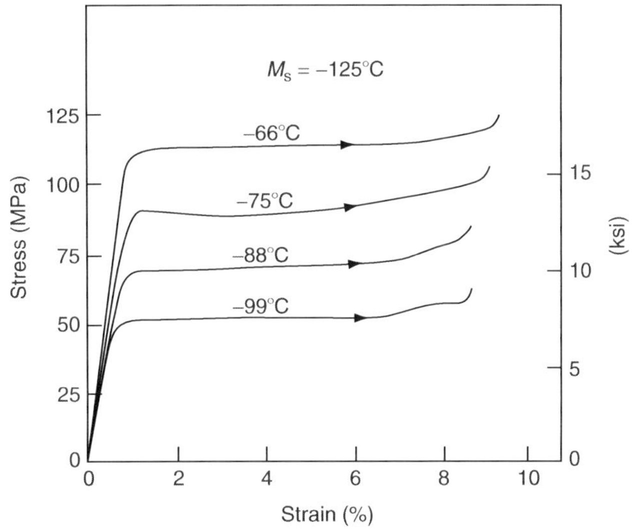

FIGURE 20.38 Stress–strain curves of a SMAs, showing a decrease in the stress plateau as the M_s temperature is approached. Martensite becomes thermally stable at that temperature, requiring less mechanical input to drive the transformation.

which happens at a lower stress than the SIM plateau (i.e., Figure 20.37b-e)) followed by elastic–plastic deformation (Figure 20.37g). This mechanism is utilized in the limiting case of shape memory (see Figure 20.40 [18]). Though not shown in this figure, the original shape will be fully restored upon heating above A_f (see Figure 20.36 [9]). At the most extreme end, it is possible to cool so far below M_f that the twin boundaries are no longer mobile, preventing detwinning. In this case, the material only demonstrates elastic–plastic deformation (see Figure 20.37h). The temperature range over which all these phenomena occur is typically from about 50°C below to about 100°C above M_s [19].

20.6.3 Summary

By covering some of the microstructural mechanisms that occur in NiTi, a systematic description of the observed macroscopic effects can be summarized in a context that allows for a better understanding of what is possible for this complex material. Although most applications typically exploit either the special case of superelasticity or shape memory, covering the additional details of all combinations of thermal and mechanical SME can illuminate additional possibilities or potential limitations, especially in highlighting the influence of temperature.

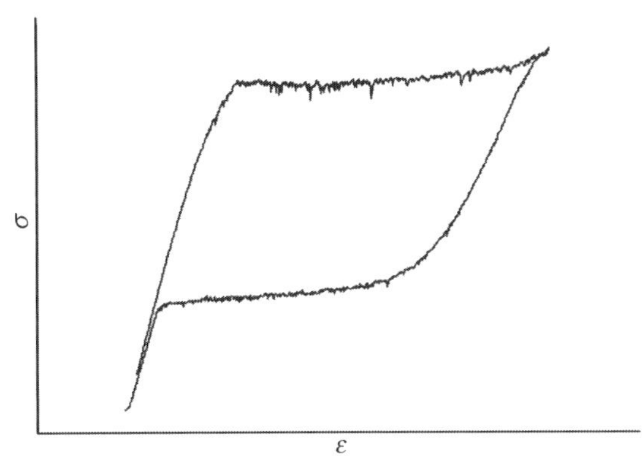

FIGURE 20.39 Stress–strain curve showing fully superelastic behavior. (From Barney, M., 2007, unpublished data. With permission.)

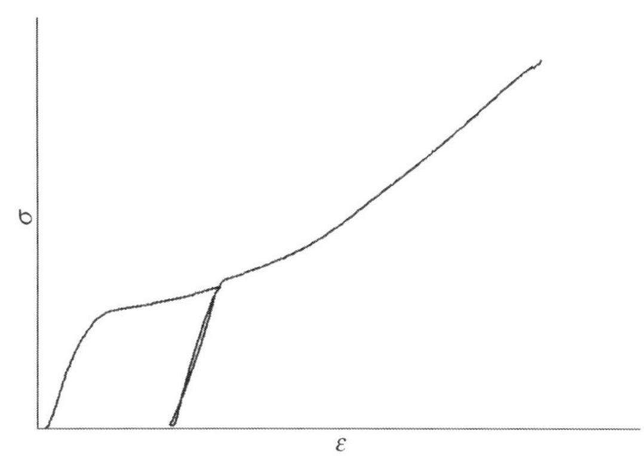

FIGURE 20.40 Stress–strain curve showing the first two steps in the shape memory mechanism. Upon unloading, the material will fully restore its original shape if it is heated above its A_f temperature. (From Barney, M., 2007, unpublished data. With permission.)

Acknowledgments

The authors would like to thank Drs T.W. Duerig, A.R. Pelton, and S.W. Robertson for the many helpful discussions.

References

1. AL McKelvey and RO Ritchie, 1999, Fatigue-crack propagation in nitinol, a shape-memory and superelastic endovascular stent material, *Journal of Biomedical Materials Research A*, **47**(3), 301.
2. K Mukherjee, S Sircar, and NB Dahotre, 1985, Thermal effects associated with stress-induced martensitic transformation in a Ti-Ni alloy, *Materials Science and Engineering*, **74**, 75.
3. JA Shaw and S Kyriakides, 1995, Thermomechanical aspects of NiTi, *Journal of the Mechanics and Physics of Solids*, **43**, 1243.
4. K Bhattacharya, 2003, *Microstructure of Martensite: Why It forms and How It Gives Rise to the Shape-Memory Effect*, Oxford University Press, Oxford.
5. K Otsuka and X Ren, 2005, Physical metallurgy of Ti-Ni based shape memory alloys, *Progress in Materials Science*, **50**, 511.
6. AR Pelton, J DiCello, and S Miyazaki, 2001, Optimization of processing and properties of medical-grade Nitinol wire, SMST-2000: The International Conference on Shape Memory and Superelastic Technologies, Pacific Grove, CA p. 361.
7. GM Michal and R Sinclair, 1981, The structure of TiNi martensite, *Acta Crystallographica B*, **B37**, 1803.
8. D Bronfenbrenner, 2006, unpublished work.
9. CM Wayman and TW Duerig, An introduction to martensite and shape memory, *Engineering Aspects of Shape Memory Alloys*, Butterworth-Heinemann, London, 1990, p. 3.
10. H-S Koo, L Chang, H Chen, and K-L Shieh, 1992, An apparatus for extinguishing a lighted cigarette, Shape-Memory Materials and Phenomena—Fundamental Aspects and Applications, MRS Proceedings, 246, , Warrendale, PA, p. 421.
11. ASTM Designation: F 2004–05, Standard Test Method for Transformation Temperature of Nickel-Titanium Alloys by Thermal Analysis.
12. ASTM Designation: F 2005–05, Standard Terminology for Nickel-Titanium Shape Memory Alloys.
13. S Miyazaki, Thermal and stress cycling effects and fatigue properties of Ni-Ti alloys, *Engineering Aspects of Shape Memory Alloys*, Butterworth-Heinemann, London, 1990, p. 394.
14. AL McKelvey, 1999, PhD dissertation, University of California, Berkeley.
15. S Miyazaki, K Otsuka, and Y Suzuki, 1981, Transformation pseudoelasticity and deformation behavior in a Ti-50.6 at% Ni alloy, *Scripta Metallurgica*, **15**, 287.
16. AL McKelvey and RO Ritchie 2000, On the temperature dependence of the superelastic strength and the prediction of the theoretical uniaxial transformation strain in Nitinol, *Philosophical Magazine A*, **80**(8), 1759.
17. TW Duerig and R Zadno, An engineer's perspective of pseudoelasticity, *Engineering Aspects of Shape Memory Alloys*, Butterworth-Heinemann, London, 1990, p. 369.
18. M Barney, 2007, unpublished data.
19. J Van Humbeeck, R Stalmans, M Chandrasekaran, and L Delaey, On the stability of shape memory alloys, *Engineering Aspects of Shape Memory Alloys*, Butterworth-Heinemann, London, 1990, p. 96.

21

Current Developments in Electrorheological Materials

Frank E. Filisko
The University of Michigan

21.1 Introduction

The history and background of electrorheological (ER) fluids are covered in the 2000 version of *Smart Materials*. This chapter focuses on those areas which might be considered a significant advancement in the field. Some of these were covered in the 2000 edition but at that time they may have been just speculation but have evolved to be more accepted within the area and more of a general concern to interested people. For the sake of space, many references were not included because they maybe redundant. However, main ones are given, which will allow interested parties to search for other relevant articles. It should be added that this includes information, which was presented at the 10th International Conference on ER/MR held in July 2006 and as such may not as yet be published except in the proceedings of the meeting. It should also be noticed that even though the title of this chapter is "current developments," many of the references are not current. The purpose of this is to give adequate recognition to those with insight who may have recognized things years ago that are just beginning to be considered.

21.2 High-Yield Strength Electrorheological Fluids

As recognized over the years as the various problems with ER fluids (ERF) were resolved, i.e., water, caking, high currents, temperature changes, etc. as discussed in the 2000 issue of this book, one of the major issues remaining is that of the yield strengths being too low, i.e., around 3–5 kPa generally. Thus major attempts have been made to increase the strengths of ER materials, most commonly by looking at various components and/or compositions of the fluids. This has been tried for the last 50 years or so and despite the almost infinite number of possibilities has resulted in virtually no consistent success, until possibly now as discussed in the giant electrorheological (GER) section. This is one of the reasons that some feel that the failures may be due to an inadequate or wrong understanding of the mechanisms involved in the phenomenon, specifically that yield occurs by chains breaking consecutively between the electrodes at high fields or strengths, which are of concern here. To further support this, studies performed to measure the modulus of ERF prior to yield have produced moduli in the MPa range [1,2]. Thus an alternate model has been proposed, which is that at high fields or strengths, yield occurs not by shear between the particle chains or columns between the electrodes but by slippage of the dense lamellar particulate structures at the electrode surfaces. There are at least two published ways that this has been tested. One method is by increasing the normal stress on the particulate structures, thus increasing the "frictional" force between the structures and the electrodes [3]. A second is by designing electrodes, which will prevent slippage, something which is not as obvious to do as it may seem [4]. It is not altogether obvious that increasing normal stresses between the electrodes or on the particulate structures can be done without moving the electrodes too close together since the particulate chains or columns (as we are generally led to view them) would apparently readily collapse or buckle. However, it was surprisingly discovered that the structures under E fields, columns if no shear or lamellae if under shear, could carry very large loads as discussed in the section on compression loading/squeeze flow. A similar study was performed by Tang et al. [5] on magnetorheological fluids (MRF) and large increases to around 800 kPa were recorded. It was suggested in this article that something similar should happen with ERF but we were unable to find any published data on this.

The second involved a rather ingenious design of devices to prevent slippage. There are two such devices, which were used and both are described in the Proceedings of the 10th International Conference on ER/MR held in Lake Tahoe, CA in July 2006. In both cases, yield stresses in the 200 kPa range were recorded at about 4 kV/mm field strengths on TiO_2/silicone oil ER formulations.

One explanation for the increased strengths by increasing normal stresses is based upon the alternative surface slip model of yield strength [6]. Another is based upon an explanation by Tang et al. [5] who also suggest surface friction but as well relate the effect to increases thickness of the columns of particles as they are compressed, i.e., increasing the force to initiating shearing as covered in the last section of this chapter. However, their studies did not involve shear, i.e., the photographs they took of the expanding columns, which if they did would have observed the lamellae. Therefore, the explanation they propose, increasing of the cross-sectional area, is identical to that proposed in the alternative model. The third type of high yield stress ERF involve the specific formulation of particles and the suspending fluids and this is separately discussed in the section of GER ERF.

21.3 ER under Compression/Squeeze Flow

ERF have not received a lot of attention in compression or squeeze flow until somewhat recently for reasons mentioned in this chapter. While these materials have been mostly studied under shear, it was pointed out by Sproston et al. [7] that they had interesting and valuable properties under compression or squeeze flow with applications in vibration damping.

The potential of ER materials in squeeze flow or compression was recognized by Stanway et al. [8] for damping control and by Williams et al. [9] who attempted to model squeeze flow for application to automotive engine mounts. In this case they considered the fluid as homogenous, modeling it as a biviscous fluid and did not consider the particle structures as acting independently to support the load although the displacements they were concerned with were small which is satisfactory for large electrode gap.

Monkman [10,11] studied ERF under uniaxial compression, which extended beyond squeeze flow to higher amounts of compression and found that certain ERF as a function of E field and gap width could support compressive stresses as high as 140 kPa, a property which he termed hardness and noted it was the same regardless of E field magnitude but at different electrode gaps. Harquing and King [12] studied an ERF under static and dynamic loading under compression and tension at constant strain rates and reported compressive stresses above 200 kPa.

Vieira and Arruda [13] did a very comprehensive study on the properties of some ERF under compression and shear but terminated their studies at low strains and thus stresses, before they reached the stage experienced by Monkman [10,11]. They reported an initial elastic region, which occurs at very low strains and which they explain as being due to compaction of the initial particle columns, but was more likely due to the "set up" of the fluids in the bleed off channels of their apparatus. This is followed by a nonlinear region in the stress–strain curves, which initiates at some critical stress, which is due to the fluid's beginning to flow. The highest critical stresses they report are around 30 kPa. They, however, very importantly recognize and discuss the consideration that the particles are flowing in/out from between the plates of the device during the tests.

Nilsson and Olson [14] reported compressive forces of 400 N and as well derived equations to model the behavior at low and medium fields under dynamic loading conditions. Recently Tian et al. [15] report compressive stresses near 250 kPa and point out that the behavior under high compressive strains rises more rapidly than predicted by any theory. They suggest this may be due to the fact that the fluids do not act as homogenous materials under all loading conditions.

More recently an extensive study by Lynch et al. [16] concentrates only on the high stress aspects of these studies as a function of particle concentration and viscosity of dispersing oil [16]. They reported compressive stresses above 1000 kPa at compressive strains of only around 30%. They find that the data generally follows a power law behavior but very interestingly the coefficients depend strongly on the viscosity of the dispersing oil. The reason for this is related to the oils pushing particles out from between the plates as the fluid is compressed.

An example of such data is shown in Figure 21.1.

Whereas most theories treat Electroheological Materials (ERM) as homogenous liquids, and model them as Bingham or biviscous fluids [9,17,18], in reality under high electric fields, the materials consist of two phases, one phase being of a series of columns or lamellae of closely packed particles extending between the electrodes in a more or less regular fashion [6] separated by a second phase of essentially particle-free dispersion oil. Depending upon particle size, concentration, and magnitude of E, the columns can be single-particle chains a few microns in diameter (very dilute suspensions) or most commonly a few hundred microns in diameter for ERF with concentrations around 10–20 vol. %. Thus under squeeze flow, stacks of particles are being compressed and a fluid phase is being squeezed out. This suggests the rheological behavior is not simple, though it is modeled as such in most cases as referenced above.

21.4 Giant ER Effect Materials

One of the more significant practical reported discoveries in the field of ER since the discovery of water-free particulate systems in the 1980's is called the giant ER effect (GER) [19–21]. These

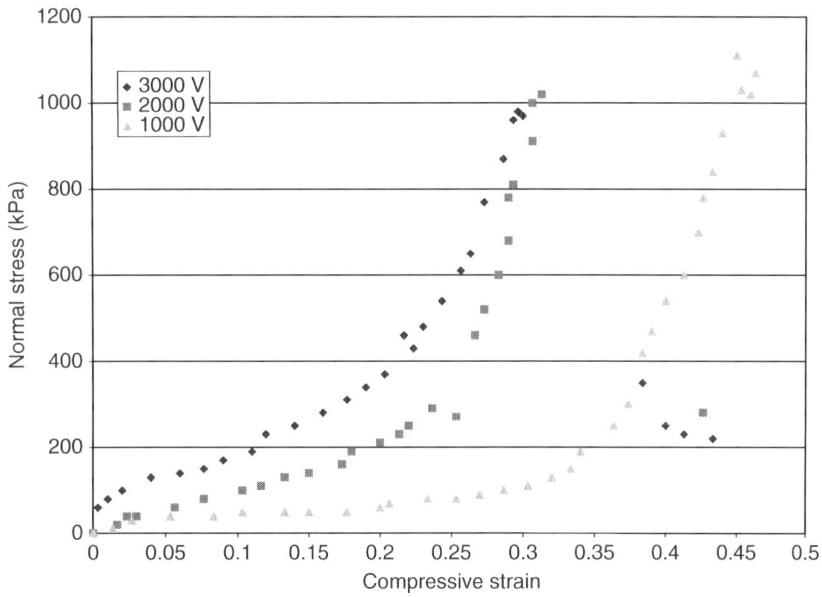

FIGURE 21.1 Normal stress vs. gap for a 40 wt% zeolite/silicone oil ERF under three electric fields.

dispersions, which can consist of a wide variety of particulate materials, and apparently dielectric fluids as well, treated in the fashion in the published papers, show yield stresses in the 100–200 kPa range consistently. Although at the time of this writing, the materials have not been licensed by any company for production or are not being produced commercially nor are samples available from the researchers. This is understandable considering the potential of these materials.

In addition to the high yield stresses, there are at least two aspects of these results, which are interesting. The first is that the results do not follow the rather elementary yet widely ascribed to views of the ER effect, i.e., an initial E^2 qualitative dependence of yield stress on electric field. Second, that the ER effect go to zero as particles get smaller, i.e., in the nanoparticle size range due to Brownian motion due to thermal effects overcoming the effects of particulate dipolar coulombic interactions.

The first reported GER effect [19–21] involved barium titanyl oxalate particles (actually nanoparticles in the 50–70 nm diameter range) coated with a thin urea layer with a dielectric constant of about 60 at 10 Hz, which are about 3–10 nm thick. The yield stresses measured on a 30 vol.% suspension of the particles in silicone oil was about 130 kPa at 5 kV/mm field strength. Other material systems have been reported by this group but in general the GER fluids are dispersions of 50–100 nm particles consisting of a core–shell structure, the core being a hard material consisting of various inorganic salts and the shell of soft urea molecules. The hard-core materials are generally of the form $M1_xM2_{2-x}TiO(C_2O_4)_2$ where M1 is Ba or Sr and M2 is Rb, Cs, or Li. A common example used much in their initial studies being barium titanyl oxalate. The procedures for synthesizing these materials are given in detail in the publications [22]. The explanation for this behavior they relate to a high degree of alignment of polar molecules within the coating and not to the bulk particle.

An extension of this behavior, which was presented at ER 10 in July 2006, was to adsorb polar molecules onto the particle surfaces, materials which were called PM-ER materials. Again very high yield stresses are reported in the range of 200 kPa [23]. The explanation for this behavior is related to polar molecules adsorbed onto the particles. These polar molecules form very strong bonds at the interface between adjacent particles, which are very strong due to the greatly attenuated E field between the particles [24].

Although only speculated on by that group, direct evidence does exist for this in the literature, but involving liquid crystalline solutions with sulfonated polystyrene particles [25]. It is clearly shown involving birefringence that the molecules are collected and oriented at the contact point of the particles. A difference being that the LC studies involved polymeric size particles in solution whereas the adsorbed molecules in their model are much shorter [25].

21.5 Lamellar Particle Structures—Effect of Coupled Electric and Shear Fields

There have been substantial reports of the appearance of unique textures or higher ordered structures (i.e., beyond strings or columns) being formed in dispersions simultaneously exposed to an electric and a shear field. Bossis et al. [26] reported the occurrence of stripes of particles aligned in the flow direction for a dispersion exposed to an electric field and under oscillatory shear. They mention that these stripes, which occur in shear fields alone, may actually be sheets due to the imposed E field.

In another study, they show that under various amplitudes of oscillation and under an E field, these sheets may be perpendicular or parallel to the flow direction. Cutillas et al. [27] compared different structures observed in ERF and MRF under field and oscillatory shear. This group refers in their papers as the phenomenon being a phase separation induced by the field, which it is, of course, in the sense that the dispersed phase, normally uniformly dispersed, is caused to associate into unique structures by either/or E/M and shear fields. Another study involving steady shear and flow through a slit, concentric cylinders, and between parallel rotating disks, Tang et al. [28] report structures consisting of parallel lines of particles in the couette and concentric rings of particles between the parallel disks. Tao and Jiang [29] also reported higher order particulate structure formation for ERF under field and shear. A subsequent extensive study by Henley and Filisko [6] using parallel disks and concentric cylinders, clearly demonstrate the electric field and shear field induced phase separation (assembly) of the dispersed particles as well as the particulate structures resulting from the coupled E and shear fields.

The occurrence of particle aggregation and structuring within dispersions exposed to electric and shear fields has not only been observed experimentally but has been shown in computer simulations [30–32]. The simulations predict the occurrence of layered structures (of particles) in the idealized situation considering point dipole interactions, Brownian, and shear forces but neglecting many-body interactions. In a more recent series of excellent articles, Klingenberg and coworkers [33] not only did extensive work on observing the structures, but developed a theory that shows that these structures must form under certain circumstances from free energy considerations (Figure 21.2)[33].

21.6 Alternative Model for Yield Strength of ER Materials

As pointed out in a previous section of this chapter, the particles in ERF and MRF as well [28] at appropriately high concentrations and under shear, upon application of an electric field, rapidly organize or regiment themselves into numerous tight packed lamellar formations, which are more or less parallel and periodic in relationship to each other. Tight packed means that the particles are crowded together as closely as possible, if not even compressed together under the influence of the E field, consistent with the irregular size and shape of the particles. Because the particles are not monodisperse spheres or any regular geometric shape, they cannot be packed into any regular lattice arrangement such as simple cubic or hexagonal close packed. Under zero shear conditions the particles agglomerate together into columns under the influence of the E field while additionally under shear these particles pack together into these continuous lamellar structures more or less equally spaced and of similar thickness. The relative arrangements and characteristics of these structures are dependent in a not as yet well determined way upon magnitude of electric field, shear rate, particle concentration, time of shear, and shear profile. These structures take the form of walls between parallel plates, short cylinders between parallel disks, and stacked washers

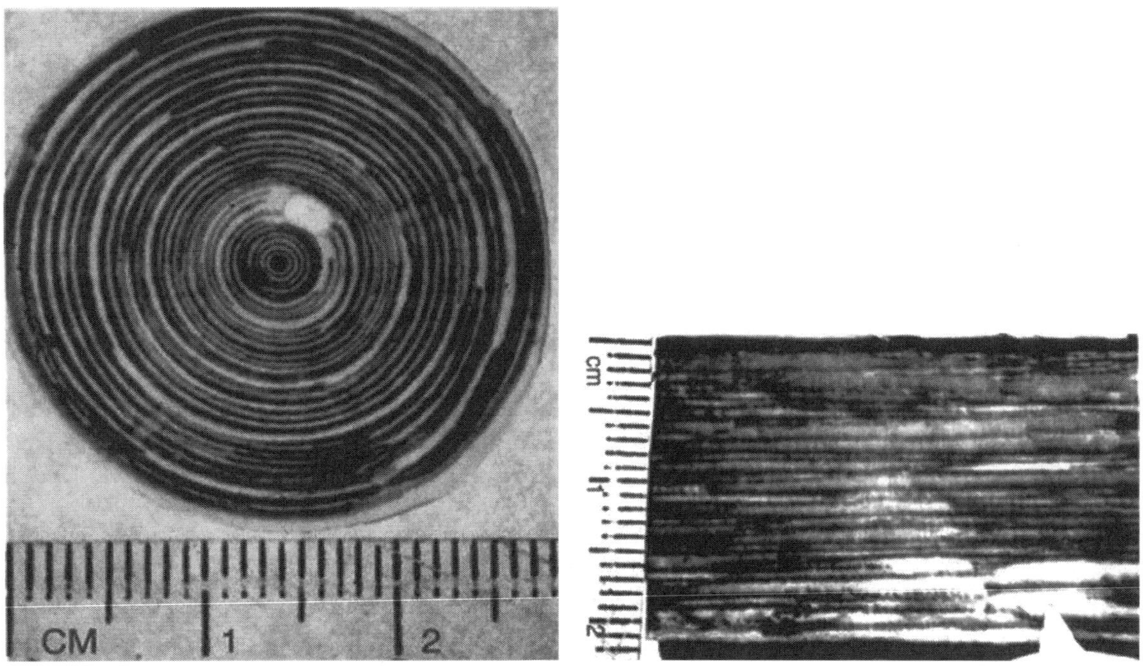

(a) (b)

FIGURE 21.2 Examples of wall and ring particulate structures, i.e., the dark line and rings are particles. The dark spots in the circular figure are the ends of columns.

between concentration cylinders. In all cases, the thickness dimension of the walls, the walls of the cylinders, and the thickness of the washers is the direction normal to plane formed by the E field and shear directions.

A major consequence of the tight packing of the particles into these structures is that it results in cooperative strengthening of the structure to failure under shear in the long dimension of the structure due simply to the increased cross-sectional area. This is because the area of the shear plane increases with lamellar length and thickness as well as the lack of an easy path for slip planes to develop within the structures due to the irregular packing and irregular size and shapes of the particles.

Thus it is hypothesized that the lamellar structures remain intact for the most part under flow (i.e., in the postyield region) and the most probable path of slippage under flow is at the interface between the particle structures and the electrodes. This being the case, it is proposed that the structures themselves do not break or shear. Because polarization forces are responsible for holding the structures together, if the structures do not fail, then these forces are irrelevant to the postyield behavior, as long as they are strong enough to maintain the structure. Thus the postyield behavior is determined primarily by the mechanics of slippage between the ends of the structures and the adjacent electrode. Shear or slippage between the ends of the lamellae and the electrodes is the more dominant source of energy dissipation in ERF under field and the source of the yield stress as well as the determining factor for the yield stress (Figure 21.3)[3].

21.6.1 Theoretical Basis

Consistent with the Bingham model approach, there is critical yield shear stress, T_y, for the onset of flow, which is of course independent of sample dimensions, and a corresponding critical yield shear strain γ_y.

$$T_y = G(E)\,\gamma_y = F_y/(Lt)$$

where

$G(E)$ is the electric field-dependent shear modulus
L is the length of lamellae
t is the thickness of lamellae
F_y is the shearing force to initiate yield

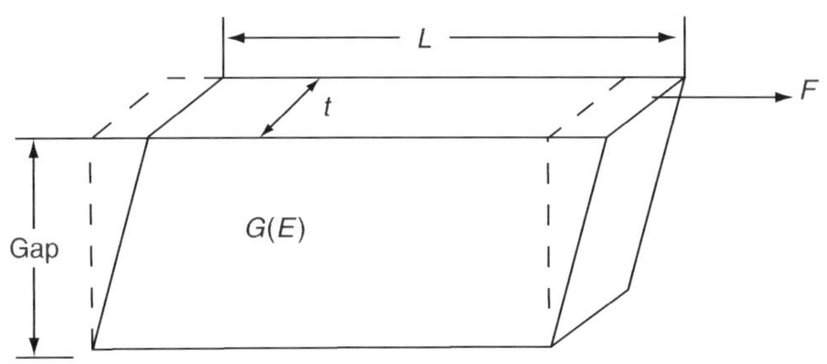

FIGURE 21.3 Model showing a portion of a particle wall under shear.

However, as the shear plane area (Lt) increases, the force(F_y) necessary to reach that yield strain increases with area and must be balanced by frictional (adhesive) forces between the lamellae and the electrodes. The frictional (adhesive) forces (F_a) between the lamellae ends and the electrodes may be partially electrical in nature or mechanical. For lack of a better understanding of the parameters and in analogy with standard frictional phenomena, the surface frictional (adhesive) forces could be described by a static frictional or adhesive coefficient, μ_S, and a normal force between a lamellae end and the electrode, N:

$$F_a = \mu_S N$$

The yield force, however, depends upon shear plane area and field-dependent modulus:

$$F_y = G(E)\gamma_y(Lt)$$

Thus, in order for a lamellae to shear internally, the surface adhesive forces must be greater than the yield force. However, if the force required for yield is greater than the frictional or adhesive forces at the surface, than slip will occur at the surface before internal shear occurs. In other words, for internal shear to occur, $F_a > F_y$. Slip at the wall occurs when $F_y > F_a$. From the above discussion, it can be seen that F_y increases with the length and thickness of the lamellae at a given field, whereas the adhesive force may be independent of the contact area. Thus it can easily be seen what these lamellar structures can accomplish increasing the force necessary for yield by virtue of the increased cross-sectional or shear plane areas.

References

1. S.O. Oyadiji, in Electrorheological fluids, in *Magnetorheological Suspensions and Associated Technology*, Ed. W.A. Bullough, World Scientific, Hong Kong, pp. 419–428, 1995.
2. D.R. Gamota and F.E. Filisko, Dynamic mechanical studies of ER materials under moderate frequencies, *J. Rheol.*, 25(3), 399–427, 1991.
3. F.E. Filisko, An alternative view for the yield stress of ER/MR materials, in *Proceedings of the 10th International Conference on Electrorheological Fluids and Magnetorheological Suspensions*, Eds. F. Gordaninejad, O.A. Graeve, A. Fuchs, Lake Tahoe, CA, July 2006.
4. R. Shen, X. Wang, Y. Lu, W. Wen, and K. Lu, *Proceedings of the 10th International Conference on Electrorheological Fluids and Magnetorheological Suspensions*, Eds. F. Gordaninejad, O.A. Graeve, and A. Fuchs, Lake Tahoe, CA, July 2006.
5. X. Tang, X. Zhang, and R. Tao, Structure enhanced yield strength of MR fluids, *J. Appl. Phys.*, 87(5), 2634–2638, 2000.
6. S. Henley and F.E. Filisko, Flow profiles of electrorheological suspensions: An alternative model for ER activity, *J. Rheol.*, 43(5), 1323–1336, 1999.

7. J.L. Sproston, R. Stanway, E.W. Wiliams, and S. Rigby, The electrorheological automotive engine mount, *J. Electrostat.*, 32(3), 253–259, 1994.
8. R. Stanway, J.L. Sproston, M.J. Pendergast, J.R. Case, and C.E. Wilne, ER fluids in the squeeze mode: An application to vibration isolation, *J. Electrostat.*, 28, 89–94, 1992.
9. E.W. Williams, S.G. Rigby, J.L. Sproston, and R. Stanway, Electrorheological fluids applied to an automotive engine mount, *J. Non-Newtonian Fluid Mech.*, 47, 221–238, 1993.
10. G.J. Monkman, The electrorheological effect under compressive stress, *J. Phys. D*, 28(3), 588–593, 1995.
11. G.J. Monkman, Exploitation of compressive stress in ER coupling, *Mechatronics*, 7(1), 27–36, 1997.
12. G. Haiqing and L-M. King, Experimental investigations on tension and compression properties of an ER material, *J. Intell. Mater. Syst. Struct.*, 7, 89–96, 1996.
13. S.L. Vieira and A.C.F. Arruda, ER fluids response under mechanical testing, *J. Intell. Mater. Syst. Struct.*, 9, 44–59, 1998.
14. M. Nilsson and N. Ohlson, An electrorheological fluid in squeeze mode, *J. Intell. Mater. Syst. Struct.*, 11, 545–554, 2000.
15. Y. Tian, Y. Meng, and S. Wen, Electrorheological fluid under elongation, compression, and shearing, *Phys. Rev. E* 65, 031507, 2002.
16. R. Lynch, Y. Meng, and F.E. Filisko, Compression of dispersions to high stress under electric fields: Effect of concentrations and dispersing oil, *J. Coll. Interf. Sci.*, 297, 322–328, 2006.
17. J. Stefan, *Akad. Wiss. Math. Natur. Wien*, 69, part 2713–2735, 1874.
18. J.R. Scott, *Trans. Inst. Rubber Ind*, 7, 1931, 169–186; 481–493, 1935.
19. W. Wen, X. Huang, S. Yang, K. Lu, and P. Shing, The giant ER effect in suspensions of nanoparticles, *Nat. Mater.*, 2, 727–730, 2003,
20. W. Wen, X. Huang, and P. Sheng, Particle size scaling of GER, *Appl. Phys. Lett.*, 85(2), 299–301, 2004.
21. K. Lu, R. Shen, X. Wang, G. Sun, and W. Wen, ER fluids with high shear stress, *Int. J. Mod. Phys, B*, 19, 1065–1070, 2005.
22. K. Lu, R. Shen, X. Weng, G. Sun, W. Wen, and J. Liu, Polar molecule type ER fluids, in *Proceedings of the 10th International Conference on Electrorheological Fluids and Magnetorheological Suspensions*, Eds. F. Gordaninejad, O.A. Graeve, and A. Fuchs, Lake Tahoe, Nevada, July 2006.
23. W. Wen, X. Huang, and P. Shing, *Appl. Phys. Lett.*, 85(2), 299–301, 2004.
24. Y. Chen, A. Sprecher, and H. Conrad, *J. Appl. Phys.*, 70(11), 6796, 1991.
25. G.P. Quist and F.E. Filisko, ER suspensions with ER active matrix liquids, *Int. J. Mod. Phys B*, 13(14–16), 1675–1681, 1999.
26. G. Bossis, Y. Grasselli, E. Lemaire, L. Petit, and J. Persello, Yield stress and field induced structure in electro and magnetorheological suspensions, in *Proceedings of the International Conference on ER fluids, Mechanisms, Properties Structure, Technology, and Applications*, Ed. R. Tao, pg. 75, World Scientific, Singapore 1992.
27. S. Cutillas, G. Bossis, A. Cebers, E. lemaire, and A. Meunier, Flow and field induced structures in relation with the rheology of ER and MR fluids, in *Proceedings of the International Conference on ERF, MRS, and Their Applications*, Eds. Koyama, K. and Nakano, M., Yonezawa, Japan, p. 43, July 22–25, 1997.
28. X. Tang, W.H. Li, X.J. Wang, and P.Q. Zhang Structure evolution of electrorheological fluids under flow conditions, *Electrorheological Fluids, Magnetorheological Suspensions and their Applications*, Eds. M. Nakano and K. Koyama, World Scientific, Singapore, pp. 164–171, 1998.
29. R. Tao and Q. Jiang, Structure formation in ER fluids, in *Progress in Electrorheology*, Eds. F.E. Filisko and K. Havelka, Plenum Press, New York, pp. 325–333, 1995.
30. J.R. Melrose and D.M. Heyes, Simulation of ER and particle mixture suspensions: Agglomeration and layered structures, *J. Chem. Phys.*, 98(7), 5873–5886, 1993.
31. M. Whittle, Computer simulation of an ER fluid, *J. Non Newtonian Fluid Mech.*, 37, 233–263, 1990.
32. J.R. Melrose, Brownian dynamics simulation of dipole suspensions under shear: The phase diagram, *Mol. Phys.*, 76(3), 635–660, 1992.
33. K. vonPfeil, M.D. Graham, D.J. Klingenberg, and J.F. Morris, Structure evolution in ER and MR suspensions from continuum perspective, *J. Appl. Phys.*, 93(9), 5769–5779, 2003.

22

Carbon Microtubes and Conical Carbon Nanotubes

Santoshrupa Dumpala
University of Louisville

Gopinath Bhimarasetti
University of Louisville

Suresh Gubbala
University of Louisville

Praveen Meduri
University of Louisville

Silpa Kona
University of Louisville

Mahendra K. Sunkara
University of Louisville

One-dimensional and conical carbon structures were reported as early as 1960 [1,2], however, no evidence of hollowness was presented at that time. The carbon nanotubes were discovered much later in 1991 [3], consisting of single or multiple layers of cylindrical graphene sheets. The morphology of all the derived structures of nanotubes was primarily cylindrical in nature with internal diameters ranging from 1 to 25 nm [4].

In the last five years, a number of new morphologies of carbon tubular structures with hollow cores have been discovered. These include the conical carbon morphologies tapering from micron-sized base to a nanometer scale tip [5–7], with a hollow core not exceeding few tens of nanometers, and micron-scale morphologies with internal diameters exceeding several hundred nanometers [8–10]. All these morphologies are unique compared to carbon nanotubes in that they provide a clearer connection between the micrometer and nanometer scales, at the same time facilitating newer applications [6]. The promising application areas include electrochemistry, biosensors, field emission, high surface area catalysts/supports, and possibly drug delivery. This chapter is devoted to providing a detailed discussion on their synthesis, possible growth mechanisms, and applications.

22.1 Introduction

In this chapter, all the new morphological forms of carbon are considered and are divided in to two main categories for ease of discussion: (a) carbon microtubes (CMTs), which include all the structures containing internal diameters greater than 100 nm and (b) (CCNTs), which include all tapered morphologies with hollow cores and internal diameters of only a few nanometers.

22.1.1 Carbon Microtubular Morphologies

Typically, carbon nanotubes are formed as a result of self-assembly of graphene sheets into tubular structures on the surface of the nanometer-sized catalyst cluster [3]. However, such self-assembly processes cannot take place readily on micron-sized catalyst clusters. Many of the earlier processes using micron-sized catalyst particles yielded solid graphite/carbon whiskers but not tubular structures [11]. So, a set of different process conditions than those used for typical self-assembly processes are required for making carbon tubular structures with large internal diameters.

The main types of morphologies of CMTs achieved are illustrated with electron microscopic images in Figure 22.1: (a) Graphite tubes with internal diameters ranging from 70 to 1300 nm having inner obstructions reminiscent of bamboo styling; (b) thin-walled, straight carbon tubular structures with internal diameters ranging from few hundred nanometers to few microns; (c) thin-walled, conical microtubular structures; and (d) micro-tubular carbonaceous tubes made by the pyrolysis of polymeric fibers.

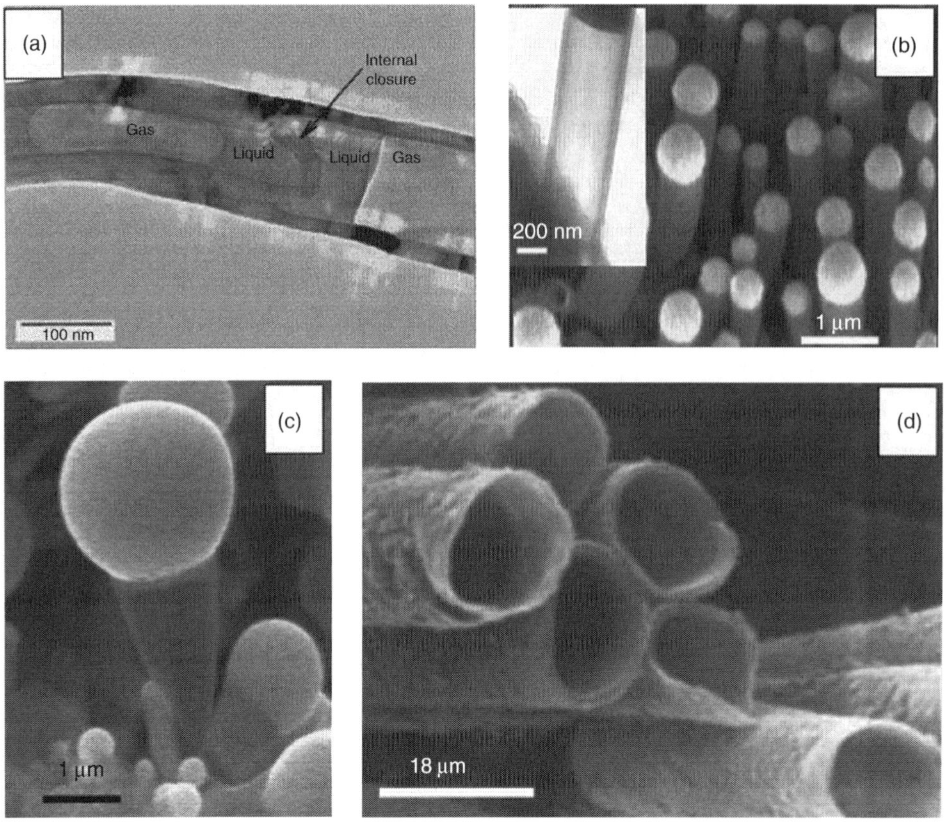

FIGURE 22.1 Various carbon microtubular morphologies. (a) CMTs with encapsulated liquids and catalyst particles using a hydrothermal method. (From Libera, J. and Gogotsi. Y., *Carbon*, 39, 1307, 2001. With permission.) (b) Straight CMTs with diameters in the range of a few microns obtained using low-melting metals such as gallium. (c) Tapered CMTs with thin walls using low-melting metals such as gallium with a different set of conditions than (b). (d) Carbon tubes made by pyrolizing polymeric fibers. (From Han, C.C. et al., *Chem. Mater.*, 11, 1806, 1999. With permission.)

The wall structures of the CMTs, shown in Figure 22.1, are all different from each other depending upon their morphology and method of synthesis. The graphite tubes shown in Figure 22.1a exhibit crystalline, graphitic walls but contain bamboo-styled internal obstructions reminiscent of the nickel catalyst used for their synthesis. The walls of the microtubes resulting from the pyrolysis of polymeric fibers typically contain disordered carbon. The thin-walled, CMTs shown in both Figures 22.2b and c contain clear internal channels and their walls are made of graphene sheets. However, the straight morphologies with uniform diameter contain walls of parallel graphene sheets similar to that of multiwalled carbon nanotubes (see Figure 22.2a). Even a slight conical angle makes the walls exhibit discontinuous regions with parallel graphene sheets. The walls within the conical microtubular structures cannot maintain any sort of ordered arrangement and so break down into an ordered arrangement of small graphitic domains within a disordered carbon matrix as shown in Figure 22.2c. Almost all of the microtubular structures with their thin walls are electron transparent even at modest accelerating voltages of 15 keV or more.

22.1.2 Conical Carbon Nanotube Morphologies

Theoretical calculations predict that a single graphene sheet can be folded through a specific arrangement of pentagonal and hexagonal rings to produce a set of conical morphologies with a limited number of discrete conical angles [12]. Some of the structures synthesized to-date include fullerene cones [13] and nanohorns [14]. All the above structures are true nanoscale morphologies. However, there is another important class of conical morphologies that extend from micron to nanometer scale. By definition, we refer to them as CCNTs in which the wall thickness tapers from micron to nanometer scale but contains a uniform hollow interior with a diameter range similar to that of multiwalled carbon nanotubes. Several morphologies of the conical carbon nanotubes were referred to earlier as tubular graphite cones (TGCs) [6], carbon nanopipettes (CNP) [5], conical graphite tubes [15], and conical nanocarbon (CNC) [7]. The morphologies of several of these CCNT structures are illustrated in Figure 22.3. All of them have micron scale bases extending conically to form nanoscale tips similar to that shown

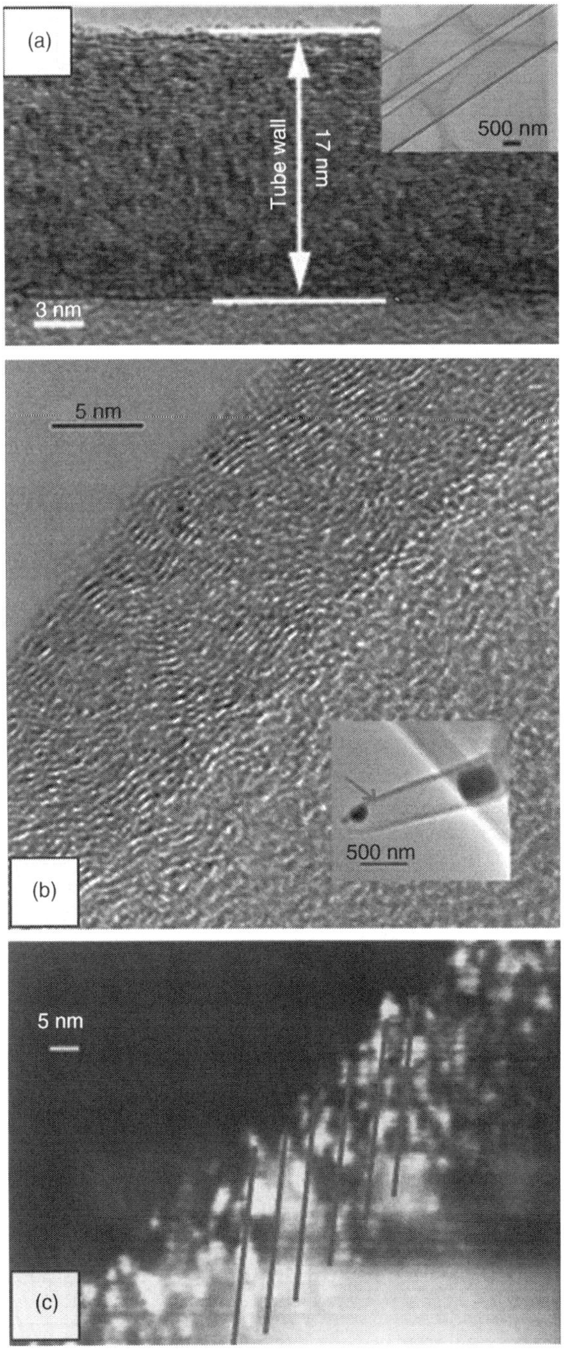

FIGURE 22.2 Variations observed in the wall structures for CMTs synthesized using low-melting metals: (a) CMT walls made of long continuous concentric graphite sheets. (From Hu, J. et al., *Adv. Mater.*, 16, 153, 2004. With permission.) (b) The walls made up of parallel, but discontinuous graphene sheets. (From Bhimarasetti, G. et al., *Adv. Mater.*, 15, 1629, 2003. With permission.) (c) TEM photograph of the wall structure showing aligned nanocrystals of graphite. (From Bhimarasetti, G., Cowley, J.M., and Sunkara, M.K., *Nanotechnology*, 16, S362, 2005. With permission.)

in Figure 22.3a. In some cases, as seen in Figure 22.3b, carbon nanotubes with extended lengths can be found at the tip of a CCNT [16]. In some cases, thick multiwalled carbon nanotubes can themselves be tapered at the tips to produce conical shape (see Figure 22.3c). In other cases, conical structures having a hollow core with the catalyst particle closing the tip have been observed (see Figure 22.3d).

Although these morphologies look similar from the exterior, the wall structures could be completely different. The CCNTs can be divided into three main types based on their wall structure as shown in Figure 22.4. One type of structure consists of a central carbon nanotube with graphene sheets helically rolled over the tube giving the conical structure [5] (Figure 22.4a). The second type also consists of a central nanotube, but its wall is made of concentric cylinders of graphene sheets with gradual reduction in the thickness along the length to yield conical shape [6] (Figure 22.4b). Neither of these structures contain any metal contamination or obstructions within their internal channels. The third structure has stacked graphene layers (similar to multiwalled carbon nanotubes) and tapers after a certain length towards its tip. The internal channel of the third type of structure may not be as perfect as that of the first two types and also contains several types of internal obstructions resulting from the catalyst particle assisted growth for extended periods of time [17] (Figure 22.4c).

22.2 Synthesis and Morphological Control of CMTs

22.2.1 Synthesis Methods and Growth Mechanisms

The synthesis of CMTs requires totally different process conditions and catalysts compared to that of multi-walled carbon nanotubes (MWNTs). CMTs with thick walls, as shown in Figure 22.1a, were produced using micron-sized nickel catalyst particles by a hydrothermal synthesis method at temperatures and pressure of ~800°C and ~100 MPa, respectively [9]. The precipitation of carbon at high pressures onto larger nickel catalyst particles suspended in water leads to the formation of highly crystalline graphite tubes with larger internal diameters. The bamboo-styled internal obstructions observed are similar to those seen in the MWNTs grown using nickel catalysts.

The most predominant morphologies of CMTs, i.e., thin-walled CMTs with no internal obstructions are made only using low-melting metals such as Ga, Sn, Zn, and In. See Table 22.1 for a concise review of all the synthesis studies reported. The first set of studies used carbothermal reduction of gallium oxide powder to produce thin-walled, straight CMTs half-filled with Ga [8,18]. Several other reports showed minor variations to produce CMTs with encapsulated metals with a range of internal diameters [19–23]. With subtle variations in the gas phase chemistry along with low-melting metals, a number of morphological shapes ranging from conical to straight tubes have been reported [10]. As shown in Table 22.1, all methods except the hydrothermal synthesis utilize chemical vapor deposition of carbon onto low-melting metals such as Zn, Sn, In, and Ga. These studies utilized various oxide, nitride, and sulfide sources for metals such as ZnS, SnS, GaN, Ga_2O_3, CdS, SnO, In, and Ga along with different gases and solid carbon sources.

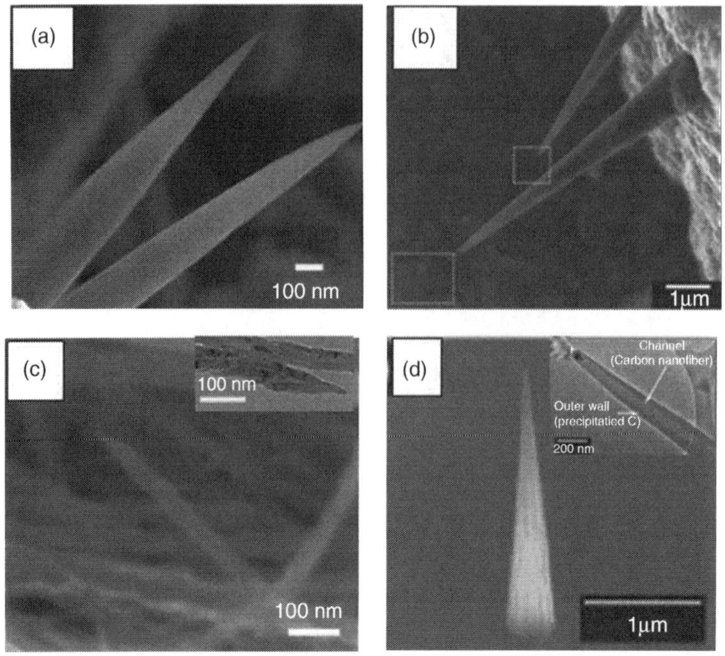

FIGURE 22.3 Various morphologies of CCNT morphologies: (a) CCNT structures termed as CNP with constant hollow cores and open tips. (From Mani, R.C. et al., *Nano Lett.*, 3, 671, 2003. With permission.) (b) A CCNT structure termed as "tubular graphitic cone" with a nanotube at the tip. (From Shang, N., Milne, W.I., and Jiang, X., *J. Am. Chem. Soc.*, 129, 8907, 2007. With permission.) (c) Long carbon nanotubes with conical tips and the inset shows the high resolution image of the conical tip. (d) A CCNT with the catalyst encapsulated at the tips. (From Merkulov, V.I. et al., *Appl. Phys. Lett.*, 79, 1178, 2001. With permission.)

All methods that employed sulfides, nitrides, and oxides for metal sources produced straight, thin-walled CMTs. Another method using a mixture of activated carbon, ZnS and SnS, produced both straight and conical carbon tubes using different process conditions [21]. The reaction was carried out in low-pressure N_2 atmosphere at 1350°C. It was suspected that the Sn nanostructures with different morphologies templated the growth of CMTs with a similar variation in morphology. When Ga_2O_3 was used as the metal source, the gallium droplets formed during dissociation assisted the growth of carbon wall around them [8,18]. In the case where Mg and Fe catalyst were used as the metal sources, growth of carbon tube at the interface of MgO and Fe was proposed [23].

Different mechanisms have been proposed for CMT growth using low-melting metals. However, only two of the suggested mechanisms are generic and can explain the growth of a variety of CMT morphologies. One of the suggested mechanisms involves vapor-liquid-solid type growth, i.e., preferential dissolution of carbon into low-melting metals such as Zn followed by precipitation as a thin wall forming the tube around the micron-sized droplets (see the schematic in Figure 22.5a). Here, ZnS was used as the source for Zn metal along with activated carbon [19]. The synthesis was carried out in N_2 atmosphere at 200 mtorr base pressure and 1400°C. However, there is no thermodynamic data available to suggest the possibility of carbon dissolution into molten gallium and other low-melting metals. On the other hand, the hydrocarbon adsorption onto the surface of low-melting metal melt is more likely (similar to that of bonding in trimethyl gallium). The adsorbed hydrocarbons could contribute to the growth of graphene wall around the molten metal droplet, leading to the growth of the CMT. The molecular level aspects of carbon precipitation on molten metal surface still need to be explored.

The droplet-led mechanism by itself does not provide any insight into the growth of various CMT morphologies. In order to understand the low-melting metal droplet led growth of CMTs, it is proposed that the addition of carbon at the molten metal–carbon interface occurs in such a way that the contact angle between the metal and carbon is constantly maintained. This mechanism is schematically illustrated in Figure 22.5b. As the carbon tube grows in length, the interface lifts the metal droplet maintaining a steady contact angle between the meniscus and the growing carbon wall, thus setting up the conical angle of the overall structure. The meniscus angle and the conical angle of the resulting carbon tube are related by the following relationship:

$$\phi = 2\theta - 180^\circ \tag{22.1}$$

where
- ϕ is the conical angle
- θ is the meniscus angle

The latter is directly related to the contact angle between the liquid metal and the carbon wall. For example, a meniscus angle of 90° should lead to a straight microtube [10].

The meniscus angle is primarily controlled by the wetting behavior of molten metal with carbon. The wetting or contact

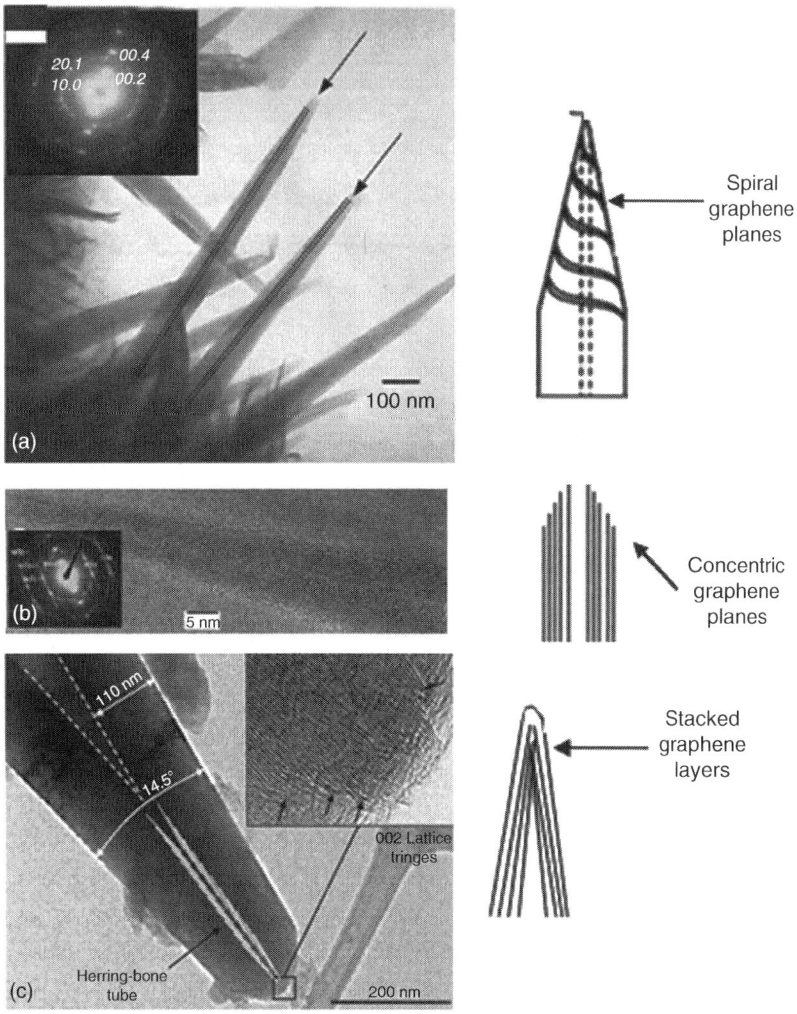

FIGURE 22.4 Different wall structures observed for CCNT structures: (a) TEM image of a CCNT with inner hallow core and diffraction pattern showing spiral graphene sheets. (From Mani, R.C. et al., *Nano Lett.*, 3, 671, 2003. With permission.) (b) HRTEM image of a CCNT with central hallow core and wall made of concentric cylinderical graphene sheets. (From Zhang, G., Jiang, X., and Wang, E., *Science*, 300, 472, 2003. With permission.) (c) TEM image of a CCNT showing the stacked gaphene layers forming a tapered inner hallow core. (From Gogotsi,Y., Dimovski, Y.S., and Libera, J.A., *Carbon*, 40, 2263, 2002. With permission.)

TABLE 22.1 Review of All Synthesis Methods Reported to-Date for CMTs

Carbon Source	Metal Source	Carrier Gas	Other Gas Phase Components	Process Conditions	Resulting Morphology	Reference
Graphite powder	ZnS, SnS	N_2	S (from dissociation of ZnS and SnS)	Thermal CVD, 1400°C, 20 Pa base pressure	Straight tube, conical horn	[19]
C_2H_2	GaN	Ar	N (from dissociation of GaN)	Thermal CVD, 1150°C, 200 torr	Tubes with very low conical angles with bulb like morphology on the top	[8]
Activated carbon	Ga_2O_3	N_2	O from dissociated Ga_2O_3	Thermal CVD, 1360°C	Straight morphology	[9,17]
Activated carbon	CdS, SnO	N_2	S and O from dissociated CdS and SnO	Thermal CVD, 1150°C, 1 atm	Straight tubes with very small conical angle	[20]
Polyethylene	Ni powder	—	—	Pyrolysis, 730°C–800°C, 90–100 MPa	Straight tubes	[54]
CH_4	Ga	H_2	Varying amounts of O_2 and N_2	PECVD, 600–800°C, 40 torr	Straight tubes, cones, tune-on-cones, funnels, n-staged morphologies capsule, dumbbells	[22]

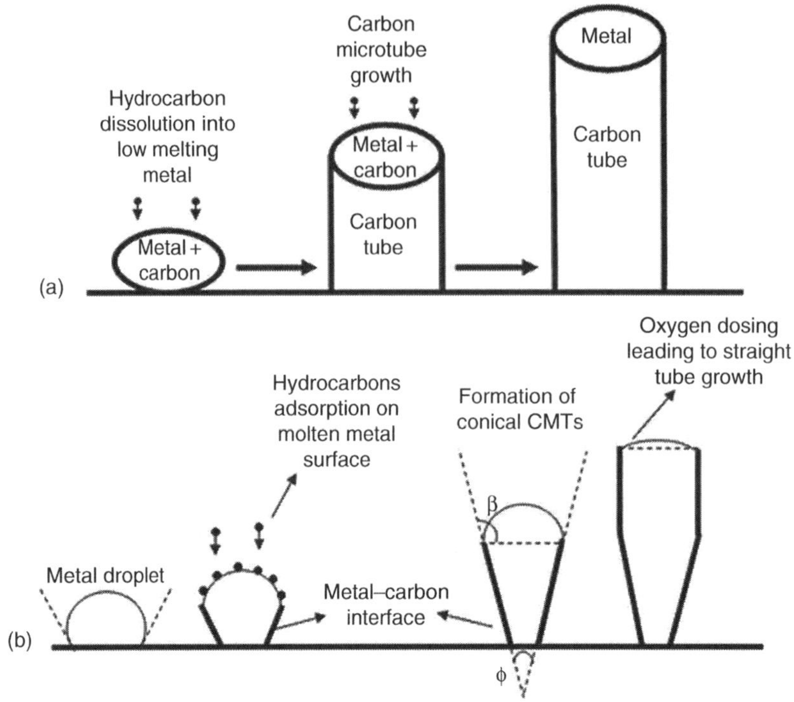

FIGURE 22.5 Schematics illustrating various suggested mechanisms for growth of CMTs using low-melting metals. (a) A schematic illustrating a mechanism in which carbon dissolves into the molten metal droplet and precipitates as CMT. (b) A schematic illustrating another mechanism, which explains the molten metal-carbon wall interface and its influence on the morphology of the resulting CMT.

angle of molten metal on carbon surfaces could vary both with the gas phase composition and the temperature. If the contact angle θ, between carbon and the low melting metal droplets, can be varied in-situ during growth, it is possible to control the morphology of the growing carbon structure. In the case of gallium, it is known that the contact angle between gallium and carbon can be reduced in the presence of oxygen or nitrogen [24,25], resulting in carbon tubular structures with very small conical angles or near straight tubes as predicted by Equation 22.1. It is also likely that the presence of sulfur in the gas phase could modify the surface of the low-melting metal melt reducing its surface tension and, in turn, improving its wetting on carbon surface [26]. This mechanism of "wetting angle controlled growth" can also be applied to almost all the methods reviewed in Table 22.1. For example, the use of gallium oxide powder provides the necessary dosing of oxygen in the gas phase and can explain the results of straight CMTs with encapsulated gallium when using Ga_2O_3 and carbon powder [8]. Similarly, straight CMT structures were obtained when using ZnS [21], which is also consistent with the proposed mechanism of the gas-phase sulfur reducing the surface tension. However, the wetting properties also depend on the temperature with contact angles being higher at lower temperatures. The observation of conical CMTs in the low-temperature regions of the furnace lead to better understanding about the wetting properties as a function of temperature.

22.2.2 Control of Morphology

According to Equation 22.1, the conical angle of the CMTs can be varied by changing the wetting behavior of molten metals with carbon, which in turn, can be controlled by changing the gas phase composition or temperature during their growth. In the absence of oxygen or nitrogen, experiments done at 600–1100 W microwave power, 40–90 torr pressure over molten gallium with molybdenum as the promoter and CH_4/H_2 plasma environments, yielded only conical morphologies with conical angles of +20° (Figure 22.6a). Similar experiments with intentional gas phase dosing of 5 sccm of oxygen have yielded straight morphologies (Figure 22.6b). Later experiments in the absence of oxygen and then followed by oxygen dosing yielded a predictable shape of tube-on-cone morphology as shown in Figure 22.6c. As predicted, the reverse sequence during CMT growth, i.e., initial dosing of oxygen followed by no dosing of oxygen produced a cone-on-tube or funnel-shaped morphology (Figure 22.6d). Repeating these dosing sequences "*n*" times, one can obtain "*n*-staged" morphologies as shown in Figure 22.6e.

Even small amounts of O_2 (5 sccm) leads to reduced surface tension of Ga, in turn reducing the conical angle of the resulting morphology significantly. However, there is a limitation to how much O_2 can be used before gallium oxidizes, resulting in Ga_2O_3 nanostructures [24]. Nitrogen on the other hand, does not readily form GaN at the process temperatures used, and therefore, can be used to further fine-tune the conical angles. In fact, the

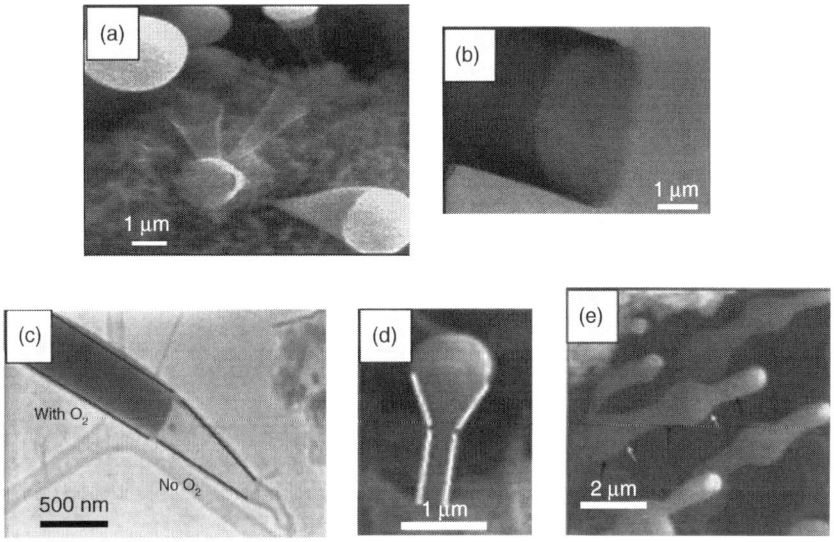

FIGURE 22.6 SEM images showing various morphologies of CMTs synthesized by controlled, in-situ oxygen dosing: (a) funnels (From Bhimarasetti, G., Cowley, J.M., and Sunkara, M.K., *Nanotechnology*, 16, S362, 2005. With permission), (b) cone-on-tube, (c) straight tubes, (d) tube-on-cone, and (e) six-stage morphology. (Figures c, d, e taken from Bhimarasetti, G. et al., *Adv. Mater.*, 15, 1629, 2003. With permission.)

experiments using nitrogen dosing with amounts ranging from 0 to 35 sccm yield morphologies with conical angles ranging from +20° to −15° (see Figure 22.7a and b). At higher nitrogen dosing, the conical angles of the resulting structures were negative, leading to converging cones compared to the observed diverging cones with no or little amounts of dosing (Figure 22.7c). The converging conical geometry limits the length of the resulting morphologies, ultimately leading to nanocapsules with high nitrogen dosing (Figure 22.7a). The ability to create converging cones could be used to create pinched morphologies using a three-step sequence of nitrogen dosing: the first step involves growth for few minutes without the presence of nitrogen; the second step involves high amounts of nitrogen during the growth for few minutes to create a converging cone; and the

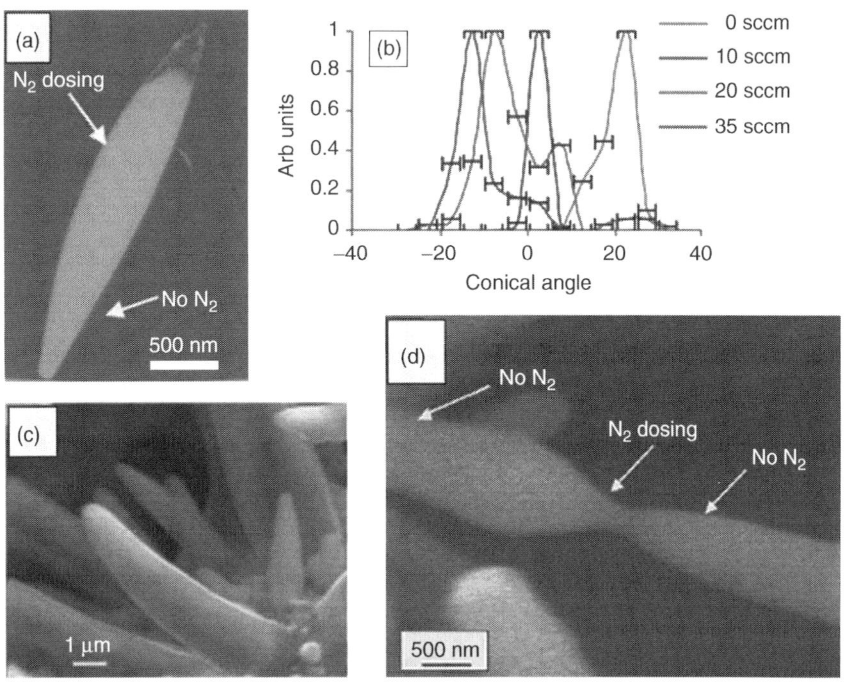

FIGURE 22.7 SEM images and a plot showing the control of conical angle of the resulting microtube morphology using nitrogen dosing during growth: (a) capsule, (b) a plot showing the effect of increasing nitrogen concentration as a function on the conical angle of the resulting CMT, (c) horn structures with converging conical angles, and (d) dumbbell-shaped morphologies. (From Bhimarasetti, G., Cowley, J.M., and Sunkara, M.K., *Nanotechnology*, 16, S362, 2005. With permission.)

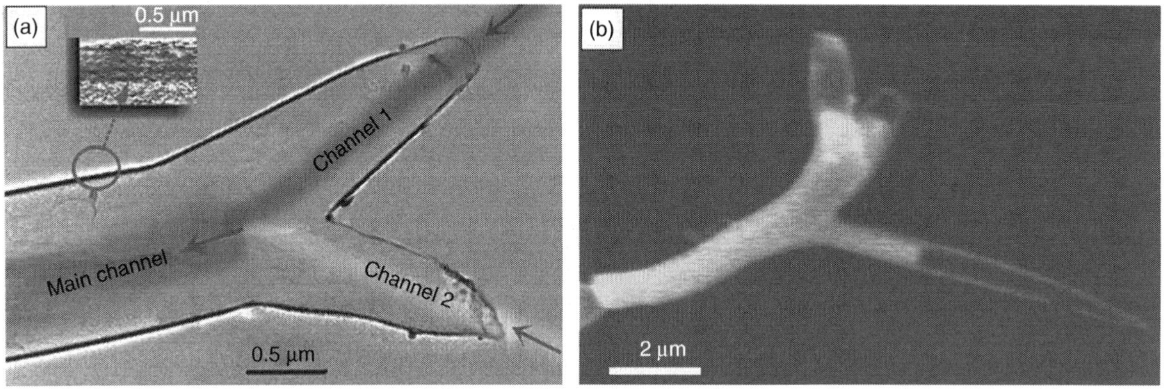

FIGURE 22.8 TEM images of branched morphologies: (a) Y-junctions formed by spontaneous coalescence of liquid metals upon physical impingement of two CMTs. (b) A CMT structure with multiple branches resulting through coalescence of several CMTs during growth.

third step involves growth without the presence of nitrogen. The net morphology will be a dumb-bell-shaped pinched structure as shown in Figure 22.7d in which the lengths of each portion are controlled by the durations used for each step. One can also control the internal diameters of the resulting structures by varying the times at which nitrogen is introduced during the CMT growth [26].

The walls of the conical CMTs contain nanocrystals of graphite whose orientation with respect to the wall depend very much on the conical angle of the tube. The specific orientation of each nanocrystal can be directly related to the contact angle between carbon and gallium during the tube formation [26]. The molten metal droplets on the tips of two or more CMTs could coalesce quickly into one larger droplet during growth upon physical impingement leading to Y- and multiple-junction CMTs for microfluidic applications as shown in Figure 22.8a and b. The as-synthesized CMT structures are partially filled with the low-melting metal used. As the CMTs have open channels, it is fairly easy to remove the low-melting metals using either acids to dissolve the metals or by heating them in vacuum.

The availability of CMT structures in large quantities will find uses as mesoporous materials, lithium ion batteries, and for other electrochemical energy conversion applications. Current efforts in the laboratories of the authors and others are aimed at large-scale production and purification of these structures from metal contamination. In addition to large-scale production, the integration of these structures into microfluidic devices will also need to be addressed.

22.3 Synthesis and Morphological Control of CCNTs

22.3.1 Synthesis Methods and Growth Mechanisms

The synthesis of the CCNT structures differs extensively from that of thin-walled, carbon microtubular structures. Firstly, the synthesis methods do not involve low-melting metals and secondly, the CCNT structures are synthesized at much higher temperatures. Most of these CCNT structures are formed with the help of transition metal catalyst particles and characteristically contain a nanometer scale hollow core or a central nanotube.

The plasma discharge assisted growth of multiwalled carbon nanotubes using catalysts resulted in the formation of conical carbon nanotubular structures or termed as CNC [7]. These structures contained the catalyst particle encapsulated at their tips while the thickness tapered from microns at the base to nanometers at the tip. The synthesis was performed using a PECVD system with acetylene and ammonia in the gas phase at 700°C and pressures of 2–5 torr. The growth of conical morphology is explained by the growth of central carbon nanofiber from the catalyst particle, and further carbon deposition from the plasma-assisted hydrocarbon decomposition. The net result is the growth in two directions, vertically (inner carbon fiber) and laterally (the deposition on the outer wall) forming the carbon nanocone. Figure 22.9a depicts the various steps in the above mechanism [7].

Two similar synthesis studies were reported for CCNT structures with open-ended tips and inner channels free of any internal obstructions: One of the studies produced conical morphologies termed as TGC by placing iron needles vertically in a microwave plasma discharge with methane and nitrogen, at 750 W, 600°C, 15 torr pressure [6]. In this case, the iron from the needle acted as the catalyst to promote the central carbon nanotube growth. In the other report, similar CCNT morphologies termed as CNPs were synthesized on platinum wires placed vertically in a microwave plasma discharge containing methane and hydrogen, at 950 W and 25 torr pressure [5]. Both types of structures are very similar, having open tips with no metal catalyst particles and no obstructions within the internal channel. In both reports, it was suggested that small amounts of methane diluted in either hydrogen or argon resulted in high surface temperature at the wire substrate. In both cases, a catalyst particle

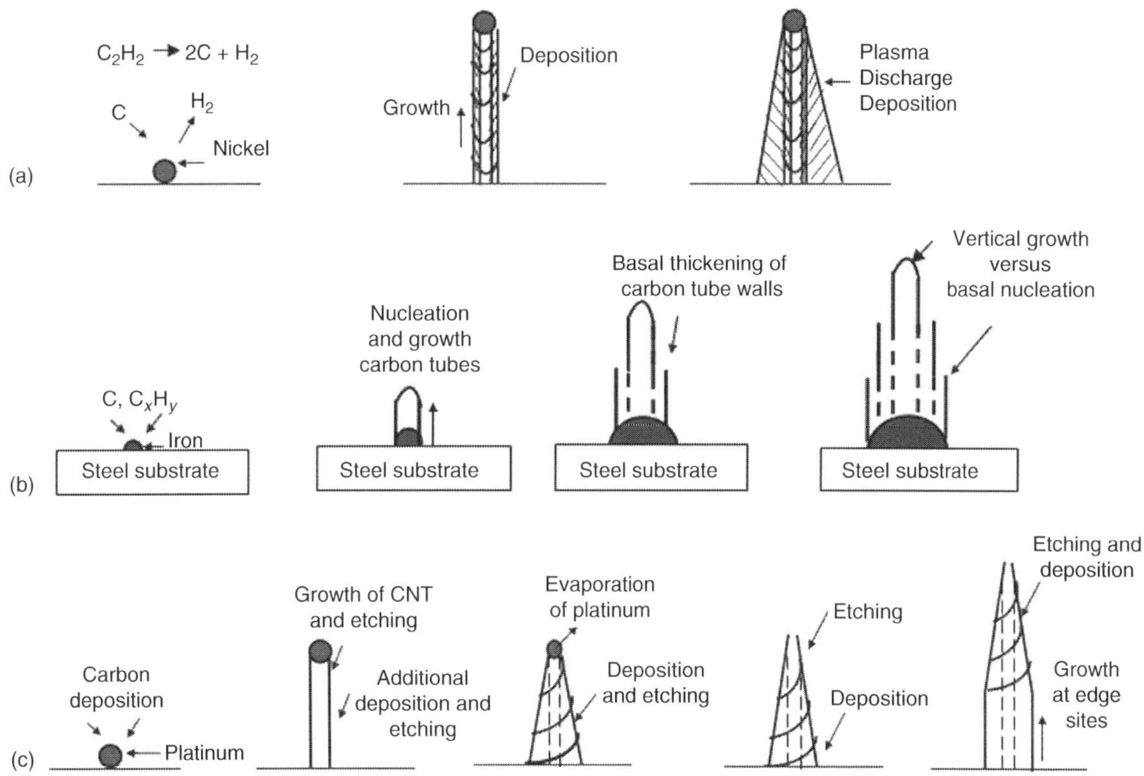

FIGURE 22.9 Schematics illustrating different growth mechanisms suggested for conical carbon nanotubular growth: (a) the catalyst assisted growth of central carbon nanotube along with uneven plasma deposition/etching enables the observed growth of conical graphite structure, (b) the catalyst-assisted basal growth of a central carbon nanotube followed by continuous nucleation of new graphene walls for thickening the base while tapering vertically in thickness forming a conical structure, and (c) catalyst assisted tip-led growth of a central carbon nanotube followed by the evaporation of the catalyst particle and simultaneous deposition and etching of the helical graphite edge planes leading to the conical morphology.

(either iron or platinum) helped with the nucleation and growth of a multiwalled carbon nanotube. However, there are two contrasting suggestions on how tapered growth occurs: The first suggested mechanism implies that the basal accumulation of iron catalyst initiates nucleation of new graphene walls thickening at the base and its competition with vertical growth leads to tapering (Figure 22.9b). The second suggested mechanism states that as the central multiwalled nanotube continues to grow axially with catalyst at its tip, it acquires a tapered morphology in the early stages as an effect of the shrinkage of the catalyst particle due to evaporation. Because of the presence of plasma discharge containing methane and hydrogen, further growth can occur with initiation of new scrolling graphene sheets and growth at the open-ended tip without the presence of catalyst (Figure 22.9c). In addition to deposition, the plasma exposure leads to greater etching near the tip, preserving the conical morphology even at longer synthesis timescales.

Experiments involving hydrocarbon decomposition on to presynthesized carbon fibers also resulted in sparse growth of conical-shaped morphologies on carbon fiber substrates [27]. The underlying mechanism is not clear but may be similar to that shown in Figure 22.9b. Conically shaped tips on multiwalled carbon nanotubes were also observed with further hydrocarbon decomposition onto iron catalyst loaded, presynthesized carbon fibers [28]. Similar structures of conical shaped tips on multiwalled carbon nanotubes shown in Figure 22.4c were synthesized using a synthesis temperature of 1325°C and bubbling of ferrocene in toluene, using hydrogen and argon gas mixtures [15]. One of the suggestions is that the conical shape is acquired from the multiwalled carbon nanotube growth basally on a conically faceted catalyst in the early stages.

Irrespective of the underlying growth mechanism, it is essential to grow CCNT arrays on to large area, flat substrates for applications. However, it is difficult to reproduce the exact conditions on the flat substrates with those existing on the wire substrates immersed in plasma discharges for forming the CCNT structures. For example, the temperature and the radical species density that exist on the immersed wire substrates are much higher than the typical conditions that can be obtained for flat substrates. Only one to two reports beyond the authors' laboratory exist on the growth of CCNTs on planar substrates and the results indicated limited control on the resulting morphology and the array density [16,29]. Much more work needs to be done in obtaining similar CCNT array structures on large area, flat substrates for various applications.

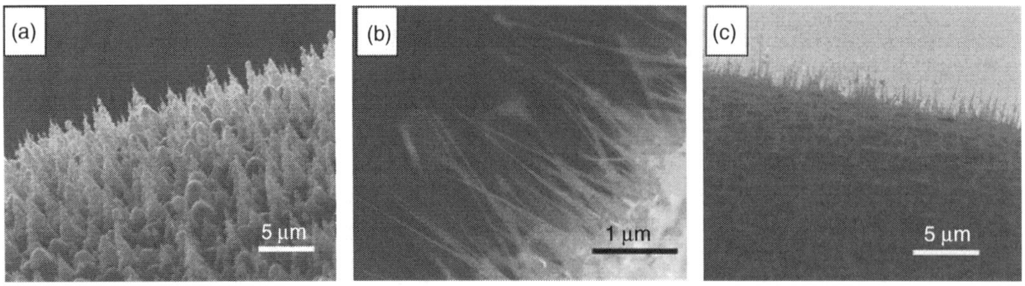

FIGURE 22.10 SEM images illustrating different morphologies of CCNT structures grown using platinum wire substrates placed vertically in the microwave plasma discharge: (a) the CCNTs with large bases containing heavy amount of loose carbon around their bases, (b) the typical morphology of a high density of CCNTs grown in an array fashion, and (c) low-density arrays of CCNTs.

22.3.2 Morphological Control

The shape of the conical carbon nanotube structures grown on planar substrates with tip-led growth were mostly affected by the gas phase composition [7]. In the case of the CCNT structures grown using direct immersion of substrates inside plasma discharges, the length, tapering, and the array density are affected by varying different process conditions such as pressure, microwave power, and the wire substrate's placement inside the plasma discharge [5,6,16,29]. The latter condition seems to be the most effective in terms of controlling the tapering geometry and the structure of the resulting tapered morphology with simultaneous etching and growth. Rapid deposition of loose carbon at bases results in low aspect ratio conical structures as shown in Figure 22.10a. The high aspect ratio CCNTs are grown where growth/etching is optimized as shown in Figure 22.10b. The aspect ratios can also be controlled by changing the composition of H_2 and CH_4 [30]. In a similar way, the density of the CCNTs can be varied by changing the process conditions as shown in Figure 22.10c.

Postsynthesis strategies could also be used for altering the tapered morphology of the presynthesized CCNT structures. The H_2 plasma etching can be used to remove the loose carbon flakes and prolonged etching can lead to straight multiwalled CNT as shown in Figure 22.11a and b [30]. Hydrogen etching can also result in carbon nanotube on CCNT tips as shown in Figure 22.11c. The oxidation of conical carbon tubular structures grown on silicon wafers [16] modified the CCNT structure to a stepped cylindrical morphology as shown in Figure 22.11d.

The tip size and inner diameter can be altered by changing the initial catalyst particle size as the latter plays a key role in determining the size of the MWNT. It is also assumed that the spinning of the catalyst droplet formed on the surface promotes the growth of helical graphite sheets surrounding the core CNTs [29]. The temperature (controlling the size of catalyst particle) and the methane concentration should be sufficient for the growth on flat substrates, where conditions are different from the wires immersed into the plasma. Though the temperature of substrate was low, self-bias around the iron needles including the strong electric field in the plasma resulted in conical structures [6].

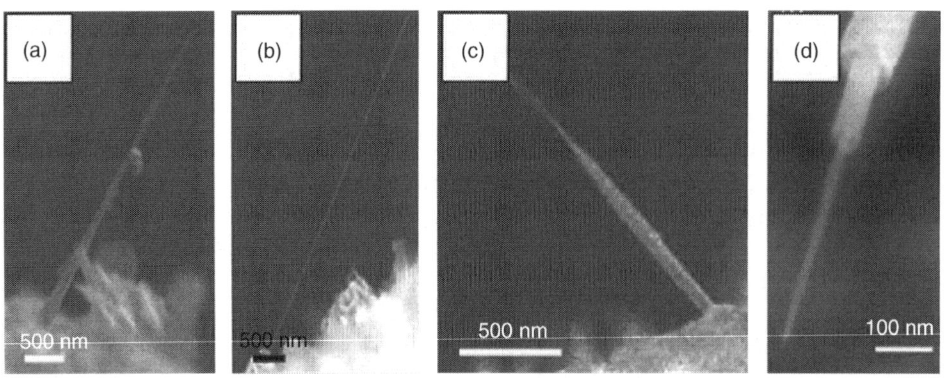

FIGURE 22.11 SEM images of the CCNTs after postsynthesis etching treatments: (a) an as-synthesized CCNT; (b) the CCNT after fully etched using hydrogen plasma exposure at 800°C–900°C temperature; (c) the CCNT structure after partial etching with hydrogen plasma exposure, and (d) the telescopic structure of the CCNT structure formed with thermal oxidation using oxygen. (From Shang, N., Milne, W.I., and Jiang, X., *J. Am. Chem. Soc.*, 129, 8907, 2007. With permission.)

Continuous supply of metal suppressed the growth of CNTs, encouraging the growth of TGCs [29]. The formation of tubes having a continuous tapered morphology, as shown in Figure 22.3c, can be explained by catalyst assisted growth, with continuous evaporation of the catalyst particle, forming ~200 nm long conical structures, before the catalyst particle evaporates completely. A steady supply of catalyst particles leads to the growth of new tapered tubes on top of the already existing ones from the new catalyst particles, ultimately forming long tubes. Similar structures, as shown in Figure 22.4b can also be explained by the same growth mechanism.

22.4 Applications

22.4.1 New Electrode Materials

All conical carbon tubular structures have unique surface structures compared to both single walled and multiwalled carbon nanotubes. The surfaces of the conical carbon tubular structures are completely made up of highly reactive graphene edge sites [5]. So, the electrochemical properties of any conical carbon material will be different from that of traditional carbon-based materials such as highly oriented pyrolytic graphite, carbon nanotubes, and diamond. Typically, several redox reactions involving neurotransmitters such as dopamine exhibit irreversibility on a number of carbon-based materials with peak separations as high as 1000 mV. However, the kinetics of the same redox reaction on the as-synthesized conical carbon tube arrays are fast and reversible with peak separations less than 150 mV. In addition, dopamine needs to be detected in the presence of high amounts of ascorbic acid. But, the oxidation peaks of dopamine and ascorbic acid compounds occur at nearly the same potential (~200 mV vs. Ag/AgCl). The as-synthesized conical carbon tube array electrodes after one-time oxidation treatment at anodic potential of 2.0 V vs. Ag/AgCl in 0.5 M H_2SO_4 shifts the ascorbic acid peak while leaving the dopamine peak intact. See Figure 22.12 showing well resolved peaks of ascorbic acid and dopamine [31]. In essence, the CCNT arrays can become an important class of electrochemical materials for sensing neurotransmitters and other important biological species.

22.4.2 Templates for Nanoelectrode Ensembles

Conical carbon nanotubular structures are also of interest for nanoelectrode applications due to their nanoscale tips. Nanoelectrodes offer several advantages compared to their micron-sized counterparts due to their small critical dimensions. Nanoelectrodes exhibit radial diffusion, rapid attainment of steady state currents, increased signal to background ratios, higher sensitivities, and decreased effects from an ohmic potential drop compared to large electrodes [31]. However, an ensemble of nanoelectrodes instead of a single nanoelectrode is required for obtaining the increased signals. In order to make such an ensemble, the spacing between each nanotip should be in the order of microns to avoid overlapping concentration boundary layers.

Conical CNP arrays, due to their conical geometry, contain tips spaced from each other by few microns depending upon their base widths. However, the as-synthesized conical carbon nanopippette arrays behave as microelectrodes. This is because the conical carbon structures touch each other at their bases.

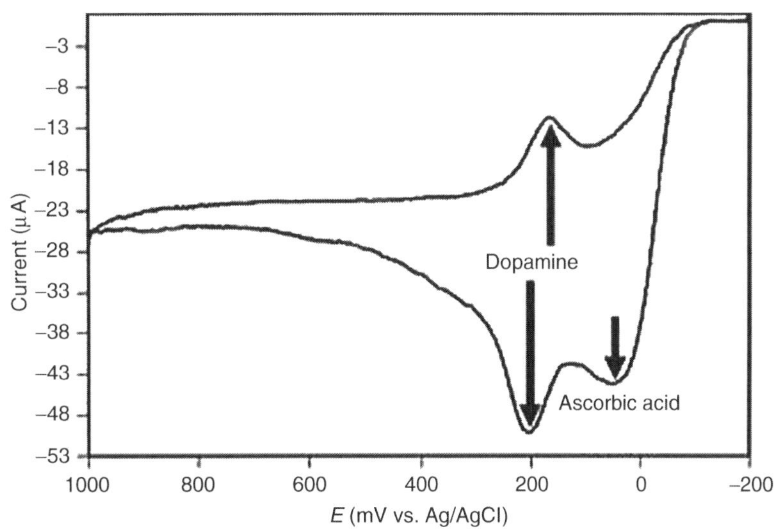

FIGURE 22.12 A cyclic voltammogram using CNP electrode at 100 mV/s scan rate in a 0.1 mM dopamine and 10 mM ascorbic acid solution in 0.1 M KCl in 0.2 M phosphate buffer solution, pH 7, after a single anodic oxidation treatment. (From Lowe, R.D. et al., *Electrochem. Solid State Lett.*, 9, H43, 2006. With permission.)

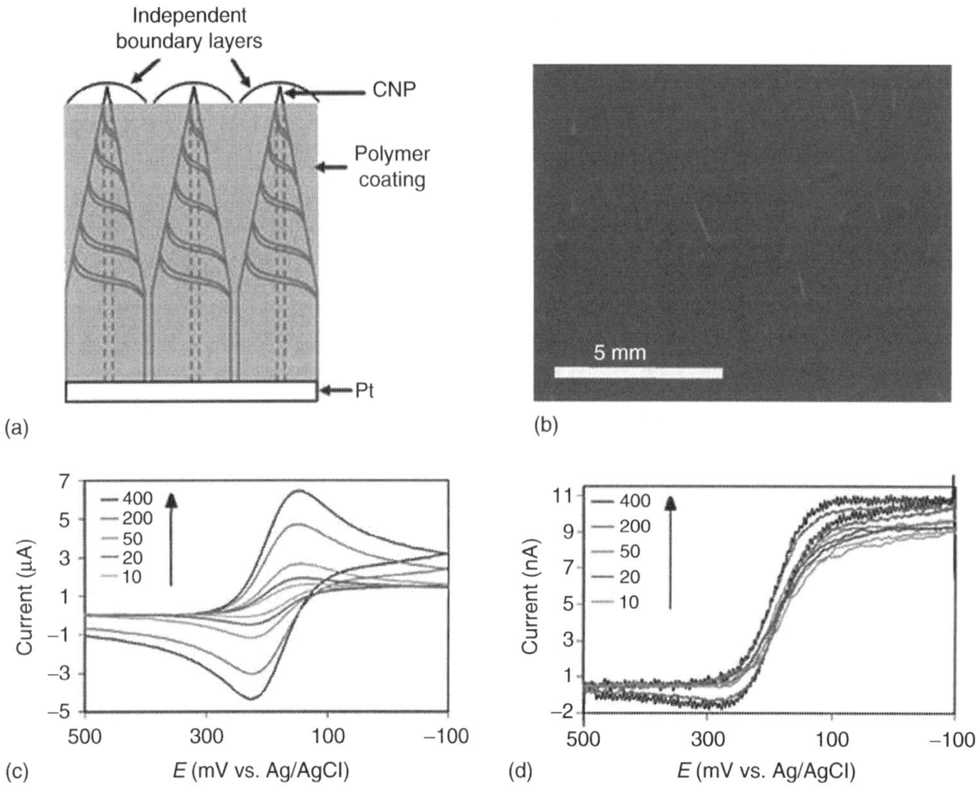

FIGURE 22.13 Fabrication and electrochemical testing of nanoelectrode ensembles using the CCNT arrays: (a) a schematic illustrating the effect of dip coating of CCNT array using a polymer for the separation of diffusion boundary layers for each nanotip exposed. Polymer coating helps to space the electrodes; (b) the SEM of CNP NEE after polymer coating showing the well-spaced nanotips; (c) the cyclic voltammograms taken at 10–400 mV/s scan rates for an uncoated CNP array exhibiting planar electrode behavior with peak-shaped responses; and (d) the cyclic voltammograms using polymer-coated CNP NEE showing the sigmoidal response indicating the nanoelectrode behavior. (From Lowe, R.D. et al., *Electrochem. Solid State Lett.*, 9, H43, 2006. With permission.)

A simple dip coating of CNP arrays using a polymer exposes only the top portion of well-separated, conical tips to electrolyte as shown in Figure 22.13a and b. Electrochemical studies performed on these electrode assemblies show a very clear shift from microelectrode behavior to nanoelectrode behavior when coated with the polymer. Cyclic voltammetry performed on these electrodes is shown in Figure 22.13c and d. The plots show that the uncoated electrodes exhibit a planar electrode behavior with a peak-shaped response and the coated electrodes exhibit a sigmoidal response, indicative of steady-state, diffusion-limited currents [31].

Even though carbon itself is an interesting material for several applications based on nanotips, conical-shaped nanotip arrays of other materials are also interesting. In this regard, the nanoelectrode ensembles of other materials could easily be fabricated by depositing materials of interest onto the conical carbon tube arrays. The diamond nanotip array electrodes are fabricated easily by depositing diamond on to conical carbon tube arrays (see Figure 22.14a and b) [30].

22.4.3 Field Emission Applications

The CCNT structures with conical geometry and nanoscale tips are interesting for field emission applications. A recent study [32] reported very low threshold voltages of 0.27 V/μm and a current density of 1 μA/cm^2. A stable emitting current density of 1.9 mA/cm^2 at only 0.6 V/μm was reported. In addition to having low work function, the conical bases allow maximum contact with the substrate for high field emission currents over long periods of time. The threshold voltage shown by these structures was three times lower than that shown by conical nanofibers [33]. The results shown in Ref. [32] are affected to some extent by the presence of encapsulated metal catalyst particles at their tips. However, the true work function and field emission properties of conical carbon nanotube structures are yet to be understood.

22.4.4 Porous Carbons

Large quantities of CMTs with controlled internal diameters and conical angles could be useful as mesoporous carbons for high surface area and catalyst support applications. In general, the applications of porous carbons can be grouped into two categories: (i) based on the architecture of their pores for structural and thermal applications and their use as templates for making ceramics and (ii) as activated carbons. Activated carbons are used extensively as catalysts and catalyst supports for water and air purification, removing color from various types of sugar syrups, separating amino acids from their solution acetic acid, separating aromatic acids, catalysts, and catalyst supports [34]. Activated carbon materials are also being investigated for possible applications in hydrogen storage [35]. Microtubular materials can offer highly interconnected porous structures enabling facile diffusion of species into and out of the carbon structures. The noncarbon-based microtubular materials are also of interest. For example, the aluminum borate microtubular materials are used in structural applications [36,37], zinc oxide microtubes show promise in gas sensing [38], and ZnO nano/microtubes are good candidates for safe biological cavities and template materials due to their innocuity and chemical characters. High surface area silicon carbide (SiC) is used as an important high-temperature structural material, such as catalyst support and hot-gas filter, and offers many advantages due to its unique properties, e.g., low thermal expansion coefficient, good thermal shock resistance, and chemical stability at elevated temperature [39] Similarly, TiO_2 microtubes for photocatalytic applications [40], diamond microtubes as reinforcement agents [41] are known applications. The CMTs can serve as templates for producing microtubes of other materials. For example, diamond can easily be deposited on to CMTs transforming them into diamond microtubes with shapes of the original CMTs as shown in Figure 22.14c and d.

22.4.5 Drug Delivery, Fluid Flow, and Other Miscellaneous Applications

Conventional syringes used for drug delivery are often painful and limit targeted drug delivery. In this regard, nano and microneedles present a suitable alternative as they are painless and can be used for highly localized drug targeting. Many different designs of microneedle-based drug delivery systems have been demonstrated. In many cases, these micropipettes are fabricated rather than synthesized. One limitation in microfabricating vertical microneedles is their maximum achievable length [42]. Also, the materials which have been used for fabricating these needles are mainly limited to silicon [43], glass [44], metal, and biodegradable polymer [45]. Carbon based nanopipettes were fabricated using chemical vapor deposition of carbon onto catalyst coated, interior surfaces of pulled quartz capillary (or nanopipette) tubes, and removing the quartz exterior [46]. Such carbon-based nanopipettes situated at the end of a long quartz capillary will be useful in both drug delivery and sensing applications.

The synthesized carbon based micro/nanopipette arrays and needles are also of immense interest to the above type of transdermal drug delivery applications if they can be integrated into macroscale systems. Moreover, recent studies on fluid flow through carbon nanotubes have shown interesting fluid behavior due to the hydrophobicity of the interior surface. The fluid flow through 7 nm diameter nanotubes was shown to be four to five orders of magnitude faster than that predicted by conventional fluid dynamics [47]. This was attributed to the almost frictionless interface at the CNT wall. Similar results were obtained using even smaller nanotubes (~2 nm) for liquids and gases [48]. Further studies showed that by attaching different end groups to nanotubes, selective liquids can be flown through them [49,50]. With open-ended tips and the mechanical rigidity of the conical structure, the fluid flow through CCNT arrays should find several applications including drug delivery.

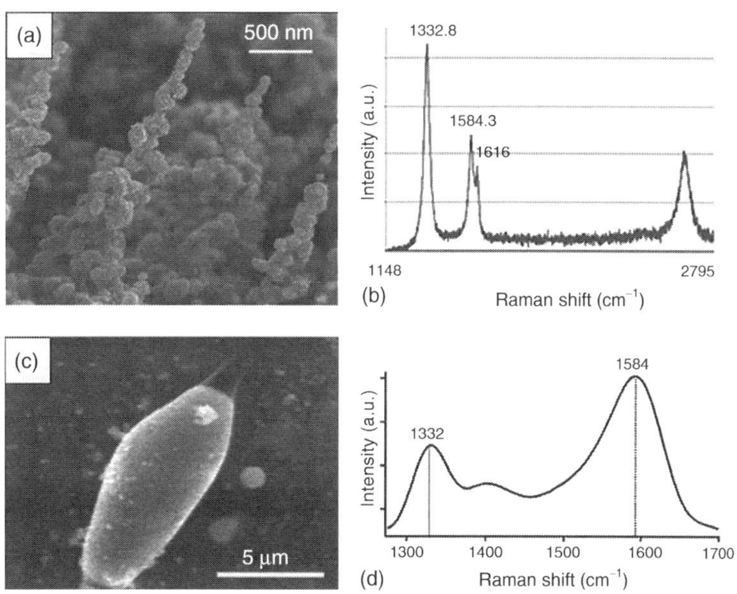

FIGURE 22.14 (a) SEM image of CCNT arrays covered with diamond, (b) visible Raman spectra obtained on the diamond coated CCNT arrays, (c) SEM image of a diamond coated CMT, and (d) UV Raman spectra obtained on the diamond coated CMT indicating the presence of ultrananocrystalline diamond.

Micropipette structures are also of interest for many new types of applications such as fountain pen chemistry when integrated with atomic force microscope's cantilever [51]. Nanopipette syringes, which can dispense fluids with a precise control of volume, are of tremendous interest. Two kinds of syringes were reported, a double barrel nanopipette [52] and an electrochemical attosyringe [53]. In the double barrel type nanopipette, an electrochemical potential is applied to two different fluid compartments, which sets up an electroosmotic flow of the fluid out of the nanopipette. By controlling parameters, such as the applied voltage and pulse duration, the volume of fluid coming out of the syringe can be controlled to attoliter precision. In the electrochemical attosyringe [54], an applied potential between electrodes immersed in the fluid inside and the fluid surrounding the nanopipette induced the fluid flow into or out of the pipette with attoliter precision.

Another interesting application for the electron transparent micro tubes, partially filled with gallium is their use as nanothermometers [8]. Gallium exists as liquid from 29.8°C to 1050°C while exhibiting low vapor pressure. Interestingly, the expansion coefficient of gallium inside the nanotube was the same as the expansion coefficient of gallium in the macroscopic state. This was unlike the melting point for nanoquantities of materials which are influenced by surface effects.

22.5 Summary

The synthesis, structure, and applications of various micron-scale and conical morphologies of carbon tubes have been described and discussed.

Acknowledgments

The authors would like to acknowledge support from U.S. Department of Energy (DE-FG02–05ER64071 and DE-FG02–07ER46375) and U.S. Army Space Missile Defense Command (W9113M-04-C-0024). The authors also acknowledge the Chemical Engineering department at the University of Louisville for support, Dr. Zhiqiang Chen for help with TEM analysis, and the Institute for Advanced Materials and Renewable Energy at University of Louisville for allowing us to access to the materials characterization facilities such TEM, SEM, and Raman spectroscopy.

References

1. Bacon, R., Growth, structure, and properties of graphite whiskers, *J. Appl. Phys.*, 31, 283, 1960.
2. Haanstra, H.B., Verspui, G., and Knippernb, W.F., Columnar growth of carbon, *J. Cryst. Growth*, 16, 71, 1972.
3. Iijima, S., Helical microtubules of graphitic carbon, *Nature*, 354, 56, 1991.
4. Ajayan, P.M., Nanotubes from carbon, *Chem. Rev.*, 99, 1787, 1999.
5. Mani, R.C. et al., Carbon nanopipettes, *Nano Lett.*, 3, 671, 2003.
6. Zhang, G., Jiang, X., and Wang, E., Tubular graphite cones, *Science*, 300, 472, 2003.
7. Merkulov, V.I. et al., Shaping carbon nanostructures by controlling the synthesis process, *Appl. Phys. Lett.*, 79, 1178, 2001.
8. Gao, Y. and Bando, Y., Carbon nanothermometer containing gallium, *Nature*, 415, 599, 2002.
9. Libera, J. and Gogotsi. Y, Hydrothermal synthesis of graphite tubes using Ni catalyst, *Carbon*, 39, 1307, 2001.
10. Bhimarasetti, G. et al., Morphological control of tapered and multi-junctioned carbon tubular stuctures, *Adv. Mater.*, 15, 1629, 2003.
11. Iwanaga, H., Kawaguchi, M., and Motojima, S., Growth mechanisms and properties of coiled whiskers of silicon-nitride and carbon, *Jpn. J. Appl. Phys.*, 32, 105, 1993.
12. Krishnan, A. et al., Graphitic cones and the nucleation of curved carbon surfaces, *Nature*, 38, 451, 1997.
13. Ge, M. and Sattler, K., Observation of fullerence cones, *Chem. Phys. Lett.*, 220, 192, 1994.
14. Iijima, S., Yusasaka, M., Yamada, R., Bandow, S., Suenaga, K., Kokai, F., and Takahashi, K., Nano-aggregates of single-walled graphitic carbon nano-horns, *Chem. Phys. Lett.*, 309, 165, 1999.
15. Kona, S., M.S. thesis, University of Louisville, 2007.
16. Shang, N., Milne, W.I., and Jiang, X., Tubular Graphite Cones with single crystal nanotips and their antioxygenic properties, *J. Am. Chem. Soc.*, 129, 8907, 2007.
17. Gogotsi, Y., Dimovski, S., and Libera, J.A., Conical crystals of graphite, *Carbon*, 40, 2263, 2002.
18. Gao, Y. and Bando, Y., Nanothermodynamic analysis of surface effect on expansion characteristics of Ga in carbon nanotubes, *Appl. Phys. Lett.*, 81, 3966, 2002.
19. Hu, J. et al., Growth and field-emission properties of crystalline, thin-walled carbon micro tubes, *Adv. Mater.*, 16, 153, 2004.
20. Pan, Z.W. et al., Gallium-mediated growth of multiwall carbon nanotubes, *Appl. Phys. Lett.*, 82, 1947, 2003
21. Shen, G. et al., Tubular carbon nano-/microstructures synthesized from graphite powders by an in situ template process, *J. Phys. Chem. B*, 110, 10714, 2006.
22. Hu, J. et al., Tapered carbon nanotubes from activated carbon powders, *Adv. Mater.*, 18, 197, 2006.
23. Sun, Z. et al., Synthesis of tubular graphite cones through a catalytically thermal reduction route, *J. Phys. Chem. B*, 108, 9811, 2004.
24. Sharma, S. and Sunkara, M.K., Direct synthesis of gallium oxide tubes, nanowires, and nanopaintbrushes, *J. Am. Chem. Soc.*, 124, 12288, 2002.
25. Graham, U.H. et al., Morphological control of tapered and multi-junctioned carbon tubular structures, *Adv. Funct. Mater.*, 13, 576, 2003.
26. Bhimarasetti, G., Cowley, J.M., and Sunkara, M.K., Carbon microtubes: tuning internal diameters and conical angles, *Nanotechnology*, 16, S362, 2005.
27. Xia, W. et al., Conical carbon filaments with axial cylindrical channels and open tips, *Adv. Mater.*, 17, 1677, 2005.

28. Muradov, N. and Schwitter, A., Formation of conical carbon structures on vapor-grown carbon filaments, *Nano Lett.*, 2, 673, 2002.
29. Shang, N.G. and Jianga, X., Large-sized tubular graphite cones with nanotube tips, *Appl. Phys. Lett.*, 87, 163102, 2005.
30. Chernomordik, B. et al., Nanodiamond tipped and coated conical carbon tubular structures, chem. vap. Deposition., 14, in press, 2008.
31. Lowe, R.D. et al., Nanoelectrode ensembles using carbon nanopipettes, *Electrochem. Solid State Lett.*, 9, H43, 2006.
32. Li, J.J. et al., Field emission from high aspect ratio tubular carbon cones grown on gold wire, *Appl. Phys. Lett.*, 87, 143107, 2005.
33. Tanemura, M. et al., Field electron emission from sputter-induced carbon nanofibers grown at room temperature, *Appl. Phys. Lett.*, 86, 113107, 2005.
34. Monacha, S., *Sadhana*, Parts 1&2, 335, 2003.
35. Thomas, K.M., *Catal. Today*, 120, 389, 2007.
36. Yang, W. et al., Polygonal single-crystal aluminum borate micro tubes, *J. Am. Ceram. Soc.*, 88, 485, 2005.
37. Ma, R. et al., Single-crystal $Al_{18}B_4O_{33}$ micro tubes, *J. Am. Chem. Soc.*, 124, 10668, 2002.
38. Fu, M. et al., L., Tetrapod-shaped ZnO microtubes synthesized from Zn/C mixtures *Mater. Res. Bull.*, doi:10.1016/j.materresbull.2007.04.035, 2007.
39. Kim, J.W. et al., Synthesis of SiC microtubes with radial morphology using biomorphic carbon template, *Mater. Sci. Eng. A*, 434, 171, 2006.
40. Motojima, S. et al., Preparation of helical TiO_2/CMC microtubes and pure helical TiO_2 microtubes, *J. Mater. Sci.*, 39, 2663, 2004.
41. Dua, A.K. et al., Formation of self-supporting hollow diamond helix and diamond sieve using jet-flow HFCVD, *Solid State Commun.*, 93, 759, 1995.
42. Reed, M.L., Microsystems for drug and gene delivery, *Proc. IEEE*, 92, 56, 2004.
43. Stoeber, B. and Liepmann, D., Design, fabrication, and Testing of a MEMS Syringe, Proceedings of Solid-State Sensor and Actuator Workshop. 2–7 June, 2002, Hilton Head Island, SC, Transducers Res. Found (TRF Cat. No. 00TRF-0001).
44. Martanto, W. et al., Mincoinfusion using hollow micro needles, *Pharmaceut. Res.*, 23, 104, 2006.
45. McAllister, D.V. et al., Microfabricated needles for transdermal delivery of macromolecules and nanoparticles: Fabrication methods and transport studies, *Proc. Natl Acad. Sci.*, 100, 13755, 2003.
46. Kim, B.M., Murray, T., and Bau, H.H., The fabrication of integrated carbon pipes with submicron diameters, *Nanotechnology*, 16, 1317, 2005.
47. Majumder, M. et al., Nanoscale hydrodynamics: Enhanced flow in carbon nanotubes. *Nature*, 438, 44, 2005.
48. Holt, J. K. et al., Fast mass transport through sub-2-nanometer carbon nanotubes. *Science*, 312, 1034, 2006.
49. Majumder, M., Chopra, N., and Hinds, B.J. Effect of tip functionalization on transport through vertically oriented carbon nanotube membranes. *J. Am. Chem. Soc.*, 127, 9062, 2005.
50. Babu, S. et al., Y. Guiding water into carbon nanopipes with the aid of bipolar electrochemistry. *Microfluid. Nanofluid.*, 1, 284–288 2005.
51. Lewis, A. et al., Fountain pen nanochemistry: Atomic force control of chrome etching, *Appl. Phys. Lett.*, 75, 2689, 1999.
52. Rodolfa, T. et al., Nanoscale pipetting for controlled chemistry in small arrayed water droplets using a double-barrel pipet kit, *Nano Lett.*, 6, 252, 2006.
53. Laforge, F.O. et al., *Proc. Natl. Acad. Sci.*, 104, 11895, 2007.
54. Han, C.C., Lee, J.T., Yang, R.W., Chang, H., and Han, C.H., A new and easy method for making well-organized micrometer-sized carbon tubes and their regularly assembled structures, *Chem. Mater.*, 11, 1806, 1999.

23

"Smart" Corrosion Protective Coatings

Patrick J. Kinlen
Crosslink

Martin Kendig
Teledyne Scientific Company

23.1 Introduction

"Smart" corrosion protective coatings have gained recent attention as a result of new developments generally in the field of smart environmentally responsive materials. For the purpose of this chapter, smart corrosion protection refers the ability of a protective coating not only to provide a barrier to corrosive compounds, but also to sense one of the following: (1) the environment, (2) changes in the environment, (3) its own condition, or (4) changes in its condition, and, on the basis of these inputs, generate an appropriate corrosion-protective response. High-performance corrosion protective coatings have always functioned more than by acting as a barrier to water, ions, and corrosive chemical species. In fact, from an historical perspective, the best corrosion protective coatings have provided some environmentally stimulated protective response. For example, zinc and tin coatings have been utilized for more than 150 years as protective coatings providing galvanic protection of defects in the presence of an electrolyte environment, and they generate corrosion-inhibiting species when a galvanic current flows. Mayne [1] pointed out that coatings serve as more than barriers since permeation by oxygen exceeds the rates necessary to sustain corrosion reactions. He also showed the reaction of lead-based pigments with an oil-based resin coating and the environment produces azelaic acid, a demonstrated corrosion inhibitor for iron. It has long been known that sparingly soluble hexavalent chromium inhibitors (e.g., Sr, Ca, and Zn chromate) release inhibiting chromates when the coating becomes wet. Results have shown that conversion coatings formed by hexavalent chromate (Cr(VI)) retain unreacted Cr(VI) that becomes active upon the formation of a pit that cathodically polarizes adjacent surfaces, releasing chromate as a result of the consequent increase in pH [2].

It is clear then that smart coatings have always provided the best inhibition. With recent developments in nanotechnology, inherently conducting polymers and composite materials, more precise engineering of smart behavior in corrosion protective coatings has started and will benefit environmental concerns about the historic use of hazardous and carcinogenic materials, such as hexavalent chromium, common to existing smart coatings. In addition, smart coatings for corrosion prevention can use the energy inherent in the metal itself (an aerated environment provides nearly 2 V of electrochemical energy) to drive a protective process such as release of an inhibitor, buffer, or passivator [3].

The scope of this chapter (corrosion protective smart materials) includes a critical discussion of recent advances where environmentally responsive materials have been considered and engineered for smart corrosion protection. We will consider only corrosion protective coatings and exclude self-lubricating, optical, nontumescent, nonskid, and tamper-resistant coatings, important topics of their own but beyond the scope of this review. Although related to corrosion protective coatings, corrosion indicating coatings such as those described by Zhang and Frankel [4] also will not be covered in great detail in this chapter.

23.2 Smart Coating Technologies for Corrosion Protection

A number of technologies currently exist for rendering protective coatings smart or responsive to the stimulus provided by corrosion. Generally the stimulus results from a chemical or electrochemical potential, but may also be provided by mechanical, thermal, or optical processes. Biological processes generally fall under the heading of chemically stimulated response. As stated in the introduction, our focus is on the intelligent response of materials to corrosion processes.

23.2.1 Chromate and Partially Soluble Pigments

In the early 1950s, as mentioned in the introduction, Mayne showed that paints containing basic lead and zinc pigments released an inhibitor in the form of heavy metal soaps to inhibit corrosion. Indeed water that had been in contact with such paints was demonstrated to inhibit the corrosion of steel. Chromates for paints include the alkaline earth chromates such as Ba, Ca, and Sr that have variable solubility as summarized by Sinko [5]. However, a critical concentration of chromate for a given pH and chloride activity is required for inhibition. Formulation of the chromate-based pigments into a coating requires a matrix that allows their slow dissolution. Formulation also requires an optimum solubility of the pigment since excessive solubility results in rapid dissolution to the detriment of the barrier property of the coating. Too little solubility or too strong binding by the matrix results in an activity of chromate below a critical concentration at defects. Often, Sr chromate exists as the pigment of choice due to its intermediate solubility [6].

In 2000, work at Ohio State University showed that hexavalent chromium adsorbed on the trivalent oxides of conversion coatings on aluminum alloys [7]. While this observation was not entirely new, the research clearly demonstrated that the hexavalent chromate deadsorbed as a function of pH and ionic strength. Elevated pH, as encountered when a pit cathodically polarized the adjacent surface, desorbs the hexavalent chromate species. In addition, the increase in ionic strength also enhanced the release of hexavalent chromium species adsorbed on the trivalent oxide. Clearly both of these chemical phenomena contribute to a smart or responsive behavior based on environmental factors. This model behavior for a traditionally, and exceptionally performing coating shows the importance of a smart response for protective coatings and has guided the way for recent investigations.

23.2.2 Environmentally Responsive Corrosion Protection

In recent years, effort has focused on using environmental conditions to stimulate release of corrosion inhibitors or imposition of a physical response (passivation) to slow corrosion. These can be divided roughly into the categories of (1) response to electrochemical potential, (2) response to chemical potential, (3) response to mechanical stress, and (4) response stimulated by living systems (electrochemical, chemical, mechanical, biological). Recognize that materials can respond to additional physical phenomena such as electromagnetic radiation and magnetic fields as a basis for the formation of "smart coatings." However, since this chapter focuses on corrosion protection, these stimuli will be of secondary importance.

23.2.2.1 Electrochemical Potential

23.2.2.1.1 Importance of Electrochemical Responsive Coatings

Response to electrochemical potential presents the most important stimulus for smart, corrosion protective coatings since corrosion is an electrochemical process requiring a galvanic potential between a cathode, typically the locus of oxygen or proton reduction, and an anode, an anodically dissolving metallic surface. A coating that connects electrically to this process to drive a corrosion-protective response represents the ultimate in smart corrosion protection since corrosion is (1) an electrical phenomenon and (2) provides sufficient electrical energy making an electrically responding coating ideal for smart corrosion inhibition. The need for anticorrosion coatings, which are pinhole and scratch tolerant, coupled with growing environmental concerns involving heavy metals, such as hexavalent chromium, has led to coating strategies employing inherently conductive polymers (ICPs) as a main focus. To date significant progress in this direction has been made primarily with ICPs as the platform for directing the signal provided by the conditions for corrosion to activate corrosion protection.

23.2.2.1.2 Smart Release of Inhibitors by ICPs

Smart coating systems are engineered to respond to electrochemical processes responsible for corrosion by providing a self-repairing system. When corrosion events are sensed, the coating automatically releases anodic and/or cathodic corrosion inhibitors to alleviate the corrosion process. Several smart coating system strategies employ ICPs such as polyaniline (PANI) and polypyrrole (PPY) [8]. ICPs have three unique properties that enable their use as coating materials for galvanically driven or electrochemical potential stimulated release of corrosion inhibitors. First, they are highly stable and inert to dissolution. Second, they are conducting, and, third, they hold and release ionic species depending on their state of charge as illustrated by the equilibria shown in Figure 23.1. The switching of the ICP redox state is triggered by local electrochemical reactions occurring on the surface of a metal. Through proper design, ICPs may be synthesized to contain inhibitor molecules or ions as dopants that release when a corrosion process is sensed.

The drug release literature provides considerable information for the potential development of smart protective coatings that release corrosion inhibitors. Much of this effort originated with the work performed at the University of Minnesota under the direction of Miller [9] and has continued over the years with many other important contributions [10].

23.2.2.1.3 Background and History of ICPs

In 1862, Letheby [11] reported that a blue substance (now known to be PANI base) was produced upon electrolysis of aniline on platinum. Twenty-nine years later, Goppelsroeder [12] discovered that a solution of aniline hydrochloride may be electrochemically oxidized first to emeraldine and then to aniline black. Othmer states that ani-

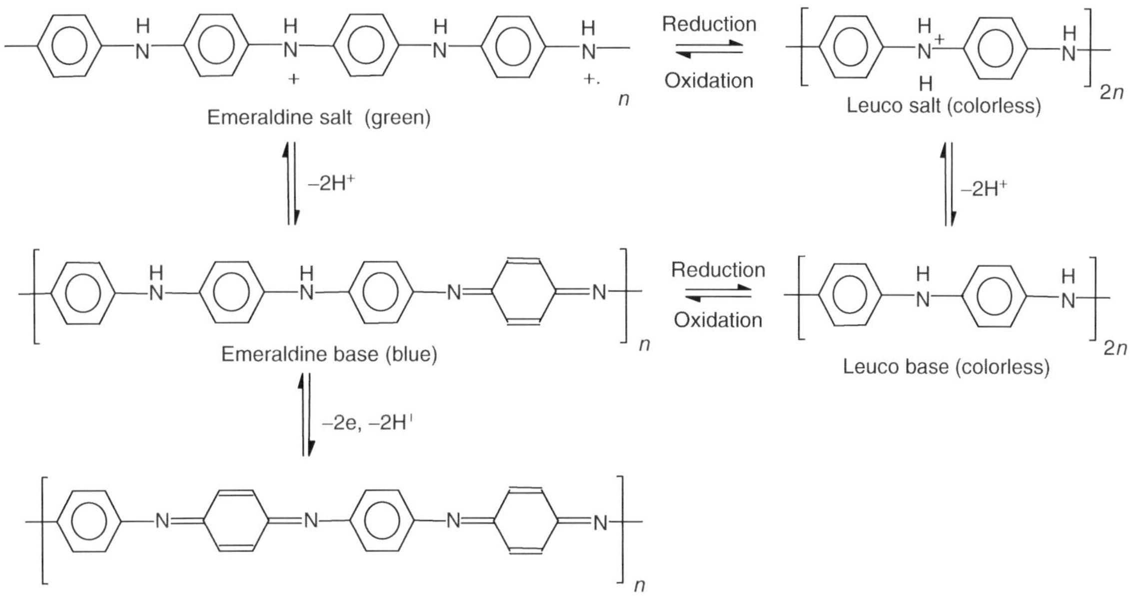

FIGURE 23.1 Redox and pH equilibria for PANI.

line bottoms were used as pigments in paints. (Note: these materials would most likely be similar to aniline black.) Hans Kuhn (formerly of Milliken) used to start his presentations with the story that crude PANI (still bottoms from preparation of aniline by reduction of nitrobenzene in Fe filings and HCl) was used to coat tank cars in Germany at the turn of the century to prevent corrosion. Hans remembers seeing the green/black painted rail cars passing on the tracks. Although Hans was told that the crude PANI was used, there are no published records on the subject. As will be discussed below, the rich redox and ion exchange chemistry of ICPs, as exemplified by PANI, governs their corrosion protection capability.

The first documented observations of corrosion protection of metals (carbon steel and titanium) using ICPs were reported by Schreiber in 1978 [13]. In Schreiber's work, polymer coatings were prepared by exposing the metals to thiophene or aniline in a large volume microwave plasma discharge. The polythiophene and PANI coatings were found to provide long-term protection from solvents and vapors. In the early 1980s, Mengoli's group [14] began a series of pioneering experiments that established definitively the applicability of electropolymerized ICPs as viable corrosion protection coatings. Since then, numerous papers have been published on the corrosion protection of carbon steel [15], stainless steel [16], iron [17], titanium [18], copper [19], and aluminum alloys [20] with ICPs such as PANI, PPY, polythiophenes, and others. Two comprehensive review articles have been published [21]. In the early 1980s, when ICPs were first considered for corrosion protection, investigators concentrated on the mechanism of corrosion protection as an anodic inhibition process. The anodic inhibition (or passivation) interpretation resulted from early experiments by DeBerry [22], who showed that conducting polymers in the oxidized form anodically protect stainless steel in dilute sulfuric acid. According to this model, ICPs stabilize the potential of the metal in a passive regime, to maintain a protective oxide layer on the metal. Oxygen reduction on the polymer film then replenishes the polymer charge consumed by metal dissolution, thereby stabilizing the potential of the exposed metal in the passive regime and minimizing the rate of metal dissolution. Equations 23.1 and 23.2 depict this concept:

ICP Reduction by metal to form metal oxide and reduced ICP:

$$\begin{aligned}(1/n)\mathrm{M} + (1/m)\mathrm{ICP}^{m+} \\ + (y/n)\mathrm{H_2O} \rightleftharpoons (1/n)\mathrm{M(OH)}_y^{(n-y)+} \\ + (1/m)\mathrm{ICP}^{\mathrm{o}} + (y/n)\mathrm{H}^+\end{aligned} \tag{23.1}$$

ICP oxidation by oxygen from air to form oxidized ICP:

$$(m/4)\mathrm{O_2} + (m/2)\mathrm{H_2O} + \mathrm{ICP}^{\mathrm{o}} \rightleftharpoons \mathrm{ICP}^{m+} + m\mathrm{OH}^- \tag{23.2}$$

Unfortunately, the oxide stabilization argument as used by DeBerry cannot be generalized, particularly for steel and aluminum in neutral chloride environments. As a result, not until the late 1990s was the importance of the dopant anion for making conducting polymers corrosion protective noticed. Kinlen et al. [23] for example, doped PANI with phosphonate corrosion inhibitors with successful inhibition of steel in acid chloride, an environment that does not lead to passivity in steel. Kendig et al. [24] used the inhibitor release mechanism to explain the inhibition of corrosion of Cu-containing aluminum alloy in neutral sodium chloride. Cook and colleagues [25] showed that release of inhibiting anions as a result of the reduction of ICPs could account for the inhibition of steel in chloride containing media where electrochemical passivation for mild steel generally does not occur. A recent review outlines the arguments for both mechanisms [26].

As far as cathodic disbonding, the ICP should quite reasonably displace the oxygen reduction reaction (ORR) away from the polymer/metal interface [27] where it does the most damage as a result of the formation of hydroxide and free radical intermediates to the ORR. However, a report using scanning Kelvin probe analysis [28] showed that this effect cannot readily be sustained, and once the ICP is reduced to a nonconducting state the disbonding proceeds at a faster rate. The report concludes that a conducting polymer interface can actually passivate sufficiently small defects. Rohwerder [29] has also reported that this mechanism only works if extended percolation networks of the conducting polymer are avoided in the coating.

In recent years, a rather considerable literature has developed around corrosion protection by ICPs. Table 23.1 provides examples of some of the more recent reports on the subject. The citations in Table 23.1 are not exhaustive, but serve to illustrate the breadth of the rather extensive work. Much of the work remains disappointing since it reports corrosion protection by ICPs usually claiming a passivation mechanism (see Table 23.1) with no data on the performance by a realistic blank, a coating with the same barrier properties as the ICP but neither conductivity nor oxidizing potential. Bare metal exposed to the macroenvironment are often considered in these reports as the blank (Table 23.1). Unfortunately, bare metal hardly represents a realistic blank when considering special protection at microenvironments of coating defects due to electrical and electrochemical properties of the ICP coating. Much of the work cited in Table 23.1 considers an apparent "enoblement" of steel by ICPs as evidence for the "passivation" mechanism (Table 23.1). The enoblement appears as a relatively high open circuit potential of the coated metal that eventually drops as the coated metal begins to rapidly corrode, explained to be a result of discharge of the electrochemically oxidizing coating. However, this argument becomes less convincing if one considers reports dating from the pre-Internet era that show nonoxidizing, nonconducting coatings such as polybutadiene on steel in NaCl to exhibit similar behavior [30]. Furthermore several comparative studies of Fe or steel coated with materials containing the oxidized emeraldine salt (ES) or emeraldine base (EB) in chloride environments show the nonconducting, nonoxidizing EB to exhibit superior performance [31] (see also Table 23.1), a result hardly supportive of the passivation model. Readers who delve into the technical literature on the subject of corrosion protection by conducting polymers should ask the following questions:

1. Does the report consider a relevant blank?
2. Does the report rely on data other than OCP transients to explain the protection mechanism?

While most of the reports in Table 23.1 propose the passivation model, as inspired from the DeBerry work for stainless steel in a nonchloride environment, a number of reports focus on other aspects of the ICPs, such as inhibitor release, as key to their corrosion protection, and of prime importance to smart protective coatings.

23.2.2.1.4 *Visual Example of Electrochemically Stimulated Release*

A visual example illustrating the smart coating phenomena can be demonstrated by incorporating a dye in an ICP film. In this case the dye, phenol red, is incorporated into PPY through electrochemical polymerization of pyrrole in the presence of phenol red, which acts as a dopant. The reaction scheme is shown in Figure 23.2. When the phenol red doped PPY is electrochemically reduced, by connecting it to a freely corroding aluminum panel, phenol red is expelled from the PPY as shown by the photographs in Figure 23.3. Clearly, the electrochemical potential of the corroding aluminum (ca. −0.5 V vs. Ag/AgCl) furnishes sufficient driving force to reduce the PPY and release the dopant (in this case phenol red). Without the driving force, phenol red is not released. In conclusion, an on-demand release system results; release will not occur unless the polymer is galvanically coupled to the corroding metal, a situation that will occur when a coating is scratched to bare metal.

23.2.2.1.5 *Case Studies in the Role of Dopants*

23.2.2.1.5.1 *PANI on Steel with Phosphonate Dopants* One of the most promising uses of ICPs for corrosion protection is PANI whose structure and oxidation states are shown in Figure 23.1. PANI in its various oxidation states mediates the anodic current between the passivated metal surface and oxygen reduction on the ICP film. In an acidic environment, the conductive emeraldine form of PANI is reduced reversibly to the leuco form, while in a basic environment, the EB form is reversibly reduced to the leuco base form.

The reactions depicted in Equations 23.1 and 23.2 and Figure 23.1 do not take into account the nature of the dopant ion. In most cases, the early work cited above utilized sulfonic acids, probably because commercially available conductive PANI salts are doped with monofunctional sulfonic acids such as *p*-toluene sulfonic acid. A major drawback to the use of sulfonic acid-doped PANIs is that sulfonic acid itself is corrosive to carbon steels with passivation only observed at high concentrations (30% w/w), which may have contributed to the somewhat contradictory results reported in the literature.

In contrast to sulfonates, phosphonates, widely used in applications for corrosion mitigation in aqueous systems, were found to be very effective as corrosion inhibitive dopants for PANI. For example, PANI doped with aminotri(methylenephosphonic acid) (ATMP) was found to be very effective for protection of carbon steel in salt fog [32]. Scanning reference electrode technique (SRET) studies have shown (Figure 23.4) that initially, PANI-ATMP-coated steel panels exhibit anodic activity in pinholes and cathodic activity on the conductive polymer surface (oxygen reduction). After ca. 92 h exposure to a corrosive environment, the control, a nonconducting PVB-coated steel panel, still

TABLE 23.1 Summary of ICP Corrosion Literature

Polymer	Substrate	Environment	Method	Imporant Observation	Authors	Reference
Numerous reviews				Mentions inhibitor release	Spinks, GM; Dominis, AJ; Wallace, GG; Tallman, DE	*J. Solid State Electrochem.*, 6(2), 85–100 (2002)
Numerous reviews					Tallman, DE; Spinks, G; Dominis, A; Wallace, GG	*J. Solid State Electrochem.*, 6(2), 73–84 (2002)
PANI	AA 2024		SVET	Controlled release of inhibitor	Kinlen, PJ; Graham, CR; Ding, Y	*ACS Symp.* 22–26, 2004, POLY-239 (2004)
PANI	AA 2024	various		Extraction of galvanically active Cu from alloy	Epstein, AJ; Smallfield, JAO; Guan, H; Fahlman, M	*Synth. Met.*, 102(1–3), 1374–6 (1999)
PANI	Fe	NaCl	EIS, raman		Bernard, MC; Deslouis, C; ElMoustafid, T; HugotLeGoff, A; Joiret, S; Tribollet, B	*Synth. Met.*, 102(1–3), 1381–1382 (1999)
PANI	Fe	various		PANI is compared to non-conducting org. coating; EB performs better than other forms	Mirmohseni, A; Oladegaragoze, A	*Synth. Met.*, 114(2), 105–108 (2000)
PANI	Fe		Dissolved [Fe], galvanic couple	Fe reduced ES; EB is superior to ES	Spinks, GM (Reprint); Dominis, A; Wallace, GG	*Corrosion*, 59(1), 2–31 (2003)
PANI	Fe	3.5% NaCl, 0.1 M HCL	EIS, galvanic couple, polarization	Passivation	Lu, WK; Elsenbaumer, RL; Wessling, B	*Synth. Met.*, 71(1–3), 2163–2166 (1995)
PANI	Fe	0.1 M NaCl, 0.1 M HCl		Inhibitor release	Cook, A; Gabriel, A; Laycock, N	*J. Electrochem. Soc.*, 151(9), B529–B535 (2004)
PANI	st	3.5% NaCl		Dopant is varied. Compared to epoxy, EB performs best	Dominis, AJ; Spinks, GM; Wallace, GG	*Prog. Organic Coatings*, 48(1): 43–49 Nov 2003
PANI	st		Polarization	Passivation mechanism	Wei, Y; Wang, JG; Jia, XR; Yeh, JM; Spellane, P	*Polymer*, 36(23), 4535–4537 (1995)
PANI	Zn		Polarization		Zhao, YP; Yin, RH; Cao, WM; Yuan, AB	*Act. Metallur. Sin.* (English Letters), 17(6), 849–855 (2004).
PANI and iodo PANI	Fe	0.5M HCl	OCP, EIS polarization	Inhibition eff. >75% after 48 h	Bereket, G; Huer, E; Sahin, Y	*Appl. Sur. Sci.*, 252(5), 1233–1244 (2005)
PANI and Poly anisidines	SS304	0.5 M HCl	EIS, OCP, polarization	Passivation mechanism	Bereket, G; Huer, E; Sahin, Y	*Prog. Org. Coat.*, 54(1), 63–72, (2005).
PANI DBSA doped	St	3.5% NaCl	EIS, polarization	Increase corrosion resistance with coating	Subrahmanya, S; Hung, VH; Holze, R	*J. Electrochem. Soc.*, 154(2), C67–C73 (2007)
PANI, Polythiophene	St		EIS		Nguyen, PT; Rammelt, U; Plieth, W	*Macromolecular Symposia*, 187, 929–938 (2002)
PANI, PPY	SS		EIS, OCP, polarization	50 nA/cm^2	Iroh, JO; Gajela, P; Cain, R; Nelson, T; Hall, S	International SAMPE Symposium and Exhibition (2005)
PANI, PPY	Fe		EIS	Adhesion promotion but no effect of passivation	Rammelt, U; Nguyen, PT; Plieth, W	*Electrochim. Acta*, 48(9): 1257–1262 Apr 20 (2003)

(continued)

TABLE 23.1(continued) Summary of ICP Corrosion Literature

Polymer	Substrate	Environment	Method	Imporant Observation	Authors	Reference
PANI/dispersed in paint	st	5% NaCl	Kelvin probe	Displace ORR from interface, observed dopant effect	Williams, G; Gabriel, A; Cook, A; Mcmurray, HN	*J. Electrochem. Soc.*, 153(10), B425–B433 (2006)
PANI/montmorillonite clay, (MMT)/polyimide	AA 2024	3.5% NaCl	EIS, OCP, polarization	Rp = 10E9 ohm cm^2	Iroh, JO; Kottarath, S; Shah, K; Rajamani, D	International SAMPE Symposium and Exhibition (2005)
PANI/Ni hexacyanoferrate dopant	Fe		RQCM, OCP		Kulesza, PJ; Miecznikowski, K; Malik, MA; Galkowski, M; Chojak, M; Caban, K; Wieckowski, A	*Electrochim. Acta*, 46(26–27), 4065–4073 Aug 24 (2001)
PANI/PMMA Blends	st			Passivation enhanced by anion release	Pereira Da Silva, JE; Córdoba De Torresi, SI; Torresi, RM	*Prog. Org. Coat.*, 58(1), 33–39
PANI/PMMA/CSA	AA 2024	0.01 M H_2SO_4	OCP, raman, scanning probe	Suppression of H2 evolution at defects	Jesse, C; Seegmiller, A; Pereira, JE; Da Silva, B; Buttry, DA; Susana, I. Córdoba De Torresi, B; Torresib, RM	*J. Electrochem. Soc.*, 152(2), B45–B53 (2005)
PANI/polyanion double strand	AA7075			Passivation mechanism proposed		*Synth. Met.*, 1997, 85(1–3) 1263–1264, (1997)
PANI-co-POA	Cu	3.5% NaCl	EIS, OCP, polarization	Passivation mechanism	Oezyilmaz, A; Tuncay; Colak, N; Sanguen, MK; Erbil, M; Yazici, B	*Prog. Org. Coat.*, 54(4), 353–359 (2005)
PDOT	St,AA2024	NaCl, dil Harrison's Soln.	SVET	Substrate anodic to ICP cathode	Jie, H; Victoria Johnston, G; Tallman Dennis, E; Bierwagen Gordon, P; Wallace Gordon, G	*J. Electrochem. Soc.*, 147(10), 3667–3672 (2000)
Poly mercaptobenzimidazole	Cu		CV EQCM	Same protection as when inhibitor is in soln.	Trachli, B; Keddam, M; Takenouti, H; Srhiri, A	*Prog. Org. Coat.*, 44(1), 17–23 (2002)
Poly o-Phenylene diammine/ phosphate	ss304	0.4 HCl, 0.1 NaCl	Potentiodynamic	Increase in OCP and Increase in exhange current density	D'elia, Luis, F; Reynaldo, L; Orti'Z; Ma'Rquez, ZOP; Ma'Rquez, J; Marti'nez, Y	*J. Electrochem. Soc.*, 148(4), C297–C300 (2001)
poly(4-vinylpyridine) with hexacyanoferrate	SS	2M H_2SO_4		Redox active anion	Galkowski, MT; Kulesza, PJ; Miecznikowski, K; Chojak, M; Bala, H	*J. Solid State Electrochem.*, 8(6), 430–434 (2004)
poly(m-toluidine)	mild Steel	3.5% NaCl	Polarization, OCP, wt. loss	Passivation mechanism	Hegazy, HS; Youssef, EAM; Abd El-Ghaffar, MA	*Egypt. J. Textile Polym. Sci. Technol.*, 8 77–88 (2004)
poly-2,5-dimethoxyaniline	SS,st	0.5 M HCl	CV, raman, OCP	Passivation mechanism	Yin, P; Kilmartin, PA	*Current Appl. Phys.*, 4(2–4), 141–143 (2004)
PPY	Fe	NaCl	EIS		Van Schaftinghen, T; Deslouis, C; Hubin, A; Terryn, H	*Electrochim. Acta*, 51(8–9), 1695–1703 (2006)
PPY	Fe	NaCl	Linear polarization, EIS, and wt. loss		Bazzaoui, M; Martins, JI; Costa, SC; Bazzaoui, EA; Reis, TC; Martins, L	*Electrochim. Acta*, 51(21), 4516–4527 (2006)

PPY	Fe	NaCl	OCP, EIS		Le HNT (REPRINT); Garcia, B; Deslouis, C; Le Xuan, Q	*Electrochim. Acta*, 46(26–27), 4259–4272 (2001)
PPY	Fe	0.1 M K_2SO_4, pH 4	Fe disk PPY ring	ICP not completely discharged before depassivation	Nguyen, Td; Keddam, M; Takenouti, H	*Electrochem. Solid State Lett.*, 6(8), B25–B28 (2003)
PPY	Fe	3% NaCl		Polyanion doped, cation selective coating	Le, Hnt; Garcia, B; Deslouis, C; Le Xuan, Q	*J. Appl. Electrochem.*, 32(1): 105–110 (2002)
PPY	Ni-plated Cu	3.5% NaCl	EIS, OCP, polarization	Passivation mechanism	Tueken, T; Yazici, B; Erbil, M	*Prog. Org. Coat.*, 54(4), 372–376 (2005)
PPY	SS304	0.1M H_2SO_4	Polarization curves	Short-term protection/No long term protection	Garcia, MA; Lucio; Smit, Mascha, A	*J. Power Sources,* 158(1), 397–402, (2006)
PPY	st		XPS, AAS	Repels chloride ion	Lamprakopoulos, S; Yfantis, DK; Depountis, S; Yfantis, CD; Schmeisser, D; Yfantis, AD	*WSEAS Trans. Environment Develop,* 2(6), 742–746 (2006)
PPY			EIS		Van Schaftinghen, T; Deslouis, C; Hubin, A; Terryn, H	*Electrochim. Acta* 51(8–9), 1695–1703 (2006)
PPY				Critical experiment on enobling mechanism	Michalik, A; Rohwerder, M	*Zeitschrift fuer Physikalische Chemie* (Muenchen, Germany), 219(11), 1547–1559 (2005)
PPY doped with DBSA	SS		CV, XPS	Reaction of dopant with interface	Prissanaroon, W; Brack, N; Pigram, PJ; Liesegang, J; Cardwell, TJ	*Surf. Interf. Anal.*, 33(8): 653–662 (2002)
PPY Oxide Composite	Fe	3% NaCl	OCP		Garcia, B; Lamzoudi, A; Pillier, F; Nguyen Thi Le, H; Deslouis, C	*J. Electrochem. Soc.*, 149(12), B560–B566 (2002)
PPY PMo12O403- and PO43-	St	3.5% NaCl pH 5.3 pH 1.9.		ICP gives 1/200 corrosion rate of bare steel	Ohtsuka, T; Iida, M; Ueda, M	*J. Solid State Electrochem.*, 10(9), 714–720 (2006)
PPY, PPY-PDAN	St	0.1 M K_2SO_4, pH 4	OCP, raman		Nguyen, TD; Pham, MC; Piro, B; Aubard, J; Takenouti, H; Keddam, M	*J. Electrochem. Soc.*, 151(6), B325–B330 (2004)
PPY, PTh	Cu	3.5% NaCl	EIS, OCP, polarization			
PPY, SDS anion	1Cr18Ni9Ti stainless steel	0.3M HCl	Polarization	Increase in pitting potential	Zhang, T; Zeng, CL	*Electrochim. Acta,* 50(24), 4721–4727 (2005)
Review				Varied dopants. EB performs best	Dominis, AJ; Spinks, GM (Reprint); Wallace, GG	*Prog. Org. Coat.*, 48(1), 43–49 (2003)
	Fe		Local EIS Raman		Nguyen, TD; Nguyen, TA; Pham, MC; Piro, B; Normand, B; Takenouti	*J. Electroanal. Chem.*, 572(2), 225–234 (2004)

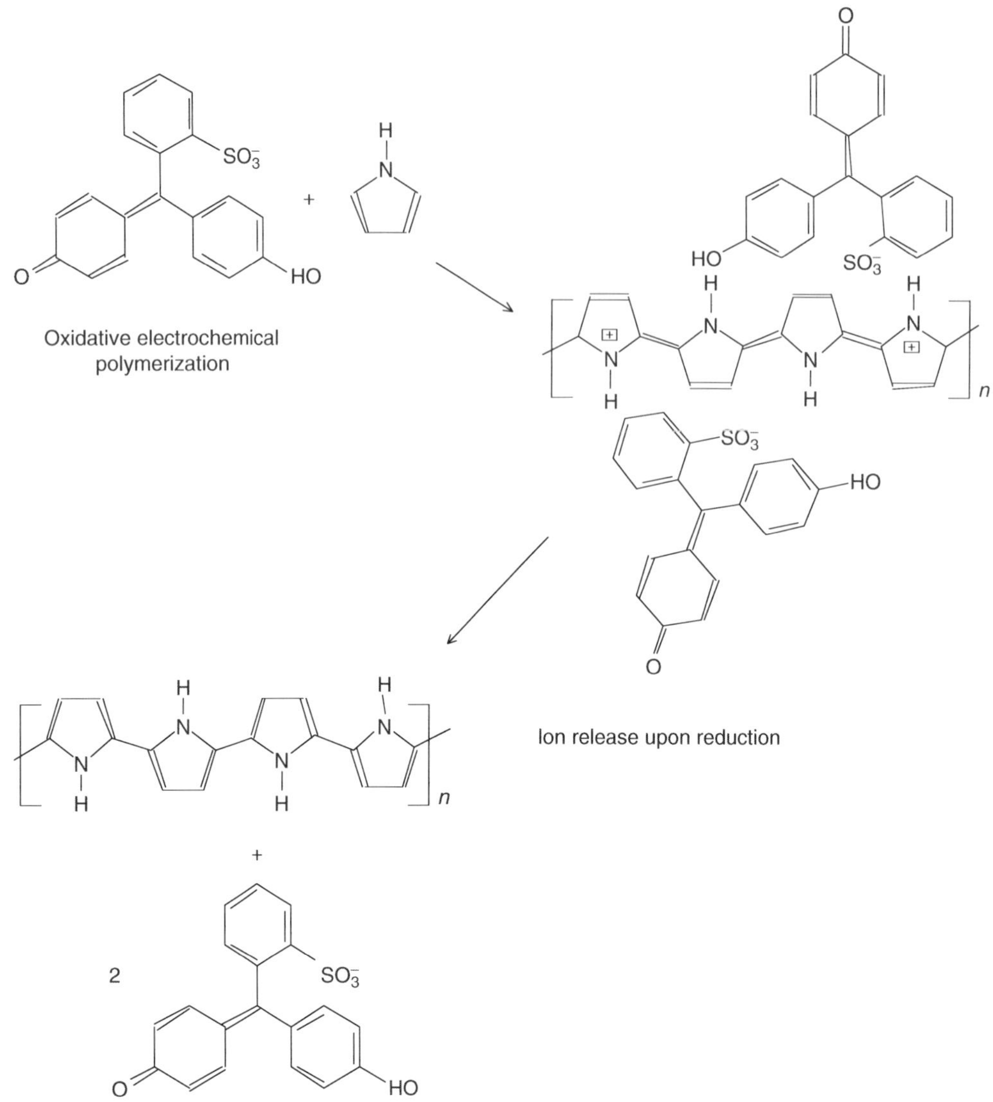

FIGURE 23.2 Reaction scheme for the release of phenol red by PPY.

FIGURE 23.3 Photograph of phenol red-doped PPY before and after contact to a reducing aluminum surface.

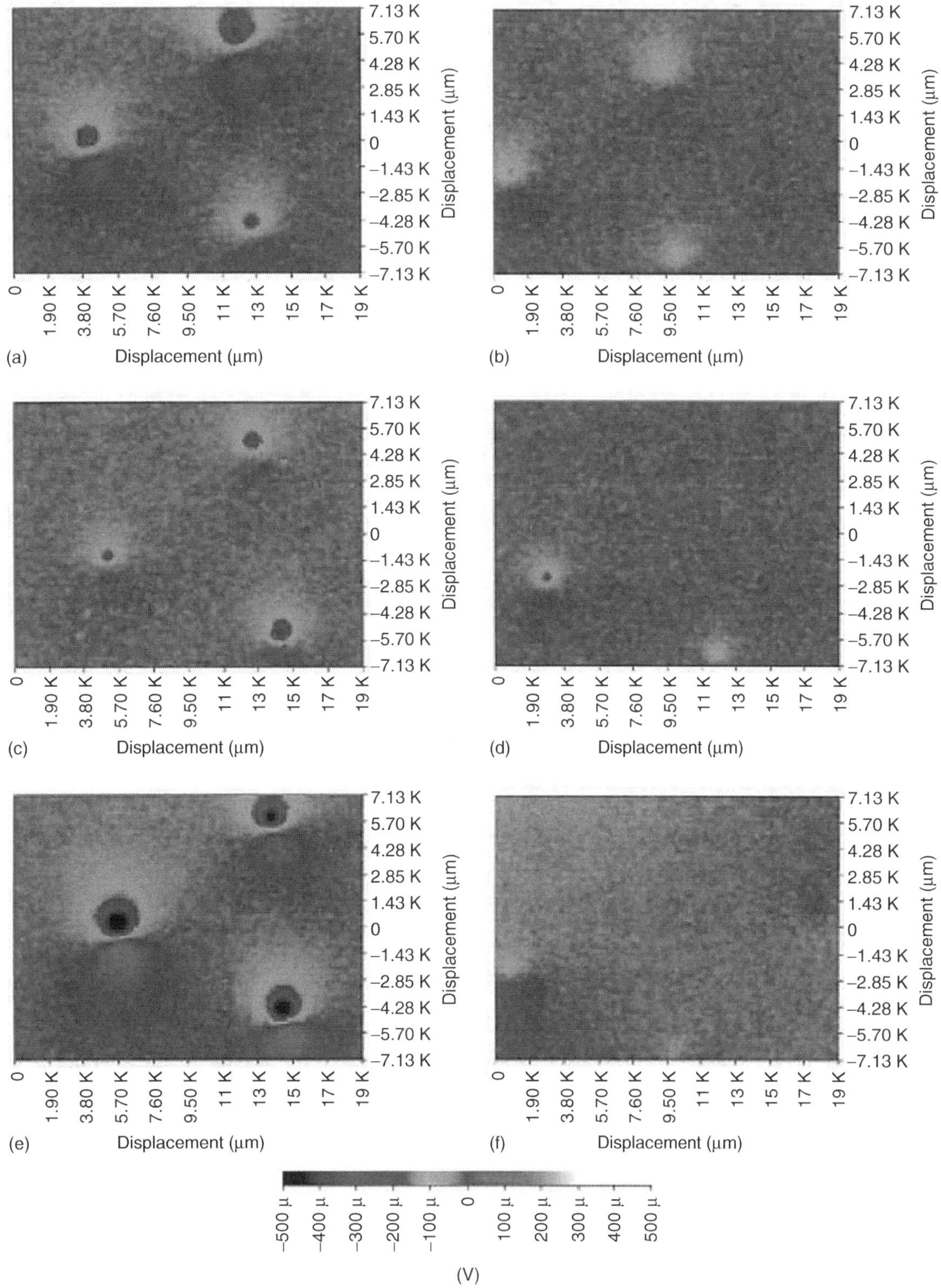

FIGURE 23.4 SRET scans over PVB coatings on carbon steel in tap water with three 1 mm diameter pinholes equidistant at 1 cm: (a) PVB first scan, (b) PVB after 92 h, (c) ATMP/PVB after 68 h, (e) PANI-ATMP/PVB, first scan, and (f) PANI-ATMP/PVB after 140 h.

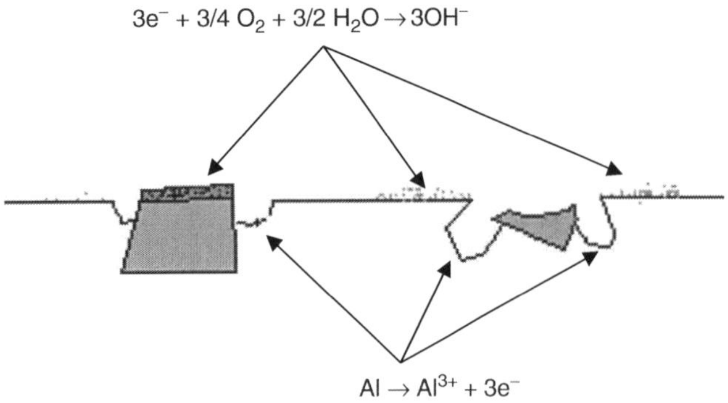

FIGURE 23.5 Schematic of an aluminum alloy surface containing intermetallic phases that catalyze the ORR.

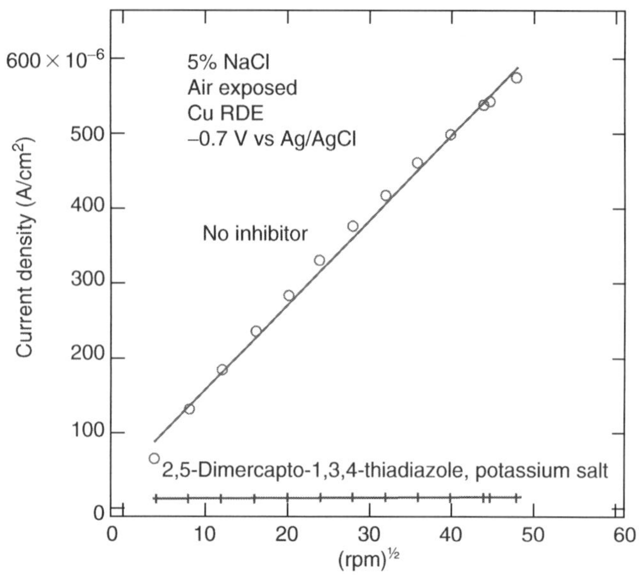

FIGURE 23.6 Oxygen reduction current at a Cu rotating disk electrode as a function of rotation rate in a sodium chloride electrolyte in the presence and absence of the di-potassium salt of DMcT. As can be seen, the addition of a low level of DMcT causes a substantial inhibition of this diffusion-limited reaction.

exhibited galvanic activity whereas the PANI-ATMP exhibited no galvanic activity. It is postulated that PANI-ATMP coating shifts the potential of the steel in a positive or noble direction, creating a combination of ferrous and ferric ions. These ions form a passivating iron/dopant complex at the coating/steel interface. For a scribed coating, Fe^{2+} is proposed to react initially with the ATMP dopant anion (supplied by the leuco-ES equilibrium in Figure 23.1 in an anodic regime). The insoluble iron-ATMP salt formed effectively passivates the exposed metal surface.

23.2.2.1.5.2 PANI on High-Strength Aluminum Containing ORR Inhibitors as Dopants Alloy 2024-T3 is an inhomogeneous aluminum/copper alloy containing intermetallic particles that are enriched in alloying elements compared to the homogeneous alloy matrix. In contrast to pure aluminum, which is resistant to pitting corrosion, the increased susceptibility of the 2024-T3 aluminum alloy to pitting corrosion is centered upon the unique electrochemistry of the intermetallic inclusions, specifically on the ORR as illustrated in Figure 23.5.

Chromate coatings inhibit through a smart release of hexavalent chromate when called upon by the presence of a corrosive environment [33]. Chromate primarily inhibits the ORR at coating defects by an irreversible electrosorption at sites otherwise catalytic to oxygen reduction [34]. Based on these results, a class of organic compounds has been found to act as inhibitors for the ORR to an extent comparable to that for chromate [35]. Application of these compounds in real coatings requires a mechanism for stabilizing them in the coating for release when needed in a fashion analogous to chromate. Fortunately the chemistry available from conducting organic polymers provides a mechanism for performing this function.

The ultimate goal is to incorporate organic ORR inhibitors into PANI coating systems that are released upon demand. 2,5-Dimercapto-1,3,4-thiadiazole (DMcT) and 1-pyrrolidine dithiocarboic acid [36] derivatives are well-known ORR inhibitors for copper and its alloys. Figure 23.6 shows the ORR currents at a Cu RDE as a function of rotation rate in the presence and absence of DMcT. DMcT virtually eliminates the ORR current. DMcT and poly(DMcT) (the polydisulfide formed by oxidative polymerization of DMcT) have been extensively studied in combination with PANI as reversible redox electrodes for lithium ion batteries.

PANI films containing the bases of 1-pyrrolidine dithiocarboic acid have enabled the release of an ORR inhibitor apparently in response to the galvanic influence of a coating defect. A PANI-coated Al 2024-T3 doped in a circular region with the 1-pyrrolidine dithiocarboic anion before exposing to the ASTM B117 salt fog environment for 48 h demonstrated corrosion inhibition around a scribe in the doped region (Figure 23.7) [37]. The SVET scan indicates an enhanced cathodic current for the scribe through the undoped region.

To summarize, conducting polymers can provide protection to aluminum alloys in chloride environment at breaks in the coating by a galvanic mechanism substantially different from anodic protection. Protection of base metals such as aluminum in chloride environments by conducting polymer coatings entails a mechanism that is similar to that provided by chromate. Rather than requiring the conducting polymer film to bias exposed substrates to a passive oxidized state, our hypothesis considers that exposed defects bias the conducting polymer, thereby causing it to (a) reduce to the nonconducting form and (b) release inhibiting species that stop corrosion in the defect.

23.2.2.1.5.3 Other Polymer Dopant Combinations Many other conducting polymer/dopant combinations have been considered and appear in Table 23.1. A typical analysis used in many investigations entails following the open circuit potential of the coated metal as a function of time. The ICP maintains the surface at a

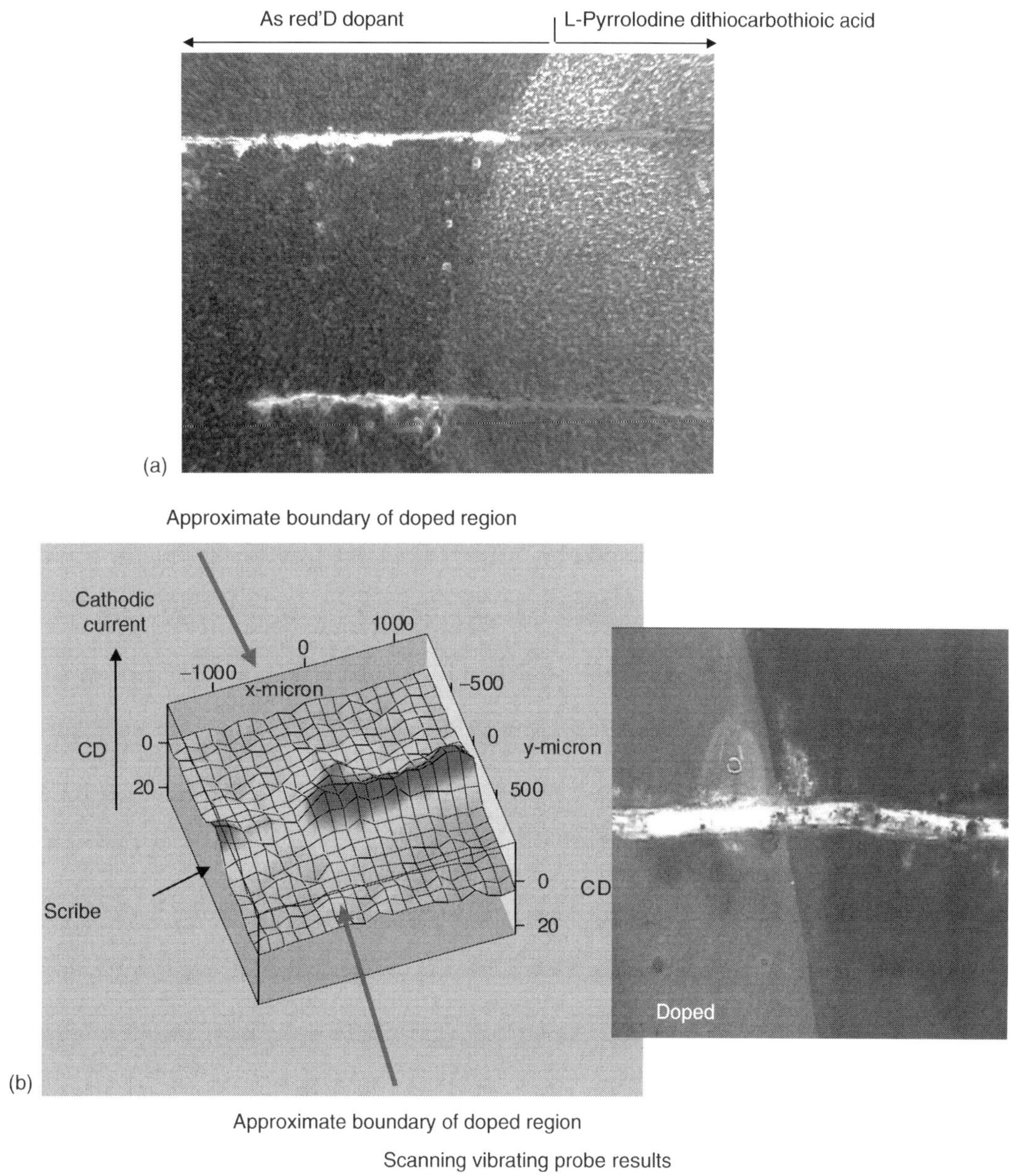

FIGURE 23.7 (a) Scribed salt fog exposure test for a portion of a PANI coating on AA 2024-T3 that was doped in a local region with 1-pyrrolidine dithiocarboic acid anion. As can be seen the anion released at the region of the scribe inhibits corrosion of the scribe. (b) SVET data for a similar specimen.

high potential until the metal depassivates with the reduction of the potential closer to the reversible potential of the metal dissolution. Bare metal exhibits a more rapid drop in potential.

23.2.2.1.6 Redox Polymers

PANI-DMcT is easily prepared by oxidative polymerization of aniline and DMcT in water using ammonium peroxydisulfate as the oxidizing agent. Size exclusion chromatograms of PANI-DMcT show that NMP-soluble fractions of PANI-DMcT are heterogeneous polymer blends of PANI and poly(DMcT), where poly(DMcT) is the product of the oxidative polymerization of DMcT (Figure 23.8a) [38]. The cyclic voltammogram in Figure 23.8b provides evidence that poly(DMcT) is reversibly reduced, expelling an anion of DMcT in the process [39]. Thus, upon electrochemical reduction, both PANI-DMct and poly(DMcT) are expected to release DMcT anions, an excellent ORR inhibitor as shown by the data in Figure 23.6.

SVET for an AA 20224-T3 surface coated with the PANI–PolyDMcT composite with pinholes appears in Figure 23.9.The three pinholes initially exhibit cathodic activity but with no corrosion. After 1 h exposure to aerated tap water, the potential of the pinholes indicates the cathode reaction has been inhibited as expected. After 42 h exposure, however, anodic activity is observed to occur. Thus although the ORR in the pin holes has been eliminated, anodic

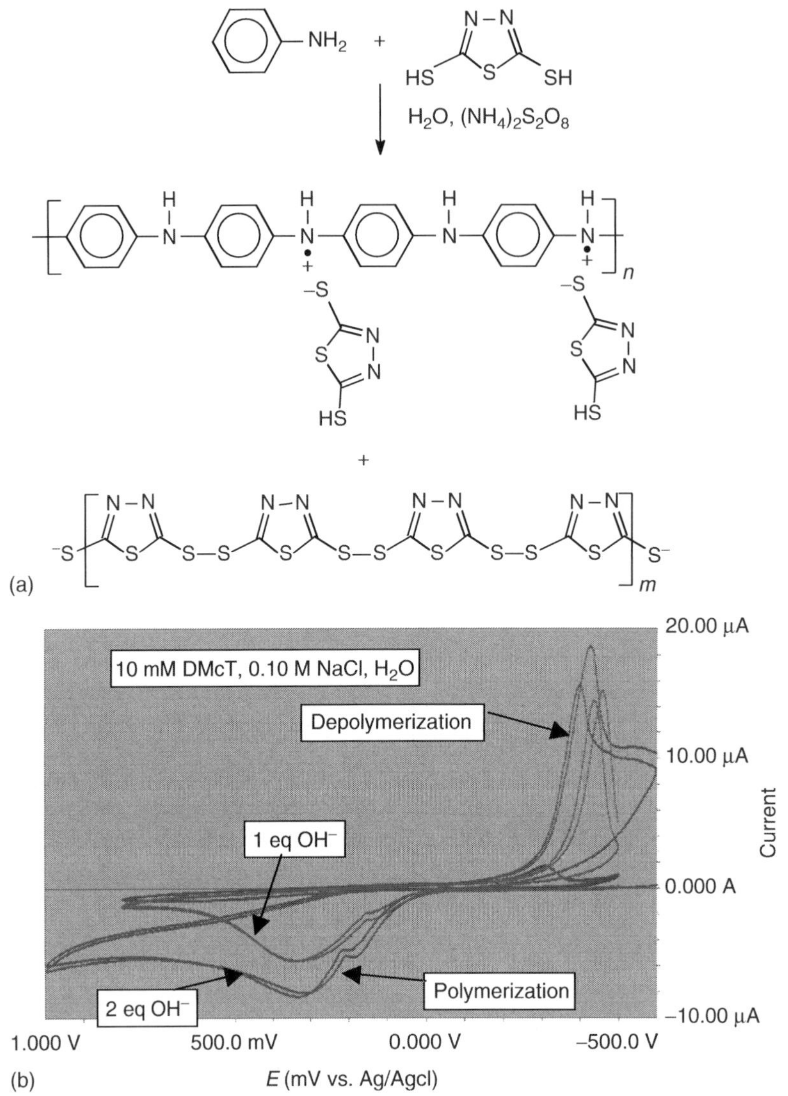

FIGURE 23.8 (a) Oxidation of DMcT to PolyDMcT and (b) cyclic voltammogram for the oxidation and reduction of DMcT to PolyDMcT.

processes in the pin hole are not completely eliminated, as might be expected since DMcT is an ORR inhibitor on copper. A potential approach to eliminate this problem would be to incorporate both anodic and ORR inhibitor as dopants into the ICP system.

More recently, the reduction of poly(dimercapto thiadiazole) (PolyDMcT) to 2,5-dimercapto thiadiazole contained in a conducting organic matrix provided a direct demonstration of galvanic release of an inhibitor, PolyDMcT, held within a carbon paste as schematically shown in the inset of Figure 23.10 [40]. A Cu RDE cathode biased at −0.7 V versus Ag/AgCl and placed above the carbon paste electrode detected the release of the ORR-inhibiting monomer as a decrease in the cathodic ORR current. Upon coupling the carbon paste to a freshly polished Al surface, the released monomer of polyDMcT was observed to lower the ORR current at the Cu RDE as shown in Figure 23.10. This observation demonstrates the feasibility of an organic coating formed from the electroactive polymer to release the inhibitor upon galvanic coupling to a defect.

23.2.2.2 Chemical Potential

For a number of years, ion exchange or host/guest compounds have shown potential as inhibiting pigments for paints. For example, the compound commercially known as Shieldex (Grace-Davison and Zin et al.) can release calcium ions as corrosion inhibitors as a result of the mass action of a high ionic strength, high corrosive ion concentration. Often these pigments have been used synergistically with other inhibitors for the corrosion inhibition of galvanized steel [41]. In other words such systems release corrosion inhibiting species as a result of mass action or a chemical potential of an ion.

More recently, McMurray and Williams [42] and many others [43] have considered the ion exchange properties of layered minerals such as hydrotalcite as sources of anion-exchangable corrosion inhibitors including chromate. Buchheit and Mahajanam used these same materials as an ionic-strength responding reservoir for vanadate inhibitors [44],

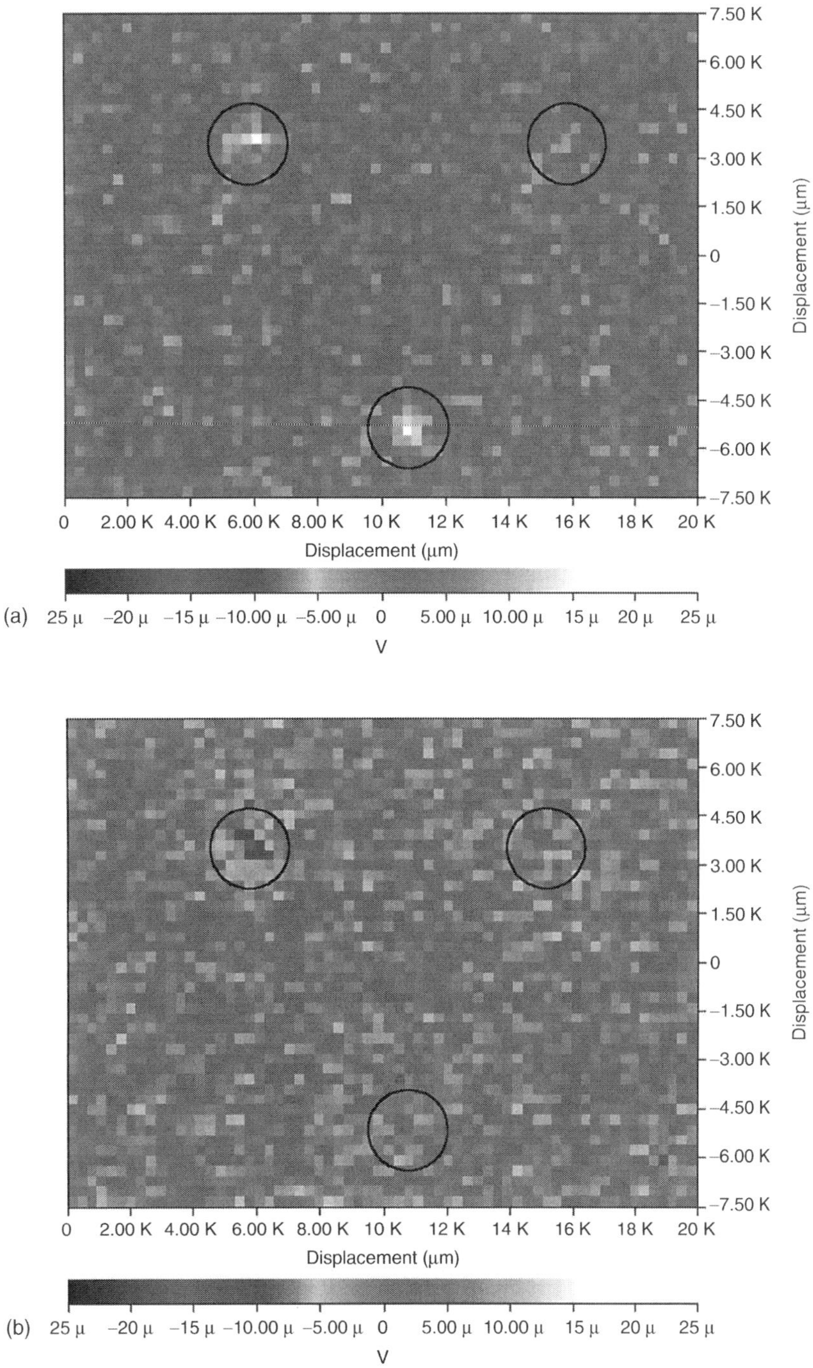

FIGURE 23.9 (a) First SVET scan of UV-cured PANI-DMcT/poly(DMcT) formulation. Circles depict pin holes (anodic areas are positive volts, cathodic negative). (b) Final SVET scan of UV cured PANI-DMcT/poly(DMcT) formulation after 48 h of exposure.

and a number of research reports have described the use of the hydrotalcites to provide controlled release of organic anion that are quite effective for slowing the ORR on Cu surfaces [45]. Similarly, ion exchange minerals can also hold exchangeable cations as in the case of Shieldex for Ca^{2+}, but also the effective rare-earth cations [46].

pH represents an important chemical potential to activate the release of corrosion inhibitors for a number of reasons. Many structural materials are amphoteric. They corrode or their passivating oxides become unstable with respect to dissolution under conditions of either high pH or low pH. As a result, local variations of pH resulting from corrosion can accelerate corrosion. For example, the alkalinity developed around noble intermetallics on Cu-containing Al alloys have been considered to promote depassivation of the alloy in the vicinity of the intermetallic. Localized crevice or pitting corrosion in stainless steels

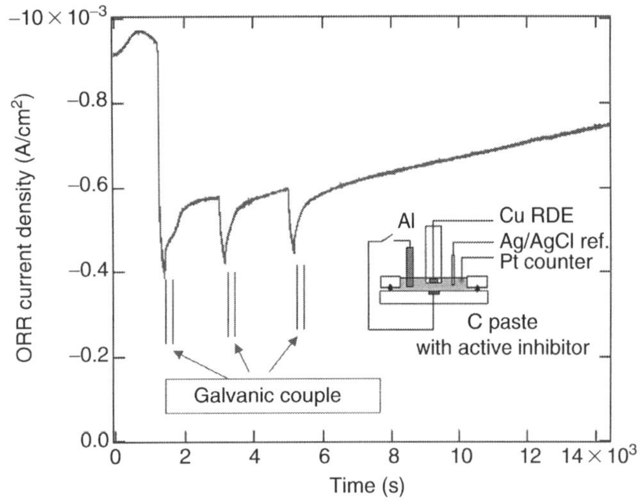

FIGURE 23.10 Demonstration of the galvanic release of the ORR inhibitor, DMcT from carbon paste electrode (inset) containing PolyDMcT.

involves the lowering of the pH in occluded regions as a result of the hydrolysis of anodic products to enhance the anodic dissolution.

$$M \rightleftharpoons M^{n+} + ne^- \quad (23.3)$$

$$nM^+ + H_2O \rightleftharpoons M(OH)^{(n-1)+} + H^+ \quad (23.4)$$

A second consideration regarding the importance of pH recognizes that the onset of localized corrosion where the anode is separated from the cathode, results in localized changes in pH. In the case of coated steel, the cathodic reduction of oxygen:

$$H_2O + 1/2\,O_2 + 2e^- \rightleftharpoons 2OH^- \quad (23.5)$$

at the coating metal interface around a defect tends to raise the pH at the interface, leading to propagation of coating failure. In such cases, the release of a corrosion inhibitor in response to increase in pH would be beneficial. Research is ongoing for devising methods for release of corrosion inhibitors as stimulated by the increase or decrease in pH. For example, Li and Calle have described an approach for microcapsules containing corrosion inhibitors that degrade to release a corrosion inhibitor as a result of alkaline hydrolysis. So far they have demonstrated nominally 10 μm microcapsules that breakdown under alkaline conditions to release pH indicators with no adverse influence on other paint properties [47]. Cook et al. reported early results on carboxylate-linked inhibitors on Boehemite. Alkaline hydroysis of the carboxylate linkage above pH 9 will release the inhibitor [48]. pH-controlled porosity of organic/inorganic composites have not yet been applied to corrosion-protective coatings, but the pH responsiveness appears promising [49].

23.2.2.3 Mechanical

Smart coatings that respond to mechanical stress resulting from abrasion, impact, and cyclic fatigue have particular engineering importance. The main contribution in this area has considered inhibitor-containing microspheres as potential or actual paint pigments. Inspired by the invention by Silver in 1970 of acrylate-copolymer microspheres that contain adhesive [50]. Bailin and Agarwala [51] developed microspheres to contain dichromate, nitrite, borate, and molybdate inhibitors as a corrosion-inhibiting pigment for paint. U.S. Army research has also demonstrated corrosion inhibition by stress release from microcapsules [52]. A recent review mentions the possibility of environmentally responsive (in the mechanical sense) polymers for smart corrosion inhibiting coatings [53]. A group at the University of Illinois has taken this generic approach several steps further with a coating that releases a two-component polymer-forming agent. The stress responsive release comes not from microcapsules, but rather from a 3D biomimetic microvascular system developed in the coating using a direct-write assembly method [54]. The vascular approach has an important advantage over the microcapsule-bourne healing agents. It enables repeated healing of a damaged region drawing material from the entire coating via the vascular system and enables use of a two-component system.

23.2.2.4 Bioproduced Inhibitors and Protective Films

Often microbial growth accelerates corrosion and general environmental degradation of materials. However, Nagiub and Mansfeld found that *Shewanella algae* and *Shewanella ana* prevented pitting of Al 2024, tarnishing of brass and rusting of mild steel in artificial seawater [55] and others report that *Bacillus subtilus* prevents pitting of Al 2024 (2002) [56]. A living material capable of inhibiting corrosion could potentially be placed in a paint or primer matrix in the form of seeds or spores only to be awakened to growth in response to adverse conditions. Kus and Mansfeld recently reported that *Shewanella oneidensis* inhibited a number of metals. Notably Cu-bearing brass and Cu appear to form a porous but somewhat insulating layer at the surface for these materials [57]. Microbes bioengineered to produce aspartate and polyphosphate inhibitors were indeed shown to inhibit corrosion in Luria Bertani growth media [58]. Although there have been no practical coatings formed using biotech-based smart inhibitors, this appears to be an important area for future development.

23.3 Important Opportunities

Protective coatings must provide a barrier to corrosive reagents, but should also be able to respond to environmental stresses in a manner to limit corrosion without depleting their reactivity or without using environmentally hazardous materials. Release of inhibitors in response to damage to the coating remains a very attractive property for protective coatings. Ideally the smart response will either not be depleted or can be continuously and inherently renewed.

Galvanic stimulation remains an important, if not the most important, signaling mechanism, given the electrochemical nature of corrosion. Therefore, electroactive and conducting polymers will continue to play a significant role in the development of smart corrosion protective coatings. Coupling the electrochemical processes responsible to corrosion to the opening of colloidal microspheres embedded in paint represents an important avenue for research.

Regarding conjugated inherently conducting polymers, there remains a need to determine the role played by their high oxidization potential. It remains hard to believe that these materials can protect base alloys in environments where they are susceptible to pitting by anodic protection alone.

Microbiological (genetic engineered) approaches relying on the growth of a living film have shown early feasibility and represent a promising future approach for certain applications. This area deserves further investigation. To render these coatings smart, however, a mechanism must exist to activate or deactivate the inhibiting process as corrosive conditions demand.

23.4 Summary

In summary, new developments in nanotechnology, biotechnology, biomimetic materials, and functional materials hold great promise for smart corrosion protective films. In all cases, the application of corrosion protection of highly energetic surfaces (high potential energy for oxidation) suggests the use of materials that exhibit some type of galvanic response or galvanic sensitivity to the bias provided by the metallic substrate in a corrosive environment. To date, conducting polymers appear to have the most promise since they can hold or release anion inhibitors depending on their oxidation state. In general, the ability of the energy released by the corrosion process itself (nearly 2 V coulomb for steel and aluminum in neutral water) is sufficient to drive corrosion-mediating processes. Their semiconductivity allows them to direct the galvanic forces that drive corrosion to a process that inhibit corrosion.

References

1. J.E.O. Mayne, *J. Oil Colour Chemists'Assoc.*, 34, 473, 1951.
2. L. Xia, E. Akiyama, G. Frankel, and R. McCreery, *J. Electrochem. Soc.*, 147, 2556, 2000.
3. M. Kendig, M. Hon, L. Warren, *Prog. Org. Coatings*, 47, 169, 2002.
4. J. Zhang and G.S. Frankel, *Corrosion*, 55, 957, 1999.
5. J. Sinko, Prog. *Organic Coatings*, 42, 267, 2001.
6. J. Sinko, Prog. *Organic Coatings*, 42, 267, 2001.
7. L. Xia, E. Akiyama, G. Frankel, and R.J. McCreery, *J. Electrochem. Soc.*, 147, 2556 2000.
8. P.J. Kinlen, Y. Ding, and D.C. Silverman, *Corrosion*, 58, 490 2002.
9. R.L. Blankespoor and L.L. Miller, *J. Chem. Soc., Chem. Commun.*, 90, 1985.
10. Reynolds et al., *J. Chem. Soc., Chem Commun.*, 620, 1987.
11. H. Letheby, *J. Chem. Soc.*, 15, 161, 1862.
12. F. Goppelsroeder, Internationale Elektrotechniche Ausstellung No. 19, 1047; No. 18, 978, 1891.
13. H.P Schreiber et al., *Ind. Eng. Chem. Prod. Res. Dev.*, 17, 27, 1978.
14. G. Mengoli, M. Munari, P. Bianco, and M. Musiani, *J. Appl. Polymer Sci.*, 26, 4247, 1981.
15. P.J. Kinlen, Y. Ding, and D.C. Silverman, *Corrosion*, 58, 490, 2002.
16. R. Gasparac and C.R. Martin, *J. Electrochem. Soc.*, 148, B138, 2001.
17. F. Beck, *Metalloberflaeche*, 46, 177 1992; F. Beck, R. Michaelis, F. Schloten, and B. Zinger, *Electrochim, Acta*, 39, 229, 1994.
18. Z. Deng, W.S. Smyrl, and H.S. White, *J. Electrochem. Soc.*, 136, 2152, 1989.
19. V. Brusic, M. Angelopulos, and T. Graham, *J. Electrochem. Soc.*, 144, 436, 1997.
20. V.J.Gelling, M.M. Wiest, D.E. Tallman, G.P. Bierwagen, and G.G. Wallace, *Prog. Organic Coatings*, 43, 149, 2001; P. McCarthy, W. Li, and S.C. Yang, *Polym. Mater. Sci. Eng.*, 83, 315, 2000.
21. T. Page McAndrew, *Trends Polym. Sci.*, 5, 7, 1997, G.M. Spinks, A.J. Dominis, and G.G. Wallace, *J. Solid State Electrochem.*, 6(2), 85, 2002; G.M. Spinks, A.J. Dominis, and G.G. Wallace, *J. Solid State Electrochem.*, 6(2), 73–84, 2002.
22. D. DeBerry, *J. Electrochem. Soc.*, 132, 1022, 1985.
23. P. Kinlen, V. Menon V, and Y. Ding, *J. Electrochem. Soc.*, 146, 3690, 1999.
24. M. Kendig, M. Hon, and L. Warren, *Prog. Org. Coatings*, 47, 169, 2002.
25. A.Cook, A. Gabriel, D. Siew, and N. Laycock, *J. Current Appl. Phys.*, 4,133, 2004; A. Cook, A. Gabriel, and N. Laycock, *J. Electrochem. Soc.*, 151(9), B529–B535, 2004.
26. Michalik and Z, Rohwerder. *Phys. Chem.*, 219, 1547, 2005.
27. P. Kinlen, D. Silverman, and C.R. Jefferys, *Synthetic Metals*, 85, 1327, 1997; T. Shaauer, A. Joos, L. Dulong, and C.D. Eisenbach, *Prog. Org. Coating*, 33, 20, 1998.
28. G. Paliwoda-Porebska, M. Stratmann, M. Rohwerder, K. Potje-Kamloth, Y. Lu, A. Z. Pich, and H.-J. Adler, *Corrosion Sci.*, 47, 3216, 2005.
29. http://www.mpg.de/bilderBerichteDokumente/dokumentation/jahrbuch/2007/eisenforschung/forschungsSchwerpunkt/index.html (2007).
30. M. Kendig and H. Leidheiser, Jr., *J. Electrochem. Soc.*, 123, 982, 1976.
31. A. Mirmohseni and A. Oladegaragoze, *Synthetic Metals*, 114, 105, 2000; G. Spinks, A. Dominis, and G. Wallace, *Corrosion*, 59, 2, 2003; A. Dominis, G. Spinks, and G. Wallace, *Prog. Organic Coatings*, 48, 43, 2003.
32. P.J. Kinlen, Y. Ding, and D.C. Silverman, *Corrosion*, 58, 490, 2002.

33. L. Xia, E. Akiyama, G. Frankel, and R.J. McCreery, *J. Electrochem. Soc.*, 147, 2556, 2000.
34. W.J. Clark and R.L. McCreery, *J. Electrochem. Soc.*, 149, B379, 2002.
35. M. Kendig, unpublished results, Rockwell Scientific Company LLC Research Study supported by Boeing Independent Research and Development.
36. J. Sinko, U.S. Patent Application 20020197468, 2002; M.W. Kendig, et al., U.S. Patent Application 20040005478, 2004; P.J. Kinlen, U.S. Patent Application 20040035498, 2004; M. Kendig, M. Hon, and J. Sinko, *Electrochem. Soc. Trans.*, 1, 119, 2006.
37. M. Kendig, M. Hon, and L. Warren, *Prog. Org. Coat.*, 47, 169, 2002.
38. P.J. Kinlen, C.R. Graham, and Y. Ding, Abstracts of Papers, 228th ACS National Meeting, Philadelphia, PA, August 22–26, 2004.
39. J.M. Pope, T. Sato, E. Shoji, N. Oyama, K.C. White, and D.A. Buttry, *J. Electrochem. Soc.*, 149, A939, 2002.
40. M. Kendig and P. Kinlen, *J. Electrochem. Soc.*, 154, C195, 2007.
41. M. Zubielewicz, W. Gnot, *Prog. Org. Coatings*, 49, 358, 2004; I.M. Zin, S.B. Lyon, V. Pokhmurskii, *Corrosion Sci.*, 45, 777, 2003; F. Deflorian, I. Felhosi, S. Rossi, L. Fedrizzi, and P.L. Bonora, *Macromolecular Symposia,* 187, 87–96, September 2002.
42. H.N. McMurray and G. Williams, *Corrosion,* 60, 219, 2004; H.N. McMurray and G. Williams, *Electrochem. Solid State Lett.*, 6, B9–B11, 2003; S. Bohm, H.N. McMurray, S. M. Powell, and D.A. Worsley, *Materials Corrosion,* 52, 896, 2001.
43. A.N. Khramov, N.N. Voevodin, V.N. Balbyshev, and R.A. Mantz, *Thin Solid Films*, 483,191, 2005; M. Kendig and M. Hon, *Corrosion* 60, 1024, 2004.
44. R. Buchheit and S. Mahajanam, *J. Electrochem. Soc.*, 502, 290, 2006.
45. G. Williams and H.N. McMurray, *J. Electrochem. Solid-State Lett.*, 7, B13, 2004; M. Kendig and M. Hon, *Electrochem. Solid-State Lett.*, 8, B10, 2005.
46. D. A. Worsley, N. McMurray, and D. Williams, *ECS Trans.*, 1(9), 153, 2006.
47. W. Li and L. Calle, Proceedings of the First International Conference on Self Healing Materials, 18–20 April 2007, Noordwijk aan Zee, The Netherlands.
48. R. Cook, R.L., Cook, J. Elliott, and A. Myers, Tri-Service Corrosion Conference 2005, Orlando, FL, November 14–18, 2005.
49. G. Garnweitner, B. Smarsly, R. Assink, W. Ruland, E. Bond, and C.J. Brinker, *J. Am. Chem. Soc.*, 125, 5626, 2003.
50. S.F. Silver, U.S. Patent 3,691,140, 1972.
51. L.J. Bailin and V.S. Agarwala, Microencapsulated DNBM Quaternary Ammonium Corrosion Inhibitors, U.S. Government Report, ADD443022, Lockheed Missiles And Space Co Inc., Palo Alto, CA, 1989.
52. S. Sarangapani, A. Kumar, C. Thies, and L.D. Stephenson, U.S. Pat, 7, 192, 993.
53. M. Urban, *J. Macromol. Sci., Part C: Polym. Rev.*, 46, 329, 2006.
54. K. Toohey, N. Sottos, J. Lewis, J. Moore, and S. White, *Nat. Mater.*, June 2007.
55. A. Nagiub and F. Mansfeld, *Electrochim. Acta*, 47, 2319, 2002.
56. D. Örnek, A. Jayaraman, T.K. Wood, Z. Sun, C.H. Hsu, and F. Mansfeld, *Corros. Sci.*, 43, 2121, 2001.
57. Corrosion Protection of Different Metals by Shewanella Oneidensis MR-1 Esra Kus, Ken Nealson, and Florian Mansfeld Meet. *Abstr. Electrochem. Soc.*, 601, 1192, 2006.
58. F. Mansfeld, H. Hsu, D. Ornek, T. Wood, and B. Syrett, *J. Electrochem. Soc.*, 1, B130–B138, 2002.

24

Smart Polymers for Biotechnology and Elastomers

24.1 Conducting Polymers

Gordon G. Wallace and C.O. Too

24.1.1 Introduction

Traditional polymers have been engineered to provide materials that are low density, mechanically strong, and environmentally stable. They have been made to be nonresponsive—in fact, lightweight dumb materials.

Intelligent polymers are engineered from the molecular level to respond to environmental stimuli in a way that enhances performance. Imagine, for example, a polymer that changes color in sunlight to stop transmission of light, or that becomes stronger in response to stress. Consider a polymer coating that releases a corrosion inhibitor in response to the onset of corrosion or even a medical device that releases anti-inflammatory drugs in response to the onset of infection.

Polymers such as polypyrrole (PPy) (see Equation 24.1) conduct electricity and can be reversibly oxidized and reduced according to the processes described by the following equation.

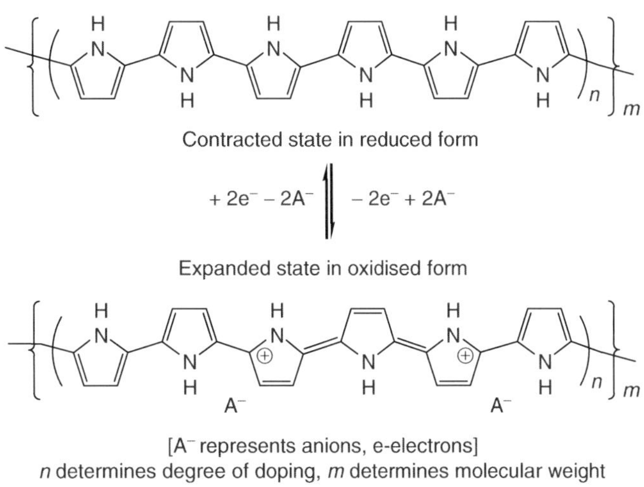

(24.1)

PPy a dynamic (electroactive) polymer structure.

Unlike traditional polymers they are not electronic insulators; they conduct electricity and also, unlike traditional polymers, they are not inert, but are dynamic (electroactive) structures.

The chemical, biological, electrical, and mechanical properties of these conducting polymers are critically dependent on the oxidation state of the structure and hence on their working environment. If this dynamic behavior can be determined by appropriate molecular engineering, truly intelligent polymer structures will emerge.

This ability of conducting polymers to respond is being coupled with an ability to sense the environment and hence produce truly intelligent systems. A simple example has involved the development of a conducting polymer trilayer system that responds to oxygen levels by opening a valve to monitor the environmental conditions needed for food packaging at an optimal level during transport [1]. As we develop our ability to molecularly engineer these systems, some fascinating systems will emerge.

Intelligent polymer systems possess the ability to sense, process information, and actuate responses. Energy is usually required to implement these functions and so energy conversion/storage capabilities are desirable if the system is to be truly autonomous. Ideally these functions would be integrated at the molecular level.

It has been clearly demonstrated over the past 20 years that inherently conducting polymers (ICPs) are capable of providing all of the above functions and as such they have a critical role to play in the development of intelligent polymer systems.

In this chapter, we briefly review the properties of ICPs and their ability to function as sensors, actuators and energy conversion/storage systems. To illustrate the ability of ICPs to provide the range of functions required for intelligent polymer systems, we will draw on examples that utilize PPy or polythiophenes. While polyanilines are also of interest, they are less amenable to use in a wide range of environments due to the need to retain protonation to ensure conductivity. Consequently, most of the examples here will focus on PPy and polythiophenes.

24.1.2 Conducting Polymer Synthesis

Each of these materials may be produced via either chemical or electrochemical oxidation of the appropriate monomer [2]. For PPy, the electrodeposition process can be described simplistically as in the following equation:

Application of a positive potential to an inert electrode substrate usually results in deposition of an insoluble conducting material on the electrode surface. Simple PPy and polythiophenes are intractable and not amenable to subsequent processing and device fabrication.

A dopant counterion (A^-) is incorporated during electrosynthesis to balance the charge on the polymer backbone. A wide range of dopants can be incorporated using this approach. The choice of dopant is critical as it is inextricably linked to the mechanical, electronic, and chemical/biological properties of the polymer. For example, the use of aromatic sulfonates is known to give rise to materials with high conductivity [3], the use of sulfonated polyethers results in excellent mechanical properties [4], and a myriad of chemical molecular recognition units such as metal complexing agents [5] or biological receptors [6] such as antibodies or biological polyelectrolytes have been incorporated. Monomer oxidation and subsequent polymerization may also be induced chemically. Common chemical oxidants such as $FeCl_3$ and $(NH_4)_2S_2O_8$, may also be used and these provide the anion from the oxidant as the dopant (A^-).

A critical feature of ICPs is that they are amenable to oxidation/reduction processes that can be initiated at moderate potentials (Equation 24.1). It is this electrochemical behavior that is the basis of their ability to sense and to respond as well as to convert and store energy. The oxidized forms exhibit good electrical conductivity (σ = 1–400 Scm^{-1}), while the reduced forms have very low conductivity ($\sigma \sim 10^{-8}$ Scm^{-1}).

If the dopant anion (A^-) is small and mobile (e.g., Cl^-) and the polymer has a high surface area to volume ratio, then upon reduction, the anion will be efficiently ejected from the polymer. However, if the dopant is large and immobile (e.g., if A^- is a polyelectrolyte such as polystyrene sulfonate), then an electrically induced cation exchange process occurs, where the cation is incorporated from the supporting electrolyte solution as the polymer is reduced and then expelled when the polymer is oxidized [7]. The exact nature of this redox process has a dramatic effect on the physical and chemical properties of the polymer. These changes are important in determining the sensing and actuation capabilities of the systems as discussed below.

24.1.3 Sensing

Nature has developed recognition systems that are able to discriminate on the basis of highly specific molecule–molecule interactions generating a unique signal. Alternatively nature utilizes arrays of less specific sensors to collect information that is deciphered using pattern recognition processes carried out in

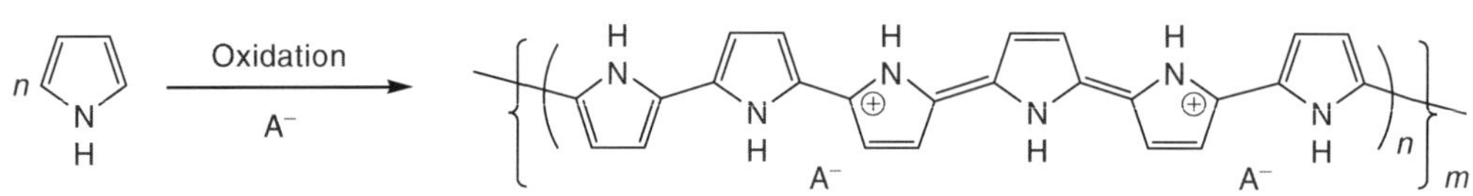

(24.2)

A⁻ molecular dopant, *n* determines degree of doping, *m* determines molecular weight

Synthesis of PPy.

the brain. Both approaches have also been pursued with ICP-based sensor systems.

Chemical sensors specificity can be induced by using molecular recognition components from nature. For example, the ICP may be used as an immobilization platform for enzymes [8–11], antibodies [12–16], oligonucleotides [17–20], or even whole living cells [21,22]. The bioactive component may be incorporated during the polymerization process. The majority of enzyme-containing ICP sensors generate a signal due to the enzymatic generation of an electroactive product (e.g., H_2O_2) or the consumption of an electroactive product (e.g., O_2). The mechanism of signal generation with antibody-containing conducting polymer sensors appears to be associated with the modification of cation movement into and out of the polymer upon oxidation–reduction in the presence of the antigen [23,24].

The selective detection of metal ions has also been achieved by the covalent attachment of molecular recognition moieties to the ICP backbone. The usual approach has been to use a monomer or dimer containing the appropriate recognition group [25]. Polythiophenes containing crown-ethers and calixarenes covalently bound to the bithiophene repeat units, which exhibit controllable selectivity toward Li^+, Na^+ and K^+ ions, have been used [26]. The incorporation of metal complexing agents such as sulfonated 8-hydroxyquinoline as dopants in PPy provides a simple route to metal ion-selective ICPs [27].

ICPs have also been assembled as the active components in microsensing arrays with a view to collecting less specific data subsequently deciphered using pattern recognition software. The so-called "electronic noses" are based on this principle [28]. A range of conducting polymers with differing molecular selectivity respond (by changes in electronic resistance of the polymer) to a complex mixture to produce a unique pattern of responses. This approach has been used to differentiate beers [29], detect microorganisms [30] for waste water management [31,32] and wine characterization [33].

The array approach has also been developed for amperometric sensing when used in solution. Change in amperometric responses in the presence of different ions are used as the signal transduction method. This has been used to discriminate between simple ions [34,35] and even proteins [36]. The approach used is similar to the "electronic nose" in that none of the sensing elements is specific, however, each polymer has a different selectivity series, giving rise to a unique pattern of responses for any given protein.

24.1.4 Actuating

ICPs are capable of undergoing transformations at the molecular through to the macromolecular level. The former can be used to build responsive molecular release systems, while the latter forms the basis of artificial muscles. A change in oxidation state of the polymer (see Equation 24.1) is accompanied by dramatic changes in chemical and physical properties. At least when the dopant (A^-) is small and mobile, reduction decreases the anion exchange capacity of the polymer and increases the hydrophobicity of the polymer backbone.

Using a membrane setup (Figure 24.1) this change in ion exchange capacity can be continually oscillated via application of a pulsed electrical stimulus. Provided a concentration or electrical field gradient is imposed, the net result is a triggered and controlled transport of ions across the membrane structure. The transport of simple cations (e.g., K^+, Na^+, Ca^{2+}) or anions (e.g., Cl^-, NO_3^-, SO_4^{2-}) [37,38], small organics [39], and even protein molecules [40,41] can be regulated in this way. The selective transport of proteins is achieved by incorporating appropriate complementary biomolecules into the conducting polymer membrane. Upon application of an appropriate potential, selective protein binding occurs and this then can be reversed by applying a different potential. This can be repetitively induced to pump proteins across the membrane as illustrated in Figure 24.1. This behavior forms the basis of controlled release devices based on conducting polymers.

For biomedical applications, the controlled release of drugs such as dexamethasone (an anti-inflammatory) [42] or the anticancer drug flouracil [43] and even growth factors that stimulate mammalian cell growth [44] has been demonstrated. Massoumi and Entezami [45,46] utilized conducting polymer bilayers to demonstrate controlled release of sulfosalicylic acid and 2-ethylhexyl phosphate. The latter is a model for a number of phosphate drug systems and has anti-inflammatory properties. The incorporation and release of oligonucleotides [47] has been achieved using polyethylene dioxy thiophene (Pedot). Pernaut and Reynolds [48] demonstrated the controlled release of adenosine triphosphate (ATP) (of interest for use as a cardiovascular therapeutic) from a PPy membrane in response to electrical stimulation or even to changes in the chemical environment (such as pH).

The incorporation of either anions or cations during redox cycling results in polymer swelling. These incorporation/expulsion events at the molecular level also result in changes in the overall volume (dimensions) of the polymer. It was these volume changes that led Baughman et al. [49] to the concept of electromechanical actuators (artificial muscles) based on conducting polymers.

By producing simple laminated membrane structures (Figure 24.2) electromechanical actuation can be induced. When one side of this membrane structure acts as the anode (oxidation), anion incorporation occurs causing swelling, and the other acts as the cathode (reduction), with anion expulsion causing contraction. The net result is movement due to electrical stimulation (Figure 24.2). The molecular dopant incorporated at the time of synthesis will play a critical role in determining the extent of this behavior as well the anions/cations present in the electrolyte solution. Using ions such as TFSI (see below), strains of up to 26% have recently been reported [50].

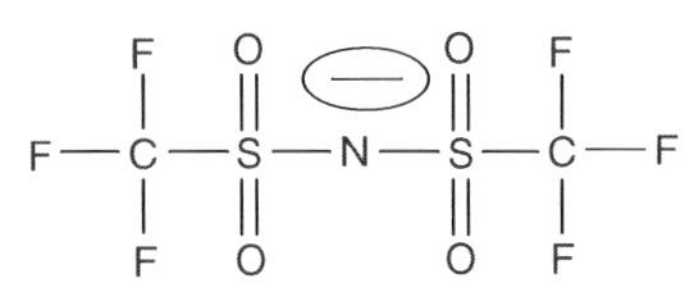

TFSI

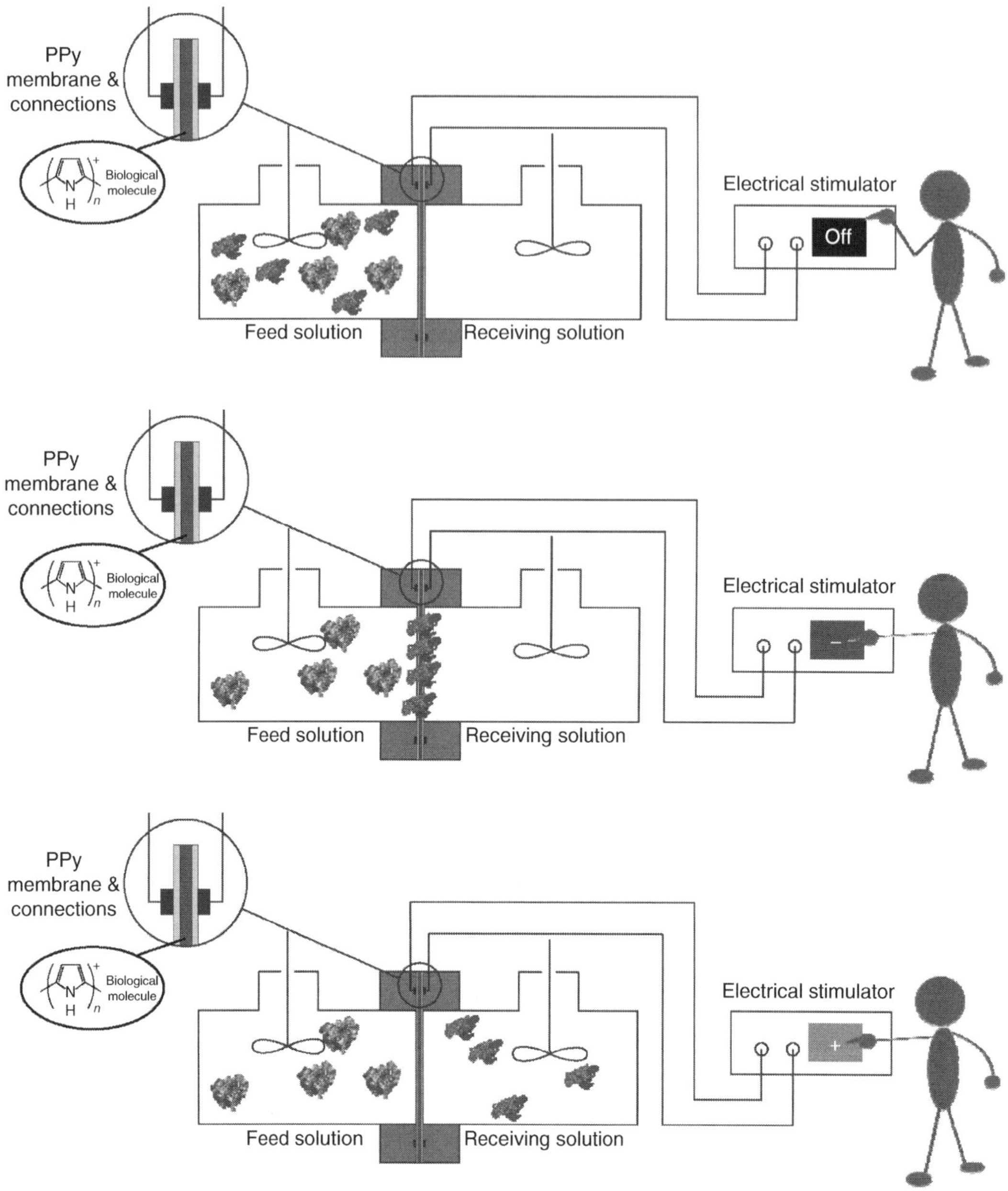

FIGURE 24.1 Stimulated transport of proteins across ICPs. Schematic of membrane transport experiment.

We have utilized electrolytes containing the TFSI anion to produce extremely fast responding actuators that operate up to 40 Hz [51].

Interestingly, such anions are also a common component of ionic liquid electrolytes [52]. Appropriate combinations of organic anions and cations produce salts molten at room temperature (considered an ionic liquid if the melting temperature is <100 °C). These salts have low vapor pressure (nonvolatile) and have extremely wide electrochemical potential windows. These benefits translate into the operation of conducting polymer-based devices (such as electromechanical actuators) that use them [53,54] with increased lifetime over multiple actuation cycles observed. In fact, not only improved stability (increasing cycles length from hundreds to hundreds of thousands of cycles) but also increased performance is observed. The latter feature arises from the inclusion of the imidazolium cation during reduction, a process that apparently preserves the inherent modulus of the materials [55]. Another interesting form of these electrolytes involves their use as solid electrolytes. By performing polymerization in the ionic liquid, a solid can be produced [56]. A highly stable, easily handled actuator can be constructed using this electrolyte.

Various applications of ICP actuators have been demonstrated. For example, conducting polymer flaps have been used as valves to open containers containing active molecules [57]. This provides a route to a novel drug delivery system wherein the drug needs to be highly water soluble as it does not need to be incorporated into the conducting polymer. In addition, using a concentric hollow fiber arrangement, we have created a micropumping system

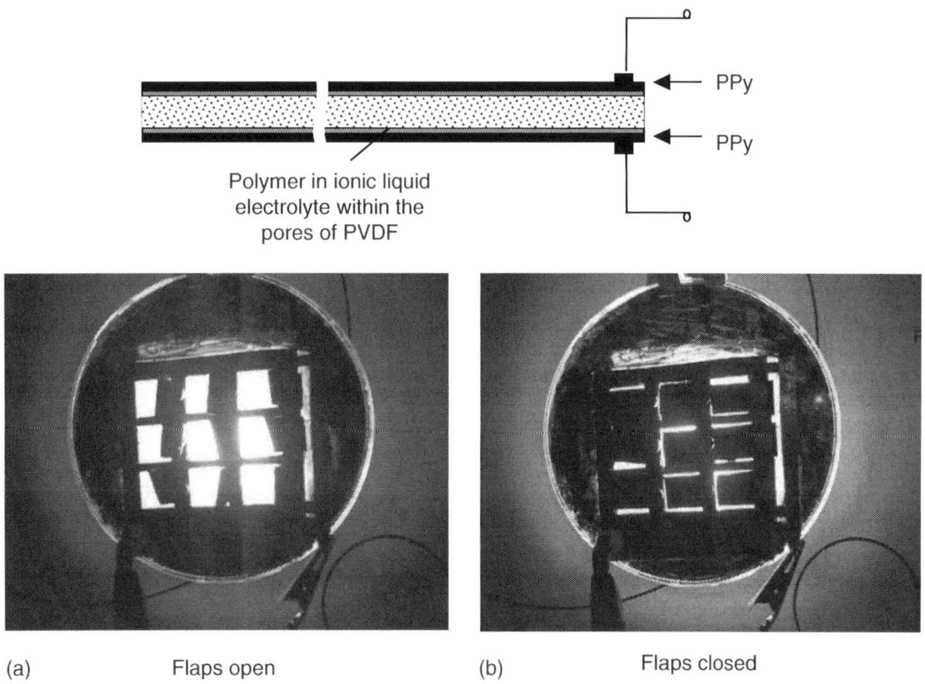

FIGURE 24.2 Cross-section of a PPy double-sided PVDF electromechanical actuator with the pores of PVDF filled with electrolyte.

(TITAN) based on the ability to produce a peristaltic action using conducting polymer actuators [58]. This also may provide an avenue for controlled delivery of active molecules. Conducting polymer actuators can also be configured to operate axially [59,60] using appropriate fiber or polymer strip geometries.

Another recent advance involves the use of carbon nanotube additives to increase the strength of conducting polymer fibers [61]. These fibers have been shown to be highly impressive artificial muscles capable of operating at stress levels in excess of 100 MPa [62].

All of the above responses (actuators) are initiated by creating an appropriate electrical potential that causes the polymer to change form. This is usually achieved by imposition of a potential from an external source although coupling to an autonomously triggered sensor is the ultimate goal.

24.1.5 Energy Conversion and Storage

Given the autonomous nature required of intelligent polymer systems, they must also be capable of converting energy from a natural source, such as sunlight, and storing it until required. ICPs and devices containing them are capable of achieving this [63,64]. Upon irradiation of the ICP, electron–hole pairs (excitons) are generated. If these excitons are transported to an interface where they can be split before recombination occurs, then electrical energy is generated. Both Schottky devices [65] and photoelectrochemical cells [66–69] have been used. Photovoltaic responses were obtained from a variety of ICPs in Schottky devices with quantum efficiencies up to 1% [68,70,71]. However, the best results have been obtained using interpenetrating blends of functionalized polyphenylenevinylenes (PPVs) [72,73] and/or polythiophenes (PThs) [74] with power conversion efficiencies up to 2%. The addition of dyes [75–77] and electron acceptors [76,78,79] to such devices has also increased the overall power conversion efficiency to around 3%. The attachment of a light-harvesting porphyrin functional group to polyterthiophene has also been considered to improve power conversion efficiency [80].

All-polymer batteries where conducting electroactive polymer comprises both the anode and cathode have been developed [81] (Figure 24.3). A specific charge capacity of 22 mA h g^{-1} at a cell potential of 0.4 V was achieved. No loss in capacity was observed after 100 cycles.

More recently an all-polymer battery based on derivatized polythiophenes supported on graphite-coated supports was described [82]. In this instance, polythiophene functioned more effectively in the n-doping region and provided an improved cell discharge voltage of 2.4 V but lower capacity. Recently, we have produced conducting polymer fibers for use in flexible electrode structures [83,84] (Figure 24.4). Capacities in the order of 10 mA h g^{-1} were observed.

An alternative energy storage approach based on conducting polymers is the use of material that is more capacitive in nature. Conducting polymer composite electrodes have proven to be extremely useful. For example, PPy has been deposited onto activated carbon to produce supercapacitor electrodes with a specific capacitance of 354 F g^{-1} [85]. Poly(3-methylthiophene) grown within a porous PVDF structure has been shown to have a capacitance of 616 F g^{-1} [86]. We recently produced an electrode structure containing PPy, carbon nanotubes, and MnO_2, with a specific capacitance of 280 F/g and this material was stable to electrochemical cycling [87].

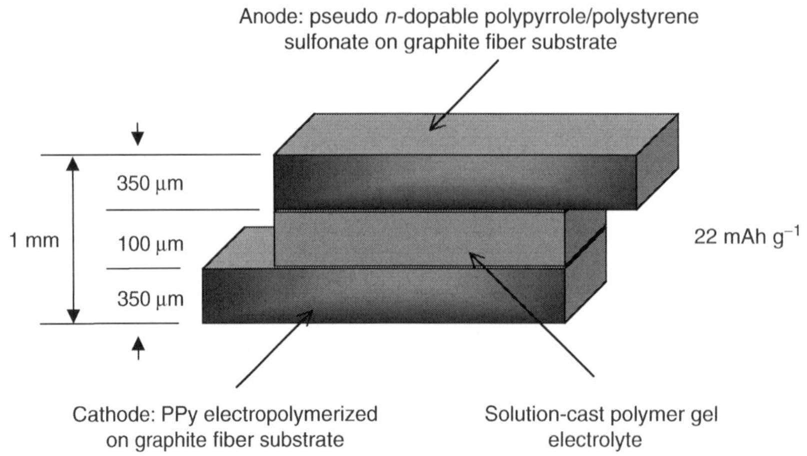

FIGURE 24.3 Schematic illustration of an assembled all-polymer battery showing the PPy cathode and the PPy-polystyrenesulfonate (PSS) anode sandwiched with a layer of solution-cast polymer gel electrolyte. The total dimensions of the assembled cell are 1 mm × 2 cm × 10 cm (8 cm of electroactive material). (From Killian, J.G., Coffey, B.M., Gao, F., Poehler, T.O., and Searson, P.C., *J. Electrochem. Soc.*, 143, 936, 1996. With permission.)

24.1.6 Polymer Processing and Device Fabrication

The integration of energy conversion and storage within a structure containing spatially arranged sensing and actuating units requires the development of fabrication/patterning systems with nanometer to micrometer resolution. Amongst the approaches to have emerged in recent years are wet spinning of fibers, inkjet printing, and screen printing. With all three approaches, it is necessary to have the polymer in solution or be able to form stable dispersions containing small particles. These solutions or dispersions are the first necessary steps in wet spinning (Figure 24.5) or inkjet printing (Figure 24.6).

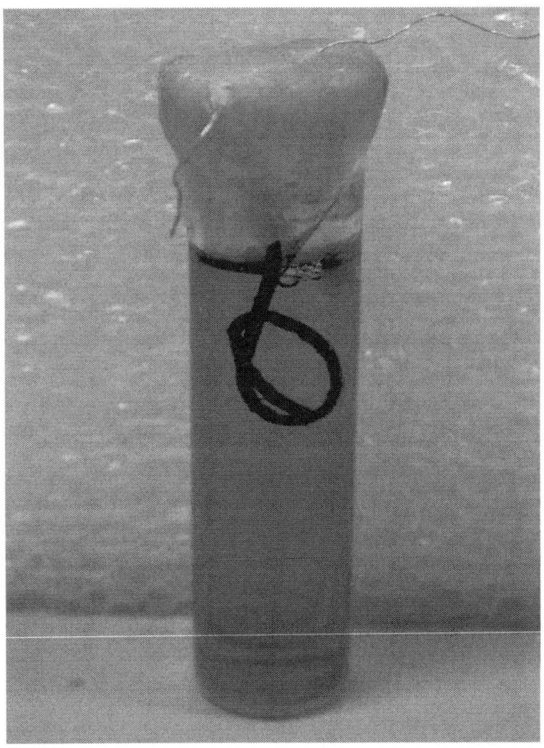

FIGURE 24.4 Photograph of knotted fiber battery.

Solubility has been induced in PPy by attaching alkyl [88,89] or alkyl sulfonate [90,91] groups to produce 3-substituted pyrrole monomers prior to polymerization. This results in markedly enhanced solubility in organic or aqueous media respectively. Polythiophenes can also be rendered either organic solvent soluble [92] or water-soluble [93] using these derivatization approaches.

Solubility has also been induced in PPy by the use of appropriate dopants (see structure below) [94]. We have recently shown that the wet spinning technique can be used to produce continuous lengths of conducting polymer fiber down to 150 μm in diameter, with conductivities in the order of 497 Scm^{-1} and tensile strength 170 MPa [95].

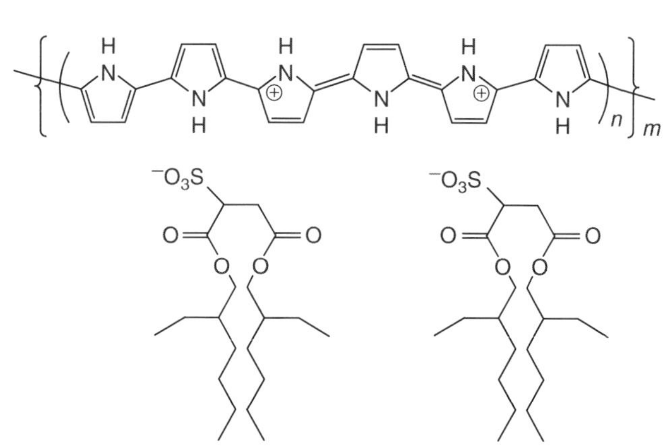

Schematic of interaction between Na^+ $DEHS^-$ and oxidised PPy

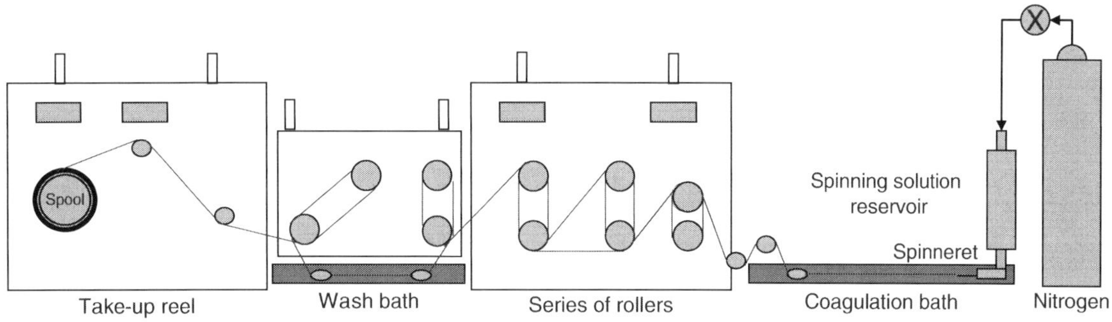

FIGURE 24.5 Schematic showing wet spinning process.

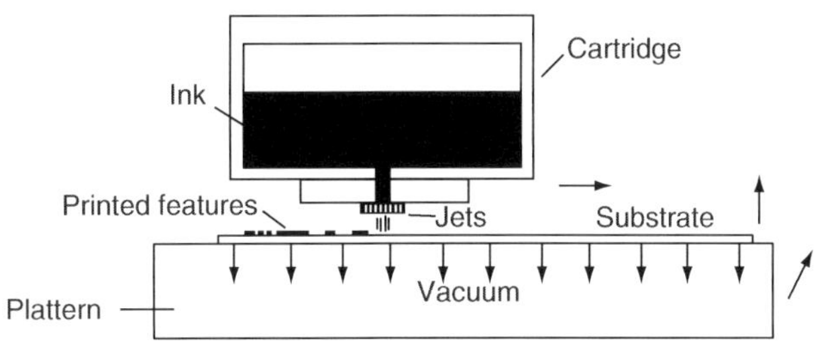

FIGURE 24.6 Schematic showing inkjet printing setup.

Due to the ability to print at higher resolution, inkjet printing has emerged as an attractive patterning technique for conjugated polymers in areas such as light-emitting diodes [96]. Patterning of organic molecules by inkjet printing can be achieved either by printing a solution of the molecule (in organic solvent) or by printing dispersions of the molecule in water or organic solvents. Sirringhaus et al. [97] developed a method to print high-resolution all-polymer transistor circuits utilizing hydrophobic surface patterning, which confines the spreading of water-based conducting polymer ink droplets upon inkjet printing. Other workers have printed PPy, Pedot, and poly[2-methoxy-5-(2-ethylhexyloxy)-1,4-(1-cyano-vinylene) phenylene] combinations to produce diodes [98]. Preparation of PPy nanodispersions and issues involved in the inkjet printing of these materials have been discussed [99]. Recently Yoshioka and Jabbour used a desktop inkjet printer to produce a patterned Pedot anode for use in an organic light-emitting diode [100].

As well as soluble conducting polymer, processable dispersions containing conducting polymer colloids or nanoparticles have also been produced for printing purposes. Emulsion polymerization is the most appropriate method for producing submicron (nano) particles due to its versatility and relative simplicity. Formation of ICP nanodispersions consisting of highly conducting nanoparticles dispersed in either organic or aqueous solution has been achieved using polyaniline [101], PPy [102], and regioregular polythiophene [103]. We have successfully prepared polyaniline nanodispersions and obtained rheological properties that make them amenable to inkjet printing [104].

Screen printing is a simple and environmentally friendly way to produce electronic circuitry and make interconnections [105]. Bao et al. [106] used screen printing to develop organic transistors with high field-effect mobilities (ca. 0.015–0.045 cm^2/V s) using regioregular poly(3-hexylthiophene). Garnier et al. [107] demonstrated the usefulness of printing in organic field effect transistor fabrication, while Gustafsson et al. [108] showed it was possible to print polyaniline to form the gate electrode. While screen printing is a highly efficient and relatively simple way to fabricate intricate electronics and circuits, this technique suffers from low resolution. In most instances, screen printing is used to pattern the parts of the circuits that do not demand high resolution [109].

24.1.7 Future Developments

Advances in the application of ICPs in sensing, actuating, energy conversion and energy storage have been reviewed. Future developments could be as follows:

(1) In the incorporation of sensors with actuators where a chemical or biochemical event is detected, leading to an actuation response, for example, in the controlled release of a drug.
(2) These combined sensors and actuators require power and this can be supplied by energy conversion devices such as plastic solar cells and/or energy storage devices such as all-polymer batteries.
(3) For these future developments to occur, advances in polymer processing and device fabrication will be crucial in order to amalgamate all these components into a single device structure. For many future applications, these new devices would need to be flexible, wearable, and even implantable.

References

1. Andrews, M.K., Jansen, M.L., Spinks, G.M., Zhou, D., and Wallace, G.G. *Sensors and Actuators A* 2004, 114, 65.
2. Wallace, G.G., Spinks, G.M., Kane-Maguire, L.A.P., and Teasdale, P.R. *Conductive Electroactive Polymers: Intelligent Material Systems* 2nd edn., CRC Press, Boca Raton, FL, 2003.

3. Price, W.E., Wallace, G.G., and Zhao, H. *Journal of Membrane Science* 1994, 87, 47.
4. Ding, J., Price, W.E., Ralph, S.F., and Wallace, G.G. *Synthetic Metals* 2002, 129, 67.
5. Reece, D.A., Ralph, S.F., and Wallace, G.G. *Journal of Membrane Science* 2005, 249, 9.
6. Kane-Maguire, L.A.P. and Wallace, G.G. *Advanced Materials* 2002, 14(13–14), 953.
7. Ren, X. and Pickup, P.G. *Journal of Physical Chemistry* 1993, 97, 5356.
8. Umana, M. and Waller, J. *Analytical Chemistry* 1986, 58, 2979.
9. Foulds, N.C. and Lave, C.R. *Analytical Chemistry* 1988, 60, 2473.
10. Adeloju, S.B., Barisci, J.N., and Wallace, G.G. *Analytica Chimica Acta* 1996, 332, 145.
11. Ma, M., Qu, L., and Shi, G. *Journal of Applied Polymer Science* 2005, 98, 2550.
12. Sadik, O.A. and Wallace, G.G. *Analytica Chimica Acta* 1993, 279, 200.
13. Barnett, D., Sadik, O.A., John, M.J., and Wallace, G.G. *Analyst* 1994, 119, 1997.
14. Bender, S. and Sadik, O.A. *Environmental Science & Technology* 1998, 32, 788.
15. Sargent, A. and Sadik, O.A. *Analytica Chimica Acta* 1998, 376, 125.
16. Sargent, A. and Sadik, O.A. *Electrochimica Acta* 1999, 44, 4667.
17. Bidan, G., Billan, M., Galasso, K., Livache, T., Mathis, G., Reget, A., Torres-Rodriguez, L.M., and Vieil, E. *Applied Biochemistry* (in press) (and references cited therein).
18. Garnier, F., Karri-Youssoufi, H.K., Srivastava, P., Mandrand, B., and Delair, T. *Synthetic Metals* 1999, 100, 89.
19. Wang, J., Jiang, M., and Kawde, A.-N. *Electroanalysis* 2001, 13, 537.
20. Xu, Y., Ye, X., Yang, L., He, P., and Fang, Y. *Electroanalysis* 2006, 18(15), 1471.
21. Campbell, T.E., Hodgson, A.J., and Wallace, G.G. *Electroanalysis* 1999, 11(4), 215.
22. Deshpande, M.V. and Hall, E.A.H. *Biosensors and Bioelectronics* 1990, 5, 431.
23. Barisci, J.N., Hughes, D., Minett, A.I., and Wallace, G.G. *Analytica Chimica Acta* 1998, 371, 39.
24. Gooding, J., Wasiowych, C., Barnett, D., Hibbert, D.B., Barisci, J.N., and Wallace, G.G. *Biosensors and Bioelectronics* 2004, 20, 260.
25. Higgins, S.J. *Chemical Society Review* 1997, 26, 247.
26. Swager, T.M. *Acc Chem Res* 1998, 31, 201 (and references cited therein).
27. Misoska, V. Chemistry Honors Thesis, University of Wollongong, 1998.
28. Lonergan, M.C., Severin, E.J., Doleman, B.J., Beaber, S.A., Grubbs, R.H., and Lewis, N.S. *Chemistry of Materials* 1996, 8, 2298.
29. Shurmer, H.V., Corcoran, P., and Gardner, J.W. *Sensors and Actuators B* 1991, 4, 29.
30. Namdev, P.K., Alroy, Y., and Singh, V. *Biotechnology Progress*1998, 14, 75.
31. Stuetz, R.M., Fenner, R.A., Hall, S.T., Stratful, I., and Loke, D. *Water Science and Technology* 2000, 41, 41.
32. Bourgeois, W., Hogben, P., Pike, A., and Stuetz, R.M. *Sensors and Actuators B* 2003, 88, 312.
33. Guadarrama, A., Fernandez, J.A., Iniguez, M., Souto, J., and de Saja, J.A., *Analytica Chimica Acta* 2000, 411, 193.
34. John, R., Ongarato, D., and Wallace, G.G. *Electroanalysis* 1996, 8, 623.
35. Nguyen, T.A., Barisci, J.N., Partridge, A., and Wallace, G.G. *Synthetic Metals* 2003, 137, 1445.
36. Lu, W., Nguyen, T., and Wallace, G.G. *Electroanalysis* 1998, 10, 1101.
37. Zhao, H., Price, W.E., Too, C.O., Wallace, G.G., and Zhou, D. *Journal of Membrane Science* 1996, 119, 199 (and references cited therein).
38. Price, W.E., Wallace, G.G., and Zhao, H. *Journal of Electroanalytical Chemistry* 1992, 334, 111.
39. Zhao, H., Price, W.E., and Wallace, G.G. *Journal of Membrane Science* 1995, 100, 239.
40. Zhao, H., Too, C.O., and Wallace, G.G. *Journal of International Material Systems and Structures* 1997, 8, 1052.
41. Zhou, D., Too, C.O., Hodges, A.M., Mau, A.W.H., and Wallace, G.G. *Reactive and Functional Polymers* 2000, 45, 217.
42. Wadhwa, R., Lagenaur, C.F., and Cui, X.T. *Journal of Controlled Release* 2006, 110, 531.
43. Huang, H., Liu, C., Liu, B., Cheng, G., and Dong, S. *Electrochimica Acta* 1998, 43, 999.
44. Thompson, B.C., Moulton, S.E., Ding, J., Richardson, R., Cameron, A., O'Leary, S., Clark, G.M., and Wallace, G.G. *Journal of Controlled Release* 2006, 116, 285.
45. Massoumi, B. and Entezami, A. *European Polymer Journal* 2001, 37, 1015.
46. Massoumi, B. and Entezami, A. *Polymer International* 2000, 51, 555.
47. Piro, B., Pham, M.C., and Ledoan, T. *Journal of Biomedical Materials Research* 1999, 46, 566.
48. Pernaut, J.M. and Reynolds, J.R. *Journal of Physical Chemistry B* 2000, 104, 4080. IPRI/7145/1.8.07
49. Baughman, R.H., Shacklette, L.W., Elsenbaumer, R.L., Plichta, E., and Becht, C. in *Conjugated Polymeric Materials: Opportunities in Electronics, Optoelectronics and Molecular Electronics* Kluwer Academic, Dordrecht, 1990, pp. 599–682.
50. Hara, S., Zama, T., Takashima, W., and Kaneto, K. *Journal of Materials Chemistry* 2004, 14, 1516.
51. Wu, Y., Alici, G., Spinks, G.M., and Wallace, G.G. *Synthetic Metals* 2006, 156, 1017.
52. Seddon, K.R., Stark, A., and Torres, M.J. *Pure and Applied Chemistry* 2000, 72, 2275

53. Lu, W., Fadeev, A.G., Qi, B., Smela, E., Mattes, B.R., Ding, J., Spinks, G.M., Mazurkiewicz, J., Zhou, D., MacFarlane, D.R., Forsyth, S.A., Forsyth, M., and Wallace, G.G. *Science* 2002, 297, 983.
54. Ding, J., Zhou, D., Spinks, G., Forsyth, S., Forsyth, M., MacFarlane, D., and Wallace, G. *Chemistry of Materials* 2003, 15, 2392.
55. Ding, J., Zhou, D., Spinks, G., Forsyth, S., Forsyth, M., MacFarlane, D., and Wallace, G. *Chemistry of Materials* 2003, 15, 2392.
56. Zhou, D., Spinks, G.M., Tiyapiboonchaiya, C., MacFarlane, D.R., Forsyth, M., Sun, J., and Wallace, G.G. *Electrochimica Acta* 2003, 48, 2355.
57. Xu, H., Wang, C., Wang, C., Zoval, J., and Madou, M. *Biosensors and Bioelectronics* 2006, 21, 2094.
58. Wu, Y., Zhou, D., Spinks, G.M., Innis, P.C., Megill, W.M., and Wallace, G.G. *Smart Materials and Structures* 2005, 14, 1511.
59. Ding, J., Liu, L., Spinks, G.M., Zhou, D., Gillespie, J., and Wallace, G.G. *Synthetic Metals* 2003, 138, 391.
60. Yamato, K. and Kaneto, K. *Analytica Chimica Acta* 2006, 568, 133.
61. Lu, W., Smela, E., Adams, P., Zuccarello, G., and Mattes, B.R. *Chemistry of Materials* 2004, 16, 1615.
62. Mottaghitalab, V., Spinks, G.M., and Wallace, G.G. *Polymer* 2006, 47, 4996.
63. Mottaghitalab, V., Xi, B., Spinks, G.M., and Wallace, G.G. *Synthetic Metals* 2006, 156, 796.
64. Dastoor, P., Officer, D.L., Too, C.O., and Wallace, G.G. *Chemical Innovation* 2000, 15.
65. Wallace, G.G., Too, C.O., Officer, D.L., and Dastoor, P.C. *MRS Bulletin* 2005, 30, 46.
66. Brabec, C.J., Sariciftci, N.S., and Hummelen, J.C. *Advanced Functional Materials* 2001, 11, 15.
67. Gazotti, W.A., Nogueira, A.F., Grotto, E.M., Gallazi, M.C., and De Paoli, M.A. *Synthetic Metals* 2000, 108, 151.
68. Yohannes, T., Solomon, T., and Inganas, O. *Synthetic Metals* 1996, 82, 215.
69. Cutler, C.A., Burrell, A.K., Collis, G.E., Officer, D.L., Too, C.O., and Wallace, G.G. *Synthetic Metals* 2001, 123(2), 225.
70. Kunugi, Y., Harima, Y., and Yamashita, K. *Journal of the Chemical Society. Chemical Communications* 1995, 787.
71. Shirakawa, H. and Ikeda, S. *Kobunshi Ronbunshu* (in Japanese) 1979, 28, 369.
72. Halls, J.J.M., Walsh, C.A., Greenham, N.C., Marseglia, E.A., Friend, R.H., Moratti, S.C., and Holmes, A.B. *Nature* 1995, 376, 498.
73. Yu, G. and Heeger, A.J. *Journal of Applied Physics* 1995, 78, 4510.
74. Granstrom, M., Petritsch, K., Arias, A.C., Lux, A., Andersson, M.R., and Friend, R.H. *Nature* 1998, 395, 257.
75. Armstrong, N.R. *Journal of Porphyrins Phthalocyanines* 2000, 4, 414.
76. Yoshino, K., Tada, K., Fujii, A., Conwell, E.M., and Zakhidov, A.A. *IEEE Transactions on Electron Devices* 1997, 44, 1315.
77. Yamaue, T., Kawai, T., Onoda, M., and Yoshino, K. *Journal of Applied Physics* 1999, 85, 1626.
78. Halls, J.J.M. and Friend, R.H. *Synthetic Metals* 1997, 85, 1307.
79. Yu, G., Gao, J., Hummelen, J.C., Wudl, F., and Heeger, A.J. *Science* 1995, 270, 1789.
80. Chen, J., Burrell, A.K., Campbell, W.M., Officer, D.L., Too, C.O., and Wallace, G.G. *Electrochimica Acta* 2004, 49, 329.
81. Killian, J.G., Coffey, B.M., Gao, F., Poehler, T.O., and Searson, P.C. *Journal of the Electrochemical Society* 1996, 143, 936.
82. Gofer, Y., Sarker, H., Killian, J.G., Giaccai, J., Poehler, T.O., and Searson, P.C. *Biomedical Institute and Technology* 1998, 32, 33.
83. Wang, J., Too, C.O., and Wallace, G.G. *Journal of Power Sources* 2005, 150, 223.
84. Wang, C.Y., Mottaghitalab, V., Too, C.O., Spinks, G.M., and Wallace, G.G. *Journal of Power Sources* 2007, 163, 1105.
85. Muthulakshmi, B., Kalpana, D., Pitchumani, S., and Renganathan, N.G. *Journal of Power Sources* 2006, 158, 1533.
86. Fonseca, C.P., Benedetti, J.E., and Neves, S. *Journal of Power Sources* 2006, 158, 789.
87. Sivakkumar, S.R., Ko, J.M., Kim, D.Y., Kim, B.C., and Wallace, G.G. *Electrochimica Acta* (in press).
88. Masuda, H. and Kaeriyama, K. *Journal of Materials Science* 1991, 26, 5637.
89. Ashraf, S.A., Chen, F., Too, C.O., and Wallace, G.G. *Polymer* 1996, 37, 2811.
90. Patil, A.O., Ikenoue, Y., Wudl, F., and Heeger, A.J. *Journal of the American Chemical Society* 1987, 109, 1858. IPRI/7145/1.8.07
91. Havinga, E.E., ten Hoeve, W., Meijer, E.W., and Wynberg, H. *Chemistry of Materials* 1989, 1, 650.
92. Bryce, M.R., Chissel, A., Kathirgamanathan, P., Parker, D., and Smith, R.M. *Journal of the Chemical Society. Chemical Communications* 1987, 466.
93. Barisci, J.N., Mansouri, J., Spinks, G.M., Wallace, G.G., Kim, C.Y., Kim, D.Y., and Kim, J.Y. *Colloids and Surfaces* 1997, 126, 129.
94. Foroughi, J., Spinks, G.M., Whitten, P.G., and Wallace, G.G. *Synthetic Metals* (submitted for publication).
95. Spinks, G.M., Mottaghitalab, V., Bahrami-Samani, M., Whitten, P.G., and Wallace, G.G. *Advanced Materials* 2006, 18, 637.
96. Hebner, T.R., Wu, C.C., Marcy, D., Lu, M.H., and Sturm, J.C. *Applied Physics Letters* 1998, 72, 519.
97. Sirringhaus, H., Kawase, T., Friend, R.H., Shimoda, T., Inbasekaran, M., Wu, W., and Woo, E.P. *Science* 2000, 290, 2123.

98. Liu, Y. and Cui, T. *Macromolecular Rapid Communications* 2005, 26, 289.
99. Winther-Jensen, B., Clark, N., Subramanian, P., Helmer, R., Ashraf, S., Wallace, G., Spiccia, L., and MacFarlane, D. *Journal of Applied Polymer Science* 2007, 104, 3938.
100. Yoshioka, Y. and Jabbour, G.E. *Synthetic Metals* 2006, 156, 779.
101. Ghosh, P., Siddhanta, S.K., Haque, S.R., and Chakrabarti, A. *Synthetic Metals* 2001, 123, 83.
102. Omastova, M., Trchova, M., Kovarova, J., and Stejskal, J. *Synthetic Metals* 2003, 138, 447.
103. Brustolin, F., Goldoni, F., Meijer, E.W., and Sommerdijk, N.A.J.M. *Macromolecules* 2002, 35, 1054.
104. Ngamna, O., Morrin, A., Killard, A., Moulton, S., Smyth, M., and Wallace, G.G. *Langmuir* (in press).
105. Gilleo, K. *Polymer Thick Films*, Van Nostrand Reinhold, New York, 1996.
106. Bao, Z. and Lovinger, A.J. *Chemistry of Materials* 1999, 11, 2607.
107. Garnier, F., Hajlaoui, R., Yassar, A., and Srivastava, P. *Science* 1994, 265, 1684.
108. Gustafsson, G., Cao, Y., Treacy, G.M., Klavetter, F., Colaneri, N., and Heeger, A.J., *Nature* 1992, 357, 477.
109. Rogers, J.A. and Bao, Z. *Journal of Polymer Science: Part A: Polymer Chemistry* 2002, 40, 3327. IPRI/7145/1.8.07.

24.2 Piezoelectricity in Polymers

Aleksandra M. Vinogradov

24.2.1 Introduction

In the past decades, rapid advances in synthetic polymer science have led to an explosive growth in the development of electroactive polymers with capabilities to detect changes in the loading or environmental conditions, decide rationally on a set of the respective actions, and implement such decisions in a controlled manner. These special qualities of electroactive polymers can be compared with biological functions that involve transformations of the sensed information into a reactive response.

The group of electroactive polymers includes piezoelectric and electrostrictive polymers, the response of which is observed either in the form of an electric charge or voltage produced by applied mechanical forces or, conversely, in the form of mechanical deformation induced by an applied electric field. These electromechanical effects have been defined, respectively, as "direct" and "converse."

It is important at this point to draw a distinction between electrostriction and piezoelectricity. Essentially, electrostriction, which is a property of all dielectrics, is characterized by a quadratic relation between mechanical deformations and the applied electric field, whereas in piezoelectric materials, this relation is linear. As a result, piezoelectric effects entail a reversal of mechanical deformations in response to reversed electric fields.

There are many polymers known to possess piezoelectric properties including, in particular, polypropylene, polystyrene, poly(methyl methacrylate), vinyl acetate, and odd number nylons. However, strong piezoelectric effects have been observed only in polyvinylidene fluoride (PVDF or PVF_2) and PVDF copolymers. At present, these polymers comprise the principal commercially produced group that plays a leading role in practical applications.

24.2.2 Properties

PVDF is a thermoplastic fluoropolymer first patented in 1948 [1]. In the 1960s, PVDF was used primarily as an electric insulator or as a base for durable long-life coatings for exterior finishes [1]. Strong piezoelectric effects in PVDF discovered by Kawai in 1969 [2] have attracted attention to the functional capabilities of the polymer stimulating qualitatively new technological applications [3–6].

PVDF is a semicrystalline polymer with typical crystallinity of approximately 50% and a molecular structure consisting of a repeated monomer unit $(–CF_2–CH_2–)_n$. The amorphous phase of the polymer has the properties of a supercooled liquid with the glass transition temperature of about −50°C and melting temperature in the range of 155–180°C. The polymer does not absorb visible light and is transparent, although its clarity is often compromised by light scattering in thicker sections as a result of crystallinity.

Permanent dipole polarization of PVDF is obtained through a technological process that involves stretching and polling of extruded thin sheets of the polymer. Typically, PVDF is produced in the form of thin films with thicknesses ranging from 9 to 800 μm (10^{-6} m). A thin layer of nickel, silver, or copper is deposited on both material surfaces to provide electrical conductivity when electric field is applied, or to allow measurements of the charge produced by mechanical deformations.

The electromechanical properties of PVDF are commonly represented by the constitutive equations of linear piezoelectricity [7]. Respectively, the piezoelectric response of the polymer is characterized by five piezoelectric coefficients d_{3i} (i = 1,2,3), d_{15}, and d_{24}, where the first subscript denotes the direction of the applied electric field and the second identifies the direction of mechanical deformation. Piezoelectric properties of PVDF are also characterized by the hydrostatic coefficient $d_h = d_{31} + d_{32} + d_{33}$, which represents the electric charge generated by hydrostatic pressure. It is important to note that the values of all piezoelectric coefficients of PVDF depend on the polarization conditions of the polymer characterized by polarization time, temperature, and polarizing field [8].

The constitutive equations of linear piezoelectricity imply that the mechanical response of PVDF is defined by the generalized Hooke's law. This formulation involves nine independent elastic constants reflecting the anisotropic nature of the polymer.

Experiments indicate that the mechanical properties of PVDF strongly depend on the orientation of the molecular chains of the polymer aligned in the stretch direction [4]. The stress–strain diagram of PVDF in the direction of the aligned molecular chains is characterized by continuous increase of stresses, culminating in sudden failure. This type of behavior is usually exhibited by brittle materials. In contrast, in the material direction normal to the orientation of molecular chains, the stress–strain response of the polymer is similar to that of ductile materials. It is characterized by an increase of stresses up to a certain maximum and followed by a sharp decrease in the load-carrying capacity of the material.

Since the electromechanical response of PVDF depends on such factors as polarization conditions, stress/strain rates, temperature and hydrostatic pressure, the piezoelectric and elastic constants of the polymer cannot be identified with absolute precision. Nevertheless, based on well documented experimental studies [4,5,9] it is possible to identify the typical values of the material characteristics of PVDF as summarized in Table 24.1, which is adopted from Ref. [4].

It is important to emphasize that the constitutive law of linear piezoelectricity tends to neglect energy dissipation, time-dependent effects, and various material nonlinearities in the electromechanical response of piezoelectric polymers. In reality, however, the electromechanical properties of PVDF are characterized by stress-dependent nonlinearities, higher order electromechanical couplings such as electrostriction, and nonlinear strain–displacement relations at large driving voltages [10–12]. Besides, the polymer tends to demonstrate piezoelectric and mechanical relaxation as well as measurable energy losses under cyclic conditions [3,8,13–15]. There are clear indications that the same molecular relaxation mechanisms that give rise to mechanical relaxation are also likely to give rise to dielectric relaxation [14,16]. Further, it has been observed that the dynamic response of PVDF under superimposed static and cyclic loading conditions is essentially nonlinear since it does not amount to a superposition of the responses to static and fully reversed cyclic loads applied separately [17]. At present, these effects in the behavior of piezoelectric polymers are not clearly understood and require further investigations.

24.2.3 Development of New Materials

To date, piezoelectric polymers have been used as active elements in transducers, sensors, and actuators. The key factors that define their functional performance include such parameters as the maximum achievable strain, stiffness, spatial resolution, and the frequency bandwidth. In the past decades, consistent efforts have been made to enhance these characteristics by modifying

TABLE 24.1 Properties of PVDF

Property	Value
Range of film thicknesses	9–800 μm (10^{-6} m)
Mass density, ρ	1.78×10^3 kg/m^3
Water absorption	0.02%
Operating temperature range	−40°C to 80°C
Glass transition temperature, T_g	−60°C to −20°C
Melting temperature, T_m	170°C to 178°C
Thermal conductivity	0.18 Wm^{-1} K^{-1}
Maximum operating voltage	30 V/μm (750 V/mil)
Breakdown voltage	100 V/μm (2000 V/mil)
Capacitance	380 pF/cm^2 for 28 μm films @1 kHZ
Piezoelectric coefficients	$d_{31} = 21.4 \times 10^{-12}$ C/N; $d_{32} = 2.3 \times 10^{-12}$ C/N; $d_{33} = -31 \times 10^{-12}$ C/N; $d_{24} = -35 \times 10^{-12}$ C/N; $d_{15} = -27 \times 10^{-12}$ C/N
Electromechanical coupling factors	$k_{31} = 12\%$; $k_{32} = 3\%$; $k_{33} = 19\%$ @1 kHZ
Permitivity	(106–113) × 10^{12} F/m
Relative permitivity	12–13
Dipole moment	2.1 D
Young's moduli	$Y_1 = 2.56 \times 10^9$ Pa; $Y_2 = 2.6 \times 10^9$ Pa
Poisson ratio	$\nu_{21} \sim 0.1$, $\nu_{31} \sim 0.8$
Yield stress	$(\sigma_y)_1 = 45$ MPa; $(\sigma_y)_2 = 39$ MPa
Yield strain	$(\varepsilon_y)_1 = 1.8\%$; $(\varepsilon_y)_2 = 1.4\%$;
Ultimate stress	$(\sigma_u)_1 = 350$ MPa; $(\sigma_u)_2 = 50$ MPa
Ultimate strain	$(\varepsilon_u)_1 = 16.9\%$; $(\varepsilon_u)_2 = 2.5\%$
Dielectric loss factor	0.015–0.25
Mechanical loss tangent	0.10
Related compounds	PVC, PTFE, PVDF–TrFE

Source: Adapted from Vinogradov, A.M., in Schwartz, M. (Ed.), *Encyclopedia of Smart Materials*, Vol. 2, John Wiley & Sons, New York, 2002, pp. 780–792. With permission.

the properties of piezoelectric polymers or creating novel polymer-based composite systems with superior properties.

In particular, strong electrostrictive response of PVDF–trifluoroethylene (PVDF–TrFE) has been obtained as a result of material irradiation [5,18]. Large electrostrictive strains in PVDF–TrFE have been also achieved by chemically increasing the degree of polymer cross-linking [19]. Certain changes in manufacturing processes have provided significant improvements in the electrostrictive response of PVDF–hexafluoropropylene (PVDF–HFP) copolymer [20]. Similar effects have been observed in PVDF–HFP as a result of special thermal treatment [21]. The use of conductive polymers in place of the traditional metallic surface layers has improved the functional performance of PVDF–TrFE over a wide range of temperatures and frequencies [22]. Enhanced performance of electroactive thermoplastic elastomers has been achieved by subjecting the materials to thermal annealing treatment. As a result, a graft-elastomer (G-elastomer) has been developed with improved functional response in terms of large electromechanical strains and a significant increase in the electromechanical output power density [23].

To date, various approaches have been developed to produce novel electroactive polymer systems by combining the properties of different materials. In particular, advanced piezoelectric composites have been produced by integrating ceramic fibers or particles of lead zirconate titanate (PZT) or calcium-modified lead titanate ($PbTiO_3$) into a polymer matrix [24,25]. Electroactive composite systems have been also synthesized using high dielectric organic compounds and electroactive polymers. A promising composite material of this type represents a system of CuPc powders used as filler and the electrostrictive polymer PVDF–TrFE as the matrix. Similarly, an all-polymer composite material has been synthesized using the conductive polymer polyaniline (PANI) as the dielectric enhancement component and the terpolymer PVDF-TrFE-CTFE as the matrix. Both composite systems are characterized by high elastic modulus, high elastic energy density, and tend to produce large strains [26,27].

A combination of an electrostrictive graft elastomer (G-elastomer) described above with a piezoelectric poly(vinylidene fluoride–trifluoroethylene) copolymer has produced several types of ferroelectric–electrostrictive molecular composite systems with dual functionality involving two types of responses, piezoelectric and electrostrictive. Such materials can form a basis for various multifunctional smart systems with the advantage of using a single film for both sensor and actuator functions [28].

24.2.4 Concluding Remarks

To date, the inherent intelligence of piezoelectric polymers has stimulated rapid progress in many modern technological fields, including aviation, space exploration, electronics, automated control, and biomedical engineering. In the immediate future, continuing progress in this field will depend on the intensity of research efforts directed toward the development of electroactive polymers with enhanced functional capabilities. Further, it is important that innovative testing methodologies be developed in order to obtain consistent material characterization of existing and emerging piezoelectric polymer systems. Renewed efforts in computational materials science are required to facilitate accurate material modeling of piezoelectric polymers as an integral part of new developments [29]. It is clear that only on this basis the unprecedented opportunities offered by piezoelectric polymer systems will continue to stimulate further technological progress.

References

1. Sperati, C.A. Fluoropolymers, Poly(vinylidene fluoride) (PVDF), in *Handbook of Plastic Materials and Technology*, I.I. Rubin, ed., John Wiley & Sons, New York, pp. 131–136, 1990.
2. Kawai, H. The Piezoelectricity of Poly(vinylidene fluoride), *Jpn. J. Appl. Phys.*, 8, 975–976, 1969.
3. Kepler, R.G. and Anderson, R.A. Ferroelectric polymers, *Adv. Phys.*, 41(10), 1–57, 1992.
4. Vinogradov, A.M. Piezoelectricity in polymers, in *Encyclopedia of Smart Materials*, Vol. 2, M. Schwartz ed., John Wiley & Sons, New York, pp. 780–792, 2002.
5. Zhang, Q.M., Bharti, V., and Kavarnos, G. Poly(vinylidene fluoride) (PVDF) and its copolymers, in *Encyclopedia of Smart Materials*, Vol. 2, M. Schwartz ed., John Wiley & Sons, New York, pp. 807–825, 2002.
6. Harrison, J.S. and Ounaies, Z. Polymers, piezoelectric, in *Encyclopedia of Smart Materials*, Vol. 2, M. Schwartz ed., John Wiley & Sons, New York, pp. 860–873, 2002.
7. Ikeda, T. *Fundamentals of Piezoelectricity*, Oxford Science, Oxford, 1990.
8. Hilczer, B. and Malecki, *J. Electrets*, Elsevier, Amsterdam, 1986.
9. Tasaka, S. and Miyata, S. The Origin of piezoelectricity in poly(vinylidene fluoride), *Ferroelectrics* 32(1), 17–23, 1981.
10. Sessler, G.M. 1982, Polymeric electrets in *Electrical Properties of Polymers*, D.A. Seanor ed., Academic Press, New York, pp. 241–284, 1982.
11. Tiersten, H.F. Electroelastic equations for electroded thin plates subject to large driving voltages, *J. Appl. Phys.*, 74(5), 3389–3393, 1993.
12. McGrum, N.G., Reid, B.E., and Williams, G. *Anelastic and Dielectric Effects in Polymeric Solids*, Dover Publications, New York, 1991.
13. Hahn, B.R. Studies of the nonlinear piezoelectric response of polyvinylidene fluoride, *J. Appl. Phys.*, 57(4), 1294–1298, 1985.
14. Vinogradov, A.M. et al. Damping and electromechanical energy losses in the piezoelectric polymer PVDF, *Mech. Mater.*, 36, 1007–1016, 2004.
15. Vinogradov, A.M. and F. Holloway, F. Electro-mechanical properties of the piezoelectric polymer PVDF, *Ferroelectrics*, 226, 169–181, 1999.
16. Lakes, R. *Viscoelastic Solids*, CRC Press, Boca Raton, FL, 1998.
17. Vinogradov A.M. and Schumacher, S.C. Cyclic creep of piezoelectric polymer polyvinylidene fluoride, *AIAA J.*, 39(11), 2227–2229, 2001.
18. Zhang, Q.M. and Scheinbeim, J.I. Electric EAP in *Electroactive Polymer (EAP) Actuators as Artificial Muscles.*

Reality, Potential and Challenges, Y. Bar-Cohen ed., SPIE Press, Washington, DC, pp. 89–120, 2001.
19. Casalini R. and Roland, C.M. Electromechanical properties of poly(vinylidene fluoride-trifluoroethylene) networks, *J. Polymer Sci.*, 40, 1975–1982, 2002.
20. Jayasuriya, A.C., Schirokauer, A., and Scheinbeim, J.I. Crystal-structure dependence of electroactive properties in differently prepared poly(vinylidene fluoride/haxafluoropropylene) copolymer films, *J. Polymer Sci., Part B, Polymer Phys.*, 39(22), 2793–2799, 2001.
21. Lu, X., Schirokauer, A., and Scheinbeim, J.I. Giant electrostrictive response in poly(vinylidene fluoride-hexafluoropropylene) copolymers, *IEEE Trans. Ultrasonics Ferroelectrics Frequency Control*, 47(6), 1291–1298, 2000.
22. Xu, H.-S. et al. All-polymer electromechanical systems consisting of electrostrictive poly(vinylidene fluoride-tryfluoroethylene) and conductive polyaniline, *J. Appl. Polym. Sci.*, 75, 945–957, 2000.
23. Su, J., Harrison, J.S., and St. Clair, T. Electrostrictive Graft Elastomers, U.S. Patent, 6,515,077, 2000.
24. Chan, H.L.W. and Unsworth, J. Mode coupling in modified lead titanate/polymer 1–3 composites, *J. Appl. Phys.*, 65(4), 1754–1758, 1989.
25. Marra, S.P., Ramesh, K.T., and Doudlas, A.S. The mechanical and electromechanical properties of calcium-modified lead-titanate/poly(vinylidene fluoride-trifluoroethylene) 0–3 composites, *Smart Mater. Struct.*, 8, 57–63, 1999.
26. Zhang, Q.M. et al. An all-organic composite actuator material with high dielectric constant, *Nature*, 419, 284–287, 2002.
27. Huang, C., Zhang, Q.M., and Su, J. High dielectric constant all-polymer percolative composites, *Appl. Phys. Lett.*, 82, 3502–3504, 2003.
28. Su, J., Harrison, J.S., and St. Clair, T. Novel polymeric elastomers for actuation, *Proc. IEEE Int. Symp. Application of Ferroelectrics*, IEEE, 2000.
29. Smith, R.C. Smart material systems. Model development, *Frontiers in Applied Mathematics*, SIAM, Philadelphia, PA, 2005.

24.3 Polymers, Biotechnology, and Medical Applications

Igor Yu Galaev and B. Mattiasson

24.3.1 Introduction

Life is polymeric in its essence. The most important components of living cell, proteins, carbohydrates, and nucleic acids are polymers. Even lipids, which have lower molecular weights, can be regarded as methylene oligomers that have a polymerization degree around 20. Nature uses polymers as constructive elements and parts of complicated cell machinery. The salient feature of functional biopolymers is their all-or-nothing or at least highly nonlinear response to external stimuli. Small changes happen in response to varying parameters until the critical point is reached; then a transition occurs in the narrow range of the varied parameter, and after the transition is completed, there is no significant further response of the system. Such nonlinear response of biopolymers is warranted by highly cooperative interactions. Despite the weakness of each particular interaction in a separate monomer unit, these interactions, when summed through hundreds and thousands of monomer units, provide significant driving forces for the processes in such systems.

Not surprisingly, understanding the mechanism of cooperative interactions in biopolymers has opened the floodgates for attempts to mimic the cooperative behavior of biopolymers in synthetic systems. Recent decades witnessed the appearance of synthetic functional polymers, which respond in some desired way to a change in temperature, pH, electric, or magnetic fields or some other parameters. These polymers were nicknamed stimuli-responsive. The name "smart polymers" was coined due to the similarity of the stimuli-responsive polymers to biopolymers [1]. We have a strong belief that nature has always striven for smart solutions in creating life. The goal of scientists is to mimic biological processes, and therefore understand them better, and also to create novel species and invent new processes.

The applications of smart polymer in biotechnology and medicine are discussed in this article. The highly nonlinear response of smart polymers to small changes in the external medium is of critical importance for the successful functioning of a system. Most applications of smart polymers in biotechnology and medicine include biorecognition and/or biocatalysis, which take place principally in aqueous solutions. Thus, only water-compatible smart polymers are considered; smart polymers in organic solvents or water/organic solvent mixtures are beyond the scope of the chapter. The systems discussed in the chapter are based on either soluble/insoluble transition of smart polymers in aqueous solution or on the conformational transition of macromolecules physically attached or chemically grafted to the surface. Systems that have covalently cross-linked networks of macromolecules, called smart hydrogels, are not considered.

One could define smart polymers used in biotechnology and medicine as macromolecules that undergo fast and reversible changes from hydrophilic to hydrophobic microstructure triggered by small changes in their environments. These microscopic changes are apparent at the macroscopic level as precipitate formation in solutions of smart polymers or changes in the wettability of a surface to which a smart polymer is grafted. The changes are reversible, and the system returns to its initial state when the trigger is removed.

24.3.2 Smart Polymers Used in Biotechnology and Medicine

The highly nonlinear transitions in smart polymers are driven by different factors, for example, neutralization of charged groups by either a pH shift [2] or the addition of an oppositely charged polymer [3], changes in the efficiency of hydrogen bonding and an increase in temperature or ionic strength [4], and critical phenomena in hydrogels and interpenetrating polymer networks [5]. The polymer systems that have highly nonlinear response can be divided into three general groups: pH-sensitive smart polymers, thermosensitive smart polymers, and reversibly cross-linked networks.

24.3.2.1 pH-Sensitive Smart Polymers

The first group of smart polymers consists of polymers whose transition between the soluble and insoluble state is created by decreasing the net charge of the polymer molecule. The net charge can be decreased by changing the pH to neutralize the charges on the macromolecule and hence to reduce the hydrophilicity (increase the hydrophobicity) of the macromolecule. Copolymers of methylmethacrylate (hydrophobic part) and methacrylic acid (hydrophilic at high pH when carboxy groups are deprotonated but more hydrophobic when carboxy groups are protonated) precipitate from aqueous solutions by acidification to pH around 5, and copolymers of methyl methacrylate (hydrophobic part) with dimethylaminethyl methacrylate (hydrophilic at low pH when amino groups are protonated but more hydrophobic when amino groups are deprotonated) are soluble at low pH but precipitate in slightly alkaline conditions [6]. Hydrophobically modified cellulose derivatives that have pending carboxy groups, for example, hydroxypropyl methyl cellulose acetate succinate are also soluble in basic conditions but precipitate in slightly acidic media [7].

The pH-induced precipitation of smart polymers is very sharp and usually requires a change in pH of not more than 0.2–0.3 units (Figure 24.7). When some carboxy groups are used to couple a biorecognition element, for example, noncharged sugar, the increased hydrophobicity of the copolymer results in precipitation at a higher pH [8]. The copolymerization of *N*-acryloyl sulfametazine with *N*, *N*-dimethylacrylamide results in a pH-sensitive polymer whose reversible transition is in the physiological range of pH 7.0–7.5 [9].

The charges on the macromolecule can also be neutralized by adding an efficient counterion, for example, a low molecular weight counterion or a polymer molecule of opposite charges. The latter systems are combined under the name of polycomplexes. The cooperative nature of interaction between two polymers of opposite charges makes polycomplexes very sensitive to changes in pH or ionic strength [10]. The complex formed by poly(methacrylic acid) (polyanion) and poly(*N*-ethyl-4-vinylpyridinium bromide) (polycation) undergoes reversible precipitation from aqueous solution at any desired pH value in the range 4.5–6.5 that depends on the ionic strength and polycation/polyanion ratio in the complex (Figure 24.8) [11]. Polyelectrolyte complexes formed by poly(ethylene imine) and poly(acrylic acid) undergo soluble–insoluble transition in an even broader pH range from pH 3–11 [12].

The pH of the transition of pH-sensitive polymers such as poly(methylmethacrylate-*co*-methacrylic acid) or poly(*N*-acryloyl sulfametazine-*co*-*N*,*N*-dimethylacrylamide) is strictly fixed for

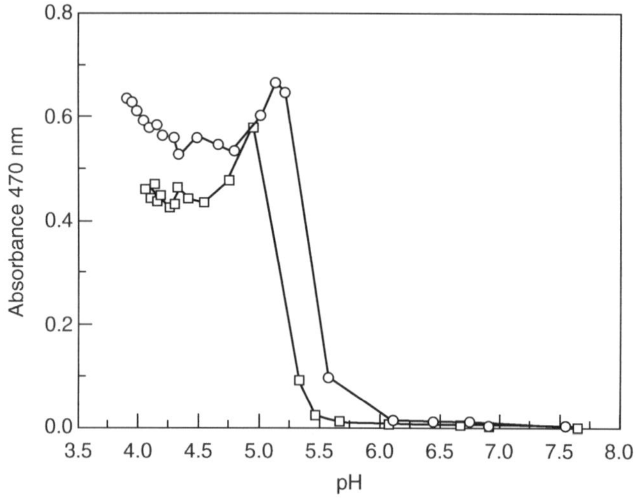

FIGURE 24.7 pH-induced precipitation of a random copolymer of methacrylic acid and methacrylate (commercialized as Eudragit S 100 by Röhm Pharma GMBH, Weiterstadt, Germany) (open squares) and *p*-amino-phenyl- α-D-glucopyranoside-modified copolymer (open circles) measured as turbidity at 470 nm. Some decrease in turbidity at lower pH values is caused by flocculation and sedimentation of the polymer precipitate. (Redrawn from Hilczer, B. and Malecki, J. *Electrets*, Elsevier, Amsterdam, 1986.)

the given composition of comonomers. Thus, a new polymer should be synthesized for each desired pH value. The advantage of polyelectrolyte complexes is that by using only two different polymers and mixing them in different ratios, reversible precipitation can be achieved at any desired pH value in a rather broad pH range.

24.3.2.2 Thermosensitive Smart Polymers

The reversible solubility of thermosensitive smart polymers is caused by changes in the hydrophobic–hydrophilic balance of uncharged polymers induced by increasing temperature or ionic strength. Uncharged polymers are soluble in water due to hydrogen bonding with water molecules. The efficiency of hydrogen bonding lessens as temperature increases. The phase separation of a polymer occurs when the efficiency of hydrogen bonding becomes insufficient for the solubility of macromolecule.

When the temperature of an aqueous solution of a smart polymer is raised above a certain critical temperature (which is often referred to as the transition temperature, lower critical solution temperature (LCST), or cloud point), phase separation takes place. An aqueous phase that contains practically no polymer and a polymer-enriched phase are formed. Both phases can be easily separated by decanting, centrifugation, or filtration. The temperature of the phase transitions depends on the polymer concentration and molecular weight (MW) (Figure 24.9) [13,14]. The phase separation is completely reversible, and the smart polymer dissolves in water when the temperature is reduced below the transition temperature.

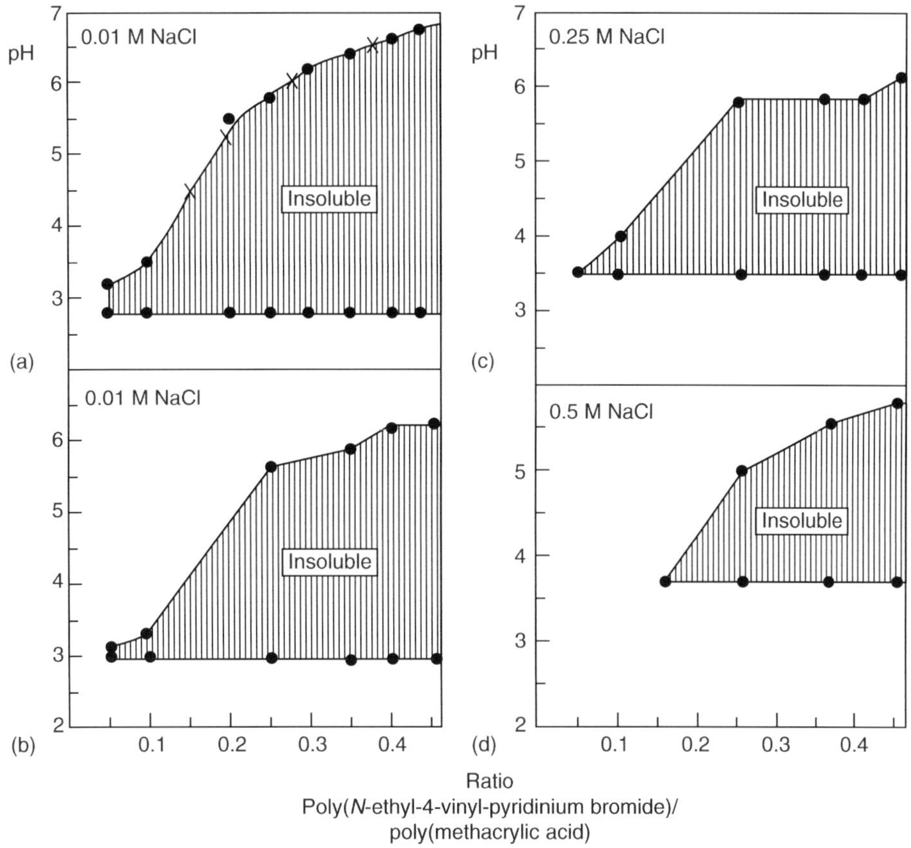

FIGURE 24.8 Phase diagram for the polyelectrolyte complex formed by poly(*N*-ethyl-4-vinyl-pyridinium bromide) (polymerization degree 530) and poly(methacrylic acid) (polymerization degree 1830). The dots present pH values at which the turbidity of the polymer solutions was first observed at 470 nm. Ionic strength was (a) 0.01 M NaCl, (b) 0.1 M NaCl, (c) 0.25 M NaCl, and (d) 0.5 M NaCl. Dashed area represents pH/composition range where the complex is insoluble. (Reproduced from tiertsen, H.F., *J. Appl. Phys.*, 74(5), 3389, 1993. With permission.)

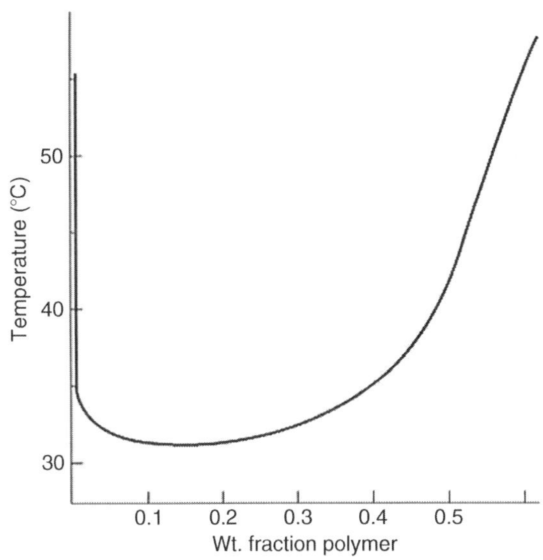

FIGURE 24.9 Phase diagram for poly(NIPAAM) in aqueous solution. The area under the binodal curve presents the range of temperatures/polymer concentrations for homogeneous solution. Separation into polymer-enriched and polymer-depleted phases takes place for any polymer concentration/temperature above the binodal curve. (Reproduced from Vinogradov, A.M. et al., *Mech. Mater.*, 36, 1007, 2004. With permission.)

Two groups of thermosensitive smart polymers are most widely studied and used:

- Poly(*N*-alkyl substituted acrylamides) and the most well-known of them, poly(*N*-isopropyl acrylamide) (poly(NIPAAM)), whose transition temperature is 32°C [14]
- Poly(*N*-vinylalkylamides) such as poly(*N*-vinylisobutyramide) whose transition temperature is 39°C [15] or poly(*N*-vinyl caprolactam) whose transition temperature is 32°C–33°C (depending on the polymer's molecular weight) [13]

A variety of polymers that have different transition temperatures from 4°C–5°C for poly(*N*-vinyl piperidine) to 100°C for poly(ethylene glycol) are available at present [16].

pH-sensitive smart polymers usually contain carboxy or amino groups that can be used for covalent coupling of biorecognition or biocatalytic elements (ligands). Thermosensitive polymers, on the contrary, do not have inherent reactive groups, which could be used for ligand coupling. Thus, copolymers that contain reactive groups can be synthesized. *N*-Acryloylhydroxysuccinimide [17] or glycidyl methacrylate [18] have often been used as active comonomers in copolymerization with NIPAAM, allowing further coupling of amino-group-containing ligands to the synthesized copolymers. The use of an initiator of polymerization [19] or chain transfer agent [20] that has an active group results in a polymer modified only at the end of the macromolecule. An alternative strategy is to incorporate a polymerizable double bond into the ligand, for example, by modification with acryoyl group, and then to copolymerize the modified ligand with NIPAAM [21,22].

An increase in the hydrophilicity of the polymer-accompanied incorporation of hydrophilic comonomers or coupling to hydrophilic ligands increases the transition temperature, whereas hydrophobic comonomers and ligands have the opposite effect [4]. The pH-induced change in ligand hydrophobicity could have a dramatic effect on the thermoseparation of the ligand–polymer conjugate. A copolymer of NIPAAM and vinyl imidazole precipitates at about 35°C at pH 8.0 where imidazole moieties are noncharged and relatively hydrophobic, but no precipitation occurs even when heating the polymer solution to 80°C at pH 6 where imidazole groups are protonated and very hydrophilic [23].

Ligand–ligand interactions in a ligand–polymer conjugate also have a significant effect on the thermoseparation. The precipitation temperature for the previously mentioned copolymers of NIPAAM and vinyl imidazole increases as the imidazole content in the copolymer increases. On the contrary, the precipitation temperature decreases as the increase of imidazole content increases, when the polymer forms a Cu(II)-complex [23]. Each Cu(II) ion interacts with two to three imidazole groups to cross-link the segments of the polymer molecule [24]. The restricted mobility of the polymer segments results in a lower precipitation temperature.

Block copolymers that have a thermosensitive "smart" part that consists of poly(NIPAAM) form reversible gels on an increase in temperature, whereas random copolymers separate from aqueous solutions by forming a concentrated polymer phase [25]. Thus, the properties of smart polymers that are important for biotechnological and medical applications could be controlled by the composition of comonomers and also by the polymer architecture.

The phase transition of thermosensitive polymers at increased temperature results from hydrophobic interactions between polymer molecules. Because hydrophobic interactions are promoted by high salt concentrations, the addition of salts shifts the cloud point to lower temperatures. When the transition temperature is below room temperature, polymer precipitation is achieved by just a salt addition without any heating. The addition of organic solvents, detergents, and chaotropic agents increases the transition temperature because these compounds deteriorate hydrophobic interactions.

24.3.2.3 Reversibly Cross-Linked Polymer Networks

Systems that have reversible noncovalent cross-linking of separate polymer molecules into a polymer network belong to the third group of smart polymers. When formed, reversibly cross-linked polymers either precipitate or form a physical gel. Polymers that have sugar ligands cross-linked by lectins with multibinding sites [26] and boronate-polyols [27–29] are the most widely used systems of this type. The reversible response in these systems is achieved by addition or removal of a low molecular weight analog of the polymer. For example, small sugars added at high concentrations compete with sugar-containing polymers for binding to lectin and destroy intrapolymer cross-links that result in disengagement of the network.

24.3.2.4 Heterogeneous Systems Using Smart Polymers

A solid surface acquires new properties when modified by adsorption or chemical grafting of smart polymers. Smart polymers that have terminal (only single-point attachment possible) or random (multipoint attachment possible) could be covalently coupled to the respective active groups on the surface [30]. Single-point attachment could also be achieved by covalent modification of the surface using an initiator of polymerization and then carrying out polymerization of monomers in the solution that surrounds the support. The growth of polymer chains occurs only at the sites where initiator was coupled [31]. Alternatively, the solid support is irradiated by light [32] or a plasma beam [33] when monomer is in the surrounding solution. Active radical sites on the surface, which appear as a result of irradiation, initiate the growth of polymer macromolecules. As a rule, irradiation methods give a higher density of grafted polymer, but polymerization is less controlled as in covalent coupling or using a covalently coupled initiator. Irradiation, especially at high monomer concentrations, could produce a cross-linked polymer gel attached to the solid support [34].

A separate group of smart polymers is represented by particulate systems. Liposomes that reversibly precipitate on salt addition and removal were prepared from a synthetic phospholipid that had a diacetylene moiety in the hydrophobic chain and an amino group in the hydrophilic head of the phospholipid, followed by polymerization of diacetylene bonds [35]. Lattices composed of thermosensitive polymers or a layer of thermosensitive polymer at the surface represent another example of insoluble but reversibly suspended particulate systems that respond to increasing or decreasing temperature [31].

24.3.3 Applications

There are numerous potential applications for smart polymers in biotechnology and medicine. The main commercial application of smart polymers is the production of "smart" pills where the shell of the smart polymer protects the pill from the harmful action of the stomach contents but allows the pill to dissolve in the intestine. There is not yet any other product in the market that applies smart polymers, but the interest in these applications is growing in both the academic community and industry. The following applications are considered in this chapter:

- Smart pills that have an enteric coating
- Smart polymers for affinity precipitation of proteins
- Aqueous two-phase polymer systems formed by smart polymers and their application for protein purification
- Smart surfaces for mild detachment of cultivated mammalian cells
- Smart chromatographic matrices that respond to temperature
- Smart polymers for controlled porosity of systems—"chemical valve"
- Liposomes that trigger the release of their contents
- Smart polymers for bioanalytical applications
- Reversibly soluble biocatalysts

24.3.3.1 Smart Pills That Have an Enteric Coating

It is common knowledge that peroral introduction of medical preparations is the most convenient method compared to subcutaneous or intravenous injection and even to nasal sprays or eye droplets. The absorption of a swallowed pill takes place predominantly in the intestine and to reach the intestine, the medicine must pass unharmed through the stomach that has a very low pH value of 1.4 and abundant hydrolytic enzymes that can degrade a broad variety of chemical structures. Many medicines are susceptible to damage in the stomach environment. The ideal condition for peroral introduction is to have a smart pill, which is insoluble in the stomach and hence passes through the stomach unaffected but easily dissolves at the higher pH in the intestine where the medicine is absorbed. Smart polymers provide the solution. Hydrophobic polymers such as poly(methylmethacrylate) or hydrophobically modified celluloses are insoluble in water per se, but the introduction of carboxy groups (either by partial hydrolysis of ester groups in methylmethacrylate or modification of cellulose HO groups by dicarboxylic acids such as succinic or phthalic acid) endows the polymers with pH-dependent solubility. The pill covered by a shell of such a polymer (enteric coating) is insoluble at low pH when the carboxy groups are protonated and uncharged, but easily soluble at a pH above 6 when carboxy groups are protonated and charged. Industrially produced polymers for enteric coating belong to two main groups, synthetic copolymers of methylmethacrylate and methacrylic acid and modified derivatives of cellulose, a natural polymer (Table 24.2). The first group of polymers is used mainly by European and U.S. manufacturers, and the second group is more popular in Japan.

Whenever the charge-bearing comonomer has an amino group instead of a carboxy group, the solubility of the polymer acquires opposite pH dependence. The polymer is soluble at low pH values but insoluble in neutral and alkaline media. Poly(diethylaminoethylmethacrylate-*co*-methylmethacrylate) (commercialized as Eudragit E) is an example of such a polymer. The shell that is composed of this polymer protects the tablet against dissolution in the neutral saliva, and the mouth is not affected by the unpleasant taste of bitter medicine, but the polymer dissolves readily in the stomach.

24.3.3.2 Bioseparation—Affinity Precipitation

All bioseparation processes include three stages: preferential partitioning of target substance and impurities between two phases (liquid–liquid or liquid–solid), mechanical separation of the phases (e.g., separation of the stationary and mobile phases in a chromatographic column), and recovery of the target substance from the enriched phase. Because smart polymers can undergo phase transitions, they could facilitate the second and the third stages of bioseparation processes.

The ability of smart polymers to form in situ heterogeneous systems is exploited in affinity precipitation (Figure 24.10). The technique is based on using a conjugate of a smart polymer that has a covalently coupled biorecognition moiety, that is, a ligand specific for a target protein. The conjugate forms a complex with the target protein but not with the other proteins in the crude extract. Phase separation of the complex is triggered by small changes in the environment, resulting in transition of the polymer backbone into an insoluble state. The target protein specifically coprecipitates with the smart polymer and the impurities in the crude remain in solution. Then, the target protein is either eluted from the insoluble macroligand–protein complex or the precipitate is dissolved. The protein is dissociated from the macroligand and the ligand–polymer conjugate is precipitated again

TABLE 24.2 Industrially Manufactured Smart Polymers for Producing Smart Pills

Polymer	Trade Name	Manufacturer	Country
Poly(methacrylic acid-*co*-methylmethacrylate) 1:1 monomer ratio, MW 135,000	Eudragit L	Röhm Pharma GmBH	United States, Germany
Poly(methacrylic acid-*co*-methylmethacrylate) 1:2 monomer ratio, MW 135,000	Eudragit S	Röhm Pharma GmBH	United States, Germany
Carboxymethylcellulose	CMEC	Freund Sangyo Co., Ltd.	Japan
Cellulose acetate phthalate	CAP	Wako Pure Chemicals Ltd.	Japan
Hydroxypropylmethyl-cellulose phthalate	HP-50, HP-55	Shin-Etsu Chemical Co., Ltd.	Japan
Hydroxypropylmethyl-cellulose acetate succinate	ASM, AS-H	Shin-Etsu Chemical Co., Ltd.	Japan
Poly(diethylaminoethyl methacrylate-*co*-methylmethacrylate) MW 150 000	Eudragit E	Röhm Pharma GmBH	United States, Germany

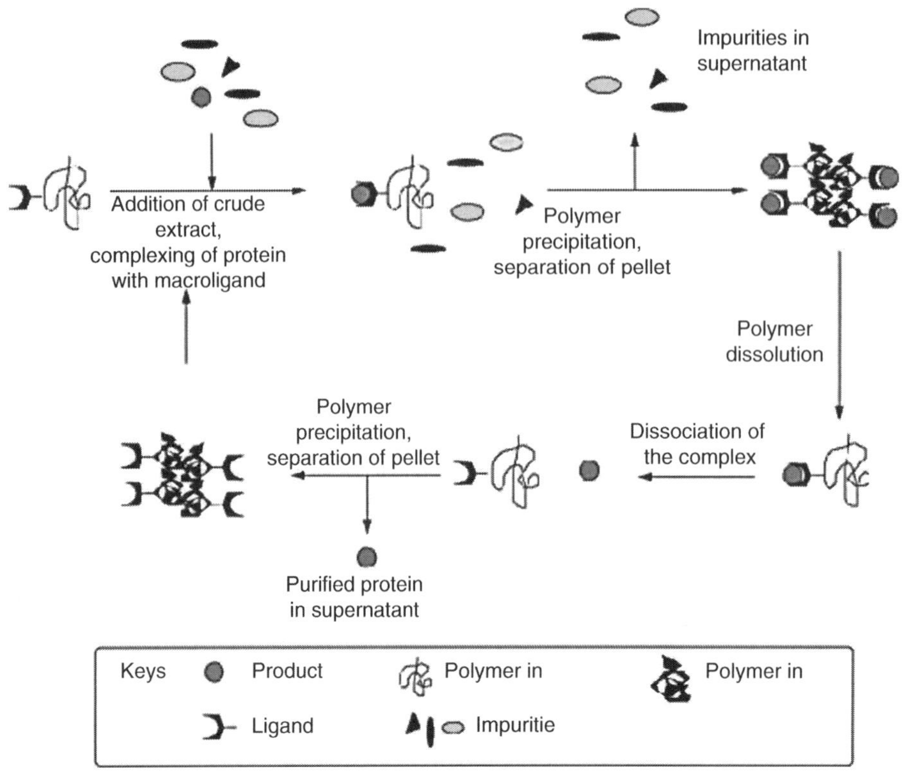

FIGURE 24.10 Schematic of affinity precipitation technique for protein purification.

now without the protein that remains in the supernatant in purified form. A variety of different ligands such as triazine dyes, sugars, protease inhibitors, antibodies, nucleotides, double-or single-stranded DNA, and chelated metal ions were successfully used for affinity precipitation [36]. After elution of the target protein, the ligand–polymer conjugate could be recovered and used in the next purification cycle [37].

Triazine dyes, robust affinity ligands for many nucleotide-dependent enzymes, were successfully used in conjugates with the pH-sensitive copolymer of methacrylic acid and methylmethacrylate, which precipitates when pH decreases (Eudragit S 100), for the purification of dehydrogenases from various sources by affinity precipitation [38,39]. Sugar ligands constitute another attractive alternative and have been used in combination with Eudragit S 100 for bioseparation of lectins [40]. Restriction endonuclease Hind III was successfully isolated using the thermosensitive conjugate of poly(NIPAAM) with phage λ DNA [21]. Human IgG was specifically precipitated with a conjugate of protein A and galactomannan. Galactomannan polymer was reversibly precipitated by adding tetraborate [41].

The efficient precipitation of Cu(II)-loaded poly(*N*-vinylimidazole-*co*-NIPAAM) by high salt concentrations at mild temperature is very convenient for metal affinity precipitation of proteins that have inherent histidine residues at the surface or for recombinant proteins artificially provided with histidine tags (usually four to six residues). High salt concentration does not interfere with protein–metal ion–chelate interaction, and, on the other hand, it reduces the possibility of nonspecific binding of foreign proteins to the polymer both in solution and when precipitated [23]. The flexibility of polymer chains in solution allows several imidazole ligands on a polymer molecule to come close enough to interact with the same Cu(II) ion and thus to provide sufficient strength of polymer–Cu(II) interactions to purify a variety of histidine-containing proteins [37].

Polyelectrolyte complexes that have pH-dependent solubility were successfully used in different bioseparation procedures. When an antigen, inactivated glyceraldehyde-3-phosphate dehydrogenase, from rabbit was covalently coupled to a polycation, the resulting complex was used to purify monoclonal antibodies specific toward inactivated glyceraldehyde-3-phosphate dehydrogenase [11]. The successful affinity precipitation of antibodies using glyceraldehyde-3-phosphate dehydrogenase bound to a polyelectrolyte complex indicates that the ligand is exposed to the solution. This fact was used to develop a new method for producing monovalent Fab fragments of antibodies. Traditionally, Fab fragments are produced by proteolytic digestion of antibodies in solution followed by isolation of Fab fragments. In the case of monoclonal antibodies against inactivated subunits of glyceraldehyde-3-phosphate dehydrogenase, digestion with papain resulted in significant damage of binding sites of the Fab fragment. Proteolysis of monoclonal antibodies in the presence of the antigen–polycation conjugate followed by (1) precipitation induced by adding polyanion, poly(methacrylic) acid, and a pH shift from 7.3 to 6.5 and (2) elution at pH 3.0 that resulted in 90%

immunologically competent Fab fragments. Moreover, the papain concentration required for proteolysis was 10 times less for antibodies bound to the antigen–polycation conjugate compared to that for free antibodies in solution [42]. Active glyceraldehyde-3-phosphate dehydrogenase from rabbit muscle was separated from the inactivated enzyme by using monoclonal antibodies specific for the inactivated enzyme covalently coupled to the polyanion component of the polyelectrolyte complex. This system can be regarded as a simplified model of chaperone action in living cells that assist in separating active protein molecules from misfolded ones [43].

Apart from specific interactions between a target protein and a ligand–polymer conjugate, nonspecific interactions of protein impurities with the polymer backbone could take place. The nonspecific interactions limit the efficiency of the affinity precipitation technique, and significant efforts were made to reduce these interactions. The advantage of polyelectrolyte complexes as carriers for affinity precipitation is low nonspecific coprecipitation of proteins when the polymer undergoes a soluble–insoluble transition [10].

Smart particles capable of reversible transition between aggregated and dispersed states were used for affinity precipitation of proteins. Thermosensitive [44] or pH-sensitive lattices [45] or salt-sensitive liposomes that have polymerized membranes [35] are examples of such systems.

Two elements are required for successful affinity precipitation. The backbone of a smart polymer provides precipitation at the desired conditions (temperature, pH, ionic strength), and the biorecognition element is responsible for selective binding of the protein of interest. By proper choice of a smart polymer, precipitation could be achieved practically at any desired pH or temperature. For example, poly(*N*-acryloylpiperidine) terminally modified with maltose has an extremely low critical temperature (soluble below 4°C and completely insoluble above 8°C) and was used to purify thermolabile α-glucosidase [46].

24.3.3.3 Bioseparation—Partitioning in Aqueous Polymer Two-Phase Systems

Two aqueous polymer solutions become mutually incompatible when the threshold concentrations of polymers are exceeded. Both of the polymer phases formed contain about 90% water and hence present a very friendly environment for proteins and other biomolecules. Proteins partition selectively between two phases depending on their size, charge, hydrophobicity, nature, and the concentration of the phase-forming polymers. The partitioning could be also directed by adding some salt or coupling an affinity ligand specific for a given protein to one of the phase-forming polymers [47]. The selective partitioning of proteins between the two phases formed has proven to be an efficient tool for purifying proteins and some low molecular weight substances. The main problem of the method—how to separate the target protein from the phase-forming polymer—has not yet been completely solved. Smart polymers provide an elegant solution to this problem—simple precipitation of the phase-forming polymer leaves protein in the supernatant (Figure 24.11): (1) The crude protein extract is mixed with the aqueous two-phase polymer system and the conditions are selected so that the protein of interest partitions into a phase formed by a smart polymer (for example by coupling

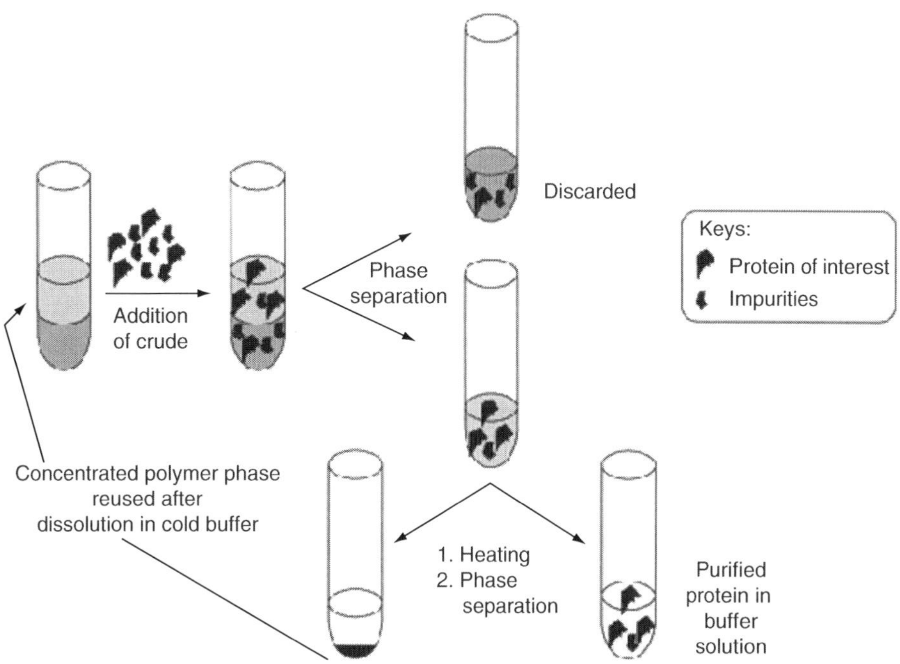

FIGURE 24.11 Schematic presentation of protein partitioning in aqueous two-phase polymer system formed by a smart (thermosensitive polymer). (Reproduced from Galaev, I.Y. and Mattiasson, B., *Trends Biotechnol.*, 17, 335, 1999. With permission.)

affinity ligand to the smart polymer), and the impurities concentrate in the other phase. (2) the phases are separated mechanically and the phase formed by the smart polymer is subjected to conditions (pH or temperature) where the polymer undergoes phase separation; (3) two new phases are formed, a polymer-enriched phase of high polymer and low water concentration, which contains practically no protein, and a polymer-depleted aqueous phase that contains most of the purified protein and minute amounts of the polymer left after phase separation.

pH-sensitive acrylic copolymers [48] or thermoresponsive polymers, poly(ethylene oxide-*co*-propylene oxide) [49,50] or poly(*N*-vinyl caprolactam-*co*-vinyl imidazole) [51], form two-phase systems from relatively hydrophilic polymers such as dextran or modified starch and have been successfully used for protein purification. The pH- or thermoprecipitated polymer opposite dextran could be regenerated by dissolution at a lower temperature. Quite recently, an aqueous two-phase polymer system was developed where both phase-forming polymers, poly(*N*-isopropylacrylamide-*co*-vinylimidazole) and poly(ethylene oxide-*co*-propylene oxide) end modified by hydrophobic $C_{14}H_{29}$ groups, are thermoresponsive and could be recycled [52].

24.3.3.4 Smart Surfaces—Cell Detachment

The driving force behind phase separation of smart polymers is a sharp increase in hydrophobicity after a small change in environmental conditions. The hydrophobic "collapsed" polymer aggregates form a separate phase. When grafted to the surface, macromolecules of the smart polymer cannot aggregate, but the conformational transition from the hydrophilic to the hydrophobic state endows the surface with regulated hydrophobicity: the surface is hydrophilic when the smart polymer is in the expanded "soluble" conformation and hydrophobic when the polymer is in the collapsed "insoluble" conformation. The change of hydrophobicity of the surface by grafted poly(*N*-isopropylacrylamide) was demonstrated by contact angle measurements [53] and water absorbency [54].

The transition temperature for adsorbed (presumably via multipoint attachment) poly(NIPAAM) molecules is lower than that in bulk solution, and the properties of the layer of collapsed macromolecules formed above the transition temperature depend strongly on the speed by which the temperature increases. At a low speed of temperature increase, the "liquid-like" polymer layer is formed, whereas at high speeds, the polymer layer has more "solid-like" properties [55]. When cooling, the collapsed polymer molecules return to the initial loopy adsorbed conformation via transitional extended conformation. The relaxation process for the extended-to-loopy adsorbed conformational transition occurs slowly and depends on the temperature observance of an Arrhenius law. Kinetic constraints, it is proposed, play an important role in this transition [56].

The change of surface properties from hydrophobic above the critical temperature of the polymer grafted to hydrophilic below it has been successfully used for detaching mammalian cells. Mammalian cells are normally cultivated on a hydrophobic solid substrate and are detached from the substrate by protease treatment, which often damages the cells by hydrolyzing various membrane-associated protein molecules. The poly(NIPAAM)-grafted surface is hydrophobic at 37°C because this temperature is above the critical temperature for the grafted polymer and that cells that are growing well on it. A decrease in temperature results in transition of the surface to the hydrophilic state, where the cells can be easily detached from the solid substrate without any damage. Poly(NIPAAM) was grafted to polystyrene culture dishes using an electron beam. Bovine hepatocytes, cells that are highly sensitive to enzymatic treatment, were cultivated for 2 days at 37°C and detached by incubation at 4°C for 1 h. Nearly 100% of the hepatocytes was detached and recovered from the poly(NIPAAM)-grafted dishes by low-temperature treatment, whereas only about 8% of the cells was detached from the control dish [57]. The technique has been extended to different cell types [58,59]. It is noteworthy that hepatocytes recovered by cooling retained their native form had numerous bulges and dips, and attach well to the hydrophobic surface again, for example, when the temperature was increased above the conformational transition of poly(NIPAAM). On the contrary, enzyme-treated cells had a smooth outer surface and had lost their ability to attach to the surface. Thus, cells recovered by a temperature shift from poly(NIPAAM)-grafted surfaces have an intact structure and maintain normal cell functions [58].

The molecular machinery involved in cell-surface detachment was investigated using temperature-responsive surfaces [60]. Poly(NIPAAM)-grafted and nongrafted surfaces showed no difference in attachment, spreading, growth, confluent cell density, or morphology of bovine aortic endothelial cells at 37°C. Stress fibers, peripheral bands, and focal contacts were established in similar ways. When the temperature was decreased to 20°C, the cells grown on poly(NIPAAM)-grafted support lost their flattened morphology and acquired a rounded appearance similar to that of cells immediately after plating. Mild agitation makes the cells float free from the surface without a trypsin treatment. Neither changes in cell morphology nor cell detachment occurred on ungrafted surfaces. Sodium azide, an ATP synthesis inhibitor, and genistein, a tyrosine kinase inhibitor, suppressed changes in cell morphology and cell detachment, whereas cycloheximide, a protein synthesis inhibitor, slightly enhanced cell detachment. Phalloidin, an actin filament stabilizer, and its depolymerizer, cytochalasin D, also inhibited cell detachment. These findings suggest that cell detachment from grafted surfaces is mediated by intracellular signal transduction and reorganization of the cytoskeleton, rather than by a simple changes in the "stickiness" of the cells to the surface when the hydrophobicity of the surface is changed.

One could imagine producing artificial organs using temperature-induced detachment of cells. Artificial skin could be produced as the cells are detached from the support not as a suspension (the usual result of protease-induced detachment) but preserving their intercellular contacts. Fibroblasts were cultured on the poly(NIPAAM)–collagen support until the cells completely covered the surface at 37°C, followed by a decrease in temperature to about 15°C. The sheets of fibroblasts detached from the dish and within about 15 min floated in the culture medium [57]. The detached cells could be transplanted to another culture surface without functional and structural changes [34]. Grafting of poly(NIPPAM) onto a polystyrene surface by photolitographic technique creates a special pattern on the surface, and by decreasing temperature, cultured mouse fibroblast STO cells are detached only from the surface area on which poly(NIPAAM) was grafted [61]. Lithographed films of smart polymer present supports for controlled interactions of cells with surfaces and can direct the attachment and spreading of cells [62]. One could envisage producing artificial cell assemblies of complex architecture using this technique.

24.3.3.5 Smart Surfaces—Temperature-Controlled Chromatography

Surfaces that have thermoresponsive hydrophobic properties have been used in chromatography. High-performance liquid chromatography (HPLC) columns with grafted poly(NIPAAM) have been used for separating steroids [63] and drugs [64]. The chromatographic retention and resolution of the solutes was strongly dependent on temperature and increased as temperature increased from 5°C to 50°C, whereas the reference column packed with nonmodified silica displayed much shorter retention times that decreased as temperature decreased. Hydrophobic interactions dominate in retaining solutes at higher temperature, and the preferential retention of hydrogen-bond acceptors was observed at low temperatures. The effect of temperature increase on the retention behavior of solutes separated on the poly(NIPAAM)-grafted silica chromatographic matrix was similar to the addition of methanol to the mobile phase at constant temperature [65].

The temperature response of the poly(NIPAAM)–silica matrices depends drastically on the architecture of the grafted polymer molecules. Surface wettability changes dramatically as temperature changes across the range 32°C–35°C (corresponding to the phase-transition temperature for NIPAAM in aqueous media) for surfaces where poly(NIPAAM) is terminally grafted either directly to the surface or to the looped chain copolymer of NIPAAM and *N*-acryloylhydroxysuccinimide, which was initially coupled to the surface. The wettability changes for the loop-grafted surface itself were relatively large but had a slightly lower transition temperature (~27°C). The restricted conformational transitions for multipoint grafted macromolecules are probably the reason for the reduced transition temperature. The largest surface free energy changes among three surfaces were observed for the combination of both loops and terminally grafted chains [30].

The introduction of a hydrophobic comonomer, buthylmethacrylate, in the polymer resulted in a decreased transition temperature of about 20°C. Retention of steroids in poly (NIPAAM-*co*-buthylmethacrylate)-grafted columns increases as column temperature increases. The capacity factors for steroids on the copolymer-modified silica beads was much larger than that on poly(NIPAAM)-grafted columns. The effect of temperature on steroid retention on poly(NIPAAM-*co*-buthylmethacrylate)-grafted stationary phases was more pronounced compared to supports modified with poly(NIPAAM). Furthermore, retention times for steroids increased remarkably as the buthylmethacrylate content increased in the copolymer. The temperature-responsive elution of steroids was strongly affected by the hydrophobicity of the grafted polymer chains on silica surfaces [63].

The mixture of polypeptides, consisting of 21–30 amino acid residues (insulin chain A, β-endorphin fragment 1–27 and insulin chain B) could not be separated at 5°C (below the transition temperature) on copolymer-grafted matrix. At this temperature, the copolymer is in an extended hydrophilic conformation that results in decreased interactions with peptides and hence short retention times insufficient to resolve them. The mixture has been easily separated at 30°C, when the copolymer is collapsed, hydrophobic interactions are more pronounced, and retention times sufficiently long for resolving polypeptides [66]. Large protein molecules such as immunoglobulin G demonstrate less pronounced changes in adsorption above and below the transition temperature. Only about 20% of the protein adsorbed on poly(NIPAAM)-grafted silica at 37°C (above the LCST) were eluted after decreasing temperature to 24°C (below the transition temperature) [67]. Quantitative elution of proteins adsorbed on the matrix via hydrophobic interactions has not yet been demonstrated, although protein adsorption on poly(NIPAAM)-grafted matrices could be somewhat controlled by a temperature shift. A successful strategy for temperature-controlled protein chromatography proved to be a combination of temperature-responsive polymeric grafts and biorecognition element, for example, affinity ligands.

The access of the protein molecules to the ligands on the surface of the matrix is affected by the transition of the polymer macromolecule grafted or attached to the chromatographic matrix. Triazine dyes, for example, Cibacron Blue, are often used as ligands for dye-affinity chromatography of various nucleotide-dependent enzymes [68]. Poly(*N*-vinyl caprolactam), a thermo-responsive polymer whose critical temperature is about 35°C, interacts efficiently with triazine dyes. Polymer molecules of 40,000 MW are capable of binding up to seven to eight dye molecules hence, the polymer binds via multipoint interaction to the dye ligands available on the chromatographic matrix. At elevated temperature, polymer molecules are in a compact globule conformation that can bind only to a few ligands on the matrix. Lactate dehydrogenase, an enzyme from porcine muscle, has good access to the ligands that are not occupied by the polymer and binds to the column. Poly(*N*-vinyl caprolactam) macromolecules undergo transition to a more expanded coil conformation as temperature decreases. Now, the polymer molecules interact with more ligands and begin to compete with the bound enzyme for the ligands. Finally, the bound enzyme is displaced by the expanded polymer chains. The temperature-induced elution was quantitative, and first reported in the literature when temperature change was used as the only eluting factor without any changes in buffer composition [69]. Small changes in temperature, as the only eluting factor, are quite promising because there is no need in this case to separate the target protein from an eluent, usually a competing nucleotide or high salt concentration in dye-affinity chromatography.

24.3.3.6 Smart Surfaces—Controlled Porosity, "Chemical Valve"

Environmentally controlled change in macromolecular size from a compact hydrophobic globule to an expanded hydrophilic coil is exploited when smart polymers are used in systems of environmentally controlled porosity, so-called "chemical valves." When a smart polymer is grafted to the surface of the pores in a porous membrane or chromatographic matrix, the transition in the macromolecule affects the total free volume of the pores available for the solvent and hence presents a means to regulate the porosity of the system.

Membranes of pH-sensitive permeability were constructed by grafting smart polymers such as poly(methacrylic acid) [70], poly(benzyl glutamate), poly(2-ethylacrylic acid) [71], poly(4-vinylpyridine) [72], which change their conformation in response to pH. Thermosensitive chemical valves have been developed by grafting poly(*N*-acryloylpyrrolidine), poly(*N*-*n*-propylacrylamide), or poly(acryloylpiperidine) [73], poly(NIPAAM) alone [33,74] or in copolymers with poly(methacrylic acid) [74] inside the pores. For example, grafted molecules of poly(benzyl glutamate) at high pH are charged and are in extended conformation. The efficient pore size is reduced, and the flow through the membrane is low ("off-state" of the membrane). As pH decreases, the macromolecules are protonated, lose their charge, and adopt a compact conformation. The efficient pore size and hence the flow through the membrane increases ("on-state" of the membrane) [71]. The fluxes of bigger molecules (dextrans of molecular weights 4400–50,600) across a temperature-sensitive, poly(NIPAAM)-grafted membrane were effectively controlled by temperature, environmental ionic strength, and degree of grafting of the membrane, while the flux of smaller molecules such as mannitol was not affected by temperature even at high degree of membrane grafting [75]. The on–off permeability ratio for different molecules (water, Cl^- ion, choline, insulin, and albumin) ranged between 3 and 10 and increased as molecular weight increased [76]. An even more abrupt change of the on–off permeability ratio was observed for a membrane that had narrow pores formed by heavy ion beams when poly(NIPAAM) or poly(acryloyl-L-proline methyl ester) were grafted [77].

Different stimuli could trigger the transition of the smart polymer, making it possible to produce membranes whose permeabilities respond to these stimuli. When a copolymer of NIPAAM with triphenylmethane leucocianide was grafted to the membrane, it acquires photosensitivity—UV irradiation increases permeation through the membrane [78]. Fully reversible, pH-switchable permselectivity for both cationic and anionic redox-active probe molecules was achieved by depositing composite films formed from multilayers of amine-terminated dendrimers and poly(maleic anhydride-*co*-methylvinyl ether) on gold-coated silicon [79].

When the smart polymer is grafted inside the pores of the chromatographic matrix for gel permeation chromatography, the transition of grafted macromolecules regulates the pore size and as a result, the elution profile of substances of different molecular weights. As the temperature is raised, the substances are eluted progressively earlier, indicating shrinking of the pores of the hydrogel beads composed of cross-linked poly(acrylamide-*co*-*N*-isopropylacrylamide) [80] or porous polymer beads with grafted poly(NIPAAM) [81].

When using a specific biorecognition element, which recognizes specific substances and translates the signal into a change of physicochemical properties, for example, pH, a smart membrane that changes its permeability in response to particular substances can be constructed. Specific insulin release in response to increasing glucose concentration, that is, an artificial pancreas, presents an everlasting challenge to bioengineers. One of the potential solutions is a "chemical valve" (Figure 24.12). The enzyme, glucose oxidase, was used as a biorecognition element, capable of specific oxidation of glucose accompanied by a decrease in pH. The enzyme was immobilized on pH-responsive poly(acrylic acid) graft on a porous polycarbonate membrane. In neutral conditions, polymer chains are densely charged and have extended conformation that prevents insulin transport through the membrane by blocking the pores. Under exposure to glucose, the pH drops as the result of glucose oxidation by the immobilized enzyme, the polymer chains adopt a more compact conformation that diminishes the blockage of the pores, and insulin is transported through the membrane [82]. Systems such as this could be used for efficient drug delivery that responds to the needs of the organism. A membrane that consists of poly(2-hydrohyethyl acrylate-*co*-*N*,*N*-diethylaminomethacrylate-*co*-4-trimethylsilylstyrene) undergoes a sharp transition from a shrunken state at pH 6.3 to a swollen state at pH 6.15. The transition between the two states changes the membrane permeability to insulin 42-fold. Copolymer capsules that contain glucose oxidase and insulin increase insulin release fivefold in response to 0.2 M glucose. After glucose removal, the rate of insulin release falls back to the initial value [83].

Alternatively, reversible cross-linking of polymer macromolecules could be used to control the porosity in a system. Two polymers, poly(*m*-acrylamidophenylboronic acid-*co*-vinylpyrrolidone) and poly(vinyl alcohol) form a gel because of strong interactions between boronate groups and the hydroxy groups of poly(vinyl alcohol). When a low molecular weight polyalcohol such as glucose is added to the gel, it competes with poly(vinyl alcohol) for boronate groups. The boronate–poly(vinyl alcohol) complex changes to a boronate–glucose complex that results in eventual dissolution of the gel [84]. In addition to a glucose oxidase-based artificial pancreas, the boronate–poly(vinyl alcohol) system has been used for constructing glucose-sensitive systems for insulin delivery [29,85–87]. The glucose-induced transition from a gel to a sol state drastically increases the release of insulin from the gel. The reversible response to glucose has also been designed using another glucose-sensitive biorecognition element, Concanavalin A, a protein that contains four sites that can bind glucose. Polymers that have glucose groups in the side

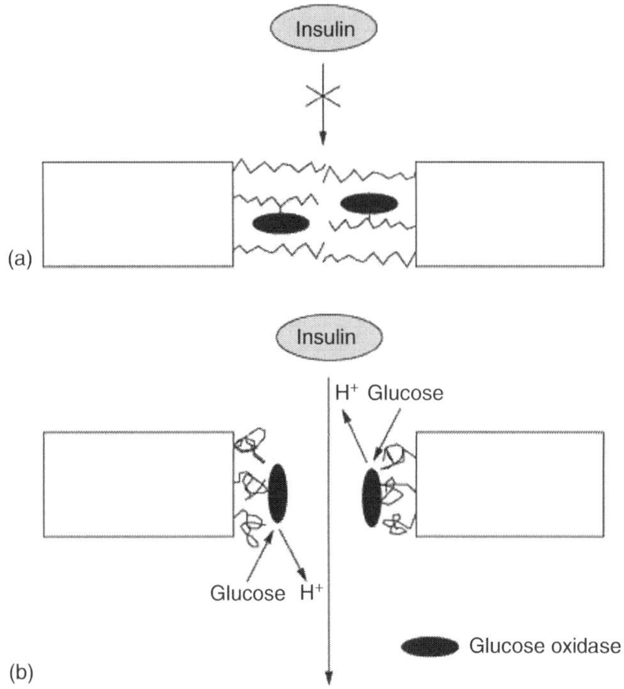

FIGURE 24.12 Schematic of a "chemical valve." Glucose oxidase is immobilized on a pH-responsive polyacrylic acid grafted onto a porous polycarbonate membrane: (a) poly(acrylic acid) is in an expanded conformation that blocks insulin transport; (b) the oxidation of glucose is accompanied by a decrease in pH and the transition of poly(acrylic cid) into a compact conformation that results in opening of the pores and transport of insulin. (Redrawn from Imanishi, Y. and Ito, Y., *Pure Appl. Chem.*, 67, 2015, 1995.)

chain such as poly(vinylpyrrolidone-*co*-allylglucose) [26] or poly(glucosyloxyethyl methacrylate) [88], are reversibly cross-linked by Concanavalin A and form a gel. The addition of glucose results in displacing the glucose-bearing polymer from the complex with Concanavalin A and dissolving the gel.

Reversible gel-formation of thermosensitive block copolymers in response to temperature could be utilized in different applications. Poly(NIPAAM) block copolymers with poly(ethylene oxide), which undergo a temperature-induced reversible gel–sol transition, were patented as the basis for cosmetics such as depilatories and bleaching agents [89]. The copolymer solution is liquid at room temperature and easily applied to the skin where it forms a gel within 1 min. Commercially available ethyl(hydroxyethyl)celluloses that have cloud points of 65°C–70°C have been used as redeposition agents in washing powders. Adsorption of the precipitated polymer on the laundry during the initial rinsing period counteracts readsorption of dirt when the detergent is diluted [90].

24.3.3.7 Liposomes that Trigger Release of the Contents

When a smart polymer is attached somehow to a lipid membrane, the transition in the macromolecule affects the properties of the membrane and renders the system sensitive to environmental changes. To attach a smart polymer to a lipid membrane, a suitable "anchor," which could be incorporated in the membrane, should be introduced into the macromolecule. This could be achieved by copolymerizing poly(NIPAAM) with comonomers that have large hydrophobic tails such as *N,N*-didodecylacrylamide [91], using a lipophilic radical initiator [92] modifying copolymers [93], or polymers that have terminally active groups [94] with a phospholipid. Alternatively, smart polymers have been covalently coupled to the active groups in the hydrophilic heads of the lipid-forming membrane [95].

Interesting and practically relevant materials for studying the behavior of smart polymers attached to lipid membranes, are liposomes, self-assembled 50–200 nm vesicles that have one or more (phospho)lipid bilayers, which encapsulate a fraction of the solvent. Liposomes are stable in aqueous suspension due to the repulsive forces that appear when two liposomes approach each other. Liposomes are widely used for drug delivery and in cosmetics [96].

The results of a temperature-induced conformational transition of a smart polymer on the liposomal surface depend significantly on the fluidity of the liposomal membrane. When the membrane is in a fluid state at temperatures both above and below the polymer transition temperature, the collapse of the polymer molecule forces anchor groups to move closer together by lateral diffusion within the membrane. The compact globules of collapsed polymer cover only a small part of the liposomal surface. Such liposomes have a low tendency to aggregate because the most of their surface is not covered by the polymer. Naked surfaces contribute to the repulsion between liposomes. On the other hand, when the liposomal membrane is in a solid state at temperatures both above and below the polymer transition temperature, the lateral diffusion of anchor groups is impossible, and the collapsed polymer cannot adopt a compact globule conformation but spreads over the most of the liposomal surface [97]. Liposomes whose surfaces are covered to a large degree by a collapsed polymer repel each other less efficiently than intact liposomes. The stability of a liposomal suspension is thereby decreased, and aggregation and fusion of liposomes takes place, which is often accompanied by the release of the liposomal content into the surrounding medium [98].

When the liposomal membrane is perturbed by the conformational transition of the polymer, both the aggregation tendency and liposomal permeability for incorporated substances are affected. Poly(ethacrylic acid) undergoes a transition from an expanded to a compact conformation in the physiological pH range of 7.4–6.5 [99]. The pH-induced transition of poly(ethacrylic acid) covalently coupled to the surface of liposomes formed from phosphatidylcholine results in liposomal reorganization into more compact micelles and concomitant release of the liposomal content into the external medium. The temperature-induced transition of poly(NIPAAM-*co*-*N*,*N*-didocecylacrylamide) [100] or poly(NIPAAM-*co*-octadecylacrylate) [101], incorporated into the liposomal membrane, enhanced the release of the fluorescent marker, calcein, encapsulated in copolymer-coated liposomes. Liposomes hardly release any marker at temperatures below 32°C (the polymer transition temperature), whereas the liposomal content is released completely within less than 1 min at 40°C. To increase the speed of liposomal response to temperature change, the smart polymer was attached to the outer and inner sides of the lipid membrane. The polymer bound only to the outer surface if the liposomes were treated with the polymer after liposomal formation. When the liposomes were formed directly from the lipid–polymer mixture, the polymer was present on both sides of the liposomal membrane [91].

Changes of liposomal surface properties caused by polymer collapse affect liposomal interaction with cells. Liposomes modified by a pH-sensitive polymer, partially succinilated poly(glycidol), deliver calcein into cultured kidney cells of the African green monkey more efficiently compared to liposomes not treated with the polymer [102]. Polymeric micelles formed by smart polymers and liposomes modified by smart polymers could be used for targeted drug delivery. Polymeric micelles have been prepared from amphiphilic block copolymers of styrene (forming a hydrophobic core) and NIPAAM (forming a thermosensitive outer shell). The polymeric micelles were very stable in aqueous media and had long blood circulation because of small diameter, unimodal size distribution (24 ± 4 nm), and, a low critical micellar concentration of around 10 μg/mL. At temperatures above the polymer transition temperature (32°C), the polymer chains that form an outer shell collapse, become more hydrophobic, and allow aggregation between micelles and favoring binding interactions with the surface of cell membranes. Thus, hydrophobic molecules incorporated into the micelles are delivered into the cell membranes. These micelles are capable of site-specific delivery of drugs to the sites as temperature changes, for example, to inflammation sites of increased temperature [103].

24.3.3.8 Smart Polymers in Bioanalytical Systems

Because smart polymers can recognize small changes in environmental properties and respond to them in a pronounced way, they could be used directly as sensors of these changes, for example, a series of polymer solutions that have different LCSTs could be used as a simple thermometer. As salts promote hydrophobic interactions and decrease the LCST, the polymer system could "sense" the salt concentration needed to decrease the LCST below room temperature. A poly(NIPAAM)-based system that can sense NaCl concentrations above 1.5% was patented [104]. The response of the polymer is controlled by a balance of hydrophilic and hydrophobic interactions in the macromolecule. Using a recognition element that can sense external stimuli and translate the signal into the changes of the hydrophilic balance of the smart polymer, the resulting system presents a sensor for the stimulus. If the conjugate of a smart polymer and a recognition element has a transition temperature T_1 in the absence and T_2 in the presence of stimuli, fixing the temperature T in the range $T_1 < T < T_2$ allows achieving the transition of a smart polymer isothermally by the external stimulus [105]. An example of such a sensor was constructed using *trans–cis* isomerization of the azobenzene chromophore when irradiated by UV light. The transition is accomplished by an increase in the dipole moment of azobenzene from 0.5 D (for the *trans*-form) to 3.1 D (for the *cis*-form) and hence a significant decrease of hydrophobicity. Irradiation with UV light results in increasing the LCST from 19.4°C to 26.0°C for the conjugate of the chromophore with poly(NIPAAM). The solution of the conjugate is turbid at 19.4°C < T < 26.0°C, but when irradiated, the conjugate dissolves because the *cis*-form is below the LCST at this temperature. The system responds to UV light by transition from a turbid to transparent solution. The termination of UV irradiation results in a slow return of the system to its initial turbid state [105]. A few other light-sensitive systems were proposed that use different chromophores: triarylmethylcyanide [106] and leuconitriles [107].

The hydrophobicity of the recognition molecule was also changed by chemical signals. Poly(NIPAAM) containing 11.6 mol% of crown ether 9 has a LCST of 31.5°C in the absence of Na^+ or K^+ ions, 32°C in the presence of Na^+, and 38.9°C in the presence of K^+. Thus, the introduction of both Na^+ and K^+ ions leads to the dissolution of the insoluble polymer at that temperature. At 37°C, this effect is achieved only by K^+ ions [108].

From better understanding of ligand–host interactions and development of new highly selective binding pairs (e.g., by using combinatorial libraries to find ligands of high affinity for particular biomolecules), one could expect that smart polymer systems will be used as "signal amplifiers" to visualize a physicochemical event, which takes place in a recognition element, by a pronounced change in the system—conversion of a transparent solution into a turbid one or vice versa.

Antibody–antigen interactions present nearly ideal analytical selectivity and sensitivity developed by nature. Not surprisingly, they are increasingly used for a broad variety of bioanalytical applications. Different analytical formats have been developed. The common feature of the most of them is the requirement for separating an antibody–antigen complex from a nonbound antibody or antigen. Traditionally, the separation is achieved by coupling one of the components of antibody–antigen pair to a solid support. The binding step is followed by washing nonbound material. Interactions of the soluble partner of the binding pair with the partner coupled to the support are often accompanied by undesired diffusional limitations, and hence, incubation times of several hours are required for analysis. Because smart polymers can undergo transition from the soluble to the insoluble state, they allow combining the advantages of homogeneous binding and, after the phase transition of the smart polymer has taken place, easy separation of the polymer precipitate from the supernatant. The essential features of an immunoassay that uses smart polymers (named PRECIPIA) are as follows. The covalent conjugate of poly(NIPAAM) with monoclonal antibodies to the κ-chain of human immunoglobulin G (IgG) are incubated for 1 h at room temperature (below the LCST of the conjugate), and the IgG solution is analyzed. Then plain poly(NIPAAM) (to facilitate thermoprecipitation of polymer–antibody conjugates) and fluorescently labeled monoclonal antibodies to the γ-chain of human IgG are added. The temperature is raised to 45°C, the precipitated polymer is separated by centrifugation, and fluorescence is measured in the supernatant [109]. Immunoassay systems that use temperature-induced precipitation of poly(NIPAAM) conjugates with monoclonal antibodies are not inferior in sensitivity to the traditional heterogenous immunoassay methods, but because the antigen–antibody interaction takes place in solution, the incubation can be shortened to about 1 h [110,111]. The limitations of PRECIPIA as an immunoassay technique are essentially the same as those of affinity precipitation, namely, nonspecific coprecipitation of analyzed protein when poly(NIPAAM) precipitates. Polyelectrolyte complexes that have a low degree of nonspecific protein coprecipitation have also been successfully used as reversibly soluble carriers for PRECIPIA-type immunoassays [112]. The conjugate of antibody and polyanion poly(methacrylic acid) binds to the antigen within a few minutes, and the polymer hardly exerts any effect on the rate of antigen–antibody binding. Subsequent addition of a polycation, poly(*N*-ethyl-4-vinyl-pyridinium bromide) in conditions where the polyelectrolyte is insoluble, results in quantitative precipitation of the antibody–polymer conjugate within 1 min. The total assay time is less than 15 min [10].

In principle, PRECIPIA-type immunoassays could be used for simultaneous assay of different analytes in one sample, provided that conjugates specific toward these analytes are coupled covalently to different smart polymers that have different precipitating conditions, for example, precipitation of one conjugate by adding a polymeric counterion followed by thermoprecipitation of the second conjugate by increasing temperature.

24.3.3.9 Reversibly Soluble Biocatalysts

The transition between the soluble and insoluble state of stimuli-responsive polymers has been used to develop reversibly soluble biocatalysts. A reversibly soluble biocatalyst catalyzes an enzymatic reaction in a soluble state and hence could be used in reactions of insoluble or poorly soluble substrates or products. As soon as the reaction is completed and the products are separated, the conditions (pH, temperature) are changed to promote precipitation of the biocatalyst. The precipitated biocatalyst is separated and can be used in the next cycle after dissolution. The reversibly soluble biocatalyst acquires the advantages of immobilized enzymes (ease of separation from the reaction mixture after the reaction is completed and the possibility for biocatalyst recovery and repeated use in many reaction cycles) but at the same time overcomes the disadvantages of enzymes immobilized onto solid matrices such as diffusional limitations and the impossibility of using them in reactions of insoluble substrates or products.

Biocatalysts that are reversibly soluble as a function of pH have been obtained by the covalent coupling of lysozyme to alginate [113]; of trypsin to poly(acrolein-*co*-acrylic acid) [114]; and of cellulase [115]; amylase [115]; α-chymotrypsin, and papain [116] to poly(methylmethacrylate-*co*-methacrylic acid). A reversibly soluble cofactor has been produced by the covalent binding of NAD to alginate [117]. Reversibly soluble α-chymotrypsin, penicillin acylase, and alcohol dehydrogenase were produced by coupling to the polycation component of polyelectrolyte complexes formed by poly(methacrylic acid) and poly(*N*-ethyl-4-vinyl-pyridinium bromide) [118].

Biocatalysts that are reversibly soluble as a function of temperature have been obtained by the covalent coupling of α-chymotrypsin and penicillin acylase to a partially hydrolysed poly(*N*-vinylcaprolactam) [119], and of trypsin [120], alkaline phosphatase [121], α-chymotrypsin [122], and thermolysin [123,124] to NIPAAM copolymers that contain active groups suitable for covalent coupling of biomolecules. Lipase was coupled to a graft copolymer composed of NIPAAM grafts on a poly(acrylamide-*co*-acrylic acid) copolymer [125]. No significant differences in biocatalytic properties were found for α-amylase coupled to poly(NIPAAM) via single-point or multipoint mode. Both enzyme preparations demonstrated increased thermostability and the absence of diffusional limitation when hydrolyzing starch, a high molecular weight substrate [126]. The temperature of a protein–ligand interaction was controlled by site-directed coupling of terminally modified poly(NIPAAM) to a specifically constructed site (close to a biotin binding site) on a genetically modified streptavidin [127].

Biocatalysts, which are reversibly soluble as a function of Ca^{2+} concentration, were produced by covalent coupling of phosphoglyceromutase, enolase, peroxidase, and pyruvate kinase to α_{s1}-casein. The enzyme casein conjugates are soluble at a Ca^{2+} concentration below 20 mM but precipitate completely at a Ca^{2+} concentration above 50 mM. The precipitate redissolves when EDTA, a strong Ca^{2+}-binding agent is added [128].

The reversible flocculation of lattices has been used to produce thermosensitive reversibly soluble (more precisely reversibly dispersible) biocatalysts using trypsin [129], papain [130], and α-amylase [131]. Lattices sensitive to a magnetic field have been used to immobilize trypsin and β-galactosidase [132]. Liposomes that have a polymerized membrane, which reversibly aggregates on changing salt concentration, have been used to immobilize α-chymotrypsin [133].

The most attractive application of reversibly soluble biocatalysts is repeated use in a reaction, which is difficult or even impossible to carry out using enzymes immobilized onto insoluble matrices, for example, hydrolysis of water-insoluble phlorizidin [134]; hydrolysis of high molecular weight substrates such as casein [123,130] and starch [115]; hydrolysis of insoluble substrates such as cellulose [135] and raw starch (corn flour) [7,134,136–138]; production of insoluble products such as peptide, benzyloxycarbonyl-L-tyrosyl-N^{ω}-nitro-L-arginine [116] and phenylglycine [139].

The hydrolytic cleavage of corn flour to glucose is an example of successfully using a reversibly soluble biocatalyst, amylase coupled to poly(methylmethacrylate-*co*-methacrylic acid), in an industrially interesting process [136]. The reaction product of the process, glucose, inhibits the hydrolysis. The use of a reversibly soluble biocatalyst improves the efficiency of the hydrolysis, which is carried out at pH 5, at which the amylase–polymer conjugate is soluble. After each 24h, the pH is reduced to 3.5, the unhydrolyzed solid residue and the precipitated conjugate are separated by centrifugation, the conjugate is resuspended in a fresh portion of the substrate at pH 5, and the hydrolysis is continued. The conversion achieved after five cycles is 67%, and the activity of the amylase after the fifth cycle was 96% of the initial value [136].

24.4 Conclusion

In the future, one looks forward to further developments and the commercial introduction of new smart polymers whose transition temperatures and pH are compatible with physiological conditions or conditions for maximal stability of target biomolecules, such as temperatures of 4–15°C and pH values of 5–8. Additional prospects will stem from a better understanding of the mechanism of cooperative interactions in polymers and increasing knowledge of structure–property correlations to enable rational synthesis of smart polymers that have predefined properties. Due to the possibility of combining a variety of biorecognition or biocatalytic systems and the unique features of smart polymers, expectations are running high in this area. Only time and more experimentation will determine whether smart polymers will live up to their generous promises.

References

1. R. Dagani, *Chem. Eng. News*, pp. 30–33, 1995.
2. H. Brønsted and J. Kopecek, in *Polyelectrolyte Gels. Properties, Preparation, and Applications*, R.S. Harland and R.K. Prud'homme, eds., American Chemical Society, Washington, DC, 1992, pp. 285–304.
3. V.A. Kabanov, *Polym. Sci.* 36: 143–156, 1994.
4. H.G. Schild, *Prog. Polym. Sci.* 17: 163–249, 1992.
5. E.Y. Kramarenko and A.R. Khokhlov, *Polym. Gels Networks* 6: 46–56, 1998.
6. Eudragit, Röhm Pharma GMBH Information Materials, 1993.
7. K. Hoshino, H. Yamasaki, C. Chida, S. Morohashi, M. Taniguchi, and M. Fujii, *J. Chem. Eng. Jpn.* 30: 30–37, 1997.
8. E. Linné-Larsson and B. Mattiasson, *Biotechnol. Tech.* 8: 51–56, 1994.
9. S.Y. Park and Y.H. Bae, *Macromol. Rapid Commun.* 20: 269–273, 1999.
10. V.A. Izumrudov, I.Yu. Galaev, and B. Mattiasson, *Bioseparation* 7: 207–220, 1999.
11. M.B. Dainiak, V.A. Izumrudov, V.I. Muronetz, I.Yu. Galaev, and B. Mattiasson, *Bioseparation* 7: 231–240, 1999.
12. U. Dissing and B. Mattiasson, *J. Biotechnol.* 52: 1–10, 1996.
13. A.A. Tager, A.P. Safronov, S.V. Sharina, and I.Yu. Galaev, *Colloid Polym. Sci.* 271, 1993.
14. M. Heskins and J.E. Guillet, *J. Macromol. Sci. Chem.* A2: 1441–1455, 1968.
15. K. Suwa, K. Morishita, A. Kishida, and M. Akashi, *J. Polym. Sci. Part A: Polym. Chem.* 35: 3087–3094, 1997.
16. I. Yu. Galaev and B. Mattiasson, *Enzyme Microb. Technol.* 15: 354–366, 1993.
17. F. Liu, F.H. Liu, R.X. Zhuo, Y. Peng, Y.Z. Deng, and Y. Zeng, *Biotechnol. Appl. Biochem.* 21: 257–264, 1995.
18. S. Mori, Y. Nakata, and H. Endo, *Protein Expression Purif.* 5: 151–156, 1994.
19. F.M. Winnik, A.D. Davidson, G.H. Hamer, and H. Kitano, *Macromolecules* 25: 1876–1880, 1992.
20. G. Chen and A.S. Hoffman, 373: 49–52, 1995.
21. M. Maeda, C. Nishimura, A. Inenaga, and M. Takagi, *Reactive Functional Polym.* 21: 27–35, 1993.
22. D. Umeno, T. Mori, and M. Maeda, *Chem. Commun.* 1433–1434, 1998.
23. B. Mattiasson, A. Kumar, and I.Yu. Galaev, *J. Mol. Recognition* 11: 211–216, 1998.
24. A. Kumar, I.Yu. Galaev, and B. Mattiasson, *Biotechnol. Bioeng.* 59: 695–704, 1998.
25. A. Durand and D. Hourdet, *Polymer* 40: 4941–4951, 1999.
26. S.J. Lee and K. Park, *J. Mol. Recognition* 9: 549–557, 1997.
27. K.-Y.A. Wu and K.D. Wisecarver, *Biotechnol. Bioeng.* 39: 447–449, 1992.
28. E. Kokufuta and S. Matsukawa, *Macromolecules* 28: 3474–3475, 1995.
29. S. Kitano, I. Hisamitsu, Y. Koyama, K. Kataoka, T. Okano, M. Yokoyama, and Y. Sakurai, Proc. 1st International Conference on Intelligent Materials, 1993, pp. 383–388.
30. T. Yakushiji, K. Sakai, A. Kikuchi, T. Aoyagi, Y. Sakurai, and T. Okano, *Langmuir* 14: 4657–4662, 1998.
31. H. Kawaguchi, in *Biomedical Functions and Biotechnology of Natural and Artificial Polymers*, M. Yalpani, ed., ATL Press, 1996, pp. 157–168.

32. H. Kubota and N. Shiobara, *Reactive Functional Polym.* 37: 219–224, 1998.
33. Y.M. Lee and J.K. Shim, *Polymer* 38: 1227–1232, 1997.
34. H.A. VonRecum, S.W. Kim, A. Kikuchi, M. Okuhara, Y. Sakurai, and T. Okano, *J. Biomed. Mater. Res.* 40: 631–639, 1998.
35. Y. Sun, K. Yu, X.H. Jin, and X.Z. Zhou, *Biotechnol. Bioeng.* 47: 20–25, 1995.
36. I. Yu. Galaev, M.N. Gupta, and B. Mattiasson, *Chemtech* 19–25, 1996.
37. A. Kumar, I.Yu. Galaev, and B. Mattiasson, *Bioseparation* 7: 185–194, 1999.
38. D. Guoqiang, M.A.N. Benhura, R. Kaul, and B. Mattiasson, *Biotechnol. Prog.* 11: 187–193, 1995.
39. D. Guoqiang, R. Batru, R. Kaul, M.N. Gupta, and B. Mattiasson, *Bioseparation* 5: 339–350, 1995.
40. E. Linné-Larsson, I.Yu. Galaev, and B. Mattiasson, *Bioseparation* 6: 283–291, 1996.
41. A.P. Bradshaw and R.J. Sturgeon, *Biotechnol Tech.* 4: 67–71, 1990.
42. M.B. Dainiak, V.A. Izumrudov, V.I. Muronetz, I.Yu. Galaev, and B. Mattiason, *Anal. Biochem.* (submitted).
43. M.B. Dainiak, V.A. Izumrudov, V.I. Muronetz, I.Yu. Galaev, and B. Mattiason, *Biochim. Biophys. Acta* 1381: 279–285, 1998.
44. A. Kondo, T. Kaneko, and K. Higashitani, *Biotechnol. Bioeng.* 44: 1–6, 1994.
45. C.S. Chen, C.K. Lee, C.Y. Chen, and M.J. Yeh, *Colloids Surf. B: Biointerfaces* 6: 37–49, 1996.
46. K. Hoshino, M. Taniguchi, T. Kitao, S. Morohashi, and T. Sasakura, *Biotechnol. Bioeng.* 60: 568–579, 1998.
47. P.-Å. Albertson, *Partition of Cell Particles and Macromolecules*. Wiley, New York, 1986.
48. P. Hughes and C.R. Lowe, *Enzyme Microb. Technol.* 10: 115–122, 1988.
49. G. Johansson and F. Tjerneld, in *Highly Selective Separations in Biotechnology*, G. Street, ed., Blackie Academic & Professional, London, 1994, pp. 55–85.
50. M. Lu, P.-Å. Albertsson, G. Johansson, and F. Tjerneld, *J. Chromatogr. B* 680: 65–70, 1996.
51. T.T. Franko, I.Yu. Galaev, R. Hatti-Kaul, N. Holmberg, L. Bülow, and B. Mattiasson, *Biotechnol. Tech.* 11: 231–235, 1997.
52. J. Persson, H.-O. Johansson, I.Yu. Galaev, B. Mattiasson, and F. Tjerneld, *Bioseparation* 9: 105–116, 2000.
53. Y.G. Takei, T. Aoki, K. Sanui, N. Ogata, Y. Sakurai, and T. Okano, *Macromolecules* 27: 6163–6166, 1994.
54. T. Kondo, M. Koyama, H. Kubota, and R. Katakai, *J. Appl. Polym. Sci.* 67: 2057–2064, 1998.
55. F.-J. Schmitt, C. Park, J. Simon, H. Ringsdorf, and J. Israelachvili, *Langmuir* 14: 2838–2845, 1998.
56. P.W. Zhu and D.H. Napper, *J. Phys. Chem. B* 101: 3155–3160, 1997.
57. N. Yamada, T. Okano, H. Sakai, F. Karikusa, Y. Sawasaki, and Y. Sakurai, *Makromol. Chem. Rapid Commun.* 11: 571–576, 1990.
58. T. Okano, N. Yamada, M. Okuhara, H. Sakai, and Y. Sakurai, *Biomaterials* 16: 297–303, 1995.
59. T. Okano, A. Kikuchi, Y. Sakurai, Y.G. Takei, T. Aoki, and N. Ogata, *J. Controlled Release* 36: 125–133, 1995.
60. M. Yamato, M. Okuhara, F. Karikusa, A. Kikuchi, Y. Sakurai, and T. Okano, *J. Biomed. Mater. Res.* 44: 44–52, 1999.
61. Y. Ito, G. Chen, Y. Guan, and Y. Imanishi, *Langmuir* 13: 2756–2759, 1997.
62. T. Bohanon, G. Elender, W. Knoll, P. Koeberle, J. Lee, A. Offenhaeusser, H. Ringsdorf, E. Sackmann, and J. Simon, *J. Biomater. Sci. Polym. Ed.*, 8: 19–39, 1996.
63. H. Kanazawa, Y. Kashiwase, K. Yamamoto, Y. Matsushima, A. Kikuchi, Y. Sakurai, and T. Okano, *Anal. Chem.* 69: 823–830, 1997.
64. K. Hosoya, K. Kimata, T. Araki, and N. Tanaka, *Anal. Chem.* 67: 1907–1911, 1995.
65. H. Go, Y. Sudo, K. Hosoya, T. Ikegami, and N. Tanaka, *Anal. Chem.* 70: 4086–4093, 1998.
66. H. Kanazawa, Y. Matsushima, and T. Okano, *Trends Anal. Chem.* 17: 435–440, 1998.
67. H. Yoshioka, M. Mikami, T. Nakai, and Y. Mori, *Polym. Adv. Technol.* 6: 418–420, 1995.
68. B. Mattiasson and I. Yu. Galaev, in *High Resolution Chromatography*, P.A. Millner, ed., IRL Press, 1996, pp. 153–190.
69. I.Yu. Galaev, C. Warrol, and B. Mattiasson, *J. Chromatogr. A* 684: 37–43, 1994.
70. Y. Ito, Y.S. Park, and Y. Imanishi, *Macromol. Rapid. Commun.* 18: 221–224, 1997.
71. Y. Ito, Y. Ochiai, Y.S. Park, and Y. Imanishi, *J. Am. Chem. Soc.* 119: 1619–1623, 1997.
72. Y. Okahata, K. Ozaki, and T. Seki, *J. Chem. Soc. Chem. Commun.* 519–521, 1984.
73. Y. Okahata, H. Noguchi, and T. Seki, *Macromolecules* 19: 493–494, 1986.
74. T. Nonaka, K. Hashimoto, and S. Kurihara, *J. Appl. Polym. Sci.* 66: 209–216, 1997.
75. S. Akerman, P. Viinikka, B. Svarfvar, K. Putkonen, K. Jarvinen, K. Kontturi, J. Nasman, A. Urtti, and P. Paronen, *Int. J. Pharm.* 164: 29–36, 1998.
76. R. Spohr, N. Reber, A. Wolf, G.M. Alder, V. Ang, C.L. Bashford, C.A. Pasternak, O. Hideki, and M. Yoshida, *J. Controlled Release* 50: 1–11, 1998.
77. H. Omichi, M. Yoshida, M. Asano, N. Nagaoka, H. Kubota, R. Katakai, R. Spohr, N. Reber, A. Wolf, G.M. Alder, V. Ang, C.L. Bashford, and C.A. Pasternak, *Nucl. Instrum. Methods Phys. Res. Sect. B* 131: 350–356, 1997.
78. S. Kurihara, Y. Ueno, and T. Nonaka, *J. Appl. Polym. Sci.* 67: 1931–1937, 1998.
79. Y. Liu, M. Zhao, D.E. Bergbreiter, and R.M. Crooks, *J. Am. Chem. Soc.* 119: 8720–8721, 1997.
80. T.G. Park and A.S. Hoffman, *Biotechnol. Prog.* 10: 82–86, 1994.
81. K. Hosoya, E. Sawada, K. Kimata, T. Araki, N. Tanaka, and J.M.J. Frechet, *Macromolecules* 27: 3973–3976, 1994.

82. Y. Imanishi and Y. Ito, *Pure Appl. Chem.* 67: 2015–2021, 1995.
83. K. Ishihara and K. Matsui, *J. Polym. Sci. Polym. Lett.* ed. 24: 413–417, 1986.
84. S. Kitano, K. Kataoka, Y. Koyama, T. Okano, and Y. Sakurai, *Makromol. Chem., Rapid Commun.* 12: 227–233, 1991.
85. K. Kataoka, H. Miyazaki, B. Miyazaki, T. Okano, and Y. Sakurai, *J. Am. Chem. Soc.* 120: 12694–12695, 1998.
86. S. Kitano, Y. Koyama, K. Kataoka, T. Okano, and Y. Sakurai, *J. Controlled Release* 19: 162–170, 1992.
87. D. Shiino, K. Kataoka, Y. Koyama, M. Yokoyama, T. Okano, and Y. Sakurai, *J. Controlled Release* 28: 317–319, 1994.
88. K. Nakamae, T. Miyata, A. Jikihara, and A.S. Hoffman, *J. Biomater. Sci. Polymer Ed.* 6: 79–90, 1994.
89. Y. Mori, M. Yoshioka, and T. Mukoyama, Cosmetic bases and cosmetics containing polymers with sol–gel transition temperature, *Jpn. Kokai Tokkyo Koho* JP 09227329 A2, Sep. 2, 1997.
90. K. Lindell, Doctoral Thesis, Lund University, Sweden, 1996.
91. H. Hayashi, K. Kono, and T. Takagishi, *Bioconjugate Chem.* 10: 412–418, 1999.
92. H. Ringsdorf, J. Simon, and F.M. Winnik, in *Colloid–Polymer Intercations. Particulate, Amphiphilic, and Biological Surfaces*, P.L. Dubin and P. Tong, eds., American Chemical Society, Washington DC, 1993, pp. 216–240.
93. X.S. Wu, A.S. Hoffman, and P. Yager, *Polymer* 33: 4659–4662, 1992.
94. K. Kono, R. Nakai, K. Morimoto, and T. Takagishi, *Biochim. Biophys. Acta* 1416: 239–250, 1999.
95. D.A. Tirrell, in *Pulsed and Self-Regulated Drug Delivery*, J. Kost, ed., CRC press, Boca Raton, FL, 1990, pp. 110–116.
96. D.D. Lasic, *Trends Biotechnol.* 16: 307–321, 1998.
97. H. Ringsdorf, J. Venzmer, and F.M. Winnik, *Angew. Chem. Int. Ed. Engl.* 30: 315–318, 1991.
98. H. Hayashi, K. Kono, and T. Takagishi, *Bioconjugate Chem.* 9: 382–389, 1998.
99. S. Sugaj, K. Nitta, N. Ohno, and H. Nakano, *Colloid Polym. Chem.* 261: 159–165, 1983.
100. K. Kono, A. Henmi, H. Yamashita, H. Hayashi, and T. Takagishi, *J. Controlled Release* 59: 63–75, 1999.
101. H. Hayashi, K. Kono, and T. Takagishi, *Biochim. Biophys. Acta* 1280: 127–134, 1996.
102. K. Kono, T. Igawa, and T. Takagishi, *Biochim. Biophys. Acta.* 1325: 143–154, 1997.
103. S. Cammas, K. Suzuki, C. Sone, Y. Sakurai, K. Kataoka, and T. Okano, *J. Controlled Rel.* 48: 157–164, 1997.
104. Y. Tomita, Sensors for concentrations of salts, *Jpn. Kokai Tokkyo Koho* JP 63206653 A2 Aug. 25, 1988.
105. M. Irie, in *Responsive Gels: Volume Transitions II*, K. Dusek, ed., Springer-Verlag, Berlin, 1993, pp. 50–65.
106. R. Kröger, H. Menzel, and M.L. Hallensleben, *Macromol. Chem. Phys.* 195: 2291–2298, 1994.
107. A. Mamada, T. Tanaka, D. Kungwatchakun, and M. Irie, *Macromolecules* 23: 1517–1519, 1990.
108. M. Irie, Y. Misumi, and T. Tanaka, *Polymer* 34: 4531–4535, 1993.
109. N. Monji and A.S. Hoffman, *Appl. Biochem. Biotechnol.* 14: 107–120, 1987.
110. Q.-Z. Zhu, F. Liu, J.-G. Xu, W. Su, and J. Huang, *Anal. Chim. Acta* 375: 177–185, 1998.
111. Y.G. Takei, M. Matsukata, T. Aoki, K. Sanui, N. Ogata, A. Kikuchi, Y. Sakurai, and T. Okano, *Bioconjugate Chem.* 5: 577–5582, 1994.
112. B.B. Dzantiev, A.N. Blintsov, A.F. Bobkova, V.A. Izumrudov, and A.B. Zezin, Dokl. *Biochem.* (in Russian) 342: 549–552, 1995.
113. M. Charles, R.W. Coughlin, and F.X. Hasselberger, *Biotechnol. Bioeng.* 16: 1553–1556, 1974.
114. E. VanLeemputten and M. Horisberger, *Biotechnol. Bioeng.* 18: 587–590, 1976.
115. M. Taniguchi, K. Hoshino, Y. Netsu, and M. Fujii, *J. Chem. Eng. Jpn.* 22: 313–314, 1989.
116. M. Fujimura, T. Mori, and T. Tosa, *Biotechnol. Bioeng.* 29: 747–752, 1987.
117. R.W. Coughlin, M. Aizawa, and M. Charles, *Biotechnol. Bioeng.* 18: 199–208, 1976.
118. A.L. Margolin, V.A. Izumrudov, V.K. Svedas, and A.B. Zezin, *Biotechnol. Bioeng.* 24: 237–240, 1982.
119. Y.E. Kirsh, I. Yu. Galaev, T.M. Karaputadze, A.L. Margolin, and V.K. Svedas, *Biotechnology* (Russian), 184–189, 1987.
120. H. Lee and T.G. Park, *Biotechnol. Prog.* 14(3): 508–516, 1999.
121. A.L. Nguyen and J.H.T. Luong, *Biotechnol. Bioeng.* 34: 1186–1190, 1989.
122. J.-P. Chen, *J. Chem. Technol. Biotechnol.* 73: 137–143, 1995.
123. F. Liu, G. Tao, and R. Zhuo, *Polymer J.* 25: 561–567, 1993.
124. K. Hoshino, M. Taniguchi, H. Kawaberi, Y. Takeda, S. Morohashi, and T. Sasakura, *J. Ferment. Bioeng.* 83: 246–252, 1997.
125. S. Takeuchi, I. Omodaka, K. Hasegawa, Y. Maeda, and H. Kitano, *Makromol. Chem.* 194: 1991–1999, 1993.
126. J.-P. Chen, D.-H. Chu, and Y.-M. Sun, *J. Chem. Technol. Biotechnol.* 69: 421–428, 1997.
127. P.S. Stayton, T. Shimoboji, C. Long, A. Chilkoti, G. Chen, J.M. Harris, and A.S. Hoffman, *Nature* 378: 472–474, 1995.
128. K. Okumura, K. Ikura, M. Yoshikawa, R. Sasaki, and H. Chiba, *Agric. Biol. Chem.* 48: 2435–2440, 1984.
129. H. Kitano, C. Yan, and K. Nakamura, *Makromol. Chem.* 192: 2915–2923, 1991.
130. A. Kondo, K. Imura, K. Nakama, and K. Higashitani, *J. Ferment. Bioeng.* 78: 241–245, 1994.
131. T. Kondo, T. Urabe, and K. Yoshinaga, *Colloids Surf. A: Physicochem. Eng. Aspects* 109: 129–136, 1996.
132. A. Kondo and H. Fukuda, *J. Ferment. Bioeng.* 84: 337–341, 1997.
133. Y. Sun, X.H. Jin, X.-Y. Dong, K. Yu, and X.Z. Zhou, *Appl. Biochem. Biotechnol.* 56: 331–339, 1996.
134. K. Hoshino, M. Taniguchi, H. Ueoka, M. Ohkuwa, C. Chida, S. Morohashi, and T. Sasakura, *J. Ferment. Bioeng.* 82: 253–258, 1996.

135. M. Taniguchi, M. Kobayashi, and M. Fujii, *Biotechnol. Bioeng.* 34: 1092–1097, 1989.
136. K. Hoshino, M. Taniguchi, Y. Netsu, and M. Fujii, *J. Chem. Eng. Jpn.* 22: 54–59, 1989.
137. K. Hoshino, M. Taniguchi, H. Marumoto, and M. Fujii, *Agric. Biol. Chem.* 53: 1961–1967, 1989.
138. K. Hoshino, M. Taniguchi, H. Marumoto, and M. Fujii, *J. Ferment. Bioeng.* 69: 228–233, 1990.
139. Yu.V. Galaev, *Prikl. Bioxim. Mikrobiol.* 30: 167–170, 1994.
140. I.Yu. Galaev and B. Mattiasson, *Trends Biotechnol.* 17: 335–340, 1999.

25

Vibration Control for Smart Structures

25.1 Vibration Control

Seung-Bok Choi and Young-Min Han

25.1.1 Introduction

The goal of vibration control is to suppress unwanted vibration of various dynamic systems. Traditional vibration control uses passive element to increase stiffness or damping. However, the insatiable demand for high performance quantified by high-speed operation, high control accuracy, and lower energy consumption has triggered vigorous researches on active and semiactive vibration control of distributed flexible structures and discrete systems. Numerous control strategies for conventional electromagnetic actuators have been proposed and implemented to suppress unwanted vibration. However, the successful realization of electromagnetic actuators maybe sometimes very difficult under certain conditions due to hardware limitations such as saturation and response speed. This difficulty can be resolved by employing smart material actuators in vibration control. As is well-known, smart material technology features actuating capability, control capability, and computational capability [1]. Therefore, these inherent capabilities of smart materials can execute specific functions autonomously in response to changing environmental stimuli. Among many smart material candidates, electrorheological (ER) fluids, magnetorheological (MR) fluids, and piezoelectric materials are effectively exploited for vibration control in various engineering applications.

A viable vibration control algorithm can be optimally synthesized by integrating control strategies, actuating technology, and sensing technology, as shown in Figure 25.1. The design philosophy presented in Figure 25.1 contains a very large number of decisions and design parameters for the characteristics of controllers, actuators, and sensors. Furthermore, the designer seeking a global optimal solution for the synthesis of a closed-loop smart structure system must also address other crucial decisions concerning the time delay of a high-voltage/current amplifier, the speed of the signal converter, and the microchip hardware of the control software. This article introduces vibration control methodology of flexible structures and various dynamic systems using smart material actuators. In Section 25.1.2, vibration control techniques for the flexible structure are presented by utilizing smart mount systems fabricated from ER fluids, MR fluids, and piezoelectric materials. In Section 25.1.3, vibration control of seat suspension and vehicle suspension systems under various road conditions is given by adopting ER damper and MR damper, respectively. In addition, piezoelectric shunt technique is presented to suppress vibration of a CD-ROM drive.

25.1.2 Vibration Control of Flexible Structures

25.1.2.1 ER Fluid-Based Mount

Utilizing ER fluids, significant progress has been made in vibration control of flexible structures. This is typically accomplished by two different methods: ER fluid-based smart structure and ER fluid-based mount. Choi et al. initiated ER-fluid-based smart

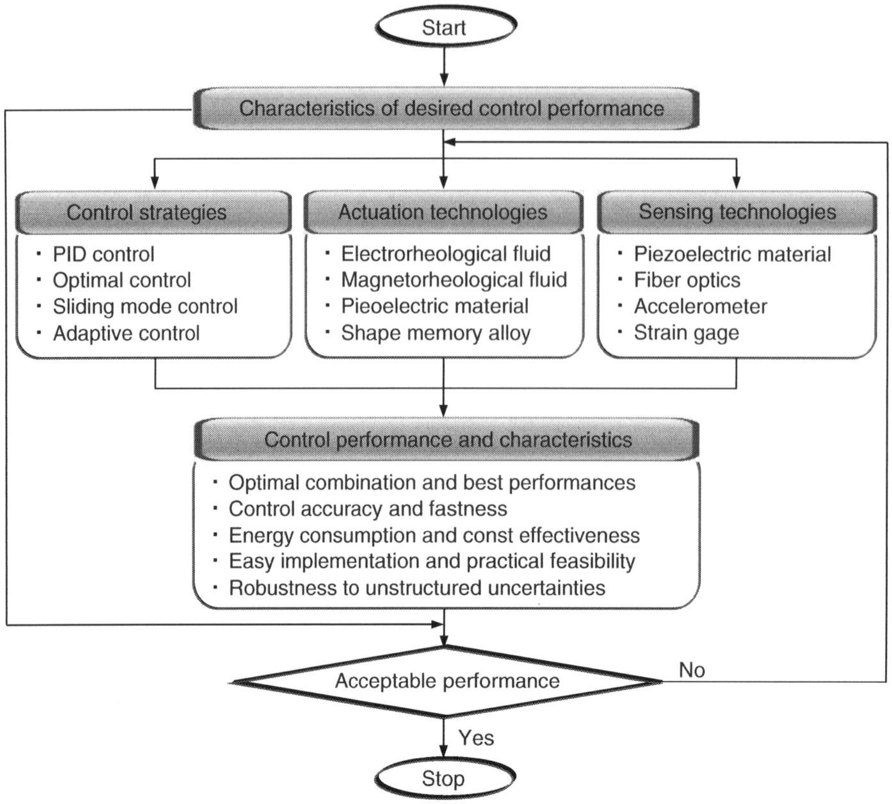

FIGURE 25.1 Flow chart for synthesizing vibration control system using smart material actuators and sensors.

structure, which can control the stiffness and energy dissipation characteristics of the structure [2]. They completed an experimental study of a variety of shear configurations based of sandwich beam structures. So far, numerous researches have been undertaken in this way [3–5]. The idea of applying the ER fluid-based mount to vibration control has been initiated in automotive engineering applications. Petek et al. proposed actively controllable damping characteristics using ER engine mount [6]. Choi et al. experimentally analyzed performances of ER engine mount [7]. This idea can be easily exploited in structural vibration control. Duclos proposed externally tunable hydraulic mount featuring ER fluid [8]. Choi et al. adopted ER mount system to vibration control of frame structure [9]. When the ER fluid-based mount is used for vibration control, operating mode of the mount can be classified by three different types: flow mode, shear mode, and squeeze mode. In the flow mode, it is assumed that two electrodes are fixed, and hence vibration control is achieved by controlling the flow motion between two fixed electrodes [10]. In the shear mode, it is usually assumed that one of two electrodes is free to translate or rotate relative to the other, and hence vibration control is achieved by controlling shear force between the two electrodes [11]. Unlike the former two modes, the electrode gap is varied in the squeeze mode and the ER fluid is squeezed by a normal force. The ER fluid-based mount using this mode is very effective for vibration control of a flexible structure subjected to small amplitude and high-frequency excitations [12].

Figure 25.2 shows the schematic configuration and photograph of the squeeze mode ER fluid-based mount. The lower electrode is fixed to the base plate, while the upper electrode is to be moved up and down. Thus, the squeeze-mode motion of the ER fluid occurs in the vertical direction. The coil spring is attached to support a static mass, which is the mass of a flexible beam structure to be controlled. The total force of the proposed ER mount can be obtained as follows:

$$F(t) = kh(t) + c_f(t)\dot{h}(t) + f_{er}(t) \tag{25.1}$$

In the above, k is the stiffness constant of the coil spring, $h(t)$ is exciting displacement, $c_f(t)$ is the damping coefficient in the absence of the electric field. $f_{er}(t)$ is the controllable damping force owing to the electric field of E and can be expressed as $f_{er}(t) = 4\pi R^3 \tau_y(E)\mathrm{sgn}(\dot{h}(t))/3(h_o + h(t))$. R and h_o are radius of the circular electrode and initial gap between lower and upper electrodes, respectively. $\tau_y(E)$ is the field-dependent yield stress, which is given by αE^β. Here, α and β are intrinsic values of the ER fluid to be experimentally determined.

Figure 25.3 shows the schematic configuration for vibration control of a flexible beam structure using the ER fluid-based mount. A continuous and uniform beam is supported by two

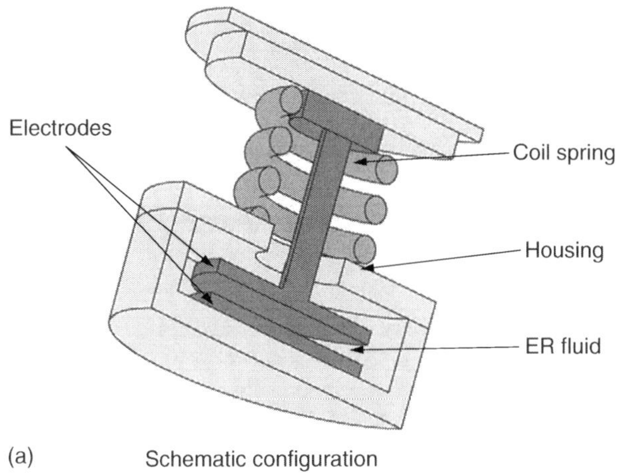

FIGURE 25.2 Squeeze mode ER fluid-based mount.

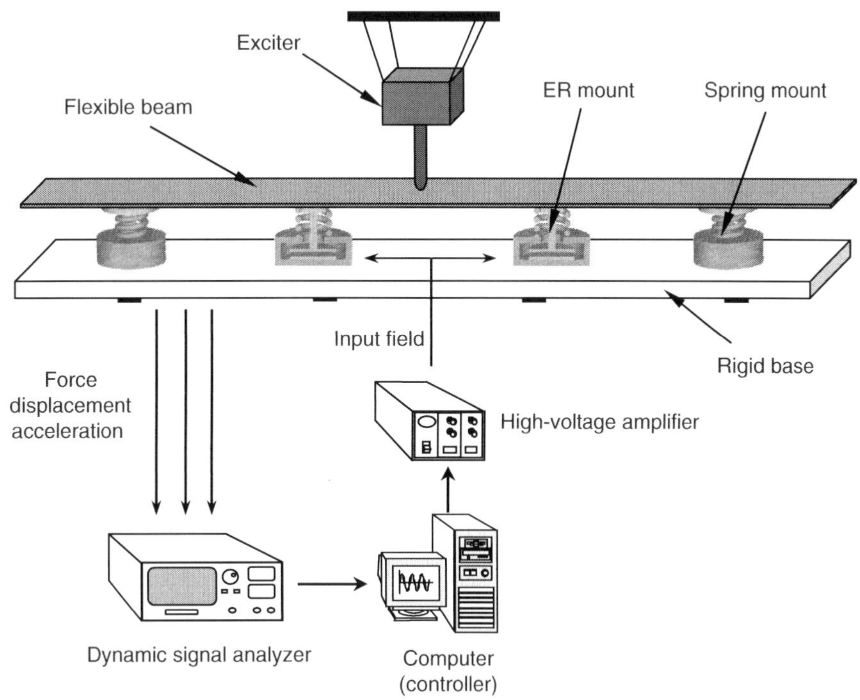

FIGURE 25.3 Experimental configuration for vibration control of a flexible beam structure using ER fluid-based mount.

spring mounts and two ER fluid-based mounts. When the Bernoulli-Eulor beam theory is applied, the kinetic energy, potential energy, and nonconservative work are obtained. By employing energy equations and Hamilton's principle, the governing equations of motion for transverse deflection $y(x,t)$ and associated boundary condition can be obtained [12]. By using the Lagrange equation, a decoupled ordinary equation for each mode is derived by

$$\ddot{q}_i(t) + 2\varsigma_i\omega_i\dot{q}_i(t) + \omega_i^2 q_i(t) = Q_i(t)/I_i + Q_{exi}(t)/I_i \tag{25.2}$$

where

- I_i is the generalized mass
- $Q_i(t)$ is the generalized force including controllable damping force
- $Q_{exi}(t)$ is the generalized force including exciting force
- q_i, ς_i, and ω_i are modal coordinate, damping ratio, and the natural frequency of the ith mode

Considering the boundary conditions of the proposed structural system, the first and second modes are chosen to be dominant for vibration characteristics. The control purpose is to regulate unwanted vibrations of the beam structure with appropriate control input voltage.

Among many potential controller candidates, an optimal controller, which is very effective for vibration control is adopted [13]. In order to implement the controller, an experimental apparatus is established as shown in Figure 25.3. The flexible steel beam is externally excited by the electromagnetic exciter, and displacement sensor (noncontact proximitor) and accelerometer are used to catch the vibration signals. The signals are fed back to the microprocessor through analog/digital (A/D) converter. Depending on the signal information, control voltage is determined by means of the optimal controller. The control voltage is then applied to the ER fluid-based mount through high-voltage amplifiers. On the other hand, the force transmissibility is measured by two force transducers: one for the input force and the other for the output force at mount position (spring or ER fluid-based mount position). Figure 25.4a presents time responses of the vibration signals when the beam structure is excited by the first mode natural frequency of 31.5 Hz [12]. It is clearly seen that displacement at the top of ER fluid-based mount is substantially reduced by activating the proposed optimal controller. It is remarked that uncontrolled response is obtained in the absence of the control voltage. Acceleration levels at the top of ER mount is measured and presented in Figure 25.4b in the frequency domain [12]. It is clearly observed that accelerations at the resonance frequencies are remarkably attenuated by activating the control voltage to the ER-fluid based mount.

25.1.2.2 MR Fluid-Based Mount

MR fluids have very similar characteristics. They are suspensions consisting of magnetizable particles in a low-permeability base fluid instead of polarized particles in nonconducting oil. Their rheological properties are changed by applying magnetic field to the fluid domain instead of electric field. Furthermore, MR fluids use low voltage and high current through a coil to generate a magnetic field whereas ER fluids use low current and high voltage to generate an electric field. The rheological change of MR fluids are primarily observed as a significant increase of the yield shear stress of the fluid. It can be continuously controlled by the intensity of applied magnetic field. By choosing an appropriate operation mode, controllable pressure or shear force can be effectively applied to various dynamic systems. Like ER fluid, MR fluid has been widely applied to vibration control fields. Choi et al. investigated the vibration control performance of MR fluid dampers for vehicle applications [14]. Wereley and Pang studied hysteresis and plug flow models of MR fluid dampers using approximate parallel plate models [15]. Stanway et al. studied various operation modes of MR fluid applications for structural vibration control [16].

In this chapter, a compact MR fluid-based mount, which consists of MR dashpot under mixed mode (flow mode and shear mode) and rubber element, is proposed for vibration control of a flexible beam structure [17]. Its schematic configuration and photograph are presented in Figure 25.5. MR fluid is filled in the gap between piston (or plunger) and outer housing. The electromagnetic coil is wired inside of the outer cylindrical housing to produce magnetic field. The housing can be fixed to the base structure, and the plunger is attached to the top end of the rubber element. This rubber element has a role to support a static mass and isolate the vibration transmission at the nonresonant regions. During the relative motion of the inner cylinder to outer cylindrical housing, the MR fluid flows through annular gap. The cross-sectional area of the plunger is regarded as the effective piston area. If a certain level of magnetic field is applied through the gap, the MR fluid-based mount produces an additional damping force due to the yield stress of the MR fluid. Figure 25.6 shows the measured damping forces in time domain according to applied current [17]. The MR fluid-based mount is placed between the load cell and electromagnetic exciter to measure the field-dependent damping force. When the shaker table moves up and down by a command signal generated from the exciter controller, the MR fluid-based mount produces the damping force and it is measured by the load cell. It is seen from the results that the damping force increases as the input current increases.

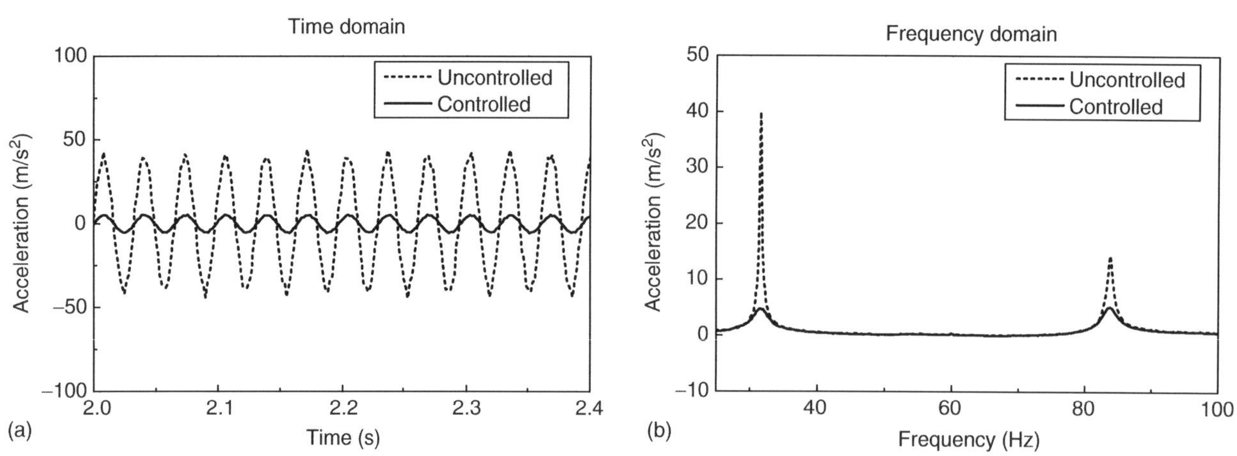

FIGURE 25.4 Vibration control responses of a flexible beam structure using ER fluid-based mount. (From Jung, W. J., Jeong, W. B., Hong, S. R., and Choi, S. B., *J. Sound Vibration*, 273, 185, 2004. With permission.)

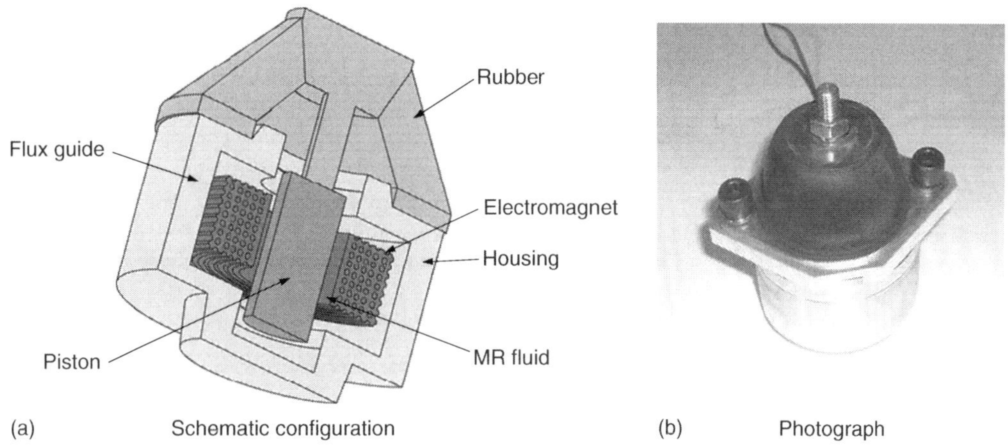

FIGURE 25.5 Mixed mode MR fluid-based mount.

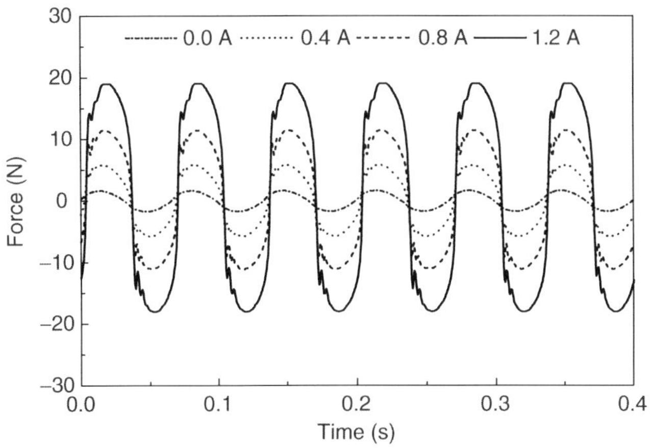

FIGURE 25.6 Field-dependent damping force of MR fluid-based mount. (From Hong, S. R. and Choi, S B., *J. Intell. Mater. Syst. Struct.*, 16, 931, 2005. With permission.)

This directly indicates that a certain desired damping force can be achieved by controlling the input current. The manufactured MR fluid-based mount is installed in a flexible beam structure, which is supported by two rubber mounts at both ends (refer to Figure 25.3). At the upside of the flexible beam, the MR fluid-based mount is placed between the beam and the exciting mass. In order to suppress vibration of the beam structure, an optimal controller integrating with Kalman filter (LQG) is synthesized as follows:

$$u(t) = -(k_1\hat{q}_1(t) + k_2\hat{\dot{q}}_1(t) + k_3\hat{q}_2(t) + k_4\hat{\dot{q}}_2(t) + k_5\hat{q}_3(t) + k_6\hat{\dot{q}}_3(t)) \tag{25.3}$$

where

k_i is the element of the state feedback gain matrix

$\hat{q}_i$ and $\hat{\dot{q}}_i$ are estimated states, which correspond to the modal displacement

Figure 25.7 presents the measured acceleration responses of the structural system with MR fluid-based mount [17]. The uncontrolled and controlled acceleration are measured at the position of the exciting mass about the MR mount and beam structure below the MR mount. The control input is determined on the basis of the LQG controller. The amplitude of excitation force is set by 1 N, and the frequency range focused

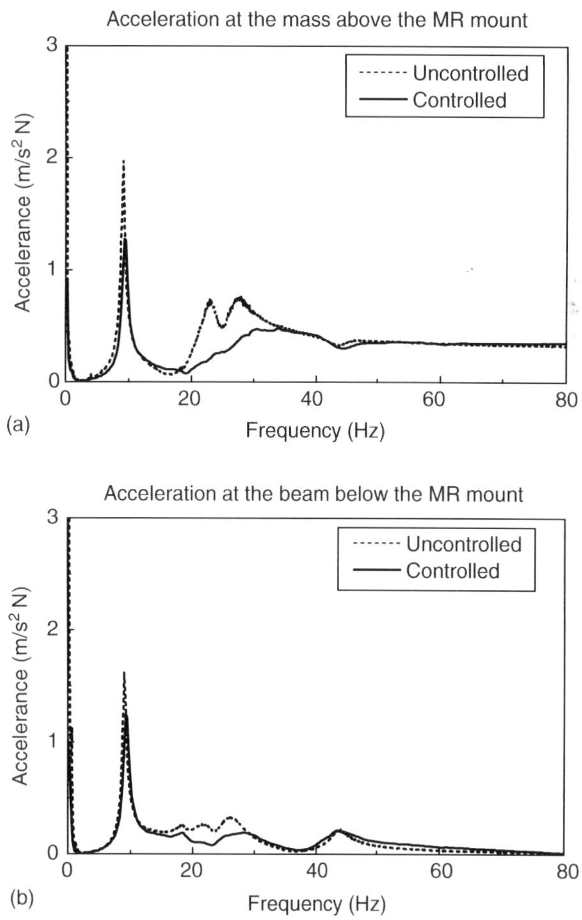

FIGURE 25.7 Vibration control responses of a flexible beam structure using MR fluid-based mount. (From Hong, S. R. and Choi, S. B., *J. Intell. Mater. Syst. Struct.*, 16, 931, 2005. With permission.)

is chosen from 5 to 80Hz. It is clearly observed that the acceleration levels of first, second, and third modes are attenuated by controlling the damping force of MR fluid-based mount.

25.1.2.3 Piezoelectric Actuator-Based Mount

Many types of passive mounts have been developed to support static load and isolate unwanted vibration of flexible structure systems and discrete systems. The rubber mount is one of the most popular and effective passive mounts applied for various vibrating systems. It is generally known that it has low damping, and hence shows efficient vibration isolation performance in the nonresonant and high-frequency excitations [18]. However, the passive mount has an intrinsic performance limitation, which leads to the study on active mounts. As well-known, the piezoactuator is featured by fast response time, small displacement, and low power consumption. Using these salient features, we can accomplish very effective vibration control performance of a flexible structure subjected to small magnitude and high-frequency resonant excitations. Active vibration control of flexible structures and discrete systems utilizing piezoelectric materials has been studied in recent years by many researchers. Premont et al. proved the robustness and effectiveness of active damping of space structure using piezoelectric actuator [19]. Choi et al. analyzed the performance of the hybrid mount, which is the combination of efficient mounts of different types, associated with semiactive ER fluid and active piezoelectric actuator [5].

In this chapter, a new type of hybrid mount featuring the passive rubber element and active piezoactuator is presented in order to achieve superior vibration control performance of a flexible beam structure at both resonant and nonresonant regions [20]. Figure 25.8 presents a schematic diagram and photograph of the proposed piezoelectric actuator-based mount. The piezostack actuator is a bipolar type and connected to the rubber mount through the intermediate mass. The intermediate mass acts like a reaction mass for the piezoelectric actuator on the beam structure. In order to evaluate vibration control performance of the proposed piezoelectric actuator-based mount, a flexible structure is established as shown in Figure 25.9. A steel beam is supported by two rubber mounts at both ends while the piezoelectric-based mount is placed between the two rubber mounts below the flexible beam. The flexible beam is excited by the electromagnetic exciter.

Among numerous control strategies, a sliding mode control (SMC) scheme, which has inherent robustness to system uncertainties and external disturbances, is adopted to isolate the vibration of the flexible beam structure. As a first step, the sliding surface $s(t)$ is defined as follows:

$$s(t) = \mathbf{G}\mathbf{x}(t) \tag{25.4}$$

where

$\mathbf{G}$ is the sliding surface gradient
$\mathbf{x}(t)$ is state variable

The existence condition of the sliding mode motion is given by

$$\frac{1}{2}\frac{d}{dt}s^2(t) \le -\eta |s(t)|$$

In this equation, η is a strictly positive constant. This condition allows the state variable $\mathbf{x}(t)$ converge to the sliding surface $s(t)$. The sliding mode controller, which satisfies the existence condition of the sliding mode motion, is obtained by [20]

$$u(t) = -(\mathbf{GB})^{-1}(\mathbf{GAx}(t) + k\,\mathrm{sgn}(s(t))) \tag{25.5}$$

where

k is the discontinuous control gain
sgn (·) is the sign function

$\mathbf{A}$ and $\mathbf{B}$ are system matrix and input matrix, respectively, which come from state space realization of the governing equation of

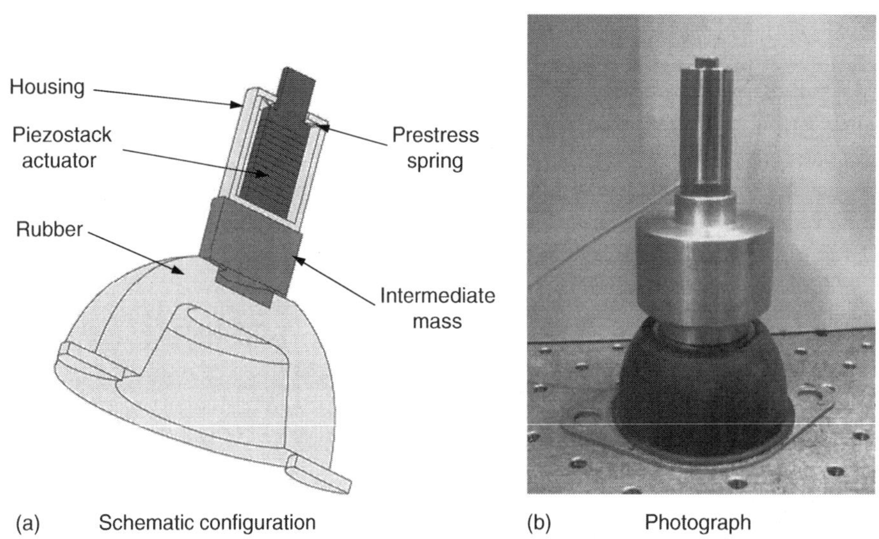

FIGURE 25.8 Piezoelectric actuator-based mount.

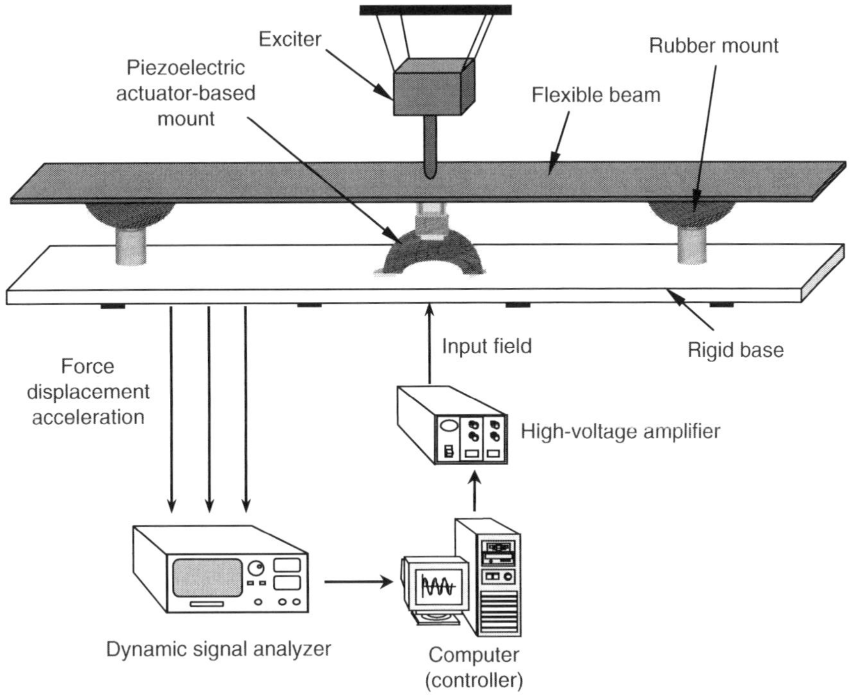

FIGURE 25.9 Experimental configuration for vibration control of a flexible beam structure using piezoelectric actuator-based mount.

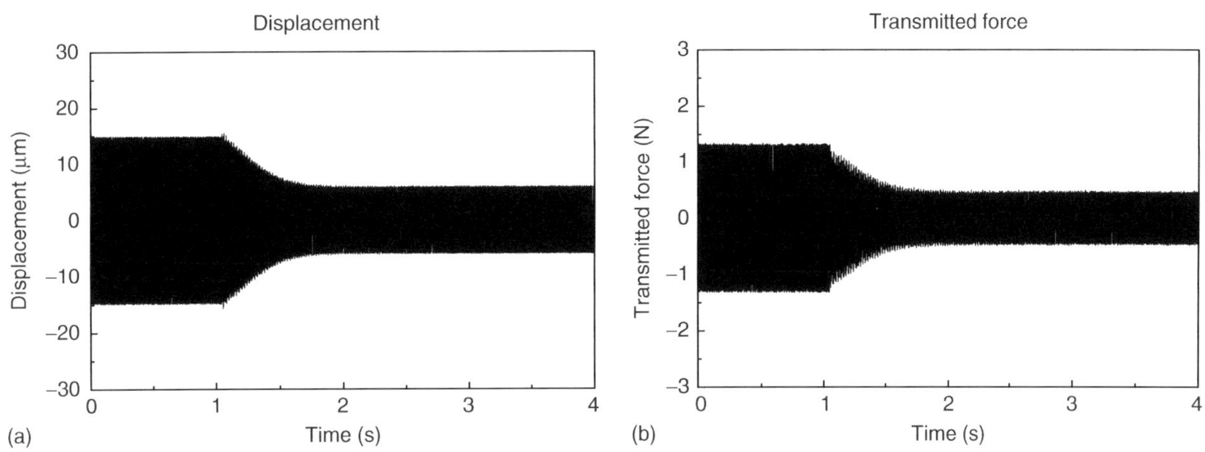

FIGURE 25.10 Time responses of a flexible beam structure using piezoelectric actuator-based mount. (From Kim, S. H., Choi, S. B., and Hong, S. R., *Int. J. Mech. Sci.*, 46, 143, 2004. With permission.)

motion. Figure 25.10 presents the measured time responses at the third mode frequency of 180 Hz [20]. We clearly observe substantial reductions of the displacement and transmitted force by activating the controller-associated piezoelectric actuator-based mount. Figure 25.11 shows the corresponding frequency response [20]. The uncontrolled displacement of the beam at mount position is 7.9 μm at the fourth mode (298 Hz). By activating the controller, the displacement is reduced to 2.1 μm. It is clear from the results that the proposed piezoelectric actuator-based mount is very effective for vibration control of a flexible structure. The force transmissibility is also evaluated as shown in Figure 25.11b [20]. From the remarkable reduction of the force transmission to the base, it is assured that the imposed vibration of the beam structure has been effectively isolated by applying the piezoelectric actuator-based mount.

25.1.3 Vibration Control of Dynamic Systems

25.1.3.1 Seat Suspension Using ER Damper

Heavy commercial vehicles are typically exposed to extended driving hours and severe working environment. Although many

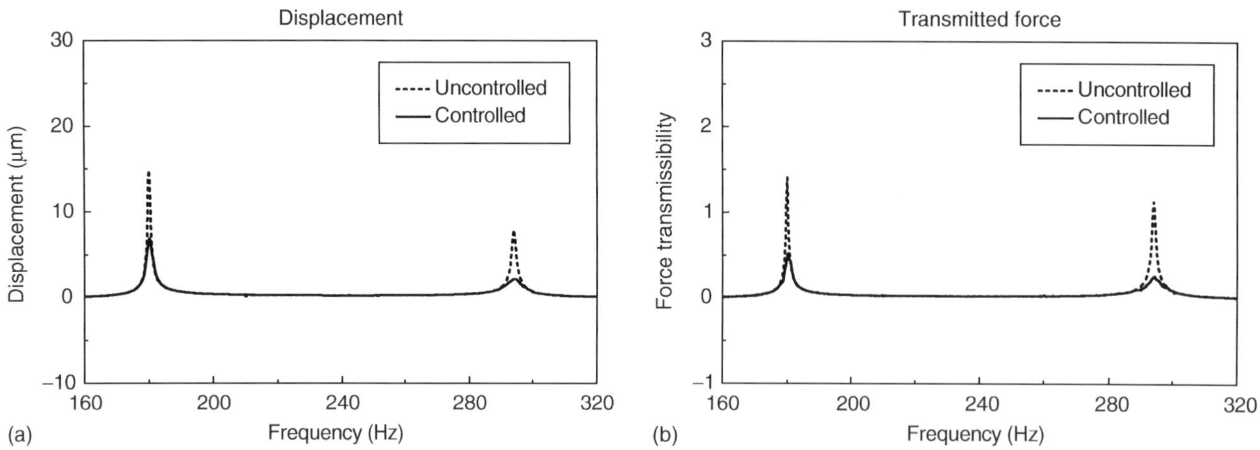

FIGURE 25.11 Frequency responses of a flexible beam structure using piezoelectric actuator-based mount. (From Kim, S. H., Choi, S. B., and Hong, S. R., *Int. J. Mech. Sci.*, 46, 143, 2004. With permission.)

efforts to improve ride quality have been made through primary and secondary (cabin) suspensions [21,22], drivers still suffer from the transmission of low-frequency and high-amplitude vibration, which is a major factor in the ride quality [23,24]. The alleviation of seat vibration is a significant problem in the heavy commercial vehicle because its ride quality is directly related to health disorders, working efficiency, and safety. A simple and effective method for attenuating unwanted vibration in a commercial vehicle is a seat suspension system that is particularly effective in the low-frequency range in which vehicle vibration energy is concentrated and a human body is sensitive [25].

Recently, a semiactive seat suspension system featuring ER fluids has been investigated for vibration isolation. This type of the semiactive seat suspension system has several advantages such as fast response time, continuously controllable damping force, and low energy consumption compared with active seat suspension systems. The schematic diagram and photograph of the ER seat damper is shown in Figure 25.12. The ER seat damper is divided into the upper and lower chambers by the piston, and is fully filled with the ER fluid. The ER fluid flows through the duct between inner and outer cylinders from one chamber to the other by the motion of the piston. A positive voltage is produced by a high-voltage supply unit connected to the outer cylinder. The gas chamber located outside of the lower chamber acts as an accumulator. If electric field is applied, the ER damper produces a damping force caused only by fluid resistance. However, if a certain level of the

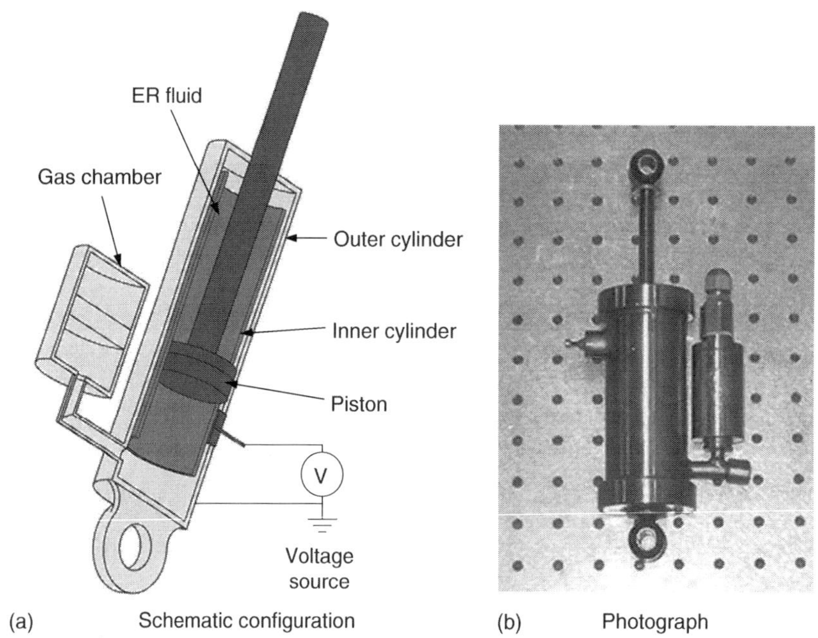

FIGURE 25.12 Controllable ER seat damper.

electric field is supplied to the ER damper, it produces an additional damping force owing to the yield stress of the ER fluid. This damping force of the ER damper can be continuously tuned by controlling the intensity of the electric field. The damping force of the ER damper can be obtained as follows [26]:

$$F = k_e x_p + c_e \dot{x}_p + F_{ER} \quad (25.6)$$

where

k_e is the effective stiffness due to gas pressure

c_e is the effective damping due to the fluid viscosity

x_p is the excitation displacement

F_{ER} is the field-dependent damping force, which is tunable by a function of applied electric field E that is typically Bingham model

On the basis of the size and required damping force level of a conventional passive oil seat damper for a commercial truck, an ER seat damper is designed and manufactured as shown in Figure 25.12b. Its field-dependent damping force is experimentally evaluated that it can be increased up to 89 N.

The most significant feature in the seat suspension system is that the seat suspension dynamics includes a seated human body. However, the seated human body exposed to vibration is a complex dynamic system. So far, various lumped parameter models have been developed to describe the human vibration motion [27]. These models consider the human body as several rigid bodies, springs, and dampers. In this chapter, a human body model is derived and integrated with the governing equations of the ER seat suspension system. For the simplification of the dynamic modeling, it is assumed that there exists only the vertical motion of the vehicle. The seat suspension system consists of the ER seat damper, linear spring, seat frame, and cushion on which the driver is seating. The spring constant is held to be constant and the damping force varies according to control input. The seat cushion is modeled by a set of linear springs and dampers whose characteristics are constant. A driver is modeled as a multisegment mass, which include head, upper torso, lower torso, and thighs [28]. The arms and legs are combined with the upper torso and thigh, respectively. Because of uncertainties in the human vibration model, one of the potential controller candidates for the ER seat suspension system is sliding mode controller, which is very effective for nonlinear and parameter uncertain system [29]. By imposing semiactive actuating conditions, the SMC is experimentally realized with a state observer.

Figure 25.13 shows an experimental setup to evaluate control performance of the seat suspension system using the ER damper [30]. In the experimental configuration, a driver directly sits on the controlled seat. The floor under the seat is excited from two types of profiles; bump and random road conditions. The excited floor displacement is calculated and then the displacement signal is converted to the hydraulic control unit for controlling the excitation input to the platform. When the vibration is transmitted to the ER seat suspension, the dynamic response signals, which are acquired from the linear variable differential transformer

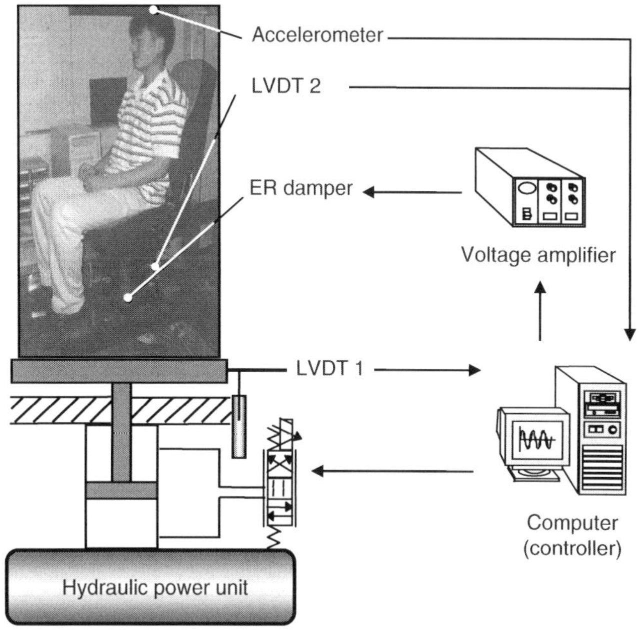

FIGURE 25.13 Experimental configuration for vibration control of a seat suspension using ER seat damper.

(LVDT) and accelerometer, are then converted to digital signals via A/D converter. The SMC is then activated to reduce the vibration level at the driver seat. Figure 25.14 presents the measured transmissibility of the proposed ER seat suspension system [30]. The natural frequency of the ER seat suspension system without electric field is 2.1 Hz, which is the same as for the passive seat suspension system for commercial truck. It is clearly observed that the peak of transmissibility is substantially reduced by activating the electric field determined from the SMC, and its frequency is slightly increased. Figure 25.15 presents the measured bump responses of the seat suspension system in the time domain [30]. As clearly observed from the results, the maximum displacement and acceleration are reduced by 40% and 30%, respectively. These results clearly indicate significant improvement of the ride quality.

25.1.3.2 Vehicle Suspension Using MR Damper

Recently, a great attention on the vibration control of a vehicle system has been significantly increased. The vehicle vibration needs to be attenuated from various road conditions. This is normally accomplished by employing suspension system. So far, three types of suspensions have been proposed and successfully implemented; passive, active, and semiactive. The passive suspension system featuring oil damper provides design simplicity and cost-effectiveness. However, performance limitations are inevitable. On the other hand, the active suspension system provides high control performance in wide frequency range. However, the active suspension requires high power sources, many sensors, servo valves and sophisticated control logic. One way to resolve these requirements of the active suspension system is to use the semiactive suspension

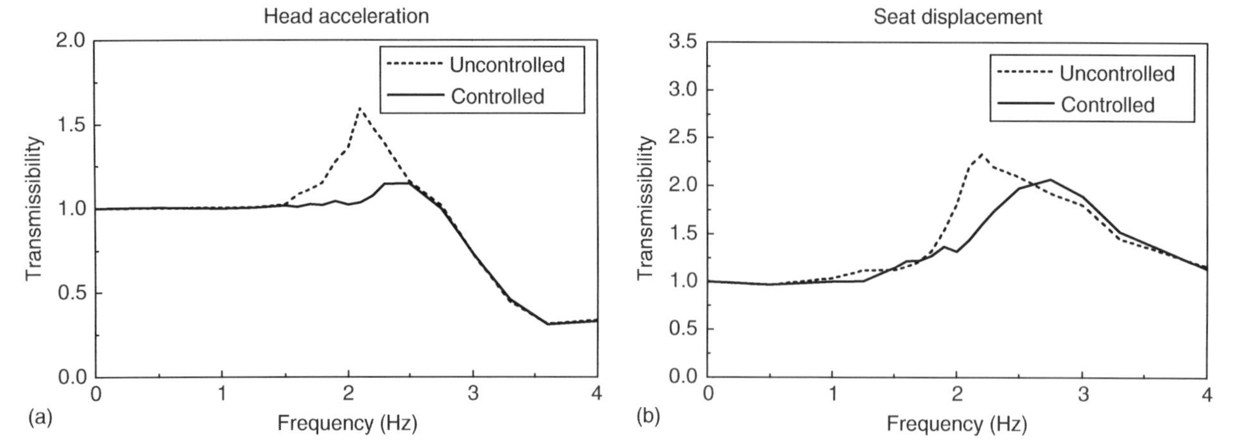

FIGURE 25.14 Transmissibility of a seat suspension using ER seat damper. (From Han, Y. M., Jung, J. Y., Choi, S. B., and Wereley, N. M., *Int. J. Mod. Phys. B*, 19, 1689, 2005. With permission.)

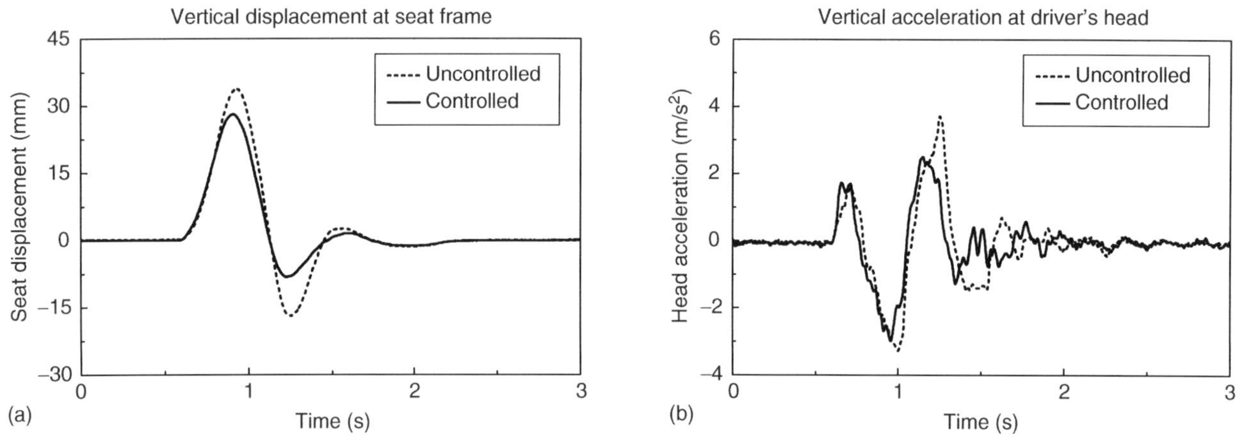

FIGURE 25.15 Bump responses of a seat suspension using ER seat damper. (From Han, Y. M., Jung, J. Y., Choi, S. B., and Wereley, N. M., *Int. J. Mod. Phys. B*, 19, 1689, 2005. With permission.)

system. The semiactive suspension system offers a desirable performance generally enhanced in the active mode without requiring large power sources and expensive hardware. Recently, very attractive and effective semiactive suspension system MR fluid has been proposed by many investigators. Carlson et al. [31] proposed a commercially available MR damper, which is applicable to on-and-off-highway vehicle suspension system. They experimentally demonstrated that sufficient levels of damping force and also superior control capability of the damping force by applying control magnetic field could be achieved. Spencer Jr. et al. [32] proposed dynamic model for prediction of damping force of a MR damper. They compared the measured damping forces with the predicted ones in time domain. Kamath et al. [33] proposed a semiactive MR lag mode damper. They proposed dynamic model and verified its validity by comparing the predicted damping force with the measured one.

In this chapter, the hardware-in-the-loop simulations (HILS) are undertaken to evaluate the performance of the full-vehicle MR suspension system [34]. It generally takes a long time and a high cost to successfully develop the new types of components for a complete system. To save the cost and time, a theoretical analysis using computer simulation method is widely used. However, since many real situations, which are difficult to be modeled and, even, cannot be modeled by analytical method, are often neglected or approximated by linearization, the theoretical method cannot precisely predict the performance of the system to occur in real field. Therefore, in order to overcome the limit of the computer simulation method, HILS method has been proposed recently. The HILS method has major advantages such as easy modification of system parameters and relatively low-cost test facilities. In addition, a wide range of operating conditions to emulate the practical situations can be investigated in the laboratory.

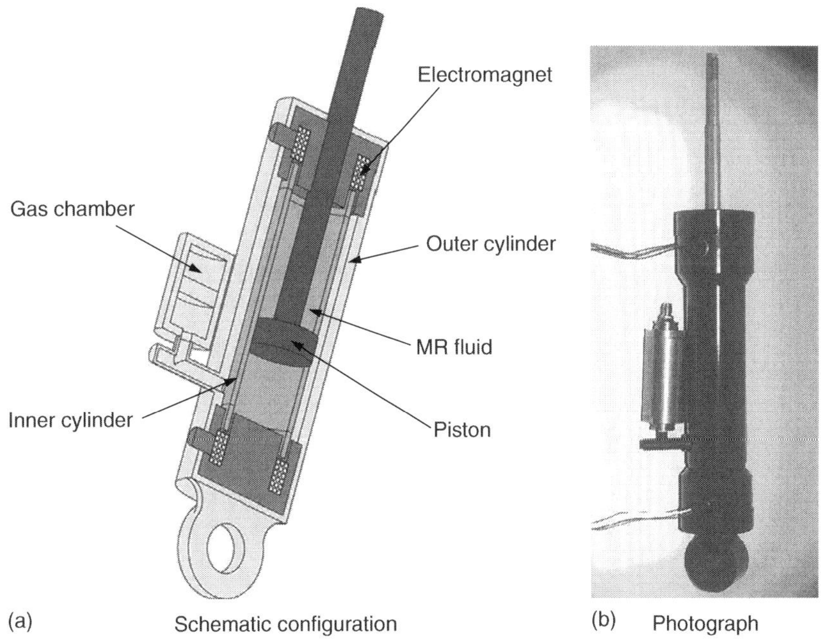

FIGURE 25.16 Controllable MR damper.

The schematic configuration and photograph of the proposed MR damper are shown in Figure 25.16. The MR damper is divided into the upper and lower chambers by the piston, and it is fully filled with the MR fluid. By the motion of the piston, the MR fluid flows through the orifice at both ends from one chamber to the other. The gas chamber located outside acts as an accumulator for absorbing sudden pressure variation of lower chamber of the MR damper induced by the rapid motion of the piston. In order to simplify the analysis of the MR damper, it is assumed that the MR fluid is incompressible and that pressure in one chamber is uniformly distributed. Furthermore, it is assumed that frictional force between oil seals and fluid inertia is negligible. Figure 25.17 shows the schematic configuration of the proposed HILS. It is divided into three parts: interface, hardware, and software. The

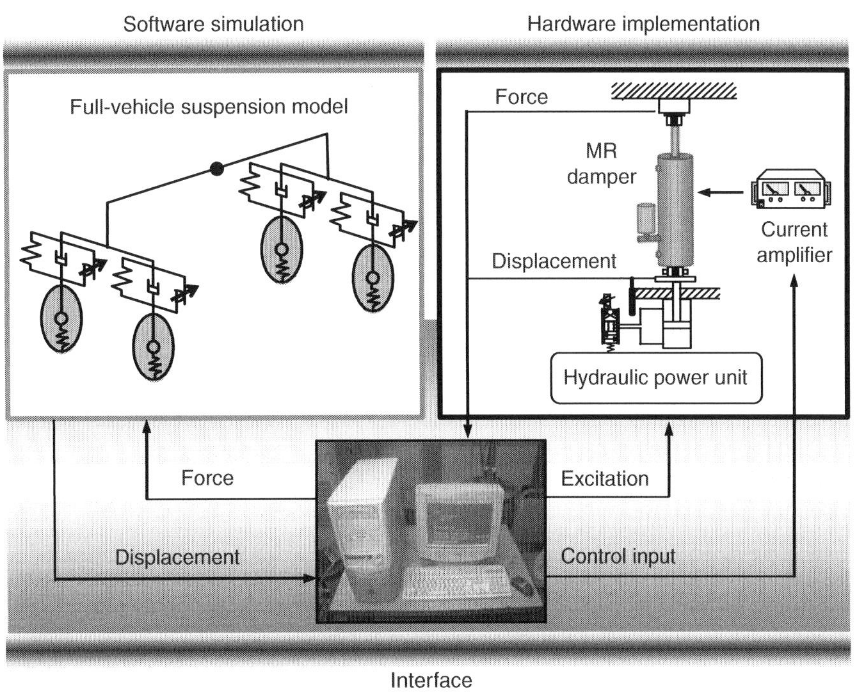

FIGURE 25.17 Schematic diagram for vibration control of a vehicle suspension using MR damper via HILS.

interface part is composed of computer on which digital signal processing (DSP) board is mounted. The hardware part is composed of the MR damper, current amplifier, and hydraulic damper tester. The software part consists of the theoretical model for the full-vehicle MR suspension system and control algorithm programmed in the computer. The vehicle body itself is assumed to be rigid and has degrees of freedom in the vertical, pitch, and roll directions. It is connected to four rigid bodies representing the wheel unsprung masses in which each has a vertical degree of freedom. In practice, the sprung mass is varied by the loading conditions such as the number of riding persons and payload. In order to take into account for these structured system uncertainties, the loop shaping design procedure (LSDP) based on H_∞ robust stabilization proposed by McFarlane and Glover [35] is adopted in this article. By choosing a suitable γ, the suboptimal controller is constituted as follows [34]:

$$\mathbf{u}=\mathbf{W}\begin{bmatrix}\mathbf{A}_{se}^{c}+\gamma^{2}\mathbf{W}_{h}^{*-1}\mathbf{Z}\mathbf{C}_{se}^{*}(\mathbf{C}_{se}+\mathbf{D}_{se}\mathbf{F}_{h}) & \gamma^{2}\mathbf{W}_{h}^{*-1}\mathbf{Z}\mathbf{C}_{se}^{*}\\ \mathbf{B}_{se}^{*}\mathbf{X} & -\mathbf{D}_{se}^{*}\end{bmatrix}\mathbf{y} \tag{25.7}$$

$$\mathbf{A}_{se}^{c}=\mathbf{A}_{se}+\mathbf{B}_{se}\mathbf{F}_{h},\quad \mathbf{W}_{h}=\mathbf{I}+(\mathbf{X}\mathbf{Z}-\gamma^{2}\mathbf{I})$$

$$\mathbf{F}_{h}=-\mathbf{S}^{-1}(\mathbf{D}_{se}^{*}\mathbf{C}_{se}+\mathbf{B}_{se}^{*}\mathbf{X}),\quad \mathbf{S}=\mathbf{I}+\mathbf{D}_{se}^{*}\mathbf{D}_{se}$$

where

$\mathbf{A}_{se}$, $\mathbf{B}_{se}$, $\mathbf{C}_{se}$ and $\mathbf{D}_{se}$ are the state space system matrices of the shaped plant

$\mathbf{G}_{se}$. $\mathbf{X}$ and $\mathbf{Z}$ are the positive definite solutions of generalized control algebraic Riccati equation (GCARE) and generalized filtering algebraic Riccati equation (GFARE)

$\mathbf{W}$ is the weighting function

As a first step for the HILS, the computer simulation of the full-vehicle MR suspension system is performed with initial value. This computer simulation is incorporated with the hydraulic damper tester, which applies the displacement (suspension travel) to the MR damper according to the demand signal obtained from the computer. It is also connected to the current amplifier, which applies control current determined from control algorithm to the MR damper. Subsequently, the damping force of the MR damper is measured from the hydraulic damper tester and the measured damping force is fed back into the computer simulation. In short, the computer simultaneously runs both the hydraulic damper tester and current amplifier during simulation loop while the computer simulation is performed based on the measured data. Figure 25.18 presents time responses of the MR suspension system for the bump excitation [34]. It is generally known that the displacement and acceleration of sprung mass and tire deflection are used to evaluate ride comfort and road holding of the vehicle, respectively. We clearly see that the vertical displacement and acceleration of sprung mass are reduced by activating the control field determined from the H_∞ controller. In addition, the pitch angular displacement and acceleration, and tire deflection are also well reduced by applying the control input. Figure 25.19 presents controlled frequency responses under random excitation [34]. The frequency responses are obtained from power spectral density (PSD) for the vertical acceleration of the sprung mass and tire deflection. As expected, the PSDs for the vertical acceleration and tire deflection are respectively reduced about 20% and 40% in the neighborhood of body resonance by applying input current. In addition, it is seen that the vertical acceleration is also reduced about 30% at the wheel resonance. The control results presented in Figures 25.18 and 25.19 indicate that both ride performance and steering stability of a vehicle can be improved by employing the proposed semi-active MR suspension system.

25.1.3.3 CD-ROM Drive Featuring Piezoelectric Shunt

The optical disk drives as the information storage device are classified into CD-ROM, CD-R/RW, DVD-ROM/RAM/RW, and so on. Especially, the CD-ROM drive has been widely used

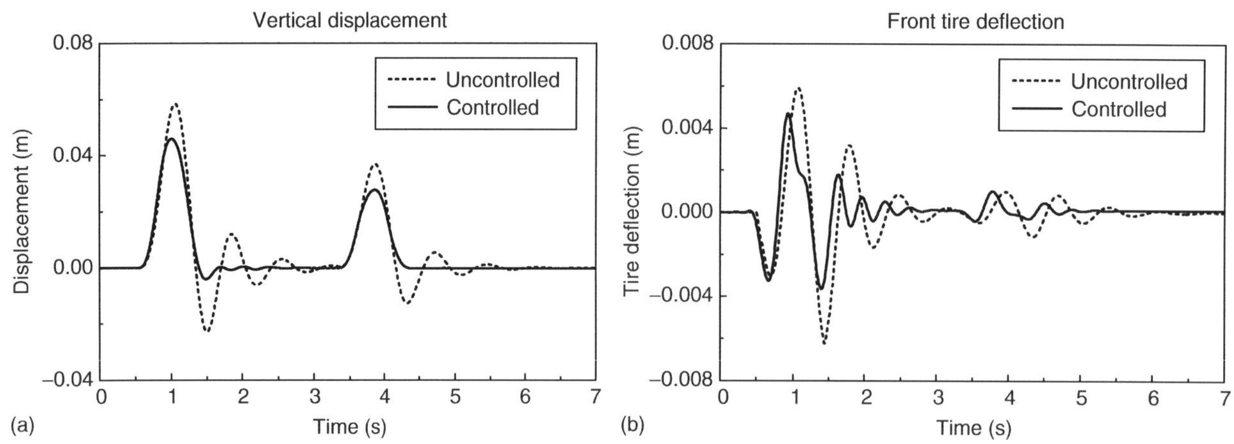

FIGURE 25.18 Bump responses of a vehicle suspension using MR damper. (From Choi, S. B., Lee, H. S., and Park, Y. P., *Vehicle Syst. Dyn.*, 38, 341, 2002. With permission.)

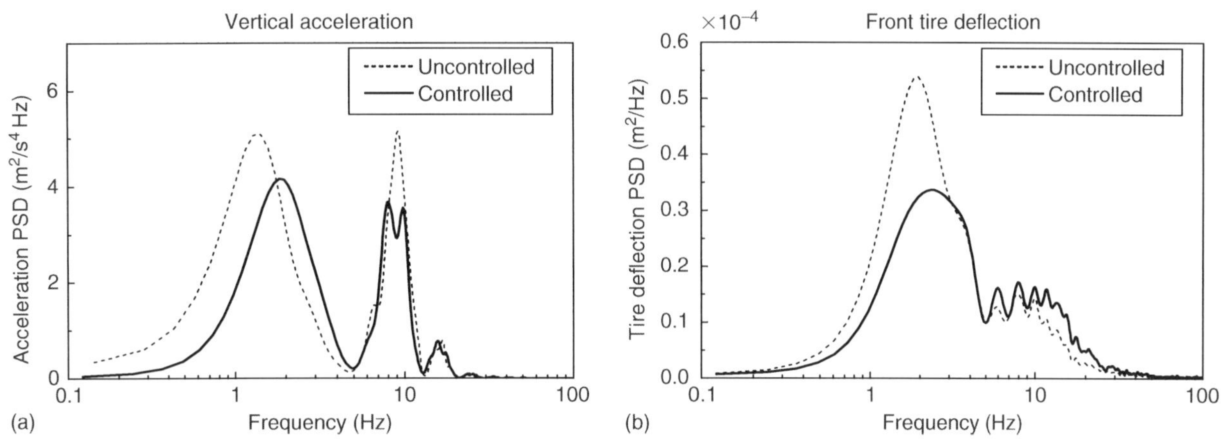

FIGURE 25.19 Random responses of a vehicle suspension using MR damper. (From Choi, S. B., Lee, H. S., and Park, Y. P., *Vehicle Syst. Dyn.*, 38, 341, 2002. With permission.)

as a secondary storage device for computer peripherals. The typical CD-ROM drive consists of the disk loading system, the feeding system including optical pick-up and spindle, the printed circuit board, and the drive base. The track density and linear density of the disk are about 16,000 tracks-per-inch (TPI) and 43,000 (BPI), respectively. In order to achieve high capacity of the CD-ROM drive, accurate position control of the optical pick-up head and vibration suppression of the feeding system are demanded. The vibration of the feeding system is generally affected by unbalanced flexible disk with high rotating speed and external excitation to the main base. Antivibration of the feeding system is an urgent problem to be solved to achieve high capability of the drive [36]. Normally, conventional drives adopt passive rubber mounts to prevent the feeding system from external excitation and the vibration of the spindle. In addition, auto ball balancer is often used [37], and a semiactive mount using ER fluid has been also studied in order to overcome the limit of the passive rubber mounts [38]. The CD-ROM drive base, which has a role of supporting the feeding system, is easily exposed to environmental vibration sources such as user's handling and high speed rotating disk. If the vibration of the main base is not effectively reduced, the robust servo control of the optical pick-up cannot be guaranteed.

Choi et al. initiated the effective vibration suppression methodologies for the CD-ROM drive by using piezoelectric damping integrated with shunt circuit [39]. It is known that each resonant mode of the smart structure integrated with the piezoelectric shunt circuit can be modeled by mechanical vibration absorber that is one of very effective vibration reduction methods [40]. The mechanical vibration absorber is then modeled by an electrical impedance model via mechanical–electrical circuit analogy. Figure 25.20 presents a schematic diagram of the proposed system that consists of the structure (CD-ROM drive base) with piezoelectric material and resonant shunt circuit [39]. A piezoceramic (PZT) is incorporated to the rear part of the drive base, which is equipped with vibration isolation workstation by fixture. The drive base is externally excited by another PZT positioned to the fixture. The vibration of the drive base is measured by an accelerometer and analyzed by dynamic signal analyzer.

The target mode and the position of the PZT can be identified by modal analysis of the CD-ROM base using finite element method (FEM). The modal analysis has been undertaken through two phases. Firstly, we investigate the modal analysis results in working frequency region for the original CD-ROM drive base, and choose a target mode shape, which exhibits the most severe vibration characteristics. In second phase, the PZT considering the deformation of the target mode shape is integrated to the drive base and modal analysis is performed again. From the analyzed mode shapes of the CD-ROM drive base, the target mode is determined to be the second mode (336 Hz) because the rear part of the base is more severely deformed than other parts at the second mode. Prior to

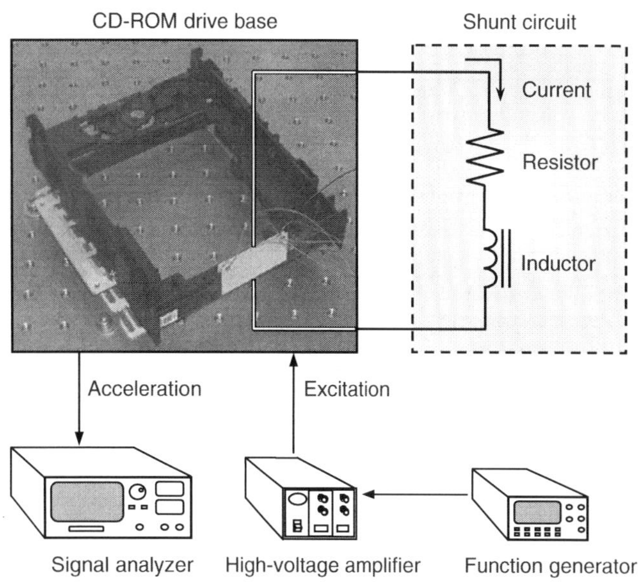

FIGURE 25.20 Experimental configuration for vibration suppression of a CD-ROM drive that features piezoelectric shunt.

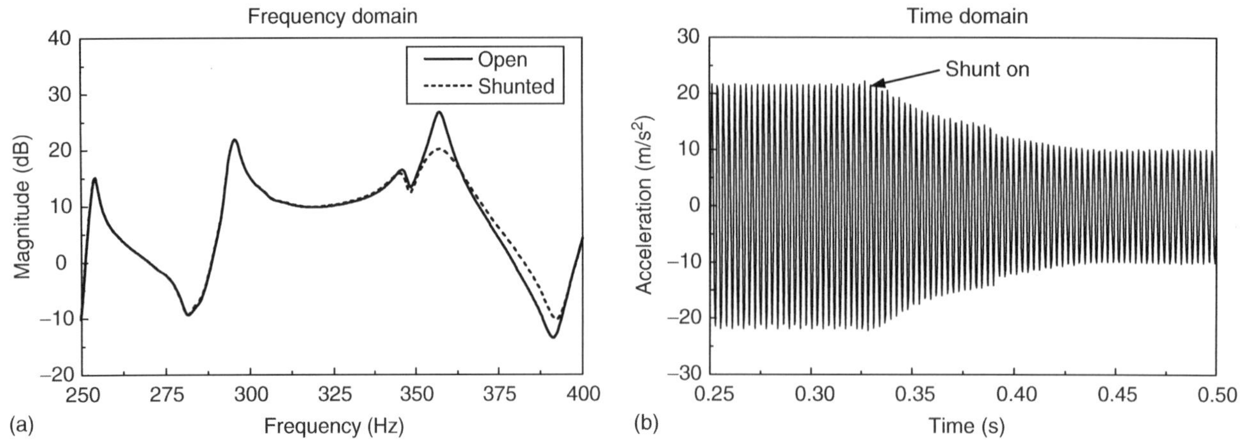

FIGURE 25.21 Vibration responses of CD-ROM drive that features piezoelectric shunt. (From Park, J.S., Lim, S. C., Choi, S. B., Kim, J. H., and Park, Y. P., *J. Sound Vibration*, 269, 1111, 2004. With permission.)

integrating the shunt circuit to the drive base, design parameters of the shunt circuit are tuned optimally by [39]

$$\frac{x}{x^{\mathrm{ST}}} = \frac{\gamma^2 + \delta^2 r\gamma + \delta^2}{(\gamma^2+1)(\gamma^2 + \delta^2 r\gamma + \delta^2) + K_{ij}^2(\gamma^2 + \delta^2 r\gamma)} \tag{25.8}$$

$$\gamma = s/\omega_{\mathrm{n}}^{\mathrm{E}}, \quad r = R_i C_{pi}^{s}\omega_{\mathrm{n}}^{\mathrm{E}}, \quad \delta = \omega_{\mathrm{e}}/\omega_{\mathrm{n}}, \quad \omega_{\mathrm{e}} = 1\Big/\sqrt{L_i C_{pi}^{s}}$$

where

L_i and R_i are the parameters of inductor and resister of the shunt circuit to be desired
C_{pi}^{s} is the inherent capacitance of the piezoelectric shunted at constant strain
ω_{n} and ω_{e} are mechanical and electrical natural frequency of the inherent structure and shunt circuit
$\omega_{\mathrm{n}}^{\mathrm{E}}$ is the natural frequencies of the structural mode of interest with short circuit piezoelectric material
K_{ij} is the generalized electromechanical coupling coefficient
x^{ST} is the static displacement of the system
s is the Laplace variable

In order to investigate performance of vibration reduction using the piezoelectric shunt damping in target frequency region, the optimal parameter values such as inductance and resistance of the shunt circuit are determined, and then set for empirical realization. The amount of dissipated energy is determined directly from the resistance, and thus a proper choice of resistance is essential. Hence, the optimal resistance value is experimentally retuned. The analytical values of optimal resistance and inductance are different from the experimentally obtained values. The optimal values for the resistance and inductance in shunt circuit are directly affected by natural frequency. In addition, the inductor in the shunt circuit can affect the optimal resistance value and vice versa. This coupling effect between the resistor and the inductor is caused by electrical structure of synthetic inductor consisting of OP amps and resistors [41]. Figure 25.21a presents the frequency response in region of target frequency of 356 Hz [39]. It is clearly observed that the vibration reduction of 6.7 dB is achieved at the resonant peak by activating the shunt circuit. This result at the target mode is represented in time domain as shown in Figure 25.21b [39]. When the shunt circuit is short, the magnitude of the acceleration is reduced from 21.7 to 10.1 m/s^2. In terms of the displacement, the vibration magnitude of 4.4 μm is reduced to 2.1 μm by activating the piezoelectric shunt circuit. Consequently, these experimental results demonstrate that the piezoelectric shunt damping can be effectively applied to vibration reduction of the CD-ROM drive base. It is expected that vibration reduction will give a high capability of the CD-ROM drive.

References

1 S. B. Choi, B. S. Thompson, and M. V. Gandhi, An experimental investigation on the active-damping characteristics of a class of ultra-advanced intelligent composite materials featuring electro-rheological fluids, Proceedings of Damping'89 Conference, 1, CAC.1–CAC.14. West Palm Beach, FL, February 1988.
2 M. V. Gandhi, B. S. Thompson, S. B. Choi, and S. Shakir, Electro-rheological-fluid-based articulating robotic systems, *ASME Journal of Mechanisms, Transmissions and Automation in Design*, 111(3), 328–336, 1989.
3 J. P. Coulter and T. G. Duclos, Applications of electrorheological materials on vibration control, Proceedings of the Second International Conference of ER Fluids, 300–325, Raleigh, NC, August 1989.
4 Y. Choi, A. F. Sprecher, and H. Conrad, Response of electrorheological fluid-filled laminate composites to forced vibration, *Journal of Intelligent Material Systems and Structures*, 3, 17–29, 1992.
5 S. B. Choi, Y. K. Park, and C. C. Cheong, Active vibration control of intelligent composite laminate structures incorporating an electro-rheological fluid, *Journal of Intelligent Material Systems and Structures*, 7(4), 411–419, 1996.

6 N. K. Petek, R. J. Goudie, and F. P. Boyle, Actively controlled damping in electrorheological fluid-filled engine mounts, SAE Technical Paper Series 881785, 1988.

7 S. H. Choi, Y. T. Choi, S. B. Choi, and C. C. Cheong, Performance analysis of an engine mount featuring ER fluids and piezoactuators, *International Journal of Modern Physics B*, 10(23), 3143–3157, 1996.

8 T. G. Duclos, An externally tunable hydraulic mount which uses electro-rheological fluid, SAE Technical Paper Series 870963, 1987.

9 S. B. Choi, S. R. Hong, and M. S. Han, Vibration control of a frame structure using electro-rheological fluid mounts, *International Journal of Mechanical Sciences*, 44, 2027–2045, 2002.

10 N. K. Petek, D. J. Romstadt, M. B. Lizell, and T. R. Weyenberg, Demonstration of an automotive semi-active suspension using electrorheological fluid, SAE Technical Paper Series 950586, 1995.

11 S. B. Choi, Vibration control of a flexible structure using ER dampers, *Journal of Dynamic Systems, Measurement and Control*, 121, 134–138, 1996.

12 W. J. Jung, W. B. Jeong, S. R. Hong, and S. B. Choi, Vibration control of a flexible beam structure using squeeze-mode ER Mount, *Journal of Sound and Vibration*, 273, 185–199, 2004.

13 S. B. Choi and W. K. Kim, Vibration control of a semi-active suspension featuring electrorheological fluid dampers, *Journal of Sound and Vibration*, 234, 537–546, 2000.

14 S. B. Choi, H. S. Lee, and Y. P. Park, H-infinity control performance of a full-vehicle suspension featuring magnetorheological dampers, *Vehicle System Dynamics*, 38(5), 341–360, 2002.

15 N. M. Wereley and L. Pang, Nondimensional analysis of semi-active electrorheological and magnetorheological dampers using approximate parallel plate models, *Smart Materials and Structures*, 7, 732–743, 1998.

16 R. Stanway, J. T. Sproston, and A. K. EI-Wahed, Applications of electro-rheological fluids in vibration control: A survey, *Smart Materials and Structures*, 5, 464–482, 1996.

17 S. R. Hong and S. B. Choi, Vibration control of a structural system using magneto-rheological fluid mount, *Journal of Intelligent Material Systems and Structures*, 16, 931–936, 2005.

18 Y. Yu, S. M. Peelamedu, N. G. Nagathan, and R. V. Dukkipati, Automotive vehicle engine mounting systems: A survey, *ASME Journal of Dynamic Systems, Measurement and Control*, 123, 186–194, 2001

19 A. Premont, J. Dufour, and C. Malékian, Active damping by a local force feedback with piezoelectric actuators, *Journal of Guidance Control and Dynamics*, 15(2), 390–395, 1992.

20 S. H. Kim, S. B. Choi, and S. R. Hong, Vibration control of a flexible structure using a hybrid mount, *International Journal of Mechanical Science*, 46(1), 143–157, 2004.

21 N. M. Wereley and L. Pang, Nondimensional analysis of semi-active electrorheological and magnetorheological dampers using approximate parallel plate models, *Smart Materials and Structures*, 7, 732–743, 1997.

22 Y. T. Choi and N. M. Wereley, Comparative analysis of the time response of electrorheological and magnetorheological dampers using nondimensional parameters, *Journal of Intelligent Material Systems and Structures*, 13(7/8), 443–451, 2002.

23 F. Amirouche, L. Palkovics, and J. Woodrooffe, Optimal driver seat suspension design for heavy tucks, *Transportation Systems ASME*, 277–291, 1994.

24 R. Andersson, The low back pain of bus drivers in an urban area of california, *Spine*, 1481–1488, 1982.

25 M. J. Griffin, *Handbook of Human Vibration*, Academic Press, London, 1990.

26 S. B. Choi, Y. T. Choi, and D. W. Park, A sliding mode control of a full-car electrorheological suspension system via hardware-in-the-loop simulation, *ASME Journal of Dynamic Systems, Measurement, and Control*, 122(1), 114–121, 2000.

27 P. E. Boileau, S. Rakheja, X. Yang, and I. Stiharu, Comparison of biodynamic response characteristics of various human body models as applied to seated vehicle drivers, *Noise & Vibration Worldwide*, 7–15, 1997.

28 P. E. Boileau and S. Rakheja, Whole-body vertical biodynamic response characteristics of the seated vehicle driver measurement and model development, *International Journal of Industrial Ergonomics*, 449–472, 1998.

29 S. B Choi, Y. T. Choi, and D. W. Park, A sliding mode control of a full-car electrorheological suspension system via hardware-in-the-loop simulation, *ASME Journal of Dynamic Systems, Measurement and Control*, 122(1), 114–121, 2000.

30 Y. M. Han, J. Y. Jung, S. B. Choi, and N. M. Wereley, Sliding mode control of ER seat suspension considering human vibration model, *International Journal of Modern Physics B*, 19(7–9), 1689–1695, 2005.

31 J. D. Carlson, D. M. Cantanzarite, and ST. K. A. Clair, Commercial magneto-rheological fluid devices, Proceedings of the 5th International Conference on ER Fluids, MR Suspensions and Associated Technology, 1995, p. 20..

32 B. F. Spencer Jr., S. J. Dyke, M. K. Sain, and J. D. Carlson, Phenomenological model for a magnetorheological damper, *Journal of Engineering Mechanics, ASCE*, 123, 230, 1997.

33 G. M. Kamath, N. M. Wereley, and M. R. Jolly, Characterization of semi-active magnetorheological fluid lag mode dampers, SPIE Conference on Smart Structures and Integrated Systems, 3329, 1998.

34 S. B. Choi, H. S. Lee, and Y. P. Park, H-infinity control performance of a full-vehicle suspension featuring magnetorheological dampers, *Vehicle System Dynamics*, 38(5), 341–360, 2002.

35 D. C. McFarlane and K. Glover, Robust controller design using normalized coprime factor plant description, Lecture Notes in Control Information Sciences, Springer-Verlag, 1989.

36 S. H. Chang, H. S. Kim, J. K. Choi, and D. G. Lee, A study on the design of vibration damper for high speed CD-ROM

drive, *Journal of Korean Society Mechanical Engineering (A)*, 22, 939–952, 1998.
37 J. Lee and W. K. Van Moorhen, Analytical and experimental analysis of a self-compensating dynamic balancer in a rotating mechanism, *ASME Journal of Dynamic System, Measurements and Control*, 118, 468–475, 1996.
38 S. C. Lim, J. S. Park, S. B. Choi, and Y. P. Park, Vibration control of a CD-ROM feeding system using electro-rheological mounts, *Journal of Intelligent Material System and Structures*, 12(9), 629–638, 2002.
39 J. S. Park, S. C. Lim, S. B. Choi, J. H. Kim, and Y. P. Park, Vibration reduction of a CD-ROM drive base using a piezoelectric shunt circuit, *Journal of Sound and Vibration*, 269(3–5), 1111–1118, 2004.
40 N. W. Hagood and A. von Flotow, Damping of structural vibrations with piezoelectric materials and passive electrical networks, *Journal of Sound and Vibration*, 146, 243–268, 1991.
41 J. Kim, Y. H. Ryu, and S. B. Choi, New shunting parameter tuning method for piezoelectric damping based on measured electrical impedance, *Smart Materials and Structures*, 9, 868–877, 2000.

25.2 Smart Materials for Sound and Vibration Control

Cai Chao, Lu Chun, Tan Xiaoming, and Zheng Hui

25.2.1 Introduction

Some materials have the ability to change shape or size simply by applying an electric potential, adding a little bit of heat, or to change their state from a liquid to a solid almost instantly when near a magnet; these materials are called smart materials. The structures with smart materials are also called smart structures. Generally, smart materials can be grouped into the following categories:

1. *Piezoelectric materials*: They can generate an electric potential in response to an applied mechanical stress (direct piezoelectric effect) and can produce the mechanical stress when subjected to an electric field (converse piezoelectric effect).
2. *Electrostrictive materials*: Electrostriction is a property of all dielectric materials. The mechanical strain is proportional to the square of the electric field. Unlike piezoelectric materials, electrostriction cannot be reversed; mechanical deformation will not induce an electric field and the reversal of the electric field does not reverse the direction of the deformation. The reason is that electrostrictive materials are not poled.
3. *Magnetostrictive materials*: Magnetostriction is the property that causes a material to change its length when subjected to an electromagnetic field. Magnetostriction properties also cause materials to generate electromagnetic fields when they are deformed by an external force.
4. *Shape memory alloys* (SMA): They are alloys that can "remember" their geometries. The materials produce shape changes when subjected to a thermal field.
5. *Optical fibers:* They use intensity, phase, frequency, or polarization of modulation to measure strain, temperature, electrical/magnetic fields, pressure, and other measurable quantities. Optical fibers can be used as various excellent sensors.
6. *Electrorheostatic* (ER) and *Magnetorheostatic* (MR) *materials*: They are smart fluids, which can experience a dramatic change in their viscosity in a millisecond when exposed to a magnetic (MR) or electric field (ER). The effect can be completely reversed just as quickly when the field is removed.

Although there exist many smart materials as mentioned above and they have been installed and are being installed into various applications from the control of aerospace structures to the development of biomechanical devices, this chapter only covers piezoelectric materials and their applications in structural sound and vibration suppression.

The Curie brothers discovered the piezoelectric effects in 1880, but it was not fully used until the 1940s when high input impedance amplifiers were developed. In the 1950s, the piezoelectric effect became commercialized with the advent of electrometer tubes. Now, piezoelectric materials have become very popular in passive and active structural control applications, both as sensors and actuators. Lead zirconate titanate (PZT) and polyvinylidene fluoride (PVDF) are two commonly available examples of piezoelectric materials in engineering applications. Generally, there are three major means to apply piezoelectric materials for structural sound and vibration control: passive, active, and hybrid.

25.2.1.1 Passive Control with Piezoelectric Materials

Piezoelectric materials are used for passive sound and vibration control in various shunt circuit techniques. The main feature of piezoelectric materials is energy transduction. This occurs when mechanical work is done on piezoelectric elements and some portion is converted to and stored as dielectric energy. In a vibrating structure, a shunting network can be configured to accomplish vibration control by modifying the dynamics of the electrical system. Lesieutre [1] described the four types of shunt circuits: resistive, inductive, capacitive, and switched. Hagood and von Flotow [2] investigated the possibility of dissipating mechanical energy with piezoelectric materials shunted with passive electrical circuits. Aldrich et al. [3] conducted research in the design of a passive control system for a space structure. Wu [4] presented the results of a piezoelectric element shunted with a parallel resistor–inductor circuit. The research interest in this area involves the shunt circuit design and the shunt circuit tuning method, etc.

25.2.1.2 Active Control with Piezoelectric Materials

Piezoelectric-enabled structures are being used increasingly in the control or cancellation of sound and vibration in the low- to mid-frequency range. They serve to bridge the limitation of the conventional method of passive damping techniques, which is only effective when used in high-frequency applications. There are many applications of piezoelectric materials for active sound and vibration control, such as the helicopter blades [5] and gun barrels [6] to reduce structural vibrations [7], an active control of sound emanating from a structure [8,9], and sound radiation control from a harmonically excited panel using multiple piezoelectric actuators [10]. There are two important issues that need to be considered in using piezoelectric materials as actuators for active control systems: (1) they usually require large amount of power for operation and (2) the complexity of the hardware involved with active control (added hardware and control law design, etc.) [11].

The research interest in this area involves how to model the dynamic properties of piezoelectric-enabled structures, optimal placement of piezoelectric actuators, optimal selection of PZT actuator numbers, controller design, controller tuning method, power management, etc.

25.2.1.3 Hybrid Control with Piezoelectric Materials

There are also some applications when active–passive hybrid systems are used to achieve the best solution for vibration suppression [12]. In a hybrid vibration control system, an active control system with a voltage source is integrated with a passive shunt circuit. The advantage of such a system lies in the increase in the overall stability of the active control system.

For whatever means above, it is necessary to have accurate and reliable numerical models for design and analysis of the piezoelectric-enabled structures. Some of the earliest models [13–15] used the induced strain by the piezoelectric actuators as an applied strain that contributed to the total strain of the nonactive structures similar to a thermal strain contribution. To find analytical solutions, electrical and mechanical equilibrium or governing equations have to be solved for a set of boundary conditions [16]. It is generally recognized that theoretical models are applicable only to a very limited range of well-defined geometries and boundary conditions. For practical applications, numerical methods such as finite element method (FEM) are often used for modeling, simulation, and analysis of different smart structure designs.

There are laminated-type and discrete-type smart structures for structural sound and vibration control in practice. The former means that piezoelectric materials cover or are embedded in the entire structure. The latter means that the piezoelectric materials occupy a relatively small area of the structure. The modeling approaches and analysis techniques differ considerably between the laminated-type and the discrete-type smart structures.

The purpose of this chapter is to provide an overview of various numerical modeling methods of the dynamic properties of piezoelectric-enabled smart structures and related issues. The review is brief and makes no attempt to be exhaustive. It is organized into the following sections: (1) linear theory of electrodynamics that outlines the fundamental governing equations of smart structure modeling; (2) methods for laminated-type smart structures that include the elastic and plate theories in addition to coupled layerwise theory; (3) methods for discrete type smart structures that include the equivalent line moment approach, the energy method together with assumed mode method and finite element approach; and (4) new research areas in smart structures, which include nonlinear characteristics of the piezoelectric materials, accuracy description of electric potentials in the piezoelectric layers, etc.

25.2.2 Linear Theory of Electrodynamics

The linear theory of electrodynamics treats all the elastic, piezoelectric, and dielectric coefficients as constants independent of the magnitude and frequency of applied mechanical stresses and electric fields. The governing equation in the absence of the body forces and body charges consists of stress equations of motion [17,18]:

$$\rho \ddot{\mathbf{U}} - \mathbf{L}_d^T \sigma = 0 \tag{25.9}$$

The charge equation of electrostatics and the electric field-electric potential relations are

$$\nabla \cdot \mathbf{D} = 0 \tag{25.10}$$

and

$$\mathbf{E} = -\nabla\phi \tag{25.11}$$

where

the symbolic dot denotes differentiation with respect to time t
the superscript T the transposition
ρ is the material density (kg/m^3)
$\mathbf{U}$ the particle displacement vector (m)
$\boldsymbol{\sigma}$ is the stress vector (N/m^2)
$\mathbf{L}_d$ the operator [17,18]
$\mathbf{D}$ the electric displacement vector (C/m^2)
$\mathbf{E}$ the electric field vector (V/m)
ϕ the scalar electric potential (V)

The constitutive relations for a piezoelectric layer are

$$\sigma = \mathbf{c}\varepsilon - \mathbf{e}^T\mathbf{E} \tag{25.12}$$

and

$$\mathbf{D} = \mathbf{e}\varepsilon + \mathbf{g}\mathbf{E} \tag{25.13}$$

where

$\mathbf{c}$ is the stiffness matrix (N/m^2) at constant electric field strength
$\mathbf{g}$ the dielectric matrix (F/m or C/Vm) at constant mechanical strain
$\mathbf{e}$ the piezoelectric stress matrix (N/mV or C/m^2)

Combination of the equations above yields the three-dimensional differential equations of the linear piezoelectric continuum in terms of the four dependent variables (three displacement components and one electric potential) as

$$\mathbf{L}_d^T \mathbf{c} \mathbf{L}_d \mathbf{U} + \mathbf{L}_d^T \mathbf{e}^T \nabla \varphi = \rho \ddot{\mathbf{U}} \tag{25.14}$$

$$\nabla \cdot \left(\mathbf{e} \mathbf{L}_d \mathbf{U} - \mathbf{g} \nabla \varphi \right) = 0 \tag{25.15}$$

In the presence of boundaries, the appropriate boundary conditions must be adjoined to Equations 25.14 and 25.15. If there is a material surface of discontinuity, across the surface there are the continuity conditions of stresses, displacements, electric displacements, and electric potentials.

25.2.3 Methods for Laminated-Type Smart Structures

The developments of piezoelectric laminate models satisfy the need for more accurate mechanic models to account for the characteristics of composite laminates and the need for more accurate electroelastic models to account for the coupling effects between the electrical and mechanical fields inside the piezoelectric components. The fundamental works by earlier researchers [17,18] provide much of the necessary theoretical basis for the modeling of static and dynamic behavior of laminated-type smart structures.

25.2.3.1 Elastic Theory

The elastic theory utilizes the elastic wave propagation method to establish the state variable relationship between upper and lower surfaces of a laminate. Elastic theory can provide an exact solution of a problem, which satisfies the governing equations at every point of the domain and the boundary conditions of the problem. There are several methods to obtain a solution of governing equations within a piezoelectric layer in the frequency domain.

25.2.3.1.1 Transfer Matrix Method

The continuity of displacements, stresses, and electric parameters at the interfaces of different layers is imposed. On the top and bottom surfaces of the laminate, normal displacements, stresses, and electric variables have to be continuous as well. The boundary condition approach constitutes the basis of transfer matrix method [20,21] and the surface impedance tensor method.

25.2.3.1.2 Surface Impedance Tensor Approach

A significant formulation to a matrix of piezoelectric composite materials appears to be the eight-dimensional state vector formalism proposed originally by Kraut [19]. At the heart of this formulation is the well-known fact that the ordinary differential equations can be transformed into a set of first-order equations. As an extension, Honein et al. [22] proposed a systematic methodology about the surface impedance tensor. The surface impedance tensor relates the components of particle displacement and the normal component of the electric displacement on a surface to the electric potential and the component of traction acting on the same surface. Once the surface impedance tensor for a single layer is obtained, a single recursive algorithm allows the evaluation of the surface impedance tensor for any number of layers. The surface impedance tensor approach is proposed to overcome a numerical difficulty, which exists when getting the solution for many layers as the product of the solutions of each layer. The method has been successfully applied to solve the acoustic reflection and transmission problem when a piezoelectric-enabled structure is subjected to an acoustic wave incidence [23,24].

The advantages of the approach lie in that it can be used not only for wave propagation (the high-order theory) in the piezoelectric layer but also for vibration analysis (the lower-order theory) of general piezoelectric laminates.

25.2.3.2 Classical Laminated Plate Theory

Classical laminated plate theory (CLPT) is an extension of the classical plate theory to composite laminates [25]. Lee and Moon [26], and Lee [27] used the assumptions of CLPT to derive a simple theory for a piezoelectric laminate. Reddy [28] presented the theoretical formulation of laminated plates with piezoelectric layers as sensors or actuators. Many investigators [29,30] used CLPT models and its variations to design piezoelectric laminates for different applications. These models used simplifying approximations in characterizing the induced strain field and electric fields generated due to an applied voltage or external load. CLPT assumes that the Kirchhoff hypothesis holds for the laminate.

CLPT implies that for a case of both plane strain and plane stress. However, from practical considerations, it uses the restriction of each layer, which is in a state of plane stress, because most laminates are typically thin compared to the in-plane dimensions. It is one pertinent assumption in establishing the constitutive relationships for the laminae of a laminated structure.

25.2.3.3 First-Order Shear Deformation Theory

To enhance the CPLT model, an alternative model, first-order shear deformation theory (FSDT) was developed [28], which belongs to the hierarchy of equivalent single-layer (ESL) laminate theories as well. In FSDT, the third Kirchhoff hypothesis is released. It means that transverse normals do not remain perpendicular to the mid-surface after deformation. It is also called as Reissner-Mindlin plate theory or a constant shear angle theory. It is noted that shear correction factors are normally adopted in computing the transverse shear force resultants to correct the discrepancy between the actual stress state and the constant stress state predicted by the FSDT.

25.2.3.4 Third-Order Shear Deformation Theory

In the FSDT model, the shear correction factors are not easy to determine. To achieve better approximation, the higher-order expansions of the displacement field have been developed. A third-order shear deformation theory (TSDT) model accommodates the vanishing of transverse shear strain on the top and bottom surfaces of a laminate [28,31]. Reddy [28] presented the Navier solutions and finite element models of CLPT and TSDT.

25.2.3.5 Coupled Layerwise Theory

Discrete layer theories [31] and layerwise theories [32] have been developed for the static and dynamic analysis of piezoelectric laminates. Unlike the simplified electric and displacement fields through the thickness of the laminate in the equivalent single-layer piezoelectric laminate theories above, the mechanical displacements and the electric potential are assumed to be piecewise continuous across the thickness of the laminate in the layerwise theories. These theories provide a much more kinematically correct representation of cross-sectional warping and capturing nonlinear variation of electric potential through the thickness associated with thick laminates. The developments of layerwise laminate theory for a laminate with embedded piezoelectric sensors and actuators are presented in the literature [33,34]. Comparisons of the predicted free vibration results from the layerwise theory with the exact solutions for a simply supported piezoelectric laminate bring out the improved accuracy and robustness of the layerwise theory over CLPT and FSDT [35] for piezoelectric laminates.

Tani and Liu [36,37] originally proposed a numerical method to investigate the surface waves in functionally gradient piezoelectric plates. Liu and Achenbach [38] named the method the strip element method (SEM). The technique has a clear advantage over the finite element method for layered structures in terms of storage requirements [39]. Recently, Han et al. [40] extended SEM to analysis of surface wave properties in a hybrid multilayered piezoelectric circular cylinder.

25.2.4 Methods for Discrete Type Smart Structures

Discrete monolithic pieces of piezoelectric material can also be used for sensing and control purposes instead of covering the entire surface in the form of layers due to weight and difficulty in fabrication considerations. The discrete patches can either be attached to the structure as in the literature [41–44] or embedded within the substrate as in the work [14].

25.2.4.1 Equivalent Line Moment Approach

In this approach, a static study of the behavior of twin, symmetric piezoelectric actuators bonded on a structure was carried out. It was demonstrated that a perfectly bonded actuator was equivalent to external line moments acting along the boundaries of the piezoelectric elements. The representative papers are from Crawley and deLuis [14] and Dimitriadis et al. [44]. In all these previous models, the usual assumption was that unless an electric field was applied, the presence of the piezoelectric material on or in the substrate does not alter the overall structural properties significantly. It was also assumed that the thickness of the bonding adhesive layer was negligible and caused negligible property changes.

Crawley and deLuis [14] presented a rigorous study of the stress–strain–voltage behavior of piezoelectric elements bonded to and embedded in one-dimensional beams. An important observation was that the effectiveness moments resulting from the piezoactuators can be seen as concentrated on the two ends of the actuators when the bonding layer is assumed infinitely thin. As an extension of the one-dimensional theory [14], Dimitriadis [44] developed the finite two-dimensional piezoelectric elements perfectly bonded to the upper and lower surfaces of elastic plate through static and dynamic analysis. The loads induced by the piezoelectric actuator to the supporting thin elastic structure are estimated in their research. Then, the equivalent magnitude of the edge moments are applied to the plate to replace the actuator patch such that the bending stress at the surface of the plate is equal to the plate's interface stress when the patch is activated.

As mentioned before, the approach used the induced strain by the piezoelectric actuators as an applied strain that contributes to the total strain of the nonactive structures, similar to a thermal strain contribution. Strictly speaking, it is not a fully coupled analysis between mechanical and piezoelectric structures. It only considers the converse piezoelectric effect. The electromechanical coefficient d_{31} is the only source of coupling between the electric and the mechanical displacement fields. Many investigators have used this analytical model for various applications where piezoelectric patches are used for controlling beams, plates, and shells. However, in many device geometries, all elements of the piezoelectric tensors couple to the strain and electric fields and the assumption that only d_{31} plays a role is no longer true.

25.2.4.2 Hamilton's Principle with a Rayleigh-Ritz Formulation

Hagood et al. [45] studied the damping of structural vibrations with piezoelectric materials and passive electric networks. They derived an analytical model for an electroelastic system with piezoelectric materials using Hamilton's principle with a Rayleigh-Ritz formulation. The requirement that all coupling between piezoelectric devices and substrate has to be included in the formulation is effectively facilitated by the Hamilton variational principle or energy methods, which is an effective method for complex structures because mechanical and electrical equilibrium equations do not need to be solved explicitly. The physics of the entire structure has been fully accounted for in the energy integrals and there is no need to derive equations based on forces and moments. The resultant equation of motion is solved by using a Rayleigh-Ritz method.

Hagood et al. [45] found that the piezoelectric energy transformation properties highly coupled to the dynamics of the electric circuit and elastic system. They studied this coupling for symmetry piezoelectric actuators to add damping to the structural modes of a cantilevered beam. Gibbs and Fuller [46] developed an analytical static model to describe the response of an infinite beam subjected to an asymmetric actuation induced by a perfectly bonded piezoelectric element. Plantier et al. [47] derived a dynamical model for a beam driven by a single asymmetric piezoelectric actuator. Their model included the effect of the adhesive bonding layer, i.e., the actuator is not assumed to be perfectly bonded to the base structure. Lam et al. [48] derived the energy equations via virtual work first. The assumed modes method was used to discretize the energy equations and Lagrange's equation was used to obtain the equations of motion for different treatments [7].

25.2.4.3 Finite Element Approach and Other Numerical Methods

For discrete piezoelectric patches embedded or bonded on beam and plate structures, investigators have mostly used numerical methods such as the finite element method since obtaining exact solutions is difficult. A potential advantage of the use of the finite element model for piezoelectric devices is that the fields in the material may be precisely described without simplifying assumptions.

Finite element methods were put into application for piezoelectric structures since Allik et al. [49], Ha et al. [43], and Rao and Sunar [50] developed a finite element method to study the dynamic as well as the static response of plates containing distributed piezoelectric ceramics based on the variational principles. Hwang [30] used a four-node quadrilateral element based on classic laminate theory with the induced strain actuation and Hamilton's principle. They assume that no stress field is applied to the actuator layer and accordingly the equivalent actuator moments per unit length are found as external excitation loads. Lam et al. [51] also developed a finite element model based on the classic laminated plate theory for the active vibration control of a composite plate containing distributed piezoelectric sensors and actuators. Zhou [31] used the same concept to establish finite element models of piezoelectric composite laminates based on Hamilton's principle as well as CLPT and FSDT and TSDT theories respectively. Shen [52] developed a one-dimensional finite element formulation for the flexural motion of a beam containing distributed piezoelectric devices. In his work, the generalized variational principle was used to include the virtual work done by the inertial and electric forces to get the functional principle. Chen et al. [53] presented the general finite element formulations for piezoelectric sensors and actuators by using the virtual work principle.

Kim et al. [54] developed a transition element to connect the three-dimensional solid elements in the piezoelectric region to the flat-shell elements used for the plate. They adopted some special techniques to overcome the disadvantages and inaccuracy of modeling a plate with three-dimensional elements. This approach has merits in terms of accuracy in modeling the piezoelectric patches and computational economy for the plate structure.

In addition, Liu [55,56] originally proposed a point interpolation method (PIM) that belongs to the mesh-free or meshless methods. It uses the nodal values in the local support domain to interpolate the shape functions. To the best knowledge of the authors, Liu et al. [57] initially dealt with piezoelectric device analysis with mechanical and electrical coupled problems using the point interpolation method. In this method, the problem domain was represented by a set of arbitrarily distributed nodes. A polynomial basis was used to construct the shape functions, which possess delta function properties. Piezoelectric structure with arbitrary shape was formulated using polynomial PIM in combination with variational principle and linear constitutive piezoelectric equations. Numerical examples [57] provide solid validations of the proposed method. It was found that the present method is easy to implement, and very flexible and stable for static and dynamic analysis of piezoelectric structures with arbitrary shape and different mechanical or electrical boundary conditions.

25.2.5 Potential Research Areas in Smart Structures

25.2.5.1 Nonlinear Characteristics of the Piezoelectric Materials

Among the currently available sensors and actuators, the smallest ones are of the order of few millimeters. However, progress toward intelligent structures requires us to develop smart material systems that are of the order of a few microns. This reduction in size has tremendous technological benefit; however clear understanding of reliability and system integrity is vital to the efficient and optimum use of these material systems. As dimensions get smaller, induced electromechanical fields get larger. Therefore, the convenience of linearity in modeling should be abandoned, and material and geometric nonlinearities should be accounted for. In addition, piezoelectric materials only exhibit a linear relationship between the electric field and strain for low field values (up to 100 V/mm). The relationship behaves nonlinear for large fields, the material exhibits hysteresis, furthermore, piezoelectric materials show dielectric aging and hence lack reproducibility of strains.

Using piezoelectric actuators, Shi and Atluri [58] conducted studies in order to control the nonlinear dynamic response of a space frame. By incorporating the effect of large deflection and in-plane loads due to the piezoactuators, Pai et al. [59] developed a fully nonlinear theory for the dynamics and active control of a laminated composite plate with integrated piezoelectric actuators and sensors undergoing large-rotation and small-strain vibration. Royston and Houston [60] presented the nonlinear vibratory behavior of a 1–3 piezoceramic composite theoretically and experimentally. In their work, the theoretical model for the electroelastic behavior of the 1–3 composite follows the conventional assumption made by prior investigators but includes nonlinear terms to account for hysteresis in the embedded PZT phase.

25.2.5.2 Accurate Description of Electric Potentials in the Piezoelectric Layers

It is noted that in much of the literature published on the mechanics model for the analysis of the coupled structure, a constant electric potential distribution of the whole piezoelectric actuator in its length dimension and a linear distribution in its thickness dimension are assumed. However, to satisfy the Maxwell equation, a constant electric potential distribution in the longitudinal direction may not be assumed, but rather obtained by solving the coupled governing equations [61].

The field inside layers of a piezoelectric laminate was previously examined by Ray et al. [62], Roh et al. [63], and Heyliger [16] using full elasticity theories, without any approximations on the mechanical and electrical fields. The exact solution obtained using the exact elasticity theory indicated that the electric and elastic field distributions were often poorly modeled using simplified theories [16]. They showed that the electric field inside the sensor layers was not zero and both the electrical and mechanical field distributions were evidently affected by the relative values of the dielectric constants of the layers in a three-layer cross-ply PVDF laminate.

To increase the order of variation of the electrical and mechanical fields inside the layers, higher order theories should be employed in the laminate models. Bisegna and Maceri [64] showed that when the ratio of thickness to width of the plate is larger than or equal to 1/5, FSDT provides results, which differ from exact solution by 20% for displacements, electric potential, and the in-plane stress components. More recently, Yang [65] included higher order (quadratic) electric potential variation through the thickness of the actuators and obtained two-dimensional equations for the bending motion of elastic plates with partially electroded piezoelectric actuators attached to the top and bottom surfaces of a thick plate. Although negligible for thin actuators, this effect of higher order electrical behavior through the thickness needs to be considered for thick actuators. Tiersten [66] derived the approximate equations for extensional and flexural motion of a thin piezoelectric plate subjected to large electric fields. Up to cubic order terms are included in the expansion of electric potential across the plate thickness to describe the higher order electrical behavior, and showed that for a very thin plate, the quadratic and cubic terms in the expansion can be ignored.

Wang and Quek [67,68] assumed the distribution in the transverse direction to be sinusoidal for the short-circuited electrodes case when considering a long and thin beam-embedded piezoelectric actuators. They concluded that the longitudinal distribution of the electric potential in the piezoelectric layer was not constant.

25.2.5.3 Nanostructured Ceramic

Nanostructured ceramic [69] has been put into the research stage because the nanoscale microstructures may result in changes in properties when the feature size is less than a particular level. Among several possible smart nanoscale materials, carbon nanotubes (CNT) have aroused great interest in the research community because of their remarkable mechanical, electrochemical, piezoresistive, and other physical properties. The potential changes in properties (improved strength and toughness etc.) are what has driven research in nanoceramics over the last decade. Kang et al. used a nanosmart material to develop a novel sensor for structural health monitoring [70].

25.2.6 Closure

Five simulation methods on laminated-type smart structures have been reviewed in this chapter. These are (1) elastic theory; (2) classic laminated plate theory; (3) first-order shear deformation theory; (4) third-order shear deformation theory; and (5) coupled layerwise theory. Three simulation methods on discrete-type smart structures are (1) equivalent line moment approach, (2) Hamilton's principle with a Rayleigh-Ritz formulation, and (3) finite element approach and other numerical methods. There are several areas of simulation methods that need attention in the years to come. These include (1) nonlinearity of the piezoelectric materials, (2) accurate description of electric fields in the piezoelectric materials, and (3) new piezoelectric materials using nanotechnologies.

References

1. Lesieutre G. A. 1998. *The Shock and Vibration Digest*, **30**(3), 187–195.
2. Hagood N. W. and von Flotow A. 1991. *Journal of Sound and Vibration*, **146**(2), 243–268.
3. Aldrich J. B., Hagood N. W., von Flotow A., and Vos D. W. 1993. *Proceedings of the SPIE—The International Society for Optical Engineering*, **1917**(2), 692–705.
4. Wu S. Y. 1996. *Proceedings of the SPIE—The International Society for Optical Engineering*, **2720**, 259–269.
5. Straub F. K., Ealey J. A., and Schetley L. M. 1997. *Proceedings of the SPIE—The International Society for Optical Engineering*, **3044**, 99–113. Society of Photo-Optical Instrumentation Engineers, Bellingham, WA.
6. Mattice M. S. and LaVigna C. 1997. *Proceedings of the SPIE—The International Society for Optical Engineering*, **3039**, 630–641, Society of Photo-Optical Instrumentation Engineers, Bellingham, WA.
7. Cai C., Zheng H., Hung K. C., and Zhang Z. J. 2006. *Smart Material and Structures*, **15**, 147–156.
8. Khan M. S., Cai C., Hung K. C., and Varadan V. K. 2002. *Smart Material and Structures*, **11**, 346–354.
9. Linder D. K., Zuonar G. A., Kirley III G. C., and Emery G. M. 1996. *Proceedings of the SPIE—The International Society for Optical Engineering*, **2717**, 543–554.
10. Fuller C. R., Dimitriadis E. K., and Bor-Tsuen W. 1991. *AIAA Journal*, **29**(11), 1802–1809.
11. McGowan A. M. R. 1999. *Proceedings of the SPIE—The International Society for Optical Engineering*, **3674**, 178–195.
12. Kahn S. P. and Wang K. W. 1994. *Active Control of Vibration and Noise—ASME, Design Engineering Division*, **75**, 187–194 ASME, New York.
13. Bailey T. and Hubbard J.E. Jr. 1985. *Journal of Guidance, Control, and Dynamics*, **8**, 605–611.
14. Crawley E. F. and deLuis J. 1987. *AIAA Journal*, **25**, 1373–1385.
15. Robbins D. H. and Reddy J. N. 1991. *Computers and Structures*, **41**, 265–279.

16. Heyliger P. 1997. *J. Appl. Mech. Trans. ASME*, **64**, 299–306.
17. Tiersten H. F. 1969. Linear Piezoelectric plate vibrations; elements of the linear theory of piezoelectricity and the vibration of piezoelectric plates, Plenum, New York.
18. Nayfeh A H. 1995. North-Holland Series in Applied Mathematics and Mechanics, Elsevier, Amsterdam.
19. Kraut E. A. 1969. *Physical Review*, **188**(3), 1450–1455.
20. Cai C., Liu G. R., and Lam K. Y. 2001a. *Journal of Sound and Vibration*, **248**(1), 71–89.
21. Cai C., Liu G. R., and Lam K. Y. 2001b. *Smart Materials and Structures*, **10**, 689–694.
22. Honein B., Braga A. M. B., Barbone P., and Herrmann G. 1991. *Journal of Intelligent Material Systems and Structures*, **2**, 542–557.
23. Barbone P. E. and Braga A. M. B. 1992. First European Conference on Smart Structures and Materials, Forte Crest Hotel, Glasgow, pp. 325–328.
24. Braga A. M. B., Honein B., Barbone P. E., and Herrmann G. 1992. *Journal of Intelligent Material Systems and Structures*, **3**, 209–223.
25. Reddy, J. N. 1997. Mechanics of Laminated Composite Plates and Shells: Theory and Analysis, CRC Press, Boca Raton.
26. Lee C. K. and Moon F. C. 1989. *JASA*, **85**, 2432–2439.
27. Lee C. K. 1990. *JASA*, **87**, 1144–1158.
28. Reddy J. N. 1999. *Engineering Structures*, **21**(7), 568–593.
29. Lam K. Y. and Ng T. Y. 1999. *Smart Materials and Structures*, **8**(2), 223–237.
30. Hwang W. -S., Park, H. C. 1993. *AIAA Journal*, **31**(5), 930–937.
31. Zhou Y. L. 1999. Master's thesis. National University of Singapore.
32. Mitchell J. A. and Reddy J. N. 1995. *International Journal of Solids and Structures*, **32**, 2345–2367.
33. Saravanos D. A., Heyliger P. R., Hopkins D. A. 1997. *International Journal of Solids and Structures*, **34**(3), 359–378.
34. Saravanos D. A. and Heyliger P. R. 1995. *Journal of Intelligent Material Systems and Structures*, **6**, 350–363.
35. Gopinathan S. V., Varadan V. V., and Varadan V. K. 2000. *Smart Materials and Structures*, **9**(1), 24–48.
36. Tani J. and Liu G. R. 1993. *JSME International Journal*, **36**, 152–155.
37. Liu G. R. and Tani J. 1994. *Transactions of the ASME Journal of Vibration and Acoustics*, **116**, 440–448.
38. Liu G. R. and Achenbach J. D. 1994. *ASME Journal of Applied Mechanics*, **61**, 270–277.
39. Liu G. R. and Xi Z. C. 2001. Elastic Waves in Anisotropic Laminates, CRC Press, pp. 235–253.
40. Han X., Liu G. R., and Ohyoshi T. 2004. *Computational Mechanics*, **33**(5), 334–344.
41. Crawley E. F. and Lazarus K. B. 1991. *AIAA Journal*, **29**, 944–951.
42. Thomson S. P. and Loughlan J. 1995. *Composite Structures*, **32**, 59–67.
43. Ha S. K., Keilers C., and Chang Fu-Kuo 1992. *AIAA Journal*, **30**(3), 772–780.
44. Dimitriadis E. K., Fuller C. R., and Rogers C. A. 1991. *Transactions of the ASME*, **113**, 100–107.
45. Hagood N. W., Chung W. H., and Von Flotow A. 1990. AIAA Pap, AIAA-90-1087-CP 2242–2256.
46. Gibbs G. P. and Fuller C. R. 1992. *JASA*, **92,** 3221–3227.
47. Plantier G., Guigou C., Nicolas J., Piaud J. B., and Charette F. 1995. *Acta Acustica*, **3,** 135–151.
48. Lam M. J., Inman D. J., and Saunders W. R. 2000. *Smart Materials and Structures*, **9**, 362–371.
49. Allik H. and Hughes T. J. R. 1970. *International Journal for Numerical Methods in Engineering*, **2,** 151–157.
50. Rao S. S. and Sunar M. 1993. *AIAA Journal*, **31**(7), 1280–1286.
51. Lam K. Y., Peng X. Q., Liu G. R., and Reddy J. N. 1997. *Smart Materials and Structures*, **6,** 583–591.
52. Shen M. –H. H. 1994. *Smart Materials and Structures*, **3**, 439–447.
53. Chen C. Q., Wang X. M., and Shen Y. P. 1996. *Computers & Structures*, **60**(3), 505–512.
54. Kim J., Varadan V. V., and Varadan V. K. 1997. *International Journal for Numerical Methods in Engineering*, **40**(5), 817–832.
55. Liu G. R. and Gu Y. T. 2001. *International Journal of Nuner. Methods and Engineering*, **50**, 937–951.
56. Liu G. R. 2002. *Meshfree Methods*. Moving beyond finite element Method, CRC Press, Boca Raton, FL, in press.
57. Liu G. R., Dai K. Y., Lim K. M., and Gu Y. T. 2002. *Computational Mechanics*, **29**(6), 510–519.
58. Shi G. and Atluri S. N. 1990. *Computer & Structures*, **34**, 549–564.
59. Pai P. F., Nayfey A. H., Oh K., and Mook D. T. 1993. *International Journal of Solids and Structures*, **30**, 1603–1630.
60. Royston T. J. and Houston B. H. 1998. *JASA*, **104**(5), 2814–2827.
61. Gopinathan S. V. 2001. PhD Thesis, The Pennsylvania State University.
62. Ray M. Ch., Rao K. M., and Samanta B. 1993. *Computers & Structures*, **47**, 1031–1042.
63. Roh Y. R., Varadan V. V., and Varadan V. K. 1996. *Smart Materials and Structures*, **5**, 369–378.
64. Bisegna P. and Maceri F. 1996. *ASME Journal of Applied Mechanics*, **63**, 628–638.
65. Yang J. S. 1999. *Smart Materials and Structures*, **8**, 73–82.
66. Tiersten H. F. 1993. *ASME AMD*, **161**, 21–34.
67. Wang Q. and Quek S. T. 2000. *Smart Materials and Structures*, **9**, 103–109.
68. Quek S. T. and Wang Q. 2000. *Smart Materials and Structures*, **9**, 859–867.
69. Cain M. and Morrell R. 2001. *Chemistry*, **15**, 321–330.
70. Kang I., Schulz M. J., Lee J. W., Choi, G. R., Jung J. Y., Choi J.-B., and Hwang S.H. 2007. *Solid State Phenomena*, **120**, 289–296.

26

Active Truss Structures

B. de Marneffe
Université Libre de Bruxelles

André Preumont
Université Libre de Bruxelles

26.1 Introduction

The concept of active trusses is quite natural: it consists of replacing one or several passive bars by active struts. Piezoelectric transducers are ideally suited for this purpose, because of their high stiffness and virtually unlimited resolution; other types of transducer, based, e.g., on electrostrictive materials can also be used but they are not investigated here. The active truss concept was first demonstrated in the late 1980s [1–3]; very effective active control strategies were proposed at that time. One of them, known as integral force feedback (IFF), is based on a collocated force sensor and has guaranteed stability [4] as long as perfect amplifier and actuator dynamics are assumed.
Although active contol can be efficient, it implies power electronics (for the actuator), a sensitive signal amplifier (for the sensor) and an analog or digital fiter (for the controller). This might be impractical in many applications and has motivated the use of the so-called *shunt* circuits, in which an electrical circuit is directly connected to a transducer embedded in the structure. The tranducer acts as an energy converter: it transforms mechanical (vibrational) energy into electrical energy, which is in turn dissipated in the *shunt* circuit [5-8]. No seperate sensor is required, and only one, generally simple electronic circuit is used. It is generally passive, i.e. made of resistors and inductors, but it can also be active as proposed in [9].

This chapter compares all these methods with each other. The benchmark truss structure used for this purpose is described in Figure 26.1. It consists of 12 bays of 140 mm each, made of steel bars of 4 mm diameter connected with plastic joints and it is clamped at the bottom. It is equipped with two active struts (piezo transducer + force sensor) as indicated in the figure.

26.2 Piezoelectric Stack Actuator

Consider the piezoelectric linear transducer of Figure 26.2: it is made of n identical slices of piezoceramic material stacked together, each of them polarized through the thickness. If one assumes that all the mechanical and electrical quantities are one-dimensional and parallel to the poling direction, the constitutive equations for the piezoelectric material are:

$$\begin{Bmatrix} D \\ S \end{Bmatrix} = \begin{bmatrix} \varepsilon^T & d_{33} \\ d_{33} & s^E \end{bmatrix} \begin{Bmatrix} E \\ T \end{Bmatrix} \tag{26.1}$$

where the standard IEEE notations have been used [10]. Integrating Equation 26.1 over the volume of the transducer gives, with the notations of Figure 26.2:

$$\begin{Bmatrix} Q \\ \Delta \end{Bmatrix} = \begin{bmatrix} C & nd_{33} \\ nd_{33} & 1/K_a \end{bmatrix} \begin{Bmatrix} V \\ f \end{Bmatrix} \tag{26.2}$$

where

$Q = nAD$ is the total electric charge on the electrodes of the transducer

$\Delta = Sl$ is the total extension ($l = nt$ is the length of the transducer)

$f = AT$ is the total force

V the voltage applied between the electrodes of the transducer, resulting in an electric field $E = V/t = nV/l$

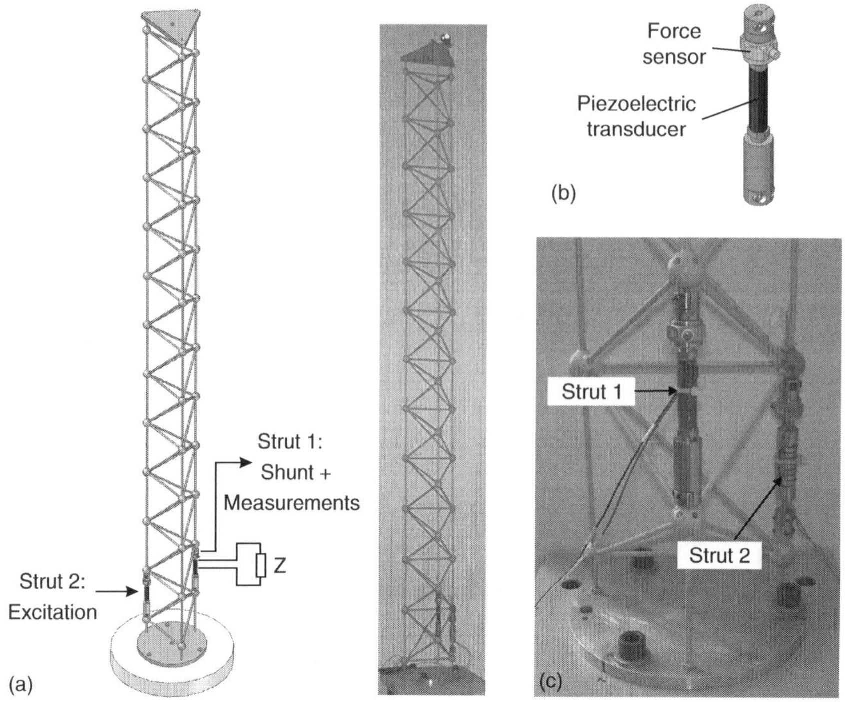

FIGURE 26.1 (a) Truss structure used in the experiment; (b) detail of an active strut; and (c) disposition of the active struts (zoom).

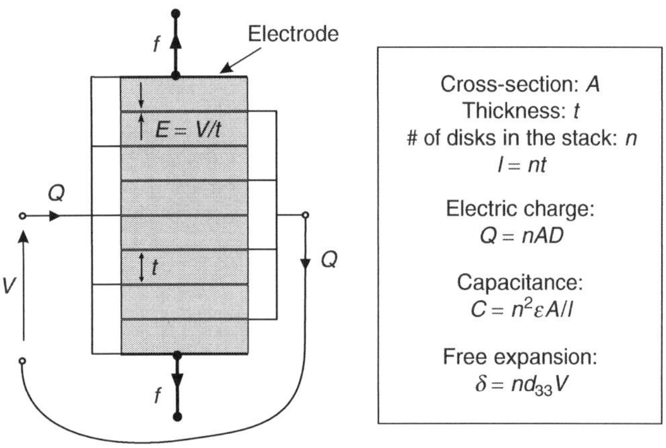

FIGURE 26.2 Piezoelectric linear transducer.

The capacitance of the transducer with no external load ($f = 0$) is $C = \varepsilon^T An^2/l$, and $K_a = A/s^E l$ is the stiffness with short-circuited electrodes ($V = 0$). The electromechanical coupling factor of the material (in the direction of polarization) is another material constant; it is defined by

$$k^2 = \frac{d_{33}^2}{s^E \varepsilon^T} = \frac{n^2 d_{33}^2 K_a}{C} \tag{26.3}$$

It measures the efficiency of the conversion of mechanical energy into electrical energy and vice versa. Finally, one finds upon inverting Equation 26.2 that the stiffness of the transducer with open electrodes ($Q = 0$) reads:

$$\left.\frac{\Delta}{f}\right|_{Q=0} = \frac{K_a}{(1-k^2)} \tag{26.4}$$

which has important consequences in the following.

Various realizations of commercially available piezoelectric uniaxial transducers are presented in Figure 26.3. The first one (Figure 26.3a) is the one used in this work. Realizations (a) and (b) follow exactly the principle of Figure 26.2. The latter is delivered with an in-built external prestressing mechanism while the former requires the introduction of an internal wire to this end. In the (c) and (d) designs, the piezoelectric stack is mounted along the main axis of an elliptical structure, which acts as a motion amplifier. The expansion of the actuator is in the direction perpendicular to that of the piezostack. This design leads to more compact (but softer) actuators. The figure also shows force sensors and flexible tips that prevent torques to be applied on the stack.

26.2.1 Structure with a Stack Transducer

Suppose that the piezoelectric transducer described in the previous section is embedded in a linear, undamped structure. A voltage V is applied across the electrodes and an electric charge Q flows into it (there is a relation between Q and V but it has no impact on the discussion). The dynamic equations of the structure alone, i.e., without the piezo, read classically:

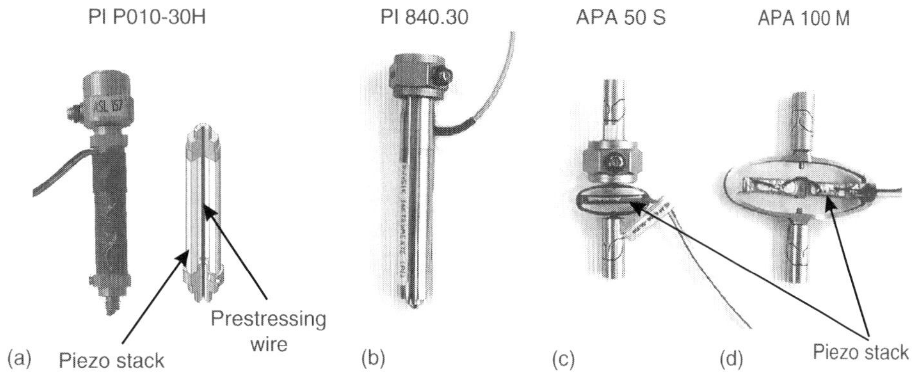

FIGURE 26.3 Realizations of uniaxial transducers (a) and (b) classical design (Physik Instrumente Gmbh), (c) and (d) amplified design (CEDRAT). In (a), (b), and (c), a force sensor is collocated with the actuator. (c) and (d) are equipped with flexible joints.

$$Mx + Kx = F \tag{26.5}$$

where K and M are the stiffness and mass matrices of the structure (without the piezo), obtained, for example, by means of a finite element model, and F is the external force vector. Here we consider that the only forces exerted on the structure come from the transducer, i.e.

$$F = bf \tag{26.6}$$

where b is the projection vector relating the end displacements of the strut to the global coordinate system, and f is the force exerted by the piezo (Equation 26.2). Taking into account that $\Delta = b^T x$, Equations 26.5 and 26.6 can be combined with Equation 26.2; one gets

$$Ms^2x + (K + K_a bb^T)x = bK_a nd_{33}V \tag{26.7}$$

$$C(1-k^2)V + nd_{33}K_a b^T x = Q \tag{26.8}$$

where $K + K_a\, bb^T$ is the global stiffness matrix (including the transducer) and s is the Laplace variable.

26.2.1.1 Various Natural Frequencies

Analyzing further Equations 26.7 and 26.8, three different eigenvalue problems can be defined, corresponding to the boundary conditions $f = 0$, $V = 0$, and $Q = 0$, respectively.

1. From Equation 26.7, the eigenvalue problem when the axial stiffness of the actuator is cancelled, i.e., $K_a = 0$, is given by:

$$(Ms^2 + kx) = 0 \tag{26.9}$$

2. If $V = 0$, i.e., if the piezo is short-circuited, the structure obeys the following relation:

$$\left(M_s^2 + K + K_a bb^T\right)x = 0 \tag{26.10}$$

3. Finally, if the structure is charge-driven instead of voltage-driven, V can be eliminated from Equations 26.7 and 26.8 and the new equation is

$$Ms^2x + \left(K + \frac{K_a}{1-k^2}bb^T\right)x = b\frac{K_a}{1-k^2}nd_{33}\frac{Q}{C} \tag{26.11}$$

and if the structure is open-circuited, i.e., $Q = 0$, it obeys the following relation:

$$\left(Ms^2 + K + \frac{K_a}{1-k^2}bb^T\right)x = 0 \tag{26.12}$$

which is the same as Equation 26.10 but with the short-circuit stiffness K_a replaced by the open-circuit one $K_a/(1-k^2)$.

The solutions of these eigenvalue problems are three different sets of natural frequencies. In this work, the natural frequencies when $K_a = 0$ are called z_i ($i = 1, \ldots, n$), those when the piezo is short-circuited are called ω_i, and those when the piezo is open-circuited are called Ω_i. Because the stiffness of the transducer increases when it is left open (Equation 26.4), one has $\Omega_i > \omega_i$.

26.3 Integral Force Feedback

This section considers the active damping of the truss based on a force sensor colinear with the piezoelectric transducer. Upon inverting Equation 26.2, the output force f is given by

$$f = K_a(b^T x - nd_{33}V) \tag{26.13}$$

where once again $b^T x = \Delta$ is the axial elongation of the transducer. The IFF consists of

$$V(s) = \frac{g}{nd_{33}K_a}\frac{y(s)}{s} \tag{26.14}$$

where g is the gain of the controller and the coefficient $1/nd_{33} K_a$ has been introduced for the purpose of normalization. Combining Equations 26.7, 26.13 and 26.14, the closed-loop poles are given by

$$\left[Ms^2 + (K + K_a bb^T) - \frac{g}{s+g}(K_a bb^T) \right] x = 0 \qquad (26.15)$$

As $g \to 0$, Equation 26.15 becomes identical to Equation 26.10, which defines the short-circuited natural frequencies ω_i. On the other hand, the asymptotic roots as $g \to \infty$ are the open-loop zeros z_i (Equation 26.9), which correspond to the situation where the axial stiffness of the active strut has been removed. If the natural frequencies of the structure are well separated, each mode is independent of the others. In these conditions, Equation 26.15 can be reduced to a set of uncoupled Equation 26.4:

$$1 + g\frac{s^2 + z_i^2}{s(s^2 + \omega_i^2)} = 0 \qquad (26.16)$$

Such a root locus (starting from ω_i and ending at z_i) is presented in Figure 26.4a when z_i and ω_i are not too far apart. In this case, the maximum modal damping is given by [4]:

$$\xi^{\max} = \frac{\omega_i - z_i}{2z_i} \left(z_i \geq \frac{\omega_i}{3} \right) \qquad (26.17)$$

When the distance between z_i and ω_i increases, the shape of the root locus changes as described in Figure 26.4b. For $z_i < \omega_i/3$, the root locus touches the real axis, which means that critical damping can be attained. Figure 26.4 indicates that the closed-loop poles are in the left half plane (and thus stable) as long as $g > 0$: this guaranteed stability is a consequence of the collocation between the actuator and the sensor. It remains valid even for multimode structures, as long as perfect actuator and sensor dynamics are assumed.

26.3.1 Extension to Cable Structures

The use of cables to achieve lightweight spacecrafts is not new: it can, for example, be found in Herman Oberth's early books on astronautics. In terms of weight, the use of guy cables is probably the most efficient way to stiffen a structure; cables can also be used to prestress a deployable structure and eliminate the geometric uncertainties due to the gaps. The dynamic modeling of such a structure and the design of a robust controller are however complex tasks, because of the nonlinear behavior of the cables and because of the complex interactions between the cables and the structure.

In this context, the use of a collocated controller with guaranteed stability such as the IFF is very appreciated. Indeed, it is easy to show that this control law is passive, i.e., it always extracts energy from the system, even in presence of large nonlinearities. Tendon control with IFF has been investigated, for example, in Refs. [11–13]. It has been found that, as long as the dynamics of the cables can be neglected, the closed-loop poles follow exactly the same root locus as that of the previous section (Equation 26.16). This root-locus is once again shown in Figure 26.5. Here ω_i is identified as the structural natural frequencies when the cable is connected, and z_i as those when the cable is removed. The theory is readily extended to the use of several decentralized active tendons with the same gain. In this case, ω_i corresponds to the structure when all the cables are connected, and the z_i to that when all the cables are removed. Experiments have shown that Equation 26.16 provides an accurate estimate of the closed-loop poles.

26.4 Passive Shunt Damping

When a piezoelectric structure is connected to an electrical circuit, the voltage V across the electrodes of the transducer and the current I that flows into them are related by:

$$I(s) = -Y_{\text{shunt}}(s)V(s) \qquad (26.18)$$

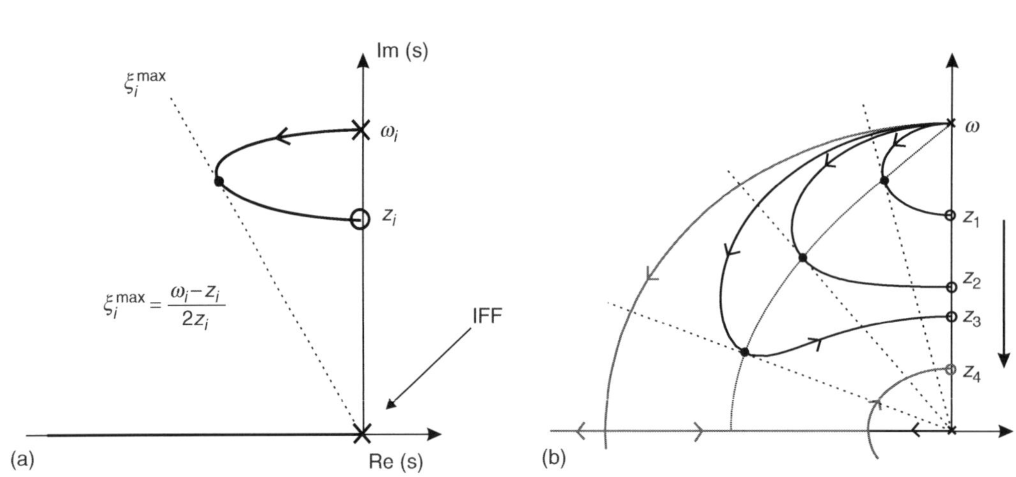

FIGURE 26.4 (a) Root locus of the IFF for a single-mode structure and (b) evolution of the locus as the zero z_i moves toward the origin. Only one half of the locus is shown, the dotted lines correspond to the maximum damping.

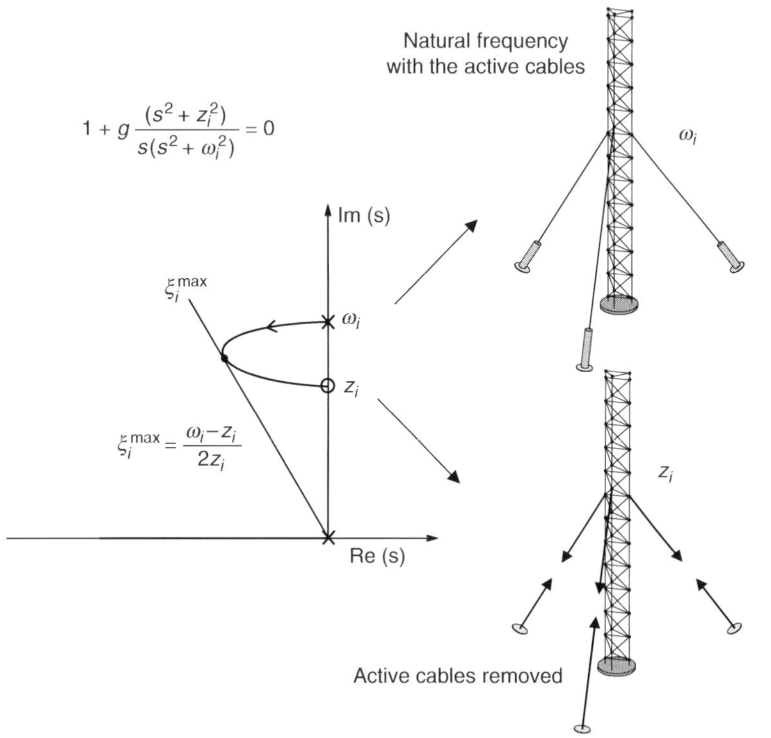

FIGURE 26.5 Open-loop poles and zeros and root locus of an active cable structure with IFF.

where Y_{shunt} is the admittance of the shunt circuit and the minus sign has been introduced to be consistent with the directions of V and I defined in Figure 26.2 (I is the derivative of Q, which in Laplace variable becomes $I = sQ$). Introducing Equation 26.18 into Equation 26.7 and 26.8 the following eigenvalue problem is obtained:

$$\left(Ms^2+K+K_abb^T+\frac{k^2}{(1-k^2)}\frac{K_abb^T}{1+Y_{shunt}/SC(1-k^2)}\right)x_i=0 \quad (26.19)$$

It becomes identical to Equation 26.10 when $Y_{shunt} \to \infty$, leading to the frequencies ω_i (short-circuited), and it is identical to Equation 26.12 when $Y_{shunt} \to 0$, leading to Ω_i (open electrodes).

26.4.1 Resistive Shunting

With $Y_{shunt} = 1/R$, Equation 26.19 becomes:

$$\left[Ms^2+(K+K_abb^T)+\frac{k^2K_abb^T}{(1-k^2)+1/sRC}\right]x=0 \quad (26.20)$$

Going into modal coordinates, all the modes uncouple into a set of independent characteristic equations:

$$1+\frac{1}{RC(1-k^2)}\frac{s^2+\omega_i^2}{s(s^2+\Omega_i^2)}=0 \quad (26.21)$$

According to Equation 26.21, the closed-loop poles follow a root locus similar to that of the IFF (Figure 26.4) when R varies from $+\infty$ to 0. The open-loop poles are in this case at $\pm j\Omega_i$ (open electrodes) and the open-loop zeros are at $\pm j\Omega_i$ (short-circuited). Just as with the IFF, the maximum achievable damping is given by

$$\xi_{i,R}^{max}=\frac{\Omega_i-\omega_i}{2\omega_i} \quad (26.22)$$

Despite the fact that the IFF and the resistive shunting share a similar root-locus, they have very different performances, as is experimentally demonstrated below.

26.4.2 Inductive Shunting

When the shunt consists of a resistance in series with an inductance, its admittance is

$$Y_{shunt}(s)=\frac{1}{Ls+R} \quad (26.23)$$

The connection of the shunt inductance with the inherent capacitance of the piezo creates an electric resonance which, if properly tuned, can increase the damping over that of a R shunt. With the same methodology as in the previous section, Equation 26.19 can once more be rearranged in a root locus form:

$$1+2\xi_e\omega_e\frac{s(s^2+\Omega_i^2)}{s^4+(\Omega_i^2+\omega_e^2)s^2+\omega_i^2\omega_e^2}=0 \quad (26.24)$$

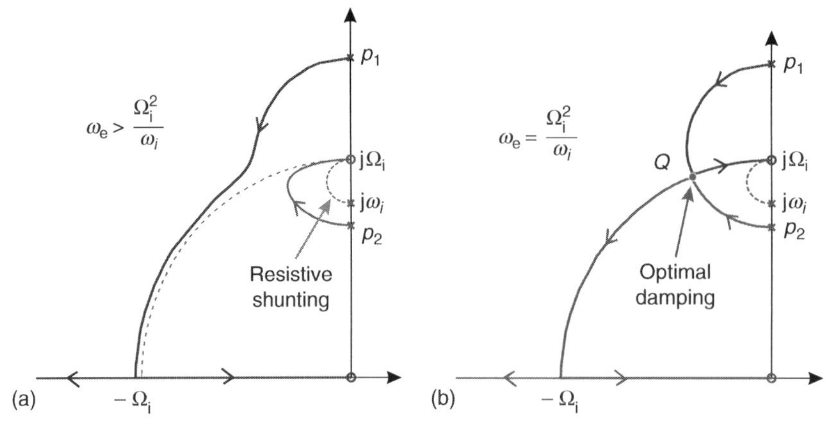

FIGURE 26.6 Root locus of the structure with a series *RL* shunt. (a) The electrical frequency ω_e is not optimally tuned and (b) Optimal tuning of ω_e.

where we have defined the electrical frequency and damping:

$$\omega_e^2 = \frac{1}{LC(1-k^2)} \qquad 2\xi_e\omega_e = \frac{R}{L} \tag{26.25}$$

Two pairs of poles p_1 and p_2 are now present; their evolution as ξ_e changes is shown in Figure 26.6 for two values of ω_e. The locus consists of two loops, starting respectively from p_1 and p_2; one of them goes to $j\Omega_i$ and the other goes to the real axis, near $-\Omega_i$. If $\omega_e > \Omega_i^2/\omega_i$ (Figure 26.6a), the upper pole is always more heavily damped than the lower one, while the opposite occurs if $\omega_e < \Omega_i^2/\omega_i$. If $\omega_e = \Omega_i^2/\omega_i$ (Figure 26.6b), the two poles are always equally damped until the two branches touch each other in *Q*. This double root can be regarded as the optimum tuning of the inductive shunting; the corresponding eigenvalues satisfy

$$s^2 + \Omega_i^2 + \Omega_i\left(\frac{\Omega_i^2}{\omega_i^2} - 1\right)^{1/2} s = 0 \tag{26.26}$$

and the corresponding damping ratio can be obtained easily by identifying the previous equation with the classical form of the damped oscillator, $s^2 + 2\xi_i\Omega_i s + \Omega_i^2 = 0$, leading to

$$\xi_i = \xi_{i,RL}^{\max} = \frac{1}{2}\sqrt{\frac{\Omega_i^2 - \omega_i^2}{\omega_i^2}} \tag{26.27}$$

This value is significantly higher than that achieved with purely resistive shunting. It is approximatively the square-root of Equation 26.22. Note, however, that it is much more sensitive to the tuning of the electrical parameters on the targeted modes.

26.4.3 Generalized Electromechanical Coupling Factor

The quantity

$$K_i^2 = \frac{\Omega_i^2 - \omega_i^2}{\omega_i^2} \tag{26.28}$$

is known as the generalized electromechanical coupling factor for mode *i*. It controls the conversion mechanism between mechanical and electrical energy. It can be shown [14] that

$$K_i^2 \approx \frac{k^2 \nu_i}{1-k^2} \tag{26.29}$$

where

k is the material electromechanical coupling factor defined in Equation 26.3

ν_i is the fraction of modal strain energy in the piezoelectric transducer for mode *i*

ν_i quantifies the capability to concentrate the vibration strain energy of mode *i* in the transducer, and *k* quantifies the conversion mechanism in the material. The performances of the *R* and *RL* shunts are directly related to K_i; indeed, Equations 26.22, 26.27, and 26.28 give

$$\xi_{i,R}^{\max} \approx \frac{K_i^2}{4} \quad \text{and} \quad \xi_{i,RL}^{\max} = \frac{K_i}{2} \tag{26.30}$$

26.5 Active Shunt Damping

Equation 26.29 highlights the importance of a high *k* in the shunt damping performances. The piezoelectric material lead zirconate titanate (PZT) typically has $k \approx 0.7$; it is currently difficult to find materials with higher values, even if promising materials such as PMN-PT with *k* as high as 0.9 have been recently reported.

k can also be artificially increased by the use of an active electronic circuit that simulates the behavior of a negative capacitance. The concept was first introduced by Forward in 1979 [9]. It remained however largely forgotten for 20 years, until some researchers reintroduced the idea [15–17]. Since then, the idea has been used in various practical applications (e.g., [18–20]). The "negative capacitance" by itself does not dissipate energy, but it does enhances the dissipation of energy in the passive (*R* or *RL*) electric circuit.

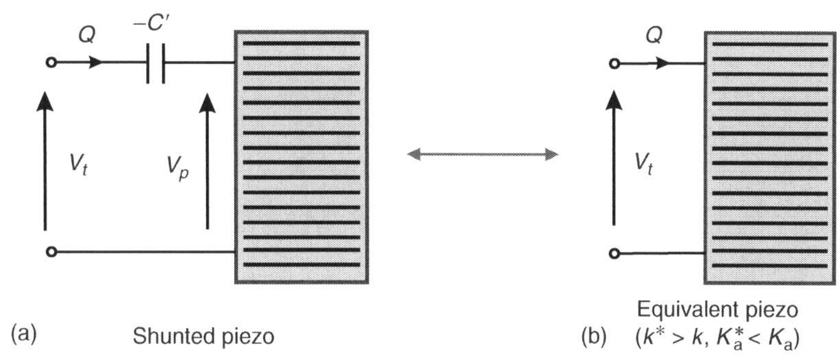

FIGURE 26.7 Piezoelectric transducer shunted on a series negative capacitance $-C'$ and equivalent transducer.

Consider a piezoelectric linear actuator shunted in series by a negative capacitance circuit of value $-C'$ (Figure 26.7a): one immediately finds

$$V_t = V_p - Q/C' \tag{26.31}$$

where
V_p is the voltage across the transducer
Q is the electrical charge flowing into it

Introducing Equation 26.31 into Equation 26.32, it is found that the shunted transducer is equivalent to a "simple" piezo transducer with a higher electromechanical coupling coefficient k^* and smaller stiffness K_a^* (Figure 26.7b). More precisely, one has:

$$K_a^* = \frac{K_a(C'-C)}{C'-C(1-k^2)} \quad k^{*2} = \frac{k^2 C'}{C'-C(1-k^2)} \tag{26.32}$$

The open-circuit stiffness (defined in Equation 26.4), on the other hand, does not change: $k_a^*/(1 - k^{*2}) = k_a/(1 - k^2)$.

A series negative capacitance thus decreases the short-circuit natural frequencies ω_i^* and has no impact on the Ω_i^*, which in turn increases the performances of the R and RL shunts (Equations 26.22 and 26.27).

The negative capacitance is an active shunt that can become unstable if improperly tuned. A stability analysis showed that a sufficient (but not necessary) stability condition reads:

$$C' > C \tag{26.33}$$

with C the constant-force capacitance of the transducer (this value of C' leads to $k^* = 1$ and $K_a^* = 0$). In practice, reasonable stability margins must be taken because of the large nonlinearities of the piezoelectric materials. Throughout this work and for the purpose of illustration, the negative capacitances were tuned as close as 90% of the stability limit. This was made possible by the accurate knowledge of the structural impedance. In realistic situations where the uncertainties on the impedance are much larger, negative circuits at 60–70% of the limit would be more advisable. Unfortunately, the performances rapidly drop when C' moves away from the stability limit. A negative capacitive circuit can be implemented as described in [21]. It is a simple circuit involving an operational amplifier. Note that the required voltage level can be quite high.

26.6 Experimental Results

In this section, the different damping techniques are quantitatively evaluated and compared with each other. To this end, they are implemented on the benchmark truss structure represented in Figure 26.1. We focus on the damping of the first structural mode, and the controller or shunt circuit is applied on strut 1 only (a modal analysis showed that this strut has a reasonable authority on the targeted mode). In this work, strut 2 is used only as a mean to excite the structure.

26.6.1 Frequency Response Function (FRF)

The FRF between the voltage at strut 2 and the force in strut 1 (see Figure 26.1a) is measured for different shunt circuits. These FRFs are shown in Figure 26.8, first when the shunt is purely resistive, then when a negative capacitance is added. The R shunt alone can raise ξ_1 from 0.18% to 0.43%. The negative capacitance shunt can raise it further to approximately 3%.

The same measurements are presented in Figure 26.9, but with a RL shunt. The damping ratio of the two pairs of poles p_1 and p_2 is measured to be 2.7% and 2.5%, respectively, while the addition of a negative capacitance brings these damping ratios to 9.3% and 5.4%, respectively. It is worth noticing that a passive RL shunt alone performs as well as an active "negative capacitance + R" shunt.

Finally, the IFF controller is implemented. The corresponding FRF is also shown in Figure 26.9. This control law was found extremely easy to implement as a result of its guaranteed stability.

26.6.2 Root Loci

The structural poles have been recorded for various values of the gain g of the IFF or of the resistance R of the shunt; these poles are presented in Figure 26.10 (each dot represents a measurement). The solid lines represent theoretical predictions based on Equation 26.16 and on the identified ω_1, Ω_1, and z_1; the agreement between theory and experiment is very good.

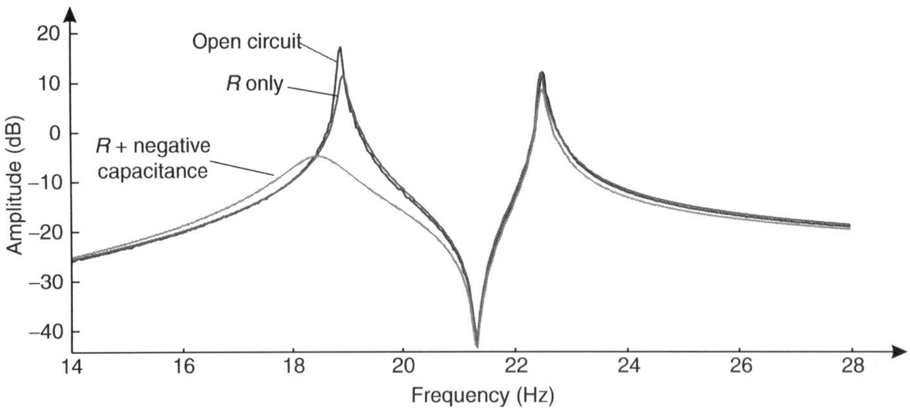

FIGURE 26.8 Structural response with an active and a passive resistive shunt.

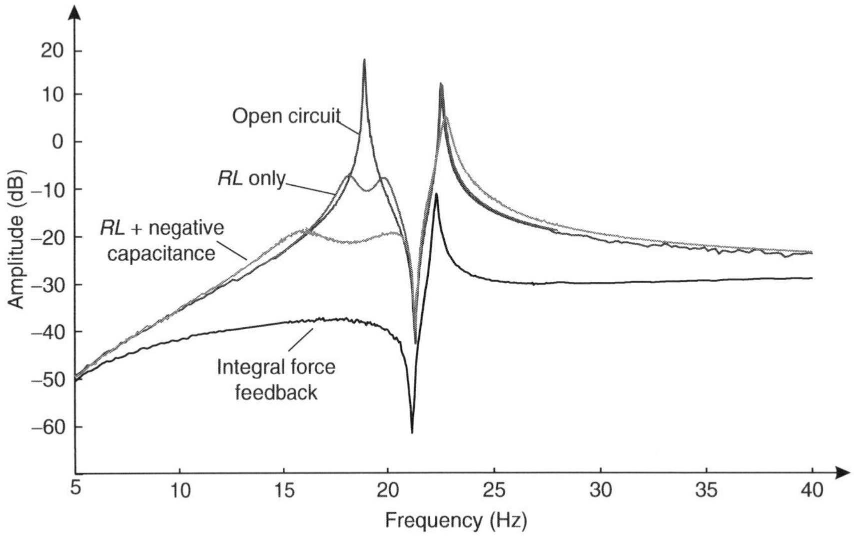

FIGURE 26.9 Structural response with an active and a passive *RL* shunt and with the active IFF controller.

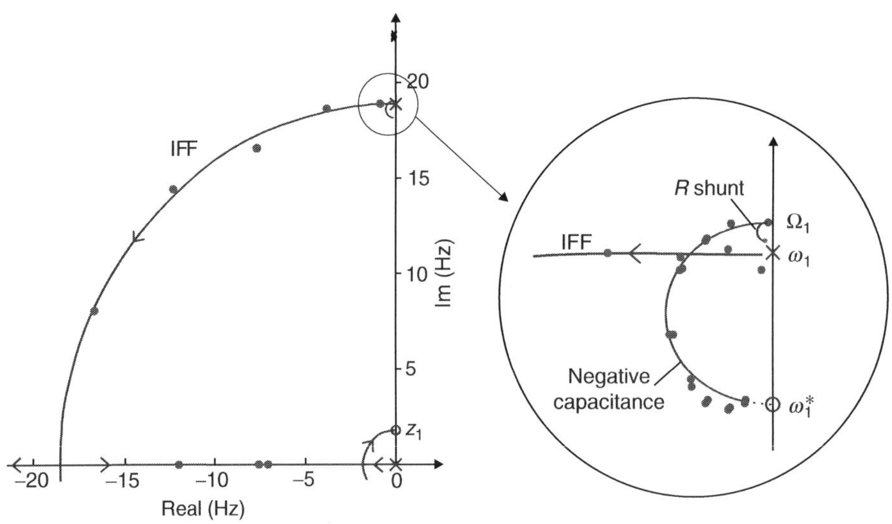

FIGURE 26.10 Identified root locus of the IFF and of the *R* shunts (with and without the negative capacitance). Each dot represents a measurement.

It is found that the root locus of the IFF is extremely large: critical damping can be attained. This is due to the physical nature of the zero z_1, which corresponds to the natural frequency of the truss when the axial stiffness of the transducer is removed, i.e., almost a mechanism. On the other hand, the loci obtained with *R* shunts (even when a negative capacitance is added) are much smaller, in spite of the relatively large values of ν_i and k^2. Figure 26.10 highlights the fact that the attainable performances with a separate collocated transducer/force sensor pair are one order of magnitude larger than those that can be attained with a shunt circuit, even active. The boundary conditions leading to the different eigenfrequencies ω_1, Ω_1, and z_1 are determinant when evaluating the potential performances of such a control system.

Acknowledgments

This study was supported by the Inter University Attraction Pole IUAP 5 on Advanced Mechatronic Systems and by EU-FW6 within the projects INMAR: Intelligent Materials for Active Noise reduction and CASSEM: Composite and Adaptative Structures: Simulations, Experimentation and Modeling.

References

1. J.L. Fanson, G.H. Blackwood, and C.C. Chu. Active-member control of precision structures. In SDM Conference, pp. 1480–1494. AIAA paper 89-1329 CP, 1989.
2. G.S. Chen, B.J. Lurie, and B.K. Wada. Experimental studies of adaptive structures for precision performance. In SDM Conference, pp. 1462–1472. AIAA paper 89-1327 CP, 1989.
3. A. Preumont, J.P. Dufour, and C. Malekian. Active damping by a local force feedback with piezoelectric actuators. *AIAA Journal of Guidance*, 15(2):390–395, 1992.
4. A. Preumont. *Vibration Control of Active Structures: an Introduction*, 2nd ed., Kluwer, Dordrecht, 2002.
5. R.L. Forward. Electronic damping of vibrations in optical structures. *Applied Optics*, 18(5):690–697, 1979.
6. N.W. Hagood and A. von Flotow. Damping of structural vibrations with piezoelectric materials and passive electrical networks. *Journal of Sound and Vibration*, 146(2):243–268, 1991.
7. J.J. Hollkamp. Multimodal passive vibration suppression with piezoelectric materials and resonant shunts. *Journal of Intelligent Material Systems and Structures*, 5:49–57, 1994.
8. S.Y. Wu. Method for multiple mode piezoelectric shunting with single PZT transducer for vibration control. *Journal of Intelligent Material Systems and Structures*, 9:991–998, 1998.
9. R.L. Forward. Electromechanical Transducer-Coupled Mechanical Structure with Negative Capacitance Compensation Circuit. U.S. patent 4,158,787, 1979.
10. IEEE Std. IEEE Standard on Piezoelectricity, 1988. ANSI/IEEE Std 176-1987.
11. Y. Achkire. Active Control of Cable-Stayed Bridges. PhD thesis, Université Libre de Bruxelles (ULB), Belgium, 1997.
12. A. Preumont, Y. Achkire, and F. Bossens. Active tendon control of large trusses. *AIAA Journal*, 38(3):493–498, 2000.
13. A. Preumont and F. Bossens. Active tendon control of vibration of truss structures: Theory and experiment. *Journal of Intelligent Material Systems and Structures*, 11(2):91–99, 2000.
14. A. Preumont. *Mechatronics: Dynamics of Electromechanical and Piezoelectric Systems*. Springer, 2006.
15. S.-Y. Wu. Broadband piezoelectric shunt for structural vibration control. U.S. Patent 6,075,303, 2000.
16. M. Date, M. Kutani, and S. Sakai. Electrically controlled elasticity utilizing piezoelectric coupling. *J. Applied Physics*, 87(2):863–868, 2000.
17. J. Tang and K.W. Wang. Active-passive hybrid piezoelectric networks for vibration control: comparisons and improvement. *Smart Materials and Structures*, 10:794–806, 2001.
18. J.S. Kim, K.W. Wang, and E.C. Smith. High-authority piezoelectric actuation system synthesis through mechanical resonance and electrical tailoring. *Journal of Intelligent Material Systems and Structures*, 16:21–31, 2005.
19. M. Neubauer, R. Oleskiewicz, K. Popp, and T. Krzyzynski. Optimization of damping and absorbing performance of shunted piezo elements utilizing negative capacitance. *Journal of Sound and Vibration*, 298:84–107, 2006.
20. H. Yua, K.W. Wang, and J. Zhang. Piezoelectric networking with enhanced electromechanical coupling for vibration delocalization of mistuned periodic structures-theory and experiment. *Journal of Sound and Vibration*, 295:246–265, 2006.
21. Philbrick Researches Inc. Application manual for computing amplifiers for modelling, measuring, manipulating & much else. Technical report, Nimrod Press, Boston, MA, 1965.

27

Application of Smart Materials and Smart Structures to the Study of Aquatic Animals

Jesse E. Purdy
Southwestern University

Alison Roberts Cohan
Pacific Whale Foundation

27.1 Introduction and Chapter Overview

27.1.1 Defining Smart Materials and Smart Structures

Smart materials sense aspects of the environment and respond appropriately. They also reverse that response when conditions change. For example, ferroelectric-based piezoelectrics generate electric polarity when subjected to mechanical stress. Conversely, if one applies a voltage to piezoelectric crystals, the crystals respond by changing shape in a linear direction. Electrostrictive and magnetostrictive are also smart materials. These materials change size in response to electric or magnetic stimuli, respectively, and produce a voltage when stretched. Rheological smart materials are fluids that change state instantly in response to electrical (electrorheological) or magnetic (magentorheological) stimuli. Electrochromic materials have the ability to change their optical properties when an electric current is applied. And smart gels shrink or swell by factors of 1000 or more a factor useful to absorb or release fluids. Similarly, smart fabrics are fabrics with a special coating that causes them to contract when strain surpasses a specified limit.

The applications for smart materials are numerous. In snow skis, piezoelectric materials detect vibrations and cause other piezoelectrics to cancel that vibration (Ashley, 1996). The resulting reduction in vibration keeps more of the ski on the snow and can be used to keep the tires of a car on the road or reduce vibrations in buildings and bridges (Gibbs, 1996; Perkins, 1997). In addition, piezoelectrics are used in sensors that cause airbags to deploy. Electrostrictive and magnetostrictive smart materials are found in pumps, wind tunnels, and landing gear and they also find application in the health sciences. Rheological smart materials are fluids also used to reduce vibration and electrochromic materials underlie liquid crystal displays (LCD). Because smart gels become increasingly stiff in response to mechanical stress, they have been used in ski outfits to reduce the chance of injury. Similarly, the smart bra is made from a fabric with a special coating that causes it to contract when strain surpasses a specified limit. The smart bra is ideal for exercise in that it stiffens and provides better support when needed (Collins, 2003).

Smart materials can be combined to produce smart structures. These structures perform sensing, controlling, and actuation activities. Or, as Cao et al. (1999) state, "a critical element of smart structures is the development of an optimized control algorithm that guides the actuators, following detection of certain sensory stimuli." Using feedback control mechanisms, smart structures can provide precise active damping functions and are used for that purpose in the space shuttle and space station. Interestingly, Cao et al. predict that future smart structures will be used to develop supersensitive noses, ears, and eyes that are smart enough to directly communicate with the human brain. Such structures could enable humans to expand their sensory input exponentially. We could hear the infrasounds produced by elephants and the ultrasounds produced by bats and many marine mammals. We could see infrared like snakes, detect ultraviolet radiation like a goldfish, and smell the world like dogs. These abilities would have a profound effect on humans and additionally could enlighten us as to what it is like to be an elephant, a bat, a goldfish, or a dog. By experiencing the sensory worlds of nonhuman animals through smart materials and smart structures, we could learn more about nonhuman animal species. And that brings us to the main topic of this chapter. What can smart materials and smart structures tell us about the behavior and physiology of animals, specifically, aquatic animals?

27.1.2 Applications for Smart Materials and Smart Structures in the Study of Animal Behavior

Smart materials and smart structures have found widespread application in the study of animal behavior. Often, these applications involve apparatuses that convert responses in the form of mechanical energy into electrical energy using ferroelectric-based piezoelectrics. Piezoelectric ceramics convert energy from one form to another and as such are transducers. Examples include microphones, hydrophones, loudspeakers, thermometers, light emitting diodes (LEDs), photocells, position and pressure sensors, and antennas. Transducers can also be used to transmit ultrasonic signals in air or water.

Scientists studying animal behavior have used smart materials and smart structures to study agonistic behavior, foraging, homing, learning, migration, and territoriality, in the field and in the laboratory. For example, Wolcott and his colleagues have developed smart structures to study the behavior of blue crabs (Shirley and Wolcott, 1991; Wolcott, 1996; Wolcott and Hines, 1996; Clark, Wolcott, T., Wolcott, D., and Hines, 1999). In one application, Clark et al. (1999) used microcontroller-based smart transmitters and receivers to study the feeding and threat behaviors of blue crabs in their natural environment. Sensors that detected and transmitted a signal indicating contraction of the mandibular (chewing) muscle were attached to the crab. Contractions indicated that the animal was feeding. In addition, sensors were attached to the both claws. The spread of both chelae a signal indicated that a meral threat display had been emitted. When population density was high, crabs expended more effort defending their territory and behaved more aggressively. This increased aggression reduced the amount of time the crabs spent feeding. Thus, high population density reduced feeding because there was a marked increase in antagonistic behavior and not because of limited availability of food. Additional studies examined molting behavior (Shirley and Wolcott, 1991). Blue crabs were tracked using ultrasonic transmitters and reed switches and magnets were used to detect molting. Under high predation pressure, crabs sought shallow water and moved to areas of less predation. Low predation pressure allowed molting to occur over a broader range.

27.1.3 Chapter Overview

Though the application of smart materials and smart structures is widespread in the study of animal behavior, the focus of this chapter is on telemetry studies with aquatic animals, specifically within the marine environment. Aquatic animal researchers have used telemetry technology to track animals to determine home range and movement patterns, quantify muscle movement, and determine energy use and requirements through measures of heart rate, physical activity, and internal temperature. In addition, telemetry studies inform how animals budget their time between foraging, fighting, resting, mating, etc. The chapter begins with specific details about the biotelemetry equipment that aquatic animal researchers use and representative companies that provide such equipment commercially. This section will be followed by examples of the methods and findings from researchers that incorporated this technology. The literature in this field is growing rapidly and this chapter is not meant to be exhaustive. Our goal is to provide representative examples of equipment and applications that the reader can then explore in greater detail.

27.2 Biotelemetry or Biologging: Introduction

Aquatic animals by virtue of camouflage and lifestyle are difficult to observe for long periods of time. However, such observations are required to understand their basic physiology and behavior. For example, does the animal expend a great deal of energy to swim long distances on a daily basis like some sharks or does it live a sedentary lifestyle like cuttlefish? Does the animal spend its entire life in one relatively small area like rockfish or does it migrate great distances like some whales? In addition, it is important to know how the environment affects the animal. For example, does the temperature range of the water animal restrict the animal's movement and is the animal confined to a narrow or wide range of depth? These questions are addressed by telemetry systems. In these systems, transmitters determine location and sense aspects of the animal and its environment. Depending on the transmitter, the data are stored or transmitted to a receiver or data scanner. Certain transmitters even provide preliminary analyses of the data. Receivers and data scanners connect to a computer that decodes, processes, and stores that information to produce a useful result. Thus, telemetric systems are smart structures and many of the transmitters and certain of the scanners/receivers utilize smart materials.

Remote observation of animals through telemetry evolved through four stages (Ropert-Coudert and Wilson, 2005). First, it was realized that animals could carry telemetric devices without compromising unduly their natural behavior. Second, devices were developed that sensed basic behavioral, environmental, or physiological data and enabled scientists to locate and track animals. This second stage, combined with acoustic and satellite telemetry, allowed scientists to study migratory movement and habitat use. The third stage saw the development of telemetry devices that placed the animal's activities on a timeline. These devices recorded telemetry data for subsequent analysis and led to studies where time budgets could be determined and energy use could be assessed. Thus, scientists could keep a record of the animal's physiology or behavior over time. Naito (2004) termed this activity "biologging." The fourth stage saw the development of devices that not only monitored physiological and behavioral data, but also monitored and stored or transmitted information about the animal's external environment. With this development, it became possible to place the animal's behavior within the context of its environment and thereby achieve an understanding of how the environment affects the animal's behavior and in turn, how the animal affects its environment.

These four stages are apparent in a telemetry system developed by Davis et al. (2004). Davis et al. developed an animal-borne video and data recorder that utilized transducers for pressure, water speed, and compass bearing. These transducers were sampled once per second and data stored on a flash memory card. The video component recorded up to 6 h of video data from the animal's perspective on Hi 8 cassettes. The two audio channels on the video recorder were also employed. One channel recorded ambient sound received by a hydrophone affixed to the animal and the second channel recorded the analog signal from an accelerometer. The accelerometer was sampled 16 times per second and determined flipper stroke frequency. A global positioning system (GPS) was attached to the head-mounted camera and could determine the seal's precise location in less than 30 s. With this apparatus, Davis et al. were able to observe the types of prey Weddell seals pursued, the strategies they used to hunt, obtain estimates of the energy used during dives, and recreate three-dimensional tracks of their dives. Davis et al. also reported plans to develop the next generation video data acquisition platform that will monitor several times per second, ambient dissolved oxygen, ambient temperature, bioluminance, compass bearing, conductivity, light level, pressure, swim speed, and tilt, pitch, and roll. In addition, a digital video recorder will record for 80 h and a GPS will acquire precise location in a few seconds. Figure 27.1 depicts a female Weddell seal with video data acquisition platform attached.

The development of sophisticated smart telemetry devices have allowed scientists to proceed from manual tracking used to study gross movement to the systematic investigation of complex behaviors including exploration, foraging, habitat selection, migration, navigation, predator–prey interactions, reproductive, and social behavior. These devices provide data from free-ranging aquatic animals of such high quality and quantity that it is becoming possible to assess cause and effect relations in the field as opposed to just the laboratory. The outcome is that scientists can now maximize internal and external validity.

The discussion begins with consideration of the equipment required to locate and track aquatic animals in a marine environment, measure physiological aspects of the organism, and record data about the external environment. This discussion is organized around representative companies that produce this equipment and equipment that is being developed by individual researchers for specific applications. The following section considers representative studies involving marine fishes, turtles, cephalopods, and marine mammals. The chapter ends with a brief discussion of the future of biotelemetry technology and its applications to the study of nonhuman animals.

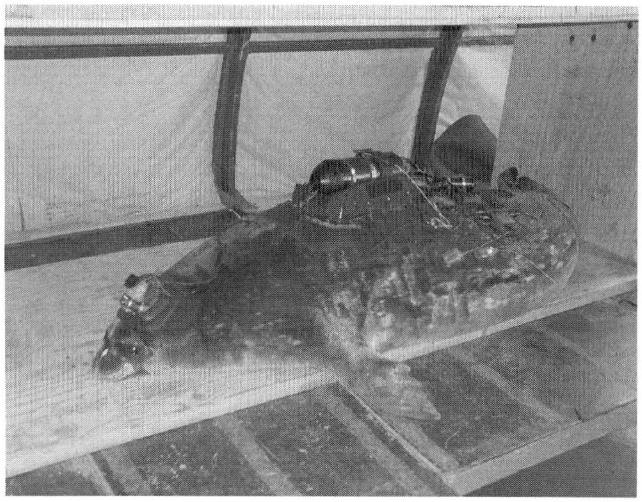

FIGURE 27.1 Weddel seal outfitted with animal-borne video and data recorder. Note camera mounted on head and video data acquisition platform attached to back of animal. Satellite and radio signals are generated by respective transmitters and the animal also carries speed sensors (mounted on shoulder) and an accelerometer mounted on tail. In addition, the seal carries an electronic compass located just behind the video acquisition platform. Photo by Don Calkins, Alaska Sea Life Center, Seward, AK. With permission.

27.3 Biotelemetry or Biologging: Equipment

Smart materials and structures are used to locate and track aquatic animals in marine environments. As radio waves quickly attenuate in salt water, researchers use ultrasonic telemetry. In a typical system, aquatic animals are fitted with transponders or pingers that emit an ultrasonic signal. One or more hydrophones sense the acoustic signal and the animal's location is determined either from the signal strength or through triangulation. In addition to locating and tracking an aquatic animal, these devices can also encode physiological and behavioral data. Receivers are then used to decode the signal and transmit or store the data. At present, it is becoming more common to use transmitters that periodically send data to a satellite and then forwarded to the user (see Figure 27.2).

27.3.1 Commercially Available Transmitters and Receivers

There are a number of companies that specialize in the construction of biotelemetry apparatus. Some of these companies provide such equipment for use in marine and freshwater systems whereas other companies specialize in just marine systems. Given the focus on marine applications, we consider companies that supply biotelemetry equipment for marine environments. A representative list of such companies includes Lotek, Sonotronics, Sony, Star-Oddi, Vemco, and Wildlife Computers.

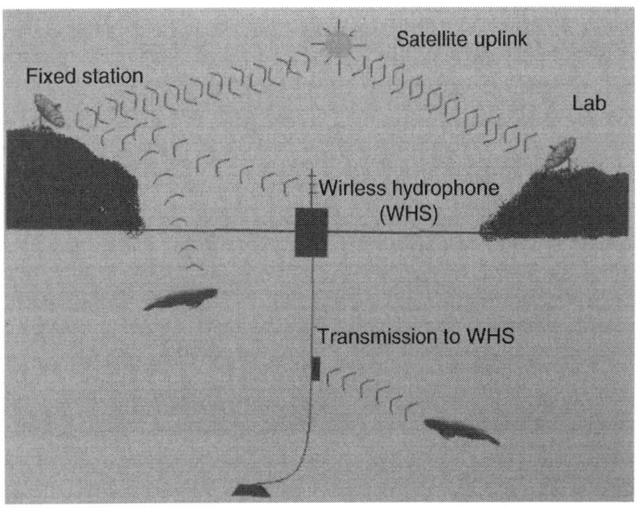

FIGURE 27.2 Typical setup for biotelemetry. Aquatic animals are outfitted with acoustic pingers that can be detected by hydrophone. In this depiction, manatees are outfitted with combined acoustic/radio transmitters that send both radio (detected by a receiver on shore) and acoustic signals (detected by wireless hydrophone) and sent by radio to receiver on short. Data are then sent to satellite and then to the laboratory. Figure modeled after Bridger, C. J., Booth, R. K., McKinley, R. S., Scruton, D. A., and Lindstrom, R. D., *J. Appl. Ichthyol.*, 17, 126, 2001. With permission.

27.3.1.1 Lotek Wireless Fish and Wildlife Systems

According to their Web site (www.lotek.com), Lotek offers a full range of biotelemetry systems for large and small marine animals. These systems typically involve the use of transmitters or archival tags to encode and produce the signal, hydrophones to receive and transduce the signal, and receivers to detect and store information. Through their MAP Series, Lotek offers coded acoustic transmitters for locating and tracking aquatic animals. These transmitters range in size from 8.5 mm in diameter and 32 mm long to 32 mm in diameter and 101 mm long. These transmitters, set at a burst rate of every 5 s, last anywhere from 16 days to greater than 5 years depending on choice. Signals from the transmitters are detected by hydrophones that can be wired directly to a receiver or can be wireless where in they send transmissions through radio waves to a land-based or mobile receiver. Lotek's MAP 600 acoustic telemetry system can support up to eight different hydrophones and is able to track hundreds of transmitters or monitor thousands of transmitters on the same frequency. The system is capable of accurate measurement of position in both a 2D and 3D format.

Lotek also offers coded acoustic transmitters through their CAFT series. Because of their unique codes, systems designed around these transmitters can distinguish up to 212 different signals on a single acoustic frequency while still maintaining individual identification. These transmitters work with the SRX 600 and SRX 400 data logging receivers and can be used to assess habitat utilization, migration, and passage data among other applications. For smaller animals, Lotek offers the NanoTag Series. These tags are claimed to be the smallest digitally encoded transmitter available. These tags utilize Lotek's proprietary coding system (CAFT series) but do not have the same duration. However, some nanotags can continue to transmit on 10 s intervals for 678 days. Nanotags are used to locate and track aquatic animals.

Lotek sensors can also detect, record, and transmit or store basic physiological and behavioral data. Sensor types include depth, motion, and temperature. These sensors can be combined into a single coded transmitter. In addition, Lotek can provide electromyogram data through their EMG Series. EMG tags are 16 mm by 53 mm in size and can be used to assess metabolic costs of tail strokes, oxygen consumption, energetic costs, and transmit temperature data, if equipped with the temperature sensor option.

Lotek also offers archival tags (LTD Archival Tags) that record and store vast quantities of data related to time, temperature, depth, and light. Through variations in sun light levels and temperature, LTD tags can provide geolocation data through onboard geolocation algorithms that give daily estimates of latitude and longitude. One can refine the geolocation data by cross-checking these data with sea temperatures and light levels available from oceanographic surveys. The real-time clock on the transmitter is temperature compensated and is accurate to within 30 s per year. Ambient light is measured with built in light sensors or a "sensor stalk." Light level data is band pass filtered to maximize the depth at which the sensor can operate. The LTD is capable of operating at depths of 2000 m with a maximum depth of 3000 m. Depending upon conditions, battery life can exceed 5 years. Archival tags are becoming the tags of choice in marine environments due to the difficulty of maintaining contact with the transmitter in the open ocean. Finally, Lotek offers its CART Series (combined acoustic radio transmitters). These tags use a conductivity sensor to determine whether the animal is in fresh water or salt water. If the animal is in saltwater, acoustic signals are transmitted and when operating in fresh water, the transmitter switches to radio signals. Locating and tracking are the primary use of these tags.

Scanning receivers (Suretrack STR-1000-W1) and combination receiver/data loggers (SRX 600, DRX-600, and SRX 400) are also available. Depending on the model, researchers can manually track aquatic animals on foot, boat, or aircraft; or both scan and log information relevant to location and tracking. In addition, the researcher can log other data including temperature, physiological, mortality, geolocation, and historical data as well as conduct 3D mapping. The SRX 400 can monitor multiple frequencies, track up to 212 individuals on a single frequency, identify transmitters as sensors, beepers or coded devices, download data to a computer, and can monitor both ultrasonic and radio frequency. The SRX 600 improves the performance of SRX 400 by offering integrated GPS capability, expanded memory, USB capability, an enhanced detection algorithm, and a dual processor. Applications include fish passage data, survival, presence/absence, site fidelity, and habitat utilization.

27.3.1.2 Sonotronics

Sonotronics (www.sonotronics.com) uses smart materials and smart structures to acoustically locate and track aquatic animals in a marine environment. Their products include a variety of ultrasonic transmitters with various combinations of lifetime, size, range, and telemetry options. Manual tracking systems for real-time determination of animal behavior are available as well as automated stations for logging animal detections, including cellular uplink and web technology. To determine fish movement, fish passage, and fish location standard coded ultrasonic transmitters are available. Arranged from smallest to largest, Pico Tags (PT series) can be detected from 300 to 750 m, Miniature Tags (IBT-96 series) from 500 m or more, Tracking Tags (CT series) have a range of 1000 m, and High Power Tags (CHP-87 series) can be detected from 3000 m. Sonotronics also offers combined acoustic and radio transmitters made in conjunction with Telonics Company. These transmitters are used in detecting aquatic animals that move from fresh water to salt water environments or from areas of low salinity to high salinity. Ultrasonic telemetry tags are also available. The Temperature Tag (CTT-83 series) measures temperature to within .5°C and can be detected from 1000 m. The Depth Tag (DT-97 series) measures depth to 5000 PSI and can be detected from 3000 m. A smaller transmitter (Mini Depth Tag, IBDT-97 series) is available with a range of 500 m and a depth range of 100 PSI. Finally, an acoustic storage transmitter (AST-05) collects temperature, pressure, and conductivity data at preprogrammed intervals. These data are archived and the code transmitted once every 3 min to allow for retrieval and data download.

Detection of the acoustic signal is via Sonotronics' directional (mod # DH-4) and omnidirectional (mod #s DH-3 and SH-1) hydrophones. The DH-4 directional hydrophone provides the greatest range in locating tags and has noise reduction capability, which improves precision. The SH-1 omnidirectional hydrophone can detect tags up to 1 km. Sonotronics also offers a towed hydrophone package (TH-1) to reduce the time necessary to transect an area. An audio amplifier (HPR-95) can be used with the piezoelectric hydrophones to allow for listening and recording of underwater sounds. Data logging receivers are available. The USR-96 narrow band receiver can scan 10 preset frequencies and a two-line LCD displays both frequency and interval. The USR-96 underlies the MANTRAK kit that has all the equipment necessary to manually track aquatic animals. Sonotronics also offers a submersible ultrasonic receiver (SUR-1) that can detect and log data to flash memory. The SUR-1 consists of an integrated hydrophone, flash memory, and a transponder that allows the user to interrogate the unit from a distance and determine if information is available. The CUB-1 cellular uplink buoy can be used automate the system and send text messages to a user's cell phone or personal computer.

27.3.1.3 Star-Oddi Marine Device Manufacturing

Star-Oddi Ltd (www.star-oddi.com) a company based in Reykjavik, Iceland, produces data storage tags or archival tags, GPS fish positioning systems, temperature recorders, and tagging equipment for marine animals. The DST-CTD and DST-CT salinity loggers detect and store for later retrieval measurements on salinity, temperature, and depth. The tags measure 15 mm × 46 mm and weigh only 12 g in water. Memory capacity is 43,582 measurements per sensor and for the salinity and temperature tag, 65,314 measurements per sensor can be obtained. Tags have a battery life of 4 years and offer 12 bit resolution. The logger can be downloaded and reprogrammed as often as needed until the battery is depleted. A communication box and software is available for the tags allowing wireless transfer of data to a personal computer. Star-Oddi also offers the DST Pitch & Roll logger. These tags measure pitch and roll, temperature, and depth. They offer a battery life of 5 years and can detect and store depth information to 2000 m. Size and weight of these tags is the same as the DST-CTD tag. The Pitch & Roll logger has 12 bit resolution and memory capacity of 174,600 measurements in total. These tags find applications for analyzing movement of marine animals and other underwater gear. The DST compass logger is a small submersible compass logger with the option of also recording temperature and depth. The electronic compass has ±15° accuracy and shows eight point compass direction or in degrees. The compass logger is more accurate in the horizontal plane. Size and weight are the same and the memory capacity is 128,000 measurements in total with 12 bit resolution.

Finally, the Star-Oddi GPS Fish Positioning system consists of a data storage fish tag (DST-GPS) and a GPS transmitting sonar/transmitter. The tag is attached externally or implanted and receives information about geographical location from satellites via Simrad sonars onboard the vessel. The vessel's GPS position is coded and transmitted underwater via the Simrad sonar signal. Tagged fish, swimming within 4 km of the GPS sonar, receive and store the geographical position as well as data on temperature, depth, and date/time (see Figure 27.1). The system is designed for medium-sized round fish (e.g., cod, saithe, salmon) and works well in marine environments. For applications in rivers, lakes, and smaller ocean areas, Star-Oddi offers a portable fish positioning sounder (FPS). Both systems can be used to determine migration patterns, habitat use, and foraging strategies among other possibilities.

In collaboration with the Icelandic Marine Research Institute, Star-Oddi has developed the underwater tagging equipment (UTE) that can be used to robotically tag deepwater fish that cannot survive large changes in temperature and pressure. The UTE is placed in the cod end of a trawl where the fish is enclosed by a grid diverting it to the tagging place. The fish is observed through a video camera and the researcher moves the tagging gun into position. A robotic arm is used to make a small incision and press the DST tag into the body cavity. In just a few seconds, the fish is released through a channel in the device. This system allows for tagging on site in the fish's natural habitat (see Figure 27.2). In October 2003, June 2004, and June 2005, successful tagging cruises were made during which nearly 2000 redfish were tagged at depths from 500 to 800 m, with tags that measured depth and temperature. As commercial and sport fisheries recapture tagged fish, activity data can be downloaded to provide valuable information for the fishing industry in the waters off Iceland and elsewhere. The UTE system is showing significant promise as a means to study deepwater fishes. Technical specifications of the UTE and a detailed discussion of how deepwater fish are tagged can be found in Sigurdsson et al. (2006).

27.3.1.4 Vemco

Vemco (www.vemco.com) specializes in the design and manufacture of acoustic telemetry systems. The company offers a wide variety of transmitters, hydrophones, receivers, and data loggers. Coded and continuous expendable pingers and data telemetry transmitters are offered to scientists who want to track and monitor aquatic animals in a marine environment. Six standard diameters ranging in size from 7 mm (V7) to 32 mm can be constructed with user-determined battery life, power output, and life expectancy options. These ultrasonic acoustic transmitters can also be equipped with sensors to measure temperature or depth.

As an example, the V16-coded transmitter measures 16 mm in diameter, with length and weight varying from 48 to 90 mm and 9 to 14 g, respectively. V16 transmitters can be equipped with silver oxide or lithium batteries, and, depending on minimum and maximum delays for transmission, required power output, and sensor type, projected battery life varies from 26 days to 7.73 years. Users have choice of sensor type (depth or temperature), battery type, power level, frequency case style, temperature range, maximum depth, coded type, code ID information, and random delay range for transmission.

The VX32TP-CHAT tag provides two-way communication between the tag and a receiver. Users can locate and track tagged marine animals and they can receive archived data from a transponder all without recapturing the animal. Data can be recorded or transmitted in real time and partial downloads are possible in the event that the animal moves out of range of the receiver. The CHAT tags have a long life span and high memory capacity; up to 5.6 years with a sample interval of 1 min and an average interval of 1 day, and multiple memory downloads are possible. Tracking receivers use reply time to estimate distance and sample rate and average interval of response is reprogrammable. These tags are equipped with depth and temperature sensors and a swim speed sensor is available on request. The tag is best suited for large and slow moving marine animals and measures 32 mm in diameter and 150 mm in length. It weighs 190 g in air and 75 g in water. Vemco will also manufacture experimental transmitters for specific applications. Examples include heart rate transmitters, differential pressure transmitters that can be used to determine swimming speed in fish or jetting in cephalopods, accelerometers, integrating accelerometers, and transmitters that detect tilt.

Vemco manufactures six different receivers, ranging from single channel receivers (VR2) that track coded transmitters automatically to receivers that measure real-time and provide detailed position information including true latitude and longitude by using three detection buoys and a two-way radio link to a base station (VRAP). Further examples include Vemco's VR3 monitoring receiver that provides remote communication to satellite, underwater modem, cell phone, or underwater observatory and can be left unattended for as long as 12 months. Satellite transmission is achieved through the ARGOS Satellite and GPS positioning system. The VR28 tracking system is connected to a four-element hydrophone array that provides 360° of coverage and all functions including frequency, gain, and echo rejection blanking interval are user determined. The VR28 is optimal for use with CHAT tags.

Vemco also offers a variety of hydrophones to be used with their receivers to detect acoustic tags. With the exception of the VHLF hydrophone that is used to monitor sounds in the audible frequency range and requires a separate system, Vemco hydrophones connect to the receivers via cable. The hydrophones are designed to be trolled through the water and are tuned to specific acoustic transmitters. The VH4x series hydrophones are designed for use with the VR28 tracking system. Finally, Vemco offers two different data loggers (Minilog-8 bit and the Minilog-12 bit). These loggers record temperature or temperature and depth information. They are waterproof, rugged, and are designed to download using the Minilog-PC interface that requires no external electrical connections. These loggers have a battery life expectancy of 5 years.

27.3.1.5 Wildlife Computers

Wildlife Computers (www.wildlifecomputers.com) offers satellite tags and archival tags. The Argos satellite system detects the transmissions from satellite tags. These tags can detect wet and dry conditions, thus providing data on haul-out behavior by seals or sea lions, and they are capable of transmitting depth, temperature, and light level data. Data are received in near real time. Three types of satellite tags are currently available. The smart position and temperature transmitting (SPOT) tag is the smallest and lightest and can be used in a marine environment. The SPOT5 tag is their latest version. This tag is available in a variety of shapes and can be deployed on cetaceans, penguins, seals, and turtles among other species. The SPOT5 tag transmits temperature data in the form of histograms to allow researchers to easily assess where the animal spends its time within specified temperature ranges and haul-out statistics are reported as the percentage of time the tag is dry for each hour of the day. Using the Argos satellite system and Cricket, the Argos-certified transmitter developed by Wildlife Computers, the SPOT5 tag provides locations with an accuracy of ±350 m.

The Argos Satellite SPLASH tag is a new satellite tag that combines the sampling and data storage functions of an archival tag with an Argos transmitter. SPLASH sensors measure depth, temperature, light level, and wet and dry periods. Temperature and depth data are collected, analyzed, summarized, and compressed for transmission. Approximately 14 Mbytes of memory are available with recovery of the tag. Available message types for transmission include histograms for dive duration, maximum dive depth, time-at-depth, time-at-temperature, percentage timelines, and 20 min timelines. Location accuracy is the same as the SPOT5 tag.

The third type of satellite tag is the Pop-up Archival Transmitting Tag (Mk 10-PAT). This tag also combines the sampling and data storage properties of an archival tag with satellite technology. The tag is designed to track large-scale movements and behaviors of pelagic fish and other animals that do not come to the surface often enough to use real-time satellite tags. Wildlife Computers contend that this tag is by far their most intelligent and complex tag. The PAT tag can provide data without the animal being recaptured, but the tag is recovered a full archival record is maintained in nonvolatile. Designed for large pelagic aquatic animals, PAT tags have been deployed on large marlin, sharks, sea turtles, swordfish, and tuna. The tag is attached with a tether that is designed to detach at a user-determined date and time. Once detached, the tag floats to the surface and transmits summarized data via Cricket to the Argos system. The tag archives depth, temperature, and light level data.

By analyzing either premature releases or when the PAT has been at a constant depth (surface or bottom) for too long, the tag can provide researchers with information on attachment failures and animal mortality. A mechanical guillotine automatically cuts the tether if the tag is dragged to depths that would crush it, thus ensuring transmission of important data. The lithium battery allows the PAT to record data for up to 1 year and over the course of 7 days, it can transmit 10,000 32-byte transmissions. Users can scale the transmitted data to meet the requirements of their deployment length. Sensors include 12 bit analog to digital converters for temperature and depth, 10 bit analog to digital converter for light level and battery voltage, and a wet and dry sensor that can detect the ideal time to transmit and provide haul out data. As with the other satellite tags, geographical positions are accurate to within 350 m. Researchers have used this tag to study migration paths, depth, and temperature preferences, and oceanographic data in the form of depth–temperature profiles.

In addition to smart satellite tags, Wildlife Computers offers a range of archival tags. Using these tags, researchers can collect and record data on the swimming behavior of marine animals. Depending on the tag, depth, environmental temperature, heart-rate, stomach temperature, and light level can be recorded. Light level data can be used to determine approximate location and time is encoded with the stored data. The Mk9 is the company's smallest archival tag and is typically configured with depth, temperature, light level, and wet and dry sensors. The tag is 67 mm long by 17 mm by 17 mm and weighs less than 30 g in air. The tag can withstand pressure to 1000 m and the 16 Mbytes of flash memory can store 8 to 16 million sensor readings. A supplied interface provides download capability to the user's PC. As mentioned a heart rate-temperature recorder (HTR), a heart rate transmitter (HRX), and a stomach temperature pill (STP) can provide data on heart rate, internal temperature, and stomach temperature. The STP tags can be used to indicate when the animal ingested cold prey, thus providing important data on foraging habits. The company indicates that they are currently developing the Mk10-AF tag that will use Fastloc technology to acquire position data in a very short time that are accurate to 10 m.

27.4 Biotelemetry and Biologging: Applications and Findings

In this section, we examine studies that have used smart materials and smart structures biotelemetry to study marine animals. We will examine representative studies and discuss briefly their findings.

27.4.1 Locating and Tracking Marine Animals: Fishes, Turtles, and Cephalopods

27.4.1.1 Fishes

Biotelemetry has been used extensively to locate and track marine fishes (Nelson et al., 1997; Lowery and Suthers, 1998; Candy and Quinn, 1999; Pepperell and Davis, 1999). More recently, Bridger et al. (2001) monitored the behavior of domestic steelhead trout (*Oncorhynchus mykiss*) in an unconstrained environment. The authors were interested in determining whether domestic fish that had escaped from an aquacultural facility would show site fidelity or would disperse. If domestic fish showed site fidelity then recapture could reduce the economic loss of large numbers of fish escaping and the environmental impact that domesticated fish might have on wild stock. Given that the salinity changed dramatically according to depth of water, the authors determined that combination acoustic and radiosmart transmitters (Lotek CART transmitters) offered the most efficient way to monitor fish location over large areas.

In total, 240 trouts were implanted with a CART transmitter. One hundred and fifty fish were released during the summer and 90 fish were released during the winter months. The authors used remote data-logging stations that were fixed and self-sustaining. The system included a Lotek SRX 400 data logger, ASP_8 (antenna switching peripheral), an ultrasonic upconverter, photocontroller, battery supply, and satellite data transceiver. In addition, a wireless hydrophone system (Lotek WHS-1000) was used to detect a signal from the transmitter. The WHS system was found to be a better solution than tethered hydrophones due to deployment and maintenance difficulties, and reliability over time. A hybrid system of wind turbines and photovoltaic cells kept the system's 400 amp hour battery charged and operational for 10 days in the absence of wind or sun light. Given the remoteness of the area (Bay d'Espoir, Newfoundland, Canada), a remote data link to the data-logging stations was required. A two way satellite-based data link was developed to allow Bridger et al. to download date and time of transmitter detection, channel, code, antenna, power level, number of events and hourly battery status. The results showed that steelhead trout released during the summer months showed strong site fidelity whereas fish released during the winter months showed decreased site fidelity. Bridger et al. concluded that attempts to recapture domestic fish that had escaped their holding pens during the summer could be successful and reduce economic loss and ecological interactions between domestic and wild stock.

To assess site fidelity on artificial reefs in the northeastern Gulf of Mexico, Szedlmayer and Schroepfer (2005) tagged and tracked 54 red snapper (*Lutjanus campechanus)* during all seasons of the year. Fish were implanted with ultrasonic transmitters and monitored with Vemco remote receivers and Sonotronics surface receivers. Manual tracking from the surface was used to follow four red snappers over night. Positions were recorded every hour for 9 or 16 h. Szedlmayer and Schroepfer found that the majority of red snappers showed strong site fidelity, with 67% of the fish staying on artificial reefs from 117 to 595 days. High residency rates were independent of season and the authors concluded that artificial reefs provided suitable habitat for red snapper residing in the northeastern Gulf of Mexico.

Estimates of fish distribution and abundance are often in error to due to acoustic dead zone loss and avoidance reactions to vessel and survey equipment (Aglen, 1994; Hjellvik et al., 2002). The acoustic dead zone refers to the area where aquatic animals can swim below the detection zone of the recording equipment and are not counted. Stensholt and Stensholt (2004) compared the estimates of abundance of Arctic cod (*Gadus morhua* L.) based on data from acoustic surveys, trawls, and data storage tags (DST) in an attempt to develop a model that could improve such estimates. The authors worked on the assumption that samples from trawls along the bottom were biased because fish higher in the water column were not counted and acoustic surveys were biased because it was possible to miss fish that were swimming on the bottom and out of the detection range of the acoustic equipment. The argument that both methods together would provide a better estimate was mitigated by the fact that water depth and temperature played a role in how much overlap there was between the two methods. In addition, prey dispersal also entered into the equation.

To develop a better model for estimating abundance, the authors began by normalizing the depth data. Then, based on estimates from acoustic surveys and trawls and data from cod tagged with data storage transmitters that provided depth and temperature information, the authors developed their model. In essence, their work resulted in modification of the fish vertical distribution due to spatial and temporal variation. This improvement in vertical distribution allowed the authors to assess fish abundance more accurately.

Holland et al. (2001) tagged tiger sharks (*Galeocerdo cuvier*) with five types of tags to evaluate the extent to which all tags provided comparable data and to assess the horizontal and vertical movement of tiger sharks off the coastline of Oahu, Hawaii. Fine scale movement of the sharks was recorded using sonic tracking methods for periods up to 50 h, whereas acoustic pingers and automated, anchored data loggers provided long-term periodicity and return frequency data. The five tag types included standard identification tags (Casey and Kohler, 1992), acoustic tags for active tracking, long-life acoustic tags detected by submerged data loggers, archiving tags, and a pop-up archival tag that transmitted stored data to a satellite. Tiger sharks were caught using long-line fishing techniques, tagged, and released less than 5 km from Waikiki beach, Oahu, Hawaii. Over a 5-year period (1994–1999), 133 tiger sharks were captured and tagged with identification tags. Seventeen of these sharks were later recaptured at the same site after intervals ranging from 14 to 336 days postcapture. A total of eight sharks were tracked actively with either internal or external tags. The tracks of the eight animals were remarkably consistent with extended periods of straight-line swimming and movement over large distances with seven of the tracks ending at Penguin Banks located 32 km from the starting site. Depth recorders revealed that the sharks stayed near the bottom when in shallow water, but moved up to between 40 and 100 m depth when in open water.

Twenty sharks were implanted with long-term pingers that were detected by data loggers in the original fishing area. Ten of these tags were later detected. Sharks returned to the area at all times of the day with some sharks showing more daytime detections than nighttime detections. There was no discernable periodicity to the sharks' movements. Four sharks received archival acoustic modem tags. Vemco CHAT tags measuring 28.8 × 6.6 cm with a nominal life span of 24 months stored depth and temperature data. A 5 min sample period with 6 h averages was used. Data were sonically downloaded via acoustic modem technology to a submerged data logger anchored to the bottom. When a tag was detected within 500 m of the data logger, downloading commenced. At the time of writing, one shark had been detected and its data downloaded. Horizontal and vertical movements were consistent with the active tracking data. One shark received an archival satellite tag developed by Wildlife Computers, Ltd. This tag recorded depth and temperature data condensing them into histogram formats. The tag was externally attached and equipped with a corrosible link that caused the tag to release and float to the surface where it transmits data to a satellite. The tag surfaced and transmitted data, but it appeared that release occurred early yielding only 13 days of data and 15 days of surface drifting. At the time of release, the tag was located 140 miles southwest of Oahu. The authors concluded that the five different tags revealed consistent data showing that tiger sharks tend to show site fidelity to a home range, which is apparently very large, with several weeks between returns to a specific site within that range. The authors also commented that more data could be collected from the acoustic and CHAT tags than from the identification and recapture method, resulting in greater accuracy.

Pacific bluefin tuna (*Thunnus thynnus orientalis*) constitute an important commercial fishery and a clear understanding of their migratory and foraging habitats is critical to ensuring the sustainability of this important resource. Domeier et al. (2005) sought to develop an automated algorithm that could provide more accurate estimates of latitude by using sea surface temperatures from satellites in combination with temperature and light level information provided by archival acoustic tags. Researchers have used light levels from such tags to provide accurate estimates of longitude, but estimates of latitude were less reliable, particularly as the subject approached the equator (Hill and Braun, 2001). To develop their algorithm, the authors tagged bluefin tuna with pop-up satellite archival tags (PSAT) supplied by Wildlife Computers, satellite-based archival tags (PTT-100 PSAT) from Microwave Telemetry, Inc. (Columbia, MD), or Lotek wireless nontransmitting archival tags (LTD2310). All tags recorded depth, water temperature, and light level. The Lotek archival tags recorded these variables and also provided data on internal fish temperature. Data from these transmitters were recorded either every 2 min (Lotek), every 2 min, and summarized hourly (Wildlife Computers), or hourly (Microwave Telemetry).

Using data from these transmitters and from satellite acquired sea surface temperatures, the authors developed an automated system for tracking called the PSAT Tracker Information System (PTIS). The key parameters of the system include time and position of tag deployment, time and position of tag recovery, light-based estimates of longitude, maximum swimming speed, and a bracketed range of latitude estimates within which the program looks for temporal matches from the sea surface temperature data. This algorithm is fully described in the paper and produced tracks that were less variable than those produced by light levels alone. The authors concluded that their algorithm showed increased accuracy in tracking particularly in those areas where there is a large north to south temperature gradient. PTIS results correlated well with estimates calculated from light-based algorithms and appeared to provide better data in those situations where light-based estimates failed (near the equator). With respect to the migratory habits of bluefin tuna, the authors found that tuna ranged from the California-Oregon border to southern Baja California, Mexico, but spent the majority of their time off the coast of central Baja, Mexico. Finally, the authors found evidence for seasonal migration with tuna preferring the central Baja location during the winter and spring and moving north toward Oregon and then back south during the summer and fall.

27.4.1.2 Cephalopods

Yatsu et al. (1999) tracked jumbo flying squid (*Dosidicus gigas*) in the Eastern Pacific Ocean using ultrasonic telemetry. Jumbo flying squid provide an important fishery in Japan and Korea that increased each year up to 1994. Perhaps due to dispersion of the species as a result of environmental changes, the catch went from a high of 165,000 metric tons in 1994 to an astonishing 80,000 metric tons in 1995. As a result, it became important to understand the behavior and movement patterns of this species. To that end, the authors deployed three squid ranging in size from 35 to 43 cm total mantle length. Ultrasonic transmitters supplied by Vemco (models V16P-1H and V16P-4H) were attached just forward of the fin with a rubber band. Transmissions were detected with an onboard hydrophone and GPS locations of the ship were determined by the NNSS-GPS navigation system. Both tag transmissions and GPS locations were transmitted at 1 s intervals. Squid were captured and released several hours after sunset and tracked between 8 (one squid) and 14 h (two squid). All squid were lost when they dived below 1020 m, a depth that exceeded the limit of the transmitters. Tracking data allowed the researchers to assess horizontal and vertical movement. Typically, squid remained above 200 m during the night and dived below 1000 m either at twilight or 5 h before sunrise. Diving speed varied between 2 and 28 m per minute. During the time of tracking, squid moved between 3 and 5 miles horizontally. The results provided important data for further work and showed that tracking the jumbo flying squid is possible through ultrasonic tagging techniques.

More recently, Jackson et al. (2005) conducted initial tests on the advantages of using hybrid acoustic/archival tags to locate and track squid and cuttlefish. Australian giant cuttlefish (*Sepia apama*) and tropical squid (*Sepioteuthis lessoniana*) were tagged with a hybrid tag developed by Vemco. The V8 acoustic tag was potted with a Vemco minilog temperature/depth tag and attached to four animals. The animals were monitored within radioacoustic positioning telemetry (RAPT) buoy system arrays. These arrays included bottom-mounted sensors that transmitted independent temperature records and reference standards for sound conductivity and position. During the course of the deployment, all four tags moved out of RAPT range, but two of the tags were recovered using a boat-mounted hydrophone and the VR60 Vemco receiver. A VUR96 diver operated receiver also aided in the recovery. Although half of the tags were lost, the authors concluded that their system provided an alternative to more expensive satellite pop-up tags and had the added advantage that it could be deployed on small species that live in or return to near-shore environments.

In a follow up paper, Aitken et al. (2005) describe the results acquired from *Sepia apama* that were tagged with acoustic transmitters and monitored with the RAPT system. The authors used position only and jet pressure transmitters. Results showed that the home range for *S. apama* was between 5300 and 23,700 square meters. S. Apama was shown to be more active during the day, spending 3.7% of their time foraging during the day and 2.1% of their time foraging at night. Over all, *S. apama* was active during the day for 32 days, but was active for only 18 nights. Interestingly, the giant cuttlefish spent 95% of its time resting, which places its behavior more in line with the octopus than the squid. The authors concluded that the cuttlefish's predisposition to quiescence allowed it to channel its energy into fast growth and later into mating.

27.4.1.3 Turtles

As indicated on the Redlist (www.redlist.org) sea turtles are endangered in all parts of the world. Greens (*Chelonia mydas*), olive ridelys (*Lepidochelys olivacea*), and loggerhead (*Caretta caretta*) are listed as endangered. More pressing, hawksbill (*Eretmochelys imbricata*), Atlantic ridley (*Lepidochelys kempii*), and leatherback (*Dermochelys coriacea*) turtles are listed as critically endangered. Biotelemetry studies can help scientists understand their behavior including foraging, migratory habits, and predator defense, and the effects of fisheries. A clear understanding of these variables and the related physiology can inform the development of conservation policies and hopefully stave off extinction.

Recent examples of biotelemetry studies with sea turtles include investigations by James et al. (2005) and Polovina et al. (2004). James et al. outfitted 38 leatherback turtles, a critically endangered species that may be facing extinction in the Pacific (Spotila et al., 2000), with one of five different satellite tags to assess their migratory habits. The authors collected data for 8288 tracking days over an average interval of 218 days. Eleven turtles were tracked for more than 1 year. Tracking data revealed that turtles spent up to 4 months off the coastlines of eastern Canada and the northeastern United States. Southward migrations typically began in October, with some turtles moving south as early as August 12, and others as late as December 15. Southern destinations were determined for 25 turtles. Eleven turtles returned to waters adjacent to nesting beaches along the north-east coast of South America, the Antilles, Panama, and Costa Rica. Other turtles migrated to pelagic waters between 5° and 23° N or to shelf waters off the coast of south-eastern United States. In February and March, turtles returned to the northwest Atlantic, north of 38° N. Typically these turtles returned within several hundred kilometers of where they occurred the previous year. Of particular interest to the authors was the finding that the turtles did not use the same paths when heading south or north, rather they spread out over much of the ocean. Also of interest was the finding that leatherback turtles in eastern Canada weighed 33% more than turtles found in the southern site of St. Croix, U.S. Virgin Islands and that turtles showed foraging site fidelity to shelf and slope waters.

Polovina et al. (2004) examined the forage and migratory habitat of 26 loggerhead and 10 olive ridley sea turtles in the central northern Pacific. Argos-linked Telonics ST-10 or ST-18 position transmitters were attached to 32 of these turtles. Four turtles were outfitted with Wildlife Computer Argos-linked satellite transmitters with dive recorder capabilities (SDR-T10). Transmissions per turtle ranged from 30 to 485 days and resulted in 2470 transmission days. In addition to position, depth, and temperature recordings, sea surface temperature, surface chlorophyll alpha concentration, and geostrophic current were obtained from the University of Miami and via satellite altimitry. Subsurface temperature and chlorophyll in the region were obtained through shipboard oceanographic transects. Genetic analyses of the turtles showed that all loggerhead turtles came from Japanese nesting beaches, six olive ridleys came from eastern Pacific nesting beaches, and three came from the western Pacific.

The results showed little overlap in the migratory paths of the two species of turtles. All loggerheads but two traveled west from their point of capture whereas olive ridleys traveled both eastward and westward. Over the course of transmissions, loggerheads moved north or south (through the region 28°–40° N), depending on season. From January to June, loggerheads favored the southern section of their range and from July to December, they were typically found in the northern section. With respect to temperature preference, loggerheads kept in waters where the surface temperatures ranged from 15°C to 25°C. Loggerheads tended to stay near the surface, spending 40% of their time at the surface and 90% of their time overall, within 40 m of the surface. In addition, for the loggerheads, it appeared that the Transition Zone Chlorophyll Front and the Kuroshio Extension Current provided important foraging and migratory habitats. Interestingly, the transmissions from one loggerhead whose recordings lasted 485 days, showed that it traveled over 9000 km in 458 days. This turtle's westward movement was assisted by the westward flow along the north side of the counterclockwise rotating eddies generated by the eastward flowing Kuroshio Extension Current. Also of interest was the finding that these areas marked regions of enhanced surface chlorophyll.

Olive ridleys, on the other hand, were found south of the loggerhead habitat occupying the region bounded by 8°–31° N in the central Pacific. There was evidence that olive ridleys also displayed a north/south preference depending on season. Olive ridleys preferred warmer waters (23–28°C) and spent more time in deeper water compared to loggerheads. On average, ridleys spent 20% of their time at the surface and 60% of their time within 40 m of the surface. It also appeared that the western stock of olive ridleys occupied different oceanic habitats from the eastern stock a fact that has implications for species formation and for conservation efforts. The longest transmission from an olive ridley showed that it traveled 7282 km in 193 days. This turtle's movements were also aided by equatorial currents, particularly the Equatorial Counter Current and the North Equatorial Current. The authors concluded that their method of using satellite remote-sensed and ship-based physical and biological data in combination with the tracking data led to a more accurate picture of the habitat needs of loggerhead and olive ridley turtles.

27.4.2 Locating and Tracking Marine Animals: Marine Mammals

Biotelemetry has been used to locate and track marine mammals for over 40 years, beginning with simple VHF radio transmitters and time-depth recorders (TDR) in the mid sixties (Kooyman, 1965). Most early biotelemetry studies focused on pinnipeds because they came ashore at predictable locations and intervals. This characteristic allowed researchers to glue the tag onto the dorsal surface of the animal and retrieve the tag later when the animal hauled out for breeding or molting (Butler and Jones, 1997; Kooyman, 2004). The incorporation of Argos satellite receivers allowed scientists to obtain information on animal location and other data from the comfort of their office, removing the need to follow animals utilizing a VHF signal. Satellite-tracking studies proliferated in the 1990s as the use of platform transmitting terminals (PTT) became more readily available and were used on an expanding range of taxa (Kooyman, 2004; see Hooker and Baird (2001) for a ranging studies in whales and dolphins). However, the amount of data transferable via satellite is limited, particularly for cetaceans (whales, dolphins, and porpoises) that only break the water surface sporadically and for seconds at a time. Thus cetacean researchers often chose archival tags that send only limited information via satellite, such as GPS location, and store the rest (Hooker and Baird, 2001). Studies of marine mammals incorporating biotechnology cover a wide array of research topics including migration routes and destinations (Mate, Lagerquist and Calambokidis, 1999); local movements related to foraging, socializing, and breeding (Chilvers et al., 2005; Mate et al., 2005; Small et al., 2005); postrelease monitoring of rehabilitated animals (Gulland, 2003); and diving behavior (Davis et al., 2003; Baird et al., 2005).

North Atlantic right whales are one of the most endangered whale populations with less than 300 individuals remaining (Moore et al., 2004). Baumgartner and Mate (2005) used satellite-monitored radio tags to track right whale movements in relation to environmental variables including depth, depth gradient, surface and bottom hydrographic properties, and remotely sensed surface temperatures and chlorophyll concentrations. The authors collected data over a number of years from 1989 to 2000. Time-depth recorders were used to record diving behavior while separate satellite-linked UHF radio tags (Telonics ST-3, ST-6, ST-15) were used to track movements. Satellite transmission was only possible when the tag was at the surface and this was determined by conductivity sensors to determine if the tag was "wet" or "dry." This methodology saved battery life when the animal was below the surface. Tags could transmit from once every 10 to 40 s. The authors expected that right whales would move from the Bay of Fundy, a favorite feeding habitat, offshore to deep basins where copepod density is high. Instead whales chose areas with low temperatures, high surface salinity, and high surface stratification. The authors speculated that the low temperatures and high vertical stratification in shallow basins caused higher densities of their primary copepod prey, *Calanus finmarchicus*. The stratification increased the shoal-layering effect that promotes *Calanus finmarchicus* survival and abundance. Right whales were able to feed longer at these prey depths (approximately 150 m) compared to whales expending energy to dive in deep oceanic basins (Baumgartner et al., 2003).

27.4.3 Assessing Energy Expenditure with Acoustic Telemetry: Fishes, Cephalopods, and Turtles

The majority of telemetry studies have involved locating and tracking animals. Such studies provide insight into the size of an animal's home range, its migratory habits of short duration (diel movement) or long duration (lunar cycles, seasonal movement or annual patterns of movement) and how it utilizes its habitat. However, this form of telemetry cannot provide fine scale data on how much energy the animal expends in a day, when and on what the animal is feeding, when the animal is mating, or how predators are avoided. Finally tracking studies do not tell us how the environment affects the animal's behavior. To address these questions more sophisticated telemetry studies are required and often this entails a greater reliance on smart materials and smart structures. As evident from the discussion above, biotelemetry companies and researchers are working together to produce "smart tags" that record aspects of the animal's environment (e.g., ambient dissolved oxygen, bioluminance, depth, light intensity, salinity, sounds, temperature, and time) and provide information from the animal itself (e.g., compass bearing, heart beat frequency, location, muscle movement, respiration rates, swim speed, tilt, pitch, and roll, tail stroke frequency, and vocalizations). These devices are more commonly used to study of marine mammals, but their application to understanding energy expenditure in fishes, cephalopods, and turtles are growing. We begin with examples of researchers using biotelemetry to detect muscle movement (used to assess heart rate, stroke frequency, and metabolic rate) and internal temperature (to assess foraging bouts and energy intake and expenditure) in fishes.

Most attempts to telemeter muscle activity have been conducted in the laboratory and in freshwater. These tags estimated energetic costs or activity levels, but they were not able to replicate the complex structure of the EMG waveform that would indicate the precise timing and extent of muscle recruitment. In addition, the older EMG tags could not telemeter in marine systems due to their reliance on radio wave technology. Dewar et al. (1999) developed an acoustic transmitter that could be used in a marine environment. This transmitter was tested under laboratory conditions with different species of fish, including toadfish (*Opsanus tau*), spiny dogfish (*Squalis acanthias*), yellowfin tuna (*Thunnus albacares*), and eastern Pacific bonito (*Sarda chiliensis*).

The EMG tag acquired muscle activity through five steps. First, bipolar electrodes sensed the EMG signal from the fast- or slow-twitch muscles of the species in question. Second, the signal was amplified 2400 times with a differential amplifier. Third, the signal was filtered with a bandpass filter that set the low pass and high pass poles at 400 and 10 Hz, respectively. Fourth, the signal was sent to a voltage-controlled oscillator that produced an FM signal whose modulations exactly mimicked the waveform of the signal. Finally, an acoustic transducer (PZT-4 ceramic tube resonance frequency of 122 kHz) sent the waveform to the recording equipment.

The authors developed three different receivers connected to one or more hydrophones to detect the FM signal and convert it back to produce the EMG waveform. Receiver 1 was abandoned because it could not provide real time display of the decoded EMG and there were problems attributed to multipath, a common problem in water acoustic systems that results from the fact that sound travels in all directions, simultaneously allowing wave reflections to arrive at the hydrophone at different times.

A second receiver improved the design, but the authors developed a third receiver to further reduce the problem of multipath and to provide finer resolution of the signal. The transmitted data were acquired from a single hydrophone, passed through an 80–260 kHz bandpass filter (AP280-5-SR, AP Circuit Corp.), and then demodulated with a Kenwood TS440 FM receiver. The signal was then low-pass filtered and sent to a speaker to provide an audible signal corresponding with the tail beat of the fish. With Receiver 3 and their transmitter tag, the authors calculated tail-beat frequency and in the absence of multipath problems replicated the EMG waveform, allowing them to measure the onset–offset duration, number of zero crossings, area under the rectified waveform, and the product of the peak and mean peak intensity. Thus, this system could lead to fine scale examination of muscle recruitment and better estimates of energy requirements for free-swimming fish. In conclusion, the authors argued that the tag's design could be combined with remote sensing devices that produce complex analog signals (strain-gauges, flow-probes) to answer more complex questions regarding animal physiology and behavior.

Recently, Campbell et al. (2005) used a data logger to record the heart rate of black cod (*Paranotothenia angustata*) over a long period of time. Heart rate is sensitive to stress (Cooke et al., 2002) and correlates with metabolic rates associated with changes in activity and feeding (Armstrong, 1986; Lucas, 1994). Consequently, heart rate can be used to estimate energy requirements and expenditure. Campbell et al. determined whether surgically implanted or externally attached electrodes provided the best measure of heart rate. The authors used a miniature electronic microprocessor controlled data logger developed at the University of Wales Bangor. This data logger provided high-resolution ECG recordings (512 Hz) from a free-swimming fish. The microprocessor analyzed the waveform and determined interbeat intervals and instantaneous heart rate. The system was programmed to sample the first 20 min of every hour and also recorded two complete ECG waves every 4000 beats to check signal quality. In addition, the data logger housed a temperature sensor that measured ambient temperature each minute. Data were stored in a nonvolatile flash memory card to protect against power failure.

Once cod had recovered from the implantation or external attachment of the electrodes and data logger, they were released into a saltwater pond where they were subject to observation. The authors showed that black cod outfitted with externally attached electrodes showed an initial heart rate of 46 beats per min for the first 24 h, but by day 20, the average heart rate had decreased to 34 beats per min. In contrast, black cod with surgically implanted electrodes had initial heart rates of 44 heart beats per min, a rate that did not decrease over the 20 day period. Campbell et al. concluded that the method of electrode attachment in EMG studies with fish could be species-specific, with some fish showing normal heart rates with surgically implanted electrodes and some responding better to external attachment techniques.

Sharks are often depicted as cold-blooded killers. Indeed the name seems to make good sense as most of us believe that all fish are cold-blooded. Yet, not all that long ago, it was discovered that a small group of lamnoid sharks (Order Lamniformes) possess vascular countercurrent heat exchangers located in the brain, musculature, and viscera. These heat exchangers allow these sharks to maintain a relatively constant internal temperature, making them something less than cold-blooded killers. Great white sharks (*Carcharodon carcharias*) are lamnoid sharks and McCosker (1987) hypothesized that white sharks could regulate their stomach temperature. Goldman et al. (1996) used telemetry to further support this ability and Goldman (1997) confirmed this ability. Goldman fed a transmitter, hidden inside a piece of elephant seal blubber, to a great white. With the use of thermistors and pressure sensors potted inside Vemco transmitters, Goldman recorded temperature and depth information from three different sharks. Whereas the ambient temperature of the water varied from 12.3°C to 16.4°C, the stomach temperature of the sharks averaged 26.0°C–27.1°C. Thus white sharks were able to maintain a relatively high and narrow range of stomach temperature even in water temperatures that were 10°C–15°C colder and more variable.

Another of the lamnoid sharks is the salmon shark (*Lamna ditropis*). Like the great white shark, these sharks also possess vascular countercurrent heat exchangers that allow for retention of metabolically generated heat. Goldman et al. (2004) tested this shark's ability to regulate internal temperature. Four free-ranging sharks in the Prince William Sound area of Southeast Alaska were induced to swallow Vemco transmitters (V22-TP) or Sonotronics transmitters (CHP-87-LT) that were hidden inside herring bait. The Vemco transmitters recorded stomach temperature and pressure. They recorded data for 9–10 days and had a range of 1 km. Sonotronics transmitters recorded only stomach temperature, transmitted for 60 days, and were detectable at half the distance of the Vemco transmitters. A directional hydrophone (Dukane Corporation, St. Charles, IL, model N30A5A) detected the signal from the transmitters and Vemco receivers (VM-10, or VR-60) or Ultrasonic Telemetry Systems, Brea, CA converted the analog signal to digital. Records were obtained over time periods ranging from 3.8 to 20.7 h. Similar to the great white shark, internal temperature ranged from 25°C to 25.7°C whereas the temperature of the water ranged from 5°C to 16°C. Goldman et al. argue that salmon are able to maintain a relatively stable internal temperature and are essentially warm blooded. This novel approach to homeothermy, according to the authors, probably underlies this shark's ability to pursue and catch highly active prey like salmon in relatively cold water.

27.4.4 Assessing Energy Expenditure with Acoustic Telemetry: Marine Mammals

Energy expenditure parameters of marine mammals were first measured unintentionally through the use of an acoustic data logger on Steller sea lions (Fletcher et al., 1996). Fletcher et al. investigated the effects of noise on Steller sea lion diving behavior and communication using digital audio tape recorder (DAT) tags. The authors were surprised to discover that the tagged animal's heartbeat could be detected when the seal was at the surface and quiet, and that swim strokes could be detected acoustically. Swim speed was measured by counting and storing the number of revolutions of a paddle wheel, and comparing the animal's rate of change in depth with the number of revolutions. Burgess et al. (1998) followed this tag design and combined the DAT with a compact acoustic probe (CAP), which together measure dive patterns, ambient and animal noise exposure, possible animal vocalizations, plus acoustic signatures of swim stroke, surface respiration, and heart rate. Burgess et al. were able to record cardiac activity at depth as well when flow noise did not obscure animal sounds.

Biotelemetry research took a large step forward upon the release of the DTAG by researchers from Woods Hole Oceanographic Institute (see Figure 27.3). The DTAG is a digital acoustic recording tag that continuously records the orientation, heading, and depth of the tagged animal in synchronicity with sounds sensed by a hydrophone (Johnson and Tyack, 2003; Figure 27.4). Orientation is calculated from pitch and roll as determined by three-axis accelerometers (Analog Devices ADXL202). A three-axis magnetometer comprising low-power magnetoresisitve bridge sensors (Honeywell HMC1021) is used to measure heading, via the direction of the earth's magnetic field relative to the tag. The audio components consist of a piezoceramic hydrophone, a preamplifier, antialias filter, and analog-to-digital converter. The DTAG utilizes a digital signal processor to combine sensor measurements (consisting of acceleration, magnetic field, and pressure sensors) with audio data to store into a memory array. Orientation can be sampled at rates up to 50 Hz. A flash memory card of up to 3 GB can be used; in the initial tag design, the DTAG had only 400 Mb of memory. The tag electronics are combined with a battery, a VHF radio beacon for tag recovery, and two suction cups for attachment to the marine mammal. This method is noninvasive and allows up to tens of hours of data collection. The suction cups usually release on their own due to natural leakage or social and surface-active behaviors on the tagged individual's part; an active release is also included via a nickel-chromium wire that corrodes and releases an air line into each suction cup (Johnson and Tyack, 2003). To date, the longest attachment of a DTAG was 21 h on a northern right whale and 10 h on a sperm whale.

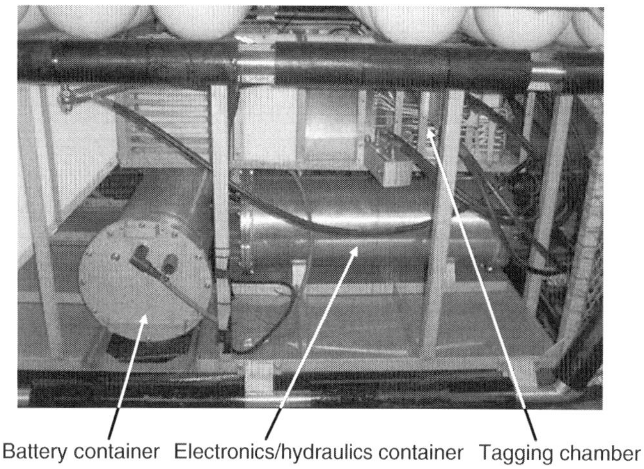

FIGURE 27.3 Underwater Tagging Equipment (UTE). Developed by Star-Oddi. Picture taken from www.star-oddi.com. With permission.

Sperm whales are champion divers that can dive at least 1200 m and hold their breath for over 1 h (Watwood et al., 2006). The DTAG system described by Johnson and Tyack (2003) was used to determine the effects of drag and buoyancy on the swimming gaits and diving patterns of sperm whales (Miller et al., 2004). The whale's length, mass, and surface area are important to estimate drag and buoyancy, and were measured via allometry

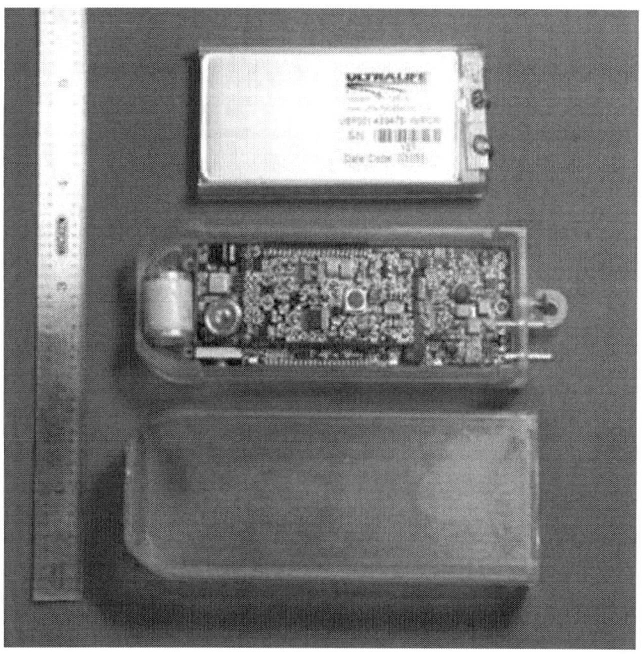

FIGURE 27.4 DTAG developed by Johnson and Tyack (2003). Top portion simplified schematic of the circuit. Bottom picture shows the internal structure of the DTAG and its size. Picture reprinted from Johnson, M.P. and Tyack, P.L., *J. Oceanic Eng.*, 28, 3, 2003. With permission.

while the research vessel was alongside the whale at the water surface. The DTAG weighed 500 g in air and consisted of a digital recorder, depth sensor, audioband acoustic recorder, and three-axis accelerometers and magnetometers. Sampling rate averaged 5.88 Hz, allowing quantification of animal pitch and depth, which allows fluking, descent, and ascent rates and speed to be calculated. Sperm whales most commonly used steady fluking on descent and stroke-and-glide fluking on ascent. Mean descent speeds were slower than ascent speeds, despite whales stroking steadily during descents. The results indicate that sperm whales make behavioral decisions based upon buoyancy, and decisions may change as buoyancy parameters change by depth, air volume, and tissue density (Miller et al., 2004).

Drag and buoyancy affect significantly the daily energy expenditure of marine mammals. Buoyancy is dependent on the amount of air in the animal's tissues and the changes in air density as an animal dives. Drag affects both horizontal and vertical movement whereas buoyancy only affects vertical movements. One way that marine mammals compensate for drag and buoyancy is to moderate their swimming "gait" or style (e.g., gliding versus steady fluking, Williams et al., 2000).

Nowacek et al. (2001) used the DTAG setup to study the effect of buoyancy on right whales. They set the sensor sampling rate at 23 Hz to track subtle and quick movements and the audio bandwidth was set at 8 kHz, within frequency range of right whales. The authors determined that volume of body cavities, the proportion of positively buoyant tissue like blubber, and the proportion of negatively buoyant tissue like bone affected tissue buoyancy. Right whales were found to be positively buoyant even when hydrostatic pressure at depth can cause negative buoyancy in some species, such as Weddell seals (Williams et al., 2000). This poses strong implications for right whale management, as positively buoyant whales on an ascent likely have reduced maneuverability. Therefore right whales may have little to no control to escape oncoming ships that pose a threat of collision. Indeed, ship strikes are a primary factor in North Atlantic right whale mortality, accounting for more than one-third of known right whale deaths (Laist et al., 2001).

Sato et al. (2003) investigated the effects of seal obesity on buoyancy and energy expenditure in Weddell seals in the wild. Swim speed, depth, two-dimensional accelerations (stroke frequency and body angle), and temperature were recorded with a data logger (Little Leonardo Corporation, Tokyo, Japan). Thinner females exhibited prolonged gliding during descent, while fatter females exhibited swim-and-glide swimming throughout their descent and ascent. Results suggested that the added positive buoyancy of fatter individuals shortened their ascent time with less energy expenditure but increased the effort required to descend. This supports the conclusions of Williams et al. (2000) that prolonged gliding is a more efficient method of swimming.

Foraging events are often inferred from two-dimensional time–depth profiles and temporal distribution of diving bouts. However, because one cannot actually observe the animal, these inferences are not always correct (R. Davis, quoted in Senkowsky, 2003). To obtain more accurate data on foraging bouts, researchers use stomach temperature transmitters (STT) to quantify the foraging frequency in large marine predators (Carey et al., 1984; Wilson et al., 1992). STTs provide better data because the stomach temperature of a warm-blooded marine mammal decreases markedly when it ingests relatively cold prey items. Andrews (1998) tested remotely releasable stomach temperature transmitters on captive Steller sea lions to determine the efficacy of the method for use in studying free-ranging pinniped feeding frequency. A small transmitter with a temperature sensor is introduced into the animal's stomach and transmits data to a logger mounted to the animal's skin or fur. The data logger package (Tattletale Fast Lite, Onset Computer Corporation, MA) records dive depth, swim speed, ambient temperature, and stomach temperature. A satellite transmitter (Telonics ST-6, Mesa, AZ) and VHF transmitter are also included in the package. When sampling dive depth, swim speed, and stomach temperature every 10 s and water temperature every 60 s, the logger's 512 kb memory could last almost 12 days (Andrews, 1998). Eventually the stomach transmitter is regurgitated or passed. The tag package atop of the seal was remotely releasable, utilizing a small heating element to melt an anchor line that attaches the tag to the animal's fur. The researchers could then retrieve the tag package via satellite transmission and VHF signal.

Austin et al. (2006) has expanded on the use of this technology, and provided the first quantitative analysis of feeding frequency in a free-ranging marine mammal (grey seals) based on stomach temperature telemetry. Austin et al. were interested in how feeding frequency would differ between males and females, and how time of day and between-meal duration affected foraging timing. A STT (Wildlife Computers) was inserted into an anesthetized gray seal's stomach via an equine intubation tube. The STT was designed to be large enough to delay passage but small enough so that over time the foam in the apparatus would break down and allow passage. A satellite-relay data logger (SRDL, Wildlife Computers, WA or ST-18-Telonics, AZ) was attached to the dorsal surface of the seal and included a time–depth recorder; the data package was removed upon the seals' return to their rookery. There were significant feeding differences between male and female seals. Males fed more often than females and spent more time feeding per day. This finding was not surprising in light of the fact that gray seal males in this study were between 30% and 45% heavier than female seals. For both sexes, the largest meal size was at dawn and the smallest meal at dusk. In addition, seals waited longer to eat their next meal after gorging themselves the previous meal. Some variability in individual feeding behavior was documented. The authors speculated that prey patchiness and individual foraging tactics between individuals likely caused some variance (Austin et al., 2006).

27.4.5 Assessing Energy Expenditure Using Video Telemetry: Fishes, Cephalopods, and Turtles

Greg Marshall (www.nationalgeographic.com/speakers/profile_marshall.html) is a pioneer in the area of video telemetry and may have been the first to attach a camera to a free-ranging marine animal and record its behavior. The idea occurred to him while he was watching a shark swimming with a host of attached remora fish. Marshall realized that it might be possible to attach a video camera to the shark to obtain a shark's eye view of the world. Such a view might allow scientists to better understand its world. With this camera, one could observe the strategies used to hunt prey, to find mates or to avoid predators under natural conditions. With those thoughts, Marshall developed and tested "crittercam."

Crittercam technology (www.nationalgeographic.com/crittercam/) incorporates sophisticated video, audio, and time–depth recording sensors that, for the first time, allow detailed examination of the factors that influence fitness and survival of marine mammals (Marshall, 1998). In addition, transmitters are also attached that record stroke frequency through accelerometers, compass bearing, depth, temperature, light levels, and velocity. This combined technology allows direct observation of swimming periods, stroke frequency, glide sequences, and allows researchers to actually observe, rather than postulate, why marine mammals are diving—whether to forage, rest, or socialize. The camera system can either face forward, in order to observe head movements, foraging and socializing, or it can face backward, in order to observe propulsive movements of flukes and flippers and perhaps social interactions. The first public viewing of images from crittercam occurred in 1993 on Geographic EXPLORER. To date, crittercam has been used on more than 30 marine species including great white sharks, emperor penguins, killer whales, sea turtles, seals, sperm whales, and walruses. Currently, this technology is being adapted for land animals with cameras being attached to bears, lions, and tigers among others.

Video telemetry is relatively new and examples of published studies with fishes, turtles, or cephalopods are rare or nonexistent. Most of the publications using video telemetry have been conducted with marine mammals, particularly seals. In the next section, we will consider representative studies from these animals. Here, we will talk about a different approach to video telemetry that is being developed by David Scheel at Alaska Pacific University and his colleagues and students from the Colorado School of Mines and the University of Arizona. Dr. Scheel, with funding from the National Science Foundation, is developing a robotic camera that could track a Giant Pacific octopus wherever it ventured. The robotic camera has been named Shadow (http://marine.alaskapacific.edu/octopus/Shadow.php). The camera will be housed in an autonomous underwater vehicle that can follow the output of an acoustic pinger attached to the octopus. This underwater vehicle is a very sophisticated smart structure requiring an ability to move through the water, navigate around obstacles, and keep the octopus in the camera's viewfinder. The project is divided into three areas with Dr. Scheel's team working on collecting and analyzing biological that can provide insight into the parameters needed to follow a moving octopus, developing a video system that provides enough scale to assess what the octopus is actually doing, and determining how the video data will be analyzed. Faculty advisor to the project, Dr. Thomas Vincent, University of Arizona, will lead the team that develops the platform and designs the thrusters, buoyancy controls, and stabilizer allowing Shadow to move through the water. Faculty advisor, Dr. Tyrone Vincent, Colorado School of Mines, will work with students to design the tracking system. This system will need to sense the acoustic signal and determine the range and direction to the octopus. Both the Navigation and Power Control teams will work together to develop the controls necessary to keep the octopus within the viewfinder of the camera and switch it on or off, depending on the physical activity of the animal. With Shadow, Scheel hopes to shed light on the life history of the Giant Pacific Octopus, which is experiencing a population decline in the Pacific Northwest. Dr. Scheel is hoping to learn more about the migration habits, foraging strategies, and predator defense strategies of the Giant Pacific Octopus (*Enteroctopus dofleini*) found in Prince William Sound among other places.

27.4.6 Assessing Energy Expenditure Using Video Telemetry: Marine Mammals

Davis et al. (1999) attached a small video imaging system to the head and a data logger to the dorsal surface of Weddell seals in the Antarctic. The data logger recorded time, depth, water speed, and compass bearing once per second; flipper stroke frequency and ambient sounds were recorded continuously with a hydrophone. The video system recorded images of the seals' heads and the environment in front of the animal. Results shed light on how Weddell seals stalk large prey in three-dimensional environments, and also pointed to the possibility that pinnipeds may use passive detection of prey as determined by their lack of emitted sounds.

Williams et al. (2000) utilized similar methods to investigate diving strategies in marine mammals. Weddell seals, an elephant seal, a blue whale, and a bottlenose dolphin were outfitted with a video imaging system and a time–depth recorder (TDR). Swimming mode, relative stroke amplitude, stroke frequency, and gliding periods were determined using a motion analysis system (Peak Performance). Data were matched to duration and depth of the associated dive. To assess effect of changes in locomotion on energetics, postdive oxygen consumption of Weddell seals was measured by having the seals breathe into a metabolic hood. Results indicated that Weddell seals and likely other marine mammals take advantage of lung compression and subsequent negative buoyancy by engaging in prolonged gliding on descent; this behavior conserves energy, allowing marine mammals to increase aerobic dive duration and achieve deeper dives.

The most recent application of this video-data logging system is the estimation of energy expended during foraging dives by Weddell seals in the Antarctic. Williams et al. (2004) measured the associated energetic costs of foraging including metabolism, locomotion, and prey digestion by measuring postdive oxygen consumption. Costs of locomotion increased with the number of flipper strokes. Also, recovery time due to oxygen consumption was increased by 45% when seals took foraging dives that the authors attribute to the necessary energy for warming and digestion of prey (Williams et al., 2004).

27.5 Predicting the Future

Predicting the near future of biotelemetry technology is pretty simple. Tags will get smaller and smarter with a wider range of sensors that will operate for longer periods of time and possess more memory. It is also likely that biotelemetry systems in the future will process more data and provide more sophisticated analyses, making easier the problem of figuring out what 1 million or more data points are telling us about the animal. In addition, we expect that behavioral and physiological ecologists and others interested in nonhuman animals will use biotelemetry to study an increasing variety of species with a great variety of applications. That is in the near term. It is more difficult to attempt to predict what the biotelemetry equipment will look like 50 or even 20 years from now or how scientists interested in aquatic animals will use such equipment. Researchers who were instrumental in the early development and promotion of telemetric systems and who today are providing important examples of this methodology have addressed this question recently. Block (2005), Cooke et al. (2004), and Ropert-Coudert and Wilson (2005) provide excellent reviews of the history of this technology, current applications, and they offer insight into what the future holds and does not hold for its use. In this section we will not paraphrase their insight, but instead we examine a new smart tag that is currently being developed and tested that may provide better data on mortality and survival of marine mammal populations. In addition, we briefly consider the possibility of telemetry systems based on nanotechnology and termed "Smart Dust" by developers. Finally we speculate on the possibility that biotelemetry could be used to directly test various theories proposed by behavioral ecologists and experimental psychologists to account for the behavior of animals.

27.5.1 Life History Transmitter

Horning and Hill (2005) describe the development of "The Life History Transmitter" that they predict will become an important tool in the study of aquatic animal population dynamics. The Life History Transmitter is designed to collect vital data from marine vertebrates for up to 10 years, which could amount to the animal's entire lifetime. The tag would record life history data from the time it is implanted until the animal dies. At this point, the transmitter is released through decay and upon reaching the surface of the ocean or exposed to air on a beach it transmits its data to a satellite. If successful, the authors contend that long-term survival rates and population dynamics of marine vertebrates could be investigated more accurately. In addition, it should be possible to obtain better estimates of the energy required to survive and reproduce.

Horning and Hill (2005) began by modeling their transmitter after Wildlife Computer's pop-up archival transmitter (PAT) tags because these tags record information until they are exposed to the air at which time they transmit their data to a satellite via the ARGOS system. The researchers then detailed the problems the tag would have to solve. For example, the tag had to discriminate between two states, mortality and extrusion. When the animal dies, a mortality event has occurred and when the tag is exposed to air either via the surface of the ocean or on a beach following the animal's death, an extrusion event has occurred. The authors decided that an internal temperature sensor could be used to determine mortality. To determine the point at which the tag was exposed to air following extrusion, the authors used a combination light and conductivity/immersion sensor. Horning and Hill then set out to build two prototypes of the Life History Tag. The first tag (Standard LHX) would house temperature, light-level, and immersion sensors only. As such, it would find application in population dynamic studies where researchers want better data on survival. The more sophisticated tag (Enhanced LHX) would record pressure and movement through motion sensors in addition to internal temperature, light-level, and conductivity/immersion. The enhanced version would provide behavioral data (total dives, total vertical displacement, stroke frequency) over the lifetime of the animal and contribute greatly to studies on energetics. To focus their efforts, Horning and Hill ranked the necessary components in terms of the potential to fail. In decreasing order of probability to fail, battery, sensors, A to D converter, housing, and software comprised the list of problems to solve.

The authors power their transmitter over a 10-year period with two primary battery systems. A BCX85PC1 lithium/bromine-chloride complex battery (Wilson Greatbatch Technologies, MA) powers the controller software and a LSH26180 lithium/thionyl-chloride battery powers the transmitter's amplifier. Both batteries lose 3% of their power annually resulting in 70% capacity after 10 years. The standard LHX tag utilizes an YSI44017 thermistor (YSI Inc., Beavercreek, OH) to measure internal temperature, a TAOS TSL257 light-to-voltage converter (TAOS Inc., Plano, TX) to measure light level, and a contact-free immersion sensor developed by Wildlife Computers, Redmond, WA. Housing requirements included positive buoyancy, pressure rated to 1000 m, a helical antenna with a cover that allowed for propagation of UHF signals, and moisture proofing for residing inside the animal for a decade or more. The LHX tag measured 42 mm in diameter and 122 mm in length. The tag weighed 115 g in air, or about .1% of the target species, stellar sea lions. The tag's buoyancy rating in salt water was −38 g. To ensure the tag remains operational for 10 years, the tag is interrogated every 30 min for sensor-data acquisition, error-check, tag-state, mortality, and extrusion events.

As mentioned briefly, mortality is measured by internal temperature. For example, if the recorded temperature falls outside the user-defined range of appropriate temperature for 96 interrogations, the animal is presumed to be dead. This positive test for TST1 triggers a heuristic algorithm to determine the highest likelihood of making a correct decision with respect to extrusion. If all sensors are operating correctly, the light level sensor indicates that the tag is in ambient light and the immersion sensor detects air the tag assigns a YES to TST2 and assumes an extrusion event has occurred. At this point, data transmission begins. The speed with which data is transmitted depends upon the confidence level of the heuristic algorithm. If the algorithm is highly confident, data are transmitted over a 10-day period. If the algorithm has detected possible problems with the sensors, it produces a moderately confident value and the transmitter reduces transmission speed and sends the data over a 20 day period. At the lowest level of confidence, transmission is reduced still further and could take up to 100 days to transmit all data. The longer periods of time to transmit are provided to allow for full extrusion in the event the tag has been only partially exposed.

To date, the Life History Transmitter has been tested on a few Steller sea lions and subjected to full simulation studies with a computer. The authors continue to develop the Enhanced LHX and plan a more extensive deployment of the Standard tag. If possible, Horning and Hill plan to implant both the standard and the enhanced version in the same animal to assess reliability. The authors contend that this tag offers a strong advance over existing methods for assessing population dynamics and could lead to substantially more and more accurate mortality and behavioral data involving fewer animals.

27.5.2 Biotelemetry and Smart Dust

In the past few years, scientists and engineers have been working on a project that has important implications for biotelemetry studies. Smart Dust (http://robotics.eecs.berkeley.edu/~pister/SmartDust/ and www.computerworld.com/mobiletopics/mobile/story/0,10801,79572,00.html) refers to devices that are very small wireless s (MEMS) that sense stimuli, process data, and provide bidirectional communication. The goal of the Smart Dust project at the University of California, Berkeley was to design sensors that could detect temperature, humidity, barometric pressure, light intensity, tilt and vibration, and magnetic fields. In addition, the device could operate for 2 years or more with a 1% duty cycle, and provide bidirectional radio communication. Finally, an onboard microprocessor could provide for preliminary analysis of the data. Amazingly, the sensors, power supply, analog circuitry, bidirectional optical communication, and programmable microprocessor would all fit within 1 cubic millimeter and data could be communicated at distances up to 1000 m.

The investigators, Kris Pister, Joe Kahn, Bernhard Boser, and Steve Morris, suggest that Smart Dust has obvious applications to the military where such devices could provide battlefield surveillance, treaty monitoring, and equipment movement. Such devices could also provide product quality monitoring, smart offices with individually tailored environments for workers, and interfaces for disabled persons. We contend that smart dust could be of significant value to the study of nonhuman animals. These devices could provide continuous data on environmental variables like temperature, humidity, salinity, light levels, magnetic fields, and movement. These data could be recorded over long or short periods of time depending upon the degree of resolution desired. Smart Dust could be used to locate and track animals, assess navigational ability and environmental constraints on daily or annual migratory behavior, determine movement, and provide estimates of energy requirements and use. Of importance, this could all be done with equipment so small as to be completely unobtrusive to the animal.

27.5.3 Methods of Testing Behavioral Hypotheses Using Biotelemetry

In 2001 and 2002, one of us (J.E.P.) spent a total of 5 months in Antarctica working with Randy Davis, Texas A&M University at Galveston, Lee Fuiman, University of Texas Marine Science Institute, Port Aransas, Terrie Williams, University of California at Santa Cruz, and Markus Horning, Hatfield Marine Mammal Center, Oregon State University studying foraging and diving characteristics of Weddell seals under the fast ice in McMurdo Sound. In analyzing certain of the data, an application for the use of biotelemetry became apparent that has not been utilized to any great extent. In their first seasons in McMurdo Sound, Davis and coworkers used an isolated hole protocol that was first developed by Kooyman (1965). In this paradigm, researchers go out on the ice and find a location where there are no cracks or holes in the ice within a circle whose diameter is 2 km or greater. Then, a hole is drilled in the center of this circle and seals are caught and transported to this location where they are outfitted with video telemetry equipment (see above description from Davis et al., 2004). Seals are released from this location where they conduct series of exploratory and foraging dives over a period of several days. Because the seal cannot get to another hole to breathe, it has return to the isolated hole. When the seal returns or when it hauls out at a location close to camp, the equipment and all physiological, environmental, and video data are retrieved.

In analyzing second by second compass bearing data from the diving seals, it was observed that seals would typically head out on a specific heading, say 90°, and return along the same heading, now 270°. This dive would typically last 16–20 min and was followed by resting in the hole for 3–5 min and diving again. Interestingly, the seal would typically head off in a different direction from the first, say 135° and return along a heading of 315°. When one looked at all of the dives together, it appeared that the seal left the hole on a heading that appeared to be chosen at random, but one that did not repeat the previous headings. This behavior has implications for the foraging strategies of the Weddell seal. If there are limited schools of prey that move around haphazardly, it is likely that the probability of finding food in the last known place is less than that of finding food in a new place. If true, the strategy of heading off in a different direction each time and not repeating directions might be the most efficient. This strategy requires that Weddell seals are able to maintain a specific heading and that they remember where they have previously been. In essence, Weddell seals must possess a good sense of direction and good spatial memory.

What makes this even more interesting is that this strategy appears to be very similar to that used by a rat in a radial arm maze. In an eight-arm radial arm maze, rats appear to choose one of eight arms at random and explore to the end looking for food. If food is found, because it is not replaced, it is to the rat's advantage to choose a different arm on the next trial. In solving the maze, rats pick one arm, go down that arm, find the food, and then return to the center. From here, the rat will pick, at random, another arm and run down it. Because food at the end of the arm is not replaced, the rat must remember which arms it has visited and which it has not. Several studies have shown that the rat is very good at this task and has shown remarkable spatial memory (Olton and Samuelson, 1976; Roberts, 1984) In fact, researchers have shown that the rat's spatial memory is such that it can remember which arms it has visited in a maze that has as many as 24 arms (Roberts, 1979).

One of the difficulties of these studies on spatial memory is that they were conducted under laboratory conditions. Thus, we know the animal is capable of the task, but we do not know if this ability is ecologically relevant. Biotelemetry may offer a way for behavioral ecologists and experimental psychologists to study these abilities in a free-ranging animal. By knowing where the animal is, the tracks that it used to navigate and when it is foraging it should be possible to judge whether the behavior observed in the laboratory reflects that observed in the field. In the end, biotelemetry might provide a means for testing in the animal's natural environment, the theories that behavioral ecologists and experimental psychologists posit to account for an animal's behavior.

27.6 Conclusions

Smart materials and smart structures serve as a foundation for a wide variety of the equipment used to study aquatic animals. This is particularly true in the area of biotelemetry where smart tags and receivers are playing an increasing role in understanding foraging, habitat selection, mating behavior, migration, navigation, population dynamics, and the ways in which the environment constrains behavior. The application of smart materials and smart materials to the study of aquatic animals has increased significantly but scientists have probably realized only a fraction of the applications that are possible. In the next 10 years, we should see biotelemetry equipment becoming increasingly smaller and more sophisticated and as a result the number of species that could be studied with this methodology will increase as well as the types of questions that can be answered. Such studies will contribute greatly to the knowledge base and will go a long way toward sustaining a great resource, aquatic animals.

References

Aitken, J. P., O'Dor, R. K., and Jackson, G. D. 2005. The secret life of the giant Australian cuttlefish *Sepia apama* (Cephalopoda): Behaviour and energetics in nature revealed through radio acoustic positioning and telemetry (RAPT). *Journal of Experimental Marine Biology and Ecology, 320,* 77–91.

Aglen, A. 1994. Source of error in acoustic estimation of fish abundance. In: *Marine fish Behaviour in Capture and Abundance Estimation.* Eds. A. Ferno and S. Olsen. Fishing News Books, Oxford, U.K., pp. 107–133.

Andrews, R. D. 1998. Remotely releasable instruments for monitoring the foraging behaviour of pinnipeds. *Mar. Ecol. Prog. Ser. 175,* 289–294.

Armstrong, J. D. 1986. Heart rate as an indicator of activity, metabolic rate, food intake and digestion in pike, *Esox lucius. Journal of Fish Biology, 29,* 207–221.

Ashley, S. 1995. Smart skis and other adaptive structures. *Mechanical Engineering, 117,* 76081.

Austin, D., Bowen, W. D., McMillan, J. I., and Boness, D. J. 2006. Stomach temperature telemetry reveals temporal patterns of foraging success in a free-ranging marine mammal. *Journal of Animal Ecology, 75,* 408–420.

Baird, R. W., Hanson, M. B., and Dill, L. M. 2005. Factors influencing the diving behaviour of fish-eating killer whales: Sex differences and diel and interannual variation in diving rates. *Canadian Journal of Zoology, 83*(2), 257–267.

Baumgartner, M.F. and Mate, B. R. 2005. Summer and fall habitat of North Atlantic right whales *(Eubalaena glacialis)* inferred from satellite telemetry. *Canadian Journal of Fisheries and Aquatic Science, 62,* 527–543.

Baumgartner, M. F., Cole, T. V. N., Clapham, P., and Mate, B. R. 2003. North Atlantic right whale habitat in the lower Bay of Fundy and on the SW Scotian Shelf during 1999–2001. *Marine Ecology Progress Series, 264,* 137–154.

Block, B. A. 2005. Physiological ecology in the 21st century: Advancements in biologging science. *Integrative and Comparative Biology, 45,* 305–320.

Bridger, C. J., Booth, R. K., McKinley, R. S., Scruton, D. A., and Lindstrom, R. D. 2001. Monitoring fish behaviour with a remote, combined acoustic/radio biotelemetry system. *Journal of Applied Ichthyology, 17,* 126–129.

Burgess, W., Tyack, P., LeBoeuf, B. J., and Costa, D. P. 1998. An intelligent acousatic recording tag first results from free-ranging northern elephant seals. *Deep-Sea Research II, 45,* 1327–1351.

Butler, P. J. and Jones, D. R. 1997. Physiology of diving of birds and mammals. *Physiol. Rev. 77,* 837–899.

Campbell, H. A., Bishop, C. M., Davies, D. A., and 7 Egginton, S. 2005. Recording long-term heart rate in *Paranotothenia angustata* using an electronic datalogger. *Journal of Fish Biology, 67,* 1150–1156.

Cao, W., Harley, C. H., and Waser, R. 1999. Smart materials and structures. *Proceedings of the National Academy of Science, USA, 96,* 8330–8331.

Candy, J. R. and Quinn, T. P. 1999. Behavior of adult chinook salmon (*Oncorhynchus tshawytscha*) in British Columbia coastal waters determined from ultrasonic telemetry. *Canadian Journal of Zoology, 77,* 1161–1169.

Casey, J. G. and Kohler, N. E. 1992. Tagging studies on the shortfin mako shark (*Insures oxrinchus*) in the Western North

Atlantic. *Australian Journal of Marine and Freshwater Research, 43*, 45–60.

Chilvers, B. L., Robertson, B. C., Wilkinson, I. S., Duignan, P. J., and Gemmell, N. J. 2005. Male harassment of female New Zealand sea lions, Phocarctos hookeri: mortality, injury, and harassment avoidance. *Canadian Journal of Zoology, 83* (5), 642–648.

Clark, M. E., Wolcott, T. G., Wolcott, D. L., and Hines, A. H. 1999. Foraging and agonistic activity co-occur in free-ranging blue crabs (*Callinectes sapidus*): observation of animals by ultrasonic telemetry. *Journal of Experimental marine Biology and Ecology, 233*, 143–160.

Collins, S. 2003. Brainy bra gives lift when you need it, 11/2/2006.

Cooke, S. J., Hinch, S. G., Wikelski, M., Andres, R. D., Kuchel, L. J., Wolcott, T. G., and Butler, P. J. 2004. Biotelemetry: a mechanistic approach to ecology. *Trends in Ecology and Evolution, 19*, 334–343.

Davis, R. W., Fuman, L., Williams, T. M., Collier, S., Hagey, W., Kanatous, S. B., Kohin, S., and Horning, M. 1999. Hunting behavior of marine mammal beneath the Antarctic fast-ice. *Science, 283*, 993–996.

Davis, R. W., Fuiman, L. A., Williams, T. M., Horning, M., and Hagey, W. P. 2003. Classification of Weddell seals dives based on three-dimensional movements and video recorded observations. *Mar. Ecol., Prog, Ser., 264*, 109–122.

Davis, R. W., Hagey, W., and Horning, M. 2004. Monitoring the behavior and multi-dimensional movements of Weddell seals using an animal-borne video and data recorder. *Mem. Natl. Inst. Polar. Res., Spec., Issue, 58*, 150–156.

Dewar, H., Deffenbaugh, M., Thurmond, G., Lashkari, K., and Block, B. 1999. Development of an acoustic telemetry tag for monitoring electromygrams in free-swimming fish. *Journal of Experimental Biology, 202*, 2693–2699.

Domeier, M. L., Kiefer, D., Nasby-Lucas, N., Wagschal, A., and O'Brien, F. 2005. Tracking Pacific bluefin tuna (*Thunnus thynnus orientalis*) in the northeastern Pacific with an automated algorithm that estimates latitude by matching sea-surface-temperature data from satellites with temperature data from tags on fish. *Fisheries Bulletin, 103*, 292–306.

Fletcher, S., LeBoeuf, B. J., Costa, D. P., Tyack, P. L., and Blackwell, S. B. 1996. Onboard acoustic recording from diving elephant seals. *Journal of the Acoustical Society of America, 100*, 2531–2539.

Gibbs, W. W. *Scientific American Explorations* (May, 1996).

Goldman, K. J. 1997. Regulation of body temperature in the white shark, *Carcharodon carcharias. Journal of Comparative Physiology B, 167*, 423–429.

Goldman, K. J., Anderson, S. D., McCosker, J. E., and Klimley, A. P. 1996. Temperature, swimming depth, and movements of a white shark at the South Farallon Islands, California. In *Great White Sharks: Ecology And Behavior,* Eds. A. P. Klimley and D. G. Ainley, San Diego, CA: Academic Press, pp. 111–120.

Goldman, K. J., Anderson, S. D., Latour, R. J., and Musick, J. A. 2004. Homeothermy in adult salmon sharks, *Lamna ditropis. Environmental Biology of Fishes, 71*, 403–411.

Hjellvik, V., Michalsen, K., Aglen, A., and Nakken, O. 2002. Examining the effective height of the bottom trawl using acoustic survey recordings. ICES Symposium on Acoustics in Fisheries and Aquatic Ecology, June 10–14, 2002. Paper no. 118. 20pp.

Hill, R. D. and Braun, M. J. 2001. Geolocation by light-level. The next step: latitude. In *Electronic Tagging and Tracking in Marine Fisheries,* Eds. J. R. Sibert and J. Nielsen, Dordrecht: Kluwer Academic, pp. 237–248.

Holland, K. N., Gush, A., Meyer, C. G., Kajiura, S., Wetherbee, B. M., and Lowe, C. G. 2001. Five tags applied to a single species in a single location: The tiger shark experience. In *Electronic Tagging and Tracking in Marine Fisheries,* Eds. J. R. Sibert and J. Nielsen, Dordrecht: Kluwer Academic, pp. 237–248.

Hooker, S. K. and R. W. Baird. 2001. Diving and ranging behaviour of odontocetes: A methodological review and critique. *Mammal Review 31*, 81–105.

Horning, M. and Hill, R. D. 2005. Designing an archival satellite transmitter for life-long deployments on oceanic vertebrates: The life history transmitter. *IEEE Journal of Oceanic Engineering, 30*, 807–817.

Jackson, G. D., O'Dor, R. K., and Andrade, Y. 2005. First tests of hybrid acoustic/archival tags on squid and cuttlefish. *Marine and Freshwater Research, 56*, 425–430.

James, M. C., Ottensmeyer, C. A., and Myers, R. A. 2005. Indentification of high-use habitat and threats to leatherback sea turtles in northern waters: New directions for conservation. *Ecology Letters, 8*, 195–201.

Johnson, M. P. and Tyack, P. L. 2003. A digital acoustic recording tag for measuring the response of wild marine mammals to sound. *Journal of Oceanic Engineering, 28*, 3–12.

Kooyman, G. L. 1965. Techniques used in measuring deving capacities of Weddell seals. *Polar Records, 12*, 391–394.

Kooyman, G. L. 2004. Genesis and evolution of bio-logging devices: 1963–2002. *Mem. Natl. Inst. Polar Res. Spec. Issue, 58*, 15–22.

Laist, D. W., Knowlton, A. R., Mead, J. G., Collet, A. S., and Podesta, M. 2001. Collisions between ships and whales. *Marine Mammal Science, 17*, 1632–1644.

Lander, M. E. and Gulland, F. M. D. 2003. Rehabilitation and post-release monitoring of Steller sea lion pups raised in captivity. *Wildlife Society Bulletin, 31*(4), 1047–1053.

Lowry, M. B. and Suthers, I. M. 1998. Home range, activity and distribution patterns of a temperate rocky-reef fish, *Cheilodactylus fuscus. Marine Biology, 132*, 569–578.

Lucas, M. C. 1994. Heart rate as an indicator of metabolic rate and activity in adult Atlantic salmon, *Salmo salar. Journal of Fish Biology, 44*, 889–903.

Marshall, G. J. 1998. Crittercam: An animal-borne imaging and data logging system. *Marine Technology Society Journal, 32*, 11–17.

Mate, B. R., Lagerquist, B. A., and Calambokidis, J. 1999. Movements of North Pacific blue whales during the feeding season off southern California and their southern fall migration. *Marine Mammal Science, 15*(4):1246–1257.

Mate, B. R., Duley, P., Lagerquist, B. A., Wenzel, F., Stimpert, A., and Clapham, P. 2005. Observations of a female North Atlantic right whale *(Eubalaena glacialis)* in simultaneous copulation with two males: supporting evidence for sperm competition. *Aquatic Mammals*, (31)(2): 157–160.

McCosker, J. E. 1987. The white shark, *Carcharodon carcharias*, has a warm stomach. *Copeia, 1*, 195–197.

Miller, P. J. O., Johnson, M. P., Tyack, P. L., and others. 2004. Swimming gaits, passive drag and buoyancy of diving sperm whales *Physeter macrocephalus*. *Journal of Experimental Biology, 207*, 1953–1967.

Moore, M. J., Knowlton, A. R., Kraus, S. D., McLellan, W. A., and Bonde, R. K. 2004. Morphometry, gross morphology and available histopathology in North Atlantic right whale (*Eubalaena glacialis*) mortalities (1970–2002). *Journal of Cetacean Research and Management, 6*(3), 199–214.

Naito, Y. 2004. Bio-logging science. *Mem Natl Inst Polar Res, Spec, Issue, 58*, 118–132.

Nelson, D. R., McKibben, J. N., Strong, Jr., W. R., Lowe, C. G., Sisneros, J. A., Schroeder, D. M., and Lavenberg, R. J. 1997. An acoustic tracking of a megamouth shark, (*Megachasma pelagios*): a crespuscular vertical migrator. *Envrionmental Biology of Fishes, 49*, 389–399.

Nowacek, D. P. et al. 2001. Buoyant balaenids: the ups and downs of buoyancy in right whales. *Proceedings of the Royal Society of London, Series B, 268*, 1811–1816.

Olton, D. S. and Samuelson, R. J. 1976. Remembrance of paces passed: Spatial memory in rats. *Journal of Experimental Psychology, Animal Behavior Processes, 2*, 97–116.

Pepperell, J. G. and Davis, T. L. O. 1999. Post-release behavior of black marlin, *Makaira indica*, caught off the Great Barrier Reef with sportfishing gear. *Marine Biology, 135*, 369–380.

Perkins, S. Science News Online (November 22, 1997).

Polovina, J. J., Balazs, G. H., Howell, E. A., Parker, D. M., Seki, M. P., and Dutton, P. H. 2004. Forage and migration habitat of loggerhead (*Caretta caretta*) and olive ridley (*Lepidochelys olivacea*) sea turtles in the central North Pacific Ocean. *Fisheries Oceanography, 13*, 36–51.

Roberts, W. A. 1979. Spatial memory in the rat on a hierarchical maze. *Learning and Motivation, 10*, 117–140.

Roberts, W. A. 1984. Some issues in animal spatial memory. In *Animal Cognition*, Eds. H. L. Roitblat, T. G. Bever, and H. S. Terace, Hillsdale, NJ: Erlbaum, pp. 425–443.

Ropert-Coudert, Y. and Wilson, R. P. 2005. Trends and perspectives in animal-attached remote sensing. *Frontiers in Ecology and the Environment, 8*, 437–444.

Sato, K., Mitani, Y., Cameron, M. F. Siniff, D. B., and Naito, Y. 2003. Factors affecting stroking patterns and body angle in diving Weddell seals under natural conditions. *The Journal of Experimental Biology, 206*, 1461–1470.

Senkowsky, S. 2003. High tech surveillance: Adapting new technology for marine studies. *BioScience, 53*, 1150–1157.

Shirley, M. A. and Wolcott, T. G. 1991. A telemetric study of microhabitat selection by premolt and molting blue crabs, *Callinectes sapidus (Rathbun)* within a subestuary of the Pamlico River. *North Carolina Marine Behavior and Physiology, 19*, 133–148.

Sigurdsson, T., Thorsteinsson, V., and Gustafsson, L. 2006. In situ tagging of deep-sea redfish: Application of an underwater, fish-tagging system. *ICES Journal of Marine Science, 63*, 523–531.

Small, R. J., Lowry, L. F., Hoef, Jay, M. V., Frost, K. J., DeLong, R. A., and Rehberg, M. J. 2005. Differential movements by harbor seal pups in contrasting Alaska environments. *Marine Mammal Science, 21*(4), 671–694.

Spotila, J. R., Reina, R. D., Steyermark, A. C., Plotkin, P. Tl, and Paladino, F. V. 2000. Pacific leatherback turtles face extinction. *Nature, 405*, 529–530.

Stensholt, B. K. and Stensholt, E. 2004. Geographical variation in the vertical distribution of cod (Gadus morhua L.) and availability to survey gears. *GIS/Spatial Analyses in Fishery and Aquatic Sciences*, (109–126).

Szedlmayer, S. T. and Schroepfer, R. L. 2005. Long-term residence of red snapper on artificial reefs in the northeastern Gulf of Mexico. *Transactions of the American Fisheries Society, 134*, 315–325.

Watwood, S. L., Miller, P. J. O., Johnson, M., and others. 2006. Deep-diving foraging behaviour of sperm whales (*Physeter macrocephalus*). *Journal of Animal Ecology 75*, 814–825.

Williams, T. M., Davis, R. W., Fuiman, L. A., and others. 2000. Sink or swim: strategies for cost-efficient diving by marine mammals. *Science, 288*, 133–136.

Williams, T. M., Fuiman, L. A., Horning, M., and Davis, R. W. 2004. The cost of foraging by a marine predator, the Weddell seal (*Leptonychotes weddellii*): Pricing by the stroke. *J. Exp. Biol, 207*, 973–982.

Wolcott, T. G. 1996. New options in physiological and behavioural ecology through multichannel telemetry. *Journal of Experimental Marine Biology and Ecology, 193*, 257–275.

Wolcott, T. G. and Hines, A. H. 1996. Advances in ultrasonic biotelemetry for animal movement and behavior: The blue crab case study. In: *Methods and Techniques of Underwater Research, Proceedings of the American Academy of Underwater Sciences Scientific Diving Symposium*, October 12–13, 1996, Eds. M. A. Lang and C. C. Baldwin, Washington, DC: Smithsonian Institution, pp. 229–236.

Yatsu, A., Yamanaka, K., and Yamashiro, C. 1999. Tracking experiments of the jumbo flying squid, Dosidicus gigas, with an ultrasonic telemetry system in the eastern Pacific Ocean. *Bulletin of the Natural Resources Institute, Far Seas Fishery, 36*, 55–60.

28

Molecular Imprinting Technology

David A. Spivak
Louisiana State University

28.1 Introduction and Applications of Molecularly Imprinted Polymers

There are many reviews on molecular imprinting technology that have chronicled the development of this field over the last two decades [1–4]. Molecular imprinting technology creates "smart materials" by integrating a template into the polymer formulation during the fabrication of these polymeric materials (Scheme 28.1). In this way, molecular level information is transferred from the template to the polymers in the form of a three-dimensional "memory" of the template's shape and other features. Because molecularly imprinted polymers (MIPs) can be tailored to specifically bind a targeted molecule or catalyze a desired reaction, these smart materials hold tremendous promise for applications such as

- Chemical and biological separations [5,6]
- Immunoassays [7,8]
- Catalysts [9]
- Sensors [10,11]
- Nanotechnology
- Biomedical, e.g., drug delivery [12–14]

Compounds currently utilized in these areas are nonspecific polymers, biological molecules, and designed small molecule receptors. MIPs are beneficial when the applications listed above require specific binding for a target molecule that is not available from nonspecific polymers such as silica gel, carbohydrates, or other polymer materials. While biological molecules exhibit good specificity toward a large range of molecules, MIP materials have improved properties such as

- stability to organic solvents, aqueous solvents, and high temperatures [15]
- ease, speed, and low cost of preparation in any volume
- tailorable binding affinities employing any functional group desired (including those not available to biological molecules)

whereas antibodies and enzymes denature under various conditions such as heat, organic solvents, acids and bases, etc. Furthermore, development of biological molecules is generally expensive and often come from animal sources and cannot be stored. A third advantage is that MIPs can be stored for several years without loss of performance unlike biological molecules. Small molecule receptors can be the rugged alternative, however, the synthesis of these compounds is often long and arduous, whereas molecular imprinting can be carried out in one day.

MIPs also have potential to create enzyme-like catalysts, specifically tailored to carry out a desired reaction with substrate-selective control of stereochemistry and regiochemistry [9]. One of the main ways to accomplish this takes advantage of transition state theory, outlined in Figure 28.1,

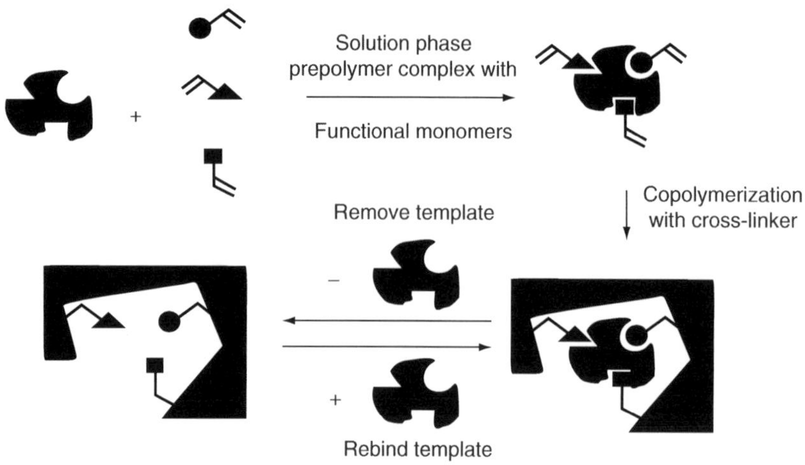

SCHEME 28.1 Outline of the molecular imprinting strategy.

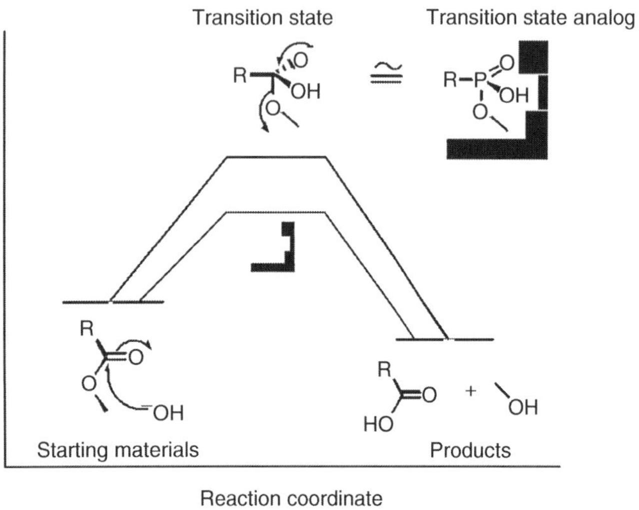

FIGURE 28.1 Energy diagram for the MIP catalyzed hydrolysis of a carboxylate ester.

the reaction coordinate diagram for hydrolysis of esters. If a MIP can be elicited to bind the transition state of a reaction (bold figure shown under the bottom reaction coordinate, and around the transition state analog), it lowers the overall energy barrier and catalyzes the reaction. However, since the transition state is too unstable to use as a template, a stable transition state analog is used for imprinting in place of the actual transition state. The MIP formed to the transition state analog can be added to the solution of starting material, which will catalyze the overall reaction as shown in Figure 28.1. MIPs that catalyze a reaction and control regiochemistry or stereochemistry, but do not have turnover (i.e., the reaction occurs once in the imprinted site) are referred to as microreactors [16].

28.2 Fabrication Methods of Molecular Imprinting Technology

The modern strategy employed for molecular imprinting (shown in Schemes 28.1 and 28.2) is much the same as introduced by Wulff in 1972, employing two types of monomers, functional monomers and cross-linkers [17]. The functional monomers incorporate a moiety that interacts with a template molecule; the template can be a small molecule of interest (e.g., drugs, hormones, sugars, etc.) or larger molecules such as proteins or other macromolecules. The functional monomer typically has a single polymerizable group that does not cross-link the material. However, cross-linking monomers that incorporate a functional group for interactions with the template have also been highly successful. The primary role of the functional monomer is to provide template–polymer interactions, referred to as the prepolymer complex, which provides rebinding interactions for the template molecule. The prepolymer complex can either be covalent or noncovalent in nature; however, noncovalent interactions are the most convenient to use and are generally preferred. An example of this method is shown in the first step of Scheme 28.2, where the functional monomer (methacrylic acid) forms a complex in solution with the template molecule (an adenine derivative).

The second type of monomers are cross-linkers, which have two or more polymerizable groups, which form a rigid network polymer that serves as a scaffold that holds the template-organized functional monomers in their positions. Some examples of cross-linkers used for molecular imprinting are shown in Figure 28.2. To form the imprinted polymer, the prepolymer complex is copolymerized with an excess of cross-linking monomer in the presence of an equal volume of inert solvent and a polymerization-initiator. Thermal- or photochemical-initiated polymerization results in a highly cross-linked polymer material, as shown in the second step of Scheme 28.2, with the template still inside.

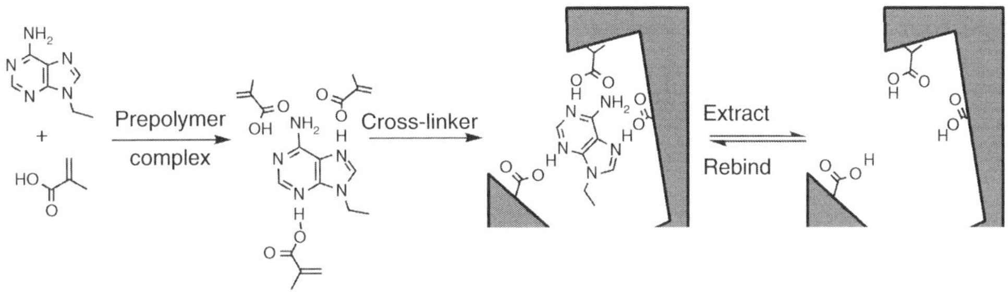

SCHEME 28.2 Molecular imprinting of a derivative of the DNA base adenine. (Adapted from Spivak, D., Gilmore, M.A., and Shea, K.J., *J. Am. Chem. Soc.*, 119, 4388, 1997. With permission.)

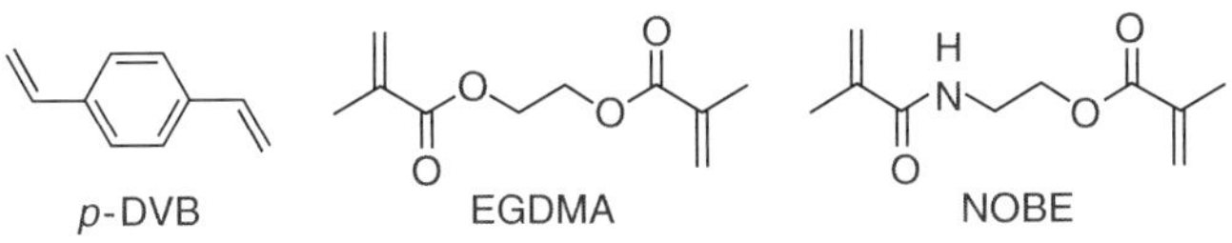

FIGURE 28.2 Examples of crosslinkers used for molecular imprinting. p-DVB, *para*-divinylbenzene; EGDMA, ethyleneglycol dimethacrylate; NOBE, *N,O*-bismethacryloyl ethanolamine. (From Sibrian-Vazquez, M. and Spivak, D.A., *J. Am. Chem. Soc.*, 126, 7827, 2004; Sibrian-Vazquez, M. and Spivak, D.A., *Macromolecules*, 36, 5105, 2003. With permission.)

Subsequent removal of the template by extraction or hydrolysis leaves the final binding sites in the polymer (step 3 of Scheme 28.2). These sites are complementary in size, shape, and functionality to the template, resembling the "lock and key" paradigm of enzymes. Rebinding of the template molecule to the molecularly imprinted sites, as shown in step 3 of Scheme 28.2, is favored for the template molecule versus any other molecule in a mixture.

28.3 Molecular Imprinting Formats

MIPs are most frequently fabricated as polymer monoliths that are ground into a powder made up of irregularly shaped particles (Scheme 28.3). The powder is generally fractionated into a specified size range, e.g., 25–45 μm, for various applications. As a powder, MIPs can be added to a solution mixture to absorb the targeted molecule selectively from the mixture; this is referred to as "batch-rebinding." MIPs are used most extensively utilized in batch rebinding mode by industry for sample preparation via solid-phase extraction of food, biological, or environmental samples. The highly cross-linked nature of the MIP materials makes them predisposed for chromatographic supports. Thus, analytical separation of molecules by MIPs is achieved by packing the MIP powder into a tube, which can be attached to liquid (or gas) chromatography apparatus. However, chromatographic applications are generally more efficient if the MIP powder is in a spherical bead format, with a narrow size range. To achieve this type of material, polymer beads can be formed from suspension, emulsion, or dispersion polymerization methods [21]. Another method of forming polymer beads is the "seed-swelling" technique, which uses a linear polystyrene latex particle as a seed particle [22]. The seed particles can absorb the MIP solution, which upon polymerization will form cross-linked MIP beads with a narrow size range. Other types of composite MIP beads can be formed, often starting with a core particle that serves as a support for another MIP coating [23]. An alternative method to grinding MIPs and packing into chromatographic columns is to polymerize directly within the column; however, the solvents needed to form the needed porous monolith limit the types of templates that can be used [24].

SCHEME 28.3 Initial polymer monolith in a test tube and the final powdered MIP product.

MIPs can also be formed in film formats [25], however, the traditional formulations for MIPs are often too brittle to make them practical for making films. There are some reports of composite polymer membranes that integrate the MIP polymer within the pores or as a coating on a stable membrane support [26,27]. In some instances, nontraditional MIP formulations have been used to directly create the membrane material [28]. Surface-imprinted polymer phases are of interest, particularly for large molecules, such as proteins, that cannot be imprinted using traditional bulk polymerization formats [29]. This is due to the fact that molecules with molecular weights larger than 1000 g/mol will remain trapped in the MIP matrix and cannot be removed. Thus, MIP binding sites formed on or near a surface can allow access to large molecules, thereby improving the mass transfer kinetics and ultimate recognition of the template molecule. Both thin-film formatted MIPs and surface-imprinted materials are often employed for sensor development toward large biomolecules [30].

28.4 Molecular Imprinting Templates

Many different types of templates have been used for molecular imprinting; some examples of these are shown in Table 28.1. MIP recognition of templates can be used to detect or separate the target compound from other compounds, or even stereoisomers of the template. To create these selective sites, there must be at least one strong binding interaction between the template and the monomers in the prepolymerization mixture. For noncovalent imprinting, these interactions are often ionic or hydrogen bonding interactions. Several other structural features have been evaluated for optimization of MIP performance such as number of interactive groups on the template, the spatial disposition of groups, and shape selectivity [31,32]. It should be noted that in general, not all of the template can be removed from MIPs. This leads to problems in trace analyses by these materials, which is solved by using a template that is not the same as (but similar to) the molecule targeted for analysis. Imprinting of large molecules, such as proteins, can be accomplished by directly imprinting the molecule using special formulation strategies (e.g., low levels of cross-linker) or novel formats (e.g., surface imprinting). Another method of imprinting large molecules is known as the "epitope approach," where only a fractional piece of the larger template is actually imprinted and recognition of the whole molecule is mediated through that piece [33]. A last remark on MIPs is that the selective recognition of molecules is an average effect of a distribution of binding sites that is due to an array of variances in the binding site structure that arise during polymerization. Therefore, the actual binding site generally behaves as an average ensemble of the different structures, rather than as a discrete structure that is often illustrated for exemplary purposes in the literature.

TABLE 28.1 Selected Examples of Templates Used in Various Molecular Imprinting Processes

Compound Class	Example	References
Carbohydrates	Mannose	[34]
	Galactose, fucose	[35]
	Cellobiose, maltose	[36]
Bioactive compounds	Morphine	[37]
	Caffeine	[38]
	Nicotine	[39,40]
	Theophylline	[41]
	Benzodiazepines	[42]
Amino acids and peptides	Phenylalanine	[43]
	Tryptophan	[44]
	Oligopeptides	[37,45–47]
Proteins	Cytochrome *c*, BSA, lysozyme, ribonuclease	[48,49]
Coenzymes	Pyridoxal	[50,51]
Nucleotide bases	Adenine	[18,52,53]
	DNA	[54]
Metal ions	Transition metals, lanthanides, group I and II metals	[55,56]

References

1 Alexander, C., Andersson, H.S., Andersson, L.I., Ansell, R.J., Kirsch, N., Nicholls, I.A., O'Mahony, J., and Whitcombe, M.J., Molecular imprinting science and technology: A survey of the literature for the years up to and including 2003. *J. Molec. Recog.*, 19, 106, 2006.

2 Yan, M. and Ramstrom, O. (Eds.), *Molecularly Imprinted Materials: Science and Technology*, Marcel Dekker, New York, 2005, 734pp.

3 Komiyama, M., Takeuchi, T., and Mukawa, H. (Eds.), *Molecular Imprinting: From Fundamentals to Applications*, Wiley-VCH, New York, 2003, 147pp.

4 Sellergren, B. (Ed.), *Molecularly Imprinted Polymers: Man Made Mimics of Antibodies and Their Application in Analytical Chemistry*, Elsevier, Amsterdam, 2001, 550pp.

5 Sellergren, B. Imprinted polymers: Stable, reusable antibody-mimics for highly selective separations, *Am. Lab.*, 14–20, 1997.

6 Pichon, V. and Haupt, K., Affinity separations on molecularly imprinted polymers with special emphasis on solid-phase extraction, *J. Liquid Chromatogr. Related Technol.*, 29, 989, 2006.

7 Haupt, K., Mayes, A.G., and Mosbach, K. Herbicide assay using an imprinted polymer based system analogous to competitive fluoroimmunoassays, *Anal. Chem.*, 70, 3936, 1998.

8 Vlatakis, G., Andersson, L.I., Muller, R., and Mosbach, K. Drug assay using antibody mimics made by molecular imprinting, *Nature*, 361, 645, 1993.

9 Wulff, G. Enzyme-like catalysis by molecularly imprinted polymers, *Chem. Rev.*, 102, 1, 2002.

10 Piletsky, S.A., Turner, and Anthony, P. F. New materials based on imprinted polymers and their application in optical sensors, *Opt. Biosensors*, 397, 2002.

11 Kriz, D., Ramstrom, O., and Mosbach, K., Molecular imprinting: New possibilities for sensor technology, *Anal. Chem.*, 345A, 1997.

12 Norell, M.C., Andersson, H.S., and Nicholls, I.A. Towards molecularly imprinted polymer based controlled release materials, *J. Mol. Recogn.*, 11, 98, 1998.

13 Allender, C.J., Brain, K.R., Heard, C.M., and Pellett, M.A. Imprinted polymers, novel excipients in controlled delivery systems, *Proc. Int. Symp. Controlled Release Bioact. Mater.* 24, 585, 1997.

14 Spivak, D.A., Optimization, evaluation, and characterization of molecularly imprinted polymers, *Adv. Drug Delivery Rev.*, 57, 1779, 2005.

15 Sellergren, B. and Shea, K.J., Influence of polymer morphology on the ability of imprinted network polymers to resolve enantiomers, *J. Chromatogr.*, 635, 31, 1993.

16 Shea, K.J. Molecular imprinting of synthetic network polymers: the de novo synthesis of macromolecular binding and catalytic sites, *Trends Polym. Sci.*, 2, 166, 1994.

17 Wulff, G. and Sarhan, A. Use of polymers with enzyme-analogous structures for the resolution of racemates., *Ang. Chem., Int. Ed.*, 11, 341, 1972.

18 Spivak, D., Gilmore, M.A., and Shea, K.J., Evaluation of binding and origins of specificity of 9-ethyladenine imprinted polymers, *J. Am. Chem. Soc.*, 119, 4388, 1997.
19 Sibrian-Vazquez, M. and Spivak, D.A., Molecular imprinting made easy, *J. Am. Chem. Soc.*, 126, 7827, 2004.
20 Sibrian-Vazquez, M. and Spivak, D.A., Enhanced enantioselectivity of imprinted polymers formulated with novel crosslinking monomers, *Macromolecules*, 36, 5105, 2003.
21 Fairhurst, R.E., Chassaing, C., Venn, R.F., and Mayes, A.G., A direct comparison of the performance of ground, beaded and silica-grafted MIPs in HPLC and turbulent flow chromatography applications, *Biosensors Bioelectronics*, 20, 1098, 2004.
22 Zhang, L., Cheng, G., Fu, C., and Liu, X., Tyrosine imprinted polymer beads with different functional monomers via seed swelling and suspension polymerization, *Polymer Eng. Sci.*, 43, 965, 2003.
23 Sulitzky, C., Rueckert, B., Hall, A.J., Lanza, F., Unger, K., and Sellergren, B., Grafting of molecularly imprinted polymer films on silica supports containing surface-bound free radical initiators, *Macromolecules*, 35, 79, 2002.
24 Schweitz, L., Andersson, L.I., and Nilsson, S., Rapid electrochromatographic enantiomer separations on short molecularly imprinted polymer monoliths, *Anal. Chim. Acta*, 435, 43, 2001.
25 Titirici, M.-M. and Sellergren, B., Thin molecularly imprinted polymer films via reversible addition-fragmentation chain transfer polymerization, *Chem. Mater.*, 18, 1773, 2006.
26 Piletsky, S.A., Panasyuk, T.L., Piletskaya, E.V., Nicholls, I.A., and Ulbricht, M., Receptor and transport properties of imprinted polymer membranes—a review, *J. Membr. Sci.*, 157, 263, 1999.
27 Takeda, K. and Kobayashi, T., Hybrid molecularly imprinted membranes for targeted bisphenol derivatives, *J. Membr. Sci.*, 275, 61, 2006.
28 Wang, H.Y., Xia, S.L., Sun, H., Liu, Y.K., Cao, S.K., and Kobayashi, T., Molecularly imprinted copolymer membranes functionalized by phase inversion imprinting for uracil recognition and permselective binding, *J. Chromatogr., B*, 804, 127, 2004.
29 Husson, S.M. and Gopireddy, D. Molecular engineering of surfaces for selective separations, *Separation Sci. Technol.*, 38, 2851, 2003.
30 Turner, N.W., Jeans, C.W., Brain, K.R., Allender, C.J., Hlady, V., and Britt, D.W., From 3D to 2D: a review of the molecular imprinting of proteins, *Biotechnol. Progr.*
31 Spivak, David, A., Selectivity in molecularly imprinted polymers, *Molecularly Imprinted Materials, Science and Technology*, Yan, M. and Ramstrom, O. (Eds.), Marcel Dekker, New York, 2005, p. 395.
32 Spivak, D.A., Simon, R., and Campbell, J. Evidence for shape selective cavity formation in molecularly imprinted polymers, *Anal. Chim. Acta*, 504, 23, 2004.
33 Minouraa, N., Rachkov, A., Higuchi, M., and Shimizu, T., Study of the factors influencing peak asymmetry on chromatography using a molecularly imprinted polymer prepared by the epitope approach, *Bioseparation*, 10, 399, 2002.
34 Wulff, G., Vesper, W., Grobe-Einsler, R., and Sarhan, A., Enzyme-analog built polymers, 4. The synthesis of polymers containing chiral cavities and their use for the resolution of racemates., *Makromol. Chem.*, 178, 2799, 1977.
35 Mayes A.G., Andersson L.I., and Mosbach K., Sugar binding polymers showing high anomeric and epimeric discrimination obtained by noncovalent molecular imprinting, *Anal. Biochem.*, 222, 483, 1994.
36 Striegler, S., Investigation of disaccharide recognition by molecularly imprinted polymers, *Bioseparation*, 10, 307, 2002.
37 Andersson, L.I., Mueller, R., Vlatakis, G., and Mosbach, K., Mimics of the binding sites of opioid receptors obtained by molecular imprinting of enkephalin and morphine. *Proc. Natl. Acad. Sci. USA*, 92, 4788, 1995.
38 Farrington, K., Magner, E., and Regan, F., Predicting the performance of molecularly imprinted polymers: Selective extraction of caffeine by molecularly imprinted solid phase extraction, *Anal. Chim. Acta*, 566, 60, 2006.
39 Spivak, D.A. and Sibrian-Vazquez, M., Development of an aspartic acid-based cross-linking monomer for improved bioseparations, *Bioseparation*, 10, 331, 2002.
40 Andersson, H.S., Karlsson, J.G., Piletsky, S.A., Koch-Schmidt, A.-C., Mosbach, K. Nicholls, and Ian, A. Study of the nature of recognition in molecularly imprinted polymers II. Influence of monomer-template ratio and sample load on retention and selectivity, *J. Chromatogr., A*, 848, 39, 1999.
41 Vlatakis, G., Andersson, L.I., Mueller, R., and Mosbach, K. Drug assay using antibody mimics made by molecular imprinting, *Nature*, 361, 645, 1993.
42 Hart, B.R., Rush, D.J., and Shea, K.J., Discrimination between enantiomers of structurally related molecules: Separation of benzodiazepines by molecularly imprinted polymers, *J. Am. Chem. Soc.*, 122, 460, 2000.
43 Sellergren, B., Ekberg, B., and Mosbach, K., Molecular imprinting of amino acid derivatives in macroporous polymers. Demonstration of substrate- and enantio-selectivity by chromatographic resolution of racemic mixtures of amino acid derivatives. *J., Chromatogr.*, 347, 1, 1985.
44 Kim, H. and Guiochon, G., Adsorption on molecularly imprinted polymers of structural analogues of a template. Single-component adsorption isotherm data, *Anal. Chem.*, 77, 6415, 2005.
45 Titirici, M.M. and Sellergren, B., Peptide recognition via hierarchical imprinting, *Anal. Bioanal. Chem.*, 378, 1913, 2004.
46 Hart, B.R. and Shea, K.J., Synthetic peptide receptors: molecularly imprinted polymers for the recognition of peptides using peptide–metal interactions, *J. Am. Chem. Soc.*, 123, 2072, 2001.
47 Nicholls, I.A., Ramstroem, O., and Mosbach, K., Insights into the role of the hydrogen bond and hydrophobic effect on recognition in molecularly imprinted polymer synthetic peptide receptor mimics, *J. Chromatogr.*, A, 691, 349, 1995.
48 Nishino, H., Huang, C.-S., and Shea, K.J., Selective protein capture by epitope imprinting, *Angew. Chem., Int. Ed.*, 45, 2392, 2006.

49 Shi, H., Tsai, W.-B., Garrison, M.D., Ferrari, S., and Ratner, B.D., Template-imprinted nanostructured surfaces for protein recognition, *Nature*, 398, 593, 1999.
50 Andersson, L.I. and Mosbach, K. Molecular imprinting of the coenzyme–substrate analog *N*-pyridoxy-L-phenylalaninanilide, *Makromol. Chem. Rapid Commun.*, 10(9), 491–495, 1989.
51 Svenson, J., Zheng, N., Nicholls, and Ian, A. A., Molecularly imprinted polymer-based synthetic transaminase, *J. Am. Chem. Soci.*, 126(27), 8554–8560, 2004.
52 Umpleby, R.J., II, Rushton, G.T., Shah, R.N., Rampey, A.M., Bradshaw, J.C., Berch, J.K., Jr., and Shimizu, K.D., Recognition directed site-selective chemical modification of molecularly imprinted polymers, *Macromolecules*, 34, 8446, 2001.
53 Matsui, J., Higashi, M., and Takeuchi, T., Molecularly imprinted polymer as 9-ethyladenine receptor having a porphyrin-based recognition center, *J. Am. Chem. Soc.*, 122, 5218, 2000.
54 Spivak, D.A. and Shea, K. J., Investigation into the scope and limitations of molecular imprinting with DNA molecules, *Anal. Chim. Acta*, 435, 65, 2001.
55 Rao, T., Prasada, K.R., and Daniel, S., Metal ion-imprinted polymers—novel materials for selective recognition of inorganics, *Anal. Chim. Acta*, 578, 105, 2006.
56 Murray, G.M. and Southard, E., Metal ion selective molecularly imprinted materials. In *Molecularly Imprinted Materials*, Yan, M. and Ramstrom, O. (Eds.), Marcel Dekker, New York, 2005, p. 579.

29

Biomedical Sensing

Dora Klara Makai
Budapest University of Technology and Economics

Gabor Harsanyi
Budapest University of Technology and Economics

29.1 Introduction

Biomedical sensing is a typical example of interdisciplinary fields that poses great challenges and desires the cooperation of experts of numerous scientific areas such as medicine, bioengineering, chemistry, electrical engineering, material sciences, nanotechnology in developing biosensors and other sensitive biomedical constructions such as drug delivery systems. The development in the field of micro- and nanotechnology and material science enabled the preparation of miniaturized biosensors that are applicable in biomedical diagnostics and monitoring.

Biosensors are special sensors using a biologically active material for measurement or indication. They consist of two main parts, the receptor and the transducer (this is not a physical separation). The receptor interacts with the analyte selectively; the transducer produces a signal (electrical or optical) as a result of the former interaction. This signal carries information about the concentration of the analyte if certain conditions are ensured.

Receptor parts contain biologically active components that are capable for specific chemical reactions with the analyte. Upon the nature of receptor types, biosensors can be classified into the following four main groups:

- Enzymatic biosensors exploit the specific reaction between enzyme and its substrate. Enzymes are biocatalysts. They enable certain biochemical reactions to proceed through their catalytic activity. Immobilizing the enzyme as receptor, its substrate as analyte can be detected.
- The operation of immunosensors is based on the formation of a specific binding between antibody and antigen.
- DNA sensors exploit the interactions between the complementary DNA sequences.
- Living biosensors use microorganisms or living tissues as receptors for selective sensing.

Biosensors can employ different transducers like calorimetric types, electrochemical types (potentiometric, amperometric, or conductometric electrochemical cells, as well as modified Ion–sensitive field effect transistor (ISFETs) fiber optic transducers, or gravimetric type resonator transducers.

Latest developments in the area of biomedical sensing are stimulated by the recent progresses in smart materials designing. Smart materials are mostly polymers that respond to external stimuli with considerable property changes. Even small variations in the environmental pH, temperature, ionic strength, electric or magnetic field can alter their structure, solubility, or can lead a sol–gel transition. These phenomena can be exploited in different internal or external biomedical applications [1].

Biomedical applications cover not only research laboratory applications or hospital instruments but home diagnostic kits commercially available for common people and applicable in everyday life such as glucose sensors for diabetic patients. The widespread availability of biosensors is preceded by long optimization processes. Mostly complicated problems have to be solved during a sensor development such as choosing the appropriate receptors, transducer types, immobilization techniques, preventing receptors from poisoning during operation, assuring appropriate stability for the receptors, compensating interferences and disturbing effects, and developing fabrication, storage, preoperation treatment, calibration, and data-processing techniques that enable the wide range mass application of low-cost, short-lifetime sensors in practice [2].

It has to be mentioned that the global biosensor market is presently dominated by medical applications but besides the glucose sensors no biosensor has yet achieved the true mass market sales [3]. After the glucose sensors, DNA chips represent the largest sector of today's biosensor market [3]. Although promising, DNA sensor technologies still need to bridge the gap between experimental status and the reality of clinical and diagnostic applications [4].

29.2 Operation of Biosensors

In the background of the apparently simple principles of biosensor operation, an immense complexity is hidden, numerous parameters should be considered in course of preparation and operation of the biosensor that are summarized below by introducing enzymatic biosensors prepared by embedding enzymes into conducting polymers.

In order to prepare a biosensor, biologically active molecules should be immobilized onto a substrate in a way that they could keep their stability and stay operable. There are different immobilization methods ranging from a "simple" adsorption or covalent binding to embedding into polymer matrices, all of them have their advantages and drawbacks.

Electrochemical polymerization is a simple, reproducible, low cost, rapid method to deposit conducting polymer matrices on electrode surfaces [5,6]. The embedding of the enzymes in a polymer layer can be achieved in one-step procedure conducting the electrodeposition from a solution containing the monomers and the enzymes as well. The thickness of the deposited film is controlled by the charge passed during the electropolymerization.

The sensor response depends on the properties of the electrodeposited polymer, which are affected by the conditions of the deposition such as the method of deposition (cyclovoltammetric, galvanostatic, potentiostatic deposition methods), the enzyme and monomer concentration in the solution, the pH of the solution, other components in the solution, speed of the deposition, and current density. The amount of enzyme in the solution has a great influence on the forming process, permeability, and morphology of the film. High enzyme concentration inhibits the nucleation process in the early stage of the deposition, results in a dense, less permeable, less porous film [7]. High monomer concentration encumbers the enzyme incorporation, which leads to lower sensor response [8].

It has to be considered that the operation of a biosensor is influenced by the sample itself as well. Other components in the sample to be analyzed should not be ignored. For example, blood contains some components that during the amperometric determination of blood glucose undergo electrode reaction at the applied potential, thus the signal derives not only from the enzymatic reaction of glucose and glucose oxidase. The substrate specificity of enzymes or the specific interactions between complementary DNA sequences or between antigen and antibody alone do not ensure the specificity of the analytical method as an analytical method is considered specific if it is perfectly selective. In the case of electrochemical biosensors, interference is usually hindered by multilayered membrane structures.

Considering this amount of the affecting parameters, it is clear that optimizing a biosensing system is not a simple task. Simulation of the impacts of these different parameters by computer is helpful and effective method for the optimization.

29.3 Smart Polymers in Biomedical Sensing

Since biomolecules are highly sensitive structures, their immobilization requires special circumstances in order to maintain their functionality and stability. The development of the material sciences has offered the possibility of choosing among a wide range of ideal matrix materials for biosensor fabrication. Besides providing a biocompatible environment to the receptor molecules, the matrix should ensure the contact with the transducer it should provide to the target molecules to get to the receptors. The use of advanced materials can improve the performance of biosensors via improving the stability and the activity of biomolecules or ensuring the operation of biosensors in extreme conditions. Such smart materials are hydrogels, sol–gel derived materials, sol–gel/hydrogel hybrids, and lipid membranes.

Hydrogels are three-dimensional polymer networks that swollen with water can provide a biocompatible environment for biomolecules. Fernandez and et al. entrapped glucose oxidase in polyacrylamide hydrogels by aqueous cross-linking copolymerization of acrylamide and *N,N'*-methylene bisacrylamide, which can be incorporated into an amperometric glucose biosensor [9].

Changes in structure of smart materials in response to different external stimuli can cause changes in functional properties of the material. Designing such smart polymeric systems has a great importance and a promising future from the perspective of biomedical application.

Zhang et al. introduced a D-glucose-sensitive hydrogel based on cross-linking carboximethyl-dextrane (CMD) with the glucose-binding lectin concanavalinA (ConA) [10]. Between ConA and terminal glucose moieties of dextrane, besides covalent bindings, affinity interactions take place as well. Increasing external glucose concentration provokes competitive displacement of affinity interactions inducing this way changes in morphology and permeability. The permeability of the hydrogel increases with increasing external D-glucose concentration.

Kim et al. studied the behavior of poly(acrylic acid)/poly(vinyl sulfonic acid) (PAAc/PVSA) copolymer hydrogel in an electric field [11]. The polymer showed reversible volume change in constant electric field indicating that this material is a good candidate for biosensor preparation.

Besides hydrogels, sol–gel-derived materials and sol–gel/hydrogel, hybrid materials provide a biocompatible environment for biomolecules. Tan et al. described an amperometric cholesterol biosensor using sol–gel chitosan/silica hybrid composite film [12]. According to the authors, the high sensitivity, good repeatability, reproducibility and selectivity, the rapid response, and long-term stability can be partly attributed to the hybrid composite film.

29.4 Enzymatic Biosensors

Enzymatic biosensors are widely applied in the medical field. Different analytes in different biological samples (blood, urine, saliva etc.) can be detected and measured via enzymatic reactions in vitro and in vivo as well. Enzymatic biosensors can also be used by production of pharmaceutical agents in bioreactors for controlling reactions.

As mentioned above, among enzymatic biosensors, glucose sensors are the most widely used. At least 171 million people suffer from diabetes worldwide and this figure may increase in the following decades [13]. One of the most important approaches of diabetes treatment is the self-monitoring of blood glucose (SMBG), which means regularly checking blood glucose levels. In this way, hypo- and hyperglycemia can be controlled in order to reduce or prevent the diabetes complications. The SMBG is recommended for all diabetic people especially for those who need to take insulin regularly [14].

Retama et al. presented an amperometric glucose sensor using polypyrrole–microgel composites [15]. For immobilizing glucose oxidase, a new immobilization matrix, water-dispersed complex of polypyrrole–polystyrensulfonate (Ppy) embedded in polyacrylamide (PA) microgel has been prepared. The polyacrylamide–polypyrrole (PAPPy) microparticles combine the conductivity of polypyrrole and the pore size control of polyacrylamide.

Besides glucose, there are other metabolites the higher or lower levels of which indicate illnesses. L-Lactate is the final product of anerobic glycolysis. Elevated L-lactate level in blood can be a marker of different diseases such as hypoxia or lactic acidosis [16]. As lactate level indicates the oxygenation state of tissues continuous monitoring of lactate level is critical during operations, intensive therapy, in critical care and sport medicine as well. For the amperometric detection of lactate, mainly NAD-dependent lactate dehydrogenase or lactate oxidase is used but L-lactate–cycochrome *c* oxidoreductase based lactate sensor has also been reported [16]. Schuvailo et al. introduced highly selective microbiosensors for glucose, lactate, and glutamate sensing that can be used in in vitro measurements [17].

There are clinically important molecules that cannot be detected amperometrically using only one type enzyme but with consecutive enzymatic reactions. The more enzyme type immobilized the more difficult the sensor preparation.

Cholesterol and its esters are precursors of different biomolecules, e.g., hormones. Monitoring of cholesterol level gains importance in diagnosis and prevention of heart diseases and arteriosclerosis. Salinas et al. introduced a multienzymatic biosensor based on the consecutive enzymatic reactions catalyzed by cholesterol esterase, cholesterol oxidase, and peroxidase for total cholesterol determination [18]. In order to eliminate the interference of L-ascorbic acid a micropacked column containing ascorbate oxidase is integrated as prereactor.

Multianalyte biosensors are able to detect and measure more analyte simultaneously. Multienzymatic amperometric sensors have more working electrodes, each functionalized with different enzymes for detection of different substrates. By fabrication of multianalyte biosensors, the cross-talk phenomenon should be considered. The molecules generated by the enzymatic reaction undergo electrode reaction in course of the sensing process. These product molecules gain the electrode surface by diffusion. Cross-talk derived from the diffusion can lead to false results. In order to prevent getting false results because of the cross-talk, the sensor construction (interelectrode distance, electrode geometry, etc.) should be rightly chosen. Simulation of processes by means of computer can help to find the most adequate construction.

29.5 DNA Sensors

The operation of DNA sensors is based on the interaction between complementary DNA sequences. The hybridization between the immobilized ssDNA strands and the DNA strands in the sample to be analyzed can be detected by various methods. Similar to enzymatic sensors, electrochemical transduction is an attractive solution due to its high sensitivity and the relatively inexpensive instruments.

By the immobilization of DNA sequences, the strong interaction between streptavidin and biotin is often exploited. Riccardi et al. presented an amperometric DNA biosensor for hepatitis C virus detection [19]. The biotinylated oligonucleotide probes have been attached to the streptavidin entrapped into siloxane–poly(propylene oxide) sol–gel films deposited onto a graphite electrode surface.

Okumura et al. also used the streptavidin–biotin interaction in the immobilization process [20]. They used hydrogel nanospheres as enhancers in surface plasmon resonance (SPR) imaging studies for highly sensitive detection of K-ras point mutation.

29.6 Immunosensors

The specific reaction between antigens and antibodies is exploited in the operation of immunosensors. Antigens are foreign substances that invade the body. Entering as unwanted invaders, foreign bioagents in living things activate the protecting functions of the immune system. Antibody production starts against the antigens in order to specifically bind them as first step of their elimination. Each antibody bears a special section, which ensures a selective fitting with a specific antigen.

Antigens can directly be detected by immobilizing specific antibodies on the sensor surface. This strategy suffers from a serious drawback namely the usually low concentration of antigens, which can lead to false-negative output. Sensitivity can be improved via indirect detection, i.e., the detection of the antibodies (instead of antigens) that are produced in huge amounts.

Pohanka et al. presented a novel method for indirect detection of *Francisella tularensis* (causative agent of tularemia, a highly infectious pathogen bacterium, potential biological warfare agent) [21]. The piezoelectric immunosensor is based on the measurement of antitularemic antibodies with covalently immobilized antigens in infected mouse serum.

Petrosova et al. reported an optical immunosensor for the detection of antibodies to Ebola virus strains Sudan and Zaire in animal and human sera [22]. The inactivated Ebola virus receptors have been photoimmobilized to the photoactivable electrogenerated polymer film modified fiber optic surface. Synthetic peptides that mimic the special immunoreactive sections provide a promising alternative to immobilized antigens.

Vaisocherova et al. developed a surface plasmon resonance biosensor for label-free detection of antibody against Epstein–Barr virus using synthetic peptide as a receptor [23]. Antibodies can be produced not only against bacteria, viruses, or other microorganisms but against "simple" molecules such as histamin. Li et al. introduced a regenerable surface plasmon resonance immunosensor for detection of antihistamine antibody [24].

29.7 Alternatives for Natural Receptors

Since biomolecules are highly sensitive materials, it is of crucial importance to handle them carefully, which practically makes difficult to work with them in practice. Replacing these molecules by alternative receptors can make biosensor preparation easier.

Molecularly imprinted polymers (MIPs) are polymers with artificial recognition sites (receptors) complementary to the target molecules, which can be proteins, DNA sequences, enzymes, antibodies but small molecules as well. During the process of molecular imprinting, functional monomers and cross-linkers form a polymer network around the templates that are eliminated at the end of the procedure. MIPs represent attractive alternatives to expensive, relatively unstable bioreceptores. They are easy to prepare with high reproducibility and their mass production is cost-effective. They can withstand to more extreme conditions than the natural receptors and show high affinity and specificity. Further advantages of MIPs over the natural receptors are that they are more compatible with the micromachining techniques and make possible creating receptors for important analytes for which no natural receptors exits. Besides these advantages MIPs usually suffer from some disadvantages such as template leakage, not easily reachable binding sites or nonspecific bindings. There is still no general procedure for MIP preparation and there are still difficulties to deal with such as the integration of transducers or the transformation of binding event to signal [25].

Numerous MIPs have been created for different clinically important analytes but mainly for small molecules. For example, Cheng et al. introduced a MIP for glucose, which can be functionalized as a part of a capacitive glucose sensor [26].

Unlike small molecules, protein imprinting is still proved to be a challenging task mainly since their 3D structure is highly sensitive to the environment, temperature, ionic strength, pH, etc. There exist more methods for imprinting proteins. The grafting method involves the formation of a thin polymer film on a solid substrate around the protein. Another method, called epitope approach, is based on that only a little part of the protein, a short peptide sequence is used as template in the MIP preparation.

Template molecules can be immobilized on a solid sacrificial substrate as well. Once the polymerization is over, the sacrificial support can be eliminated by different chemical treatments. This method enables preparing MIPs with templates that are insoluble in the polymerization mix.

Peptide nucleic acids (PNAs) are oligonucleotide analogs having uncharged pseudeopeptide backbone instead of sugar–phosphate backbone [27]. PNAs can hybridize with complementary DNA, RNA, or PNA strands. The PNA–DNA duplexes show higher thermal stability than the respective DNA–DNA duplexes [28]. The stability of PNA–DNA complexes, unlike that of DNA double helices, is barely affected by the ionic strength of the medium. PNAs show great resistance against enzymatic degradation [28]. Brandt et al. introduced PNA microarrays combined with time-of-flight secondary mass spectrometry (ToF-SIMS) as a sensitive method for label-free DNA diagnostic on a chip [29].

29.8 Drug Delivery

Drug delivery by smart materials can also be considered as biomedical sensing process although the way they are working is markedly different from the aforementioned biosensors. The use of smart materials (such as nanosystems, microgels, MIPs) in the field of drug delivery has a significant impact on the efficiency of treatment improving the clinical and the commercial value of the products. The increased efficiency of smart materials in drug delivery is mainly due to the targeted drug delivery and the intelligent drug release. Targeting and intelligent, controlled release of drugs ensures that the drug can reach the desired site of the body and the desired way (sustained or pulsative). Drug targeting reduces the side-effects of medications as the drug is carried to the desired site of action, minimizing this way the exposure of other tissues and organs to the drug. Active targeting can be achieved by using carriers bearing a motif that can bind only with receptors, which are specific for the target cells. Intelligent drug release can be realized by drug carriers, which respond in a predictable way to specific stimuli (e.g., changes of pH).

Molecularly imprinted polymers can function as drug carriers as they can be prepared to able to selectively bind to the target motifs. After binding to the target, they can release their payload. Nanosystems are systems containing submicron-sized dispersed particles as nanoparticles, nanocapsules, polymeric self-assemblies, or dendrimers (nanometer-sized synthetic polymer macromolecules) [30]. The major advantages of nanoparticles in drug delivery derive from the size and their nature. Their size makes possible to penetrate small capillaries and to be taken up by cells while their biodegradable nature ensures a sustained drug release over a longer period after administration [31]. By the aid of nanosystems, water solubility and stability of drugs can be improved [30]. Drug delivery by nanoparticles plays an important role in cancer therapy, gene therapy, AIDS therapy, delivery of protein, antibiotics, and vaccines.

The surface of nanoparticles can be modified in order to further improve the efficiency of drug delivery. Surface modification gain importance, for example, in intravenous administered drugs as it can reduce the opsonization (interaction with special blood proteins) of nanoparticles and increase the circulation time (stealth nanoparticles). These effect were observed with poly(ethylene-glycol)-modified nanoparticles [31].

Delivering of insulin is a highly investigated field in the drug delivery searching. In case of insulin as most of the other drugs, the oral route of administration is the most desirable way of administration due to its high patient compliance. As insulin is a protein, its oral administration poses great challenges that have to be overcome by developing intelligent drug carrier systems. Oral administration of proteins is not efficient as the majority of the molecules suffer degradation from the enzymes of gastrointestinal tract before reaching the blood stream. Peppas et al. developed a hydrogen-bonding copolymer network of poly(methacrylic acid) grafted with poly(ethylene-glycol) for oral delivery of peptides and proteins [32]. In the acidic environment of the stomach, the proteins remain entrapped in the delivery system unreachable for the enzymes, safe from degradation. The proteins can reach this way the intestine where they should be absorbed and can be released from this delivery "vehicle." The mechanism behind this protection and release phenomenon can be explained by the changes in mesh size of the hydrogel.

29.9 Future Prospects

The development of biomedical sensing systems is still on the rise. Advances in material sciences and in nanotechnology can be exploited in biosensor preparation in order to fabricating more and more sophisticated biosensing systems using smart materials. Experts still face challenge of combining the existing and the newly developed materials and methods and optimizing the biosensing systems to get biosensors that can be widely used.

References

1. Jeong, B. and Gutowska, A., Lessons from nature: Stimuli-responsive polymers and their biomedical applications, *Trends Biotechnol.*, 20, 305, 2002.
2. Harsanyi, G., *Sensors in Biomedical Applications*, Technomic Publishing Company, Lancaster, Basel, 2000, 226.
3. Bogue, R., Developments in biosensors—where are tomorrow's markets? *Sens. Rev.*, 25, 180, 2005.
4. Hahn, S. et al., Nucleic acid based biosensors: The desires of the user, *Bioelectrochemistry*, 67, 151, 2005.
5. Cosnier, S., Biomolecule immobilization on electrode surfaces by entrapment or attachment to electrochemically polymerized films, *Biosens. Bioelectron.*, 14, 443, 1999.
6. Thanachasai, S., Yoshida, S., and Watanabe, T., Effects of fabrication parameters on the enzyme loading and sensor response of enzyme-carrying conductive polymer electrodes, *Anal. Sci.*, 19, 665, 2003.
7. Shin, M.C. and Kim, H.S., Electrochemical characterization of polypyrrole/glucose oxidase biosensor: Part I. Influence of enzyme concentration on the growth and properties of the film, *Biosens. Bioelectron.*, 11, 161, 1996.
8. Shin, M.C. and Kim, H.S., Electrochemical characterization of polypyrrole/glucose oxidase biosensor: Part II. Optimal preparation conditions for the biosensor, *Biosens. Bioelectron.*, 11, 171, 1996.
9. Fernandez, E. et al., Viscoelastic and swelling properties of glucose oxidase loaded polyacrylamide hydrogels and the evaluation of their properties as glucose sensors, *Polymer*, 46, 2211, 2005.
10. Zhang, R. et al., Synthesis and characterization of a D-glucose sensitive hydrogel based on CM-dextran and concanavalin A, *React. Funct. Polym.*, 66, 757, 2006.
11. Kim, H. I. et al., Electroactive polymer hydrogels composed of polyacrylic acid and poly(vinyl sulfonic acid) copolymer for application of biomaterial, *Synth. Met.*, 155, 674, 2005.
12. Tan, X. et al., An amperometric cholesterol biosensor based on multiwalled carbon nanotubes and organically modified sol-gel/chitosan hybrid composite film, *Anal. Biochem.*, 337, 111, 2005.
13. http://www.who.int/diabetes
14. http://www.fda.gov/diabetes/glucose.html#2
15. Retama, J.R. et al., Design of an amperometric biosensor using polypyrrole–microgel composites containing glucose oxidase, *Biosens. Bioelectron.*, 20, 1111, 2004.
16. Smutok, O. et al., A novel L-lactate-selective biosensor based on flavocytochrome b2 from methylotrophic yeast *Hansenula polymorpha*, *Biosens. Bioelectron.*, 20, 1285, 2005.
17. Schuvailo, O.M. et al., Highly selective microbiosensors for in vivo measurement of glucose, lactate and glutamate, *Anal. Chim. Acta*, 573, 110, 2006.
18. Salinas, E. et al., Multienzymatic-rotating biosensor for total cholesterol determination in an FIA system, *Talanta*, 70, 244, 2006.
19. Dos Santos Riccardi, C. et al., Immobilization of streptavidin in sol–gel films: Application on the diagnosis of hepatitis C virus, *Talanta*, 70, 637, 2006.
20. Okumura, A. et al., Point mutation detection with the sandwich method employing hydrogel nanospheres by the surface plasmon resonance imaging technique, *Anal. Biochem.*, 339, 328, 2005.

21. Pohanka, M., Pavlis, O., and Skladal, P., Diagnosis of tularemia using piezoelectric biosensor technology, *Talanta*, 71, 981, 2007.
22. Petrosova, A. et al., Development of a highly sensitive, field operable biosensor for serological studies of Ebola virus in central Africa, *Sens. Actuators, B*, 122, 578, 2007.
23. Vaisocherova, H. et al., Surface plasmon resonance biosensor for direct detection of antibody against Epstein-Barr virus, *Biosens. Bioelectron*, 22, 1020, 2007.
24. Li, Y. et al., Surface plasmon resonance immunosensor for histamine based on an indirect competitive immunoreaction, *Anal. Chim. Acta*, 576, 77, 2006.
25. Piletsky, S.A., Turner, N.W., and Laitenberger, P., Molecularly imprinted polymers in clinical diagnostics—Future potential and existing problems, *Med. Eng. Phys*, 28, 971, 2006.
26. Cheng, Z., Wang, E., and Yang, X., Capacitive detection of glucose using molecularly imprinted polymers, *Biosens. Bioelectron.*, 16, 179, 2001.
27. Nielsen, P.E. et al., Sequence-selective recognition of DNA by strand displacement with a thymine-substituted polyamide, *Science*, 254, 1497, 1991.
28. Egholm, M. et al., PNA hybridizes to complementary oligonucleotides obeying the Watson–Crick hydrogen-bonding rules, *Nature*, 365, 566, 1993.
29. Brandt, O. et al., Development towards label- and amplification-free genotyping of genomic DNA, *Appl. Surf. Sci.*, 252, 6935, 2006.
30. Kidane, A. and Bhatt, P.P., Recent advances in small molecule drug delivery, *Curr. Opin. Chem. Biol.*, 9, 347, 2005.
31. Xu, T., et al., Modification of nanostructured materials for biomedical applications, *Mater. Sci. Eng., C*, 27, 579, 2007.
32. Peppas, N.A. and Kavimandan, N.J., Nanoscale analysis of protein and peptide absorption: Insulin absorption using complexation and pH-sensitive hydrogels as delivery vehicles, *Eur. J. Pharm. Sci.* (in press); corrected proof available online May 10, 2006.

30

Intelligent Chemical Indicators

Christopher O. Oriakhi
Hewlett-Packard Company

30.1 Introduction

Most people remember a chemical indicator from their high school chemistry. This kind of indicator is a material which changes color to signify the endpoint of a titration reaction or to provide a relative indication of the acidity or alkalinity of a chemical substance. The use of indicators extends far beyond this. For example, food, cosmetic, pharmaceutical, and other chemical formulations undergo complex chemical, enzymatic, and microbial interactions when they are exposed to UV light or temperature fluctuations over time. Consequently, product quality may be degraded, and may lead to additional safety concerns. The challenges facing the produce industry include successful implementation of good manufacturing practices (GMP), hazard analysis and critical control point (HACCP), total quality management (TQM) programs, and other regulations demanding compliance [1].

To address these issues and increase consumers' confidence on product quality, safety, and authenticity, many manufacturers incorporate inexpensive indicator monitoring devices into their products during production, packaging, or storage. A large number of consumer-readable indicators are available commercially. Some examples of these types of indicators are tags, labels, seals, and thermometers. Some are designed to give a visual color change in response to degradation of product quality, tampering, or to detect a counterfeit. They are used extensively in the chemical, food, and pharmaceutical industries where consumers need assurance of product integrity, quality, and safety during postmanufacture handling.

Generally, chemical indicators maybe defined as stimulus responsive materials that can provide useful information about changes in their environment. Organic dyes, hydrogel or "smart polymers," shape memory alloys, thermochromic or photochromic inks, and liquid crystals are some examples. They may function by forming structurally altered ionic or molecular complexes with species in their environment through chemical or physical interactions involving proton exchange, chelation, hydrogen bonding, dipole–dipole interactions or van der Waal forces [1]. The resulting characteristic biochemical, chemical, optical, magnetic, thermal, or mechanical changes can be tailored to provide the desired indication response.

This chapter will focus on inexpensive disposal chemical indicating devices such as pH indicators, temperature indicators, time–temperature indicators (TTI), tampering and counterfeit indicators. The temperature and TTIs are widely used in the food and pharmaceutical products where date coding on a package may sometimes be inadequate.

30.2 Chemical Indicating Devices Are Smart

Smart materials or devices maybe defined as materials that produce strong visually perceptible changes in physical or chemical property in response to small physical or chemical stimuli in the medium. The material properties measured may include pH, concentration, composition, solubility, humidity, pressure, temperature, light intensity, electric and magnetic field, shape, air velocity, heat capacity, thermal conductivity, melting point, or reaction rates [2–6]. Chemical-indicating devices can respond reversibly or irreversibly to small changes in the physical or chemical properties in their environment in a predictable manner. They may be regarded as smart materials because of the range of material properties they encompass. Typical materials include shape memory alloys, piezoelectric materials, magnetostrictive substances, electrorheological and magnetorheological fluids, hydrogel polymers, photo and thermoresponsive dyes [2–6].

30.2.1 Classification

Indicators may be classified on the basis of the response mechanism, or operating principles, and application. Thus there are chemical, biological, biochemical, electrical, magnetic, or mechanical indicators according to the response mechanism. Based on the intended application, indicators may be classified as temperature indicator, time–temperature indicator, pH indicator, counterfeit indicator, tamper indicator, freeze and thaw indicator, or freshness indicator.

30.2.2 General Operating Principles

The response mechanism of most indicators includes one or more of the following: physical, chemical, physicochemical, electrochemical, and biochemical. The physical mechanisms are based on photophysical processes, phase transition, or other critical material properties such as melting, glass transition, crystallization, boiling, swelling, or changes in specific volume. In most cases, these transitions are driven by changes in the interaction forces (e.g., hydrophilic–hydrophobic forces) within or around the indicator material. Indicator response mechanism may also be based on chemical, biochemical, and electrochemical reactions. Examples include acid–base, oxidation–reduction, photochemical, polymerization, enzymatic, and microbial reactions. Many of these changes are irreversible and the onset or termination can be observed visually as color change, color movement, or mechanical distortion [1–6].

30.2.3 Choice of Indicators

Some factors governing the selection of a given indicator device include the following:

Cost: It must be relatively inexpensive. The indicator should not be more expensive than the product it is protecting.

Application: Easy to attach to a variety of containers or packages. Once installed, the device must remain intact and readable during the service life of the packages.

Response: The response mechanism must have fast kinetics on the order of seconds to a few hours and be reproducible. There should be no time delay in response between reactions involving solid, liquid, or gas. Most applications require the response in the indicator to be irreversible to preserve the needed indication record.

Sensitivity: The indicator must be highly sensitive, accurate, and easily activated. A user-friendly indicator that can provide useful information when needed will make both the product manufacturer and the consumer happy.

Shelf-life: The indicator must have a shelf-life equal to or longer than that of the product it is monitoring.

It should be technically difficult to duplicate or counterfeit the indicator's response. In this case, the indicator is acting as a "smart" locking mechanism.

30.2.4 pH Indicators

The pH indicator is probably the oldest and simplest smart chemical indicating device known. The chemistry of acid–base indicators is well documented [7] and involves proton or electron exchange reactions. It will be mentioned only briefly here. By Brönsted and Ostwald's definition, indicators are weak acids or bases. Upon dissociation, they exhibit a structural and color change that is different from the undissociated form. Indicators commonly used in acid–base and redox titrations are conjugated organic dyes containing one or more light-absorbing groups called chromophores. The electronic structure of these dyes can be changed by redox or proton exchange reactions. This also changes the absorption energy in the visible region of the electromagnetic spectrum, and consequently, the color. The color change therefore provides a visual indication of an endpoint of a titration [1,7].

30.2.4.1 Indicator Materials

Indicators for pH measurement may exist as a solution, emulsion, colloidal gel, paper, or electrode. Methods for preparing them are described in the literature [1,7,8]. Table 30.1 lists examples of common indicators used in solution form or as indicator papers.

There are also nondye pH-sensitive smart polymers, which exhibit critical material changes in response to changes in the concentration of hydrogen or other ions. Typically, there are polymer electrolytes derived from homo- or copolymerization of acidic or basic functionalized monomers. Examples include poly(ethylene oxide), poly(dimethylsiloxane); copolymers of *N,N*-dimethylacrylamide, *N-t*-butylacrylamide, and acrylic acid; mixtures of methacrylic acid and methyl methacrylate; mixtures of diethylaminoethyl methacrylate and butyl methacrylate; polystyrenesulfonate; polyacrylamide, functionalized cellulose copolymers, etc. [8,9]. In the presence of specific ions, these polymer gels can swell or shrink over a range of pH at room temperature, resulting in a visible color change. Indicators based on them are highly selective, sensitive, and inexpensive [8].

30.2.5 Temperature and Time–Temperature Indicators

Recently, there has been a growing interest in the use of critical temperature indicators and TTIs in intelligent packaging technology. The production, processing, distribution, storage, and point-of-use of temperature-sensitive products pose a great challenge to industries. Temperature plays a leading role in the growth of microorganisms in foods and thus in the incidence of food poisoning. As a result, many perishable food and nonfood products are prone to temperature-induced quality loss and degradation.

Pharmaceutical products, medical devices, chemical reagents, industrial formulations such as inks, paints and coatings,

TABLE 30.1 Selected Indicators for pH Measurement

Common Name (nm)	Formula	pK_a[a]	pH Range	Color in Acid	Color in Alkaline	λ_{max}[a]
Bromothymol blue	$C_{27}H_{28}O_5Br_2S$	7.3	6.0–7.6	Yellow	Blue	617
Bromocresol green	$C_{21}H_{14}O_5Br_4S$	4.9	3.8–5.4	Yellow	Blue	618
Bromocresol purple	$C_{21}H_{16}O_5Br_2S$	6.3	5.2–6.8	Yellow	Purple	588
o-Cresol red	$C_{21}H_{15}O_5S$	8.2	0.2–1.8	Red	Yellow	572
Chlorophenol red	$C_{19}H_{12}O_5Cl_2S$	6.25	4.8–6.4	Yellow	Red	572
Methyl red	$C_{15}H_{15}N_3O_2$	5.0	4.4–6.3	Red	Yellow	526
Phenol red	$C_{19}H_{14}O_5S$	8.0	6.4–8.2	Yellow	Red	559
Phenolphthalein	$C_{20}H_{14}O_4$	9.5	8.0–9.8	Colorless	Red–Violet	552
Tetrabromophenol blue	$C_{19}H_6O_5Br_8S$	3.56	3.0–4.6	Yellow	Blue	605
a-Naphtholphthalein	$C_{28}H_{18}O_4$	8.4	7.4–8.8	Colorless	Green–Blue	660
Thymol blue	$C_{27}H_{30}O_5S$	9.2	1.2–2.8	Red	Yellow	598

[a] Measurements made in solution at zero ionic strength and for a temperature ranging from 15°C to 30°C.

photographic materials, and other related items are highly sensitive to temperature abuse. Therefore, it is useful to monitor the temperature history from production to point of consumption for such products.

There has been a greater need to monitor and prevent overheating-related failure of machine and electronic parts. Routine quality control and preventive maintenance of automobile and aircraft engines system against overheating may help avoid catastrophic and costly failures.

To ensure product safety, freshness, minimize losses and retain customer's confidence, dedicated temperature indicators or TTI are attached to many commercial products. They show the temperature or record temperature history of the product and indicate if abuse has occurred, thus allowing the consumer to make judgments about product quality or safety. The indicator may incorporate a color-changing dye, polymeric substance, enzyme, or time–temperature integrating materials, which must be thermally activated to function [10].

30.2.5.1 Critical Temperature Indicators

Critial temperature indicators (CTIs) visually indicate that a material or product has been exposed to an undesirable temperature below or above a reference critical temperature for a time long enough to alter product quality or cause safety concerns [11]. Several temperature indicators have been described in the patent literature although only a few have been commercialized. Their principle of operation depends on temperature changes resulting from freezing, melting transition, liquid crystal formation, polymerization, enzyme reactions, or electrochemical corrosion [12]. Some formulations are derived from a dye and a chemical blend or a thermoresponsive hydrogel polymer with a specific freezing or melting transition. Some commercial indicators are described below.

30.2.5.2 High-Temperature Indicating Crayons

These are used mainly for plant maintenance, quality assurance, engine monitoring, heat treatment of metals and alloys, diagnosis, and troubleshooting. For example, a crayon would be useful where a knowledge of the exact temperature of engine parts is desirable [13]. Such crayons are made of materials with known melting points calibrated to have a response time of a few seconds with an accuracy of ±1%. Two commercial examples include the Telatemp temperature-indicating crayon (made by Telatemp Corporation) and Tempilstik temperature indicators (made by TEMPIL Division, Air Liquide America, South Plainfield, NJ). They are available for use in temperature ratings between 125°C and 1200°C. In an application, the indicating crayon is used to mark a desired surface of known operating temperatures such as the exhaust manifold, cylinder heads, or fuel injection pump. An instantaneous color change upon marking indicates the rated temperature has been exceeded. Alternatively, if a color change is observed after 1 or 2 s, the rated temperature of the indicator has been reached. When the material is operating below the rated temperature, a color change may not be observed or take much longer time to occur.

30.2.5.3 Irreversible Temperature Labels

Temperature labels are designed to permanently and irreversibly respond to and record overheating of surfaces. The indicator labels are made by sealing one or more temperature-sensitive materials in a highly stable self-adhesive polymeric or paper strip [14]. Temperature ratings are printed on, below, and above each indicator window in degree centigrade (°C) and Fahrenheit (°F). Several models in custom sizes and shapes and customer-selected temperature ranges are available from Telatemp Corporation. To use indicator labels, they are installed onto a desired surface. When the calibrated and rated temperature is exceeded, the indicator window of the label undergoes a noticeable color change (from silver to black in most Telatemp products). Response time is usually less than 1 s with a guaranteed tolerance of ±1%. Some uses include monitoring overheating in engine components, electronics, transformers, chemicals, foods, and pharmaceutical items.

30.2.5.4 Defrost Temperature Indicators

These are color-change indicators that are mostly used to monitor chill and frozen temperatures of chemical and food products during storage or distribution [15]. They are go/no go without delay or go/no go with some delay time devices. The go/no go without delay indicators provide information that the temperature has risen above a threshold value over a short time period. The go/no go with delay indicators show the indicator has been exposed to a predetermined temperature for a specified length of time. Some examples are as follows:

1. ColdMark indicators (IntroTech, Inc.) are used to show if a product has been exposed to critical cold temperature conditions with an accuracy of ±1°C. Several designs have been described in the patent literature [16]. They consist essentially of a glass bulb filled with a colorless fluid and a tube partially filled with a colorless and a violet colored fluid separated by a green barrier fluid. The tube is enveloped in a transparent plastic casing provided with a pressure-sensitive adhesive backing for attachment onto the exterior or interior surface of a desired package [17]. For a freeze-monitor indicator, the response temperature is 32°F/0°C. Upon exposure to temperatures at or below the response temperature for about 30 min, the colorless fluid in the bulb freeze and contract, causing the colored fluid in the tube to flow downward, coloring the bulb. When warmed above the response temperature, the bulb remains permanently colored. When not in use, the indicator is stored above the response temperature.
2. TwinMark indicator (IntroTech, Inc.) is a dual purpose indicator that can respond to cold and hot temperatures. It works like the ColdMark and shows irreversible color change in the bulb when the cold side or hot side of the indicator is exposed to temperatures below or above predetermined values [17].
3. ColdSNAP indicator (Telatemp Corporation) is a temperature recorder that contains a bimetallic sensing element [14]. When attached to a product, a safety tab is pulled to activate the sensor. A clear window in the indicator means safe product storage temperature. When exposed to a damaging critical temperature, the bimetallic sensor snaps into the indicator window permanently changing it from clear to red. The accuracy of the indicator is ±2°C and customers can select snapping point temperatures ranging from −20°C to +40°C.

30.2.5.4.1 Liquid Crystal Display Labels

Liquid crystal displays (LCDs) are miniature reversible thermometers engineered on a label. They are provided with an indicator window and a self-adhesive backing. A liquid crystalline material or polymer is placed behind the indicator window. When installed, the material changes from opaque to transparent at rated temperatures [12,14]. It is possible to select the indicator's color change of interest. LCDs can be designed to continuously monitor surface temperatures covering various temperature ranges or indicate selected temperatures. For example, Telatemp markets LCD reversible temperature decals covering a range of −30°C to 120°C in 2–5°C increments and with visual tan/green/blue color changes permitting readings to 1°C.

30.2.5.5 Time–Temperature Integrator or Indicators

A time–temperature integrator/indicator is a continuous monitoring device or tag that measures both the cumulative exposure time and temperature of perishable products from production, distribution, storage, and even point of use. When a TTI is attached to a product and activated, the time–temperature history to which the product has been exposed is recorded and integrated into a single visual result (such as color change). In theory, TTI can be used as an informational, monitoring, and decision-making protocol at any point in the distribution chain of product quality and safety [11,18].

The response mechanism of a TTI is based on the irreversible physical, chemical, or biochemical changes such as oxidation and reduction reactions, protein denaturation, enzymatic browning, sublimation of ice, and recrystallization of ice that occurs from the time of activation. The rate of these changes increases with temperature. The application and reliability of TTIs have been studied extensively by several workers [19–22]. Applications of TTI include food produce, pharmaceutical and blood products, cosmetics, adhesives, paints and coating, photographic and film products, avionics, plastics, as well as shipment of live plants and animals.

30.2.5.6 Chemical Kinetic Basis for TTI Application

Major factors responsible for the deterioration of food and related chemical products include microbial growth, physical changes and biochemical or chemical reactions occurring within the perishable products. Some examples of spoilage reactions include acid–base catalysis, enzymatic reactions, free radical processes, hydrolytic reactions, lipid oxidation, meat pigments oxidation, nonenzymatic browning reactions, and polymerization or cross-linking processes [23]. Like any chemical reaction, the rates of degradation or quality loss of perishables are influenced by environmental factors such as temperature, concentration (e.g., composition of internal gaseous phases), and water activity. Increasing the temperature, for instance, increases the rate of these reactions.

The response of TTIs and changes in food quality (or other perishables) can both be modeled by chemical kinetic equations. It is critical to gather accurate and reliable kinetic data for TTI to be successful and reliable. Several studies are available on the subject [23–25]. Briefly, studies on several frozen and refrigerated foods indicated that the response of TTIs correlates with storage-related quality changes [26]. In another study, the sensory changes in frozen hamburger were reported to correlate with the response of a commercially available TTI [27].

Several investigators have studied the influence of temperature on the rate of quality loss. It is generally accepted that the rate of quality loss behaves as an exponential function of the reciprocal of the absolute temperature. This is the Arrhenius equation shown below:

TABLE 30.2 Activation Energies of Some Selected Reactions Resulting in Food Quality Losses

	Activation Energy Range	
Type of Reaction	(kJ/mol)	(kcal/mol)
Diffusion controlled	0–60	0–14.3
Acid/base catalysis	80–120	19.1–28.6
Enzyme reaction	40–60	9.5–14.3
Hydrolysis	50–60	11.9–14.3
Lipid oxidation	80–100	19.1–23.9
Nutrient losses	20–120	4.8–28.6
Maillard reaction	100–180	23.9–43
Protein denaturation	300–500	71.6–119.3
Spore destruction	250–350	59.7–83.5
Vegetable cell destruction	200–600	47.7–143.2
Microbial growth	60–200	14.3–47.7

Source: From Taoukis, P. S. and Labuza, T. P., *J. Food Sci.*, 1989, 54(4), 783; Fu, B., Taoukis, P. S., and Labuza, T. P., *Drug Dev. Indu. Pharm.*, 1992, 18(8), 829. With permission.

$$k = Z \exp\left(\frac{-EA}{RT}\right)$$

where
- k is the reaction rate constant
- Z is the temperature independent preexponential factor
- E_A is the activation energy, which describes the temperature sensitivity of the quality loss reaction
- R is the universal gas constant
- T is the absolute temperature in Kelvin (K)

To measure loss of quality or shelf-life in perishables, one or more characteristic measurable quality factor, denoted X are selected. The quality factor X can be a chemical, microbiological, or physical parameter. The rate of change of the quality factor, with time under isothermal storage conditions is given by the following:

$$\frac{d[X]}{dt} = -k[X]^n$$

where
- t is the reaction time
- n is the reaction order

For most food quality losses, reaction order of zero or one is typical for a simple rate constant. By considering the kinetics of changes in food quality at various temperatures the activation energy E_A for the quality loss reaction can be obtained [19,24,25]. Table 30.2 lists typical values of $E_{A(Food)}$ for quality losses. From the TTI kinetics, the activation energy of the indicator ($E_{A(TTI)}$) can be obtained.

30.2.5.7 Correlation of TTI Response with Food Shelf Life

Taoukis and Labuza have developed a correlation scheme for predicting the shelf life of a product based on the TTI response [19,20]. This scheme is based on a series of kinetic equations describing quality loss and TTI response (Figure 30.1).

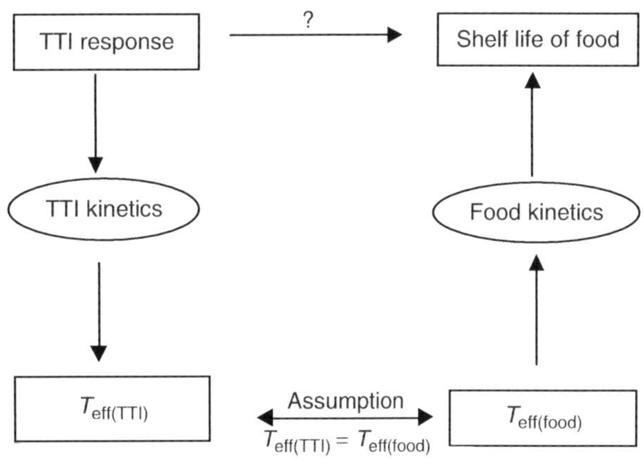

FIGURE 30.1 Application of TTI as food quality monitors. (From Fu, B. and Labuza, T. P., *J. Food Distr. Res.*, 1992, 23(1), 9. With permission.)

The effective temperature for a variable time–temperature distribution is determined from the TTI response kinetics. This value is used in conjunction with the food kinetics to estimate the remaining shelf life or quality loss. The scheme shown in Figure 30.1 assumes that the effective temperature and the activation energies of both the TTI device and the food are the same. However, this is not always the case. For practical applications, a difference between $E_{A(food)}$ and $E_{A(TTI)}$ of ±8.4 kJ/mol (2 kcal/mol) generally gives a good prediction [19,20].

30.2.5.8 Commercially Available TTIs

There are several types of TTI devices that have been described in the patent literature, but only three have continued to receive significant attention from the industrial and scientific community for the past 15 years [11,18]. These include the 3M Monitormark, Lifeline, and VITSAB TTI. They are available as labels, tags, pallet sticks, or pallet with sheathes. Each of these is described below.

30.2.5.9 3M Monitormark TTI

The principles, characteristics, and operation of the 3M Monitormark TTI (3M Identification and Converter Systems Division, St. Paul, MN) is well documented [29,30]. It is designed to give a visual response only after a critical threshold temperature has been exceeded. The indicator consists of an assemblage of porous paper wick, and a rectangular paper pad saturated with a blue colored fatty acid ester and phthalate mix that has a specific melting point, a polyester barrier layer, which separates the wick from the reservoir pad (paper pad) and adhesives. The response critical temperature is usually the melting point of the specialty chemical. The prepared TTI tag remains inactive until it is activated.

The indicator is activated when a polyester pull-strip is removed allowing contact between the porous paper and the reservoir. When exposed to temperatures higher than the threshold temperature, the chemical in the reservoir melts and migrates along the porous wick. The indicator has five windows that progressively change color as the migrating chemical runs through.

The observation of color in the first window implies that the indicator has been exposed to a temperature higher than the preset threshold temperature. The extent of color movement through these windows indicates the cumulative exposure time spent above the threshold temperature. Response cards provided by the manufacturer are used to correlate the time–temperature relationship of each indicator. When all five windows turn blue, the product is interpreted to be unacceptable. If stored according to manufacturer's guidelines, the tag shelf life is 2 year from the date of manufacture.

A plot of the square of the run-out distance against time is linear for a Monitormark TTI and obeys the Arrhenius equation [20]. The design and operation of this indicator relies on a diffusion mechanism so the activation energy of this TTI is limited to 0–60 kJ/mol (0–14.3 kcal/mol). Therefore, it is recommended for use in many enzyme and diffusion-controlled spoilage reactions.

30.2.5.10 Lifeline Fresh-Scan and Fresh-Check TTI

Lifeline Technologies (Lifeline Technologies, Morristown, NJ) offers two "full history" TTIs that monitor the freshness of perishables independent of temperature threshold. These products are the Fresh-Scan and Fresh-Check indicators and are based on the color change resulting from solid state polymerization of a diacetylenic monomer. The construction, operation, and characteristics of the device is described in detail by Patel and his coworkers [31,32] as well as by Field and Prusik [33].

The Fresh-Scan indicator consists of the following: a standard bar code containing product information and a diacetylenic monomer deposited on a pressure-sensitive band; a portable microcomputer with a laser scanner, which measures changes in reflectance of the indicator; and a data analysis work station. The indicator band initially shows about 100% reflectance. As the diacetylenic monomer undergoes a time–temperature dependent solid state polymerization during storage, the indicator band darkens, causing a decrease in the measured reflectance. The rate of color change follows Arrhenius behavior so higher temperatures enhance color development rate. The microcomputer combines this time–temperature characteristic with the product information on the bar code to predict the shelf life of the product.

The Fresh-Check indicator is a consumer-readable visual label, which is also based on the color change of an incorporated polymerizable monomer, but with a different design and configuration [15]. The circular device consists of an inner polymer shell, containing the monomer and an outer nonpolymer shell painted with a reference color. When exposed to some time–temperature storage conditions, the monomer at the center part converts to polymer, causing a progressive color development at a rate that increases with temperature. If the polymer center become darker than the outer reference, the consumer is advised to discard the product, regardless of the printed expiration [15].

These two Lifeline indicators are active as soon as they are manufactured and must be refrigerated at very low temperatures (≤ -24°C) to preserve high initial reflectance. Wells and Singh have investigated the use of these indicators as a quality change monitor for perishable and semiperishable products such as tomatos, lettuce, canned fruitcake, and UHT-sterilized milk [22]. It was shown that the response kinetics of these indicators correlates with quality loss in the products studied. The indicators studied have activation energies of 84–105 kJ/mol (20–25 kcal/mol). Major limitations in these indicators are that the polymerization reaction is photosensitive and also their inability to respond to short time–temperature history.

30.2.5.11 VITSAB TTI

Numerous studies have been carried out on the time–temperature-dependent enzyme reaction on which the Vitsab TTI (formally known as I-Point) is based [34,35]. It is a "full history" indicator that records the temperature conditions and time history of a product independently of threshold temperature. The device uses two color-coded compartments to store the chemicals. The green part houses a mixture of enzyme solution and a pH-indicating dye. The gray part contains a lipase substrate suspension. The line separating the two chambers represents a pressure-sensitive barrier. The device is activated by breaking the barrier between the two chambers. This causes the enzyme solution and the substrate to mix to form the indicating solution. As the reaction progresses, the lipase substrates (triglycerides) hydrolyze into their component fatty acids, causing the pH to drop. A change in the pH causes the dye to change color from an initial green color to a final yellow color at a rate that is temperature-dependent. A variety of VITSAB indicators can be custom-designed for specific temperature-sensitive commodities by judicious selection of enzyme concentration and enzyme–substrate combination.

At present, there are four different types of indicators, which come in one and three dots standard configurations depending on the level of product safety risk to be communicated. The single dot indicator will change color from green to yellow above a predetermined temperature. The three-dot indicator conveys the degree of temperature exposure with high reliability.

The activation energies of the major TTI products are summarized in Table 30.3. By matching the E_A of these TTIs to the E_A values for some common deterioration reactions in food and pharmaceutical products (Table 30.2), current TTI technology should provide a range of selections that will meet manufacturer's growing interest and demand for these products.

30.2.6 Anticounterfeiting and Tamper Indicator Devices

Counterfeiting, forgery, tampering, and piracy of valuable documents and products have existed from time immemorial. Counterfeiting has become a multibillion dollar business that is prospering more than many of the victim companies. This highly organized trade results in loss revenue to both companies and governments, as well as loss of jobs in private and public sectors. The motivation for the professional counterfeiters is the ease in making huge profit with very low risk of being caught and

TABLE 30.3 Activation Energy Values of Some Common Commercial TTIs

Manufacturer	TTI Model	TTI Activation Energy (kJ/mol)	TTI Activation Energy (kcal/mol)	Reference
Visual Indicator Tag Systems, AB-Malmo, Sweden	VITSAB Type 1	57.4	13.7	[36]
	VITSAB Type 2	94.7	22.6	[36]
	VITSAB Type 3	126.1	30.1	[36]
	VITSAB Type 4	194.8	46.5	[36]
	I-Point 3014[a]	47.8	11.4	[19]
	I-Point 4007[a]	137.0	32.7	[19]
	I-Point 4014[a]	101.8	24.3	[19]
	I-Point 4021[a]	141.2	33.7	[19,20]
3M Packaging System Div., St. Paul, MN	All types	41.1	9.8	[19,36]
Lifeline Technology Inc., Morris, Plain	Fresh-Check A7	155.0	37.0	[36]
	Fresh-Check A20	81.3	19.4	[19]
	Fresh-Check A40	81.7	19.5	[19]
	Fresh-Check B21	88.0	21.0	[19]
	Fresh-Scan-18	113.1	27.0	[19]
	Fresh-Scan-41	85.8	20.5	[19,20]
	Fresh-Scan-68	82.5	19.7	[19,20]

[a] Vitsab was formally known as I-Point.

penalized. It is impossible to enumerate products or documents that are counterfeited or quantify the impact of these goods on vulnerable consumers. Activities of counterfeiters have increased in developed countries. The incidence of counterfeiting is even greater in developing countries, which have become a dumping ground for counterfeit products. A partial list of goods that are commonly faked include antiquities, currency, credit letters, credit cards, jewelry, chemicals, processed food, pharmaceuticals, and cosmetic products, product identification marks, videos, compact disks (CDs), clothing and apparel accessories, electronics, auto and airplane parts, children's toys, watches, sporting goods, travel documents such as passports, identity cards and driving permits, and textile materials [37–39].

Advances in technology such as computers, lasers, scanners, charge-coupled devices (CCD), color printers, optical holograms, and analytical tools have enhanced the state of the art for anticounterfeiting security and encryption. Ironically, these advances are also being exploited by the adversaries to promote high-tech forgery. To combat these activities, research efforts have continued to develop tamper-resistant security mechanisms that will make counterfeiting too expensive to be lucrative. This subject has been reviewed recently [37,38]. A summary of some of the available technologies relevant to this chapter will be presented.

30.2.6.1 Holographic Technology

An optical hologram is a three-dimensional photo created on a photoresist film or plate by the reflection and refraction pattern of laser light incident on an object. Two of the holograms widely used in security applications are the Embossed and the Denisyuk hologram. The embossed hologram can be readily mass-produced as a mechanically tough and durable thin film at a very low cost.

The relatively thicker Denisyuk hologram requires a more stringent manufacturing process. In addition to being more expensive to produce, it is not as mechanically robust as the embossed hologram. This hologram is the more easily counterfeited optical hologram.

Holographic Dimensions, Inc. offers the Verigram™ security system that consists of two elements, the display hologram and the machine readable Verigram™ hologram, which afford a completely secured device. Although the display component may be counterfeited, the Verigram™ element is claimed to be impossible to forge [40].

One of the leading manufacturers of anticounterfeiting holograms is American Bank Note Holographics (ABNH) Inc. Their products are used for various commercial and security applications. They include holographic stripe, holomagnetic stripe, holographic thread, transparent holographic laminate, hot stamp foil, hologuard, frangible vinyl and pressure-sensitive and tamper-evident labels. ABNH can manufacture holograms with magnetic stripe, visible, and invisible bar codes, or microprinting for added security, product tracking, audits, and authentication. The list of producers and vendors of commercial and security holograms is endless. Only two are mentioned here as examples.

30.2.6.2 Microtaggants Technology

Taggants are microscopic color-coded particles derived from several layers of a highly crossed-linked melamine polymer and are used extensively as markers [41]. Layers of magnetic and fluorescent materials are added to the taggants' particles during formulation to simplify detection and decoding. The technology, originally developed for explosives, finds applications in antiterrorism, authentication, piracy, counterfeiting, and product identification [41]. Taggant particles can be coded with information such as the product manufacturer, production date and batch number as well as the distributor. When used in antiterrorism, taggants are thoroughly mixed with the explosives. When a bomb containing tagged explosive is detonated, remnant taggants maybe collected from debris with the aid of a UV light and a magnet. The coded color sequence information can be retrieved with an optical microscope. This reliably helps in tracing the origin of the explosive.

Tagged particles can also be added to printing ink for security printing. This provides a way to combat counterfeit, forgery, and product diversion and assist product/property identification. Microtaggants can be added to many products either directly during formulation or by thermal transfer, films, laminates, spraying, and so on. Some of the companies that market Microtaggant identification particles are Microtrace Inc., Minneapolis, MN; MICOT Corporation (St. Paul, MN); SW Blasting (BÜLACH, Switzerland); and Plast Labor (Bulle, Switzerland). Patronage comes from government agencies, law enforcement, and industry.

30.2.6.3 Smart Ink Technologies

Smart inks or coatings change color predictably in response to alteration of their environment. Color change may or may not be reversible and smart inks or coatings are finding applications in areas such as security printing, lenses, and other optical devices. The two main classes of smart inks include those derived from photochromic and thermochromic materials.

Photochromic inks are formulated from light-sensitive dyes, pigments, polymers, or other colorless chemical compounds. These inks change color when exposed to ultraviolet light. Examples include extrusion lamination inks, invisible ink security markers, tagged inks, and inks based on copper-free metallics.

Thermochromic inks incorporate temperature-sensitive polymers, dyes, or pigments. When exposed to a predefined temperature, the ink undergoes a color change. Typical color-sensitive components include diacetylenic compounds, hydrogel polymers, and thermochromic liquid crystals.

Security inks incorporates fluorescent dyes, photochromic dyes, thermochromic dyes, or organic pigments in their formulation. These dyes or pigments reversibly change color when exposed to ultraviolet light or heat of predefined intensity or temperature. This forms a basis for their use in security printing, counterfeit or forgery indicators, tamper-evident, and other applications. Several companies around the world manufacture and distribute counterfeit detector pens or specialty fluorescent light bulbs for quick authentication of currency bills, gift certificates, and other documents.

30.2.6.4 Tamper-Indicating Devices

Tamper indicators are security seals or labels designed to detect product or document tampering. Indicator devices available are based on fiber optic active seal technologies, optical thin film, holograms, and chemical reactions. A major requirement for these materials is that they provide unambiguous evidence of tampering. They must be resistant to environmental conditions such as heat or solvents and relatively cheap. Several low-cost tamper indicator products are commercially available to meet various customers' needs.

Telatemp (Telatemp Corporation) sells TelaSEAL security labels constructed of polyester film, high strength instant-bond adhesive, and other proprietary chemicals [14]. A repeating geometrical pattern is printed onto the TelaSEAL during construction for additional security. The seal can be applied to a variety of surfaces including plastics, metals, wood, and painted surfaces. A disrupted repeating pattern with a custom message "OPENED" is transferred to the applied surface when the device is peeled from the substrate. This provides permanent and unambiguous evidence of tampering.

Meyercord (Sentinel Division, Carol Stream, IL) supplies several tamper-evident adhesive-base materials that leave a "VOID" message on surfaces, thus providing indication of unauthorized opening or tampering of products. The "VOID" message can be customized to include company's name, trade mark, logos, or serial number.

3M has developed several authenticating devices based on optical thin film technology for counterfeit detection and tamper indication. A proprietary metallic surface coating is formulated and deposited on a pressure-sensitive material. Specialized imaging technologies are used to generate unique covert graphics and print product information or serial number on the device. When the device is peeled from the substrate, the covert image is revealed followed by an irreversible color change, which provides a sure indication of tampering. These devices find uses in asset control, package sealing and identification, parts identification, serial numbering, and consumer information.

30.2.7 Indicating Device Issues and Limitations

Interest in indicating devices continues to grow and numerous potential applications have been proposed. Despite significant advances made in chemical indicator technology during the last 25 years, only a few devices have been commercialized when compared to the number of patents granted each year. There are several reasons why many manufacturers, governments, and consumers are reluctant to adopt chemical indicating devices. Major considerations on the part of manufacturers include device reliability, cost, and education. Manufacturers fear the consumers will be reluctant to embrace products containing these indicators for reasons of additional cost. The addition of a leak indicator to the headspace of packaged food or pharmaceutical products is likely to be greeted with stiff resistance on the part of the consumer for safety reasons. To gain acceptance, commercial indicators must be relatively inexpensive compared to the products to which they are applied. Also, they must be technically competent to perform their intended functions.

Response mechanisms of many indicators are poorly understood [42]. For example, to produce a reliable time–temperature indicator, the quality index, kinetics, and mechanism of food or chemical product degradation under various environmental conditions must be understood. Some degradation reactions may not always follow simple Arrhenius kinetics. Therefore, predictions based on them may be faulty and thus affect product reliability. Consumer surveys [12,30] on the use of TTI on food products on a good note indicate that the public reception of these indicators is growing. It also raised awareness on the need to educate the consumers on the usefulness, reliability, and application of indicator devices in order to increase the current level of acceptance.

Even though most of the counterfeit and tamper indicator technologies currently available are reliable, they are quite expensive. For this reason, many industries are reluctant to implement them. Some the indicators are limited in their scope of application and are easily forged or copied as is the case for many optical holograms [37–40]. This has been a major problem with many credit and various identification cards. The challenge is to produce indicators that are reliable, technically difficult to copy or fake, and are available at a relatively low price.

Recently, several indicating devices have made inroads into the market place with unsubstantiated claims of what the product can offer. In some cases, instructions for consumers on the application and interpretation of indicator response or color change are inadequate. Such consumers appear to be confused. Information about the safety of the indicator should be provided too. Efforts should be made to design indicators that are childproof to avoid ingestion and other accidents. A general specification for the various classes of indicators needs to be established. Such a specification should provide performance criteria that each indicator design must meet for a given application. It should also set standards for acceptable qualification testing. This will ensure that commercial indicating devices meet the needs of industry.

30.3 Conclusion

The application of chemical indicating devices has come to stay and now available. In many cases, they are living up to expectations in reducing the incidence of counterfeiting, forgery, tampering or piracy, and losses of perishable products. Temperature and TTIs are now being used in routine packaging of some critical food and chemical products. For TTI, only limited information on the kinetics and response mechanism is currently available in the literature. New indicators covering a large range of activation energies for many spoilage reactions need to be designed. This will allow the scope of application of TTIs to be extended. More research and development effort could and should be directed toward addressing the present technical and manufacturing challenges hindering commercialization of many patented inventions. Public education will accelerate interest and ultimate acceptance of these products. Such acceptance may create sufficient volumes of use, which would lead to lower costs for manufacturing these smart devices.

Acknowledgments

The author wish to thank Dr. Ted Labuza for providing key references on TTI, Dr. Annapoorna Akella for help with patent literature search, Dr. John Evans, Dr. Jim Harvey and, Dr. David Landman, for helpful comments; and to companies who graciously provided relevant information.

References

1. Boside, G. and A. Harmer., *Chemical and Biochemical Sensing With Optical Fibers and Waveguides*, Boston, MA: Artech House, Inc. 1996.
2. Nagasaki, Y. and K. Kataoka, Polysilamines as intelligent materials, *CHEMTECH*, 1997, 3.
3. Hoffman, A. S., Intelligent polymers in medicine and biotechnology, *Artf. Organs*, 1995, 19, 458–467.
4. Galaev, I. Y., Smart polymers in biotechnology and medicine, *Rus. Chem. Rev.*, 1995, 64, 471.
5. Mucklich, F. and H. Janocha,Smart materials—The IQ of materials in systems, *Z. Metallkd.*, 1997, 87, 357.
6. Snowden, M. J., M. J. Murry, and B. Z. Chowdry, *Chem. Ind.*, 1996, 531.
7. Kolthoff, I. M. and C. Rosenblum, *Acid-Base Indicators*, London: Macmillan, 1937.
8. Bromberg, L. and G. Levin, *Macromol. Rapid Commun.*, 1996, 17, 169.
9. Dagani, R., Intelligent gels, *Chem. Eng. News*, 1997, 26.
10. Bryne, C. H., Temperature indicators—the state of the art, *Food Technol.*, 1976, 6, 66.
11. Taoukis, P. S., B. Fu, and T. P. Labuza, Time–temperature indicators, *Food Technol.*, 1991, 45(10), 70 and references therein.
12. Woolfe, M. L., Temperature monitoring and measurement. In: *Chilled Foods, A Comprehensive Guide*, Dennis, C. and M. Stringer (eds.), New York: Ellis Horwood Ltd, 1992, Ch. 5.
13. Anon. Temperature indicators find many applications in engine equipment maintenance, *Diesel Prog. Eng. Drives*, 1996, 8, 36.
14. Telatemp Corporation, Fullerton, CA Product catalog, 1997.
15. Selman, J. D., Time/temp' indicators: how they work, *Food Manufacture*, 1990, 65(8), 30 and references therein.
16. Manske, W. J., Critical Temperature Indicator, U.S. Patent No. 4,457,252, 1984.
17. Introtech, Inc., St. Paul, MN Product catalog, 1992.
18. Schoen, H. M. and C. H. Byrne, Defrost indicators: many designs have been patented yet there is no ideal indicator, *Food Technol.*, 1972, 26(10), 46.
19. Fu, B. and T. P. Labuza, Considerations for the application of time–temperature integrators in food distribution, *J. Food Distr. Res.*, 1992, 23(1), 9.
20. Taoukis, P. S. and T. P. Labuza, Applicability of time–temperature indicators as food quality monitors under nonisothermal conditions, *J. Food Sci.*, 1989, 54(4), 783.

21. Taoukis, P. S. and T. P. Labuza, Reliability of time–temperature indicators as food quality monitors under isothermal conditions, *J. Food Sci.*, 1989, 54(4), 789.
22. Wells, J. H. and R. P. Singh, Application of time–temperature indicators in monitoring changes in quality attributes of perishables and semi-perishable foods, *J. Food Sci.*, 1988, 53(1), 148.
23. Labuza, T. P., Application of chemical kinetics to deterioration of foods, *J. Chem. Edu.*, 1984, 61(4), 348.
24. Hendrickx, M., G. Maesmans, S. De Cordt, J. Noronha, A. Van Loey, and P. Tobback, Evaluation of the integrated time-temperature effect in thermal processing of foods, *Crit. Rev. Food Sci. Nutrir*, 1995, 35(3), 231 and references therein.
25. Villota, R. and J. G. Hawkes, Reaction kinetics in food systems. In: *Hand Book of Food Engineering*, Heldman, D. R. and D. B. Lund, (Eds.), New York: Marcel Dekker, 1992, p. 39.
26. Wells, J. H. and R. P. Singh, A kinetic approach to food quality prediction using full-history time-temperature indicators, *J. Food Sci.*, 1988, 53(6), 1866.
27. Grisius, R., J. H. Well, E. L. Barrett, and R. P. Singh, Correlation of time–temperature indicator response with microbial growth in pasturized milk, *J. Food Proc. Preserv.*, 1987, 11, 309.
28. Fu, B., P. S. Taoukis, and T. P. Labuza, Theoretical design of a variable activation energy time–temperature integrator for prediction of food or drug shelf life, *Drug Dev. Indu. Pharm.*, 1992, 18(8), 829.
29. Manske, W. J., Selective Time Interval Indicating Device, U.S. Patent No. 3,954,011, 1976.
30. Selman, J. D., Time–Temperature Indicators, In: *Active Food Packaging*, Rooney, M. L., (Ed), New York: Blackie Academic & Professional, 1995, Ch. 10 and references therein.
31. Patel, G. N. and K. C. Yee, Diacetylene Time–Temperature Indicator, U.S. Patent No. 4,228,126, 1980.
32. Patel, G. N., A. F. Preziosi, and R. H. Baughman, Time–Temperature History Indicators, U.S. Patent No. 3,999,946, 1977.
33. Field, S. C. and T. Prusik, Shelf life estimation of beverage and food products using bar coded time–temperature indicator labels. In: *The Shelf Life of Foods and Beverages*, Charalambous, G., (Ed.), Amsterdam: Elsevier, 1986, p. 23.
34. Blixt, K., S. I. A. Tornmarck, R. Juhlin, K. R. Salenstedt, and M. Tiru, Enzymatic Substrate Composition Adsorbed on a Carrier. U.S. Patent No. 4,043,871, 1976.
35. Blixt, K. and M. Tiru, An enzymatic time/temperature device for monitoring the handling of perishable commodities, *Dev. Biol. Std.*, 1977, 36, 237.
36. Vitsab TTI Product catalog, 1997, @ www.vitsab.com.
37. van Renesse, R. L., (ed.) *Optical Document Security*, Boston, MA: Artech House, 1994.
38. Lancaster, I., Progress in counterfeit deterrence; the contribution of information exchange, *Proc. SPIE*, 1996, 2659, 2.
39. McGrew, S. P., Countermeasures against hologram counterfeiting, Proc. Optical Security Systems Symposium, Zurich, Oct. 14–16, 1987.
40. Brown, K. G., J. Weil, and M. O. Woontner, Counterfeit Proof and Machine Readable Holographic Technology Verigram Systems, @ http://www.hgrm.com/counterf.htm.
41. Rouhi, A. M., Government, industry efforts yield array of tools to combat terrorism, *Chem. Eng. News*, 1995, July 24.
42. Selman, J. D., Time–temperature indicators: do they work? *Food Manufacture*, 1988, 63(12), 36 and references therein.

31

Piezoelectric Polymer PVDF Microactuators

Yao Fu
Silverbrook Research Pty. Ltd.

Erol C. Harvey
Swinburne University of Technology

Muralidhar K. Ghantasala
Western Michigan University

31.1 Introduction

Microactuators are one of the most important components in microelectromechanical systems (MEMS). The simplest and earliest actuating elements are microcantilevers. In the late 1980s, microactuators received increasing attention when electrostatically driven micromotors started to appear [1,2]. In the last 15 years, many types of microactuators utilizing various driving forces (e.g., electrostatic, electromagnetic, piezoelectric, shape memory alloy (SMA) etc.) have been developed. Piezoelectric actuation is one of the most popular actuation principles used for microactuators.

Piezoelectric microactuators generate large forces but only small displacements by applying a voltage to the piezoelectric materials. Typically used constructions are the bimorph [3,4] and multilayer structures. In addition, more and more unimorph composite beams [5–8] are employed to meet the increasing requirements of special applications.

The piezoelectric actuation is employed in many applications, such as electric fans [4], hydrophones [9], microphones [10,11], inkjet printers, control valves [12], micropumps [13], tactile sensor [14], acoustic control [15], and micromotors [16], etc. The main advantages of this actuation are its high precision, speed, and mechanical power.

In these applications, piezoelectric ceramic materials such as zinc oxide [17] and PZT [7,8] are most commonly used, as they exhibit large piezoelectric coefficients. However, the major difficulty associated with their use in many applications is the requirement of advanced deposition technologies and facilities to prepare stoichiometric thin films. Further, these are usually brittle, and have a relatively large Young's modulus, limiting the achievable strain. Some composite active polymers [18] comprised of piezoelectric ceramics and epoxies were fabricated to compensate for these disadvantages. Piezoelectric polymers such as polyvinylidene fluoride (PVDF) and its copolymers can overcome some of these difficulties even though they have a relatively low piezoelectric coefficient. Low numerical Young's modulus values of these polymers [19,20] have the potential for enabling relatively large strain piezoelectric actuators.

The special characteristics of piezoelectric polymers have attracted the attention of many researchers from many different disciplines. A lot of devices were designed from these materials since the 1980s. A long list of papers and patents can be found in the piezofilm sensors technical manual from Measurement Specialist [21]. In some of these devices, PVDF has been used mostly in the form of a thin foil, either stretched or mounted. The ability to pattern PVDF into required shapes and sizes is highly desirable in order to increase the range of applications, particularly in the millimeter to micrometer scale. Several different methods have been employed to pattern PVDF, including the use of an excimer laser [22], UV light source [23], x-rays [24,25], and hot embossing [26]. Each method has its own strengths and problems. Excimer laser ablation will change the UV-VIS photoabsorption spectrum of PVDF polymer [22]. Heat diffusion can easily damage the piezoelectric effect of PVDF and can also burn the edge of the pattern [27]. A direct pattern transfer into PVDF is possible through x-ray-induced etching [24]. However, the x-ray fabrication is much more expensive than laser ablation. Hot embossing PVDF will have the risk of damaging the piezoelectric effect in PVDF polymer as it typically requires molding temperatures in the range of 175°C to 185°C. One group in Japan [28] used an electrospray deposition method to deposit thin-layer PVDF on the substrate biased with a voltage of 8–15 kV.

A piezoelectric polymer composite cantilever was designed by using finite element method simulation. The aim of the design was to find the optimal parameters and dimensions of a piezoelectric cantilever to achieve large tip deflection and force. The influence of the parameters of the cantilever on tip deflection was analyzed based on the simulations. The designed piezoelectric unimorph actuator was fabricated using punching and electroplating techniques. The microstructures on the top of the piezoelectric polymer PVDF were realized by a low-temperature hot embossing process, thus avoiding the depoling of the PVDF polymer. The material property piezocoefficient of PVDF was characterized. Finally, the experimental test results of the composite cantilever are explained.

31.2 Piezoelectric Polymer Cantilever

A unimorph cantilever beam is chosen as the basic element. The cantilever is made out of a commercially available piezoelectric polymer PVDF sheet, with a thin permalloy layer electroplated onto one side in our laboratory. The schematic cross-section of the cantilever is shown in Figure 31.1. The thickness of the original PVDF sheet is 28 μm and is covered with a 40 nm thick nickel-copper alloy metal electrode layers on both sides. The PVDF polymer is polarized in a direction parallel to its thickness. The nonpiezoelectric layer of the unimorph cantilever is fabricated by electroplating permalloy to a thickness of approximately 5 μm. The length of the cantilever beam is 6 mm and the width is 1 mm.

31.3 Modeling and Simulation

Simulation and modeling using numerical methods employing available finite element software is an efficient way to design and optimize the microstructures. We performed the simulation using ANSYS and CoventorWare. The 3-D model

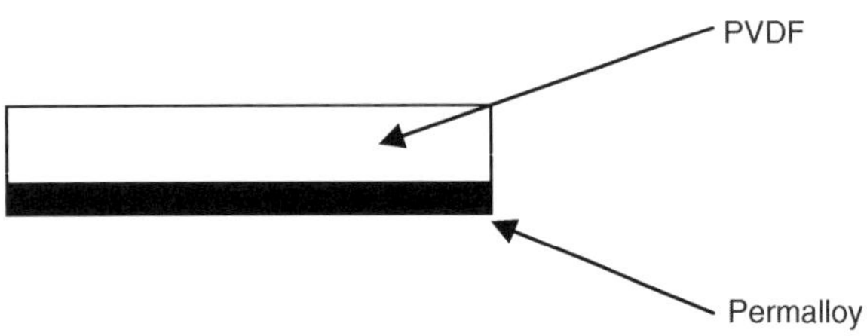

FIGURE 31.1 Cross-section of the piezoelectric polymer cantilever.

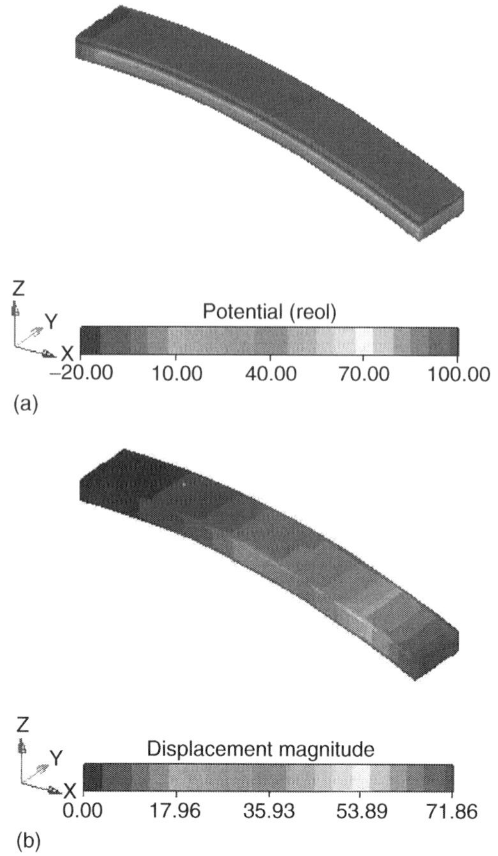

FIGURE 31.2 Simulation of a cantilever beam using CoventorWare. (a) Distribution of the electrical potential across the PVDF polymer. (b) Displacement of the cantilever caused by piezoelectric effect.

of a cantilever beam simulated using CoventorWare package is shown in Figure 31.2. It is showing two layers, one for PVDF and the other for permalloy. The beam dimensions with a length of 5 mm, a width of 1 mm, and a thickness of 28 μm for the PVDF layer with the thickness of nonpiezo permalloy layer of 5 μm. The Young's moduli of PVDF and that of permalloy were input as 2.8 [21] and 150 GPa [29], respectively. The tip deflection at an applied potential of 100 V is found to be 71.9 μm. Figure 31.2a shows the geometric distribution of the electrical potential over the thickness of piezoelectric element and Figure 31.2b the beam deflection under this potential.

31.3.1 Parametric Analysis for the Cantilever Deflection

The tip deflection of a unimorph cantilever is determined by its geometrical dimensions, material properties, as well as by the external electrical field. In order to design an effective microactuator with maximized deflection, it is necessary to optimize all these parameters. A detailed analysis of the effect of individual parameters on the tip deflection of the cantilevers was performed.

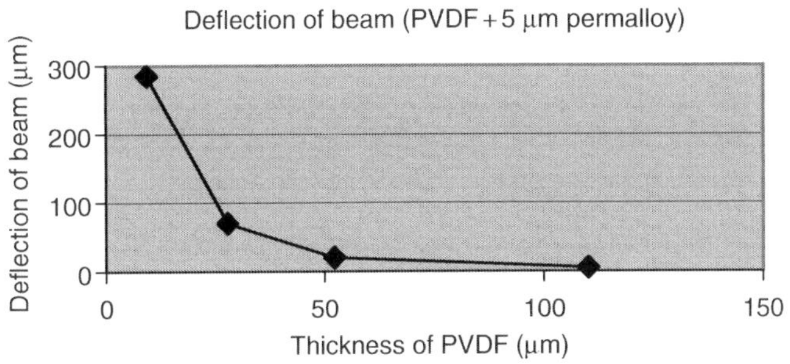

FIGURE 31.3 Tip deflection versus PVDF thickness.

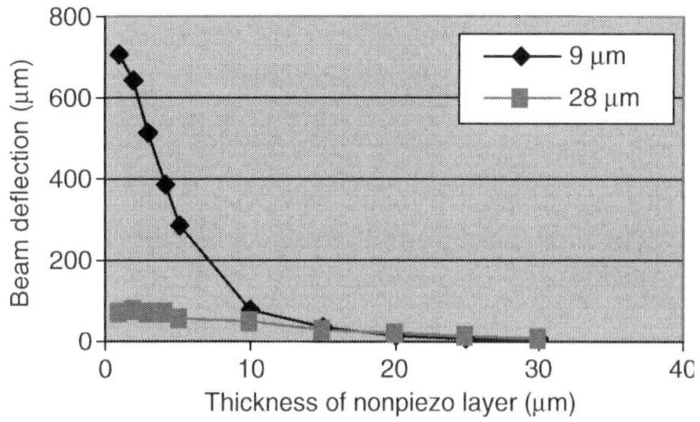

FIGURE 31.4 Tip deflection versus the thickness of nonpiezoelectric layer.

As expected, the tip deflection increases proportionally to the square of the beam length and is proportional to the applied voltage. The width of the cantilever has little effect on the beam deflection. Therefore in the simulation, the width of the cantilever is fixed at 1 mm and applied voltage on the cantilever was set at 100 V.

The tip deflection of a composite cantilever beam depends on the thickness of both of its piezo and nonpiezo layers. The effect of the piezo layer thickness is first studied. Figure 31.3 shows the variation of the beam deflection, as the thickness of the PVDF layer is varied in the range 9–110 μm. This graph clearly shows that the thinner the PVDF layer, the larger the deflection of the beam. This kind of variation in beam deflection with the piezolayer thickness can be explained by the fact that the induced strain on the thinner PVDF under the same electrical potential is larger due to the large electrical field (voltage/thickness of the beam) and also the thinner cross-section of the cantilever beam has less moment of inertia, therefore the deflection of the cantilever increases with the decrease of the PVDF thickness. Even though the 9 μm thick PVDF layer provides the largest deflection, it is not practical for use in most applications; as such, a structure is very flexible and provides little force. Therefore, a 28 μm thick PVDF layer was is used in this design.

It is equally important to determine the effect of the nonpiezo layer. Figure 31.4 shows the variation of the beam deflection with the thickness of the nonpiezo permalloy layer, when all other parameters are kept constant. It can be seen from the graph that for a 28 μm thick PVDF layer the deflection has a maximum of around 75 μm with a permalloy layer thickness of about 2 μm, but decreases very little up to a thickness of 5 μm. When the thickness is increased to 30 μm, the deflection decreases to 9 μm. For a 9 μm thick piezo layer, the beam deflection can reach up to 700 μm with a nonpiezo layer thickness of 1 μm, but decreases to 50 μm with a nonpiezo layer thickness of 15 μm. Then the two curves converge after the thickness of the nonpiezo layer increases to values above 20 μm. This means that when the nonpiezo layer thickness exceeds 20 μm, the beam deflection is mainly affected by the nonpiezo layer only. In this design, a thickness of 5 μm was chosen as an optimal value for a 28 μm piezoelectric polymer, as a thinner permalloy layer might get damaged during actuation. This analysis shows that with certain material the thickness of the nonpiezoelectric layer can be optimized to achieve the maximum deflection.

Young's modulus of the nonpiezoelectric layer has been identified as another important parameter with respect to the beam deflection. In order to achieve the maximum tip deflection of the cantilever, an optimal material of the nonpiezoelectric layer can be worked out from this simulation. It is observed from the simulation that the useful range of Young's modulus from a practical material point of view is in the range of 10–1000 GPa.

The numerical results obtained from the finite element analysis (FEA) simulations were compared and confirmed with the theoretical analysis based on bimetal theory. The optimal parameters for the piezoelectric composite cantilever are determined by the thickness and the Young's modulus of the nonpiezoelectric layer. For a given material of the nonpiezo layer, an optimum thickness can be calculated for obtaining the maximum tip deflection. Similarly, a suitable material can be selected from this kind of simulation depending on the chosen actuator configuration and the thickness of the layer. The potential materials for the nonpiezoelectric layer can be metals (e.g., aluminum) or polymers, such as polyimide and Mylar. The use of a magnetic material as the nonpiezoelectric layer facilitated the formation of a hybrid microactuator. Since the simulation studies indicated a required nonpiezo layer modulus in the range of 10 to 1000 GPa, we selected a soft magnetic permalloy ($Ni_{80}Fe_{20}$) with a modulus of 150 GPa for this purpose.

This simulation and modeling work helped us in arriving at some of the optimal parameters of the beam length of 5 mm (depending on the required deflection), width of 1 mm, and thicknesses of the PVDF and permalloy layers of 28 and 5 μm respectively.

31.4 Fabrication of Piezoelectric Composite Cantilevers

Even though PVDF can potentially be used in many microactuator applications, it is still a challenge for MEMS researchers to pattern and form the polymer foils into the required shapes and sizes with high accuracy. The present work demonstrates a novel route for this purpose using laser micromachining, electroplating, and punching (microembossing) techniques.

31.4.1 Excimer Laser Ablation

Excimer laser micromachining has been used as a tool for machining and/or patterning many polymers or their layers on different substrates. Laser micromachining has already been demonstrated as an excellent tool for patterning different photoresists, polyimides, polycarbonate, PET, and similar materials. The availability of an excimer laser based micromachining system (Exitech Ltd., United Kingdom) has prompted us to explore the possibility of applying this method for machining the cantilevers using PVDF material.

A typical machined pattern in these experiments is shown in Figure 31.5. The square hole is approximately 300 × 300 μm, machined using 64 shots at a laser fluence of 1135 mJ/cm^2. Unfortunately, the edges turned out to be quite rough and could not be improved significantly, at a range of laser machining conditions tried. As the number of shots increased, the edges became worse. This may mainly be attributed to the laser absorption properties (@ 248 nm) and the thermal characteristics of the PVDF polymer. This led to the conclusion that the direct excimer laser micromachining (@ 248 nm) is not suitable for machining or patterning PVDF polymer foils.

31.4.2 Punching of the PVDF Cantilever

Punching is one of the most commonly used processing methods for thermoplastics [30,31]. In conventional punching, a pair of preheated male and female punch tools is used to cut the polymer by shear forces. A polarized piezoelectric polymer can be exposed to no more than its Curie temperature (70–90°C for PVDF), beyond which the material loses its piezoelectric characteristics. Therefore, conventional punching process cannot be directly used in this case. Instead of using male and female plug tools, a nickel shim with the required punch profile was used to cut the PVDF polymer cantilever in a home-made embossing system. This facilitated a low-temperature punching process, forming the designed profile of the piezoelectric polymer cantilever well below its Curie temperature. The nickel shim was

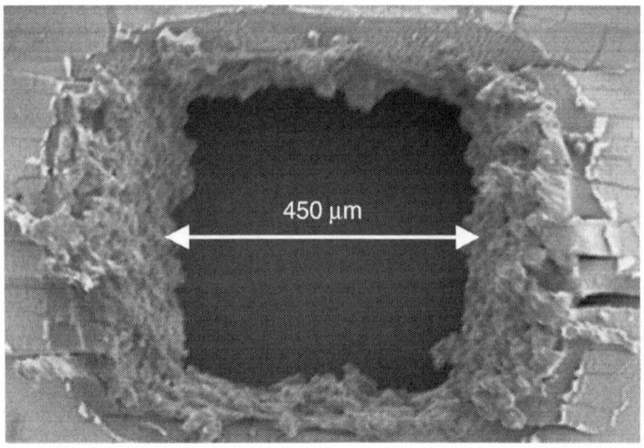

FIGURE 31.5 SEM images of metallized PVDF patterned with an excimer laser at energy density of 1135 mJ/cm^2.

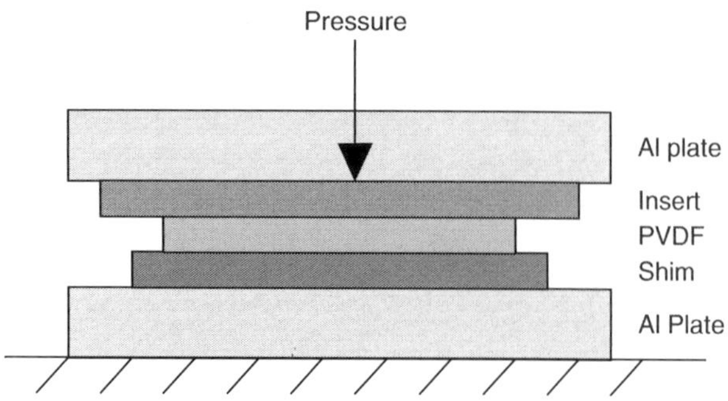

FIGURE 31.6 Hot embossing system used for the punching process.

fabricated using laser micromachining of a mold in a given polymer substrate and subsequent electroforming.

The punching of the cantilever from a PVDF polymer sheet was performed using a microembossing system at room temperature. The procedure used for this purpose is similar to the embossing technique described by Becker et al. A nickel shim, PVDF film, and the insert were fixed in a sandwich structure between the two aluminum plates shown in Figure 31.6. This arrangement is enclosed in a vacuum chamber inside the hot embossing system. A hydraulic pump was used to apply the force on the chamber.

A force of 0.27 bar/mm^2 was applied onto the sample for one minute. PVDF cantilevers with the designed shape were punched out along the PVDF stretch direction.

31.4.3 Electroplating Permalloy on the Metallized PVDF

As the composite cantilever requires a nonpiezo layer on the PVDF cantilever structure, the next logical step would be to metallize, the PVDF cantilever to form the required bilayer. Permalloy is chosen for this purpose to build a hybrid actuator combining piezoelectric and electromagnetic principles. In this case, the permalloy serves the purpose of a soft magnetic layer, which can assist in providing additional force. It is deposited by employing electroplating techniques. The bath composition used for this purpose is presented by Ref. [32]. A basic sulfate bath with a pH of 3.5 was used. The electroplating was carried out at room temperature (25°C) at a current density of 5 mA/cm^2. With a plating rate of about 6 μm/h, it takes less than an hour to plate 5 μm.

31.5 Testing and Evaluation of Cantilevers

The testing and evaluation of the fabricated cantilevers consisted of two parts. The first part of the testing used a bimorph piezocantilever to determine the piezocoefficient, which ensured that the value used for the PVDF is realistic. The second part of the evaluation mainly compared the deflection obtained by the analytical and numerical (FEA simulation) methods with those obtained experimentally using a fabricated unimorph composite cantilever.

31.5.1 Determination of Piezocoefficient of PVDF Using a Bimorph Cantilever

The piezocoefficient of the PVDF polymer was verified using a bimorph cantilever having a length of 6 mm and a width of 1 mm. The bimorph constitutes two PVDF punched cantilevers from a 28 μm thick PVDF sheet and attached together. When an AC voltage was applied to this test structure, the bimorph vibrated like a fan. The measured resonance frequency was found to be about 600 Hz. The tip deflections of the cantilever were measured under an optical microscope. The cantilever was excited by an oscillator. The variation of tip deflection with the applied voltage was measured in this experiment is shown in Figure 31.7. The bottom curve gives the tip deflection of the cantilever as a function of the applied voltage at 50 Hz, while the upper pink curve indicates the deflection when exciting at its resonance frequency. It maybe noted that the deflection at 50 Hz excitation is similar to the values obtained with pure DC excitation. Due to a Q-value larger than 1, the amplitude at the resonance frequency is about five times higher for small voltages. At high voltages, the amplitude does not show a linear dependence any more, which may be due to a shift of the resonance frequency for high deflections. When using the linear dependence of the deflection on the voltage in the low-frequency case, the piezo strain coefficient d_{31} of the PVDF can be calculated using the following equation:

$$\delta = \frac{3}{4} d_{31} \frac{l^2}{t^2} V$$

Where

l is the cantilever length
t is the thickness of a single PVDF sheet
V is the applied voltage

By substituting the deflection observed in the above equation, d_{31} was obtained by equation as 25×10^{-12} m/V, which is exactly the same value as the one used in the simulation.

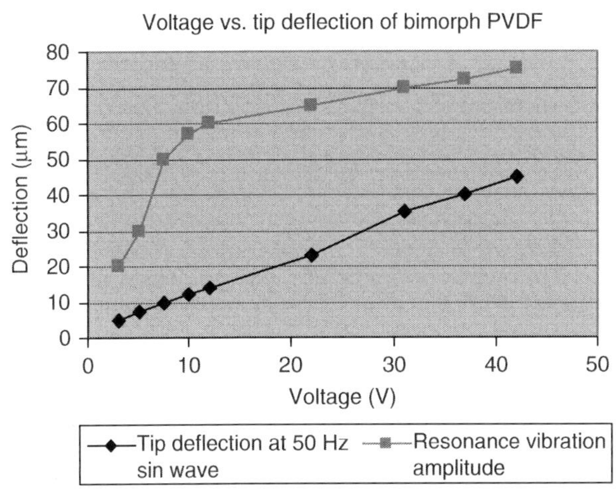

FIGURE 31.7 Bimorph cantilever deflection as a function of applied voltage for different frequencies.

31.5.2 Testing of Unimorph Cantilever

The dimensions of the unimorph cantilever used in this experiment were 6 × 1 mm with a PVDF sheet thickness of 28 μm having a 5 μm thick electroplated permalloy layer on one side. The cantilever was actuated by a DC voltage in the range from 60 to 180 V. The tip deflection of the cantilever was observed in each case under an optical microscope mounted with a micrometer. Three samples with the same nominal dimensions were measured to ensure the repeatability of the results. The deflections observed with each of these beams are shown in Table 31.1.

To ensure the reproducibility of the readings, deflections were measured in each case at least twice (switching the voltage off and on) and the range of the values obtained is provided in Table 31.1. It may be noted that the observed deflection increases four times when the applied voltage is tripled. Also the values from beam to beam were very consistent.

The variation of the average tip deflection as a function of the applied voltage is plotted in Figure 31.8, along with the deflection values calculated analytically and using finite element method (FEM)s analysis for a beam length of 6 mm, width of 1 mm, and thicknesses of PVDF and permalloy 28 and 5 μm, respectively.

In general, the deflection of the cantilever is linearly proportional to the applied voltage. These results showed that the values of the deflection obtained through analytical calculations are in excellent agreement with the numerical values obtained through CoventorWare simulation using finite element methods. However, the deflections observed experimentally were 20% less than the values obtained through analytical numerical computations. This difference between the experimental and computed values maybe attributed to a combination of uncertainties in the thickness measurement, nonuniformity of the electroplated permalloy layer, residual stresses within the electroplated permalloy film, and a difference in the Young's modulus of the electroplated permalloy compared to the bulk value, which is used in the simulation.

The electroplated layers are, in general, known to contain residual stress essentially due to the electrochemical kinematics occurring during the plating process [33]. The plating parameters such as current density, bath concentration, pH, and temperature are known to contribute to the residual stress

TABLE 31.1 Measurement of Tip Deflections of a Composite Cantilever (μm)

Potential (V)	Beam 1	Beam 2	Beam 3
60	35–40	40–45	40–45
100	90–100	85–95	90–95
120	105–115	100–110	105–115
150	130–135	130–145	135–140
180	145–155	150–160	150–160

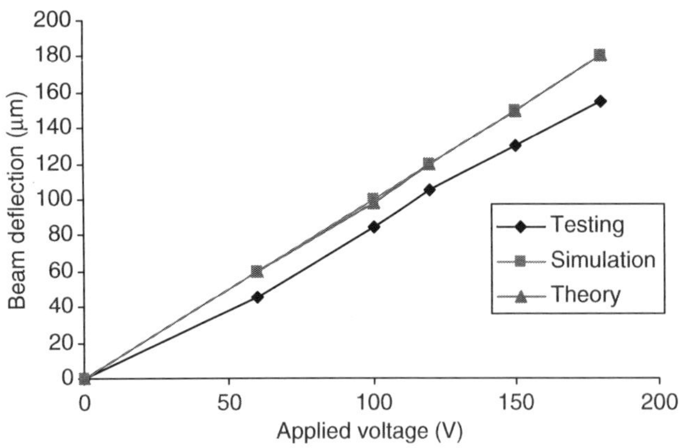

FIGURE 31.8 Comparison of experiment and simulation.

developed during the deposition process. In addition, residual stress in electroplated permalloy films is also sensitive to their iron content, beyond certain values of the nickel to iron ratio. Different techniques have been used to measure the internal stress during electroplating [33,34]. However, it is hard to find a process variable that does not influence internal stresses of the deposited layers. In this work, we found that current density is a critical parameter with respect to the reduction of residual stress. As the current density increased the plating rate also increased at the same time increasing the surface internal stress. Therefore, in order to minimize the stresses a lower current density (5 mA/cm^2) was used instead of 10 mA/cm^2 during the electroplating of the permalloy layer. However, it is extremely difficult to deposit completely stress-free films. These residual stresses within the electroformed permalloy layer might have contributed to some extent to the beam deflection observed in the experiment. It maybe noted that both the simulation and the analytical calculations assumed a stress-free metallic layer.

The Young's modulus of the nonpiezo layer used in the simulation might be another possible cause for the difference in deflection observed in the experiment. The Young's modulus of the nonpiezo layer used in the simulation is mainly a bulk value for permalloy with a nominal composition of 81% Ni and 19% iron. The Young's modulus of the electroplated could be different from the bulk value reported in the literature. All of these factors might have together contributed to the deviation of the experimental cantilever deflection from the simulation. This certainly requires further investigations to determine and identify the specific reasons for such a large difference.

31.6 Summary

A piezoelectric polymer composite unimorph cantilever was designed and optimized by finite element method modeling. The unimorph cantilever consists of a piezoelectric polymer PVDF and a nonpiezo elastic layer. The effect of various parameters on the tip deflection of the cantilever was analyzed. These studies showed that an increase in the thickness of the piezoelectric layer reduced the tip deflections of the composite cantilever. Thickness of the nonpiezo layer affected the tip deflection dramatically. In order to apply this unimorph cantilever as a component of a later hybrid actuator, which combines electromagnetic actuation along with the piezoelectric cantilever, permalloy was chosen as the material for the nonpiezo layer of the unimorph cantilever. Considering the above results and the different aspects of fabrication, a 28 μm thick PVDF layer with a 5 μm permalloy has been selected for this work. The tip deflection of the cantilever is found to be proportional to the square of its length.

The details of the fabrication routes followed in making the PVDF cantilevers are described. Two types of cantilevers, unimorph and bimorph, were fabricated and tested. The cantilevers were punched using electroformed nickel shims, which was laser micromachined. The bimorph cantilevers were formed by attaching two PVDF cantilever together, whereas the unimorph cantilevers constituted a single PVDF cantilever with one side having a 5 μm thick electroformed permalloy layer. The effect of microstructures on the deflection of cantilever beams was also investigated by fabricating them using microembossing techniques. The unimorph composite cantilever has generated a 70 μm tip deflection at an excitation voltage of 100 V. The measured results of the cantilever have shown reasonable agreement with the values obtained by analytical and FEA simulation.

Punching is a promising way to shape and fabricate polymer films, which are flexible and sensitive to temperature. Furthermore, it can be used favorably to shape a surface microstructure on a PVDF polymer film.

References

1. Fan, L., Tai, Y., and Muller, R.S., IC-processed electrostatic micromotors, *Sensors and Actuators* 20: 41–47, 1989.
2. Fujita, H., Microactuators and micromachines, *Proceedings of the IEEE* 86: 1721–1732, 1998.
3. Robertson, C., Bimorph drivers-an electric solution, Pennwalt (Europe) 79: 1–8, 1979.
4. Todals, M. et al., A new electromotional device, *RCA Engineer* 25: 24–27, 1979.
5. DeVoe, D.L., Piezoelectric thin film micromechanical beam resonators, *Sensors and Actuators A* 88: 263–272, 2001.
6. Elvin, N.G., Elvin, A.A., and Spector, M., A self-powered mechanical strain energy sensor, *Smart Materials and Structures* 10: 293–299, 2001.
7. Kueppers, H. and Leuerer, T., PZT thin films for piezoelectric microactuator applications, *Sensors and Actuators A* 97–98: 680–684, 2002.
8. Zurn, S. et al., Fabrication and structural characterization of a resonant frequency PZT microcantilevers, *Smart Materials and Structures* 10: 252–263, 2001.
9. Lau, S.T. et al., Piezoelectric composite hydrophone array, *Sensors and Actuators A* 3145: 1–7, 2001.

10. Lee, G. and Chen, S.H., Microfabricated plastic chips by hot embossing methods and their applications for DNA separation and detection, *Sensors and Actuators B* 75: 142–148, 2001.
11. Schellin, R. and Hess, G., A silicon subminiature microphone based on piezoresistive polysilicon strain gauges, *Sensors and Actuators A* 32: 555–559, 1992.
12. Roberts, D.C. and Li, H., A piezoelectric microvalve for compact high-frequency, high-differential pressure hydraulic micropumping systems, *Journal of Microelectromechanical Systems* 12: 81–92, 2003.
13. Yoseph, B.C. and Chang, Z., Piezoelectrically actuated miniature peristaltic pump, Proceedings of SPIE's 7th Annual International Symposium on Smart Structures and Materials, Newport, March 1–5, 2000.
14. Dargahi, J., Parameswaran, M., and Payandeh, S., A micromachined piezoelectric tactile sensor for an endoscopic grasper—theory, fabrication and experiments, *Journal of Microelectromechanical Systems* 9: 329–335, 2000.
15. Li, L.X. and Shen, Y.P., The optimal design of piezoelectric actuators for acoustic control, *Smart Materials and Structures* 10: 421–426, 2001.
16. Ruprecht, R. et al., Molding of LIGA microstructures from fluorinated polymers, *Microsystem Technologies* 2: 182–185, 1996.
17. Jenkins, D.F.L. et al., The use of sputtered ZnO piezoelectric thin films as broad-band microactuators, *Sensors and Actuators A* 63: 135–139, 1997.
18. Friese, C., Goldschmidtboing, F., and Woias, P., Piezoelectric microactuators in polymer-composite technology, Proceedings of Transducers'03, The 12th International Conference on Solid State Sensors, Actuators and Microsystems, Boston, MA, June 8–12, pp. 1007–1010, 2003.
19. Sun, D. and Mills, J.K., Control of a rotating cantilever beam using a torque actuator and a distributed piezoelectric polymer actuator, *Applied Acoustics* 63: 885–899, 2002.
20. Xu, T.B., Cheng, Z.Y., and Zhang, Q.M., High performance micromachined unimorph actuators based on electrostrictive poly (vinylidene fluoride-trifluoroethylene) copolymer, *Applied Physics Letters* 80, 2002.
21. www.msiusa.com
22. Izumi, Y. et al., Irradiation effects of excimer laser light on poly(vinylidene fluoride) (PVDF) film, *Bulletin of the Chemical Society of Japan* 71: 2721–2725, 1998.
23. Katan, E., Narkis, M., and Siegmann, A., The effect of some fluoropolymers' structures on their response to UV irradiation, *Journal of Applied Polymer Science* 70: 1471–1481, 1998.
24. Manohara, H.M. and Morikawa, E., Pattern transfer by direct photo etching of poly(vinylidene fluoride) using x-rays, *Journal of Microelectromechanical Systems* 8: 417–422, 1999.
25. Duca, M.D., Plosceanu, C. L., and Pop, T., Effect of x-rays on poly(vinylidene fluoride) in x-ray photoelectron spectroscopy, *Journal of Applied Polymer Science* 67: 2125–2129, 1998.
26. Ruprecht, R. et al., Molding of LIGA microstructures from fluorinated polymers, *Microsystem Technologies* 2: 182–185, 1996.
27. Fu, Y., Harvey, E.C., Ghantasala, M.K., and Spinks, G.M., Design, fabrication and testing of piezoelectric polymer PVDF microactuators, *Smart Materials and Structures* 15: S141–S146, 2006.
28. Fujitsuka, N., et al., Monolithic pyroelectric infrared image sensor using PVDF thin film, *Sensors and Actuators A* 66: 237–243, 1998.
29. Khoo, M. and Liu, C., Micro magnetic silicone elastomer membrane actuator, *Sensors and Actuators A* 89: 259–266, 2001.
30. Becker, H. and Gartner, C., Polymer microfabrication methods for microfludic analytical applications, *Electrophoresis* 21: 12–26, 2000.
31. Simdekova, I. et al., A study of hot embossed microchannels using confocal microscopy, Proceedings of SPIE Nano- and Microtechnology: Materials, Processes, Packaging, and Systems, Melbourne, Australia, Dec. 2002, pp. 82–92.
32. Ghantasala, M.K. et al., Design and fabrication of a micro magnetic bearing, *Smart Materials and Structures* 9: 235–240, 2000.
33. Stein, B., A practical guide to understanding, measuring and controlling stress in electroformed metals, AESF Electroforming Symposium, Las Vegas, March 27–29, 1996.
34. Hyoung, J.C. and Ahn, H.C., Magnetically-driven bi-directional optical microscanner, *Journal of Micromechanics and Microengineering* 13: 383–389, 2001.

32

Ultrasonic Nondestructive Testing and Materials Characterization

John A. Brunk
National Nuclear Security Agency's Kansas City Plant

32.1 Scope and Purpose

Sound and ultrasound have a tremendous variety of industrial and scientific applications in materials characterization, flaw detection, and measurement. Although the same basic principles always apply, there are huge differences in the practical details. There is continuous improvement in equipment, and new techniques are always being developed to the point of being practical for routine use. In 2006, it is impossible to give a broad view of the technology that will be up-to-date for even a few months. There are many good sources of basic information, such as *The NDT Handbook* [1] Rather than duplicate available training material, this chapter will describe a few things that can be accomplished with ultrasound, with some explanation of why they need to be worked with some caution. Some differences between flaw detection and material characterization methods are described. Ultrasonic evaluation is only one of many nondestructive methods for these applications. It is sometimes necessary to utilize two or more to obtain all of the desired information. An example of using x-ray radiography as a complementary technique is included.

32.2 Sound Frequency Range for Materials Characterization and Testing

One definition of sound is mechanical vibrations transmitted by an elastic medium. We generally think of sound as what we can hear, from 15 or 20 Hz or cycles/second up to as high as 20,000 Hz (20 kHz) for normal young people. The upper limit tends to decrease with age, more for men than for women. A middle-aged person may have an upper limit of 12 to 14 kHz. "Ultrasound," above the frequency of human hearing, is generally considered to be anything above about 20 kHz.

Applications for the lower frequencies are limited or indicated by the fact that some animals hear (or transmit and detect) higher frequencies than humans. These include dolphins, porpoises, whales, bats, and some species of frogs. Ultrasonic burglar alarms that operate in the 20–30 kHz range are not suitable for homes with dogs. Elephants communicate using infrasound (frequencies too low for humans to hear) down to as low as about 14 Hz.

Frequencies from as low as 3 kHz (sonic) up to at least 250 MHz are used for flaw detection, material property evaluation, measuring, and process monitoring. The most common industrial application maybe testing structural welds using frequencies from 1 to 5 MHz.

32.3 Flaw Detection Practices

Ultrasonic flaw detection is required by many industrial codes and specifications, such as the ASME Boiler and Pressure Vessel Code [2] and the American Welding Society's D1.1/D1.1M Structural Welding Code-Steel [3]. These documents provide inspection and acceptance criteria for specific product forms. They also include requirements for training, experience, examination, and certification of personnel who perform ultrasonic and other nondestructive tests. Personnel requirements are made by reference to other publications. In the United States, these are usually SNT-TC-1A Personnel Qualification and Certification in Nondestructive Testing [4] or NAS 410, NAS Certification & Qualification of Nondestructive Test Personnel [5].

Inspector qualification is beyond the scope of this discussion. More information can be obtained from the referenced documents and their publishers. Appropriate and sufficient training is essential for successful evaluation of product forms and materials to their specifications. However, training directed toward specific types of test objects does not necessarily cover information important for other applications.

32.4 Flaw Detection References

Flaw detection to determine whether a product or material meets acceptance criteria requires some definitions before any testing is initiated:

1. "Flaw" must be defined. "Discontinuity" is a term often used for an inhomogeneity that has been detected and must be evaluated in accordance with requirements. Definitions maybe general such as "linear indications" and "rounded indications" or specific such as (in welds) "slag inclusions" and "lack of fusion."
2. Smallest discontinuity or test signal to be evaluated must be stated. This maybe an actual dimension, but in ultrasonic testing, it is usually an indication that gives a certain signal amplitude compared to a reference reflector.
3. Portion of the test object to be evaluated should be stated. This maybe the entire volume, or only part, such as what will remain after subsequent machining.

Most references are not real flaws. One reason for this is a need for standardization. No two real flaws would be alike. The most common references used to simulate defects are side-drilled holes (with the sidewall of the hole as the reflector) and flat-bottom holes (with the flat hole bottom as the reflector). Others include v-notches and u-notches in the material surface. Many specifications define reference requirements [6].

Some applications require special references to provide a reasonably realistic approximation of defects that are of concern. There are specifications covering fabrication of references in solid composite laminates [7] and in composite honeycombs [8].

An important limitation to real flaws is that they often can be fully characterized only by cross-sectioning or some other destructive means, leaving nothing to be used as a reference in the future. There are exceptions, such as in Example 1 below.

32.5 Basic Equations

Ultrasound can be used to determine variations in bulk properties or localized changes within a material sample. Wave propagation is related to the elastic properties of the propagating medium. Longitudinal waves are those where particle displacement is in the direction of propagation. They may exist in solids, liquids, and gases.

Shear or transverse waves are those where particle motion is transverse to the direction of propagation. They may exist in solids, and in a few very viscous liquids, but not in gases. For bulk materials (when the dimensions of the material are many times larger than the wavelength of the ultrasound) the relationships are

Young's Modulus of Elasticity, $E = [\rho v_s^2 (3v_l^2 - 4v_s^2)]/(v_l^2 - v_s^2)$

where

ρ = density, kg/m^3 [or lb/in.^{3}]
v_l = longitudinal velocity, m/s [or in./s]
v_s = transverse velocity, m/s [or in./s]
E = Young's modulus of elasticity, N/m^2 [or lb/in.^{2}]

Poisson's ratio, $\sigma = [1 - 2(v_s/v_l)^2/2[1 - (v_s/v_l)_2]$

where

σ = Poisson's ratio
v_s = ultrasonic transverse velocity, m/s [or in./s]
v_l = ultrasonic longitudinal velocity, m/s [or in./s]

Shear modulus, $G = \rho v_s^2$

Bulk modulus, $K = \rho[v_l^2 - (4/3)v_s^2]$

Surface waves may travel on the surfaces of bulk materials, and also may exist in thin layers. There are several types of surfaces waves, but the type most often used in ultrasonic testing is a Rayleigh wave. These have an elliptical particle motion in the vertical plane. Their velocity is typically about 90% of the longitudinal wave velocity in the material.

From these relationships, it can be seen that measurements of ultrasonic velocity can be used to indicate variations in material properties. A very important parameter affecting the transmission and reflection of ultrasound at boundaries where material properties change is "acoustic impedance." This is the product of density and wave velocity:

Acoustic impedance, $z = \rho v_l$

where z is the acoustic impedance (kg/m^2 s [or lb/in.^{2} s]).

The percentage of incident power that is reflected from and transmitted across a boundary is given by

Reflection coefficient for energy (R): $R = (Z_2 - Z_1)^2/(Z_2 + Z_1)^2$

where
Z_1 is the acoustic impedance in medium 1
Z_2 is the acoustic impedance in medium 2

Transmission coefficient for energy (T): $T = (4Z_2Z_1)/(Z_2 + Z_1)^2$

The change in acoustic impedance at a boundary allows detection of discrete flaws such as gas porosity or cracks in metals, measurement of layers such as stainless steel weld cladding applied to low-alloy steel for corrosion resistance, and other types of discrete material changes within test objects.

32.6 Ultrasonic Velocity and Attenuation Measurements

There is an ASTM document covering ultrasonic velocity measurements with conventional pulse-echo ultrasonic flaw detectors [9]. It also includes methods that require more specialized equipment and can produce more accurate measurements. There are other documents covering specialized methods; for example, for measuring velocity and measuring attenuation in advanced ceramics [10,11].

32.7 Sources of Ultrasound

Most ultrasonic transducers use piezoelectric elements to convert electrical impulses into mechanical vibrations and vice-versa. Reference [1] includes a discussion of piezoelectric transducers and also of alternatives such as electromagnetic acoustic transducers and laser-generated ultrasound. In general, piezoelectric transducers used with immersion techniques or in contact with some sort of liquid coupling are used unless there is some special reason to do otherwise. For example, when it is desired to test metal while it is very hot or when a material is so porous, its properties would be altered by liquid intrusion [12].

Piezoelectric transducers are made from several types of materials and constructed in various ways for different applications. A good transducer catalog is full of useful information, and some manufacturers can provide a great deal of assistance in selecting the right transducer for a specific application. This is too complex a subject to try to cover here. In Example 1, a set of self-coupling contact transducers was especially made for the application. In Example 2, it can be seen that two different immersion transducers produced very different signals from a steel sample.

Some recent developments include high-efficiency air/gas coupled transducers that can eliminate the need for any other couplant for some applications [13] and phased arrays where selective sequential pulsing of individual elements in an array can be used to "steer" the resulting ultrasound beam [14]. NDT.net is an excellent resource for following new developments in ultrasonic and other NDT methods. It gives access to papers from many technical conferences.

32.8 Couplants

Some sort of material must be used between the transducer face and the entry surface of the test object. Most contact transducers have hard surface wear plates. The purpose of a couplant is to reduce the acoustic attenuation mismatch that would exist with an air gap (although air-coupled testing is definitely possible for some applications). Suitable couplants for most contact tests include light-grade oil and glycerine. For transverse wave testing, there are two types of methods. Direct contact shear wave transducers produce a transverse wave traveling perpendicular to the transducer face. This is the same path followed by a longitudinal wave from the same sample surface location. The best practice is to have both beams follow the same path, especially for any materials that may be anisotropic. Transverse waves may be coupled with high-viscosity liquids such as honey, or with slightly compressible solids. For either type of wave, liquid intrusion into porous materials must be avoided. Dry couplings may be an integral part of the transducer (instead of a hard face) [15], and air-coupling is possible for some materials [16].

Transverse waves may also be produced by refraction. When the angle of incidence at the test surface is not zero (the center of the incoming beam is not perpendicular to the test surface) the direction of ultrasound entering the test material is given by Snell's Law:

$$\frac{\sin \phi}{\sin \phi} = \frac{VLW}{VM}$$

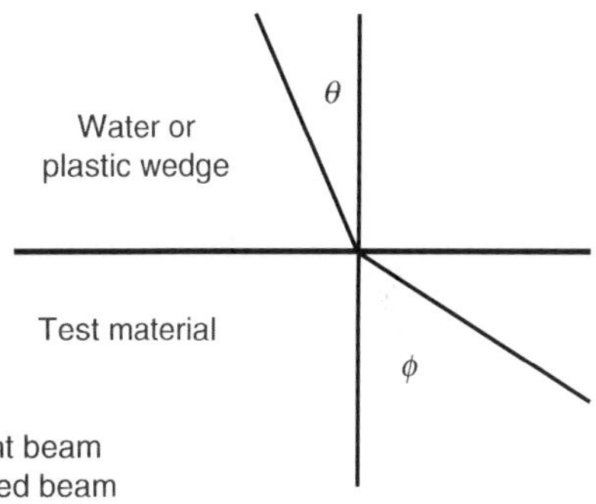

where
θ = angle of incident beam
ϕ = angle of refracted beam
VLM = velocity of longitudinal waves in water or wedge
VSM = velocity of shear or longitudinal waves in material under examination

The shear wave angle is always smaller than the longitudinal wave angle. When the calculated angle increases to 90°, the longitudinal wave becomes a surface wave. By varying the incident angle, data can be taken in multiple directions. This is easier with water coupling compared to using a number of plastic wedges giving different angles.

Water coupling is also more convenient for scanning large test objects and collecting data during the scan. Scanners are available for contact testing. When using longitudinal waves with a thin layer of liquid couplant, actual scanning is possible. For shear wave transducers requiring a very viscous couplant, and with self-coupling transducers, the transducer must be picked up and put down for each data point.

32.9 Complications

Characterization of materials must be approached with caution because the most readily measured quantities, velocity, and attenuation, are functions of several variables. In addition, the relationships to variables change under different conditions. For example, "all else being equal" the attenuation of ultrasound passing through a particular material increases with increasing frequency. That is, higher frequencies are attenuated more than lower. But the relationship between attenuation and frequency is not a simple one. In a polycrystalline material, where the wavelength of ultrasound is much greater than the average grain diameter, attenuation is proportional to the fourth power of the frequency. When the wavelength is approximately equal to the grain diameter attenuation is proportional to the square of the frequency. In between these two, there is a continuous change. To further complicate matters, "test frequency" is not a single value but a frequency range. Transducers may produce broader or narrow bandwidths according to their construction and how they are excited, and receivers may be broadbanded or tuned. In many cases, the received signal has lost relatively more of its higher frequency components. Beam spread is also a part of total attenuation, and this is dependent upon transducer size and frequency as well as material properties. A very good discussion of the complexities of ultrasonic materials characterization is given by Papadakis [17].

Ultrasound can still be a very valuable tool for characterizing or comparing materials. Attenuation as a function of frequency can be compared for samples of the same alloy with different grain sizes even though there is not one simple correlation function for the entire range of sizes studied. Ultrasonic measurements are effectively used for a very wide assortment of applications. To name just a few:

- Characterization of processes for freezing foods [18]
- Evaluation of cheese as it matures [19]
- Characterization of graphite particles in cast iron [20]
- Evaluation of density variations in green and sintered ceramics [21]
- Monitoring grain size while producing seamless metal tubing [22]

It is important to note that material characterizations are comparative, just as are flaw detection methods. That is, quantitative results are produced only when samples are compared with materials with known properties determined by some other method.

32.10 Simplifications

Useful applications require limiting the numbers of variables and/or unknowns in the test objects in addition to proper selection of equipment. Because mathematical models can become extremely complicated and must take some variables as constants, they can at best give the investigator a good start in the right direction. It is also important to avoid adding unexpected variables with the equipment used.

It is well understood that ultrasonic attenuation and grain scattering in metals is influenced by grain size and shape, and will vary with the propagation direction in samples that have been worked in a manner that produces a preferred grain orientation. Anisotropy may also occur in materials where it might not be expected, such as in blocks of polystyrene. Whenever possible, it is a good idea to prepare test references from material that is as close as possible to the samples to be evaluated, and to consider that ultrasonic properties may be directional. Howard and Enzukewich reported an excellent practical example of dealing with the effects of microstructure on ultrasonic attenuation of alloy steels such as 4330 and 4340 [23]. Normalizing a hardened alloy steel reference was found to effectively reduce ultrasonic attenuation and "noise" to about that from fully annealed material to be tested.

AWS D1.1/D1.1M covers ultrasonic examination of welds in a limited range of structural steels. Material characterization is not an issue, so several test simplifications are possible. Flaw detectors are required to have certain operating characteristics. Dimensions and nominal frequencies of transducers are prescribed. Standard test blocks are defined for distance and beam angle calibration. The same blocks have holes to provide reference signals for discontinuity evaluation. With specific material and transducer characteristics, attenuation of ultrasound by the material is taken to be a constant 2 dB/in. after the first inch. Only one reference hole is used, and this attenuation factor is used to compensate for actual metal paths. The intent is that everyone using approved equipment and procedures will be able to accept welds that meet code requirements and reject any indications that do not.

The ASME Boiler and Pressure Vessel Code covers a much wider range of materials and the requirements for ultrasonic examinations apply to low-alloy steels, various stainless steels, and alloys such as Monel and Inconel. Calibration blocks are required to be of the same material to be tested and where applicable in the same heat treat condition. Instead of calculating equivalent hole signal amplitudes for different distances, several side-drilled holes are used to determine the distance–amplitude relationship for the same size hole in different locations. There is more latitude in the choice of transducers, but the distance–amplitude relationship must be determined for each transducer used. The appropriate setup for material characterization can be quite different from one optimized for flaw detection When straightforward flaw detection procedures are performed as specified, there can be times when material property changes other than the "defects" of interest create unusual results that must be explained.

32.11 Example 1: Real Defects Used to Establish and NDT Procedure

This application was unusual because it employed through-transmission over a path with two 90° corners, and special dry-coupling 4 MHz transducers. It was developed to test small U-shaped ferrite cores to determine if they had cracked in the course of being potted and then incorporated into an assembly with a stainless steel housing. They were ground flush with the steel surface of the housing. After potting, it was still possible to radiograph the cores for cracks, but x-ray was ineffective for the assemblies. Figure 32.1 shows parts before and after potting.

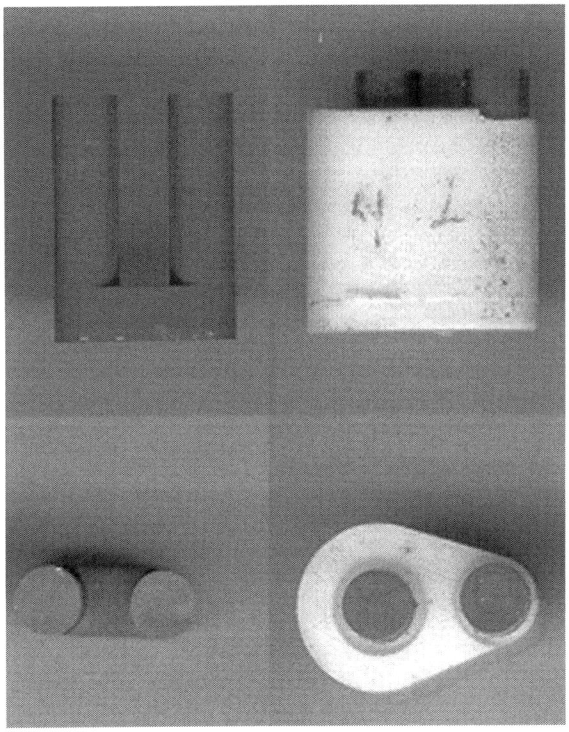

FIGURE 32.1 Parts before and after potting.

The circular ends are 0.200 in. (5.1 mm) in diameter with a center-to-center spacing of 0.350 in. (8.9 mm).

A number of potted cores were radiographed until some with cracks were identified. Figure 32.2 is a radiograph that shows a completely fractured leg. Figure 32.3 shows a less severe leg crack near its junction with the bridge between the legs. Cracks in these locations should be detectable by a pulse-echo ultrasonic test with a transducer coupled to the round end of each leg in turn. Figure 32.4 shows a crack in the bridge. This was of particular interest because this crack could not be detected by a simple pulse-echo test. It was used to develop a through-transmission test to detect a crack in the bridge. A fixture was made to hold a pair of dry-coupling contact transducers for a through-transmission test. Transmission

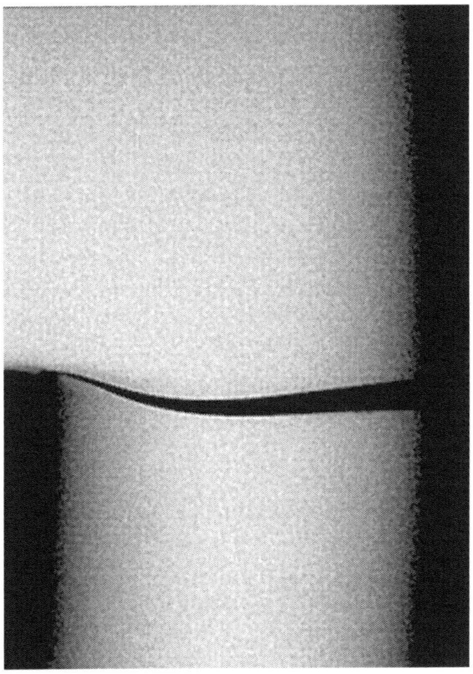

FIGURE 32.2 Radiograph shows fractured leg.

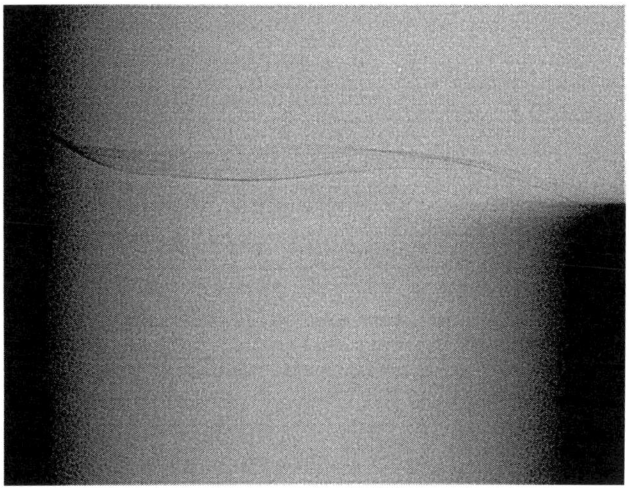

FIGURE 32.3 Less severe crack shown in radiograph.

FIGURE 32.4 Cracks in the bridge shown by through-transmission test.

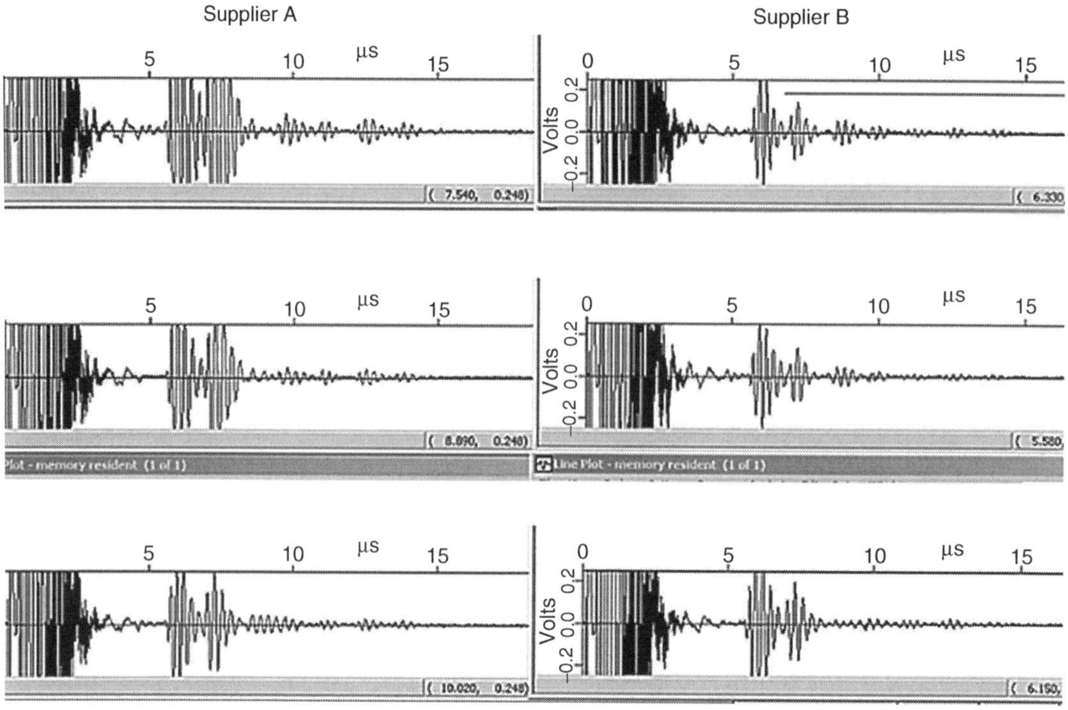

FIGURE 32.5 New cores not potted.

through an undamaged core was easy. The reflections and redirections between the transmitter to the receiver would be extremely difficult if not impossible to model. It was sufficient to demonstrate that it could be done with repeatable results in the absence of a defect.

To establish repeatability, a number of similar parts from two suppliers were examined. Figure 32.5 shows some typical through-transmission time–amplitude displays of cores that had not been potted. Cores from supplier B gave consistent responses. Those from supplier A were substantially more variable. They were also found to vary more in density and in other properties of interest.

A number of parts were tested after potting and then put into assemblies. Being in assemblies did not significantly change the ultrasonic response, indicating the cores had not been damaged. Some assemblies were subjected to mechanical shocks that might crack the cores and tested again. A few were found to have cracked. Figure 32.6 shows the difference between cores considered good and the sample shown in Figure 32.4.

32.12 Example 2: Material Property Variation Shown during "Flaw Detection"

A 304L stainless steel round bar stock was tested with a water immersion pulse-echo technique with the ultrasound beam directed radially into the bar. Reference bars with flat-bottom holes (drilled radially to various depths from the beam entry surface with flat bottoms parallel to a plane tangent to the entry point) are made for each lot of material. Discontinuity indications are compared with a calibration curve of amplitude versus depth for a specific flat-bottom hole diameter. The amplitude of the back surface reflection is also monitored. Unfavorably oriented flaws could reflect energy so that it is not received by the transducer, but still reduce back surface reflection. References for larger bars are often made with holes drilled no deeper than the radius so the "back half" of the sound path through the bar is tested.

The upper part of Figure 32.7 shows a time–amplitude signal typical of most material in a particular lot of 4 in. (102 m) diameter bar. The lower part shows a higher "noise" signal level and lower back reflection that aroused suspicion. The transducer used had been found to be suitable based on detection of reference holes in a bar from the same lot. It had a nominal center frequency of 5 MHz, 0.75 in. (19 mm) diameter, spherical beam focus with a focal length in water of 6 in. (152 mm).

The suspicious region was reexamined using a transducer with a nominal center frequency of 10 MHz. Compared to the 5 MHz unit, it had a much lower energy output and a much broader bandwidth. It had an element diameter of 0.4 in. (10 mm), spherical focus with a focal length in water of 1.4 in. (35 mm). The resulting display is shown in Figure 32.8. This clearly shows a reflector at a depth where the lower part of the previous figure has what appears to be part of the front surface "noise." This is an example of a large difference in response with a test frequency change.

A slab with a length of about 0.2 in. (5 mm) was cut from the suspicious part of the bar and scanned with ultrasound directed

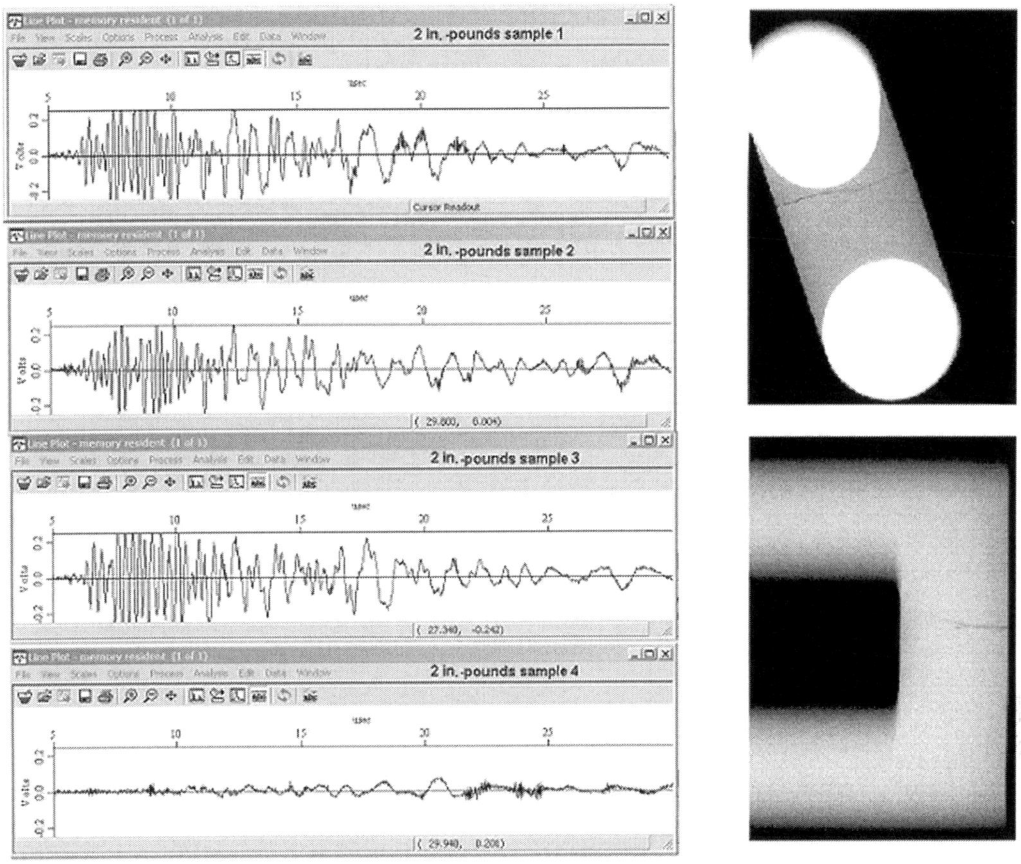

FIGURE 32.6 Top three are good cores in assemblies. Bottom is the core in Figure 32.4.

in the axial direction. Figure 32.9 shows a plan view plot of the back reflection amplitude. This scan was made with a 20 MHz transducer to increase amplitude variations. There is a very definite ring of lower amplitude dividing the sample into what appears to be an inner and an outer section.

Figure 32.10 is a photograph of the beam entry surface after it was etched to bring out the grain structure. The original ultrasonic test was responding to the well-defined change in grain structure. The original setup for detecting "real" defects, as represented by flat- bottom holes, was not capable of indicating this change except by increased "noise" and loss of back reflection. If the initial test had utilized a lower test frequency, the condition might not have been detected at all. If a "cored" structure had been suspected and testing for it had been part of the requirements, the material would probably have been tested at two different frequencies with different transducers. In fact, a uniform grain structure was required for this material.

32.13 Example 3: Filtering to Separate Discontinuities from Material Variations

Figure 32.11 shows responses from a 7.5 in. (190.5 mm) diameter 304 stainless steel bar with a cored grain structure similar to the one in the previous example. Longitudinal wave contact tests were done with a 2.25 MHz ceramic composite transducer, 0.375 in. (9.5 mm) in diameter. The upper part of the figure shows the response using a spike pulser and a broadband (0.1 to 10 MHz) amplifier. The front surface reflection noise obscures anything else within the first third of the sound path. The lower part of the figure was produced with the same transducer in the same location. The same pulser- receiver was used, but a tunable preamplifier/bandpass filter was added between the transducer and the receiver input. The filter was tuned to 2.25 MHz and the preamplifier set to 30 dB. The main receiver gain was reduced until the back reflection amplitude was the same as before. This reduced the front surface noise greatly, and a discontinuity within Gate 1 was clearly resolved. Frequency spectra of the back surface reflection indicate that removing frequencies below about 1.5 MHz eliminated most of the front surface noise. Tuning to a frequency much higher than 4 MHz resulted in excessive loss of back reflection. The back surface reflection shows that sufficient ultrasonic energy did make the round trip through the material. Proper tuning made it possible to detect side-drilled 0.020 in. (0.5 mm) reference holes throughout the material at distances from 0.5 in. (12.7 mm) to 7.0 in. (178 mm).

Pulse tuning, using a square wave or toneburst pulser can produce similar results by causing a broadband transducer to produce a narrow frequency spectrum with higher power

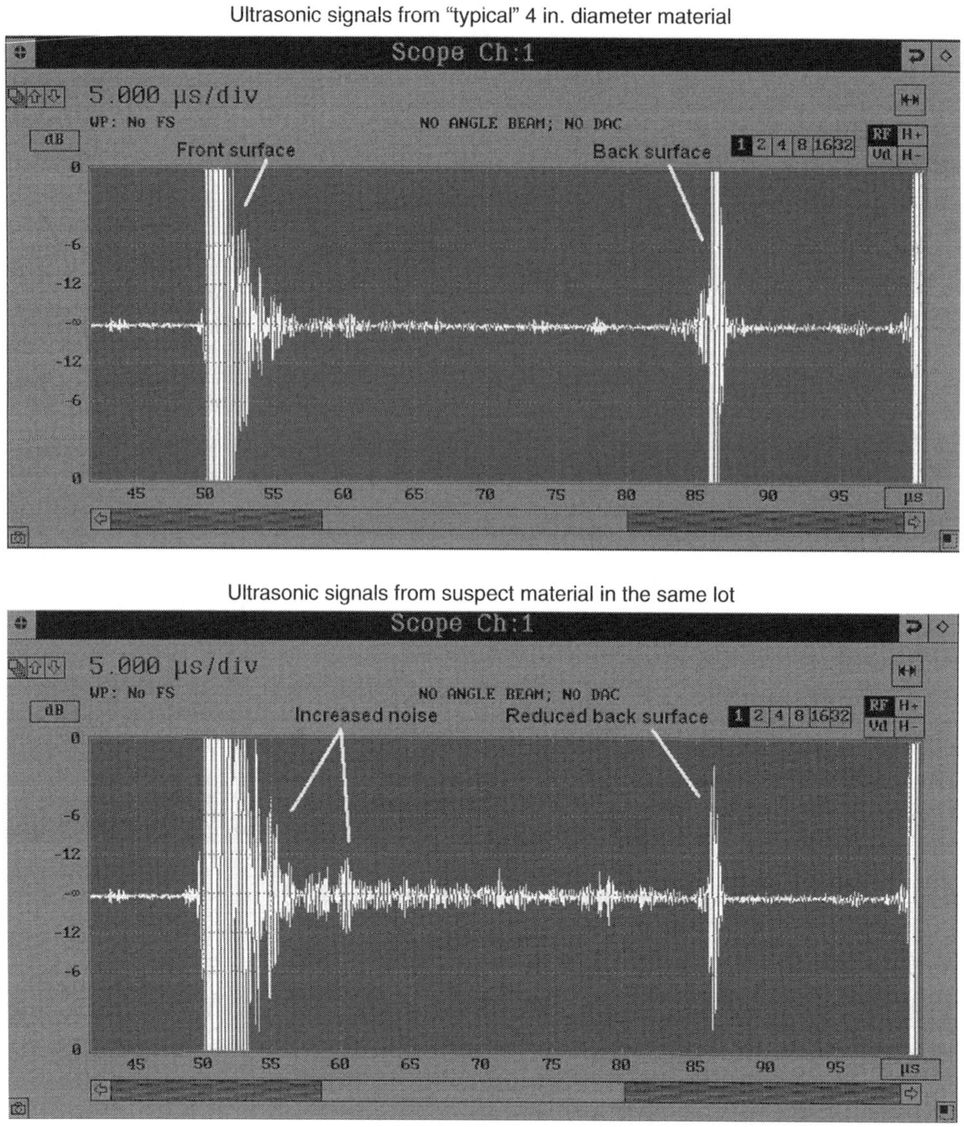

FIGURE 32.7 Time–amplitude signals from typical and suspect material, tested with the usual procedure using a medium-bandwith 5 MHz transducer.

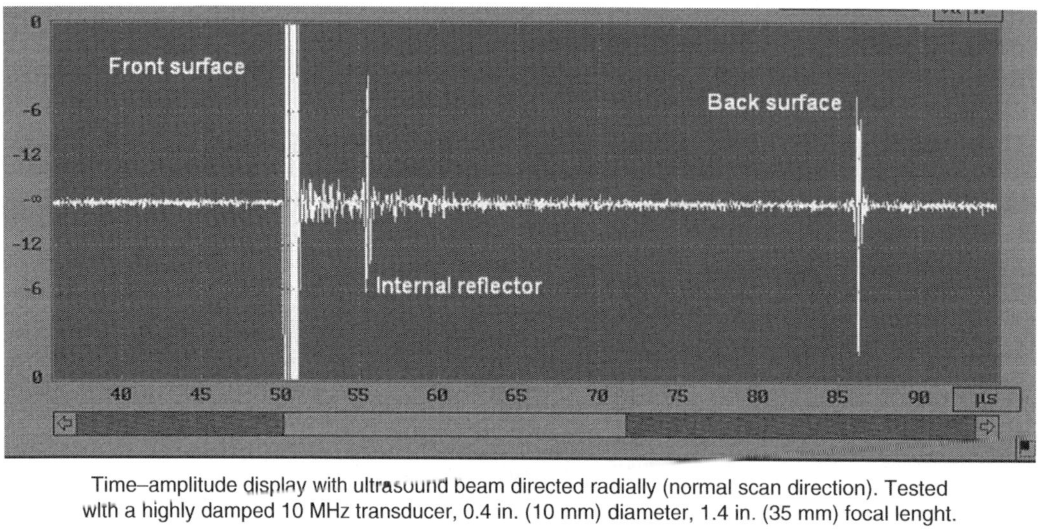

FIGURE 32.8 Display from same region of suspect material using a 10 MHz broadband transducer.

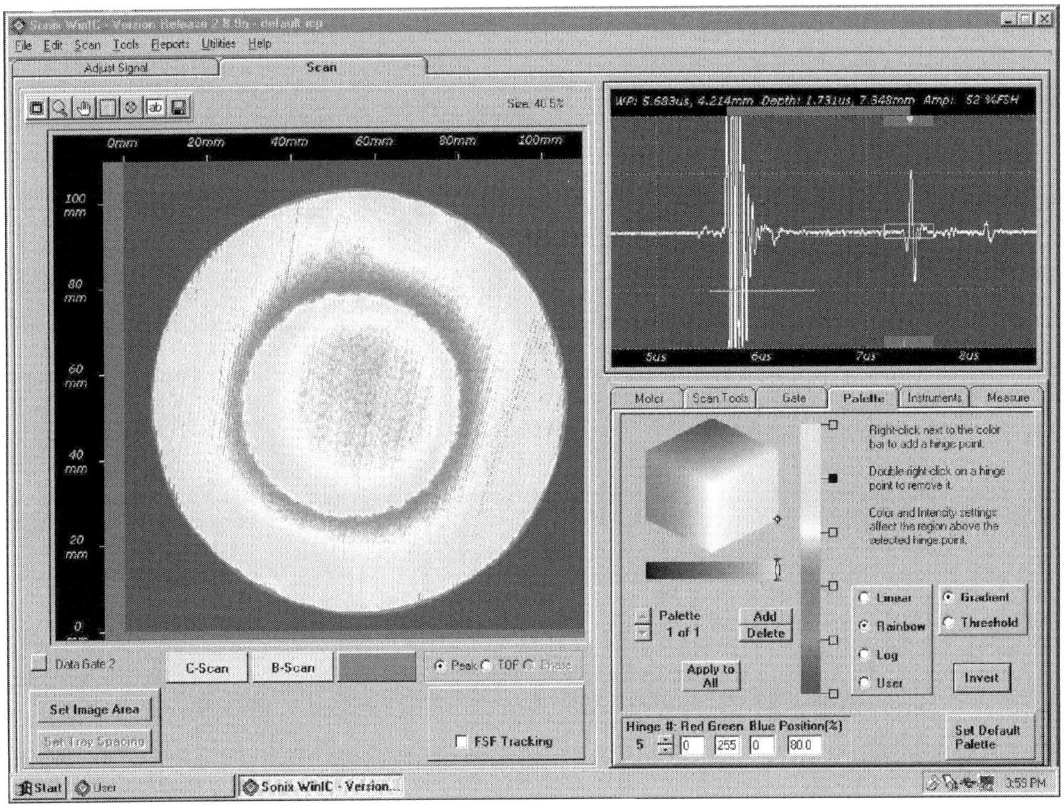

FIGURE 32.9 Plan view plot.

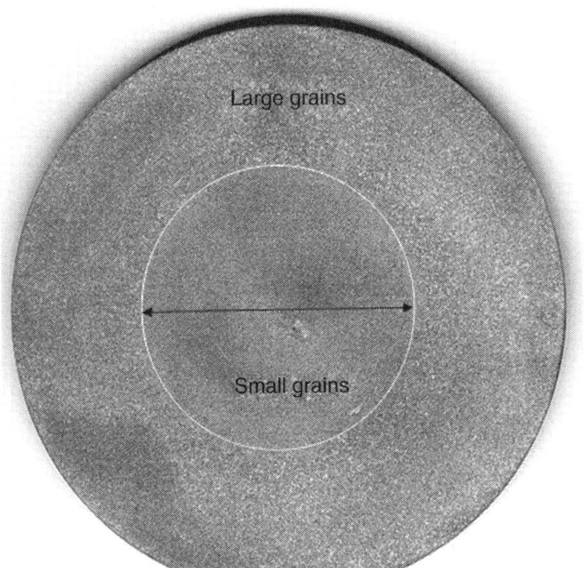

FIGURE 32.10 Photograph of beam entry surface.

compared to when a spike pulser is used. A broadband transducer used this way still has its full normal bandwidth as a receiver. Pulse tuning and receiver tuning can be used simultaneously. It is important to remember that techniques for removing unwanted noise signals may also remove information of interest when the purpose is material characterization instead of flaw detection.

32.14 Conclusion

Ultrasound provides a tremendous assortment of useful tools for flaw detection and material property characterization. It can be easy to go astray because there is no simple universal tool kit. Fortunately, a very large amount of useful information is readily available. It is important to understand how the characteristics of all the components of a test system can influence the results. These include transducer frequency, size, damping characteristics (bandwidth), and focusing. Pulser variables include type (spike, square wave, toneburst), and the number and type of controls. For example, spike pulsers may have continuously variable pulse damping, a limited number of damping settings, or no operator adjustment for damping. Some have two or more settings for pulse voltage. Receivers may be

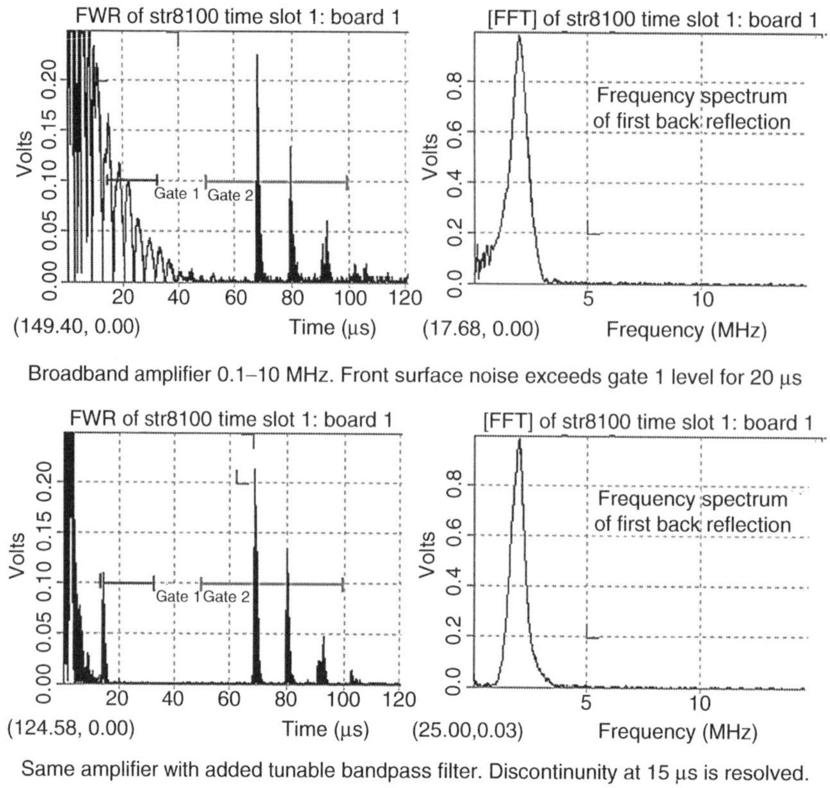

FIGURE 32.11 Response in steel bar with cored grain structure.

broadband with ranges up to perhaps 10, 50 or 100 MHz. Tuned reception frequencies may include a limited number of specific passbands, but a tuneable bandpass filter is often more useful. Portable instruments designed for a limited number of applications often have fewer adjustments in the interest of reliability and repeatability with less chance for operator error. Having the right equipment is as important as the right knowledge. Understanding how test object geometry and possible material property variations may interact with ultrasound is essential for understanding test data.

References

1. *The NDT Handbook*, 2nd ed., Vol. 7 Ultrasonic Testing. 1991 The American Society for Nondestructive Testing, 1711 Arlingate Lane, Columbus, OH 43228-0518. (The third edition is expected in late 2006 or early 2007.)
2. Published by The American Society of Mechanical Engineers, Three Park Avenue, New York, NY 10016-5990.
3. Published by American Welding Society, 550 N.W. LeJeune Rd., Miami, FL 33126.
4. Published by The American Society for Nondestructive Testing, Inc., 1711 Arlingate Lane, Columbus, OH 43228-0518.
5. Published by Aerospace Industries Association of America, Inc., 1250 Eye St. N. W., Washington, DC, 2005.
6. ASTM E 137-06. Standard Practice for Fabrication and Checking Aluminum Alloy Ultrasonic Standard Reference Blocks. ASTM International, 100 Barr Harbor Drive, West Conshohocken, PA 19428.
7. SAE ARP 5605 "Solid Composite Laminate NDI Reference Standards", SAE International, 400 Commonwealth Drive, Warrendale, PA 15096.
8. SAE ARP 5606 "Composite Honeycomb NDI Reference Standards", SAE International.
9. ASTM E 494-05. Standard Practice for Measuring Ultrasonic Velocity in Materials. ASTM International, 100 Barr Harbor Drive, West Conshohoken, PA 19428–2959.
10. ASTM C 1331-96. Standard Test Method for Measuring Ultrasonic Velocity in Advanced Ceramics with Broadband Pulse-Echo Cross-Correlation Method.
11. ASTM C 1332-96. Standard Test Method for Measuring Ultrasonic Attenuation Coefficients in Advanced Ceramics by Pulse-Echo Contact Techniques.
12. Brunk, J. Ultrasonic examination of porous materials. Materials Evaluation, April 1988. The American Society for Nondestructive Testing.
13. Hillger, W. et al. Inspection of CFRP components by ultraonic imaging with air coupling. NDT.net, October 2002, Vol. 7, No. 10.
14. Poguet, J. and Ciorau, P., Reproducibility and reliability of NDT phased array probes, Paper presented at the 16th

World Conference of Nondestructive Testing, Montreal, 2004. Available at NDT.net, Oct 2004, Vol. 9, No. 10.
15. Brunk, J. Applications and advantages of self-coupling transducers for materials characterization and inspection, *Acousto-Ultrasonics Theory and Application*. J. Dule (ed.). Plenum Press, New York, 1988.
16. Bhardwaj, M. and Stead, G. "Introduction to contact-free ultrasonic characterization of consolidated materials", Application of Non-Destructive Evaluation in Powder Metals Seminar, Iowa State University, Ames, IA, April 2000.
17. Papadakis, E. The inverse problem in materials characterization through ultrasonic attenuation and velocity measurements. Nondestructive Methods for Material Property Determination. C. Ruud and R. Green, Jr. (eds.), Plenum Press, New York, 1984.
18. Jivanuwong, S. et al. Design and development of an on-line system to study ultrasonic properties of model food systems during freezing. Paper presented at the 2002 Annual Meeting and Food Expo, Anaheim, CA.
19. Graham-Rowe, D. That cheese sounds great, *New Scientist*, August 1998, p. 20.
20. Kruger, S.E. et al. Ultrasonic backscattering formulation applied to cast iron characterization. Proceedings of the 7th European Conference on Nondestructive Testing, Copenhagen, 1998.
21. Bhardwaj, M. "Non-contact ultrasonic characterization of ceramics and composites", Proceedings of the American Ceramic Society, Vol. 89, 1998.
22. Levesque, D. et al. Thickness and grain size monitoring in seamless tube making process using laser ultrasonics", Proceedings of the Fifth International Workshop, Advances in Signal Processing for Nondestructive Evaluation of Materials, Quebec City, August 2005.
23. Howard, Q. and Enzukewich, S. The effects of microstructure on the ultrasonic testing of alloy steels. Materials Evaluation, December, 1997.

33

Lipid Membranes on Highly Ordered Porous Alumina Substrates

Andreas Janshoff
Johannes Gutenberg Universität

Claudia Steinem
Georg-August Universität

33.1 Introduction

For a living cell, the main barrier between the inside and outside world is a 4–6 nm thick layer mainly composed of lipids and proteins, the plasma membrane. The lipids, arranged in a bilayer, form a barrier, impermeable for water-soluble ions and molecules. Membrane spanning proteins are embedded, which are in part responsible for the selective transport of otherwise impermeable ions and molecules. The fundamental understanding of these membrane proteins as well as their applications in biosensors and screening assays make the development of membrane structures attached to a surface, which allow for the functional insertion of transmembrane proteins, very attractive. Such bilayers provide, however, only an appropriate environment for membrane proteins if several requirements are fulfilled. Each leaflet of the lipid bilayer should be in the fluid state and the entire membrane should be surrounded by an aqueous phase. Several different bilayer systems have been developed over the past 30 years to fulfill these requirements, two of which will be briefly discussed here: (1) Solid supported membranes, first described by Brian and McConnell in 1984 on glass surfaces [1], allow to functionalize conducting as well as nonconducting surfaces [2–8]. The resulting membranes have, however, some major drawbacks. The membrane is in direct contact to the surface, which hampers the insertion and functionality of transmembrane proteins. Due to the lack of a second aqueous compartment, the transport of ions and small molecules mediated by ion channels, transporters, and pumps cannot be followed easily. To monitor the transport of ions or molecules from one compartment to another, a lipid bilayer needs to be prepared that spans a small aperture, thus separating two aqueous compartments. Such membranes have been developed in the 1970s by Müller and Rudin [9]. However, they are very fragile, and automation and parallization as required for the design of biosensors and automated screening systems, is not possible. Hence, attempts have been made to suspend lipid bilayers across small apertures, manufactured in glass or silicon [10–14]. In recent years, a hybrid system has been established, which is based on highly ordered mesoporous substrates, on which lipid bilayers are prepared in a way that they suspend the pores. These membranes are to be expected to combine the merits of solid supported and freestanding lipid bilayers suspending a 1–100 μm sized single pore.

33.2 Preparation of Highly Ordered Porous Alumina Substrates

The substrate that is used is highly ordered porous alumina. Anodized porous alumina has been extensively investigated over the last five decades [15,16]. The material is sufficiently stable in aqueous solution and hydrophilic so that the pores are filled with water. It can be prepared in various pore sizes ranging from around 20 to 400 nm, making it an ideal mesoporous/macroporous material for the development of a new bilayer system. To obtain a sieve-like structure of hexagonally ordered pores with diameters of about 60 nm, the following procedure is pursued (Figure 33.1). Aluminum foils (thickness 0.5 mm, purity 99.999%) are cleaned with ethanol, electropolished, and anodized in

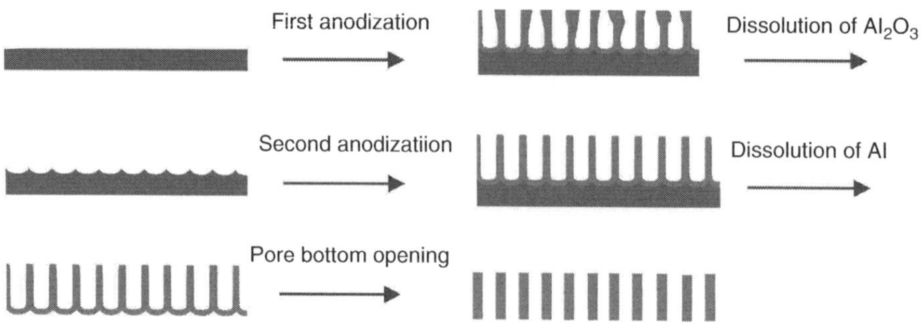

FIGURE 33.1 Schematic drawing of the procedure to generate a sieve like structure of hexagonally ordered pores in alumina. Further details can be found in the text.

aqueous 0.3 M oxalic acid solution at $U = 40\,V$ and $T = 2°C$ for 3 h. The formed porous oxide layer is removed by wet chemical etching in a mixture of phosphoric acid (6 wt%) and chromium(VI) oxide (1.8 wt%) at 60°C for at least 3 h [17–19]. The remaining pattern on the aluminum substrate serves as a mask for the second anodization process using the same parameters as in the first step. The second anodization is performed for several days to achieve a porous layer of reasonable thickness, which guarantees mechanical stability. Under the chosen conditions, the pores grow around 1–2 μm in depth per hour. The resulting porous alumina substrates are then incubated in a saturated $HgCl_2$ solution to remove the underlying aluminum layer. Finally, pore bottoms are removed by chemical etching at $T = 30°C$ with 10 wt% phosphoric acid solution. The process of pore opening can be monitored online by impedance spectroscopy [20]. The resulting porous substrates are analyzed by scanning electron microscopy.

33.3 Preparation of Pore-Suspending Lipid Membranes

The basic idea is to span lipid bilayers on the mesoporous alumina substrate. A prerequisite for such a preparation is that the membrane covers the upper part of the porous material, while the inner walls are noncovered. To achieve this, the bottom surface of the sieve like alumina structure is coated with a thin 25 nm gold layer, allowing for the chemisorption of thiol compounds. As yet, two different strategies were realized to obtain pore-suspending lipid membranes on porous alumina substrates dependent on the requirements of the membrane.

33.3.1 Nano–Black Lipid Membranes

If highly insulating lipid bilayers are required for the insertion of peptidic and proteinacous ion channels to monitor their activity down to the single channel level, membranes are prepared in analogy to the technique to obtain black lipid membranes (BLMs) developed by Müller and Rudin (Figure 33.2). First, the gold-coated substrate is functionalized with 1,2-dipalmitoyl-*sn*-glycero 3-phosphothioethanol (DPPTE) to render it hydrophobic

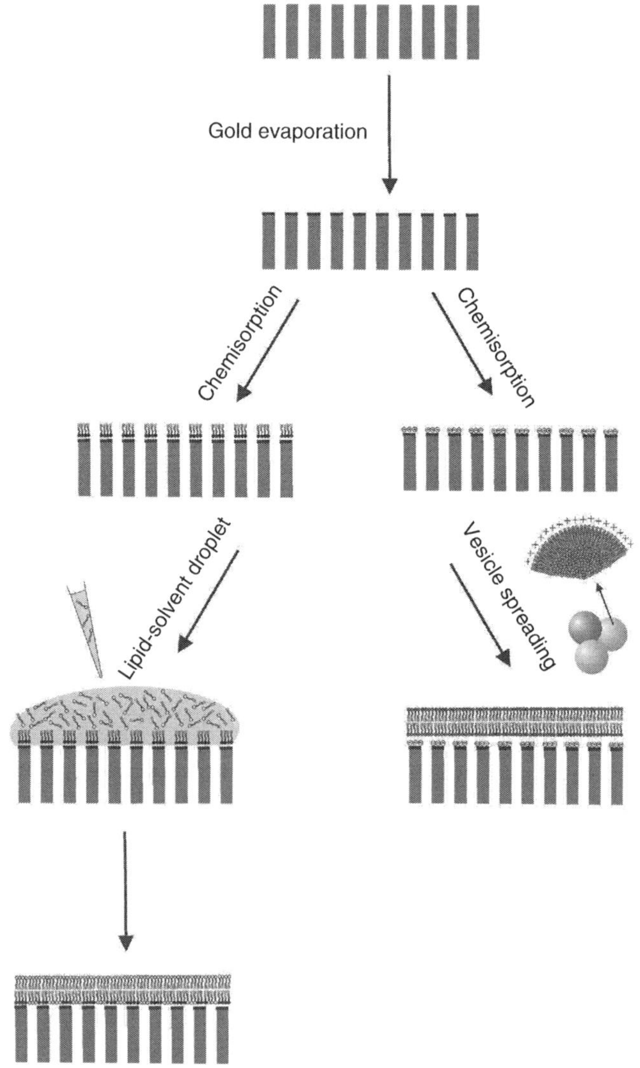

FIGURE 33.2 Schematic drawing of the formation of pore suspending membranes. On the left-hand side, the formation of a nano-BLM is illustrated, while on the right-hand side the fusion and spreading of unilamellar vesicles onto functionalized porous alumina substrates is shown leading to solvent-free pore suspending lipid membranes.

and to provide a monolayer on the solid part of the support, which needs to be completed by a second lipid monolayer. The formation of a full lipid bilayer covering the solid part and suspending the pore part is achieved by applying a solution of 1,2-diphytanoyl-*sn*-glycero-3-phosphocholine (DPhPC) (2% w/v) in *n*-decane to the surface of the functionalized porous substrate [20–22]. The process of lipid bilayer formation as well as the characteristic electrical parameters of the membrane are investigated by impedance spectroscopy. To model the impedance data, an equivalent circuit is required, which in this case is composed of a parallel connection of a capacitance C_m and a resistance R_m representing the pore-suspending membranes in series to an Ohmic resistance representing the electrolyte resistance. In general, a good accordance between data and fit is achieved, resulting in a mean specific membrane capacitance of $C_m = (0.4 \pm 0.1)$ μF/cm² taking the porous area into account [20]. This value is in good agreement with specific capacitance values obtained from classical BLMs generated by the Müller-Rudin technique, which are reported to be around 0.5 μF/cm² [23]. To attribute to this similarity, we called the membranes nano-BLMs. The membrane capacitance is an invaluable parameter to control and evaluate the process of nano-BLM formation. The second characteristic parameter is the membrane resistance R_m, which is determined at very low frequencies in the impedance spectrum. A nano-BLM with a membrane resistance of >1 GΩ is essential for its application in single-channel recordings. In almost all preparations, the resistance of the pore-spanning membranes exceeds the critical value of around 1 GΩ, which is sufficient for low conductance measurements and stays at this value for the experimental time period of 8–10 h. The membrane resistance then slowly and continuously decreases owing to the individual rupturing of some pore suspending membranes.

33.3.2 Solvent-Free Pore Suspending Membranes

It is desirable to create pore spanning lipid bilayers that are entirely free of solvent due to unwanted electrical and mechanical contribution from solvent molecules interacting with the membrane constituents. A straightforward way to obtain solvent-free membranes is to spread and fuse unilamellar vesicles on solid supports forming planar supported lipid bilayers free of detergents and solvents. The success of this preparation method depends strongly on the adhesion between the attached vesicles and the substrate. Hence, in most cases, strong attractive electrostatic interactions are employed to overcome the threshold adhesion energy and to enforce rupture of vesicles leading to planar lipid bilayers. Solvent-free membranes on porous substrates are particularly challenging due to the reduced contact area of the vesicle with the porous matrix, which results in an inherently reduced adhesion energy. One way to create planar pore spanning bilayers is to chemisorb thiols with charged end groups, which allow to deposit oppositely charged vesicles (Figure 33.2). The gold surface is first functionalized with a monolayer of 3-mercaptopropionic acid

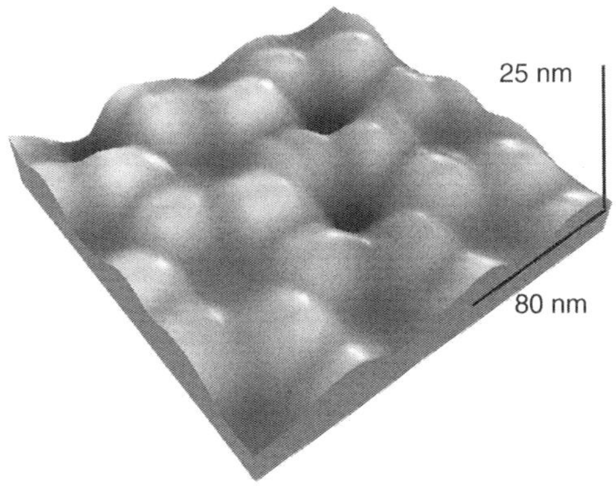

FIGURE 33.3 Scanning force microscopy image of a lipid bilayer suspending the pores of a porous alumina substrate. Two pores are not covered by a membrane.

followed by an incubation for 4 h at $T = 55°C$ with large unilamellar vesicles with a diameter of 100 nm composed of either positively charged *N,N,*-dimethyl-*N,N,*-dioctadecylammonium bromide (DODAB) or 1,2-dioleoyl-3-trimethylammonium-propane chloride (DOTAP). In situ atomic force microscopy (AFM) images clearly reveal the presence of a single lipid bilayer (DODAB) covering the porous alumina matrix next to uncovered pores (Figure 33.3).

Pore spanning bilayers offer the unique possibility to challenge their electrical properties on a local scale. Recently, Steltenkamp et al. [24]. managed to directly measure the restoring forces of free-standing lipid membranes composed of DODAB and DOTAP covering a nanoscopic hole of selected pore size, which is displaced by an AFM tip with a defined normal force. With this technique, it is feasible to infer elastic properties of lipid bilayers, i.e., the bending modulus and lateral tension from only a few lipid molecules (5000–30,000 lipids). The indentation experiments were quantified by means of variational calculus describing the indentation of a pore-spanning bilayer with a finite-sized tip.

Typical force–distance curves obtained from fluid phase DOTAP and gel phase DODAB membranes are depicted in Figure 33.4. All curves were taken in the center of the pores. Figure 33.4 shows trace (indentation) and retrace (relaxation) curves taken on gel-phase DODAB and fluid-phase DOTAP bilayers covering pores with an average radius of 90 nm (insets). Strikingly, the indentation curves exhibit a linear dependency of the restoring force on the penetration depth for DODAB as well as for DOTAP membranes over the full range. The mean slope representing the apparent "spring constant" of the membrane is a function of the physical state of the bilayer. Fluid DOTAP bilayers show an apparent spring constant of $\bar{k}_{\mathrm{DOTAP}}^{90\mathrm{nm}} = (0.0039 \pm 0.0008)$ N m^{-1}, while DODAB bilayers in the gel-state are considerably stiffer, displaying a spring constant of $\bar{k}_{\mathrm{DODAB}}^{90\mathrm{nm}} = (0.015 \pm 0.004)$ N m^{-1}. Importantly, both trace and retrace curve lie on top of each other except for the snap-on and snap-off positions, illustrating that under these conditions, it is possible to assess static, elastic properties. It is further possible to

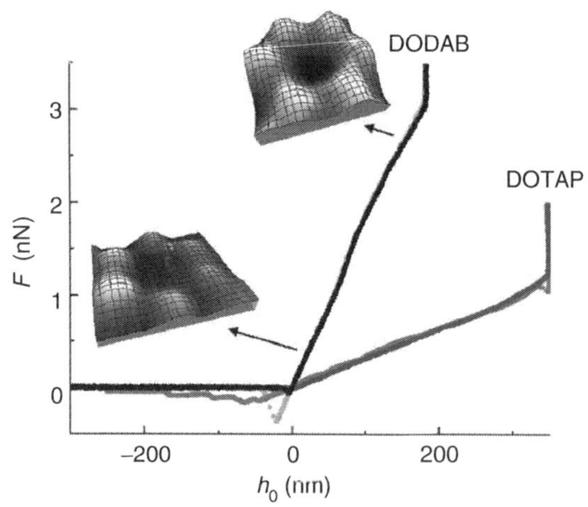

FIGURE 33.4 Force-distance curves (black line: indentation; grey line: retraction) taken in the center of gel-phase DODAB and fluid-phase DOTAP bilayers covering alumina pores with a diameter of 180 nm. The insets show AFM images taken at low and high normal forces illustrating the situation on the surface.

extract the bending modulus of the bilayer and the lateral tension representing the adhesive force of the membrane attached to the pore rims by solving the Euler-Lagrange equation yielding the shape of the indented membrane. This method is the first one to assess bending rigidity of membranes on a nanoscopic scale.

33.4 Insertion of Ion Channel Forming Peptides and Proteins

Pore-suspending membranes have been invented with the aim to insert ion channel peptides and proteins and monitor their activity. If the resistance of such a membrane is sufficiently high, namely in the Gigaohm regime, it is possible to monitor the ion flow through a single ion channel using high gain current amplifiers. The general functionality of nano-BLMs was first demonstrated using gramicidin as a channel-forming peptide [20]. The peptide is added to the aqueous phase. With time, a peptide spontaneously partitions into the bilayer, which can be monitored by the resulting ion-current that is driven by a holding potential of $V_h = +70$ mV across the bilayer in 0.5 M symmetric KCl solution. The peptide activity is characterized by stepwise current changes, which can be attributed to single conductance states and multiples of those. A single open state O_1 exhibits a current flow of (4.2 ± 0.15) pA under the chosen conditions, which translates in a gated conductance state of (60 ± 2) pS. In general, gramicidin is a good indicator of the formation of single lipid bilayers, since gramicidin is, with a length of 2.6 nm, only capable of spanning the hydrophobic part of one bilayer leading to the observed current steps.

Besides peptides, also large proteins can be inserted in membranes suspending pores with diameters of 60 nm. A good example is the reconstitution of the outer membrane protein F (OmpF) from *Escherichia coli*. It is composed of 16 antiparallel aligned β-sheets (β-barrel), connected by amino acid sequences referred to as loops and turns, building up a water-filled pore. Three of these monomeric units with a molecular weight of 37.1 kDa [25] and a length of 5 nm [26] are arranged around a threefold molecular axis. Loop 3 (L3) folds into the barrel forming a constriction zone of (11 × 7) Å^2 at approximately half the height of the channel. The constriction zone is assumed to be a decisive factor regarding conductivity and ion selectivity as elucidated by means of computer simulations and mutant studies. To achieve the insertion of OmpF in preformed nano-BLMs, the protein is added from a detergent solution to the aqueous phase [27]. The reconstitution of the trimeric protein into the membrane causes both a three-step increase in conductance and a step decrease in ionic current due to the opening and closing of the channel subunits (Figure 33.5A). From all-point histogram analysis of such current traces, the three different conductance levels with G_1 = (1700 ± 80) pS, G_2 = (3360 ± 80) pS, and

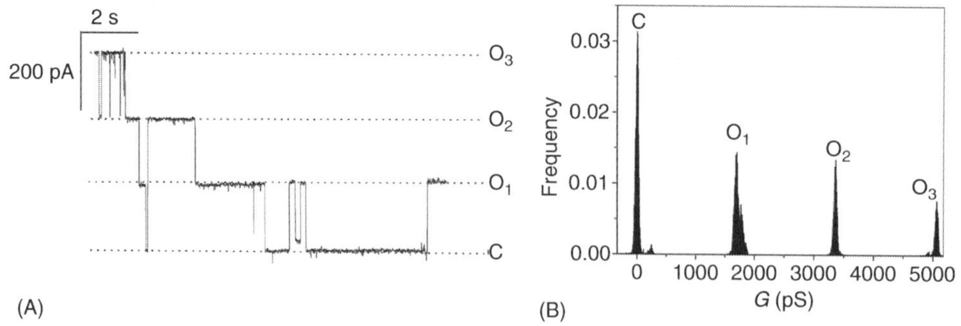

FIGURE 33.5 (A) Time trace of a characteristic channel activity (change in current) of one OmpF-trimer reconstituted in nano-BLMs at a holding potential of $V_h = -100$ mV. Three different current levels of the different opening states (O_1, O_2, O_3) as well as the closed state "C" can be distinguished. (B) All-point histogram of the current trace.

$G_3 = (5060 \pm 50)$ pS (Figure 33.5B) can be determined. The conductance of one subunit monomer calculated as the difference between the conductance states averages $G_m = (1690 \pm 20)$ pS. The current traces unambiguously demonstrate that even large proteins can be inserted into nano-BLMs. To give an impression about the sizes, one can calculate that one pore of the porous alumina holds about 4100 lipid molecules, taking the molecular surface area of one DPhPC molecule of 69 Å^2 into account [28]. The OmpF protein occupies an area of around 80 nm^2 [26], which is around 3% of the total membrane area on one pore.

33.5 Conclusions

Pore-suspending lipid membranes on highly ordered porous alumina substrate are developed as a versatile alternative to classical freestanding and solid-supported membranes. It has been shown that nano-BLMs on porous alumina substrates are well suited for the functional insertion of ion channels, allowing the analysis of their activities down to the single channel level. Owing to the coupling of the membranes to a support, these membranes are in principle suited for the development of biosensors and screening devices.

References

1. Brian, A. A., McConnell, H. M., Allogeneic stimulation of cytotoxic T cells by supported planar membranes, *Proc. Natl. Acad. Sci. USA*, 81, 6159, 1984.
2. Hillebrandt, H., Wiegand, G., Tanaka, M., Sackmann, E., High electric resistance polymer/lipid composite films on indium-tin-oxide electrodes, *Langmuir*, 15, 8451, 1999.
3. Janshoff, A., Steinem, C., Transport across artificial membranes—an analytical perspective, *Anal. Bioanal. Chem.*, 385, 433, 2006.
4. Kiessling, V., Tamm, L. K., Measuring distances in supported bilayers by fluorescence interference-contrast microscopy: Polymer supports and SNARE proteins, *Biophys. J.*, 84, 408, 2003.
5. Knoll, W., Morigaki, K., Naumann, R., Sacca, B., Schiller, S., Sinner, E.-K., Functional tethered bilayer lipid membranes, in *Ultrathin Electrochemical Chemo- and Biosensors*, V. M. Mirsky, ed., Springer, Berlin, 2004, p. 239.
6. Sackmann, E., Supported membranes: Scientific and practical applications, *Science*, 271, 43, 1996.
7. Steinem, C., Janshoff, A., Ulrich, W.-P., Sieber, M., Galla, H.-J., Impedance analysis of supported lipid bilayer membranes: A scrutiny of different preparation techniques, *Biochim. Biophys. Acta*, 1279, 169, 1996.
8. Tanaka, M., Sackmann, E., Polymer-supported membranes as models of the cell surface, *Nature*, 437, 656, 2005.
9. Müller, P., Rudin, D. O., Action potentials induced in biomolecular lipid membranes, *Nature*, 217, 713, 1968.
10. Cheng, Y., Bushby, R. J., Evans, S. D., Knowles, P. F., Miles, R. E., Ogier, S. D., Single ion channel sensitivity in suspended bilayers on micromachined supports, *Langmuir*, 17, 1240, 2001.
11. Fertig, N., Klau, M., George, M., Blick, R. H., Behrends, J. C., Activity of single ion channel proteins detected with a planar microstructure, *Appl. Phys. Lett.*, 81, 4865, 2002.
12. Ogier, S. D., Bushby, R. J., Cheng, Y., Evans, S. D., Evans, S. W., Jenkins, T. A., Knowles, P. F., Miles, R. E., Suspended planar phospholipid bilayers on micromachined supports, *Langmuir*, 16, 5696, 2000.
13. Osborn, T. D., Yager, P., Formation of planar solvent-free phospholipid bilayers by Langmuir-Blodgett transfer of monolayers to micromachined apertures in silicon, *Langmuir*, 11, 8, 1995.
14. Schmidt, C., Mayer, M., Vogel, H., A chip-based biosensor for the functional analysis of single ion channels, *Angew. Chem. Int. Ed.*, 39, 3137, 2000.
15. Masuda, H., Fukuda, K., Ordered metal nanohole arrays made by a two-step replication of honeycomb structures of anodic alumina, *Science*, 268, 1466, 1995.
16. Thompson, G. E., Wood, G. C., Anodic films on aluminum, in *Treatise on Materials Science and Technology*, Vol. 23, J. C. Scully, ed., Academic Press, New York, 1983, p. 203.
17. Li, A. P., Müller, F., Birner, A., Nielsch, K., Gösele, U., Hexagonal pore arrays with a 50–420 nm interpore distance formed by self-organization in anodic alumina, *J. Appl. Phys.*, 84, 6023, 1998.
18. Li, A. P., Müller, F., Birner, A., Nielsch, K., Gösele, U., Polycrystalline nanopore arrays with hexagonal ordering on aluminum, *J. Vac. Sci. Technol. A*, 17, 1428, 1999.
19. Li, A. P., Müller, F., Birner, A., Nielsch, K., Gösele, U., Fabrication and microstructuring of hexagonally ordered two-dimensional nanopore arrays in anodic alumina, *Adv. Mater.*, 11, 483, 1999.
20. Römer, W., Steinem, C., Impedance analysis and single-channel recordings on nano-black lipid membranes based on porous alumina, *Biophys. J.*, 86, 955, 2004.
21. Horn, C., Steinem, C., Photocurrents generated by bacteriorhodopsin adsorbed on nano-black lipid membranes, *Biophys. J.*, 89, 1046, 2005.
22. Römer, W., Lam, Y. H., Fischer, D., Watts, A., Fischer, W. B., Göring, P., Wehrspohn, R. B., Gösele, U., Steinem, C., Channel activity of a viral transmembrane peptide in micro-BLMs: Vpu_{1-32} from HIV-1, *J. Am. Chem. Soc.*, 126, 16267, 2004.
23. Benz, R., Fröhlich, O., Läuger, P., Montal, M., Electrical capacity of black lipid films and of lipid bilayers made from monolayers, *Biochim. Biophys. Acta*, 394, 323, 1975.
24. Steltenkamp, S., Müller, M. M., Deserno, M., Hennesthal, C., Steinem, C., Janshoff, A., Mechanical properties of pore-spanning lipid bilayers probed by atomic force microscopy, *Biophys. J.*, 91, 217, 2006.

25. Buehler, L. K., Kusumoto, S., Zhang, H., Rosenbusch, J. P., Plasticity of Escherichia coli porin channels. Dependence of their conductance on strain and lipid environment, *J. Biol. Chem.*, 266, 24446, 1991.
26. Cowan, S. W., Schirmer, T., Rummel, G., Steiert, M., Ghosh, R., Pauptit, R. A., Jansonius, J. N., Rosenbusch, J. P., Crystal structures explain functional properties of two *E. coli* porins, *Nature*, 358, 727, 1992.
27. Schmitt, E. K., Vrouenraets, M., Steinem, C., Channel activity of OmpF monitored in nano-BLMs, *Biophys. J.*, 91, 2163, 2006.
28. Pownall, H. J., Pao, Q., Brockman, H. L., Massey, J. B., Inhibition of lecithin-cholesterol acyltransferase by diphytanoyl phosphatidylcholine, *J. Biol. Chem.*, 262, 9033, 1987.

Index

A

B

C

T

U

V

W

X

Y